MW00845567

# Essential Biomaterials Science

This groundbreaking single-authored textbook equips students with everything they need to know to truly understand this hugely topical field, including essential background on the clinical necessity of biomaterials, relevant concepts in biology and materials science, comprehensive and up-to-date coverage of all existing clinical and experimental biomaterials, and the fundamental principles of biocompatibility. Drawing on the author's 40 years' experience in biomaterials, this is an indispensable resource for students of biomaterials science and engineering studying these life-saving technological advances.

Featuring

- A new classification system for biomaterials, a systematized framework for biocompatibility theory, new concepts for tissue engineering templates, and applications for nanomaterials in medicine.
- Extensive case studies from a wide range of clinical disciplines, interweaved with the theory, equipping students with a practical understanding of the phenomena and mechanisms of biomaterials performance.
- A whole chapter dedicated to the biomaterials industry itself, including guidance on regulations, standards and guidelines, litigation, and ethical issues, to prepare students for industry.
- Informative glossaries of key terms, engaging end-of-chapter exercises, and up-to-date lists of recommended reading.

**David Williams** is a Professor at the Wake Forest Institute for Regenerative Medicine, North Carolina, with over 40 years' experience in biomaterials science. He is Editor-in-Chief of the international journal *Biomaterials*, the President of the Tissue Engineering and Regenerative Medicine International Society, and a former Director of the UK Centre for Tissue Engineering at Liverpool University, where he is now an Emeritus Professor. In addition he is Advisory Professor of Shanghai Jiao Tong University, Visiting Chair Professor of Biomedical Materials, Taipei Medical University, a Visiting Professor of the Christiaan Barnard Department of Cardiothoracic Surgery, University of Cape Town, of the National University of Singapore, Tsinghua University, Beijing, Beihang University, Bejing, the University of New South Wales, Australia, and the Sree Chitra Tirunal Institute for Medical Science and Technology, Thiruvanthapuram, India. He has travelled extensively to promote excellence in scientific research and writing. He is a Fellow of the Royal Academy of Engineering, and has received numerous awards, including the 2012 Acta Biomaterialia Gold Medal.

"This is the long overdue single-author compendium students, scientists and clinicians were waiting for. Anyone expecting a dry scientific compilation will be pleasantly surprised by the wonderfully lively style in which Prof. Williams takes the reader on an exciting journey into the world of modern biomaterials and the opportunities it offers to patients. In a field long plagued by self-sustained paradigms, wrong models and wrong questions, this book boldly introduces each chapter on the basis of true clinical needs, taking the captivated reader into the deepest depths of material science and biology and eventually leaving him in a position where his own understanding and judgment has undergone a quantum leap."

**Peter Zilla**
**University of Cape Town, South Africa**

"This revolutionary book provides a coherent synthesis of the entire field of biomaterials, from the underlying sciences, to its practical applications. The book is the culmination of thought from one of the leading pioneers in the field, David F. Williams, who has been active for over 45 years, and is able to bring together not only the importance of the subject matter, but also its historical perspective, and future trends. With a strategic focus of thought, this unique text is a seminal contribution that provides an invaluable and thorough resource for anyone interested in the biomaterials field, not just for students, but also for scientists, and government and industry personnel."

**Anthony Atala**
**Wake Forest University School of Medicine, USA**

"This book distils the wide-ranging field of biomaterials down to critical topics, and presents them in an accessible and user-friendly way. In writing the book, the author applies his innovative ideas, vast knowledge, and rich experience to adroitly tackle the challenge of 'less is more' in processing a wealth of subject matter, placing a special focus on dynamic interactions between various biomaterials with complex biological systems, and translation of tissue engineering products to the clinic. Another valuable feature of this book is the pedagogical implications contained in each topic, which begins with a clear, simple diagram to introduce the reader to the core information, and ends with a number of questions to help the reader to integrate basic concepts into practice. Accordingly, this book provides a great reference for graduate students, researchers, and doctors specializing in biomaterials science. Such empowerment will inevitably lead to advancing the state of the art in the field."

**Xiaosong Gu**
**Nantong University, China**

"David Williams is one of the leading international authorities in biomaterials. Drawing on his vast multidisciplinary experience in the field, Prof. Williams presents in this attractive textbook not only a comprehensive view of biomaterials in their various facets, but also innovative ideas, along with the clarity of thought and precision of expression that those who know him well have come to expect of him. Although written primarily for students in biomaterials curricula, I see this book as 'a must' for the personal and institutional library."

**C. James Kirkpatrick**
**Johannes Gutenberg University of Mainz, Germany**

"This book provides the reader with the most up-to-date information on the ground-breaking revolutions in biomaterials sciences, and huge application potentials to overcome the most acute clinical challenges in the 21st-century. Reading this book is an academic enjoyment!"

Yan Li
Zhongnan Hospital of Wuhan University, China

"It is a remarkable achievement for any one individual, even if that individual is David Williams, to construct such an accomplished and authoritative text. Based on a lifetime spent in the field, this book is comprehensive, thought-provoking, and forward-looking, and is beautifully written and illustrated. While intended, primarily, as a student text, it is certain that there will be biocompatibility between this work and academics, clinicians, regulators, and industry practitioners alike, and it is destined to become a definitive biomaterials science text."

Keith McLean
CSIRO, Australia

"As the advancement of medical science cures various diseases, the role of biomaterials and their applications in medicine is recognized as growing. Almost every week, new biomaterials are announced and launched in the market. This book is composed of several chapters containing important information with many beautiful illustrations and photographs, which help students to understand biomaterials from very basic to near clinical applications. As one of the unique points of this book, each chapter has a brief glossary of biological and medical terms, which may be unfamiliar for students."

Teruo Okano
Tokyo Women's Medical University, Japan

"Williams' *Essential Biomaterials Science* combines comprehensive scope, single-authored consistency, and contemporary translational practicality in this novel textbook on biomaterials. The book clusters detailed considerations of materials, pathobiology, applications, regenerative therapeutics, and considerations of commercialization and clinical implementation, with an overriding focus on biocompatibility and concepts of biomaterial–tissue interactions, a key theme of Williams' many contributions to and leadership in this field. Well-illustrated, particularly with conceptual graphics, well-referenced with suggested readings, and with end-of-chapter questions, the book is most likely to be most useful to university students at an advanced undergraduate or graduate level, and nicely complements other available references in adding to the richness and usefulness of literature in the field."

Frederick Schoen
Brigham and Women's Hospital, Harvard Medical School, USA

"This is an extraordinary, impressively thorough reference source and textbook. David Williams has a rare knack for clear communication. He draws on a unique combination of outstanding knowledge, remarkable experience, and a rare appreciation of the key concepts. This book is an absolutely essential, superbly comprehensive, and valuable resource for anyone who wants to truly understand the field of biomaterials."

Tony Weiss
University of Sydney, Australia

## Cambridge Texts in Biomedical Engineering



*Cambridge Texts in Biomedical Engineering* provide a forum for high-quality textbooks targeted at undergraduate and graduate courses in biomedical engineering. It covers a broad range of biomedical engineering topics from introductory texts to advanced topics, including biomechanics, physiology, biomedical instrumentation, imaging, signals and systems, cell engineering, and bioinformatics, as well as other relevant subjects, with a blending of theory and practice. While aiming primarily at biomedical engineering students, this series is also suitable for courses in broader disciplines in engineering, the life sciences and medicine.

# Essential Biomaterials Science

DAVID WILLIAMS

Wake Forest Institute for Regenerative Medicine

CAMBRIDGE
UNIVERSITY PRESS

University Printing House, Cambridge CB2 8BS, United Kingdom

Cambridge University Press is part of the University of Cambridge.

It furthers the University's mission by disseminating knowledge in the pursuit of education, learning and research at the highest international levels of excellence.

www.cambridge.org
Information on this title: www.cambridge.org/9780521899086



First published 2014

Printing in the United Kingdom by TJ International Ltd. Padstow Cornwall

*A catalog record for this publication is available from the British Library*

*Library of Congress Cataloging in Publication data*
Williams, D. F. (David Franklyn), author.
Essential biomaterials science / David Williams.
p. ; cm. – (Cambridge texts in biomedical engineering)
Includes bibliographical references.
ISBN 978-0-521-89908-6 (Hardback)
I. Title. II. Series: Cambridge texts in biomedical engineering.
[DNLM: 1. Biocompatible Materials. QT 37]
R856
610.28–dc23 2013032183

ISBN 978-0-521-89908-6 Hardback

Additional resources for this publication at www.cambridge.org/biomaterialsscience

# CONTENTS

*The colour plates are situated between pp. 396 and 397*

# PREFACE

Biomaterials are crucial components of many health care products. They are in the news daily as we hear of new devices that allow deaf people to hear, and of techniques to return patients to a near normal life after a heart attack. Headlines tell us of titanium dental implants, ceramic artificial hips, carbon heart valves, collagen cosmetic injections and clear plastic lenses in the eye. Science magazines talk of new drug–biomaterial combinations that are radically altering cancer chemotherapy and immunotherapy and of nanoscale contrast agents that give far more power to MRI and CT imaging systems for better and earlier disease diagnosis. Biomaterials save lives and improve the quality of life for millions of people.

The science that underpins these advances is, not surprisingly, called biomaterials science. It has grown and developed from tentative beginnings half a century ago into a major academic and clinical discipline today. This science is both multidisciplinary and interdisciplinary since it brings together many classical disciplines of science, engineering and medicine, but also adds new knowledge that fits within the gaps between the classical subjects.

Biomaterials science is one of the most attractive subjects in any curriculum and is taught in colleges and universities across the globe. Students of biomaterials science go on to become research scientists at the forefront of medical technologies, clinicians who actually use biomaterials-based products on a daily basis, industrialists who manufacture the products, regulators who decide what can and can't be used in clinical practice or a practitioner of any one of the other contributory professions. It is obvious that these students need a textbook to guide them through the complexities of the individual components that make up biomaterials science. That is the rationale for this book, *Essential Biomaterials Science.*

It will be evident from the above comments that the compilation of all of the essential features of biomaterials science in a single book is not a trivial task, since these essentials cover so many different themes. For this reason, most books on this subject are multi-author books, with individual contributions from many different scientists and clinicians, brought together by a panel of editors. *Essential Biomaterials Science* is different. It is the work of a single author. It brings together, with one style, the various components of biomaterials science and integrates them, hopefully with little repetition and few gaps, into a logical story. It covers the essential underlying sciences, both materials/engineering sciences and biological sciences, and the clinical applications.

I have spent over 40 years working in the area of biomaterials science, developing the understanding of the subject that allows such a book to be written. I wrote one of the first textbooks on biomaterials and medical devices in the early 1970s (Williams and Roaf, *Implants in Surgery*, W.B. Saunders) and my own research work, writing and teaching has covered many of the scientific areas that are discussed in this book. In addition I have been the Editor-in-Chief of the premier research journal in this field,

*Biomaterials*, since 2001, a position that has allowed me to monitor and influence the developments on a global basis.

*Essential Biomaterials Science* is primarily intended as a textbook for students who are studying biomaterials at senior undergraduate and postgraduate levels. It should also serve as a reference source for anyone, at any level, who utilizes biomaterials in the course of their professional work. This especially includes those at post-doctoral or early faculty levels whose major disciplines have not been materials science, but also those in industry, regulatory or legal professions who regularly deal with health care products.

The book starts with an introduction to the world of biomaterials and medical devices served through a series of exemplars of current clinical practices that employ health care products routinely. Chapter 2 covers the essentials of materials science that underpin these clinical uses. In many ways Chapter 3 can be considered as the heart of the book since it deals with the mechanisms whereby the materials science and biology intersect, that is within the subject of biocompatibility; here a new unified framework of biocompatibility mechanisms is presented. Chapters 4 to 7 deal with the four main clinical applications, of implantable medical devices and artificial organs, tissue engineering and regenerative medicine, drug and gene delivery, and imaging and diagnostic systems. Chapter 8 comprehensively guides the reader through the array of current and potential biomaterials, with a new system for their classification. The final chapter discusses some of the infrastructure issues, including ethics, regulation and litigation.

In each chapter there are summaries of learning outcomes, glossaries, lists of recommended further reading and sample questions. The reading matter includes some other textbooks (marked *) but mainly provides citations to major review and opinion papers that take the reader more deeply into critical issues. All of these have been carefully chosen, mostly from the current literature and each citation includes a brief summary of the content of the work. Since this is a textbook, there are no references to support individual statements. I have also resisted the temptation to refer to commercial products and trade names since these are often ephemeral and become out-dated quickly. Occasionally a trade name becomes a generic descriptor and these are sparingly introduced. The glossaries are included in order to supply definitions of key individual terms, whose meanings may not be intuitively obvious.

The views expressed in this book mostly reflect my own beliefs, philosophies and prejudices and I therefore take full responsibility for the whole of the contents. Naturally I have been influenced by the writing and lectures of very many individuals, in many different countries and cultures. It is inappropriate to single out any such individuals here, but hopefully many of them will read or glance at this book and recognize where their thoughtful contributions have had an impact. Several of these colleagues have generously provided original artwork for illustrations; their collective support is gratefully acknowledged here and full credit is given to each in the relevant captions. I am very grateful to the staff at Cambridge University Press, especially Michelle Carey and Elizabeth Horne, for their guidance during the five years it has taken to prepare the book.

Most people who work in the area of biomaterials science will be aware that I have been supported in all of my endeavors by my wife, Peggy. Contributors to the journal *Biomaterials* will recognize her role, as Managing Editor of the journal, in coordinating much of the work that is published on this subject. Countless others have met her on our many biomaterials-related journeys around the world. Without her constant and unwavering support this book could not have been written. Too many days and nights have been spent, in North Carolina, South Africa and elsewhere, researching, writing and compiling the book, and her support, both practical and emotional, has been invaluable. My debt to Peggy is publicly acknowledged here.

# 1 The clinical necessity of biomaterials in the twenty-first century

In this opening chapter you will be introduced to the extent to which health care products contribute to the delivery of therapeutic and diagnostic procedures across a massive array of clinical problems and solutions. Included here are examples of long-term implantable devices, procedures of regenerative medicine, the diagnosis of disease and injury, and the specialized delivery of drugs and genes. You will then see how biomaterials science has evolved in order to optimize the performance of these products. The concepts of biomaterials science are introduced, along with a general discussion of the requirements of biomaterials and their essential characteristics.

## 1.1 Health care products in medical practice

You are an observer in a busy doctor's clinic on a Monday morning during a cold wet month of the winter. This is a large polyclinic, which includes not only primary care physicians but a plethora of specialists, who deal with the diagnosis and uncomplicated treatments for a variety of conditions, ranging from dental and ophthalmological conditions, to neonatal care, trauma, geriatric complaints and common infectious diseases. A few hundred meters away is a major teaching hospital, able to deal with virtually every acute and chronic condition that is likely to be seen in this mid-size industrial city, which encompasses people of all ages and genetic background.

The doctors and their colleagues deal with regular, routine patients who have regular routine non-emergency, but increasingly complex, chronic conditions. They also deal with patients who present with recently acquired illnesses and are in urgent need of a diagnosis and referral, with those who have minor and moderate trauma, those who suffer from degenerative diseases that either before, after or instead of major therapeutic or surgical intervention require ameliorative or rehabilitation therapies. There are those who have acquired (by whatever route, perhaps while travelling overseas, through lifestyle decisions or just by living next door to someone) infections by bacteria, viruses, fungi or prions. They see children and adolescents who need ongoing support for the corrective therapies that changed their physical appearance when they were young. There are those who have no illness at all, but just think that they have, or who might have in the future if something is not done now. And of course they have to deal with those who are sensible enough to undertake regimes of preventive medicine, perhaps through vaccines, prophylactic drugs or physiotherapy.

The medical practitioners in this clinic clearly wish to provide the best service to these patients and in order to do this they need the best tools of their trade. First and foremost, of course, are the skills and knowledge that they themselves bring, a point which we shall return to several times in this book. In addition, however, they need a variety of products, often called health care products, which allow them to deliver the required therapy or make the correct diagnosis. These could be drugs and vaccines, simple clinical sundries such as needles, scalpels and bandages, or more complex instruments or devices.

Let us follow some of these patients as they pass through the various offices in the clinic, often being referred onwards to the specialist clinics of the adjacent hospital.

### 1.1.1 The ubiquitous low back pain

The first patient is one of those individuals who experiences low back pain. The primary care physician sees many such people; the diagnosis may be complex since pain is difficult to quantify and there may be few indications on an X-ray that there is anything present other than normal ageing tissues. A prescription of anti-inflammatory agents and pain medication may be all that is necessary. This particular patient has been to the doctor before, however, and the condition is getting worse; there is chronic continuous pain with frequent spasms that leave him immobile and scared to try any movement. The time has come for more extensive diagnostic tests, where CT and MRI scans reveal that there are irreversible changes to one of the intervertebral discs in the lower, thoracic, part of the spine. The disc is a flat cylindrical piece of cartilage, which has an outer tough membrane, the annulus fibrosus, which encloses a viscous gel-like mass, the nucleus pulposus. This disc separates adjacent vertebrae and allows for bending and rotational movements of the spine. It is subject to degenerative changes over time and may, suddenly or gradually, be displaced from the intervertebral space, causing intense pain as two vertebrae now come into contact and interfere with nerves. This cartilaginous mass does not heal very well, and although the spinal doctors and neurologists know that efforts are being made to use stem cells to regenerate such discs, they are only too aware that, at this time, they have limited options.

The preferred treatment involves total removal of the disc and the intentional fusion of the two adjacent vertebrae. This results in some loss of mobility at that part of the spine, but with a total of 24 movable bones in the cervical, thoracic and lumbar sections of the spinal column, this is a small price to pay for the relief of the pain that accompanies the elimination of bone-to-bone sliding. The obvious practical question is how to produce this fusion. The surgeon has available to him/her a few devices which he/she can place, or implant, in the affected area to produce this fusion. The most common is a metal or plastic cage that is filled with bone grafts and secured between the vertebrae; this bone remodels over a short period of time to produce bony continuity and fusion (Figure 1.1). In some cases, this process is speeded up or enhanced through the use of a growth factor that is added to the bone graft. Our patient here is scheduled for this procedure a few weeks later. The doctors know he has at least a 75% chance of a good fusion and relatively pain-free back movements over the next 10 years.

**Figure 1.1** Device for spinal fusion surgery. The illustration shows an Infuse® Bone Graft with an LT-Cage® used in the lower part of the spine. This device consists of titanium cages that contain collagen sponge mixed with recombinant Bone Morphogenetic Protein-2 (BMP-2); this system incorporates technology developed by Gary K Michelson MD. Image provided by Medtronic, Inc.

### 1.1.2 Sporting injuries and arthritis

Also in the doctor's office that morning are several patients who received sporting injuries over the previous weekend. Knees are common sites of injury, especially to the cartilage or ligaments. Small cartilage lesions and ligament tears will heal over time without too much intervention, but assistance may be required, especially if the injury is extensive or if there is an overwhelming reason for the patient to be repaired as quickly as possible, as with valuable sportsmen and women. Cartilage is a difficult tissue as far as healing is concerned and quite small lesions can lead to progressive destruction. One patient has a relatively small lesion but it is in a central area of the knee joint, on one of the condyles at the end of the femur, and the doctors agree to carry out a relatively new technique called mosaicplasty. This is suitable for small but painful injuries. It attempts to encourage the cartilage to heal by drilling a small hole at the site of the lesion through to the bone, and taking a graft from an undamaged, peripheral area of the joint and inserting this into the drilled hole. The drilling causes some local bleeding which helps with the healing process. The procedure is carried out with minimally invasive techniques rather than major surgery and it is very convenient. The doctors are very confident with this process, as they have done it many times, and they know from published studies that over 90% of patients have very good outcomes, without any recurring problems over 10 or more years.

With another patient, the doctor knew, from previous experience, that one knee joint did show signs of localized arthritis, but the patient still carried on playing various sports and now had some acute damage in the area of the lesion. The doctor did not believe that this patient was a good candidate for mosaicplasty, but instead thought that it was worth trying an alternative method, which had been used since the mid 1990s in such patients; it is, however, considerably more expensive and involves far longer rehabilitation. This is the technique of autologous chondrocyte implantation (ACI). This is a type of cell therapy, for it involves harvesting some of the patient's own cartilage cells, the chondrocytes, culturing them in a special laboratory such that they expand and form a significant mass of cells, which can then be re-injected into the site of the lesion, where they generate new cartilage. The first stage with this patient had been completed several weeks ago and now, in the hospital, he was to receive this injection of cells. This is a major surgical procedure that has to be carried out with great care since these chondrocytes are very sensitive and easily damaged. The surgeon prepares the site of the injection in the cartilage very carefully and uses a layer of the patient's own periosteum, the thin outer lining of bone, to cover the site, using fine sutures and a special surgical adhesive to make a tight seal. They are all aware that success is far from guaranteed with this procedure and experimental work is still under way with different materials and techniques that will minimize chondrocyte damage and leakage, and maximize new cartilage generation. Around 80% of patients report good results after 5 years, but this is very much dependent on the co-operation of the patient over a full year of rehabilitation.

Still on the subject of cartilage and joints, since in some countries more than half the population over the age of 60 have signs and symptoms of osteoarthritis, it is not surprising that the doctors in this clinic will see quite a few arthritic patients in one session. They will be at different stages of the disease. Today there are a couple of patients, in their seventies, with fairly early stage arthritis, one of the hip and the other of the knee, who are going through the routine of trying the various pain and anti-inflammatory medications, under medical supervision, in order to achieve the right balance of good results without too many side effects. Another patient, although under 60 years of age, has a more advanced stage of arthritis in one hip. He has had the usual radiographic diagnostic examinations and is now visiting the primary care physician for the last time before being admitted to the orthopedic clinic for the surgical solution to his problem – total hip replacement. Next week, he will receive a metal-on-plastic combination that will replace the top end of his femur and a small part of the pelvis. This is a routine operation, the orthopedic surgeons carrying out this procedure on a daily basis, expecting to get a lifetime of 15–20 years with the device in 95% of patients.

That is a very good success rate, but does, of course, imply that 5% will not be so successful and also that after 20 years of function we might expect more problems. Unfortunately, on this particular day one of that 5% turns up at the hospital, in a wheelchair. Her device, after only 8 years, is clearly not working well and has become loose, giving extreme pain. She had not received this device in this hospital, having only moved to the neighborhood recently. The surgeons cannot readily identify the device on X-rays. The lady is obese and not in good shape. There is only one option, and that is to carry out a revision procedure, in which the old device is removed and a new one inserted, but the surgeons are well aware that, since the bones and joint have already been damaged, the chances of a really good outcome are not high.

### Box 1.1 Glossary of terms

In each chapter, a brief glossary of (mainly) biological or clinical terms that have been introduced but not explained in detail will be provided. In some cases, the opportunity is used to define related or similar terms that have not specifically been introduced here, but will be in later chapters. The majority of definitions are taken from *The Williams Dictionary of Biomaterials*, published by Liverpool University Press, 1999.

**agonist** a drug that has an affinity for, and stimulates physiological activity at, cell receptors.

**angina** paroxysmal pain in the chest usually due to interference with the supply of oxygen to the heart muscle.

**antibody** an immunoglobulin protein with a site able to combine specifically with the antigenic determinant on an antigen; an *antigen* is any molecule that is capable of being recognized by the immune system, for example T-cell receptors.

**artery** any vessel in which blood flows away from the heart; a *vein* is any vessel in which blood flows towards the heart in the major circulation.

**arthritis** inflammation of a joint; *osteoarthritis* is primarily a degenerative disease, while *rheumatoid arthritis* has immunological origins.

**atrium** a chamber affording entrance to another structure or organ, especially the upper, smaller cavities on either side of the heart; *ventricles* are the lower chambers in the heart.

**autograft** a graft taken from a source in the individual who receives it; an *allograft* is a graft taken from another individual of the same species as the recipient and a xenograft is taken from a different species altogether.

**cancer** a general term used to refer to a malignant tumor.

**cardiology** that part of medical science that is concerned with the function and diseases of the heart.

**cartilage** the hard, compliant form of connective tissue, in which cells are embedded in chondroitin, and which is extensively distributed around the musculoskeletal system.

**cataract** opacity of the lens sufficient to cause visual impairment.

**clinical trial** a controlled study involving human subjects, designed to evaluate prospectively the safety and effectiveness of new drugs or devices.

**Computerized Axial Tomography (CAT or CT)** an imaging method in which the cross-sectional image of the structures in a body plane is reconstructed by a computer from X-ray absorption data.

**condyle** the smooth, rounded protuberance at the end of a bone.

**disease** the condition in which the normal function of some part of the body is disturbed.

**echocardiography** examination of the structure and function of the heart using reflected pulsed ultrasound.

**etiology** study of the causes of medical conditions.

***ex vivo*** performed outside of the body; *in vivo* signifies something that takes place inside the body and *in vitro* refers to activities in solution or culture outside of the body.

**growth factor** one of a family of polypeptide hormones that regulate the growth, division and maturation of cells.

**immunohistochemistry** pathological technique for the microscopic examination of cells or tissues that have been stained by specific immunological markers in order to identify discrete cellular features.

**lesion** any pathological or traumatic discontinuity of tissues or loss of function of part of the body.

**ligament** bands of fibrous tissue that connect bones and cartilage and serve to support joints.

**Magnetic Resonance Imaging (MRI)** the use of nuclear magnetic resonance of molecules to produce images of the human body.

**mutation** sudden random change in the genetic material of a cell that may cause it to differ in appearance or behavior from the normal type.

**neurotransmitter** chemical that mediates the transmission of a nerve impulse across a synapse; a *synapse* is the site of functional apposition between neurons.

**plaque** a superficial, solid, elevated lesion, including the plaque that forms on the inner wall of arteries or the bacterial plaque that forms on teeth.

**prophylaxis** the use of methods to prevent the occurrence of disease, including *prophylactic drugs*.

**receptor** a molecule on the surface or within a cell that recognizes and binds with specific molecules, producing a specific effect in the cell.

**sphincter** a ring-like muscle that closes a natural orifice or passage.

**stenosis** the narrowing or contraction of a duct or canal.

**thalamus** either of the two large ovoid masses, consisting mostly of gray matter, forming part of the lateral wall of the third ventricle.

**trauma** injury.

**vaccine** a suspension of attenuated or killed microorganisms, or antigenic proteins derived from them, administered for the prevention, amelioration or treatment of an infectious disease.

### 1.1.3 The end of the road for cataracts

The next lady has a much better story. She is the same age, around 70, and is visiting the ophthalmologist. A few months ago, she was very depressed because of failing eyesight. Gradually she was losing the ability to see clearly, either near or far, and even recognising the faces of her five grandchildren had become impossible. She was otherwise very healthy, and never needed doctors and it took a great deal of persuasion by her children to seek help. She was petrified of needles and other medical paraphernalia and simply would not consider anyone going near her eyes with a surgical instrument. But now she had no choice, she had cataracts in both eyes; her natural lenses had become very cloudy and she was nearly blind. She could not believe it when, a few days after her first appointment she had a minor procedure under local anaesthetic, hardly noticing what was going on, and a piece of plastic was placed in each eye, in the form of intraocular lenses. Although such lenses had been used for a few decades, the ones she was fitted with were the relatively new self-accommodating lenses, and her vision was now truly excellent. Today the ophthalmologist was telling her that the results of the procedure were ideal and she could resume swimming and taking her grandchildren out.

### 1.1.4 The challenge of Parkinson's disease

Females around the age of 70 are frequent visitors to the clinic, and not all are so lucky as the lady with cataracts. Neurodegeneration is a massive problem in many countries. Although Alzheimer's disease is usually the major factor here, Parkinson's disease is also of huge significance. This disease has a complex aetiology, is progressive, incurable and difficult to treat. The incidence rises with age, from roughly 20 in 100 000 person years between 50 and 59 years of age, to over 90 in 100 000 person years between 70 and 79 years. Naturally as life expectancy in many countries is increasing, so the incidence of Parkinson's is increasing, and it will shortly affect around 1 in 800 people. The impact of the disease is therefore profound. It is caused by the failure of cells in that part of the brain known as the substantia nigra to produce adequate levels of dopamine, a neurotransmitter, with the result that the body experiences tremors, stiffness, slow movements and a loss of balance. These symptoms are not themselves life-threatening but collectively have a significant effect on the quality of life. Patients with Parkinson's have a median life expectancy of around 15 years following diagnosis, death often being due to complications of these physical limitations, such as pneumonia.

Why these dopaminergic neurons fail is still not clear. Most cases of Parkinson's disease are considered to be sporadic, occurring in individuals with no apparent history of the disease in their family, probably as the result of a complex interaction of environmental and genetic factors. Certain drugs may cause Parkinson-like symptoms, and various environmental toxins have been claimed to be associated with the disease. Approximately 15% of those diagnosed with Parkinson's have a family history of the disease, these familial cases often being caused by mutations in several genes.

Two patients, at different stages of disease progression attend the clinic. One, in the early stages, is solely on medication. It is not possible to supply dopamine directly to the body by a drug, but the molecule levadopa can be given orally, which is converted to dopamine within the brain. Levadopa can be converted too rapidly to dopamine outside of the brain, and has some unpleasant side effects, and so is usually combined with carbidopa. This combination is the major Parkinson's drug and it is this that has been prescribed for this first patient. However, it loses effectiveness over time and symptoms gradually reappear, even with increased dose levels. Other drugs include dopamine agonists, which mimic the function of dopamine, monoamine oxidase B inhibitors, which slow down the degradation of dopamine, and anticholinergics, which help to control the tremors. Her doctors know that she has only a few years left during which the drug will be really effective.

As with many conditions, doctors know that their remedies are not always beneficial over the long term, but they await new developments that will take over and give new hope. The slow progressive loss of dopamine-producing neurons suggests that an alternative strategy may be based on neuroprotection through the delivery of neurotrophic factors, which could protect the neurons from cell death and promote their regeneration. Neurotrophic factors derived from glial cell lines do have the ability to promote such repair, but limited clinical trials have not been very successful as yet, since there are problems with the optimal delivery of the molecules and with the apparent development of anti-neurotrophic factor antibodies in recipients. The concept is still valid, however, and we should expect to see better results when superior delivery systems involving the incorporation of these factors into biomaterial-based systems, possibly in the form of nanoparticles, are developed.

A further alternative is to alter the cells such that they themselves produce such a neurotrophic factor *in situ*. This is the basis of gene therapy, where the delivery of the appropriate gene would allow the cells to constantly express that factor. There are still serious issues to resolve as far as safety is concerned. Again, alternatives may eventually be found through the use of new materials, either in the form of non-viral polymeric vectors for the genes, or through the use of microencapsulated, *ex vivo* genetically modified cells that have been programmed to express a neurotrophic factor. These developments will probably not occur in a time frame that will help this particular lady.

The next approach involves the transplantation of immature dopamine-producing cells into the brain, a type of cell therapy. This was actually first started over 20 years ago, through the use of embryonic or fetal neurons obtained at abortion. Leaving aside the ethical issues, there have been many problems with this, and although some patients have considerably improved following this transplantation, success is very difficult to achieve because so many of the cells are lost during sourcing and implantation procedures. It is possible that types of stem cells may be beneficial in this therapeutic approach in the future.

The second Parkinson's patient is being treated by a relatively new method that is proving effective. This involves the deep brain stimulator, and is generally used as an adjunct to medication. The goal of the procedure is primarily to improve the condition in the "off" medication state. Patients with Parkinson's disease have both "on" and "off" states. The "on" state is when the medications appear to be working and when the patient is reasonably mobile. The "off" state is when the patient is slower and stiffer. Deep brain stimulation does not usually improve the "on" state, which is the best condition, but improves the patient when they are in their worst state.

This technology requires the placement of an electrode deep inside the brain, which is attached to a subcutaneously placed electronic device that generates electrical signals that stimulate the brain and counter those effects which produce the Parkinson's symptoms (Figure 1.2). There are several locations in which the electrode can be placed, most involving either the thalamus or the sub-thalamic region. The procedure, which is more complex for the sub-thalamic stimulation device, is usually performed in two stages, the first involving the placement of the electrode and the optimization of its position, and the second involving implantation of the stimulator and connections. The procedure appears to have significant beneficial effects in addition to those achieved with medication by itself, and there is good evidence that the benefits last at least five years. There are indications that these benefits may be reduced at longer periods of time and it is clear that not all patients will benefit from this treatment. Obviously it is not inexpensive.
This particular lady had the device fitted one year previously and this routine examination today shows that the treatment is going well. It took a few months for her to adjust to the "feel" of the system, but her quality of life has been immeasurably improved.

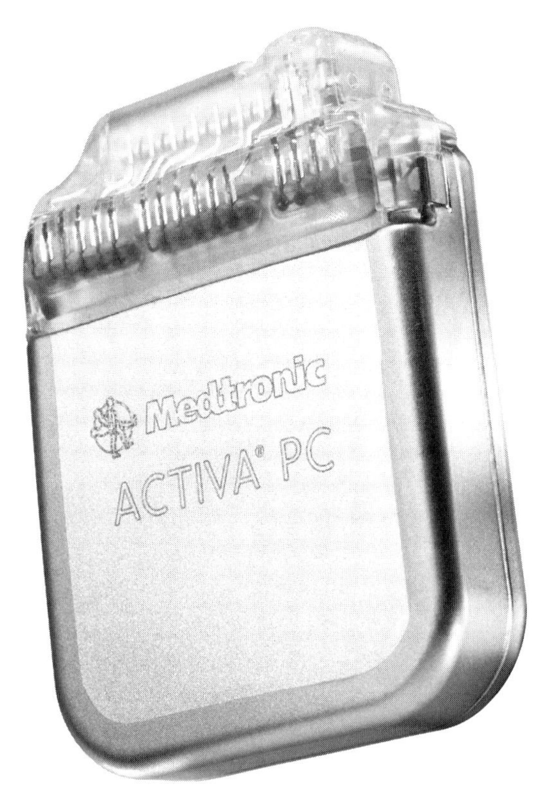


**Figure 1.2** Deep brain stimulator. The deep brain stimulator used in the treatment of Parkinson's disease. The device has an outer casing of titanium and provides stimulation to specific parts of the brain by stereotactically implanted electrodes. Image courtesy of Medtronic, Inc.

She is aware that this may not last, but prefers to be very positive for her future.

## 1.1.5 The cath lab and beyond

Most people are fully aware that the major cause of death in developed countries is heart disease. In the USA there are close to half a million deaths a year from it, which accounts for one in every six deaths. No wonder there is such an emphasis on the prevention, diagnosis and treatment of this disease, and our clinic is heavily involved in all such aspects.

The heart is a magnificent engineering and physiological structure, but there is much to go wrong, especially as individuals get older, when changes to the structure of valves, coronary arteries and other parts cause loss of function. Sometimes the symptoms are dramatic, with sudden collapse from a heart attack. Often there are early signs of disease, usually ignored by the individual, which lead to discomfort, difficulties with breathing and movement, and chest pain, but most people will seek some form of help sooner or later. A whole variety of non-invasive tests are available to give some indication of the presence of heart disease, including exercise stress tests, echocardiograms, MRI scans and nuclear imaging. If there are signs of serious disease, more advanced tests are required and several treatment options may be considered. This hospital, as with most major hospitals, has full facilities to conduct such diagnostic tests and deliver various therapies. The main facility is the catheterization laboratory, known as the cath lab, in which the state of blood flow in arteries can be assessed, especially in the coronary arteries which are themselves supplying blood to the muscular walls of the heart. If it is obvious that one or more of the coronary arteries are blocked, either following the onset of certain symptoms or immediately following a heart attack, consideration may be given to the use of a percutaneous coronary intervention, often referred to as angioplasty. Here a catheter is placed, under sedation, through an incision in the groin and along the femoral artery into the area of the blockage. The catheter has a balloon attached to its end and once it reaches the blockage, the balloon is expanded and compresses or dislodges the plaque that has slowly built up on the inner surface of the artery, opening the artery up again.

One such patient attends the cath lab this morning. He is in his sixties, has had angina for some time, and tests have determined that he has coronary artery disease that should be amenable to angioplasty. The cardiologists have, as always, been very truthful, and have said that although the angioplasty works very well, the effects may not last and the blockage may return, perhaps as soon as within one year. They do have an alternative, which doctors have been using for some 20 years. This involves using the

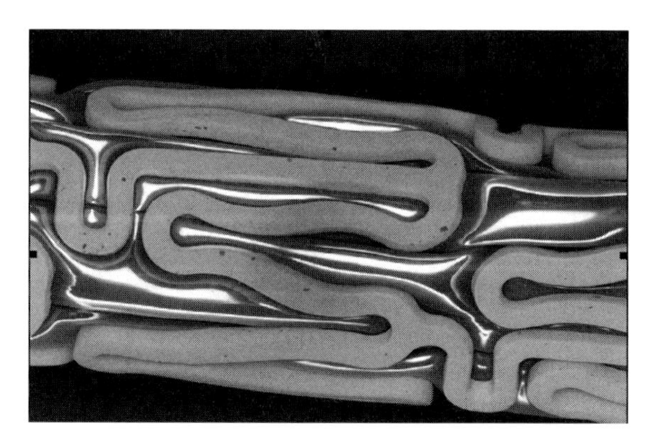

**Figure 1.3** Intravascular stent. Scanning electron micrograph (original magnification ×200) of a stainless steel intravascular stent mounted on an expandable balloon.

same type of procedure but instead of simply compressing the plaque and then leaving it alone, a small metal device, known as a stent, is left behind, in the hope that it can hold the vessel open permanently (Figure 1.3). The patient is asked to give permission for the cardiologists to use a stent if they think that is the best thing to do. The story is a little more complex. It has become obvious that even with a stent in place there is no guarantee that the vessel will not close up again. Many of the stents now available have an additional component that is meant to address this problem. This component is a thin layer of a polymer on the surface of the metal; this polymer contains a minute amount of a very powerful drug, in fact a chemotherapy drug, which stops the cells in the blood vessel wall from proliferating (just as it normally stops cancer cells from proliferating). Our clinic is taking part in a clinical trial to assess the latest design of a drug-eluting stent, and our patient is asked to give informed consent to take part. He will not know before he is sedated whether he will receive the original type of stent (the bare metal stent), a new drug-eluting stent, or no stent at all.

One other patient in the hospital wasn't quite so lucky. He had ignored the warning signs and went straight to the heart attack. He had too many arteries blocked, in poor configurations, and was not suitable for the angioplasty and stent. He was admitted to the emergency room at dawn, a common time for heart attacks, and it was decided that he had to have a coronary bypass. In this, the blocked parts of the coronary arteries are bypassed by a segment of a tube, which allows the blood to flow freely.

In this situation, there is no magic piece of synthetic material for the tube. The coronary arteries are rather small, usually less than 5 mm in diameter, and no one has yet made a synthetic tube that works in this position – they would block rather quickly. Interestingly it is different when arteries in some other parts of the body become diseased, for example in the upper part of the leg, for these arteries are larger, and synthetic replacements made out of common textiles such as polyesters work quite well in these positions. A number of patients in the vascular ward of this hospital have had such procedures this week, most of them suffering from the consequences of lifetimes of smoking, since atherosclerosis, which is the blocking of arteries by plaque, is one of the major sequelae of this habit. Instead, with the coronary arteries, the surgeon will use one of the patient's own veins for the bypass, having determined that he has a good-quality greater saphenous vein in one of his legs that can be used. This is a relatively superficial vein that runs from just above the ankle to the top of the thigh. Removing this vein is not a problem for the long-term survival of the leg since there are several other veins that compensate as far as the major circulation is concerned, but it is not a trivial process removing it, usually with an incision running the whole of the length of the leg. The best sections of the vein are cut and trimmed and used for the bypass; in the case of this patient three of the coronary arteries are involved. When he comes around from the operation and recovers from the ordeal over the next few days, the patient will be concerned about his future, but the cardiac surgeons are very confident and positive. Many famous and active individuals have gone through this procedure and had very successful and rewarding lives afterwards. The British explorer

Sir Ranulph Fiennes had a double bypass in 2003 and went on to climb Mount Everest in 2009 at the age of 65. Well-known figures in the USA, such as Bill Clinton and Larry King, have had successful bypasses, so our patient is in good company. This area poses several difficult questions in relation to the preferred treatment modality. At one time drug-eluting stents appeared to offer the best prognosis, but as the results of long-term follow-ups are analyzed, the outcomes may not be better that coronary artery bypass surgery.

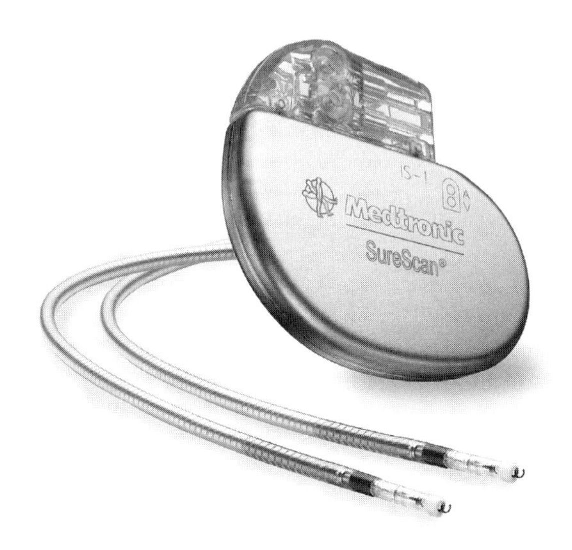


Figure 1.4 Cardiac pacemaker. Structure of an implantable cardiac pacemaker. The device has a chamber encased with titanium, two leads with electrode tips and a header through which the leads enter the chamber. This particular device has been designed to be MRI-compatible since many patients with pacemakers often require imaging and conventional pacemakers are not compatible with the magnetic fields used in MRI. Image courtesy of Medtronic Inc.

## 1.1.6 Pulses and shocks in the heart

The heart is not only about blood flow through tubes and valves; it needs to be driven in some way. The driving force is based on electrical signals; these are generated in the sinus node at the top of the right atrium of the heart and in healthy individuals the pulses produced trigger regular movements of the heart wall muscles, with a frequency (the heart rate) that is responsive to conditions such as exercise. One of our patients in attendance this morning had experienced some difficulties with his heart rate and rhythm a number of years ago and it was clear that his natural pacemaker was not functioning correctly. He was fitted with an artificial pacemaker (Figure 1.4).

This implantable device has two main components. One is the pulse generator, which is a small hermetically sealed metal container that is placed under the skin in the region of the shoulder. It contains the generator circuitry and the power supply. It is connected to electrodes in the heart by fine metallic wires that conduct and deliver the impulses to the muscle. Pacemakers are configured in different ways depending on the precise condition of the patient. This patient has a dual chamber pacemaker with leads going both to atria and ventricles. It is a demand pacemaker that does not work all the time; it senses the heart rhythm and switches itself on when the heart rate drops below a pre-determined level. These patients are routinely monitored and all have a device at home, linked through a computer, which can send information about the state of the pacemaker over the phone to the clinic every three months. Once a year our patient comes into the clinic for a physical check and for inspection of the battery condition. Depending on the amount of time that the pacemaker is actually generating signals, a battery may last between 5 and 15 years; our patient had had the pacemaker for 7 years and today the tests show that he will need the device changed sometime in the near future and so he arranges the necessary appointment. The leads and electrodes look fine and so this will be a simple outpatient procedure.

Another patient with heart rate problems is not so confident. He has a history of arrhythmic episodes, where his heart beats irregularly, and he has been on anti-arrhythmic medication for a couple of years. The cardiologists are worried, however, and he does seem to be at risk of ventricular fibrillation, in which the ventricles quiver instead of contracting, so that no blood is pumped from the heart. It is fatal unless treated immediately. There is one type of treatment involving an implantable device rather similar to a pacemaker, but this has been much more controversial. This is the implantable cardioverter-defibrillator (ICD). Most people are familiar with the normal defibrillator, the device with two big pads that can be placed on the chest of a victim of acute ventricular fibrillation and which deliver a massive electrical shock that stops the irregular beat and returns the heart to normal – it is often portrayed as the exemplary dramatic medical intervention and defibrillators are installed in many public places such as airports. The chances of being close to a defibrillator when an attack happens are usually rather small, however, and the possibility of personalizing this device was behind the development of the ICD. It does look similar to a pacemaker in terms of combinations of generator and electrodes, and has some similar functions, but with a big difference. The ICD can deliver pacing signals if required. It can also deliver cardioversion, which controls a heart beat that has become too fast – this is a mild electrical shock, detected by the patient as equivalent to being thumped in the chest. If ventricular fibrillation is detected, then a much stronger shock is delivered, this being powerful and unpleasant. In the early days, ICD was seen as a treatment of last resort, one of the issues being that they were not 100% reliable in the detection of ventricular fibrillation, and the powerful shock could be given when it wasn't required. Living with a system that could, without warning, deliver a massive shock – equivalent, some people say, to a horse kicking you in the chest – was psychologically disturbing to many people, and even more so when you were aware that often the horse had no reason at all to kick you. Depression and anxiety are not uncommon experiences in these patients. The situation has become far more reliable in recent years and almost all relevant medical authorities worldwide have concluded that ICD therapy is effective. The cardiologists here have concluded that this patient should be placed on the list for this treatment, and he now awaits this with just a little trepidation.

### 1.1.7 Care of the newborn

Not all patients in this medical facility are old and in need of care because of disease; many are very young and in need of care because of just that, they are too young to care for themselves. The Neonatal Intensive Care Unit (NICU) in this hospital is there to assist in supporting extra young individuals who have been born under less than optimal conditions. Prematurity and low birth weights have become significant problems. Prematurity is usually defined as delivery at less than 37 weeks pregnancy, and very premature at less than 32 weeks. Low birth weight is considered to be less than 2500 grams (around 5½ pounds) and very low birth weight is around 1500 grams (3½ pounds). Over 12% of American babies are born prematurely, with figures elsewhere ranging from 5% in Ireland to 17% in southern Africa. Risk factors for prematurity include diabetes, smoking, drug habits and infertility treatments in the mother. Just being born early is not necessarily a problem, it is more the health status of the baby, often reflected in the birth weight. According to the Center for Disease Control in Atlanta, 7% of babies are of low birth weight and 1% of very low birth weight.

Our maternity unit has its fair share of premature and low birth weight babies, some of whom have to spend time in the NICU – about 2000 per year, either premature or with life-threatening conditions, spend their first few days of life here. They are supported in their fight for life by a variety of machines and devices that allow them to breathe and feed.

Much of the equipment involves small disposable but life-saving items such as tubes, sensors and routine monitoring equipment. Babies that cannot be fed have catheters to assist in nutrition – we have one very low birth weight infant that needs all its nutrition delivered this way, a procedure known as total parenteral nutrition (TPN), and has a catheter placed directly into a central vein for this purpose. It may take several weeks for him to be strong enough to be weaned off this system. The lungs and respiratory system are also vulnerable in premature infants and respiratory stress syndrome is the most common cause of death. These infants have to be assisted in their breathing by use of a ventilator, with constant monitoring of oxygen levels and other vital signs.

One of the infants is in quite a bad way, suffering from persistent pulmonary hypertension (PPHN), which arises from a failure of the babies' own circulation to successfully take over from that of its mother. This is life-threatening and the pediatricians have already placed the baby on the best ventilator for this purpose, the high frequency oscillation ventilator, but the results have not been good enough and now they have decided to place the baby on extracorporeal membrane oxygenation (ECMO). This is the equivalent to the so-called heart–lung machine used in major cardiac operations in adults, and takes the babies' blood from the heart and lungs and passes it through a machine where oxygen is added to the blood and carbon dioxide is removed, and then the blood is passed back to the major circulation. This baby will need to stay on ECMO for a few days before it is well enough to take over the heart and lung functions itself. The introduction of neonatal ECMO in the 1980s saved many infant lives, although fewer babies need it now because of advances in ventilation, including the use of nitric oxide in the ventilator, which improves the response from the lungs.

### 1.1.8 Diagnosing and treating breast cancer

According to the World Health Organization (WHO), there are around 8 million deaths annually worldwide from cancer, and this is increasing. Developed countries have a high prevalence of prostate, breast and colon cancer, while developing countries see high rates of liver, stomach and cervical cancer. With the very mixed population in our city, we see many different forms of cancer, and rather too many cases altogether. Since many cancers are preventable through lifestyle choices, there is a big emphasis in the community on prevention, while within the hospital itself early detection of cancer and the various forms of therapy that can be delivered after diagnosis are very important.

Breast cancer is a major problem, but there have been significant advances recently and our hospital is at the forefront of progress. Breast cancer screening is a politically sensitive issue in many parts of the world since the success rates of detection through screening processes such as mammography are controversial and, of course, there are difficult balances to make with respect to the health economic issues involved. We are more concerned here with what to do once a tumor is suspected. Of particular relevance is the fact that it is the nature of the tumor that may be as important as its presence.

Of special relevance here is the so-called HER-2 type of breast cancer. In some breast cancers, the cancer cells have an unusually large number of Human Epidermal Growth Factor 2 receptors on their surface. This occurs because of a gene mutation and causes the tumor to be more aggressive. Fortunately, there are some treatments specifically directed at HER-2 positive tumors, but the tumors need to be identified in the first place. Usually the front-line treatment will be surgical removal of the tumor and at that time a biopsy will be taken. There are two good tests for HER-2 positivity, one being immunohistochemical and the other involving the technique of fluorescence *in situ* hybridization (or FISH). Our laboratory uses the latter, and it is very effective, but it does take a little time. The oncologists are aware, however, of some new techniques that are in advanced stages of development that should be able to give real-time assessment of the presence of these receptors, especially using highly fluorescent nanoparticles called quantum dots. There is an international effort to develop these tests, and our group is liaising with groups in China, Japan and Singapore on these.

Of course, identifying the type of tumor is only really helpful if there is a specific treatment for it. Fortunately there is for HER-2 positive tumors. The extracellular domain of this receptor (that is the part of the surface receptor molecule that sits on the outside of the surface membrane) can be targeted by a molecule that can attach to it, and effectively inactivate it; usually this will be a monoclonal anti-HER-2 antibody. One of these drugs is Herceptin, which is proving remarkably effective (although there others, such as Lapatinib, which has a slightly different target site on the cell membrane), as shown in Figure 1.5. Herceptin is very expensive and a great deal will depend on the availability of finance, private or public, to pay for the cost of the treatment, which is in the region of 50 000 US dollars or euros or their equivalent elsewhere.

Many questions then arise over whether there should be other types of concurrent treatment. Radiotherapy is usually advised, but the role of chemotherapy is of some importance, since it is unclear whether chemotherapy adds much to the effects of this immunotherapy (or biological therapy) associated with Herceptin. Again, however, we have to look at the future here, which is what our oncologists are always doing. Everyone knows that chemotherapy is quite effective in general, but has significant side effects. This arises because in most cases the chemotherapeutic drug, which kills cancer cells, also has effects on other cells. The goal of most work in this area is to design the drug so that it specifically targets the cancer cells and leaves other cells alone. The oncologists here are already using some recently approved drugs in which the drug is combined to some other structure, which is essentially harmless in the body until it is taken up by cancer cells, the toxic part then being released. Two of the most powerful chemotherapy drugs have been formulated in this way, doxorubicin being structured in the form of lysosomes for use in myeloma, and paclitaxel being attached to albumin nanoparticles for the treatment of non-HER-2 breast cancers. Even more enticing is the possibility to take multifunctional drug-containing, nanoparticles that simultaneously have diagnostic and therapeutic roles.

### 1.1.9 The challenge of diabetes

There are over 180 million people in the world suffering from diabetes, a staggering 32 million of these being in India, 34 million in Europe, 20 million in China and 18 million in the USA. The WHO anticipates that the global figure will be 366 million by 2030. Type I diabetes accounts for 10–15% of cases; it is an autoimmune disease where the body is destroying the insulin-producing beta cells. Treatment involves regular injections of insulin and control of diet. Type II, usually referred to as late-onset diabetes, involves insulin resistance and insufficiency, and is far more common. One of the real concerns is the rapid increase in Type II diabetic adolescents around the world. If this disease is diagnosed late in life it is usually managed by control of diet and exercise, but there are complications that take some time to develop, including significantly higher risks of cardiovascular disease, retinal neuropathy that leads to blindness, kidney disease and the development of leg ulceration. There are a number of obese young people in our clinic that already have signs of Type II diabetes, who are being followed carefully. There are also several elderly patients with serious leg ulcers associated with their diabetes (Figure 1.6). This is a somewhat controversial area since there are several types of wound dressing available, including polyurethane foams and alginate colloids. However, not all authorities agree that the more expensive dressings recently introduced give better overall results than cheap simple dressings. Our nurses are well aware that they have to carefully monitor each patient and prescribe the most appropriate dressing for them, and also that great care with cleanliness and sterility is very important. Many attempts have been made to accelerate or enhance the repair of chronic ulcers through regenerative medicine procedures; for example, the patients' own skin cells can be harvested and grown in culture to produce a layer

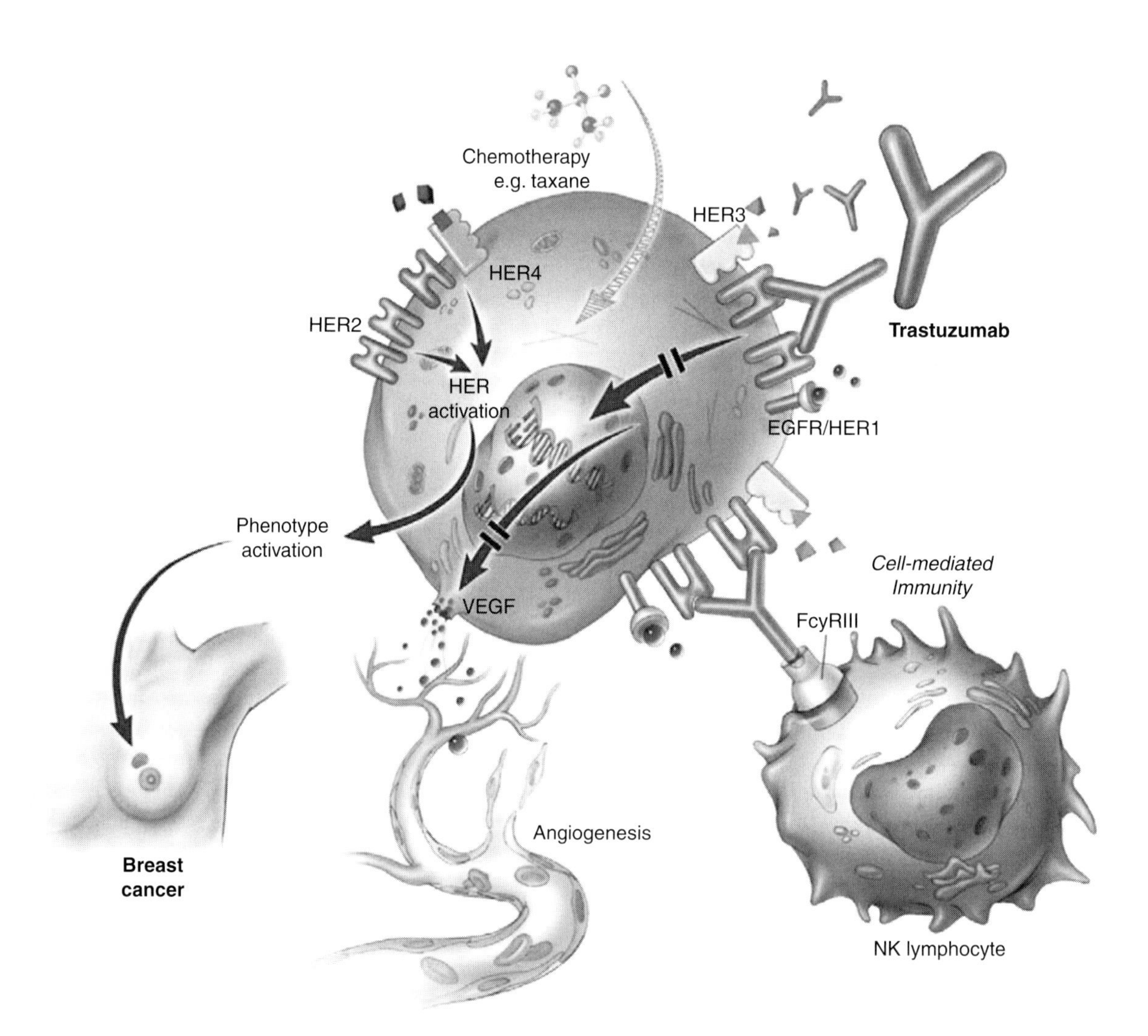


**Figure 1.5** Herceptin and breast cancer. Trastuzumab, best known by the trade name Herceptin™ (Genentech), is a recombinant humanized monoclonal antibody that is directed against the extracellular domain of the HER-2 protein. In HER-2 positive individuals, this protein activates multiple signaling pathways leading to aggressive tumor formation. Herceptin binds to the protein and diminishes the receptor signaling and induces cell cycle arrest and accelerates the death of cancer cells. Although this monoclonal antibody has good effectiveness, new mechanisms involving the conjugation of some biomaterials to the antibody are being developed to improve performance. This provides an example of how biomaterials can be employed to enhance drug delivery, as discussed in detail in Chapter 6. Reprinted from *Cancer Letters*, 232(2), Nahta R and Esteva FJ, Herceptin: mechanisms of action and resistance, 123, © 2006, with permission from Elsevier.

of skin. The main problem with this approach is that it doesn't address the underlying problems affecting the poor blood supply; it could also be a very expensive alternative to traditional wound care and bandages.

Young patients with Type I diabetes have to learn how to self-inject their insulin and monitor their glucose levels to keep on top of this delicate balance. Not all manage this successfully and real-time continuous glucose monitors have been in use for

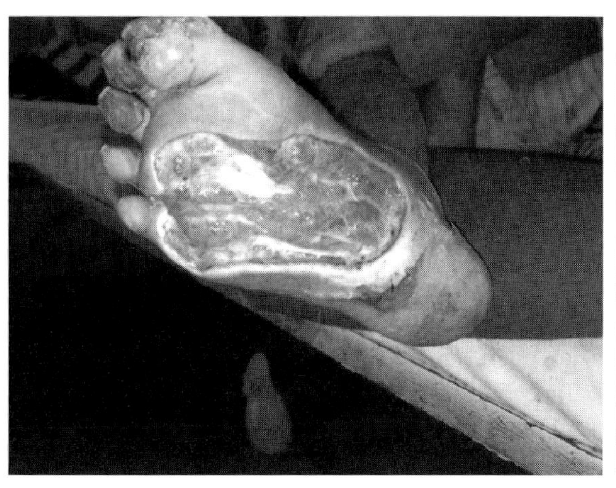
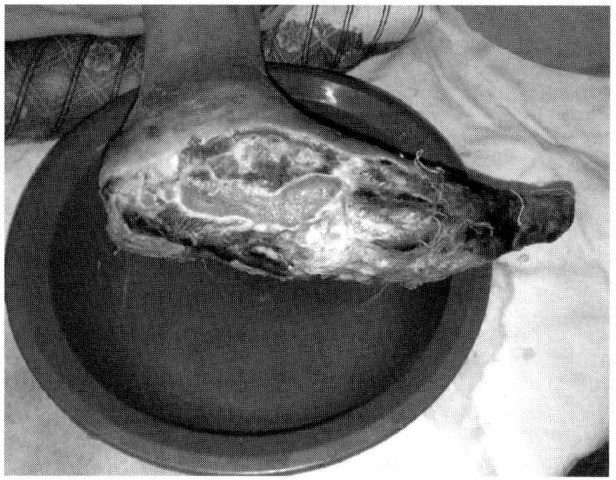

**Figure 1.6** Diabetic foot ulcer. One of the more severe consequences of diabetes is a compromised blood supply to the lower limbs, which results in the progressive development of ulcers. Examples here are severe, and may lead to limb amputation. Ulcers are very difficult to treat once they become established. Simple dressings may no longer suffice. Attempts are being made to develop techniques to regenerate the skin and underlying tissue, as discussed in Chapter 5. Reprinted from *The Foot*, 22(1), Zubair M *et al.*, Incidence, risk factors for amputation among patients with diabetic foot ulcer in a Northern India tertiary care hospital, 24, © 2012, with permission from Elsevier. See plate section for color version.

selected patients for a few years. One of the attendees this morning is particularly difficult. However, this clinic is taking part in a clinical trial of a continuous monitor that is linked to an insulin infusion pump. This subcutaneous closed loop system is still under evaluation, and monitoring this 18-year-old boy, who was diagnosed 3 years ago, should provide valuable information on this type of system.

### 1.1.10 The messy business of incontinence

This is one of the more embarrassing parts of the clinic; it shouldn't be this way. If the sphincter between stomach and duodenum, the pyloric sphincter, is not working properly, you get a duodenogastric reflux, with excess bile going back into the stomach. You are referred to a gastroenterologist and get on with the treatment. If, however, it is the urethral or anal sphincter that is in trouble, you may experience either urinary or fecal incontinence, which is where the embarrassment comes in for this is quite difficult to hide, and lifestyles change radically. We needn't go into the unpleasant aspects here except to consider one patient in the clinic who is having trouble with urinary incontinence, and specifically stress urinary incontinence. The International Continence Society estimates that about 200 million people worldwide suffer from some form of urinary incontinence, with stress urinary incontinence in women forming a large part of this. Although often associated with old age, this is not necessarily so, and many otherwise very healthy women between ages 35 to 55 suffer this to some degree. The effects of childbirth and obesity are two major factors. In these patients, the urinary sphincter lacks sufficient strength to prevent urinary leakage when abdominal pressure exceeds a certain level, which may happen with exercise, for example when playing tennis, or even under the more trivial of conditions such as a cough or a sneeze, or lifting a bag of shopping. No wonder there is embarrassment, which itself often leaves the patient unwilling to see a doctor about it. Treatments vary from protective pads and other clothes – the market for these in the USA alone is $500 million per year – right up to surgery.

It is the surgical option that is of interest here, since one of our patients is about to undergo one of the relatively new procedures for stress urinary incontinence. She is 45 years old, obese and smokes. The doctors, specifically the urogynecologists, have deliberated with great care. She is intractable to more conservative treatments and is rather insistent on surgery as she sees this as her only option – she has seen this on the internet and wants the procedure. The treatment involves a procedure in which a special surgical tape is inserted, via the vagina, into the soft tissues and placed under the mid-section of the urethra with the intention of pulling it upwards, thereby making it much more difficult for urinary leakage to take place. The technique works pretty well – clinical studies from all around the world show a success rate of 85–90%, which can be maintained for many years (or at least for the duration of the studies in this rather new technique). The doctors have stressed that success cannot be guaranteed and a lot will depend on the patient – lose some weight, cut back or stop smoking, follow rigorous personal hygiene to avoid urinary tract infection, and so on. They are encouraged by the fact that at least a million women have had this treatment, but that doesn't stop them being a little nervous.

## 1.2 The ubiquity of materials in clinical medicine

In all of these cases the clinician has chosen to use some form of modern health care technology to effect the therapy or determine the diagnosis. Often this is carried out with the parallel administration of one or more drugs that may be directly targeting the same condition, or dealing with the associated issues for which pain relief, anti-infective, diuretic, anti-inflammatory agents and so on are required.

The extent of the usage of these technologies is probably obvious from some of the figures already given, but it is worth putting this into some perspective. The medical technology market is in the region of US$100 billion annual sales, around 75 billion euros in the European Union and the equivalent of US$25 billion in Japan. Out of the US $2000 annual total spend per capita in health care, some $150 goes on this technology in the USA. Many other countries of course do not spend anything like these amounts but some are catching up. In the so-called BRICS (Brazil, Russia, India, China, South Africa) countries, Brazil spends US$13 per capita on health care, Russia US$17, India US$2 and China US$3, but these have high growth rates and many countries have their own medical technology industries. The markets for some individual products are themselves huge, the pacemaker market being over US$4 billion per year, for example. Some three million cataract procedures are performed each year in the USA, with lenses costing between US$1500 and US$3000 each. And of course it is not just the major devices themselves that we have to consider. Each surgical operation, of which there may be more than 15 million performed annually in both the USA and Europe, will require packs of sutures, drains and catheters, swabs and gauzes, dressings, surgical gloves and so on.

There is, therefore, a major impact on the delivery of health care to be derived from these products, and on both manufacturing and service industries.

The products of these technologies, which include the joint replacement prostheses, the cardiac pacemaker, the intravascular stent, the cartilage regeneration system, the intraocular lens and the incontinence device, are almost always manufactured devices that have to be made of one or more materials. The requirements of these materials will naturally vary from one situation to another, but these can be distilled into a small number of crucial issues: they have to perform the desired function efficiently and for the intended duration; they must do the patient no harm; and they have

Box 1.2 **The generic requirements for biomaterials**

| | | |
|---|---|---|
| Functionality – *optional, depending on application* | *General* | Volume, size, shape<br>Surface characteristics |
| | *Mechanical* | Elasticity/rigidity<br>Strength/fracture resistance<br>Tribological properties (wear, friction)<br>Stress transfer to cells/tissues |
| | *Physical* | Electrical properties<br>Thermal properties<br>Optical/optoelectronic properties<br>Magnetic properties |
| | *Chemical* | Control of biostability/biodegradation |
| | *Biological and pharmacological* | Control of cell phenotype<br>Control of molecular targeting<br>Pharmacokinetics/pharmacodynamics |
| Safety – *mandatory* | *Intrinsic biocompatibility* | Appropriate local host response<br>Control of cytotoxicity<br>Absence of remote or systemic adverse effects |
| | *Clinical application* | Technique sensitivity<br>Patient sensitivity |
| Practical features – *variable importance* | *Supply* | Suitability for quality manufacturing<br>Sterilization and infection control |
| | *Economics* | Acceptable costs of goods<br>Appropriate business models |
| | *Regulatory* | Absence of insurmountable hurdles |
| | *Ethical* | Absence of insurmountable hurdles |

to be deliverable to the patient in an efficient and cost-effective way (see Box 1.2). It is, hopefully, obvious that we cannot just use any readily available household material for these purposes; they have to be chosen and handled with great care. The word we use for these special substances is "biomaterial." Biomaterials are special and, obviously, immensely important. The scientific, medical and technological details that underpin their performance constitute a major subject in its own right. We call this area "*biomaterials science*" and this book is concerned with all of these aspects.

## 1.3 The biomaterials concept

It is worth considering why we use the term "biomaterial" and what are the implications of so doing. There are several potential meanings of the word, since the prefix "bio" simply means pertaining to life and the noun "material" is usually taken to mean a substance useful for making things. Biomaterial could therefore, quite logically, mean *either* a living substance useful for making things *or* a substance used for making things that are relevant to life or living systems. As with many recently introduced technical terms, popular usage more often determines the meaning of a word than strict logic. Although some people refer to biomaterials with the former meaning, it is almost universally accepted that it is the latter meaning, involving substances that are used for making things that are relevant to living systems, that is correct, as we discussed earlier in this chapter. Even this, however, is rather vague, and it would be very helpful to have a little more focus on which living systems and what manner of relevance should be involved. Again popular usage has provided some answers here, and it generally accepted that the living systems are primarily humans and that it is the maintenance of the health of humans that is relevant to the use of the word, again as we have alluded to earlier.

This book is concerned with those materials that are used in this context of making things that directly impact on human health. We have to be even more precise, however, if we wish to define an area about which a textbook can be written, since it is easy to contemplate some objects, and the materials from which they are made, which impact on human health but which we do not consider to relate to biomaterials. The glass from which a pair of spectacles is made is obviously beneficial to health, but we do not consider this to be a biomaterial. Similarly, the materials used to construct high-performance footwear for athletes give protection against injury, and therefore assist in maintaining health, but they are not biomaterials. Over the last several decades this issue has been partially resolved by the restriction of the word biomaterial to those situations in which the material exerts its beneficial effect on human health by specific, direct and intentional contact with the tissues of the human body, and where the nature of that contact is instrumental in determining the performance of the object that is intended to impact on health. The glass of the spectacles does not contact the eye and is not therefore a biomaterial. In contrast, however, the material used to make a contact lens, by definition, contacts the eye, as does the material of the intraocular lens discussed previously, and these are, therefore, biomaterials. We shall see in this book just why this difference is important.

A number of years ago, biomaterials scientists at a Consensus Conference tried to solve the conundrum of exactly what is a biomaterial, and concluded that a biomaterial was "*a material intended to interface with biological systems to evaluate, treat, augment or replace any tissue, organ or function of the body.*" We shall see in the next section that our views have changed a little in recent years, but this is a good starting point.

It should be noted that "biomaterial" is often used interchangeably with the phrase "biomedical material." As we shall see later, there is a little more logic with the term biomaterial, since "biomedical material" is more restrictive and does not have the flexibility to deal with some emerging technologies, but this is not of any great concern as long as both are used with common sense. It is also notable that there has been little logic to the development of terms that are subsidiary to biomaterial. The term "bioceramic" has a directly analogous meaning to biomaterial and is used when the material in question is a ceramic. However, the same usage does not apply to "biopolymer," which instead means a naturally derived polymer, with no implied direct connection to health; either of the phrases "biomedical polymer" or "polymeric biomaterial" are usually reserved for the analogy to biomaterial. To add to the confusion, there is little usage of the word "biometal," while "biocomposite" is used to mean either a composite biomaterial or a naturally derived composite material such as bone.

## 1.4 The evolution of biomaterials science and the changing concepts

To explain the subject matter of this book in more detail, we should consider how biomaterials science has evolved and in which direction it is now heading. The most significant uses for biomaterials have always been in the construction of devices for placing within the human body. Initially, before the word biomaterial was introduced, these were objects that were intended to assist in a healing process, such as a suture, ligature or bone fracture plate. They were simple structures that utilized a variety of common materials and the performance was very limited.

### 1.4.1 Early implantable devices

It was in the 1940s and 1950s that the potential for biomaterials to replace parts of the body, or replace some function of the body, was realized, as the medical profession saw that the quality of life of many individuals was seriously compromised by the loss of function of parts of the body through structural changes to the tissues as they age. Orthopedic surgeons had been struggling with the increasing incidence of arthritis, especially of hips and knees, and the concomitant pain and reduced mobility as both cartilage and bone degenerated. Cardiac surgeons witnessed increasing levels of heart valve malfunction as the tissue leaflets failed to open or close properly on account of their poor condition, leading to major difficulties with even moderate exercise. Their colleagues, the vascular surgeons, saw significantly raised levels of vascular disease as vessels closed up and became stenotic due to the plaque built up within their lumens, leading to vascular insufficiency and, frequently, the need for amputation. Ophthalmic surgeons noted the effects of cataracts in the ageing population, leading to blindness as the lens in the eye degenerated and clouded over.

Many of these surgeons and physicians appreciated that if the offending degenerated part of the body, that is, a hip or knee joint, a heart valve, an artery, a lens, could be removed and replaced with some appropriate material or structure, the quality of life could be profoundly improved. It should be noted that the actual function of these diseased parts was quite simple in concept and always either mechanical or physical, and never biological. The question was always the same: what material and what structure could be developed to satisfactorily perform in a situation where nature's own material, optimized over hundreds of thousands of years of evolution, had ultimately failed? It was at this point that biomaterials science became established and the pathway to the development of specially designed biomaterials, and exquisitely engineered medical devices, was initiated. Soon along came the total hip replacement, the artificial heart valve, the synthetic vascular graft and intraocular lens. There then followed an extensive period in which the materials and designs improved, clinical outcomes, generally, but not always, were better, and areas of application were extended. Some 50 years after the creation of biomaterials science and implant surgery, a wide array of implantable devices had become available to a wide range of physicians, aimed at an increasingly diverse range of medical conditions, and often with exceptionally good performance and durability, as discussed in detail in Chapter 4. It will become clear that the optimal route to success as far as the biomaterial is concerned involves the selection of those materials that have the greatest resistance to attack by the very aggressive physiological environment of the human body, coupled with the appropriate mechanical and physical functionality, taking into account a number of pragmatic factors such as ease of manufacturing and sterilization. Much of this book is concerned with the materials, the devices and the clinical conditions and outcomes associated with the traditional concepts of biomaterials for long-term implantable devices.

However, the story is not solely about such devices and the reality is that as medicine itself has evolved beyond all recognition in the last 15–20 years, so has the role of biomaterials. It is no longer just within the provenance of long-term implantable devices but has extended to many new areas of medical technology. In this book we will discuss the enormous benefits that have been, and will continue to be, derived from these medical technologies and devices, as well as the difficulties that have either been overcome or which remain. Whatever the situation, however, and whatever success has been achieved, one simple fact will always remind us of the limitations of these devices. As noted earlier, the functions of the first implantable devices were mechanical or physical, and not biological. That situation remains today. The treatment of a disease by the replacement of living tissue with a synthetic non-living material will always be limited to those situations where the loss of function is structural rather than chemical/biochemical. We may use implants to sustain and transmit mechanical forces, control fluid flow, apply electrical stimuli, conduct electromagnetic radiation, fill spaces and alter contours, but we cannot use them to permanently control active biochemical pathways or cellular activity.

### 1.4.2 The emergence of regenerative medicine

In treating diseases, we can now consider that we have several different options, mostly concerned with the technologies of regenerative medicine. As we shall discuss in detail in Chapter 5, adult humans have very limited powers of tissue regeneration. Several species of animal, including the salamander and newt, have the capacity to regenerate some of their tissues, especially limbs. During the evolution of mammals, such capacity appears to have been lost, at the expense of vastly increased brain power, and we have been left with the possibility of limited repair of some tissues that might be regularly compromised. Skin is often breached and we do regenerate skin, otherwise we would be continually prone to infection. Bones often break and we are able to regenerate bone tissue, otherwise our mobility would be continuously restrained. With most other tissues of the body, spontaneous regeneration is not possible. We do not spontaneously regenerate cartilage to compensate for arthritic changes, nor do we regenerate most organs, vessels, bladders, kidneys, eyes, teeth or most other tissues when they become diseased or injured. Nevertheless we did have the ability to generate those tissues and organs in the first place, as embryos and fetuses, and to grow those tissues during infancy and adolescence, so the mechanisms were available. Regenerative medicine is concerned with recapitulating those mechanisms for the purposes of forming new tissues and organs in adults, when required. It should be immediately obvious that the process of tissue regeneration is very different to tissue replacement by synthetic structures, and offers the possibility of treating diseases that are not amenable to replacement by implants and the possibility of superior performance in some circumstances.

Regenerative medicine is in its infancy and does not have the same history as implantable medical devices. It is conceivable that it may not fulfill the potential that many people claim at this time. However, many of the underlying concepts have been demonstrated to be correct and feasible, and clinical indications and applications are increasing at a significant rate. The question naturally arises as to whether the advent of regenerative medicine will bring with it a reduced use of, or interest in, biomaterials and implantable devices. Time will tell, but we should note that the very considerable successes achieved with many types of implanted device, and the accompanying cost effectiveness of these established procedures, suggest that regenerative medicine will not be able to displace them in clinical practice, at least for a long while. We did see earlier in this chapter that there are some techniques to facilitate small areas of cartilage to regenerate, and this is a beginning. The treatment of Parkinson's disease may well be changed if we can actually persuade the dopaminergic cells to regenerate, and it should be eventually possible to cure diabetic patients of their difficult decubitus ulcers by regeneration of the skin and the underlying vascular tissue.

More importantly, many of the routes towards regeneration themselves involve the use of biomaterials. There are two important routes, one that is referred to as cell therapy and a second that is termed tissue engineering. In order for the regenerative process to take place, selected cells, at specific sites, have to be activated in order to produce new tissue or tissue functions. As the names imply, cell therapies involve the use of donor cells that are placed at the appropriate site where they are able to carry out the function previously supplied by the now diseased or injured tissue. Biomaterials are not necessarily integral to cell therapy. On the other hand, tissue engineering involves the development of macroscopic pieces of tissue. Such a process still needs the appropriate cells, but they have to be organized and stimulated, and the new tissue has to be formed or shaped. Typically a biomaterial is used as a template within which the new tissue is generated, usually involving a porous structure or a gel, and this process may be achieved external to the body in some form of bioreactor, or within the body itself. The stimulation of the cells may be achieved in a number of ways, the methods broadly being either molecular or mechanical, and it is usually a requirement that the template facilitates the delivery of these stimuli. It is self-evident that in the process of tissue regeneration, the material template should simultaneously degrade. In contrast, therefore, to the specifications for a biomaterial for a long-term implantable device that involve inertness and, generally, the lack of interactivity with the tissues of the body, the specifications for a biomaterial in regenerative medicine are associated with positive interactions with cells and deliberate but safe degradation. It is no wonder that the science and scope of biomaterials has changed so much recently.

### 1.4.3 Nanotechnology, diagnostics and drug delivery

The story of biomaterials science in the twenty-first century does not stop here, however, and we have to consider how recent developments have caused us to reassess even what constitutes a material within medical technologies, especially in the context of the nanotechnologies. Pharmaceutical products have not been considered as biomaterials since they are essentially soluble molecules that are metabolized in the body, rather than structural materials. Although many drugs can be administered to patients as solutions, many are prescribed in the form of tablets, but, even though the tablet is in solid form, we do not normally consider this to be a biomaterial since the tableting constituents are purely ancillary to the drug itself.

It is well known that the optimal dosing of patients with pharmaceuticals is a difficult process since the periodic administration of a tablet, say by mouth, neither directs the drug to the site of need or effect, nor does it result in a consistent systemic level of the drug because of the continuous metabolic processes that occur between doses. Biomaterials have long been used in techniques of controlled drug release and drug targeting that are intended to overcome these problems. As discussed in Chapter 6, traditional drug delivery devices have generally been solid objects that contain a drug, the latter being released from the former by one of a variety of mechanisms, for example, erosion, diffusion, biodegradation, osmosis or electromechanical pumping. However, as the emphasis has been moving more towards drug targeting, in which the molecules of the pharmaceutical are directed specifically to the tissues or cells where they will exert their specified effect, for example as discussed earlier with a chemotherapeutic agent targeted to cancer cells, these molecules are being combined with agents that have an affinity for the targets. Increasingly commonly, these agents are in the form of nanoparticles or nanostructured materials. As well as delivering drugs more accurately and effectively to the body, these new pharmaceutical technologies are also being used to deliver or suppress genes in the quest to treat some very important diseases.

In addition the techniques that doctors use to diagnose disease and injury have become far more sophisticated, with magnetic resonance imaging (MRI), positron emission tomography (PET), ultrasound, computed tomography (CT) and others being added to the list of essential diagnostic instrumentation. In many of these cases, the information is obtained through the use of substances, usually referred to as contrast agents, that enhance the interaction of the energetic system of the machine with the relevant tissues. Such substances may be administered to the patient prior to the test.

The question arises as to whether the agents used for targeted drug or gene delivery, and those used in imaging and diagnostic systems, can be considered as materials and whether, therefore, these processes involve biomaterials, and biomaterials science. Can a liquid, or a bubble, be an object; can a nanoparticle be a material; is a quantum dot used for cell marking a medical device? These are the fundamental questions that we have to answer in any quest to determine and explain the science of biomaterials. In this book, we shall be as expansive as possible in answering these types of question.

### 1.4.4 The current definition of a biomaterial

The concept of biomaterials science, and the range of situations that may be covered by the word "biomaterial" have been recently re-examined; this book will follow and discuss the definition that arose out of this analysis:

> A biomaterial is a substance that has been engineered to take a form which, alone or as part of a complex system, is used to direct, by control of interactions with components of living systems, the course of any therapeutic or diagnostic procedure.

## 1.5 The classification of biomaterials applications

At this stage it will be useful to summarize the applications in which we use biomaterials and the general types of specifications we may wish to identify for our optimal biomaterials when used in these applications. The different classes of biomaterials applications are given in Box 1.3. It will become obvious that the structure of this book is partly based on this classification.

It is important to reflect on this classification at this stage because it reinforces the fact that some biomaterials have only one significant clinical application but others are used in multiple applications. Thus, titanium alloys are used in Classes 1.1, 1.2, 1.3, 1.4, 1.5, 2.1, 4.1 and 4.2. Polyurethanes may be used in Classes 3.1, 4.1, 5.1 and 7.3. Biodegradable polyesters are used in Classes 2.1, 2.2, 6.4, 6.5 and 7.3. It is a major intention of the style and content of this book that the properties of individual biomaterials are set out to explain why and how they are, or could be, used in these specific and quite often disparate applications.

## 1.6 The components of biomaterials science

We noted above that the requirements of these biomaterials will naturally vary from one situation to another, but these can be distilled into a small number of crucial issues; they have to perform the desired function efficiently and for the intended duration, they must do the patient no harm and they have to be deliverable to the patient in an efficient and cost-effective way.

In order to explain the science behind the ability of materials to carry out these functions, we need to consider the essential features of the materials themselves, the essential features of the tissues and biological systems they come into contact with, and the possible ways in which materials and tissues can interact. These issues can be summarized as follows, shown schematically in Box 1.4.

Looking back at Box 1.2, we know that biomaterials have major functional requirements with respect to the applications referred to in the previous section. The properties that allow these materials to meet these requirements are largely determined by the structures determined by conventional materials science, as we

### Box 1.3 The classification of biomaterials applications

Class 1 Permanent (long-term) implantable devices
Class 1.1 Permanent implantable devices for the anatomical replacement of parts of the body that have undergone some form of degenerative disease
Class 1.2 Permanent implantable devices for the anatomical replacement of parts of the body that have undergone surgical removal of cancerous tissue
Class 1.3 Permanent implantable devices for the correction of congenital or developmental deformities
Class 1.4 Permanent implantable devices for the restoration or correction of function after injury
Class 1.5 Permanent implantable devices for the restoration or correction of function as a consequence of disease
Class 1.6 Permanent implantable devices for cosmetic purposes

Class 2 Short-term implantable devices
Class 2.1 Implantable devices to assist in the repair of broken bones
Class 2.2 Implantable devices to assist in the repair of soft tissue

Class 3 Invasive but removable devices
Class 3.1 Indwelling catheters and shunts
Class 3.2 Contraceptive devices

Class 4 External artificial organs/organ assist devices
Class 4.1 Devices attached to the patient that deliver short-term support
Class 4.2 Devices attached to the patient that act as a bridge to transplant or life-long support

Class 5 Surgical and clinical accessories
Class 5.1 Wound dressings
Class 5.2 Short-term catheters and drains

Class 6 Drug and gene delivery systems
Class 6.1 Oral drug delivery systems
Class 6.2 Infusion systems
Class 6.3 Systems for delivery across epithelial/mucosal surfaces
Class 6.4 Monolithic implantable devices
Class 6.5 Microparticulate and nanoparticulate systems
Class 6.6 Prodrugs and polymer therapeutics
Class 6.7 Antimicrobial systems
Class 6.8 Immunotherapy and chemotherapy hybrids
Class 6.9 Non-viral gene vectors

Class 6.10 Engineered viral vectors
Class 6.11 Vaccine delivery systems
Class 6.12 Theranostic systems

Class 7 Tissue engineering systems
Class 7.1 Engineered cell therapy products for regenerative medicine purposes
Class 7.2 Engineered gene therapy products for regenerative medicine purposes
Class 7.3 *Ex vivo*/bioreactor generated tissue constructs
Class 7.4 Cell seeded implanted scaffolds or templates
Class 7.5 Cell-free implanted scaffolds or templates
Class 7.6 Injectable cell seeded products
Class 7.7 Injectable cell-free products
Class 7.8 Cell sheet engineered constructs
Class 7.9 Engineered systems for drug discovery and testing
Class 7.10 Engineered tumor models

Class 8 *In vivo* diagnostic systems
Class 8.1 MRI contrast agents
Class 8.2 Ultrasound contrast agents
Class 8.3 Fluorescence and bioluminescene imaging systems
Class 8.4 Contrast enhanced CT systems
Class 8.5 Implantable biosenors

shall see in the next chapter. The critical issues of biomaterials science are based on the need to take these functional materials and place them into the human body or components of the body.

Let us consider the simplest of all situations in which a biomaterial comes into contact with tissues of the body, and then consider how they could interact once the biomaterial and tissue are in contact. This simplest configuration involves on the biomaterial side a single-phase, isotropic, homogeneous, chemically defined substance. The significance of these terms will become clearer in the next chapter but together they mean that this material has one chemical structure and configuration only. It is like saying we have a piece of absolutely pure gold, or a single crystal of diamond, or a slab of pure uncontaminated ice. For the tissue, we can simply assume it is normal, healthy and uncontaminated, such as a piece of muscle, fat or bone. We can ignore for the moment just how the material found its way into the tissue – the complications that arise from this process of entry will be considered later.

Exactly what would happen in this hypothetical situation will depend on a number of factors; these events constitute the phenomena of biocompatibility and are discussed in detail in Chapter 3. Listed in Box 1.4 are the events that are likely to occur in the short term with all materials and all tissues to a greater or lesser extent. The phrase "in the short term" is used here as a caveat since biocompatibility phenomena are interactive and time dependent. We start this discussion with the assumptions that they are largely independent

## Box 1.4 Critical issues in biomaterials–tissue interactions

**The baseline biomaterial configuration:**
**The material is a single-phase, isotropic, homogeneous, chemically defined substance**
(i.e the material has one simple chemical composition and one configuration)

Interaction 1
There is a thermodynamically driven adsorption of tissue components onto the material surface – *components of blood, extracellular fluid, urine, saliva, tears, etc. immediately attach themselves to the biomaterial surface.*

Interaction 2
The tissue responds to the presence of the material – *a non-specific response of the body to invasion by a foreign object, mediated by size, shape and surface characteristics.*

Interaction 3
The tissue responds to the physical characteristics of the material – *biophysical processes influence the relationship between the biomaterial surface and the tissues.*

Interaction 4
The tissue and material interact mechanically – *mechanical forces significantly influence the development of the longer term response from the tissue and its components and also the response of the material to these components.*

Interaction 5
The material responds to the fluid tissue environment – *the time-dependent response of the material to the aggressive tissue fluids, where deviations from the baseline biomaterial configuration (such as the presence of complex microstructures and additives/contaminants) start to influence the material response, either beneficially or deleteriously.*

Interaction 6
The tissue responds to the chemical nature of the material and any released components – *this is the ultimate determinant of the performance and safety of the biomaterial, and is significantly influenced by compositional and structural deviations from the simple, baseline configuration.*

variables and that we need not consider long-term time dependence just yet.

These six events are broadly speaking in some chronological order. Interaction 1 is the thermodynamically driven adsorption of tissue components onto the material surface. That this should happen is no surprise; we just have to reflect on the ease with which components of blood, sweat and tears stick to material surfaces in everyday life to appreciate these phenomena.

Interaction 2 is the response of the tissue just to the presence of a foreign surface. The internal tissues of the body do not normally come into contact with surfaces that are foreign to them and will respond, by processes that are mediated by surface topographical parameters, and the overall shape of the object. As we shall see in Chapter 3, the tissues of the body are exquisitely sensitive to invasion by anything foreign and it is likely that their normal defensive mechanisms, finely tuned to deal with microorganisms, will be adapted in order to respond to this intrusion.

Interaction 3, probably the least well understood, concerns the response of the tissues to physical characteristics of the material. Tissue fluids are electrolytes and physiological processes may involve bioelectrical phenomena. Materials surfaces may have electrical charge and other electrical properties. It should not be surprising that electrically driven processes, for example in attracting cells to surfaces, play some role. Interaction 4 involves mechanical phenomena. Most tissues need some form of mechanical stimulus to survive and function properly, but only up to a point. Excessive forces or repeated forces may cause tissue to respond in some adverse way, rather like the way a blister or a callus forms on skin due to excessive rubbing. The main issue with a biomaterial in contact with tissues is that they will rarely have equivalent stiffness. A piece of metal in muscle will have a considerable disparity in stiffness such that when the material–tissue complex is subject to motion, there will be relative movement at the interface and the tissues are likely to respond, essentially with their equivalent of a callus.

Interactions 5 and 6 are different mechanisms, but these two are related. They are also time dependent to some extent. Interaction 5 concerns the manner in which the material responds to the presence of the fluid tissue environment. Several processes can take place. The material may absorb components into itself. The material may be soluble and dissolve in these fluids. Depending on the nature of the material it may be corroded or degraded by them; the electrolytic and biologically active nature of tissue fluids make them surprisingly hostile to most synthetic materials. The consequences of these processes may be total or partial destruction of the material, which of course may be intentional in some medical devices such as surgical sutures. In addition, the consequences may contribute to those phenomena included in Interaction 6, which are those effects that the products of these degradation processes have on the tissues.

The essence of biomaterials science is contained within the mechanisms involved in Interactions 1–6, and the derivatives of these interactions that come into play when we take into account the complexities of materials and tissues in the real world, and add interdependence and time factors into the equation. In order to understand these phenomena, we have to consider in some detail the structure and composition of real materials (covered in Chapter 2) and the mechanisms of biocompatibility (in Chapter 3).

## Summary and learning objectives

1. Biomaterials science is concerned with the clinical, scientific and engineering implications of the applications of biomaterials in health care, the word "biomaterial" specifically being defined as "*a substance that has been engineered to take a form which, alone or as part of a complex system, is used to direct, by control of interactions with components of living systems, the course of any therapeutic or diagnostic procedure.*"
2. These procedures cover very many clinical disciplines, ranging from orthopedic and maxillofacial surgery, to cardiovascular

interventions, the treatment of sensory organs and diseases of urinary, gastrointestinal, reproductive and respiratory systems. The procedures, and the materials used within them, may involve the replacement, repair or regeneration of tissues and organs, the control of the delivery of active molecules to patients, or the technologies of diagnosis and imaging.

3. The materials that are encompassed by biomaterials science include metals, polymers, ceramics, synthetic composite, natural tissues, recombinant tissue molecules, and any combinations of these. They may be macroscopic or nanoparticulate, they may be solid or liquid, they may be manufactured or self-assembled *in situ*. There are no chemical or structural limitations to what may constitute a biomaterial.
4. The performance of biomaterials and associated health care products is largely determined by the interactions between the materials and the tissues of the host, within the phenomena of biocompatibility, and the properties that allow the material to perform the intended function.
5. The use of biomaterials and associated health care products is governed by systems that encourage disease prevention, early stage diagnosis of conditions, and judicious use of drugs and rehabilitation therapies. Their use should be seen in the context of complementary approaches to the maintenance of health in patients. The skill and knowledge of health care professionals is of paramount importance to the successful and effective performance of biomaterials.

## Questions

1. Describe current treatment options for patients with a degenerating intervertebral disc and the relative roles of biomaterials-based devices, drugs and cell therapies.
2. Sports injuries and arthritis are the main conditions affecting knee joints. Describe the causation of these conditions and the need for regeneration or replacement technologies in each case.
3. Discuss the influence of drugs, implantable devices and cell therapies during the different stages of Parkinson's disease.
4. What are the essential biomaterials characteristics required for intravascular stents?
5. Describe the current technologies used for the diagnosis and treatment of breast cancer and discuss how these could be improved or replaced in the future.
6. How do health economic considerations affect the options for treating patients suffering from incontinence?
7. Discuss the evolutionary trends that have taken place over the last decade in biomaterials science.
8. Describe how nanotechnology has influenced the scope of biomaterials in health care applications.

## Recommended reading

(* Indicates a textbook.)

* Ducheyne P, Healey K E, Hutmacher D, Grainger D E, Kirkpatrick C J (eds.) *Comprehensive Biomaterials*, Elsevier, Amsterdam, 2012. *Six-volume, 4000-page reference work on biomaterials.*

Huebsch N, Mooney D J, Inspiration and application in the evolution of biomaterials, *Nature* 2009;462:426–32. *Insightful article on the increasing sophistication of biomaterials.*

* Park J, Lakes R S, *Biomaterials: An Introduction*, Springer, New York, 3rd edition, 2007. *Basic level textbook.*

* Ratner B D, Hoffman A J, Schoen F J, Lemons J E (eds.) *Biomaterials Science: An Introduction to Materials in Medicine*, Elsevier, Amsterdam, 3rd edition, 2012. *Multi-author collection of reviews and technical chapters covering biomaterials and their applications.*

* Temenoff J S, Mikos A G, *Biomaterials: The Intersection of Biology and Materials Science*, Prentice-Hall, 2008. *Textbook aimed at second- and third-year undergraduates.*

Williams D F, On the nature of biomaterials, *Biomaterials* 2009;30:5897–909. *Essay on the changing concepts, scope and applications of biomaterials.*

In addition, the contents of the following journals should be routinely monitored, listed in order of relevance:

*Biomaterials*, www.journals.elsevier.com/biomaterials/

*Acta Biomaterialia*, www.journals.elsevier.com/acta-biomaterialia/

*Nature Materials*, www.nature.com/nmat/

*Advanced Materials*, http://onlinelibrary.wiley.com/journal/10.1002/ (ISSN)1521-4095

*Biomacromolecules*, http://pubs.acs.org/journal/bomaf6/

*Journal of Biomedical Materials Research, Part A*, http://onlinelibrary.wiley.com/journal/10.1002/ (ISSN) 1552-4965

# 2 Essential materials science

In this chapter you will be introduced to the principles of the structure and characteristics of materials in general, and the specific features of the main classes of practical materials. These are metals and alloys, the different forms of polymer-based materials, ceramics and glasses, composite materials and natural materials. This is followed by discussions on how these structures give rise to the specific properties of these material classes, with an emphasis on the mechanical and physical properties and the chemical stability of materials in various environments, especially the physiological environment. Attention is given to many different specialized materials that are now used in health care, including nanocomposites, quantum dots, polymeric micelles, dendrimers, hydrogels and biopolymers. The objective is to allow you to understand why these different materials have their own properties and how biomaterials can be designed to meet the very critical performance specifications required in medical technology.

## 2.1 Introduction

It is important to start reading this chapter with no preconceived ideas of what a material should look like or how it should perform. In order to understand and appreciate the high-performance, esoteric materials that are used in advanced engineering applications of the twenty-first century, including medical engineering, it does not help to have fixed in our minds the idea that a material has to look and behave as if it were a macroscopic, solid object that is made by some conventional manufacturing process and which we can hold in our hand and examine visually. Most of today's sophisticated materials do not behave in a similar manner to the more traditional steel, plastic, textile, glass, concrete or wooden structures that have been the mainstay of materials engineering for many decades. We should not be constrained by concepts of state (materials do not have to be solid), of size (they may be macroscopic, microscopic or of nanoscale dimensions), of activity (they do not have to be inert but may be intentionally active, or even living) or of permanence (they may be intentionally biodegradable). They do not have to be manufactured by conventional means but may be formed *in situ* by self-assembly. In other words, a collection of intensively active, macromolecular self-assembled nanoparticles is just as much a material as the piece of forged titanium that constitutes the bulk of a total hip replacement prosthesis.

This does not mean to say that we should ignore or forget the basic rules of materials science, and indeed it is important that we build on the basic fundamentals and modify these as we describe the nature of the materials in use today, especially in the form of biomaterials. This chapter is concerned with the essentials of materials science and how these are adapted to the special and specific applications in medicine. To do this, following on from the end of the previous chapter, we start with a very simple model of a material, describing the basic options for their atomic/molecular structure and set the scene for the description of how these structural features determine the properties of the material. We then consider how these basic structures can be modified and adapted in order that these properties can be controlled to enhance and optimize their performance, especially in health care products.

Crucial to our understanding of the current direction of biomaterials science is the explanation of the marked shift in emphasis in the requirements of biomaterials. For several decades, as biomaterials were used in increasing numbers of implantable devices, the demands on these materials in terms of mechanical properties and corrosion/degradation resistance were considerable, leading to the use of so-called high-performance engineering materials such as the highest quality stainless steels and titanium alloys, strong and ductile plastics, carbon fiber composites and toughened ceramics. We will explain here how these evolved and how, from structural and mechanistic perspectives, they achieved their properties and their status.

The world of biomaterials applications then started to change, with much more emphasis being placed on non-mechanical objectives, such as chemical and biochemical functionality and environmental responsiveness, and sophisticated electronic, optoelectronic, fluorescence and magnetic properties. Moreover, the use of these new materials is not necessarily predicated on conventional surgical procedures for them to be placed in the body; it is far more likely that they will be delivered through minimally invasive techniques or by transport across intact epithelial membranes or even by self-assembly or curing *in situ*. We need to use the same basic model of materials structure but adapted to the highly specific needs of medical technology in order to explain how these properties and characteristics can be achieved.

## 2.2 The basic materials model

The previous chapter ended with a discussion about the simplest model of a material, the "single-phase, isotropic, homogeneous, chemically defined material" and analogies were given to this as a material that has one chemical structure and configuration only, such as a piece of absolutely pure gold, or a single crystal of diamond or a slab of pure uncontaminated ice. Let us take the gold as an example of how this very simple structure can be modeled. Gold is an element. The expression "absolutely pure gold" explicitly states that the piece of gold we are looking at consists solely of atoms of gold, an atom being the smallest component of an element having the chemical properties of that element. Obviously, since the radius of a gold atom is around 0.15 nm, there are very many of these atoms in a tangible or visible piece of gold, and the first level of the materials model is concerned with how all of these atoms are arranged. As it happens, with gold this is an easy matter, for these gold atoms are arranged in a very regular manner, with considerable symmetry, in a three-dimensional lattice.

I borrow the phrase "chemically defined" from the technology of tissue or cell culture, where a chemically defined medium is that solution where the chemical nature of all the ingredients and their amounts are well known and characterized. There are many substances and materials where this is not necessarily so, and being chemically defined is a useful concept when dealing with many aspects of biomaterials science. As this is pure gold, we have to consider it to be chemically defined. Homogeneous means uniform in nature; its use with pure gold is unnecessary but it is important to recognize that

## Box 2.1 Glossary of terms

Most of these terms are discussed at some length in this chapter, but are included here for completeness and ease of reference. The definitions have been derived from a number of sources.

**additive** a substance added to polymers to alter certain properties.

**amorphous** description of a material without the periodic ordered structure of crystalline solids.

**amphiphilic** relating to one or more molecules in a biological membrane having a polar water-soluble terminal group attached to a water-insoluble hydrocarbon chain.

**atom** smallest particle of an element that can take part in a chemical reaction; the smallest unit of an element, having all the characteristics of that element and consisting of a dense, central, positively charged nucleus surrounded by a system of electrons.

**biodegradation** the breakdown of a material mediated by a biological system.

**biostable** said of a material, usually a polymer, which resists chemical or structural degradation within a biological environment.

**combinatorial chemistry** an approach to the synthesis and characterization of materials using automated techniques to generate large libraries of elemental combinations on a microscopic scale.

**composite material** structural material made of two or more structurally different materials, where each component contributes positively to the final properties.

**covalent bond** chemical bond in which two or more atoms are held together by the sharing of some of their outer electrons.

**crystal** (1) a solid with a regular polyhedral shape; (2) homogeneous solid displaying an orderly and repetitive steric arrangement of its atoms.

**dipole** pair of separated opposite electric charges.

**electron** fundamental particle with negative electric charge that is a basic constituent of the atom.

**electrophoresis** technique for the analysis and separation of colloids based on the movement of charged colloidal particles in an electric field.

**fluorophore** component of a molecule that causes that molecule to fluoresce.

**glass** a hard, amorphous, brittle substance, made by fusing together one or more of the oxides of silicon and certain other oxides, and cooling their product rapidly to prevent crystallization.

**glass-ceramic** ceramic produced to final shape as a glass and then induced to crystallize by controlled thermal treatment.

**glass transition temperature** the temperature at which an amorphous solid becomes soft upon heating or brittle upon cooling.

**hydrophilic** having an affinity for water.

**hydrophobic** not readily interacting with water.

**interface** sharp contact boundary between two substances, materials or phases, either or both of which may be solid, liquid or gases.

**ion** any atom or molecule that has resultant electric charge due to loss or gain of valence electrons.

**ionic bond** chemical bond between two or more oppositely charged ions.

**isotope** one of two or more atoms of the same element that have the same number of protons in their nucleus but different numbers of neutrons.

**isotropic** having the same value of a property in all directions; *anisotropic* refers to a material whose properties vary with direction or orientation.

**luminescence** emission of light by a substance for any reason other than a rise in its temperature.

**macromolecule** a single, very large molecule.

**metabolism** the sum of all the physical and chemical processes by which living, organized substance is produced and maintained.

**metallic bond** the force that holds like ions together in a metallic structure, derived by the attraction between the positively charged ions and the negatively charged free electrons.

**molecule** finite group of atoms that is capable of independent existence and has properties characteristic of the substance of which it is the unit.

**optoelectronics** the branch of physics that deals with the interconversion of electricity and light.

**orbital** the wave function of an electron in an atom or molecule, indicating the electron's probable location.

**plastic** generic name for a material based on an organic substance that, under heat and pressure, can be shaped or cast into moulds, extruded or used in a variety of other processes.

**quantum dot** a quantized electronic structure in which electrons are confined with respect to motion in all three dimensions.

**resorption** the assimilation of a substance in the body.

**sol–gel process** chemical route to glass formation, using reactions in solution to produce a gel precursor, which yields a glass on drying out, thus avoiding the melting stage.

**steric** pertaining to the arrangement of atoms in space.

**stoichiometry** anything associated with the quantities of substances that enter and are produced during chemical reactions.

**supramolecular** relating to or consisting of more than one molecule.

**tension** the condition of being stretched.

**textile** term used to describe any fiber, filament or yarn, and the products produced from them.

**topography** delineation of the natural and artificial features of an area of a surface.

**ultrasound** sound waves that have a frequency above the limit of human perception.

**upconversion luminescence** the non-linear optical process that converts two or more low-energy pump protons from the near infrared spectral region to a higher energy/shorter wavelength output.

most real materials are far from homogeneous in the sense that they combine many different components that are non-uniformly, or heterogeneously, mixed. Isotropic means invariant with respect to direction. This will not necessarily be so with real materials, in which case they are said to be anisotropic. This is very important in biomaterials science since the vast majority of tissues, which we may be trying to emulate, regenerate or interrogate, are very complex heterogeneous anisotropic structures. This allows tissues to possess some of their unique properties, for example the mechanical characteristics of tendons and heart valves, and the electrophysiological properties of the retina and peripheral nerves.

Very, very few materials are as simple structurally as a piece of pure gold. Let us look at this two-dimensionally and consider each identical X unit in Figure 2.1(a) to be the smallest identifiable component of a material. We shall see in the next section that X could be an atom or a molecule. Here a number of these X units are arranged in one pattern. One would expect the characteristics of that pattern to depend on the precise characteristics of X and their physical arrangement within it. Now consider that we have another similar pattern, still with the same X units, which comes alongside, as in Figure 2.1(b). If there are many of them, as in Figure 2.1(c), a higher level structure is formed, the overall characteristics now depending not only on the nature of X and the relationships between X units, but also on the relationships between each pattern. This structure will only be strong if all parts are strong; weak relationships between different parts usually lead to weak structures. The two-dimensional schematic of Figure 2.1(c) defines the second level of the materials model, a series of individual patterns, linked together. Depending on the type of material that is involved, these patterns can be called names such as grains, domains or crystals.

Our simple example of a piece of gold was described as a single phase, that is there is only one pattern in the whole of the object; the gold atoms stretch from all sides with the one simple geometrical arrangement, and this is called a crystal because of its

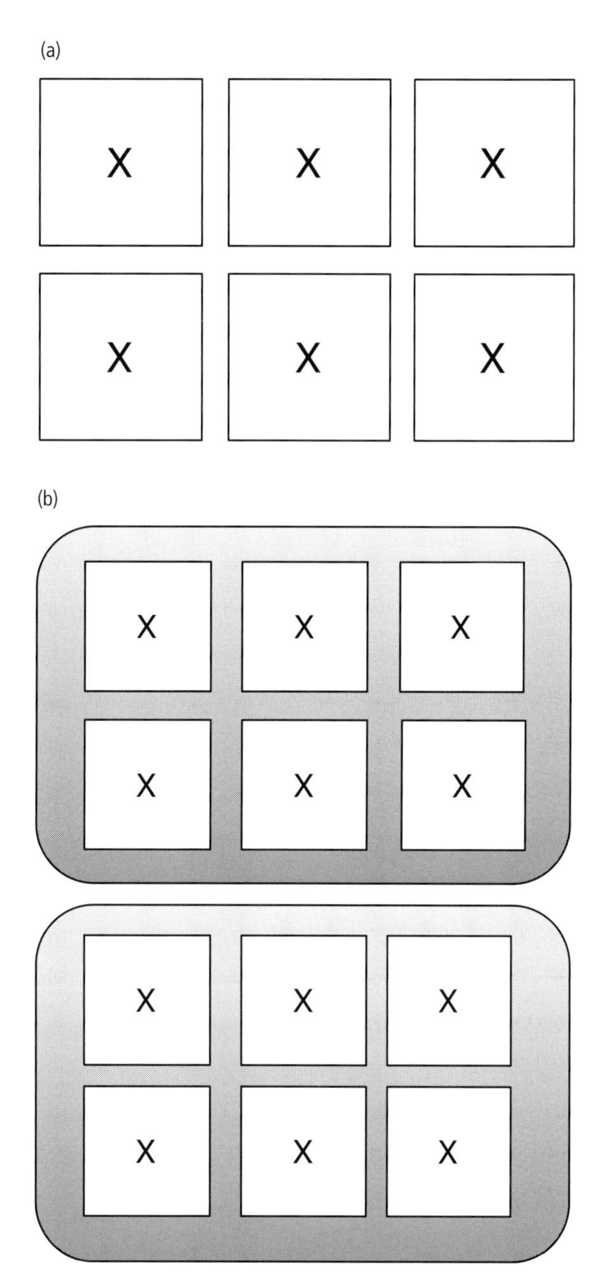


**Figure 2.1**
**(a)** Relationships between structural units in material structure. X is the smallest identifiable component, which could be an atom or molecule – the arrangement of a number of X units gives one aggregate.
**(b)** Two identical aggregates of X in close proximity.

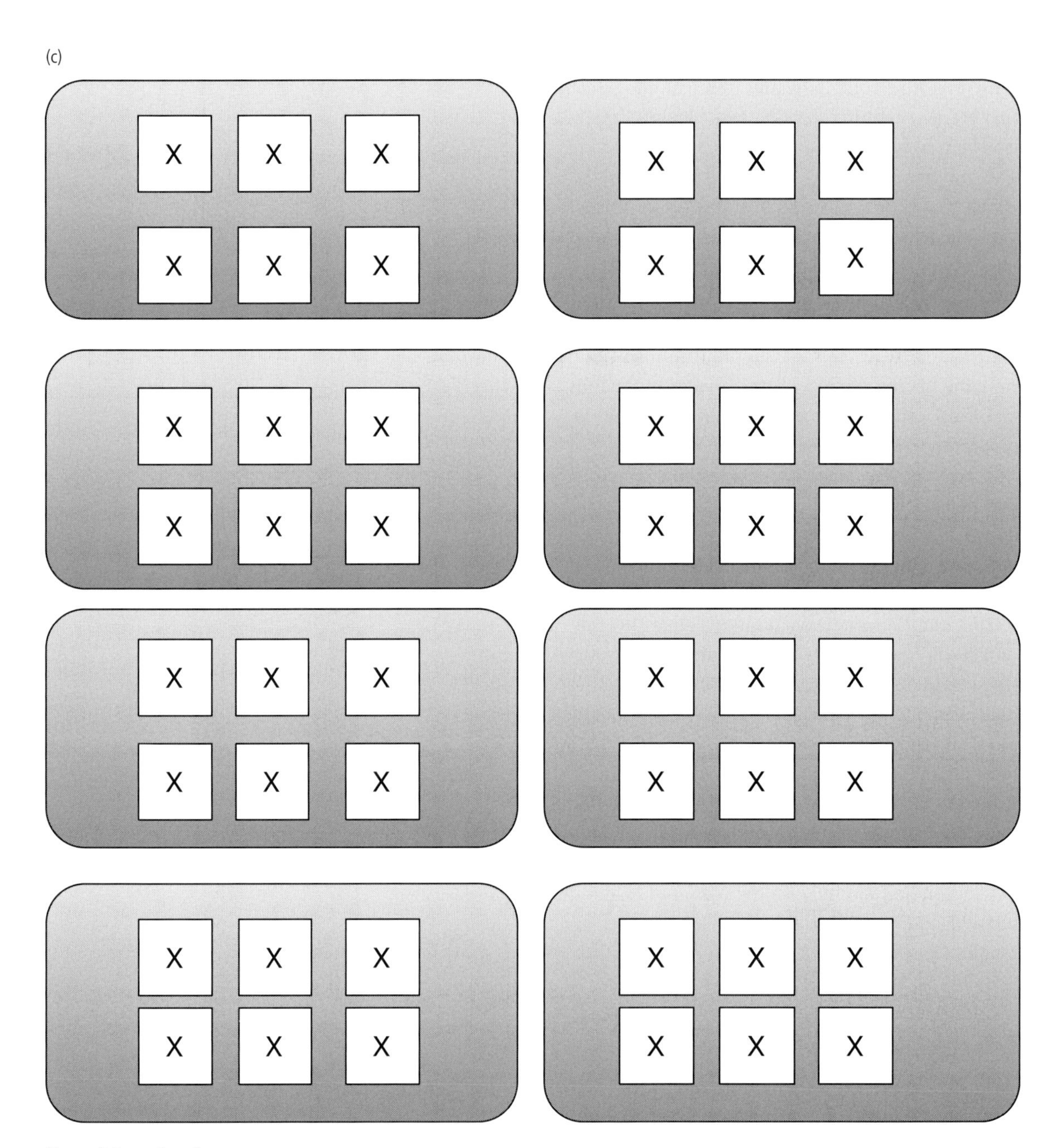


**Figure 2.1** *continued*
(c) A community of aggregates of X.

symmetry. This could be as simple as the situation given in Figure 2.1(a), where X is an atom and rows stretch from one side to another in all three directions. However, most pieces of gold, and indeed most pieces of almost anything, consist of more than one crystal or domain. This is most easily visualized by taking some pure water and allowing it to freeze. Individual crystals of ice form in the water

**Figure 2.1** *continued*
**(d)** Creation of individual crystals in solid ice. At the top left, as water cools, small crystals of ice form; these grow (top right, bottom left) until they impinge and form a solid mass of ice, comprised of a large number of individual crystals.
**(e)** A section through solid pure metal, showing individual crystals, or grains.
**(f)** A micrograph of a section through a titanium alloy, with grains of two different phases being visible.

(g)

**Figure 2.1** *continued*
(g) An anisotropic fiber reinforced composite.

(Figure 2.1(d)) and each grow until they meet up and the whole of the volume of water is solid. Each ice crystal is the same solid water, but each may have a different orientation, morphology and size. It is not so easy to visualize this in gold since it is opaque, but the same thing happens as liquid gold cools down after casting. A two-dimensional (2D) micrograph of a section of a metal such as gold, Figure 2.1(e), shows that it consists of many crystals.

It is also rare for any material to consist of only one chemical component. Most everyday materials, including plastics, ceramics and metals, consist of multiple components, either because it is not possible (or too expensive) to obtain absolutely pure substances, or because better properties are obtained by using mixtures. In the context of their structure, unless the components are mutually soluble in each other, this means that the material will consist of more than one type of microscopic pattern; these are usually referred to as phases. Figure 2.1(f) shows a micrograph of one of the most popular biomedical materials, a titanium alloy, where crystals of two separate and easily identifiable phases can be seen. Such a structure will be microscopically heterogeneous, although if, as in this figure, the phases are essentially randomly distributed and oriented, the material may be isotropic. If one of the phases is distributed with a clear orientation, as with the fiber reinforced composite in Figure 2.1(g), then it will have anisotropy.

So here we have the basic model of a material structure. At one level we have atoms or molecules as the essential fundamental unit. These are brought together to form second level units, the characteristics of which are controlled by the interatomic and/or intermolecular forces. There may be multiple types of second level units and the characteristics of the structure as a whole will depend on the nature of each unit (as crystals, phases, domains, etc.), the relationships between these units in terms of relative proportions, phase sizes, shapes and orientations, and the nature of the boundaries between these units.

So where does this lead us with respect to material properties? If we consider the traditional biomaterials mentioned at the beginning of this chapter, which are analogous to most materials encountered in other engineering disciplines, the main attributes are good mechanical properties such as strength and toughness, combined with good chemical (or environmental) stability, and, where necessary, good thermal properties. All three attributes imply that the structure of the material has to be resistant to applied energy; respectively, mechanical, chemical and thermal energy. Since energy is associated with all bonds between atoms and molecules, and with all interfaces between phases, crystals and domains, it follows that a material will remain stable as long as any applied energy is lower than these components of internal energy. As soon as the external or applied energy becomes greater than the internal energy associated with any one of these features, the material becomes susceptible to instability (Figure 2.2). If sufficient thermal energy is applied to a material it may soften and then melt as this energy overcomes the internal

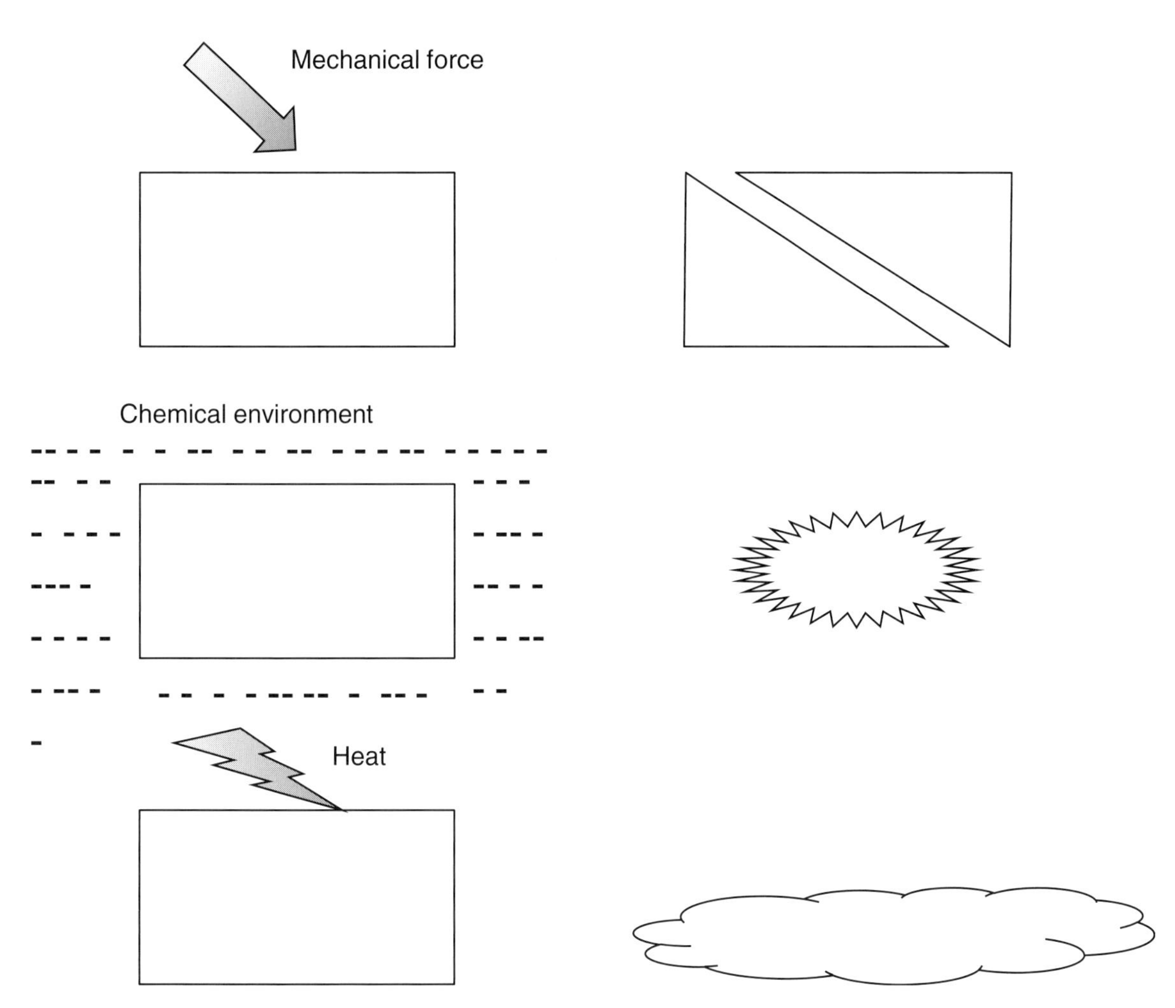


**Figure 2.2** Schematic of the effects of applied energy on a material; mechanical energy may cause the material to fracture, chemical energy may cause it to degrade or dissolve, thermal energy usually causes it to melt.

energy. Chemical energy may cause a material to dissolve or degrade. Mechanical energy may cause deformation and fracture as the applied energy overcomes the energy holding some part of the structure together. These are, of course, broad generalizations, and there is a myriad of mechanisms and individual features that control the behavior of materials in practice. Nevertheless it has been the goal of the materials scientist to create structures that are best able to resist all forms of external energy that they will encounter in service.

As biomaterials science has evolved, different types of attribute have become as, or even more, important. Many applications do not require such stability and inertness, and some functions in diagnostics and drug or gene therapies may be transient. Far from wanting to resist chemical changes, these may be actively desired in many applications that have biological functions in mind. The essential performance characteristics may require subtle electrical, magnetic and optical properties rather than mechanical strength. In many of these situations, it is far more likely that the properties of the first level units will be critical, and the purpose of the biomaterials scientist is to optimize the performance of those units within the broader structural context, more often now at the nanoscale than at the macroscale.

## 2.3 Atoms, molecules, interatomic and intermolecular bonds

### 2.3.1 Atoms and the nature of elements

It is not necessary to describe basic chemical and physical principles of atomic structure here. An atom, in isolation, consists of a nucleus and a series of orbiting electrons. The gold atom mentioned above has a nucleus that normally consists of 118 neutrons and 79 protons, and a total of 79 orbiting electrons. Neutrons are uncharged, but protons are positively charged and electrons negatively charged, each atom being neutral since there are equivalent numbers of protons and electrons. The atomic number of an element is defined by the number of protons. Since the mass of the protons and neutrons is essentially the same, and the mass of each electron is very much smaller, the atomic mass is equivalent to the combined numbers of protons and neutrons. There are many other types of sub-atomic particle contained within the nucleus, but we need not consider them in relation to biomaterials science.

While it is tempting to consider each element as having a single, well-defined atomic structure based on this precise number of protons and neutrons, this is not necessarily the case. The number of neutrons may vary, giving a series of slightly different structural forms with the same atomic number but different atomic masses. These different forms are called isotopes. This is rarely of relevance to biomaterials science, with one exception. Some isotopes are unstable and may change over time, this change being associated with the release of energy, in the form of radiation. Whilst radiation is potentially very harmful to health, it may be controlled and directed to specific biological sites for highly specific beneficial effects. This is the basis of the techniques of nuclear medicine, and, as we shall see, may be associated with the uses of some biomaterials.

Since each element is defined by its atomic number, they can be arranged in a sequence of ascending numbers, starting with hydrogen of atomic number 1, with its most common isotope having one proton, one neutron and one electron, to the heaviest natural element, uranium with atomic number 92. There are some even heavier elements but these are highly unstable and need not be considered here. It is well known that there are many similarities between some elements, and this is best seen when they are arranged in the Periodic Table of elements, shown in Figure 2.3.

The Periodic Table has rows and columns, and generally it is elements within the same columns that have the major similarities. This is not just coincidental but arises from the manner in which the electrons are arranged in their orbits around the nucleus. Just as we have seen with the brief discussion about material stability and the relationship between the energy of interatomic or intermolecular bonds and externally applied energy, structures and properties are determined by the need of substances to adopt minimum energy configurations whenever possible. The positions that electrons take is controlled by energy considerations. Electrons inhabit regions known as orbitals. With hydrogen the single electron occupies a spherical orbital by itself. This is the 1s orbital. The next element in the table is helium, which has two electrons. These two electrons also occupy the 1s orbital. This structure is energetically very favorable. However, an orbital cannot contain more than two electrons, so that in the next element, lithium, the three electrons take up a configuration of two 1s orbital electrons and one single electron in the second orbital, 2s.

Orbitals have different shapes. At the first level, there is only the spherical orbital, 1s. At the second level, there is the spherical 2s orbital, but also three p orbitals, which are double-lobed shapes, each with a different orientation, x, y and z. As noted above, the maximum number of electrons in the first energy

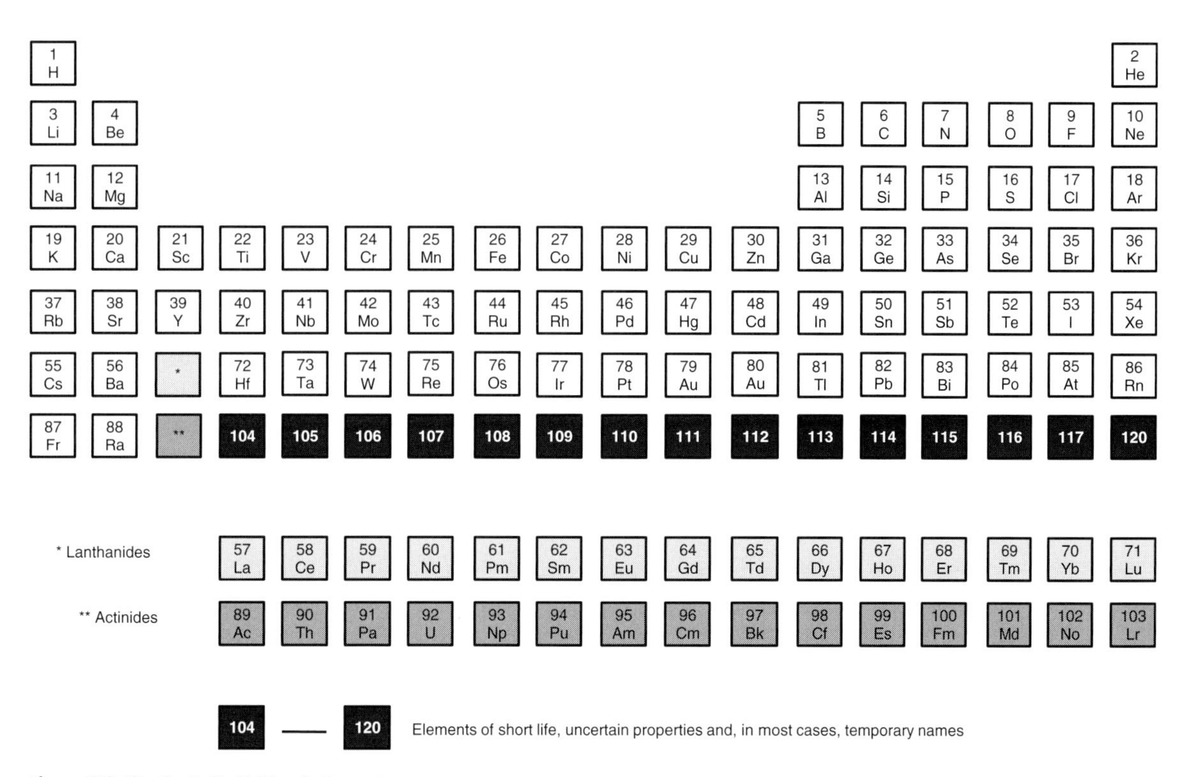


**Figure 2.3** The Periodic Table of elements.

level, sometimes called a shell, is 2. For the second energy level it is 8, involving two in each of the 2s, $2p_x$, $2p_y$ and $2p_z$ orbitals. For the third energy level, the maximum number of electrons is 18, with two electrons in each of the one s, three p and five d orbitals, the latter having yet a different shape. Higher level shells have different combinations of orbitals. We shall see later how some subtleties in the energy levels of orbitals lead to important properties of metallic elements.

The structure of the Periodic Table is determined by the number of shells (the rows) and the number of electrons in the outer shell (the columns). On the left-hand side of the table in the first column, there is just one electron in the outer shell, and it will be seen that these elements, lithium, sodium, potassium and so on are very similar, collectively being known as the alkali metals. On the right-hand side, column 8, in each case the outer shell is full. Again these elements are similar, helium, neon, argon, krypton, xenon and radon being the inert gases. The Periodic Table is not simple. The complexity largely arises from the fact that although the energy levels broadly increase as we go from first shell, to second, third and so on, there are differences between the energy levels in s, p, d and other orbitals, such that the energy level of some orbitals within one shell can be lower than some orbitals in a lower shell. Thus, the energy level of 4s electrons is lower than that for 3d electrons. This means that the 4s orbitals will fill before the 3d orbitals. This accounts for the grouping of some elements in the Periodic Table, which have very similar properties on account of their equivalent electronic makeup in the outer shell but with minor differences associated with the variation of numbers of electrons in orbitals in lower shells. This gives rise to the groupings of elements within the Transition Metals, Rare Earths and so on. These groupings will be discussed later when they become relevant to biomaterials applications.

### 2.3.2 Interactions between atoms

All of the above discussion has been based on single atoms of each element. In practice, of course, and as mentioned earlier, any sample of any element consists of billions of atoms, and it is necessary to consider how they interact with each other. There are two factors to take into account. First, in a hypothetical situation, if we have two atoms of the same element and allow them to move towards each other, there will both attractive and repulsive forces associated with unlike and like charges and these forces are associated with energy. As we see in Figure 2.4(a), we can plot these energies as a function of interatomic spacing. There are different forms of the energy–distance curves for the different forces, the overall energy being given by their sum. We can see that there is an interatomic distance of minimum energy, and atoms will attempt to adopt this. Secondly, since an atom is in its own lowest energy state when its outer shell of electrons is full, when atoms come together they will try to rearrange their electron configuration so that they obtain full outer shells.

These points are best considered by reference to specific examples of the atoms in the columns of the Periodic Table (Figure 2.3). On the far right we have, as noted above, atoms that are all in their minimal energy configuration by virtue of their full outer shells. These atoms have no energetic incentive to interact with anything. This is why these are the inert gases; inert because they will not interact with any other element, and gases because their own atoms do not want to interact amongst themselves and so they are free to move independent of each other (Figure 2.4(b)).

In the column just to the left we have elements that have nearly full outer shells, in fact each atom needs only one more to acquire a full outer shell. When such atoms come together they conspire to achieve this structure by sharing one each of their outer electrons, both of them effectively becoming full and stable. In other words they form a diatomic molecule, as seen in Figure 2.4(c). The bond that holds these two atoms together is called the covalent bond. It is one of the primary interatomic bonds and is very strong. This column includes fluorine, bromine, chlorine and iodine. As each of these molecules is stable they now have limited affinity for each other and they may also exist as gases or liquids. Similar situations are seen with column 6, where each atom needs two further electrons in their outer shells, and they also achieve this by sharing two outer electrons, giving diatomic molecules such as $O_2$.

Moving across to the left-hand side of the table, in column 1 we have elements with just one electron in their outer shell. In order to achieve their stable state they each need to lose this one electron per atom. When masses of these atoms come together they do this by simply giving up this spare electron. The resulting "cloud of free electrons" envelops the mass of residual atoms, which by virtue of their lost electron, now become positively charged ions (Figure 2.4(d)). There is now a type of communal bond between all of the negatively charged free electrons and the positively charged ions. This is the second primary interatomic bond, and is called the metallic bond. This is also very strong. Elements in column 2 will do the same by donating two outer electrons and those in column 3 by giving three electrons. It is in this situation where we see the effect of the minimal energy interatomic spacing, since it will be necessary for each of these atoms to adopt a position with a maximum number of neighboring atoms at this spacing. This will not be achieved by a random association of the atoms but rather by a highly regular arrangement. The very significant consequence of this is discussed in the next section, but the implication here is that these collections of atoms form very strong materials, the metals. In addition, the cloud of free electrons, provides the metals with unique powers of electrical conductivity.

For elements in the columns in the center of the Periodic Table, the situation is not quite so clear, and we shall deal with a few of these cases individually later. At this point, however, it is essential to introduce

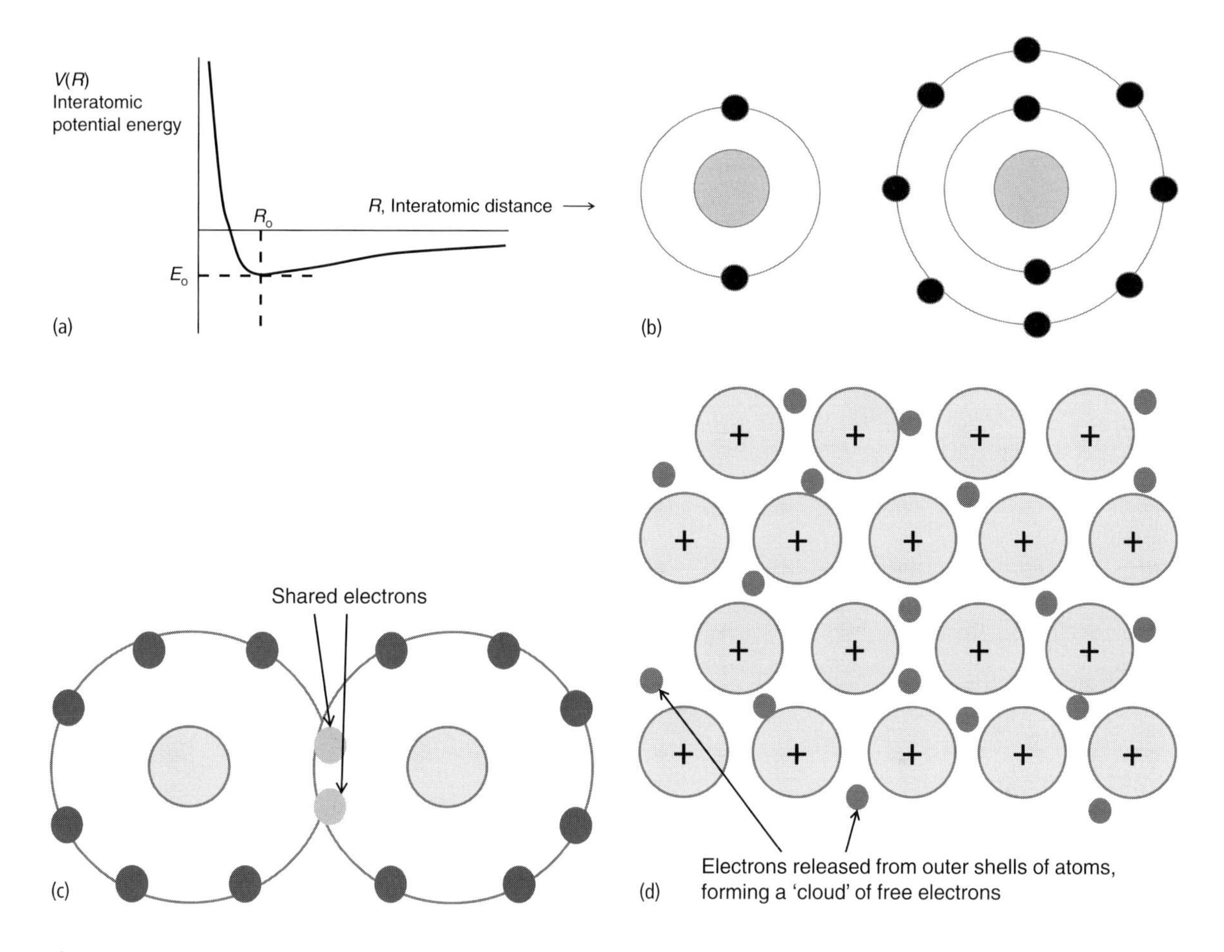


**Figure 2.4** Interatomic forces and interactions between atoms.
**(a)** Representation of attractive and repulsive forces (given as interatomic potential energy $V(R)$ as a function of interatomic distance, with an indication of the position of minimal energy $E_0$ and the ideal interatomic distance $R_0$.
**(b)** Atomic configuration of inert gases, where outer electron shells are full; on the left is helium, on the right, neon.
**(c)** Electronic configuration of covalent bonds; both atoms are one electron short of a full outer shell, so they each share one electron with the other atom, both effectively becoming stable.
**(d)** Representation of the metallic bond, with an array of positively charged ions and a "cloud" of free electrons, each of which has been released from the outer shells of the atoms; the overall electrostatic attraction between free electrons and the array of positively charged ions gives a very strong metallic bond.

the third type of primary interatomic bond and we have to consider what happens when atoms of two separate elements are brought together. Let us consider sodium, in column 1, with a single outer electron, and chlorine, column 7, with seven outer electrons. The "reaction" between these two atoms is straightforward: the sodium atom donates its outer electron to the chlorine atom, both thereby achieving a stable structure and becoming charged ions, $Na^+$ and $Cl^-$, in the process (Figure 2.4(e)). This is the ionic bond and it, too, can be strong. As with the metal ions, it is energetically favorable for these ions to adopt a regular pattern, and strong solids may arise from this type of structure. One very important point to note here, however, is that while in the solid metallic structure each ion is indistinguishable from another, with an ionic solid we have both positive and negative charged ions and their placing with

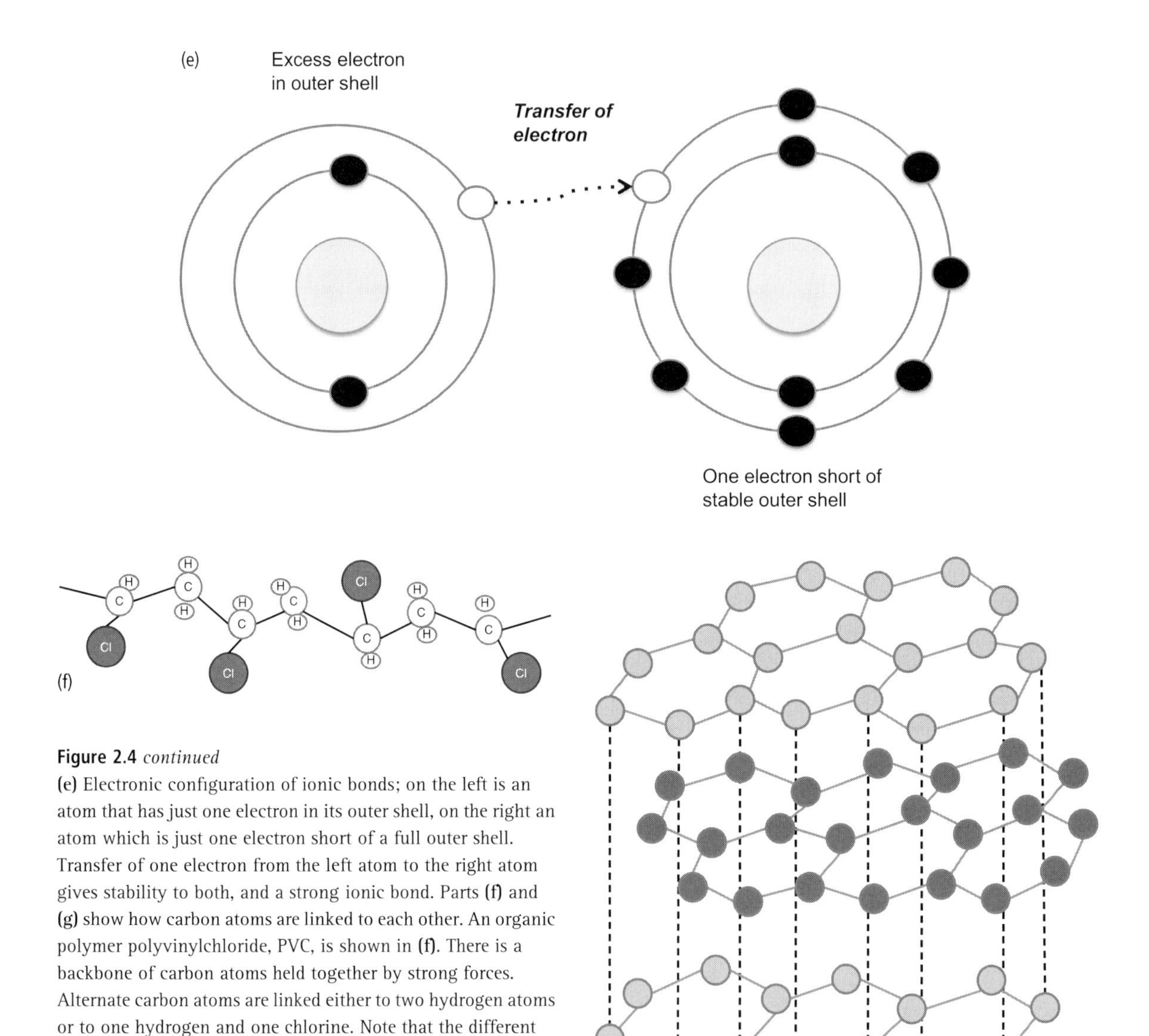


**Figure 2.4** *continued*
**(e)** Electronic configuration of ionic bonds; on the left is an atom that has just one electron in its outer shell, on the right an atom which is just one electron short of a full outer shell. Transfer of one electron from the left atom to the right atom gives stability to both, and a strong ionic bond. Parts **(f)** and **(g)** show how carbon atoms are linked to each other. An organic polymer polyvinylchloride, PVC, is shown in **(f)**. There is a backbone of carbon atoms held together by strong forces. Alternate carbon atoms are linked either to two hydrogen atoms or to one hydrogen and one chlorine. Note that the different sizes of these atoms will affect the manner in which the overall molecule forms.
**(g)** Shows the structure of graphite, where all atoms are carbon. These atoms are arranged in sheets with an hexagonal arrangement. In this diagram, one sheet at the top and one sheet at the bottom are arranged in an identical manner. The intermediate sheet has the same hexagonal structure but it is displaced to the right. Although the carbon–carbon bonds within the sheets are strong, the distance between adjacent sheets is too large for these primary bonds, and the sheets are held together by weaker secondary forces. This allows the sheets to slide over each other on application of forces, giving rise to lubrication properties.

respect to each other is critical. As we will see, this leads to enormous differences between these two types of solid, especially in relation to mechanical properties.

These primary interatomic bonds involve, therefore, the sharing or transfer of electrons. The number of electrons that is involved per atom defines the valence of that element. An Na atom loses one electron to form metallic Na and transfers one to Cl to give sodium chloride; Na has a valence of one. A Cl atom similarly gains one electron from the Na or shares one electron with another Cl atom to give Cl gas; Cl has a valence of one. Oxygen has a valence of two, aluminum of three, and so on. Naturally, not all reactions occur between two elements of identical valence. If we are able to combine this O with the Al, it will require three of the former and two of the latter to give $Al_2O_3$. This is aluminum oxide, or alumina, a very hard ceramic used frequently as a biomaterial. Nor do the reactions occur between just two species. Many metallic elements and many non-metallic elements may react to form some highly complex glasses for example.

Returning for a moment to some of the more complex parts of the Periodic Table, we need to discuss electronic configuration of the elements, especially those of rows three and higher, in some more detail. Whilst the concept of the outer electrons, and the need to maintain a complete outer shell, is quite straightforward, the situation often arises, as we have seen, when there are electrons in the outer shell but one or more of the inner shells is incomplete. This allows for the possibility of inter-shell movement of electrons, which can give rise to multiple valences for one element. This also gives rise to phenomena where the bonding between atoms is not exclusively ionic or covalent but can be a mixture of the two, again discussed later.

The one significant discussion point in the center of the table is the position of carbon. The C atom has two 1s, two 2s and two 2p electrons. One consequence of this structure is that C can form very many molecules by bonding both with itself and other elements, with possibilities for both single and double bonds between them. C gives rise to a number of covalently bound molecules, for example between one C and two O atoms gives carbon dioxide. A more complex arrangement between one C and one O gives carbon monoxide. From the materials science perspective, we can visualize these opportunities by considering a C atom having a valence of four and establishing covalent bonds between it and other atoms through the sharing of these four outer electrons. By linking C atoms to each other (Figure 2.4(f) and (g)), there is the possibility of establishing chains, with each C atom being free to bond to other, different, atoms, or even to further carbons. These molecular structures give rise, in the first case, to the organic polymers and in the second case to carbon materials such as graphite, diamond, carbon fibers and carbon nanotubes. The one other element with a valence of four that is worth commenting on is silicon, which also gives rise to many interesting materials, as we shall see.

### 2.3.3 Interactions between molecules

As implied earlier in the case of either oxygen or chlorine molecules, there is little inducement for the molecules to interact with each other, so that, at normal temperatures, these exist as gases – in the gas phase there are no effective intermolecular forces. With most substances, however, some structural features of the molecules permit some degree of intermolecular attraction. There are three broad types of intermolecular force; these vary in magnitude but are all typically less than 5% of the strength of the primary intramolecular forces holding the atoms together in the molecules.

Dipole–dipole interactions arise from an uneven distribution of charge in molecules, for example because molecules are asymmetrical or have unbalanced electronegativities. Hydrogen bonding occurs between molecules that have a hydrogen atom bonded to a more electronegative atom such as fluorine, oxygen or nitrogen. Hydrogen bonds are quite strong and give rise, for example, to the stability and high boiling point of water. London dispersion forces arise from the temporary and rapidly changing dipoles that are formed as electrons move within their orbitals. These forces are very weak. In general, intermolecular forces increase with the size of the molecules. Significant intermolecular forces give rise to liquids and, in some cases, especially with very large molecules, to solids.

## 2.4 The organization of atoms and molecules: states of matter, crystalline and amorphous materials

### 2.4.1 The states of matter

In the previous section we have defined the three primary interatomic bonds, that is the covalent, ionic and metallic bonds and the much weaker intermolecular bonds, and have discussed how the nature of these bonds determine the essential characteristics of the substances that arise from them. Now we have to move on to describe how practical, useful materials can be produced on the basis of these bonds.

First we shall consider the states of matter, that is the gases, liquids and solids. As noted above, at one level this is a straightforward discussion, for the distinction between them is largely dependent on the relationship between the internal bond energies and external energy, especially thermal energy. If there is any bond within a substance that has an energy that is much lower than the thermal energy at any one temperature, then atoms, or more likely molecules, will be essentially free of each other and the substance will exist either as a gas, that is it will have no defined volume or shape, or a liquid, which has volume but no shape. This is especially seen in those substances that have small molecules and weak intermolecular forces. However, if the substance has uniformly strong bonds that all exceed thermal energy at a given temperature, then it will be a solid, with both volume and shape. Generally, as the temperature rises, a substance will transform from a solid to a liquid at the melting point and then to a gas at the boiling point.

Metals are nearly always solid at ambient temperatures, generally with melting points in the range 500–1000°C and boiling points in the range 1000–5000°C. The exceptions include mercury, with a melting point at -38.8°C, and gallium, cesium and francium, all of which have melting points just under 30°C.

For the reasons mentioned above, many covalently bound substances are gases or liquids. In the halogen series, fluorine and chlorine are gases, their small molecules having little affinity for each other. The bromine atom is larger, and so therefore is the bromine molecule. The intermolecular bond energy is higher, and the element exists as a liquid. With iodine, this relationship is taken one step further and the element can be a solid at ambient temperatures.

This is also an important matter with the properties of polymers. It was noted above that C may form long chain molecules because of its valence of four. Polymers may have a small number of C atoms in the chain, but it is more usual for them to have hundreds or thousands of them. As the molecules get larger so do the intermolecular forces, and many very useful solid polymeric materials can be formed. In addition, the large molecules may show some degree of entanglement, further increasing their mechanical properties. The chemical functionality associated with some of the side atoms or groups on these chains may also allow for covalent bonds to be established between adjacent chains, which have very significant effects on the material properties.

### 2.4.2 Crystallinity

One of the most important structural features of solid materials is the manner in which the atoms or molecules are arranged in order to provide for the lower energy interatomic spacing.

#### Metallic crystals

Metals provide the simplest format to discuss this point here; we shall call the individual entities "atoms" although they are technically ions. Consider the 2D representation of a plane of atoms M shown in Figure 2.5(a). We can see here that the structure is very regular, each atom being equidistant to six other atoms. This has to be extended into the third dimension. Let us have another row of atoms with an

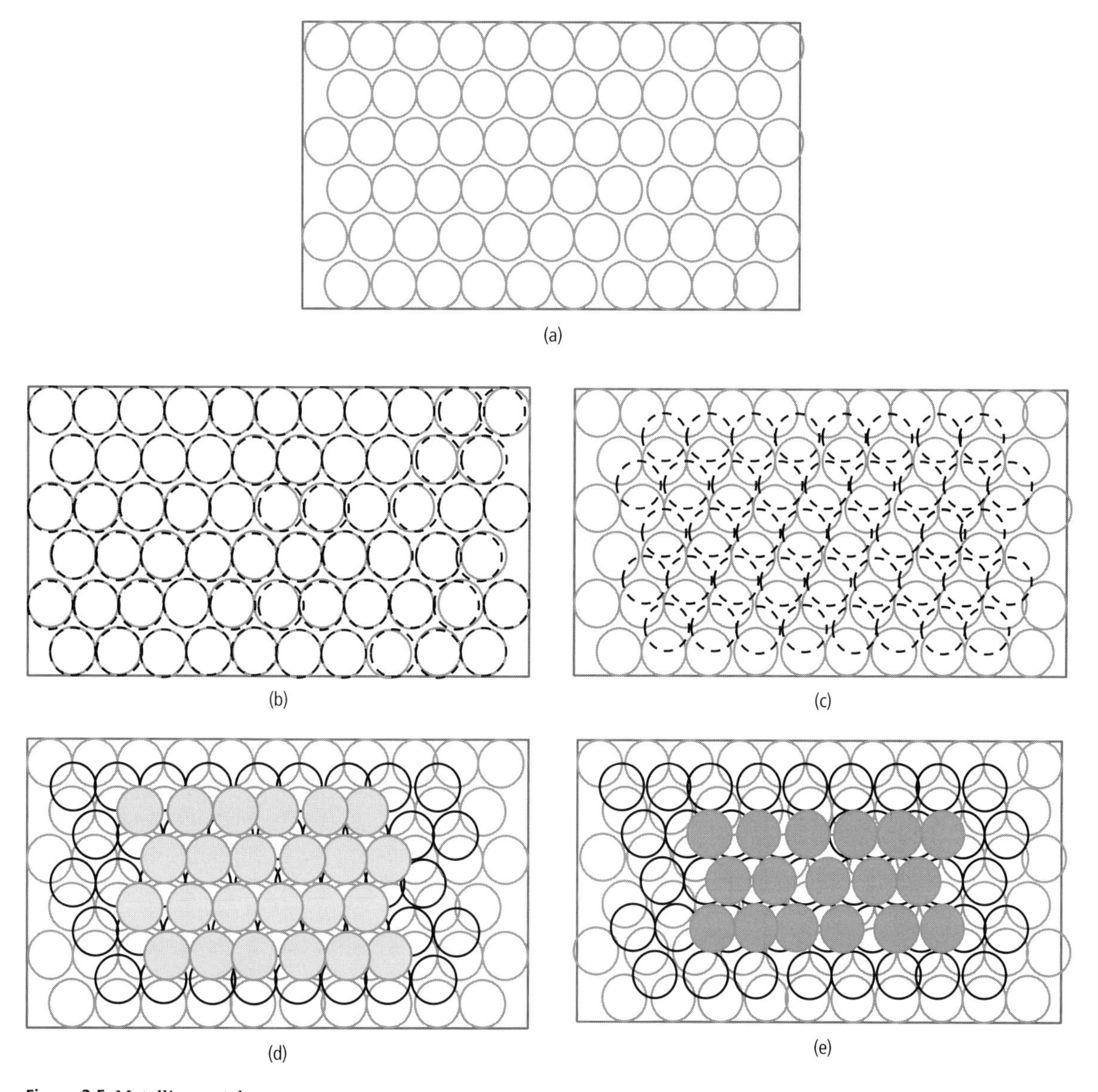


**Figure 2.5** Metallic crystals.
**(a)** A two-dimensional representation of a single plane of atoms of a metal. The atoms are arranged in a very regular sequence.
**(b)** A second layer of atoms is placed over the first, each atom being placed directly over identical atoms in the first layer.
**(c)** The second layer could be displaced from the first such that each atom fits into a depression between adjacent atoms of the first row.
**(d)** A third layer of atoms is placed over the second layer.
**(e)** Same as **(d)** but with the third row atoms being placed in new positions.

identical format. This could be placed directly over the first layer, as in Figure 2.5(b). Alternatively it could be placed with the atoms of the second row now directly over the spaces between the atoms of the first (Figure 2.5(c)). In either case, the regularity is extended into this third dimension, although we can see that there are different ways of arranging the planes. There are even more opportunities for variation with a third layer, whose atoms may be placed directly over the atoms of the first plane (Figure 2.5(d)) or in new positions (Figure 2.5(e)).

A solid material in which the atoms are arranged with this three-dimensional (3D) regularity is defined as being crystalline. There are many ways in which the planes of atoms in a crystal can be arranged, which means that there are different crystal structures and these will have different degrees of regularity, or symmetry. Ideally crystals will try to adopt formats that maximize the number of atoms that have neighbors at optimal distances; such structures are said to be "close-packed." In these, various planes of atoms with well-defined regularity can be seen, such as the parallel planes of Figure 2.6; these are described as crystallographic planes. The best way to visualize the structure of crystals is draw a 3D image with dots representing the center of atoms. With metals there are three common structures, shown in Figure 2.6. The 3D pattern that is formed from the layering of atoms in Figure 2.5(d) is shown in Figure 2.6(a); this is called a face centered cubic structure – the form is that of a cube with atoms at all corners and with atoms at the center of each face. The 3D pattern that is formed from the layering of atoms in Figure 2.5(e) is shown in Figure 2.6(b); this is called a close packed hexagonal structure. The third common structure is shown in Figure 2.6(c) and is called the body centered cubic structure – it is possible to discern the overall cubic form but here there is an additional atom in the center of each cube. There are a few other crystal forms that metals can adopt, but they have less symmetry and are more rare. Some metals exist in different crystal structures at different temperatures; this is due to the fact that thermal energy influences interatomic distances such that it may be preferable to adopt a different structure on changing the temperature in order to maintain the lowest energy configuration. This is a matter of immense importance in determining the structure and properties of metal systems, as we shall see later. The crystal structures of common metallic elements are given in Table 2.1.

The 3D arrangements shown in Figure 2.6 obviously represent only an infinitesimally small volume of a piece of metal. We can imagine that the crystallographic planes extend in all three dimensions in order to establish a metal crystal, just as we saw with the example of ice in Section 2.2. Although tangible pieces of metal can contain just one crystal, this rarely happens in practice. Almost every example of metals that we come across will consist of multiple crystals. Why this happens is best seen by reference to the solidification of molten metal. Metals are virtually always prepared by

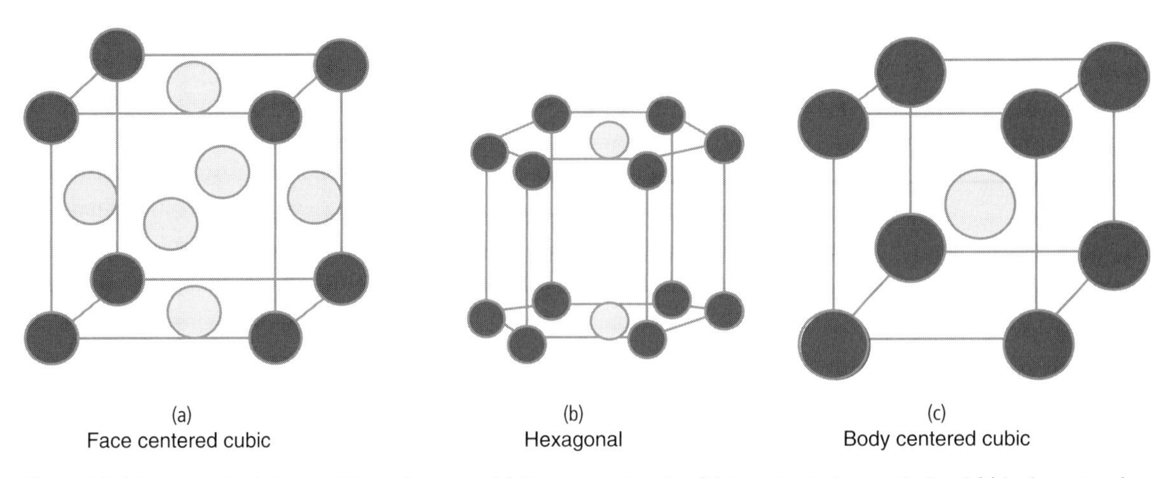


**Figure 2.6** Schematic of relative positions of atoms in (a) face centered cubic, (b) hexagonal close packed and (c) body centered cubic crystal lattices.

Table 2.1 **The crystal structures of common metallic elements**

| Element | Structure |
|---|---|
| Aluminum, Al | fcc |
| Chromium, Cr | bcc |
| Cobalt, Co | hcp below 460°C; fcc above 460°C |
| Copper, Cu | fcc |
| Gold, Au | fcc |
| Iridium, Ir | fcc |
| Iron, Fe | bcc up to 912°C; fcc 912–1394°C; bcc above 1394°C |
| Magnesium, Mg | hcp |
| Molybdenum, Mo | bcc |
| Nickel, Ni | fcc |
| Niobium, Nb | bcc |
| Palladium, Pd | fcc |
| Platinum, Pt | fcc |
| Ruthenium Ru | hcp |
| Silver, Ag | fcc |
| Tantalum, Ta | bcc |
| Titanium, Ti | hcp below 900°C; bcc above 900°C |
| Tungsten W | bcc |
| Vanadium, V | bcc |
| Zinc, Zn | hcp |
| Zirconium, Zr | hcp below 862°C; bcc above 862°C |

bcc – body centered cubic; fcc – face centered cubic; hcp – hexagonal close packed

processes that include a molten stage, in which there is no regularity at all. As the metal cools down past the melting point it will solidify. The solidification process takes place by nucleation and propagation phases. At some point in the liquid, the energy level falls to a point where the solid form is favored, and a few atoms get together to establish the nucleus of a crystal. In quick succession, more and more atoms are deposited onto this nucleus and the crystal grows. However, the formation of that nucleus is a random event and is responsive to fluctuations in thermal energy, or thermal gradients, in the liquid, and in practice very large numbers of nuclei will form. The orientation of these nuclei is also random, so that although the crystal structure in each propagating crystal is the same, the orientation of lattices in each will be different. Thus a given volume of solid metal will consist of very many individual crystals. These are often referred to as grains. The individual grains in a cross section of a metal are shown in Figure 2.7(a). Because the orientation of individual grains is different, there will be discontinuities at the places where the grains meet, as shown in Figure 2.7(b); these are called grain boundaries. In some metals, the individual grains may be large enough to be visible to the naked eye but usually they are much smaller, typically being in the range 10–100 μm.

### Ceramic structures

With ceramic and other ionically bound solids, the formation of crystals is still possible, although with greater complexity. If we have a simple chemical compound such as sodium chloride, we can see how the ionic bond is established between the positive and negative ions to form a molecule (Figure 2.8(a)). If we now place another, identical, molecule alongside (Figure 2.8(b)) it too will have its own directional ionic bond. However, the two sodium ions are identical to each other and the two chloride ions are identical to each other and there will be an identical electrostatic attraction between sodium and chloride ions if they are at the same interatomic distance (Figure 2.8(b)). We can therefore establish a highly regular and symmetrical arrangement of these ions, in 3D, provided that they alternate in the way shown in Figure 2.8(b, left). We therefore have a crystal of sodium chloride. This process, however, gets more complex when we have different valence atoms involved. Oxide ceramics, such as aluminum oxide, $Al_2O_3$, do form crystals, but the atomic

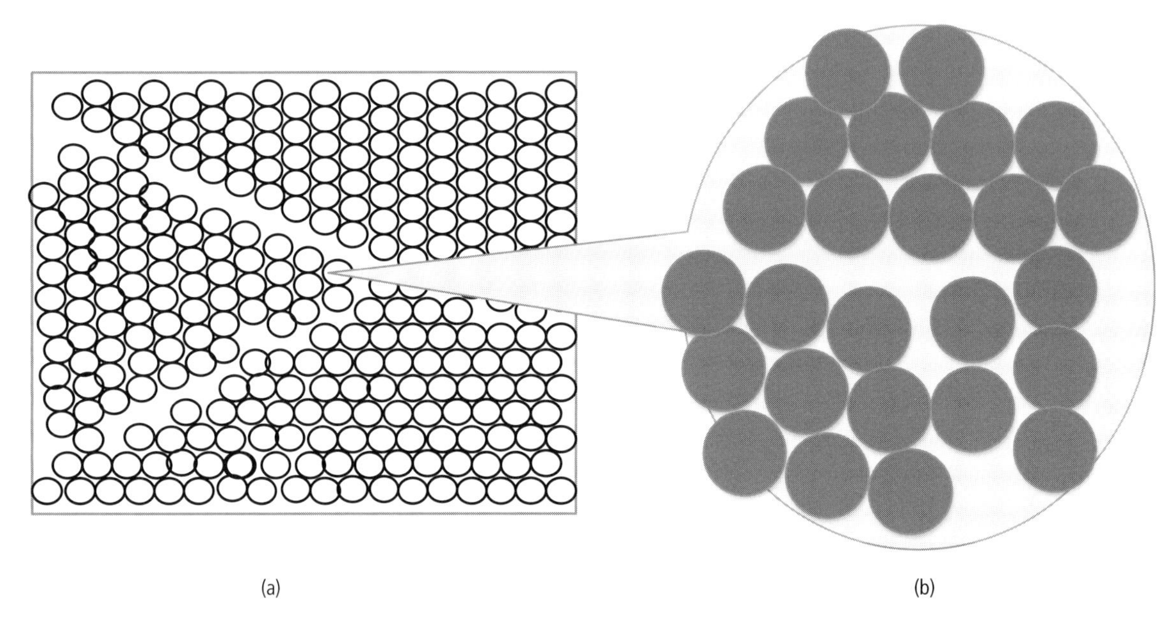


**Figure 2.7** Grains and grain boundaries. In a metallic material there will usually be very many small three-dimensional crystals, usually referred to as grains. In a pure metal these grains will be of identical crystal structure, but each having their own orientation, as shown in **(a)**. There will inevitably be a zone of atomic disarray where grains meet, as in **(b)**. These are known as grain boundaries.

configuration is more complex, and rarely of cubic form. It is often the case that such oxides are able to exist in more than one form. Silica, $SiO_2$, for example, can exist as α quartz, which has a rhombohedral form, β quartz, which is hexagonal, α tridymite, which is orthorhombic and α cristobalite, which is tetragonal (and indeed several other forms), as shown in Figure 2.8(d). Aluminum oxide exists as corundum, a nearly close packed hexagonal structure, but it can also exist as the gemstones sapphire and ruby, which are of trigonal structure, these crystals having characteristic colors depending on traces of impurities of other metal ions.

## Polymer structures

**Solid polymers** It is a little more difficult to visualize how the large and often irregular-shaped molecules of polymers can form crystals, but they sometimes do, at least in part. If the molecules are uniform in terms of a regular backbone of carbon atoms and uniform side atoms or groups, then it is possible for these molecular chains to be arranged in a parallel manner, in bundles, which constitute a crystalline structure. Polyethylene and polytetrafluoroethylene are the best examples of crystalline polymers since they both have an all C backbone and just one type of side atom, H and F respectively, as seen in Figure 2.9(a, b). Polymethylmethacrylate, which has two large and disparate side groups, cannot form crystals and has irregular disposition of molecules (Figure 2.9(d)). Some polymers, such as polypropylene, can form crystals, but with varying options for where side groups are placed and, in consequence, different types of crystallinity. Isotactic polypropylene has all the methyl groups on one side of the molecule while syndiotactic polypropylene has them alternating on both sides (Figure 2.9(c)).

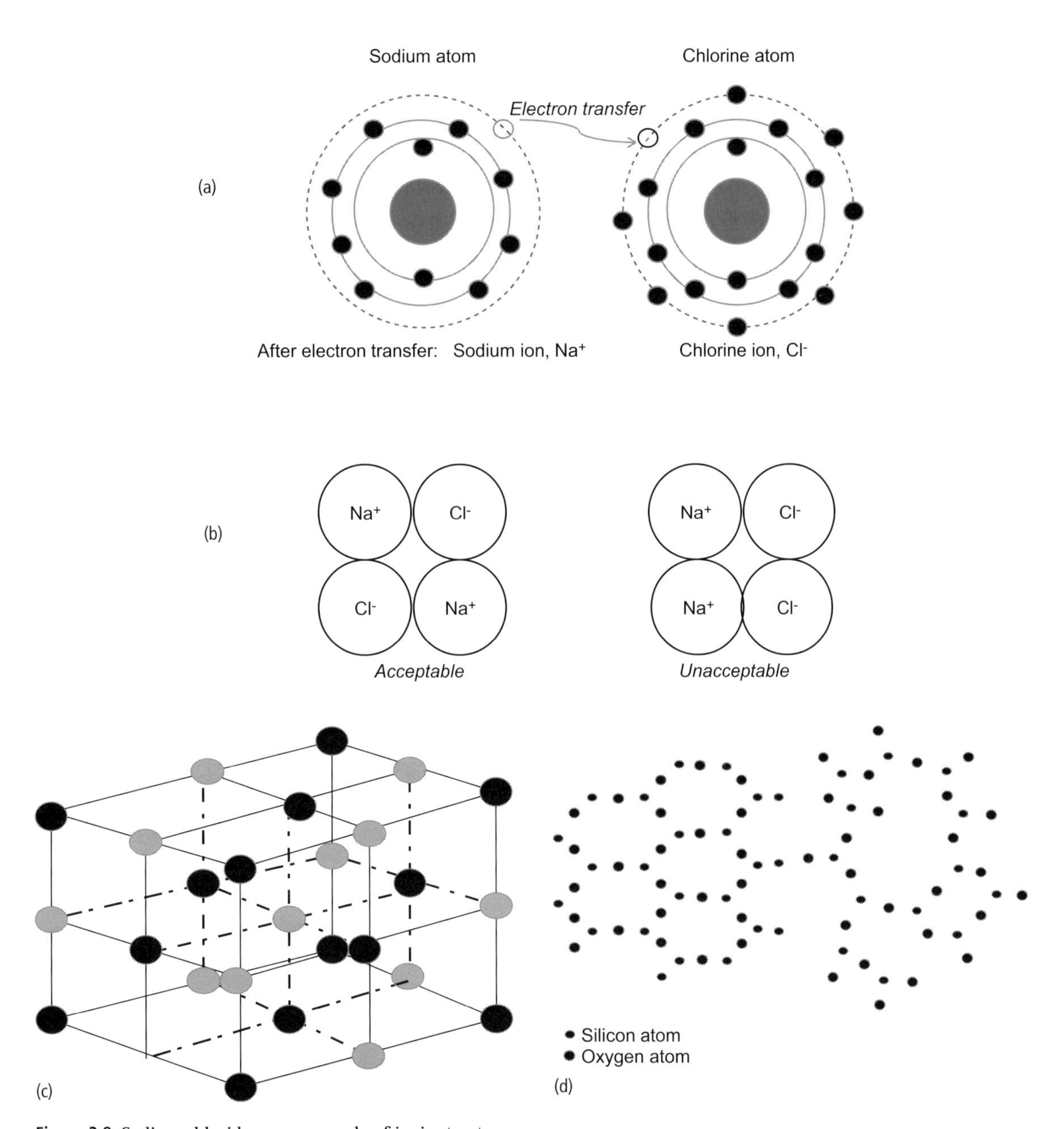


**Figure 2.8** Sodium chloride as an example of ionic structure.

**(a)** Transfer of electron from one atom of sodium to one atom of chlorine, giving two stable ions, held together by the strong ionic bond between them.

**(b)** Energetically acceptable and unacceptable arrangement of ions.

**(c)** Three-dimensional structure of sodium chloride.

**(d)** Structure of silica, $SiO_2$, on the left with both short-range and long-range order, giving a crystalline structure, on the right with short-range order but no long-range order, giving an amorphous structure.

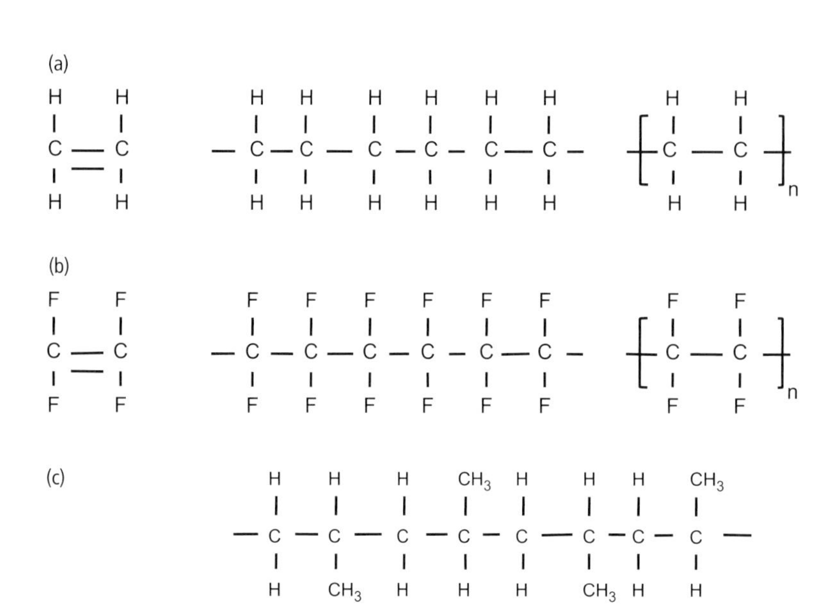


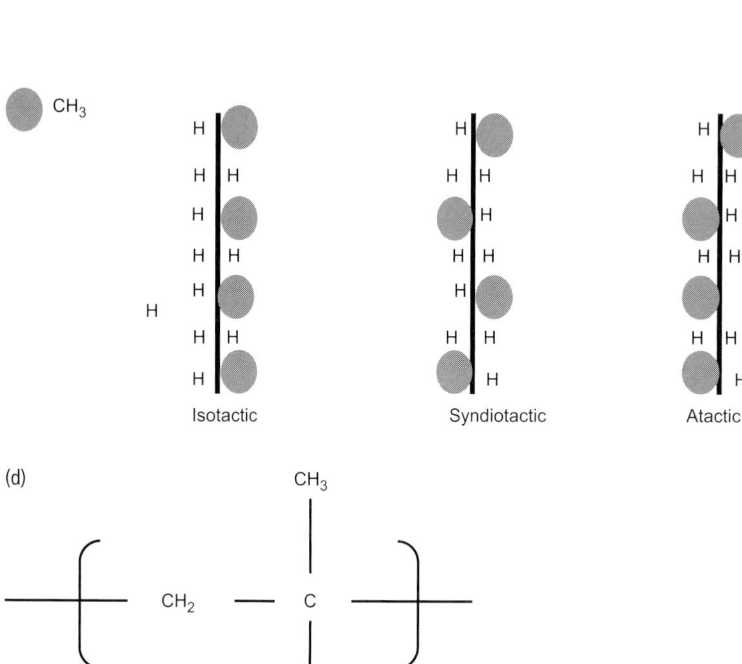


(d)

$CH_3$

$CH_2$ — C

n

C = O

$OCH_3$

**Figure 2.9** Basic polymer structure. **(a)** Polyethylene. On the left the monomer ethylene, in the center the arrangement of carbon and hydrogen in a linear molecule, on the right, a short-hand notation for the molecule. **(b)** Polytetrafluoroethylene, where the hydrogen atoms are replaced by fluorine. **(c)** Polypropylene, similar to polyethylene, with one-in-four hydrogen atoms replaced by the $CH_3$ group, the polymer being derived from the propylene monomer. The $CH_3$ group can be placed in different positions, giving different structures. On the left, all of these groups are on the same side of the polymer backbone (Isotactic polypropylene), in the center, they are arranged symmetrically on either side (Syndiotactic) and on the right, in a random fashion (Atactic). Note that "tactic" means arrangement or order, "iso" means the same and "a-" means without. **(d)** Polymethylmethacrylate, with an all-carbon backbone but methyl and methacrylate side groups. The sizes of these groups, and especially the disparity of sizes, makes it very difficult to achieve regularity and symmetry in the structure, and these polymers are generally amorphous. Since amorphous materials are usually transparent (e.g., in glasses), these acrylic polymers usually have excellent optical properties and are used in lenses.

**Soluble polymers** Structural, solid materials were the mainstay of the polymer catalog for the vast majority of medical applications until quite recently. For a variety of reasons, however, there has been a trend towards soluble polymers, and especially water-soluble polymers in many industrial sectors, these reasons often being concerned with the need to avoid solvent-based technologies for environmental reasons. In several situations, including a number that involve medical technology, there are more important functional advantages of utilizing water-soluble rather than insoluble polymers, especially when these materials are based on reversible supramolecular organizations. These applications do not require the mechanical characteristics of the structural polymers; their self-assembly is not dependent on covalent bonds along the length of large molecules but rather on the weak non-covalent interactions such as the hydrogen bonding and van der Waals forces. There is often a delicate balance between attractive and repulsive forces, which permits the structure to vary with environmental conditions.

The structures formed by soluble polymers are of significant interest. When the balance of forces determine that they self-assemble, a wide variety of supramolecular structures is possible depending on the nature of these forces and the topographical nature of the interfaces between molecules. These structures have very considerable architectural variety. They include, for example, dendrimers, which are highly branched macromolecules that have a tree-like architecture, usually with multiple branching levels (Figure 2.10(a)). These normally have a globular character, with diameters ranging from 1 to 100 nm, and are monodisperse and spherical. They have varying shaped internal cavities and usually have a large number of functional groups at their periphery. Many soluble polymers are amphiphilic, that is they comprise of a mixture of hydrophilic and hydrophobic segments. This can give rise to the formation of micelles. In aqueous solutions these substances form as nanoaggregates with a hydrophobic core and a hydrophilic shell (Figure 2.10(b)). The characteristics of these core and shell components determine the overall properties. Both dendrimers and micelles are very attractive as drug or gene carriers since the internal cavities of the former and the hydrophobic core of the latter can be designed to function as a molecular depot. The wide-ranging functionality of the surfaces of these aggregates can also be manipulated to facilitate control over the biological properties. We shall come across these structures later in this chapter in the context of nanoparticulate and environmentally responsive materials.

#### Amorphous solids

The above discussion of the crystallinity of polymers begs the question, if a solid is not a crystal, what is it? Those materials that are not crystalline are said to be amorphous, which simply means that there is no regularity or order to the arrangement of atoms or molecules. The majority of polymers exist as partly crystalline solids, in which there are some crystalline and some amorphous regions. Ionically bound solids that cannot form crystals because of their complex molecular structure also exist in the amorphous state. Such substances are usually referred to as glasses. Glasses (and amorphous polymers) are usually transparent because the lack of any regular crystal planes means that light cannot be reflected or refracted.

**Liquid crystals** For completeness, it should be noted that there are some substances that exist in an intermediate state between the liquid and crystalline solid states. These are the liquid crystals that have the ability to flow like liquids but have symmetrical structures similar to those in crystalline solids. Some liquid crystals have significant optical properties and are used in optoelectronic displays. In addition, many biological materials form liquid crystals, in particular cholesterol and myelin.

### 2.4.3 Macro-, micro- and nanostructures

The scale at which we examine and control the structure of a material is taking on increasing significance.

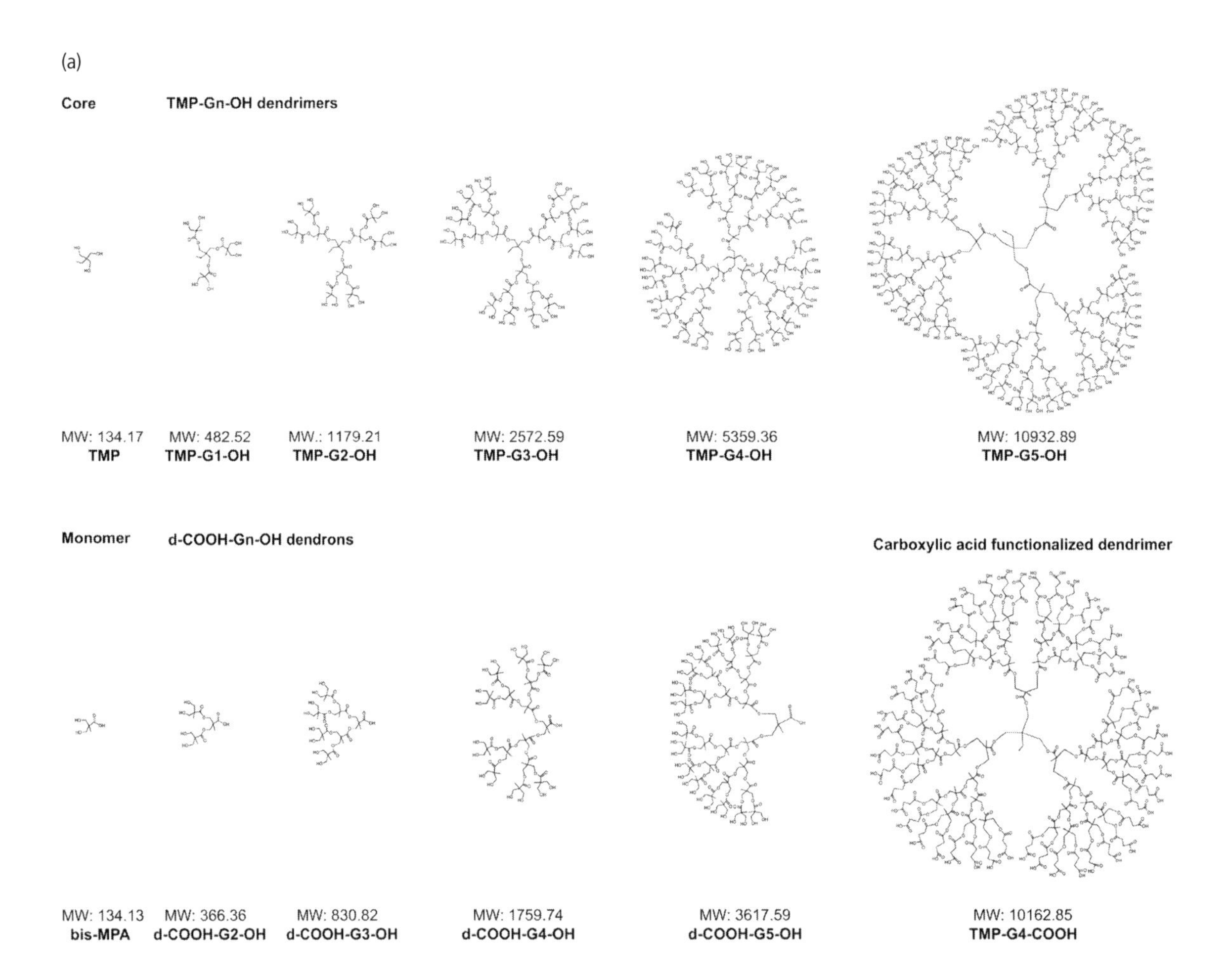


**Figure 2.10**
(a) Dendrimer structure. Dendrimers are highly branched macromolecules that have a tree-like architecture, usually with multiple branching levels. The complexity of dendrimers is reflected by the description of their structure in terms of "generations." The examples in the top row are based on the trimethylolpropane (TMP) monomer, which forms the core. TMP-Gn-OH dendrimers of generations 1 to 5 are shown. In the bottom row the monomer is 2,2-bis(methylol)proprionic acid (bis-MPA) and there are G2–5 generations. The structure at the bottom right is a carboxylic acid functionalized dendrimer, showing the versatility of these structures. Reprinted from *Biomaterials*, 33(7), Feliu N *et al.*, Stability and biocompatibility of a library of polyester dendrimers in comparison to polyamidoamine dendrimers, 1970–1981, © 2012, with permission from Elsevier.

The macrostructure of a material or product is that which we can discern with the naked eye. There is very little we can say about this that is not obvious. Below this level there is the microscale, with the associated microstructure. This is usually taken to mean dimensions of around 1 μm up to 1 mm.

Whatever the material, the microstructure plays a very important role in determining the properties, and frequently we find that "structure–property relationships" determine the way in which materials can be manipulated in order to optimize performance. Among the features which determine these properties are the intrinsic crystal/grain/domain characteristics (e.g., grain size, phase distribution) and defect structures, which include porosity and the presence of impurities.

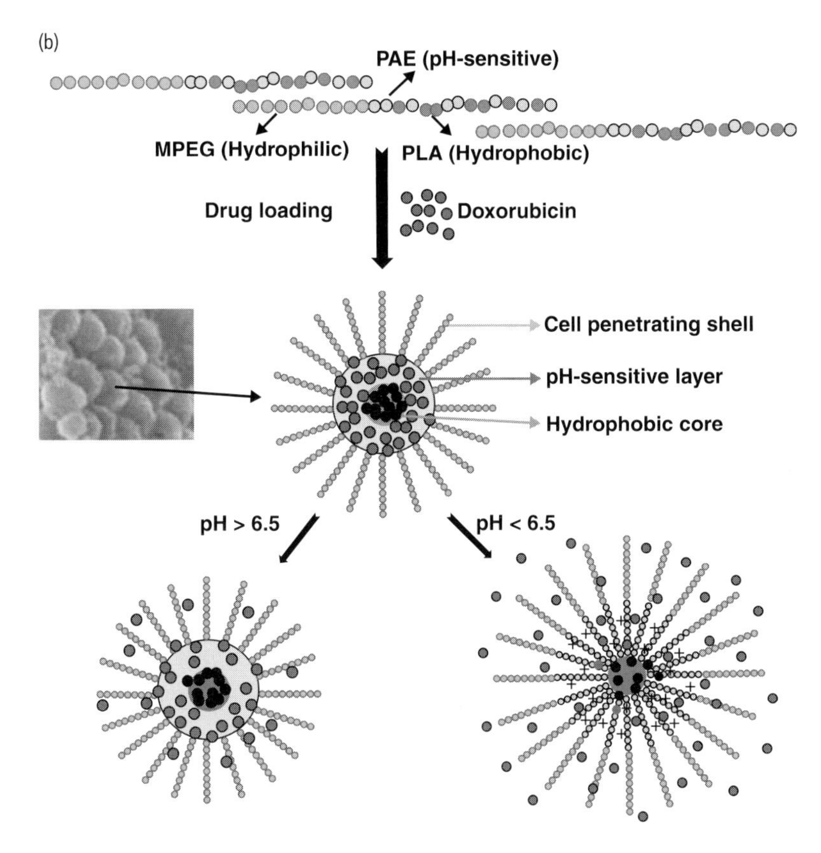


**Figure 2.10** *continued*
**(b)** Micelle structure and self-assembly. This shows an example of the structure of a micelle and how it can be used for drug loading and pH-dependent release. This is an amphiphilic poly(ethylene glycol) methyl ether-b-(polylactic acid-co-poly(amino ester)), MPEG-b-(PLA-co-PAE). The micelle has a hydrophobic core, a pH-sensitive layer and a cell penetrating shell. Drugs may be loaded into the micelle at high pH, and then released at pH higher than 6.5. Reprinted from *Biomaterials*, 33(26), Zhang C Y *et al.*, Self-assembled pH-responsive MPEG-b-(PLA-co-PAE) block copolymer micelles for anticancer drug delivery, 6273–83, © 2012, with permission of Elsevier.

The nanoscale has been much discussed in recent years. The prefix nano- implies dimensions around the nanometer, 1 nm. There is a huge difference between 1 nm and 1 μm, and this is now being exploited in various forms of nanoscience and nanotechnology. Most authorities agree that the nanoscale does not extend up to 1 μm, but rather covers the range up to around 100 nm. It is preferable to consider the range from 100 nm to 1μm as being sub-micron. The reason for this is that quite different properties can be expected below 100 nm, especially involving the activity of particles with the very high surface area to volume ratio and quantum effects.

In terms of biomaterials applications there are several important features of the nanoscale. First, these features have far more chance of influencing biological events than those at the microscale, bearing in mind that these dimensions are approaching those of critical molecules in biological systems, including DNA and cell surface receptors. Secondly, nanoparticles containing or attached to drugs can be better targeted to relevant sites, potentially allowing for safer and more effective therapies. Thirdly, quantum effects associated with certain very small particles can be relevant to the detection of physical events that can be used in diagnostic and related activities.

It should be noted here that the prefix meso- is sometimes used in the context of nanotechnology and this can cause some confusion. Traditionally meso- refers to an intermediate scale and in materials science has implied some feature that is intermediate between the microscale and the macroscale. However, in chemistry the term is now used for something between the nanoscale and the microscale. This is only important in biomaterials science when it refers to porosity, where "microporous" implies pores that are less than 2 nm, "mesoporous" means between 2 and 50 nm and "macroporous" means greater than 50 nm. Mesoporous materials, especially mesoporous silica-based materials, are of interest in drug delivery systems.

## 2.5 Surface properties of materials

The surface of a solid piece of a material has different characteristics and different properties to the interior of that material. These differences are based on two factors. First, the internal structure is predicated on the fact that individual entities such as atoms or molecules are surrounded by other, similar entities, and forces are distributed equally in a 3D matrix. At the surface, this is not the case. In the outermost layer of atoms in a metal, forces of attraction exist on one side, directed to the interior of the sample, but not on the other, external, side. The energy balance of the surface is, therefore, quite different to that in the interior, and we should therefore expect the properties to be different. There is no universally applicable definition of what constitutes a surface since this will vary from material to material; the distance through which the properties vary from the exterior to that of the bulk interior may range from a few nanometers in the case of simple metals through to a hundred nanometers for complex polymers. The second factor, which is consequential on the first, is that, depending on the external environment in which the material exists, extraneous atoms or molecules from that environment may adsorb onto and react with the atoms or molecules of the surface. Thus, the chemical nature of the surface will reflect the nature of the adsorbed species as well as the intrinsic composition of the material. This could be disadvantageous since the desirable properties of the material may be masked by the layer of adsorbed species. On the other hand, this phenomenon may be put to good use if the surface layer can be controlled, or manipulated, to confer special properties to the material; in biomaterials this is particularly important since surface modifications can be directed towards the control of biological functionality of engineering materials.

We shall discuss surfaces in terms of the energy considerations, the surface chemistry and the surface topography. These discussions include mention of the main techniques to evaluate surface characteristics. We will then describe the principles of surface modification.

### 2.5.1 Surface energy: the hydrophilic/hydrophobic balance

As noted above, the energy of a material at the surface is different to that in the interior. The imbalance of forces at the outermost layer implies that the surface layer will have a higher energy than the bulk material; this excess of energy is referred to as the surface energy, measured in units of energy per units of area, such as Joules per square meter ($J\ m^{-2}$). Of course, unless the material is in a vacuum, it will be in contact with some other medium, usually a liquid or gas, and one might expect that the nature of that medium will have some influence on the surface energy of the materials. In view of this, it would be better to refer to the interfacial free energy rather than surface energy of the material when discussing the behavior of a material in a specific situation. Irrespective of the material chemistry, this surface/interface energy will have a significant influence on reactions that take place at the surface. Since all materials will tend to minimize their energy, a material with a high surface energy will tend to encourage adsorption of substances from their surroundings if they are able to reduce the energy.

The surface energy is usually discussed in terms of the behavior in water and this brings us to considerations of hydrophobicity and hydrophilicity. This discussion may be based on observations of a drop of water that is placed on the surface. Depending on the surface energy, that water may form a spherical droplet or it may flow over the surface as an even layer. In the former case, the surface energy of the material is very low and there is no incentive for the water to spread since that would increase the total energy. In the latter case, the surface energy is high, and spreading of the water will reduce the total energy. The former material is described as being hydrophobic (literally water-hating) and the latter as hydrophilic (water-loving); alternatively the former is described as non-wettable and the latter as wettable. Most materials will show behavior in between these extremes and the water drop will take on a shape that depends on the energy. In particular, an angle is established by the tangent of the droplet surface and the material surface, as shown in Figure 2.11(a), this angle, $\theta$, being known as the contact angle. This angle is determined by the balance of forces at this point, in particular by the balance of attractive forces between the water molecules (i.e., cohesive forces) and the attraction of water molecules to the surface molecules (i.e., adhesive forces). The surface energy of the material, that is interfacial free energy between the solid surface s and the vapor v, $\gamma_{sv}$, is determined by the relationship:

$$\gamma_{sv} = \gamma_{sl} + \gamma_{lv} \cos\,\theta,$$

where $\gamma_{sl}$ is the interfacial energy of solid and liquid (i.e., water), and $\gamma_{lv}$ is the interfacial energy of liquid and vapor. It is generally considered that solids interact with liquids largely through dispersion and polar forces and it is possible to develop extensive analyses of the basis of surface energy by consideration of these different forces.

Clearly the interfacial free energy is a function of the characteristics of all three phases (s, l and v) and is not a unique fundamental property of a material. The most common way of determining a quantifiable parameter for the surface energy of an individual material is to measure the contact angle formed by several standard liquids on that material and plot a graph of the cosine of the measured contact angle against the $\gamma_{lv}$ (i.e., the surface tension) of the liquids. This is known as a Zisman plot, shown in Figure 2.11 (b). The point at which the extrapolated straight line drawn through the points on this graph intersects the horizontal line at $\cos\,\theta = 1$ is known as the critical surface tension of the material. A liquid will only wet, or flow over, a material surface if its surface tension is less than the critical surface tension of the material. Thus a material with a very low critical surface tension will resist being covered by most liquids, and, when that liquid is water, will be very hydrophobic. This characteristic is very important in adhesion science since the best adhesives tend to have low surface tensions, and those materials with very low critical surface tensions will be very difficult to bond to adhesives. This is exemplified by the behavior of polytetrafluoroethylene (PTFE), which has a critical surface tension of less than 20 mN/m (equivalent to 20 dynes/cm) and is well known for not sticking to anything. Polymers generally have critical surface tensions in the range 20–50 mN/m, while metals, glasses and ceramics exhibit much higher values, in some cases exceeding 1000 mN/m.

It is tempting to assume that the critical surface tension of a material will have a major influence on the interactions between materials and biological environments. In practice, it has not been easy to define such correlations. There are many reasons for this, which will be explored in Chapter 3. On the material side, the critical surface tension refers to ideal conditions using very clean surfaces, but in practice, as noted above, biomaterials surfaces will usually be adorned with environmental contaminants, such as hydrocarbons, and their surface energy in practice may be very different. On the tissue side there are many different components, including cells and physiological macromolecules as well as water and various ions, all of which have their own non-specific surface energy or tension characteristics together with their own highly specific features that control binding to other substances, such that the competition for the surface is far from simple.

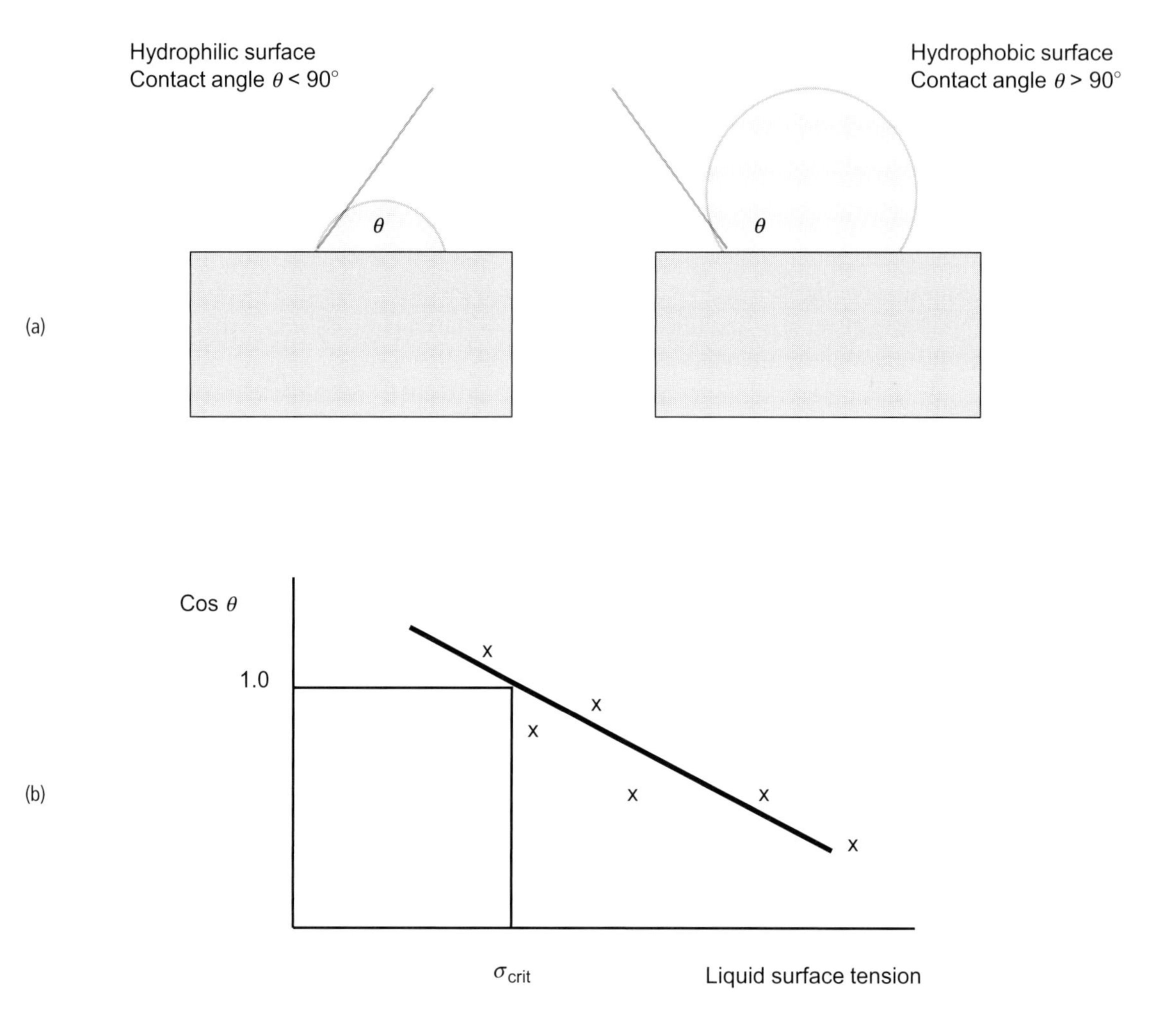


**Figure 2.11** Surfaces.
**(a)** Contact angles. On the left is a hydrophilic surface, where the contact angle is less than 90°. With a hydrophobic surface, on the right, the contact angle is greater than 90°.
**(b)** A Zisman plot. A series of different liquids, with different surface tensions, are placed separately on the test surface and their contact angles measured. The cosine of the contact angle, is plotted against the liquid surface tensions (data points given as x). The critical surface tension of the solid, $\sigma_{crit}$, is that which corresponds to $\cos \theta = 1.0$.

## 2.5.2 Surface chemistry

It has been noted above that the unique properties of surfaces are derived from the imbalance of forces on atoms and molecules at that surface. It may well be the case that the actual chemical composition at the surface is no different to that in the interior (this is essentially the case with gold placed in an inert environment) and the change in energy is the only difference. In most practical situations, however, there may be more substantial differences and the chemical composition and morphology at the surface can be radically altered. There are few common principles here and each material–environment combination has to be considered separately. Among the factors that

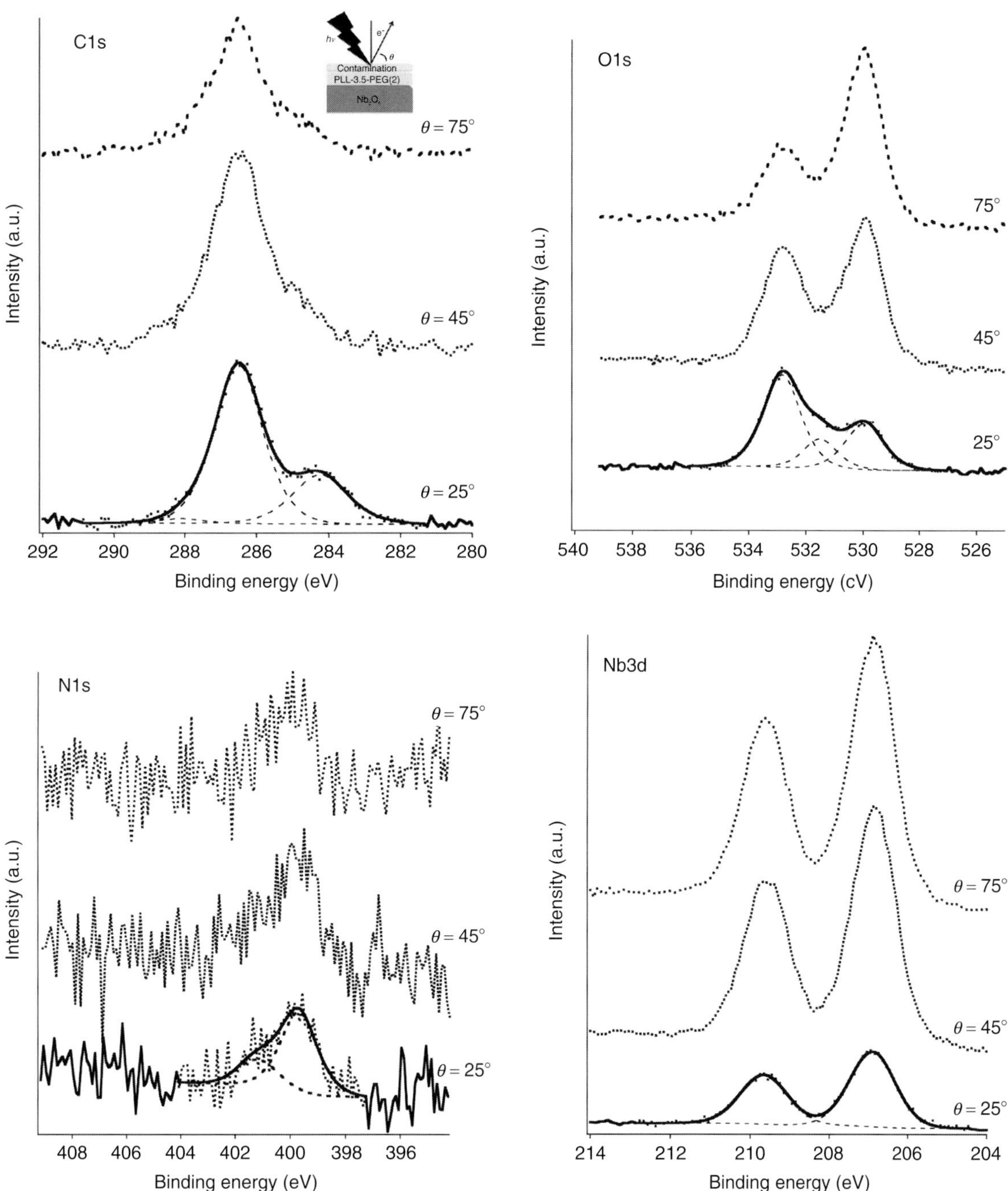


**Figure 2.11** *continued*

**(c)** Example of angle-resolved X-ray photoelectron spectroscopy, for the protein-resistant polymer poly(L-lysine)(20)-g[3.5]-poly (ethylene glycol)(2) (PLL-g-PEG) adsorbed on $Nb_2O_5$. The spectra show the C1s (top left), O1s (top right), N1s (bottom left) and $Nb3d_{5/2}$ and $Nb3d_{3/2}$ peaks (bottom right) – dots experimental, full line reconstructed by curve fitting – at electron take off angles (ToA) of 75°, 45° and 25°. The ToA is the angle, *q*, between the direction of the emitted electrons and the sample surface (see insert) exploiting the different sampling depths. The attenuation of the nitrogen signal at 75° indicates that it is associated with the interface; the O1s peak at 530.5 eV belongs to the $Nb_2O_5$ and is more intense at 75°, the other two components (more pronounced at the grazing angle of 25°) correspond to the adsorbed molecule. Image courtesy of Professor Antonella Rossi and Professor Nic Spencer, Laboratory of Surface Science and Technology ETH, Zurich, Switzerland.

(d)

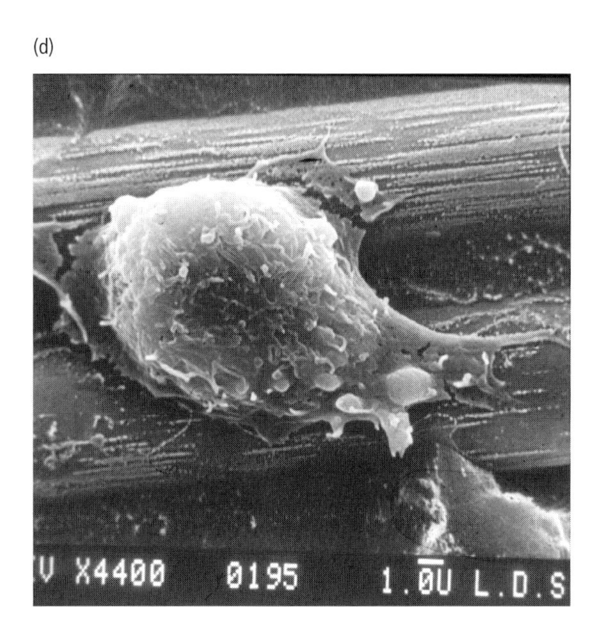


(e)

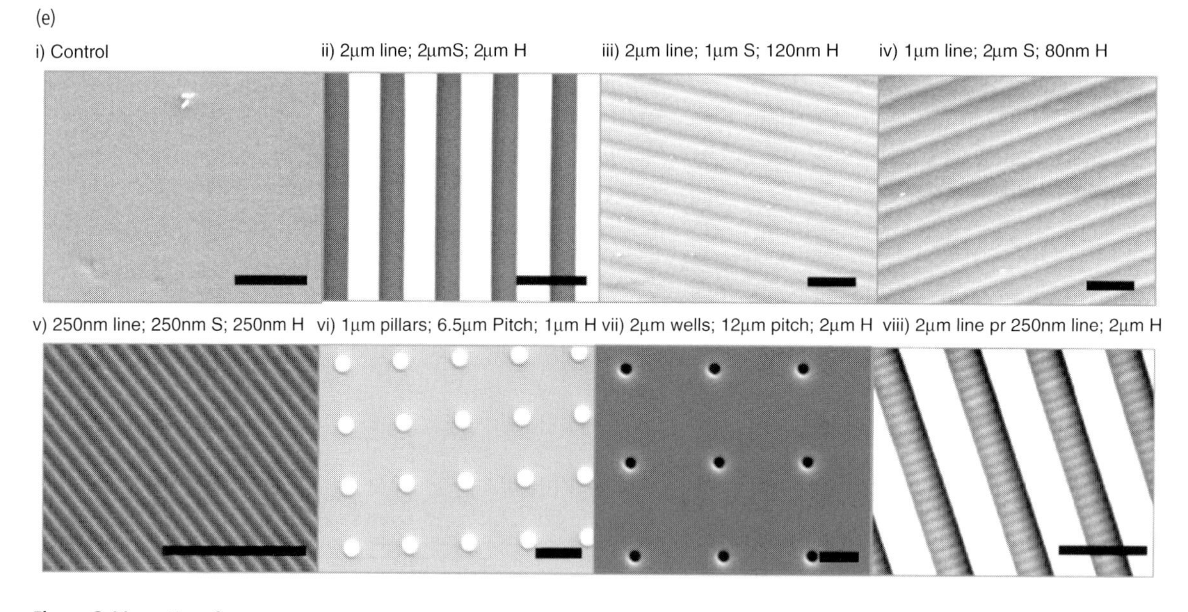


**Figure 2.11** *continued*

**(d)** Scanning electron micrograph of textured surface, showing a cell aligning itself with the oriented fibrous structure.

**(e)** Scanning electron micrographs of patterns generated on polydimethylsiloxane using soft lithography, used for culturing stem cells. *S* is the spacing between gratings and *H* the height of the topography. Reprinted from *Acta Biomaterialia*, 9, Ankam S *et al.*, Substrate topography and size determine the fate of human embryonic stem cells to neural or glial lineage, 4535–45, © 2013, with permission of Elsevier.

(f)

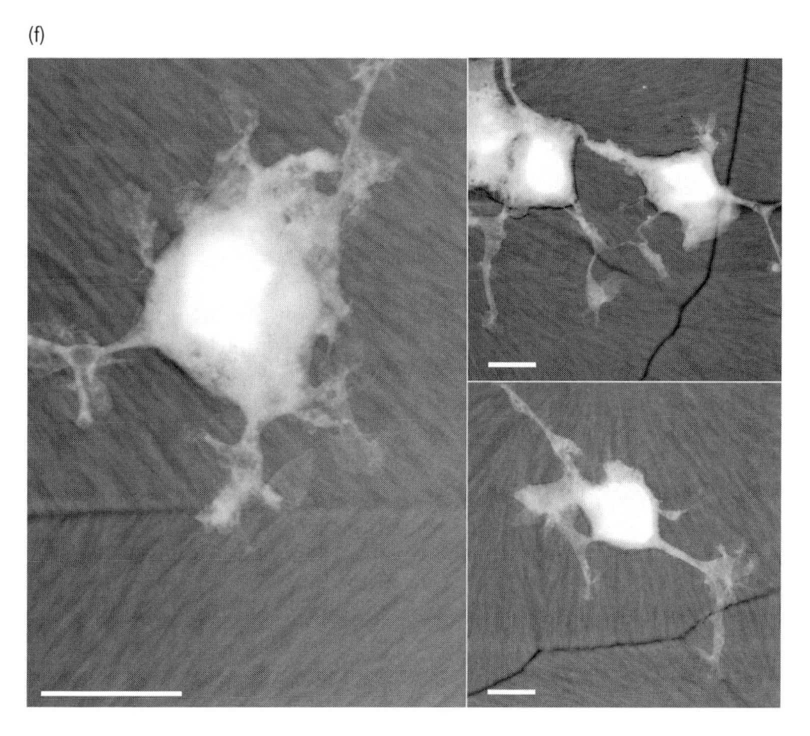

**Figure 2.11** *continued*
**(f)** AFM micrographs (tapping mode) depicting MC3T3 osteoblasts adhering to and branching on poly(caprolactone) spherulitic and lamellar organizations. The micrograph was recorded under dry conditions after 4 hours of adhesion and following cell fixation. The white scale-bar is 20 microns. Images courtesy of Dr. Edmondo Maria Benetti, Materials Science and Technology of Polymers, University of Twente, The Netherlands and Professor Nic Spencer, Laboratory of Surface Science and Technology (LSST), ETH Zurich, Switzerland.

control these effects, which are often interrelated, are:

- the readiness of the material to be oxidized on contact with air;
- the ease with which contaminants such as hydrocarbon molecules are adsorbed;
- the orientation of polymer side groups;
- the mobility or flexibility of polymer side groups;
- the surface charge density;
- the diffusion to the surface of additives and other minor components of the material.

Many of these factors will be discussed in terms of specific biomaterial applications later in the book.

### Characterization of surface chemistry

A range of sophisticated methods is available to characterize surface chemistry. It has to be emphasized that these methods depend on the interrogation of surfaces by radiation or ions and require high vacuum conditions. This means that these techniques do not necessarily provide information of biomaterials surfaces that is directly relevant to the performance in biological environments.

Electron spectroscopy for chemical analysis (ESCA), also known as X-ray photoelectron spectroscopy (XPS), involves a focused beam of X-rays directed at the sample surface; the interaction of these X-rays with the atoms of the material surface causes the emission of inner shell electrons. The energy levels of the emitted electrons are measured and these provide information about the atoms that are present. Typically a wide scan is carried out first in which the energies of all electrons within a wide range are detected, followed by narrow scans in regions of interest at higher resolution (Figure 2.11 (c)). All elements of atomic number 3–92 can be detected when present at levels greater than 0.1 atom %, with a semi-quantitative assessment of the whole elemental surface composition (note that *H* cannot be detected). The spatial resolution is usually between 10 and 150 μm and information can be obtained up to a depth of 50 Å. Angular-dependent ESCA allows assessments to be made of surface heterogeneity, oxidation states, organic groups, etc.

Secondary ion mass spectroscopy (SIMS) involves a beam of accelerated ions (of xenon or argon for example) directed at the surface. Detection of positive and negative secondary ions takes place, giving information that is complementary to that obtained by ESCA. There are several modes of operation, primarily based on static and dynamic modes. Static SIMS interrogates the outermost atom layers. Dynamic SIMS uses a higher ion dose and is able to yield information from depth profiling, up to 1 μm into the sample with a spatial resolution less than 500 Å.

### 2.5.3 Surface topography

Topography is a general term that denotes the shape of a surface. This could refer to geographical, architectural or many other features, but in materials science it is used to describe how the surface of a material object varies with respect to smoothness and roughness. For macroscopic objects such as implantable devices, this topography is usually determined by the fabrication process. Machining and casting may result in significant roughness, which can be reduced by various finishing techniques including electropolishing. A large number of parameters are used within engineering manufacture to characterize the nature of the surface finish, including the average roughness, $R_a$, the root mean square roughness, $R_q$, the maximum peak height, $R_p$, and the maximum peak-valley height, $R_{max}$. In principle, the surface finish can control tribological properties such as the friction and wear at joint replacement bearing surfaces, corrosion and degradation processes and cell–material interactions.

Because biological components such as cells and proteins are able to interact with specific features of material surfaces, a great deal of attention has been paid recently to the deliberate introduction of surface topographical features that may be able to control these interactions. When the topography is deliberately introduced, and has a regular form, it is usually referred to as a texture; alternatively the material surface may be described as patterned. Bearing in mind that cells normally interact with their environment at the molecular level, most of this interest has been directed to microscale or, especially, nanoscale textures and patterns. These interactions will be discussed in Chapter 3.

It is necessary to distinguish here the techniques to produce a texture on the material surface itself and those that produce a pattern of another substance on the material surface. The latter really come under the heading of surface modification and are discussed in Section 2.5.4 below, although there are many similarities in the processes. Textures on the materials themselves may be produced by a variety of techniques, mostly based on lithography, including electron beam lithography and laser interference lithography. Examples of the textures that may be produced are shown in Figure 2.11(d)–(f).

#### Techniques for characterizing surface topography

Probably the most widely used technique to explore the surface topography of biomaterials is the scanning electron microscope (SEM). This can accommodate quite large specimens and can provide images of the surface with such a depth of field that these are quasi-three-dimensional images. A focused beam of electrons is swept across the surface; electrons escaping from the surface are captured and counted by a detector. The counts at each point of the raster are compiled into a digital image. The images of Figure 2.11(d) and (e) were obtained by SEM. It is possible to couple an energy-dispersive X-ray spectroscope to the SEM (SEM EDX or SEM EDAX) in order to provide chemical composition data alongside the topographical image. There are some limitations with a conventional SEM. As with most electron beam techniques it is necessary to place the sample in a high vacuum. The surface also has to be clean, dry and electrically conductive. Non-conducting materials can be imaged but they have to be coated with an ultrathin layer of a conducting medium such as gold or carbon. Because of these extensive limitations, the environmental scanning electron microscope (ESEM) was developed. This uses multiple pressure limiting apertures that separate the specimen chamber from the vacuum column, allowing wet surfaces to be examined, a major factor in studying biomaterials that have been in contact with tissues.

The second major technique is atomic force microscopy (AFM). A 3D image of the surface is generated by scanning a tip that is mounted on a flexible cantilever across the surface and measuring the tip deflection that is caused by interactions between it and the surface. Since the deflections are magnified by the cantilever arm, the vertical resolution may be less than 100 nm. However, the sensitivity and spatial resolution is limited by the size of the tip and the properties of the cantilever. There are many variations on this theme. Contact mode AFM uses a very low force on the tip and is typically used for metals, ceramics and hard polymers. In non-contact mode AFM, the scanning tip is oscillated perpendicular to, and just above, the sample surface and a photodiode detector monitors the deflection of a laser beam reflected from the cantilever, giving a measure of attractive forces, again these being converted to an image. This may be used with softer hydrophobic polymers. In tapping mode AFM, the cantilever is vibrated close to its resonance frequency. As the scanner moves across the surface, a feedback loop maintains constant oscillation frequency or amplitude. This technique may be used for a wide range of polymeric materials. A typical AFM image is shown in Figure 2.11(f).

### 2.5.4 Techniques for surface modification

A very wide range of techniques is available for the surface modification of biomaterials, some of which are used in commercial products and the majority have been the subject of experimental analysis. The modification process may aim to cover the basic biomaterial with a totally different material, for example covering a polymer with a layer of a metal, or a metal with a layer of a bioceramic. Alternatively, it may aim to introduce different chemical groups into the surface, for example by graft polymerization or oxidation. It may intend to introduce discrete ions into the surface layer, or to build up multilayer films.

This range of techniques covers a spectrum rather than a discrete list of individual methods, since there may be considerable overlap between the various methods. We will discuss some of these methods at various points in this book, but we should expect to see the following principles and concepts used in many of them.

**Wet chemistry** As the name implies, this involves placing the material in question into a reaction vessel, where a chemical reaction results in the attachment of specific groups to the surface molecules.

**Hydrothermal methods** The above process may be assisted by high temperature and pressure, commonly used for the modification of bioceramics.

**Electrophoretic methods** The wet chemistry process can also involve the use of DC electrical fields that drive charged particles towards the surface to be modified, also used for bioceramics.

**Sol–gel techniques** This is also a wet chemical method in which a substance in the state of a sol, which has fluid-like properties, is prepared and the material in question is dipped into it and slowly

(a)

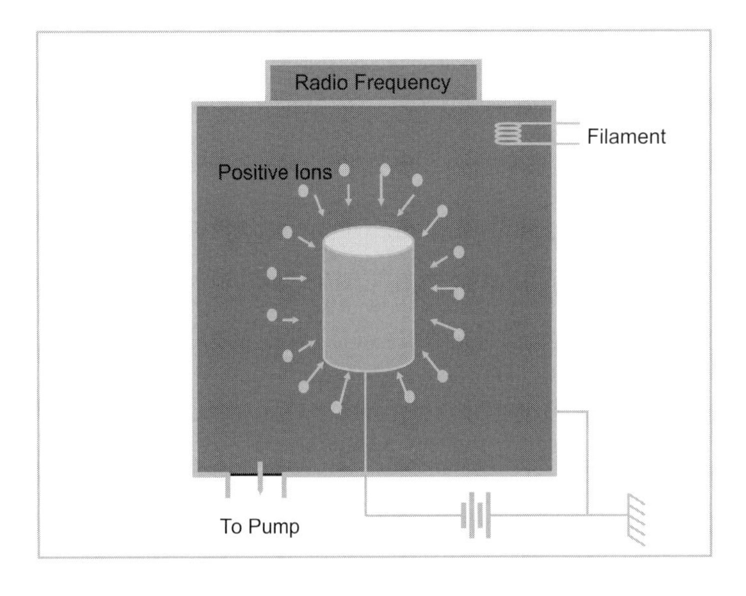


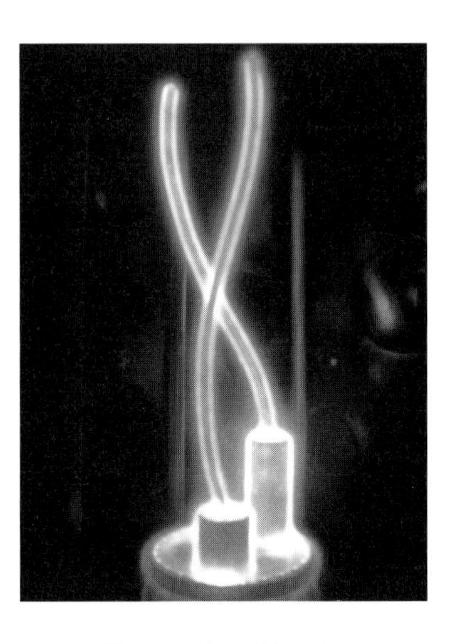

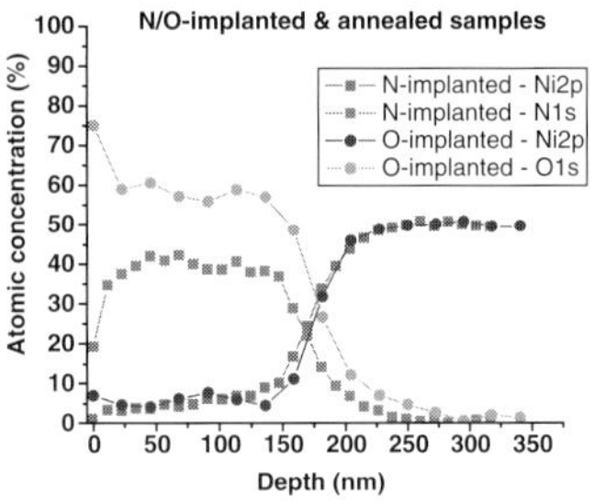


**Figure 2.12** Surface modifications.
(a) Plasma immersion ion implantation and deposition. Left, a schematic of the process. Upper right, 3D plasma treatment of scoliosis correction rod made of NiTi to mitigate leaching of Ni ions into tissues. Bottom right, surface composition after treatment. Image courtesy of Professor Paul Chu, Plasma Laboratory of City University of Hong Kong. See plate section for color version.

withdrawn, when it transforms into a gel, with more solid-like properties, through the loss of water. Again this may be used to coat some metals with bioceramics.

**Ion implantation** Ion implantation involves a technique in which ions are accelerated to high energy levels and directed to the surface of the material to be modified. This is widely used to inject small ions, such as those of nitrogen, carbon or boron, into the surface of metals, which results in increases in hardness and wear resistance. The surface and sub-surface changes may be quite profound in view of the energies involved.

**Plasma processes** These are probably the most widely used techniques (Figure 2.12(a)). Plasmas are dissociated gaseous substances that may contain electrons, positive or negative ions, atoms, molecules, free radicals and photons. The plasma environment is typically generated in a temperature range up to 60°C, with an electron energy between 1 and 10 eV and a pressure between 0.025 and 1 torr. These conditions allow plasmas to be used on sensitive substrates; for example, the temperatures are low enough to allow the treatment of most polymers. Depending on the conditions, the plasma may produce etching, ablation, deposition or molecular immobilization effects. The latter two

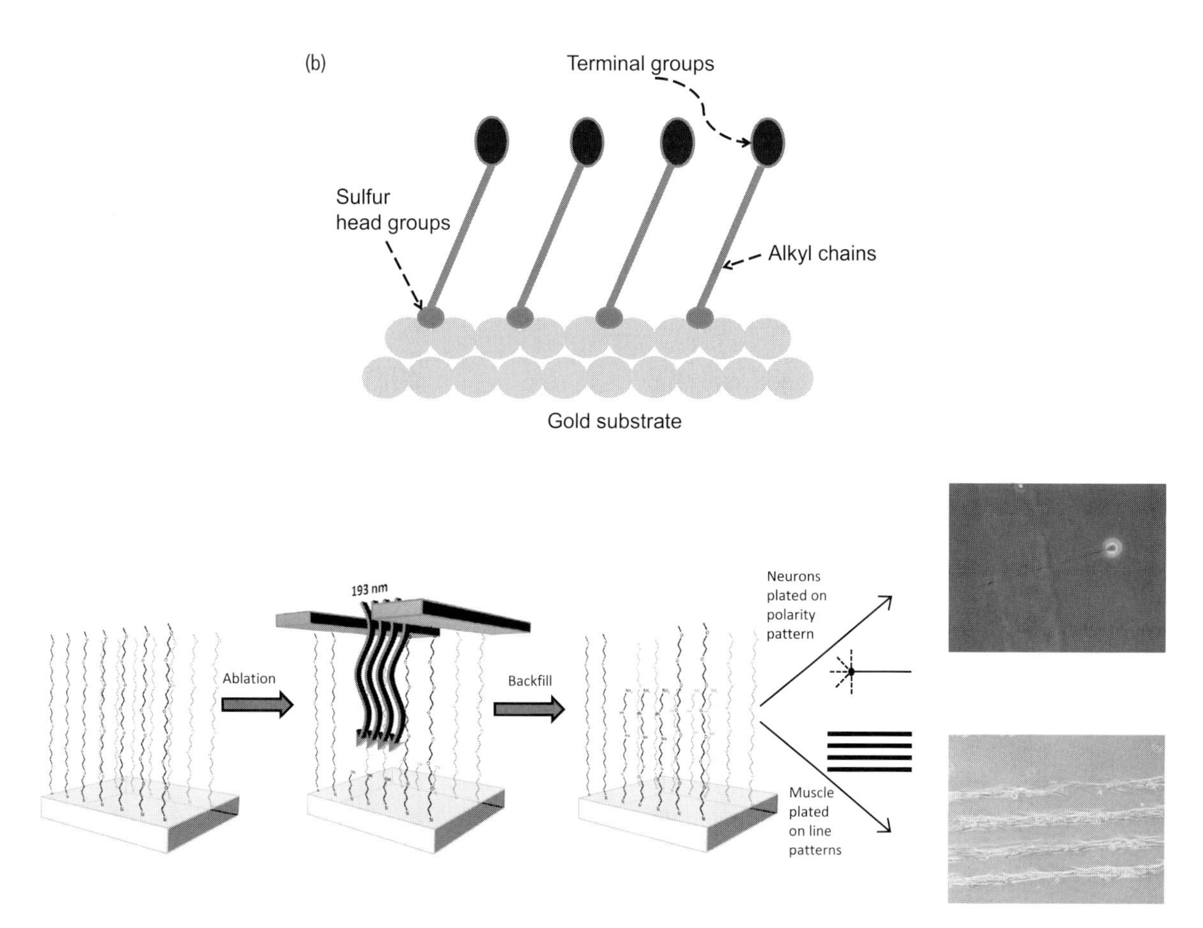


**Figure 2.12** *continued*
**(b)** Self-assembled molecules. In the top image, molecules based on alkyl chains self-assemble on a gold surface, sulfur head groups of the chains binding to gold atoms, and terminal groups at the other end of the chains providing functionality. A practical example is given in the lower image, courtesy of Dr. James Hickman, NanoScience Technology Center, University of Central Florida, Orlando, Florida, USA. Light is passed through a wavelength selective mirror and reflected onto the sample, which has been placed under a patterned chrome photomask. The exposed regions of the poly(ethylene glycol) (PEG) monolayer are then photoablated by the incident laser light leaving a substrate suitable for reaction with another silane, in this case (3-trimethoxysilyl propyl) diethylenetriamine (DETA). The use of a cell repulsive substrate, PEG, and a cell supporting substrate, DETA, promotes differential adhesion and growth of cells. This system ultimately results in cell patterns that promote organized and directed cell adhesion and maturation.

effects allow surface chemical modifications where the surface layer is usually conformal, defect-free and adhesive, although the chemistry itself will be complex, and indeed the resulting surface may be unique without there being any corresponding structures created by conventional chemistry. The processes allow controlled modification of wettability, hardness, chemical reactivity and so on.

**Photopolymerization** Photopolymerization at a material surface, sometimes called photografting, may be achieved with either UV or visible light and is used to achieve covalent coupling to the surface. It is particularly advantageous since this coupling may be produced without much prior surface preparation. For example, some reactive species can simply be introduced to the polymer substrate under irradiation conditions when its molecules are excited to a reactive triplet state that facilitates hydrogen abstraction, leading to cross-linking with the polymer surface. These can be simple and inexpensive processes. These are sometimes referred to as living radical polymerization. They are particularly effective in the production of microfluidic devices, as with the technique of contact lithography. Photopolymerization can be used to prepare devices with variable chemistry substrates for high throughput screening procedures for cell culture.

**Radiation grafting** Radiation-induced graft polymerization involves the use of ionizing $\gamma$ radiation or high-energy electron beams to break chemical bonds on the surface of the substrate, forming reactive species such as free radicals or peroxides, which are then reacted with a monomer. The subsequent reaction proceeds as a free radical chain reaction with this and, if necessary, other monomers. A wide variety of surface chemistries can be created. The substrate may be limited by the possibility that it may be damaged by the irradiation.

**Self-assembled monolayers (SAMs)** These are assemblies derived from the adsorption of molecular species from solution or gas onto a solid or liquid surface. They are often prepared from small amphiphilic molecules on which the head group has a specific affinity for the solid surface. These molecules adsorb and pack laterally on the substrate with a high degree of order. Obviously the binding of these head groups to the substrate has to be strong, especially filling every possible site on the surface and displacing contaminants, and the ordered structure will only effectively materialize if there are reasonably strong secondary interactive forces between the adsorbed molecules. Many useful SAMs are based on n-alkyl chain assembly or thiols (Figure 2.12(b)). For example, n-alkyl silanes can be adsorbed to form monolayers on hydroxylated surfaces and disulfides can be assembled on the surfaces of metals such as gold.

It is also possible to prepare SAMs with macromolecular chains instead of small molecules. These offer enhanced stability and easier processing. The molecules tend to form coils rather than fully extended chains. Increased functionality may be derived from the use of random copolymers. One further variant is the development of so-called layer-by-layer techniques to create nanoscale thick layers of alternating charged molecules. This may be particularly interesting for controlling some biocompatibility events through polyelectrolyte coatings.

**Micropatterning** Micropatterning involves the chemical and spatial design of surfaces in order to control complex environmental reactions at these surfaces. This is particularly important in the control of cell and protein behavior, and different forms of micropatterning have become very important in biotechnology and regenerative medicine applications. There are several options for the micropatterning process. The principle of these techniques is that a pattern needs to be developed on a surface with spatial resolution at the microscale such that some areas are amenable to reaction with the cells or molecules in question, while other areas are not. This implies that it is crucial to use a non-fouling surface for the latter part in order to limit non-specific adsorption in those areas. Many systems use poly(ethylene glycol), PEG, or PEG derivatives, for this purpose, while some lipid bilayer structures and thermoresponsive polymers have also been used.

The question then arises as to how a pattern of specific interactive areas and non-fouling areas can be generated with the resolution and preciseness that is required. A well-tried group of processes are those of soft-lithography, where an elastomeric material, such as polydimethylsiloxane, usually known as silicone elastomer, is used to create stamps that replicate the microstructured master, which is usually made of silicon. In micro-contact printing, this process involves casting the liquid phase silicone over the master. The stamps are "inked" with a solution containing the biomolecules to be printed, which are then transferred to the required substrate. This is generally followed by back-filling with the second, non-fouling molecule. Many surfaces can be patterned in this way; sometimes this is assisted by microfluidic processes. Alternative, more complex, techniques have been developed, including photolithography, ink-jet patterning and laser-guided cell writing.

### 2.5.5 Special considerations at the nanoscale

At several places in this book we refer to the influence of the nanoscale on biomaterials-related phenomena. From a purely materials science perspective, the finer the structure, for example with grain size, usually the better the properties. From a biocompatibility or toxicological view, we will see that nanoparticles are associated with different properties than microparticles or visible particles, sometime for the good, other times for the worse. As far as the surface of materials is concerned, topography can be an important mediator of a number of properties with special emphasis on nanoscale topography. There are some effects on rates, and even mechanisms, of metallic corrosion and polymer degradation, and of some mechanical and physical properties, derived from the nature of the surface topography. It is, however, the biological properties that are most affected. Bear in mind that most cells of the human body have diameters of a few microns and have critical surface features that operate at the molecular or supramolecular level, so there should be no surprise that both micro- and nanoscale features of the surface should have some influence on the tissue's response to the material.

It is clear that surface nanotopography affects cell adhesion and proliferation, but definitive relationships between surface metrics and cell behavior are not yet understood. It is likely that there will be optimal surface roughness, which, all other things being the same, will define the influence on cells, and this optimum will not be ultra-smooth nor ultra-rough.

## 2.6 Practical materials

### 2.6.1 Metals and alloys

#### Simple solid solutions

In Section 2.4, the basic crystal structure of metals was introduced, and that discussion centered on pure metals. In reality, we rarely use pure metals for practical applications and so it is necessary to describe the really useful metallic materials known as alloys. An alloy is a metallic structure (i.e., based on the metallic bond) that involves two or more elements. In most cases there will be between two and six metallic elements intentionally involved in this mixture, but there may be some non-metallic elements, especially C, O and N, and there will almost inevitably be some impurities, which are usually derived from the ores from which metallic elements are refined, or picked up during a processing phase, and which are difficult to remove.

In the simplest case we can consider one element A and another element B. If these elements have atoms of roughly equivalent size, and if the two elements form the same type of crystal structure, and if they are of approximately the same melting point, then they are likely to be mutually soluble. Adding atoms of element B into a lattice of element A (Figure 2.13(a)) will result in the direct substitution of one for the other. This results in crystals of the solid solution A+B. This is a single phase, with all crystals or grains of identical composition; the microstructure will be indistinguishable from that of either pure metal. This process is known as solid substitution.

If we plot a graph of the melting points of alloys of different composition across the range 100%A, 0%B to 0%A, 100%B, we will get a diagram similar to that seen in Figure 2.13(b). Note that while for each pure element there is a single well-defined melting point, for intermediate concentrations there are two points, and hence two curves. For any one composition, the alloy will start to melt as the temperature is raised to one defined point and then it will be partly liquid and partly solid until it reaches a higher temperature, when is becomes completely liquid. The lower line representing the boundary between solid and solid–liquid phases is known as the solidus and the upper line, the liquidus. This diagram is known as a phase diagram, which is of immense importance for understanding the structure and properties of alloys. In this simple phase diagram, the single solid phase that is stable is given a Greek symbol, usually α.

With most combinations of metallic elements, the differences between the constituent atoms are sufficiently great to prevent complete miscibility across the composition range. If now we have elements A and C, adding some C to A may result in a solid solution, perhaps called the α phase, but there will be a limit to the amount that can dissolve. If we add A to C, then another phase, perhaps called the β phase, will form, again there being a limit. At compositions in between these two limits of solid solution, the two phases α and β will co-exist. These can be distinguished on examination of the microstructure (Figure 2.13(c)) and will give rise to a phase diagram such as in Figure 2.13(d). It will be seen that the liquidus has two forms and that most alloy compositions have melting points lower than those of either A or C. There is one composition, known as the eutectic composition, which has the lowest melting point; this is useful when designing low melting point soldering and brazing materials, the melting points of some eutectic mixtures being only a few hundred degrees centigrade.

(a)

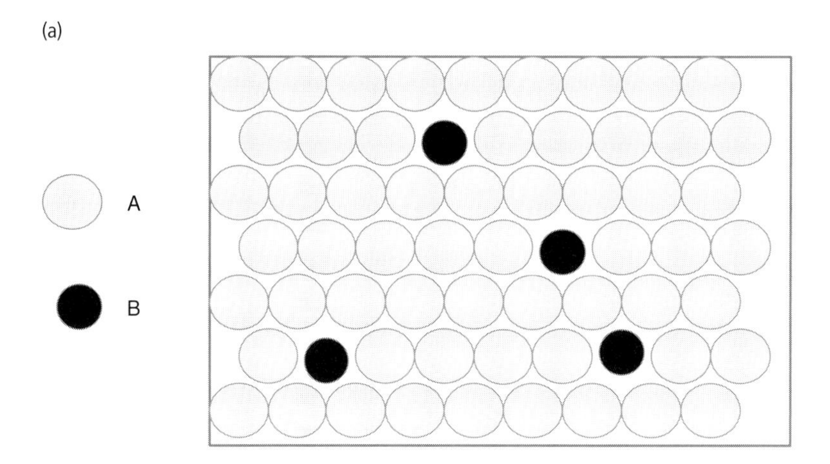


(b)

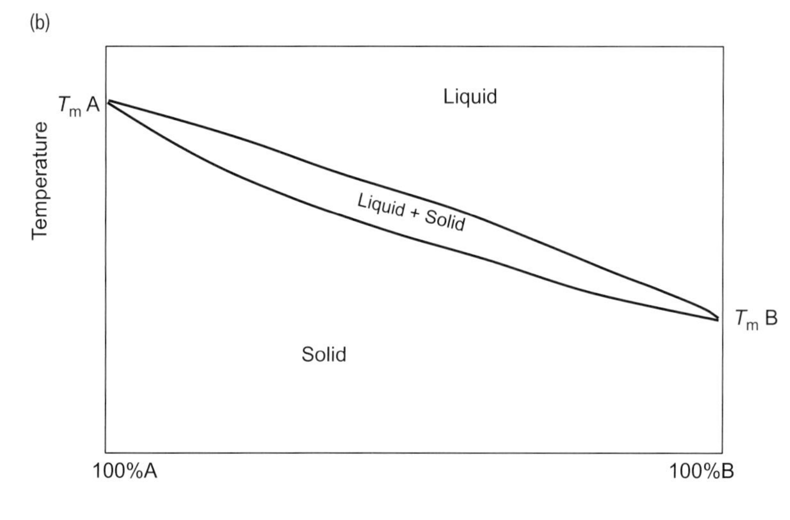


**Figure 2.13** Metallic structures.
**(a)** Two-dimensional representation of a substitutional solid solution involving two metallic elements A and B. The atoms of B are a little smaller than those of A and essentially dissolve on the lattice of A.
**(b)** A phase diagram of a solid solution. $T_m$ A is the melting point of element A, $T_m$ B is the melting point of B. On heating any alloy, melting will take place over a range of temperatures, the structure between the start and finish of melting consisting of a combination of liquid and solid.

(c)

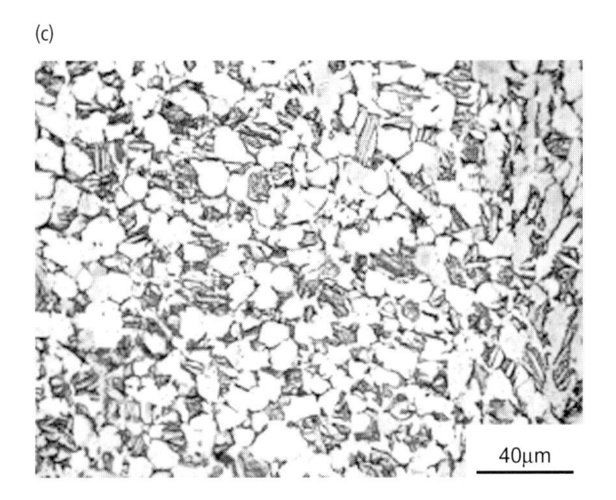


(d)

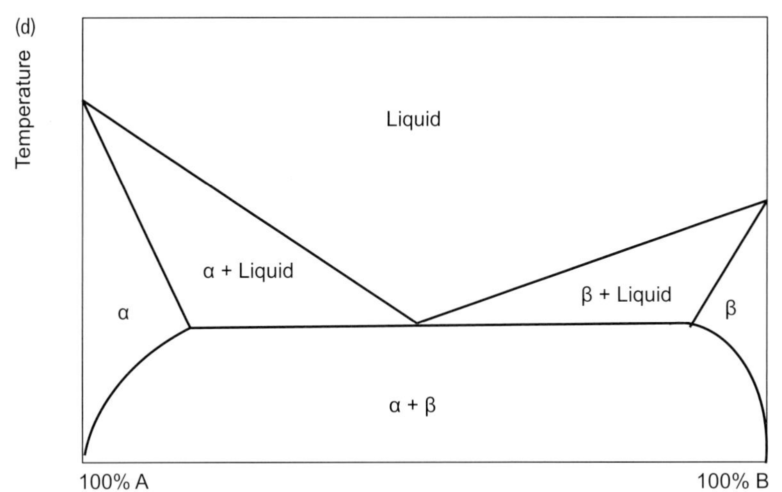


(e)

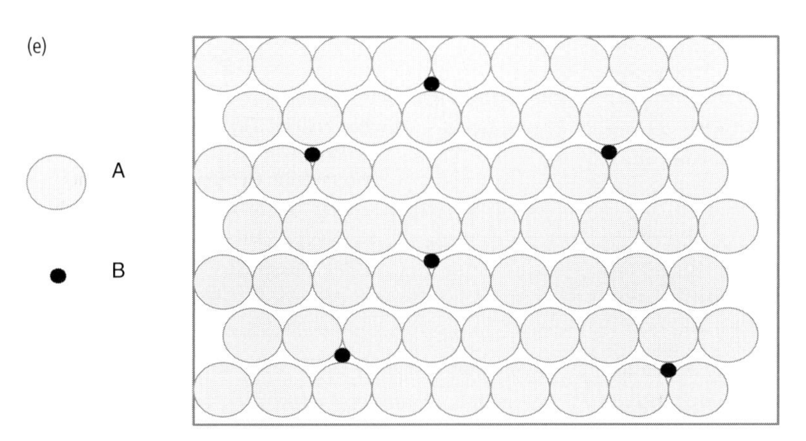


**Figure 2.13** *continued*
**(c)** An optical micrograph of two-phase Ti-6Al-4V alloy. Reprinted from *Mechanics of Materials*, 42(2), Luo J *et al.*, Constitutive model for high-temperature deformation of titanium alloys using internal state variables, 157–65, © 2010, with permission of Elsevier. **(d)** A phase diagram of a two phase alloy system. A small amount of element B may dissolve in A (the amount increasing with temperature), giving the α phase. A similar situation is seen to the right of the diagram, with the β phase. Between the limits of solubility, alloy structures involve mixtures of α and β. Melting takes place over a range of temperatures, with the exception of one point, where the melting point is the lowest, this usually being lower that the melting points of both pure metals.
**(e)** Two-dimensional representation of an interstitial solid solution involving two elements A (a metal) and B (usually a non-metal). The atoms of B are much smaller than those of A and are located in small spaces between the atoms in the lattice of A. The limit of solubility is usually very small. A good example is the solution of very small amounts of carbon in iron, which is the basis of steels.

### Complex solid solutions and intermetallic phases

The ability to form solid solutions is obviously dependent on the relative sizes of the atoms. If the two elements are of vastly different size, substitutional solid solution may be impossible. That does not mean to say that any form of solution is impossible, but it will not occur by substitution and will be very limited in terms of the composition range. In this situation, the atoms of the smaller element will occupy spaces between the larger atoms. These spaces are known as interstices and the result is an interstitial solid solution, shown in Figure 2.13(e). The best example is that of iron and carbon, the latter atoms occupying interstices between the large iron atoms. This is the basis of all steels. As we shall see later, the benefit in terms of strength arising from interstitial solutions is very considerable.

Even more important in the case of steels is the nature of the phase that forms when the limit of carbon-in-iron solubility is reached. This limit is less than 0.01%. At greater levels of carbon, an intermetallic phase, of stoichiometric ratio $Fe_3C$, known as cementite, forms. This phase has an orthorhombic crystal structure, with 12 atoms of iron to 4 of carbon, and with a weight percentage of 6.67% carbon. This phase is very brittle but can impart great strength to the steel.

It is at this point that we should return to the matter of multiple crystal forms for some metallic elements. The crystal structure adopted by a metal is determined by a balance of thermodynamic factors and the equilibrium structure will depend on the prevailing conditions. Foremost among the factors that determine this structure is temperature. Raising the temperature of a pure metal may mean that, although one phase is stable at room temperature, a different phase becomes stable at a higher temperature. As noted in Table 2.1, different crystal structures of the same substance are called allotropes. This is of considerable relevance to metallic biomaterials, since both iron and titanium exist in more than one crystal state. Iron is body centered cubic up to 912°C (as the $\alpha$ phase), between 912°C and 1394°C it is face centered cubic (the $\gamma$ phase) and then it reverts to a body centered cubic structure at 1394°C (the $\delta$ phase), which is stable up to the melting point at 1538°C. Titanium exists as the body centered cubic $\alpha$ phase up until 900°C, when it changes to the close packed hexagonal $\beta$ phase. We shall see later just how significant this is in biomaterials science; suffice it to say here that the transition point can be controlled by the addition of alloying elements, which gives powerful control over mechanical properties.

### Multi-elemental alloys

The main principles of alloying have been discussed above in the context of systems involving two elements. In practice, alloy systems usually have more than two elements. It is difficult to make generic arguments to explain the principles here, partly because phase diagrams become very difficult to construct and interpret – three elements require a three-dimensional diagram, and four or more elements make diagrams impossible. One of the main points here is that the benefits of alloying in terms of improved mechanical properties (see Section 2.8 below) are often enhanced in an amplified way when two or more phase modifications are introduced, and that alloying additions may be used to simultaneously enhance different types of property. For example, the carbon added to iron gives good mechanical properties to the resulting steel, but if we wish the steel to be corrosion resistant we have to add chromium, or better still a combination of chromium and molybdenum. We will reserve discussion of these points until specific metallic biomaterials are covered in Chapter 8.

## 2.6.2 Synthetic structural polymers, plastics, elastomers and textiles

Let us now turn our attention to those solid materials that are based on the covalent bond. Since, as we have seen, polymer molecules may contain many thousands of atoms, including C by definition but also, possibly, H, O, Cl, S, P and some others, there is an almost infinite library of polymers available to us. There are many ways in which they may be classified; we shall do so here on the basis of some important characteristics that control the properties of the materials.

### The physical nature of solid polymeric materials

Let us first deal with the major factors that control the overall nature of the polymer. Since the polymer consists of many long chain molecules, we can consider a structure in which these molecules are relatively unconstrained by each other, the intermolecular bonds being much weaker than the intramolecular bonds. As the polymer is heated, so the imposed thermal energy gives these molecules even more freedom and the material will become softer and softer until it eventually becomes liquid at its melting point. For reasons that will become clearer later in this chapter, the material is said to become more plastic as it gets hotter, and, essentially, the material starts to flow. Such a polymer is said to be thermoplastic; this is very useful since it means that shaping of the polymer is achieved relatively easily in molding, extrusion, blowing and other mechanical processes at elevated, but not really high, temperatures.

On the other hand, we may have a polymer in which strong cross-links are established between molecular chains, and these have the profound effect of tying the molecules down. These cannot flow at elevated temperatures; heating them has no effect on plasticity until they effectively decompose. Most cross-linked polymers are prepared by taking a low molecular weight resin and applying a curing agent/catalyst which causes the cross-links to form by chemical reaction, which is often exothermic. As such the polymers are "cured" or "set," and they are known as thermosetting resins; epoxy resins are the best-known examples.

The mechanical properties of thermoplastic and thermosetting polymers can vary quite widely. Some are very soft, some quite hard; some are very strong, some very weak. There is one group of polymers, however, that has some very special characteristics in the sense that they can deform very extensively. We shall discuss the phenomenon of elasticity later in the chapter, but the word "elastic" has a well-known common meaning, and we all know that a piece of elastic will stretch and stretch. Although there are a few chemical entities that we can call elastic, they are all polymers and have a unique characteristic that their molecules are coiled and the stretching of the material is associated with the molecular uncoiling. These materials are called elastomers. Their relevance here is that several natural tissues have elastomeric characteristics, which we often wish to emulate in a synthetic material.

Next we come to fibers and textiles. Consider the thermoplastic material, previously shown in Figure 2.9, in which we have crystalline areas and amorphous regions. If we take a part of a crystalline region, we can see that the properties in a direction parallel to the molecular orientation will largely be controlled by the interatomic bonds along the backbone. The properties in a perpendicular direction will be controlled by the much weaker intermolecular bonds. The overall properties of the material will reflect the different contributions of amorphous and crystalline regions, and the differences between these orthogonal directions within the crystalline regions. If, however, we prefer the properties that are associated with the long axis of the molecules in the crystalline region, we may be able to separate out some of these regions, and build our material around this structure. This means that we could produce the structure seen in Figure 2.14(a). If this is very thin and very long, then we have a fiber. This, of course, is not very original and nature got there a long time ago, for example with silk fibers, which are immensely strong. Their molecules are not simple linear carbon-only backbones, but are far more complex and have biological functionality as well, which is why there is so much interest in them for regenerative medicine applications.

Spiders naturally realized there were limited applications of infinitely long thin fibers, and developed the technology to weave them into complex and useful shapes. The same has happened with synthetic polymer chemists, who copied spiders and other creatures to produce synthetic fibers and structures, which we now call textiles, based on these fibers (Figure 2.14(b)). The first really useful textile was developed in the 1940s when nylon (and nylon fabrics) were produced during attempts to emulate the sheep's capacity to produce wool.

Finally in this section let us take a look at polymers that contain water. Many polymers, on exposure to moisture will absorb a little of the water, perhaps up to a few percent weight/weight. This is important in biomaterials science, but not of concern here. Instead let us consider a structure of polymer molecules, as shown in Figure 2.14(c), where there are large spaces between the chains. What would happen if we immersed such a structure in water? The answer is that we might expect a large influx of water molecules, and indeed there are some such materials where their equilibrium water content is over 70% of the polymer. Now we would expect that many of the properties would depend as much on the water as on the polymer chains. This is not a bad idea in

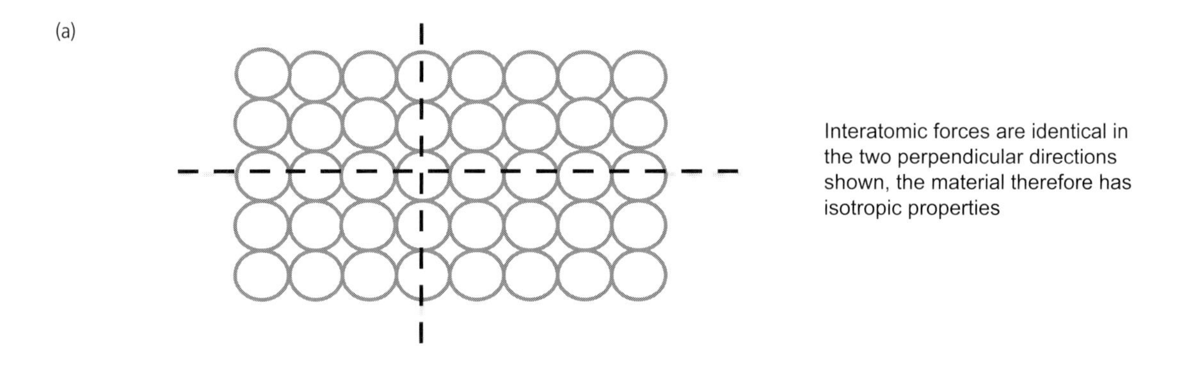


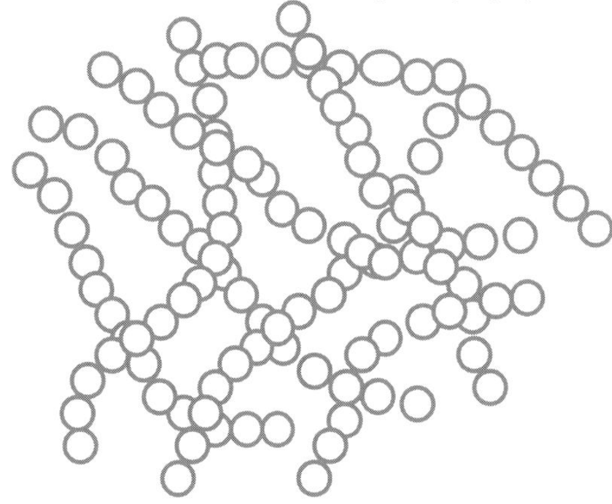


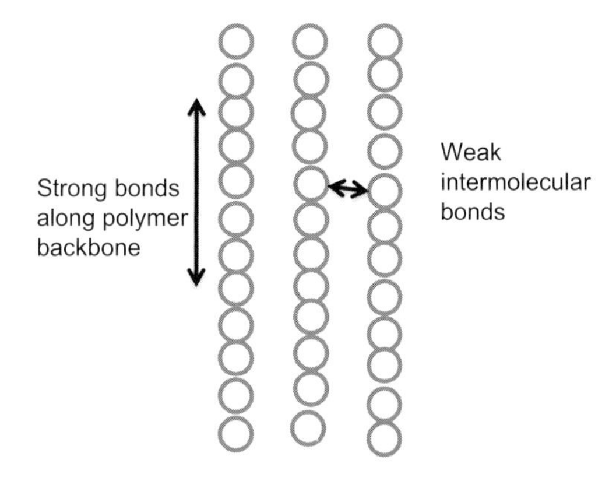


**Figure 2.14** Textiles and hydrogels.
**(a)** Scheme showing the dependence of mechanical properties on arrangements of atoms and molecules. Top left shows a highly crystalline structure (metallic or ionic) where interatomic forces are equivalent in mutually perpendicular directions, giving isotropic properties. With a polymeric structure, bottom left, if the molecules are randomly arranged, there will again be isotropic behavior. However, at bottom right, if the molecules are arranged essentially parallel, then the strength of the primary intramolecular forces is maximized, giving highly anisotropic properties. This is the basis of polymeric fibers, which have maximum strength along their axes.

(b)

(c)

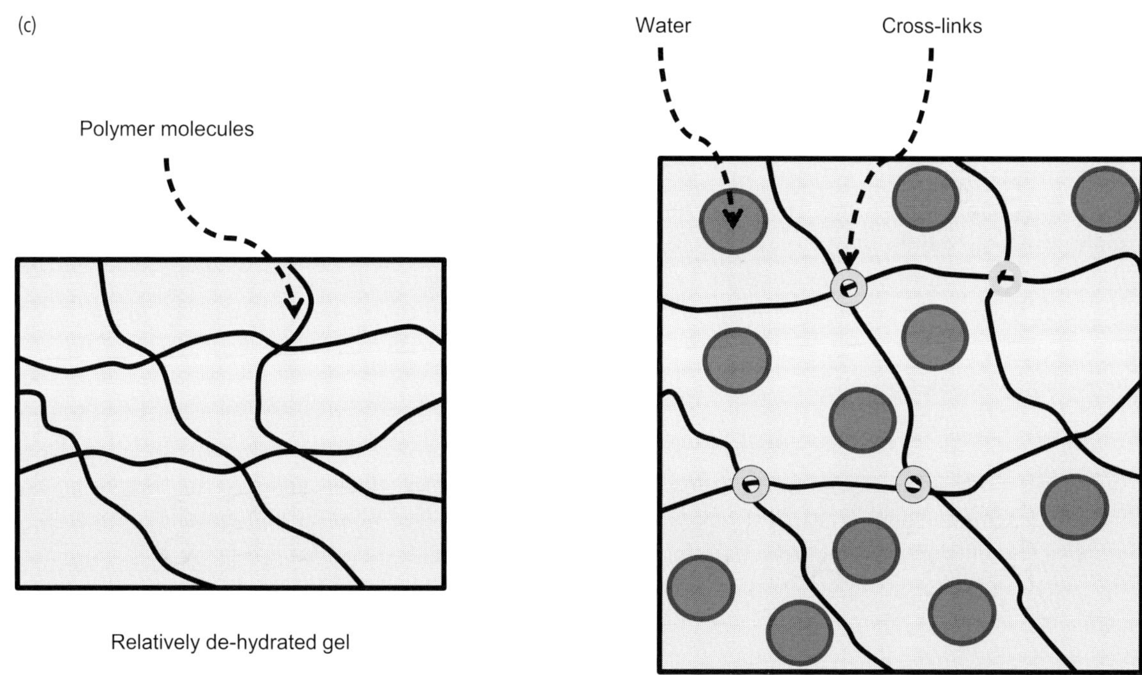


**Figure 2.14** *continued*
**(b)** Fibers can be arranged in many ways in order to give textiles. **(c)** On the left are the polymer molecules of a gel in a relatively de-hydrated state. Upon hydration, water molecules penetrate the spaces between the polymer molecules. A cross-linking reaction may also be initiated during the swelling, for example using photopolymerization, to give a strong hydrogel.

biomaterials science since many natural tissues similarly contain large amounts of water. These materials are called hydrogels.

We therefore have a wide range of solid polymer concepts from which to choose our biomaterials, from thermoplastics to thermosetting resins, and elastomers, hydrogels, fibers and textiles. Much of this book is concerned with the varied performance of these different types.

### 2.6.3 Biopolymers

Although synthetic polymers have been used very extensively in medical applications over the last 40–50 years, we have witnessed a major increase in interest in natural polymers, often referred to as biopolymers. This is not surprising in the light of the comment above concerning the attempts to emulate the properties of silk. For a long time, the use of natural products in medical devices was limited by the inherent variability in natural materials, this variability being inconsistent with the requirements of product quality control. Attempts were made to use natural substances such as wood and ivory in some prostheses, but the only sustained examples for many years were the natural products such as silk, cotton and, especially, catgut used in surgical sutures. Even their use has declined as better and more consistent synthetic alternatives have become available. At the same time the development of materials based on specific components of natural substances has become important. These primarily include proteins and polysaccharides. Among the proteins are collagens, elastin, keratin, fibrin/fibrinogen and silk. Among the polysaccharides are alginates, chitosan, hyaluronan and heparin. These will be discussed individually in Chapter 8 and elsewhere in the book.

### 2.6.4 Ceramics, glasses and glass-ceramics

#### General features

We noted in Section 2.4 that ceramics are based on the ionic bond and that simple compounds such as aluminum oxide can exist as single crystals or polycrystalline materials with wide-ranging characteristics. However, the majority of ceramic materials are highly complex, involving not only multiple elements but also combinations of ionic and covalent bonds. The directionality of the bonds between atoms and groups often makes the formation of ordered crystal structures difficult and so ceramic materials may either be crystalline, amorphous or both. Amorphous ceramics are referred to as glasses. A glassy substance, nearly always transparent, is essentially metastable and can slowly transform to a crystalline form, a process known as devitrification, often seen as old glass windows slowly become opaque. Silica, which is the oxide of silicon, is a material that exists in natural form; there are many versions of this rather simple material including both crystalline and amorphous forms, which have quite different properties, including toxicity.

As we shall see later, ceramic structures tend to be very brittle, which limits their engineering applications. They can be used in massive bulk form, such as concrete and cement. They are, however, often very hard and can be used for bearing surfaces and cutting or abrading materials, and in addition can be very environmentally resistant, hence their use in decorative objects such as porcelain and various glasses.

A major limitation to ceramics, imposed by their inherent brittleness, is the difficulty of being shaped by mechanical means. Ceramics are therefore often cast or sprayed from the molten state or sintered by applying intense heat and pressure to powders. This also contributes to poor properties since these techniques often leave residual porosity.

Structural ceramics have very few biomedical applications, except for the wear-resistant oxide ceramics used in joint prostheses and the so-called bioactive ceramics discussed in the next section. However, nanostructured ceramics, especially nanoparticles, have considerable potential. One of the best examples is iron oxide $Fe_2O_3$, which can be easily produced in nanoparticulate form within the

20–50 nm range, when it has superparamagnetic properties and can be used as a contrast agent in imaging techniques. An electron microscope image of such particles can be seen in Figure 2.15(a).

### Bulk metallic glasses

Bulk metallic glasses (BMGs) are, from a compositional point of view, metallic alloys, but they are able to avoid crystallization during solidification. They retain the amorphous structure of the parent liquid alloy and form glasses, which are chemically homogeneous. Many alloys based on Zr, Fe, Cu, Ni, Ti, Mg and Pd can be prepared as BMGs. The most important are those that involve Mg, Ca and Zn. The lack of crystalline lattice defects means that strengths can be very high, albeit with limited ductility.

### Nanoscale ceramics – semiconductors

A quite different family of materials based on ceramic crystals has been developed over the last half-century, which have become very important at the nanoscale. These are the semiconductors. They will be considered in more detail in Section 2.7.2; as we shall see, metallic materials are conductors of electricity, but ceramics are insulators. By taking a material with a defined crystal structure and adding very small amounts of specific elements, some conductivity can be introduced. Silicon and germanium, both from Group IV of the Periodic Table, provide the basis of well-known semiconductors that have widespread use in electronics, their properties being defined by minute levels of so-called doping agents such as boron and arsenic. Other semiconductors are based on ceramic structures derived from combination of elements such as gallium arsenide and indium antimonide. These conventional semiconductors do not have direct biomaterials applications, but they do when produced at the nanoscale in the form known as quantum dots. The best example is cadmium selenide, an equiatomic compound that crystallizes in hexagonal form. This can be produced in colloidal form where the particles are in the size range of up to 10 nm. The number of atoms in a quantum dot ranges from 10 000 up to 100 000 and the properties depend on their size. In particular, because of their electronic structure, they are able to fluoresce very strongly, thus finding use in imaging technologies. An image of a quantum dot is given in Figure 2.15(b).

## 2.6.5 Bioceramics

Just as with polymers and the use of biopolymers, ceramics have their natural counterparts, the bioceramics. There is a slight difference in that although some of the materials that are collectively called bioceramics are naturally occurring, some are entirely synthetic. The defining character of a bioceramic is that it is a ceramic, glass or glass-ceramic that has specific biological activity that can be exploited in biomaterials applications. The bases of the majority of bioceramics are the various forms of the compound calcium phosphate, which occurs as the mineral phase of hard tissues in animals, especially bones and teeth. The compounds that are formed by the combination of calcium and one of the different forms of phosphate ions are numerous. Some may be pure forms of calcium phosphate such as $Ca_3(PO_4)_2$ or they may have some small additions of substitute elements such as silicon, strontium or fluorine, or may be hydrated to some extent. In nature, calcium phosphate usually exists as a form of hydroxyapatite, $Ca_{10}(PO_4)_6(OH)_2$; again there are natural variations on this related to substitutions and stoichiometry. Many different forms of calcium phosphate, including different hydroxyapatites have been used as biomaterials. They are usually crystalline solids that are prepared from powders, often sintered, deposited from solution, or melt sprayed onto substrates.

## 2.6.6 Composites

As their name implies, composites are materials made up from combinations of different types of substance. We should be clear, however, that they are not random mixtures, nor are they combinations of quite similar but not exactly the same substances. Alloys

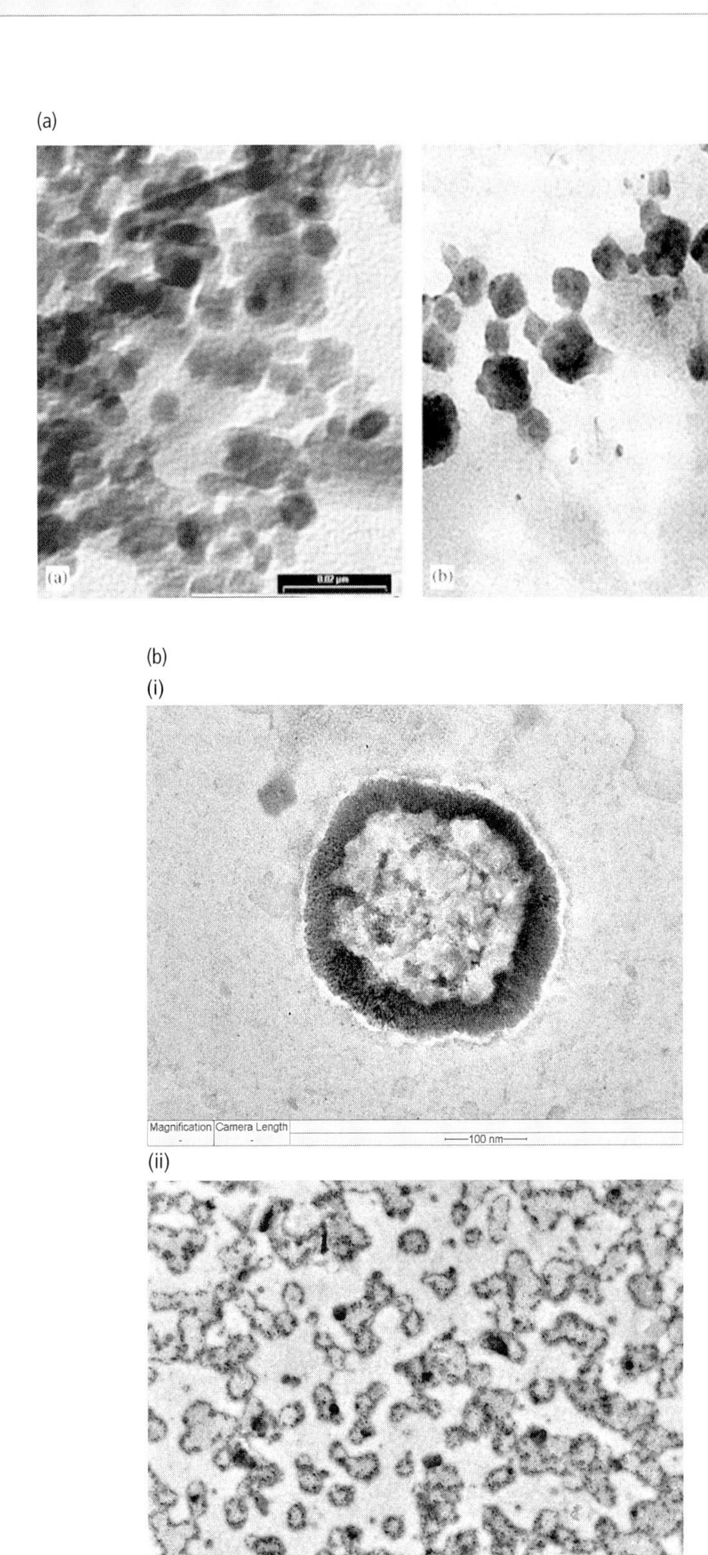


**Figure 2.15** Nanoparticles.
**(a)** Iron oxide nanoparticles. Transmission electron micrographs of magnetite nanoparticles prepared in bulk solution (left) and in water-oil microemulsion. Reprinted from *Biomaterials*, 26(18), Gupta A K and Gupta M, Synthesis and surface engineering of iron oxide nanoparticles for biomedical applications, 3995–4021, © 2005, with permission from Elsevier.
**(b)** Quantum dots. Field-emission transmission electron micrograph of (i) an individual quantum dot-loaded liposome and (ii) multiple quantum dot loaded liposomes. Reprinted from *Biomaterials*, 33(12), Muthu M S *et al.*, Theranostic liposomes of TPGS coating for targeted co-delivery of docetaxel and quantum dots, 3494–3501, © 2012, with permission of Elsevier.

involving different elements and phases are not composites, nor are polymer blends and copolymers. A commodity plastic that contains fillers just to reduce costs is not a composite. Composite materials are also bulk materials rather than standard materials with surface coatings – these may be composite structures but not composite materials.

The real significance is that composites involve two, or occasionally three, different classes of materials where the structure of the composite has been designed to maximize the benefits of each component and, if at all possible, minimize their disadvantages. The majority of useful composites combine a ceramic phase with a polymer phase, although polymer–metal and metal–ceramic composites are also possible. Usually the composite structure can be identified with one continuous phase and one dispersed phase. For example, the continuous phase may be a polymer, while the discontinuous phase may be a dispersion of ceramic particles. Such a composite in which the ceramic particles are uniformly dispersed in the polymer will tend to have isotropic properties. A very good example in the biomedical field is the composite dental filling material. The polymer will be a clear resin such as urethane dimethacrylate and the ceramic could be silica. The resin, effectively a thermosetting material, is able to polymerize and cross-link within the tooth while the ceramic gives the material toughness and abrasion resistance. Optimal properties are usually obtained by maximizing the amount of ceramic, which can be achieved through the use of multimodal size distributions.

Other composite materials can be obtained by using different physical forms of the dispersed phase, the best example being seen with the use of fibers. Again usually involving a thermosetting resin such as an epoxy or phenolic resin, these may contain high percentages of long fibers, usually made of a glass (fiberglass) or carbon. Carbon fiber composites are amongst the strongest and toughest synthetic materials, especially on a strength-to-weight ratio basis; these are obviously highly anisotropic. It is difficult to process thermoplastic materials with very long fibers, but very good properties can be achieved by cutting the fibers into short lengths. Some very good engineering thermoplastics, such as polyetheretherketone, which is used as a biomaterial in several areas, can have certain properties substantially improved by the incorporation of chopped carbon fibers.

Nanotechnology has also had a significant impact on composite materials. Some properties are even more enhanced when the dispersed phase involves sub-100 nm particles and some new concepts have been introduced, for example the incorporation of silver nanoparticles into polymers to confer anti-bacterial properties.

### 2.6.7 Natural composites

The use of natural biopolymers and bioceramics has already been discussed in Sections 2.6.3 and 2.6.5, where it was implied that most naturally occurring substances, especially tissues, are mixtures; indeed nature determined very early in the evolutionary process that composites, as defined above, give the best combination of properties. Bone is an exquisite composite that is based on a distribution of nanoscale/microscale apatites in a largely collagen-based proteinaceous matrix. The precise composition and stoichiometry of the apatites can vary, as can the amount and nature of the non-collagenous proteins and other organic substances present. Moreover, there is a microarchitecture to bone, involving, for example, channels for blood vessels, which yields a highly multifunctional material. In teeth, there are similar composites; however, the amount of apatite in the outer enamel layer is maximized in order to give hardness and abrasion resistance, in contrast to the inner layer of dentin, which has much less apatite and more protein to give better toughness. There are several other natural, animal- or plant-derived composites which also have significant properties. The ivory of the tusks of elephants and other animals, the shells of crustaceans, and plant material such as bamboo or wood are good examples.

The importance of natural composites to biomaterials science lies in the way that they can demonstrate how we can emulate tissues when we need to replace them. The terms biomimesis and bioinspired are often used here. Biomimesis is the noun that denotes the process by which we may mimic, or replicate, a natural substance. Bioinspired is the adjective often used to describe how this is done. A bioinspired approach to bone replacement may involve a hydroxyapatite–synthetic polymer composite coated with a deposit of an apatite.

## 2.7 Physical properties of materials

Having dealt with the structure of materials, we now turn to their properties, which refer to the performance of that material when used in an application. We use titanium alloys in aeroframes, nickel alloys in aeroengines, reinforced concrete in civil engineering structures, carbon fiber composites in golf clubs and tennis racquets and so on. These choices have not been random; they have been based on an understanding of the structure of the materials in question and of how these structures determine their behavior in real-life situations.

It is most instructive to consider these properties in the context of the applications. In this section we deal with physical phenomena, that is the response of materials to physical energy in the form of thermal, electrical, optical and magnetic phenomena. We will then deal with mechanical, environmental and biological processes.

### 2.7.1 Thermal properties

There are few situations in which the thermal properties are of significance, bearing in mind that the majority of biomaterials applications are anticipated to take place at body temperatures. Those properties we must mention are conductivity, expansion and exothermic reactions, with an additional brief comment on cryogenics and hyperthermia in tumor ablation.

#### Thermal conductivity

Thermal conductivity, which refers to the ability to conduct heat, is largely irrelevant in biomaterials science. In metals it is associated with the ability of freely moving valence electrons to transmit energy. Silver, copper and gold have conductivities in the region of 300–400 W/m K (watts per Kelvin meter) and steel around 20. Polymers, ceramics and tissues generally have conductivities less than 1. Diamond may be as high as 2000 and graphene over 5000.

#### Thermal expansion

Thermal expansion is also relatively unimportant. It does have some relevance in restorative dentistry where small changes in temperature intra-orally can give rise to volume changes in filling materials, which can open up gaps at their interface with teeth, allowing bacterial ingress. The tendency for solids to expand on raising the temperature reflects interatomic and intermolecular bond strengths such that ionic solids have the lowest coefficients of expansion (typically $1–3 \times 10^{-6}$/°C for the linear coefficient of expansion), metals somewhat higher, between 10 and $40 \times 10^{-6}$/°C and polymers higher at $50–100 \times 10^{-6}$/°C. The expansion of gases and liquids will be much higher. This may be used to advantage with heat sensitive microbubbles in the monitoring of thermal ablation therapies, discussed below, since changes in microbubble volume on raising the temperature can provide ultrasound contrast. Such agents include a perfluorocarbon liquid contained in a polymer shell.

#### Exothermic reactions

Exothermic reactions are those reactions that release energy in the form of heat. This has no clinical relevance for conventionally manufactured materials and devices but could be important for those systems that depend on *in situ* polymerization or setting. This was a particular concern in the early uses of acrylic bone cements in total joint replacement; the polymerization of methylmethacrylate is strongly exothermic and there is no doubt that some formulations resulted in temperature rises within the hip to 50°C or greater. This can cause thermal necrosis of tissues and may influence the porosity and residual stresses in the resulting cement. These problems have been resolved through different formulations and techniques but consideration to the exotherm will always be necessary with such polymers, especially when used in areas of sensitive tissue, such as in vertebroplasty, where cements are used close to the spinal cord. Calcium phosphate cements are only mildly exothermic.

### Hyperthermia, thermotherapy and thermoablation

Cells respond to heat in a variety of ways and these responses may be used to kill tumor cells. When heated to between 41 and 46°C, apoptosis, a form of cell death, is enhanced; the cytoskeleton is disorganized, there is irreparable damage to organelles and apoptosis-inducing proteins are activated (see Chapter 3). At higher temperatures, proteins are denatured, lipids melt and the cell lyses. If tumor cells can be targeted such that their temperature is raised, then either apoptosis or cell lysis may occur, giving rise to the possibility of thermotherapy and thermoablation respectively. It is obviously necessary to heat the tumor cells and not surrounding healthy cells. This may be achieved by targeting nanoparticles to the tumor cells where they can be selectively heated. There are two main candidates, iron oxide and gold nanoparticles, and some minor ones such as carbon nanotubes. The iron oxide nanoparticles, of which there are several types, are magnetic and are heated when they are placed in an alternating magnetic field caused by rotational friction and the dissipation of magnetic energy. Radiofrequencies that are otherwise harmless to human tissue may be used to create the magnetic field. Gold nanoparticles have plasmon resonance properties, which allow the generation of heat under near infrared radiation (NIR). Carbon nanotubes can also be heated by NIR.

## 2.7.2 Electrical properties

It is necessary to consider electrical properties from the perspective of those biomaterials that are used in devices that deliver electrical power or detect electrical signals.

### Electrical conductivity

In the vast majority of conventional implantable medical devices, electrical conductivity is of no relevance. Conductivity ($\sigma$) is a measure of the ability of a material to permit conduction of an electrical current; it is measured in units of Siemens per unit of distance, typically S/cm. Resistivity ($\rho$) is the reverse, a measure of resistance to a current, measured in Ohms ($\Omega$). There are of course some implantable devices whose function is to deliver electrical charge, in which case these parameters become crucial. A cardiac pacemaker, for example, has to deliver an electrical charge to heart muscle. The current has to flow from the implanted pulse generator to that muscle, which implies that there must be electrical connectivity between the two, achieved by virtue of one or more electrodes, with adequately high electrical conductivity. However, that current must not be dissipated to the surrounding tissues and fluids in between these locations, which implies that the electrode has to be surrounded by a material that does not conduct electrical current, such a material being known as an insulator. Fortunately there is a very large margin between good conductors and insulators. Platinum alloys are often used as conducting electrodes and they have a resistivity of around $10^{-7}\ \Omega\,m$. Elastomers such as silicones or polyurethanes can be used as insulators, with resistivities around $10^{20}\ \Omega\,m$. This massive difference is again associated with the ability of free electrons in metals to carry electrical charge. Different types of stimulator may require variations in the metals used. For cardiac pacemakers, which may operate with a pulse amplitude up to 7 V, platinum alloys are usually preferred. With implantable defibrillators, where the energy pulse during defibrillation may involve 700–800 V, it may be necessary to use even better conducting metals, for example silver, which has a resistivity of around $1.6 \times 10^{-8}\ \Omega\,m$.

**Electrically conducting polymers** Although, as noted above, there is a very large difference between the electrical conductivities of metals and polymers, there is one special class of material in which this difference is substantially eliminated. This refers to the group known as conducting polymers. Metals largely conduct electricity through the movement of their free valence electrons. In the majority of polymers, the valence electrons are associated with $sp^3$ hybridized covalent bonds and are essentially unable to contribute to conductivity. However, there are a few polymers that have, in the molecular chains, alternating double- and single-bonded $sp^2$ hybridized atoms, which gives the materials some semiconducting character. This in itself is not so significant. However, these neutral polymers can be doped with anions or cations to give charge carriers, which are able to migrate along the polymer chains under the influence of an electrical field. Some of these doped polymers have considerable conductivity. The best example is polypyrrole, but there are several others, most of them being of polyheterocyclic structure. The attractive combination of electrical conductivity coupled to the possibility of molecular functionalization, has led to a number of biomaterials applications, for example in biosensors, bioactuators and neural probes.

#### Dielectric constants

The dielectric constant of a material has often been considered to be a determinant of its biocompatibility, especially in the context of the response of tissues to the oxidized surfaces of metals (see below). There is no proof of this, and indeed it would appear to be unlikely, but it is worth mentioning what type of material a dielectric is, and what is the dielectric constant. A dielectric is essentially an insulator. When placed in an electric field, the positive and negative charges of a dielectric are slightly separated and the material is electrically polarized. The extent of this polarization is quantified by the dielectric constant of a material. The oxide layer on a metal surface will be a dielectric, in contrast to the conducting metal and so there may be a considerable difference in charge distribution at their interface. Dielectric phenomena are most important in the design of devices that store and distribute electrical charge, such as capacitors. The dielectric constant does not have units and varies from 2–4 for most polymers to 100–1000 for glasses and ceramics. The dielectric constant is likely to be important in techniques of thermotherapy where electrical energy is used to heat tissues; values for tissues are usually in the region of 30–50.

### 2.7.3 Electronic and optoelectronic properties

We have seen that electrons in solids are either in localized or delocalized states. The inner or core states are localized, and are very similar to the states in free atoms. The outer or valence electrons are delocalized since they can extend outside the region of just one atom. In metals, the delocalization of the valence electron(s) means that they are shared by all atoms in the solid.

The electronic properties of a material are largely controlled by the band gap, which refers to the energy difference between the top of the valence band and the bottom of the conduction band. Electrons are able to jump from one band to another. However, in order for an electron to jump from a valence band to a conduction band, it requires a specific minimum amount of energy, known as the binding energy. This energy differs with different materials. Electrons can gain enough energy to jump to the conduction band by absorbing either heat or light energy.

The band gap in semiconductors and insulators results from the interplay between the electron wavelengths and the symmetric arrangement of atoms in the solid. A semiconductor is a material with a small but non-zero band gap which behaves as an insulator at absolute zero but allows thermal excitation of electrons into its conduction band at temperatures which are below its melting point. In contrast, a material with a large band gap is an insulator. In conductors the valence and conduction

Table 2.2 **Some common fluorophores**

| Dye | Typical molecular target | Wavelength (nm)/color |
|---|---|---|
| Polymethine cyanines<br>Cy 5<br>Cy 5.5<br>Cy 7 | Folate receptors<br>Apoptotic cels<br>Cancer cell integrins | Abs 650, Em 670, red<br>Abs 675, Em 694, red<br>Abs 743, Em 770, red |
| Squaraine derivatives | Various proteins | Abs 630–670, Em 650–700 |
| BODIPY FL (boron dipyrromethene) | Nucleotides | Abs 503, Em 512, green |
| Phthalocyanines | Malignant and dysplastic tissues | Abs 650–800, Em 700–1000 |
| Texas red | Various proteins, antibodies | Abs 589, Em 615, Red |
| Fluorescein isothiocyanate (FTIC) | Antibodies | Abs 495, Em 518, Green |
| Green fluorescent protein | Marker of gene expression | Abs 395, Em 509, Green |
| DAPI (4',6-diamidino-2-phenylindole | Nuclei | Abs 345, Em 455, Blue |

Abs = absorption; Em = emission.

bands may overlap, so they may not have a band gap. Band-gap engineering is the process of altering the band gap of a material by controlling the composition of the material.

Although electronic properties of materials control the performance of many health care products and medical devices, these materials do not often come into direct contact with tissues and cannot usually be considered as biomaterials. They are not, therefore, discussed at length here. However, there is one related aspect that is taking on increasing significance and this concerns the inter-connectivity between electrical and optical energy, in the area known as optoelectronics, and the role of the various phenomena in imaging systems. Most of these systems rely on fluorescence or bioluminescence.

### Fluorescence

Fluorescence is the emission of light by substances that have absorbed electromagnetic radiation of a different wavelength. Fluorescence in medical technology is used as a non-destructive way of tracking or imaging biological molecules by means of the fluorescent emission at a specific frequency, particularly where there is no background from the excitation light, relatively few cellular components being naturally fluorescent. A protein or other component can be labeled with a extrinsic fluorescent dye, known as a fluorophore which can be a small molecule or protein. Some examples of medically and biologically relevant fluorescent materials are given in Table 2.2.

Until recently, fluorophors could be grouped as either organic dyes or biological molecules. In the former group, fluorescein, a synthetic organic dye, was one of the first fluorescent compounds used, along with derivatives such as fluorescein isothiocyanate (FITC). The small size of the derivatives gives them an advantage since they can be linked to macromolecules like antibodies without interfering with their biological function. Biological fluorophores were first used in the 1990s when a green fluorescent protein, known as GFP, was synthesized from the jellyfish *Aequorea victoria* and applied as a gene expression reporter. Several other biological fluorophores have been developed and they may be used in cells or whole organisms to study biological processes. In general they do not have such good photostability as synthetic fluorophores and they may also change the function of the process in which they are being studied.

It was because of the deficiencies of both organic dyes and biological fluorophores that other agents, which are now included in the scope of biomaterials, have been introduced. These primarily include quantum dots and unconversion nanoparticles. As we discuss in more detail elsewhere, quantum dots are 2–50 nm sized semiconductors which emit fluorescence when excited at a wavelength that is dependent on the size of the particle. They are characterized by large absorption spectra but narrow and symmetric emission bands that can span the light spectrum from ultraviolet to infrared. At the lower end of the size range, the band gap is larger so that the dots emit at higher energy. The color of the emission changes from blue to red as the size increases. These are considerably more photostable than fluorophores and also give high quantum yield, which is the ratio of emitted to absorbed photons.

### Upconversion luminescence

Upconversion luminescence is the non-linear optical process that converts two or more low-energy pump protons from the near infrared (NIR) spectral region to a higher energy/shorter wavelength output. These may show considerable reduced autofluorescence, enhanced photostability and enhanced tissue penetration compared to conventional fluorophores. This latter point is particularly important in the technique of photodynamic therapy (PDT), especially of cancer. In this, photosensitizer molecules are administered to the patient and exposed to electromagnetic radiation, where they transfer the absorbed photon energy to oxygen molecules in the tissue, generating cytotoxic reactive oxygen species (discussed in Chapter 3), thereby killing cancer cells. Visible and UV light have limited tissue penetration but NIR radiation has much greater penetration ability, and upconversion nanoparticles are able to convert the NIR radiation to visible photons which activate photosensitizers adsorbed onto their surface to generate the reactive oxygen species. A similar process, referred to as fluorescence resonance energy transfer (FRET), may be used to attach DNA derivatives to nanoparticles for NIR-mediated gene delivery. Several types of nanoparticle may exhibit the upconversion effect, mostly involving rare earth doped (ytterbium, erbium), silica-coated materials such as $NaYF_4$.

### Bioluminescence

Bioluminescence imaging is based on the process whereby light is produced by a chemical reaction that originates in an organism. Specifically, two chemical reagents are required, one that produces the light (the luciferin) and one which catalyzes the process (the luciferase). Absorbed light drives an electron from its stable orbital into a higher orbital, but it will only be stable there for a short time, when it returns to its lower energy orbital, emitting energy as a longer wavelength photon. Most luciferins are found in marine organisms, although one is also found, and functions, in the firefly.

Mammalian cells do not produce the luciferase substrates that are necessary for this process. However, cells (for example, stem cells) may be genetically modified with luciferin derived from the firefly which, with appropriate detection systems, allows imaging of whole animals that indicate cell distribution, as shown in Figure 2.16. Since some bacteria can also be rendered bioluminescent, these techniques may be used to track infections.

**Molecular beacons** One special form of photoluminescent particle is the so-called molecular beacon. These are used in intracellular RNA imaging, discussed in Chapter 7. They are hairpin-shaped probes that contain photoluminescent species, usually a fluorophore as mentioned above, at one end and a quencher at the other end. The loop structure of the hairpin is designed such that fluorescence only occurs when the probe binds to a complementary target structure in the nucleic acid.

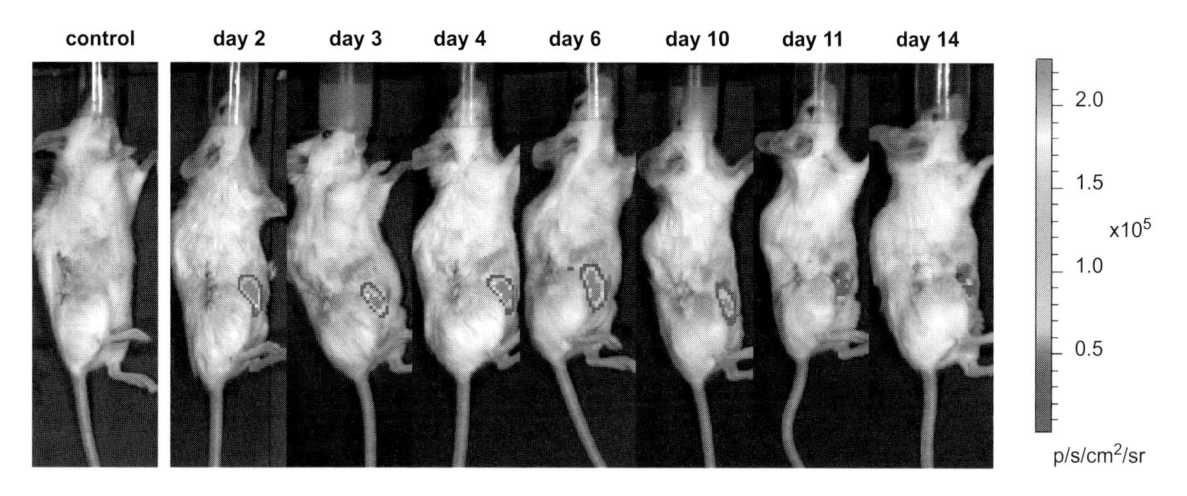


**Figure 2.16** Bioluminescence and the monitoring of biomaterial-related infection.
A pseudo-color representation of the bioluminescent radiance from a surgical mesh, pre-contaminated with *S. aureus* Xen29 bacteria implanted subcutaneously in mice during a 15-day follow-up. The images show that the amount of viable bacteria contributing to the bioluminescence signal decreases in time during the course of infection. Bioluminescence was imaged with a CCD camera 2 days following implantation of the mesh and subsequently on selected days. Image was produced by Seyedmojtaba Daghighi in the laboratory of Dr. Jelmer Sjollema and Dr. Henk Busscher, University Medical Center Groningen, Netherlands. See plate section for color version.

## 2.7.4 Magnetic properties

As with the other physical properties discussed above, the magnetic properties of materials have not played a significant role in the development of conventional biomaterials and medical devices. However, the position has changed in the last decade or so with the deployment of magnetic nanoparticles in a variety of systems for the targeted delivery of active molecules to patients, especially in cancer therapy and in systems for imaging, particularly contrast agents for magnetic resonance imaging. It is necessary therefore to discuss some principles that underpin the magnetic properties of materials, and the structure of the various classes of magnetic materials and also describe the significance of the magnetic properties at the nanoscale.

### Magnetic materials

We discussed the electronic structure of materials in Section 2.3. The origins of the magnetic properties of materials can be found in the way in which electrons interact with each other with respect to their existence within the orbital structure, and especially whether there are paired or unpaired electrons in partially filled orbitals. When a material is placed within a magnetic field, the electrons will be affected in some way, largely depending on their spin. In most atoms, electrons are arranged in pairs, and within one pair the electrons spin in opposite directions. These opposite spins cause their magnetic fields to cancel each other. If there are unpaired electrons there will be a net magnetic field. The magnetic properties of a material, which are concerned with the behavior of the material in an external magnetic field, depend on these internal magnetic fields. There are four main types of magnetic behavior, associated with diamagnetic, paramagnetic, ferromagnetic and ferrimagnetic materials.

**Diamagnetic materials** In diamagnetic materials, all the electrons are paired so that there is no net magnetic moment for each atom. There is, however, a slight realignment of electron paths under the influence of an external magnetic field; they are slightly repelled by the field and are considered to have a negative magnetic susceptibility. They do not retain any magnetic properties once the external field has been removed. Most elements are diamagnetic.

**Paramagnetic materials** Paramagnetic materials, in contrast, have a small positive magnetic susceptibility. They have some unpaired electrons and are attracted by a magnetic field, although again they do not retain magnetic properties once a field has been removed. A few elements, including magnesium and tantalum, and many minerals containing iron, are paramagnetic.

**Ferromagnetic materials** Ferromagnetic materials also have unpaired electrons in partially filled orbitals, but the atomic moments are very strong and these materials are able to retain their magnetic properties after the external field has been removed. The strength of this magnetic effect is related to the existence of magnetic domains, in which large numbers of atomic moments are aligned so that the magnetic force within a domain is very strong. In the unmagnetized state, the near random organization of domains means that net magnetic field is close to zero; when a magnetic field is applied, the domains are aligned to produce a strong magnetic field within the material. Iron is the best example, although some other metals such as cobalt and nickel are also ferromagnetic. If ferromagnetic materials are heated, thermal energy may eventually overcome the electronic exchange forces and the magnetization is lost. The temperature at which this happens is known as the Curie temperature.

**Ferrimagnetic materials** Ferrimagnetic materials tend to be ionic compounds, especially including some oxides of iron such as magnetite, where the crystal structure involves sub-lattices. The magnetic properties are derived from the anti-parallel alignment of spins between two different sub-lattices.

#### Magnetic nanoparticles

The recent emphasis on the medical uses of nanoparticles is discussed in several places in this book. Nanoparticles have several attractive properties, including the quantum effects seen at the lower end of their size range, and we also find some very interesting magnetic properties. Several materials that would be considered as normal paramagnetic materials in bulk have quite different properties when they exist as nanoparticles since the nanoparticle size may be within the size range of the domains of ferromagnetic materials. This gives rise to two related effects, the single domain ferromagnetic and superparamagnetic effects. We shall consider the latter here; the array of nanoparticles, each with its own magnetic moment, can be magnetically saturated in a field because of their alignment, exactly as with the domains, but the magnetization returns to zero on removal of the field through the effects of thermal energy. This is superparamagnetism; the behavior of the small magnetic moment of a single paramagnetic atom is considerably amplified in the much larger magnetic moment of a magnetic nanoparticle because of the coupling of many atomic spins. The most widely used magnetic nanoparticles are those that have a core of iron oxide (superparamagnetic iron oxide nanoparticles (SPION)). These particles can be functionalized in order to give targeting and molecule delivery features for *in vivo* use, as discussed in Chapter 7.

### 2.7.5 Ultrasound

Ultrasound refers to sound waves that are above the audible range of human hearing, which is around 20 000 Hz. Ultrasound techniques, often referred to as ultrasonography, have been extensively used for many years in medical imaging and diagnostic procedures, including echocardiography for cardiac conditions, obstetric and neonatal sonography for the examination of a fetus, abdominal sonography for imaging of solid organs in the abdomen, and arterial sonography and venosonography for detecting abnormalities in blood vessels.

These techniques do not involve biomaterials, but the relatively recent innovation of contrast-enhanced ultrasound does. Ultrasound contrast agents rely on the different ways in which sound waves are reflected from interfaces between substances. This may be the surface of a small air bubble or a more complex structure. Most clinically used contrast media are gas-filled microbubbles that are administered intravenously. These have a high degree of echogenicity, the ability of a substance to reflect the ultrasound waves, which is very different to that of soft tissue. Contrast-enhanced ultrasound can be used to image blood perfusion in organs and to measure blood flow. Typical products include air bubbles with a galactose–palmitic acid surfactant, octafluoropropane with an albumin shell and sulfur hexafluoride with a phospholipid shell.

These contrast agents are essentially hydrophobic gases within a stabilizing shell; they have some advantages over other imaging modalities, including low cost, safety and ease of use. Functionality is enhanced by causing oscillation and fragmentation. They can survive passage through the pulmonary vasculature and show contrast persistence over a long time. Recent developments in this technology, extending applications to cancer detection outside of the vascular space, include a combination of ultrasound with NIR imaging where the nanobubbles are conjugated to antibodies to over-expressed antigens for highly specific tumor location.

## 2.8 Mechanical properties of materials

The mechanical properties of materials are concerned with the behavior of materials when they are subjected to mechanical forces. Many health care products that utilize biomaterials will be subjected to mechanical forces at some point and it is important to understand how they will respond to these forces. There are two significant general phenomena that control the response. The first is that of deformation; whenever a force is applied to a material it will deform. The applied external energy causes changes in interatomic or intermolecular distances or causes atoms or molecules to move to new positions, the result being a change of shape of the material. Depending on the magnitude of the force, this change may be very small and difficult to detect, or may be very large such that the material distorts very visibly. It may be transient such that the deformation is reversible when the force is removed, or it may be permanent. The deformation may be instantaneous and constant or it may change over time with prolonged or repeated application of the force. The second phenomenon is that of fracture. Here the applied force reaches a level where the external energy exceeds the energy holding atoms or molecules together, such that they separate. In any one device, the fracture may occur in one place such that it breaks into two pieces. Alternatively, the device may shatter into many pieces, or the forces may be confined to surfaces, in which case the material abrades, or wears, with small fragments separating from the bulk material.

We shall concentrate here on the principles of the general deformation and fracture mechanisms and then discuss those that are most relevant to biomaterials. This largely refers to the medical devices that are used for implantation in the body for structural purposes and scaffolds for regenerative medicine, with minimal relevance to the applications of particulate products used in drug and gene delivery or in imaging and diagnostic procedures.

### 2.8.1 Forces and stress: deformation and strain

If a patient who has a total hip replacement walks on a road, it is obvious that mechanical forces will be applied to the prosthesis. Because of the complexity of the design of the prosthesis and the multiplicity of muscles that will be acting in the vicinity of the hip, it would seem that these forces are also complex, with some elements of bending, twisting and so on. Equally, a tubular artificial blood vessel will be subjected to complex forces as the blood flows through the vessel under sinusoidally varying pressure and unpredictable movements of the surrounding tissue. As complex as these situations are, however, we can always rationalize the forces into three distinct types. These

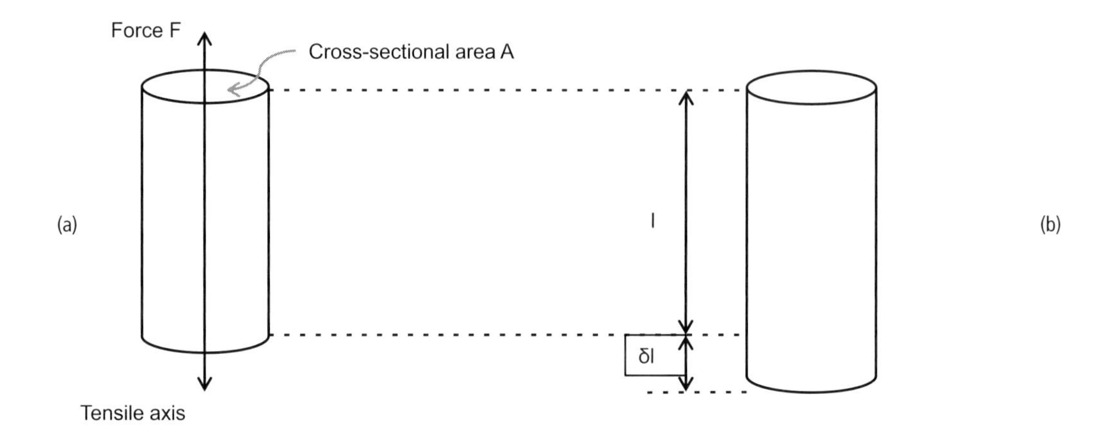


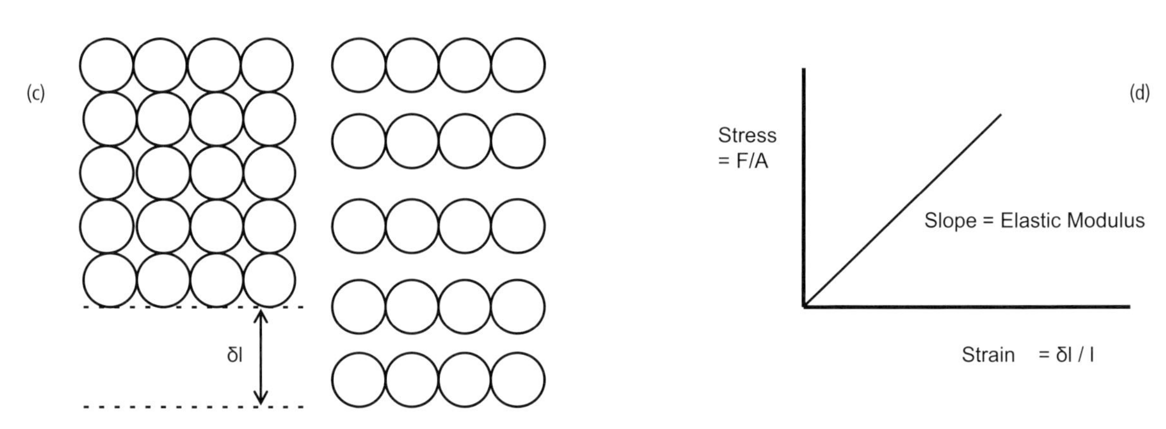


**Figure 2.17** Elastic deformation.
**(a)** Application of a tensile force $F$ to a cylinder of cross-sectional area $A$. The tensile stress is given by $F/A$.
**(b)** The application of this stress results in deformation, in this case tensile elongation. The elongation is $\delta l$, and the original length is $l$. The strain is given by $\delta l/l \times 100$, measured as a percentage.
**(c)** This shows the movement of atoms that results in this deformation. Up to a limit, known as the elastic limit, this deformation is reversible, the sample recovering its original length on removal of the stress.
**(d)** The stress–strain curve for elastic deformation.

are tensile forces, which cause the material to stretch, compressive forces, which squeeze the material, and shear forces. Pure tensile and compressive forces are applied parallel to the long axis of the material. A shear force is applied perpendicular to the long axis. It will be obvious that very few devices will experience only pure tension, compression or shear, but the forces can always be resolved into these separate components. Bending and torsion are practical variations of shear.

The effect of a force on a material will depend on the magnitude and type of the force, but also on the shape and size of the material. In the case of a tensile force acting on a uniform cylindrical specimen (shown in Figure 2.17(a)) the effect of the force will depend on the cross-sectional area of the cylinder. The force acting per unit cross-sectional area is called the stress, in this case the tensile stress being $F/A$. A compressive stress is defined in the same manner, whilst a shear stress will also be given by $F/A$ where $F$

is the force perpendicular to the long axis and $A$ is the cross-section. Stress is often given the Greek symbol $\sigma$; the units of stress are $N/m^2$, referred to as pascals, Pa.

As noted, the application of a force results in deformation. In the case of pure tension acting on a cylindrical specimen, we can imagine a section of a structure that experiences this force (Figure 2.17(b)), in which the atoms move apart, making the section longer. This is tensile deformation. Again, the amount of deformation will depend on size and shape. The extension is usually normalized to the original length, this ratio being called the strain, and given the Greek symbol $\epsilon$. If $\delta l$ is the extension and $l$ the original length, then $\epsilon = \delta l/l$. Being a ratio, it has no units.

## 2.8.2 Elasticity and plasticity

Having defined the nature of mechanical forces and the parameters by which force and deformation are measured, we now consider how deformation occurs in practice. We shall distinguish two types of deformation here, elastic and plastic deformation. The distinction between these two is critically important.

### Elastic deformation and elastic modulus

Consider the simple situation of a single crystal of pure metal (Figure 2.17(c)) subjected to an increasing tensile stress. As the stress increases, atoms are stretched apart, giving the resulting strain. If the stress is removed, the atoms will revert to their previous positions. There is no permanent evidence that the deformation took place – it is entirely recoverable or reversible. Deformation that is reversible in this way is described as elastic deformation. If a graph of applied stress against resulting strain is plotted, as in Figure 2.17(d), then a straight line will be produced. This is manifested as Hooke's law, which states that, under these conditions, stress is proportional to strain. The slope of this straight line is one of the most important fundamental properties of a material. It is called the elastic modulus; in the case of tensile and compressive stresses, this is specifically called Young's modulus. If it is a shear stress, this modulus is called the *shear modulus*, or the *modulus of rigidity*. The units of an elastic modulus are the same as those for stress, i.e., pascals.

We would expect the value of an elastic modulus to depend on the ease with which the atoms or molecules can be stretched, which is itself dependent on the magnitude of the interatomic and intermolecular forces. Thus, for metals and ceramics, which have uniform metallic or ionic bonds, the modulus should be high. For polymers, which may have quite strong interatomic bonds, but only weak bonds holding the molecules together, the modulus should be quite low. If cross-links are introduced into polymer structures, then the modulus will increase. If the molecular structure in a polymer is such that individual molecules are coiled or wavy, then significant elastic deformation may occur by uncoiling or straightening out of chains even at very low stresses, so that the modulus will be correspondingly low.

A material with a high elastic modulus can be considered to be rigid. One with a low modulus can be considered to be flexible. Table 2.3 gives values for the elastic modulus of a range on materials.

When a material stretches longitudinally under the application of a tensile stress, then we would expect its cross-sectional area to reduce as it elongates. The ratio of the transverse strain to the axial strain is known as Poisson's ratio. Most materials have a Poisson's ratio between 0 and 0.5. There are some materials that have a negative Poisson's ratio, that is they become thicker as they extend. These are mostly polymer foams, including some of those used as biomaterials.

**Superelasticity** Superelasticity is closely associated with shape memory, discussed later in this chapter and elsewhere, and is considered only briefly here. Conventional elasticity is derived, as we have seen, from the reversible movement of atoms or molecules within a material where the state of that material, determined by phase structure, is stable. Under some conditions, usually involving temperatures very close to the transition temperature between two phases,

Table 2.3 **Elastic (Young's) modulus for various materials**

| Material | Young's modulus (GPa) |
|---|---|
| Natural rubber | 0.001–0.005 |
| Silicone elastomer | 0.001–0.005 |
| Plasticized PVC | 0.02–0.05 |
| Low density polyethylene | 0.20–0.40 |
| PTFE | 0.41–0.75 |
| Polypropylene homopolymer | 0.80–1.30 |
| High density polyethylene | 0.60–1.40 |
| Polylactic acid | 0.35–2.80 |
| Polymethylmethacrylate | 2.4–3.3 |
| PVC | 2.4–4.1 |
| Polyethylene terephthalate | 2.5–3.0 |
| PEEK | 2.7–12 |
| Carbon fiber PEEK | 3.0–138 |
| Cortical bone | 14 |
| Calcium phosphates | 30–60 |
| Magnesium | 45 |
| Nickel–titanium, martensitic | 28–45 |
| Nickel–titanium, austenitic | 83 |
| Tooth enamel | 83 |
| Titanium alloys | 105–120 |
| Stainless steel | 200 |
| Cobalt–chromium alloys | 215 |
| Alumina | 350–400 |
| Tungsten | 400 |
| Tungsten carbide | 650 |
| Single wall carbon nanotube | 1000 |
| Diamond | 1220 |

the application of stress stimulates a diffusionless, shear, transition, which is accompanied by major deformation. This is reversible since it will revert once the stress is removed and hence it is deemed to be elastic, and here the effective modulus can be very low as these major changes in dimensions occur with very little increase in stress. It is, however, described as superelasticity, or pseudoelasticity, to indicate its unusual nature. It is best seen in nickel–titanium alloys and is used in some health care applications, especially in orthodontics where it is effective in producing tooth movement at low applied forces.

### Plastic deformation

In the above examples of elastic deformation, we would not expect materials to continue to deform for ever as the stress increases – intuitively we would expect that there would be a limit to the extent to which atoms could be reversibly pulled apart. In practice this is the case, but one of two things may happen once that limit is reached. The first is that the bonds between atoms are irretrievably broken such that the material fractures or falls apart. We discuss this in the next section. The second is that the atoms accommodate the stress by moving to new positions. This is indicated in Figure 2.18, which again represents atoms of a single crystal of a pure metal. In (a), the metal is at rest, unstressed. In (b) an applied stress is accommodated by elastic deformation. In (c), an increased level of stress is still accommodated elastically, but it is close to the limit of what can be achieved. In (d), the atoms have now moved along one crystal plane. The strain has been increased by this movement (the sample appears longer) rather than by further stretching of the bonds.

This movement of atoms to new positions is not reversible, as they would stay in their new positions once the stress was removed. This irrecoverable, irreversible deformation is called plastic deformation. If we plot stress against strain for this situation, as in Figure 2.18(e), we see a discontinuity. Elastic deformation has initially take place, but at the elastic

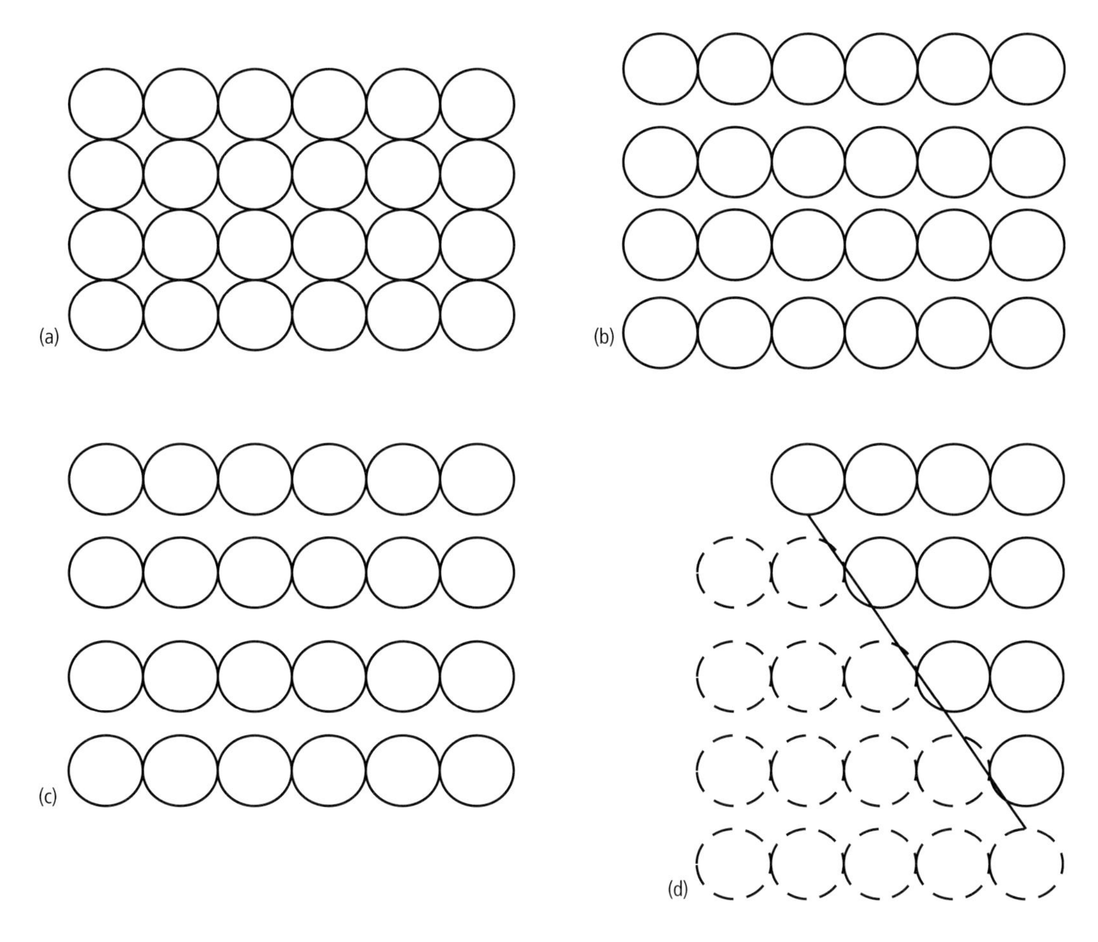


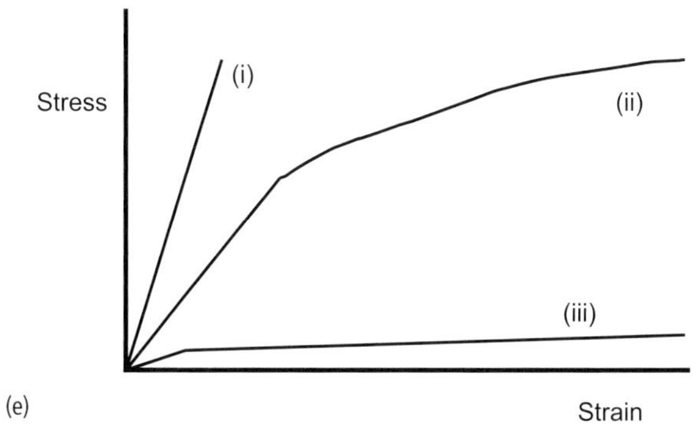


**Figure 2.18** Plastic deformation.
**(a)** Unstressed crystal lattice.
**(b)** Movement of atoms on application of a small stress.
**(c)** Further movement of atoms, now at the elastic limit.
**(d)** Atoms slide to new positions, recreating the same crystal structure but now considerably greater elongation along the tensile axis. This is plastic deformation, which is not recoverable on removal of the force.
**(e)** Stress–strain curves: (i) a material showing only elastic deformation and no plasticity; (ii) a material which undergoes extensive deformation after the elastic limit – note that increments of strain are achieved with smaller and smaller increments of stress; (iii) a material that has a very low elastic limit and extends plastically extensively at very low stresses.

limit the relationship no longer remains linear, more deformation being produced for incremental increases in stress. The nature of the transition from elastic to plastic deformation varies; it may be a gradual transition, barely perceivable, or it may be abrupt, where suddenly a large amount of strain is produced for very little increase in stress. This process is known as yield, and the point at which it occurs is the yield point, this being especially important in metals.

A number of factors control the ability of a material to deform plastically. The example of the single crystal of a pure metal is a good place to start this discussion. The metallic bond is strong but it is not directional. Each atom is exactly the same and the energetics will be the same wherever they are placed in the crystal lattice. Thus there are no energetic constraints to this movement and we would expect the process of plastic deformation to be easy once it has been initiated. It is well known, for example, that pure gold deforms plastically very easily. The only issue here is at what stress does the plastic deformation start. This is not a trivial matter. It might seem that there should be a correlation between the stress at the elastic limit and the strength of the relevant interatomic bonds. In reality plastic deformation starts to occur at stress levels much lower than the theoretical levels based on bond strengths. The theoretical argument would suggest that the movement of atoms along the designated plane would occur when the applied energy exceeded the aggregated energy of all bonds between atoms along that plane. In reality, the movement of atoms does not occur simultaneously, but starts at one point, which creates a local discontinuity, and it is this discontinuity, known as a dislocation, which effectively moves along this plane. There will of course be very many planes of atoms taking part, and very many dislocations; the secret to the control of the mechanical properties of metals and alloys is based on the manner in which dislocation creation and propagation is controlled. Almost the whole of the metallurgical industry is based on this fact.

With ionic materials the situation is quite different. As we see from Figure 2.19(a) on the left, instead of all the atoms being identical as with a pure metal, there will be more than one type of atom, each type having a different charge and usually with strong directionality of the bonds. The complexity of these atomic arrangements means that it will be very difficult, and often impossible, for the movement of atoms to take place. Movement of atoms along a slip plane, represented in Figure 2.19(a) on the right, shows that this would result in the juxtaposition of like charges and massive repulsive forces, which effectively prohibits this plastic deformation. Ceramics and glasses are notorious for their inability to be deformed. With covalently bound materials, and especially polymers, the situation is again different. The deformation will be focused on the weak intermolecular bonds. In a crystalline non-cross-linked polymer, plastic deformation may be quite easy, the movement of molecules over each other being assisted by small increases in temperature. Increasing levels of cross-linking will reduce this ability. Many amorphous polymers have very limited plasticity.

### 2.8.3 Fracture, ductility and toughness

With all materials, a point is reached where deformation of any type is no longer possible as the stress is increased. It may be that the elastic limit has been reached, but there is no available mechanism for plastic deformation. Alternatively, there may have been extensive elastic deformation up to the yield point and then extensive plastic deformation, but the level of disruption that has been sustained within the atomic and molecular structure means that the movement of dislocations becomes too difficult. In either case, the applied energy causes the material to fracture. This usually happens when the stress level at one or more locations exceeds bond strengths and causes cracks to form. A material may sustain the formation of multiple small cracks without outward signs of failure, but sooner or later, one crack will open up, or propagate catastrophically and the material, or device, breaks into two or more pieces. This process is called fracture and the stress at which it occurs is often called the fracture stress.

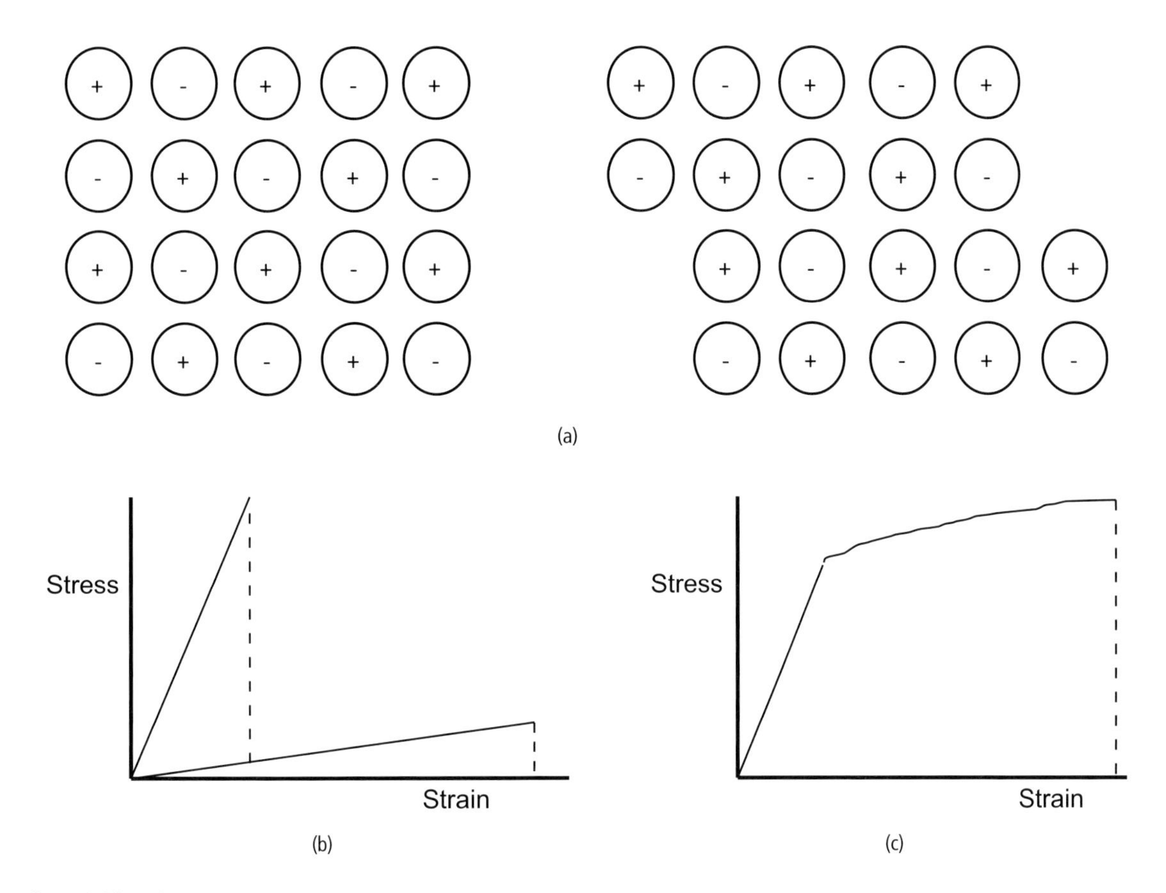


**Figure 2.19** Deformation of ionic structures.
**(a)** On the left, a 2D representation of a simple ionic structure, with two elements of the same valency, and the ions with equal and opposite charge. As seen in Figure 2.8, similarly charged ions cannot exist next to each other. If slip of atoms were to take place in the horizontal plane, as shown on the right, an unacceptable juxtaposition of charges would result, making this deformation more or less impossible. These materials are very brittle.
**(b)** Stress–strain curves for a material that can withstand high stresses but not high strains as with a ceramic, and for a material that can withstand high strains but not high stresses, as with many polymers.
**(c)** Stress–strain curve for a material that displays considerable plasticity, sustaining both high stress and high strain. This behavior is seen with many alloys, such materials being described as "tough."

One important practical point follows on from this concept of crack initiation and crack propagation. Most materials contain inherent flaws. Some processing routes inevitably result in defective structures. The casting of metals and injection molding of polymers usually leads to some degree of porosity, for example. Any such defect can act as micro-crack, which can propagate once stresses are applied. Fracture would be expected at much lower stresses than in defect-free materials. Since many biomaterials are used in porous form, for example porous surfaces for bone ingrowth and porous tissue engineering scaffolds, it is important that we do not inadvertently provide incipient cracks within the materials through this porosity.

We have to introduce the term "strength" at this stage. We often talk of strong materials, that is those materials with a high strength, but it is not intuitively obvious what we mean. The first point to make is that strength is not the same as rigidity, although these

Table 2.4 **Yield strength, ultimate strength and ductility for various materials**

| Material | Yield strength (MPa) | Ultimate strength (MPa) | Ductility (%) |
|---|---|---|---|
| Silicone elastomer | | 5 | 300–800 |
| PTFE | | 25 | 600 |
| High density polyethylene | 20–25 | 40–50 | 800 |
| Polypropylene | 30–35 | 30–40 | 100–600 |
| Polylactic acid | | 10–60 | 1.5–300 |
| PMMA | 50–70 | 50–70 | 2–5 |
| PEEK | 50–90 | 70–100 | 30–150 |
| Carbon fiber PEEK | | 300 | 2 |
| Carbon fiber-epoxy | | 750 | < 1 |
| Carbon fiber | | 3800–4200 | < 2 |
| Carbon nanotube | | 10 000–60 000 | Variable |
| Kevlar | | 3500–4000 | 3 |
| Alumina | | 300–550 | < 1 |
| Toughened zirconia | | 800–1500 | < 5 |
| Stainless steel | 205–310 | 515–620 | 30–40 |
| Titanium | > 170 | > 240 | 30 |
| Ti-Al-V alloy | 830–1100 | 900–1150 | 10–15 |
| Spider silk | | 1000 | > 20 |

are often confused. A material with a high elastic modulus is rigid but not necessarily strong. Strength refers to the point at which a material starts to deform plastically or the point at which it breaks, and usually it is sensible to be clear which we are using. In the former case, it is the yield point that we are considering and we may describe this as the yield strength. In the latter case, it is the stress at fracture, and this is usually referred to as the ultimate strength.

If a material fractures without any appreciable plastic deformation, it is described as brittle. As noted above, most ceramics, glasses and amorphous polymers are brittle. If a material fractures after appreciable plastic deformation, it is described as ductile. The amount of plastic strain that occurs before fracture is the ductility of the material, and since this is usually measured in tension, it is often referred to as the elongation at break, or simply the elongation. Metals often have a ductility in the range of 10–50%. Some polymers can go much higher than this. Values of ultimate strength and ductility for a range of relevant materials are given in Table 2.4.

The range of stress over which plastic deformation takes place is important. In many products, we require both a suitable technology for fabrication and also suitable mechanical properties for performance. Many fabrication technologies depend on plastic deformation, such as forging, extrusion, wire drawing and so on. It is very helpful if these can be carried out at relatively low stresses. However, quite often we would also like these materials to perform in their application without exhibiting any plastic deformation or fracture. An alloy for a joint replacement prosthesis would be of no value if it deformed plastically once the patient started walking.

In Figure 2.19(b) we see the stress–strain curves for a rigid but brittle material and for a very flexible but ductile material. Clearly the former can sustain high stresses but not high strains, whereas the latter can sustain high strains but not high stresses. The energy that these materials absorb before they fracture is represented by the area of their stress–strain curve, which in neither case is very high. For high-performance situations, which include many medical device applications, it would be better to have materials with stress–strain curves more like that of Figure 2.19(c). This material can sustain both high stresses and high strains, and has a high energy absorbance. This property is referred to as toughness. The area under the stress–strain curve is a good indicator of toughness. Strong tough materials are exceptionally valuable and several technologies have been developed to produce such materials.

### Strengthening and toughening of alloys

As noted above, a single crystal of a pure metal will normally have a low yield point and extensive ductility, primarily because dislocations move easily through the structure. Using methods that impede dislocation movement can produce stronger and tougher metallic materials. The first, and simplest, method is through the use of grain boundaries; that is by using very small grain sizes. If dislocations are to propagate in polycrystalline materials, they have to cross grain boundaries, which is not always easy, Typically the yield strength of an alloy will double by decreasing the grain size from 100 μm to 10 μm and triple by decreasing it to 2 μm. The second method is through the use of multi-phase structures, as discussed in Section 2.4.3. This can be achieved simply with two phase structures as in the alpha–beta titanium alloys, or through the use of fine second phase precipitates, as seen in many forms of steel. Each time a phase boundary is encountered, so more energy is necessary to move the dislocations across.

The third method is described as work hardening. In Figure 2.20(a) we see a representation of an alloy in the un-deformed state. If stress is applied, and the material deforms, the structure becomes distorted through the generation of very many dislocations; Figure 2.20(b) shows this distorted structure. The more deformation that takes place, the more distorted it becomes, and the more difficult it is to drive the dislocations through the structure (Figure 2.20(c)). This is the reason why the stress–strain curve for ductile metals does not take on a horizontal shape after the yield point, which would occur if dislocations were free to move unimpeded once they had been generated and released at the yield point, but rather the stress level necessary to sustain plasticity continuously increases. It is when so much internal distortion has taken place that cracks develop in regions of high internal strain and fracture ensues. We can make use of this work hardening phenomenon to produce tough and useful materials. Starting with the un-deformed material, perhaps as a cast ingot, the shaping process, say rolling or forging, results in a distorted structure. This will obviously be stronger than the starting materials. The extent of work performed during manufacture can be adjusted to give the desired final properties. It may well be that the distortion produced during manufacturing will be too much, and the remaining ductility reduced too far. In this case, the alloy can be given a heat treatment, or annealing treatment; if it is heated to a pre-determined temperature and held there for a specified time, the atoms will acquire thermal energy which will allow them to readjust to a certain extent and reduce the distortion level, Several intermediate annealing treatments may be given during a fabrication process.

### Toughening of ceramics

We have noted that ceramics tend to be brittle; they usually have no intrinsic mechanisms for plastic deformation to take place, such that the only process

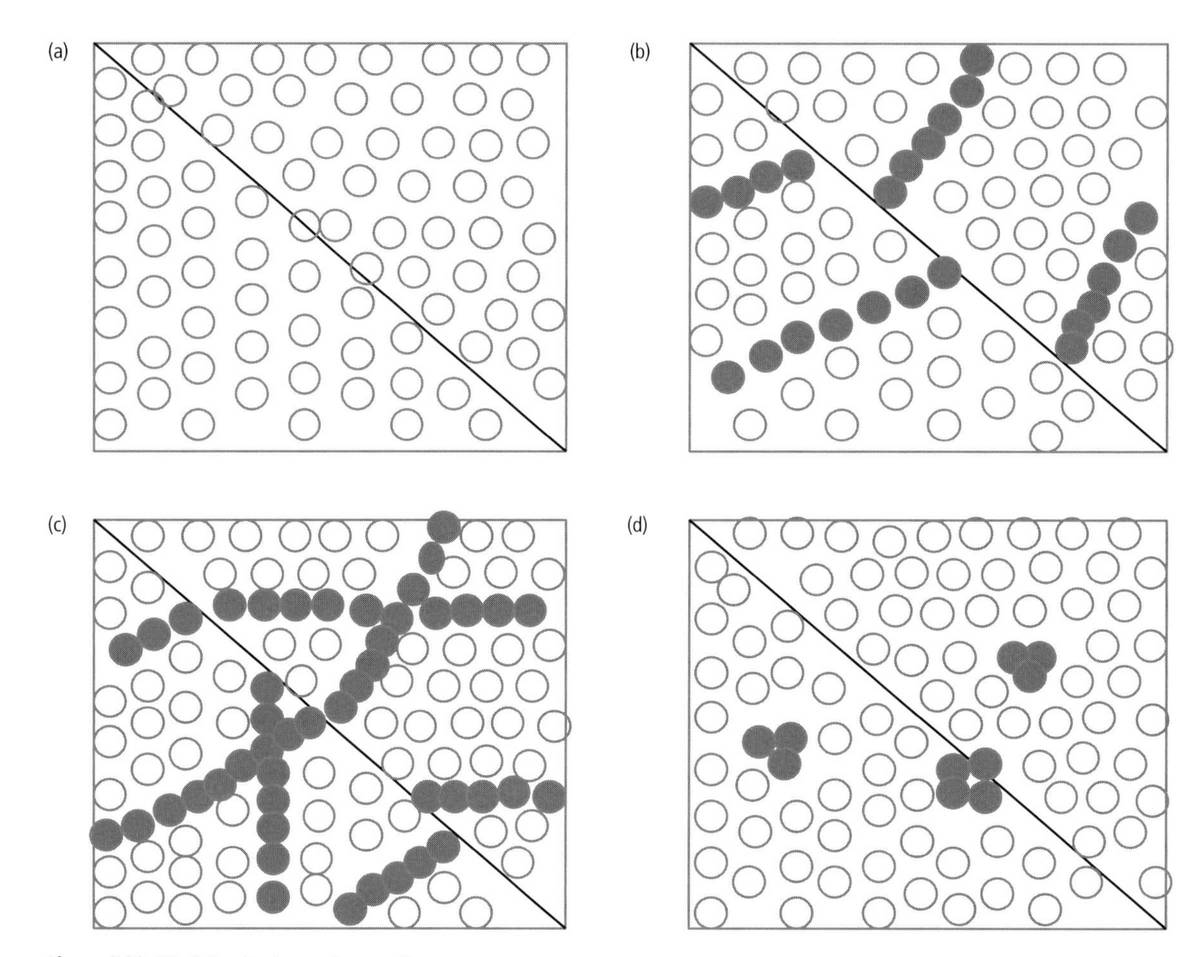


**Figure 2.20** Work hardening and annealing.
**(a)** Two-dimensional representation of atomic positions in an unstressed metal in two adjacent crystals.
**(b)** When a stress above the elastic limit is applied, plastic deformation takes place by slipping of atoms in close-packed crystal planes. In the direction of movement, disturbances of atoms known as dislocations move along the planes (shaded atoms). They will meet grain boundaries, where they usually pile up, with considerable atomic disruption.
**(c)** At higher stresses, several slip planes may operate, resulting in further atomic disarray where they intersect. The energy associated with these interaction zones means that it gets progressively more difficult for the dislocations to move. The material therefore becomes harder and stronger as this internal energy and disruption increase. This process is known as work hardening and may be used to produce high-performance alloys, the stress being applied during fabrication methods such as extrusion and rolling.
**(d)** One disadvantage with the work hardening arises from the reduced ductility that is associated with the increasing difficulty of plastic deformation. Heating the alloy may give the atoms sufficient thermal energy to move to lower energy positions, re-creating the original structure, albeit with a different shape. Some areas of disturbance may still be present. This process is known as annealing.

by which the material can accommodate excessive stress is for cracks to propagate, often catastrophically, leading to fracture. There are general ways in which the fracture strength can be elevated, such as reducing the grain size and avoiding internal cracks as much as possible. Under some special circumstances, a different mechanism may be available; this is referred to as transformation toughening.

The vast majority of work on transformation toughening of ceramics has concerned various forms of zirconium oxide (zirconia); the biomaterials aspects of this are covered in Chapter 8. This is achieved by exploiting the tetragonal to martensitic phase transformation of discrete zirconia particles within the ceramic matrix. Zirconia exists in three polymorphic forms, a high-temperature cubic phase, a tetragonal structure that is normally stable above 1167°C and a monoclinic phase, which can be produced by a stress-induced, martensitic-like, transformation of the tetragonal phase. The addition of different oxide species to the zirconia is able to profoundly alter the transformation temperatures and different phases may be stable at room temperature with different compositions. Several mechanisms contribute to the overall toughening effect. The main mechanisms involve the stress-induced transformation that takes place when a small crack propagates in the tetragonal phase and transforms into the monoclinic phase, which changes the stress field around the crack and effectively blunts and arrests it.

### Composite materials

Fracture occurs when cracks propagate through the bulk of a material. There are two basic ways of preventing fracture; the first involves preventing crack initiation, the second is preventing crack propagation, or at least preventing crack propagation throughout the material. Structural composite materials have been developed primarily to generate strong and tough materials by increasing the resistance to crack propagation. If we consider a composite of parallel ceramic fibers within a polymer matrix, as in Figure 2.21(a), and apply a stress to this structure, we would expect a significant difference in the stress fields within these two phases. The ceramic fibers, which have a high elastic modulus, will sustain a much higher stress than the lower modulus polymer matrix. The ceramic, however, will be far more brittle and unable to sustain any plastic deformation, while the polymer is able to deform plastically, but will do so at low stress. We might expect, therefore, to see some cracks develop in the fibers (Figure 2.21(b)), but as soon as they reach the interface with the polymer, the latter will deform plastically at the point of interception with the crack, and dissipate the energy of that crack, preventing its further propagation. It is important that there is substantial interfacial adhesion to prevent the rapid propagation along the interface. This principle is used in glass fiber and carbon fiber composites. They are strong and tough, although there is very limited overall ductility.

It is not essential to use long fibers in composites, and in many complex shapes it is impossible to do so. Some composite materials may be made by incorporating short fibers in a matrix, which may even be injection molded, although the percentage of fibers in the product will be much less than in a long fiber composite, and the improvement in strength will normally be less. In other situations equiaxed particles may be used; it is these types of composites that are used in many contemporary dental restorative materials. Very good mechanical properties, including good wear resistance, can be achieved, especially if care is taken over the design, with multimodal distributions of particle size in order to maximize the total quantity of particles and the highest particle-matrix interfacial surface area (Figure 2.21(c)).

To a first approximation, the physical and mechanical properties can be represented by a simple mixtures law, where $P_c$ is the composite property and will be given by

$$P_c = P_f V_f + P_m V_m,$$

where $P_f$ and $P_m$ are the properties of the fiber f and matrix m, and $V_f$ and $V_m$ are the respective volume fractions. One manifestation of this is that the elastic modulus of a composite can be finely tuned by adjusting volume fractions, which has been important in the development of biomimetic composites.

## 2.8.4 Time-dependent deformation: fatigue, creep and viscoelasticity

The discussion of mechanical properties so far has assumed that stress–strain relationships are

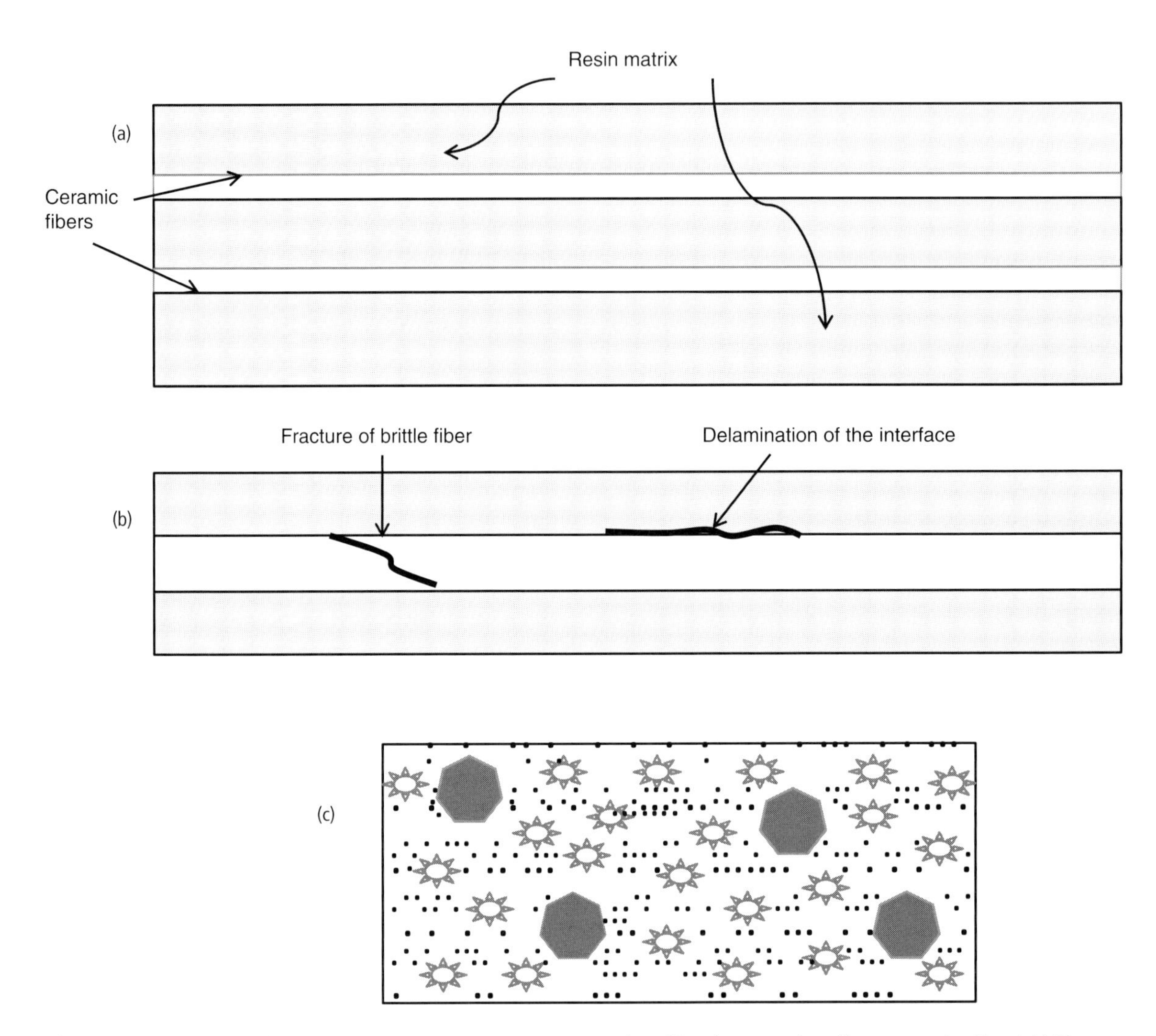


**Figure 2.21** Composites. (a) A long-fiber ceramic–resin composite, such as fiberglass or carbon fiber composite. The rigid fibers sustain most of the stresses, leaving the resin matrix relatively lightly stressed. (b) One consequence of the stresses in fibers is the occasional fracture of a fiber. However, the ductile matrix usually will not allow the crack to propagate any further – many fibers can fracture without too much detriment to the material. Mechanical forces may also cause delamination between fiber and matrix – most composites use a coupling agent to maximize interfacial bonding. (c) Particulate composites may also give good properties, including isotropic behavior which is difficult with long-fiber composites. With a soft matrix and hard particles, the amount of particles present, usually measured as their volume fraction, is an important determinant of properties. This volume fraction can be maximized by using particles of different sizes – in this example there are three sizes, which could be of micron, sub-micron and nanometer diameter respectively.

instantaneous and independent of time. This is not necessarily so and we often have to take into account the role of time-dependent processes. Two main mechanisms are involved here, one concerning fatigue and the other creep or viscoelastic behavior.

## Fatigue

With fatigue, the time variable is associated with the number of times a stress is applied. We might assume that if the maximum stress applied to a material is below the elastic limit, then only elastic deformation occurs. That is essentially true if we are considering macroscopic deformation. However, since the stress distribution is not necessarily uniform on a microscopic scale, being influenced by porosity, impurities and other discontinuities, it is possible for there to be localized regions of microplasticity. If stress is applied and removed repeatedly, then the effects of repeated microplasticity will be cumulative, often resulting in the nucleation of small cracks, which can then propagate. The result is that the material may fracture under the influence of large numbers of stress cycles, even though the maximum stress is less than the normal ultimate strength of the material. This phenomenon is known as fatigue. If we repeatedly apply a stress at, say 90% of the normal elastic limit of the material, we can count the number of times that stress has to be applied before fracture occurs. We may do the same for 85%, 80%, 75% of the elastic limit, and so on, and plot a graph, known as an *S–N* curve, which shows the number of stress cycles, *N*, that will cause fatigue fracture at any stress amplitude level *S*. Since the whole process of fatigue depends on the ability of microplasticity to occur and the ability of fatigue cracks to nucleate and propagate, there can be no absolute certainty about how one specimen, or one product, will perform, and the *S–N* behavior has been to considered approximate, and with statistical variability.

Different materials display different types of *S–N* curves, but they can be divided into two types. In the first, a material will display fatigue fractures as long as the stress amplitude exceeds a certain level. Below this stress level, known as the fatigue limit, which will typically be around 50% UTS, fracture should not occur, again taking into account the statistical variability; critical product design should always provide a good safety margin in order to accommodate this variability. The second type of behavior indicates that there is no fatigue limit and that fatigue failure is inevitable if we apply a stress a sufficient number of times. It should be noted that fatigue failures occur in practice after tens or hundreds of millions of cycles.

Since many implantable devices operate under physiological conditions, where heart beats or walking cycles constitute repetitive stress situations, we may assume that fatigue is an important biomedical engineering phenomenon. A heart valve, intravascular stent or artificial artery will be subjected to, on average, $80 \times 60 \times 24 \times 365$ stress cycles per year, or 40 million per year. A device in the musculoskeletal system may experience half a million cycles per year. As we shall see in Section 2.10, the fatigue behavior may be influenced by environmental factors such that, in implanted devices, corrosion fatigue becomes the defining mechanism.

### Creep

Creep is a process whereby deformation continues to occur under constant stress over prolonged periods of time. If a material is stressed to a level above the elastic limit, it will experience an instantaneous level of plastic deformation. If that stress level is sustained, we should expect the amount of deformation to slowly increase, as thermal energy may allow dislocations to keep on moving. This creep behavior is usually infinitesimally small at low temperatures, so that with metals it would not be observed at body temperatures. With polymers, however, body temperature may not be too far away from melting points so that creep may be a clinically relevant problem. Polymers used in the musculoskeletal system with constant applied stresses may suffer from creep.

### Viscoelasticity

The creep behavior of polymers is best considered in the context of viscoelasticity. Thermoplastics above their glass transition temperature may be regarded as very viscous fluids. Time-dependent deformation of polymers is therefore very important; creep occurs readily in the elastic region, the strain being almost wholly recoverable, given sufficient time, after removal of the load. This behavior is known as viscoelasticity. Although polymers are often characterized, for convenience, by Young's modulus, it is more relevant to take time into account. A tensile creep modulus may be defined as the ratio of stress to strain at a given time, say at 100 seconds. This modulus may vary, not only with time but also possibly with stress level, in which case it is described as a non-linear viscoelastic material. The tensile creep modulus also varies with temperature. Below the glass transition temperature, when the material is a glassy solid, deformation is almost wholly due to the stretching of interatomic and intermolecular bonds, there being very little time-dependent deformation. As the temperature is raised through the glass transition region, there is an increasing contribution from viscoelastic phenomena. The stress tends to alter the configuration of the molecular chains, especially straightening kinked or bent chains, a process known as rubbery elasticity. In addition, the thermal energy allows molecular chains to slide past each other, analogous to viscous flow. Greater levels of crystallinity reduce the contributions from viscoelasticity.

### 2.8.5 Hardness

The hardness of a material is a measure of the resistance to abrasion or indentation. A very rough comparison of hardness is made by the ability of materials to scratch each other. In the Mohs' scale, a series of standard materials are arranged in order, with talc as No 1 and diamond as No 10. More conveniently for conventional engineering materials, the hardness is measured by the amount of deformation induced in a specimen by a standard indentation. There are several variants on the type of indenter, the Brinell, Vickers and Rockwell being the most common. In all cases, the size of the indentation produced under a known load is measured and converted to a hardness value. High values indicate hard materials and low values, soft materials.

Such methods are relatively crude and more sophisticated methods are required for complex structures. Nanoindentation techniques have been introduced for this purpose. These use depth-sensing methods to determine the hardness of the outermost surface and are being increasingly used in biomaterials characterization. Typically the properties are measured to a depth of less than a micron, with high spatial resolution. Several variations are possible, with both quasi-static and dynamic methods widely used. With dynamic nanoindentation, a small oscillatory load of known maximum amplitude and frequency is applied to a specimen and the displacement response measured. This response has components due to the nanoindenter itself as well as the material. Software is used to separate these components, and the output may be used to calculate a number of material properties including the elastic modulus.

### 2.8.6 Friction and wear

Many biomaterials applications involve different components in contact with each other or in contact with tissues, and where there is relative movement between them. The mechanical properties of materials under sliding conditions, collectively discussed under the term tribology, are very important; two features of the interactions at sliding surfaces are particularly relevant – the frictional forces that are generated and the wear or destruction of the surfaces during the movement.

#### Friction

If two solids are placed in contact, there will be some forces of attraction between their surface atoms or molecules. This means that there will be a degree of bonding at the interface, which implies that a force will be necessary to break the bonds if the materials are to slide over each other. This adhesion accounts for the frictional forces involved in surface movement. At the interface, however good the surface finish of the components is on the microscale, they will be atomically rough. Exaggerating the situation slightly, Figure 2.22(a) represents the interface, showing that there is actual contact in only a few places. These are called junctions and the sum of all the contact areas of these junctions constitutes

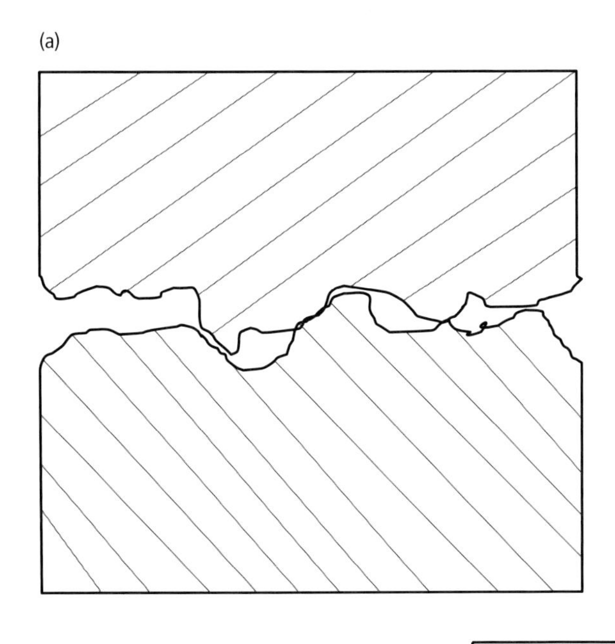


**Figure 2.22** Contact at surfaces.
**(a)** Although two surfaces may appear to be quite smooth, on the atomic scale they will usually be undulating or irregular. The real area of contact may represent just a small fraction of the apparent area of contact.
**(b)** One consequence of this very small area of contact is the interfacial stresses may be very high at these points since they have to sustain all of the external stresses. Under the influence of these high localized stresses, strong adhesive forces may arise; with some materials, this could result in a degree of localized welding. When attempts are made to separate the surfaces, in sliding for example, the strength of the adhesion will determine whether the separation takes place exactly along the original interface (on the left) or by fracturing of some of the asperities (on the right), which yield particles. These are adhesive wear particles, resulting in material loss from one or more surfaces.

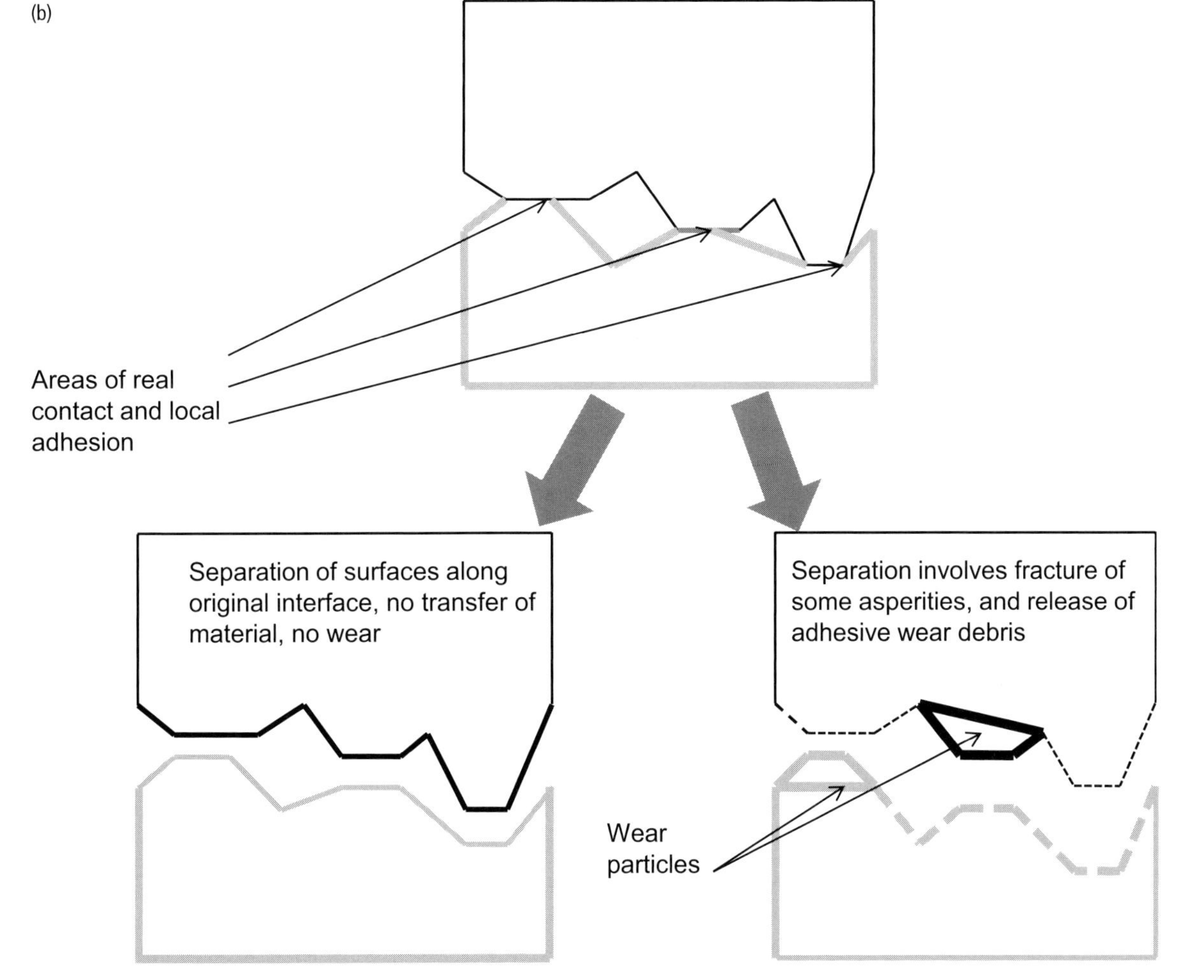

the real area of contact. The total macroscopic area of contact is the apparent area. Because the forces between atoms are short range, the only contribution to the adhesion comes from the real area. The magnitude of the frictional force, $F$, resisting the motion of one solid over another will be equal to $\mu L$, where $L$ is the normal force acting between the objects and $\mu$ is the coefficient of friction.

The real area of contact is governed by the normal force acting on the surface and on the strength of the material. For two identical materials in contact, the real area will be proportional to the ration of $L$ to the hardness and the frictional force will be proportional to the product of the shear stress and the real area. With two materials of quite different mechanical properties, the frictional behavior will be controlled by the weaker of the two for it is this which will yield plastically in compression to give a larger real area, and this will also shear to produce the sliding. In the absence of lubrication, the coefficient of friction is basically independent of load, sliding speed, surface geometry and roughness. The other feature to note is that since the friction is dependent on the ratio of hardness and shear strength, and since these two properties are related to the same basic material properties, forces of friction should be very similar for most materials. There is therefore a remarkably narrow range for the coefficients of friction for clean materials in sliding contact. For clean metal surfaces, the coefficient of friction is usually in the range of 0.3 to 1.5; in practice the range may be higher because of contamination and other factors. Polymers generally have coefficients of friction in the range 0.4 to 0.6. There are some exceptions, most notably polytetrafluoroethylene, which is so chemically inert that forces of adhesion at an interface are minimal and hence the coefficient of friction is very low, around 0.04. A further type of low friction material involves structures that are close-packed with strongly bound atoms within layers but very low binding forces between layers. Shear in a direction within the layers is therefore easy but compression normal to those planes is difficult, resulting in abnormally low friction. Examples are graphite and molybdenum disulfide.

### Wear

Wear is defined as the removal and relocation of material arising from the contact of two solids. There are four distinct types of wear, namely adhesive, abrasive, corrosive and surface fatigue wear. Corrosive wear involves the wear of surface oxide films that would otherwise inhibit corrosion, thereby accelerating corrosion. Surface fatigue wear occurs under repeated loading and sliding cycles, which results in the formation of surface cracks and eventual breakdown of the surface.

Abrasive wear occurs when a rough hard surface slides over a much softer surface and ploughs a series of grooves in it. The material from the grooves is displaced in the form of wear particles. Adhesive wear occurs when two surfaces slide over each other and fragments are sheared off at points within the real area of contact (Figure 2.22(b)). During sliding there are several possibilities for the adhesive forces to be overcome. This could possibly happen along the interface itself, resulting in no loss of material on either side. Alternatively shear could take place within the asperities of either side resulting in the release of fragments, or wear particles. Generally the amount of wear is directly proportional to the load applied across the interface and to the distance of sliding, and is inversely proportional to the hardness of the surface being worn away.

### Lubrication

Both friction and wear properties may be significantly altered if a lubricant is present between the surfaces. There are several different types of lubrication, ranging from situations where a liquid film is thick enough to give complete separation of the components to those where the film is only of atomic dimensions. In many engineering situations, the lubricating film will be relatively thin, and the mode of boundary lubrication will apply. The sliding may also be accompanied by elastic deformation of the asperities, giving elastohydrodynamic lubrication.

## 2.9 Corrosion and degradation mechanisms

The environment to which biomaterials are exposed during use can be considered to be an aqueous medium containing various anions, cations, organic substances and dissolved oxygen, as detailed in Box 2.2. The anions are mainly chloride, phosphate, and bicarbonate ions. The principal cations are $Na^+$, $K^+$, $Ca^{2+}$ and $Mg^{2+}$. The concentration of dissolved oxygen also influences the aggressive nature of the environment and in venous blood is approximately a quarter of that in air. The organic substances include low molecular weight species as well as relatively high molecular weight proteins and lipids. The protein content of the environment is known to have a significant influence on the corrosive nature of body fluids. The temperature is constant around 37°C and the pH is around 7.4, although inflammation may cause it to change to 4 or 5 for short periods. In some situations, very active molecular species, including intracellular enzymes, free radicals and superoxide ions may be released into extracellular spaces.

### Box 2.2 The composition of physiological fluids

These non-exclusive lists give the major components and typical levels; they are included here to indicate the complex and aggressive nature of the physiological fluids that biomaterials will encounter.

(A) Inorganic concentrations of fluids (mM)

| | Whole blood | Other extracellular fluids |
|---|---|---|
| Sodium | 130–140 | 5–160 |
| Potassium | 3.5–5.0 | 2–40 |
| Magnesium | 0.7–2.0 | 0.1–2 |
| Calcium | 1.9–3.0 | 1–3 |
| Bicarbonate | 16–31 | 2–40 |
| Chloride | 95–110 | 6–140 |
| Total phosphorus (mg/dl) | 2.5–4.5 | 0.5–10 |

(B) Organic concentrations of fluids (mg/dl)

| | Whole blood | Other extracellular fluids |
|---|---|---|
| Amino acids | 4.5–7.5 | 0–7 |
| Cholesterol | 115–125 | 0.1–200 |
| Fatty acids | 250–400 | 0–500 |
| Glucose | 80–100 | 10–110 |

(*cont.*)

| | Whole blood | Other extracellular fluids |
|---|---|---|
| Total lipids | 450–600 | 0–850 |
| Phospholipid | 225–285 | 0–300 |
| Urea | 20–40 | 0–40 |

(C) Protein concentrations of fluids (mg/dl)

| | Plasma | Other extracellular fluids |
|---|---|---|
| Albumin | 4000–5500 | 0–2500 |
| Fibrinogen | 200–400 | negligible |
| Fibronectin | 1.5–3 | $< 1.5$ |
| IgG | 650–1600 | $< 700$ |
| Lysozyme | negligible | 100–280 (tears) |
| Macroglobulin | 150–450 | $< 50$ |
| Transferrin | 200–320 | negligible |

(D) Chemicals released from cells or activated during inflammation

| | |
|---|---|
| Free radicals | Hydrogen peroxide, superoxide anions |
| Vasoactive agents | Prostacyclin, histamines, serotonin |
| Lysosomal protease enzymes | Collagenase, elastase |
| Cytokines | Interleukins, tumor necrosis factor |
| Plasma proteases | Bradykinin, kallikrein, complement fragments, coagulation factors |
| Growth factors | PDGF, FGF, TGFs, EGF |

It is necessary here to consider how biomaterials interact with this varying but potentially very aggressive environment, bearing in mind that on many occasions we wish the material to be stable and as inert as possible but yet in other situations we wish them to degrade, safely, and often rapidly. With metals we discuss the various processes of corrosion. With polymers, there are several potential mechanisms of degradation. With ceramics, the performance may range from simple solubility to slow degradation.

Of course, we cannot expect the biological environment to be totally constant; there are variations with time, location, activity, health status, etc. Moreover, mechanical stress plays a very important role in the corrosion and degradation processes, both potentiating existing effects and initiating others. Under these circumstances, corrosion and degradation are not necessarily progressive homogeneous reactions with zero-order kinetics; they can be quiescent but then become activated, or they can be active but then become inactive, with transient fluctuations in conditions playing a part in these variations. The discussion takes into account the effect of the corrosion/degradation process on the properties of the material over time and the potential effect of the products of the process on the host tissues.

## 2.9.1 Metallic corrosion

On the basis of their structure we should predict that metals, as a generic group, should be relatively susceptible to reactions in the human physiological environment.

### General principles

We shall discuss first the behavior of metallic biomaterials in a pure aqueous environment and then consider how other species in that physiological environment affect that behavior. There are always two reactions that occur with metals in water – the anodic reaction, which yields metallic ions, for example:

$$M \rightarrow M^{(n+)} + n \text{ (electrons)},$$

and the cathodic reaction, in which these electrons are consumed, for example by the reduction of hydrogen:

$$2H^+ + 2e^- \rightarrow H_2,$$

or the reduction of dissolved oxygen:

$$O_2 + 4H^+ + 4e^- \rightarrow 2H_2O \text{ (in acidic solutions) or}$$

$$O_2 + 2H_2O + 4e^- \rightarrow 4OH^- \text{ (in neutral or basic solutions)}.$$

The rate of the anodic or oxidation reaction must equal the rate of the cathodic or reduction reaction. Variations in the local environment can affect the overall rate of corrosion by influencing either the anodic or cathodic reactions, and preventing either of them can stop the corrosion process. As we have seen, a pure metal consists of positive ions closely surrounded by free electrons. When the metal is placed in a solution, there will be a net dissolution of metal ions since the free energy for the dissolution reaction is less than for the reverse reaction. The metal therefore develops a net negative charge, thus making it increasingly difficult for the positive ions to leave the surface. There will come a point when the free energy for the dissolution reaction will equal that for the reverse reaction and a dynamic equilibrium is established, a potential difference being set up across the charged double layer at the metal surface. This potential difference will be characteristic of the metal; when measured against a standard hydrogen electrode in a 1 N solution of its salt at 25°C, it is defined as the standard electrode potential for that metal. Metals can be placed in a series, known as the *electrochemical series*, on the basis of these standard electrode potentials (Table 2.5). The position of a metal in this series

Table 2.5 **The electrochemical series**

| Element | Electrode potential (volts) |
|---|---|
| Lithium | −3.04 |
| Magnesium | −2.37 |
| Aluminum | −1.67 |
| Titanium | −1.63 |
| Chromium | −0.74 |
| Iron | −0.44 |
| Cobalt | −0.28 |
| Nickel | −0.24 |
| Molybdenum | −0.20 |
| Hydrogen | 0.00 |
| Copper | +0.34 |
| Silver | +0.80 |
| Platinum | +1.20 |
| Gold | +1.45 |

indicates the order with which metals displace each other from compounds and also gives a general guide to reactivity in aqueous solutions. Those at the bottom are the noble, relatively unreactive metals, whereas those at the top are the more reactive.

In this situation of a homogeneous pure metal existing within an unchanging environment, equilibrium is reached and no further net movement of ions takes place. In other words, the corrosion process takes place only transiently, but is effectively stopped once equilibrium is attained. In reality, we do not usually have entirely homogeneous surfaces or solutions, or complete isolation of the metal from other parts of the environment, and this equilibrium is easily upset. If the conditions are such that the equilibrium is displaced, the metal is said to be polarized. Two main factors determine the extent of corrosion in practice. The first concerns the driving force for continued corrosion (the reasons why the equilibrium is upset), and the second concerns the ability of the metal to respond to this driving force.

If either the accumulating positive metal ions in the surrounding media or the accumulating electrons in the metal are removed, the balance between the dissolution and the replacement of the ions will be upset. For example, this may occur in the physiological environment due to the interaction of proteins with the metal ions. Metal ions can form complexes with proteins and these complexes can be transported away from the immediate vicinity. This removes the metal ions from the charged double layer at the interface allowing further release of metal ions to re-establish the equilibrium. The equilibrium is established because of the imbalance of charge, so that if the charge balance is disturbed, further corrosion will occur in an attempt to re-establish the balance. The result will be continued dissolution as the system attempts to achieve this equilibrium, giving sustained corrosion. An environment that allows the removal of electrons in contact with the metal or stirring of the electrolyte will achieve this.

The process of galvanic corrosion is important here. Consider a single homogeneous pure metal placed within an electrolyte. The metal will develop its own potential with respect to the electrolyte. If a different metal is placed into the same electrolyte, but without contact with the first metal, it will develop its own potential. If these two potentials are not equal, there will be a difference in the numbers of excess free electrons in each. This is of no consequence if the two metals are isolated from each other, but should they be placed in electrical contact, electrons will flow from that metal with the greater potential in an attempt to make the two potentials equal. This upsets the equilibrium and causes continued and accelerated corrosion of the more active metal (anodic dissolution) and protects the less active (cathodic protection).

Galvanic corrosion may be seen whenever two different metals are placed in contact in an electrolyte. It is not necessary for the components to be macroscopic, monolithic electrodes for this to happen, and the same effect can be seen when there are different microstructural features within one alloy, for example different phases or inclusions. In practice, it is the regional variations in electrode potential over an alloy surface that are responsible for much of the generalized surface corrosion that takes place in metallic components.

Most metallic biomaterials are alloys that contain highly reactive metals (i.e., with high negative electrode potentials), such as Ti, Al and Cr. Because of this high reactivity, they will react with oxygen upon initial exposure to the atmosphere. This initial oxidation leaves an impervious oxide layer firmly adherent to the metal surface; thus all other forms of corrosion may be significantly reduced because the oxide layer acts as a protective barrier – the metal is then said to be passivated. The manufacturing process for alloys in medical devices usually includes a passivating step to enhance the oxide layer prior to implantation, for example nitric acid treatment of stainless steel.

The presence of biological macromolecules should not cause completely different corrosion mechanisms, since the electrochemical events described above are universally applicable and follow sound basic laws. However, they can influence the rate of corrosion by interfering in some way with the anodic or cathodic reactions. For example, biological molecules could consume one or other of the products of the anodic or cathodic reactions; proteins can bind to metal ions and transport them away from the implant surface. This will upset the equilibrium across the charged double layer and allow further dissolution of the metal.

The stability of the oxide layer also depends on the electrode potential and the pH of the solution. Proteins and cells can interact with the charges formed at the interface and thus affect the electrode potential. Inflammatory cells and microorganisms can alter the pH of the local environment through the generation of acidic metabolic products that may alter the equilibrium. Moreover, the stability of the oxide layer is also dependent on the availability of oxygen, and adsorbed macromolecules and cells could limit the diffusion of oxygen to regions of the surface, causing preferential corrosion of the oxygen-deficient regions and breakdown of the passive layer. In addition, the cathodic reaction often results in the formation of hydrogen and the buildup of hydrogen in a confined locality tends to inhibit the cathodic reaction, restricting the corrosion process. If the hydrogen were eliminated, for example by the activity of bacteria, then the active corrosion could proceed. It is known that the presence of proteins and cells can influence the rate of corrosion of some metals by these types of effect. With some pure metals under simulated physiological conditions, the presence of proteins can significantly enhance corrosion rates while with some other metals there may be a decrease in the rate. One of the reasons for this variability may be related to the wide range of protein adsorption characteristics.

### Corrosion in physiological environments

Examples of corrosion of metallic biomaterials are shown in Figure 2.23.

The need to ensure minimal corrosion has been a major determining factor in the selection of metallic biomaterials, and two broad approaches have been developed. The first involves the use of noble metals, that is, those metals where the electrochemical series indicates excellent corrosion resistance. Examples are gold, silver, and the platinum group of metals. Because of cost and relatively poor mechanical properties, these are not used for major structural applications; silver is sometimes used for its antibacterial activity and platinum-group metals (platinum, palladium, iridium and rhodium) are used in electrodes.

The second approach involves the use of the passivated metals. Three metallic elements are strongly passivated, Al, Cr and Ti. Al cannot be used on its own because of toxicity problems; however, it has an important role in several titanium alloys. Cr is very effectively protected but cannot be used in bulk. It is widely used in alloys, especially in stainless steels and in the cobalt–chromium-based alloys, where it is normally considered that a level of above 12% gives good corrosion resistance and about 18% provides excellent resistance; the passivating layer is essentially chromium oxide. Ti is the most strongly passivated, through the formation of titanium dioxide, and is used as a pure metal or as the major constituent of alloys. Other alloying elements may be present in the surface oxide and have an influence on the passivity of the layer. The molybdenum in some stainless steels enhances the performance of the chromium oxide layer. Production procedures need to be controlled because of their influence on the surface oxides, for example the cleaning and sterilization procedures.

Although these metals and alloys have been selected for their excellent corrosion resistance, corrosion will still take place when they are implanted in the body. All metals will suffer a slow removal of ions from the surface, largely because of local and temporal variations in microstructure and environment. The rate may either increase or decrease with time, but metal ions will be inevitably released into that environment, which is particularly important since it is the effect of these potentially toxic or irritant ions that largely control the response of tissues to the materials, as discussed in Chapter 3.

(a)

(b)

(c)

**Figure 2.23** Metallic corrosion.
**(a)** Generalized corrosion in stainless steel component of dialysis machine. **(b)** Galvanic corrosion on stainless steel fracture plate that was used on the proximal part of a femur in contact with a titanium plate placed at the distal end. **(c)** Crevice corrosion at the interface between two components of a femoral osteotomy device.

Even with a strongly passivated metal, there will be a finite rate of diffusion of ions through the oxide layer, and possibly dissolution of the layer itself. Even Ti is steadily released into the tissue from titanium implants.

### Specific corrosion mechanisms

In addition, some specific mechanisms of corrosion may be superimposed on this general behavior as seen below.

**Pitting corrosion** The chromium oxide at a surface passivates stainless steel biomaterials. In a physiological saline environment, however, the driving force for repassivation of the surface is not high, so that if the passive layer is broken down, it may not repassivate and active corrosion can occur. Localized corrosion takes place as a result of imperfections in the oxide layer, producing small areas in which the protective surface is absent. These localized spots will actively corrode and pits will form in the surface, resulting in a large degree of localized damage since the small areas of active corrosion become anodic and the entire remaining surface becomes cathodic. Since the rate of the anodic and cathodic reactions must be equal, a small area of the surface will initiate a relatively large amount of metal dissolution, and large pits may form.

**Fretting corrosion** The passive layer may also be removed by a mechanical process, such as with a continuous cyclic process of abrasion. Any attempt at spontaneous repassivation may result in removal of the new surface by the next cycle of abrasion. This is known as fretting corrosion. The continual removal of the oxide film obviously increases the corrosion rate. The process may also induce localized plastic deformation of the contact area, producing high strain fatigue and enhancing corrosion fatigue. With titanium, there is also the possibility that fretting enhances hydrogen absorption into the metal, which causes hydrogen embrittlement and a decrease in mechanical properties.

**Crevice corrosion** Accelerated corrosion can also be initiated in a crevice in a metallic device, for example at the interface between two components. Crevices may restrict diffusion of oxygen at the interface. Initially, the anodic and cathodic reactions occur uniformly over the surface, including within the crevice. As the crevice becomes depleted of oxygen, the reaction is limited to metal oxidation balanced by the cathodic reaction on the remainder of the surface. The buildup of metal ions within the crevice causes the influx of other ions (for example, chloride ions) to balance the charge by forming, in this case, metal chlorides, which are the corrosion products. In the presence of water, the chloride will dissociate to its insoluble hydroxide and acid. This is a rapidly accelerating process since the decrease in pH causes further metal oxidation.

**Stress corrosion cracking** Stress corrosion cracking is an insidious form of corrosion since an applied stress and a corrosive environment work synergistically and may cause complete failure in situations where neither the stress nor the environment alone would be problematic. The stress level may be very low and the corrosion may be initiated at a microscopic crack tip that does not repassivate rapidly. Incremental crack growth may then occur, resulting in fracture. This is a risk in some stainless steels in saline environments.

**Galvanic corrosion** As we have seen above, if two metals are independently placed within the same solution, each will establish its own electrode potential with respect to the solution. If these two metals are placed in electrical contact, then a potential difference will be established between them, electrons passing from the more anodic to the more cathodic metal. A continuous process of dissolution from the more anodic metal will then take place; this is galvanic corrosion. This is important if two different alloys are used together in an implantable device when the more reactive may corrode freely.

Whenever stainless steel is joined to another implant alloy, it may suffer from galvanic corrosion. If both alloys remain within their passive region when coupled in this way, the additional corrosion may be minimal. Some modular orthopedic systems are made of titanium alloys and cobalt-based alloys on the basis that both should remain passive, but evidence of corrosion has been reported. Titanium stems of modular prostheses can exhibit corrosion.

**Intentional corrosion and resorption of implantable alloys** The vast majority of applications of metallic systems in medical devices are predicated on the assumption that the device will remain in the patient's body for their full life or until it is intentionally removed surgically. There are some situations where permanence is not required, and indeed could be disadvantageous and an intentionally degradable and resorbable material may be considered as an alternative. There is increasing interest in the possibility of using magnesium alloys that are known to degrade in the body, and are potentially resorbable. Mg itself corrodes readily, and samples can disappear in a short time. It would appear that the Mg ions are relatively well tolerated, Mg being an essential element in humans. Developments with this element have concentrated on alloying additions that reduce the corrosion rate without inducing significant cytotoxicity or radically altering the mechanical properties. Alloys of Mg-Mn-Zn and Mg-Ca, where all alloying additions are also essential elements, may be considered here.

## 2.9.2 Polymer degradation

### General principles and definitions

With polymeric materials, the nature of the individual atoms and groups, and the manner in which they are arranged within these molecular and crystalline structures, control most of the properties of the resulting material, including the susceptibility to degradation. Of crucial importance to the discussion of degradable and resorbable polymers is the environmental stability of the structures. It is appropriate to carefully define the relevant terms here – many terms used within the field of biomedical polymer degradation and resorption have widely used generic meanings as well as precise meanings. "Degradation" is a general term used in materials science that means "deleterious changes in the chemical structure, physical properties or appearance of a material." "Biodegradation" means "the breakdown of a material mediated by a biological system."

These are very general terms and are not sufficiently precise or focused to allow a clear understanding of what is required, or achieved, in intentionally degradable medical devices. Neither term mentions at all the nature of any degradation products or their fate, either within the body or associated with their elimination from the body. In many situations we require the materials to be as stable as possible. The intention with other polymer biomaterials is that they are fully and harmlessly eliminated from the body as the end-stage of the degradation process. Such polymers were initially described as absorbable polymers but this term is not entirely appropriate since "absorb" is a simple verb that describes the process of taking in or assimilating a substance. Common usage has determined that "absorbable" and "bioabsorbable" are still used in this area, and the latter term has come to be associated with the phenomenon of being "capable of being degraded or dissolved or subsequently metabolized within an organism." The terms "resorbable" and "bioresorbable" are more appropriate although there is no clear consensus of the distinction between them. The noun "bioresorption" and its adjective "bioresorbable" are preferred, the former being defined as "the process of removal by cellular activity and/or dissolution of a material in a biological system." When the term "resorbable polymer" is used in the context of medical applications, it de facto becomes synonymous with "bioresorbable polymer," which can be taken to mean "a polymer that is capable of removal by cellular activity and/or dissolution in a biological system such as the human body."

A polymer that degrades in an aqueous environment must have linkages between atoms and between molecules that can be broken in that environment. The simple polymers with all-carbon backbones are usually stable because the C-C bond is very strong and not easily broken by thermal or chemical energy. On the other hand, there are several bonds that are quite readily broken by such energies. In polymers, wherever carboxyl groups are found there will be susceptibility to degradation, as there will in polymers based on ester and amide structures. Inherently biodegradable polymers will usually contain one or more of these structures, shown in Figure 2.24.

Polymers in general are susceptible to degradation through the effects of heat, light and ionizing radiations and in certain chemical environments. Within the human body, we are not concerned with excesses of heat or exposure to light or radiation, but implanted biomaterials will be constantly bathed in tissue fluids. Their biodegradation is therefore largely dependent on interactions with these fluids and the ions and molecules they contain. Water is the main constituent of these fluids, and hydrolysis, which is the breakdown of any substance caused by interaction with water, is the main mechanism by which degradable polymers break down in the body. As a general rule, hydrolysis of organic substances takes place minimally under neutral conditions, that is when the pH of the environment is around 7.0, but proceeds much faster under acidic (low pH) or

*Polymers resistant to hydrolysis*

All-carbon-backbone polymers

- C – C – C – C – C – C – C – C --

Siloxane polymers

- Si – O – Si – O – Si – O – Si – O -

*Polymer groups susceptible to hydrolysis*

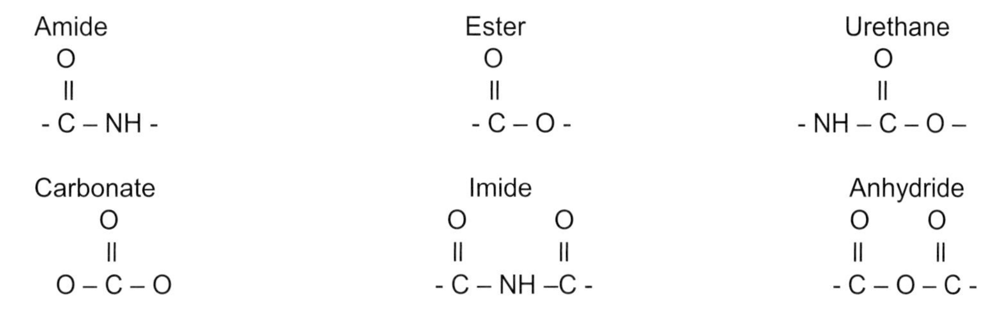


**Figure 2.24** Hydrolysis of polymers. Hydrolysis is the most significant form of degradation of polymers within the body. Polymers with an all-carbon backbone are the most resistant; siloxane polymers with alternating Si and O atoms along the backbone are also highly resistant. A number of groups, especially involving NH and O, that are found in heterochain polymers are susceptible to hydrolysis. This figure gives some prominent examples.

alkaline (high pH) conditions, a fact which becomes very important in discussion of both *in vivo* and *in vitro* degradation and the design of biodegradable biomedical polymers.

In order for the water of tissue fluids to cause the molecules of a polymer to degrade, it has to gain access to the susceptible bonds. Within any one type of polymer system, much will depend on the ability of the water to diffuse into the polymer. For a polymer that is very resistant to water absorption, there will be limited opportunity for hydrolysis to take place and it may well be that the degradation process is confined to the surface. In other words, the polymer degrades slowly by a process of continual erosion of the surface, the polymer essentially degrading from the outside towards the center. For polymers that readily take up water, we would expect the degradation process to be more uniform and faster. This characteristic of absorbing water is controlled by the structure and chemical form of the molecules themselves and the molecular arrangement. The degree of crystallinity is one important factor, as is the overall hydrophilic/hydrophobic nature of the material. Some polymers such as PTFE, poly(tetrafluoroethylene), are extremely hydrophobic, giving the non-stick quality to domestic products made of Teflon®, and do not absorb water at all. It is interesting to note that this polymer has many applications in long-term implantable devices because it is so biologically inert through the combination of this hydrophobicity and the presence within the molecular structure of only high strength C-C and C-F bonds. The intentionally biodegradable polymers are those that are both hydrophilic and contain hydrolysable bonds.

It is also necessary to take into account the nature of end groups of the polymer chains. The repeating units in the polymer backbone control overall properties, but each molecule have ends that will usually be of a different structure to these repeating units, and these will play some role in the initiation of the degradation. Manipulation of polymer end group chemistry is one way that biodegradable polymers can be fine-tuned with respect to degradation rates. In addition, since all medical devices have to be sterilized, the effect of the sterilization process (often $\gamma$ irradiation or electron beam) on the polymer has to be taken into account. Such processes often cause up to 50% drop in molecular weight of the polymer, with a consequent effect on the degradation rate.

As noted earlier, there are also many types of cell and macromolecules within the tissues, which can be highly active. Many of these entities can contribute to polymer degradation processes, and it is important to note that the constitution and activity of this tissue environment are changeable, both with time and from one location to another. For example, we will see in Chapter 3 that inflammation is the response of the body to injury or irritation, which is mediated by a series of cells and molecules that are present in the tissue fluids. Thus the tissues in contact with biomaterials will contain, at any one place and at any one time, a myriad of very active and aggressive molecules and cells, which may assist the hydrolysis process, or introduce another mechanism of degradation such as oxidation.

We therefore have an immensely interesting situation where the implantation of a polymer into the body may stimulate an inflammatory reaction and where that inflammatory reaction may influence the way in which the polymer degrades. Polymer degradation within the body may be autocatalytic, which makes the understanding of the degradation process difficult and adds considerable uncertainty to the prediction of *in vivo* performance.

To add to this complexity, there are some additional factors that influence polymer degradation mechanisms and their kinetics. First, there are some molecular species other than the products of cell activation that may accelerate the degradation process, principal amongst these being metal ions such as iron and cobalt. Secondly, the degradation process is likely to be influenced by the mechanical stresses applied to the polymer. Stress will influence the molecular conformation, and the stretching of bonds within and between molecular chains may increase susceptibility to environmental attack. This may happen with applied stress, for example associated with tying the knot of a suture, or with residual stress following a manufacturing process.

### The selection of resorbable polymers

Based on the comments in the section above, the essential requirements for a resorbable biomedical polymer can be summarized as follows. The polymer should have:

- appropriate mechanical and physical properties suitable for the desired function;
- a molecular structure that is capable of hydrolysis at pH 7.4 with the desired kinetics;
- a hydrophilicity that allows for the required degree and kinetics of water absorption;
- a degradation process that results in the predictable formation of degradation products;
- a chemical composition that reliably results in degradation products that are readily assimilated and/or metabolized within the body without any significant pro-inflammatory or cytotoxic characteristics;
- a structure and form that can be conveniently fabricated without the need for any processing additives that would compromise the biocompatibility of the polymer as it degrades.

The polymeric materials that are able to satisfy the above requirements may be grouped as either naturally derived or synthetic. This is not an entirely accurate division since some polymers that are based on structures found in nature may actually be prepared by synthetic routes, including recombinant technology, but it is useful for present purposes. These are the biopolymers, which include proteins and peptides, such as collagen, elastin, keratin and silk, and polysaccharides such as chitosan, alginate and hyaluronan, and lipids.

Over the last few decades, several groups of biodegradable/resorbable polymers have been developed or adapted for medical use. There is no straightforward way of classifying these and it is better to consider each group as a separate entity, although there will be some similarities between many of the groups. First and foremost of these groups are the polyesters based on lactic acid and glycolic acid and the closely related polycaprolactone. In no particular order, the other clinically used or experimental degradable biomedical polymers are the polyanhydrides, polydioxanone, polyorthoesters, polyglycerols, sebacate polymers, fumarate polymers, cyanoacrylate polymers, degradable polyurethanes and the polyhydroxyalkanoates. These are all discussed in Chapter 8.

### Biostable polymers

As noted above, virtually all polymers should eventually succumb to some form of degradation. In the previous section we discussed those that have been designed to be intentionally degradable; now we have to mention those that were intended to be non-degradable, or biostable, but which have shown some degree of degradation in the human body. Taking forward the arguments presented earlier, we may assume that some polymers will have bonds that are potentially hydrolysable under certain circumstances and other polymers that will absorb water in some conditions. It is convenient, therefore, to rank groups of polymers according to their sensitivity to these processes:

- Hydrophilic and hydrolysable – bulk degradation.
- Hydrophilic, no hydrolysable bonds – may swell, but little or no degradation.
- Hydrophobic, hydrolysable – surface activity only.
- Hydrophobic, no hydrolysable bonds – most biostable.

This is a fair, but only fair, representation of what happens in practice. It should be expected that all homochain polymers, that is with a backbone solely of carbon atoms or rings, should be essentially biostable. These include polyethylene, polypropylene, polytetrafluoroethylene, polyetheretherketone, polyvinylchloride and polymethylmethacrylate, which should all survive in the body, under normal circumstances, for many years. Polymers that have some degree of hydrophilicity and hydrolysability include families of aromatic polyesters, polyurethanes and polyamides. Some of these can be used in certain situations, for example the aliphatic polyester, polyethylene terephthalate is widely used as an implantable textile, but there are situations in which degradation will eventually occur.

The reasons why this is only a fair representation of the *in vivo* experience are based on three factors. First, as implied above, the process of conversion of a monomer to a polymer has to be considered reversible given the right circumstances. Secondly, hydrolysis is not the only available mechanism for polymer degradation, even if the most obvious one under the water-based, ambient temperature environment of the human body or cell culture. Thirdly, factors other than the environmental chemistry may play a part. Other than these generalities, there are few common threads by which we may characterize the degradation of putatively biostable polymers. Examples of such degradation are found in various chapters of this book. We will list a few examples here that indicate the nature of the phenomena.

**Chain scission of homochain polymers** There is a general process of degradation, known as chain scission, which can cause the breakdown of most polymers under specified conditions. The process involves initiation, propagation and termination stages. During initiation, energy is absorbed from the external source causing the cleavage of a covalent bond (either the primary chain bond or a cross-link) and the formation of active radicals. Thermal degradation occurs when the vibrational, rotational or translational energy exceeds the activation energy required to break a C-C bond on increasing the temperature. With polyethylene there is a random chain scission event (called random since it could occur anywhere along the polymer chain), which is followed by propagation, leading to the elimination of an ethylene molecule, or termination, where free radical transfer takes place that stops the process. The result of this is a mixture of chain fragments of varying size and some released monomer. Clearly the activation energy for carbon bond scission is the controlling factor, and this will vary with a number of factors, including the nature of the side groups and the molecular weight. Polytetrafluoroethylene has one of the highest activation energies and is one of the most stable of all polymers.

Polymethylmethacrylate can also undergo thermal degradation but this involves so-called unzipping where the polymer molecule effectively depolymerizes from an end. Steric hindrance from the $CH_3$ and $COOCH_3$ side groups prevents free radical transfer and so, once initiated, the process proceeds rapidly to full depolymerization.

The purpose of this discussion is not to show that we should expect extensive chain scission and depolymerization in homochain polymers under ambient conditions, but rather to point out that such mechanisms are available. Whenever a thermally activated chemical mechanism exists, and when activation energies can be measured, it should be assumed that statistically some parts of the mechanism could take place, with a greater frequency at higher temperatures. It is usually possible to make estimates of the lifetime of materials under ambient conditions on the basis of the kinetics of the processes at higher temperature; such estimates for homochain biomaterials at body temperature usually exceed 20 years.

Radiation, involving either UV (giving photolysis) or high-energy radiation (radiolysis) can also cause degradation. Biomaterials are not normally susceptible to photolysis since they do not absorb radiation energy in the UV range, but as noted in a later section, they can suffer radiolysis during gamma ray sterilization.

**The effect of mechanical forces** Since many biomaterials applications involve mechanical forces, it is necessary to consider whether such forces could influence the degradation. There is no evidence that mechanical stress causes degradation in materials for which no known degradation mechanism exists, but it is clear that it can contribute to the rate of degradation of susceptible biomaterials. All intentionally degradable biomaterials will be influenced in this way, with the rate of degradation being increased significantly when high tensile stresses are applied. Most polyurethanes are also quite susceptible to so-called stress-cracking. This has been clinically problematic in cardiac pacemaker leads that are subjected to repeated flexing, although this does depend on the precise chemical structure and morphology.

**Effect of tissue components** As with the effect of mechanical forces, it is known that the presence of certain ions and molecules in the biomaterial environment can influence the degradation of susceptible polymers. This primarily refers to the effect of metal ions such as iron and cobalt. The other factor of clinical note is that some polymers may be influenced by lipids. This is certainly the case with polydimethylsiloxanes, which absorb lipids and are degraded by them.

### 2.9.3 Ceramic degradation

Ceramics are basically structures in which anions and cations are bound together by ionic bonds. The degradation process of ceramics is, however, vastly different to that of the metallic elements on which they are based, partly because of the nature and energetics of these bonds and, conversely, the lack of free electrons, which prevents the same electrochemical processes taking place. The rate of degradation of ceramics within the body can vary considerably and ceramics may be highly degradation resistant or highly soluble.

As a general rule, we should expect to see a very significant resistance to degradation with ceramics and glasses. Since the corrosion process in metals is one of a conversion of a metal to ceramic structure (i.e., metal to a metal oxide, hydroxide, chloride, etc.) we must intuitively conclude that the ceramic structure represents a lower energy state, in which there would be less driving force for further structural degradation. The interatomic bonds in a ceramic, are, as we have seen, strongly directional and large amounts of energy are required for their disruption. By way of example, it takes a great deal of energy to extract aluminum metal from the ore aluminum oxide, but the reverse process takes place readily by surface oxidation. Thus, we should expect ceramics such as $Al_2O_3$, $ZrO_2$, $TiO_2$, $SiO_2$ and TiN to be stable under normal conditions. This is what is observed in clinical practice. There is limited evidence to show that some of these ceramics (e.g., polycrystalline $Al_2O_3$ and $ZrO_2$) do show "ageing" phenomena, with reductions in some mechanical properties, but the significance of this is unclear.

With some variations in structure, there is evidence that problems may arise if incorrect structural forms are used. Zirconia is often used in the transformation toughened state; as with most oxide ceramics, the material is inherently brittle but this can be ameliorated by the use of doping elements which cause the formation and retention of a metastable state. This is achieved by addition of small amounts of the rare earth oxide yttria to the zirconia. The toughening arises from the transformation of crystallographic structure at a crack tip under strain, which causes blunting of the crack and prevents its catastrophic propagation. However, this metastable state can transform under other conditions. Many batches of improperly heat-treated zirconia hip replacement components underwent such changes over a short time after implantation, causing significant structural failure under load.

Alternatively, there will be many ceramic structures that, although stable in the air, will dissolve in aqueous environments. Consideration of the classic fully ionic ceramic structure NaCl and its dissolution in water demonstrates this point. It is possible, therefore, on the basis of the chemical structure, to identify ceramics that will dissolve or degrade in the body, and the opportunity exists for the production of structural materials with controlled degradation. Since any material that degrades in the body will release its constituents into the tissue, it is necessary to select anions and cations that are readily and harmlessly incorporated into metabolic processes and utilized or eliminated. For this reason, it is compounds of sodium, and especially calcium, including calcium phosphates and calcium carbonates, that are primarily used.

The degradation of such compounds will depend on chemical composition and microstructure. For example, tricalcium phosphate, $Ca_3(PO_4)_2$, is degraded fairly rapidly while calcium hydroxyapatite, $Ca_{10}(PO_4)_6(OH)_2$, is relatively stable. Within this general behavior, however, porosity will influence the rates so that a fully dense material will degrade slowly, while a microporous material will be susceptible to more rapid degradation. Dissolution rates of these ceramics *in vivo* can generally be predicted from behavior in simple aqueous solution. However, there will be some differences in detail within the body, especially with variations in degradation rate seen with different implantation sites. It is possible that cellular activity, either by phagocytosis or the release of free radicals, could be responsible for such variations.

In between the extremes of stability and intentional degradability lie a small group of materials in which there may be limited activity. This is particularly seen with a number of glasses and glass-ceramics, based on Ca, Si, Na, P and O, in which there is selective dissolution on the surface involving the release of Ca and P, but in which the reaction then ceases because of the stable $SiO_2$-rich layer that remains on the surface. This is of considerable interest because of the ability of such surfaces to bond to bone, and this subject is dealt with elsewhere in this book. On the basis of this behavior, bioceramics are normally classified under three headings: inert or "nearly inert" ceramics; resorbable ceramics; and ceramics of controlled surface reactivity.

### 2.9.4 Performance of composites

It is difficult to make general statements about the degradation of composite materials since the behavior will depend on the individual constituents. With ceramic reinforced polymers, the main controlling factor will be the diffusion of water through the continuous polymer matrix and its susceptibility to hydrolysis. If the degradation rate of this polymer is higher than that of the ceramic filler, the fragments of the latter are likely to be released into their environment as individual entities as the polymer disappears. With soluble fillers, the dissolution products may well have some influence on the polymer degradation, especially if they substantially alter the pH or if they contain metal ions or other species that catalyze the degradation. Particular emphasis may have to be placed on the interface between filler and matrix, since if this is very susceptible to hydrolysis, fragmentation of the composite may readily occur.

## 2.10 Synthesis, fabrication and manufacturing considerations

In this section we very briefly refer to factors associated with processing and manufacturing. In many cases, well-defined and understood processes are used, which are covered in specialized textbooks and there is no need to discuss them in detail here since they are not biomaterials specific.

### 2.10.1 Metallurgical processes

The vast majority of metals used in biomedical devices can be manufactured by traditional metallurgical processing routes. Inevitably, since metals have to be refined from ores, and since alloys usually have to be prepared by mixing elements under liquid conditions, the starting point for manufacture is that of the casting of a liquid metallic material. Under certain conditions, the liquid metal may be cast directly into the required shape for the final component; this is usually only done when the alloy in question is very difficult to process mechanically or if the component is of extremely complex shape. Certain cobalt-based alloys used in orthopedics and dentistry come into the first category, and some dental components such as bridges and inlays come into the latter. The properties of the resulting structure depend on a variety of parameters, including the rate of solidification and cooling; the grain size and shape and the presence of porosity are very important factors.

More commonly, the alloy is cast into a regular-shaped ingot, and this is processed further by mechanical shaping and forming. The most important reason for this is the cast structures tend to be brittle, and the processing can radically improve the mechanical properties, especially when they involve the combined methods of mechanical force and elevated temperature (thermomechanical processes) which result in the most appropriate microstructures. We shall see in Chapter 8 that the critical properties of many metallic biomaterials depend on the precise nature of these processes.

There is an alternative procedure that can be used with a few metallic materials when specialized structures and properties are required. This is the process of sintering, which is especially useful for hard brittle materials that are not readily shaped mechanically and for structures where a certain degree of porosity is required. Sintering comes within the group of processes referred to as powder metallurgy. It is particularly useful for cobalt-based alloys. Fine powders of these alloys may be prepared by atomization techniques, in which each particle retains the structure equivalent to a rapidly cooled casting. These powders may be consolidated to full density within a suitable evacuated vessel, when pressure is applied at high temperature, for example in excess of 100 MPa pressure above 1000°C. This process may also be used to prepare porous surfaces on alloys used in contact with bone, as discussed in Chapters 3 and 4. In such cases, the particles may be quite coarse, adjusted to the required parameters of the porosity, say of the order of a few hundred microns. It is important to achieve sintering conditions that allow effective particle–particle and particle–substrate bonding.

## 2.10.2 Polymer processing

The preparation of polymeric materials involves two phases, which are often distinct, although they do not have to be. The first involves the synthesis of the polymer and the second the processing of the polymer into useful objects. The subjects of polymer synthesis and polymer processing are far too large to discuss here and we shall not even attempt to summarize the conventional methods. We shall, however, mention some examples of recently introduced concepts that are having a specific influence of the development of new polymeric biomaterials. Other individual synthetic routes will be considered at appropriate places in other chapters.

### Click chemistry

At the beginning of the twenty-first century many people argued that methodologies of organic chemistry had not moved forward very much in recent times and were not entirely appropriate for the requirements of new chemical applications, especially those of drug discovery. At that time a new concept emerged, called "click chemistry." This is now having a significant effect, not only on drug discovery but also on biomaterials synthesis. The underlying philosophy is that polymeric molecules derived by nature have preference for carbon–non-carbon bonds over carbon–carbon bonds, as evidenced by nucleic acids and proteins which are condensation polymers of sub-units linked through these carbon–heteroatom bonds. This has led to the strategy of making large oligomers from simple building blocks which can be linked by reversible heterocarbon bonds. It was intended that click chemistry would involve processes that were modular and high yielding, with inoffensive by-products that could be easily eliminated and easily removed solvents. A variety of click reactions have already been used to prepare, for example, hydrogels with far better functionality and mechanical properties than can be achieved by more conventional routes.

### Reversible addition-fragmentation chain transfer (RAFT)

While conventional polymerization processes have been used for many years with very successful outcomes, they do have limitations with respect to the control of molecular weight and molecular weight distribution. Much better control of such features are required for the more sophisticated applications of polymers, especially in drug and gene delivery applications. With normal free radical polymerization, widely used for many structural polymers, the growth of molecular chains may not be easily controlled because of termination and transfer reactions that take place. So-called living radical polymerization techniques have been designed in order to minimize these reactions and give greater control over complex architecture and site-specific functionality. One of the more important of these methods is reversible addition-fragmentation chain transfer (RAFT). In this process, the polymerization of one or more monomers is highly controlled by chain transfer agents, which are organic substances containing a thiocarbonylthio moiety, which has sulfur–carbonyl groups. We shall see in Chapter 5 just how important such methods are in the production of dendrimer- and micelle-based biomaterials.

### Combinatorial or high throughput methods

It is obvious that the rational design of polymers with controlled structure and properties that meet defined specifications is a time-consuming process if carried out by conventional means, involving one experiment at a time. The combinatorial approach to molecular design was initiated by the difficulties inherent in this trial-and-error process. This approach has a number of steps which include the development of a discrete or continuous combinatorial library in which the properties are varied systematically in one or more directions. High throughput techniques are then used to investigate the structure–property relationships within the parameters set for the library. Large data-sets are obtained, which are analyzed and validated. This process has been used to identify and prepare libraries of several series of biomedical polymers, in which the dependence on precise polymer characteristics for cell behavior can be demonstrated.

## 2.10.3 Ceramic processing

Since many ceramics are brittle, high-melting-point materials, they are not amenable to most of the processing techniques available to metals and thermoplastic polymers. The vast majority of ceramics cannot be formed by plastic deformation, nor does the application of heat make much difference. They are, instead, usually prepared as powders, which are processed and formed into shapes and then fired at high temperature in order to consolidate the powder particles. Additional post-firing processes may be required, including machining, surface treatments and special techniques to impart specific properties such as electrical and magnetic properties. In traditional methods, powder processing is obviously important if not technically sophisticated, since particle size and contamination levels have to be strictly controlled for high-performance engineering ceramics. These steps usually involve milling and mixing steps. Since ceramics have no intrinsic plasticity, organic binders are added to the powder. Forming of the required shape may be achieved in a number of ways. Powder, binder and solvent can be mixed and spray dried, or formed as slurries. The resulting substances may then be cold-formed using die compaction, isostatic compression, extrusion, calendaring, tape casting or injection molding. These vary in complexity and cost. The formed objects are then fired, using appropriate combinations of time, temperature and atmospheres. The first effect of heating is the removal of the binders and residual moisture; this has to be done carefully and slowly in order to avoid decomposition of components and cracking. Following this there is a period of soaking, where the temperature is kept constant at the highest point of the firing cycle, before slow cooling. There are many refinements on these techniques to allow for optimal property development.

Of course, as with the other major groups of materials, the processing strategies for recently developed forms of ceramics have had to change. This mostly relates here to the use of ceramic-based nanoparticles, which are discussed briefly below in Section 2.12.1 and elsewhere in this book. These strategies largely rest with the synthesis of the nanoparticles themselves rather than with consolidation of material into monolithic objects. Monodisperse nanoparticles are usually prepared in a solution phase, where it is usual to separate out initiation and growth stages of the particles. This is necessary in order to control size and concentration. It is often arranged for there to be a burst initiation phase with control over the number of nuclei. As the conditions are changed, growth of the particles proceeds, where strict control means that they grow at the same rate so that a monodisperse (i.e., of the same size) collection is produced. With iron oxide nanoparticles, for example, stoichiometric mixtures of inorganic salts are reacted in aqueous media under controlled conditions of pH and temperature. Alternatively, high-temperature decomposition of metal complexes, together with surfactants and a high boiling point solvent, may be used. More complex structures, such as hollow nanoparticles, require more sophisticated methods, most of which are tailored to the specific chemistry and morphology.

## 2.10.4 Composites

Many important composites are fiber reinforced thermosetting resins. Usually the component will be made from prepregs, or preimpregnated fibers, in which bundles of filaments are impregnated with the resin, which is partly cured, and then these prepreg forms are laid-up within a mold, for final impregnation by the resins and cure completion. The key is to avoid entrapment of air at the interfaces since the interlaminar shear strength will decrease by around 7% for every 1% of porosity. Molding under pressure (press molding) enhances this process. Several other, more complex processes may be used for high-performance composites. Resin-transfer molding is a low-cost process in which the fibers are enclosed in a mold and the resin injected under a vacuum. Reaction injection molding (RIM), instead of using pre-catalyzed resin, brings two fast-reacting components together and mixes them just prior to injection into the mold; this is widely used for polyurethane composites. Pultrusion is the equivalent of the extrusion of unreinforced polymers; bundles, or rovings, of the fiber, are passed through a bath of the resin, and then through dies when resin completely penetrates the bundles, where it gels and cures. Reinforced thermoplastics may be processed by methods analogous to those used for unreinforced polymers, with adjustments made according to the properties of the ceramic, glass or carbon filler.

## 2.10.5 Sterilization

Any device that is inserted into the human body, whatever the intended duration, must be sterile; that is, it should not introduce any risk of infection to the patient. There are several implications of this simple statement. The first is that an acceptable definition of sterility is required. Secondly, there must be adequate methods to assure that devices meet the acceptable standards of sterility during production. Thirdly, these methods should not leave behind residues that, although not living, could still act as antigens and cause an immune response. Fourthly, the methods used to remove infection risk should not damage the materials used in the device.

Sterility is the absence of all living organisms. This is essentially concerned with microorganisms, that is bacteria, fungi and viruses. A single microorganism on a medical device could, in theory, be responsible for an infection, but in practical terms it may not be possible to remove each organism let alone be possible to give an assurance that all have been killed. Sterilization validation, which is an essential part of quality control, works through the use of a sterility assurance level (SAL). This is based on statistical probability; the sterilization operates by reducing the number of viable microorganisms logarithmically, and the result depends on the original bioburden and the amount of energy applied during the process. The SAL is the probability that a product will be non-sterile after exposure to a specified sterilization process. The accepted SAL for an implantable device is $10^{-6}$. There are several standards that are used to inform the technical aspects of the sterilization of health care products. With respect to residual components of microorganisms following sterilization, these are known as endotoxins; they and their effects are discussed in Chapter 3.

### Moist heat sterilization

This is the oldest method to be used and simply involves exposing the device to saturated steam, typically at 121°C. These conditions kill microorganisms by destroying structural and metabolic parts of the organism provided all surfaces are maintained at this temperature for 30 minutes. It is a simple process and leaves no residues but many materials are unable to withstand these conditions without undergoing unacceptable changes.

### Ethylene oxide

Ethylene oxide is a gas (above its boiling point of 11°C) that affects the nucleic acids of microorganisms and is a very effective sterilization medium. Objects are placed in gas permeable packages, which are themselves placed in vessels with a relative humidity between 60 and 80% and a temperature around 40–50°C. The ethylene oxide is used either pure or with some reasonably inert gas to reduce explosion risk. Exposure times range from a few hours up to 15–20 hours depending on the volume of packages and the required SAL. The process is compatible with many materials and devices and is very efficient. It does have one major drawback, which is toxicity. Both ethylene oxide itself and the by-product ethylene chlorohydrin have toxicity and carcinogenicity profiles so that the processes have to be strictly controlled with respect to environmental contamination, workforce exposure and residuals in or on the products. Strict regulations apply to all post-sterilization handling and storage procedures.

#### Gamma irradiation

Gamma rays cause ionization of cellular components such as DNA, which readily causes cell death. The rays are highly penetrating and provide a very effective sterilization mechanism. Usually the rays are generated by a $^{60}$Co isotope source, which decays to $^{60}$Ni, emitting the rays and electrons, and a dose of 25 kGy is sufficient to satisfy the implantable device SAL under most circumstances. Many materials, including most polymers, can be sterilized by this method, although some are too sensitive to radiation for this to be used. Fluorocarbon polymers such as polytetrafluoroethylene come into this category and care has to be taken with any material being subjected to gamma irradiation for the first time to validate compatibility. Particular attention has to be paid to the atmosphere used inside the packaging since the oxygen can synergistically work with the irradiation to produce degradative effects. Polyethylene can oxidize under these circumstances and irradiation may have to be undertaken *in vacuo* or under a gas such as nitrogen.

#### Electron beam radiation

Electron beams provide an alternative method of applying radiation energy for sterilization purposes, operating by similar mechanisms and with similar doses. Penetration into objects is not so good and applications are limited. Nevertheless, many devices including surgical dressings, wound care products, IV administration kits and cardiac catheters are routinely sterilized by electrons. The process, involving an accelerator, is simpler and, in many circumstances, more cost-effective than gamma irradiation.

## 2.11 Environmental responsiveness

Traditional or conventional materials generally respond in a reasonably linear fashion to external stimuli, such as mechanical forces or changes in temperature. However, many natural materials show considerable sensitivity to such stimuli, often showing little variation in properties over wide ranges of conditions but then exhibiting abrupt changes over a very narrow range. This is very useful as proteins, polysaccharides and nucleic acids are required to undergo profound changes in their properties under certain well-defined conditions. These usually involve conformational changes and reversible phase transitions, resulting in significant changes to solubility or hydrophobicity. It is not surprising that materials scientists have tried to replicate these phenomena in synthetic materials, usually employing structural and property changes that are induced by changes in temperature or pH, or, under very special circumstances, mechanical stress. The majority, although not all, of these materials are hydrogels.

### 2.11.1 Thermally responsive polymers

We have seen that hydrogels are soft materials that contain a great deal of water; they usually involve the cross-linking of hydrophilic macromolecules, which allows absorption of water and swelling, but not dissolution. The key to using such materials in environmentally responsive modes is to enable the macromolecular structure to switch from a

Lower temperature

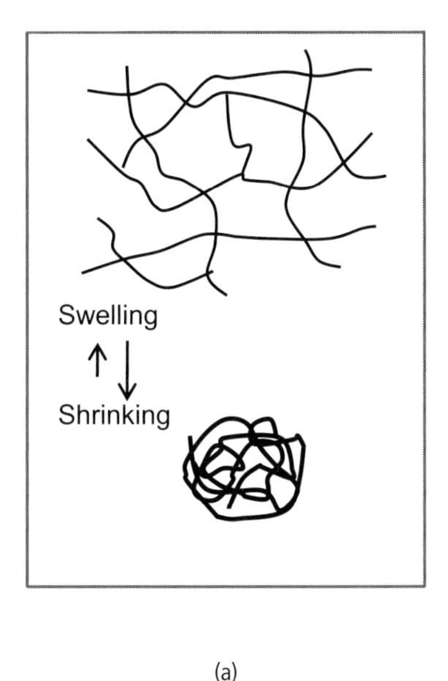


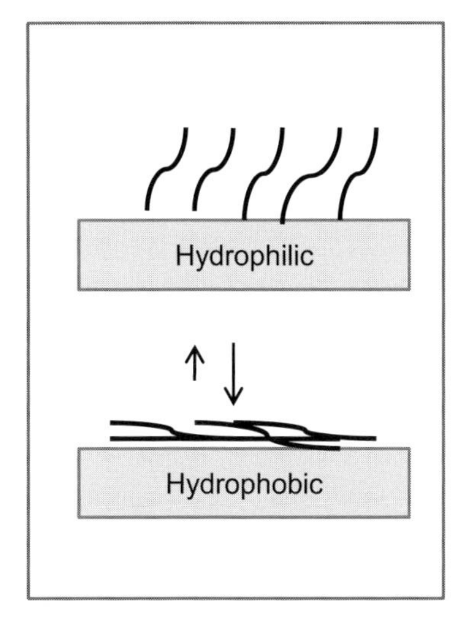


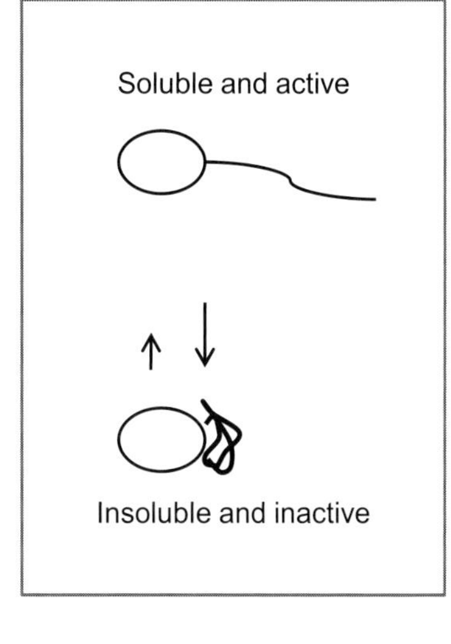


(a) (b) (c)

Higher temperature

**Figure 2.25** Thermoresponsive polymers. A small change in temperature in certain polymers can produce significant changes. **(a)** Decreasing the temperature causes the polymer to swell; **(b)** the hydrophobic surface changes to a hydrophilic surface, and **(c)** an insoluble and inactive polymer becomes soluble and active.

hydrophilic state to a hydrophobic state on a slight change to an environmental parameter, so that the material can reversibly swell and de-swell, or change its surface energy. Usually this occurs where a homopolymer has one repeat unit that is hydrophobic and one that is hydrophilic.

This phenomenon is best seen with poly(N-isopropylacrylamide), PNIPAAm, which has hydrophilic amide groups and hydrophobic isopropyl groups. This has a lower-critical solution temperature (LCST); below this temperature the polymer is fully hydrated with an extended coil structure, but above the LCST it is dehydrated, compact and globular. The driving force for this is the disruption of hydrogen bonding between water molecules and the polymer chains. This is depicted in Figure 2.25. The fact that the LCST is around 32°C, which can be adjusted by copolymerization and is reversible, means that this profound change can be used to advantage in medical technology applications. The kinetics of the change will be governed by diffusion and may be quite slow. If it is surface changes that are important, for example a change in surface energy or wettability, which is key to the process of cell sheet engineering discussed in Chapter 5, then surface grafting of PNIPAAm to gels may be used. Also, the effects may be transferred to other systems by the conjugation of PNIPAAm or similar polymers to, for example, proteins. Many thermoresponsive copolymers have been produced, which may be random or block copolymers and may be either cationic or anionic depending on whether basic or acidic comonomers are employed.

### 2.11.2 pH responsive materials

Polymers that have acidic or basic functional groups may also react strongly to changes in pH. The ionization level of these groups may change profoundly around the $pK_a$, the logarithmic measure of the acid dissociation constant, which then changes the solubility of the polymer chains. If a pendant group accepts protons at low pH, and then releases them at high pH, the hydrogels will show a significant increase in hydrodynamic volume and swelling capacity on a change in pH. Some quite simple synthetic polymers such as poly(acrylic acid) show pH responsiveness, as do many polysaccharide derivatives such as some of those based on chitosan, hyaluronic acid and alginates. Bearing in mind that the pH changes quite considerably along the gastro-intestinal tract, this phenomenon has led to the development of pH-dependent oral drug delivery systems. This option is even more attractive when thermoresponsive nanoscale structures are considered.

### 2.11.3 Shape memory materials

The two types of environmental responsiveness discussed above refer to structural changes that take place when a material experiences a change in either pH or temperature; these are fundamental properties that are inherent to the materials. Shape memory is a little different since it is not an intrinsic property but depends on a combination of material properties and specific processing conditions. Whatever the material (and there are not many that qualify), shape memory is established by a processing method that gives its initial permanent shape, which we can denote A. It is then deformed such that it takes up a different, temporary, shape, B. Then, on application of an external stimulus, it recovers, or remembers, its initial shape A. This may be done just once, or repeatedly, depending on the application.

There are examples of shape memory in both metal and polymer classes. The former were utilized first, but applications of the latter are increasing. Although a few metals exhibit this effect, there is one dominant material in this class, which is a nickel–titanium alloy, referred to as Nitinol, which is close to equiatomic composition, the most commonly used being 50.8% Ni–49.2% Ti. The shape memory property here is based on the diffusionless transformation between a high-temperature austenitic-like phase and a lower-temperature martensite-like phase. The transition temperature can be adjusted by thermomechanical means and may be set at temperatures close to body temperature (Figure 2.26). Control of composition is extremely important as a 1% move either way in the Ni–Ti balance can alter the transition temperature by 100°C. A wide range of medical devices that need to experience shape change, either for delivery to the body as with intravascular stents, or for function as with orthodontic wires, are made from Nitinol.

There may be a certain amount of hysteresis between thermal and mechanical behavior so that the shape change may not be instantaneous. Some shape memory polymers have much less hysteresis, which has been one of the main stimuli for their development. These are elastic polymeric networks that contain stimuli-sensitive switches and netpoints. The permanent shape is determined by either or both covalent bonds or intermolecular interactions, and the morphology has at least two segregated domains. Hard segment domains associated with the higher thermal transition act as specific netpoints, while chain segments in other domains of the lower transition serve as molecular switches. With flexible switching domains, different entropic elastic behavior is seen on moving through the transition temperature. The shape-memory functionality is obtained by temporarily setting the polymer network in the deformed state under the temperature conditions required by the application. The chain segments are reversibly prevented from recoiling through the introduction of reversible netpoints, such as reversible covalent cross-links. Switching between shapes is then achieved by cycling through points (usually determined by temperature) when the cross-links establish and disestablish. The best effects are seen with linear block copolymers such as certain polyurethanes, some polyesters and urethane-isoprene copolymers.

## 2.12 Special considerations

In this final section, a few recent developments in biomaterials science are introduced in order to place some detailed discussions in later chapters into perspective. These are at the heart of the change in emphasis in many applications away from the classically manufactured engineering materials to those that attempt to resemble, in structure, synthesis and/or properties, the tissues of the body. Inevitably the mechanisms involved are usually taking place at the nanoscale.

### 2.12.1 Nanoparticles as biomaterials

With respect to the nanoscale, there are several factors that we have to take into account when

(a)

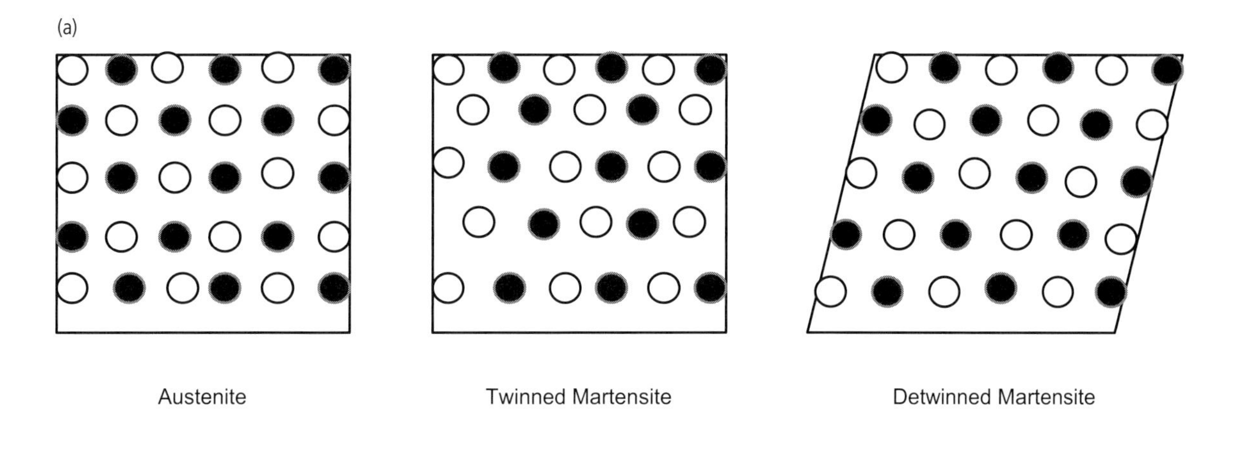


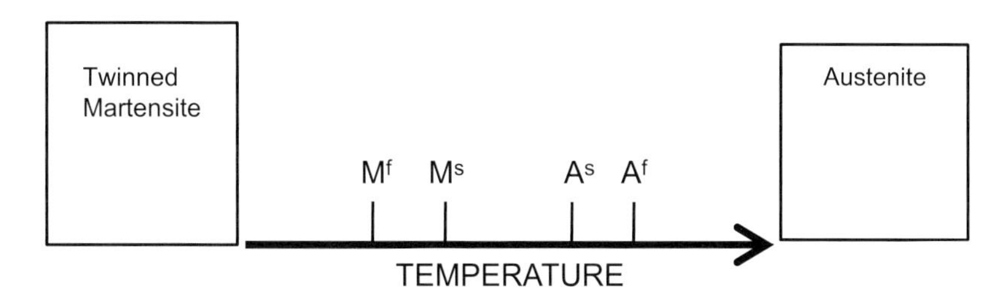


**Figure 2.26** Shape memory effect.
(a) In an alloy that displays shape memory, such as Nitinol (nickel–titanium), transformations take place between crystal structures as a function of stress and temperature. At the top are shown schematically the relevant crystal structures austenite (cubic structure, stable at high temperature) and martensite (the low-temperature phase of monoclinic structure); the martensite undergoes shear transformation between twinned and detwinned states. Twinned martensite transforms, without mechanical stress, to austenite on heating, shown in the lower part. There are four specific temperatures associated with the transformation. $M^s$ is the martensitic start temperature and $M^f$ the martensitic finish temperature. $A^s$ is the austenite start temperature and $A^f$ the austenite finish temperature.

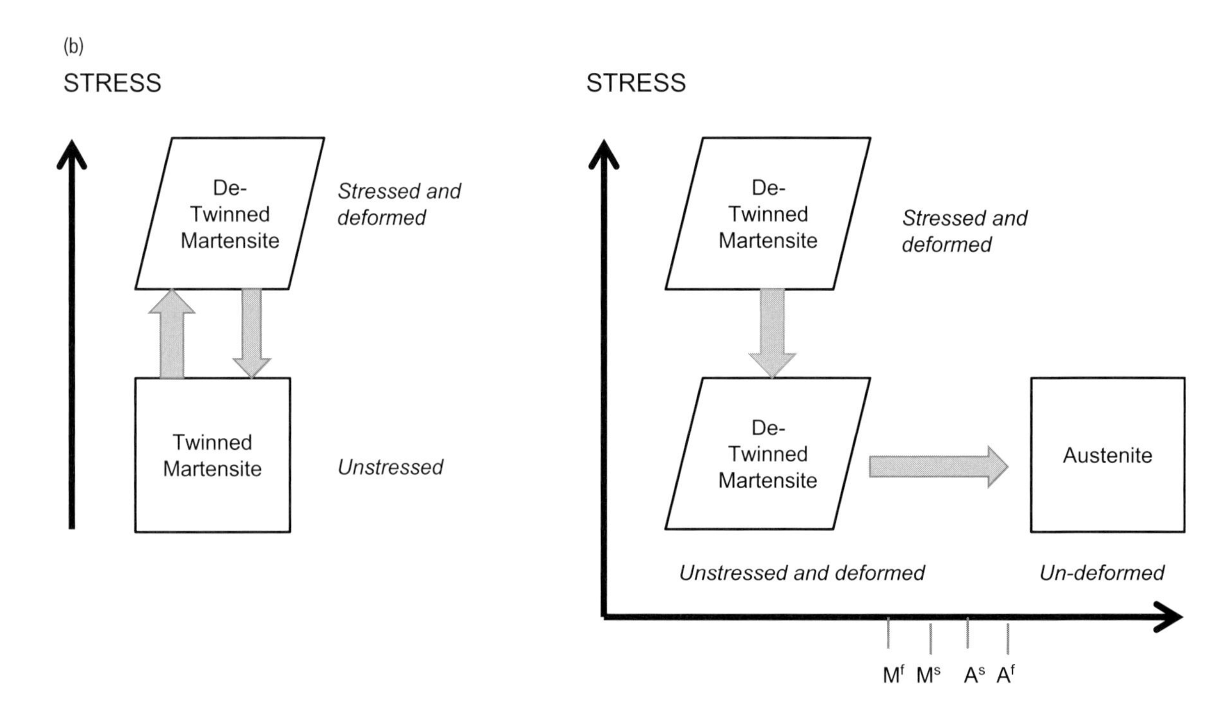


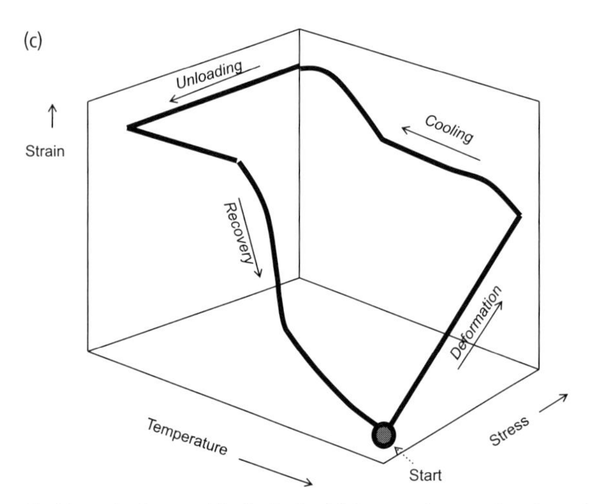


**Figure 2.26** *continued*
**(b)** Shape memory is achieved with combined stress and temperature. On the left, unstressed twinned martensite is transformed to detwinned martensite on application of stress at low temperature and undergoes shape change. On the right, removing the stress keeps the same detwinned structure which also retains the acquired shape change. Heating the material causes a transformation to the austenite, which recovers the initial shape, and it is unstressed. In an implanted device such as a stent, the device is manufactured in the unstressed state at the required final shape. The material is then stressed, for example crimping the stent inside a narrow catheter, when it changes its shape, and retains this shape until there is a temperature rise. If the transformation temperatures are between ambient and body temperatures, the original required shape is then recovered when that temperature is reached, that is on deployment in the body. **(c)** Some polymers also show shape memory. With such a polymer, in the heated unstressed state, it will deform on application of a stress. If it is cooled, it will undergo more strain and on reaching ambient temperatures it will be in its fixed state, retaining this new shape. This shape will be retained on reducing stress in the unloading phase, but will then recover the original shape on raising the temperature again.

considering its relevance to biomaterials. In materials science in general, nanocomposites with nanoscale dispersed phases, and nanocrystalline materials in which the very fine grain size affords quite different mechanical properties to conventional microstructures, are already in use. In surface science and surface engineering, nanotopographies offer substantially different properties related to adhesion, tribology, optics and electronic behavior. In biological sciences, the fundamental understanding of molecular motors and molecular functional entities at the nanometer scale has been responsible for advances in drug design and drug targeting and nanoscale functionalized entities and devices are in development for analytical and instrumental applications in biology and medicine, including tissue engineering and imaging. Although nanostructure and nanotopography are clearly important in biomaterials applications, it is the properties of nanoparticles that are the most relevant.

The principal parameters of nanoparticles are their shape (including aspect ratios where appropriate), size, and the morphological sub-structure of the substance. Nanoparticles are presented as either an aerosol (mostly solid or liquid phase in air), a suspension (mostly solid in liquids) or an emulsion (two liquid phases). The interactions between individual nanoparticles are also very important. In the presence of chemical agents such as surfactants, the surface and interfacial properties of particles may be modified, which influences these reactions. Indirectly such agents can stabilize against coagulation or aggregation by conserving particle charge and by modifying the outmost layer of the particle.

At some point between the angstrom level and the micrometer scale, the simple picture of a nanoparticle as a ball or droplet changes. Both physical and chemical properties are derived from atomic and molecular origin in a complex way. For example, the electronic and optical properties and the chemical reactivity of small clusters are completely different from the better-known property of each component in the bulk form. At the nanoscale, particle–particle interactions are either dominated by weak van der Waals forces, stronger polar and electrostatic interactions or covalent interactions. By the modification of the surface layer, the tendency of a colloid to coagulate can be enhanced or hindered. The details of nanoparticle–nanoparticle interaction forces and nanoparticle–fluid interactions are of key importance to describe physical and chemical processes, and the temporal evolution of free nanoparticles.

### Nanotechnology definitions

In order to discuss the role of nanoparticles in biomaterials science it is important that precise definitions are used. In many ways the selection of the size limits associated with the prefix "nano-" in all aspects of nanoscience is somewhat arbitrary. From the scientific evidence so far available, there does not appear to be any sharp change in biological properties, including either toxicokinetic or toxicodynamic properties, of substances at any particular size, which has to be taken into account when considering the preciseness of any definitions.

The majority of terms that need to be considered in the context of nanoscience and nanotechnology are those that start with the prefix "nano-," followed by a noun, such as in nanoscience and nanotechnology themselves. The prefix "nano-" specifically means a measure of $10^{-9}$ units, the nature of this unit being determined by the word that follows. Thus a nanosecond is $10^{-9}$ seconds and a nanometer is $10^{-9}$ meters. There are certain situations in which explanations are required in the development of the framework for the terminology, and where the use of the prefix "nano-" is not intuitively obvious or understandable. For example, clarification may be needed on how precisely the term refers to the $10^{-9}$ measure. Also the addition of the prefix "nano-" to a noun may not adequately explain exactly to what the $10^{-9}$ measure refers. For instance, with the word "nanomaterial," to what structural feature of a material should the prefix "nano-" refer? Are crystal size, grain size, domain size, surface topology or any other feature included?

It is widely accepted that although the prefix nano specifically refers to $10^{-9}$ units, in the context of nanoscience the units should only be those of dimensions, rather than of any other unit of scientific measurement, such as for time, energy or power. Moreover, it is unrealistic, for practical purposes, to consider the prefix "nano-" to solely and precisely refer to $10^{-9}$ meters, just as it is not considered that "micro-" specifically and solely concerns something with a dimension of precisely $10^{-6}$ meters. It is further widely agreed that one of the characteristics that confers special properties to products of nanotechnologies is the large surface area to volume ratio that is encountered at very small dimensions and that those of the order of 100 nm and below are most likely to be associated with such properties.

Based on the considerations mentioned above, the main word in the hierarchy of terminology in nanotechnology and nanoscience related to size is "nanoscale." It follows that the definition of nanoscale should be any feature characterized by dimensions of the order of 100 nm or less. Since the changes in characteristics that are seen on reducing dimensions do not occur uniquely at the 100 nm dimension, some of the derivatives of the nanoscale have to allow for a range of dimensions at this level. The term "structure" is generally held to mean "a complex entity composed of many parts." It is logical, therefore, to consider that "nanostructure" implies a complex entity composed of discrete functional parts, many of which will be at the nanometer scale. Reference here is made to the functionality of the component parts. This is required since it is not the mere presence of very small entities that determines that a substance is nanostructured (all substances consist of atoms that exist at the nanoscale), but rather the existence of such entities that control the properties and functions that are unique to the nanoscale for that substance.

It follows that most substances will have internal structures that individually could be considered as being at the nanoscale, for example molecules, crystals or domains, but these do not, a priori, qualify for classification as nanostructures. For example, simply because a polymeric material may consist of individual molecules of nanometer dimensions does not necessarily confer nanostructure status on that substance.

Some terms have been introduced into nanotechnology in order to describe some general types of process or product. The only ones of significance are those that relate to engineered or manufactured nanostructures. An engineered nanomaterial is any material that is deliberately created such that it is composed of discrete functional parts, either internally or at the surface, many of which will have one or more dimensions of the order of 100 nm or less.

In practice, these products of nanotechnology rarely consist of a single entity with one or more dimensions of 100 nm or less, or of large numbers of identical entities with identical sizes. Rather, they consist of large numbers of similar but non-identical entities, as, for example, in a powder where rarely will the sample be monodisperse or homogeneous.

It is helpful to have terms that differentiate between discrete entities having either one, two or three dimensions in the nanoscale. In nanotechnology the term "nanoparticle" is used as a collective term for any material consisting of discrete entities with one, two or three dimensions of the order of 100 nm or less. It is preferable to consider a nanoparticle to have a comparable scale in all three dimensions. In other situations, a nanosheet is a discrete entity which has one dimension of the order of 100 nm or less and two long dimensions, while a nanorod (or a nanowire or nanowhisker) is a discrete entity that has two dimensions that are of the order of 100 nm or less, and one long dimension. A nanotube is a discrete hollow entity that has two dimensions of the order of 100 nm or less and one long dimension.

With these concepts and definitions in mind we will be able to consider the impact of nanotechnology, nanostructures and nanoparticles in most of the remaining chapters of this book.

## 2.12.2 Injectable materials

The traditional approach to the design of implantable devices has, of necessity, utilized pre-manufactured products that have been placed in the body by surgical methods. This practice has been very successful and will no doubt remain as the optimal procedure in many situations. There are, however, several drawbacks, including the lack of specific customization for individual patients and the need for open surgery. In a few situations, these procedures have involved, in whole or in part, a liquid component that has set *in situ*, including those situations where an adhesive or attachment reaction was required, as in bone cements. These have been useful but hardly ideal. In many situations, therefore, consideration has been given to materials that can be administered minimally invasively and then undergo a liquid to solid type of transition once in place, where it can adapt to the required form in the body. This approach has become more important with tissue engineering and cell therapy applications, where the material may have to incorporate active biomolecules and even live cells. As we will see in Chapter 3, the potential for the body to respond to a mobile, reactive, liquid has to be different to that involving a pre-formed solid object, so that injectable, *in situ*-curing materials have some additional hurdles to overcome.

*In situ*-curing materials have to be designed so that the curing reaction does not have harmful side effects, such as an unacceptable exotherm or the generation of harmful by-products. The reaction has to be achieved in the absence of toxic or irritant catalysts. There has been interest in injectable calcium phosphate-based substances as bone substitutes for use in spinal surgery, where the setting reaction involves hydration and re-precipitation of phosphate phases. Most attention, however, has focused on polymeric systems, and especially those that are of hydrogel structure. The interest here is based on the ease with which some of the materials can undergo a sol–gel transition under ambient conditions without harmful residues or by-products and also can incorporate drugs, proteins and cells. They can also have highly suitable rheological characteristics for simple injection procedures. Included here are some thermosensitive block copolymer hydrogels such as those based on poly (ethylene glycol) and thermally gelling chitosan and polypeptides, which undergo a transition on raising the temperature to physiological levels. There are also pH-sensitive hydrogel systems that contain pH-sensitive moieties that control gelation. These avoid some of the disadvantages of temperature-responsive materials, such as the premature gelation in the injection needle.

### 2.12.3 Self-assembly

Here we are concerned with the mechanisms by which complex tissues are formed naturally and the attempts to replicate these mechanisms in biomaterials. These natural processes are best described as molecular assembly, in which the nanostructured tissue components are synthesized and assembled by cells to give specific shapes, dimensions, internal structures and, in consequence, highly specific properties. Complex objects are built up by a series of supramolecular self-assembly processes that involve their own signaling and control mechanisms. Recapitulation of this process with man-made precursors is not a trivial process, not least because many of these natural mechanisms have yet to be fully elucidated. A start has been made, however, and a number of approaches to self-assembled biomaterials have been explored. Most of these employ peptide nanostructures, especially the spherical, planar and cylindrical shapes of peptide amphiphiles. Peptides are good choices for these innovations since they should allow the incorporation of known biological signaling structures within them. This is important since if these materials are ever able to replicate cell-driven self-assembly, they have to take part in signaling pathways. These aspects will be taken up again in Chapters 3 and 5.

### 2.12.4 Biomimetic materials

As a final comment in this chapter, which is associated with the self-assembly discussed above, we should emphasize the role of biomimicry, which is the process of emulating nature in the design of products, in the future of biomaterials. The vast majority of biomaterials applications involve either replacing or regenerating tissues, or delivering agents to the tissues. While considerable success has been achieved with classical, synthetic materials in many areas, this approach, by definition, has serious limitations, quite simply because pieces of metal, glasses or solid polymers do not look like, feel like or act like tissues. For long, the hope has been that we can emulate both the structure and function of natural tissues in the new biomaterials. This is not a trivial task because of the complexity of these tissues but now, especially with the innovations in nanotechnology, significant inroads to biomimetic materials are being made. It should be emphasized that this is more than making a composite material with the same elastic modulus as bone or a synthetic polymer that is degradable by enzymes; indeed there have been many false dawns in this area, often described as bioinspired materials. The true inspiration is coming from molecular or nanoscale structure and organization and the processes of self-assembly discussed above.

## Summary and learning objectives

1. The sophisticated biomaterials that we use today do not behave in a similar manner to the more traditional steel, plastic, textile, glass, concrete or wooden structures that have been the mainstay of materials engineering for many decades. We are not now constrained by concepts of state (materials do not have to be solid), of size (they may be macroscopic, microscopic or of nanoscale dimensions), of activity (they do not have to be inert but may be intentionally active, or even living) or of permanence (they may be intentionally biodegradable). They do not have to be manufactured by conventional means but may be formed *in situ* by self-assembly. This chapter places these current concepts of biomaterials into the overarching understanding of structure–property relationships of materials in general.
2. The first part of the chapter deals with the basic materials model, and discusses how the structure of materials evolves in a hierarchical manner, starting with atoms and then molecules, with atom–atom interactions and bonds and molecule–molecule interactions and bonds, and the spatial organization of these entities. This leads to discussions about the states of matter, of crystalline and amorphous conditions, and the identification of the main classes of synthetic materials, that is metals, ceramics and polymers, and the main morphological or architectural forms such as defined on the macro-, micro- and nanoscales. The main learning objective here is to facilitate the understanding of how the structure of a piece of gold differs from that of a piece of glass, or a plastic material and so on.
3. Special attention is given to the surfaces of materials since, in most biomedical situations, it is the surface that interacts with the physiological environment. We include here the features of surfaces that control these interactions, including surface energy, the hydrophilic–hydrophobic balance, surface morphology and surface chemistry. It is important to recognize just how these features, individually or collectively, control many material properties.
4. A large part of this chapter then places these discussions about bulk and surface structures into the context of real, practical materials. We deal here with the metallic system, including pure metals such as titanium or silver, simple alloys such as platinum alloys used for medical electrodes, and multi-elemental alloys such as Fe, Cr, Ni, Mo steels and Co–Cr–Mo–W alloys for joint replacements. This is followed by a description of synthetic structural polymers,

including medical plastics, textiles and elastomers, of soluble polymers and of natural biopolymers. Ceramics, glasses and glass-ceramics, including those that display bioactivity, follow, and then synthetic and natural composites. Special attention is given to the nanoscale where important properties ensue, such as in nanoscale polymers and semiconductors. This section is intended to allow the reader to understand the structures of materials they may have heard about, or used. This is important as the basis for the following sections on material properties.

5. The properties of materials are discussed under three main headings and a few minor ones. These properties really concern the behavior of a material when it is subjected to external energy. First there are physical forms of energy, and we discuss thermal, electrical and magnetic properties, Then there are mechanical forms of energy and we discuss elasticity, strength, hardness, toughness and so on. Thirdly there are environmental energies, giving rise to discussions of corrosion and degradation mechanisms. In each case, the properties are discussed in relation to the structures described previously and bearing the special conditions that exist inside the human body clearly in mind.

## Questions

1. Discuss the simple model of a "single-phase, isotropic, homogeneous, chemically defined material."
2. Describe the differences between metallic and ionic structures.
3. Give examples of how some ceramic structures exist in partly ionic and partly covalent states.
4. Why is argon an inert gas and magnesium a reactive metal?
5. How does the electronic structure of the metallic state allow platinum to be used for pacemaker electrodes?
6. Discuss what is the difference between crystalline and amorphous states.
7. How does the molecular configuration differ between polyethylene, polypropylene and polyvinylchloride and how is this reflected in their properties?
8. What factors control the surface reactivity of biomaterials?
9. Explain the difference between a substitutional solid solution and an interstitial solid solution and give examples of metallic biomaterials that have these characteristics.
10. Discuss the differences between diamagnetic, paramagnetic and ferromagnetic states and how these differences affect biomaterials applications.
11. The elastic modulus is often confused with the strength of a material. Explain why these are quite different properties.
12. Compare medical-grade alumina with transformation-toughened zirconia with respect to toughness.

13. Discuss the difference between the corrosion behavior of gold and titanium.
14. Describe the design specifications for intentionally degradable synthetic polymers.
15. Contrast the conventional production of thermoplastic components with the synthesis of polymeric nanostructures.
16. Discuss the merits of nanoscale terminology.
17. What will be the significance of self-assembly in the future applications of biomaterials?
18. Discuss the relevance of sterilization processes to health care products.

## Recommended reading

Behl M, Lendlein A, Shape memory polymers, *Materials Today* 2007;10:20–5. *Discussion of the concepts and mechanisms of shape memory polymeric materials.*

Bettinger CJ, Synthesis and microfabrication of biomaterials for soft tissue engineering, *Pure and Applied Chemistry* 2009;81:2183–2201. *Review of methods for biomaterials synthesis and biomaterials system design for tissue engineering scaffolds.*

Borden MA, Zhang H, Gillies RJ, Dayton PA, Ferrara KW, A stimulus-responsive contrast agent for ultrasound molecular imaging, *Biomaterials* 2008;29:597–606. *Discussion of the development of stimulus-responsive microbubbles for highly specific molecular imaging.*

Chu PK, Chen JY, Wang LP, Huang N, Plasma-surface modification of biomaterials, *Materials Science and Engineering R* 2002;36:143–206. *Very detailed description of plasma surface modification techniques and applications to biomaterials.*

Falconnet D, Csucs G, Grandin HM, Textor M, Surface engineering approaches to micropattern surfaces for cell-based assays, *Biomaterials* 2006;27:3044–63. *Review of biologically motivated micropatterning of two-dimensional surfaces.*

Garcia AJ, Interfaces to control cell-biomaterial adhesive interactions, *Advances in Polymer Science* 2006;203:171–90. *Discussion of the mechanisms by which cells attach to biomaterials surfaces.*

Gentile F, Tirinato L, Battista E, *et al.*, Cells preferentially grow on rough substrates, *Biomaterials* 2010;31:7205–12. *An essay on the control of cell behavior by surface nanotopography.*

Hao R, Xing R, Xu Z, *et al.*, Synthesis, functionalization, and biomedical applications of multifunctional magnetic nanoparticles, *Advanced Materials* 2010;22:2729–42. *Summary of the strategies to nanoparticle fabrication and functionalization.*

* Hummel RE, *Electronic Properties of Materials*, Springer, New York, 2001. *Major textbook on electronic and physical properties of materials.*

* Hummel RE, *Understanding Materials Science*, Springer New York, 2nd edition, 2004. *Introduction to the principles of materials science.*

Khademhosseini A, Bettinger C, Karp JM, *et al.*, Interplay of biomaterials and micro-scale technologies for advancing biomedical applications, *Journal of Biomaterials Science – Polymer Edition* 2006;17:1221–40. *Overview of the merger of micro-scale technologies and biomaterials in two-dimensional surface patterning, device fabrication and three-dimensional tissue-engineering applications.*

Liu R, Fraylich M, Saunders BR, Thermoresponsive copolymers: from fundamental studies to applications, *Colloid and Polymer Science* 2009; 287:627–43. *Extensive review of thermally triggered conformational changes in polymers.*

Lutholf MP, Hubbell JA, Synthetic biomaterials as instructive extracellular microenvironments for morphogenesis in tissue engineering, *Nature Biotechnology* 2005;23:47–55. *Explanation of the way in which symbiosis of materials engineering and cell biology may ultimately result in synthetic materials that contain the signals to recapitulate developmental processes in tissue- and organ-specific differentiation and morphogenesis.*

Ma E, Xu, J, The glass window of opportunities, *Nature Materials* 2009;8:855–7. *Short review of the properties of metallic glasses.*

Ma P X, Biomimetic materials for tissue engineering, *Advanced Drug Delivery Reviews* 2008;60:184–98. *Review of biomimetic approaches to tissue engineering scaffolds and methods to produce nanofibrous materials.*

Mano J F, Stimuli-responsive polymeric systems for biomedical applications, *Advanced Engineering Materials* 2008;10:515–27. *Review of polymers and polymer surfaces that can reversibly alter their physico-chemical characteristics in response to environmental factors.*

Merrett K, Cornelius R M, McClung W G, Unsworth L D, Sheardown H, Surface analysis methods for characterizing polymeric biomaterials, *Journal of Biomaterials Science – Polymer Edition* 2002;13:593–621. *Detailed review of methods available to analysis the surface properties of polymeric biomaterials.*

Mitragotri S, Lahann J, Physical approaches to biomaterial design, *Nature Materials* 2008;8:15–23. *Discussion of the role of the physical properties of materials in the regulation of biological performance.*

Moses J E, Moorhouse A D, The growing applications of click chemistry, *Chemical Society Reviews* 2007;36:1249–62. *Tutorial review of the click chemistry approach to the assembly of new molecular entities.*

Nguyen M K, Lee D S, Injectable biodegradable hydrogels, *Macromolecular Bioscience* 2010;10:563–79. *Discussion about the requirements of injectable biomaterials and the options currently available.*

Park J-W, Kim H, Han M, Polymeric self-assembled monolayers derived from surface-active copolymers: a modular approach to functionalized surfaces, *Chemical Society Reviews* 2010;39:2935–47. *A discussion of the design principles for self-assembled monolayers and the resulting chemical and morphological structures.*

Pelton A R, Russell S M, DiCello J, The physical metallurgy of Nitinol for medical applications. *JOM: Journal of Minerals, Metals and Materials Society* 2003;55: 33–7. *Review of processes and properties that relate to shape memory alloys and their medical applications.*

Petersen L K, Narasimhan B, Combinatorial design of biomaterials for drug delivery: opportunities and challenges, *Expert Opinion on Drug Delivery* 2008;5:837–46. *Review of the concept of combinatorial chemistry and its application to polymeric biomaterials design.*

Pinaud F, Michalet X, Bentolila L A, *et al.*, Advances in fluorescence imaging with quantum dot probes, *Biomaterials* 2006;27:1679–87, *Essay on the potential for quantum dot technology in fluorescence imaging.*

* Ratner B D, Hoffman A J, Schoen F J, Lemons J E (eds.) *Biomaterials Science: An Introduction to Materials in Medicine*, Elsevier, Amsterdam, 3rd edn, 2012. *Major multi-editor, multi-author book that covers the science and clinical applications of biomaterials.*

Roy D, Brooks W L A, Sumerlin B S, New directions in thermoresponsive polymers, *Chemical Society Reviews* 2013;42:7214–43. *Review of strategies to develop thermoresponsive polymers.*

Simms H M, Bowman C M, Anseth K S, Using living radical polymerization to enable facile incorporation of materials in microfluidics cell culture devices, *Biomaterials* 2008;29:2228–36. *Description of contact lithographic photopolymerization techniques for preparation of surfaces for high throughput screening applications.*

Sjollema J, Sharma P K, Dijkstra R J B, *et al.*, The potential for bio-optical imaging of biomaterial-associated infection in vivo, *Biomaterials* 2010;31:1984–95. *Review of bioluminescence and luminescence imaging technologies, especially as applied to biomaterials-related infections.*

* Smallman R E, Ngan A H W, *Physical Metallurgy and Advanced Materials*, Butterworth-Heineman, New York, 2007. *Major textbook on structure and properties of advanced materials.*

Somorjai G S, Li Y, Impact of surface chemistry, *Proceedings of the National Academy of Sciences* 2011;108:917–24. *Essay on surface chemistry at the molecular level and its relevance to medical technology.*

* Sperling L H, *Introduction to Physical Polymer Engineering*, John Wiley, New Jersey, 2006. *Major textbook dealing with the engineering of polymer systems.*

Stupp S I, Self assembly and biomaterials, *Nano Letters* 2010;10:4783–6. *Brief perspective on the role of self-assembled materials, especially those based on peptide amphiphiles, in medical technology.*

Uskokovic V, Challenges for the modern science in its descent towards nano-scale, *Current Nanoscience*

2009;5:1–18. *Extensive discussion of the challenges with nanotechnologies and nanostructured materials.*

Wang Y-X, Robertson J L, Spillman W B, Claus R O, Effects of the chemical structure and the surface properties of polymeric biomaterials on their biocompatibility, *Pharmaceutical Research* 2004;21:1362–73. *Discussion of the ways in which surface structure and properties affect the performance of biomaterials.*

Williams D F, On the nature of biomaterials, *Biomaterials* 2009;30:5897–909. *Essay on the radical changes in the concepts of biomaterials and their practical applications.*

Williams D F, The scientific basis for regulating nanotechnologies, in Hodge G S, Bowman D M, Maynard A D, *International Handbook on Regulating Nanotechnologies*, Edward Elgar, Cheltenham, 2010:107–23. *An explanation of the principles of risk assessment with nanoparticles and presentation of a series of nanotechnology definitions.*

# 3 Biocompatibility pathways

Biocompatibility is the most critical factor that controls the success of biomaterials and those health care products that incorporate biomaterials. It is concerned with the mechanisms of interaction between biomaterials and the human body, and the consequences of these interactions. This chapter first introduces the concept of biocompatibility and then provides you with a series of scenarios that cover the whole range of situations in which biomaterials come into contact with tissues. In each case, critical mechanisms are explained and discussed, leading to the presentation of a unified framework of the sequence of events that constitute biocompatibility. This is based on the simple concept that in biocompatibility there are *causative events* within the biomaterial–host interactions that lead, through a variety of different but interconnected *pathways*, to physiological or pathological *effects* and then to their *clinical consequences*. This framework is then used to explain a variety of situations in which biocompatibility has proved to be so important.

## 3.1 Introduction

We discuss here a wide variety of situations in which interactions take place between biomaterials and the patients in which they are placed, and where the nature of that interaction determines both the level of satisfaction and risk that the patient receives or perceives.

Consider the following examples, which are represented in the images of Figure 3.1:

- The performance of a prosthetic heart valve depends on the ability to control the inherent susceptibility of blood to clot once it contacts the surface of the valve. If the blood does clot, there is a high risk of valve malfunction since the clot impedes blood flow and also a high risk of part of the clot being released into the bloodstream, where it could cause a stroke should it lodge in a small vessel in the brain. Either way, the consequences are very serious, and possibly fatal, for the patient.
- The ability to differentiate stem cells into the required phenotype for specific tissue regeneration depends on the interaction between those cells and the material surfaces on which they are seeded. The precise chemical and topographical characteristics of those surfaces control the potential for stem cells to regenerate new tissue.
- The lifetime of a hip replacement prosthesis is largely determined by the response of the surrounding bone to the minute wear particles that are released from the prosthesis during normal

walking activity. These highly successful medical devices may eventually fail because these particles, which may be of nanoscale dimensions, interact with bone resorbing cells and inflammatory cells, causing the bone at the interface with the prostheses to soften and recede.

- Maintenance of clarity in intraocular lenses is dependent on avoiding disruptive tissue changes at the interface between the lens polymer and the eye tissue. These are also very successful in the majority of cases, but in patients where there may be other diseases present, such as diabetes, the tissues may be unable to accommodate the lens without adverse changes, resulting in a lack of visual clarity.
- The ability of a stent to maintain patency of the coronary artery is controlled by the interaction between the metal stent and the endothelium. These stents mechanically hold the arteries open in patients who are diagnosed with coronary artery disease, but they do irritate the lining of the vessel, which may cause it to become blocked again. The result is that the procedure may have to be repeated at regular intervals.
- The safe use of a nanoparticle-based contrast agent in magnetic resonance imaging depends on the level of cytotoxicity of these potent but potentially harmful particles. Contrast agents, which provide exquisitely sensitive and accurate information to assist in the diagnosis and monitoring of disease, are often, of necessity, made of potentially toxic chemicals, and it is necessary to balance the risk associated with these particles against the benefits of accurate and timely diagnosis.
- The ability of a material to carry DNA into a target cell for gene therapy depends on how the material–DNA conjugate is able to penetrate the cell membrane and then avoid destruction by the cell's defensive mechanisms. One of the reasons why it has been so difficult to translate the technology of gene transfer into clinically successful gene therapies, has been the difficulty of achieving efficiency of gene transfer through the use of so-called vectors, or carriers, of the gene, without introducing unacceptable toxic risk.

(a)

**Figure 3.1** Examples of biocompatibility phenomena. **(a)** Thrombogenicity: a blood clot within a mechanical heart valve. Perturbations to blood flow hemodynamics within the valve orifice and the inherent thrombogenicity of valve materials leads to a tendency to clot formation. Patients are anticoagulated in order to minimize this possibility. Image courtesy of Frederick J. Schoen, M.D. Professor of Pathology, Brigham and Women's Hospital and Harvard Medical School, Boston, MA, USA.

At this stage we do not need to know the details of each of the above types of material or product to appreciate just how important these interactions are. Ever since biomaterials were first used within the human body, it has been accepted that interactions take place between them and the components of the body, and that these interactions have a significant impact on the ability of that material to satisfactorily perform the desired function for the desired length of time.

The above examples give a hint that the nature of these interactions varies from one clinical

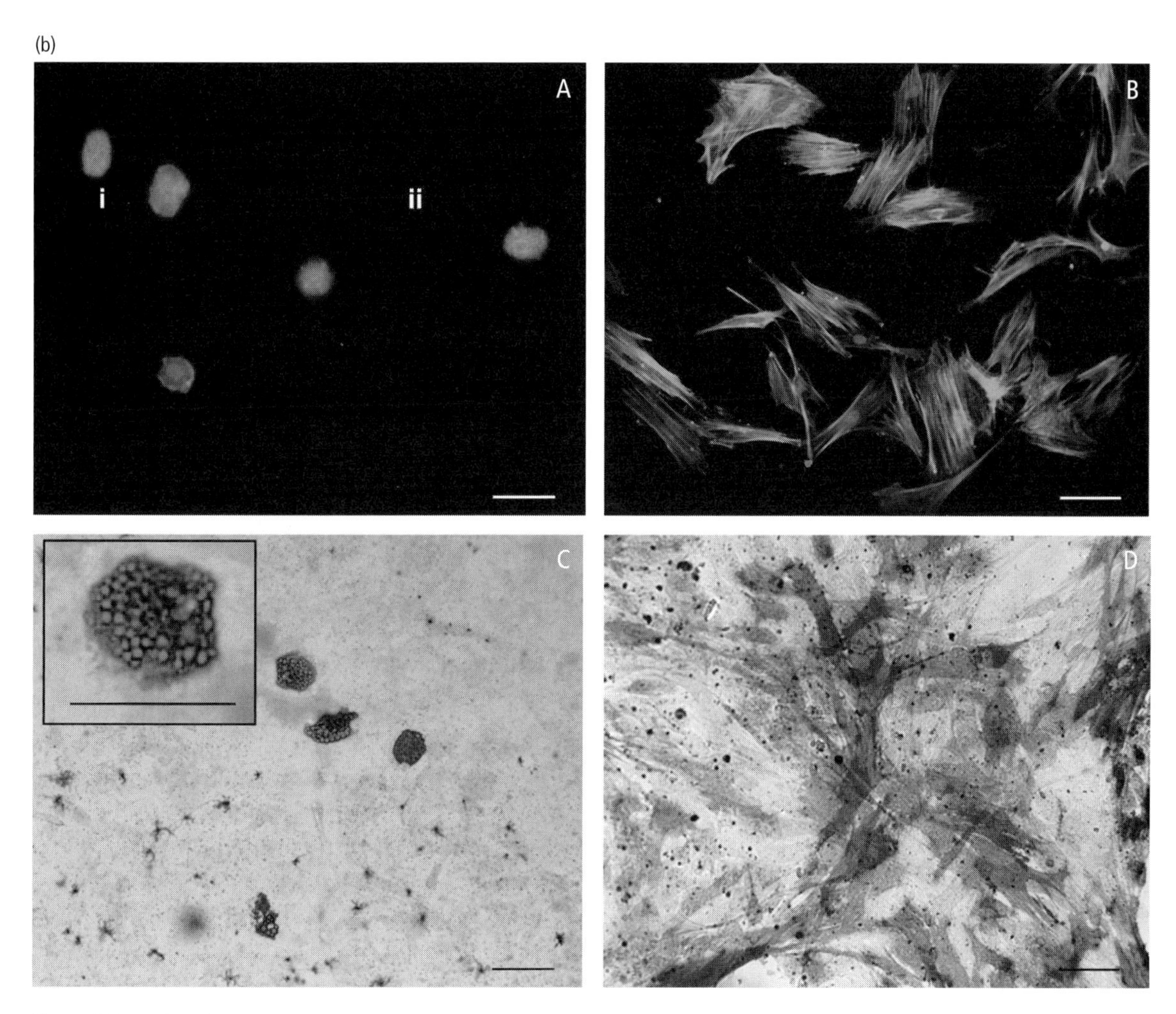


**Figure 3.1** *continued*
**(b)** The effect of substrate stiffness on stem cell behavior. Fluorescent images of human mesenchymal stem cells cultured on (A) soft 3 kPa or (B) stiff 30 kPa hyaluronic acid-based hydrogels for 14 days in mixed adipogenic/osteogenic induction media and stained for F-actin (green) and nuclei (blue). Corresponding brightfield images (C) soft and (D) stiff of hMSCs stained for lipid droplets (red, adipogenic marker) and alkaline phosphatase (blue, osteogenic marker). Scale bars are 100 microns and 50 microns (inset). Image courtesy of Murat Guvendiren, PhD and Jason A. Burdick, PhD Department of Bioengineering University of Pennsylvania, USA. See plate section for color version.

situation to another. In reality, there is not just one overarching type of interaction that could take place; there are very many of them. There is not just one phenomenon taking place here, but a myriad of individual and interconnected reactions. For many years, these different phenomena have been discussed collectively under the broad heading of biocompatibility. For much of that time, few people had any idea of what biocompatibility really meant and we are still largely ignorant of the precise details of these mechanisms; it has been, however, a convenient and recognizable term that encompasses all reactions between biomaterials and the human body.

(c)

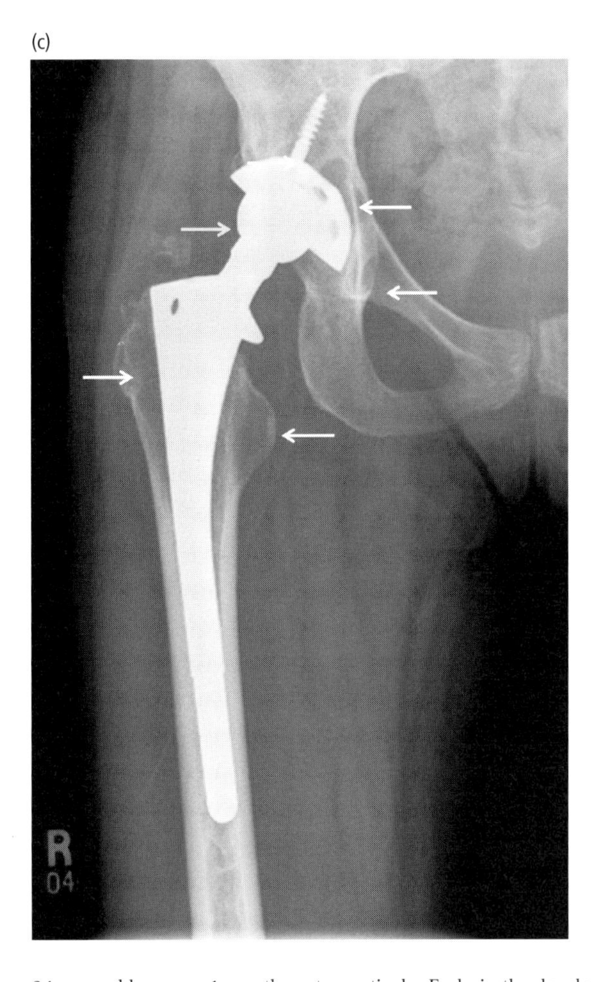


(d)

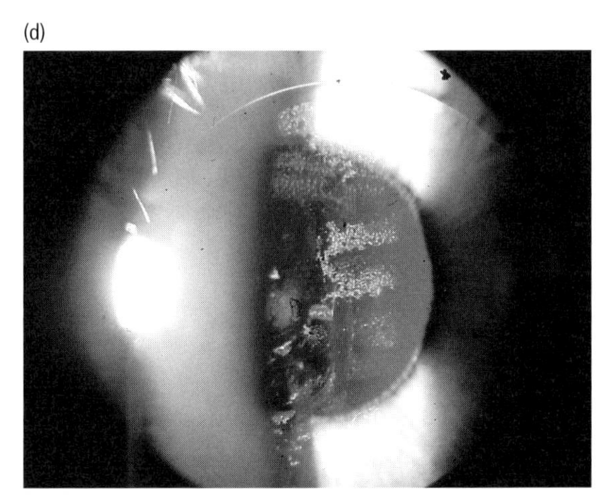

**Figure 3.1** *continued*
**(c)** Loosening of hip replacements. A radiograph of a cementless total hip arthroplasty with polyethylene wear and periprosthetic osteolysis (bone destruction). The arrow at top left points to eccentric positioning of the femoral head within the plastic insert of the acetabular component, indicative of polyethylene wear. The other arrows point to bone destruction associated with the chronic inflammatory reaction to polyethylene wear particles. Progressive undermining of the acetabular component by this process eventually leads to loosening of the cup, requiring revision surgery. Image courtesy of Dr. Stuart Goodman, Department of Orthopaedic Surgery, Stanford University, California, USA.
**(d)** Protein deposits on intraocular lenses. Protein deposits are seen here on an anterior chamber IOL in a 34-year-old woman 1 month postoperatively. Early in the development of IOLs these were clinically important as they could negatively affect visual acuity. Improvements in lenses and techniques have markedly reduced this problem. Reprinted from *Journal of Cataract and Refractive Surgery*, 36, Kohnen T *et al.*, Phakic intraocular lenses, 12–39, © 2013, with permission of Elsevier.

Biocompatibility was first discussed in the context of implantable medical devices and early definitions of the term reflected this situation. Probably the most widely accepted definition of biocompatibility was derived in the 1980s, and states that biocompatibility "refers to the ability of a material to perform with an appropriate host response in a specific application." This, of course, is a conceptual definition rather than one of immediate practical usefulness, but indicates three important factors:

- the biomaterial has to perform and not simply exist;
- out of all the possible reactions and responses from the tissues of the body, the response that is the most appropriate for the application in question is required; and
- we always have to define biocompatibility in the context of the specific application.

The last point is extremely important since it implies that, for a given material, the biocompatibility

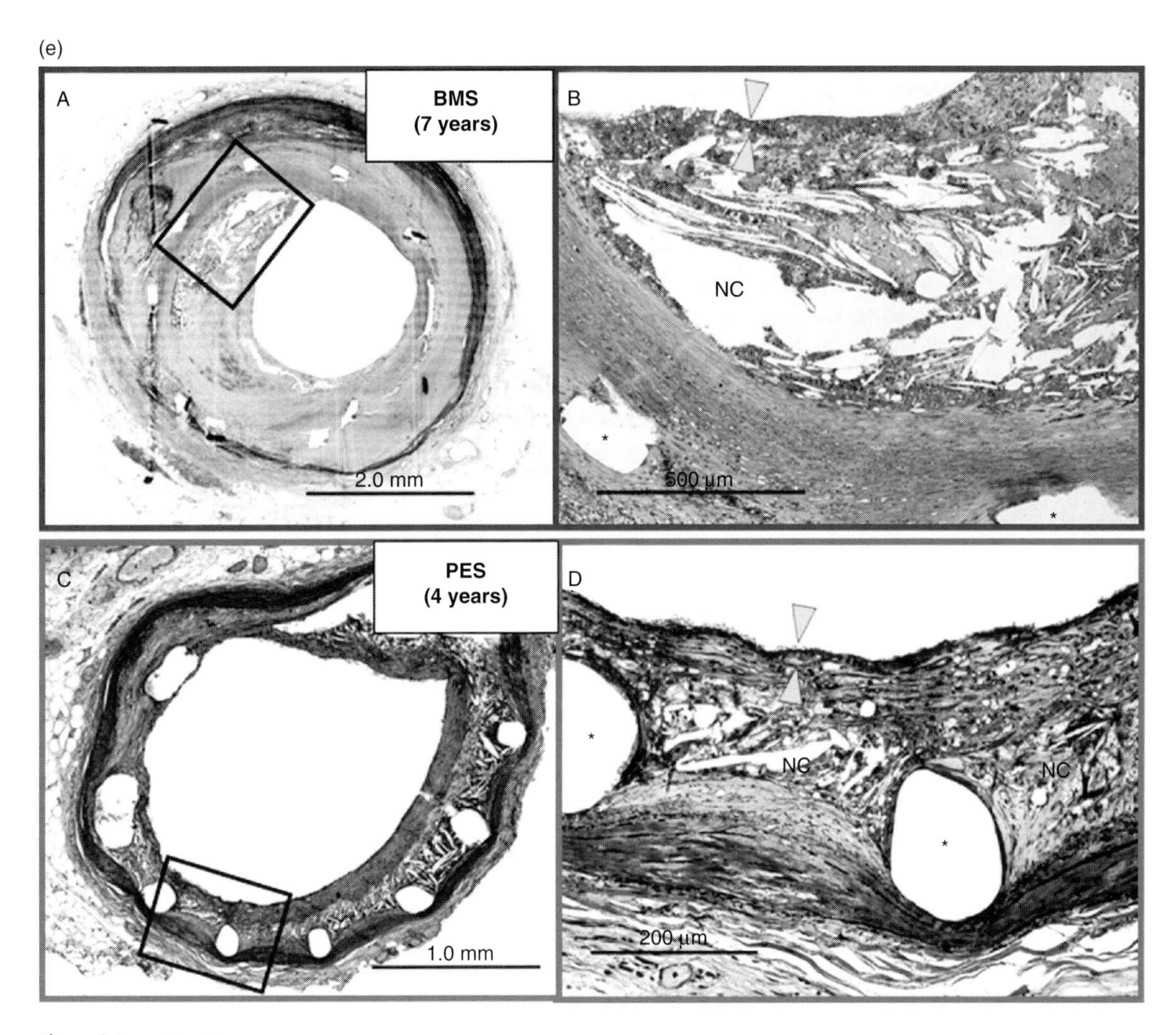


**Figure 3.1** *continued*
**(e)** In-stent restenosis. Intravascular stents, used to open blocked arteries, may themselves become blocked again in a process known as restenosis. (A) Cross-sectional histology of bare-metal stent (BMS) implanted in the coronary artery for 7 years. (B) High-power image of the box in A (×100). A large necrotic core (NC) containing cholesterol crystals is identified within the neointima. The fibrous cap overlying the NC is infiltrated by numerous foamy macrophages and is markedly thinned (yellow arrowheads point to thinnest portion), which resembles vulnerable plaque encountered in native coronary arteries. The asterisks represent metal struts. (C) Cross-sectional histology of paclitaxel-eluting stent (PES) implanted in the coronary artery. (D) High-power image of the box in C (×200). A relatively small NC containing cholesterol crystals is formed around metal struts (asterisk). The fibrous cap is infiltrated by numerous foamy macrophages and is markedly thinned (yellow arrowheads point to thinnest portion). Reprinted from *Journal of the American College of Cardiology*, 59, Park S-J *et al.*, In-stent neoatherosclerosis, 2051–7, © 2012, with permission of Elsevier. See plate section for color version.

characteristics vary from situation to situation. Biocompatibility is not, therefore, a property of a biomaterial; it is a characteristic of a material–tissue system. The corollary of this is that there is no such thing as a biocompatible material, a fact that is still lost on the majority of biomaterials scientists and which is at the heart of a number of biomaterial failures.

(f)

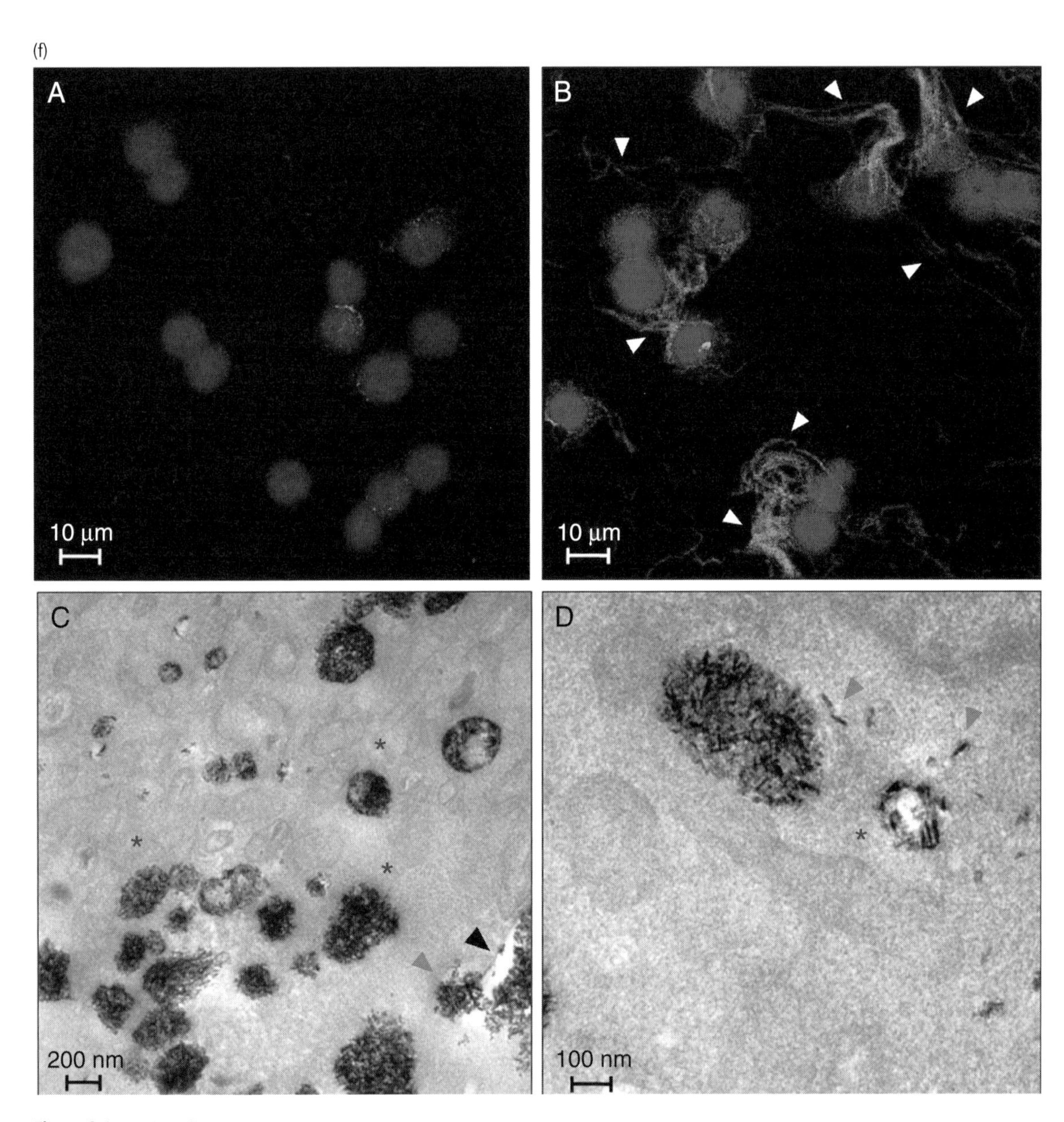


**Figure 3.1** *continued*

**(f)** Biocompatibility of contrast agents. Cultured mouse hepatocytes (NCTC 1469) were treated with a MRI contrast agent (ferucarbotran, Resovist), superparamagnetic iron oxide nanoparticle (SPION, 0.2 mM) alone (A) or concomitantly treated with the SPION and static magnetic field (SMF, 0.4 Tesla) for 1 h. In (B), the fibrillar SPION aggregates (green, arrowheads) were notably observed by confocal fluorescence microscopy in SMF-exposed cells, which were stained with anti-dextran antibody. Nuclei (blue) were stained with DAPI. The electron micrograph (C) of internalized SPION in the SMF-exposed cells show at the red arrowhead, an emerging phagocytic vesicle encapsulating SPION aggregates and at the surface of the cell membrane (black arrowhead). The asterisks show some of the trafficking vesicles containing SPION aggregates. The electron micrograph (D) show cytosolic SPION spilled by trafficking SPION-harboring vesicle(s) (asterisks). Note the spike-formed SPION aggregates in the vesicles and the cytosol. Image courtesy of Professor Kwon-Seok Chae, Kyungpook National University and Professor Myung-Jin Moon, Dankook University, Republic of Korea. See plate section for color version.

(g)

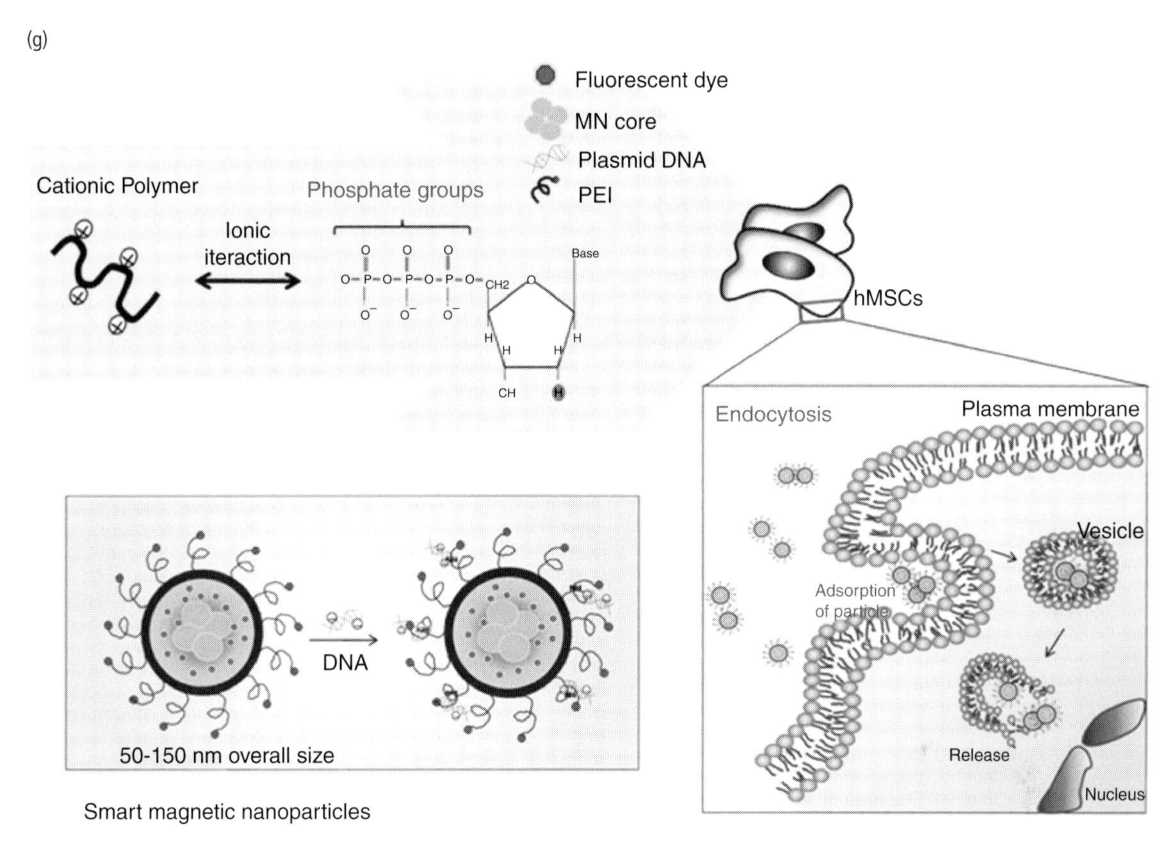


**Figure 3.1** *continued*
(g) Intracellular release of plasmid DNA. Schematic diagram of the endosomal escape and gene release of PEI-modified nanoparticles. Mixing PEI-modified nanoparticles with negatively charged plasmid DNA (pDNA) results in the spontaneous electrostatic formation of stable complexes. In the endosomal compartment, these interact with negatively charged lipid bilayers and induce the influx of water and ions into endosomes, thus causing endosome destabilization and gene release into the cytosol. Reprinted from *Biomaterials*, 31, Park J S *et al.*, Non-viral gene delivery of DNA polyplexed with nanoparticles transfected into human mesenchymal stem cells, 124–32, © 2010, with permission of Elsevier.

This basic concept of biocompatibility is a good place to start, although we shall see very soon that many of our ideas about biocompatibility have had to change in recent years with new and varied biomaterials applications. New definitions have had to be introduced, and a discussion about the mechanisms of biocompatibility today has to be quite different to those of just a decade ago. The goal of this discussion here is the development of a generic framework of biocompatibility mechanisms that provides an integrated approach to our understanding of these phenomena.

## 3.2 The fundamental biocompatibility paradigm

We start this discussion with two clear statements of fact. First, with very few exceptions, when man-made, engineering or commodity materials are used as biomaterials they are not intrinsically compatible with physiological systems, nor have they been designed to be so. Secondly, the tissues of the human body have not evolved in order to benignly accommodate these materials within their midst and they will be treated as "foreign" and

## Box 3.1 Glossary of terms

**actin** muscle protein located within myofibrils, responsible for contraction and relaxation of muscle.

**activation** the process whereby certain cells, such as lymphocytes and macrophages, differentiate from a resting state to an active state, acquiring new capabilities and functions.

**allergy** reaction of the body to a substance to which it has become sensitive, characterized by edema, inflammation and destruction of tissue.

**cell adhesion** process by which cells become attached to a substrate, including a biomaterial surface.

**cell adhesion molecule** molecule on the surface of a cell that is involved in intercellular adhesion.

**cell differentiation** process in which a less specialized cell becomes a more specialized cell in order to perform a specific function.

**chromatin** substance of which eukaryotic chromosomes are composed, consisting of proteins, DNA and small amounts of RNA.

**coagulation** process in which colloidal particles come together to form larger masses.

**conformation** the shapes or arrangements in three-dimensional space that an organic molecule assumes by rotating carbon atoms or their substituents around single covalent bonds.

**cytokine** a protein released by a cell that stimulates or inhibits the differentiation, proliferation or function of immune cells.

**cytoplasm** protoplasm of a cell, exclusive of that of the nucleus.

**cytoskeleton** network of fibers permeating the matrix of living cells that provides a supporting framework for organelles, anchors the cell membrane, facilitates cellular movement and provides a suitable surface for chemical reactions to take place.

**cytosol** semi-fluid soluble part of the cytoplasm of cells.

**embryonic stem cells** stem cells that are found in the inner cell mass of blastocysts.

**endocrine system** system of ductless glands and organs that secretes substances directly into the blood to produce a specific response from another part of the body.

**endocytosis** cellular ingestion of macromolecules by invagination of the plasma membrane to produce an intracellular vesicle which encloses the ingested material.

**enzyme** protein that catalyses chemical reactions.

**epitope** that part of an antigen to which an antibody attached itself.

**exocytosis** discharge from a cell of particles that are too large to diffuse through the cell wall.

**exogenous** originating outside, or caused by factors outside, an organism.

**extracellular matrix** the non-cellular matrix of proteins and glycoproteins that surrounds cells within tissues.

**gene** a hereditary unit consisting of a sequence of DNA that occupies a specific location on a chromosome and which provides coded instruction for the synthesis of RNA.

**gene expression** any of the processes by which nuclear, cytoplasmic or intracellular factors influence the differential control of gene action at the level of transcription or translation.

**hematopoietic stem cells** adult stem cells that give rise to all kinds of blood cell.

**hyperplasia** over-development of an organ or tissue, due to increased production of cells.

**hypersensitivity** excessive immune response that leads to undesirable consequences.

**hypertrophy** enlargement or overgrowth of an organ or tissue part, due to an increased amount of work of that organ or part.

**interstitial fluid** extracellular fluid within tissue spaces that is not lymph, plasma or transcellular fluid.

**killer cells** cells that are morphologically indistinguishable from small lymphocytes without T or B cell surface markers but which have cytotoxic activity against target cells coated with specific IgG antibodies.

**kinase** an enzyme that catalyzes the phosphorylation of its substrate by ATP.

**mesenchymal stem cell** multipotent stem cells that can differentiate into a variety of cell types such as osteoblasts, chondrocytes and adipocytes.

**mesenchyme** loosely organized undifferentiated cells that give rise to structures such as connective tissue, blood and the lymphatic system.

**mitochondria** mobile cytoplasmic organelles whose main function is the generation of ATP by aerobic respiration.

**mitosis** method of indirect cell division in which two daughter nuclei normally receive identical complements of the number of chromosomes characteristic of the somatic cells of the species.

**necrosis** death of a cell, or group of cells, while still part of the living body.

**opsonin** substance such as an antibody or complement fragment that promotes adhesion of an antigen to a phagocytic cell.

**opsonization** coating of antigen with opsonin in order to enhance phagocytosis.

**osteogenesis** the formation of bone.

**pathogenic** disease-producing.

**phenotype** the set of observable physical or biochemical characteristics of an organism, derived both genetically and environmentally.

**phosphorylation** chemical reaction that combines phosphorus with an organic compound.

**progenitor cell** similar to stem cell but already somewhat differentiated down a lineage with limited capacity for self-renewal.

**receptor** molecule on the surface of, or within, a cell that recognizes and binds with specific molecules, producing a specific effect in that cell.

**somatic cells** cells of the body, as distinct from germ cells, sperm and eggs.

**stem cell** a cell from a multicellular organism that is capable of self-renewal and of division and differentiation into diverse specialized cells.

**systemic** pertaining to or affecting the body as a whole.

**transcription factor** a protein that binds to specific DNA sequences, thereby controlling the movement of genetic information from DNA to mRNA.

"potentially harmful." The default position, therefore, is that there is inherent incompatibility between these two compartments, the biomaterial and the tissue.

It is, however, much more serious than that. The problem with this default position is that the human body has evolved in such a way as to have exquisite detection mechanisms that readily identify foreign objects (historically in the form of microorganisms), and equally important, there are exquisite defensive mechanisms that deal with such objects once they have been detected; we are confronted, therefore, with an active incompatibility and not a passive one. These mechanisms evolved naturally to deal with bacteria and viruses, but they are often capable of diversion towards any synthetic material that might find its way into the body or any type of biological stress that may arise with this use. This becomes especially important when we design biomaterial components that have some similarities with the bacteria and viruses, both in size and chemistry, so that we are inviting these reactions to take place and have to devise means to avoid them. It is useful to bear in mind that the introduction of a biomaterial into the human body normally represents a physiologically stressful event and we should expect the body to present some adaptive response.

We have already identified in Chapter 2 the fact that many synthetic materials are susceptible to degradation within aqueous environments. As we have seen in Box 2.2, the tissues of the body are aqueous-based, and in addition, have a collection of species, both cellular and molecular, that are mobile and aggressive, so that the already corrosive environment is powerfully enriched by these active agents. It should be of no surprise that aggressive host–biomaterial interaction mechanisms are readily available as soon as the physician or technologist exposes the former to the latter. The role of the biomaterials scientist is to design materials that are best able to resist or accommodate these mechanisms.

Responses will also evolve over time, depending on the physical characteristics and biostability of the material, and may change with the circumstances within the body. The evolving response has often been referred to as the foreign body response, especially with implantable devices, although this is a rather imprecise term. Clearly the nature of this response will depend on the specific characteristics of the material behavior and of the host tissue behavior, which we shall discuss before addressing the totality of the foreign body response.

### 3.2.1 Some general principles

There are a few general principles that we have to bear in mind; these are summarized in Box 3.2. These principles can be considered in terms of the chronology and location of events and the controlling factors:

- *Chronology*: we should not necessarily expect the mechanisms of biocompatibility to show linear progression with time. In many situations, usually on the side of the host, one event may be triggered spontaneously, at any time, the effect of which can be powerfully amplified by one or more mechanisms, changing the whole nature of the response in a short space of time.
- *Location*: the consequences of the interactions between material and host may be localized to the vicinity of the material, giving the local foreign body response. Alternatively, or additionally, the effects may be remote, affecting either the whole body (a systemic response), or affecting a specific discrete remote site, for example the site where a corrosion or degradation product is eventually stored. Local foreign body responses are very important in the area of implanted medical devices. Systemic effects can occur with any biomaterial but have taken on greater significance as molecular or nanoscale biomaterials have been introduced for drug and gene delivery or for imaging contrast agents, where the products themselves are highly mobile and often delivered by injection rather than surgically.

**Box 3.2 The basic ground rules of biocompatibility**

1. The introduction of a biomaterial into the human body represents a physiologically stressful event and we must expect the body to present some adaptive response.
2. The default position is that there is inherent incompatibility between foreign materials and the tissues of the body.
3. Biocompatibility is not a property of a material; it is a characteristic of a material – tissue system.
4. In biocompatibility, an event may be triggered spontaneously, at any time, the effect of which can be powerfully amplified by one or more mechanisms, changing the whole nature of the response in a short space of time. Mechanisms of biocompatibility do not necessarily show a linear progression with time.
5. The consequences of interactions may be localized to the vicinity of the material, often referred to as the foreign body response, or they may be seen at some distant site, or they may be truly systemic.
6. Biocompatibility phenomena vary from patient to patient and may vary with the techniques used to administer the biomaterial to the patient.

- *Controlling factors*: although biocompatibility is obviously controlled by the nature of the material, device or agent, it is clearly influenced by many other factors. Biocompatibility phenomena vary from patient to patient, and they vary with the techniques used to administer the biomaterial to the patient. These issues must not be overlooked.

### 3.2.2 The components of biocompatibility

The story of biocompatibility is one of a series of individual phenomena, which combine to generate the overall host response. Because there are so many variables that may affect each of these components and the mechanisms that are associated with them, it should not be surprising that there is an almost infinite number of possible features of the host response. As noted above, biocompatibility is not a precise characteristic of a material; we can only discuss the biocompatibility of individual materials in very general terms. We should not expect similar experiences with all patients receiving identical medical devices, nor should we expect similar outcomes when one material is used in different circumstances. We should, however, anticipate that there is a common generic pattern around which these mechanisms are arranged.

We shall conceptually consider the totality of biocompatibility by identifying those physiological processes that may be perturbed by the presence of biomaterials, and describing the mechanisms by which these perturbations may take place. For example, how do cells that normally interact with a specialized extracellular matrix respond to the surface of a non-physiological biomaterial? How does a plasma protein respond to a metallic intravascular stent when it normally interacts with an intact endothelium? How does a cell respond to a nanoscale non-viral gene vector that is able to circumvent the normal defense process and pass through the cell wall and into the endoplasm?

It has been common practice in the development of explanations of biocompatibility to separate out different materials, different applications and different phenomena. Thus interactions with blood have been considered separately from interactions with solid tissues. Bone and soft tissue biocompatibility have usually been considered separately and the biocompatibility of implant materials has been characterized separately from that of materials used in tissue engineering scaffolds, drug delivery and imaging systems. In this book we take another approach.

### 3.2.3 A systems approach

A systems approach may be used to assess how these individual mechanisms contribute, in general terms, to the overall host response. These processes are discussed in terms of physical/mechanical systems (that is, the physical presence of a biomaterial in the host tissue), of chemical systems (that is, the release of chemical products from biomaterials into the tissue), of pharmacological systems and of external systems, such as clinical factors, patient variability and so on. We also do this in terms of the scale of the biomaterial, for example with a macroscopic implantable medical device, a microscopic distribution or a nanoscale material.

We will then develop the framework of biocompatibility mechanisms. After the systems have been identified, explained and consolidated, specific examples of the biocompatibility of clinically relevant devices and materials will be shown.

In developing the hierarchy of mechanisms, a basic scenario of the biomaterial–tissue system can be established and the material and system variables that can give rise to the various and varied responses identified. This simple situation is depicted in Figure 3.2. A biomaterial exists within a volume of tissue. We can ignore for the moment exactly how the material came to be within the tissue – this is an extremely important point, and is a major feature that distinguishes the approach to biocompatibility discussed here to virtually all previous approaches. The traditional approaches have started with the assumption that the biomaterial gains access to tissues by means of a surgical incision and that the host response can be considered to be a perturbation of normal wound healing. This is one possibility, but only one; it shall be taken into account later but it will not underpin the basic mechanisms. We should also note here that although we would normally expect this tissue to be fully formed functional tissue, this is not necessarily the case: applications of biomaterials in many forms of regenerative medicine involve the effect of materials on tissue that is in the process of being formed.

In Figure 3.2 we start with the assumption that the biomaterial is a solid object that is immobile, chemically non-reactive with physiological components, and unchanging with time. The tissue is generic; it consists, to varying extents and with varying individual characteristics, of cells and their extracellular matrix (ECM). It could be solid tissue, blood, a collection of cells or a tissue engineering construct. In contrast to most forms of the biomaterial, however, the tissue provides a dynamic environment in which the cells and their ECM change over time. Solid tissue is also permeated by circulating cells, such as blood and inflammatory cells, and by circulating molecules, such as proteins. In this simplest of conditions, the biocompatibility of the system is concerned with the effect that the physical presence of the material has on the dynamic response of the tissue components. This is the conceptual starting point, from which a series of positions emerge where the biomaterial is presented to the physiological environment in different forms or in different situations. Each of these takes the basic inert scene and adds, in turn, the complexities of chemical reactions with solid surfaces or soluble components, reactions with solid microscale entities, reactions with nanoscale entities and those influenced by pharmacological agents. Note that we do not differentiate tissue types, material types or clinical applications. There is no special section on blood or bone compatibility. The nanoscale can be

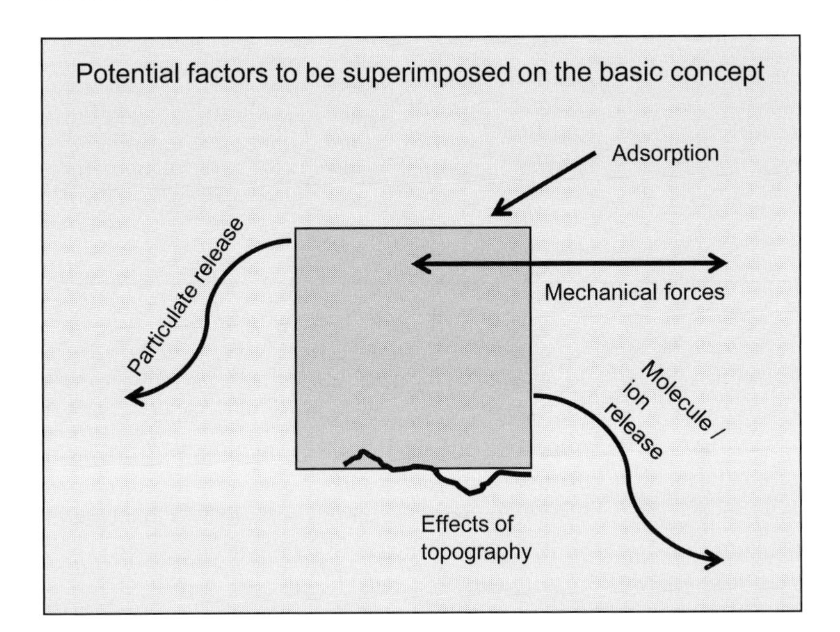


**Figure 3.2** A systems approach to biocompatibility. The conceptual starting point is the understanding of the effects of the physical presence of a material on the dynamic response of the tissues; superimposed on this concept are several, varied factors that ultimately influence the outcomes.

considered generically whether it is encountered in contrast agents or regenerative scaffolds.

## 3.3 Biocompatibility scenarios

### 3.3.1 Introduction to biocompatibility mechanisms and pathways

Having identified the basic biocompatibility paradigm, we now discuss the specific mechanisms by which biocompatibility interactions take place by reference to a series of individual scenarios and to an overview of the totality of biocompatibility phenomena. These scenarios take into account the following criteria and questions:

- Is the host response determined by direct chemical interaction between a macroscale biomaterial and the tissue?
- Is the host response determined by mechanical forces?

### Box 3.3 The consequences of biocompatibility failures

| | |
|---|---|
| Death of the patient | Thromboembolic complications<br>Anaphylactic shock<br>Organ failure |
| Major disease | Cancer<br>Central nervous system damage<br>Autoimmunity<br>Incontinence<br>Peripheral nerve damage<br>Reproductive errors<br>Loss of sensory function<br>Viral disease transmission |
| Minor–major complaints | Allergy<br>Inflammation/edema<br>Hyperplasia<br>Hypoplasia/resorption |
| Failure to achieve desired outcome | Osteolysis<br>Loss of patency of vessel<br>Infection |

- Is the host response substantially determined by the release of particulate or soluble substances from the biomaterial?
- Is the host response significantly influenced by the presence of a pharmaceutical agent in the biomaterial?
- Is the host response substantially determined by the generation of new tissue within a biomaterial?
- Is the host response substantially determined by the microscale features of the biomaterial?
- Is the host response substantially determined by the nanoscale features of the biomaterial?
- Does the host response involve direct contact between the biomaterial and the major blood circulatory system?
- Are there significant systemic or remote site components of the host response?

Obviously with any specific material and/or device, more than one scenario may apply; these scenarios are applied to mechanisms and not discrete devices. Also some specific mechanisms may be operative in more than one scenario; they are discussed here in the most appropriate places.

It is also important at this point to bear in mind the potential consequences of failing to achieve an appropriate, or acceptable, host response in clinical practice. After all, the purpose of using biomaterials in health care products is to provide clinical benefit to patients and one of the main detractors of benefit is the lack of biocompatibility. These are summarized in Box 3.3. They range from death of the patient, through the acquisition of serious disease associated with the material, minor irritations and to the failure of the device to perform the intended function. The last point is an important one. There are, usually, several options for the treatment of a specific condition. If a doctor chooses one option that includes a biomaterial component, he/she may forgo an alternative option,

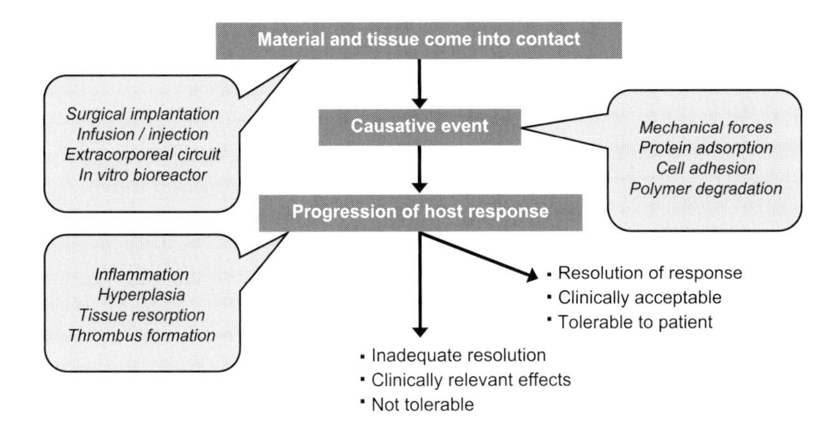


**Figure 3.3** The essential biocompatibility paradigm. The process starts with the initial contact between material and the tissues, from which arise one or more causative events, which lead to the progression of the host response and, ultimately, to acceptable resolution or an unacceptable clinical outcome.

perhaps involving a drug or physical therapy so that the failure to perform optimally may suggest that this was the wrong option.

Each of the individual clinical consequences may be considered to be the manifestations of the *effects* of biocompatibility processes. In establishing the framework of biocompatibility mechanisms, we aim to identify the *causes* and link them to these effects by a series of *pathways*. Although these pathways will, at molecular and cellular levels, be different, we will identify common or generic features associated with them.

At its very simplest, the essential biocompatibility paradigm is seen in Figure 3.3. This will be a common theme as the nature of the pathways emerges. During the discussion of this paradigm, a series of cells and extracellular matrix proteins will be introduced. The most important of these are briefly described in Boxes 3.4 and 3.5, along with definitions of some of the relevant terms.

## 3.3.2 Scenario A: Mechanisms of biocompatibility that do not rely on direct chemical interactions between the biomaterial and the host

It may be argued that no material is inert and all biomaterials will interact chemically with the body. This is largely true, and I have already implied that this is the case earlier in this chapter. However, it is impossible to develop an argument about mechanisms of biocompatibility in real life if we do not start with the basic, fundamental position, that all such real phenomena have to be interpreted in the context of a host response to the presence of an inert object. We shall build on this basic position to consider those scenarios where other factors come into play, as indicated in Figure 3.2.

It should be emphasized here that this baseline scenario does not state, nor does it imply, that chemical reactions are not involved in the processes that take place. What it does say is that the cause of the events that take place is not a chemical process. Once the process has been initiated, then the reaction may be propagated by chemical processes, especially within and between cells.

There are two major considerations here and several minor ones. The major ones are associated with the role of mechanical factors and the adsorption or adhesion of tissue components onto biomaterials surfaces.

### Mechanical interactions

Mechanical forces play a significant role in homeostasis, especially within connective tissues. It is well known that physical activity promotes tissue development and maintenance of tissue condition, but a lack of activity, for example with enforced rest or under zero gravity conditions, may cause atrophy of both muscles and the skeleton. This implies that

## Box 3.4 Essential cells in biocompatibility phenomena

The cells listed here are those that might be found within the active zone of the host response to a biomaterial. They do not include the target cells for any particular therapy (see Section 3.5.2)

**Cells of inflammation and the immune response**

Basophils
B-lymphocytes
Dendritic cells
Eosinophils
Foreign body giant cells
Macrophages
Mast cells
Monocytes
NK cells
Plasma cells
Platelets
Polymorphonuclear leukocytes (neutrophils)
T-lymphocytes

**Cells of reparative and hyperplastic responses**

Endothelial cells
Epithelial cells
Fibroblasts
Myoblasts
Myofibroblasts
Odontoblasts
Osteoblasts
Smooth muscle cells

**Cells of tissue resorption**

Chondroclasts
Odontoclasts
Osteoclasts

the cells of these tissues must be able to sense strain within the ECM and translate the characteristic features of strain into some form of adaptive response. It should be no surprise, therefore, that when the physical presence of a biomaterial in tissues is able to cause some perturbation to the strain system within the tissue, it will itself invoke some such adaptive response. In a similar manner, we should expect that cells in a culture dish or bioreactor should respond to applied stresses and consequential

## Box 3.5 Essential biological molecules in biocompatibility phenomena

There are very many molecules that are released from cells during the inflammatory and reparative processes that constitute biocompatibility phenomena. They have been classified in a number of ways, for example according to the cells that secrete them, by their functions or by the receptors on which they act. None of these classifications are very relevant to our discussion of biocompatibility mechanisms. Therefore, the more important of these biological molecules are listed here alphabetically, often in groups, with a brief statement of their characteristics.

| | |
|---|---|
| Bradykinin | Polypeptide, a powerful vasodilator |
| Chemokines | Small cytokine molecules that regulate chemotaxis, especially the extravasation of leukocytes |
| Complement | System of proteins that facilitates antibody and phagocytic response to microorganisms (Section 3.3.2) |
| Histamine | Organic nitrogen compound, an important mediator of inflammation, with effects on vasodilation and smooth muscle contraction |
| Immunoglobulins | Key molecules of the immune response (Section 3.3.3) |
| Interferons | Family of glycoproteins that have anti-viral activity |
| Interleukins | Family of cytokines released by many cell types including leukocytes, that regulate the immune response and inflammation. More than 20 molecules have been characterized, with effects on NK cells (IL-1), proliferation of T cells (IL-4) and neutrophil chemotaxis (IL-8) |
| Lectins | Family of sugar-binding proteins that regulate cell adhesion and control tissue glycoprotein levels |
| Lymphokines | Cytokines produced by lymphocytes that stimulate further lymphocyte and macrophage production |
| Prostacyclins | Lipid molecules involved in hemostasis through inhibition of platelet activation and vasodilation |
| Prostaglandins | Autocrine or paracrine lipid molecules that act on mast cells, platelets and the endothelium |
| Tumor necrosis factor | Group of cytokines that regulate apoptosis; also can stimulate cell proliferation and differentiation under some conditions |
| Tyrosine kinase | Enzyme, one of many protein kinases, that controls phosphorylation of proteins and regulates many cell functions and signal transduction |

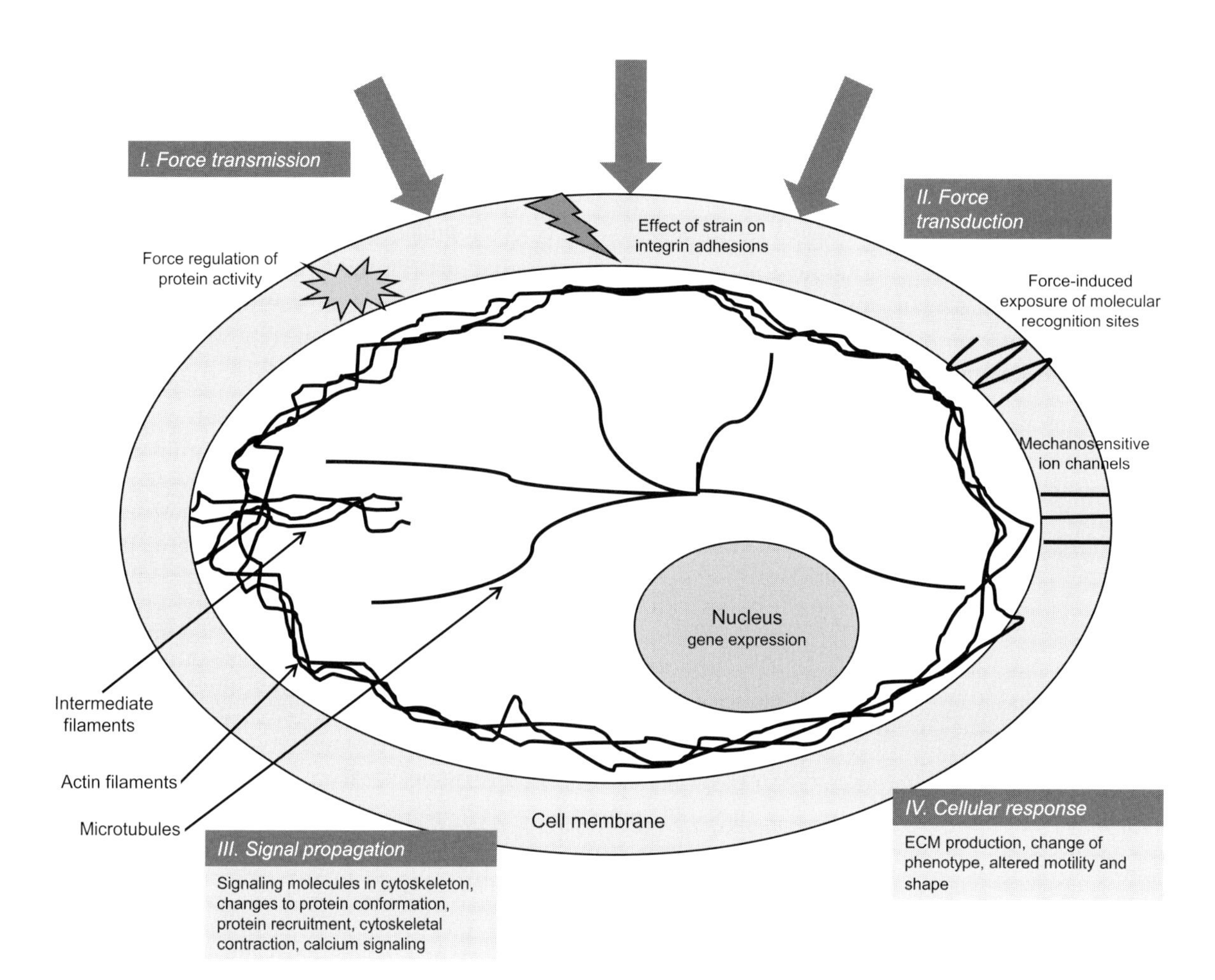


**Figure 3.4** Mechanotransduction. The four discrete stages of mechanotransduction: force transmission, force transduction, signal propagation and cellular response.

strains in structural or fluid flow situations by also adapting their behavior.

The various mechanisms by which cells are able to convert mechanical forces into chemical activity are collectively described by the term "mechanotransduction." We can easily see here why it is important to have this baseline of biocompatibility mechanisms that do not rely on direct chemical interactions between biomaterial and tissue, since these events associated with mechanotransduction are entirely independent of the chemical composition of the biomaterial.

Mechanotransduction is generally considered to involve four discrete stages: force transmission, force transduction, signal propagation and cellular response, as shown in Figure 3.4. The first concerns the transmission of force from some external point to the cell surface. Force transduction involves the sensory action of the cell that transduces the mechanical stimulus to a chemical signal, which is transmitted within the cell in the third phase. The cell then responds to this transduced signal by modulating gene expression. There are several specific mechanisms by which these events can occur. For example, the transduction of mechanical forces may occur through changes in protein conformation, which influences enzyme activity and protein interactions.

Increasing tension within the lipid bilayer of a cell membrane may affect channel-opening processes.

Incidentally, the four-stage sequence involving cause, transduction, propagation and effect, is consistent with the paradigm of Figure 3.3. Several examples of mechanical stress induced biocompatibility phenomena may be identified.

**Implantable devices: bone** With invasive devices, it would be exceptional for the device to have exactly the same elastic properties as the surrounding tissue. The virtually inevitable mismatch of the material's elastic modulus and structural flexural rigidity with those of the tissue means that there will be differential stresses and strains in these components, the nature of which depends to some extent on the degree of attachment between them.

If a high modulus metal is placed within bone, in general the metal will sustain by far the greatest stress, leaving the bone with less than normal physiological loading, resulting in resorption of bone. This is observed clinically in some total hip and knee replacements and can be especially troublesome when revision surgery has to be performed. As osteoblasts deposit the mineralized Type I collagen matrix that primarily constitutes bone, they become trapped and transform into osteocytes, which are considered to be the main mechanotransduction cells of bone. These exist in unmineralized pockets, the lacunae, and interconnect with each other via protrusions, the canaliculi. Under mechanical stimulation, the network of osteocytes express multiple factors that stimulate osteoblast function, including prostaglandin E2 and nitric oxide, while at the same time releasing negative factors that decrease osteoclast activity. Without sufficient stimulation, the reverse will be true. It is generally agreed that dynamic rather than static forces are required, and that compressive stress per se is not directly involved. Instead, compressive forces generate pressure gradients that stimulate flow of interstitial fluid and develop high shear stresses. Shear stress stimulates transient increases in intracellular calcium, which stimulates the osteoblast function.

**Implantable devices: blood vessels** In the cardiovascular system, there are different scenarios, and different specific mechanisms, but with the same type of outcome. In the healthy individual, the blood pressure is very carefully regulated so that blood flow is optimal under all physiological conditions. Within the process known as the myogenic response, arterial smooth muscle cells regulate local flow in a very time sensitive manner. If the response is not sufficiently effective, and blood pressure rises on a chronic basis, the cardiovascular system will remodel, with induction of atherosclerosis at sites of high flow disturbance and cardiac hypertrophy. The responses of the cells are partly mediated through calcium signaling changes. Vascular smooth muscle cells, when stretched, open up non-specific cation channels, which induces calcium entry and regulation of vessel constriction. Transmission of mechanical strain to cells in arteries occurs through integrins on their cell membrane, and strain on integrin adhesions is known to influence their signaling pathways. In the intact vascular system, in the absence of any surgical or pharmacological intervention, changes to the morphology and elasticity of vessels that result in flow disturbances will cause clinically discernable intimal hyperplasia and/or atherosclerotic lesions; that is, de facto, changes to shear stresses in disturbed flowing blood cause pathological changes (Figure 3.5). It would appear very obvious that any intervention that involved the placing of a biomaterial in this system should produce changes to that pathology, and these changes have to be considered as the baseline for the biocompatibility of that biomaterial system. Thus, the placing of a stent within an artery, by definition, will alter flow conditions, and the new flow pattern will affect the host response. Whenever a blood vessel is replaced or bypassed by a graft, the same situation will apply. It is well known that the region around the anastomosis of graft to natural tissue is the most susceptible, and most prosthetic grafts of less than 6 mm diameter will ultimately fail because of

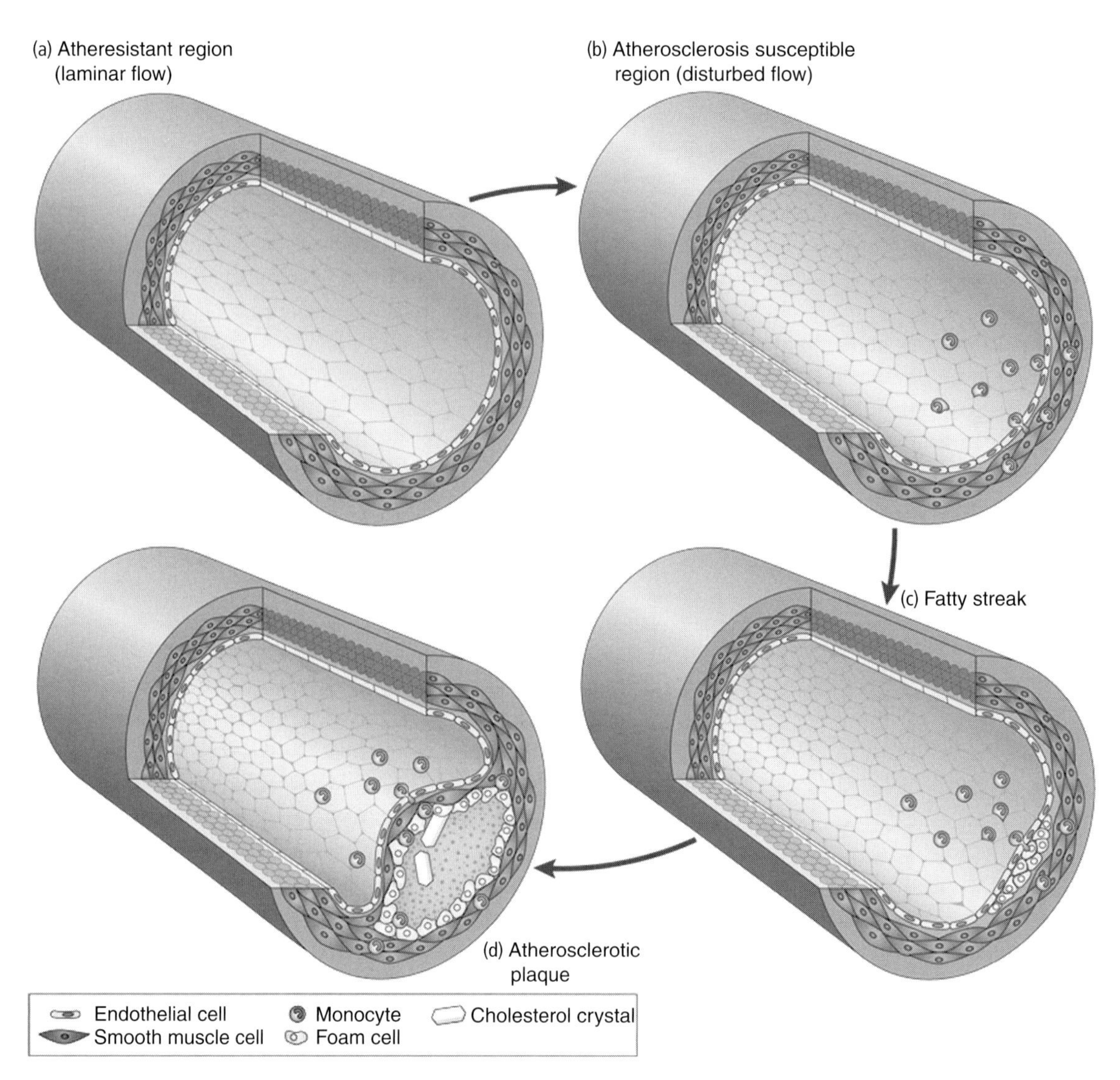


**Figure 3.5** Mechanotransduction in blood vessels. In a healthy artery (a) with high laminar shear stress, endothelial cells are aligned in the direction of blood flow. When flow is disturbed (b), the cells are poorly aligned and become activated, with expression of leukocyte adhesion receptors monocytes then binding to the endothelium and migrating into the vessel wall. Fatty streaks form in the regions of disturbed flow (c), where monocytes differentiate into macrophages which take up low density lipoproteins, varying cholesterol to the tissues. These streaks progress to form atherosclerotic plaques (d). The role of disturbed flow and the influence of mechanical factors seen here in natural atherogenesis are replicated in situations where the flow is disturbed by medical devices. Reprinted from *Nature Reviews Molecular Cell Biology*, 10(1), Hahn C and Schwartz M A, Mechanotransduction in vascular physiology and atherogenesis, 53–62 © 2009, with permission of Elsevier.

hyperplasia just downstream of the proximal anastomosis. Matching the compliance of the graft to that of the tissue should have some beneficial effect on minimizing the problem.

**Implantable devices: soft tissue** In soft tissue, we may expect to see similar phenomena, although not necessarily so dramatic. In most situations it will be the fibroblast that will be the cell most affected. Fibroblasts are certainly able to sense mechanical forces. As with previously mentioned mechanotransduction systems, two types of transmembrane structure are potentially involved, the integrins and ion channels, although most attention has been paid to the former. The transmission of ECM forces to the interior of fibroblasts via integrins involves the assembly and growth of focal adhesion complexes and a consequent increase in cytoskeletal traction and triggering of MAP kinase and NF-κB pathways. There are three possible ways by which genes for ECM components could be regulated in this way. First, a transduction pathway could activate an already available transcription factor that binds to a regulatory element in the gene. Secondly, a mechanical signal could induce the synthesis of a nuclear factor that transactivates a gene. Thirdly, the stress could directly induce the synthesis or secretion of a growth factor that regulates the expression of the ECM gene. Growth factors affected by mechanical stress may induce the tenascin-C gene that can play a significant role in ECM development. The significant endpoint should be the control of the expression of many ECM components by the stress, resulting in greater levels of fibrous tissue. Again it should be expected that the presence of a biomaterial in soft connective tissue should disturb the local biomechanical environment and, therefore, have an influence on the host response directly. If a hollow silicone tube is placed in the spinal musculature of a rat, the fibrous capsule around that tube should, in equilibrium, be of minimal thickness. If a rod of titanium is placed in that tube, and the ends sealed, the resulting fibrous capsule will normally be much thicker, this being regulated by the now considerable difference in rigidity between tube and muscle, and the greater mechanical stimulus experienced by the fibroblasts at the interface. This situation is also seen with breast implants: these are large devices, which, although designed to replicate the flexibility and compliance of breast tissue, are often unable to do so, and the large relative movement between implant and tissue often results in excessive fibrosis (Figure 3.6).

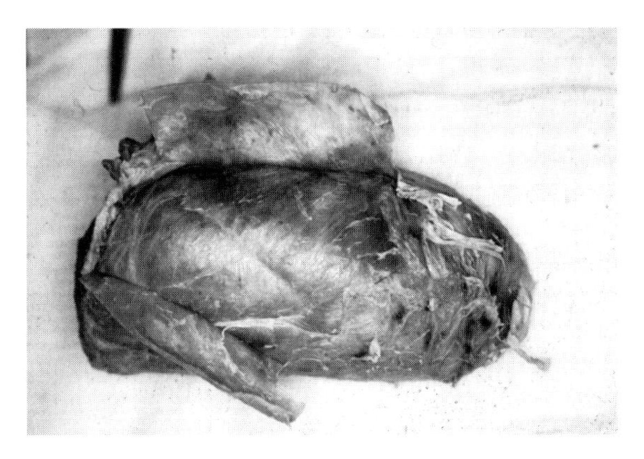

**Figure 3.6** Mechanotransduction in breast implants. These implants can be very large and, through normal everyday activity, move relative to the surrounding breast tissue, the mechanical stresses associated with this movement stimulating fibroblasts and myofibroblasts that can generate excessive fibrous tissue.

The fibrous tissue that forms in this latter situation is usually dense and in contracted form. This arises because in the tissue around the implant, which is effectively injured tissue, the normal fibroblasts, both resident and recruited, may acquire smooth muscle features, expressing smooth muscle actin and forming contractile stress fibers. These myofibroblasts contract and express large amounts of ECM, resulting in this dense contracted capsule.

**Implantable and extracorporeal devices: hemolysis** Erythrocytes, the red blood cells, are susceptible to damage when contacted by certain types of medical device. These are biconcave-shaped cells that lack a nucleus and most organelles. The cytoplasm is rich in hemoglobin, which picks up oxygen in the lungs and delivers it to tissues and organs during blood circulation. These cells are regularly exposed to stress, including oxidative stress in the lungs, osmotic stress in the kidneys and mechanical stress within capillaries. Red cells age and have a programmed lifetime of 100–120 days. During this natural ageing process, the cell membrane changes and it becomes more rigid and less able to deform when trying to squeeze through capillaries; their biconcave shape is optimal for such shape changes because it maximizes the cell surface to cell volume ratio, but the ageing of the membrane makes this increasingly difficult towards the end of their lives. Should cells rupture, the hemoglobin is released into the extracellular fluid, which will be filtered in the glomerula of the kidneys, precipitating in and damaging the tubules. In order to avoid this happening routinely, a mechanism has evolved that allows disposal of erythrocytes without this harmful release of hemoglobin. This process is a variation of apoptosis (discussed later in this chapter), although it is different from apoptosis in nucleated cells. Exposure to cellular stress activates a $Ca^{2+}$ permeable cation channel in the cell membrane. The subsequent entry of $Ca^{2+}$ into the cell leads to the exposure of phosphatidylserine at the cell membrane surface, which is recognized by macrophages, especially those in the spleen. These rapidly engulf the affected cells and ingest them before there is any hemoglobin release.

The premature ageing and destruction of erythrocytes is termed hemolysis. This is highly relevant to some biomaterials and medical devices since hemolysis can be induced by repetitive mechanical damage. This can occur with prosthetic heart valves and some extracorporeal circuits. Specifically a few types of mechanical prosthetic valves, some ventricular assist devices and some membrane oxygenators have been known to produce hemolysis in patients. This may be of clinical or sub-clinical consequence. In the latter case the effects may be seen in altered rheological properties of the blood, but in the former case there are increasing levels of plasma-free hemoglobin and other unscavenged cell constituents and decreased levels of haptoglobin, which can lead to neurologic and renal sequelae.

It should be emphasized that these are primarily device- not biomaterial-related components of the host response. It is often seen that hemolysis tests are carried out on biomaterials to assess their overall biocompatibility. These tests have nothing to do with the blood compatibility of materials but can be used as a surrogate for *in vitro* cytotoxicity since they have good endpoints for the assessment of the effects of material components on cell membranes.

**Stem cells and mechanical signals** It is inevitable that during embryonic and fetal development *in utero*, the shape and activity of pluripotent stem cells will be influenced by mechanical forces, both within the relevant matrix and at the cell–matrix interface. These forces combine with many other factors, especially including soluble factors, to control the differentiation of these cells. It should not be surprising, therefore, that when stem cells are used in tissue engineering or cell therapy procedures, the physical and mechanical nature of the cell's environment will also have an influence on these processes. Since the developments in these therapies are in their infancy, knowledge about the mechanical influences and their mechanisms is relatively unclear.

We do know that the elastic modulus of the matrix, whether the natural matrix or a synthetic gel matrix, plays a crucial role in determining stem cell fate. In the natural state, muscle cells require a compliant matrix so that they can deform it during contraction, while osteoblasts want a more rigid matrix, which they can mineralize. These requirements are reflected in the way that the matrix controls stem cell fate because a very soft matrix will favor neurogenic differentiation since it mimics brain tissue; a soft but stiffer matrix should be myogenic as it mimics the muscle, while a much stiffer matrix would be osteogenic. Generally an embryonic stem cell has high self-renewal capability during the first days of differentiation, but as the surrounding matrix gets stiffer, so the self-renewal genes may be downregulated and matrix expression genes are upregulated. For example, the transcription factor *oct*3/4, which is critical for embryonic stem cell self-renewal, is downregulated by cyclic stress, transmitted through focal adhesions, as the cells start to differentiate (Figure 3.7).

The exposure of mesenchymal stem cells (MSCs) to shape-changing stresses can result in the

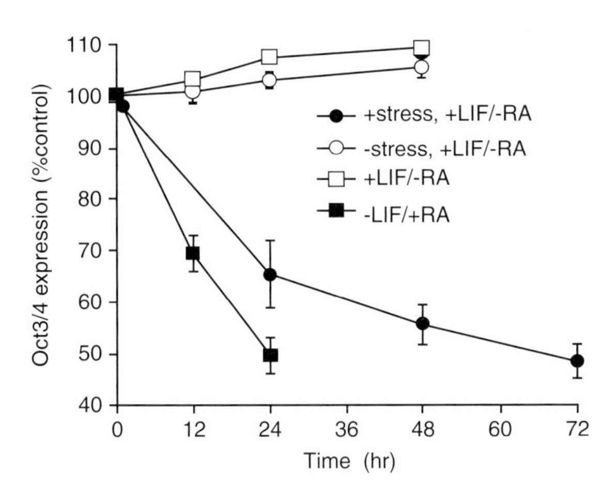


**Figure 3.7** The effect of stress on gene expression in stem cells. Loss of the expression of the genes oct3/4 in embryonic stem cells is a hallmark for differentiation. After continuous application of a 17.5 Pa local cyclic stress at 0.3 Hz for only 60 min, oct3/4 expression in these stem cells was downregulated by around 35% within 24 h, and by 50% within 72 h, whereas control cells a few micrometres away in the same dish without stress continued to express oct3/4. The culture was performed in the presence of LIF, leukemia inhibitory factor. Reprinted by permission from Macmillan Publishers Ltd: *Nature Materials*, Chowdhury F *et al.*, Material properties of the cell dictate stress-induced spreading and differentiation in embryonic stem cells, 9, 82–8, © 2009.

upregulation of genes associated with the critical phase, known as condensation, when the lineage commitment is determined. Mechanical forces are transduced to the nucleus. The duration of exposure appears to be a critical factor in determining the balance of power between mechanical effects and the biochemical environment on stem cell differentiation fate. Mechanical signals control differentiation of MSCs down osteogenic and chondrogenic pathways through the regulation of transcription factors, particularly c-Fos, the Fos family of genes being particularly important in osteogenesis.

Hemopoietic stem and progenitor cells in culture are profoundly affected by the stiffness of their culture substrate. Highly elastic (i.e., low modulus) materials lead to a significant increase in cell population, apparently through activity of membrane receptors such as integrins and of the actin–myosin cytoskeleton.

Shear stress plays an important role in the fate of blood vessel cells, as we have seen above, and it would not be unreasonable to assume that they would influence endothelial cells in culture or bioreactors. Endothelial cells respond to shear stress by activating mechanosensitive ion channels, inducing cytoskeletal re-arrangements of F-actin filaments and altering shape to align with the direction of strain. Embryonic stem cells can be influenced to differentiate into endothelial cells under the influence of shear stress through the VEGF signaling pathway.

In all of these situations it is clear that the totality of the mechanical environment of stem cells plays a role in the determination of their fate. This is independent of the chemical nature of that environment. Of course there are many biochemical species that are also involved, and indeed may have a more significant role, but we cannot consider the biocompatibility of engineered support systems for stem cell manipulation without this mechanical basis. The same is true in general terms for the manipulation of any cells within tissue engineering.

**The essence of mechanical biocompatibility mechanisms** Taking the paradigm of the cause and effect connection through biocompatibility pathways discussed in the introduction to Section 3.3 and Figure 3.3, we may summarize these mechanical phenomena by a series of interconnected pathways, shown in Figure 3.8. These start with the cause, which is the application (*in vitro*) or the perturbation (*in vivo*) of mechanical forces that are transmitted to the system that involves both biomaterial and host. They end with the effect, which is a specific cellular or tissue level event(s), which has physiological and, perhaps, clinical consequences. There are several pathways by which force transduction and signal propagation connect these two different ends. Note that the resulting effects may be tissue or cell destructive (bone loss, premature red cell ageing), tissue hyperplasia (fibrosis, intimal hyperplasia) or direction changing (stem cell differentiation).

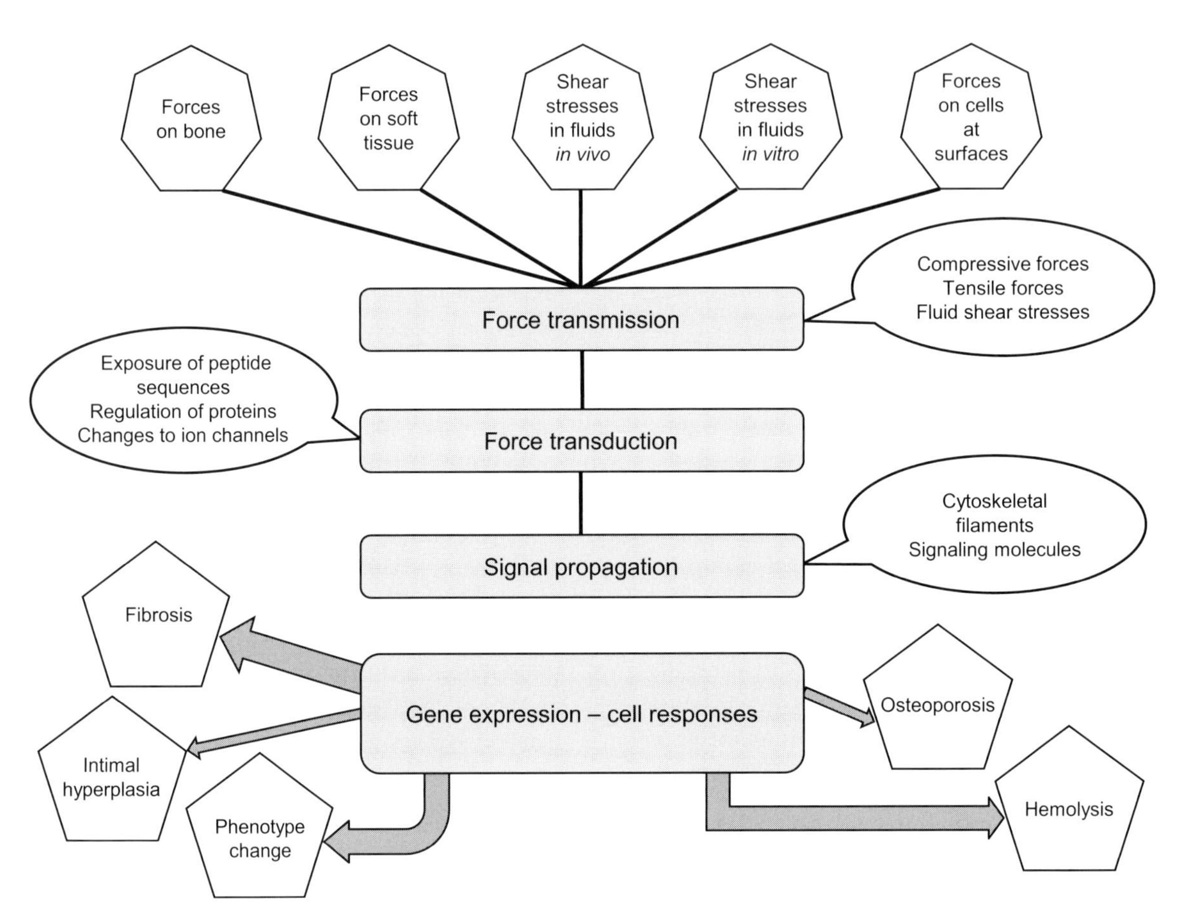


**Figure 3.8** The essence of mechanical biocompatibility mechanisms. This scheme summarizes the pathways involved in the processes of mechanotransduction that may be established following contact between biomaterials and tissues or cells.

## Absorption, adsorption and adhesion processes: events at surfaces

It will be obvious that the surface of a biomaterial plays an enormously significant role in biocompatibility processes. It is worth considering at this stage some general characteristics of the surface and the interfacial region that will control the nature of this role. The interfacial region, whether we are considering an implanted device or a tissue-engineering scaffold, usually represents only a small part of the biomaterial–host complex and it is likely that the vast majority of tissue components never come into contact with the material surface. The significance, however, is that once some of these components do come into contact with the surface, they may undergo changes that activate sequences of events that have far-reaching consequences. Included here are conformational changes in proteins once they are adsorbed on the surface and the activation of cells once they adhere to surfaces.

It is difficult to clearly separate out the various material properties that control, or even influence, these events. A great deal of experimental evidence has been accumulated during attempts to correlate material surface variables with molecular and cellular responses, but few clear, unequivocal relationships have emerged. The reasons for this can be found on both sides of the interface. The more important material surface variables are surface chemical composition, surface topography and biophysical parameters such as surface energy, surface charge, degree of hydration and molecular mobility. It is, however, difficult to vary just one of these parameters without introducing changes in one or more of the others. Surface chemistry may be altered by using a range of polymers that vary within one feature, for example, the length or nature of side groups, but this will usually have consequential effects on surface energy or molecular mobility. On the tissue side of the interface it is very difficult to investigate either macromolecular or cellular events alone under realistic conditions since they operate collectively. It is difficult to study cell behavior without the presence of proteins since cells need proteins for survival.

Notwithstanding the above comments, this section attempts to identify those surface parameters, apart from chemistry and topography (both discussed later), that are significant factors. This will be done first with a discussion of macromolecular adsorption on biomaterials surfaces, both in general terms and with respect to the one major consequential effect of adsorption, which is the contact phase activation of blood coagulation. This will be followed by a discussion of the effect of these biophysical parameters on cell adhesion, again including the important process of platelet activity at a material surface as a major exemplar.

**Macromolecular adsorption** Immediately after contact with tissues, proteins (and possibly other macromolecules) from blood and interstitial fluids adsorb onto biomaterials surfaces. This adsorption is a complex process since there are so many types of molecule competing for the surface, each with its own affinity for the surface, its own concentration in the fluid, its own kinetic parameters of diffusion and so on. The consequences of adsorption of different species also vary, as does their ability to remain on the surface and the tendency to undergo conformational change.

This is a subject of considerable potential importance but also controversy. There is a fundamental problem here since we know that proteins adsorb on surfaces and we know that adsorbed proteins have the potential to mediate subsequent events at or near the surface, but the evidence that they do have such effects and the mechanisms by which they may do so are far from clear. In view of this, we shall only summarize the essential facts here and provide the rationale for the putative consequences of protein adsorption for biocompatibility.

There are several reasons for this difficulty in understanding protein adsorption, and these are worth mentioning. At the heart of the problem is the inherent uncertainty over what constitutes adsorption and the adsorbed layer. How strongly does a protein have to bind to the surface to be considered adsorbed? Do we consider an adsorbed layer to be a monolayer or can it be multilayered and if the latter is the case, which parts of the multilayered structure are the most relevant to subsequent events. Related to these points are the effects of chronology and dynamics at the interface; the composition and structure of the layer are likely to change over time, especially in the short term, as proteins may spontaneously desorb or be actively displaced by other proteins or simply be covered by them. From a practical point of view, serious questions arise over the methods that are used to interrogate the processes of adsorption and desorption since many of the methods may actually interfere with adsorption process, for example those methods that involve radiolabels. The inability to resolve temporal and compositional variations in adsorption profiles with complex biological fluids such as blood and plasma has led most investigations to reply on observations with single solutions. This is problematic since the adsorption behavior in complex fluids will be controlled by the competition between proteins on the basis of diffusion and affinity parameters.

The adsorption process is governed by energy considerations and by the presence of water, which competes with protein molecules for the surface through hydrogen bonding. There is reasonable evidence to conclude that hydrophobic surfaces generally adsorb more protein than hydrophilic surfaces. In order for proteins to adsorb, the surface has to dehydrate to some extent; this is thermodynamically favorable for hydrophobic surfaces but not for hydrophilic surfaces. In a fluid such as blood, the dominant proteins such as albumin and fibrinogen adsorb in greater amounts than those proteins that are present in much lower concentrations. The composition of this adsorbed layer will then change as the faster diffusing molecules are displaced by proteins that have a higher affinity for the surface, such as vitronectin and fibronectin. This phenomenon is often referred to as the Vroman Effect (Figure 3.9). Proteins are usually less tightly bound to hydrophilic surfaces so that this displacement takes place more readily on these surfaces. In single protein solutions the amount of protein adsorbed tends to increase with increasing concentration until it reaches a peak, after which increases in concentration usually cause a decrease in adsorption.

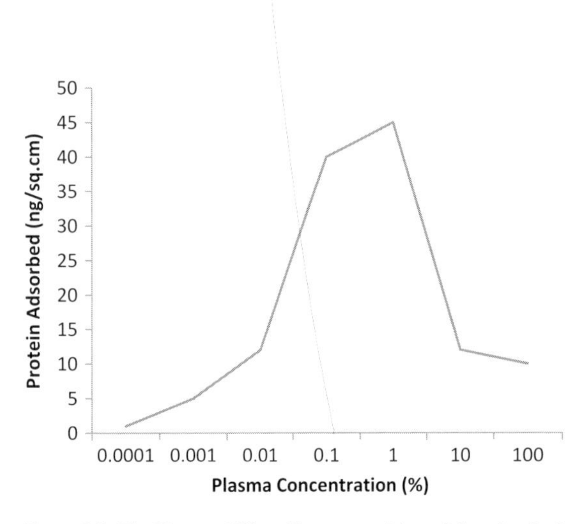


**Figure 3.9** The Vroman Effect. The composition of the adsorbed layer of proteins on a biomaterial changes as faster diffusing molecules are displaced by those that have a higher affinity for the surface. This is the Vroman Effect. In single protein solutions the amount adsorbed tends to increase with increasing concentration until it reaches a peak, after which increases in concentration usually cause a decrease in adsorption.

One potentially important aspect of the adsorption process is the scope for conformational change in the adsorbed layer. There is likely to be a tendency to lose secondary structure within the molecules, facilitating increased contact between the protein and the surface, with obvious implications for the strength of the attachment and likelihood of displacement. Proteins vary considerably in their ability to retain their natural structure when adsorbed and, should they undergo change after adsorption, to regain natural structure should they be displaced. It is quite possible for the protein layer to show spatial heterogeneity with respect to conformation.

**Cell adhesion and activity mediated by adsorbed species** Based upon the observations discussed above it may be assumed that whenever cells approach a macroscale biomaterial surface, they will not directly interface with that surface but with a film of proteins and therefore the cell behavior should be determined by that film. If we consider implantable devices, and in the absence of mechanical effects already discussed and chemical reactions and topographical effects that will be discussed shortly, the adsorbed film that develops over time will vary very little between different surfaces, so that we should not expect much difference in cell behavior. This is consistent with clinical experiences. The surface wettability discussed above may play some slight role, as cell activity in general may be greater on protein layers on hydrophilic substrates.

The situation with tissue engineering scaffolds *in vitro* may be somewhat different, although the situation is far from clear. Cells are usually cultured in medium containing proteins, in the form of fetal bovine serum for example, so similar generic principles may apply. Under culture conditions it is possible to have greater control of the environment and many approaches have been used to modify surfaces so that they are more compatible with cells. This may include functionalizing the surface with proteins or amino acid motifs, as will be discussed in Chapter 5. It is also possible to treat the surfaces with one of the techniques discussed in the previous chapter in order to modify wettability, specifically to increase hydrophilicity. However, while some general correlations may be made, there are far too many variables and contradictory performances for specific rules to be written. For example, some surface modifications may increase cell adhesion but have the opposite effect on cell spreading or other cell properties. The evidence so far would suggest that both material surface characteristics and the adsorbed protein characteristics together control cell behavior; the adsorbed proteins need to interact with surface characteristics such that the proteins on the outermost layer have to be in the right conformation to optimally promote cell function. At the same time, once a cell adheres to a protein-covered surface it may be able to reorganize that protein layer locally and secrete its own distinctive protein environment.

**Contact phase activation of blood coagulation** We now come to discuss blood and interactions with material surfaces and medical devices. As noted earlier, blood compatibility is usually considered separately from tissue biocompatibility. This is not necessarily helpful since it is better to regard biocompatibility as one set of related phenomena, where interactions with blood are a specific subset. Similar types of reaction occur and indeed thrombosis, which is a major endpoint of interactions with blood, has been considered to be a special case of inflammation. Blood compatibility is considered in this section, since although there may be some involvement of chemical activity, the processes are dominated by biophysical/thermodynamic events, where the precise chemical constitution of the material is largely irrelevant. The discussion considers contact phase activation first, since this is controlled by proteins, followed by interactions with platelets, which are the specific cells that control events. These two processes are brought together in the context of the formation of blood clots.

There are four major components of blood that we have to consider. These are the platelets, the leukocytes, the proteins of coagulation system and complement. Platelets are disc-shaped cells of 3–4 microns diameter (Figure 3.10). They have no nucleus. In the resting, non-activated, state, they have a dense system of tubules and a circumferential array of microtubules that maintain the shape of the cell. There is also a substantial amount of muscle proteins such as myosin and actin that facilitate the contraction that is required when they are activated. The external surface of the cell membrane contains many membrane-bound receptors, including the glycoproteins known as Ib and IIb/IIIa, as well as phospholipids and an open canalicular network structure that can expand and selectively adsorb molecules. The cells contain three types of storage granules in their cytoplasm. The $\alpha$ granules contain

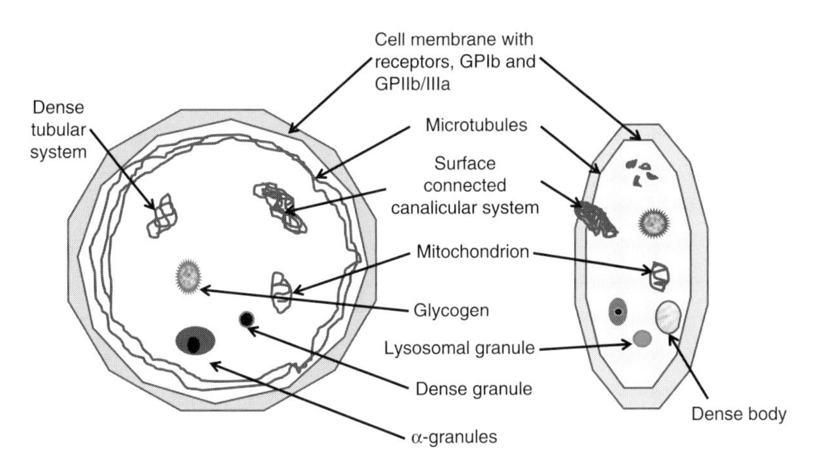


**Figure 3.10** The general structure of platelets. This is a schematic representation (plan and side views) of a platelet, showing the cell membrane, the microtubules, the dense tubular system and the surface connected canalicular system, and the various intracellular structures, including granules and dense bodies.

the proteins platelet factor 4 (PF4) and β thromboglobulin (βTG) as well as some plasma proteins. Secondly, there are dense granules that contain adenosine diphosphate (ADP), serotonin and $Ca^{2+}$ ions. Thirdly, there are lysosomal granules containing acid hydrolase enzymes.

There are several different types of circulating leukocytes but with respect to blood compatibility, it is the monocytes and neutrophils that are important. Neutrophils are in fact the most abundant white blood cells, comprising between 40 and 60% of total white cell count, the monocytes being less numerous at around 5%. These cells have short half-lives when circulating in blood, with maxima at 20 and 100 hours respectively. They both have receptors for complement products (discussed below) and other molecules. Leukocytes are significantly involved in inflammatory responses in the circulation system.

**The coagulation cascade** A zymogen is an inactive enzyme precursor. A number of proteins in the plasma are zymogens and they constitute the system of plasma coagulation factors (listed in Table 3.1). The human body has a defense mechanism for the arrest of bleeding after trauma and this powerfully depends on this system. Put simply, this mechanism becomes operative as soon as tissue sustains damage that results in escape of blood from the vasculature, and results in the formation of a blood clot that mechanically prevents further bleeding. This coagulation process functions through a sequence of self-amplifying conversions of zymogens into enzymes, which results in the formation of thrombin, a serine protease that is able to convert fibrinogen into fibrin. This forms as a mesh and becomes the basis of the clot.

This series of reactions is referred to as the coagulation cascade, which has two separate branches (see Figure 3.11). One is the extrinsic pathway and the other is the intrinsic pathway. Although these pathways are initiated separately, they merge into a common pathway that results in the formation of thrombin. These pathways are complex and, although the principles are well established, some details are still controversial; only a simplified form is presented here. The extrinsic pathway is the dominant natural process that starts with the expression of tissue factor (TF) on cells that have been damaged at the site of vascular injury. Factor VII, known as plasma factor, binds to TF. Once complexed in this way, FVII is activated by a number of circulating factors, generating FVIIa. TF-FVIIa cleaves FX into FXa and FXI into FXIa in the presence of calcium. Factor Xa plays a critical role here since it cleaves FII, which is prothrombin, into FIIa, which is thrombin. As noted above, thrombin converts fibrinogen into fibrin.

Table 3.1 **Plasma coagulation factors and other relevant molecules**

| | |
|---|---|
| Factor I | Fibrinogen |
| Factor II | Prothrombin |
| Factor III | Tissue factor *or* thromboplastin |
| Factor IV | Calcium |
| Factor V | Proaccelerin *or* labile factor |
| Factor VII | Proconvertin *or* stable factor |
| Factor VIII | Antihemophilic factor A |
| Factor IX | Antihemophilic factor B *or* Christmas factor |
| Factor X | Stuart–Prower factor |
| Factor XI | Plasma thromboplastin antecedent |
| Factor XII | Hageman factor |
| Factor XIII | Fibrin-stabilizing factor<br>Von Willebrand factor<br>Prekallikrein<br>High molecular weight kininogen<br>Tissue plasminogen activator |

The conventional view of the intrinsic pathway is that it is initiated by the contact of blood with an unrecognized surface. Factor XII, known as the Hageman factor, is activated on adsorption to a surface. FXIIa then converts prekallikrein to kallikrein, which together with high molecular

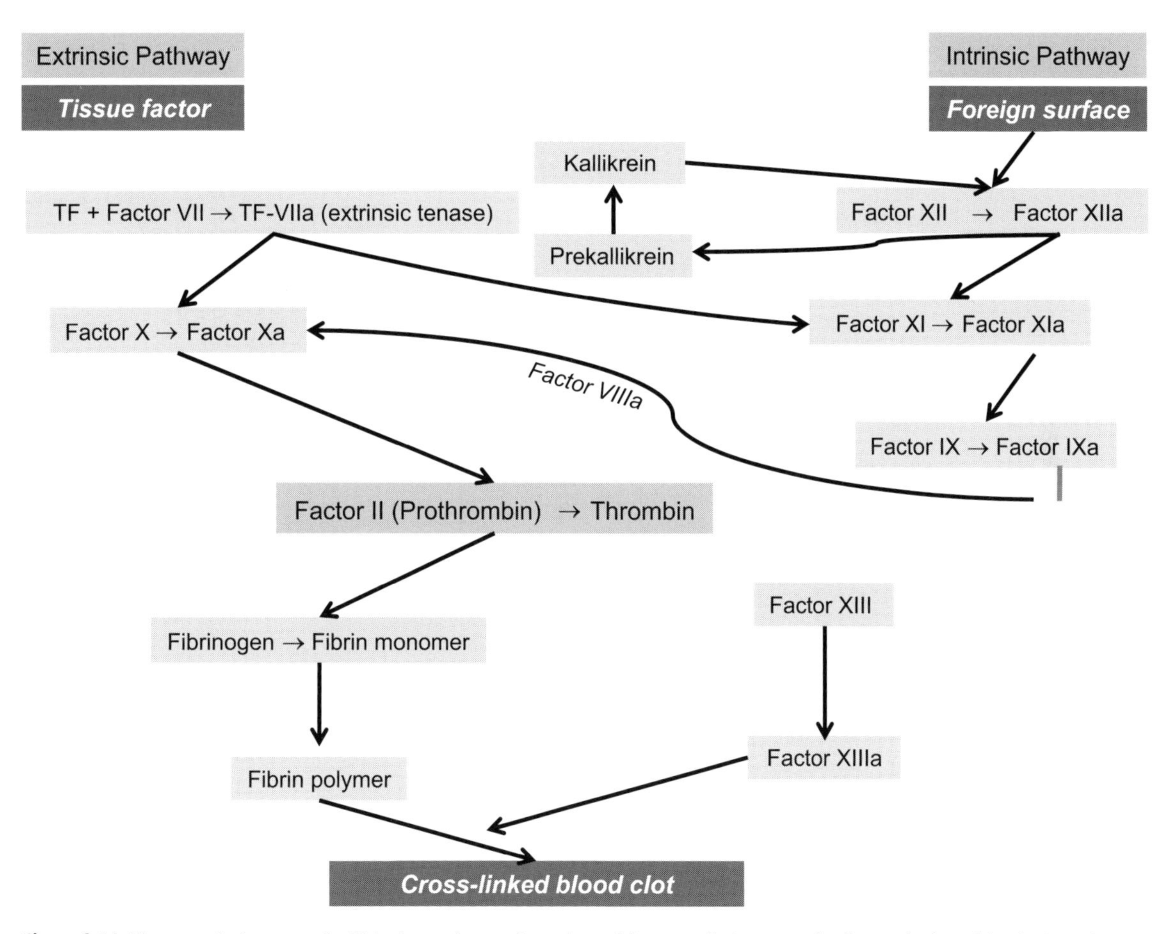


**Figure 3.11** The coagulation cascade. This shows the two branches of the coagulation cascade, the extrinsic and intrinsic pathways. Although these pathways are initiated separately, they merge into a common pathway that results in the formation of thrombin. These pathways are complex and still controversial; this is a simplified form.

weight kininogen as a co-factor, converts FXI to FXIa, which activates FIX to FIXa, leading into the common pathway. The surfaces that appear to be strongly activating of FXII tend to be anionic or hydrophilic, but this does not imply that these characteristics are essential. In fact, it is quite likely that both intrinsic and extrinsic pathways could be involved with foreign surfaces. It is possible that blood contact with a material surface could itself induce TF expression on white cells.

**Complement** The complement system is relevant both to blood coagulation and inflammation and needs to be discussed in some detail. The term "complement" is used to describe a series of components in blood plasma. The system is composed of more than 30 plasma and membrane-bound proteins which function as either enzymes or binding proteins; it essentially acts as a system to discriminate between self and non-self molecules. Some complement proteins, including C3 and C4, serve to identify and opsonize pathogens. C3a and C5a recruit and activate inflammatory cells. Some, such as C5b-9, assist in the lysis of pathogens while others, including C3, C3a and C5a augment cellular immune response, which we shall discuss later.

Just as with the coagulation system, the complement system functions as a pathway or

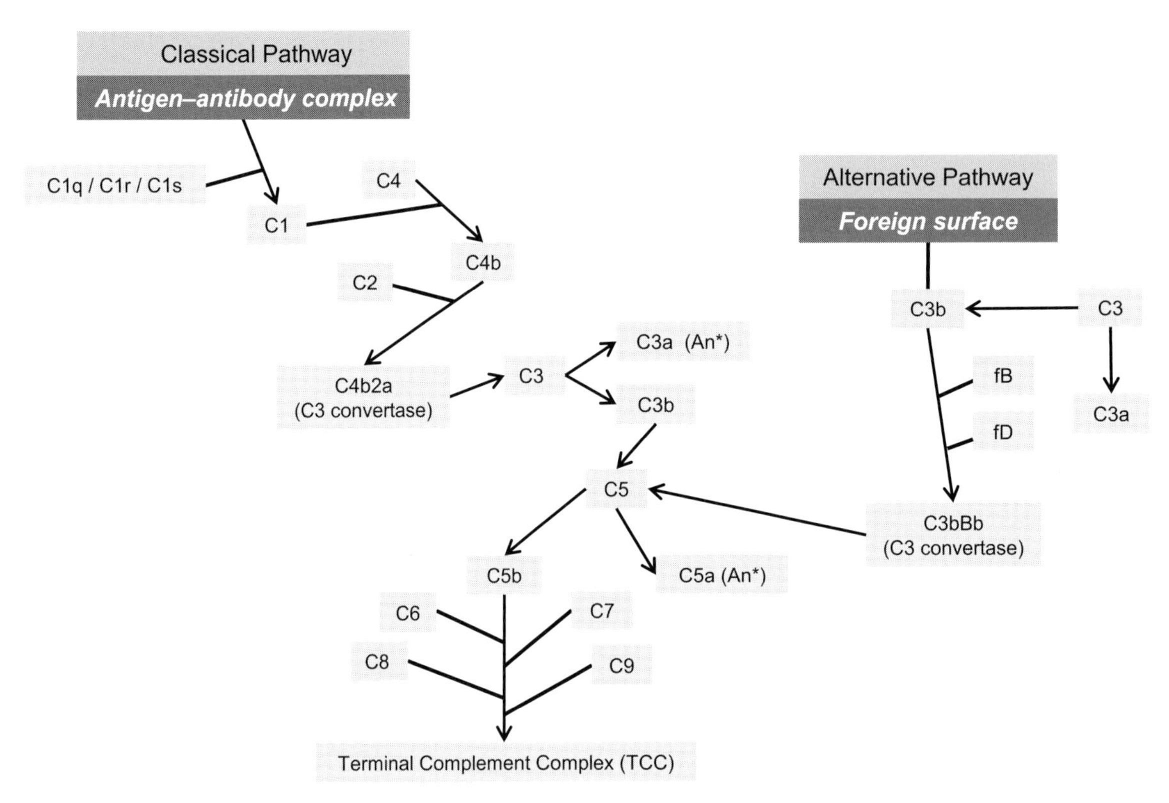


**Figure 3.12** The complement pathways. There are two complement pathways, the classical (initiated by an antigen–antibody complex) and alternative (initiated by a foreign surface) pathways, both of which finish with the formation of the terminal complement complex.

cascade of activation. One complement protein is activated, following which a series of enzyme-mediated cleavage reactions takes place, resulting in the release of fragments that have their own specific function. The significance to biocompatibility is that some biomaterials are able to activate complement and the process is involved in a number of biomaterials-related phenomena. There is, in fact, more than one pathway, two important ones being relevant here.

These two pathways are the classical and alternative pathways, both of which finish with the formation of the terminal complement complex (TCC). The classical pathway operates within the immune response since it is activated by antigen–antibody complexes. The proteins of this pathway are C1, C2, C4, C1inhibitor and C4 binding protein. As shown in Figure 3.12, this is initiated when C1 binds to the antigen–antibody complex. C1 has three subunits C1q, C1r and C1s. The binding takes place with one end of C1q, which activates C1s, which in turn cleaves C4 into C4a and C4b. C2 then binds to C4b, which is cleaved by C1s, releasing C2a. This produces the C3 convertase C4bC2a, which is able to cleave C3 into C3a and C3b. The former is an anaphylatoxin, while C4bC2b and C3b combine to give the C5 convertase, which cleaves C5 into C5a, also an anaphylatoxin, and C5b. The C5b is part of the TCC.

The alternative pathway does not need the antibody–antigen complex for activation, which can be initiated by microorganisms directly or by foreign surfaces. The proteins involved here are C3, B, D, H, I and P. Initiation is a spontaneous process where parts of C3 in the plasma undergo conformational change and hydrolysis of the internal thioester group, producing an activated form of C3, that binds to Factor B, which itself cleaves another C3 molecule into C3a and C3b. At this point, a foreign surface may become involved since C3b binds covalently to hydroxyl or amine groups on that surface. This allows binding of Factors B and D to C3b. Factor B is cleaved into Ba and Bb and so the C3 convertase of the alternative pathway, C3bBb, is formed, providing an amplification loop. The continued production of C3b yields the convertase C3bBbC3b, which cleaves C5.

In both cases the result is the formation of C5b, which has an affinity for C6 in the TCC. The C5bC6 binds C7, C8 and C9, which forms the TCC C5b-9, the membrane attack complex, which inserts itself into lipid layers in microorganism membranes.

The role of complement in immune responses is discussed later. Here there are two significant points. First there is cross-talk between complement and coagulation cascades. Complement effectors directly influence coagulation and may interfere with anticoagulation factors, while coagulation enzymes are able to activate complement components. Thrombin activates several complement proteins while many complement components, such as C3bBb, cause proteolysis of coagulation cascade molecules. Secondly, it is clear that under some circumstances, biomaterials are able to activate complement themselves. It is known that this can occur when polymers with hydroxyl and amino groups are used in large surface area devices, especially in extracorporeal circulation. From the above discussion it would appear that this occurs by the alternative pathway, but the evidence is not totally supportive of this position and the classical pathway may be involved in some situations. The role of conformational changes in adsorbed proteins on the capacity to bind complement fragments has to be considered.

**Platelet behavior at surfaces** Platelets are very sensitive cells and we should not be surprised that they play a role in blood compatibility processes. We need to consider here platelet adhesion to surfaces, platelet activation and their aggregation. When a blood vessel is injured, platelets adhere to tissue surfaces, especially collagen, by the interaction with platelet glycoprotein Ib (GPIb), a reaction that requires the presence of von Willebrand factor (vWF). The platelets are then activated following the stimulation of the platelet surface. Plasma proteins of the coagulation system, collagen and some complement fragments may be potent platelet activators. The activation of platelets leads to a significant shape change and restructuring of the cell membrane, but also the release of a number of components. (Figure 3.13). The granules mentioned earlier will release their soluble factors, including PF4 and βTG, into their microenvironment and also so-called platelet microparticles, which are rich in coagulation factors. The protein P-selectin is also expressed on the platelet membrane, which plays a major role in mediating cell–cell interactions. Significant platelet adhesion receptors are GPIb and GPIIb/IIIa, the latter being a major integrin receptor. Activation of the platelet leads to conformational changes that expose these binding sites, with a high affinity for fibrinogen. The result is that activation leads to significant recruitment of more platelets and the aggregation of platelets through their cell–cell interactions that enhances clot formation and structure.

With respect to biomaterials, there is no doubt that platelets play a major role in blood compatibility but it is far from clear how mechanisms of platelet behavior control how different types of surfaces fare in relation to platelet adhesion, aggregation, consumption and overall thrombogenicity.

**Thrombosis and embolism** Clearly, one major consequence of the interplay between protein adsorption on biomaterials surfaces and the subsequent attraction of cells to the adsorbed layer, and their own activation, is the formation of a blood clot. The activation of the clotting cascade and the activation of platelets yields a fibrin–platelet matrix, which then traps red blood cells, and establishes the clot, which is designed to arrest bleeding. This is a very effective defensive measure against major blood

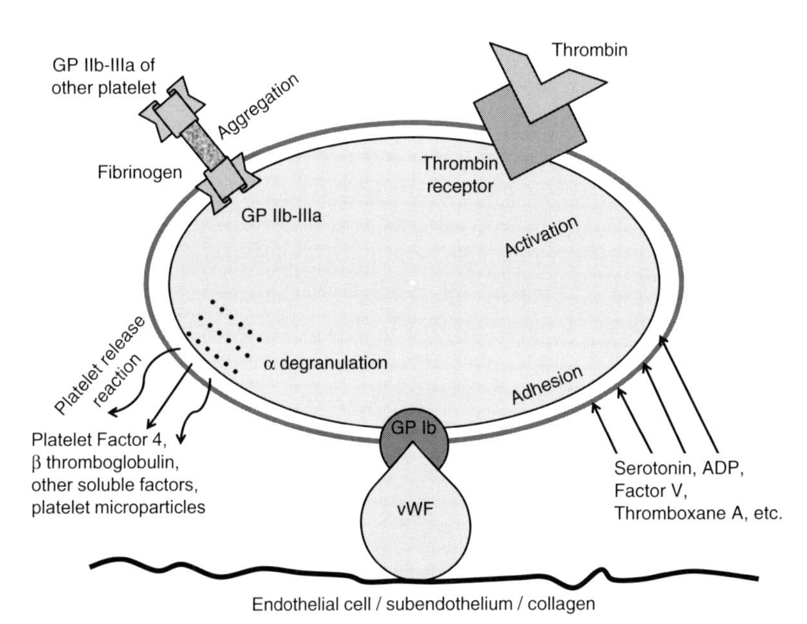


**Figure 3.13** Platelet activation. The activation of platelets leads to a significant restructuring of the cell membrane and the release of a number of molecules. The granules release their soluble factors, including PF4 and βTG, into their microenvironment and also platelet microparticles, which are rich in coagulation factors.

loss following injury. However, when this occurs subsequent to placing a biomaterial within the circulatory system, the resulting arrest of blood flow within that system can have a devastating effect both on the performance of the device and the health of the patient. The clot is known as a thrombus and the process is often termed thrombosis and the phenomenon as thrombogenicity.

A few further important points should be made here. First, there is a natural process by which a blood clot dissolves. This is called thrombolysis and it results in the restoration of blood flow once healing has occurred. One method that is available to minimize the effect of a blood clot that forms on a medical device involves the administration of drugs that have thrombolytic properties.

Secondly, one potentially very harmful additional effect of a device-related thrombus is the breaking off from the main thrombus of a fragment of the clot that can circulate in the bloodstream, eventually lodging in a small vessel, for example within the brain, causing a blockage at that site. This fragment is called an embolus. This is a cause of minor strokes in patients that have medical devices within the central circulatory system.

Thirdly, we come to the question of if and how biomaterials surfaces can be rendered non-thrombogenic. The unfortunate answer is that this is very difficult to do. In the vast majority of cases, the most significant factor that controls whether a clot forms is the hemodynamic state. If the device radically alters the blood flow mechanics, causing either areas of stasis or reduced blood flow, or areas of high turbulence and high shear stresses, there is a far higher chance of initiating a thrombus than with relatively unperturbed flow regimes. Modifying surface chemistry generally has little effect although microtopography can have some influence if it changes wall shear stresses on a localized microscale. Some surfaces have been engineered to resist protein adsorption and these may have somewhat better resistance to thrombosis; these include phospholipid surfaces that mimic cell membranes. Surfaces that have been modified in order to resist platelet activation or to interfere with phases of the coagulation cascade may appear to be effective under experimental conditions but this rarely transforms into good clinical benefit.

The one procedure that we do have available to minimize the risk of thrombosis following the use of medical devices is that of systemic anticoagulation. These regimes include the anti-thrombosis drugs such as warfarin and anti-platelet drugs such as plavix, and aspirin, which are generally known to reduce blood clotting tendencies. These drugs are used with mechanical heart valves and intravascular stents. There is one potential problem with such drugs, which relates to the fact that artificially reducing the tendency for clot formation around a device through the systemic administration of a drug will simultaneously increase the risk of excessive bleeding elsewhere in the body; a very careful watch on this balance will be necessary.

**Fluid absorption** The final process to be discussed briefly in this section is fluid absorption. Some biomaterials may absorb water and associated ions or macromolecules during use. This is not a biocompatibility issue per se but it may influence some processes that influence biocompatibility mechanisms and/or material performance. The following processes should be considered:

- The performance of hydrogels is obviously dependent on water absorption; this is discussed in several places in this book.
- The intentional degradation of polymers is also controlled to a large extent by the absorption of water and other molecules.
- The stability of some putatively biostable polymers may be affected by absorption of some physiological molecules. Silicone elastomers, although very stable in aqueous solutions, are affected by lipids; in the past this has led to significant problems with mechanical heart valves that utilized silicones in the ball component since this suffered swelling and cracking. This was also a problem with some breast implants that utilized a lipid-based gel within a silicone elastomer shell.
- Some elastomeric materials used for indwelling catheters may absorb and concentrate systemically distributed drugs. This can be a problem with certain polyurethane suprapubic catheters since that part of the catheter that is in continuous contact with urine inside the bladder may take up drugs or their metabolites over time.

**The essence of surface biocompatibility mechanisms** In this subset of Scenario A phenomena, the initiation processes are those of macromolecular adsorption onto the biomaterial surface. These are controlled by the compositional characteristics of the biological fluid environment and the prevailing hydrodynamic (including hemodynamic) parameters in that environment. The surface chemistry of the material is of minor importance although hydrophilicity and microtopography may have some effect. The main consequences of these adsorption processes are two-fold. First, the process may allow activation of one or other of the defensive cascade systems, of clotting and complement, which although very effective in dealing with normal injuries, can here result in the initiation of thrombosis. In this situation, surface characteristics such as charge may play a role in the initiation of the cascade since surface groups may interact with key proteins in the cascade. Secondly, the nature of the adsorbed layer, and its variation over time, may control the behavior of cells that come into contact with the surface. These can include platelets, the activation of which significantly affects the blood clotting process, or they could be cells in culture, where the nature of the protein layer can influence the development and maintenance of cell phenotype. Again here we have to consider the interactions between molecular structures of the adsorbed layer and receptors on the cell surface and the subsequent signal propagation that leads to events such as platelet degranulation or stem cell differentiation. These events are summarized in Figure 3.14.

### 3.3.3 Scenario B: Mechanisms of biocompatibility involving chemical reactions between macroscale biomaterials and their soluble derivatives with the host

This scenario takes into account the response of the tissues to the presence of a foreign

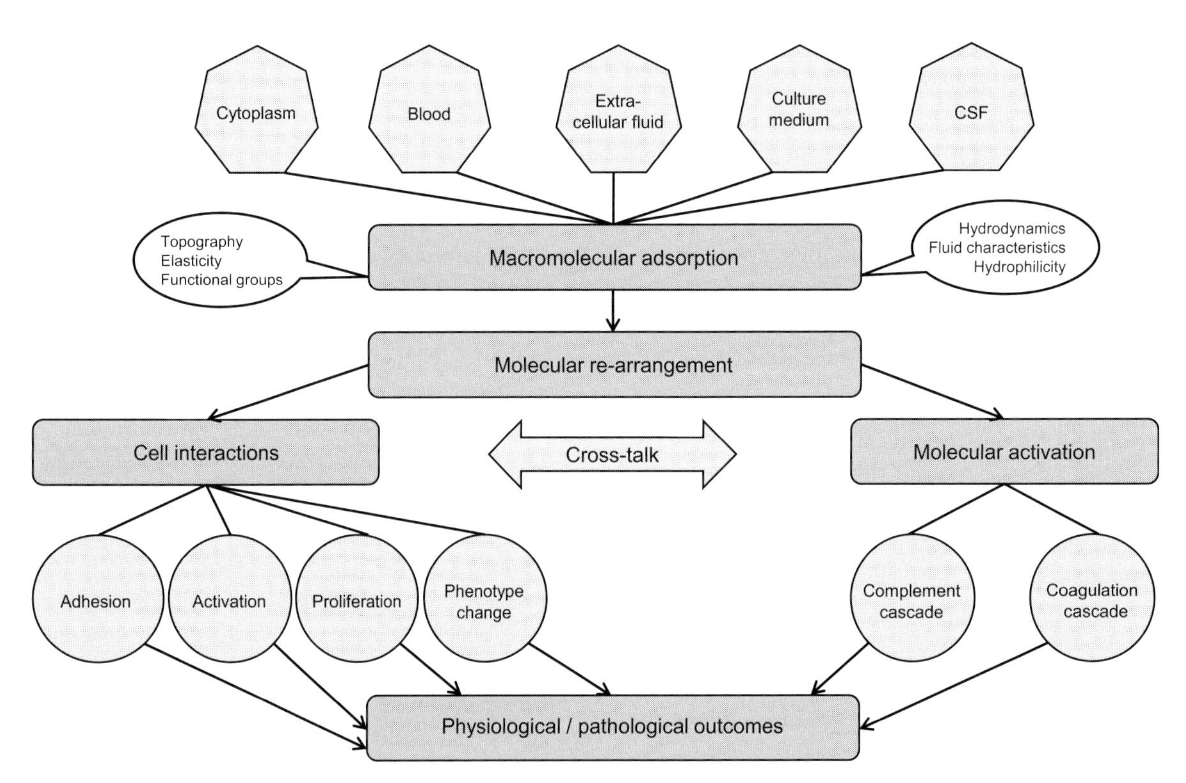


**Figure 3.14** The essence of surface biocompatibility mechanisms. This schematic summarizes the generic pathways for biocompatibility phenomena that are initiated by events at the interface between biomaterials and tissues and controlled by subsequent molecular and cellular interactions. Note that molecular re-arrangement on the material surface is likely to influence cellular events such as adhesion and phenotype change, that molecular activation may involve powerful amplification events and that there is cross-talk between many events. All processes will have physiological outcomes and possibly detectable pathological outcomes.

object that has the potential for influencing the components of the host via direct chemical effects. We do not take into account any mechanical or physical factors here, nor do we consider responses that are mediated by microscopic or nanoscopic topographical features, nor particulate release per se. The chemical effects may be initiated at the material surface, but may also arise from the release of soluble chemical entities into the tissues from the material. These could be anions or cations, monomers, soluble oligomers and polymers, additives or contaminants and similar species. We shall deal first with the potential influence of these species on cell signaling and then go on to discuss specific phenomena.

## Biomaterial surface–cell interactions: adhesion and cellular signaling

The majority of host response characteristics involve interactions between one or more biomaterial components and one or more cell types. In their normal environment, cells interact with other cells and with the extracellular matrix. In this situation, cells are able to communicate with each other, to adhere to each other, and also to interact with the matrix. It should not be surprising that biomaterials components may take part in, or interfere with, these cellular interactions. These processes may influence how materials are able to bond to some tissues, for example in the so-called osseointegration between prostheses and bone. They should be able to determine how a material is able to control cell phenotype, for example in orchestrating the differentiation of stem cells in tissue engineering processes. They may influence how an implanted material causes hyperplastic responses within the vascular endothelium. In short, the cell–biomaterial interactions that are built upon normal cell–cell and cell–ECM interactions may be pivotal in many biocompatibility processes.

**Receptors** Cells communicate by molecular mechanisms that are collectively termed cellular signaling processes. Crucial to this process is the presence of receptors on cell membranes. These are molecular structures that selectively receive and bind to specific substances. Most receptors are transmembrane receptors, which have three parts or domains (Figure 3.15). The extracellular domain is located on the external surface of the cell; its main function is to recognize, respond to, and often bind to a specific ligand. The transmembrane domains pass through the cell membrane, with varying degrees of tortuosity and often have the structure of an alpha helix. Upon activation of the extracellular domain, pores within the domain become accessible to ions, constituting an ion channel, or undergo some form of conformational change. The intracellular domain interacts with the interior of the cell, relaying the signal via protein–protein interactions or via enzyme activity.

The discussion of the last paragraph implied that these receptors vary in their structure and in their mechanism of operation. Generally three classes of receptor can be identified, as shown in Table 3.2. The first are the channel-linked receptors, sometimes called ligand-gated ion channels, which have the receptor and transducing functions within the same protein molecule. Specifically, interaction between the chemical signal and the binding site on the receptor opens or closes ion channel pores and the resulting change in ion flux causes changes in the membrane potential of the cell. The second class is the enzyme-linked receptor where the intracellular domain is an enzyme that is regulated by the signal. These receptors are usually protein kinases, which are enzymes that modify other proteins by chemically adding phosphate groups to them (in the process of phosphorylation), resulting in functional change in the target protein. One major type here are the tyrosine kinases, which transfer the phosphate group from ADP to the protein, effectively acting as an on-off switch. The third group involves the G-protein coupled receptors. These possess seven membrane-spanning transmembrane helices. The signal activates a G-protein, a guanine nucleotide-binding protein, which sets off a cascade of events directed towards the cell interior.

**Cell signaling pathways** The processes that take part in these cascades that follow recognition by

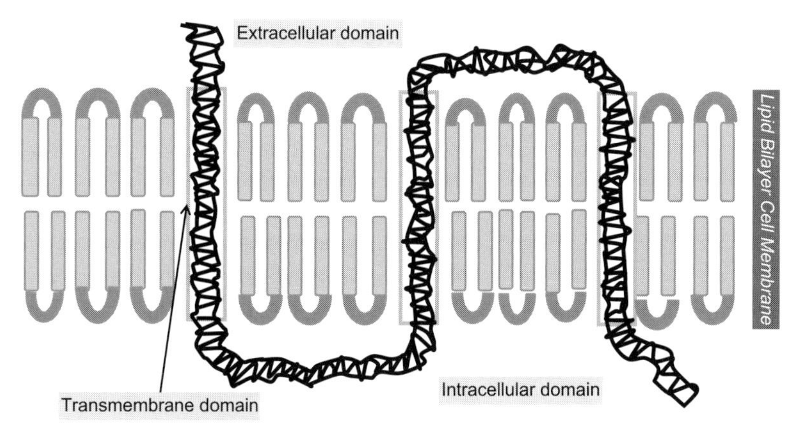


**Figure 3.15** Transmembrane receptors. Crucial to the processes of cell signaling and communication is the presence of receptors on cell membranes. Most are transmembrane receptors. The extracellular domain of the receptors is located on the external surface and recognizes and often binds to specific ligands. The transmembrane domains pass through the cell membrane. The intracellular domain relays the signal to the interior of the cell.

Table 3.2 **Types of receptor**

| | |
|---|---|
| Channel-linked receptors | Also called ligand-gated ion channels. These have both receptor and transducing functions. Interaction between the binding site on the receptor and the chemical signal allows ion channel pores within the molecule to open or close, modulating ion flux and the membrane potential. Neurotransmitter receptors operate in this way |
| Enzyme-linked receptors | These receptors are usually protein kinases, including tyrosine kinases (Box 3.5). Their intracellular domain is an enzyme, the catalytic activity of which is controlled by the binding of a signal to the extracellular domain of the receptor. The enzyme activity is usually phosphorylation of the intracellular target proteins |
| G-protein coupled receptors | These receptors operate via an intermediate transducing molecule, known as GTP-binding proteins (G-proteins). The receptors have an N-terminus extracellular domain and a C-terminus intracellular domain, the molecules traversing the cell membrane seven times. The activating signals can be chemical or physical and their action is determined by mechanical effects on the G-protein |

receptors constitute the cell signaling pathways. Whatever molecules or structures are involved, these pathways have three components, signal initiation, signal transduction and gene activation, again consistent with Figure 3.16. The sequences are collectively described as cell signaling pathways, of which there are very many, some of which are operative in a number of situations but some of which are much more specific.

Signal initiation is the interaction of a ligand with its receptor. In general terms we can distinguish three categories of molecule that can serve as ligands, those that are freely diffusible, the ECM proteins and membrane bound proteins. Within the first group are proteins, such as growth factors, and hormones, which are produced in originating cells and are distributed with the intention of locating and interacting with their target cells, the distance between originating and target cells being quite variable. Long-range endocrine action may involve large intracellular distances while some are short range and the target cell has to be very close to the originator. In addition, some common physiological species act as ligands to trigger highly specific responses; these include calcium and oxygen. It is easy to see here how the presence of xenobiotic compounds derived from biomaterials, including other metal ions, monomers, oligomers and intrinsically biologically active molecules, can take part, either positively or negatively, in cell signaling processes.

**Integrins** As far as the ECM is concerned, many of the ECM proteins and proteoglycans act as ligands for specific receptors and are crucial for the adhesion between cells and the ECM. This adhesion is pivotal to many physiological processes, especially embryonic development, the organization and repair of many types of tissue, and wound healing. Without cell–ECM adhesion there is no tissue structure. Many cellular processes, including muscle cell contraction and cell spreading and migration, simply cannot occur. Critical to this process are the family of glycosylated transmembrane receptors called integrins (Figure 3.17(a)). These are $\alpha\beta$ heterodimers; a heterodimer is a protein made of paired polypeptides that differ in amino acid sequences. There are several different $\alpha$ and $\beta$ sub-units; the different integrins are defined by the specific units they consist of, for example $\alpha_1\beta_1$, $\alpha_6\beta_1$ and $\alpha_M\beta_2$. Many of the significant integrins, with their respective ligands and binding sites, are shown in Table 3.3. These binding sites are usually specific short peptide/amino acid sequences of the ECM proteins, such as vitronectin, fibronectin, various collagens and laminin, as well as plasma proteins such as thrombospondin and von Willebrand factor.

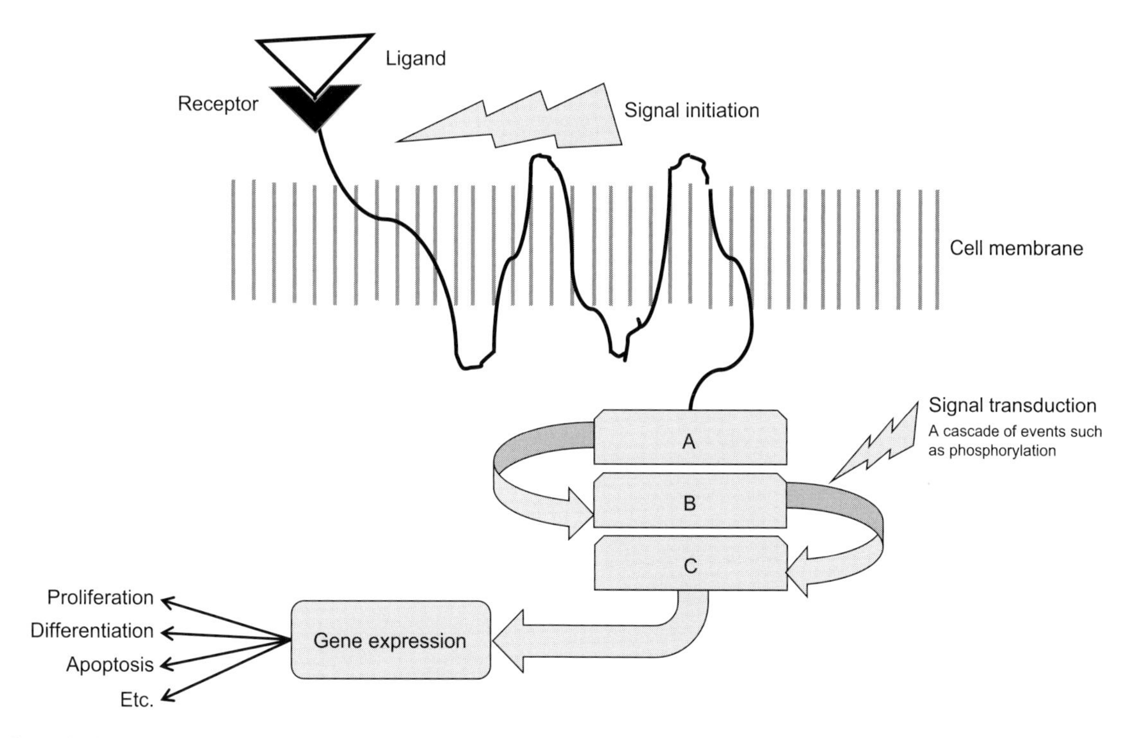


**Figure 3.16** Schematic of cell signaling pathways. Signal initiation follows interaction between a ligand and receptor. Signal transduction usually involves a cascade of reactions, leading to gene activation, controlled by the rate of transcription of a gene, and the resulting fate of the cell.

These sequences include the arginine–glycine–aspartic acid motif, referred to as RGD.

The integrin receptor has a large extracellular domain formed by both α and β subunits, with a single transmembrane pass and two short non-enzymatic intracellular domains. The extracellular domains also contain divalent metal ion binding sites. In the signal transduction phase, upon ligand binding, the integrins associate with the actin cytoskeleton of the cell and form clusters and focal adhesions that provide the structural links between the cytoskeleton and the ECM. These clusters consist of structural proteins such as vinculin and some signaling molecules. The transduction tales place through various signaling pathways, shown in Figure 3.17(b). Included here are the MAPK (mitogen-activated protein kinase) and FAK (focal adhesion kinase) pathways. The end stage of the process is gene activation, largely controlled by the rate of transcription of the gene. Transcription factors, the sequence-specific DNA binding proteins, may be activated by the signal and bind to the promoter region of the target gene, which in turn activates the RNA polymerase complex and directs the cell function.

**The influence of biomaterials on cell signaling** Here again we can readily see the potential to influence these signaling pathways, including the role of integrins, when biomaterial components come into contact with cells. We shall consider the potential role for surface topography to influence cell adhesion processes, especially with bone, in later scenarios; here we discuss the implications for surface chemistry. It is recognized that a clinically used biomaterial will rarely come into contact with cells without the presence of some physiological fluid, usually blood or extracellular fluid. In most *in vitro/ex vivo* tissue engineering processes, some protein-containing medium will also be present. It is not expected,

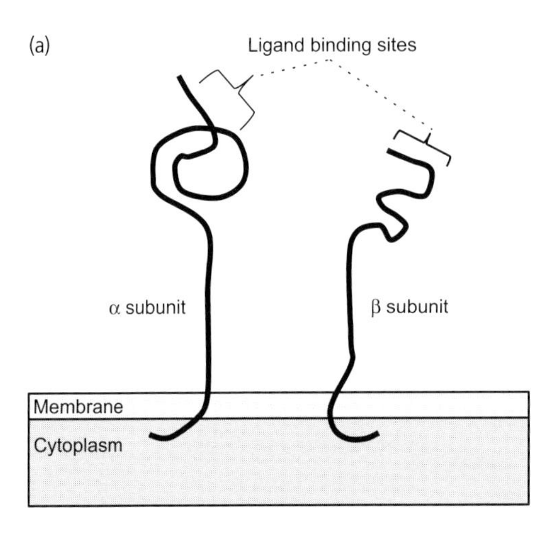


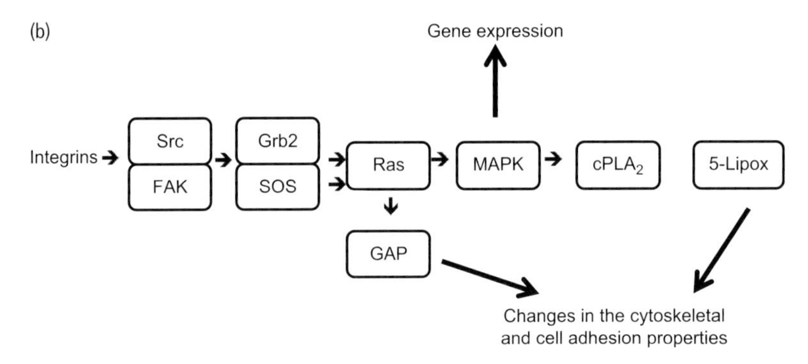


**Figure 3.17** Integrins.
**(a)** Integrins are families of transmembrane heterodimers, that integrate the function of cells with the extracellular matrix and molecules. They are referred to as αβ heterodimers, each with a pair of polypeptide chains known as the α subunit and the β subunit respectively. These chains terminate with ligand binding sites or domains, the nature of which vary with the specific integrin. The mammalian genome contains 18 α subunit and 8 β subunit genes. The α domains determine the ligand specificity of the integrin, whilst the β domain affects multiple signaling pathways.
**(b)** On the binding of a ligand to a binding site on the integrin, structural links are formed between the ECM and the cytoskeleton through clusters of signaling molecules. Signal transduction and propagation takes place via the pathways established by these clusters. This scheme is a simplified version of a pathway that involves FAK (focal adhesion kinase), which is a tyrosine kinase, and MAPK (mitogen-activated protein kinase). The pathway involves Src, which is also a tyrosine kinase, Grb2 and SOS (Son of Sevenless, an activator of Ras, which is itself a small molecular weight guanine nucleotide-binding protein). Activation of MAPK leads to the phospholipid cPLA2 and 5-Lipoxygenase. The outcomes of this pathway are changes to the cytoskeleton and, in consequence, the cell adhesion properties (mediated through the activation of Ras via GAP and 5-Lipox, and also to gene expression following activation of MAPK). Note that several other molecules such as phosphorus and arachidonic acid are involved in various places. This is a simplified scheme of one simple pathway; there are many more pathways.

therefore, that cells will contact a biomaterial directly, but rather will contact it indirectly via a layer of proteins or other biological macromolecules. The effect on adhesion processes and cell signaling will therefore be mediated by the adsorbed protein, which is likely to be conformationally altered during the process, and possibly by biomaterial-derived components that have diffused to the surface through

Table 3.3 **Significant integrins with their ligands**

| | |
|---|---|
| $\alpha_1\beta_1$ | collagen, laminin |
| $\alpha_2\beta_1$ | collagen, laminin |
| $\alpha_2\beta_3$ | fibrinogen |
| $\alpha_4\beta_1$ | fibronectin |
| $\alpha_5\beta_1$ | fibronectin |
| $\alpha_6\beta_1$ | laminin |
| $\alpha_6\beta_4$ | laminin |
| $\alpha_7\beta_1$ | laminin |
| $\alpha_L\beta_2$ | I-CAM 1 |
| $\alpha_V\beta_3$ | vitronectin |

this protein layer. It is also possible for any cells that contact the material surface to secrete their own ligand-bearing macromolecules onto the surface.

This is not a simple process and many factors have to be taken into account when assessing how a biomaterial surface, through surface chemistry, can influence cell signaling, integrin expression and cell adhesion. The complexity is manifest by the intensely dynamic nature of the interactions and the fact that there are few logical and reproducible sequences of events that lead a cell to recognize a specific material chemistry and cause it to perform specific functions. The adhesion mechanisms will change from material to material and from cell to cell, and the principal adhesion mechanism may change over time. Individual dominant adhesive ligands may be displaced and replaced. Moreover, many cells, once having made contact with a protein-covered material surface, may rapidly degrade and/or reorganize the layer. As noted in Table 3.3, most cells have several integrins specific for the same ligand and the binding to receptors may rapidly change with different integrin conformation.

A further difficulty arises in the determination of mechanisms of cell adhesion to biomaterial surfaces with the notable differences between *in vitro* and *in vivo* behavior; this is particularly unfortunate since it is difficult to follow cell adhesion phenomena *in vivo*, but that is where most (but not all) interest lies. We can consider, for example, osteoblast adhesion since this is an important step in the process of the attachment of bone to orthopedic prostheses. *In vitro*, osteoblasts depend on vitronectin and fibronectin for initial adhesion to surfaces such as titanium and hydroxyapatite. This implies that materials that are able to preferentially adsorb these proteins from their immediate environment are more likely to be successful in initial bone bonding. The osteoblasts express a wide variety of integrins, especially with subunits $\alpha1$, $\alpha2$, $\alpha3$, $\alpha4$, $\alpha5$, $\beta1$, $\beta3$ and $\beta5$, and this expression varies with the age of the cell. There is evidence that osteoblasts express different integrin subunits on different metal or ceramic surfaces, and that the strength of attachment varies with the chemical nature of the substrate.

### Inflammation and the immune response

We deal here with the role of biomaterials and biomaterials-derived species on inflammation and the immune response. There has always been confusion about these two related components of the host response, especially as the terms themselves "inflammatory reaction" and "immune response" conjure images of harmful events as if they were to be avoided at all costs. These are normal features of the human defense system and, in many situations, not only are they not harmful but they are actively beneficial to the body's health and indeed self-preservation. The key to understanding their role in biocompatibility is to understand the broad mechanisms that are involved and determine those parts that are beneficial to the development of the appropriate response and those that should be avoided if at all possible. We also repeat here that this discussion does not impress considerations of inflammation over wound healing phenomena. Wound healing is discussed later but here we consider inflammation/immunity outside of healing in order to get our baseline.

The immune system has two components, the innate, non-specific immune system and the adaptive, specific immune system. Each has both cellular and humoral components through which they carry out their dedicated functions. A couple of key features suggest which system should be more relevant to biomaterials. The innate system has features that are constitutively present and ready to be activated or mobilized immediately on exposure to the invading agent, usually a microorganism. This system is not antigen-specific and can respond equally well to a variety of stimuli. The adaptive system, however, is antigen-specific: it takes some time to respond to a new stimulus but has immunological memory such that the response is far more rapid upon re-exposure.

These facts make it very clear that in the majority of situations, biomaterials will be associated with the innate response. Generally the adaptive response will only be involved when they contain antigens that will be recognized by the immune cells, which essentially implies that this will be confined to materials based on foreign proteins, including xenogeneic or allogeneic systems.

**Innate immunity** It is preferable to consider innate immunity as a system of defense that has primarily evolved in order to destroy microorganisms but is equally able to respond to tissue injury. Inflammation is an integral part of the innate immune response. It is a coordinated sequence of events that is aimed at resolving infection and repairing damage, effectively returning the body to a state of homeostasis. The emphasis is on destroying bacteria in the case of infection and repairing damage in the case of tissue injury. The key to an effective response is the correct balance between these events since the powerful inflammatory response can do more damage than good.

The principal components of innate immunity that are relevant to our discussion are phagocytic cells, natural killer cells, circulating proteins such as complement and coagulation factors, and the cytokines that regulate the overall cellular response. There are other components, such as physical barriers, but these do not concern us here. It is logical to start with the process of cellular recognition of signals that inform of the presence of microbes or tissue damage. It should be emphasized here that the tissue damage can take the form of many different types of stress stimuli, including heat, hydrostatic pressure, oxidative stress, various forms of radiation and exposure to exogenous chemicals. With infection, the signals are provided by a number of structures in the microbial cell wall, such as bacterial peptides, oligosaccharides and lipopolysaccharides, or viral/bacterial RNA or DNA. With non-microbial sources of damage, the signals may be derived from denatured connective tissue elements produced during trauma or fragments of molecules generated during coagulation. With foreign materials the most likely source is the denatured proteins that will be adsorbed to the material or conjugated to material-derived components.

In contrast to the situation with adaptive immunity, which we shall discuss later, the specificity of innate receptors is rather limited; this evolutionary trait has ensured that many different classes of potential pathogens can be recognized easily and quickly. The recognition process targets a number of highly conserved features, or patterns, of the microbial surface. This effect has been termed pattern recognition and the relevant receptors as pattern-recognition receptors (PRRs). Whether or not these function in the same way for signals of tissue damage is not really clear, but here are several important features that are likely to be relevant.

**Inflammasomes** The PRRs are expressed by cells of the innate immune system, including macrophages monocytes, dendritic cells, neutrophils and epithelial cells. They primarily include the membrane-bound Toll-like receptors (TLRs) and C-type lectins (CTLs), which probe the extracellular space and endosomal compartments for pathogen-associated molecular patterns (Figure 3.18(a)). There are also intracellular DNA sensing PRRs which cover the cytosol. Signal transduction from these receptors target a group of signaling molecules, including NF-κB, which promote the production of pro-inflammatory

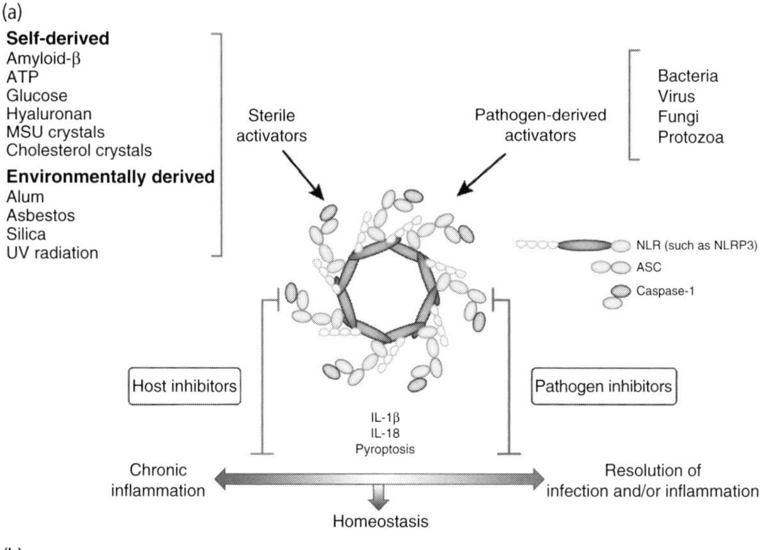


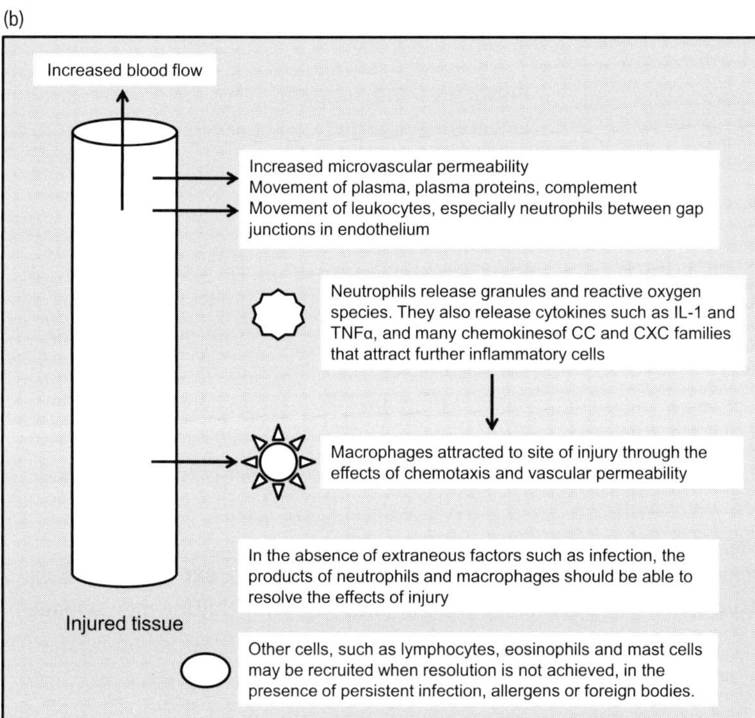


**Figure 3.18**
**(a)** Inflammasomes and mediators of inflammation. Inflammasome activity regulates homeostatic processes and inflammation during infection and tissue injury, being activated by a wide array of danger-associated molecular patterns. The initial event leads to activation of caspase-1 and the release of IL-1β and IL-18, which results in the recruitment of effector cell populations of the immune response and tissue repair. Under normal circumstances, activation of the inflammasomes culminates in the resolution of infection or inflammation and contributes to homeostatic processes. However, perpetuation of inflammasome activation can lead to chronic inflammatory diseases. Reprinted by permission from Macmillan Publishers Ltd: *Nature Immunology*, Henao-Mejia *et al.*, Inflammasomes: far beyond inflammation, 13(4), 321–4, © 2012. **(b)** This scheme demonstrates the factors and agents involved in inflammation. Blood flow is increased in the small vessels, and their walls become more permeable. White cells, along with plasma components rapidly pass through the endothelium. Neutrophils dominate the response, releasing cytokines, including chemokines which attract more cells, and agents such as granules and reactive oxygen species which attempt to clear the area of the products of injury. Within a few days, numerous macrophages will dominate, resolving this clearing process and preparing the way for tissue repair.

cytokines and chemokines. A different group of PRRs are the NOD-like receptors that recognize both microbial and non-microbial molecular patterns. This latter group constitutes a platform of caspase-1 activating molecules that are collectively known as inflammasomes. The net result of these processes of molecule recognition, signal transduction and gene expression is, therefore the activation of resident macrophages and other cells and the release of cytokines and other mediators of inflammation.

**Inflammatory mediators** Figure 3.18(b) indicates the scheme of cellular involvement in this process. The release of these pro-inflammatory mediators induces a number of changes in the local tissue, especially in the endothelial cells of the microvasculature, giving it the characteristics of inflamed tissue. Through increased permeability of the vessel walls and the effects of molecules such as TNF$\alpha$ and IL-1 there will be an influx of leukocytes and plasma. In particular, neutrophils arrive very early, followed by monocytes. The neutrophils release protease-laden granules and ROS, both aimed at destroying bacteria should they be the cause of the injury, and then phagocytose any debris. Interestingly, if the neutrophils do not directly encounter microbial matter, they will still release their granules into the extracellular space. Thus, should the injury be non-microbial, as with, for example, biomaterial-derived particles, qualitatively the same type of response will be seen, and both the ECM and resident cells can be destroyed. These granules contain powerful enzymes such as broad-spectrum serine proteinases (e.g., elastase and cathepsins). In evolutionary terms this process is intended to create a hostile environment should bacteria subsequently invade, but it clearly has other consequences. After releasing their granules the neutrophils die through apoptosis and are subsequently cleared by macrophages, which enter the inflamed site a little later.

Even if there are no bacteria to initiate the entry of the very first neutrophils, this type of sequence can still occur in response to tissue injury, with every indication that TLRs associated with resident cells are very much involved. It is important to note that once initial recognition of tissue stress has taken place, a variety of endogenous molecules released from the first cells to die by necrosis play a role in the prolongation, acceleration or resolution of the inflammatory response. Principal amongst these are the so-called heat shock proteins, which are able to protect against further tissue damage.

If the innate immune response to bacteria has been effective, it is then oriented towards resolution and the generation of anti-inflammatory signals. The mechanisms for this are not clear but a number of lipid-derived mediators, such as lipoxins, appear to play a role.

**Adaptive immunity** Since innate immunity is not effective against all microbial invasions, an alternative and more powerful strategy is necessary. Processes of evolution derived this strategy by producing a type of molecule that could bind to the invading microorganisms *and* activate humoral and cellular components *and* do so with an acquired memory such that the defensive process was both specific and timely. This process is adaptive or acquired immunity.

**Antibodies and antigens** This adaptor molecule is called an antibody (see Figure 3.19). Since there are so many possible microbial structures that could invade the tissues, there have to be recognition segments of antibodies that could bind to each of them, and there have to be enough carriers of these antibodies to deal with hundreds of thousands of identical but rapidly replicating microbial cells. Antibodies are made and carried by lymphocytes. Resting lymphocytes are small cells with little cytoplasm and few mitochondria. There are two subsets of lymphocytes, the B-lymphocytes and T-lymphocytes. The part of the microorganism that evokes a response from antibodies is known as an antigen. B-lymphocytes are programmed to make a single type of antibody, which is located on the outer surface and acts as a receptor. Microbial bound antigens are confronted with a massive array of B-lymphocytes when they enter the body, all of these cells bearing different antibodies. The antigen will only bind to those receptors to which it is matched. Once a lymphocyte has an antigen bound to its surface receptor antibody, it receives a signal that causes it to develop into an antibody-forming plasma cell, which then secretes identical antibodies. The lymphocytes undergo successive waves of

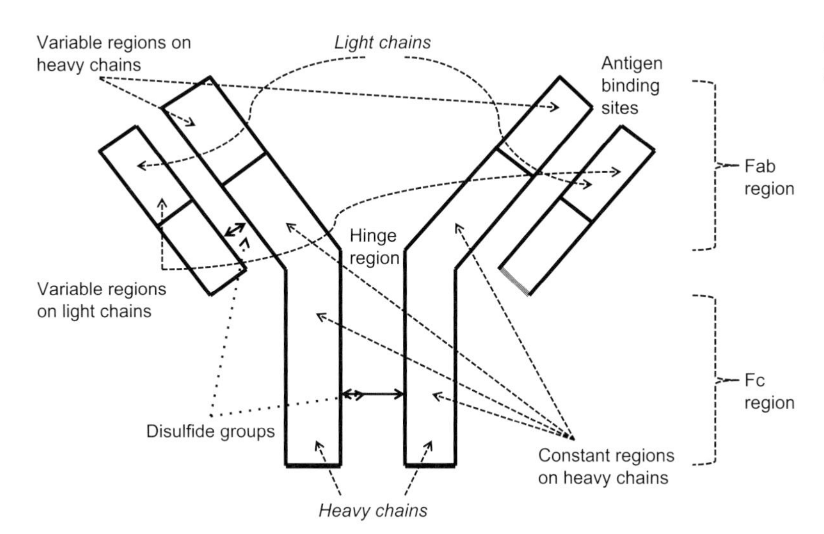


**Figure 3.19** The general structure of antibodies.

proliferation, which builds up a large clone of antibody-producing plasma cells that collectively combat the infection.

**Complement, antigen-presenting cells and the major histocompatibility complex** There are other mechanisms that contribute to the effectiveness of this process of dealing with infection. The first is antigen-presentation. Dendritic cells play a key role here. Immature dendritic cells originate in the bone marrow and migrate through tissues where they tend to remain dormant. Their function is to find and capture antigens. Once captured, the antigen is processed by one of several different pathways and presented to lymphocytes. The second concerns the activation of complement, which we have already introduced. An antibody that is bound to the microbial antigen links to a molecule in the classical complement sequence (C1q), which triggers proteolytic activity of the C1 complex and activates the various cascades, resulting in the formation of a variety of complement fragments. Some of these, such as C5a, are potent chemotactic agents for polymorphonuclear cells. Others, such as C3a, act on mast cells causing the release of other mediators. C3b and C5b link covalently to the microbial surface, facilitating phagocytic cell recognition.

Many microorganisms reside inside host cells, out of the reach of humoral antibodies. These include viruses, which have to replicate inside cells. In this case a different form of acquired immunity process, based on T-lymphocytes, operates. In this, T cell surface receptors recognize an antigen that is coupled to a molecule on the cell surface, this binding leading to phagocytosis; these molecules form part of the major histocompatibility complex (MHC). These are cell surface molecules encoded by a large gene family; they mediate interactions between leukocytes and other cells. In addition, a subset of the T-lymphocytes called T-helper cells can recognize an antigen bound to MHC molecules and release powerful cytokines, including $\gamma$ interferon, which assist in macrophage activity.

**The role of biomaterials in adaptive immunity** It has generally been assumed that synthetic materials such as polymers, ceramics and metal systems should not be involved in adaptive immunity since they do not constitute antigens as described above. However, lymphocytes, plasma cells and dendritic cells may be seen in the tissue surrounding implants and *in vitro* evidence suggests that there are some interactions between biomaterial components and both humoral and cellular components of acquired immunity.

The dendritic cells mentioned in the previous section, when quiescent and inactive are described as immature, but when exposed to a stimulus terminally differentiate into mature cells, this process being described as maturation. Some biomaterials initiate maturation and some inhibit it, and indeed those cells that appear to be activated by a biomaterial surface develop an immunogenic phenotype, characterized by increased expression of MHC molecules and other markers. The state of maturation may be determined by integrin-mediated adhesion. It would seem that several Toll-like receptor signaling pathways are involved, with increased expression of pro-inflammatory cytokines and activation of antigen specific T cells. It is also known that T cells can adhere to biomaterials surfaces but the significance of this is not clear.

In addition to the direct effects of material surfaces, it is necessary to consider the role of adsorbed organic molecules. Protein and carbohydrate moieties adsorbed onto a material surface may serve as activating ligands and interact with dendritic cell receptors, again Toll-like receptors being significantly involved.

**Animal tissue-derived biomaterials** The situation is potentially different when the biomaterial is derived from the tissues of animals. This occurs when the ECM from various species is used as the source of materials, including collagen, pericardium and small intestine submucosa. Although the subject of vital organ transplantation from human donor to human recipient is beyond the scope of this book, it will be clearly understood that a major limitation with this technology is immunological rejection, where reactions based on the mechanisms of adaptive immunity discussed above take place. The antigens responsible for rejection are histocompatibility antigens, especially those located on the MHC, and specifically the human leukocyte antigen system. These antigens are expressed by MHC genes, the inheritance of which leads to so-called tissue matching between identical siblings, and the possibility of immunological risk-free transplantation. T cells are central to rejection mechanisms, as we would expect. With unrelated individuals, the course of rejection can be minimized by matching of histocompatibility genes and blood groups and cross-matching to minimize the effects of prior sensitization to HLA antigens.

Xenotransplantation is an alternative to human transplantation, either with whole vital organs or specific tissues. Whole organ transplantation is currently too difficult because this genetically encoded barrier is even greater; there are also significant threats from viral transmission, such as PERV, porcine endogenous retrovirus, should pig to human xenotransplantation be considered. It may well be possible to genetically engineer animals to reduce risks of xenotransplantation, through either or both the removal of particular pig characteristics, especially the gal epitope (discussed below), or the addition of human characteristics, such as human complement regulators. By current definitions, such engineered organs would be considered as biomaterials, but we are not there yet. We do have, however, treated animal tissues to consider.

The gal epitope now becomes important in biomaterials science. The full name is galactosyl ($\alpha$1,3) galactose epitope. It is a terminal carbohydrate residue present on carbohydrates and glycolipids in all animal species up to New World primates but is absent from Old World primates and humans. Its absence in humans is due to the development of anti-gal antibodies. It is present in many but not all porcine tissues. The immune response to porcine tissues depends to a large extent on the presence and activity of the gal epitope. The T-helper cell is also relevant since a Th-1 response is likely to lead to rejection, while a Th-2 response allows for greater accommodation, With the SIS material, porcine small intestine submucosa, widely used in regenerative medicine, the lack of evidence of an immune response has been associated with both minimal gal epitope presence and a Th-2 dominated response.

**Allergies and hypersensitivity** The immune response described above has clearly evolved to deal with microorganisms. However, it is possible for the immune system to be triggered by invading molecules, quite often when they present no threat to the host. The events that happen once this response has been triggered can be very serious, even fatal, and the amounts of the foreign substance necessary to cause the effect may be extremely small. These phenomena are collectively described under the heading of hypersensitivity and the clinical consequences as allergies. Hypersensitivities are known to occur with some biomaterials and their components, although mechanisms are far from clear. In general, an allergy is an acquired condition where the response requires both a genetic disposition and prior exposure and sensitization to the substance in question. Clinical consequences include eczema, asthma, urticaria and, in the extreme, anaphylaxis. It is also possible for there to be an intolerance reaction, which is similar to an allergy but does not require prior sensitization. This may occur with some drugs and organic molecules.

It is generally considered that there are four types of hypersensitivity (immunologists are currently discussing refinements to this classification but these need not concern us here):

1. Type I reactions are known as immediate hypersensitivity reactions and involve antibody-mediated release of histamine and other molecules from mast cells and basophils.
2. Type II reactions are described as cytotoxic hypersensitivity reactions and involve the binding of antibodies to cell surface antigens and fixation, or inactivation, of complement.
3. Type III reactions are immune-complex reactions and involve circulating antigen–antibody immune complexes that deposit in blood capillaries.
4. Type IV reactions are known as delayed hypersensitivity reactions, or cell-mediated immunity; these are mediated by T lymphocytes.

Both Type I and Type IV reactions can be seen with biomaterials. It is not known whether there is any significant involvement of biomaterials in Classes II and III. Dealing first with Types I–III, it is necessary to consider the nature of these antibodies, otherwise known as the immunoglobulins. These are globular proteins that are composed of four polypeptide chains (Figure 3.20). Two of these chains are identical and described as light chains and two are identical heavy chains; these are held together by inter-chain disulfide bonds. The light chains are composed of 220 amino acid residues, with two equal-sized domains known as the variable (V) and constant (C) regions. The heavy chains have between 440 and 550 amino acids, the nature of which determines the class of immunoglobulin. These chains also have V and C regions. The arms of the molecule combine in the Y-shaped hinge region, which provides some flexibility. Each immunoglobulin has two distinct types of fragments. These are known as the Fab, or fragment antigen binding, and Fc, or fragment crystallizable, fragments. There are two Fab fragments, each of which is monovalent and consist of a light chain and parts of a heavy chain. These, as the name implies, contain the antigen binding sites, which are found within the V regions. There is one Fc fragment, which consists of the remaining heavy chains.

There are five classes of immunoglobulin, IgG (gamma), IgM (mu), IgA (alpha), IgD (delta) and IgE (epsilon). The differences are based on the amino acid sequences in the constant region of the heavy chain. The different immunoglobulins have different functions based on their structures. For example, IgM and IgG activate complement. IgD is a B-cell receptor and IgG also binds to phagocytic cell surfaces.

Of importance here is IgE which binds to the surfaces of mast cells, basophils and eosinophils, and is primarily involved in Type I hypersensitivity responses. The binding of IgE to an intruding molecule triggers the release of active mediators, especially vasoactive substances such as histamine. These immediate hypersensitivity reactions are usually triggered by skin contact (for example, following contact with latex products such as gloves), inhalation (e.g., pollen) and ingested food (e.g., peanuts). Quite often a substance is

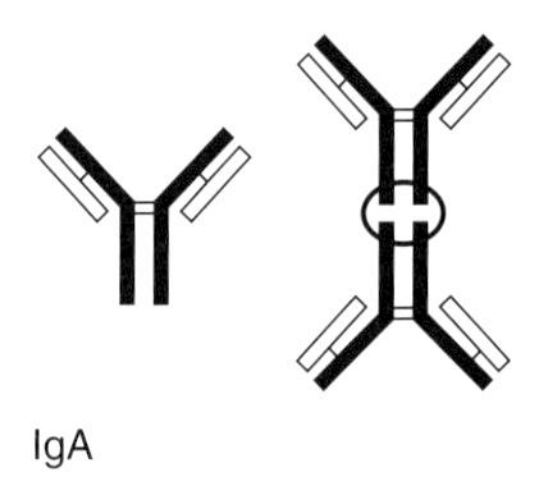

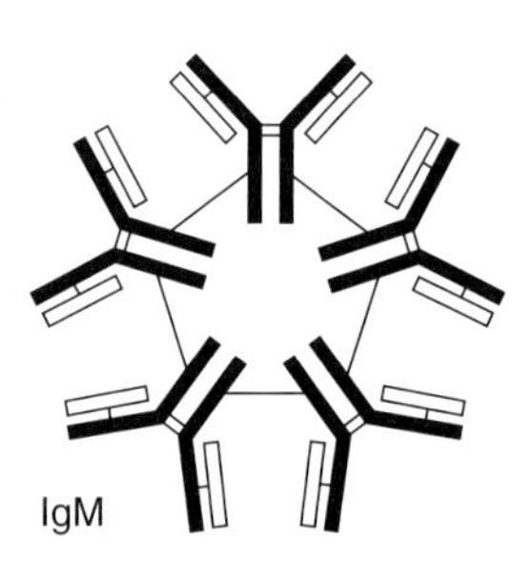


**Figure 3.20** Essential features and properties of the five immunoglobulins, IgG, IgD,IgE, IgA and IgM.

capable of invoking both Types I and IV responses. Some individuals are known to have a particularly significant predisposition to IgE-based hypersensitivity; this phenomenon is known as atopy and the individuals are described as atopics.

Type IV hypersensitivity is referred to as delayed because the response may not be seen for many hours or indeed days. The T-lymphocytes that are involved will have been previously sensitized to the substance in question and act in conjunction with other cells, especially phagocytic cells in order to generate a type of persistent granulomatous response. Skin contact is normally considered to be the most important route of exposure, giving rise to the condition of allergic contact dermatitis.

Biomaterials themselves are rarely the cause of Type IV hypersensitivity. However, biomaterials may release low molecular weight components that combine with proteins and the resulting complex may initiate the response. These substances are known as haptens and include organic materials derived from polymers and metal ions. The effect of hypersensitivity initiation here may be local but also systemic. The hapten binds with selected proteins to form a complete antigen, which is brought into contact with regional lymph nodes where it activates T cells that pass into the general circulation. This is the sensitization phase and upon further exposure the T cells release inflammatory mediators, setting up a prolonged inflammatory reaction. Of considerable importance to the potential for implanted biomaterials to initiate this type of hypersensitivity is the fact that the material may constantly release these haptens so that the distinction between sensitization and allergic response is a moot point. Moreover, the chronic inflammation that ensues is likely to have similar clinical signs as a non-hypersensitivity inflammation, and indeed the histological patterns determined by conventional pathological examination will have many similarities.

Allergic contact dermatitis is frequently seen in dentistry following prolonged contact between some polymeric and metallic fixtures and the oral mucosa.

This should be expected as the oral mucosa represents a front line of defense against the oral influx of bacteria, either air or food borne. The oral epithelium contains cells, the Langerhans cells, which constitutively express a high affinity IgE receptor and the Toll-like receptor TLR4. Dendritic cells in the epithelial layer process and present antigens to T-cells. Mechanisms similar to those used in defense against microbes will be effective against hapten-producing dental materials, contributing, for example, to the development of denture stomatitis.

With reference to implanted devices, the possibility of Type IV hypersensitivity has most commonly been linked to orthopedic devices, especially joint replacement where there may be extensive release of individual molecules or micro-/nanoparticles from articulating surfaces. It has to be said that the role of the immune response, including delayed hypersensitivity, is not clear in the development of inflammatory responses to these devices. These prostheses often involve a metallic component, however, from which several types of ion, including nickel, cobalt and chromium, are known both to be released from the implants and to have records of inducing hypersensitivity in other situations. It is unlikely that polymer-derived components involve significant numbers of T-cells in the responses they develop, but with metals, and especially those devices that have metal-on-metal bearing surfaces, there is often a clear T-cell accumulation and it is possible that there is a T-cell mediated Type IV hypersensitivity component of the responses that are seen in a subset of patients.

### Cell death: apoptosis and cytotoxicity

Cells of the human body die. They may do so naturally in what could be regarded as a programmed process but they may also die prematurely through the influence of endogenous or exogenous factors. It is quite possible, and maybe inevitable, that the presence of biomaterials in tissues influences these processes.

There are different types of cell death that may be defined by morphological, biochemical or functional criteria (see Figure 3.21(a)). Generally, from a mechanistic point of view, cells first take part in a process of dying that is initially reversible until a step is reached that is irreversible. There is no single event that defines this point of no return but rather a group of possible events. A cell may be considered to be dead when either it has lost the integrity of its plasma membrane, as defined by the incorporation of vital dyes, or has undergone complete fragmentation into discrete bodies, or has been engulfed by an adjacent cell *in vivo*. These criteria differentiate a dead cell from one that has been arrested in its normal cell cycle.

**Apoptosis** There are several discrete processes of cell death. Of relevance here is apoptosis. Apoptosis has often been termed “programmed cell death,” a term which certainly conveys the meaning that this is natural phenomenon although they are not truly synonymous as there are other processes of programmed cell death that are not apoptotic. Apoptosis is best defined by the morphological features that accompany cell death; these features include rounding-up of the cell, retraction of pseudopods, reduction of cell volume, chromatin condensation, nuclear fragmentation and plasma blebbing. Mechanisms are complex, but they can be considered as belonging to two cascades of molecular events. These are the extrinsic and intrinsic pathways. (Figure 3.21(b)) The extrinsic signaling pathway involves transmembrane receptor-mediated interactions which include death receptors, part of the TNF group of receptors, and the activation of caspase 8, which is one of the family of cysteine-aspartic proteases. The intrinsic pathway starts with an array of non-receptor-mediated stimuli, which include toxins, radiation energy and conditions of hypoxia. This stimulates mitochondrial changes and activation of caspase 3. Both of these pathways lead to one common pathway, the execution pathway, which results in DNA fragmentation, the degradation of cytoskeletal and nuclear proteins, the expression of ligands for phagocytic cells and eventual

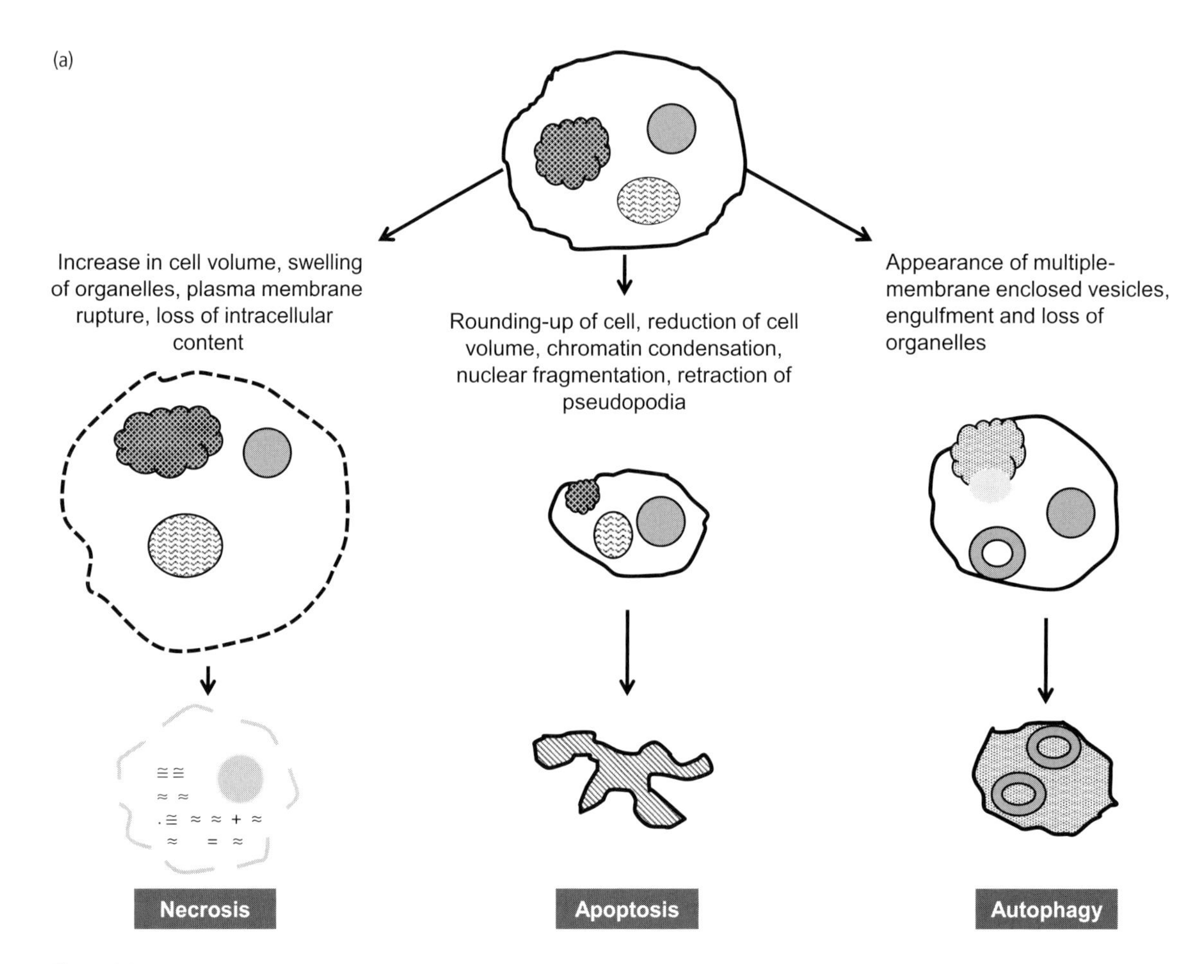


**Figure 3.21** Cell death.
**(a)** Characteristics of cells during the main processes of cell death, necrosis, apoptosis and autophagy, consistent with the recommendations of the Nomenclature Committee on Cell Death 2009 (*Cell Death Diff* 2009, 16(1) 3–11).

phagocytosis. The involvement of mitochondria in the intrinsic apoptosis pathway is important since cytotoxic species are able to stimulate the mitochondria to release cytochrome c, which enhances the activation of caspase 3.

Thus the mechanism of programmed cell death, which allows for the natural removal of old cells by phagocytosis, incorporates a mechanism for exogenous agents, potentially including components derived from biomaterials, to initiate a different pathway towards death that ends in the same final phagocytic events. There are several other forms of cell death, including autophagy and necrosis. The latter is characterized by a gain in cell volume, the swelling of organelles and rupture of the cell membrane. It is likely that such components from biomaterials will influence these processes as well, especially necrosis.

It is difficult, if not impossible, to define a single pathway by which soluble entities derived from biomaterials cause or influence cell death. There are indeed many possible ways since there are so many variations in the nature of these products. The chemical agents that are potentially released include monomers and oligomers, residual catalysis, additives in polymer products, polycations and metal

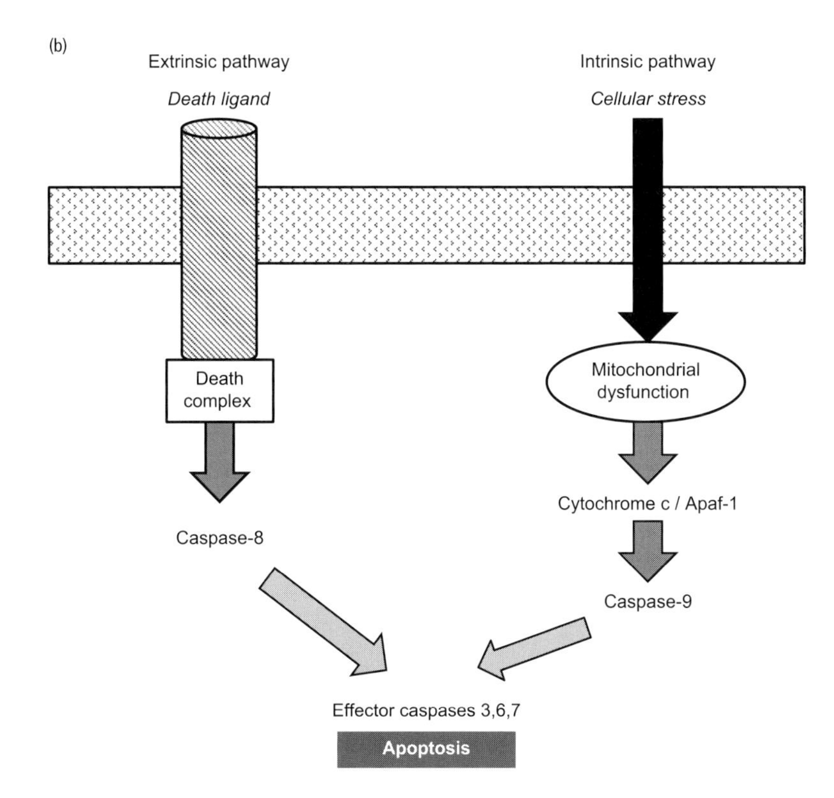


**Figure 3.21** *continued*
**(b)** Pathways of apoptosis. There are two pathways, the extrinsic and intrinsic pathways, in which caspases, cysteine-aspartic proteases, play prominent roles. On the left side, death ligands prompt the formation of a death complex, which activates caspase-8. On the right, cellular stress causes disruption of mitochondrial function and the release of cytochrome c into the cytoplasm. This is a heme protein associated with the inner membrane of the mitochondrion that is involved in electron transport chains. It forms a complex with apoptotic-protein activation factor-1 and caspase-9. Both pathways combine with the activation of effector caspases, which induce the final apoptotic change.

ions. Cationic macromolecules interact with cell membrane proteins and phospholipids through electrostatic interactions, disturbing both structure and function. Both charge density and molecular weight are significant factors. Metal ions can influence apoptosis and necrosis mechanisms, through the generation of reactive oxygen species, upregulation of caspase 3, effects on nuclear transcription factors and so on. Some metal ions exert their effects by competing with and displacing essential metal ions, especially those associated with enzyme function. In addition some metal ions have significant effects on gene expression, especially those involved with the regulation of heat shock proteins and metallothionein. Monomers such as methylmethacrylate influence apoptosis. Other monomers, including triethylene glycoldimethacrylate and hydroxyethyl methacrylate, reduce the level of natural radical scavengers such as glutathione that protect cells from the effects of reactive oxygen species.

## Material surfaces and cell phenotype

In several situations, the ability of biomaterial surfaces to influence the phenotype of cells in the vicinity of the surface is of major interest. There are two main reasons for this, the first being that the course of the host response could be modified if the phenotype of critical cells were changed, the second being that it may be possible to proactively produce cell differentiation within culture to facilitate tissue regeneration.

There are not too many examples of the former scenario, but this possibility has taken on some significance with the potential for biologic scaffold materials to influence macrophage phenotype. Although the discussions in this chapter have assumed that all macrophages are the same, which is a perfectly satisfactory basis for the phenomena included here, in recent years it has been recognized that there are some variations within macrophages, based on their functional properties and cytokine profiles. Macrophages may now be considered as being of either M1 or M2 types, or even as a continuum of subtypes between these. M1 are classically activated pro-inflammatory macrophages that are activated by IFNγ and secrete a variety of interleukins and other cytokines. M2 are induced by exposure to a variety of different signals and then secrete a series of different molecules. It is not clear whether the presence of synthetic biomaterials influences the polarization of macrophages towards M1 or M2. However, biologic scaffolds appear to have effects that depend on whether the tissue is cellular or acellular. Polarization towards M1 appears to result in the formation of more dense connective tissue, while the M2 phenotype gives more constructive remodeling. Pathways for phenotype changes are shown in Figure 3.22.

The situation with scaffold surfaces and cell differentiation is complex, although it is clear that different surface features are able to influence stem cell differentiation. There are several material characteristics that could be responsible, including surface chemistry, surface topography, substrate stiffness and surface functionalization by active molecules such as growth factors. Some of these effects are summarized in Table 3.4.

Stem cells will be discussed in much greater detail in Chapter 5. Here it is sufficient to note that stem cells are undifferentiated cells that are characterized

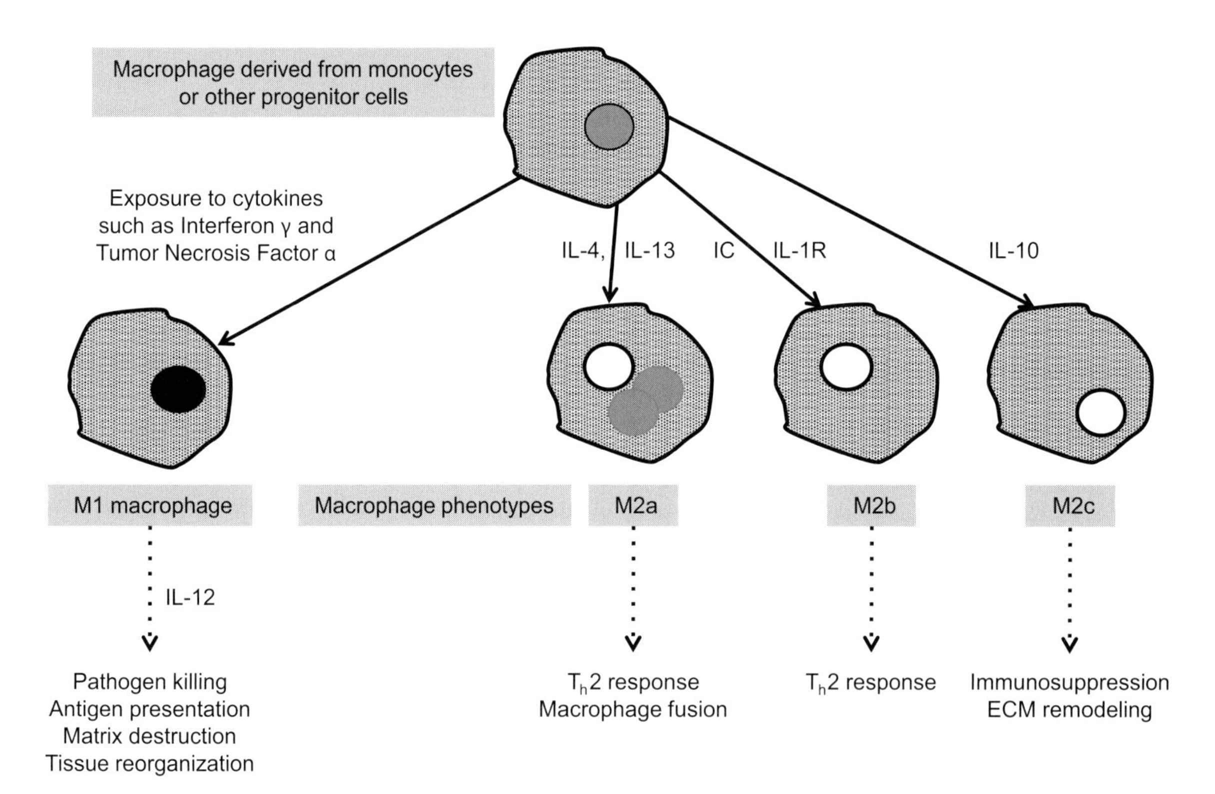


**Figure 3.22** Macrophage phenotype. It has become clear that not all macrophages are the same. They may be derived from monocytes or other progenitors but can change their phenotype depending on stimuli. There are two main types, M1 and M2, although the latter have three sub-types, M2a, M2b and M2c. The M1 phenotype arises following exposure to cytokines such as IFN-γ and TNF-α, and is then responsible for pathogen destruction. The M2 phenotypes are influenced by different interleukins and are responsible for immunosuppression and ECM remodeling. Understanding of macrophage has only developed recently and the simple scheme presented here is likely to be more complex. Scheme based on information from Kou and Babensee, *Journal of Biomedical Research* 2011, 96A, 239–60.

by the ability of both self-renewal and multipotential differentiation. The self-renewal process arises from the cell division that takes place within the microenvironment in which the cells reside, this being known as the stem cell niche. The stem cell division could result in a daughter cell that remains a stem cell and a progenitor cell, or in two daughter stem cells, the balance between these two outcomes leading to the correct replacement of stem cells and the replacement of progenitor cells that generate a differentiated progeny when subjected to different signals. In different stages of an individual's development, from embryo to fully mature adult, stem cells have different roles and differing characteristics of their niches. In adult stem cell niches, cytoarchitectural organizational features are maintained by the relationship between the stem cells and their somatic cell neighbors. Inside the niche, the stem cells are exposed to a variety of complex temporally and spatially controlled molecular mixtures, involving chemokines, cytokines and growth factors and ECM molecules. The combination of this molecular mix and the cytoarchitectural organization controls both the maintenance of stem cells within that niche and the progress of progenitor cell migration and differentiation.

Within the biomaterials context, the critical features of the stem cell niche have to be recreated in an essentially artificial environment, which is typically a bioreactor, where this balance is not automatically present. An *ex vivo* stem cell culture may be created with the types of molecule that should induce the required differentiation, but the physical environment that recapitulates the architecture of the niche will be provided by a biomaterial construct within the bioreactor. It is not yet clear which are the most important biomaterial characteristics for control of stem cell differentiation, but all of those listed in Table 3.4 have been shown under experimental conditions to have some influence.

Table 3.4 **Cell differentiation and material surfaces. The surface features listed are able to influence stem cell fate, including differentiation. The effects may be different with different stem cell types and can depend on other conditions, such as soluble molecules and oxygen levels in the media**

| | |
|---|---|
| Surface topography | Microscale roughness<br>Nanoscale roughness<br>Surface porosity<br>Surface features; grooves, pillars, etc.<br>Alignment/orientation of features |
| Material characteristics | Surface energy<br>Stiffness, elasticity<br>Molecular mobility |
| Mechanical forces | Tension<br>Compression<br>Cyclic strain<br>Hydrostatic pressure<br>Fluid flow; fluid shear stress |
| Surface functionalization | Plasma treatment<br>Peptide attachment<br>Growth factor attachment |

## Wound healing

It was noted earlier that many previous descriptions of biocompatibility phenomena started with the premise that a biomaterial was introduced into the body by a traumatic process, usually a surgical incision, and that the host response to the material was dominated by the response to the wounding process, that is wound healing. The biocompatibility paradigm presented here is not based on this assumption, but we do have to take wound healing into account, and this is best achieved at this point, although this shall be done briefly and only to point out where the chemical interactions in biocompatibility could be relevant.

Most surgical wounds involve the skin, although interventions increasingly have been designed to avoid this, as other mucosal surfaces may be used. The wound healing process is usually considered to have four overlapping phases of hemostasis, inflammation, proliferation and remodeling. It is easy

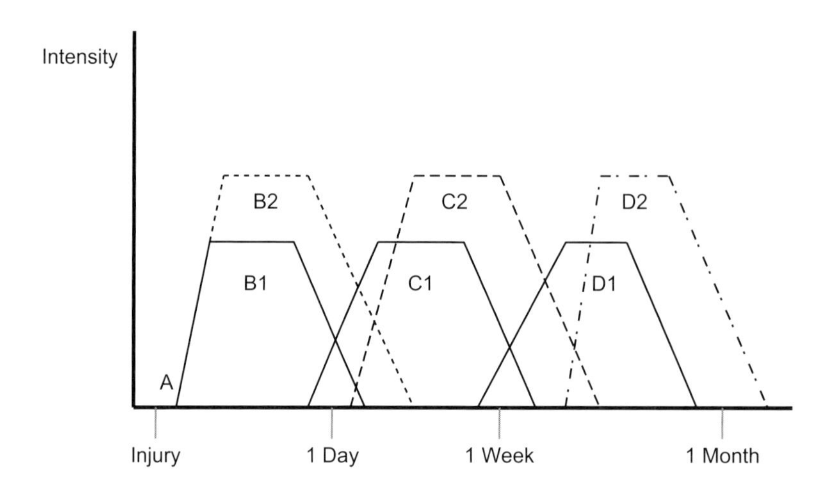


**Figure 3.23** Wound healing. This is a schematic of the phases of wound healing in soft tissue. (A) is the immediate phase of hemostasis, followed very quickly by (B) the acute phase of inflammation. Under simple conditions (B1) the intensity and duration are reasonably constant, but both intensity and duration can be increased (B2) with increasing severity of the injury. This phase merges with chronic inflammation, under the same variable conditions (C1 and C2), and then repair processes, leading to scar tissue (D1 and D2), which may take between 1 and 2 months to reach maturity.

to see why the foreign body response may be considered to be a variation of wound healing since they have many of the same features.

The timescale of wound healing is important (Figure 3.23). The hemostasis phase should be activated immediately and completed over a short period of time. Inflammation is also activated quickly and polymorphonuclear leukocytes (neutrophils, PMNs) arrive within a short time. Macrophages may take a day or so to arrive but quickly become the dominant inflammatory cell. Depending on the extent of the injury, the inflammatory phase will start to subside from a few days later, to be replaced by the proliferative phase, primarily involving fibroblasts. Revascularization should take place through the process of angiogenesis. In an open wound, granulation tissue, which consists of the inflammatory cells, proliferative cells and the beginnings of new extracellular matrix, will form at the same time. This allows re-epithelialization to commence at the surface over the granulation tissue. Maturation of the wound and its remodeling usually begin after a week, and this will be accompanied by some degree of contraction, and some form of scarring.

The hemostasis phase involves the immediate contact between platelets and exposed extracellular matrix molecules (especially collagen). As discussed previously with respect to blood compatibility in general, the platelets release coagulation factors and cytokines such as TGFβ and platelet-derived growth factor (PDGF), resulting in the formation of a clot and the release of vasoactive mediators that regulate vascular permeability. The neutrophils and then the macrophages initiate phagocytosis of any foreign matter, bacteria and fragments of damaged tissue. Neutrophils play a crucial role as they coordinate recruitment of inflammatory cells. They do so through a four-stage process:

- Activation of members of the selectin family of adhesion molecules (specifically L-, P- and E-selectin) that facilitate rolling along the vascular endothelium.
- Triggering of signals that activate and upregulate leukocyte integrins mediated by cytokines and leukocyte-activating molecules.
- Immobilization of neutrophils on the surface of the endothelium through $\alpha_4\beta_1$ and $\alpha_4\beta_7$ integrins

binding to endothelial cell adhesion molecule-1 (VCAM-1) and MadCAM-1.

- Transmigration through the endothelium to the site of injury facilitated by molecules such as matrix metalloproteinases.

As soon as the wound is reasonably cleared by the inflammatory cells, the proliferative phase starts. This is dominated by fibroblasts, which are attracted by the various cytokines already released. Within the fibroblasts, procollagen molecules are synthesized and secreted into the extracellular space, where they are processed into collagen. Initially this is rather immature but rapidly gains strength through cross-linking processes. Other growth factors, including VEGF, are released and they promote angiogenesis or revascularization, while the surface area re-epithelializes.

Under normal circumstances, the repair process is self-limiting. The production of collagen and other extracellular matrix molecules will cease, although there should be a continuing increase in collagen maturity and some remodeling of the tissue architecture. The formation of this new ECM may not, of course, be consistent with the structure and composition of tissue that has been damaged, in which case a zone of fibrous scar tissue will remain, which will be functionally different and may impair normal properties. Because the maturation of collagen depends on cross-linking, there will inevitably be some contraction associated with this healing process. Under some conditions, pathological responses may be superimposed onto this normal, uneventful, process. The most significant of these is fibrosis, which is the replacement of the normal structure of the tissue by excessive, distorted scar tissue. This results in keloid or hypertrophic scars on the skin, tendon adhesions, urethral strictures and so on. It is possible that some implanted devices will be associated with fibrosis, the most obvious of which is the constrictive fibrosis seen around breast implants. It is not at all clear whether such breast implant fibrosis is related to keloid and hypertrophic scar processes, which are associated with increased production of TGFβ.

#### Chronic exposure to biomaterials, the resolution of inflammation and the fibrous capsule

We now come to the critical question. A material has been implanted or injected into the tissue. It has been subjected to the immune response, most probably innate but also possibly adaptive. It, or its released components, has variously reacted with fluids, cells and molecules. In the short- and medium-term the results of these reactions could range from a barely detectable and clinically irrelevant inflammatory response through to extensive and clinically significant inflammation and either the proliferation or resorption of tissue or the development of some type of hypersensitivity. Some of these responses can be sufficiently significant as to cause failure of the device to function correctly, necessitating its removal, as with stents and joint replacements, or produce pain and general ill-health. The question is, what happens in those situations where there is no obviously implant related adverse effect requiring clinical intervention during these phases but where the implant is still present for a long time, perhaps continually releasing molecular, ionic or particulate species.

The evidence would suggest that, in the absence of a dramatic inflammatory, necrotic or hyperplastic response, the host response pathway is a continuum, albeit a complex continuum, up to a certain point, at which several different pathways emerge (Figure 3.24). Up to this point this corresponds to a quiescent state, perhaps even an equilibrium, where the pathological state is one of a mixture of minor chronic inflammation and minor fibrosis. With a solid implant this corresponds to the co-called classical foreign body capsule. This equilibrium reflects the possibility of a continuing but decreasing slow release of species from the surface and the protective, isolating effect of the capsule. This is the *chronic quiescent pathway of benign acceptance*. This contrasts with the dramatic effects discussed in the previous paragraph, where the pathway may be described as the *destructive pathway*.

The next possibility is that the tissues will recognize that continued inflammation is potentially damaging and shut it down. The immune system

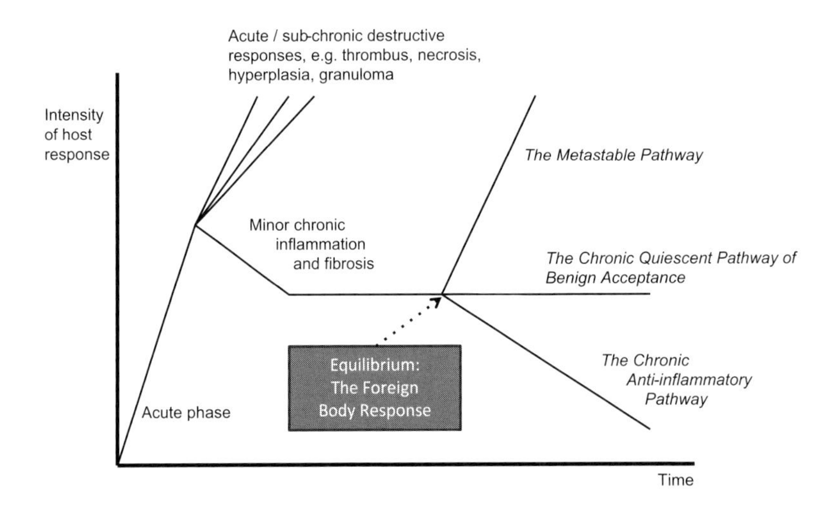


**Figure 3.24** Chronic exposure to biomaterials . The intensity of the host response varies considerably with time and conditions. There is an initial natural acute response. It is possible for there to be causes of serious and destructive responses during the acute and sub-chronic phases. In the absence of these, the response will subside, giving a combination of chronic inflammation and repair, effectively producing an equilibrium condition, which is usually referred to as the Foreign Body Response. The continued presence of a biomaterial in the tissue may affect the response. At any time, the quiescence may be disturbed since it can be regarded as a metastable state, and any of the early described destructive events may occur. In most cases, the equilibrium may be continued, and this may be considered as the chronic quiescent pathway of benign acceptance. It is also possible for there to be some form of reduced inflammation, giving the chronic anti-inflammatory pathway.

has several mechanisms to resolve an inflammatory condition, which requires the termination of pro-inflammatory signaling pathways and the clearance of inflammatory cells, allowing the return to normal tissue function. Many of the genes that are induced by pro-inflammatory stimuli are also required to inhibit excessive inflammation, and macrophages may be re-programmed to take on anti-inflammatory roles and promote normal wound healing. This may be considered to be the *chronic anti-inflammatory pathway*.

Finally we have to consider the possibility of a discontinuous pathway. This is best explained by a couple of examples. A prosthetic heart valve may perform exceptionally well for many years with apparently good biocompatibility, that is with good healing of the sewing ring, no hemolysis, no hyper-plastic tissue overgrowth and no thrombo-embolic complications. Then, without warning, perhaps after ten years, a blood clot forms on the valve and the patient dies. In another situation, a slowly degrading polymer is used as a resorbable plate to assist in the reconstruction of the craniofacial skeleton. The bone heals well, the plate ceases to be palpable under the skin indicating it is resorbing, but then, several years later, just as the degradation process is terminating, there is a massive inflammatory reaction, with swelling and pain. In the first case, the problem arose because the patient forgot to take anticoagulation medicine; in the latter case the final degradation products of the polymer were pro-inflammatory, whereas the polymer itself was not. These examples serve to emphasize that the equilibrium of biocom-patibility is not necessarily stable. It is better to refer to the equilibrium as metastable. This term implies that a physical state should not really exist if all the rules were followed, but it is allowed to because of local or current circumstances. It will, however, revert to the normal state as soon as the conditions are changed. This term was used to describe the state of transformation-toughened zirconia in Chapter 2. We may consider this phenomenon as metastable biocompatibility, and the pathway to be the *metastable pathway.*

### The essence of chemical biocompatibility mechanisms

In their normal environment, cells interact with other cells and with the extracellular matrix. Cells communicate by molecular mechanisms, and receptors on cell membranes are crucial to this process of cell signaling. Cell signaling involves signal initiation, signal transduction and gene activation. The initiation phase is very important to biocompatibility since, of the three general categories of molecule that can serve as ligands, two of them, the freely diffusible molecules and ECM proteins can both consist of, or be associated with, biomaterial components. Xenobiotic compounds derived from biomaterials, including metal ions, monomers, oligomers and active molecules contained within ECM biomaterials can all take part, either positively or negatively, in cell signaling processes. These potential effects on cell communications arising from the chemical nature of the surface should be seen as additional to the mechanical and biophysical effects already discussed, and topographical effects which will be seen later.

Specifically in relation to chemical events and cell signaling in biocompatibility, we have to recognize their role in inflammation and both innate and adaptive immunity. Critical to these roles are the processes where the key components of innate immunity, which are the phagocytic cells, natural killer cells and circulating proteins, control the overall response. Tissue damage that results in inflammation can take the form of many different types of stress stimuli, including heat, oxidative stress and radiation; exogenous chemicals associated with biomaterials form just one group of stressful stimuli, and we have to consider the role of such chemicals within the various forms or phases of the responses, including apoptosis and wound healing in the context of the available pathways of cell signaling in response to the stress stimuli. These are summarized in Figure 3.25.

Notwithstanding these observations that biomaterials-induced innate immunity responses have to follow general pathological mechanisms, there is one major difference, and that in contrast to most other forms of stress stimuli, biomaterials may well exert their effect within tissue for very long periods of time, and we have to consider the mechanisms of the response to chronic exposure and the eventual resolution of inflammation. Here we incorporate those features of Figure 3.24 that deal with these resolution pathways into the general scheme of Figure 3.25.

## 3.3.4 Scenario C: Mechanisms of biocompatibility involving microscale biomaterials and microparticles

The description of biocompatibility phenomena in the previous sections has referred to macroscale materials and devices and has not taken into account the mechanisms that are influenced by features at the microscale and nanoscale. We turn to these issues in this and the following section. These characteristics are consistent, rather than inconsistent, with those discussed in Section 3.3.3, including the inflammation and immune response components, but with different emphases, with variations in the influence of material parameters on the response, and with upregulation and downregulation of certain events that ensure that the overall balance is perturbed and that there are different clinical manifestations of the collective biocompatibility phenomena. In both micro- and nano- situations, biomaterials science has been engaged with the objective of optimizing the desirable features that may come with the scale while minimizing, or hopefully avoiding, those events that could be harmful. Also in both cases, we deal with the twin phenomena of topographical control of biocompatibility and the host responses to particles.

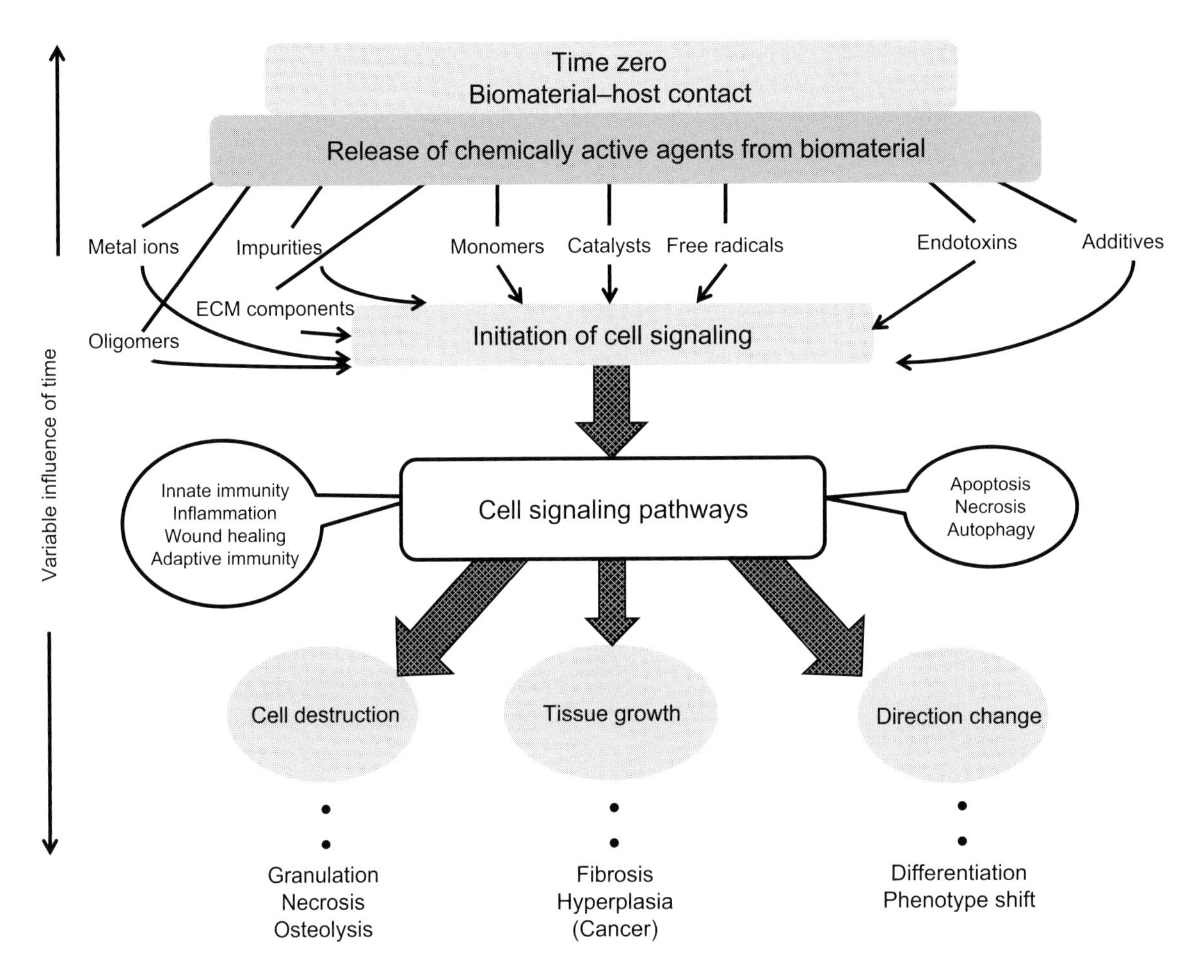


**Figure 3.25** Essence of chemically controlled biocompatibility pathways. This scheme shows the sequence of events that follow the release of chemically active agents from biomaterials. Note that there are multiple and variable sources of these agents, and there will be a variable influence of these agents with time, depending on release rates and overall biological activity. The presence of molecules and ions in the tissue initiates cell signaling and the events follow one or more cell signaling pathways. The outcomes tend to be related to cell destruction, tissue growth and direction change.

## Microscale topography

It has been known for some time that surface topography may have a significant influence on cell behavior. In this section we shall consider the generic issues of changing surface morphology, including features at the microscale, on cell behavior; the special case of the nanoscale is considered later.

It is first necessary to describe the features of surface morphology. In the case of what might be called accidental topography, such as those features left after manufacturing processes, including abrasion, machining marks and casting or molding surfaces, there are likely to be random features that may be realistic as far as medical devices are concerned but difficult to study and interpret. A basic paradigm does emerge from observations on accidental roughness, which is that in general rough surfaces enhance cell adhesion. On the other hand, most information and theories are derived from carefully and deliberately introduced surface features where many different parameters can be compared with those of cells.

These classically include grooves or ridges, but also pits, cliffs, dots, spikes and meshes or fibers.

In most situations, the responses vary with cell type, which is to be expected since some are anchorage-dependent and some are not, and they have different types of receptors that may respond differently to mechanical surfaces. Fibroblasts, leukocytes, neurons, chondrocytes and oligodendrocytes are among the cells known to respond to topography. Cells appear to respond significantly to single cliffs on surfaces, especially when they have a height of up to 20 μm. Cells recognize grooves of the same type of dimension and will show orientation effects. They tend not to respond to grooves that are wider than cells but effects are seen when the grooves are of the same magnitude of the cells. *In vitro* experiments reveal that these surface features, especially the curvature on discontinuities, can affect cell surface receptors and the cytoskeleton, resulting in variations in gene expression much as described for mechanotransduction. The use of topography is clearly one of the features that has been used in attempts to control cell behavior, as shown in Figure 3.26. A major problem has been concerned with identifying correlations with *in vivo* and clinical experience since it is not clear how significant this topography is for the practical application in humans.

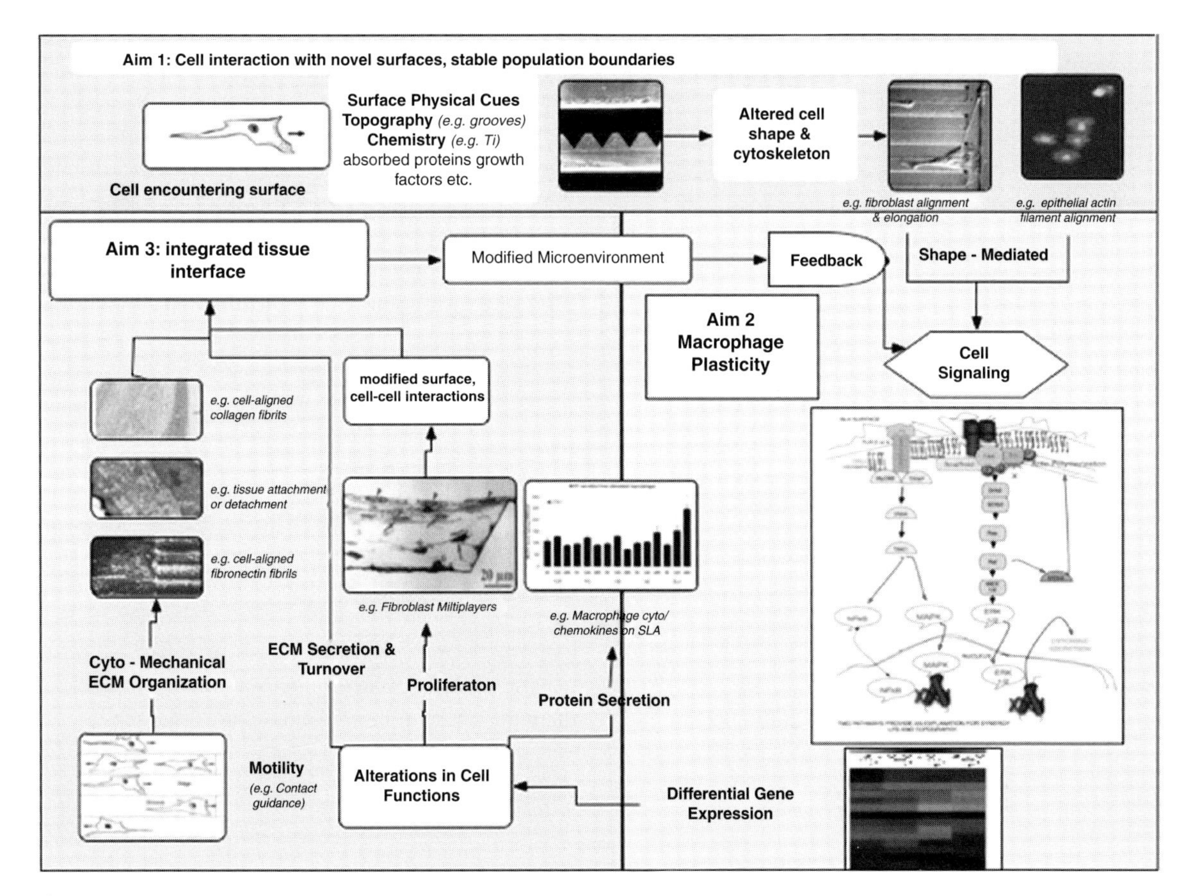


**Figure 3.26** Microscale surface features. Several research groups have used surface topography along with other characteristics to influence cell and tissue responses. Shown here is the strategy of one group to use surface topography and adsorbed molecules to influence cell phenotype and tissue integration. Image courtesy of Dr. Donald Brunette, Oral, Biological and Medical Sciences, University of British Columbia, Vancouver, Canada.

## Microparticles

Microparticles are important from two perspectives, the response to microparticles inadvertently or unavoidably released into the host environment, and the effect of microparticles deliberately delivered to the body in attempts to deliver drugs or other agents. Attention was given to microparticles long before nanoparticles and we deal with these separately because there are some significant differences notwithstanding some similarities. One thing that these different scale particles do have in common is the reliance on experience with lung pathology to get some initial ideas about mechanisms. In the microparticle area, this arises from the long known phenomenon associated with the inhalation of particles during environmental or occupational exposure. We start our discussion of microparticle-induced host responses by considering the lessons learnt from asbestos.

**Lessons from pneumoconiosis** Asbestos fibers are known to cause a number of lung pathologies, including pulmonary fibrosis, pleural effusions and plaques, and malignancies such as malignant mesothelioma and bronchogenic carcinoma. These conditions have arisen following long-term inhalation exposure. There are two forms of asbestos fiber, both of which are naturally occurring hydrated silicate compounds, one being the serpentine fibers that are curled strands, and the other the amphibole fibers, which are rod-like. The diseases are largely concerned with uptake by pulmonary epithelial cells and susceptibility varies with the chemistry, morphology and persistence of the fibers. The overall pattern of events involves early damage to the alveolar epithelium and the formation of foci of fibrosis in areas of damage (Figure 3.27). There is an accumulation of alveolar macrophages and the development of diffuse pulmonary inflammation. Specifically, there are several mechanisms and pathways involved here, in particular asbestos-induced reactive oxygen species production, DNA damage, mitochondrial dysfunction and intrinsic apoptosis. It should be noted that many other inhaled particles, such as coal dust and silica, are associated with similar types of pulmonary diseases, collectively called pneumoconiosis, but they have their own clinical courses and pathological features. Asbestos

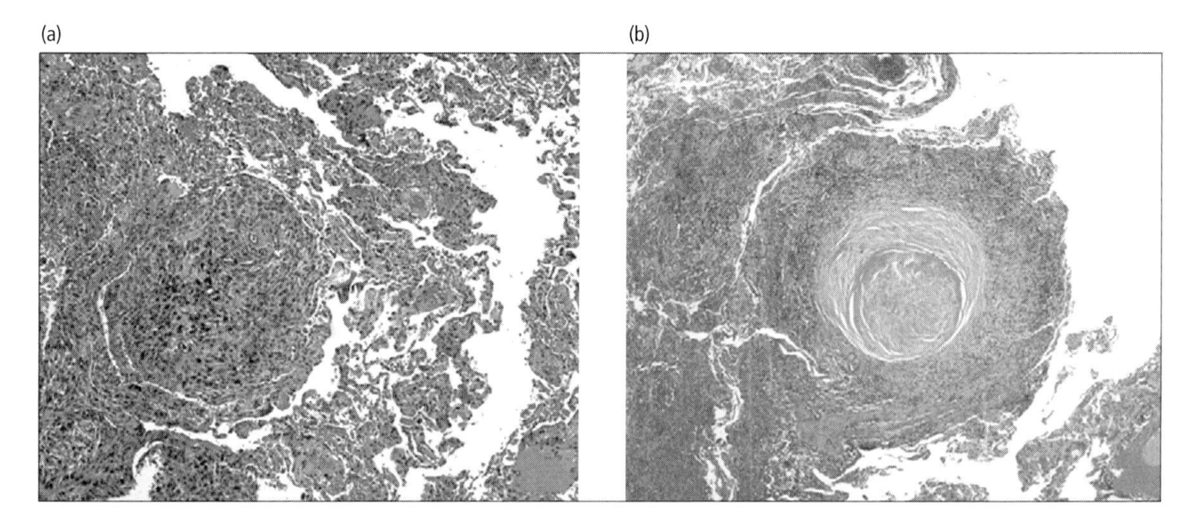


**Figure 3.27** Fibrosis of the lung. Histological sections of lung with silicotic lesions:
**(a)** early silicotic lesion as cellular nodule of dust-laden macrophages (×100).
**(b)** Chronic silicotic nodule with concentric fibrosis in the center and peripheral dust-laden macrophages (×40)
Reprinted from *The Lancet*, 379, Leung C C, Silicosis, 2008-18, 2012, with permission from Elsevier. See plate section for color version.

has the more significant history and we can associate these events with the particle size, shape and surface chemistry. At its very simplest, we can note that the individual fibers are often a few microns in diameter and of varying length depending on how they were separated from the long parent fibers. It would not be unusual for these fibers to be 2–20 microns in length, giving highly favorable conditions for frustrated phagocytosis, where an alveolar macrophage can attempt to phagocytose a particle, partially but not completely engulfing it. The now damaged cell releases pro-inflammatory molecules, such as intracellular enzymes, cytokines and a variety of chemotactic agents, such that there is a proliferation and accumulation of inflammatory cells, following which any one of the above disease conditions can emanate. The precise chemical nature of the asbestos particle is of less importance. The amphibole asbestos fibers are more fibrogenic and carcinogenic, mostly because their rigidity is more destructive to the macrophages and because they have greater biopersistence. It is possible that the leaching of metal ions from the surfaces of the fibers plays a contributory but not significant role.

Clearly the inhalation of asbestos fibers is not representative of either the normal route of patient exposure to biomaterials or the physical and chemical form of products derived from biomaterials, but there are many lessons, and the host response to microparticles released from hip joint prostheses or injected drug loaded microparticles, may follow similar pathways. To understand this, we have to consider some of the specific mechanisms involved, building on the earlier generic discussion of inflammation.

**Oxidative stress** We start with oxidative stress. This is defined as the imbalance between the production of reactive oxygen species, ROS (or in some cases reactive nitrogen species) and antioxidant defenses. This is an immensely important subject, with implications far wider than the biocompatibility of biomaterials. This balance is crucial for cell survival and growth. Cells have to be able to remove ROS efficiently to maintain vitality. Oxidative stress is involved in the initiation and progression of many diseases, including cardiovascular diseases, hypertension, viral pathogenesis, inflammatory diseases and cancer. Most cells in the human body are capable of generating ROS, which by definition are more reactive than molecular oxygen. They are beneficially involved in many physiological reactions, including fighting infection. They are produced by the mitochondria as part of their metabolic activity, but are also released by inflammatory cells such as eosinophils, neutrophils and macrophages through the activity of the NADPH oxidase that is in the cytoplasm. ROS include the superoxide anion ($O_2^-$), hydrogen peroxide ($H_2O_2$) and the hydroxyl radical ($OH^-$). Under normal quiescent conditions, cells are able to cope with the production of ROS by anti-oxidant activity. For example, anti-oxidant enzymes such as superoxide dismutase (SOD) reduce $O_2^-$ to $H_2O_2$, which can be reduced by catalase to $H_2O$. Glutathione peroxidase is able to neutralize many ROS. Several therapies directed towards oxidative stress in general involve the use of anti-oxidants, including vitamins. It is unclear how and why particles have such a significant effect on oxidative stress. There have been suggestions that the effects are material specific, for example involving metal ions that take part in redox reactions or inhibit the glutathione peroxidase, but the evidence is far from clear. It is evident, however, that the ROS can directly affect both inflammation and fibrosis, damage DNA, enhance lipid peroxidation and modify calcium homeostasis.

We have seen in Section 3.3.3 that several signaling cascades are involved in the response of cells to foreign surfaces, which are mediated by kinases of the MAPK family, including ERKs and JNKs. It would seem that under some situations phagocytic cells respond to microparticles via MAPK cascades and transcription factors. Particulate matter can activate JNK and p38 MAPK. Importantly, activation of p38 MAPK can lead to activation of NF-κB (nuclear factor kappa-light chain enhancer of B cells), which is a protein that controls the transcription of DNA; this has implications for the control of apoptosis, inflammation and fibrosis.

**Microparticles generated *in situ* within tissues** Let us turn now to the situation of microparticles generated within tissues; the best example involves the release of debris from articulating surfaces but there are others, including the particles generated during the *in vivo* degradation of polymers. The role of macrophages has been discussed earlier. Here we focus on the role of the macrophage in phagocytosis of particles. Phagocytosis is generally considered to be the process where particles of diameter greater than around 0.2 micron are actively taken into cells. This process is restricted to a few cells, of which macrophages and monocytes are the principal examples. The process involves the binding of particles to the cell membrane via receptor molecules or non-specific electrostatic forces and the activation of receptors. This initiates intracellular signaling, which leads to actin polymerization and coordinated cytoskeletal movements and internalization of the particle. The particle is then contained within a phagosome, which is a membrane bound vesicle, and this will fuse with a lysosome to form a secondary lysosome, sometimes called a phagolysosome. This is rich in lysosomal enzymes; should the particle be a bacterium, then these enzymes will rapidly degrade it. With biomaterial-derived particles, a great deal will depend on the chemical composition and whether the lysosomal enzymes are able to initiate any degradation of the material.

As far as the ability of a macrophage to take up different types of particle is concerned, we have to take into account the availability of receptors for different material surfaces and the physical characteristics of the particle that control the internalization process. There are indeed many macrophage receptors so that cell attachment is not normally a problem, especially considering that the particle will usually be coated with a layer of molecules during opsonization. Many of these are receptors for complement fragments and immunoglobulins, especially IgE and IgG. There are also scavenger receptors that may bind certain macromolecules and mannose-containing molecules. Naturally interference with these receptors, such as in association with disease states or pharmacological activity, can influence phagocytosis. The effect of the physical characteristics of the particles is related to the fact that some physical resistance by the particle is required for the cell membrane to deform around its surface (see Figure 3.28). Rigid particles are more easily deformed than soft ones. Size also has an influence, with greater rates of phagocytosis being found with particles between 1 and 5 microns.

It is now necessary to address the question of the significance of the macrophage behavior with respect to particles, again using the discussion of inflammation as the starting point. The endpoint of the phagocytic ingestion of bacteria is the destruction of the bacteria and, with the help of some other mechanisms, the resolution of infection. When the invading particle is a foreign material instead of a bacterium, it is far less readily destroyed and may be very persistent. The consequences may be localized or systemic. Just as macrophages that are activated by macroscale biomaterial surfaces release a variety of inflammatory mediators, so do macrophages that encounter, or are attracted to, free microparticles. It is clear that a number of genes are either upregulated or downregulated as a result of the phagocytic process. Many of these genes follow the same pattern as seen when bacteria are ingested but some are more specific to the foreign material, being determined by size, shape and chemistry of the latter. Insufficient evidence currently exists to allow a complete picture to emerge, but some features stand out. Genes responsible for cytokine production feature strongly in those that are upregulated, including the interleukins IL-1$\beta$ and IL-6 and the tumor necrosis factor TNF$\alpha$. Equally, certain genes that influence ECM turnover, such as the tissue inhibitor of metalloproteinase (TIMP-1) and some matrix metalloproteinases such as MMP-14 may be upregulated. The gene expression profiles suggest that there is a significant increase in

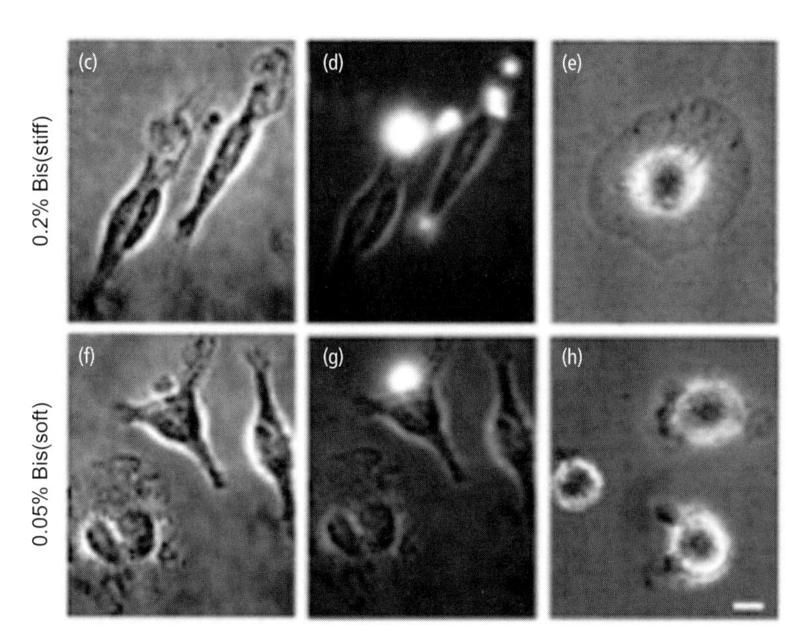


**Figure 3.28** Phagocytosis and the mechanical properties of the target particle. Phagocytosis is dependent on actin-based processes in order to ingest particles, usually greater than 0.5 μm diameter. The mechanical properties of the particles significantly affect the efficiency of the process. In the upper row macrophages are able to phagocytose stiff polyacrylamide beads and sheets. In the lower row the cells were unable to phagocytose soft particles of similar chemistry. Reprinted from *Journal of Cell Science*, Beningo K A and Wang Y-L, Fc-receptor-mediated phagocytosis is regulated by mechanical properties of the target, 2002, 115(4), 849–55, courtesy of the Company of Biologists Ltd.

cytokine release, which is not surprising, and leads to migration and recruitment of additional inflammatory cells to the region, and disruption of the ECM. Naturally we would expect the gene expression to change with time. With bacteria, the significant early up-regulation of genes that are important to the destruction is transient, and decreases as the microorganism is digested. With non-degradable material particles, this may not be the case, and the upregulation and activity may be prolonged; the early significant release of TNF$\alpha$ quite quickly slows down in the former case but remains high when the macrophage has ingested non-degradable polymeric particles.

These processes indicate that the presence of microparticles can result in prolonged inflammation. This may or may not be clinically significant. If the particles are corrosion products of, say, stainless steel, then a significant and noticeable chronic edema can develop; in the early days of metallic devices in orthopedics, when corrosion resistance was not optimized, this was a clinical problem, possibly necessitating implant removal (see Figure 3.29(a)). With degradable polymers, macrophage-dominated inflammation may be seen as the degradation profile leads to the release of small fragments (Figure 3.29(b)). It is noticeable that the nature of the macrophage may change during these processes; a significant change in phenotype is not unusual. Immunostaining of macrophages observed in the transition from acute to chronic stages of inflammation show that the dominant macrophage markers change with time, especially showing variations in the strength of ED1, ED2 and MHC class II markers.

This change in macrophage phenotype and activity is a prelude to the later consequences. A transient inflammation may not be very significant if it were the only result in particle-mediated macrophage activation. To understand the other consequences, we have to discuss the process of macrophage fusion. This was mentioned briefly before in the context of foreign body giant cells. Cell–cell fusion is essential in embryonic development but is not a widespread phenomenon in adults, being restricted to a small number of physiological processes. It is, however, important in the case of macrophages. Macrophage fusion involves a number of stages, starting with the induction of a fusion-competent status. Chemotaxis then allows directional cell migration, followed by

(a)

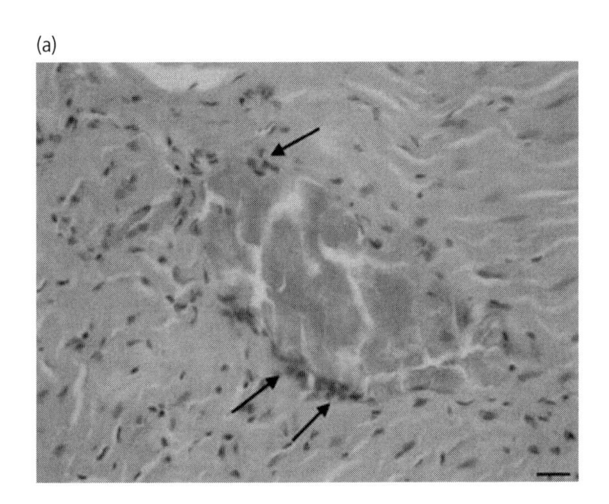

(b)

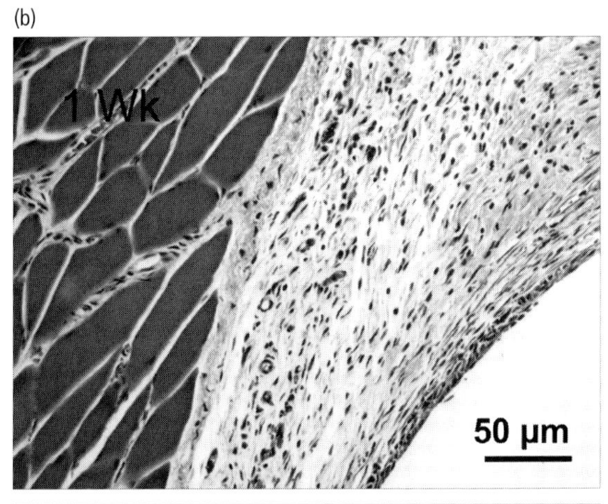


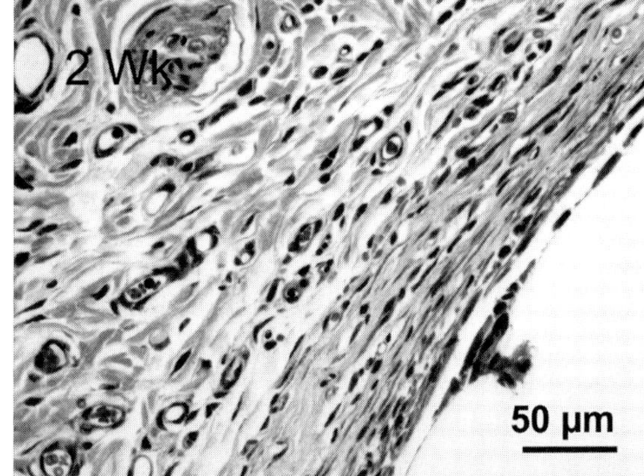


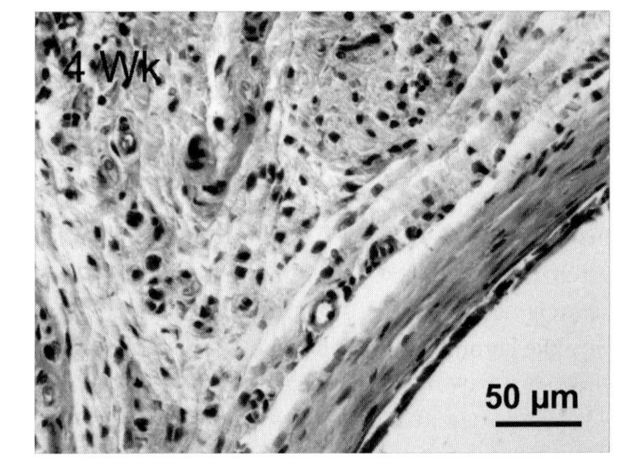


**Figure 3.29**
**(a)** Phagocytosis of corrosion products. Histological section showing dense fibrous tissue surrounding a corrosion deposit, demarcated by foreign body giant cells. Reprinted from *Acta Biomaterialia*, 5, Huber M *et al.*, Presence of corrosion products and hypersensitivity-associated reactions in peri-prosthetic tissue after aseptic loosening of total hip replacements with metal bearing surfaces, 172–80, © 2009, with permission of Elsevier. See plate section for color version.
**(b)** The initial tissue response to polymers is usually seen in terms of an inflammatory response and fibrotic tissue adjacent to the polymer. When polymer particles of any form are released into the tissue they may be associated with focal points of inflammatory cells. These sections show the tissue at 1, 2 and 4 weeks after the intramuscular implantation of elastomer cylinders. Images courtesy of Dr. Brian Amsden, Chemical Engineering, Queens University, Kingston, Ontario, Canada. See plate section for color version.

cell–cell attachment and the cytoskeletal re-arrangement that results in fusion. A variety of exogenous and endogenous factors are required for re-programming the cells, including IL-4 and E-cadherin. The transmembrane protein DC-STAMP appears to play a significant role here. A variety of chemokines are effective chemotactic agents for fusion-competent macrophages, and E-cadherin is an

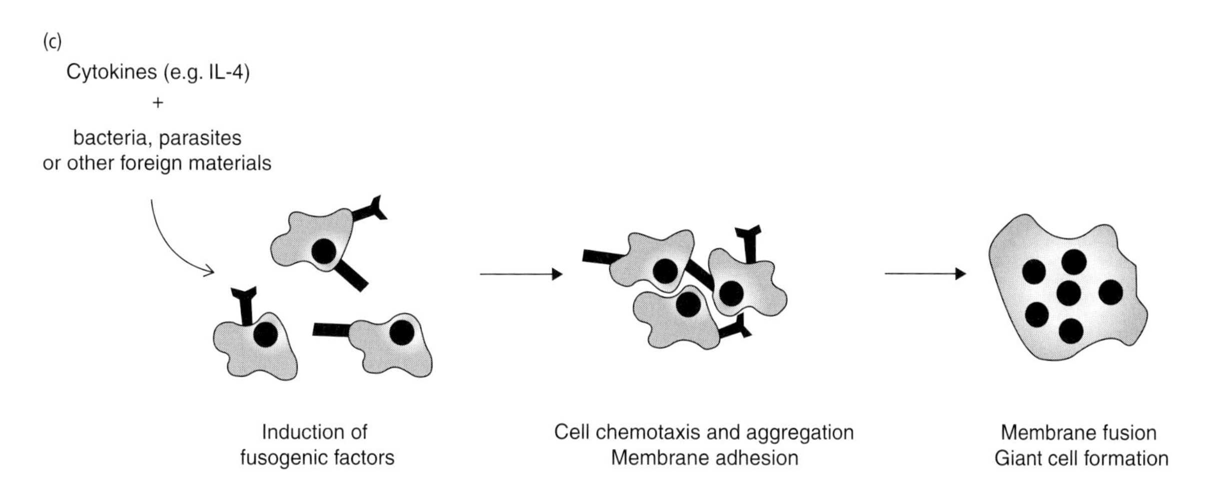


**Figure 3.29** *continued*
(c) Fusion of macrophages. The local microenvironment of macrophages, in the presence of bacteria or foreign bodies, which includes many types of cytokine, induces the expression of fusogenic factors on the cells. Chemotactic factors mediate cell aggregation and membrane adhesion of adjacent cells. Reorganization of the cell membranes is influenced by multiple factors, including membrane proteins, biophysical factors and intracellular signaling, the result being the formation of a single, multinucleated giant cell. Reprinted from *Immunobiology*, 212, Helming L and Gordon S, The molecular basis of macrophage fusion, 785–93, © 2008, with permission of Elsevier.

important mediator of adhesion. The result of this is that, under certain conditions, macrophages are able to fuse together (Figure 3.29(c)).

There are two important consequences here, one generic and one highly specific. The generic one involves the formation of the multinucleated foreign body giant cells mentioned before. This does not solely relate to particles since they can form within the capsule of a large implant. The process is, however, a result of the inability of macrophages to ingest and digest large objects, and the giant cell, which may be in excess of 50 microns in diameter with many nuclei, represents an attempt to deal with the foreign object more successfully. It is not unusual to see the chronic granulomatous response to multiple particles dominated by these giant cells.

**Osteolysis (Figure 3.30)** The specific situation is one that involves bone, and is responsible for one of the most intractable problems of biomaterials-related host responses. Osteoclasts are the cells of bone that are responsible for bone resorption. Under normal circumstances, they work in concert with bone-forming osteoblasts to ensure effective and continuous bone turnover. Osteoclasts are formed from hematopoietic cells of the mononuclear phagocytic lineage that includes macrophages, and are multinucleated cells formed by cell–cell fusion, but the process and the reasons are different to those associated with foreign body giant cells. The osteoclasts become polarized when attached to bone. A ruffled membrane surrounded by an actin ring controls attachment to the bone but the nuclei are found at the other end of the cell, away from the resorptive surface. The resorption takes place by acidification of the interface between cell and bone through the activity of a proton pump and the secretion of the acid protease cathepsin K. In this situation, the macrophage fusion is induced by the two mediators RANKL and M-CSF. A crucial glycoprotein involved in bone turnover is osteoprotegerin (OPG). RANKL, receptor activator of nuclear kappa B ligand, is a ligand for OPG and binds it to osteoclast precursor cells. Macrophage-colony stimulating factor, M-CSF, is a molecule that is able to induce the differentiation of osteoclast precursor into mature osteoclast.

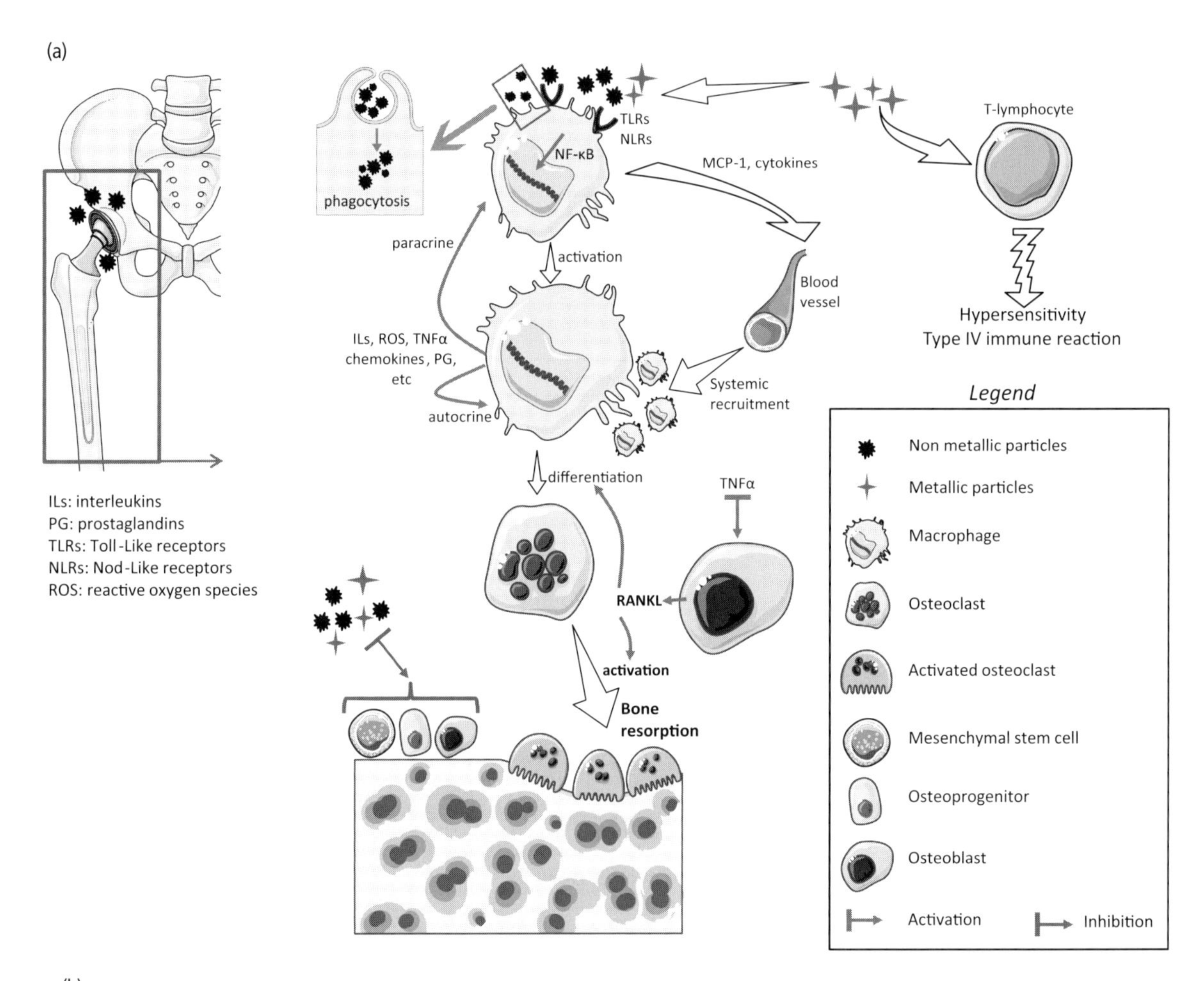


(b)

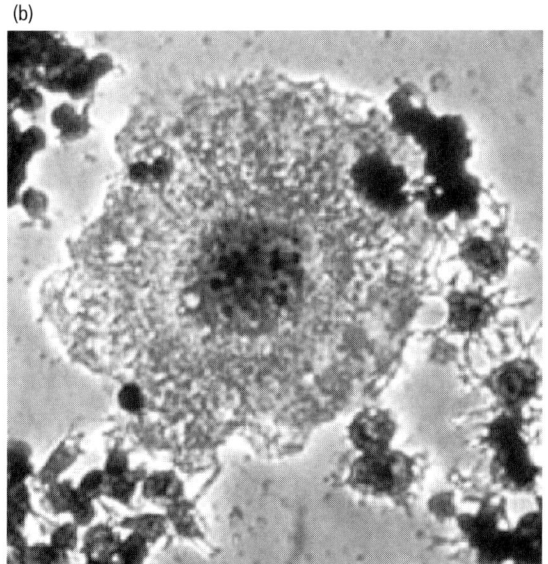

**Figure 3.30** Osteolysis.
**(a)** Mechanism of bone resorption following release of wear debris from hip replacement prosthesis. Image courtesy of Dr. Stuart Goodman, Orthopaedic Surgery, Stanford University, California, USA.
**(b)** Osteoclasts. This image shows a multinucleated osteoclast, observed with tartrate-resistant acid phosphatase (TRAP) staining, that has been stimulated by endogenous macrophage-colony stimulating factor (MCSF) and RANKL. Reprinted from *Pharmaceutical Research*, 28(5), Newa, M *et al.*, Antibody-mediated "universal" osteoclast targeting platform using calcitonin as a model drug, 1131–43, © 2011, with kind permission from Springer Science and Business Media. See plate section for color version.

Working together, these two molecules play vital roles in bone resorption. Unfortunately for joint replacement prostheses that generate wear particles, macrophages that are activated by these particles secrete chemotactic cytokines, such as TNFα and IL-1, which appear to play a role in the differentiation of precursors into mature osteoblasts by stimulating RANKL and M-CSF production in stromal cells, hence upsetting the osteoblast–osteoclast balance in favor of the latter. One major clinical situation where this happens regularly is in the disease of osteoporosis. This situation with wear debris recapitulates the osteoporotic environment and causes loss of bone at the prosthesis surface. This is known as osteolysis and is a major cause of prosthesis loosening.

The final comment on the response to microparticles relates to the systemic consequences. Concerns have often been expressed about the eventual fate of biomaterial-released microparticles. Interest in this topic has increased with the deliberate use of particles, of both micro- and nano- scale, to deliver drugs and genes systemically or targeted to remote sites. These latter situations are discussed elsewhere. Considering the fate of microparticles released into tissue, although some other distribution routes cannot be totally excluded, the one process that solely commands attention here is that of dissemination through the lymphatic system. There is no doubt that, as nature intended, phagocytic cells containing particles can migrate to lymph vessels and accumulate in the lymph nodes. Wear debris from joint replacements has been identified in regional lymph nodes, especially in axillary nodes and those in the pelvis. Polymer fragments derived from degrading polymers are similarly disseminated. Often these lymph nodes will contain an abundance of mono- and multinucleated cells. The clinical consequences of these observations are unclear but there is no strong evidence to show any significant adverse effects. It is likely that any slowly degrading or corroding particles will remain in the nodes, releasing oligomers, monomers, metal ions or other products slowly, these being eventually cleared in the liver or kidneys as the lymphoreticular system drains into the central venous system. Distribution of some of these entities, especially metal ions, to distant organs, may occur during this process. Those particles, or their remnants, that cannot be degraded most likely stay in lymph nodes indefinitely.

**The essence of microscale biocompatibility mechanisms** The most important aspect here is the influence of microparticles that either collectively constitute a biomaterial or are derived from macroscale biomaterial devices. This is a size, shape and chemistry controlled process but the pathway is relatively straightforward. With equiaxed microparticles of size just below and just above the micron, the particles can be phagocytosed by a number of cells, especially including macrophages. Once internalized, the particles may initiate ROS production, DNA damage, mitochondrial dysfunction and intrinsic apoptosis. The extent of these events will vary. If the engulfed material is biodegradable, then it will be degraded and eliminated; essentially the macrophages are doing the work they were designed to do and there should not be any lasting effect. If, however, the particles are of a shape and/or size that cannot be conveniently engulfed or they are non-degradable, bio-persistent and chemically irritant, the process is not so smooth and the cells will release into the surrounding tissue a variety of pro-inflammatory molecules, including lysosomal enzymes, chemokines and cytokines. This process then initiates the innate immunity pathways discussed above. In some situations, such as with osteolysis, specific consequences in relation to the host response may develop. This scenario of microparticle controlled host responses is summarized in Figure 3.31.

### 3.3.5 Scenario D: Mechanisms of biocompatibility involving nanoscale biomaterials

As with the previous section, we are concerned here with the interactions that occur between biological

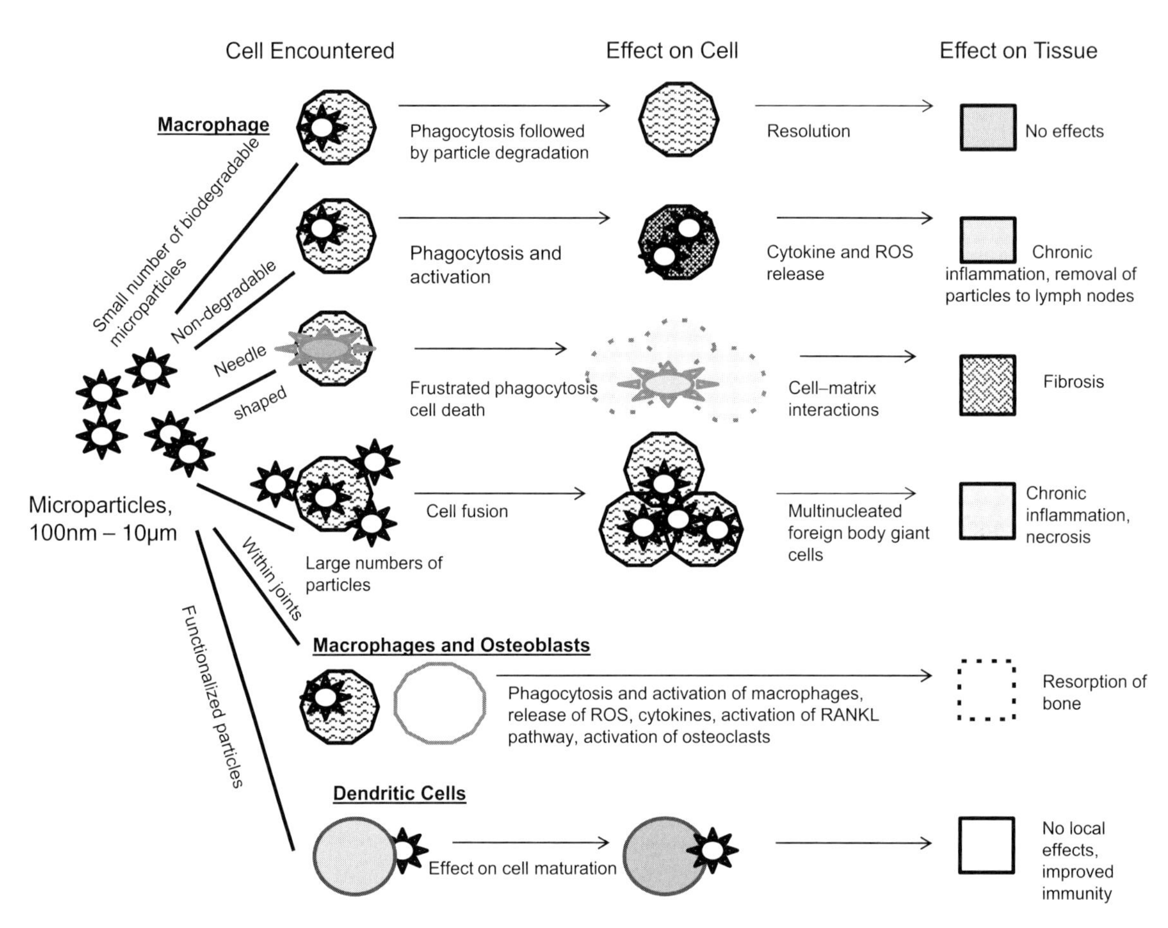


**Figure 3.31** Microparticle controlled biocompatibility pathways. This scheme shows the potential interactions between microparticles and cells and tissues. Microparticles will be in the size range 100 nm–10 μm and could be wear debris, corrosion or degradation products or therapeutic or diagnostic agents injected into the body. The macrophage dominates the response. Phagocytosis may result in rapid degradation of the particles leading to resolution and no long-term effects. An inability to degrade the particle may result in the release of cytokines, reactive oxygen species, enzymes, etc., giving chronic inflammation. Particles may be removed and deposited in the lymphatic system. Some particles, e.g., some forms of asbestos fibers, may resist phagocytosis, resulting in cell death and profound local responses including harmful fibrosis. Large numbers of particles may cause fusion of macrophages to give multi-nucleated foreign body giant cells and chronic inflammation and, possibly, necrosis. If the particles encounter macrophages within or near joints, there may be interactions with osteoclasts, resulting in bone resorption. The particles may encounter cells of the immune system and may influence dendritic cell maturation.

systems and biomaterial features that are specific to the scale parameters of the biomaterial. And again it is appropriate to deal with these in the context of how the tissues responds to surface topography at this scale, in this case the nanoscale, and how the human body responds to particles whose dimensions relate to that scale, in this case nanoparticles.

## Nanoscale topography

This discussion follows on from that on microtopography in the previous section. The question arises as to whether any effects on biocompatibility seen at the microscale are enhanced or negated if nanoscale topographical features are superimposed. From the implantable device

(a)

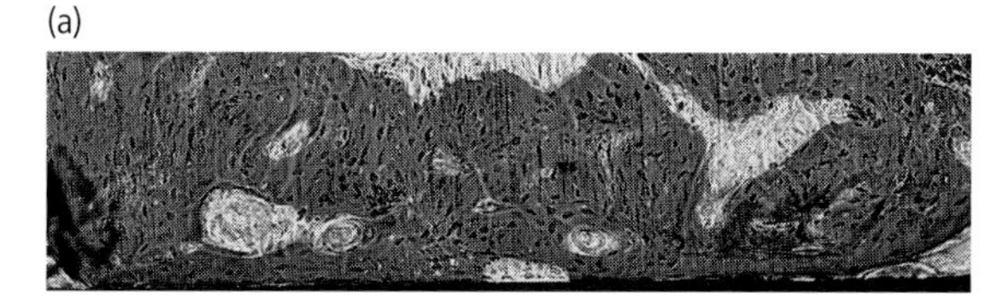

(b)

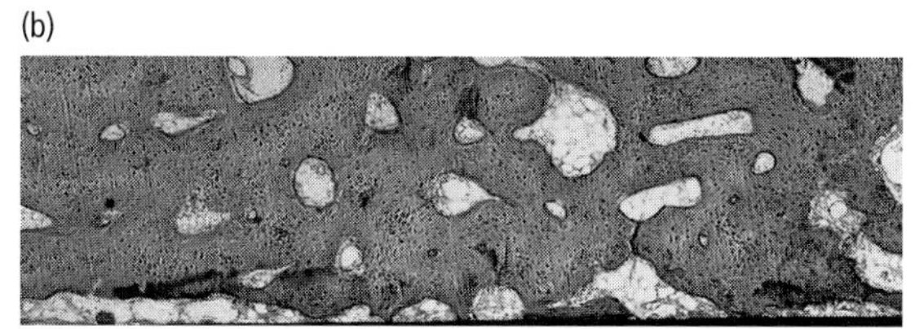

**Figure 3.32** Effects of nanotopography on bone attachment. Digital photographs of stained (methylene blue/basic fuchsine) histological sections at 4 weeks after implantation into minipig skull.
(a) Implant is titanium with a coating of 70 nm $TiO_2$ nanotubes. (b) Implant is machined titanium. Images courtesy of Professor Zhenting Zhang, Capital Medical University, Beijing, China. See plate section for color versions.

perspective, there is ample evidence to show that cellular and protein responses can be modified by nanoscale topographical features under experimental conditions, but whether this is clinically significant has yet to be demonstrated.

Most of the interest here has been with implants in contact with bone, especially dental implants, in view of the necessity to optimize the quality and speed of bone attachment (see Figure 3.32). Facilitation of positive osteoblastic response is a specific goal of this approach. Conventional implant materials such as titanium may be surface treated in order to produce this nanoscale topography. Such techniques include acid etching, alkali treatments, anodization and peroxidation. Under *in vitro* conditions, several features in the range 10 to 100 nm have been shown to produce changes in the osteoblast cytoskeleton, and in adhesion and proliferation. Once again, the clinical relevance here is not clear.

With respect to tissue engineering, there is a tremendous interest in identifying whether nanotopographical features can be used to control cell behavior within scaffolds and matrices. We shall reserve the majority of this discussion to Chapter 5. There is an interesting dichotomy here. If the material surface within a scaffold is influential in controlling cell behavior, for example the differentiation of stem cells, then it is intuitively obvious that the amount of surface area should be maximized. However, in order to do that, the size of the scaffold pores has to be decreased, which then makes cell seeding and nutrient flow more difficult. Combinations of microscale and nanoscale features have been used in order to circumvent some difficulties. It is also very difficult to differentiate the effects of material topography from all the other variables, including substrate stiffness and total surface area.

## Nanoparticles

Nanoparticles are relevant to biocompatibility in two main ways. First, components released from solid biomaterials, such as wear particles, are often of nanometer dimensions and consideration has to be given to the influence of particle size on the response. Secondly, the biomaterial may be deliberately introduced into the body in nanoparticulate form, for example to deliver a drug or a gene, or to track and label cells, and it is necessary to determine mechanisms whereby the particles are best delivered to their intended site of action and to estimate the potential for these particles to cause toxicity. This introduces the relatively new field of nanotoxicology into biocompatibility. Some of the issues discussed here are also touched upon in other parts of the book, dealing specifically with drug delivery and contrast agents for example. It is also important to recognize that much remains to be discovered here and it is necessary to take evidence of nanotoxicology in non-human health situations into account when attempting to understand mechanisms.

In some ways, interactions between the tissues of the body and nanoparticles mirror those with microparticles discussed in the previous section, but just different in scale. In other ways, however, they are very different. Cell surfaces are very heterogeneous, both with respect to chemistry and charge. At length scales from 10–50 nm, interactions may be associated with variation in surface protein structures. Particles of microscale dimension interacting with this heterogeneity tend to experience an averaging effect with respect to energy levels. Nanoparticles, however, of dimensions equivalent to these discrete patches interact quite differently because they may experience quite different specific environments.

We may analyze the biological performance and/or fate of nanoparticles within biological systems by considering the separate components of the overall system and their interactions. These components are the nanoparticle itself, the liquid biological medium (cell culture medium, extracellular fluid, plasma, blood, serum and so on), the solid–liquid interfacial region, and the interface between the latter region and the cell surface. These are indicated in Figure 3.33. The significance of these multiple interfacial regions is that the nanoparticle will interact with the liquid medium and adsorb components from that medium onto the particle surface, and it is this coated particle that will interact with the cells.

Not surprisingly, these interfaces, and the events that occur within them, are complex, being controlled as they are by multiple dependent and independent variables. On the nanoparticle side, these variables include size, shape, surface area, surface charge and energy, functional groups and ligands, valence state, crystallinity and the hydrophobic/hydrophilic balance. Within the human body the fluid varies in composition and character. In culture media, as will be described in Chapter 5, a wide variety of chemicals may be present. Within the solid–liquid interfacial region, it is largely surface hydration, free energy changes and charge variations due to ion adsorption, charge neutralization,

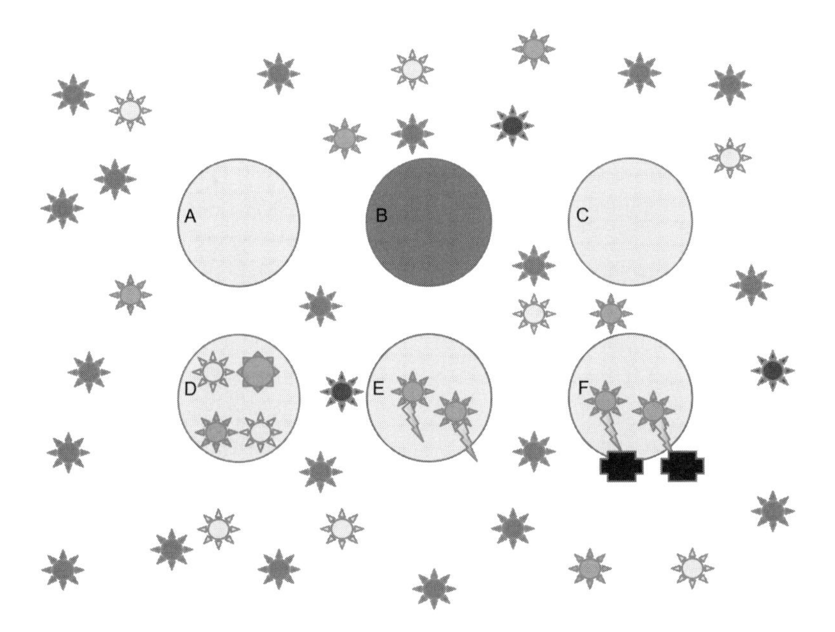


**Figure 3.33** The nanoparticle environment. Nanoparticles are associated with a protein corona once exposed to a biological environment. In A, the nanoparticle is introduced into the protein-rich environment, different protein species being colored red, yellow, blue and green. Protein red is the most abundant and will be initially adsorbed onto the nanoparticle surface preferentially (B). However, protein yellow has much higher affinity and may displace protein red (C). Protein blue has little affinity for the nanoparticle surface but does have affinity for protein yellow, and may attach to parts of that protein layer (D). Note that protein green is in low concentration and has little affinity and takes no part in the corona. In E we see that some epitopes on protein blue are exposed, which could combine with antibodies and other molecules. Based on diagrams in Monopoli *et al.*, *Nature Nanotechnology*, 2012, 7, 779–86. See plate section for color version.

electrostatic and electrosteric interactions that characterize the dynamic environment. Factors that control interactions between this layer and cells include receptor–ligand binding interactions, specific and non-specific interactions with membranes, and interactions within biomolecules, such as conformational changes and energy transfer.

A very important part of the nanoparticle–tissue interactions is that of protein coating of the nanoparticle. Proteins will adsorb rapidly to the nanoparticle surface, which may undergo conformational change and expose new epitopes, and display new functionality. This layer is sometimes referred to as the protein corona of the nanoparticle. As we would expect from knowledge of protein adsorption to material surfaces in other situations, the nature of this corona is a dynamic one and there will be exchanges between proteins over time, perhaps with the initial adsorption of abundant but non-adhesive proteins being superseded by the later adsorption of those which have a greater affinity for the surface but are less abundant (Figure 3.33).

Nanoparticles, with their protein corona, may interact with a cell membrane via adhesion to the cell surface lipid bilayer, involving both specific and non-specific interactions. The specific interactions are those between surface ligands and cell membrane receptors, potentially resulting in receptor-mediated endocytosis. Obviously the ligand has to meet the receptor for this to occur and the receptor may diffuse along the cell membrane in order to optimize this. Non-specific interactions arise from surface charge and hydrophobicity characteristics; for example, charged nanoparticles may be attracted to the charged phospholipid head groups in the bilayer. There are several different pathways of endocytosis, which are characterized on the basis of the endocytic proteins that opsonize the particles. The classical route for entry is dependent on clathrin, a fibrous protein that forms a polyhedral coat around a particle. The particle coated with clathrin becomes engulfed in a pit on the cell membrane, which is then pinched off by enzymatic activity. This is known as clathrin-mediated endocytosis. The secondary pathway is clathrin-independent endocytosis, or caveolae-mediated endocytosis, where caveolae form a lipid-based raft with the particles in the membrane.

The surface topography of the cell may also play a role, as nanoparticles interact with small radii depressions or protrusions. As noted earlier, nanoparticles should penetrate cell membranes rather more easily than microparticles. This does not mean, however, that the smaller the nanoparticle the easier will be the process. With extremely small nanoparticles there may be too few contact sites for there to be an energetically favorable entry; there appear to be optimal particle size radii of between 15 and 30 nm for endocytosis. It might be expected that particle shape would have some influence, but reproducible data has not emerged in this area.

In the context of biomaterials, an interesting situation arises once a particle has been endocytosed, since a nanoparticle has a number of opportunities to cause damage but also the possibility of doing good. We may consider these effects in terms of the cytoplasm and the nucleus. In the cytoplasm, the particle may enter the mitochondria where it can cause oxidative stress. Within the cytosol it can cause an increase in calcium, which also causes oxidative stress. The results of the oxidative stress include the release of inflammatory mediators, such as pro-inflammatory transcription factors, and the promotion of apoptosis and necrosis. At this stage, the toxicological pattern has to be determined on a case-by-case basis. Cationic polymers, for example, have been investigated for drug and gene delivery, but at nanoscale several of these have significant toxicity profiles. Polyethyleneimine induces lactate dehydrogenase leakage, induction of mitochondrial apoptosis and lysosomal disruption. Cholesterol cationic lipids inhibit protein kinase C. Quantum dots can produce subtle alterations in cell function. Quantum dots based on cadmium can release free cadmium that has its own toxicological profile, including the release of free radicals. Fullerenes may be pro-inflammatory, increasing IL-8 and TNF$\alpha$ release, but there is also evidence of anti-oxidative effects. These effects are summarized in Table 3.5.

Table 3.5 **Toxicological effects of nanoparticles. This list represents the possible effects of nanoparticles on the human body. There is clear experimental and clinical evidence for some of them; in other cases the effects are theoretical and not necessarily proven**

| |
|---|
| Generation of reactive oxygen species, leading to oxidative stress |
| Pro-inflammatory responses, including granuloma formation and acute phase protein expression |
| Effects on phagocytic function, especially prolongation of chronic inflammation and inhibition of clearance of microorganisms and tissue debris |
| Effects on proteins, denaturation and loss of enzyme activity |
| Effects on mitochondria, including disruption of energy processes |
| Uptake by reticulo-endothelial system, sequestration in liver and spleen |
| Nuclear uptake, leading to DNA damage and mutagenesis |
| Uptake in neuronal tissue, damage to central and peripheral nerve systems |
| Cardiovascular effects, including thrombosis, myocardial infarction and stroke |
| Effects on cell cycle, including proliferation, senescence and apoptosis |

Even though this evidence of mechanisms of nanoparticle toxicity may be inconsistent, there are some clear observations of harmful *in vivo* effects, usually arising from inhalation exposure, which indicate that caution has to be exercised. Nanoparticles can certainly cause pulmonary inflammation, with effects covering many forms of particle and generally related to dose as measured by surface area. They can also affect the immune response to common allergies.

With respect to the cell nucleus, nanoparticles have considerable potential for genotoxicity, as we discuss in the next section. On the other hand, in gene therapy or gene transfer, it is essential for the nanoscale vector to reach the nucleus. This will be discussed more in Chapter 6. The important point here is that there are defensive processes within cells that normally prevent foreign molecules and particles gaining access to the nucleus. These involve endosomes. The vesicles associated with endocytosis transfer the nanoparticles to endosomes. Endosomes located at the periphery of the cell have a mildly acidic pH. Internalized matter is then transferred to late endosomes where the pH is around 5.5, following which molecules and particles are sorted and passed to lysosomes with a pH of 4.8 and an abundance of enzymes. These normally degrade particles and return breakdown products to the cytoplasm. It is possible to design systems that avoid this process, giving the so-called endosomal escape, again discussed in Chapter 6. If the DNA-carrying nanoparticulate vector gets through to the nucleus, it may be able to cross the nuclear membrane.

**Genotoxic potential of engineered nanoparticles** As we have seen, nanoparticles may be able to enter cells by a variety of routes, and, once inside the cell may be able to promote DNA damage. If they are able to penetrate the nucleus, direct interactions between them and DNA molecules or related proteins may lead to damage, perhaps by causing intranuclear protein aggregation which can inhibit replication and transcription processes. In addition, indirect damage may occur, not by interaction with the DNA molecule itself, but by their effect on other cellular proteins which influences the environment within which the DNA operates (Figure 3.34).

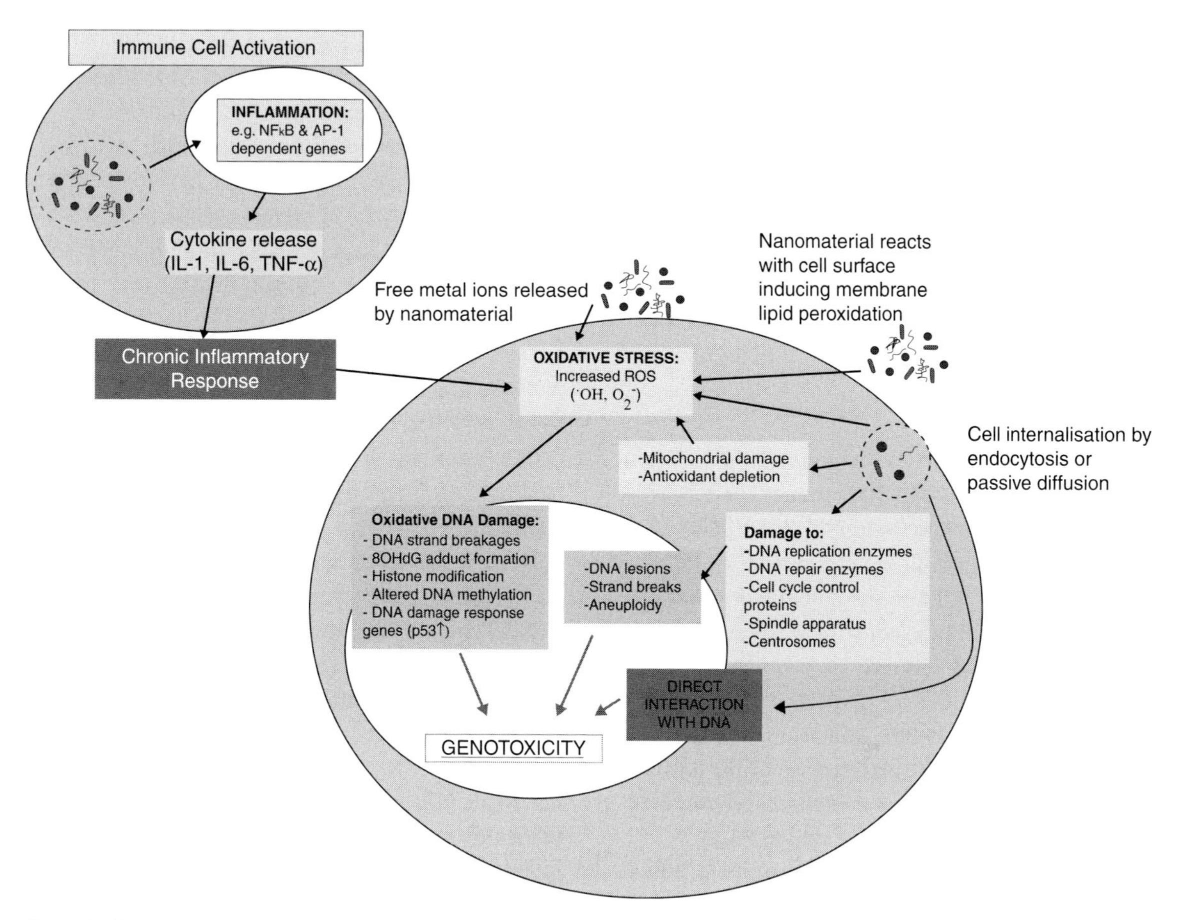


**Figure 3.34** Nanomaterial genotoxicity can arise by primary or secondary mechanisms. Primary genotoxicity is imparted by the nanomaterials themselves and can be induced by direct interaction with the DNA to cause damage or by indirect mechanisms. Secondary genotoxicity is induced by chronic inflammation as a result of inflammatory mediators released by immune cells, which induce oxidative stress in the cells of surrounding tissue. Courtesy of Dr. Shareen H. Doak, Institute of Life Science, Swansea University, SA2 8PP, Wales, UK.

One of the more important mechanisms of genotoxicity here is that which involves compromising DNA repair. DNA repair mechanisms are essential for preventing genetic damage, which occurs frequently, leading to permanent mutations. The tumor suppression gene p53 is a key molecule in effecting DNA repair, but it also is able to trigger apoptosis if the DNA damage is extensive. Nanoparticles may be able to increase the activity of p53 and cause a significant amount of apoptosis and cell death.

As might be expected, there are several material-specific effects on genotoxic potential and several generic contributory factors. Gold nanoparticles may cause DNA damage through oxidative stress and decreased expression of DNA repair genes. Cobalt ions released from cobalt nanoparticles may cause single strand breaks and chromosomal aberrations. Titanium dioxide nanoparticles can act as catalysts for the generation of ROS, especially in the presence of UV light and cause chromosomal damage. CdSe/ZnS quantum dots can cause DNA fragmentation under some circumstances. Fullerenes can form complexes with DNA inducing strand breakages, while carbon nanotubes have the ability to cause chromosomal damage and point mutations. These are also summarized in Table 3.5.

With respect to the generic characteristics, size and shape, which together determine surface area, are very important since they influence both the capacity of the cell for internalization and interactivity with the nucleus. Related to these factors is the tendency towards agglomeration, which is heavily dependent on hydrophobicity. In addition, the surface charge plays a significant role in internalization processes and in DNA interactions.

**Immunological properties of nanoparticles** This is an area of particular importance since there is a potential for nanoparticles to initiate an immune response that is harmful with respect to overall biocompatibility, but also the potential for them to augment the immune response to antigens when used as an adjuvant in a vaccine.

The immunostimulatory properties of nanoparticles are complex in nature, and depend on their molecular structure, the architecture of the particles, their chemical reactivity and degradation products and stability. Once again these responses are material-specific. One aspect that appears to influence the inflammatory response is the maintenance of the balance between T-helper 1 and T-helper 2 responses. While large (microparticle) sizes tend to induce Th1 responses, small nanoparticles induce the Th2, inducing cytokine and immunoglobulin production. There is also evidence that nanoparticles may stimulate adaptive immunity, promoting CD8 and CD4 T-cell responses.

Several types of nanoparticle have been shown to act as adjuvants, giving enhanced immunoglobulin production compared to standard adjuvants. This has been seen with PAMAM dendrimers, lipid-coated nanoparticles and fullerenes. Generally it is assumed that these nanoparticles enhance antigen uptake and stimulate antigen-presenting cells.

**The essence of nanoscale biocompatibility mechanisms** Dealing first with nanoparticles, the general principles on which nanoparticle toxicology is based can be built upon the discussion about microparticles but with some significant differences. These are based on the different mechanisms of passage across the cell membrane, which involve endocytic processes rather than phagocytosis, and then the mechanisms of passage of the particle through the cytoplasm, possibly gaining access to mitochondria where significant damage may be caused through ROS generation. The normal pathway for nanoparticles that have been internalized involves a series of transfer processes through endosomes, resulting in their degradation and inactivation within lysosomes. It is possible, through design of the nanoparticle chemistry, to avoid this fate, through endosomal escape and pass directly towards the nucleus, which could under adverse circumstances lead to genotoxicity, or under favorable conditions with DNA-carrying nanoconstructs, to intentional and beneficial gene transfer. These mechanisms for nanoparticle-mediated biocompatibility are shown in Figure 3.35(a).

With respect to topography, at both micro- and nanoscale, it is very difficult with current knowledge to identify their roles in biocompatibility. We attempt in Figure 3.35(b) to summarize the current situation.

### 3.3.6 Scenario E: Mechanisms of biocompatibility involving delivery of pharmaceutical agents from macroscale biomaterials to host

In some situations, the mere presence of an implantable device does not guarantee ideal clinical outcomes. It may be that in achieving one desirable result the device and/or material induces a different but undesirable effect. Alternatively the device itself may not yield the optimal effect, either in the quality of the result or the speed at which it is produced. In these types of situation, it may be beneficial to assist the material by the simultaneous delivery of a pharmaceutical agent; note that we are not discussing here the use of biomaterials to deliver drugs to the body with improved characteristics, as will be discussed in Chapter 6, but those situations where the drug is an integral part of the device and is present solely to improve the performance of the device. The combination of biomaterial and drug is invariably intended to beneficially modify the biocompatibility of the device. A few important examples demonstrate the significance of this approach.

### Anti-proliferative drugs – drug-eluting stents

Intravascular stents will be discussed in Chapter 4. These are used to address the massive clinical problem of atherosclerosis, particularly in the coronary arteries. The problem faced here may be considered as a simple series of events, where good, life-saving effects are seen through the use of an implantable device, but which is associated with unfortunate biocompatibility sequelae. The sequence starts with the progressive development of atherosclerotic plaque in the artery, then moves to the eventual clinically detectable consequences of the stenotic vessel and the minimally invasive procedure of angioplasty in which a catheter-delivered balloon enlarges the vessel lumen, reducing the stenosis. Unfortunately there is then the return of stenosis because the effects of the angioplasty are only transient. A stent, which is a metal tubular structure, is then implanted in the affected area in order to keep the vessel open. Even more unfortunately, the so-called bare-metal stent does not always keep the vessel open since it is itself able to irritate the endothelium and, in order to minimize this effect, a layer of a powerful drug-containing material is placed on the surface of the stent in order to control the in-stent restenosis. In practice, because of the apparent clinical success of the drug-eluting stents (DES), the clinician usually circumvents the middle stages and implants one or more DES as the front-line treatment. Let us now consider the mechanisms involved and implications for biocompatibility.

It should be emphasized here that this medical device is being placed in an area of active disease and the host response has to be considered in the context of the damage that has already been done and the changes that have occurred, both reversible and irreversible, to the tissues (Figure 3.36). The development of the atherosclerosis follows a pattern in which the initial lesion, which is populated by isolated macrophages, leads to a partially fatty lesion with intracellular lipid accumulation. This extends to an intermediate stage with small extracellular lipid pools, which may consolidate in core areas to give an atheroma, which develops into a fibroatheroma as it becomes fibrotic and calcific, and finally the full lesion, which may be associated with hematoma and thrombus. This is a process that involves inflammation, and granuloma and matrix formation. The inflammatory phase involves platelets, macrophages and a variety of leukocytes, releasing a range of molecules such as PGDF and transcription factors c-fos and c-myc. With granuloma formation, various growth factors, including PDGF, bFGF, EGF and TNF are released, and matrix formation significantly involves smooth muscle cells, and especially their migration and proliferation, with effects on collagen, proteoglycans and elastin, as the plaque develops. It is into this environment that a stent is placed; it is not surprising that material stents alone are not equipped with the means to routinely and effectively arrest this process.

As noted above, restenosis will occur after angioplasty alone. Immediately after balloon dilatation, elastic recoil of the arterial wall will take place as the elastic fibers react to overstretching and a form of contraction occurs, rather like the contraction seen in wound healing. A stent will prevent this, but now we have a foreign material that is compressing on the inner lumen of the vessel. This causes the so-called in-stent restenosis (ISR). The stent struts contribute to endothelial denudation and deeper injury of the vessel wall; the importance of the mechanical influence on the endothelium is seen from the fact that stents with thin struts produce less damage and are associated with lower incidences of ISR in many situations. Restenosis is seen in Figure 3.1. The ISR is predominantly associated with neointima formation, as smooth muscle cells (SMCs) and myofibroblasts migrate from the media and adventitia of the vessel into the lumen, where there is then

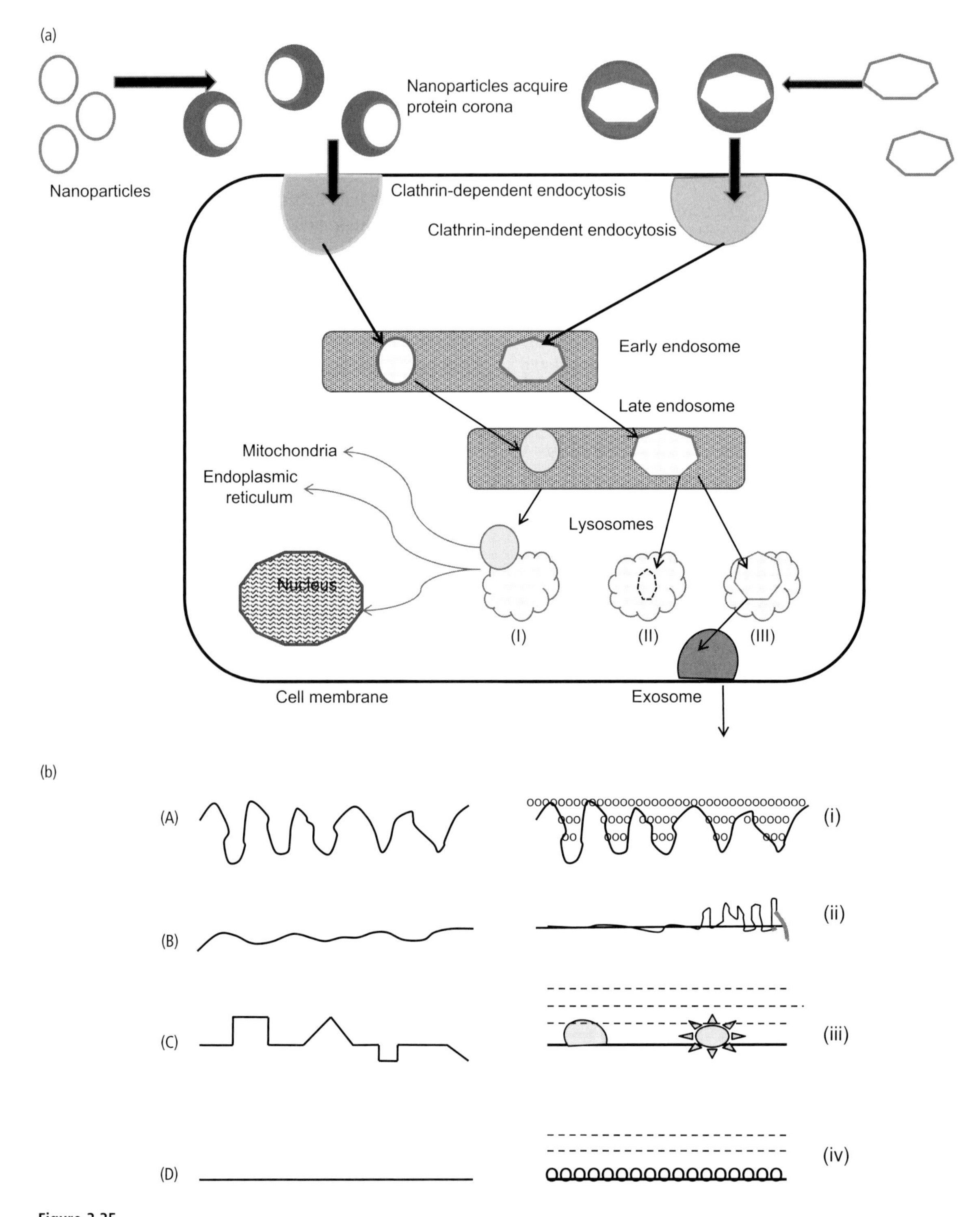


**Figure 3.35**

**(a)** Nanoparticle-controlled biocompatibility pathways. Nanoparticles, by definition with diameters up to 100 nm, will have varying shape and reactivity. They will acquire a protein corona. Particles may be internalized through clathrin-dependent or proliferation and secretion of ECM and remodeling of the intimal surface. Numerous growth factors and cytokines are released from the inflammatory cells, platelets and endothelial cells. From the white cells, PDGF, FGF and TGF β1 stimulate SMC proliferation and production of ECM and leucotrienes cause vessel spasm. From the platelets, serotonin enhances aggregation and thromboxane A2 again enhances SMC proliferation. This is also assisted by the release of many factors from injured endothelial cells. Crucial to this discussion of neointima formation, or intimal hyperplasia, is the fact the release of growth factors and cytokines affect cell cycle entry genes and kinase inhibitors. Such events are responsible for SMC proliferation and hyperplasia. This leads to the possibility that inhibition of SMC proliferation, for example by disruption of certain cell cycle stages, could reduce hyperplasia and ISR.

This is where the pharmaceutical agents suggest themselves as possible solutions. Drugs that have anti-proliferative and anti-inflammatory properties, as well as pro-healing activity would be desirable. Several drugs have been used clinically in DES, the three most common being sirolimus, everolimus and paclitaxel (Table 3.6). Sirolimus (also known as rapamycin) is primarily used as an immunosuppressive drug in organ transplantation patients, with a mechanism that inhibits the response to IL-2 and blocks the activity of T- and B-cells. Everolimus is quite similar, having anti-proliferative and anticancer activity derived from blocking growth-driven transduction signaling in T-cells. Paclitaxel is a chemotherapeutic agent that affects microtubules and blocks cells in the G2/M phase such that they are unable to undergo mitotis. Stents with any of these three drugs incorporated into their surfaces have been shown to significantly reduce ISR, although with some differences in clinical outcomes. Their mechanisms of action are also a little different in detail but they function in some type of anti-proliferative manner. With sirolimus for example, the mechanism appears to be a receptor-based antagonism of the intracellular enzyme known as mTOR, mammalian target of rapamycin, which is a mediator of a phosphorylation signaling pathway, inhibition of which leads to cell cycle arrest in late G1 to S phases.

The lesson here is that through an understanding of the molecular and cellular bases of biocompatibility, it is possible to determine the critical event and address the point at which the host response takes a turn in the wrong direction. This involves interventions that cause a different route to be taken, converting a destructive pathway into a chronic pathway of benign acceptance, or, more probably, a metastable pathway. It has to be admitted, of course, that the use of these very powerful agents, with their own toxicity profiles, is not necessarily a good precedent

clathrin-independent endocytosis, following which they will pass through cytoplasm and may be taken up within early endosomes and then late endosomes and then transferred to lysosomes, which contain highly reactive chemical species. Route (II) involves rapid destruction of the nanoparticles by these species. Particles that are resistant to this may be transferred to exosomes and released, probably after modification, into the extracellular space (Route III). It may be possible to design the particle surface such that take up by lysosomes is avoided, and they may be directed to the nucleus or organelles. This so-called endosomal escape allows delivery of nanoparticle content to specific sites, for example DNA to the nucleus in gene transfer as in Route I.

**(b)** Surface topography and biocompatibility. It is unclear exactly how surface topography influences biocompatibility, and there is much contradictory evidence. On the left are the generic topographical features, which could exist at an length scale: (A) is a porous surface, (B) represents random topographical features, such as produced by machining or grit-blasting, (C) represents controlled features or patterning and (D) shows a surface without any recognizable feature. On the right are the generic responses, including (i) tissue ingrowth, (ii) protein adsorption with varying degrees of conformational change and strength of attachment, (iii) shows random cell attachment, with varying degrees of adhesion, activation and phenotype change, (iv) shows regular cell patterning on the surface, potentially controlled by surface patterning of the material, for example cell alignment.

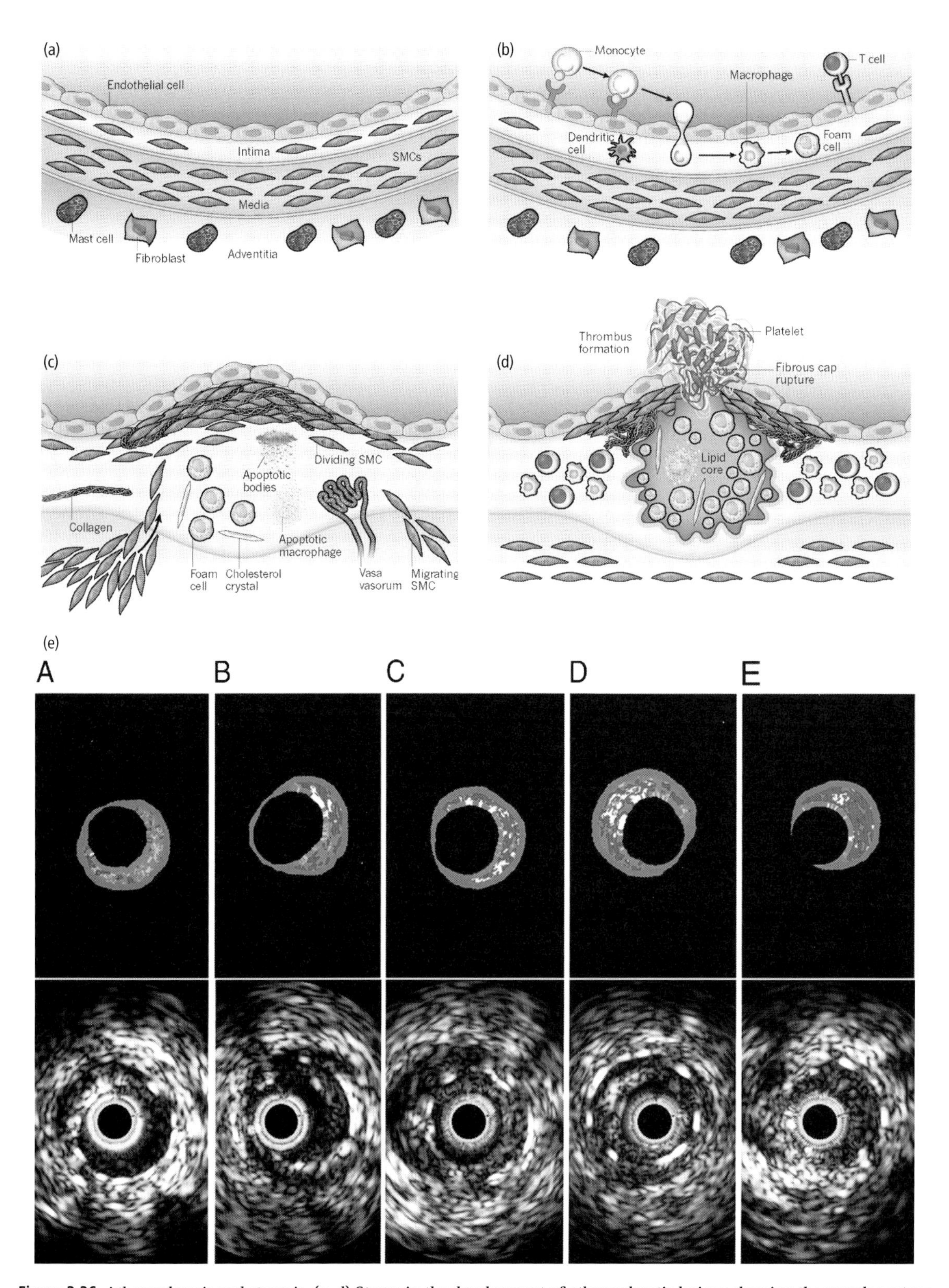


Figure 3.36 Atherosclerosis and stenosis. (a–d) Stages in the development of atherosclerotic lesions showing the musclar artery and the cell changes that occur during disease progression are shown.

and they constitute a high risk solution to a biocompatibility problem. Moreover, it has to be admitted that complete knowledge of the pharmacokinetic and pharmacodynamic properties is only emerging with clinical experience. Fundamentally it cannot be assumed that these properties will be the same when the drugs are released from surfaces of medical devices as when they are delivered by other routes, for example by intravenous injection. These factors have become rather obvious with the evidence that the beneficial effect with respect to ISR may be countered by some adverse effects. This particularly concerns the observations of so-called late stent thrombosis, in which a higher incidence of thrombo-embolic complications are seen with DES than with bare-metal stents. The reasons for this, or indeed the reproducibility of the data, are not clear, but evidence does suggest that it is associated with delayed arterial healing; in other words it may be difficult to obtain the ideal balance as reduced endothelial hyperplasia may be achieved to the detriment of normal endothelial healing. This problem is counteracted by the systemic delivery of appropriate anti-platelet agents, usually a combination of aspirin and clopidogrel, but even then controversies arise on the duration of treatment. It does seem that some patients are at risk as soon as the treatment ceases, say after 6 or 12 months, but continuation after this time may lead to unacceptable risks of bleeding.

### Bone morphogenetic proteins in spinal surgery

Implantable devices used in spinal surgery will be discussed in Chapter 4. One of the principal areas here involves spinal fusion, in which a material is placed between two vertebrae in order to encourage fusion. Although this reduces mobility of that part of the spine, it gives enormous relief of pain once a degenerated intervertebral disc has been removed. The most widely used material here is autologous iliac crest bone, which gives high rates of fusion. However, the process does involve a second operative site with associated risks of morbidity and pain. One alternative relies on a biomaterial cage that contains so-called bone-substitute materials, often calcium phosphate granules and/or allografts. In recent years these have been supplemented with biologically active agents that are intended to increase the osteogenic potential of the device. These agents include bone morphogenetic proteins (BMPs), and specifically BMP-2 and BMP-7. These are growth factors that naturally exist within the bone matrix and act as pleiotropic regulators (that is, a single gene that can influence multiple phenotypic traits) of extracellular matrix synthesis, chemotaxis, mitosis and other events. They bind to membrane-bound receptors and profoundly control intracellular cascades and are responsible for osteoblast differentiation and bone matrix formation. BMP-2 is one of this family that possesses osteoinductive properties and it has been widely investigated as a

(a) Normal artery.
(b) Initial steps of atherosclerosis.
(c) Lesion progression.
(d) Thrombosis. Shown is a fracture of the plaque's fibrous cap, which has enabled blood coagulation components to come into contact with tissue factors in the plaque's interior, triggering the thrombus that extends into the vessel lumen, where it can impede blood flow. Reprinted by permission from Macmillan Publishers Ltd: *Nature*, May 18, 2011, 473(7347), Libby P *et al.*, Progress and challenges in translating the biology of atherosclerosis.
(e) Appearance of cross-section of arteries after atherosclerosis and stenting, using virtual histology intravascular ultrasonography. (A) Six-month follow-up of paclitaxel-eluting stent (percent necrotic core [%NC] 10%, percent dense calcium [%DC] 2%); (B) 9-month follow-up of paclitaxel-eluting stent (%NC 28%, %DC 8%); (C) 22-month follow-up of paclitaxel-eluting stent (%NC 39%, %DC 20%); (D) 48-month follow-up of bare-metal stent (%NC 40%, %DC 25%); and (E) 57-month follow-up of bare-metal stent (%NC 57%, %DC 15%). Reprinted from *Journal of the American College of Cardiology*, 59, Park S-J *et al.*, In-stent neoatherosclerosis, 2051–7, © 2012, with permission of Elsevier. See plate section for color version.

Table 3.6 **Drugs in DES: a selection of clinically used drug-eluting stents**

| Drug | Stent material | Polymer |
|---|---|---|
| Sirolimus | Stainless steel | Polyethylene-co-vinyl acetate and poly-n-butyl methacrylate |
| Sirolimus | Stainless steel | Polylactic acid |
| Sirolimus | Cobalt–chromium | Polylactide-co-glycolide |
| Sirolimus | Stainless steel | No polymer/microporous surface |
| Paclitaxel | Stainless steel | Poly(styrene-isobutylene-styrene) |
| Paclitaxel | Platinum–chromium | Poly(styrene-isobutylene-styrene) |
| Paclitaxel | Stainless steel | Polylactic acid |
| Zotarolimus | Cobalt–chromium | Phosphorylcholine |
| Everolimus | Cobalt–chromium | Poly-n-butyl methacrylate |
| Biolimus A9 | Nickel–titanium | Polylactic acid |
| Biolimus A9 | Cobalt–chromium | Polylactic acid |

possible clinically acceptable enhancer of bone formation.

Initially BMP-2 was derived from bovine bone matrix but is now available in a human recombinant form, rhBMP-2. A rhBMP-7 is also used, although not so extensively. It is clear that rhBMP-2 is able to enhance bone formation in clinical situations. It binds to kinase receptors on a cell surface, which then transduce signals via Smad proteins, and activate bone-forming genes, as shown in Figure 3.37. The real issues with these recombinant proteins have concerned dosage and the related pharmacokinetics and pharmacodynamics. Clinically they are used in situations where they are added to a simple biomaterial that has potential to release the molecule over time. Collagen sponge is one such material. Matching the dose with the known pharmacokinetic properties associated, for example, with intravenous administration, is very difficult. Uptake of rhBMP-2 is rapid in most tissues but residence time is usually very short, in a matter of minutes in most situations. The residence time for rhBMP-2 within a collagen sponge *in vivo* is less than 10 days and the half-life for local delivery is 2–3 days. Bone induction is observed at the site of delivery and is limited to the time period when the rhBMP-2 is present. The quantity of new bone is dose-related up to a threshold, above which no further bone formation occurs. The question arises as how to best control BMP dosage so that optimal enhancement of bone formation occurs in an acceptable time without inducing any undesirable effects.

Such undesirable effects can and do occur. Complications arising from its use include osteolysis, swelling, heterotopic bone formation and antibody reactions. The osteolysis has occurred in adjacent bones just away from the cage and the heterotopic bone formation in adjacent soft tissue sites. These events, the former of which tends to be transient but the latter being chronic and troublesome, suggest that this complex, multifunctional molecule is difficult to control and may have significant, and seemingly competing, effects on bone remodeling. The swelling or edematous response is also of concern. This may occur in the vertebral body, where it may be asymptomatic or associated with persistent pain, but may occur some distance away. This

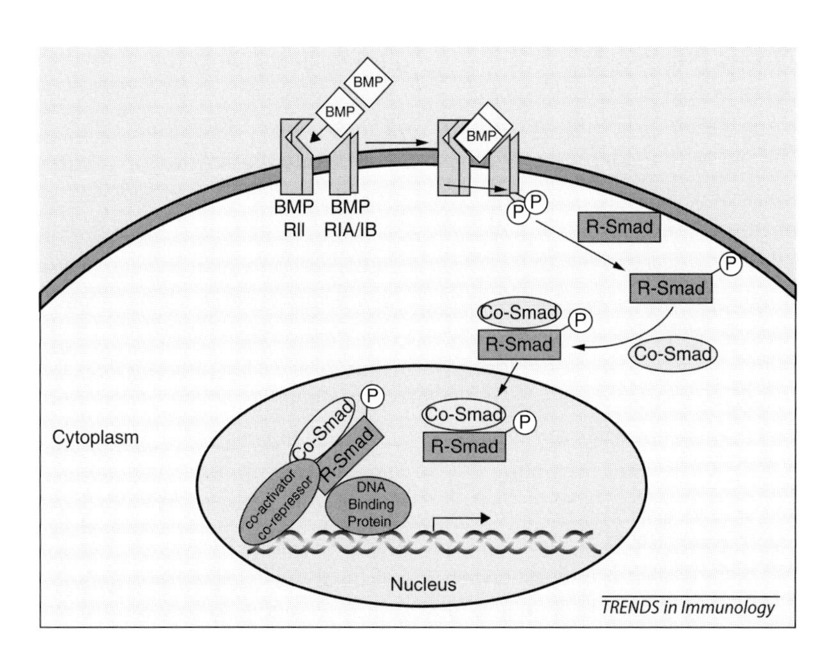


**Figure 3.37** Bone morphogenetic protein signaling pathways. BMP2/4 ligands attach to type II BMP receptor (BMPR-II), which then recruits and phosphorylates (P) type I BMP receptors (BMPR-IA or BMPR-IB). The phosphorylated BMPR-IA and/or BMPR-IB then phosphorylate the receptor-regulated R-Smads (Smad-1, Smad-5 and Smad-8), which recruit the common mediator Co-Smad (Smad-4) and translocate to the nucleus. In the nucleus, the R-Smad–Co-Smad heterodimers recruit DNA-binding proteins and co-activator or co-repressor proteins, and activate or repress the expression of target genes. Reprinted from *Trends in Immunology*, 24, Varas A *et al.*, The role of morphogens in T-cell development, 197–206, © 2003, with permission of Elsevier.

became obvious in some cases where rhBMP-2 products were used in anterior cervical spine locations, where swelling in the region of the patient's airway can produce significant problems. It is also known that resident cells of the CNS are responsive to BMP signaling, which could have an impact on the spinal cord.

## Bisphosphonates

As noted elsewhere in this chapter and in Chapter 4, the major limitation of joint replacements is the loosening associated with osteolysis. It will be recalled that activated macrophages influence the osteoblast/osteoclast balance that results in bone resorption. Considerable attention has been paid to the possibility of some pharmacological intervention to slow down or reverse this process. One of the molecules that is involved in peri-prosthetic bone resorption is prostaglandin $E_2$ ($PGE_2$). This is synthesized by activated macrophages. One group of drugs is theoretically able to suppress $PGE_2$ synthesis. These are molecules of the bisphosphonate family, which are close analogues of natural pyrophosphates and are known to decrease bone resorption and have general anti-inflammatory properties in certain conditions, including osteoporosis. Bisphosphonates have been used in attempts to reduce osteolysis. It has been shown that the drug acts directly on osteoclasts by enhancing their apoptosis and indirectly by inhibiting $PGE_2$ activity. An interesting question arises as to how bisphosphonates could be administered to patients. Both local delivery through incorporation of the molecules into devices (in cements or adsorbed to implant surfaces) or systemic delivery have been used. The evidence shows that some small difference may be seen with both approaches.

### Growth factors in tissue engineering

As we shall see in Chapter 5, tissue engineering is concerned with the generation of new tissue, usually through a combination of cells and biomaterials, and the stimulation of those cells by combinations of molecular and mechanical signals. One of the more important groups of molecules that achieve this signaling is the family of growth factors. Growth factor delivery in tissue engineering is not the same as conventional drug delivery, which is discussed in Chapter 6, since in tissue engineering the material is not simply a vehicle that delivers drugs but it works synergistically with the molecules of the drug.

Growth factors play a central role in transferring information between cells and the ECM and stimulating mechanisms of tissue formation and repair. Their modes of action include the stimulation of chemotaxis, mitogenesis, cell differentiation, metabolism and apoptosis. In order to be effective, the growth factor molecules have to reach the target sites and remain there long enough to exert their effect. This is a critical point as the simple infusion of growth factors in solution into the target area will not generally be effective since they diffuse away and are inactivated. Hence the growth factor has to be provided with good spatiotemporal control through an appropriate materials-based technology. Among the growth factors relevant to tissue engineering are the bone morphogenetic proteins (BMPs), vascular endothelial growth factor (VEGF), epidermal growth factor (EGF) and fibroblast growth factor (FGF).

These processes are discussed in detail in Chapter 5 and are not considered further here.

### The essence of pharmacological biocompatibility mechanisms

Referring back to Figure 3.24, it will be recalled that several pathways associated with the chronic response to implanted devices were identified. Within the destructive pathway were events and mechanisms that could lead to undesirable long-term effects including excessive tissue proliferation or tissue resorption. One major factor in the pharmacological biocompatibility scenario is the delivery of active molecules that alter this pathway and return it back to the quiescent pathway of benign acceptance. This is shown in Figure 3.38(a).

In an alternative version of this scenario, the initial stages of the host response may not be ideal in relation to the required result, either in terms of the nature of the tissue that is being formed or the speed at which this is happening. Under *in vivo* conditions this may be addressed by the delivery of active molecules that re-direct this process, as indicated in Figure 3.38(b).

Under *in vitro* conditions, as with tissue engineering processes, active molecules may be required to assist in the control of cellular processes, including stem cell differentiation and maintenance of phenotype. Since the regeneration of tissue under these conditions is determined by both molecular and mechanical signaling, it is important to have the right balance, with synergistic effects of active molecules and material surfaces, as indicated in Figure 3.38(c).

## 3.4 Systemic or remote site biocompatibility

### 3.4.1 Reproductive toxicology

As the title suggests, there are some implications for reproductive processes arising from the use of biomaterials. Although potentially of major significance should these occur, the possibilities are generally remote and may be considered only briefly. The major consideration here is the possibility of the systemic distribution of chemicals in females, their passage across the placenta and into a developing fetus. Control of reproductive function is accomplished by the coordinated activity of the hypothalamus, the pituitary gland and the gonads, known collectively as the HPG axis, within the endocrine system. The HPG axis forms *in utero*, and the associated hormonal milieu is extremely important for fetal development, especially with respect to testosterone and estrogen. Exposure to

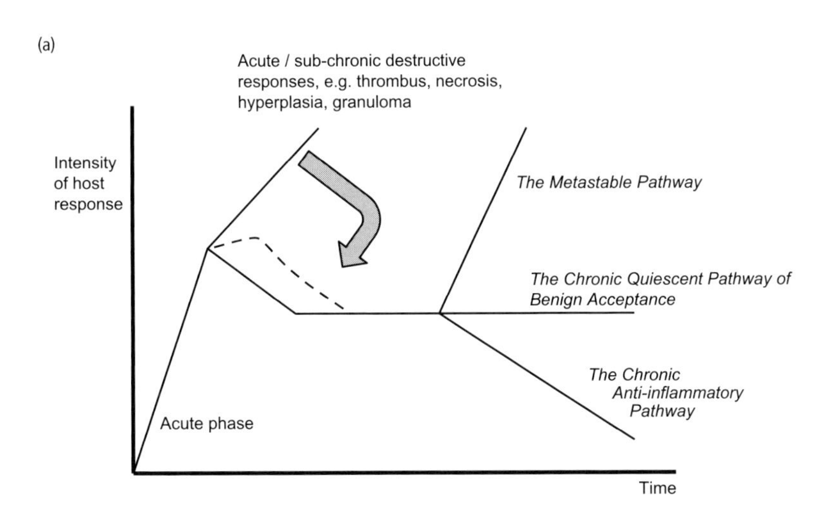


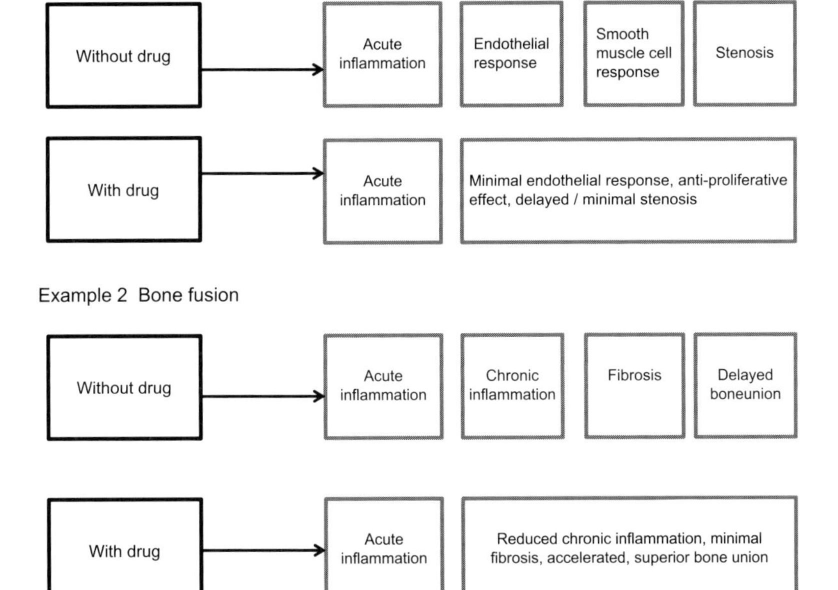


**Figure 3.38** Pharmacological control of biocompatibility pathways.
**(a)** The use of a drug to alter, suppress or ameliorate the acute/sub-chronic destructive response so that it approached the desirable chronic quiescent pathway.
**(b)** Two examples of how the addition of a drug to a biomaterial, and the sustained release of that drug into the tissue, can modulate the long-term response.

chemicals at this time can adversely affect these hormones, by inhibiting their synthesis or signaling mechanisms, affecting the development of the whole of the neuroendocrine system and potentially altering reproductive function and fertility.

Within the area of environmental exposure and toxicology, a number of industrially used chemicals are known to be endocrine disrupters. These include the major toxins polychlorinated biphenyls (PCBs) and polycyclic aromatic hydrocarbons (PAHs). These are not important as far as biomaterials are concerned, but there are some other substances that do require consideration. Bisphenol A is widely used in the manufacture of medically used plastics, including polycarbonates and many dental preparations. Di-ethylhexyl phthalate (DEHP) is used

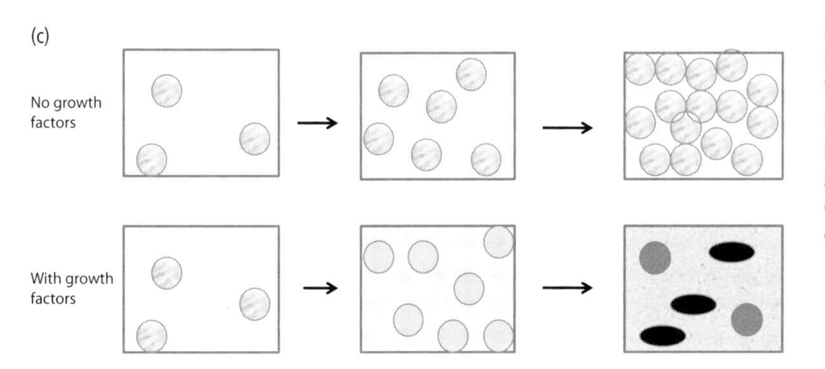


**Figure 3.38** *continued*
(c) Growth factors in cell culture. Without specific growth factors, cells may undergo normal expansion and produce confluence. With specific growth factors, the cells may be directed to change phenotype and/or express extracellular matrix.

as a plasticizer in polyvinylchloride, which is extensively used in catheters and blood bags. In both cases there is evidence from experimental animals that some effects on the reproductive organs are possible, but effects in humans are unclear. As always, a risk–benefit analysis has to be undertaken in order to assess the clinical significance of these possibilities, but these facts do indicate that it is prudent to consider the potential for reproductive toxicity in any newly developed biomaterials.

### 3.4.2 Carcinogenicity

Carcinogenicity is the term used to describe the phenomenon of cancer initiation by an external agent. In the context of biomaterials it is traditionally used to denote the situation in which cancer is caused directly by the material. Strictly speaking, this generic use of the term is incorrect since a carcinoma is only one of many types of tumor: tumorogenicity would be a better term. Cancer is the uncontrolled growth of abnormal cells. A carcinoma originates in the skin or the lining of internal organs. A sarcoma originates in connective tissues – there are many different types depending on the nature of the tissue, for example an osteosarcoma originating in bone. Leukemia originates in blood-forming tissue and lymphomas in tissue of the immune system.

There are several causes of cancer, including ionizing radiations, but of specific relevance here is the fact that many chemicals are either actual or potential carcinogens. Under some circumstances, some of the components of biomaterials, including certain monomers and metal ions, can be associated with the initiation or promotion of cancer and so it is not surprising that attempts have been made to establish causal links between the uses of biomaterials and cancer in patients. This discussion can be divided into the theoretical risks of cancer and the clinical evidence concerning such causation.

The theoretical risk is not so much concerned with overt carcinogenicity of biomaterial components but rather the indirect way in which materials and devices can initiate events that lead to tumors when they are present in tissues for a protracted period of time. The evidence of the potential risk comes from studies in animals, and especially rodents, and this evidence first came to light in the 1960s and 1970s. It is clear that the implantation of many materials in rats can produce tumors; indeed it could be argued that all biomaterials can produce tumors in such animals under some circumstances. The incidence can vary with different materials and implant designs. Implants of pure nickel will invariably produce tumors from what might be assumed to be a chemical carcinogenic effect. Discs of many essentially inert and definitely non-carcinogenic polymers may also cause tumors, with solid discs being more prone that porous ones. These effects have been described as solid-state carcinogenicity and it has been assumed that the mechanism of tumor formation is related to the development of the foreign body response.

With respect to the latter point, there has been an increasing emphasis in recent years on the role of inflammation in cancer. With viral or chemical carcinogenesis, the agent may initiate a sub-threshold neoplastic state where there is irreversible DNA damage but the effect is latent until it is exposed to a promoter, which may be a product of inflammation, especially the chemokines. Promotion then leads to cell proliferation and the generation of ROS, giving oxidative DNA damage and reduced DNA repair. Whilst normal inflammation is self-limiting, chronic inflammation such as that seen with the prolonged presence of an implant, may result in this sustained promotion effect. It is well known that UV light can lead to skin inflammation and eventually melanoma, while the asbestosis described previously is an inflammatory process that can lead to the cancer mesothelioma. Experiments have shown that the implantation of plastic discs in mice that are deficient in the p53 tumor suppressor gene significantly enhances sarcoma formation associated with oxidative stress in the foreign body response.

We now turn to the human clinical evidence. In spite of many epidemiological studies, there is very little evidence to suggest that patients receiving implants are at greater risk of developing cancer. Examination of orthopedic implant registries and comparisons with tumor registries show no consistent pattern and no statistically significant differences between incidence of tumors in patients with joint prostheses and the general population. There are, of course, individual case histories of sarcomas found in association with large implants, but these may have occurred spontaneously and independent of the device. There are also reported cases of tumors found adjacent to breast implants, typically a mesenchymal tumor such as fibromatoses, but again direct causal relationships have not been found. There is some circumstantial evidence that spinal cord injury patients with chronic incontinence may have an elevated risk of bladder cancer possibly associated with the increased infection and irritation associated with a chronic indwelling urinary catheter.

Altogether, therefore, there is good reason to believe that solid-state carcinogenicity is a real phenomena but it is likely to be restricted to certain animal species with little or no clinical relevance.

### 3.4.3 Autoimmunity

Autoimmunity is the reaction of components of the immune system with constituents of the body's own tissues that leads to clinical symptoms and demonstrable pathological effects. The resulting conditions can be systemic or organ-specific. Systemic autoimmune diseases are primarily connective tissue diseases, including systemic lupus erythematosus, scleroderma and Sjögren's disease. Organ-specific diseases can affect the thyroid, the liver and the ovaries. These conditions are associated with expansion of self-reactive T and B cells, the production of autoantibodies and general tissue damage. The significance of autoimmunity in biocompatibility is that various chemicals are known to promote the development of this condition and there have been concerns that some components of biomaterials may be included here.

Most of the evidence about chemically induced autoimmunity comes from the use of certain types of drugs. In addition, some well-known toxins such as dioxins and asbestos and some oils have been implicated. So far there is no specific evidence of biomaterial involvement although there is evidence that certain metals now being used in nanoparticulate form, such as gold and silver, can produce or influence the development of antinuclear antibodies in some animal species. It is likely that autoimmunity can be induced by chemicals via a number of different mechanisms, although they all involve factors such as chronic inflammatory responses and associated cell death, and the suppression of immunoregulation. Once a toxin does promote inflammation and aberrant cell death, the dying or dead cells release cellular debris that are able to activate receptors such as the Toll-like receptors, which may lead to inflammasome activation and the release of pro-inflammatory cytokines. The activation of elements of the innate immune system is then accompanied by antigen presentation to T cells and the activation of autoantibody-producing B cells. These processes are similar in part to those of the chronic inflammation seen with the host response to biomaterials, such that autoimmunity has to be considered as a potential factor in biomaterials biocompatibility.

## 3.5 The unified framework of biocompatibility

It is now time to take all of the separate, apparently discrete, mechanisms described in the above scenarios and translate then into a generic framework that can be used to explain in a logical manner the events that we observe during the clinical use of biomaterials.

### 3.5.1 The generic biocompatibility pathway

For the moment, let us ignore details of scale, location and chronology, and simply consider a generic biomaterial that it to be used in some form of therapy. As indicated in Figure 3.39(a), there is a simple generic pathway, which starts with the presentation of a clinical condition and leads to the decision to use a therapy that involves a biomaterial. In order for that therapy to yield the optimal clinical outcome, the biomaterial may interact with cells in the determination of the appropriate host response. In some situations, for example with an intravenous delivery catheter, where the required function is very simple and transient, there is no specific target cell that determines the outcome, but there will be cells present that can interfere with the outcome, for example platelets that can adhere to the catheter and cause a blood clot. In other, and indeed most, situations the desired clinical outcome can only be achieved through a combination of effects on certain critical cells and the avoidance of effects on certain other cells.

The part of the pathway between biomaterial and cells constitutes the generic biocompatibility pathway. We emphasize that the material could be a massive implantable device, a collection of quantum dots or a tissue engineering scaffold. The critical cells could be embryonic stem cells, platelets or osteoblasts. The location could be an *in vitro* bioreactor, an extracorporeal support system, an intravascular catheter or the central nervous system. The timescale could be a matter of minutes, hours, days or years. The clinical outcomes could be tissue replacement, functional support, tissue regeneration or a diagnosis. It is the intention here to identify the common elements of the biocompatibility pathways in each case.

The general schematic of these pathways is shown in Figure 3.39(b). To the left we have the biomaterial, which will influence the events within the biological environment by either mechanical or molecular signaling processes, or more commonly by both; it should be emphasized that more than one type of signaling process should be anticipated in most situations. The biomaterial will inevitably encounter macromolecules in these environments. The biomaterial itself and most biomaterial-derived components will become coated or enveloped by an adsorbed layer, typically of proteins. Even biomaterial-derived ions or molecules may become coupled to proteins at this stage. All subsequent interactions will take place between the macromolecule-coated biomaterial and its environment, although the extent to which this layer influences these interactions is variable. At this point we see one possibility of controlling biocompatibility phenomena through the design of surfaces that influence the nature of the adsorbed layer, although again in practice it has been difficult to achieve measurable success with this strategy.

### 3.5.2 Target cells, defensive cells and interfering cells

We can now entertain the prospect of the macromolecule-coated biomaterial or biomaterial component interacting with cells. For the purpose of

(a)

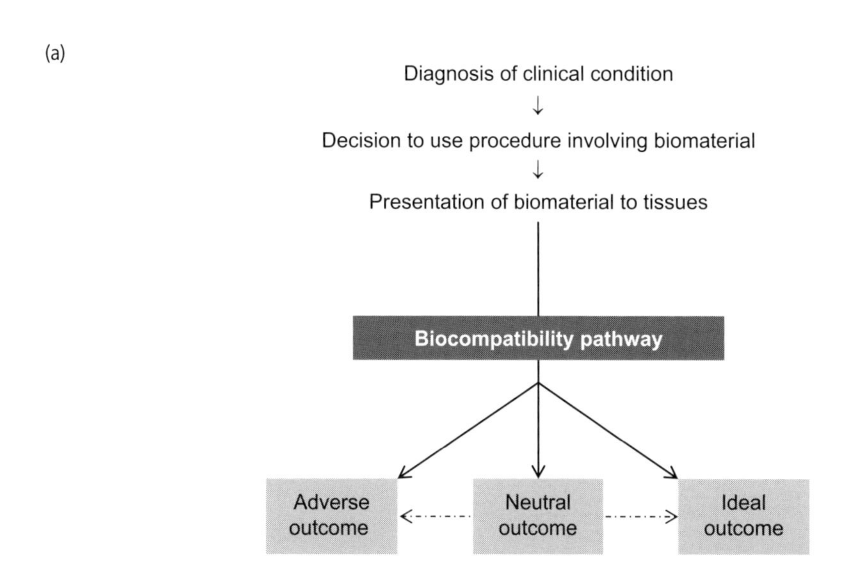


(b)

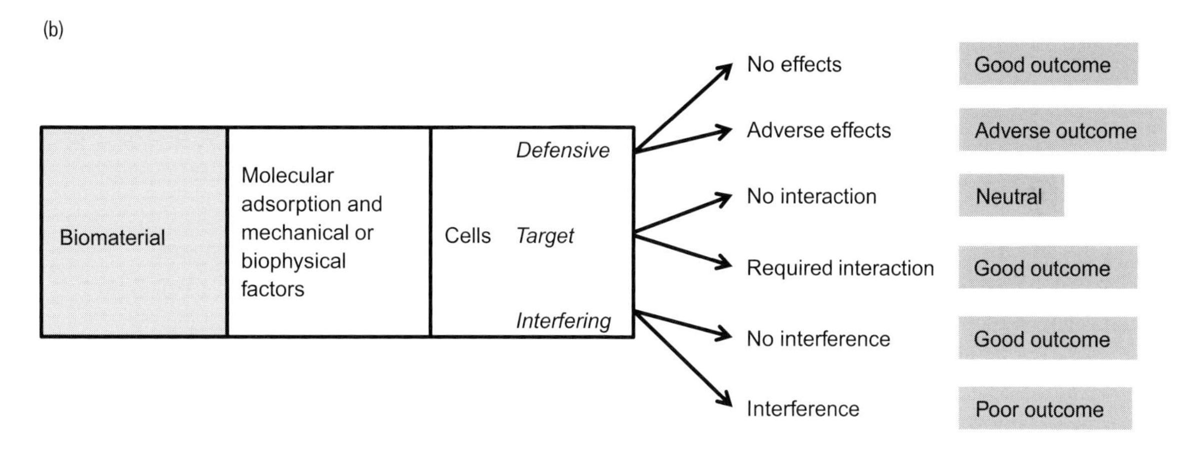


**Figure 3.39** Summary of biocompatibility pathways.
**(a)** The general scheme.
**(b)** The pathways from biomaterial contact to clinical outcome, with specific reference to the generic cell types that will be encountered.

this framework, we can divide cells into three groups. First there are the *target cells*, these being the cells at which the therapy is aimed. These could be stem cells in a tissue engineering bioreactor, osteoblasts in a bone-contacting device, cancer cells in a polymer-chemotherapeutic agent or those cells that a contrast agent has to enter in order for a specific imaging functionality to work.

Secondly, there are the *defensive cells*. These primarily include the cells of innate and adaptive immunity and platelets, whose very existence is based on the need to repel and remove injurious external agents. In some circumstances the biomaterial component has many of the features of the microorganisms that the immune system was designed to fight and the response of such cells to these biomaterial components is crucial to the overall biocompatibility phenomena.

Thirdly there are *interfering cells*. Usually these are cells that are in their natural habitat and essentially

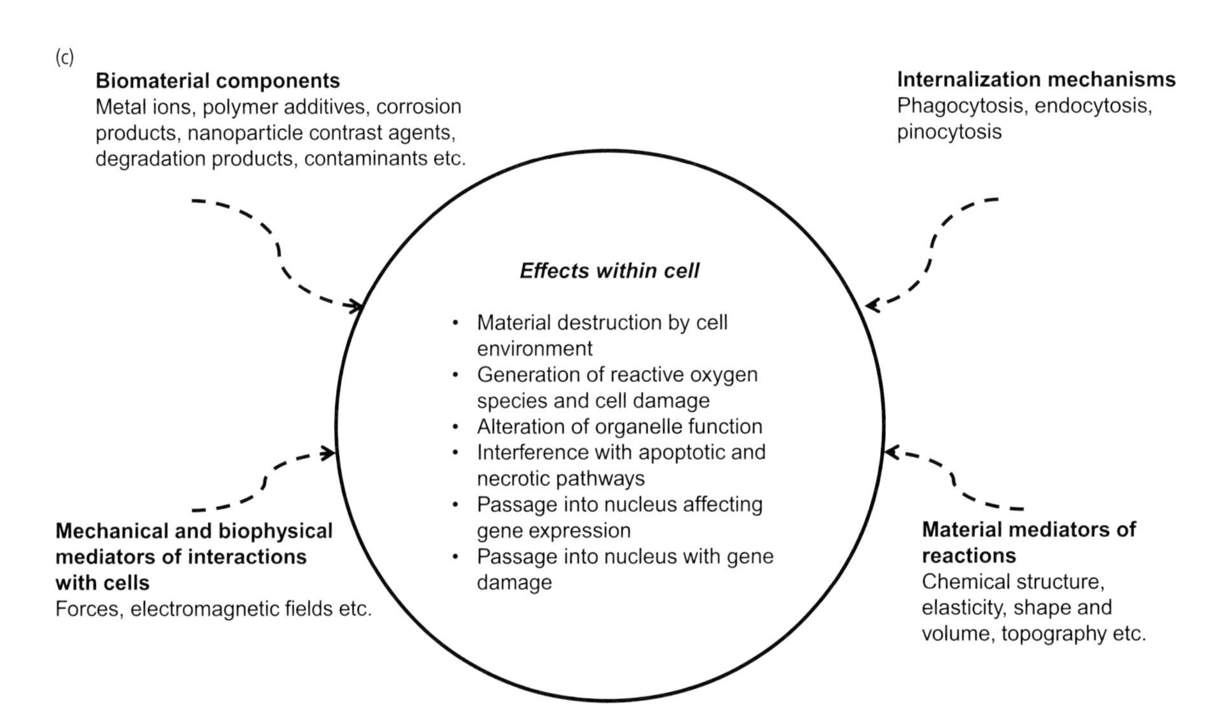


**Figure 3.39** *continued*
(c) Summary of the mediators of biocompatibility and critical cellular responses.

get in the way. In so doing, they interfere with the response that we are seeking, for example smooth muscle cells in the vasculature, fibroblasts in soft connective tissue and osteoclasts in bone. The activity of these cells can lead to hyperplasia, or tissue resorption or other undesirable events.

## Processes with target cells

The biocompatibility pathway that dominates any particular situation will be determined by the events within these three groups of cells. With the target cells we have to consider those events that lead to the desired result and those that lead to undesirable effects. The desired result may simply be maintenance of phenotype and a healthy status of the cell; this is usually required in long-term implantable devices. In tissue engineering constructs, the desired result may be the differentiation of cells down a pre-determined lineage. In a quantum dot-based agent, the desired response may be the internalization of the particle in the target cells and its elimination soon after the imaging process is complete. In a polymer-chemotherapeutic agent, the desired result is the internalization within the cancer cell and the destruction of that cell. With a non-viral gene vector, the desired result is the internalization of the vector in the target cells, their avoidance of the destructive lysosomes, the delivery of the DNA to the nucleus and the elimination of the DNA-depleted vector afterwards. Obviously in some of these situations the biomaterial component is actually carrying an active molecule. In other situations, the achievement of the desired result may be positively influenced by the application of exogenous active molecules, as with the effect of growth factors or transcription on cells in tissue culture or the enhancement of bone regeneration in implanted devices by the presence of BMPs.

With respect to undesirable effects on target cells, and with the exception of those chemotherapeutic agents that are intended to kill cancer cells, we should note that the biomaterial component should not cause any stress state that would lead to apoptosis or necrosis. Such a state could be caused by coincidental mechanical or chemical mechanisms, which usually involve the generation of reactive oxygen species. The key to an appropriate host response, that is excellent biocompatibility, is the dominance of the desirable effects over undesirable effects on the target cells, coupled with the avoidance of unacceptable responses via the defensive and interfering cells, which we shall now discuss.

#### Processes with defensive cells

The involvement of defensive cells is inevitable. The critical question is whether the responses here are controlled or uncontrolled. By uncontrolled here we mean that those cells of the immune response react to the presence of the biomaterial component in such as a way that process follows that destructive pathway defined back in Figure 3.27. In this, the initial effects of the biomaterial component on these cells, especially the cells of the innate immune response, result in activation of these cells and the release of a variety of pro-inflammatory mediators. The combined cellular and humoral compartments of inflammation may lead to an accelerating, aggressive process, which can be destructive of both biomaterial and tissue. The continuing presence of the irritant biomaterial, which itself may corrode or degrade faster in the now enzyme- and free radical-rich environment, may lead to foreign body giant cell formation and granulation tissue. Often the domination of this response over anything else will diminish any positive effects on target cells and cause failure of the therapy. A tissue engineering construct that generates the required tissue through the activity of the target cells may eventually fail if the polymer scaffold is pro-inflammatory as it degrades, resulting in a mass of macrophages and giant cells instead of the desired connective tissue.

We should bear in mind that the defensive cells of the immune systems are assisted by a variety of macromolecules, some of which are involved in cascade processes, as with complement for example, such that a relatively quiescent situation may suddenly change and become destructive.

It should also be noted that this pathway could be controlled, meaning that it does not dominate the effects on target cells and that it may also resolve naturally, as discussed in Section 3.4. There is also the possibility that the application of exogenous active molecules, such as anti-inflammatory agents, could assist in the minimization of these effects, although this is not widely used.

#### Processes with interfering cells

As noted above these cells form part of the normal anatomical structure into which a biomaterial may be inserted. They are not the target cells for the intervention and just happen to be there, but their influence can have a profound effect on the outcome. Unless the biomaterial components have been designed with specific cell targeting mechanisms, these components will usually be non-specific as to the cells they meet. As with defensive cells, the response of these interfering cells may be uncontrolled, often leading to excessive tissue growth, to tissue loss and possibly calcification of tissue, because of the perturbation to normal homeostasis. As we have seen in Section 3.4, it is possible to intervene here with active molecules that are able to inhibit these mechanisms. This does not necessarily lead to resolution and it is probably better to consider this position in the context of metastable biocompatibility.

With most implantable devices it is inevitable that biomaterial components will meet and interact with interfering cells. With biomaterial systems that are delivered systemically, as in some drug and gene delivery systems and contrast agents, the probability of meeting with non-target cells is obviously much higher. It is for this reason that these systems are being designed with specific targeting mechanisms (to be discussed in Chapter 6) so that harmful effects on non-target cells are minimized.

### 3.5.3 Cellular mechanisms in biocompatibility pathways

We now turn our attention away from the schematic view of the generic pathway towards the focal point of the pathway, which is the manner in which the biomaterial components actually interact with the cell, indicated in Figure 3.39(c). We can assume for the moment that the biomaterial component arrives at the cell membrane. Through a variety of mechanisms, which include phagocytosis, pinocytosis and endocytosis, the component may pass through the cell membrane and into the cell. Whichever mechanism operates, it will be controlled by physical and chemical parameters including the size and chemical nature of the component, as discussed in the various parts of Section 3.4. Once inside the cell the component may follow one of several pathways through the cytoplasm, which could lead to:

- its removal from the cell by the action of endosomes and lysosomes;
- the generation of reactive oxygen species and cell damage;
- alteration of organelle function and cell damage;
- interference with apoptotic and necrotic pathways;
- passage into the nucleus with effects on gene expression;
- passage into the nucleus with gene damage.

With several of these individual processes, active molecules generated within the cells, including chemokines and cytokines, may be released into the extracellular space, where they then can have effects on surrounding cells, thus potentially accelerating the process.

In addition to these pathways for chemically mediated mechanisms, there is also the possibility of mechanically induced effects on the cell. Here the biomaterial component may come into direct contact with the cell, or mechanical forces generated by the biomaterial may be transmitted to the cell by an intermediary substance such as the ECM. The mechanotransduction mechanisms here can be considered as equivalent to the chemically mediated mechanisms since the mechanical signals are transduced into chemical signals passing from the cell membrane to the nucleus, the stress system here being responsible for similar effects on gene expression, apoptosis, cell damage and so on.

## 3.6 Specific clinical examples of biocompatibility phenomena

We now consider a small number of actual clinical examples to see how this framework of biocompatibility pathways operates in practice. These are clinical situations involving implantable devices since these are the areas where we have greatest experience and evidence. We shall cover specific areas in regenerative medicine, drug delivery and imaging at appropriate places in later chapters.

### 3.6.1 Osteoinduction, osteoconduction and osseointegration

The biocompatibility of materials and devices in contact with bone is of crucial significance to many clinical applications, including joint prostheses, dental implants and spinal fusion. With permanently implantable devices there is a very specific requirement that, after insertion of the device into the relevant site, new bone forms within the interfacial region between existing bone and biomaterial, and without any intervening soft tissue. It is usually clinically desirable that this takes place in as short a period of time as possible. It is a further requirement that this integrated bone-material system is maintained throughout the lifetime of the device, which usually means the lifetime of the patient.

#### Bone healing

Fundamental to the mechanisms involved here is the normal bone healing process. Contrary to many other biocompatibility scenarios, these phenomena are based upon wound healing since the devices, by definition, are implanted by a surgical technique, which specifically involves mechanical disturbance of, and direct trauma to, vital bone. For very good evolutionary reasons, bone has developed a sound healing process, which means that, under optimal conditions, bone heals or regenerates spontaneously, one of the few tissues to do so. The process of new bone formation is termed osteogenesis. Among those factors that determine the optimal conditions of healing are those that govern the geometry of the wound. After a fracture of bone, healing can take place spontaneously but only if the severed parts can be placed in close proximity and protected from too much movement in the early stages, hence the use of splints, casts or plates. Similarly, if a hole is drilled into bone it may heal spontaneously but only if it is not too big. The maximum size hole that can heal completely is referred to as a critical size defect, the size of which will vary from one species of animal to another and with the anatomical site.

Fracture healing, the process that restores the bone to its original physical and mechanical properties occurs in three distinct but overlapping stages: the early inflammatory stage, the repair stage and the remodeling stage. In the inflammatory stage, a hematoma develops within the fracture site during the first few hours and days. Inflammatory cells, including neutrophils, macrophages, monocytes and lymphocytes, and fibroblasts infiltrate the site, resulting in the formation of granulation tissue, the ingrowth of vascular tissue, and migration of mesenchymal cells. During the repair stage, fibroblasts lay down a stroma that supports vascular ingrowth. As vascular ingrowth progresses, a collagen matrix is laid down and osteoid, an unmineralized tissue, is secreted. This is subsequently mineralized, leading to the formation of a soft form of bone called callus around the repair site. This callus is initially very weak and requires adequate protection. Eventually, the callus ossifies, forming a bridge of woven bone between the fragments. If optimal immobilization is not used, ossification of the callus may not occur, and an unstable fibrous union may develop instead. Fracture healing is completed during the remodeling stage in which the healing bone is restored to its original shape, structure, and mechanical strength. Remodeling of the bone occurs slowly over months to years and is facilitated by mechanical stress placed on the bone. Adequate strength is typically achieved in 3 to 6 months.

One extremely important point here is that bone healing takes place by well-defined mechanisms and it is very difficult to interfere with the natural process in order to speed it up or produce better results. In other words, in an otherwise healthy patient there is little that can be done to improve on the natural process. In some patients the fracture healing process may be impaired, producing non-unions or delayed unions and here it may be possible to redress this deficiency. Treatments may involve the application of physical cues such as ultrasound or electrical stimulation; these effects may be beneficial but results are not guaranteed. It is also possible to positively influence a delayed healing process with the application of some active molecules such as bone morphogenetic proteins or platelet-rich plasma.

Turning now to the attachment of implanted devices to bone, the preparation of the bone, through drilling for example, is equivalent to the trauma of fracture and we should expect a similar type of healing process to take place. There are, in fact, two processes by which a biomaterial can influence this response from the bone: one is osteoinduction and the other is osteoconduction.

Osteoinduction means that primitive, undifferentiated and pluripotent cells are stimulated to develop into the bone-forming cell lineage, from which they induce the process of osteogenesis. These undifferentiated cells support the differentiated bone cells, including osteoblasts and are of utmost importance for healing or anchorage of an implant, since they can be recruited to form osteoprogenitor cells. Osteoconduction, on the other hand, refers to bone that grows on a surface. An osteoconductive surface is one that permits bone growth on its surface or down into pores, channels or pipes, rather than one which actively induces new bone. It is usually difficult to identify whether the gap between an implant and pre-existing bone is filled by osteoinduction or osteoconduction. The important factor is whether it fills sufficiently well with new bone, by whichever mechanism, to give a functional connection, as opposed to being filled substantially with soft tissue, which does not constitute a good functional connection. The situation in which there is a direct structural and functional connection between the pre-existing, ordered, bone and the surface of a load-bearing implant is described as osseointegration.

### Osseointegration

A great deal has been written about osseointegration, with much speculation about mechanisms and controlling factors. The most important issues are described below.

Osseointegration is largely controlled by chemical inertness. There is no doubt that commercially pure titanium is the most effective metal from the point of view of osseointegration, being the most corrosion-resistant structural metal, but, from a pathological perspective, equivalent bone formation can be seen with alumina and polyetheretherketone, two highly inert non-metals. Stainless steel, which is less corrosion resistant, does not produce such a good result as titanium.

There is some evidence that surface topography can influence the rate at which bone growth occurs, but not necessarily the end result (i.e., the eventual amount of bone-material contact or the mechanical strength or rigidity of the system). Whether this is because different topographies influence the osteogenesis mechanism itself or merely facilitate better connectivity of the bone that does form is not too clear. It does seem that microroughness is a factor in determining the strength of attachment, for example as measured by the torque necessary to remove a dental implant, but this could be simply due to the increased friction compared to that with an ultra-smooth surface. The surface topography varies with manufacturing processes, with machining, grit blasting, acid etching and anodizing all having their characteristic morphology. There is some evidence that nanoscale features superimposed on microtopography make have some beneficial effect.

A major discussion point has been the role of so-called bioactive surfaces. In this context, bioactivity refers to the ability of a material surface to positively enhance the rate or quality of bone formation. As discussed in both Chapters 4 and 8, there are several possibilities here, including bioactive glasses that actually form a chemical bond to bone. We shall concentrate here on the role of calcium phosphates. We shall see later that there are several different forms of calcium phosphate that can be considered as potential bioactive substances that can be applied to the surfaces of implants intended to interface with bone. These have varying stoichiometries (specifically calcium to phosphate ratio) and solubility/degradation characteristics. It has been argued that there are three essential constituents for bone formation, these being the presence of collagen, the presence of phosphate and the regulation of inhibitors of mineralization. It would seem intuitive that a surface of calcium phosphate that is a source of phosphate ions should be beneficial to the process of bone formation. The use of such surfaces in general does result in a clinically faster achievement of functional osseointegration, but long-term results are not necessarily any better than with uncoated surfaces and the eventual resorption/dissolution/debonding of these layers cast doubt on their value for good chronic biocompatibility.

## 3.6.2 Heart valves

As discussed in Chapter 4, prosthetic heart valves play a major role in the treatment of valve disease. There are two major types of valve replacement, mechanical and bioprosthetic valves, and these have some biocompatibility issues in common and some that are valve specific. The common issues largely concern the healing process around the sewing ring. The valve-specific issues are thromboembolic events with mechanical valves and calcification/structural deterioration in bioprosthetic valves.

### Healing in sewing rings

Sewing rings are discoid-shaped fabrics that cover the annulus of prostheses. They primarily function by providing an anchoring base for sutures that are placed through the heart muscle of the valve annulus. However, it is important that tissue from this muscle grows into the interstices of the fabric to provide an effective seal. Failure to do so can result in paravalvular leakage, that is the flow of blood around rather than through the valve, which is a major cause of valve dysfunction. On the other hand, this ingrowth of tissue should not proceed to gross proliferation, which could result in overgrowth of tissue, called pannus, into the valve orifice, causing obstruction. Thus the functional performance of the valve is critically dependent on this tissue ingrowth, which is essentially a modification of wound healing within the vascular compartment. The process is very similar to that which occurs in vascular grafts. Blood rapidly penetrates the interstices, where it clots because it is static, following the mechanisms discussed earlier in this chapter, and the clot becomes reorganized into fibrous tissue. It is unclear exactly what controls pannus overgrowth but obviously this is a modified healing process that results in hyperplastic growth.

### Thromboembolic events in mechanical valves

It is well known that hemodynamic forces are powerful mediators of thrombosis. Flow disturbances may produce both areas of stasis and abnormally high shear stresses, both of which influence the equilibrium between platelets and material surfaces. Mechanical heart valves produce greater flow disturbances than bioprosthetic valves since they have leaflets or other structures that interfere with normal blood flow. All clinically used mechanical valves are associated with elevated risks of thrombosis and the consequential risks of released emboli and it is a clinical necessity that patients are routinely anticoagulated. The lesser disturbances seen with bioprosthetic valves yield lower but not insignificant risks. The biocompatibility of mechanical heart valves is therefore dominated by the mechanical force induced thrombogenicity and the anticoagulation therapy regime. This has implications for the role of patient variables in clinical biocompatibility phenomena. Not all patients are at the same risk. Medical risk factors here include the presence of atrial fibrillation, hypertension, increased left ventricular size, reduced ejection fraction of the heart, intrinsic hypercoagulability and post-operative infection. Non-medical risk factors include smoking. The risk of thromboembolic events is reflected by the parameter known as INR, or institutional normalized ratio, which is determined by tests on the patient's blood. Typical target INRs are in the region of 2.5. Obviously there is a limited to the amount of anticoagulant that can be prescribed because of the risk of bleeding or hemorrhage in high-risk patients. Warfarin is the most widely used anticoagulant, which can be delivered by oral administration, although self-administered low molecular weight heparin and aspirin are also used. This is the classical example of metastable biocompatibility since good performance can be achieved for long periods of time, only to be upset by instability in the anticoagulation regime.

### Calcification and structural degeneration in bioprosthetic valves

Bioprosthetic valves are derived either from whole porcine valves or fashioned from bovine pericardium. Although they clearly have far greater similarities with natural valve tissue than do the carbon/metal mechanical valves, they still have some structural differences. The manufacturing sequence has to involve a chemical process that sterilizes the tissue and removes antigenicity, primarily through cross-linking processes. As well as cross-linking the proteins, both collagen and elastin, these processes remove endothelium (of porcine valves) or mesothelium (of pericardium) and reduce interstitial cell viability and cause their fragmentation and degeneration. Collagen bundles are loosened, with loss of amorphous extracellular matrix and, as a consequence, loss of compliance. A series of autolytic, chemical and physical changes take place between tissue harvesting and implantation, while a further series of pathologic changes will take place following implantation. Cuspal surfaces become covered in fibrin and cells, both platelets and inflammatory cells. Fluids are absorbed into the tissue, facilitating both cell and protein infiltration. Endothelialization of the surfaces takes place only sporadically and usually only small parts of the surface will be covered by endothelium.

Valves eventually become dysfunctional because of the deterioration of the cuspal tissues, mediated by two separate but synergistic processes, calcification and degradation. Calcification is initiated deep within the tissue with distinct regional variations. The calcific deposits predominate at commissures adjacent to attachment areas where deformation is at a maximum. Evidence strongly points to the significant effect of mechanical stress on this process. Obviously other factors are involved since the rate of calcification depends on many non-mechanical factors. The calcification process has initiation and propagation phases, the initiation being controlled by the ability of calcium derived from the plasma being able to react with the phosphorus released from the non-viable interstitial cells. Mechanical forces are partly responsible for the movement of these ions. Structural degradation can take place independent of calcification but tends to occur in the same highly stressed regions. The chemical treatment of the valve material alters mechanical properties such that there are increased flexural stresses in parts of the cusp, which cause structural damage of the collagen. These changes in the collagen also are associated with biochemical changes, including depletion of glycosaminoglycans, which causes alteration to the viscoelastic properties and the inherent ability to accommodate mechanical strain.

Since these tissues are xenogeneic, it is natural to speculate on the role of the immune response in these valve failures, although this has been a controversial matter. There is clinical evidence that shows that patients in whom valves have failed have higher levels of circulating antibodies to the tissue components compared to patients where valves have not failed. There is also good animal data to show that fixed xenogeneic tissue calcifies much faster than fixed autologous tissue. When this data is considered alongside the dramatic experience of cryopreserved decellularized xenogeneic valves for pediatric patients, in which rapid failures occurred through immediate neutrophil and macrophage dominated inflammation, followed by extensive lymphocytic responses and calcification, it seems fairly certain that the immune system has to be involved with these processes.

We see, therefore, that the primary mechanisms of bioprosthetic heart valve failure are directly related to the biocompatibility of the chemically treated collagen-based tissue, which in turn are mediated by mechanotransduction processes that significantly control the movement of chemical species that react to initiate calcific deposits and structural deterioration of this tissue. Intimately associated with these mechanically initiated events, it is likely that immune responses contribute to the pathologic changes that occur.

## 3.6.3 Blood vessels

Synthetic vascular grafts, as will be seen in detail in Chapter 4, consist of tubular structures whose walls are composed of microporous polymers. These are attached to the existing blood vessels using fine sutures, these junctions being referred to as anastomoses. Ideally, once the graft has been attached to the vessels and the clamps removed from the vascular system to allow blood to flow through the graft, blood oozes through the interstices of the graft wall where it clots. The clot is then reorganized over time, being replaced by smooth muscle cells within the wall and covered on the inner lumen with a thin layer of endothelium. Both the muscular and endothelial layer are important for functionality and biocompatibility, the endothelium being particularly important for the biocompatibility. Endothelialization can in theory take place by transanastomotic migration (i.e., lateral extension from the pre-existing vessel into the graft), by transmural growth (i.e., derived from growth through the graft wall) or by blood-borne cellular deposition.

### Transanastomotic endothelialization and hyperplasia

A major problem with the assessment of this performance is that in most animal models, the transanastomotic endothelialization (TAE) takes place quite readily, and short- to medium-length grafts can eventually be fully endothelialized. This does not take place in humans; experience shows that TAE at most extends 1–2 cm into the graft, even after many years. In addition to the effect of species, the age of the host has a significant effect and even in the most favorable animals the rate of endothelialization decreases significantly with age. In most situations with human patients, the peri-anastomotic endothelium is unstable and will tend to thicken to give a region of intimal hyperplasia.

Although several factors contribute to this phenomenon, the major factor is mechanical, where the change in compliance across the junction between graft and natural vessel leads to flow disturbances and raised wall shear stresses, which cause thickening of the endothelium. This is an excellent example of mechanotransduction, and potential signaling pathways have been identified. For example, insulin growth factor-1, IGF-1, is a growth factor that participates in multiple pathophysiological pathways including those involved in vascular smooth muscle cell migration and proliferation, and hence the development of atherosclerosis and implantable device hyperplasia. Egr-1 (early growth response-1) is a mechanosensitive transcription factor that binds to the IGF-1 receptor and promotes its activity. Mechanical forces increase both mRNA and protein expression of Egr-1 in vascular smooth muscle cells, and perturbations to these forces leads to intimal hyperplasia.

### Mid-graft response

In the mid-graft region, a different response is seen. With low porosity PTFE grafts in humans, a 10–20 μm thick fibrin layer develops during the first two weeks post-implantation. This remains with this thickness but changes in structure, such that in the long term it consists of a compacted acellular fibrin layer. There is no transmural tissue ingrowth in the early stages. Slowly there is some sporadic invasion of connective tissue from the outside, accompanied by macrophages and foreign body giant cells. With higher porosity grafts, the fibrin may penetrate the pores and become infiltrated by inflammatory cells, fibroblasts and sprouting capillaries. In knitted Dacron grafts, where the porosity is even greater, the intimal surface meshwork of fibrin, white cells, erythrocytes and platelets rapidly penetrates into the thickness. For the majority of the length of the prosthesis compacted fibrin remains the major component of the luminal surface. Some transmural growth will take place, although not with extensive capillary content, but does not reach the inner surface because of this fibrin layer. It is also possible for there to be a few islands of endothelium established by blood-borne cells.

### Implications for the overall biocompatibility of vascular grafts

The implications of these events at anastomses and mid-graft sites is that there are serious biocompatibility limitations with these devices. Failure occurs either as intimal hyperplasia at anastomotic sites or thrombus formation in the mid-graft regions, or possibly both. These grafts, when used in the major circulation, eventually become non-patent for either of these reasons. Moreover, these materials cannot be used in vessels of small diameter, including the coronary arteries, because this loss of patency occurs very early. These mechanisms of failure are related to the mechanotransduction effects in intimal hyperplasia, and the lack of good healing along most of the graft length. It is interesting to note that the compacted fibrin that forms on the luminal surface is unusual. It tends to have a high level of fibrinogen that is absorbed from the flowing blood and also is rich in TGFβ, released from platelet granules, which inhibits angiogenesis, explaining the lack of capillaries within the vessel wall. In addition, macrophages that are initially present within the interstices play their usual role in healing, but in later stages, especially as the tissue within the fabric tends to become hypoxic, macrophages may well change their phenotype which, together with an increase of foreign body giant cells, negatively affects the healing process.

## 3.6.4 Nerve repair

Nerves are often injured and their repair is a matter of tremendous importance. This topic is usually considered in two separate sections, dealing with the peripheral nervous system (PNS) on the one hand and the central nervous system (CNS), especially the spinal cord, on the other. There are many differences between these systems, the two that concern us here being that, from a biological perspective, the CNS is an immunopriviledged site whereas the PNS is not, and from a functional perspective, the PNS can spontaneously regenerate under some conditions, whereas the CNS cannot.

### The peripheral nervous system

The damage here can be traumatic when a nerve is severed or damaged by compression, or it may be surgically induced through operative procedures that require sacrifice of one or more nerves. There are varying rates of success in terms of repair since the greater the trauma to the nerve and surrounding tissue, and the longer the period between injury and attempted repair, the more extensive will be the inflammation and the more difficult the regeneration. Peripheral nerves can be repaired through autografts. Although quite successful, they do have significant disadvantages, including donor site morbidity and the matching of size and structure to the nerve that is to be repaired. The autograft does supply a type of template, or scaffold, that facilitates the regeneration and incorporation of nerve, as well as supplying cells, the growth factors that these cells secrete and ECM molecules. Since the autografts are not ideal, there have been many attempts to produce and use synthetic equivalents. In essence, these try to recapitulate as many of these features of the autograft as possible, which means in theory they could consist of a scaffold alone, or a scaffold that incorporates cells, growth factors or ECM molecules.

Over the last 25–30 years, many different scaffold materials have been used for this purpose. They include natural materials such as autologous veins or pieces of muscle, non-degradable polymers, especially including silicone elastomers, degradable synthetic polymers including a range of polyesters, and biopolymers such as collagen and chitosan. Graft lengths are usually in the range 5–30 mm. The fact that all of these disparate materials work to some extent suggests that they possess no intrinsic, highly specialized, biocompatibility characteristics. There are some variations, which will depend on how the material influences the course of events following implantation of the material, which is typically in the form of a tube, often referred to as a conduit or nerve guidance channel.

If we consider a non-degradable, relatively inert conduit material such as the silicone elastomer, within a few hours of implantation the tube will fill with serum, which will contain neurotrophic factors, cytokines and inflammatory cells. Macrophages will remove cellular debris from the injury and the fibrin will form a type of clot that should extend from one end of the tube to the other. The next stage is critical, since this fibrin mass, sometimes called a fibrin cable, becomes invaded by cells from both proximal and distal ends of the conduit. These include inflammatory cells, fibroblasts, endothelial cells and the principal structural cells of peripheral nerves, the Schwann cells. If the system here is significantly pro-inflammatory, then the Schwann cells will find it hard to compete for the space. If fibroblasts dominate, then scar rather than nerve tissue will be favored. The axons, which are the functional nerve cells, migrate into this fibrin matrix, and they rely on the cytoplasm of the Schwann cells for their progress. Once the axons from the proximal end reach the distal end and become myelinated, functional nerve repair is established.

It has become clear that the most essential, basic property required of the conduit material is that it should not stimulate inflammation during this critical phase. In terms of the biocompatibility pathways, we are seeking the quiescent pathway of benign acceptance. There are some features of the material or adjunct components that could help this process. The structure of the conduit wall may be tuned in order to provide greater passage of essential molecules through the wall, for example neurotrophic factors, and permeability and porosity can be varied for this purpose. It is also possible that the inner surface of the tube may be patterned with longitudinal features that may guide nerve cell growth; under experimental conditions *in vitro* it is clear that axons and Schwann cells do respond to topographical cues so that there is the possibility of micro- or nanoscale influences on biocompatibility. This may also be enhanced by the use of longitudinally oriented degradable fibers within the tubes.

In addition, the tubes may be pre-filled with ECM proteins, including collagen and laminin, and fibrin in order to facilitate the repair process and these may contain neurotrophic factors, including nerve growth factor (NGF) and brain-derived growth factor (BDGF). The addition of Schwann cells to the mix may also be beneficial, although outcomes of such practices are not yet clear.

### Central nervous system: spinal cord injury

As noted above, the CNS is an immunoprivileged site and we should therefore expect that the response to injury, including the inflammatory response, should be different. In the spinal cord, an injury, either compression or resection, initially causes uncontrolled necrotic cell death, inflammation, hypoxia and ischemia, driving neurons and oligodendrocytes into apoptosis. The injured tissue quickly becomes isolated through the activity of astrocytes and the formation of a glial scar. This gliotic response is very important. Under normal conditions, glial cells help to protect CNS tissue from damage. Injury results in a response from glial cells, especially a major proliferation of these cells, that tends to protect uninjured tissue and minimize further damage, but in doing so they form the glial scar which contains many inhibitors of axonal growth.

Under these conditions, the use of a simple biomaterial conduit is of very limited value. It is not good enough to have a material that minimizes inflammation and fibrosis since the spinal cord has its own intrinsic mechanisms of promoting inflammation and scarring. There has been good progress, however, with the development of complementary strategies that include the targeting of anti-inflammatory agents to the site of injury and the local delivery of stem cells and neurotrophic factors, as discussed in Chapter 5. It should be noted here that biomaterials may or may not form part of such strategies; they probably will with the drugs and factors but not with the cells. It has already been demonstrated that chitosan tubes can act as delivery vehicles for factors such as neurotrophin-3 (NT-3). This is a powerful neurotrophic factor in fetal development and facilitates spinal cord regeneration, but has a very short half-life. When released slowly from chitosan it can allow cord regeneration over several centimeters under experimental conditions. The most promising biomaterial option, and biocompatibility pathway, here would appear to involve the delivery of pharmaceutical agents from macroscale devices in order to control the local host response.

## 3.7 Influences of bacteria, fungi, viruses, prions and endotoxins

Much of the discussion of biocompatibility has been predicated on the assumption that we can consider the processes and mechanisms of biocompatibility to be determined by the reactants, that is the biomaterial and the host. This is not necessarily always the case, however, and it is necessary to introduce some other variables. In this section we consider the role that microorganisms and related agents may play. As noted in several places in the book, microorganisms frequently gain access to the tissues of the body and our defense mechanisms have almost wholly evolved to deal with the invasion of microorganisms. The invasion and multiplication of microorganisms into the body is associated with the phenomenon of infection, in which this invasion causes local cellular injury, through a variety of cellular and toxic mechanisms. There are several groups of microorganism that could be involved, primarily including bacteria, fungi (especially yeasts and molds) and viruses. Bacteria are unicellular prokaryotic organisms.

Fungi are multicellular organisms. Viruses are non-cellular but are parasitic within cells, and consist of a nucleic acid core and a protein coat. Although not being considered strictly as microorganisms we shall also deal briefly here with prions, which are infectious misfolded proteins, and endotoxins, which are fragments of dead microorganisms. This section will concentrate on bacterial infections.

Microorganisms are important players in biocompatibility phenomena for two main reasons. First, biomaterials may alter the pathogenicity and virulence of microorganisms, especially bacteria. Secondly, the microorganisms may influence the stability and performance of the biomaterials. This section starts with an explanation of the mechanisms of biomaterial–bacteria interactions before reviewing the evidence of clinical significance.

### 3.7.1 Biofilms

The process of protein adsorption on biomaterials surfaces has been discussed earlier; this is a potentially important stage in many individual biocompatibility processes. The film of proteins and other matter that form on medical device surfaces plays a significant role in biomaterial–bacterial interactions, especially as bacteria can have a significant role in the development of this film, which is generally referred to as a biofilm. The impact of a biofilm can be seen by the possibility that it can facilitate bacterial colonization and provide protection to the bacteria from more normal environmental agents, including phagocytic cells and anti-bacterial molecules.

Biofilm formation may be seen as a four-stage process, involving bacterial transport to the surface, bacterial adhesion to the surface, biofilm matrix synthesis and film maturation. Adhesion is critical and follows a sequence of reversible, unstable physico-chemical interactive forces followed by essentially irreversible receptor–ligand binding; it is here that the initial protein layer on the biomaterial surface comes into play because of opportunities for binding provided by conformationally changed biomolecules. The biofilm matrix proliferates through the species-specific substances such as exopolysaccharides.

The formation of biofilms on medical devices depends on many factors, including the locality (i.e., flowing blood or urine, solid tissues, etc.) and the nature of the material surface, where topography (at macro-, micro- and nanoscale) and chemistry are both likely to be significant. Clearly the films also vary with the bacterial strain. The membrane surface macromolecules, collectively called adhesin molecules, control adhesion behavior. These include pili, which are filamentous proteins, the larger flagella proteins, lectins that are able to bind to carbohydrates and some polysaccharides.

### 3.7.2 Material variables and bacterial adhesion

Material surfaces vary quite considerably in their ability to be colonized by bacteria; the relationships may reflect their role in initial biofilm formation or the actual adhesion process itself, it being difficult to separate these stages mechanistically. As with mammalian cell adhesion, the surface topography and chemical nature are both possible mediators. It is very evident that bacteria respond to topographical features, since bacterial colonies are often preferentially found in areas of defects, such as pits, rather than flat defect-free areas of surfaces. With extended wear contact lenses, there is a tendency for greater risk of bacterial contamination if the hydrogel surfaces show signs of wear. On the other hand, the scientific evidence that bacteria respond unequivocally to certain topographical features is lacking. It would seem intuitively obvious the bacteria could preferentially adhere to the bottom of grooves on surfaces because of protection against shear stresses. It might also be expected that surface irregularities promote bacterial adhesion through greater surface area of contact. However, it is far from clear if and how either microtopography or nanotopography influence biofilm formation and bacterial adhesion. It might be expected that bacterial cells would be different to eukaryotic cells in this respect since they have different cell membrane characteristics. Experimental evidence does support the view that nanoscale topography by itself can change the morphological, genetic and proteomic profile of many bacteria. Some *in vitro* work suggests that nanotopographical features lower bacterial adhesion whilst increasing osteoblast adhesion, which would favor better bone biocompatibility and better resistance to infection, but there must be significant doubt as to whether this is of practical value.

With respect to chemical and biophysical variables, the hydrophobic/hydrophilic balance should be a major factor, but again there is equivocal evidence. In the literature there are simplistic stances that determine that hydrophilic materials are more resistant to bacterial adhesion than hydrophobic surfaces and equally positions that state the opposite. The reality is that the situation is far from simple, and that the hydrophobic/hydrophilic nature of the bacterial wall and the nature of the substances immediately released from bacterial cells on contact are major complicating factors. It is known, for example, that the extracellular DNA in bacterial cell walls and in the biofilm can enhance adhesion through involvement with acid–base interactions and alterations to the thermodynamic conditions within the system.

It should be noted that almost all bacteria have cell walls that are negatively charged at physiological pH and that attractive interactions may take place with material surfaces that are positively charged. Chemical functionality of the material surface may also have an effect, especially when they mediate receptor–ligand type interactions. The general cell-repulsive effects of the hydrogel poly(ethylene glycol) (PEG), produced by the effects of the stable interfacial water layer on inhibiting direct cell-material contact, are responsible for bacterial adhesion-resistance properties of PEG-coated surfaces.

From a mechanistic perspective, the most important factors are the processes whereby bacterial cells anchor themselves to host tissue and protein-covered biomaterials. These involve a variety of host protein-binding receptors, which are commonly called adhesins. In some well-known highly pathogenic bacteria, the bacteria are capable of synthesizing the polysaccharide intercellular adhesin (PIA), a fibrinogen-binding (Fbe) adhesin and a collagen-binding protein known as GehD lipase. Analysis of bacterial strains in fluids around infected implants frequently reveal the presence of the adhesins and those genes encoding for these proteins.

### 3.7.3 Clinical aspects

Infection is clearly a major risk factor in the clinical use of biomaterials. In the extreme, infected devices are life threatening; should a prosthetic heart valve become infected, giving the condition known as bacterial endocarditis, the consequences are frequently fatal. With major prosthetic reconstruction, as with joint replacement, an infection usually requires removal of the device, followed by a difficult period in which attempts are made to disinfect the residual tissue, and then re-operation to provide another prosthesis. With contact lenses, an infection can have serious long-term effects on eye tissues, including the sensitive cornea. Any chronic placement of an externally connecting tube or wire in the body, for example an intravenous catheter, a suprapubic urinary catheter, or a connection between an implanted device and an external accessory (for example, a power source) is at risk of infection, which is usually very serious. In cell therapy and tissue engineering, *in vitro* bioreactors are at risk of contamination, in which case there can be serious functional and economic consequences.

Infection can occur in the immediate perioperative time frame, when bacteria could be introduced at the time of the clinical intervention or just afterwards. It is also possible for infection to occur much later, typically involving a metastatic spread of bacteria from an unrelated infected site, for example a urinary tract infection.

It is worth mentioning here that, in discussions concerning the causes of biomaterial-related failure, it is frequently stated that a device or a material caused an infection. This is manifestly not the case; bacteria cause infections and not biomaterials. We must be careful, however, with these arguments. First, as discussed in Chapter 2, all medical devices have to be sterilized before use. The sterilization process, if properly validated, should exclude an infection risk, but the process itself and the validation system are determined on a statistical basis and there cannot be a 100% guarantee that absolutely no bacteria are present. With the increasing use of tissue-derived products, even greater attention to decontamination and sterilization processes are necessary.

In addition, there is significant evidence that the presence of biomaterials within tissue can affect any bacteria that happen to be present. Experience with surgical skin wounds shows that the number of bacteria necessary to cause a clinically significant wound infection is markedly lower when materials (e.g., sutures) are present. Also, certain strains of bacteria that are normally of low pathogenicity and/or virulence can become far more pathogenic or virulent in the presence of a biomaterial. We should also mention here the serious issues that have arisen in recent years with the emergence of antibiotic-resistant bacteria such as methicillin-resistant *Staphylococcus aureus* (MRSA), which is increasing the difficulty of combating infections, including those related to medical devices.

### 3.7.4 Avoidance and treatment of bacterial infection

In view of the major significance of biomaterial-related infections, it is essential that every effort is made to avoid infection. This includes the sterilization process mentioned above as well as meticulous attention to sterile fields during clinical procedures. Attention may also be focused on antibacterial properties of the biomaterials themselves. Control of bacterial activity on or near the material surface can be attempted by several different processes, which involve either inhibition of bacterial adhesion or direct bacterial killing, or a combination. Anti-adhesive surfaces include the PEG modified surfaces mentioned above as well as a series of polymer chains that can be grafted onto surfaces to inhibit biofilm formation. Direct effects on bacteria are usually associated with antimicrobial agents that include antibiotics, oxidants and biocides. Some forms of bone cement used with joint prostheses contain antibiotics that leach out over time. Silver, as a type of biocide, is increasingly being used to coat medical devices in order to confer antibacterial activity, especially using nanocrystalline forms of silver.

One major problem with antimicrobial agents involves the need to supply inhibitory effects on bacterial cells without inducing any harmful effects on host cells. Often there will only be a small window of concentration when the former effects can be achieved without the latter. One solution to this lies with the permanent attachment of molecules to surfaces that do not leach out, minimizing effects on host cells, but preventing any bacterial activity. Cationic antimicrobial agents such as molecules with quaternary amine groups at very high surface charge density are capable of very effective bacterial killing whilst remaining strongly attached to biomaterials surfaces.

An alternative strategy involves the exploitation of antimicrobial peptides (AmPs). These are broad-spectrum peptides that constitute part of the innate immune system in mammals, being defensively effective against a wide variety of gram-positive and gram-negative bacteria. They are often able to block biofilm development. They may be coated onto biomaterials surfaces with a controlled release rate in order to provide local antibacterial effects.

### 3.7.5 Fungi

Fungal infections on medical devices are much more rare but nevertheless they can pose problems. Candida are the most common organisms involved, with *Candida albicans* being the most common species. Probably the better known examples are seen on dental products, denture stomatitis, characterized by chronic erythema and edema of the palatal mucosa, being particularly troublesome. *C. albicans* may also colonize contact lenses and intrauterine devices. These fungi may also be associated with the biofilms on catheters, and systemic, blood-borne infections may arise. Catheter-related candidiasis may be treated by anti-fungal agents such as amphotericin B, although resistance to such agents can occur.

### 3.7.6 Viruses

Viral infections associated with implantable and extracorporeal medical devices are also uncommon. There is one potentially important issue with increased susceptibility to bacteremia in patients on hemodialysis when those patients have a hepatitis C infection. This is especially seen with those cases involving tunneled catheters. The viral infection may inhibit phagocytosis or directly depress humoral immune responses. Viral infections are obviously a risk with allogeneic and xenogeneic products used in tissue engineering, these issues being discussed in Chapter 5.

### 3.7.7 Prions

The transfer of prions to humans from contaminated animals became a serious issue in the 1990s, especially in the UK. This arose from the consumption of bovine spongiform encephalopathy (BSE) contaminated meat products. The possibility of this type of infectivity spreading to other situations, including those involving biomaterials, raised some significant concerns. As with viruses, these issues have to be addressed whenever xenogeneic materials are used but there are also concerns with contaminated medical devices. The human equivalent of BSE is Creutzfeldt–Jakob disease. Some tissues of individuals with this disease are known to be infectious, as the protease-resistant pathological form of the prion protein may be detectable in lymphoid tissues such as the spleen and tonsils. The main problem here is that these prion proteins are extremely difficult to eliminate, and may resist decontamination procedures used with re-usable surgical instruments. Cases of disease transmission have been recorded following the use of instruments in procedures such as tonsillectomies. More appropriate decontamination procedures have been developed.

### 3.7.8 Endotoxins

In contrast to the infectious agents in the above three sections, the risks with endotoxins may be very widespread. Endotoxins, also known as lipopolysaccharides, form an integral part of the outer cell membrane of gram-negative bacteria. Endotoxins are released on bacterial cell death, which means they are released in large numbers when medical devices are sterilized. They consist of a lipid part, a core oligopolysaccharide and a heteropolysaccharide segment, which constitutes a surface antigen. The endotoxins are hydrophobic and attach to a range of biomaterials, especially those that are hydrophobic and/or cationic. Endotoxins can bind to some serum proteins and there is evidence that they can activate complement. It is reasonable to assume that the presence of endotoxins can influence the host response to biomaterials through a number of mechanisms. It is also clear that endotoxins present in xenogeneic tissues used in regenerative medicine can also influence the immune response to them.

## 3.8 Clinical and patient variables

The point has been made several times in this chapter that biocompatibility is a characteristic of a material–host system and that factors other than the material may play a significant role. It is difficult to present a systematic profile of these effects, but we may enumerate some of the facts.

### 3.8.1 Clinical technique

These factors relate more to implantable devices since these often require considerable surgical skills. For the procedures requiring the highest levels of skill, including heart valve replacement, vascular grafts, urogynecological surgery, total joint replacement and spinal surgery, the best clinical outcomes are obtained in hospitals that carry out large numbers of procedures on a daily basis and with highly qualified surgeons who carry out the procedures routinely. In many of these situations, the consequence of less-than-optimal clinical skills will be manifest by poor biocompatibility outcomes. These are some of the facts:

- In joint replacement, the alignment of prostheses in the respective bones determines both the overall stress system and the forces exerted at the bearing surfaces. These have consequential effects on the wear process and hence on the potential for loosening.
- With any procedure that involves the preparation of a bone bed for implant attachment, the mechanical forces that are applied to the bone, for example by drilling alveolar bone for dental implants, should not damage the remaining bone since that will impair the process of bone healing – unnecessarily high temperatures produced during drilling denatures bone proteins.
- In vascular surgery, it is primarily the quality of the suturing at the anastomoses between vascular grafts and existing vessels that controls flow disturbances at these sites, with consequences on intimal hyperplasia.
- Similarly in heart valve replacement the preparation of the tissue bed and the care with which the sewing ring is attached to that bed determines the course of post-operative healing and the avoidance of paravalvular leaks and thrombosis.
- Catheter-delivered stents are so common now that it is easy to diminish the importance of clinical skills but the precise location and mode of delivery of the stent have a great influence on the initial damage to the endothelium and subsequent restenosis.

### 3.8.2 Patient variables

There is even more evidence to implicate patient variables in the determination of biocompatibility and clinical outcomes. There are two generic effects here, one group being related to lifestyle/compliance issues with the patients and one with genetic factors. These are some of the facts:

- Avoidance of thromboembolic events with current prosthetic heart valves is mostly dependent on the compliance of the patient with anticoagulation medication.
- The best outcomes for hip and knee replacements are achieved when the patient follows careful and rigorous physiotherapy regimes very soon after surgery.
- Complications with endosseous dental implants related to poor-quality soft and hard tissue responses are twice as high with smokers compared to non-smokers.
- Smoking is also the single most important factor that adversely affects outcomes of the treatment of diabetic foot ulcers.
- Recreational drugs impair heart rhythm and can make cardiac pacing more difficult.
- Obesity is a major factor in the causation of pelvic organ prolapse and in resistance to therapies involving biomaterials.
- Women are at greater risk of implant-related complications (such as hematoma) than men after pacemaker implantation.
- Genetic factors contribute to long-term performance of coronary artery bypass grafts.

### 3.8.3 General comments

It is appropriate to conclude this chapter on biocompatibility with a brief discussion of clinical and patient variables. While we are obviously concerned with the mechanisms by which the materials themselves influence the host response, they may not ultimately control that response. In many clinical situations, the performance of the clinicians and the subsequent behavior of the patient, may totally denominate the outcome. One of the more significant consequences of understanding biocompatibility mechanisms will be the development of materials and techniques that are more forgiving of these factors.

## Summary and learning objectives

1. Interactions take place between the materials and the components of the body, and these have a significant impact on the ability of that material to satisfactorily perform the desired function for the desired length of time in that body. These interactions are collectively described under the heading of biocompatibility.
2. With very few exceptions, materials are not intrinsically compatible with physiological systems. Moreover, the tissues of the human body have not evolved in order to benignly accommodate these materials and they will be treated as "foreign" and "potentially harmful." The human body has evolved to have exquisite detection mechanisms that readily identify foreign objects and there are exquisite defensive mechanisms that deal with such objects once they have been detected; there is, therefore, active and not passive incompatibility.
3. We should not expect the mechanisms of biocompatibility to show linear progression with time; one event may be triggered spontaneously, at any time, the effect of which can be powerfully amplified by one or more mechanisms. Also, the consequences of the interactions between material and host may be localized to the vicinity of the material, giving the local foreign body response, or they may be remote, affecting either the whole body (a systemic response), or a specific discrete remote site.
4. Each of the individual clinical consequences may be considered to be the manifestations of the *effects* of biocompatibility processes. The framework of biocompatibility mechanisms identifies the *causes* and link them to these effects by a series of *pathways*, which will be different at molecular and cellular levels, but have common or generic features associated with them.
5. In this book we start with a baseline position that a biomaterial is a solid object that is immobile, chemically unreactive with physiological components, and unchanging with time. Furthermore, we assume that the tissue with which it is in contact provides a dynamic environment in which the cells and their ECM change over time. In this baseline scenario, the biocompatibility of the system is concerned with the effect that the physical presence of the material has on the dynamic response of the tissue components. This is the conceptual starting point of biocompatibility mechanisms. Specifically this states that the baseline mechanisms of biocompatibility do not rely for their initiation on direct chemical interactions between the biomaterial and the host; the cause of the events that take place is not a chemical process, although once the process has been initiated, then the reaction may be propagated by chemical processes, especially within and between cells.
6. There are five biocompatibility scenarios based upon whether they are determined by mechanical/biophysical processes, chemical processes, pharmacological events and either microscale or nanoscale controlled events.
7. In the mechanical/biophysical scenarios, we can distinguish two main types of phenomena, those associated with mechanotransduction and those associated with biomolecular adsorption to biomaterial surfaces. Mechanical forces play a

significant role in homeostasis, especially within connective tissues. The various mechanisms by which cells are able to convert mechanical forces into chemical activity are collectively described by the term mechanotransduction. In this part of the first scenario of biocompatibility, we discuss how mechanical forces are primarily responsible for the response of bone, blood vessels and soft connective tissue to implanted devices, for damaging effects on red blood cells and the control of cell differentiation in tissue engineering bioreactors.

8. With macromolecular adsorption on biomaterials surfaces, we show how the processes of adsorption are governed by energy considerations, the presence of water and the competition for the surface by proteins that have differing compositions and differing affinities for surfaces. The nature of the protein layer changes over time, as initially adsorbed molecules are covered, displaced or conformationally changed. We show here that although much is known about protein adsorption under experimental conditions, its influence on real-time, clinically relevant biocompatibility is unclear and probably of minimal importance. It is certainly true that cells interacting with biomaterials surfaces have to do so through a protein layer but exactly what influence the layer has is not known in most circumstances.
9. The one situation where these events are important is that of the interaction of biomaterials surfaces with blood, where there is complex interplay between material surface, the proteins of both clotting and complement cascades and blood cells, especially platelets. These mechanisms are explained in this chapter in terms of the biophysical processes of macromolecular adsorption and cell activation for although the cascades and activation processes are clearly chemically driven, and although material surface chemistry does play a minor role, it is these biophysical events, especially including the fluid mechanical environment, that initiate events. These aspects of blood compatibility are considered to be integrated into the collective biocompatibility mechanisms and not separate from them.
10. In considering the role of chemical mechanisms of biocompatibility, the second of our scenarios, we recognize that in their normal environment, cells interact with other cells and with the extracellular matrix. Cells communicate by molecular mechanisms that are integral to processes of cell signaling. Xenobiotic compounds derived from biomaterials, including metal ions, monomers, oligomers and active molecules contained within ECM biomaterials can all take part, either positively or negatively, in cell signaling processes. These potential effects on cell communications arising from the chemical nature of the surface should be seen as additional to the mechanical and biophysical effects already discussed, and topographical effects which will be seen later. Specifically in relation to chemical events and cell signaling in biocompatibility, we emphasize and explain their role in inflammation and both innate and adaptive immunity. Critical to these roles are the processes where the key components of innate immunity, which are the phagocytic cells, natural killer cells and circulating proteins, control the overall response. Tissue damage that results in inflammation can take the form of many different types of stress stimuli, including heat, oxidative stress and radiation; exogenous chemicals associated with biomaterials form just one group of stressful stimuli, and we have to consider the role of such chemicals within the various forms or phases of the responses, including apoptosis and wound healing, in the context of the available pathways of cell signaling in response to the stress stimuli.
11. There is one major difference between biomaterial-induced inflammation, and that is, in contrast to most other forms of stress stimuli, biomaterials may well exert their effect within tissue for very long periods of time, and we have to consider the mechanisms of the response to chronic exposure and the eventual resolution of inflammation. The different pathways to destructive biocompatibility events and resolution are presented here.
12. We then deal with both microscale and nanoscale driven processes, involving the

influences of biomaterials surface topography and micro/nanoparticles on the host responses. Although much has been written in the literature on the role of topography in biocompatibility phenomena, we are rather circumspect here as there is little evidence to support a role in clinical situations. With respect to particles, that is certainly not the case and much is known about the mechanisms whereby small entities gain access to cells and interact with intracellular components including organelles and nuclei. We discuss these mechanisms in some detail since they are of crucial importance to the biocompatibility of nanoscale products in drug and gene delivery and imaging systems.

13. When it is clear that really desirable and appropriate host responses are difficult to achieve in practice, and where the mechanisms by which the host response deviates from the ideal are understood, it may be possible to incorporate active molecules into the biomaterial construct in order for them to modulate the response and even re-direct it. These processes are discussed here under the general heading of pharmacological biocompatibility mechanisms.
14. Having discussed the five biocompatibility scenarios, we briefly address systemic or remote site phenomena such as reproductive toxicity, autoimmunity and carcinogenesis. Although these phenomena are rarely encountered, they cannot be ignored since they are very serious effects and all new biomaterials have to be rigorously tested in order to avoid their occurrence. We also address issues of contamination and infection, which can also be serious but are far more common.
15. Combining all of the individual mechanisms and individual pathways into one over-arching scheme is the focal point of this chapter. Section 3.5 does this by text and graphics. This is a unified biocompatibility theory, or a presentation of biocompatibility pathways under a single scheme, into which each separate biomaterial experience can fit and be explained. It revolves around the identification of three generic cell types, the target cells, the defensive cells and the interfering cells, and the balance between the power and influence that each of these displays under specific conditions. Going right back to the original definition of biocompatibility and the appropriateness of the host response, that appropriateness is dependent on achieving optimal pathways to and effects within the target cells, avoiding uncontrolled responses from defensive cells, and controlling, often by pharmacological processes, the undesirable and coincidental responses by interfering cells.

## Questions

1. Discuss the reasons why synthetic materials are intrinsically incompatible with mammalian physiological systems.
2. Describe how biocompatibility phenomena vary between local sites, remote sites and systemically.
3. Give examples of how failure to achieve desired biocompatibility characteristics can lead to (a) fatal, (b) severe and (c) mild consequences.
4. Discuss the different stages of mechanotransduction.
5. Describe the influence of mechanical forces on response of bone to implanted devices.
6. Describe the process of hemolysis and its significance for medical devices.

7. Discuss the structure of platelets and the processes involved with platelet adhesion to surfaces and their activation.
8. Describe the role of material surfaces in blood clotting.
9. Explain the phenomena of thrombogenicity and embolization with respect to medical devices.
10. Describe the general structure and function of integrins.
11. Discuss the essential differences between the innate and adaptive immune responses.
12. Discuss the potential roles of biomaterials in adaptive immunity and hypersensitivity.
13. Describe the essential structure and function of the immunoglobulins.
14. Compare and contrast the processes of apoptosis and necrosis in relation to cell death.
15. Explain the time course for wound healing in soft connective tissue.
16. Give and explain your views on the pathway of metastable biocompatibility.
17. Discuss the overall host response to biomaterials-derived chemical components in the context of the body's response to stress stimuli.
18. Discuss the lessons that have been learnt from the pathology of the pneumoconioses with respect to the host response to microparticles derived from biomaterials.
19. Describe the process of osteolysis and the loosening of joint prostheses.
20. Explain the over-arching presentation of biocompatibility pathways and give your views on its usefulness.

## Recommended reading

Allori A C, Sailon A M, Warren S M, Biological basis of bone formation, remodeling and repair – part 1: Biochemical signaling molecules, *Tissue Engineering* 2008;14:259–69. *Review of signaling molecules that contribute to the biochemical system associated with bone healing.*

Anderson J M, Rodriguez A, Chang D T, Foreign body reaction to biomaterials, *Seminars in Immunology* 2008;20:86–100. *Discussion of some classical concepts of the development of the host response to biomaterials.*

Badylak S F, The extracellular matrix as a biologic scaffold material, *Biomaterials* 2007;28:3587–93. *Opinion paper on the potential and limitations of ECM-derived materials in therapeutic applications.*

Barton G M, A calculated response: control of inflammation by the innate immune system, *Journal of Clinical Investigation* 2008;118:413–20. *An overview of innate responses to microbial and non-microbial tissue damage and the interplay between innate immunity and inflammation.*

Bavry A A, Bhatt D L, Appropriate use of drug-eluting stents: balancing the reduction in restenosis with the concern of late thrombosis, *Lancet* 2008;371: 2134–43. *A discussion of the evidence concerning the biological responses to drug-eluting stents.*

Brown B N, Valentin J E, Stewart-Akers A M, McCabe G P, Badylak S F, Macrophage phenotype and remodeling outcomes in response to biologic scaffolds with and without a cellular component, *Biomaterials* 2009;30:1482–91. *Explanation of the polarization of macrophage phenotype induced by biologic scaffolds.*

Caplan M R, Shah M M, Translating biomaterial properties to intracellular signaling, *Cell Biochemistry and Biophysics* 2009;54:1–10. *A review of the manner in which materials can influence cell behavior through cell-surface interactions and variations in cell signaling phenomena.*

Chen F-M, Zhang M, Wu Z-F, Toward delivery of multiple growth factors in tissue engineering, *Biomaterials* 2010;31:6279–308. *Review of the concept of growth factor delivery in tissue engineering and the methods to achieve this delivery.*

Chiquet M, Renedo A S, Huber F, Fluck M, How do fibroblasts translate mechanical signals into changes in extracellular matrix production? *Matrix Biology* 2003;22:73–80. *Review of mechanotransduction processes in fibroblasts.*

Chiu J-J, Chien S, Effects of disturbed flow on vascular endothelium: pathophysiology basis and clinical perspectives, *Physiological Review* 2011;91:327–87. *Extensive review of the mechanisms by which mechanical forces involved within blood flow disturbances affect the endothelium.*

Chowdhury F, Na S, Li D, *et al.*, Material properties of the cell dictate stress-induced spreading and differentiation in embryonic stem cells, *Nature Materials* 2010;9:82–8. *Explanation of theories of the effect of mechanical stress on gene expression in stem cell differentiation.*

Coussens L M, Werb Z, Inflammation and cancer, *Nature* 2002;420:860–7. *New insights into the way in which inflammatory cells orchestrate the tumor microenvironment.*

Dalby M J, Topographically induced direct cell mechanotransduction, *Medical Engineering & Physics* 2005;27:730–42. *Review of the influence of topographical features on mechanisms of mechanotransduction.*

De Jong W H, Borm P J A, Drug delivery and nanoparticles: applications and hazards, *International Journal of Nanomedicine* 2008;3:133–49. *An overview of mechanisms of toxicity with nanoparticles used in drug delivery.*

Dobrovolskaia M A, McNeil S E, Immunological properties of engineered nanomaterials, *Nature Nanotechnology* 2007;2:469–78. *Review of immunotoxicity of nanoparticles including antigenicity, adjuvants and inflammation.*

Eckes B, Zigrino P, Kessler D, *et al.*, Fibroblast-matrix interactions in wound healing and fibrosis, *Matrix Biology* 2000;19:325–32.

Elmore S, Apoptosis; a review of programmed cell death, *Toxicologic Pathology* 2007;35:45–516. *A review of mechanisms of apoptosis, including the role of chemicals in the initiation of the intrinsic pathway of apoptosis.*

Franz S, Rammelt S, Scharnweber D, Simon J C, Immune responses to implants – a review of the implications for the design of immuomodulatory biomaterials, *Biomaterials* 2011;32:6692–709. *Review of mechanisms of the immune response to biomaterials and a discussion of strategies aimed at triggering appropriate responses.*

Glassman S D, Howard J M, Sweet A, Carreon L Y, Complications and concerns with osteobiologics for spine fusion in clinical practice, *Spine* 2010;35:1621–8. *Discussion of the potential complications associated with the use of active agents such as BMPs in the spine.*

Gorbet M B, Sefton M V, Biomaterials-associated thrombosis: roles of coagulation factors, complement, platelets and leukocytes, *Biomaterials* 2004;25:5681–703. *Review of the role of interactions between proteins and cells in thrombosis.*

Helming L, Gordon S, Molecular mediators of macrophage fusion, *Trends in Cell Biology* 2009;19:514–22. *Review of the mechanisms of giant cell formation and the related signaling processes.*

Holst J, Watson S, Lord M S, *et al.*, Substrate elasticity provides mechanical signals for the expansion of hemopoietic stem and progenitor cells, *Nature Biotechnology* 2010;28:1123–9. *Demonstration of the role of substrate elasticity, provided by tropoelastin, on stem cell expansion.*

Hori K, Matsumoto S, Bacterial adhesion; from mechanism to control, *Biochemical Engineering Journal* 2010;48:424–34. *Review of physicochemical models of bacterial adhesion and potential mechanisms for the control of biofilm formation and cell adhesion.*

Human P, Zilla P, Characterization of the immune response to valve bioprostheses and its role in primary tissue failure, *Annals of Thoracic Surgery* 2001;71:S385–8. *Experimental observations on the role of the immune response to bioprosthetic valves.*

Ingham E, Fisher J, The role of macrophages in osteolysis of total joint replacement, *Biomaterials* 2005;26: 1271–86. *Review of the relationships between wear particles, macrophages and osteolysis.*

Khan W, Farah S, Domb A J, Drug eluting stents: developments and current status, *Journal of*

*Controlled Release* 2012;161:703–12. *Review of stents used for coronary artery disease management.*

Kloxin A M, Benton J S, Anseth K S, In situ elasticity modulation with dynamic substrates to direct phenotype, *Biomaterials* 2010;31:1–8. *Discussion of the effects of microenvironment elasticity on critical cell functions.*

Kroemer G, Galluzzi L, Vandenabeele P, *et al.*, Classification of cell death, *Cell Death and Differentiation* 2009; 16:3–11. *Recommendations of the Nomenclature Committee on Cell Death on the criteria for definitions of cell death.*

Le Guehennec L, Soueidan A, Layrolle P, Amouriq Y, Surface treatment of titanium dental implants for rapid osseointegration, *Dental Materials* 2007;23: 844–54. *Explanation of the potential for different surface treatments to control the speed and quality of osseointegration.*

Liu G, Beri R, Mueller A, Kamp D W, Molecular mechanisms of asbestos-induced lung epithelial cell apoptosis, *Chemico-Biological Interactions* 2010;188:309–18. *Discussion of asbestos-induced apoptosis, oxidative stress and mitochondrial dysfunction.*

Mallander V, Landfester K, Interaction of nanoparticles with cells, *Biomacromolecules* 2009;10:2379–400. *Review of uptake of nanoparticles by cells and mechanisms of intracellular effects.*

Martino S, D'Angelo F, Armentano I, Kenny J M, Orlacchio A, Stem cell-biomaterial interactions for regenerative medicine, *Biotechnology Advances* 2012;30:338–51. *Review paper that discusses the translation of stem cell biology into therapeutic applications.*

Maurer-Jones M A, Bantz K C, Love S A, Marquis B J, Haynes C L, Toxicity of therapeutic nanoparticles, *Nanomedicine* 2009;4:219–41. *Introduction to the applications of therapeutic nanoparticles, their potential toxicity and methods of evaluation.*

Mazzoli-Rocha F, Fernandes S, Einiker-Lamas M, Zin W A, Roles of oxidative stress in signaling and inflammation induced by particulate matter, *Cell Biology and Toxicology* 2010;26:481–98. *Description of reactive oxygen species and their potential role in the host response to particulate materials.*

McBride S H, Falls T, Knothe Tate M L, Modulation of stem cell shape and fate; Mechanical modulation of cell shape and gene expression, *Tissue Engineering A* 2008;14:1573–80. *Detailed analysis of the role of the mechanical environment on stem cell condensation.*

Mei Y, Saha K, Bogatyrev S R, *et al.*, Combinatorial development of biomaterials for clonal growth of human pluripotent stem cells, *Nature Materials* 2010;9:768–78. *A description of structure-function analysis for stem cell activity on polymer surfaces.*

Moller P, Jacobsen N R, Folkmann J K, *et al.*, Role of oxidative damage in toxicity of particulates, *Free Radical Research* 2010;44:1–46. *Very detailed analysis of the evidence concerning reactive oxygen species and their role in the toxicology of particles.*

Monopoli M P, Aberg C, Salvati, Dawson K A, Biomolecular coronas provide the biological identity of nanosized materials, *Nature Nanotechnology* 2012;7:779–86. *A critical assessment of the role of the nanoparticle corona in determining the particle fate.*

Montanaro L, Campoccia D, Arciola C R, Advancements in molecular epidemiology of implant infections and future perspectives, *Biomaterials* 2007;28:5155–68. *Review of the application of molecular epidemiological techniques to the analysis of bacterial adhesion and colonization of biomaterial surfaces.*

Murata H, Koepsel R R, Matyjaszewski K, Russell A J, Permanent non-leaching antibacterial cells – 2: How high density cationic surfaces kill bacterial cells, *Biomaterials* 2007;28:4870–9. *Description of polymer brushes that can be tuned for bacterial killing while attached to biomaterial surfaces.*

Nel A E, Madler L, Velegol D, *et al.* Understanding biophysicochemical interactions at the nano-bio interface, *Nature Materials* 2009;8:543–57. *Review of the interactions between cells and nanoparticles, especially in relation to the dynamic forces and molecular interactions that are involved.*

Orr A W, Helmke B P, Blackman B R, Schwartz M A, Mechanisms of mechanotransduction, *Developmental Cell* 2006;10:11–20. *Review of specific mechanisms that control the effects of mechanical cues on cells.*

Park S-J, Kang S-J, Virmani R, Nakano M, Ueda Y, In-stent restenosis: a final common pathway of late stent failures, *Journal of the American College of Cardiology* 2012;59:2051–7. *A well-illustrated discussion of the pathology of in-stent restenosis.*

Ploux L, Ponche A, Anselme K, Bacteria/material interfaces: role of the material and cell wall properties, *Journal of Adhesion Science and Technology* 2010;24:2165–201. *Description of the development of biofilms and the mechanisms of bacterial adhesion.*

Pollard K M, Hultman P, Kono D H, Toxicology of autoimmune disease, *Chemistry Research Toxicology* 2010;23:455–66. *Discussion of mechanisms of autoimmune disease with an emphasis on the effects of extrinsic factors such as exposure to materials and chemicals,*

Rogers T H, Babensee J E, The role of integrins in the recognition and response of dendritic cells to biomaterials, *Biomaterials* 2011;32:1270–9. *Experimental demonstration of the role of biomaterials surfaces in dendritic maturation on initiation of adaptive immunity processes.*

Sahay G, Alakhova D Y, Kabanov A V, Endocytosis of nanomedicines, *Journal of Controlled Release* 2010;145:182–95. *Review of mechanisms of endocytosis and variables which influence cell entry by nanoparticles.*

Shukla A, Fleming K E, Chuang H F, *et al.*, Controlling the release of peptide antimicrobial agents from surfaces, *Biomaterials* 2010;31:2348–57. *Description of antimicrobial peptides and their mechanism of activity against antibiotic-resistant bacteria.*

Singh N, Manshian B, Jenkins G J S, *et al.* NanoGenotoxicology: the DNA damaging potential of engineered nanoparticles, *Biomaterials* 2009;30:3891–914. *A review of the genotoxic potential of engineered nanomaterials and methods of risk assessment.*

Tesfamariam B, Local vascular toxicokinetics of stent-based drug delivery, *Toxicology Letters* 2007;168:93–102. *A review of the relationship between local drug delivery from stents and the release profile.*

Varkouhi A K, Scholte M, Storm G, Haisma H J, Endosomal escape pathways for delivery of biologicals, *Journal of Controlled Release* 2011;151:220–8. *Review of mechanisms by which substances escape from endosomes within the endocytic pathway.*

Vogler E A, Siedlecki C A, Contact activation of blood-plasma coagulation, *Biomaterials* 2009;30:1857–69. *Essay on the biochemical mechanisms of contact activation of blood-plasma coagulation.*

Wilson C J, Clegg R E, Leavesley D I, Pearcy M J, Mediation of biomaterial-cell interactions by adsorbed proteins; a review, *Tissue Engineering* 2005;11:1–18. *Discussion of the pivotal role of extracellular adhesion proteins in cell adhesion, morphology and migration.*

Wu P, Grainger D W, Drug/device combinations for local drug therapies and infection prophylaxis, *Biomaterials* 2006;27:2450–67. *Review of combination devices, including interactions between material and drug components.*

Wu X, Cheng J, Li P, *et al.*, Mechano-sensitive transcriptional factor Egr-1 regulates insulin-like growth factor-1 receptor expression and contributes to neointima formation in vein grafts. *Arteriosclerosis, Thrombosis and Vascular Biology* 2010;30:471–6.

Zilla P, Bezuidenhout D, Human P, Prosthetic vascular grafts: wrong models, wrong questions and no healing, *Biomaterials* 2007;28:5009–27. *Critical appraisal of the biocompatibility and performance of vascular grafts.*

# 4 Implantable medical devices and artificial organs

After studying this chapter you will be able to identify all of the significant applications of biomaterials in those devices that replace the structure and/or function of tissues and organs by the use of medical devices. These may be implanted within the patients, usually for the remainder of their lives, or connected to the patient for some short-term assistance; these applications were summarized at the end of Chapter 1. This discussion covers all of the clinical disciplines. It includes implantable devices that have been in use for decades, and you will be able to understand the reasons for their success, and the reasons for failures where they have occasionally occurred. It also covers the implantable and support systems that have recently been developed and introduced into clinical practice so that you can appreciate where the technology of the twenty-first century is leading us in health care products.

As noted earlier, implantable medical devices were, for many years, the main focus of attention within biomaterials science. The rationale and performance of such devices are discussed in this chapter. Each application and each situation is different and it is not possible to deal with this in an entirely satisfactory systematic manner, but the major headings given in Chapter 1 are covered and dealt with in relation to the clinical discipline that is involved. This includes permanent (or long-term) devices, short-term devices, invasive but removable devices and artificial organs or assist devices that are attached to the body. We will conclude the chapter with an assessment of the overall performance of implantable devices and the lessons learned.

## 4.1 Orthopedics

The rationale for the use of permanently implantable devices, as given in Chapter 1, includes the anatomical replacement of parts of the body that have undergone some form of degenerative disease, the anatomical replacement of parts of the body that have undergone surgical removal of cancerous tissue, the correction of congenital deformities, the restoration of function as a consequence of disease, and cosmetic purposes. The devices that are used in orthopedics may well be associated with more than

Box 4.1 | **Glossary of terms**

**acetabulum** cup-shaped socket in the pelvis in which rests the head of the femur.

**approach** specific surgical procedure by which an organ or part is exposed.

**arrhythmia** variation from the normal rhythm of the heartbeat.

**arthropathy** any joint disease.

**arthroplasty** replacement of a joint, in whole or in part, with a prosthesis.

**articular joint** junction between two bones in a skeleton.

**atrophy** wasting away; a diminution in the size of a cell, tissue, organ or part.

**bioactive material** biomaterial that is designed to elicit or modulate biological activity.

**bioprosthesis** implantable prosthesis that consists totally or substantially of non-viable, treated, donor tissue.

**bone cement** material intended for use in arthroplasty procedures for the fixation of prosthetic components to bone.

**bradykinesia** abnormal slowness of the movements of the body.

**cerebrospinal fluid (CSF)** fluid contained within the ventricles, the subarachnoid space and the central canal of the spinal cord.

**cochlea** spiral tube forming part of the inner ear, which is the essential organ of hearing.

**congenital deformities** malformations that are present at birth.

**craniofacial** relating to the cranium and the face.

**cryopreservation** maintenance of the viability of excised tissue or organs by storing them at very low temperatures.

**diabetes mellitus** condition characterized by a raised concentration of glucose in the blood because of a deficiency in the production or action of insulin.

**dialysis** method by which molecules in solution may be separated by selective diffusion through a semi-permeable membrane.

**distal** remote or farther from a point of reference; proximal is nearest to a point of reference.

**extravasation** escape of fluids, such as blood, from the vessels that contain them.

**fibrillation** uncoordinated contraction of individual muscle fibers of the heart, giving rise to an irregular and inefficient action of the heart.

**hernia** protrusion of a portion of an organ or tissue through an abnormal opening.

**infarct** localized area of ischemic necrosis produced by occlusion of the arterial supply or the venous drainage from the part.

**Kaplan–Meier analysis** analysis of the relationship between the percentage of a population that still survives and the time after a specified event, derived from clinical follow-up studies.

**laparascopic** of any technique of examination or treatment of internal tissue, especially the peritoneal cavity, by means of an endoscope.

**mesh** expanded metal or polymer used as a reinforcement of tissue.

**modular prosthesis** any prosthesis that is constructed from several components, the size and nature of which may be varied during assembly to suit the particular requirements of patients.

**mucosa** a mucous membrane, a layer of tissue comprising an epithelium supported on connective tissue.

**occlusion** the act of closure, or the state of being closed by an obstruction.

**ossicle** small bone, especially one in the middle ear.

**patent** description of a tubular structure that is open and unobstructed.

**percutaneous** performed through the skin.

**periosteum** specialized connective tissue covering the bones and having bone-forming potential.

**prion** abnormally folded protein of unknown function that undergoes rapid turnover in the brain, being responsible for spongiform encephalopathy.

**prophylaxis** the use of mechanical or medical means to prevent the occurrence of disease.

**retina** light sensitive tissue of the eye.

**sphincter** ring-like muscle that closes a natural orifice or passage.

**thrombectomy** surgical removal of a blood clot from a blood vessel.

**vasodilation** the process by which an increase in the caliber of blood vessels is produced.

one of these reasons and the subject is best treated on a regional basis. Surgery of the joints is covered first, followed by ligaments, tendons and muscle, then the spinal column and various types of bone defect, including cancer. This section closes with a discussion of short- and medium-term devices intended to assist in bone fracture repair. For ease of reference and clarity, the main bones and joints of the human body that are relevant to implant-related reconstruction are given in Box 4.2.

### 4.1.1 Joint replacement

As noted in Chapter 1, the articular joints of the body are susceptible to disease and trauma, with considerable effects on mobility and quality of life. The major disease that affects the hip, knee, elbow, ankle and shoulder joints is osteoarthritis. All of these joints can also be affected by rheumatoid arthritis although much less frequently. The finger joints are the most susceptible to the latter disease. There are some other minor diseases that affect all of these joints.

The treatment options for osteoarthritis are limited (see Box 4.3). The condition largely arises from the mechanical destruction of the cartilage and bone of the joints. There are some natural remedies, including food supplements such as glucosamine, which may be used to alleviate the symptoms, although they have very limited effects. The condition is progressive and is not amenable to any cure or even significant amelioration through the use of drugs; anti-inflammatory agents and pain relief medication may help most patients for some time but surgical intervention is pretty well inevitable.

Early surgical attempts to relieve the pain and limited mobility of arthritic patients, primarily involving the hip, included an osteotomy, in which the bones of the joints were surgically re-aligned, and arthrodesis, in which the bones of the joint were

## Box 4.2 The major bones and joints of the human body

Diseases and injuries within the skeletal system of the human body vary in character with the different parts of the system. The main bones and joints of the system that may require treatment are listed here, using a basic form of classification.

### The skeleton

It is convenient to divide the skeleton into two parts:

- The *axial skeleton* consists of the bones of the body axis, including the skull (cranium), vertebral column (cervical, thoracic, lumbar vertebrae, sacrum, coccyx), ribs and sternum.
- The *appendicular skeleton* consists of the bones of the limbs and the supporting thoracic and pelvic girdles. In the upper limbs are the humerus, ulna, radius, carpals, metacarpals and phalanges. In the lower limbs are the femur, tibia, fibula, patella, tarsals, metatarsals and phalanges. The thoracic (or shoulder) girdle consists of the clavicle and scapula. There are two pelvic girdles, consisting of the ileum, ischium and pubis, these two girdles being joined at the symphysis pubis.

### The joints

- There are some joints in which the surfaces of the bones are in very close contact and joined by connective tissue such that there is essentially no movement. These are known as synarthroses and are found in the skull.
- There are some joints that are only slightly movable, such as at the articulation between tibia and fibula and between vertebral bodies, usually there being fibrocartilage or intraosseous ligament holding them together.
- Most joints are freely movable, being known as diarthroses. They have different forms of motion, including rotation and sliding. The bony surfaces are covered with articular cartilage and they are connected by ligaments. These include:
  - ball and socket joints, e.g., hip and shoulder joints;
  - hinge joints, e.g., interphalangeal joints, knee and ankle joints;
  - rotary joints, e.g., radio-ulnar articulation;
  - condyloid articulations, e.g., wrist joints;
  - gliding joints, e.g., carpal and tarsal joints.

fused together, giving no relative movement at that joint but pain-free overall movement of the limb. These procedures rarely gave truly satisfactory long-term outcomes and attempts at replacing the tissues of arthritic joints with synthetic engineered structures were made in the first half of the twentieth century, initially by replacing one of the articulating surfaces (a hemiarthroplasty) or by interposing a material between the defective surfaces (an interposition arthroplasty) and finally involving all moving surfaces (total joint replacement or resurfacing).

## Box 4.3 Options for the treatment of osteoarthritis

There is no "cure" for osteoarthritis; the condition has to be managed through a variety of methods ranging from simple lifestyle changes to different types of surgical procedure. Not all options work in all patients.

| | |
|---|---|
| Lifestyle changes | Weight loss<br>Controlled rest and exercise<br>Use of aids, such as braces, shoe inserts |
| Dietary supplements | Vitamins<br>Glucosamine<br>Traditional Chinese medicines |
| Alternative therapies | Acupuncture |
| Medication | Pain relief, e.g., acetaminophen drugs<br>Anti-inflammatory (NSAIDs)<br>Pain/anti-inflammatory with narcotics<br>Corticosteroid injections |
| Interventional | Intra-articular injections, e.g., hyaluronic acid<br>Joint realignment (osteotomy)<br>Joint fusion (arthrodesis)<br>Joint replacement (arthroplasty), either total or partial |

With joint replacement, a small number of generic technical issues have confronted the bioengineer and orthopedic surgeon over the years and these have, it is probably fair to say, not yet fully been resolved. Three such issues may be discussed here in the context of the evolution of these materials and devices. These are the preferred mechanisms by which the prosthetic joints should operate, the materials used for the construction of the devices themselves, and the method of attachment of a device to the remaining parts of the skeletal system.

### Mechanisms of joint replacement prostheses

With respect to the mechanism of operation, the choice depends on the joint in question. With the hip joint, for many years there was no doubt as to how it should be replaced and how the new joint should operate; the replacement of a simple ball and socket joint by a direct synthetic equivalent is intuitively obvious and this probably explains why total hip replacement was achieved earlier and with greater success than the replacement of other joints. One issue with this process is the amount of bone that has to be removed to accommodate the components, which has led to some developments in which it is primarily the surfaces that are replaced rather than large amounts of the bone, but it is still a ball-and-socket action. In the knee, the situation is quite different since the mechanical operation of the knee, involving both sliding and rotation, is more complex, as are the arrangements of bones, cartilaginous surfaces, ligaments and tendons. Designers of knee replacements have struggled with the decision as to whether the femoral and tibial components should be totally independent of each other, which leads to difficulties of joint stability, or whether they should be constrained by some interconnecting component. This has led over the years to a multitude of different designs ranging from a fully hinged prosthesis to one in which the opposing surfaces only are replaced.

### Material choices for joint replacement prostheses

With respect to the materials of construction, the major issue has been the wear resistance of the articulating surfaces, and here the principles of tribology and biocompatibility have often given contradictory advice to the designer. The essential problem here is that healthy natural joints display a coefficient of friction that is lower than has ever been achieved in an engineering construct (Table 4.1), arising from the unique combination of lubrication regimes, including elastohydrodynamic lubrication, that are associated with two cartilage surfaces operating within an environment of synovial fluid. It has been impossible to replicate this combination so that joint replacements have had to rely upon combinations of synthetic materials that give the lowest friction and especially the lowest wear rates when in sliding contact. At the very beginning of the era of total hip replacement, an argument arose as to whether a metal-on-metal bearing or a metal-on-plastic bearing would be better. The originator of total hip replacements, John Charnley, firmly believed in the latter approach and rather naively assumed that the low coefficient of friction observed with polytetrafluoroethylene (PTFE) under lightly stressed circumstances would be translated into good performance within a hip joint. This was not the case, but the situation was recovered through the later choice of ultra-high-molecular-weight polyethylene (UHMWPE) and metal-on-plastic bearings have been in use ever since.

Table 4.1 **Coefficients of friction**

| | |
|---|---|
| Steel on steel | 0.80 |
| Cobalt–chromium alloy on cobalt–chromium alloy | 0.25 |
| Titanium alloy on polyethylene | 0.25 |
| Steel on polyethylene | 0.07 |
| Alumina on alumina | 0.05 |
| Alumina on polyethylene | 0.05 |
| Steel on PTFE | 0.04 |
| PTFE on PTFE | 0.04 |
| Human articular joints | 0.01 |

The metal-on-metal combination also suffered some early difficulties. Several prostheses involving articulating cobalt–chromium alloy components were introduced clinically at approximately the same time as Charnley was working with his polyethylene-based prosthesis, and whilst there is no doubt that some patients have survived 20–30 years of successful hip function with these all-metal devices, the adverse publicity associated with a few of them led to their decline. A number of these patients suffered significant tissue response and bone loss, which was attributed (although not really proven) to the host's immunological response to the nickel contained in these alloys. Since it is impossible to totally eliminate the release of metal from the surfaces of a metal-on-metal combination, and since the principles of electrochemistry determine that the tribologically favoured combination of dissimilar metals could never be used *in vivo*, there seemed little way forward with such devices. However, superior engineering techniques have led to better designs, and it is metal-on-metal combinations that are used in resurfacing. As we shall see, however, problems with the host response have not gone away, and some patients have experienced difficulties in this respect.

Other combinations have been utilized in this search for reduced wear. Early attempts to use ceramics were not very successful because of their inherent brittleness, but now the concept that alumina-on-alumina should give even lower wear rates because of the extreme hardness of both surfaces is well accepted.

### Mechanisms of attachment of prostheses to bone

With respect to the attachment of prostheses to bone, the story again starts with the work of

Charnley. Although he was not the first to utilise cement in orthopedic surgery, it was he who developed the use of acrylic cements for the fixation of joint prostheses. Acrylics are very useful for *in situ* polymerization and the ability to produce a material that would set within minutes of mixing and placing in the bony cavity around prostheses was the critical property of this material. Not all problems with cementing were resolved immediately, the exotherm associated with the polymerization causing some bone necrosis and the inherent brittleness of this amorphous polymer being two of the issues that had to be addressed. Even more critical was the toxicity of methylmethacrylate, which caused the death of several early patients through its significant effect on the respiratory system.

It has to be said that the mechanism of fixation by cement is not ideal since there is no chemical bond formed between the cement and either the bone or prosthetic component. The cement functions simply by filling this space at the interface. This observation caused a number of surgeons and prosthesis designers to question the wisdom of using cement and a trend towards so-called cementless fixation was started. At least three different types of procedure have been used for this, including highly accurate machining of the bone to achieve an interference fit, the use of porous surfaces that would encourage bone ingrowth and the use of bioactive surfaces on the components that would result in a chemical bond with the bone. With porous surfaces, it was recognized in the 1960s that bone would grow into a microporous surface, almost independent of the material of construction, depending only on the pore size and geometry (Figure 4.1). Although much of the early work was performed on ceramics, most attention was paid to the porous metals, sintering and other metallurgical techniques being used to create porous titanium and cobalt–chromium alloy components. In some situations this has worked quite well, although the fact that bone inside a porous structure may not necessarily sustain any mechanical loads means that the bone, once integrated into the porosity, may suffer atrophy. Thus bone that grows in may also grow out of the prosthesis.

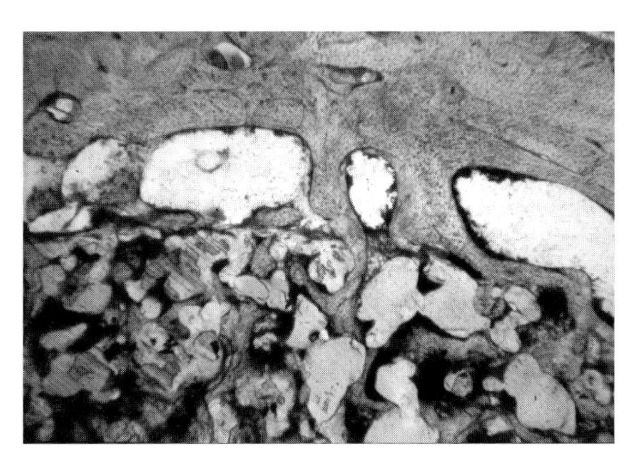

**Figure 4.1** Bone ingrowth into porous surfaces. Bone can grow into pores on the surfaces of most biomaterials, including metallic systems, both degradable and non-degradable bioceramics, and polymers, especially thermoplastics. In this example bone has invaded the pores of a high density polyethylene. The size of pores and of the interconnections between them are important determinants of the rate, extent and quality of the ingrown bone. See plate section for color version.

Adhesion to bone has been the subject of much discussion over the years and indeed this remains a very controversial aspect. Much of the attention has been focused on the various calcium phosphate materials that show a degree of chemical similarity to the apatite phase of bone itself. In particular calcium hydroxyapatite has been utilized by many joint prostheses manufacturers for the coating of components, especially the intramedullary components of hips and knees, and the outer surface of acetabular components in hips, in order to achieve fixation. Again some success has been achieved, although it is by no means guaranteed. It should also be said that claims have been made with respect to the putatively special properties of titanium concerning attachment to bone and the so-called phenomenon of osseointegration has been utilized in some products. These aspects have been discussed in Chapter 3 and will be considered again in Chapter 8.

## Hip replacement

Currently, the surgeon has a number of generic choices for the procedure and device for hip replacement. Within each of these categories, there are several specific designs produced, globally, by many manufacturers.

The variations in hip replacement design are based on the stem and head of the femoral component, the acetabular cup and the method of fixation of both the stem and the cup to the bone. As noted above, the principal design used over several decades involved an all-metal femoral component, a polymeric cup, with fixation of both components by bone cement. A typical example is given in Figure 4.2(a).

**Femoral stems** The stem may be forged or cast from one type of alloy, usually either stainless steel or preferably a cobalt–chromium alloy. Because of the requirement to have a range of sizes of prosthesis to accommodate all patients, in particular with respect to overall length and the diameter of the head, many manufacturers prefer a modular system, in which the stem and the head are produced separately and fitted together at the time of surgery. This allows for a smaller number of devices to be kept in inventory at the hospital. This also allows for the use of different metallic materials to be used for the stem and the head. This is a controversial issue. It may be argued that a lower modulus alloy is required for the stem in

(a)

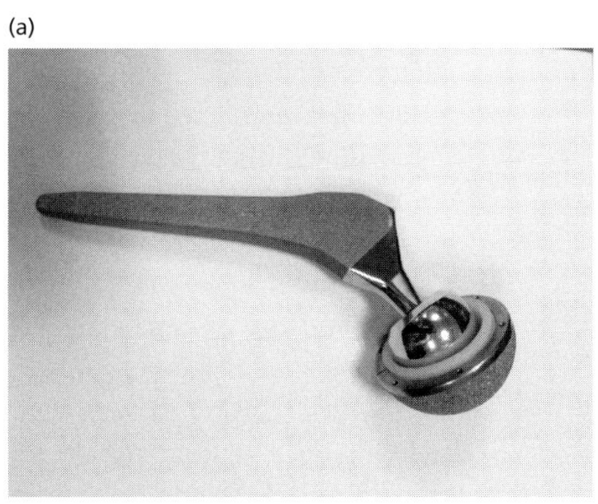

(b)

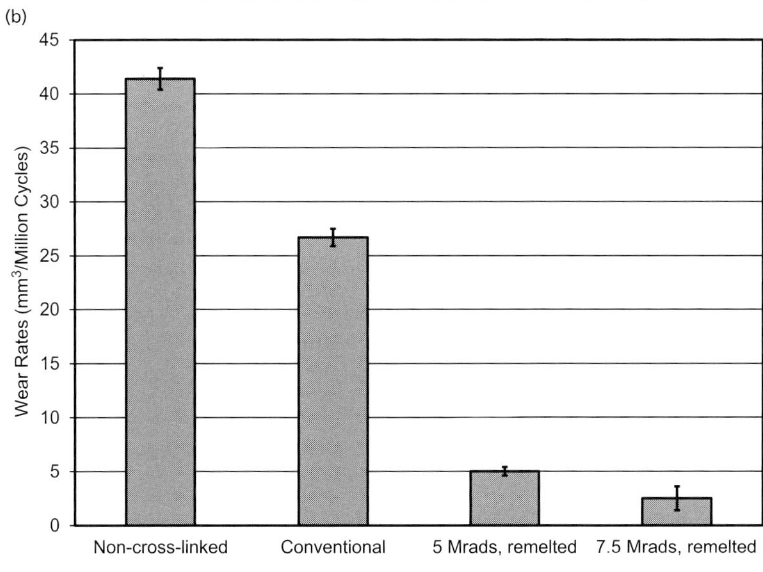


**Figure 4.2** Total hip replacements. Metal-on-polyethylene total hip replacement prostheses. Images courtesy of Dr. Harry McKellop, The J. Vernon Luck Sr MD Orthopaedic Surgery Research Center at the Orthopaedic Hospital, Los Angeles, USA.
(a) A conventional prosthesis with metallic femoral stem and femoral head, interfacing with a cross-linked polyethylene liner and a porous coated metallic acetabular cup.
(b) The effect of cross-linking and radiation on polyethylene wear rates, determined in a hip simulator to 5 million cycles: non-cross-linked – polymer sterilized by ethylene oxide; conventional – moderately cross-linked, sterilized with gamma radiation; 5 Mrads, remelted – gamma irradiated at 5Mrads for cross-linking, melted to remove free radicals, machined and gas plasma sterilized; 7.5 Mrads, remelted, as above with 7.5 Mrads irradiation.

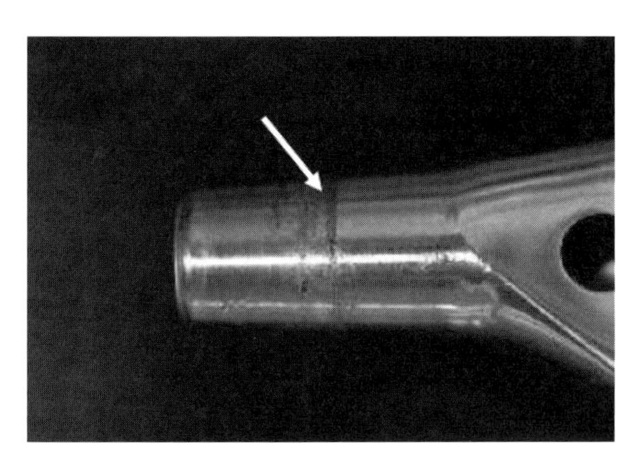

**Figure 4.3** Interfacial corrosion. Corrosion at the interface between taper neck and head of a modular cobalt–chromium hip replacement prosthesis, caused by a combination of fretting and crevice mechanisms (see Chapter 2). Reprinted from *Acta Biomaterialia*, 4, Virtanen S *et al.*, Special modes of corrosion under physiological and simulated physiological conditions, 468–76, © 2008, with permission from Elsevier.

order to minimize stress shielding discussed in Chapter 3, while the main specification for the head is maximum hardness in order to reduce the wear rate. The optimal materials for these components would therefore seem to be a titanium alloy for the stem and a cobalt–chromium alloy for the head. The counter argument to this is that, as discussed in Chapter 2, the mixing of different metals within the same device should be avoided because of the increased risk of galvanic corrosion. This can certainly occur (see Figure 4.3); however, whether the increased corrosion becomes a matter of clinical concern, and whether these risks are outweighed by the lower wear rate, is dependent on precise design, manufacturing quality and surgical technique. There have been reports of corrosion fatigue, surface cracking and fracture as a result of micro-motion and fretting corrosion.

**Polymeric acetabular cups** Almost exclusively, the polymer used for the acetabular cup is a type of polyethylene, as noted earlier. Over the 40 or so years since it was introduced, there have been some refinements to this polyethylene. In order to achieve the best mechanical properties, the polymer should be of the high density form rather than low density (see Chapter 8) and the molecular weight should be very high, typically around 5 million daltons. The polymer is usually described as ultra-high-molecular-weight polyethylene, or simply UHMWPE. As we shall see below, the main limitation of these joint prostheses is the rate of wear of the materials at the bearing surface, since the rate of release of wear particles, and their morphological and chemical characteristics, determine the response of the surrounding tissues to the prosthesis. Clinically, high wear rates result in loss of bone, known as osteolysis. It is imperative, therefore, that the rate of wear of the polyethylene is kept to an absolute minimum. This has created some difficulties over the years. Because of the need to control the risk of infection, the devices have to be sterilized before use, and the industry standard for sterilization procedures with medical devices moved towards gamma ray sterilization during the 1980s. However, polyethylene is somewhat sensitive to gamma irradiation if it is carried out in the presence of air or oxygen, resulting in increased rates of degradation and wear. The performance of these prostheses therefore became a significant issue clinically. This matter has largely been resolved through the use of sterilization under nitrogen and through the introduction of cross-linking techniques, which improve resistance to the degradation. These effects are shown in Figure 4.2(b). Further attempts to minimize the oxidative degradation have involved the use of antioxidants within the polyethylene. The materials science of polyethylene is discussed in Chapter 8. One further problem can occur with the polyethylene acetabular cups and that is creep deformation. The cups may slowly change shape as a consequence of the static and dynamic stresses occurring through normal activity over a period of time. This is largely overcome by the use of a metallic backing to the cup, as seen in the example of Figure 4.2(a).

**Ceramic prostheses** In view of the need to minimize wear, attention has been given to alternative materials. Ceramics provide one possibility since some of them have extremely high hardness. The same structural features that give this hardness, however, also tend to result in significant brittleness. First attempts to introduce ceramics into hip replacements involved aluminum oxide femoral heads. Although resulting in lower wear rates, they did suffer from some cases of brittle fracture. Over the years significant improvements both to material specifications, especially grain size and purity, and to manufacturing quality, particularly through eliminating porosity, have led to much better performance. Designs have evolved towards ceramic-on-ceramic, again with alumina. An example of such a prosthesis is seen in Figure 4.4. Thus, while the linear wear rate for a cobalt–chromium alloy against polyethylene can be around 200 μm per year, it is less than 100 for alumina on polyethylene and less than 5 for alumina on alumina. Clinical performance is very good, with very few mechanical failures reported. There is a regional variation in the use of ceramic prostheses, with continental Europe favoring their use, but not the USA. One factor in this selection is the occasional case where a ceramic-on-ceramic prosthesis emits an audible noise, or squeak, when the patient walks.

One final note here refers to the possibility of using alternative ceramics. Aluminum oxide was an obvious first choice because of its hardness and chemical inertness. There are some other oxide ceramics that could be used, with a very attractive option of using one that could be much tougher, that is less brittle than the alumina. This distinction falls to zirconium oxide, or zirconia. Zirconia can exist in more than one phase and a transition between the two can occur under the influence of mechanical stress, this being used to toughen the material. The formation of the relevant structures is facilitated by doping the zirconia with a rare earth oxide such as yttrium oxide or magnesium oxide. This mechanism of transformation toughening had been discussed in Chapter 2 and the specific features of zirconia will be discussed in Chapter 8. The significance here is that much tougher ceramic components can be produced for hip replacements using this material and indeed it became the preferred ceramic for many manufacturers. The fracture toughness and compressive strength of transformation-toughened zirconia can both be double those of alumina. However, the stability of these phases does depend on the precise manufacturing process, especially any thermal treatments that the prostheses undergo. A major problem emerged when one manufacturer changed from a batch process to a continuous process in which the temperature regime was not sufficiently well controlled, resulting in much inferior mechanical properties and premature failure. A further alternative is oxidized zirconium, in which the

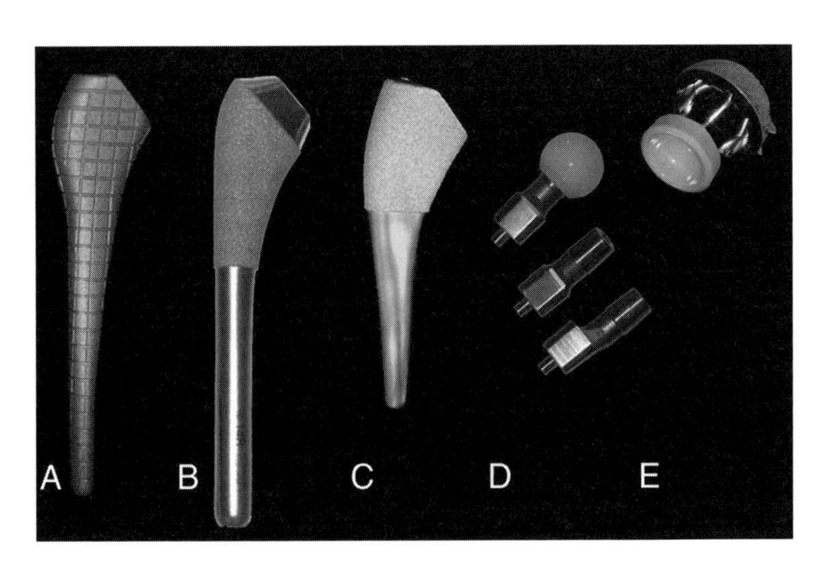


**Figure 4.4** Ceramic-on-ceramic hip replacement prostheses. Examples of alumina bearing surfaces on femoral head and acetabular components. Reprinted from *Journal of Arthroplasty*, 27, Sugano *et al.*, Eleven-to-14 year results of cementless total hip arthoplasty using a third-generation alumina ceramic-on-ceramic bearing, 736–41, © 2008, with permission from Elsevier.

surface of zirconium metal is oxidized to give a very hard surface of a tough substrate.

**Metal-on-metal replacement and resurfacing** An X-ray of conventional metal-on-polymer hip replacement is shown in Figure 4.5. It will be obvious that the metallic stem occupies much of the medullary cavity in the femur, and a great deal of bone has had to be removed. The bone involved in the arthritic changes is largely confined to the area just below the cartilage (the sub-chondral bone) so that much of the bone removed is quite healthy; this is not an ideal maneuver and it would be preferred that joint replacement involved removal of much less bone. This is the rationale for the so-called resurfacing procedure, in which much less bone is removed and, as the name implies, only the surfaces of femoral head and acetabulum are replaced. The mechanical properties of polymers and ceramics are not suited to the relatively thin section components of resurfacing, and the preferred materials are metals – hence the alternative name for these prostheses, metal-on-metal. An example is seen in Figure 4.6. In addition, some manufacturers have introduced metal-on-metal combinations that are full joint replacements rather than resurfacing devices.

This is still a somewhat controversial area. Suitability for younger patients is clearly a big factor, and the procedure has become popular in some sectors; it has been estimated that up to one-third of all hip replacements in the USA between 2005 and 2010 involved metal-on-metal devices. Overall failure rates have been reported to be higher than with conventional prostheses, for example 6% compared to 4% at 8 years. Risk factors include older patients, small femoral head size, some designs and some co-existing conditions such as dysplasia. With some designs, females have been associated with greater rates of problems.

Attention has been focused on the tissue response to the metals and their wear debris. Aseptic loosening of the acetabular cup and reactions to the metals may occur. Reactions may be of an inflammatory nature, the delayed-type hypersensitivity response being termed aseptic lymphocytic vasculitis-associated lesions (ALVALs). A variety of soft tissue changes, including necrotic changes, have been reported. Effusions may be seen intra-operatively at revision surgery, sometimes with fluid described as creamy or milk colored, or sometimes yellowish-brown. These reactions may occur within the first one or two years of implantation. There are reports of elevated cobalt

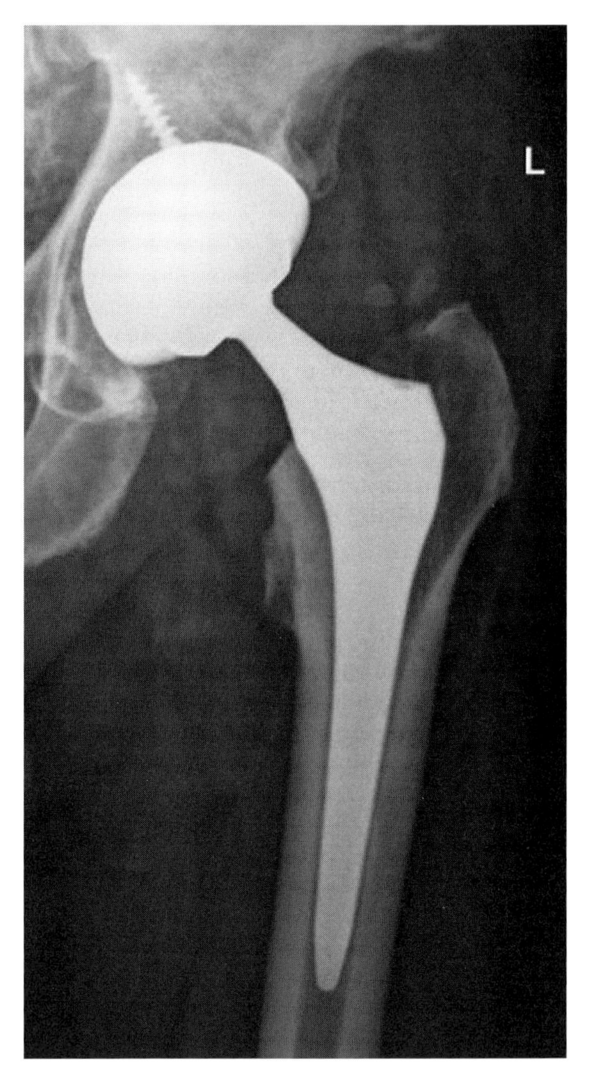


**Figure 4.5** Cementless total hip arthroplasty. Radiograph shows the large size of the prosthesis and the significant sacrifice of bone necessary for its placement and fixation. Image courtesy of Dr. Harry McKellop, The J. Vernon Luck Sr MD Orthopaedic Surgery Research Center at the Orthopaedic Hospital, Los Angeles, USA.

**Figure 4.6** Hip resurfacing. The components of the Birmingham Hip Replacement (BHR) System, image courtesy of Smith & Nephew Inc.

and chromium levels in both the serum and hip aspirate fluid in some patients but the relevance of this is not clear. Mechanical problems also occur, in particular, techniques of resurfacing may result in femoral neck fractures associated with iatrogenic notching of the bone.

**Bone cements** There are many factors that control the performance of hip replacements, but a most critical factor is the ability to achieve and maintain close adaptation of the prosthetic components to the bone. Put quite simply, we can attempt to produce this fixation by either arranging for some form of chemical bond to be established at the interface, to allow for ingrowth of bone into the surface, or to use a cement. It was the last of these options that was used first, and remains quite popular today. The cement most widely used is based on polymethylmethacrylate, usually referred to as acrylic cement, or simply bone cement. The idea for this material came from dentistry where the material was used to fabricate dental prostheses. The fundamental requirement is that the material in paste form should be capable of being placed in the space between bone and prosthesis where it should set and establish a bond between the two surfaces. Methylmethacrylate is one of the few monomers that can polymerize at low temperature and is therefore a preferred material for these types of application.

This is not a trivial process, and there are many issues we have to face. The chemistry of the acrylic polymers is discussed in Chapter 8; we deal here with the practical issues of the cement, which can be summarized as follows:

- The polymerization process is exothermic and the temperature rise can damage tissue.
- The monomer, methylmethacrylate has some degree of toxicity, and care has to be taken to minimize exposure of the patient.
- The polymerization also results in contraction and may lead to porosity.
- The appropriate degree of force is required in order to achieve filling of all the space.
- The cement does not chemically bond to bone; the attachment occurs via penetration into the interstices of the bone, especially cancellous bone.
- The polymer is not radiopaque and is difficult to see on X-rays.
- The polymer is brittle and prone to micro-cracking.

Based on these factors, the composition of current bone cements are given in Table 4.2. It should be noted that many commercial bone cements contain an antibiotic, which is intended to address potential infections that may arise, and radiopaque fillers.

Notwithstanding the less than ideal combination of properties, the acrylic bone cements are still widely used, and although over the years several alternatives have been proposed and indeed used, none have replaced the acrylic cement as the preferred choice.

**Porous surfaced components** Bone is a dynamic structure and, when damaged, has the capacity to remodel and regenerate. If an ultra-smooth surface of, say, a metal, is placed within a bony cavity, the bone is likely to grow up to the surface, and remain in intimate contact, but will not chemically bond to that surface. If that surface has a porous structure, the same bone growth may continue, with new bone penetrating the porosity, giving a degree of

Table 4.2 **Typical formulations of acrylic bone cement**

| | Product A | Product B | Product C | Product D |
|---|---|---|---|---|
| *Powder* | | | | |
| PMMA | 82.8 | 65.3 | | |
| MMA/styrene copolymer | | 18.6 | | |
| PMMA/MMA copolymer | | | 82.5 | 82.1 |
| Benzoyl peroxide | 3.0 | 1.85 | 0.75 | 0.78 |
| Barium sulfate | 10.0 | 10.0 | | |
| Zirconium dioxide | | | 10.0 | 15.0 |
| Gentamicin sulfate | 4.2 | 4.2 | 3.8 | 2.1 |
| Clindamycin hydrochloride | | | 2.8 | |
| *Liquid* | | | | |
| MMA | 98.2 | 98.0 | 98.0 | 98.0 |
| Dimethyl-p-toluidine | 1.8 | 2.0 | 2.0 | 2.5 |
| Hydroquinone | 75ppm | 75ppm | 75ppm | 75ppm |
| Figures are wt/wt% except for hydroquinone | | | | |

attachment. The following factors influence the nature of the ingrown bone and quality of attachment:

- The size and geometry of the pores and of the interconnection between the pores have to be such to allow osteoblast invasion and the protrusion of blood capillaries into the structure. Typically a pore size of 150 μm is required.
- Good fixation can be achieved with a small depth of porosity – too great a depth may leave spaces filled with fluid rather than tissue, and may compromise the mechanical properties of the device.
- Obviously the ingrowth of bone is not a spontaneous event and the process will take some time, with allowances made during patient rehabilitation.
- The choice of material is not critical since ingrowth can occur in metals, ceramics and polymers, although biocompatibility (for example, associated with greater corrosion due to higher surface area and poor oxygenation) should not be compromised.
- The porous layer should have good attachment to the substrate and not be subject to delamination.

Based on these factors, the majority of porous surfaces on hip replacements are produced from titanium, cobalt–chromium alloys or hydroxyapatite. An example of a porous surface is shown in Figure 4.7.

**Bioactive surfaces** The phenomena of bioactive biomaterials has been discussed in general terms in Chapter 3. In the context of orthopedic prostheses, bioactivity is considered to be associated with the chemical bond that can form between a material and bone. This possibility has been discussed for many years, and although bioactive surfaces are applied to a variety of prosthetic components, the practice is still somewhat controversial. There are two major groups of material that are known to have the ability to bond to bone in this way, those based on calcium phosphate and those based on biologically active glasses or glass-ceramics. Although the latter

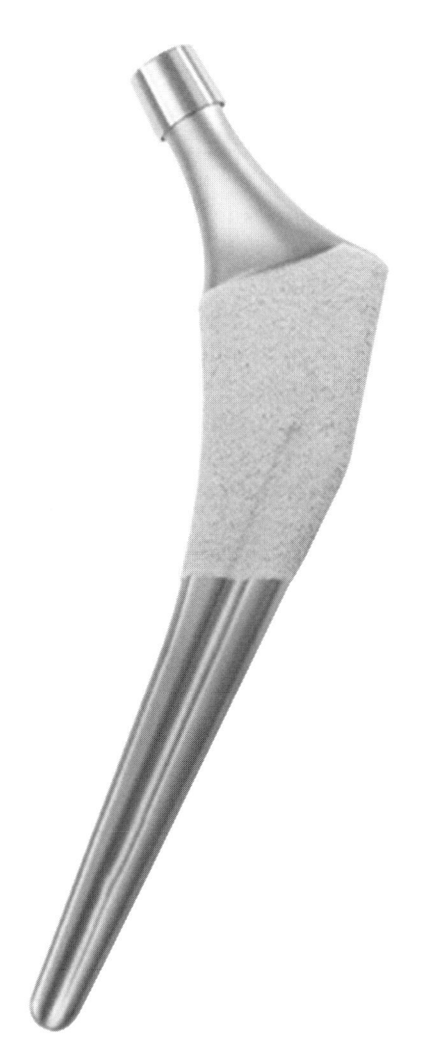

**Figure 4.7** Porous surfaces on hips. The illustration shows the femoral stem of the Accolade TMZF femoral hip system, which is made from a β titanium alloy and has a circumferential section with a plasma-sprayed hydroxyapatite coating. Image courtesy of Stryker Orthopaedics, New Jersey, USA.

certainly can be used in some clinical situations, they are not used in joint prostheses because of their poor mechanical properties. The emphasis is therefore on the calcium phosphates. The characteristics of these are discussed in Chapter 8.

The important questions relating to these calcium phosphates relevant to their performance include the following:

- How does the precise chemical composition influence the strength of the bond?
- How does the stability/degradability influence the overall performance?
- How good are their mechanical properties?
- What are the optimal coating procedures?

In general, after several decades of experience, the orthopedic device industry have settled on hydroxyapatite as the preferred option, using a small number of equally effective coating techniques and with sufficient stability and mechanical properties to allow safe clinical performance. As we shall see in the next section, however, the real issues are whether these devices give as good performance as cemented prostheses, and the answer seems to be no.

**Clinical performance of hip joint replacements** Since total hip replacements have been in use clinically for around 50 years, it would be expected that good information would be available about clinical performance and survival rates. This is the case to some extent, but the continual evolution of designs and techniques, and the considerable variation in surgical skills and patient characteristics make comparisons and general statements rather difficult. Data from various national registries, especially in Scandinavia and Australasia, suggest that 10-year survival rates should be around 95% and 20-year survival rates at around 90%. One major factor in the variability concerns the use of cement for fixation. Although the difference has been decreasing in time, cemented components have tended to give better results than uncemented ones, especially with the cup. Kaplan–Meier analyses show that the 90% survival at 15 years for cemented devices falls to 70% in uncemented prostheses. Failures are largely associated with component loosening as a result of wear and osteolysis. There are occasional fractures of components and dislocation of prostheses. A major cause of concern is infection. The general factors relating biomaterials and infection are dealt with elsewhere in the book, but it should be noted that infection of total hips is a significant issue as it very often leads to the need to remove the prosthesis. *Staphylococcus aureus* and *Staphylococcus epidermidis* are two important polysaccharide biofilm-forming bacteria that frequently colonize hip prostheses and great care has to be taken to minimize this risk. Also venous thromboembolism is a significant concern. Hospital procedure-acquired conditions, especially deep vein thrombosis and pulmonary embolism, can be life-threatening, with total hip replacement procedures being major risk factors. Prevention of these conditions through prophylaxis measures and rapid treatment should they occur are very important.

### Knee replacement

The knee is a more complex joint than the hip; it involves three bones, the femur, the tibia and the patella, and the major articulation between the femoral condyles and the tibial plateau involves two compartments (inner and outer condyles), which may suffer degeneration and wear at different rates. The movement is also a combination of rotation and sliding. Therefore, while many of the biomaterials concepts with devices to treat degenerative changes are the same as with hip replacements, there are some significant differences in design concepts.

Some of the early designs of total knee replacement involved two long stem components that were constrained to each other through a hinge. These may still be used in patients who have highly unstable knees with very poor ligament function but they have largely been displaced by non-constrained implants. These typically will have a metal tibial plate and short stem that is placed within the tibia. These will be of titanium or cobalt–chromium alloys that can be fixed by either bone cement or through bone ingrowth into porous surfaces. A polyethylene plate is secured to the tibial plate and acts as a bearing surface. Opposed to this will be the femoral component, which is usually a contoured metal structure that replicates the two condyles and is attached to the bone (Figure 4.8). There are several variations on these designs. Because of the complexity of the movement at the bearing surfaces, it may be considered important to replicate sliding and rotation components more accurately; some products have a mobile polymer component that can itself slide within the tibial plate. The surgeon also has the option to retain the patella or to resurface it; some devices have surface features, such as grooves, that allow easier movement of the patella.

One important issue with knee replacement is the fate of the posterior cruciate ligament. This plays a crucial role in providing stability of the joint. Some designs allow for retention of this ligament, whilst others anticipate loss of the posterior cruciate ligament and have features, such as sloping surfaces or posts, that attempt to compensate for this loss. As noted earlier, it may also be the case that only one condyle of the femur (usually the medial condyle) is involved in arthritic changes, and it may be desirable to surgically address that compartment alone. A number of products are available for this process of unicompartmental (or unicondylar) knee replacement. These tend to give overall better results than total knee replacement, and may have the added advantage of less invasive surgical approaches.

### Elbow replacement

The elbow is a complex joint anatomically and biomechanically, and injury or disease can have profound effects. Up to 60% of patients with rheumatoid arthritis have elbow involvement and it is also a target for primary degenerative osteoarthritis. It is frequently injured, with both fracture and dislocation being common. Early attempts to replace the elbow joint were largely unsuccessful, but improved designs have led to better success.

Prostheses are generally of two varieties, those where humeral and ulnar components are linked and those where they are not (Figure 4.9). Implants in the former group vary in terms of the degree of constraint, some being rather loose and some being rigidly constrained. Linked designs tend to have more widespread use, the unlinked devices requiring good integrity in the surrounding capsular and ligamentous structures and being more prone to dislocation. Some designs are available in either linked or

(a)

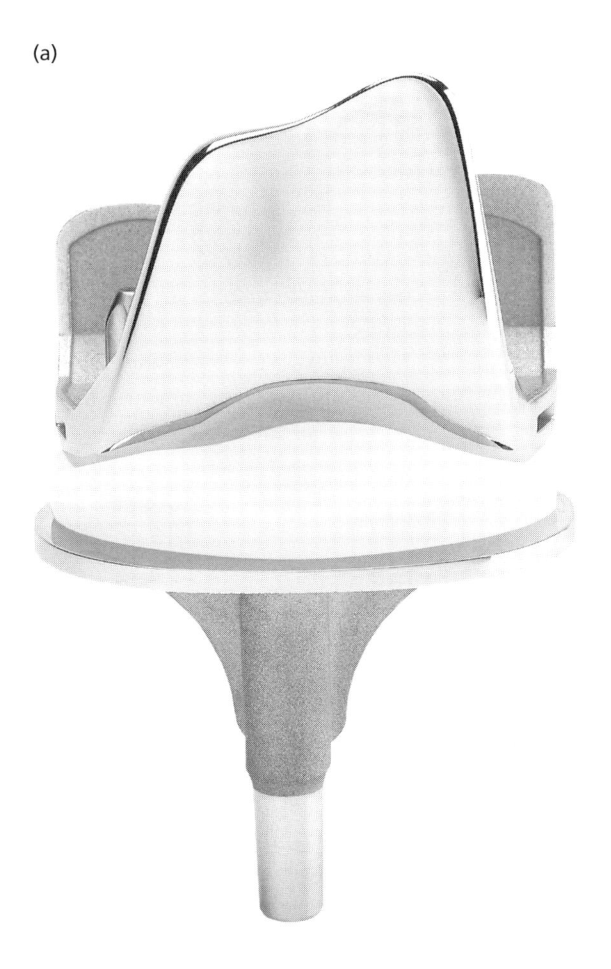

(b)

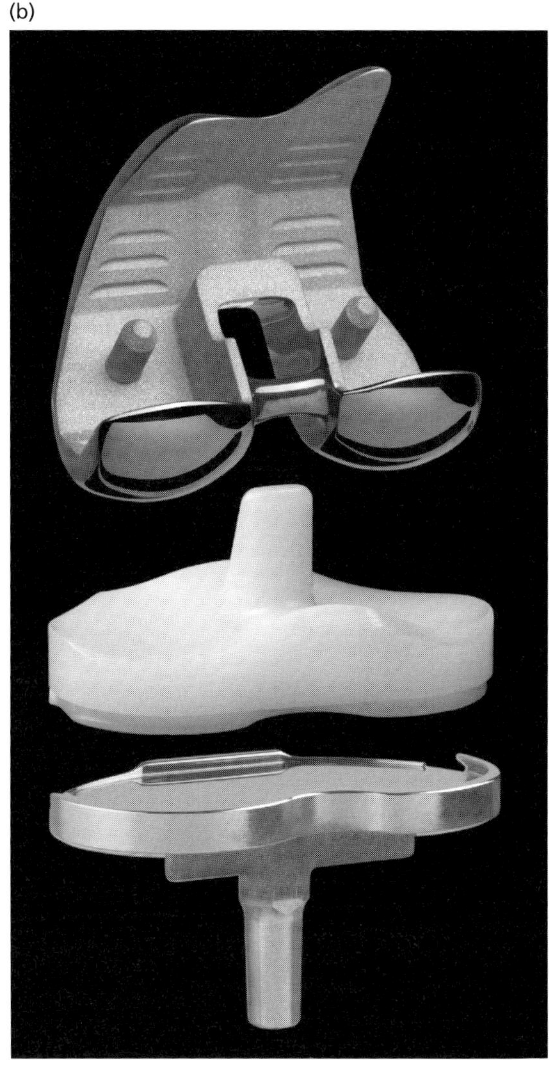

**Figure 4.8** Total knee replacement systems.
**(a)** Ox Apparatus Knee;
**(b)** Genesis II Total Knee System. Both images courtesy of Smith & Nephew Inc.

unlinked configurations. Although the radius forms an important part of the elbow joint, it has proved very difficult to incorporate a radial component into total elbow prostheses. Most prostheses have metallic stems for both humeral and ulnar components, utilizing either cobalt–chromium or titanium alloys, these stems usually being fixed by bone cement. The linked prostheses usually have some type of metal pin that joins them in a hinge configuration, with a polyethylene bush. The unlinked prostheses employ a polyethylene cap on a metal stem as the ulnar component and an all-metal humeral component.

Loosening of the prostheses is a major complication, especially with the unlinked devices, with some 10–15% failing by 5 years.

## Ankle replacement

The ankle is a relatively difficult joint to replace. Indications for total ankle arthroplasty primarily include pain associated with traumatic arthritis following a fracture and generalized inflammatory arthropathy. Some early designs had two components that were constrained, that is linked together, and were cemented in place. The rather poor

(c)

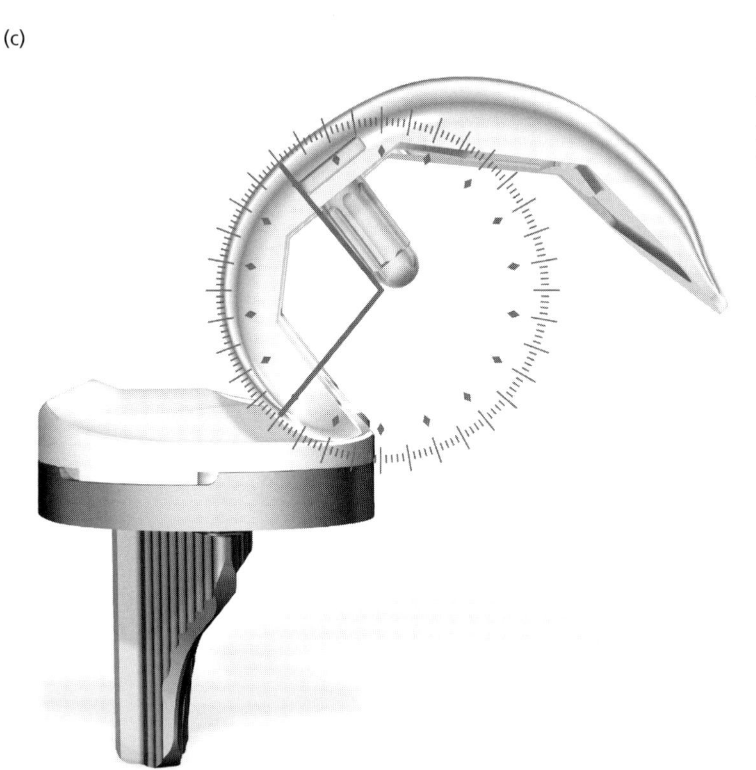

**Figure 4.8** *continued*
(c) The Triathlon Total Knee System, with a tibial-bearing insert of highly cross-linked polyethylene. Image courtesy of Stryker Orthopaedics, New Jersey, USA.

performance led to different designs. One of the more popular has a cobalt–chromium talar component that sits on the talar bone in the foot and a tibial component that has both titanium and polyethylene parts, the former allowing attachment to both tibia and fibula and the latter acting as the bearing surface opposite the talar component. Porous surfaces of the metal allow for fixation by bone ingrowth. The overall performance is still not very good, and considerably inferior to both hip and knee. According to some studies, roughly 50% of patients will require some revision surgery within 5 years, although other reports give survival of prostheses at 5 years of around 75–85%. An example of an ankle joint prosthesis is seen in Figure 4.9(a).

## Shoulder replacement

The shoulder (or glenohumeral) joint consists of a ball-shaped head of the humerus, which sits inside the glenoid fossa. The stability and range of movement of the joint are controlled by the rotator cuff, a collection of muscles that surround the joint and are connected to the bones. The rotator cuff is a frequent site of sporting injury, especially involving a cuff tear. This can be treated fairly effectively but frequently leads to cuff tear arthropathy. The shoulder may also suffer arthritic changes, but not so frequently as the hip or knee. It is often possible to treat this arthritis by hemi-arthroplasty, that is replacing only one of the sides of the joint, almost exclusively this being a metal humeral head and stem, which can be press fit into the humerus. If the glenoid cartilage is seriously affected, a total shoulder arthroplasty may be necessary. Traditionally this involves the same type of humeral head and a polyethylene glenoid component. An alternative, which became quite popular, involved the reverse, a cemented polyethylene humeral cup and a metallic glenoid component. Variations on these themes exist, with varying material combinations and varying methods of fixation, including screw fixation of the glenoid component. The techniques are reasonably

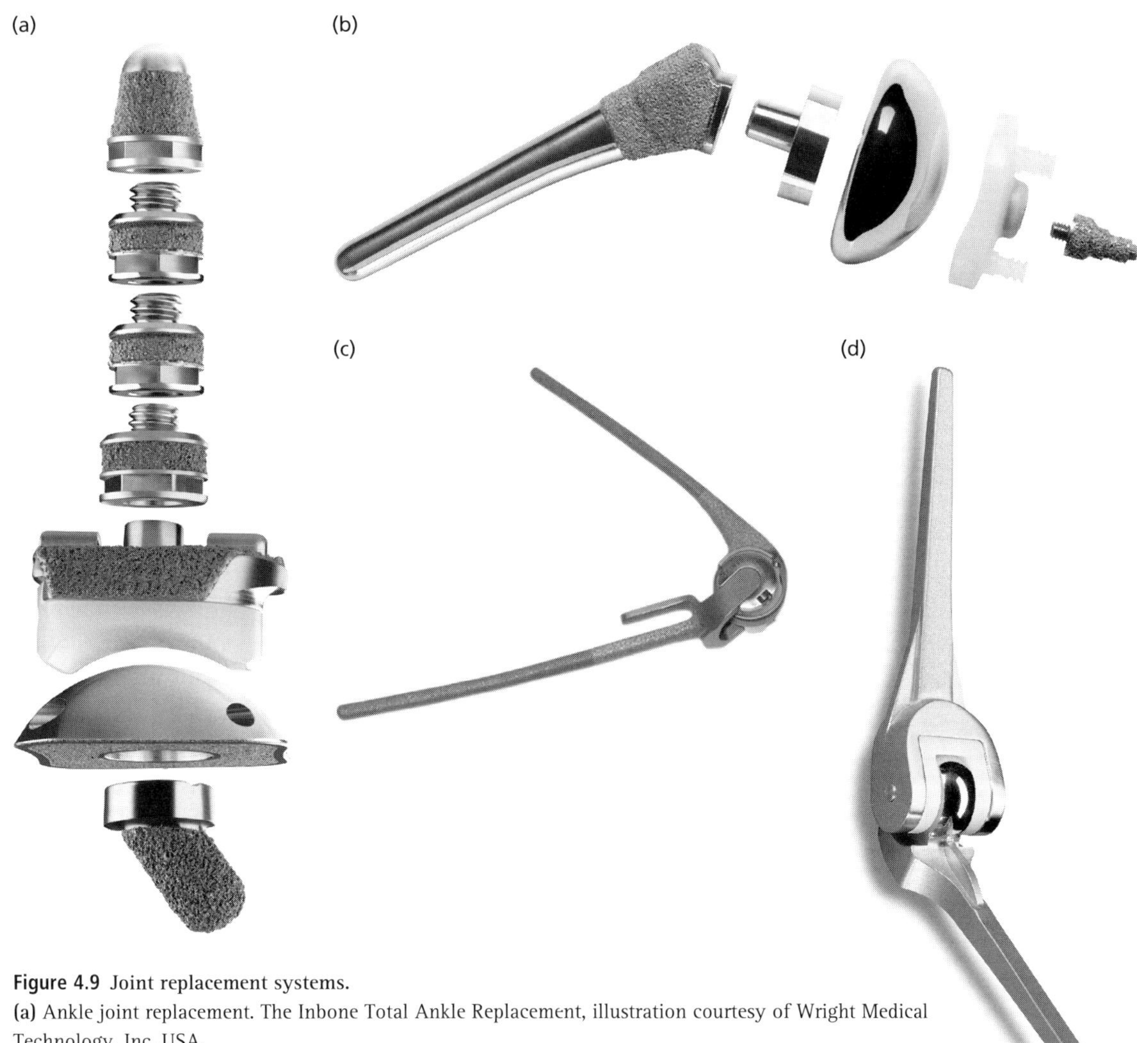


**Figure 4.9** Joint replacement systems.
**(a)** Ankle joint replacement. The Inbone Total Ankle Replacement, illustration courtesy of Wright Medical Technology, Inc. USA.
**(b)** Shoulder joint replacement. The Comprehensive® Shoulder System, illustration courtesy of Biomet Inc, Warsaw, Indian, USA.
**(c)** Elbow joint replacement. The Discovery® Elbow System, illustration courtesy of Biomet Inc, Warsaw, Indiana, USA.
**(d)** Elbow joint replacement. The Solar Elbow System, designed to restore the normal center of rotation of the elbow joint, illustration courtesy of Stryker Orthopaedics, New Jersey, USA.

successful, with complication rates usually between 10 and 20%. An example of a total shoulder replacement joint is given in Figure 4.9 (b).

## Finger joints

The proximal interphalangeal (PIP) joints of the hand represent a primary target for rheumatoid arthritis and some degenerative and post-traumatic conditions. The functional impairment and pain associated with these conditions suggest that replacement of the joints is a priority. Metallic hinge prostheses were first used for these joints over 50 years ago but with limited success. As a consequence, integral hinge prostheses, made of a single component elastomeric material, were introduced. Some devices used polypropylene, but the most widely used

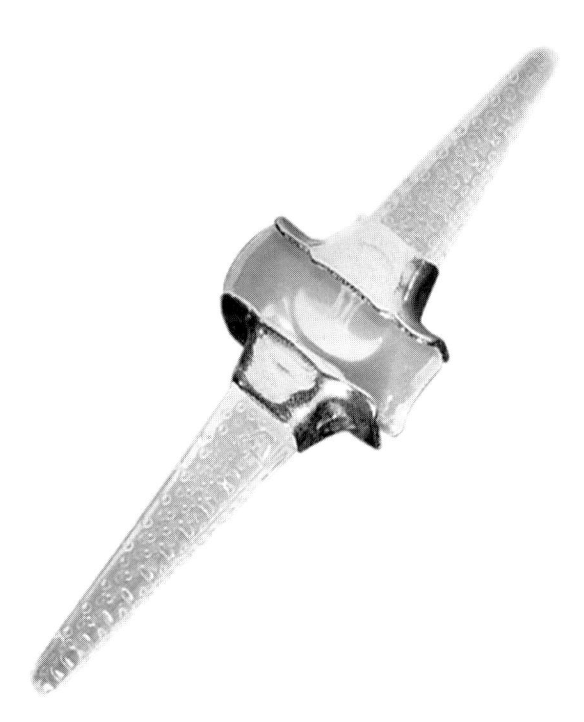

**Figure 4.10** The Swanson Finger Joint Prosthesis, made of silicone elastomer, illustration courtesy of Wright Medical Technology, Inc.

PIP prostheses have been made of silicone elastomer (Figure 4.10). The two stems are placed in intramedullary positions without cement. These prostheses do provide pain relief in the majority of patients, but improvements to the range of motion of the fingers, and the overall mechanical functioning of the hand, are rather limited. Survivorship of the implants may be 75% at 12 years and as low as 50% at 15 years. Fractures are quite common. Some alternative materials and designs have been introduced but, as yet, without evidence of better success.

## 4.1.2 Ligaments and tendons

Ligaments and tendons represent one of the biggest challenges for medical technologies and biomaterials. On the one hand, there is an enormous need for effective repair strategies simply because of the frequency with which these tissues become damaged. On the other hand, these are very difficult tissues to treat, whether that is by repair, regenerative or replacement methods; this is partly because they are highly stressed and partly because they are attached to muscle and bone in situations where the recapitulation of the natural integration of tissues in these regions is not at all easy. We shall discuss tissue engineering approaches to the treatment of injuries to these tissues in the next chapter. Here we will consider those situations in which materials are used to assist in repair of the tissues. These procedures do not involve the use of total replacement devices but rather structures that can carry load but which encourage the formation of reparative tissue, usually within their porosity.

The need for repair of these tissues is largely associated with sporting or occupational injuries. Different parts of the body suffer differentially, with the anterior cruciate ligament of the knee, the rotator cuff of the shoulder, the Achilles tendon in the ankle and the extensor carpi radialis brevis of the elbow being commonly affected. The prevalence of these injuries is staggering, with millions of individuals in the USA alone being affected each year. The slow and often ineffective repair processes means major loss of quality of life and significant economic sequelae.

In the knee, which suffers more than most joints, injuries to the medial collateral ligament are treated conservatively since this ligament will spontaneously heal, perhaps assisted by an external brace. On the other hand, injuries to the anterior cruciate ligament (ACL) and, to a lesser extent the posterior cruciate ligament, are much more difficult to heal, especially if they involve the site of attachment to the bone, and they frequently require surgical intervention. The ACL provides a good reference point. Repair/reconstruction may be achieved with autograft, allograft, xenograft or synthetic material (Figure 4.11). Two main sources of autograft tissue have emerged, the patella tendon and the hamstring tendon. In the former case bone can be harvested along with a section of the tendon and used as a so-called "bone-patellar tendon-bone autograft" (BPTB). The pieces of bone allow for fixation to the existing bone with

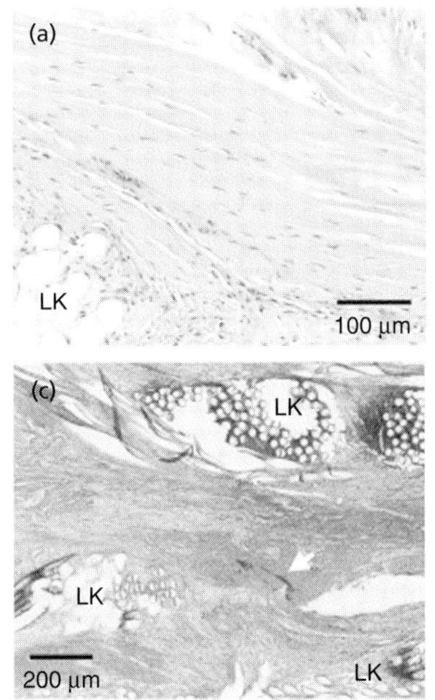


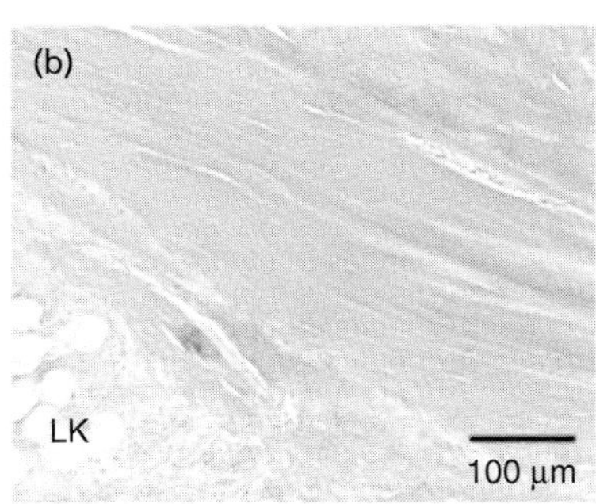


**Figure 4.11** Ligament prostheses. These histological images relate to the Leeds–Keio polyester fabric prosthesis used in reconstruction of the medial patellofemoral ligament. The micrographs show the tissue that has formed inside the prosthesis. In (a) and (b) derived from a patient 48 months after implantation where there are regularly aligned collagen bundles but little elastin. More elastic fibers are seen in (c) in a patient after 70 months. In some situations significant inflammatory cells will be seen between the fibers, seen at the lower left corner in (a) and (b) and upper center in (c). Reprinted from *Biomaterials*, 26(15), Nomura E *et al.*, Histological evaluation of medial patellofemoral ligament reconstructed using the Leeds–Keio ligament prosthesis, 2663–670, © 2005, with permission of Elsevier.

screws. The hamstring tendon is preferred by some because of a somewhat lower morbidity, the location of the patellar tendon graft source often being associated with long-term discomfort. Allografts obtained from tissue bank cadaveric sources are mainly used when autograft is unavailable. Overall they give inferior performance and are associated with a finite infectivity risk. Some products are xenogeneic, for example being fabricated from porcine small intestine submucosa (especially useful in rotator cuff injuries), bovine pericardium and bovine or porcine dermis.

There has been a long history of using synthetic materials for ligament and tendon repair; at one time long strands of carbon fiber were used. A relatively small number of materials have proved successful; these include polyethylene terephthalate textile and microporous PTFE. These fabrics are stapled to the bone and slowly allow tissue ingrowth to occur. There are still concerns about the long-term effectiveness of these devices, mechanical failures associated with creep and fatigue being common, and success rates may be less than 60% at four years.

### 4.1.3 The spinal column

The spinal column is critical to the quality of life, as alluded to in the introductory chapter. We have to face three crucial facts here. First, the spinal nerve is essential for all sensory and locomotor function from the head downwards. If it is damaged, even slightly, daily functions may be seriously compromised. Secondly, the intervertebral discs, which separate the individual vertebrae, are prone to serious degeneration, which can be extremely painful and debilitating. Thirdly, infants may be borne with abnormal anatomical features of the spine, especially unusual curvatures, which often become exacerbated during skeletal growth. These can be very unsightly but also can have an impact on the positioning and functioning of internal organs.

At this stage, implantable devices by themselves have no role in therapies for spinal cord injury. This is an immensely difficult area and we shall touch on future opportunities in the next chapter in relation to regenerative medicine techniques. We can discuss, however, spinal deformities and the treatment of degenerative disease.

## Spinal deformity

The spine may be deformed in the sense that it does not follow the normal geometrical curvatures. There are several possible causes. The first is congenital, being associated with problems that arise with the formation of the vertebrae during development within the womb. Closely related to this are developmental causes that arise early in life, for example, from abnormal fusion of the ribs. Neuromuscular-derived curvatures are caused by muscle weakness or imbalance, or as a consequence of some other diseases such as spina bifida or cerebral palsy. Many cases, indeed most cases in adolescents, have no apparent cause, and are known as idiopathic conditions. There are three main forms: scoliosis, which involves sideways curvature; kyphosis, which is an outward curve of the thoracic spine; and lordosis, which is inward curvature of the lumbar spine.

Minor curvatures may require no treatment or the use of externally worn braces. However, with increasing degrees of curvature there may be significant aesthetic consequences and, more importantly, pressure on internal organs, with, for example, reduction of lung function. A wide variety of physiotherapy options are available but with curvatures above 50° it may be necessary to resort to surgery. Surgery usually involves a combination of the fusion of vertebrae and the use of implantable rods, plates, screws and other devices that aim to alter the degree of curvature. The types of device are similar for all three types of deformity, although their configuration will vary. The placement of the instrumentation may be either anterior or posterior. A typical device will consist of rods that are connected to screws, which are inserted into the vertebrae, as shown in Figure 4.12. The hardware is usually made of titanium alloys or stainless steel.

## Spinal fusion

The option of spinal fusion for the treatment of disc problems was introduced in Chapter 1.

The intervertebral discs are fibrocartilaginous structures that separate the vertebral bodies and allow some extension and flexion movement. They are composed of two components, the annulus fibrosus and a nucleus pulposus. The annulus fibrosus is a strong radial structure made of concentric sheets of collagen fibers connected to the vertebral end-plates. These sheets are orientated at various angles. The annulus fibrosus encloses the nucleus pulposus. This consists of collagen fibers within a polysaccharide/proteoglycan-rich gel. When healthy, this gel is strongly hydrated, but this reduces with age and disease. The healthy adult disc has few blood vessels, but it has some nerves, mostly restricted to the outer lamellae. The hyaline cartilage end-plate is normally totally avascular and aneural.

With increasing age the disc changes in morphology, becoming more and more disorganized. As it degenerates, the annular lamellae become irregular and interdigitating. These changes may lead to herniation or prolapse of the disc when it bulges or ruptures posteriorly or posterolaterally, and press on the nerve roots in the spinal canal. The herniation results from the migration of isolated degenerate fragments of nucleus pulposus through pre-existing tears in the annulus fibrosus. The condition is most frequently seen in the lumbar spine.

The most commonly used treatment for serious herniation is spinal fusion, in which the degenerated disc is excised and the two adjacent vertebral bodies fused together. There are several different techniques and surgical approaches, including the anterior approach (from the front), the posterior approach (from the back) and from the side, the lateral approach. Minimally invasive techniques may also be used.

All spinal fusions use some type of bone graft, to promote the fusion. For this, bone may be harvested from the patient's hip and used as an autograft. Alternatively an allograft, that is cadaver bone acquired through a bone bank, may be used. Some bone substitutes may be used to augment the bone graft, including demineralized bone matrices and calcium phosphates.

There are several types of implantable device that can be used as adjuncts to the fusion process. Pedicle

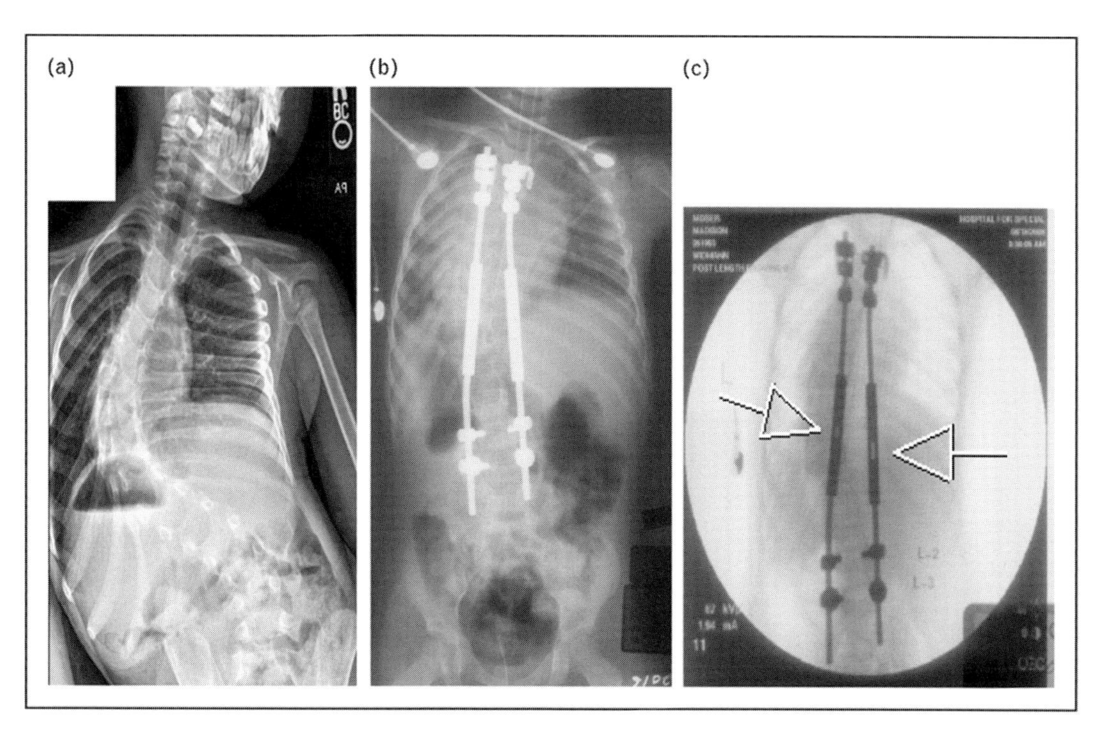


**Figure 4.12** Treatment of scoliosis. This is a 4-year-old patient with juvenile idiopathic scoliosis.
(a) An anteroposterior radiograph before surgery.
(b) The situation just after surgey using two expandable rods and laminar hooks to hold the fused segments.
(c) The patient 6 months later when the rods were lengthened to accommodate the growing spine. Reprinted from *Current Opinion in Pediatrics*, 21 (1), Han K *et al.*, Update on the management of idiopathic scoliosis, 55–64, © 2009, With permission from Wolters Kluwer Health.

screws are screws that are attached to the vertebral bone and may be connected by rods; they provide for stability during the fusion process. It is possible to remove these screws once fusion has taken place although not all surgeons advocate this. At one time this procedure was criticized in view of excessive complications, including screw fracture and spinal cord compression but improved technology and techniques have significantly improved outcomes. Perhaps the most widely used adjunct is the spinal cage, which is inserted into the intervertebral space. This assists in maintaining bone grafts in place and gives good stability. Some examples are shown in Figure 4.13. Initially cages were almost universally made of titanium but more recently the thermoplastic material polyetheretherketone has been introduced with good success; being compatible with MRI is an attractive feature. Carbon fiber composites have also been introduced for the same reason.

Somewhat controversially, an additional component has been introduced to spinal fusion technology, and that is the incorporation of a growth factor, bone morphogenetic protein 2 (BMP-2), into the product, for example as a solution contained within a collagen sponge held in a cage. This approach has been discussed in the previous chapter.

## Intervertebral disc replacement

The alternative to using interbody fusion techniques to reduce back pain associated with disc degeneration involves replacement of the disc. As we will see in Chapter 5 there is significant interest in the use of tissue engineering techniques to generate new disc material but at this point it is the use of synthetic

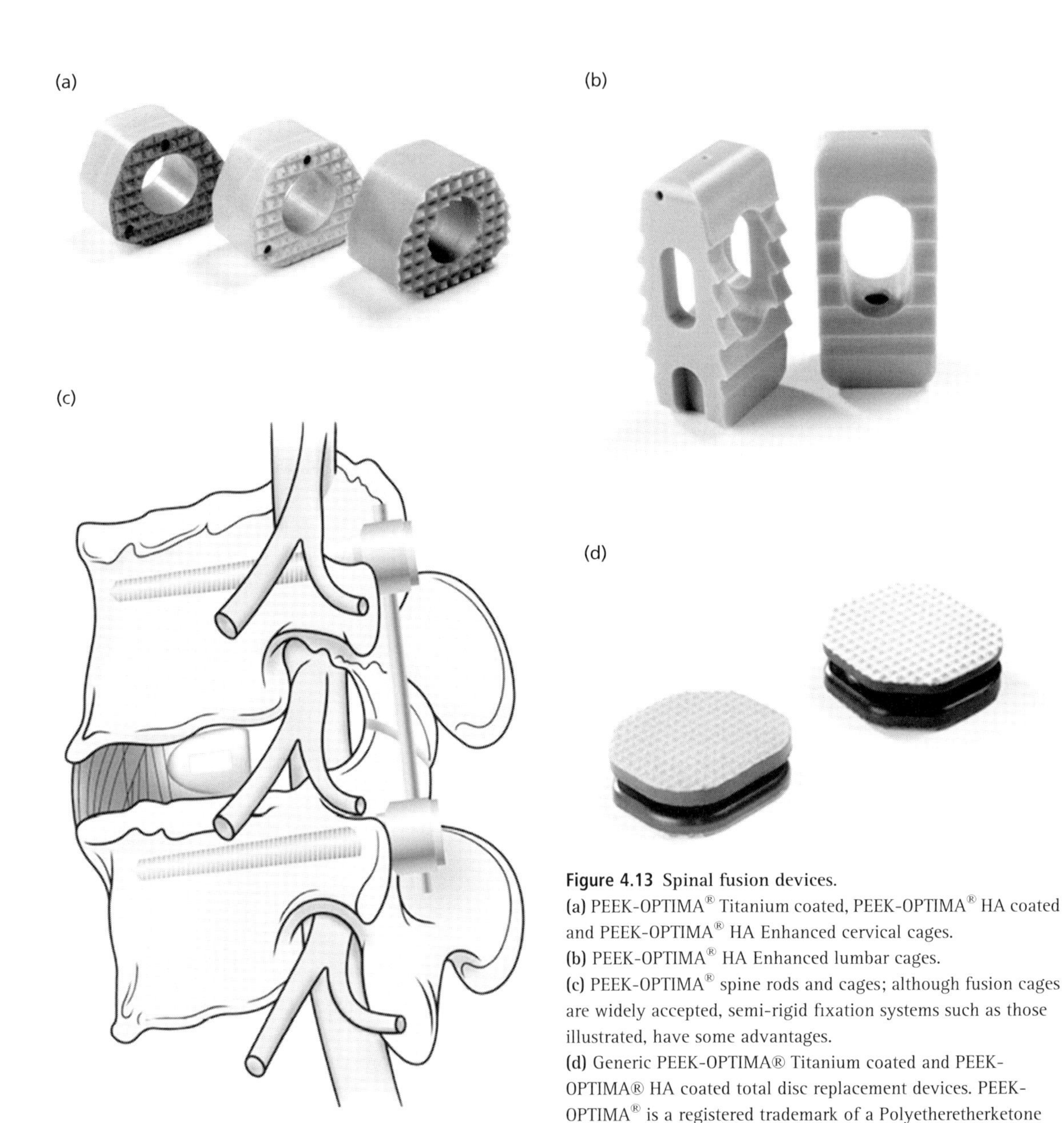


**Figure 4.13** Spinal fusion devices.
**(a)** PEEK-OPTIMA® Titanium coated, PEEK-OPTIMA® HA coated and PEEK-OPTIMA® HA Enhanced cervical cages.
**(b)** PEEK-OPTIMA® HA Enhanced lumbar cages.
**(c)** PEEK-OPTIMA® spine rods and cages; although fusion cages are widely accepted, semi-rigid fixation systems such as those illustrated, have some advantages.
**(d)** Generic PEEK-OPTIMA® Titanium coated and PEEK-OPTIMA® HA coated total disc replacement devices. PEEK-OPTIMA® is a registered trademark of a Polyetheretherketone thermoplastic material (see Chapter 8); HA is hydroxyapatite. Images courtesy of Invibio® Biomaterial Solutions.

materials to replace the disc that is used clinically. Indeed, total disc replacement has been gaining popularity for the treatment of degenerative disc disease, particularly in the lumbar region. There are several types of design of prosthesis, which give varying degrees of articulation. One type involves two metallic components, usually of cobalt–chromium alloys, which are attached to the surfaces of adjacent vertebra and constitute articulation by having concave and convex features on facing surfaces. There has been an interest in using PEEK or carbon-fiber-reinforced PEEK for these components

(Figure 4.13(d)), partly to reduce risks of tissue responses to wear debris from the metal-on-metal joints and partly to allow MRI imaging. Other devices have metal end-plates attached to the vertebrae and an intervening UHMW polyethylene insert between them to provide articulation against one of these plates. Some prostheses have metal end points with an intervening elastomeric component to give some flexibility. There is little clear evidence that disc replacement will be better than fusion. Whilst it is limited to situations where the disc degeneration is confined to a single level, between L4 and S1 in non-osteoporotic patients without severe instability and neural element compression, the results appear to be about the same; whether complication rates up to 30%, which appears to be common, is ultimately acceptable, remains to be seen. Failures appear to be technique rather than material/device factor related, and freedom from pain in the long term is the major concern.

It should also be noted that it may not be necessary to remove and replace the whole of the intervertebral disc. For example, it may be possible to excise and replace the nucleus pulposus. The concept here is to remove the anterior part of the disc, where most of the pain arises, and replace it with a device that gives some, but very limited, motion; this effectively places tension on the annulus fibrosus rather than the anterior segment and minimizes irritation to the sinuvertebral nervous network. The nucleus replacement could be an injectable, *in situ* curing polymer, or hydrogel that absorbs water following setting, or a preformed polymer.

## 4.1.4 Bone defects

Bone defects may arise from a number of causes, particularly being associated with congenital conditions, trauma and tumors. With congenital bone defects, there is clearly no possibility of spontaneous resolution, and physician and patient have to decide whether the defect is sufficiently serious, functionally or cosmetically, to warrant any surgical intervention. With tumors, surgical excision may or may not have to be accompanied by post-surgical reconstruction, again depending on function and appearance. With trauma, as we have seen, small defects may spontaneously heal and remodel, but there is a limit to the size of the defect that can be resolved without intervention.

### Bone tumors

In most situations where a bone tumor has been diagnosed, the prospects are not good for the patient and aggressive reconstruction is not often an option. We discuss here one situation in which reconstruction may be possible, that is in the lower limbs, and also one where it is rarely possible, that is metastatic spinal tumors.

**Lower limbs** The conventional approach to malignant tumors in the lower limbs was, for many years, amputation of the affected limb, offering the patient a lower risk of metastases. Since the 1980s there has been a major move towards salvage of the affected limb and reconstruction with what have become known as megaprostheses. Because individual patients vary so much with respect to their anatomy and the extent of the tissues affected by the tumor, these devices are usually custom made or assembled from modular components. Different groups have worked with varying materials systems, but these do not differ greatly from the materials and design concepts of traditional hip and knee prostheses, that is involving titanium or cobalt–chromium alloys, polyethylene, bone cements and hydroxyapatite surfaces. Single axis or rotating hinge designs are usually used in the knee. It is difficult to assess the overall performance of these devices, partly because low numbers of procedures are carried out in the centers that treat these patients and partly because of the additional effects of the cancer on the patients. Leaving aside those patients who die from their underlying malignancy, the major complications are infection (a much higher risk than normal arthritic patients), loosening and dislocations at the knee.

**Metastatic spine tumors** The spine is the commonest site for metastasis of tumors to bone; with some types of cancer up to 70% of patients will eventually have spinal metastases, often with significant compression of the spinal cord. Treatment options are rather limited; radiation without surgery is often preferred to surgery, but this rarely improves life expectancy over the normal 1 year following diagnosis. Surgical procedures are available but obviously they diminish the quality of life in the immediate post-surgical period and are not considered effective if the life-expectancy is less that 3–6 months. Surgery may consist of cord decompression for pain relief but this may be assisted by some form of stabilization. Several forms of stabilization device have been used. These include different types of cement, either polymethylmethacrylate or calcium phosphate, sometimes accompanied by silicone spacers. More complex devices, including adjustable metallic cages and alumina ceramic inserts have also been used. The relative rarity of these procedures and the usual short-term prognosis has not allowed reliable data on significant numbers of patients with these devices to be obtained.

## 4.1.5 Bone fracture

Bone fractures occur readily, at all ages. In most cases they are caused by trauma and very often the individuals who experience the fracture are otherwise healthy. Fracture healing can occur spontaneously, or with just minimal help. In other cases the fractures occur because the patient is not healthy and the bones do not have the ability to withstand forces that are just outside the normal range, for example in post-menopausal women with severe osteoporosis.

Bone heals readily under the right circumstances, discussed in Chapter 3; the minimal help that may be required usually takes the form of external splinting, through plaster, which holds the fracture surfaces together so they heal with proper alignment. There are several situations in which this spontaneous healing cannot take place and additional techniques are required. This may be because the fracture has taken place in a location which is difficult to stabilize, for example within the condyles, or if the fracture is complex, yielding multiple pieces, or if it is a serious open fracture with much damage to and loss of surrounding tissue. Under these circumstances, internal fixation of the fracture, or possibly external fixation, using appropriate instrumentation, may be indicated. These techniques have been used for many decades and little has changed in either the principles or the materials, and so we can deal with this subject quite briefly.

### Internal fixation

Internal fixation involves the use of a range of devices, including plates, screws and rods, which stabilize the fracture region. This can be achieved by the use of a rod placed in the medullary space, which has a mechanical advantage because the rod can be essentially co-linear with the long axis of the bone. The major alternative for a long bone fracture is the plate, which is attached to the cortical bone by means of screws. Small fragments of bone can be reattached by means of screws or wires. Examples are shown in Figure 4.14.

**Metallic systems** Conventional internal fixation devices are made from either titanium alloys or stainless steel. Clearly fracture devices have to be strong enough to resist fracture themselves, in particular under fatigue conditions, and both of these types of alloy satisfy this requirement; although many years ago some plates and rods experienced fracture, this is not an issue with current materials. Additionally, however, the yield strength should not be too high since it is preferable that the plates and rods can be manipulated intra-operatively to allow the surgeon to match the bone contour. It is also necessary that the screws have high torsional strength to allow for insertion and withdrawal without shear failure. There is no significant evidence to suggest that either alloy is superior in relation to performance.

The one issue of interest here concerns the desirability of leaving a fracture fixation device in a

(a)

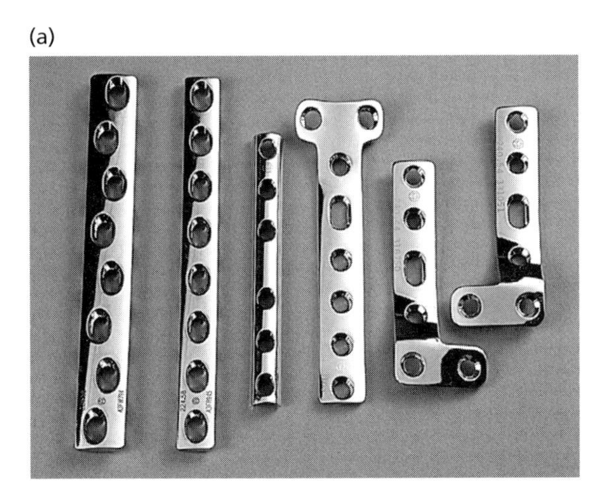

(b)

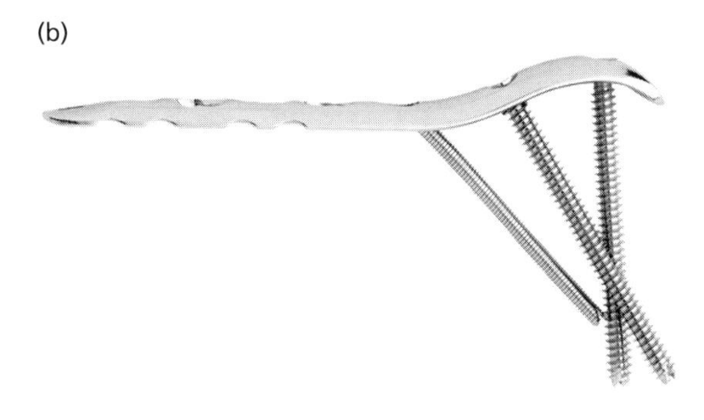

(c)

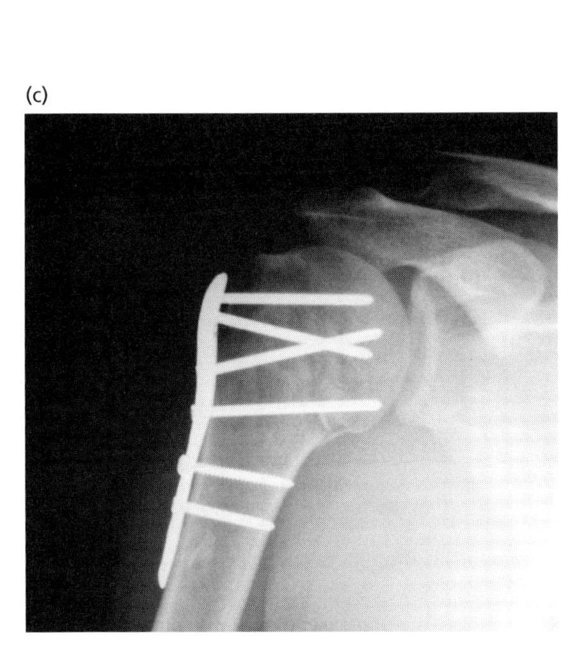

(d)

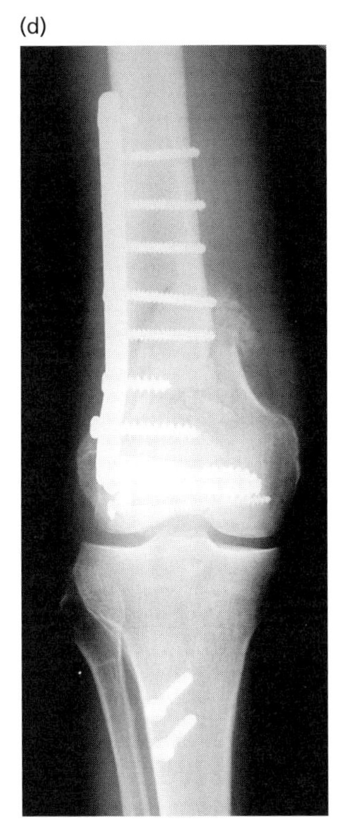

**Figure 4.14** Examples of internal fracture fixation devices. Images copyright by AO Foundation, Switzerland, provided courtesy of Dr. Geoff Richards and Urs Ruetschi, AO Foundation.
**(a)** Metallic plates for long bones;
**(b)** plate and screws for fractured neck of femur;
**(c)** radiograph of fractured neck of femur with device *in situ*;
**(d)** radiograph of complex fracture at distal end of the femur and proximal end of the tibia.

patient after the healing has taken place. The two concerns here include the potential for chronic adverse effects, both locally and systemically, over many years during which there is no additional benefit from the presence of the device; some have argued about the putative risks of hypersensitivity and carcinogenicity for example. In addition, the presence of the metallic device, which has a much higher elastic modulus than cortical bone will inevitably alter the stress fields within the bone and may result in so-called stress shielding. It is difficult to know how important these possibilities are, but many surgeons will routinely remove the implants after, say, a year, to minimize the risks of these effects. The alternative is to use devices that are made of degradable materials.

**Resorbable polymers** In the light of the above comments, degradable polymers have been used for bone fracture plates. Obviously the mechanical properties are different, which limits their use. A typical biodegradable polymer may have only 10% of the yield and ultimate tensile strengths of a titanium alloy, for example, which essentially rules them out of consideration for highly stressed long bones such as the femur or tibia. Most attention has therefore been paid to less highly stressed bones, including the proximal humerus, the olecranon, the radial head, the talus, the metatarsal bones and phalanges of the toes, and facial fractures.

Various polylactide-based materials have been used in maxillofacial trauma. Although patients of all ages may be suitable for resorbable plates, young children are especially suitable since rigid fixation using metals may adversely affect skeletal growth and development. The plates tend to be thicker than metal ones because of the need for greater strength, although even after taking this into account they may not be strong enough for the mandible, and their use is preferred in the mid-face. The devices may also be used for craniofacial reconstruction in addition to traumatic fractures, including the treatment of encephalocele, which is a congenital gap in the skull, and craniosynostosis, which is the premature fusion of cranial sutures that prevents normal cranial growth.

The resorbable materials can be used, with care, in other parts of the body, again especially in children. Screws may be used to treat flexible flatfoot, where there is excessive flexibility at the sub-taler joint and pins have been used in the management of osteochondral fractures of the knee that arise from high levels of sporting activity in children.

#### External fixation

The alternative to internal fixation is external fixation, which has been used in one form or another for over a century. This is essentially a minimally invasive procedure, which may be used in two situations. The first is where there is a severe open fracture that is not amenable to conventional internal fixation procedures; this is usually called the damage-control regime. The second is the definitive-treatment regime, in which it is used even if internal methods are available but where it is considered to be the preferred option. In either technique, the methodology uses pins that transverse the bone on either side of the fracture and exit the skin in multiple sites where they are attached to an external frame.

There are many different designs of pins and frames, and several different views on where they should be placed, and indeed how they should be placed. The materials used are not irrelevant but it is interesting to note that most publications dealing with these devices do not even mention the materials. Titanium alloys are most widely used for the pins. Because of the significant risk of infection along the percutaneous pins there has been interest in applying silver coatings to them. There have also been attempts to coat the bone-interfacing segments with a bioactive material such as hydroxyapatite in order to achieve a more favorable bone response, although how this matches the requirement to remove the pins after union is not clear. One of the more significant aspects is the benefit of having materials for the frame that can withstand MRI assessments, and for this reason, aluminum, carbon and carbon fiber composites have been used for the frames.

### 4.1.6 Cartilage defects

Not all compartments of the skeletal system are amenable to either repair or replacement by biomaterials-based devices. Cartilage is the prime example. It is avascular but frequently suffers from trauma and disease and options for self-renewal or replacement by synthetic materials are very limited. This is compounded by the fact that the mechanical properties of cartilage are very difficult to replicate in a synthetic, or even biomimetic, material, and also because the critical cartilage tissues from a functional perspective are usually seamlessly attached to bone, such a complex system being even more difficult to replicate synthetically. We shall return to cartilage repair in the context of regenerative medicine in the next chapter, but as far as we are concerned here, the replacement of cartilage by implantable devices is not an attractive option.

## 4.2 The cardiovascular system

In this section we shall cover arteries and veins of the major circulation, the blood supply to the heart, the muscular structure of the heart, the valves of the heart, the electrical apparatus of the heart, and therapy options for heart failure. For convenience the major anatomical structures encountered in this discussion of the cardiovascular system are given in Box 4.4.

### 4.2.1 Arteries and veins

Diseases of the cardiovascular system lead to a variety of problems but they can be distilled into a few major clinical issues. First, over time there may be an accumulation of deposits on the luminal surface of an artery, as described in Chapter 3. This leads to elevated blood pressure as circulation becomes more difficult, and to localized effects on the tissue, which is affected by the reduced blood supply. These effects are controlled to some extent by individual patient variables and the biomechanical/hemodynamic characteristics of the blood flow in the arteries. Secondly, the walls of the arteries may become thinner and unable to withstand the pressure of the flowing blood, such that the wall balloons outward, eventually rupturing; this condition is known as an aneurysm, and most often affects the aorta. Thirdly, within the venous system, where the flow of blood returning to the heart is controlled by vein valves, the valves may lose their competence, leading to poor venous return and the condition of varicosity.

#### Artery replacement and bypass in the major circulation

The deposition of atherosclerotic plaque in the major circulation most commonly affects the arteries providing oxygenated blood to the legs, starting with the descending aorta, and on to the femoral and popliteal arteries. The main effects are seen with the compromised tissues in the lower parts of the leg, often leading to diminished blood supply in the feet and toes, and eventual gangrene. The treatment of choice is the replacement or circumvention of that part of the arterial system that is blocked. Synthetic materials have been used in this procedure for many years, with considerable success. The choice of material is very important, but so is the architecture of the tubular structure used. The natural artery has a complex microstructure with an inner layer, the endothelium, which controls the interactions of the vessel with the circulating blood, and a structural layer composed of oriented collagen and elastin which allows the vessel to accommodate the sinusoidal flow of blood, that is the radial expansion and compression following systolic and diastolic cycles of the heart. Replacing the vessel with a solid, impervious tube would not be able to replicate this activity. Moreover, arteries have to be able to flex and change shape with the movement of the limb or abdomen – the femoropopliteal artery descends the leg at the back of the knee and it would be of no use at all if the artery or its replacement kinked or squashed every time the knee was bent. We therefore need to have compliant, flexible, kink-resistant structures for the tubular prosthesis. The prosthesis also has to be attached to the remaining parts of the artery. The actual join, or anastomosis, is crucial in determining the overall performance of the prosthesis, and usually a layer of fine but strong sutures is used for this, placing a further requirement on the material structure since very few materials can be

## Box 4.4 The major anatomical features of the cardiovascular system

The cardiovascular system pumps blood around a network of vessels in order to provide oxygen and key biological molecules to all parts of the body. Diseases of this system are obviously critical to the overall health of individuals. They may be associated with the pumping mechanism itself, with the systems that control the blood flow, and with the network of blood vessels.

### The heart

The heart is a four-chambered muscular pump. The upper two chambers (the atria) receive blood from the major and pulmonary circuits, the lower two (the ventricles) pump blood into these circuits. The *mitral valve* controls flow of blood between the left atrium to the left ventricle and the *tricuspid valve* controls flow between the right atrium and right ventricle. The *pulmonary valve* opens to let blood flow to the pulmonary artery and the *aortic valve* controls flow from the left ventricle to the aorta and then to the major circulation. Any of these valves may become diseased but the aortic and mitral valves, which operate at higher stresses, are the main candidates.

The pumping action of the heart is delivered by the *myocardium*, a powerful contractile muscle that occupies much of the heart wall. This is surrounded by the *epicardium*, which is the inner layer of the protective sheath, known as the *pericardium*, which envelops the heart. The inner surface of the muscle, which contacts the blood is the *endocardium*. The heart wall receives blood from a network of *coronary arteries*, which branch off the ascending aorta. These arteries are very prone to disease. If they become narrow and blocked, the myocardium does not receive sufficient oxygenated blood and becomes damaged, ultimately leading to a heart attack, or *myocardial infarction*.

Control of the pumping action is the responsibility of the electrical network that guides impulses according to the needs of the body. This is primarily organized by the *sinoatrial node*, which is located in the right atrium. Irregularities of this activity lead to arrhythmias, which have to be controlled by medical devices such as pacemakers.

### The circulation

Oxygenated blood that is pumped from the left ventricle through the aortic valve passes into the major blood vessel, the *aorta*. This artery is about 2 cm in diameter. As with all arteries, it has an inner lining, *the endothelium*, a muscular middle layer and an outer layer, the *tunica media*. As we go further away from the heart, the arteries get smaller and smaller. Major arteries include the *iliac*, *femoral* and *popliteal* arteries in the legs, the carotid artery leading to the brain, the *brachiocephalic*, *subclavian*, *ulna* and *radial* arteries of the upper limbs and those arteries that lead to organs, such as the *renal artery* leading to the kidneys. These arteries are also subject to disease, especially atherosclerosis that causes blockage, particularly in the legs, and thinning of the vessel wall, causing aneurysms, mainly in the aorta. The arteries eventually lead to the delivery of blood into generalized tissue, within the arterioles of microvascularity. This blood is taken up in venules and then the veins, which return the blood to the lungs for re-oxygenation. The venous system is far less susceptible to disease.

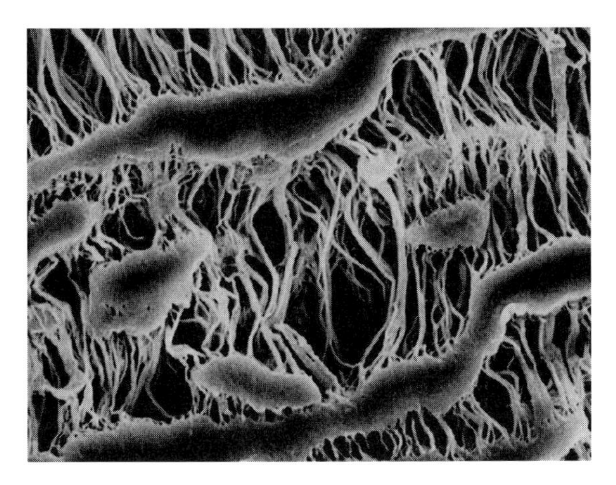

**Figure 4.15** Vascular graft. The surface of an expanded-PTFE vascular graft. Image courtesy of Professor Peter Zilla, Cardiovascular Research Unit, University of Cape Town, South Africa.

easily pierced with a suture needle and hold the tightened thread without it pulling out.

The result of this rather difficult set of specifications is that tubular textiles are used for these prostheses. In reality, two forms of material are used, a polyester textile and a polytetrafluoroethylene (PTFE) fabric, with some minor interest in polyurethanes. Both these polyester and PTFE materials were developed for other non-medical applications, typically in clothing. The polyester is essentially Dacron®, used for shirts and other garments, while the PTFE is essentially the microporous expanded PTFE of GoreTex®, well known for waterproof clothing and other applications. An example of prostheses made from the last material is shown in Figure 4.15. We should note in passing that these devices are usually referred to as vascular grafts, even though they are technically not grafts but synthetic devices.

Some prostheses are straight tubes while others have bifurcations, the use of which depends on the nature of the arterial structure to be replaced. These prostheses may be sutured to the existing vessels by end-to-end or by end-to-side anastomoses.

The essential feature of these synthetic arteries is that they are microporous, giving the flexibility and compliance of the tube and also the possibility to suture it to the adjacent vessel. It will be obvious, however, that a microporous structure should be permeable to the flowing blood. While this might intuitively sound contradictory to the required function, which is to allow blood to flow through the tube and not to escape or seep through the side walls of the tube, this is actually an advantage since the flowing blood, as it enters the interstices of the wall, clots rather rapidly. The clot within the walls is reorganized during the healing process, providing a tissue seal that has mechanical characteristics quite similar to the natural vessel. It is, of course, a major requirement that the blood clotting does not extend into the lumen of the vessel. This is controlled by the nature of the tissue that forms on this surface. Ideally, a new endothelial layer should form along the length of the prosthesis. Whilst this is sometimes seen in certain experimental animals, it does not happen in humans. Endothelium may grow inwards from the anastomosis for a short way, but typically the luminal surface will be comprised of a layer of compacted fibrin. This is obviously not ideal but does give good clinical outcomes in many circumstances and in many patients. The mechanisms involved here are shown in Box 4.5.

## Box 4.5 Mechanisms of healing of vascular grafts

Large diameter prostheses appear to work fairly well in humans. However, in arteries that have diameters less than 7 mm, which includes coronary arteries and those in the lower leg, there are significant limitations. The coronary arteries still cannot be replaced with synthetic devices and vascular grafts in lower leg arteries suffer from anastomotic intimal hyperplasia and thrombogenicity.

The critical feature required in the healing of vascular grafts is that of complete endothelialization along the length of the prosthesis. There are three possible mechanisms for this. The first is transanastomotic growth of endothelium, that is growth of an endothelial layer from the retained parts of the vessels, across the anastomotic area and along the blood-contacting surface of the prosthesis. The second is transmural endothelialization, which is the formation of endothelium following ingrowth of tissue through the wall of the graft. The third is blood-borne endothelialization, where endothelial cells spontaneously deposit on the graft inner surface and nucleate areas of endothelium.

In reality, none of these happen clinically to any significant extent. Although transanastomotic growth has been seen in some animal models, which usually involve only short grafts, this does not happen in humans. The growth will normally only extend 1–2 cm, which is irrelevant in prostheses of 10–50 cm in length. Most vascular grafts are too dense and too impervious to allow transmural growth. None are known to spontaneously develop a neo-intima through blood-borne deposition to any significant extent, although occasional endothelial islands may be seen.

The two main components of the tissue response to vascular grafts are fibrin and macrophage infiltration. The majority of the inner lumen of grafts is covered with compact fibrin, hundreds of micron thick. This does not appear to be conducive to endothelial cell colonization. The interstices of the graft become populated by macrophages. These become a chronic feature, the persistence of which inhibit full healing. Instead of the formation of a full endothelium, therefore, the inner surface of a graft becomes covered with a layer of fibrin. The failure of grafts, which with current prostheses is more-or-less inevitable, eventually occurs through peri-anastomotic intimal hyperplasia and mid-graft thrombosis.

There are variations in the tissue response, depending on the patient, the material/architecture combination and the position in the graft. The transanastomotic endothelialization (extension of endothelium from natural vessel into the graft across the anastomosis) and anastomotic hyperplasia (the hyperplasia seen in the peri-anastomotic region) are important. Since the intimal proliferation can only occur on top of existing tissue, and since the endothelialization occurs to such a limited extent, the hyperplasia, and consequential reduced patency, is confined to the anastomotic regions, this narrowing and occlusion being one of the main causes of clinical failure. In the middle section of the graft, a persistent type of foreign body response within the graft interstices dominates such that healing does not occur and there is a constant threat of thrombosis. With the PTFE grafts, the histological appearance varies with the porosity. With low porosity material, a thin (10–20 micron) layer of fibrin forms, which remains acellular in the vast majority of cases, and there is no transmural tissue ingrowth. With higher porosity, the fibrin within the interstices becomes progressively populated with macrophages and leukocytes and sprouting capillaries can be seen in the outer layers of the graft along with some smooth muscle cells. With woven polyester, there is limited tissue ingrowth and the compact fibrin layer tends to remain intact. In knitted polyester there is greater tissue ingrowth from the exterior.

There have been some changes to vascular grafts over the years, but the principles remain essentially the same. Autologous *in vitro* endothelialization of PTFE grafts, with cells derived from subcutaneous veins, has been shown to improve patency; for example, patency rates at 5 years can be as high as 70%. These aspects will be discussed further in the next chapter. Some devices are now coated with heparin in order to minimize thrombosis and some are coated with nanoparticulate silver to minimize infection. Other approaches include the attachment of nitric oxide generating substances, such as S-nitrothiols, to the graft in order to reduce intimal hyperplasia.

In general, vascular grafts used in above-knee situations, such as in aortoiliac positions, have high patency rates, but below the knee patency rates may be much lower, with only 50% survival at 5 years.

## Abdominal aortic aneurysms

The aorta, especially that part of the aorta within the abdomen, is susceptible to aneurysms, which are areas where the vessel wall thins and bulges outwards. In the early stage of disease progression, this is asymptomatic and not particularly harmful. However, it is usually progressive and carries increasing risks of blood leaking through the wall, and ultimately rupture. It is generally considered that an aorta with a diameter of less than 4 cm is not at risk, but the risk of rupture, which has extremely serious consequences, increases with diameter; at 8 cm the risk of rupture may be 50%.

If the aorta ruptures, or if it is diagnosed at a point where the aneurysm is far advanced and the risk is very high, the treatment involves replacing the diseased section with a prosthesis, in much the same way as discussed in the previous section. At earlier stages, where the disease is progressing but not yet causing significant effects, that is, without bleeding and with manageable pain, the treatment may be carried out through an endovascular stent graft. This is a catheter-based process in which a device that consists of a mesh prosthetic graft reinforced by a nitinol stent is inserted at the affected site (Figure 4.16). The technology of stents is discussed in the next section, and this will not be pre-empted here. Over the last decade there have been many

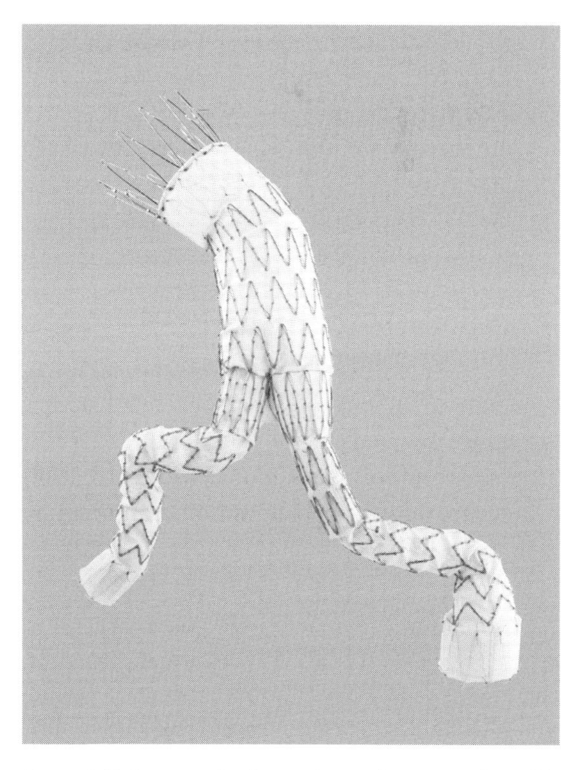

**Figure 4.16** An example of an endovascular stent graft used in the treatment of abdominal aortic aneurysms. Permission for use granted by Cook Medical Incorporated, Bloomington, Indiana, USA.

studies of the performance of these grafts and the outcomes of the procedures. There have certainly been a number of problems both with the design and the procedure, but these have evolved and the technique is widely considered now to be the standard of care. Adverse events still occur, including structural failures, such as fatigue of the nitinol, migration of the stent, and a failure to prevent the rupture. There is a risk of peri-operative mortality of around 20% and of long-term complications and morbidity of around 40%, and so clearly improvements are still required. The use of stent grafts is most complex within the aortic arch because of the branches, and complications of mis-alignment, leaks and retrograde dissections occur.

## The coronary arteries

The coronary arteries provide a somewhat different situation to the vessels of the major circulation, both in terms of therapeutic options and outcomes.

**Coronary artery disease** The coronary arteries are derived from the aortic arch and deliver oxygenated blood to the muscle of the heart wall. Their disposition is shown in Figure 4.17. These arteries have a small diameter, typically around 3 mm, and are rather tortuous. They are particularly prone to plaque deposition. Risk factors include smoking, high blood pressure, and high levels of fat, cholesterol and/or sugar in the blood. This may simultaneously occur in several of the arteries and may be asymptomatic for a long time. The plaque may be hard and continuous, causing the artery to become narrow and stiff, or may be soft and prone to fragment. In the former case the blood flow to the heart muscle is restricted, eventually causing angina pain, referred to as stable angina, most noticeably during exercise or emotional stress. The latter is more likely to become disrupted and cause blood clots, which can completely block the artery, resulting in a heart attack.

A range of drugs is available to treat the symptoms and reduce the progress of the stable angina, including aspirin, nitrates, beta-blockers, calcium channel blockers and statins. Nevertheless, the progress of coronary artery disease, of either form, is likely to reach a level, seen either through chronic

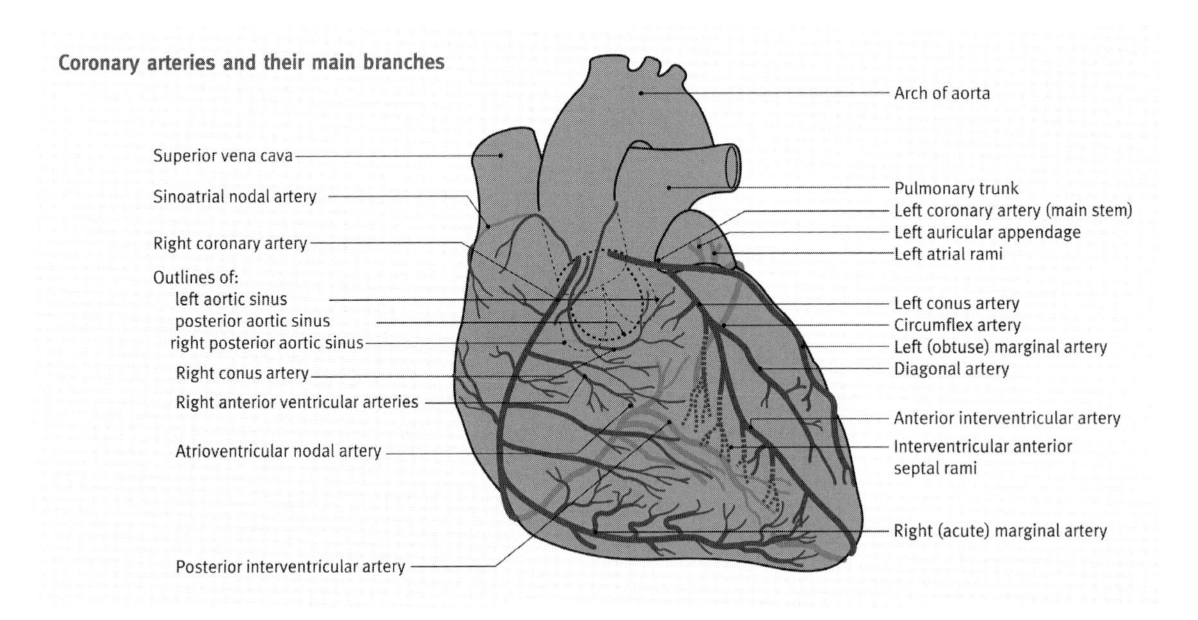


**Figure 4.17** The coronary arteries and their main branches. Reprinted from *Surgery* (Oxford), 30 (1), Mahadevan V, Anatomy of the heart, 5–8, © 2012, with permission from Elsevier.

observation or acute events, which require some intervention. There are two common options, the minimally invasive procedures that attempt to remove or counteract the plaque deposition, or the surgical implantation of a device to bypass the blockage, We shall deal with them in this order, even though they entered clinical use in reverse order.

**Angioplasty and stents** There are several techniques that are used to diagnose coronary artery disease, one of these involving the insertion of a catheter into the vessels and the use of a radiopaque dye that can allow imaging of the blood flow and areas of blockage. This technique is known as angiography. A very widely used therapy involves an extension of this diagnostic procedure, referred to as angioplasty, in which the catheter is used to physically displace or compress the plaque. The catheter is assisted in this process by an inflatable balloon that is threaded along its lumen and expanded at the site of the lesion. Angioplasty does provide immediate relief for the majority of patients and is minimally invasive. However, since the process of plaque compression or disruption inevitably affects the underlying endothelium, it should not be surprising that this responds with the formation of more proliferative tissue, usually resulting in further blockages after a few months or years. The blockage of a vessel is referred to as a stenosis, and the secondary type of blockage is known as restenosis. A few other more aggressive techniques were introduced in attempts to reduce the incidence of restenosis, including atherectomy, in which the plaque was removed by a rotating bur threaded down the catheter, or brachytherapy, in which radioactive materials were placed into the catheter in an attempt to suppress the proliferation, but these are not widely used.

A critical event in the development of these minimally invasive interventions was the introduction of the intravascular stent, which is probably now the most widely used of all major implantable medical devices. The concept is simple: after compression or disruption of the plaque, why not try to keep the vessel open to its fullest extent by placing a synthetic tube inside it? At its simplest, a stent is just such a tube. This can be passed down the same catheter and released into the artery at the site of the lesion. The locating of the device in the right place should not be too difficult since it can be done under radiographic control. The real design issues start, however, with the structure of this tube. The first point is that it cannot be a solid walled tube as this would be too damaging to the endothelium and would not have sufficient flexibility. It is much better to have an open structure such as a mesh. However, there are two conflicting requirements here. First the stent has to be threaded down the catheter that is placed within the rather tortuous coronary artery. A device that was rigid and inelastic could not traverse this passage – we need something that can bend elastically with respect to the longitudinal axis. Once at the site of the legion, however, it has to be forced to expand from its small initial diameter, which may be around 50% of the diameter of the vessel, in order to contact and restrain the vessel wall. This has to be a permanent expansion, since elastic recoil would be of no value; that is, the tube has to deform plastically. In other words the stent material and design must be such that the device can flex elastically one minute and deform plastically the next.

The history of stent design shows how stents have evolved on the basis of this requirement. Some of the early designs were essentially thin tubes with slots or coils. Later versions had combinations of very short tubes joined together by flexible segments. These evolved into complex structures such as that seen in Figure 4.18 in which there are elements that provide for longitudinal flexibility and those that provide for radial expansion. Not surprisingly the precise designs here have been crucial for clinical effectiveness and have been the subject of many patent disputes. These stents are usually manufactured by laser-cutting techniques.

The method by which the stent is deployed to achieve its radial expansion is of significant interest. Two techniques are available. The first involves the use of a strong balloon inside the stent, which can be pressurized once the stent is in the right place and

**Figure 4.18** The intravascular stent. The Taxus® Express paclitaxel-eluting coronary stent system. Image courtesy of the Boston Scientific Corporation.

the catheter withdrawn in order to release it. The material used for these balloon expandable stents is usually stainless steel or a cobalt–chromium alloy. The alternative technique involves a shape memory alloy such as nitinol, discussed in both Chapters 2 and 8. This is prepared by designing the stent in its fully deployed state, then compressing it within the catheter. The alloy will have a transition temperature close to that of the blood so that when the stent is released from the catheter it will revert to its expanded state.

These stents are very effective in relieving symptoms, and the effect usually lasts much longer than with angioplasty alone. However, restenosis still occurs with these bare-metal stents, as discussed in Chapter 3. There have been attempts to minimize these effects by the use of surface coatings, but these are not very effective in relation to avoiding restenosis. This is clearly a situation where innovations produce some improvements but without giving the ideal biocompatibility characteristics that avoid the chronic irritation to the endothelium and the tendency towards further intimal hyperplasia and restenosis. The next event in stent evolution, however, appeared to go a long way towards this goal. The control of proliferation of the endothelium has been achieved through the use of anti-proliferative drugs. These cannot be delivered systemically because of their powerful effects that would be seen over the whole body. Instead they are applied to the surface of the stents, from which they are slowly released. The most common drugs used are paclitaxel and sirolimus, both of which are powerful chemotherapeutic agents. They are typically dispersed in a thin layer of polymer, which is usually non-degradable, such as SIBS, or could be resorbable, such as a polyester. A number of clinical trials have been performed and regulatory approval obtained for many of these commercial products. The incidence of restenosis has been reduced. There is a small downside and that is an increased risk of thrombus formation, which occurs in a few percent of patients, the cause of which is unclear.

The final episode in the stent story, at least to date, concerns the need for permanency of the stent. It has been argued that the benefit of the stent is achieved during the first few months and that the problems of proliferation, hyperplasia and restenosis arise only because of the chronic irritation associated with the persistence of the stent. If this is the case, then one solution may involve the use of resorbable stents. Much experimental work has been done, both with resorbable polymers such as those of the lactide–glycolide family, and also with magnesium alloys which corrode in the body. This is clearly an interesting approach, although significant questions need to be resolved over the influence of a degrading stent, and the inflammation that is almost inevitably associated with this degradation, on the endothelium.

**Coronary artery bypass** The coronary artery bypass procedure, commonly known to the layman as heart bypass and to the clinician as CABG, or cabbage, is one of the most widely discussed medical procedures. In many ways this follows the same principles as a bypass in the arteries of the legs, but with one significant difference. The diameter of the coronary arteries is much smaller than those of the arteries in the legs and as a result, synthetic materials are unsuitable; they would clot and stenose far too readily. The surgeon therefore has to use a transplanted blood vessel. In the most common procedure, the surgeon removes either the internal mammary artery or the saphenous vein from one of the patient's legs. In the latter case, depending on how many of the coronary arteries are diseased and require bypass, the surgical team will take that vein and locate the best parts, avoiding valves and bifurcations, and use these for the bypass.

This procedure is very successful, although obviously it is far more invasive and disruptive than angioplasty and stenting. Typically the patient is placed on cardiopulmonary bypass, in which the heart is stopped to allow for the delicate surgery and the blood is pumped through an oxygenator (see later in this chapter). This all requires a large clinical team and usually the patient is hospitalized for 4–6 days afterwards. In an attempt to reduce the overall level of complications, especially those associated with prolonged cardiopulmonary bypass, some surgeons started to use so-called off-pump techniques, in which the heart was left beating by itself and special instruments used to stabilize the relevant area to allow for the surgery. This does not appear to be as popular as once thought.

For many years the saphenous vein source was the first choice, but this is no longer the case. Early patency rates for the internal mammary artery are very high and these are maintained at over 90% at 10 years, compared to 60% for saphenous vein grafts. This has been linked to the effects of exposure of the veins to arterial pressures and to the structure of the artery that decreases the risk of atherosclerosis.

**Overall outlook for coronary artery disease** Obviously surgical and minimally invasive interventions have had a major impact on the treatment options for coronary artery disease and the prognosis for those who are diagnosed with the condition has considerably improved in recent years. With this condition, possibly far more than in many others, the susceptibility to the disease may, to a significant extent, be controlled by the individual. Of course there are some un-modifiable risk factors such as age and gender, but many can be modified, including lifestyle issues such as smoking, obesity and physical fitness, and others that can be kept under control, such as high blood pressure and diabetes.

The current uses of medical technologies are generally divided into those of surgical bypass and interventional angioplasty and stenting. The evidence oscillates between these two with respect to best options and outcomes and this is now largely a matter of clinical planning and expediency depending on the severity of the condition when first diagnosed or detected. Both types of technology, although giving very good results, could be improved. This requires a greater understanding of the biology of vascular healing and the influence of medical devices on these healing processes. We shall see in the next chapter whether tissue engineering and cell therapy will provide even better options.

## 4.2.2 Valves

The heart has four valves, which are responsible for controlling the flow of blood into and out of the chambers of the heart. These valves are susceptible to disease, the consequences of which are either reduced flow of blood through them (as with arteries, this restricted flow is called stenosis) or to reduced control of forward flow, usually by failing to close properly, giving a condition of incompetence or regurgitation. When the effects of valve dysfunction become severe, the affected valve(s) may need replacing or repairing.

### Valve disease

Heart valve disease may be congenital or acquired. Congenital defects mostly affect the pulmonary and aortic valves and are associated with an insufficient number of leaflets, malformed leaflets or parts that are of the wrong size and/or shape. Acquired heart valve disease are usually associated with calcific deposits that cause leaflets to thicken or become more rigid, interfering with their movement, or with infection. Rheumatic heart disease is a consequence of rheumatic fever, in which autoimmunity accompanies the body's defense against the bacteria, resulting in damage to heart valve tissue. At one time this was the major cause of heart valve disease. Although largely eradicated in many parts of the world through the use of antibiotics, it is still prevalent in many developing countries. Heart valves can also be damaged through other infections, infective endocarditis being the migration to and accumulation on the valves of bacteria that are associated with specific infections elsewhere, for example in the urinary tract or a tooth abscess. The mitral and aortic valves, being on the left side of the heart and subject to greater workloads, are far more susceptible to acquired disease than pulmonary or tricuspid valves. The most common forms of heart valve disease are aortic stenosis as a result of calcification, aortic regurgitation as a result of age-related dilation of the aorta and mitral prolapse, in which the valve leaflets lose structural form and become floppy. Valve disease places significant stress on the heart and may lead to congestive heart failure. There are no effective pharmacological treatments although some degree of relief may be afforded by vasodilators, anticoagulants and diuretics.

### Surgical valve replacement and repair

Without any effective drug treatments, it was realized many years ago that the only solution to valve disease was to surgically intervene in order to either replace or repair the defective valve.

**Surgical replacement** It has been possible to surgically replace diseased heart valves since the 1960s. The requirements for such valves are fairly straightforward. They must have an opening and closing mechanism that allows for the control of blood flow without too much variation from normal physiological parameters, that is with physiologically acceptable flow rates and energetics. They must not damage the blood components to any clinically significant extent, nor must they interact with blood clotting mechanisms so as to cause thrombosis or thromboembolic events. They should be mechanically robust and retain that robustness for an acceptable length of time, which implies that they possess fatigue, creep, degradation and corrosion resistance, and they should not interact with blood in any way that causes significant structural change. They must be capable of attachment to heart tissues. The materials used should obviously be non-toxic and have no deleterious biocompatibility properties. Ideally the valves should be close to normal anatomic size, especially not taking up excessive space within heart chambers.

Within these rather broad but simple specifications are hidden a number of crucial engineering and biological nuances. No valve designed to date can achieve all of the requirements and there have had to be a number of compromises.

We see from Box 4.6 that the function of the natural valve is dependent on very complex architectural and structural properties such that it is very difficult to replicate this function in a synthetic engineering device. Moreover, whilst flowing blood is able to interact with healthy natural valves without any adverse consequences, this does not apply with engineered devices and blood compatibility with prosthetic valves is a serious issue. Furthermore, the complexity of the conjoint mechanical–biological environment is such that structural changes to the prosthetic valve may occur over time.

The first prosthetic heart valves were made wholly of synthetic materials and functioned as typical engineered valves, with one or more moving parts contained in a housing. As we see from Figure 4.19, which gives profiles of early stage clinical valves, it was impossible to provide for normal physiological flow of blood through the devices, such that hemodynamic parameters including pressure gradients and energy losses were less than optimal and unfavorable interactions between rigid materials and flowing blood were likely to lead to clotting or hemolysis. Developments over the last 20 years have involved either improvements to the design or material selection of these so-called mechanical valves and the clinical handling of the recipients, or a move towards a different type of structure altogether, with the

## Box 4.6 Crucial engineering and biological issues with prosthetic heart valves

| Mechanical valves | |
|---|---|
| Design principles: | A structural synthetic device that has an annular component enclosing a circular orifice, which can be secured to the heart muscle via sutures and a sewing ring, and one or more moving parts that are either fixed to the annulus or move within some cage structure that is also integral with the annulus. |
| Potential failure modes: | Thromboembolic events<br>Hemodynamics, energy loss<br>Structural stability, fracture<br>Endocarditis<br>Paravalvular leakage |
| **Bioprosthetic valves** | |
| Design principles: | A device that consists mainly of biological tissue that formed as multiple leaflets, which are attached to an annular component enclosing a circular orifice, which can be secured to the heart muscle via sutures and a sewing ring. |
| Potential failure modes: | Calcification<br>Tissue degradation<br>Endocarditis<br>Paravalvular leakage |
| **Typical algorithm for selection of heart valve** | |
| Aged over 60 years (aortic) or 65 years (mitral) | |
| with no risk factors for thromboembolism | *Bioprosthesis* |
| with risk factors for thromboembolism | *Mechanical valve + anticoagulation* |
| with risk factors for thromboembolism and difficulties with anticoagulation and life expectancy $< 10$ years | *Bioprosthesis* |
| Aged under 60 years (aortic) or 65 years (mitral) | |
| with no risk factors for thromboembolism, life expectancy very short | *Bioprosthesis* |
| with no risk factors for thromboembolism, life expectancy long | *Mechanical valve + anticoagulation* |
| with risk factors for thromboembolism | *Mechanical valve + anticoagulation* |

so-called bioprostheses, which are based on animal tissue rather than synthetic materials.

**Mechanical heart valves** The majority of mechanical heart valves today are of the bi-leaflet design, as shown in Figure 4.20. These valves have an annulus, which is usually made of titanium or a cobalt–chromium alloy. Within this annulus are two thin leaflets which are attached to the annulus at diametrically opposing hinges and which are free to rotate such that they constitute an effective planar disc covering the whole of the annular diameter at closure, and two parallel half-circular discs perpendicular to the annular orifice at full opening. These leaflets are made of pyrolytic carbon or an equivalent material. Different manufacturers have slightly different hinge and leaflet designs, which control ease of movement and influence blood flow disturbances. The annulus is covered with a sewing ring, made of either a polyester textile or microporous PTFE, which, as the name implies, allows sutures to be used to attach the valve to the heart wall.

These valves give very good performance in the majority of patients, but care still has to be taken and problems and failures can still arise. The most important issue here is the possibility of thrombus formation. As noted in Chapter 3, foreign surfaces have an inherent tendency to cause blood to clot and one significant risk factor involves flow disturbances that alter normal mass transport and rheological characteristics in the blood within a device. Forcing blood to pass through a mechanical valve almost inevitably causes such flow disturbances, as seen in Figure 4.21, the result being areas of turbulence and

**Figure 4.19** Mechanical heart valve profiles. Images of various mechanical valves, showing different profiles with different types of closure (ball, disc, leaflets). Image courtesy of Professor Peter Zilla, Cardiovascular Research Department, University of Cape Town, South Africa.

(a)

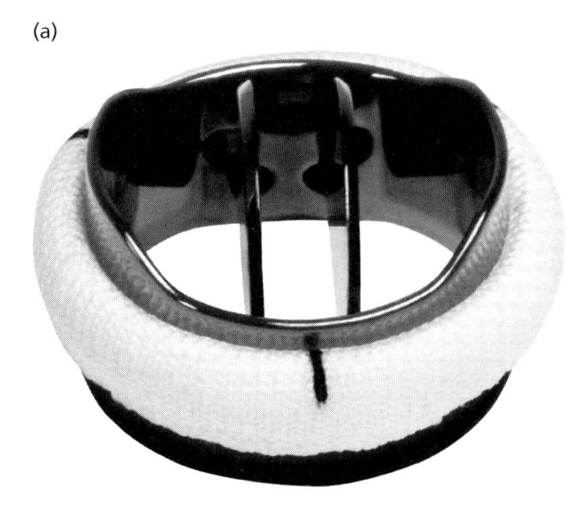

(b)

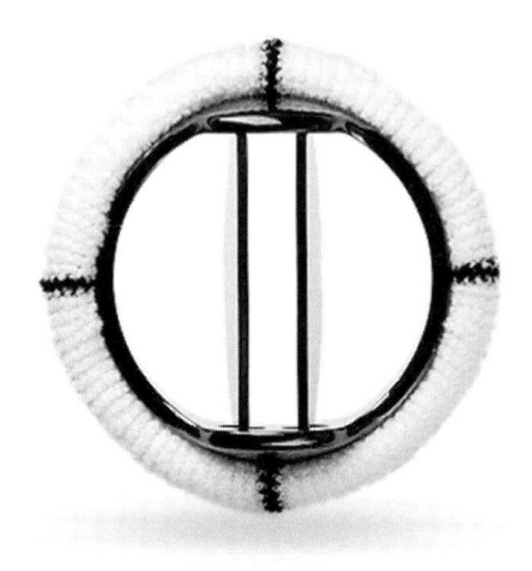

**Figure 4.20** Current mechanical heart valves.
(a) The On-X standard aortic valve prosthesis. Image courtesy of On-X Life Technologies Inc, Austin, Texas, USA.
(b) The St Jude Medical® Masters HP Series valve. Image courtesy of St Jude Medical, St Paul, Minnesota, USA.

areas of stasis. In the latter areas, there is a greater chance of clot formation. Such clots can result in malfunction of the valve through failure to open or close, and can also lead to the release of micro-emboli. These emboli, within the greater circulation, can become lodged in arterioles, the small blood vessels in the brain being especially susceptible, leading to the possibility of strokes. In view of these risks, and independent of the materials used in valve construction, it is necessary to place these patients on an anticoagulation regime. Adverse events from blood clots still occur; in the main these are caused by lack of patient compliance with the strict anti-coagulation medication process. The anticoagulation levels of individual patients, measured as the INR (the international normalized ratio), are easily determined during routine clinic visits.

Other biological/physiological complications arising from mechanical heart valves include hemolysis, paravalvular leaks, pannus overgrowth and infection. Hemolysis involves accelerated ageing of red blood cells following repetitive contact with the hard, rigid materials of the valve, and can lead to clinically significant anemia. Paravalvular leaks occur when there is inconsistent healing around the sewing ring, and involves flow of blood around the outside rather than through the valve. On the contrary, if there is excessive healing and tissue growth, an excess of reparative tissue, called pannus, can form, leading to obstruction of valve components. As noted earlier, natural heart valves are susceptible to bacterial colonization during any bacteremia episode; the same is true for mechanical valves and it is imperative for patients to minimize these chances,

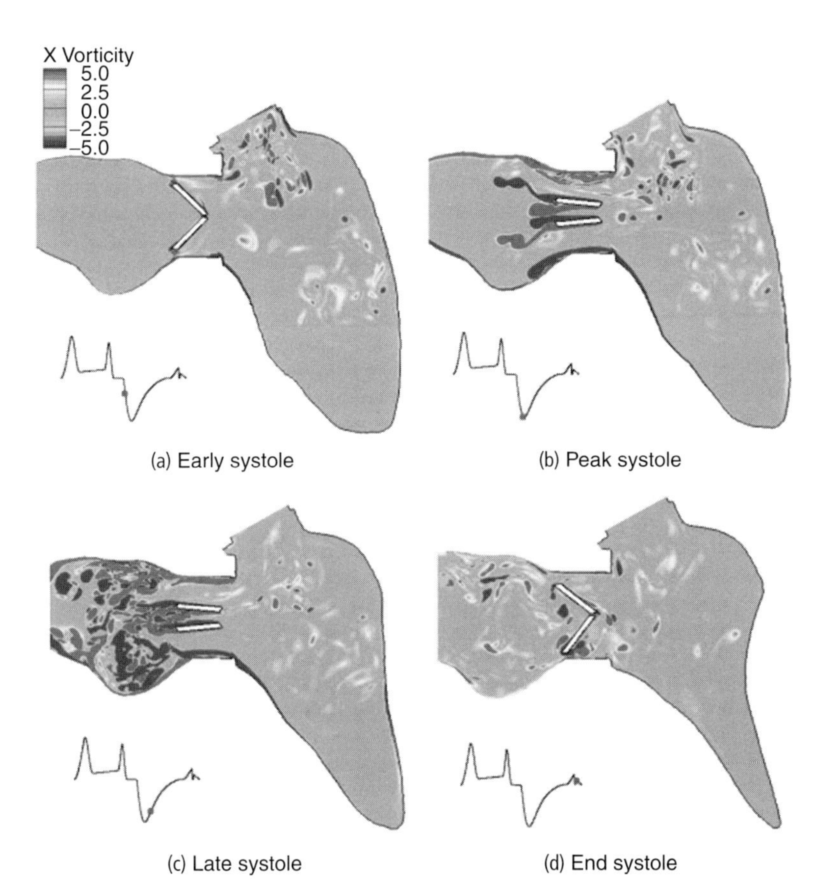


**Figure 4.21** Blood flow in heart valve prostheses. Blood flow through valves may be simulated in a kinematic model. In (a) at the early phase of systole, there is reasonably coherent flow. As the bileaflet valve opens at peak systole (b), an unstable shear layer forms at the leaflet surfaces, then complex 3D flow structures form at the end of systole at the aortic root (c). At closure (d) the leaflets vary slightly in closure velocity and there is some leakage back through the valve. Reprinted from *Journal of Computational Physics*, 244, 41–62, Le *et al.*, Fluid–structure interaction of an aortic heart valve prosthesis driven by an animated anatomic left ventricle, 2013, with permission from Elsevier. See plate section for color version.

primarily through the use of antibiotics during any activity with elevated risk, such as tooth extractions.

Mechanical failures can also occur in these valves. Struts or hinges in the housing can fracture under the imposed fatigue conditions. The brittle carbon materials can fracture or fragment, and cavitation conditions within the turbulent blood flow may contribute to this. Some extremely important issues arise when defects are suspected in valves after many patients have been treated. This aspect is discussed in detail, with a case study, in Chapter 9.

**Bioprosthetic heart valves** The need for long-term coagulation of patients with mechanical heart valves led to the development of alternative valve substitutes. By far the most widely used are the xenogeneic bioprostheses, an example being shown in Figure 4.22. Although at one time bovine sources were used for these prostheses, the outbreaks of bovine spongiform encephalitis, a potentially devastating transmissible prion disease, in certain parts of the world in the 1990s, led to the emergence of porcine sources as the preferred option. As discussed in Chapters 3 and 8, systems were put in place to minimize risks of disease transmission in bovine products. Today, porcine aortic valves and bovine pericardium that is fashioned into valves, are both extensively used. The need to render the tissue sterile (whether with either porcine or bovine tissues, or concerned with bacteria, viruses or prions) has led to the introduction of rigid processing treatments. In addition, the collagen component of these tissues is weakly antigenic so that the risk of immunogenicity has to be minimized, also through a processing treatment. For both of these purposes, glutaraldehyde solutions have been used to process these tissues. Because the tissue valves have a structure closely resembling the natural human valve, especially with blood flow through the center of the valve rather than around a mechanical disc or leaflet, and because the tissue is less thrombogenic than the synthetic materials, these valves may be used in patients without anticoagulation, and without any significant risk of thrombus formation. This is a major advantage.

However, the valves do have one disadvantage: the tissue degenerates and slowly calcifies; thus the valves have a built-in design life, in the early years this being around 10 years. By the year 2000, current valves were expected to give 75% functional life by 10 years, but over 50% were expected to fail by 12–15 years. Porcine valves in the mitral position gave poorer performance than those in the aortic position because of the higher closing pressures. Calcification is the major factor in this degradation, which leads to regurgitation through tears in the

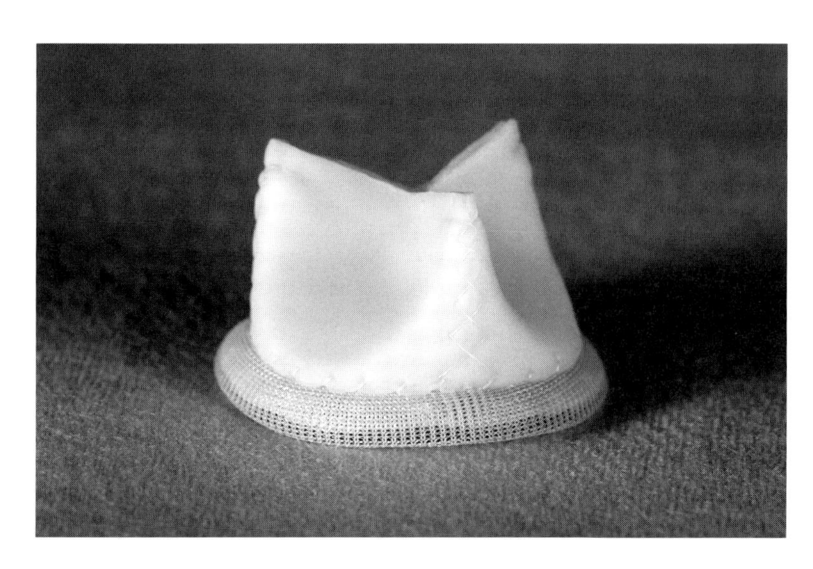

**Figure 4.22** Bioprosthetic heart valve. The Mitroflow Aortic Pericardial Heart Valve. Image courtesy of the Sorin Group.

calcified cusps. Early bovine heart valves had the tissue mounted on the outside of a polymeric stent; these tended to fail through tears in the vicinity of the attachment points. In view of this, designs changed so that the pericardial tissue was suspended inside a flexible low-profile stent.

**Valve homografts** A homograft valve is one that is obtained from a human donor. The use of these valves actually pre-dates the use of mechanical prostheses, but although they can work very well, they have never been used extensively, partly as a result of donor supply. It was the problem of donor supply for fresh homograft valves that led quite quickly to the development of preservation techniques, especially involving the use of cryoprotectants and storage at −180°C. In many ways, in spite of increasing sophistication with the cryopreservation technology, it is a surprise that these valves work at all. There are two factors that predispose to failure. First, early phase cell survival is low. Cryopreserved cusps are essentially acellular within a few months and valvular surfaces become covered in a fibrin sheath. Secondly, the stress of reimplantation coupled with the new abnormal environment triggers apoptosis in any remaining cells. The fact that some valves do work quite well, and success rates increase with better harvesting and cryopreservation techniques, appears to be due to the fact that these valves act as a scaffold for *in vivo* fibrosis, which can lead, in some patients, to a functioning valve.

**Mitral valve repair** The etiology of mitral valve disease is a little different to that of aortic valves and in some cases different treatment modalities are available. The mitral valve is composed of leaflets, the chordae tendinaea, the annulus and papillary muscles, damage to any of these structures potentially causing mitral regurgitation. This primarily involves fibroelastic deficiency or myxomatous degeneration. This can lead to myocardial damage and heart failure. It is possible to replace such diseased valves, but increasingly cardiac surgeons prefer to repair the valve. The repair process usually utilizes an annuloplasty ring that reduces the size of the mitral annulus and helps brings the leaflets into better coaptation. There are several forms of ring although they are typically made of a combination of expanded PTFE and polyester sutures. Excess tissue may also be removed. In an alternative technique, the two leaflets are sutured together in their mid-section in order to restrict prolapse and achieve competence. Increasingly the repair process is being carried out percutaneously, similar to those processes discussed in the next section.

### Minimally invasive techniques

As noted above, open surgical aortic replacement has really been the only treatment modality available for aortic stenosis for many years. However, this is associated with a degree of peri-operative morbidity, which may be 5%. In addition, or perhaps even more importantly, many patients are denied this open-heart surgery because they are considered to be too high risk for the procedure, especially when there is significant co-morbidity. In these situations, attention has turned towards the possibility of using minimally invasive, non-surgical techniques.

**Percutaneous valve delivery** The main option for minimally invasive replacement of heart valves involves the percutaneous route. This avoids open-heart surgery and the need for cardiopulmonary bypass, which is highly advantageous for high-risk patients, but the technique still requires a skillful clinical team and sophisticated equipment. These percutaneous techniques are sometimes called catheter-based or transcatheter heart valve replacements. There are several anatomical routes involving either the femoral or subclavian arteries or possibly the femoral vein. The valve is usually made of bovine or porcine pericardium and it is mounted on a collapsible stent, usually this is made of stainless steel and will be balloon-expandable. This is introduced by a catheter and deployed across the aortic valve (Figure 4.23).

There are several limitations and potential complications of this technique. The delivery catheter has

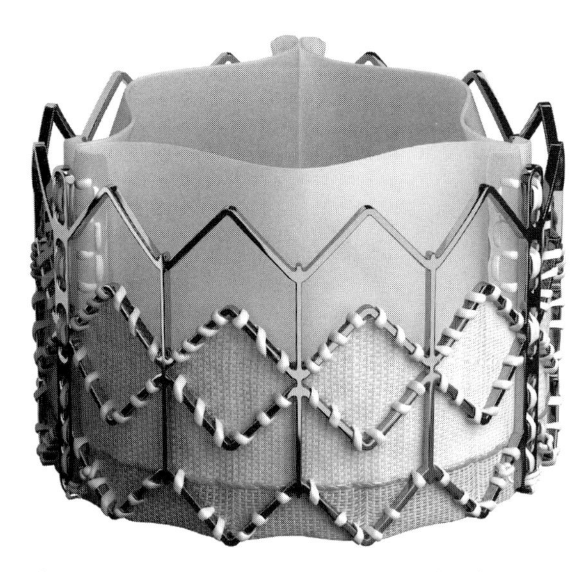

**Figure 4.23** A transcatheter heart valve. The Edwards SAPIEN valve. Image courtesy of Edwards Lifesciences LLC, Irvine, CA. Edwards SAPIEN and SAPIEN are trademarks of Edwards Lifesciences Corporation.

to be large enough to accommodate the valve and itself has to be accommodated by the vessel along which it passes. This may be complicated by the tortuosity and atherosclerotic nature of the vessels. Angiography has to be performed to assess the feasibility of the approach and many hopeful patients have to be excluded because of anatomical limitations. These techniques were introduced around 2006 and evidence of outcomes is limited. Most studies show a very high success rate of implantation but with concern over survival of patients in the short and medium term. Approximately 10% fail to survive for one month; this is perhaps not surprising considering the state of health of the patients and the evolving nature of both the devices and the clinical technique. One long-term concern relates to the performance of the pericardium, which has to be considerably deformed during compression of the device within the catheter and which may not possess adequate mechanical properties and stability after this sequence. The process of valve delivery does sometimes adversely affect the electrical activity of the heart, many patients requiring simultaneous pacemaker installation.

**Transapical valve delivery** There is an alternative minimally invasive approach that avoids the need to guide the valve along an artery. This is the transapical approach, in which the device is inserted directly through the apex of the heart following an incision in the thorax. This has several major advantages and is now in limited clinical use. A major disadvantage in most systems is the need to place the heart in some form of arrhythmia during insertion in order to temporarily minimize cardiac output.

### Overview of prosthetic heart valve outcomes and criteria for selection

The number of patients that can be considered for percutaneous or transapical valve delivery is very small, so the choice for a patient is largely based on the mechanical-stented bioprosthesis characteristics. Overall, in the USA and with the majority of valve types, the mechanical prostheses give more than 20 years lifetime, with little to choose between these types. With bioprosthetic valves, the optimal design and type depends on whether aortic root replacement is required; where root reconstruction is required, the operative mortality is usually much higher. Also, since the failure mode with bioprosthetic valves is structural deterioration, and since the younger the patient the higher the risk of this deterioration, comparisons are more difficult. The rate of deterioration after 10 years implantation may be three times higher for patients under 65 years of age than for those over 65 years. The ultimate cause of death in patients with prosthetic heart valves is more often unrelated to valve disease than valve related. With aortic valve replacement, valve-related causes of death amount to 37% with mechanical valves and 41% for bioprosthetic valves, corresponding figures for mitral valve replacement being 44 and 57%. Overall, patient survival is around 70% at 10 years.

Randomized clinical trials indicate that there are no statistically significant differences in outcomes between mechanical and stented bioprostheses for mitral valve replacement for up to 15–20 years implantation and for aortic valves at all ages. These outcomes are based on systemic embolism, valve thrombosis, prosthetic endocarditis and valve-related complications. The most important determining factor is the personal preference for the avoidance of anticoagulation therapy and the avoidance of structural deterioration, and this balance clearly depends on age.

A generalized algorithm for the choice of prosthetic heart valve may be constructed on the basis of clinical observations and outcomes. This can be seen in Figure 4.24.

### 4.2.3 Congenital heart defects

There are several ways in which the heart fails to develop correctly in the fetus, which may have serious consequences. Several of these conditions, or defects, concern abnormal openings in the structures, or septae, between chambers. For example, there may be an opening in the wall that separates the left and right upper chambers, or atria, which is known as an atrial septal defect. Similarly a ventricular septal defect is an opening between the right and left ventricles.

An atrioventricular septal defect is a deficiency in the atrioventricular septum caused by an inadequate fusion of the superior and inferior endocardial cushions in the mid-portion of the atrial septum and the muscular portion of the ventricular septum. Quite often these defects give rise to no signs or symptoms after birth and they may heal spontaneously. Serious defects will require immediate intervention to correct the hemodynamic imbalance. In the lesser, untreated cases, the individual may be symptomless for several decades, until signs of shortage of breath or fibrillation, eventually leading to the diagnosis of the condition.

In those patients where the hemodynamic disturbance is significant and warrants treatment, there are surgical and transcatheter methods available. In the former case, open-heart surgery is required and the defect is closed with a textile patch under direct visualization. Transcatheter methods are now given preference in the majority of cases. The precise method depends on the site of the defect. A very popular device, referred to as a septal occluder, consists of two self-expandable nickel–titanium mesh discs that are connected to each other, with a polyester fabric enclosed within

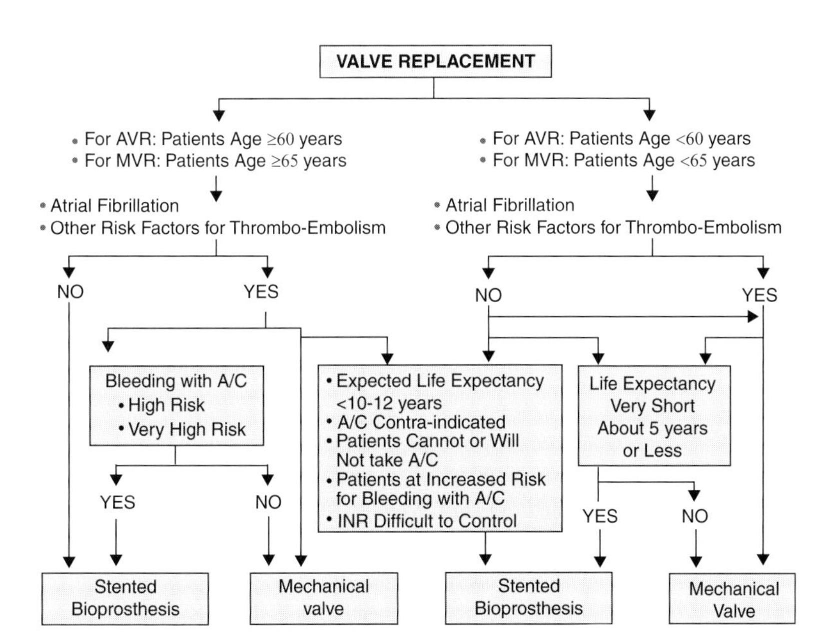


**Figure 4.24** Algorithm for choice of prosthetic heart valve. Key: A/C = anticoagulation; AVR = aortic valve replacement; INR = international normalized ratio; MVR = mitral valve replacement. Reprinted from *The Journal of the American College of Cardiology*, 55, Rahimtoola S H, Choice of prosthetic heart valve in adults: an update, 2413–26, © 2010, with permission of Elsevier.

the mesh. The device is deployed from a catheter such that each of the discs, which have been chosen with a diameter just larger than that of the defect, is placed on opposite sides of the septum. The polyester facilitates clot formation within the mesh and tissue ingrowth and reorganization readily occur.

A closely related defect is patent ductus arteriosus. The ductus arteriosus is a blood vessel that allows blood to bypass the lungs of a fetus by connecting the pulmonary arteries to the aorta. After birth, this blood vessel is no longer needed since the baby requires full pulmonary circulation, and the vessel closes naturally. In patent ductus arteriosus, the vessel fails to close so that there is abnormal blood circulation between the heart and the lungs. This opening can be closed using a transcatheter approach and a similar type of occluder.

## 4.2.4 Myocardial infarction

Acute myocardial infarction, generally known as a heart attack, is caused by a blockage of blood flow to the heart and the resulting damage to heart muscle. Each year, close to 10 million people worldwide have a heart attack, and this is increasing. We have seen earlier that patients surviving a heart attack have a few options for restoring blood flow, but there remains a critical point about keeping patients alive during the time before they can be treated in an intensive care unit. It is a significant goal to minimize the time from the onset of symptoms to reperfusion and to maximize the effect of early stage procedures to protect the myocardium during this time.
These procedures may involve a combination of pharmacological agents and biomaterial-based devices; in the latter case, devices may be deployed via catheters in order to remove thrombus deposits. These thrombectomy devices include filters, high pressure saline jets and dual lumen catheters that have torque cables with helical cutters and vacuum suction facilities. It is not clear just how well these current devices assist in the protection of these patients, but they are likely to be further developed in the future.

## 4.2.5 Arrhythmias

The functioning of the heart is controlled by a cycle of electrical impulses. These cycles can be disturbed on both acute and chronic bases, causing irregularities of the cardiac function, which are known as arrhythmias. These may be very serious and life threatening, and different forms of therapy have emerged for their treatment.

The cycle starts with an impulse initiated by the sino-atrial node within the right atrium. This passes through both left and right atria, causing atrial contraction because of depolarization of the cardiomyocytes. The impulse proceeds to the atrioventricular node, which is at the posterior right atrium, and on to the bundle of His before spreading to right and left ventricles, causing ventricular contraction. Irregularities can arise because of disturbance to either initiation or conduction of the impulses. For example, there may be sino-atrial node dysfunction affecting the impulse generation, or conduction blocks can occur along these pathways. A wide range of antiarrhythmic drugs is available. These tend to act on sodium channels (fast channel blockers) or calcium channels (slow channel blockers) or may have blocking effects on catecholamines at the β-adrenergic receptors, the so-called β-blockers. If the cardiac arrhythmias cannot be controlled by these drugs, several surgical interventions are possible, primarily including implanted devices that provide electrical therapy. There are two major types of device, the pacemakers and cardioverter-defibrillator.

### Cardiac pacemakers

Cardiac pacing is used for the treatment of conduction block, which is manifest by bradycardia, an abnormally low heart rate. A cardiac pacemaker is comprised of the pulse generator, including a power source and circuitry, and one or more conducting leads, which have electrodes at their distal ends. Pacemakers have many different forms, the use of which depends on the patient's condition. At its simplest, the device may be a single-chamber pacemaker, which gives a preprogrammed stimulus. In many patients, the stimulus is not required continuously and the most appropriate pacemaker is a "demand" type that incorporates a sensor that detects intrinsic cardiac activity and only sends signals when they are required. Many pacemakers are dual-chamber that have electrodes in both atrium and ventricle for sequential stimulation and synchrony of the chambers.

There are a number of important biomaterials considerations with pacemakers. The pulse generator is fairly straightforward. In the early stages of pacemaker development, the pulse generator was encased in an epoxy resin. Since the batteries in use then were short-lived, the resin was sufficient for the duration of implantation before the generator had to be replaced, but the evolution of longer-life batteries meant that the epoxy, through which water diffuses, was insufficient. Pacemaker pulse generators today are hermetically sealed in laser or electron beam welded titanium cans (Figure 4.25). The can has to be electrically insulted from the conducting lead; this is normally achieved by the use of a ceramic plug (usually alumina) at the feed-through, which separates the can from the lead.

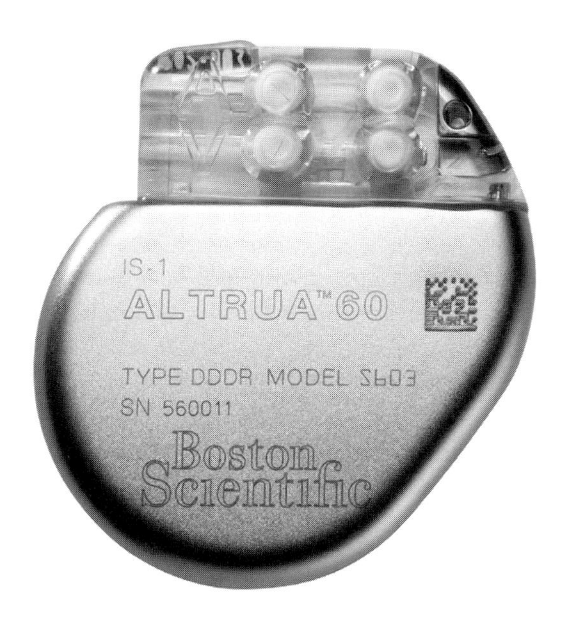


**Figure 4.25** Cardiac pacemaker. The Altrua™ 60 Pacemaker. Image courtesy of the Boston Scientific Corporation.

The leads have to be good electrical conductors with excellent fatigue resistance; the lead terminates with the electrode, which has a combination of design and materials to optimize charge transfer and is capable of attachment to the myocardial tissue. There have been many changes to lead and electrode materials over the years. The early uses of stainless steel and simple cobalt–chromium alloys have been superseded by better performance materials and their incorporation into more complex designs including co-axial systems. Lead materials now include platinum–iridium alloys, platinum coated with platinized titanium and Elgiloy. A recent trend has involved the use of the so-called drawn-brazed-strand (DBS) concept, which has a central silver core surrounded by a number, usually six, of nickel alloy wires, usually MP35N, the wire being formed by simultaneously drawing and heating. These give very good combinations of conductivity and mechanical properties. Electrode materials may be very similar, but with a trend to achieve better charge delivery characteristics through surface coating, for example with vitreous carbon, titanium nitride or iridium oxide. Ideally the electrode tip should have a small radius to maximize charge density and a large surface area (achieved through surface porosity) to reduce polarization. The electrodes may have geometrical features such as screws or helices to facilitate tissue attachment; these helices may be electrically active or inactive. Because the electrodes will irritate the tissues to some extent, through the effects of the intrinsic properties of the material and the charge that they are constantly delivering, it is desirable to minimize the associated inflammation. Chronic inflammation leads to an increasingly thicker layer of fibrous tissue around the electrode, which decreases the efficiency of charge transfer. In some systems, an anti-inflammatory agent is incorporated in the electrode, for example dexamethasone may be incorporated into the porous surface of a silicone insert.

Obviously the electrical impulse must be delivered to the precise part of the myocardium, and should not be transferred to the tissue between the pulse generator can and the electrode. The lead therefore has to be insulated. Usually this is achieved by use of a flexible insulating sleeve, usually made of silicone or polyurethane elastomers. There have been serious problems with some polyurethane insulated leads because of polymer degradation, which eventually results in loss of insulating capacity. Superior polyurethanes give much better performance. The silicones and polyurethanes in current use have their own advantages and disadvantages but both give very adequate performance.

## Defibrillation

Pacing involves, as we have seen, superimposition of a regular heart rhythm when the heart is beating too slowly or irregularly. The other major rhythm dysfunction occurs when the heart beats too fast, with a quivering or twitching action, and with such a frequency that it is unable to pump blood. This fibrillation can affect the atria, which is manageable by several methods, or the ventricles, which is far more severe. We shall deal with atrial fibrillation first.

Atrial fibrillation is the most common sustained cardiac-rhythm condition, with substantial consequences for stroke, thromboembolism and, ultimately, heart failure. The incidence of this condition is increasing worldwide, both as a result of general demographic changes and also improvements in the management of myocardial infarction, which leaves more people susceptible to age-related cardiac-rhythm problems. At 40 years of age, the lifetime risks for developing atrial fibrillation are around 24%, being slightly higher for men than women.

Atrial fibrillation is usually easily diagnosed, using surface electrocardiograms and clinical symptoms. The arrhythmia may be a single event but is usually recurrent; paroxysmal fibrillation stops spontaneously within a few days, but persistent fibrillation continues and needs either electrical or pharmaceutical intervention, or cardioversion, to stop. It is usually considered preferable to manage fibrillation by pharmaceutical methods, which involve either rate-control or rhythm-control strategies. Rate control aims at producing a target resting heart rate by a drug, either intravenously or orally, using a nodal blocking drug such as β-blockers or digoxin. In rhythm control, the objective is the reduction in paroxysms and the return to long-term maintenance of sinus rhythm. Several antiarrhythmic drugs are available, although there is clinical evidence that they have low efficacy and are associated with several adverse side effects. If atrial fibrillation persists for more than 7 days, the effectiveness of these pharmaceutical interventions is low, and electrical cardioversion may be needed to restore sinus rhythm. The most readily available electrical cardioversion method aims to deliver external transthoracic synchronized shock. Successful cardioversion has to be followed by maintenance, and this may be compromised by concomitant disease.

Difficulties with the pharmaceutical approaches have led to significant developments with non-pharmaceutical methods. These include variations on interventions to physically alter conduction pathways and barriers, and variations on electrical manipulation of rhythm. The former methods include surgical maze procedures in which patients who are undergoing concomitant open-heart surgery undergo surgical ablation and restructuring of critical areas of right and left atrial surfaces. More common are catheter ablation techniques that do not require such major surgery; these isolate electrically the pulmonary veins from the left atrium since these veins are a common source of rapidly depolarizing arrhythmogenic foci that cause the paroxysms. Atrial pacing may be beneficial in patients that have conventional indications for pacemakers and who have a history of fibrillation, but there is no real procedure for spontaneous defibrillation, as we see for ventricular fibrillation below.

Ventricular fibrillation is far more serious since no blood will flow from the heart during a period of fibrillation, which is likely to lead to syncope (fainting) and cardiac arrest. This is clearly a medical emergency and is usually only treatable by the use of an external defibrillator, which delivers an electrical shock through the chest wall. Obviously such a device has to be readily available, which is not always the case. For patients who survive a period of ventricular fibrillation and who are likely to have repeat episodes, an implantable device is available. This is the implantable cardioverter-defibrillator (ICD). This has the combined functions of converting some episodes of tachyarrhythmias into normal mode but then providing a shock, of up to 30 joules, when serious arrhythmias occur. Many of the materials and components used here are similar to those of the pacemaker. Storage capacitors are placed inside the pulse generator can to provide the shock capability, and the leads are more substantial with better conductivity.

It is by no means certain that the ICD is sufficiently effective to justify its widespread use. There are views that clinical benefits have been over-estimated, adverse effects on morbidity and quality of life under-estimated and cost-effectiveness is poor. It is likely that anticipated benefits will have to be assessed on a patient-to-patient basis.

### 4.2.6 Heart failure

Heart failure is a progressive disease and, in spite of many significant advances in our understanding of the causes, and in both diagnosis and prevention, it has become the most common cause of morbidity and mortality in people over the age of 65 years in many countries. There are few treatment options. Drugs are available, including the positive inotropes that increase the force of cardiac contractions, but these have significant limitations. Elsewhere in this book we discuss the use and limitations of heart transplants. It is clear that there is a major role for mechanical devices that are able to give support to the heart and the circulatory system, either on a short-term basis or permanently.

Since the major circulatory system is controlled by the activity of the left ventricle, it is this part of the heart that usually needs assistance. The ventricular assist device (VAD) and more specifically the left ventricular assist device (LVAD) is used to protect the failing heart and help maintain cardiac output and vital organ perfusion. The use of an LVAD is considered under three headings, as a bridge to recovery, as a bridge to transplantation or as "destination therapy," which is the position where the device is intentionally used as a permanent solution. The bridge to recovery is used where the ventricular condition is considered reversible and only in need of temporary support over a matter of days or weeks, for example after a severe but non-fatal myocardial infarction. The bridge to transplant, by definition, is used for patients who have no alternatives to transplantation and have to wait for a donor heart to become available. Isolated right ventricular dysfunction is rare, but can occur secondary to failure of the left ventricle, in which case biventricular support is needed.

There are two generic types of ventricular support device, and several commercial products within each (Figure 4.26). The first is a volume-displacement pump that has a pulsatile chamber that fills with blood, either passively or by suction, and then is compressed by a pusher plate. This produces pulsatile flow that mimics the systole and diastole of the natural heart. The devices have inflow and outflow conduits, unidirectional valves and a pumping chamber. It may operate either pneumatically or electrically. The second type, which appears to becoming more popular, is a continuous rotary pump. These have no valves and produce constant non-pulsatile flow, which is either driven by a spinning rotor around a central shaft or by a suspended spinning rotor. The specific materials used in these devices are often proprietary, but usually they are standard biomaterials used in other cardiovascular devices. Titanium alloys are widely used for the structural parts. Sometimes the blood contacting surfaces are coated with a porous polyurethane. Biostable polyurethane–silicone elastomers, to be discussed in Chapter 8, are used in some flexible sacs. Ceramics such as zirconia are used for journal bearings; conduits are made from low porosity polyester and PTFE fabrics. Not surprisingly, all devices require

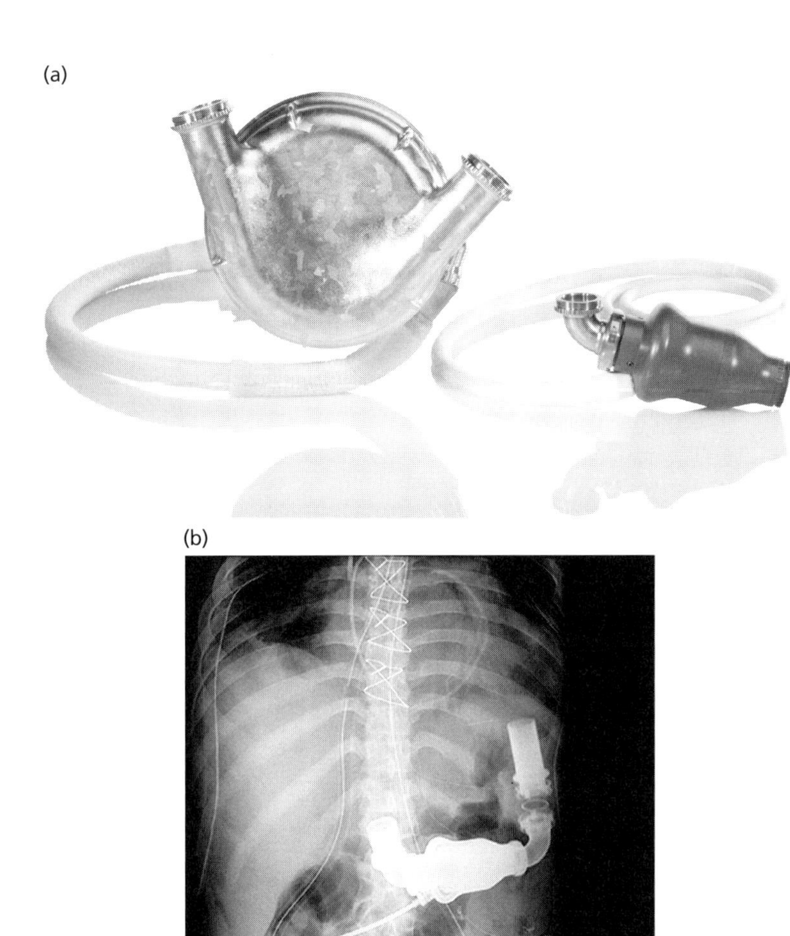


**Figure 4.26** The HeartMate II® left ventricular assist device. This is attached to the heart and is designed to assist the pumping function of the left ventricle. The device is placed just below the diaphragm in the abdomen. It is attached to the left ventricle, and the aorta. An external, wearable system that includes a small controller and two batteries is attached by an external driveline. The wearable system is either worn under or on top of clothing.
**(a)** The components of the device;
**(b)** a radiograph of the system in place in a patient. Images courtesy of Thoratec Corporation, Pleasanton, California, USA.

patients to be anticoagulated, either by warfarin or anti-platelet therapies.

Good success has been achieved with many of these devices, although there are many possible peri-operative complications and long-term problems. Included in the former groups are hemorrhage, infection, right ventricle failure, conduit kinking and air embolism. Device malfunction does occur, although the incidence decreased after design improvements were introduced. They may also be associated with thromboembolic events, and transient ischemic attacks are not uncommon. The entire LVAD is susceptible to infection, with a special focus on the percutaneous driveline; sepsis is the leading cause of death with most LVADs and may result in multi-organ failure.

## 4.3 The major organs: liver, kidney, lungs, pancreas

The heart is not the only major organ to experience dysfunction and chronic disease. Complex as it is, however, the heart is actually easier to assist when it is in difficulty than most of the other organs. In this section we cover some of the techniques that may be used in attempts to assist failing or temporarily disabled organs.

## 4.3.1 The kidney: hemodialysis

If the kidneys fail, a full kidney transplant is the ideal solution, but donor supply is a serious impediment. In the absence of a transplant, the only effective treatment of chronic kidney failure at this stage is renal dialysis. This involves taking blood from the patient and passing this through the dialysis machine in which waste products and extra fluids are removed from the blood by membrane dialysis and the purified blood returned to the body (Figure 4.27). This has to be done regularly, perhaps three days apart, each treatment lasting several hours. The passive dialysis mechanism may be assisted by convection processes (hemodiafiltration).

Over the half-century since hemodialysis was introduced into routine clinical practice, which now involves 1–2 million patients worldwide, the process and the biomaterials have changed considerably. The basic principle is that the blood is passed through a system where it contacts a semi-permeable membrane, on the other side of which is a dialysis solution, or dialysate. The dialysate is an electrolyte that is very similar to body fluids with respect to ionic species. The concentration difference across the membrane between blood and dialysate with respect to small molecules allows them to diffuse through the membrane, but larger molecules such as proteins and cells are unable to do so.

It has become clear that the efficiency and safety of hemodialysis depend on many material and configurational variables. These include the surface roughness and good mechanical stability,

(a)

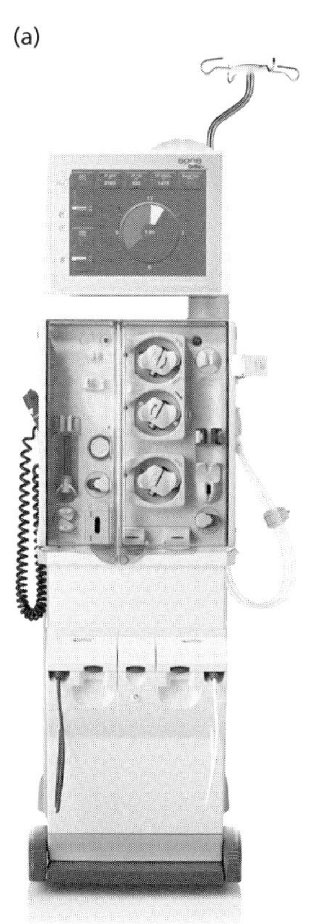

(b)

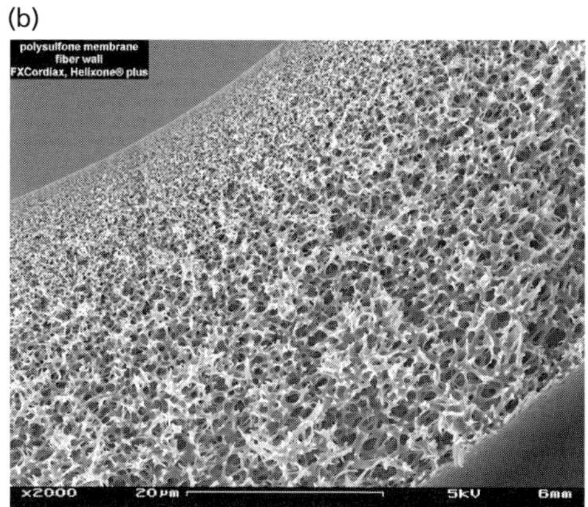


(c)

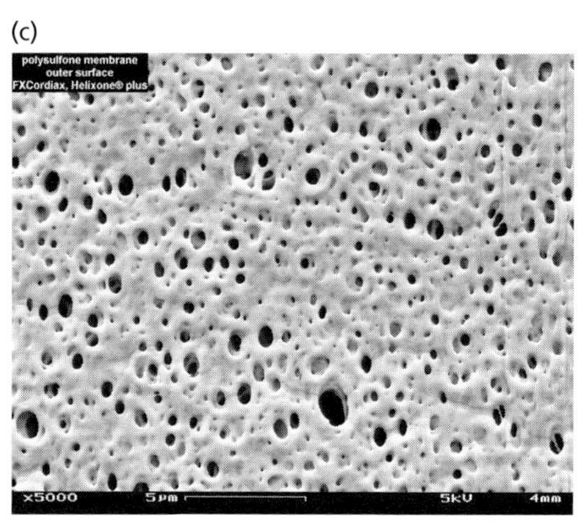


**Figure 4.27** Hemodialysis equipment.
**(a)** The 5008 CorDiax generation of dialysis machines by Fresenius Medical Care.
**(b)** Cross-sectional view through an asymmetric polysulfone dialyser membrane (Helixone® plus), with sponge-like structure gradually increasing in porosity.
**(c)** Pore structure at the outer, dialysate, surface of the same polysulfone dialysis membrane. All images courtesy of Fresenius Medical Care.

along with high porosity and narrow pore size distribution. The membranes may be characterized as of either high flux, with a water permeability greater than 0.15 mL/(hm$^2$ Pa) and those of low flux, with a water permeability of 0.03–0.09 mL/(hm$^2$ Pa).

For many years, the most prominent hemodialysis membrane material was regenerated cellulose. This provided for very effective dialysis but had poor blood compatibility, especially because of the reactivity of free hydroxyl groups, which caused complement activation and other effects. Some methods to alleviate these problems involved surface coatings for example with poly(ethylene glycol), or substituting the material with cellulose acetate. More recently, synthetic polymers such as polysulfone, polyarylethersulfones and polyacrylonitrile have been used. Many of the membranes are now fabricated as hollow fibers rather than the early flat sheets or tubes. Several studies and reviews indicate that there are few indications that synthetic materials produce significantly better clinical outcomes than cellulosic materials, although they do appear to be associated with fewer incidences of dialysis-related complications such as amyloidosis.

In this context, the future of hemodialysis membranes has to be based on better control of membrane architecture and biocompatibility. The technology of fabrication of polysulfone hollow fibers, for example, is now able to generate microstructures such as that seen in Figure 4.27(b), this morphology allowing clearance to be based on micro-convection as well as concentration differences, giving control over hydraulic permeability. These high flux membranes may also allow some clearance of low molecular weight proteins. There is also a conceptual, if not yet practical, trend towards more sophisticated engineering and biological solutions, involving, for example, microfluidics and extracellular matrix molecule coating of membranes. This leads towards the so-called bioartifical kidney, which is discussed in the next chapter.

### 4.3.2 The lungs: extracorporeal membrane oxygenation

Biomaterials used in the context of lung support are, at this stage, not involved in chronic lung failure but rather in temporary support during surgery. In particular, this involves procedures where the heart is stopped (e.g., heart valve insertion) and where efficient gas exchange has to be artificially provided for several hours. This may have to deal with up to 5 L blood per minute, with 95–100% hemoglobin saturation, and an outlet of around 40 mmHg $CO_2$. Although there have been several types of oxygenators developed and used, the only relevant type today is the membrane oxygenator, where there is no direct contact between the blood and the oxygen, but diffusion of oxygen and $CO_2$ through a suitable membrane.

The solubility of both $O_2$ and $CO_2$ in blood is low so that high blood flow (2–4 L/min) through the device is required. In the oxygenator, the driving force for $O_2$ is generally much higher than for $CO_2$, which is the reverse of what happens in the lung. Maximizing $CO_2$ transport is therefore critical. Silicones have been used for oxygenation because of their high $O_2$ permeability and good biocompatibility, but better all-round performance is now seen with some polyolefins and olefin copolymers, such as poly(4-methyl-1-pentene). Polypropylene microporous hollow fibers have also gained a great deal of attention.

### 4.3.3 The liver

Just as chronic kidney failure is a life-threatening condition and preferably treated by transplantation, so is liver failure very dangerous, with few treatment options. The function of the liver is very complex and there is no equivalent of hemodialysis for this organ. There have been excellent developments in the direction of so-called bioartifical livers but nothing is available at this stage for chronic disease. Liver failure can be caused by short-term exposure to toxins and this acute failure may be amenable to detoxification systems in which the toxins in question can be sequestered from the blood. Activated charcoal has been used in such perfusion columns.

Far more interest is now being paid to so-called "metabolically active" extracorporeal liver assistance devices, in which hemodialysis type equipment incorporates cells that are able to remove certain toxins and other products. One such product incorporates cloned, immortalized human liver cells. These approaches are more relevant to regenerative medicine and are discussed in the next chapter.

### 4.3.4 The pancreas: insulin pumps

The treatment of diabetes is clearly of immense importance. The ability to monitor and treat diabetes has improved over the years, but this progress has been outpaced by the rise in the incidence of the disease. Some aspects of the diagnosis and treatment of diabetes are addressed in other chapters of this book. In relation to medical devices, we shall concentrate here on insulin pumps. These have been available for many years, especially for patients with Type 1 diabetes but they have not found a major place in treatment modalities. The principle is straightforward; an insulin pump should be able to provide a patient with a continuous infusion of the required insulin, rather than depending on multiple daily injections. There have been attempts to produce totally implantable closed-loop pumps that deliver insulin from its reservoir to the body on the basis of continuous glucose monitoring. This has proved very difficult, not least because of problems with the long-term stability of insulin within the reservoir and catheters. Instead of this complete solution, insulin pumps in clinical use are worn externally and provide insulin to subcutaneous tissues via a fine catheter and short metal cannula. The devices are not automatic in the sense that they do not have integrated glucose sensors, but most have a system that allows for a basal rate of delivery, which can be overridden by patient-initiated bolus or correction delivery. Some devices can be operated by wireless signals of periodically monitored glucose levels. The materials issues here are not too difficult since the subcutaneous cannula is no different to other needle-like devices (i.e., stainless steel will suffice) and since the insulin cartridge only has to carry the insulin for a short time, as they are disposable and changed frequently.

## 4.4 Eyes and ears

Implantable devices within the sensory organs such as the eyes and ears represent a tremendous opportunity to improve the quality of life. The potential number of patients that could be beneficially treated is huge, as is, therefore, the commercial market. In some cases deficiency in hearing or sight is apparent in the early stages of infancy. In other cases, problems arise with ageing, more often these days in middle rather than advanced age. There are situations in which the deficiency is readily compensated for, as with spectacles or external hearing aids. In other situations the disease is more profound and often really effective solutions are not yet available. As always, there are intermediate positions, and it is here that most technologies, to date, have found their role.

### 4.4.1 Contact lenses

Throughout the world, well over 100 million people wear contact lenses. These are simple alternatives to spectacles, and they correct vision in the same way by altering the diffraction of light. The majority of users prefer daily wear lenses, which have to be cleaned overnight, and are replaced monthly. However, an increasing number now use daily disposable lenses, extended wear, which are changed weekly, or continuous wear, which are worn without removal for up to a month before being replaced.

A contact lens material should obviously be optically clear, but should also be hydrated and permeable to oxygen and ions, be resistant to protein and lipid adsorption and should maintain a continuous tear film. Although some lenses in use are hard or rigid, typically made of polymethylmethacrylate, the vast majority is now described as soft lenses. The first soft material to be used, and indeed the first material to be specifically formulated solely for medical use, was poly(2-hydroxyethylmethacrylate). The oxygen permeability of this hydrogel is not very high, and has been improved over the years by use of co-monomers such as methacrylic acid. Polydimethylsiloxane, a silicone elastomer, has also be extensively used, having very high oxygen permeability, but it is non-wettable with respect to tears and is prone to lipid deposition. Increasingly common are the hybrid silicone hydrogels that combine the advantages of the two varieties. There are several different versions of this hybrid, which variously incorporate agents to increase wettability or are surface treated to increase resistance to deposition.

Contact lenses are very successful and have improved the quality of everyday life for vast numbers of individuals. They may be adapted for personal preference with cosmetic features, such as tints, and forms of lens have been prepared which offer the possibility of using them for prolonged drug delivery. The increased use of extended or continuous wear lenses has resulted in a small but finite risk of infection. Microbial keratitis (or ulcerative keratitis), for example associated with *Pseudomonas aeruginosa* or *Staphylococcus epidermidis*, can have serious consequences, with the loss of corneal epithelium and inflammation of the underlying stroma. The risk may be as high as 20 per 10 000 extended wear users annually in most populations, significantly increasing in communities with lower standards of hygiene. There is no doubt that the risks of infection increase with extended and continuous wear, as the production and maintenance of the tear film are compromised, especially during sleep when lack of flow increases mucin production and reduces lysozyme concentration, an enzyme with antibacterial properties. Protozoas such as *Acanthamoeba* may also cause keratitis. Prolonged antimicrobial treatment is required for all infections.

### 4.4.2 Intraocular lenses

The lens of the eye may suffer age-related deterioration, especially through the formation of cataracts. The proteins that constitute the structure of the lens may distort, diminishing the optical transparency. Usually this starts in the center of the eye with a small cloudy spot, which then grows until the whole lens becomes opaque. Patients with diabetes are particularly at risk and the cloudy area may start at the outer edge and grow inwards. There is no medical treatment for cataracts, but fortunately there is a simple operative procedure, utilizing simple biomaterials, that provide very effective treatment; as many as 3 million Americans undergo this treatment each year. The device is an intraocular lens (IOL). As seen in Figure 4.28, an IOL has a central optic, with peripheral haptics that assist its retention in the eye. The cloudy lens is removed and replaced by an IOL.

IOLs are generally made from silicones or acrylics. The silicones are essentially polydimethylsiloxane. There are two broad types of acrylic lens. Rigid acrylic lenses are made from polymethylmethacrylate. Softer, foldable lenses are made from acrylic hydrogels (hydrophilic lenses) or flexible acrylic copolymers (hydrophobic). The trend has certainly been towards foldable lenses, since these are inserted less traumatically, but several factors determine the characteristics of the different varieties.

The biocompatibility of IOLs has been a matter of debate for decades, and the pattern of the tissue response, with the phenomena of protein adsorption and cellular infiltration well documented. In reality, although cellular responses can be detected, they are rarely clinically significant and the long-term success of IOL implantation is very good. The main complication is opacification, either in the anterior capsule or more commonly in the posterior capsule, which is basically a secondary cataract forming behind the lens.

Conventional IOLs have been monofocal, meaning that they can only offer correction at one distance, far, near or intermediate. In recent years, much more sophistication has been introduced, such that

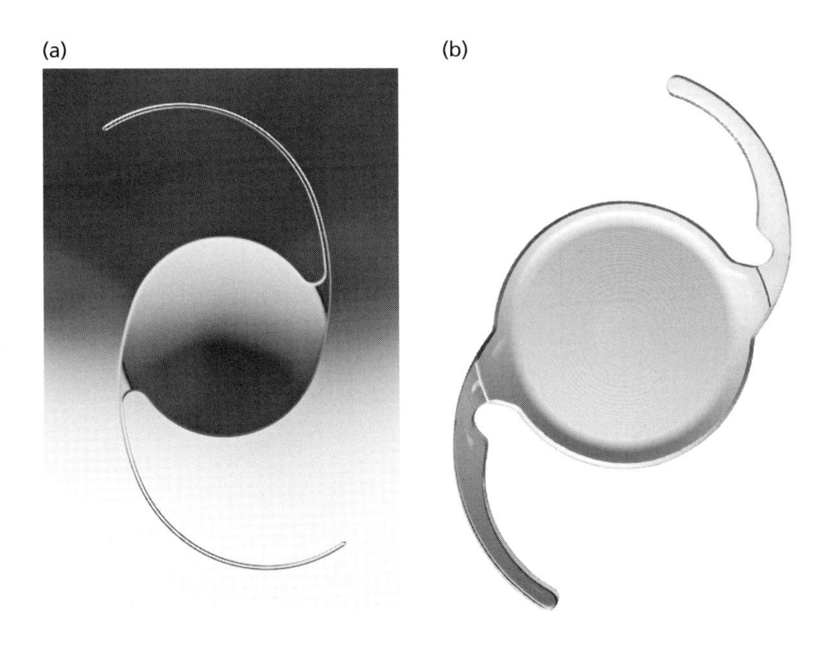


**Figure 4.28** Intraocular lenses. Two examples of hydrophobic acrylic lenses.
**(a)** Courtesy of Aaren Scientific, California, USA.
**(b)** The Technis® One Piece Lens, courtesy of Abbott Laboratories, Illinois, USA.

multifocal and accommodating lenses are available. Lenses are now able to correct astigmatism, and can be made with blue light, or UV/violet filtering chromophores to protect the retina from radiation.

### 4.4.3 Glaucoma

Glaucoma is a major cause of blindness, especially among African-Americans, the incidence markedly increasing with age. It involves a group of diseases that affect the optic nerve and loss of retinal ganglion cells. The disease is asymptomatic for a long time and usually significant damage to the optic nerve occurs before symptoms are present, starting with a loss of peripheral vision. Glaucoma is caused by a rise in intraocular pressure. The aqueous humor that nourishes the lens, cornea and iris is continuously produced in the ciliary tissues and, under normal circumstances, it drains away through the trabecular network and drainage channels. If there is difficulty in drainage, especially caused by structural changes in the trabecular network, then pressure builds up, eventually damaging the optic nerve. Glaucoma involves irreversible changes and thus cannot be cured, but progress of the disease can be reduced by medication, especially with pilocarpine, which opens drainage canals, beta blockers, which decrease the rate of aqueous humor flow, and prostaglandin analogues which increase uveoscleral flow, that is the escape of fluid through the front of the eye.

If the glaucoma is infractory to medication alone, the blockage in the trabecular network may be treated surgically with the process of trabeculotomy, in which a part of the network underneath the sclera is excised. This works fairly well but there are many potential complications, including fibrosis or infection. Lasers can also be used to selectively alter the structure of the trabeculae without physical penetration. This is known as laser trabeculoplasty; it reduces the complications of trabeculotomy, but has to be repeated frequently. Because of the difficulties with these techniques, implantable devices, generally referred to as glaucoma drainage devices, have been used to relieve the intraocular pressure by directing aqueous humor through an anterior chamber tube to the subconjunctival space or equivalent location. These tubes, with attached endplate are usually made of a silicone elastomer, although some use stainless steel. Examples are shown in Figure 4.29. Success rates are difficult to define at this stage since many

(a)

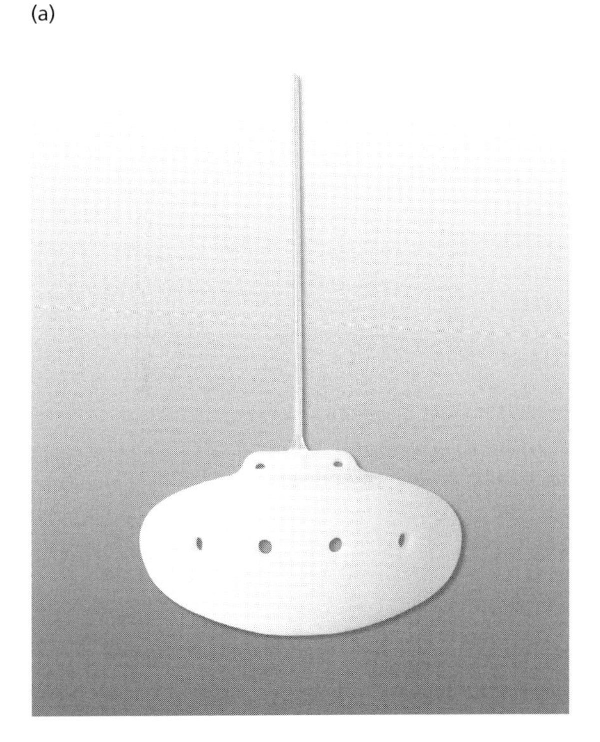

(b)

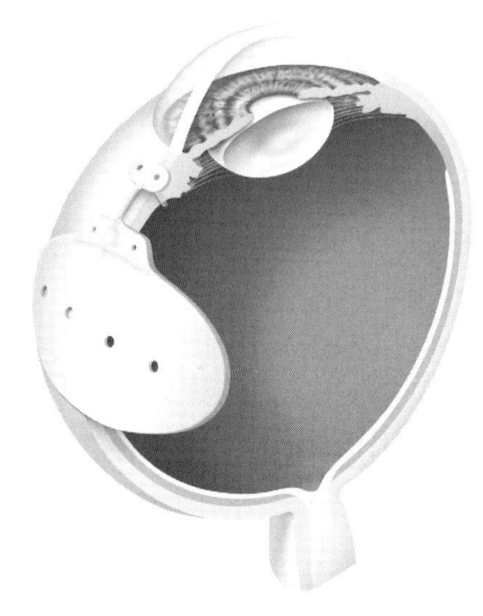

**Figure 4.29** Glaucoma drainage devices.
**(a)** The Baerveldt® device;
**(b)** *in situ* in the eye. Images courtesy of Abbott Medical Optics, a division of Abbott Laboratories, Illinois, USA.

devices have only recently been introduced. Complications are primarily related to fibrosis.

## 4.4.4 Retina and artificial vision

Diseases of the retina are serious and sight threatening, with few types of therapy available. Among these diseases are forms of macular degeneration and retinosa pigmentosa. Dry macular degeneration is probably the most significant. The macula is a 5 mm diameter area in the center of the retina; it consists of large numbers of light sensitive cells, the rods and cones, and is responsible for the clarity of vision. These cells produce the proteins that are sensitive to light, but these proteins age and are replaced. Immediately behind the retina is a layer of cells, the retinal pigment epithelium, which is responsible for the removal of the waste protein fragments. However, with age, these epithelial cells deteriorate, begin to lose their pigment and become less efficient in removing these waste proteins, This results in fat deposits, known as drusen, forming in the retinal cells, which blur vision, especially central vision. Dry macular degeneration progresses slowly. There is no cure at the present time, and it may transform into wet macular degeneration, which is much faster progressing. Clinical trials are underway with some forms of cell therapy, but otherwise there are few options.

Nevertheless, attempts to develop retinal implants have been gaining increasing attention; over the last decade significant progress has been made with techniques for the electrical stimulation of the inner surface of the retina, based on the hypothesis that such stimulation will force neurons in the visual system to change their membrane potential and fire action potentials, circumventing the damaged normal route. The work is largely experimental with a small number of limited human experiences being reported. The main emphasis has been on the retina, usually with either an epiretinal or subretinal approach. With the epiretinal approach, the systems typically consist of an intraocular array of stimulation electrodes and an extra-ocular device which

(c)

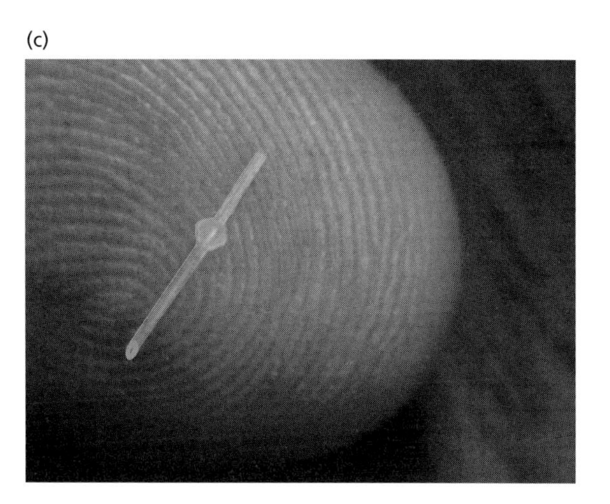

(d)

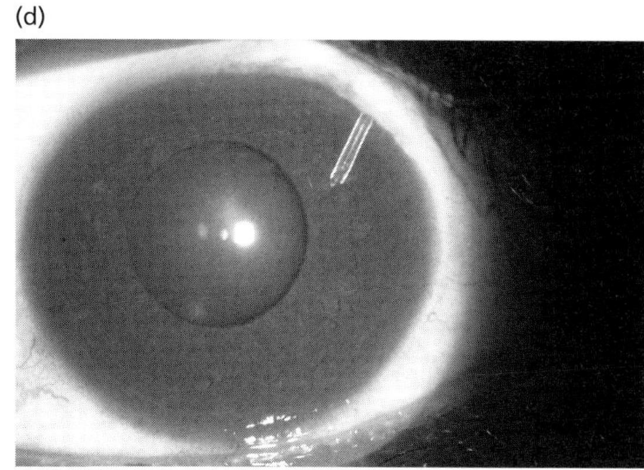

**Figure 4.29** *continued*
(c) The InnFocus MicroShunt™ and
(d) after two years implantation in the human eye, shunting aqueous humor from the anterior chamber to under the conjunctiva. The shunt is made from SIBS polymer, see Chapter 8. Images courtesy of Dr. Len Pinchuk, Innovia, Florida USA and Dr. Juan Batlle, Centro laser, Santo Domingo, Dominican Republic.

serves for image acquisition, data processing and encoding and data transfer. The electrodes vary in number (in some systems up to 100) and in materials; polyimide is used in some devices. Surface microtopographical features enhance transfer of signals to retinal cells, seen in Figure 4.30. Data transfer takes place either by cable or wireless techniques. The subretinal approach typically involves the use of arrays of thousands of microphotodiodes underneath the retina which convert light into electrical current and activate postsynaptic neurons. The array has to be connected to an external energy source which passes through the sclera. In order to avoid potential complications of surgical intervention close to the retina, other teams are working with superchoroidal approaches which, for example, activate the visual cortex using episcleral electrodes.

## 4.4.5 External ear

The ability to reconstruct the complete external ear remains elusive, but significant advances have been made in recent years. The need for such reconstruction arises from either congenital microtia or traumatic loss of the ear. For several decades, costal cartilage, obtained from the patient's ribs has been used, but with varying success, problems arising from harvesting the cartilage and crafting the graft. It is possible to produce an external prosthesis and attach this to the tissues, for example with screws, but this is not very popular. Some success has been achieved with porous polymers such as polyethylene, which are secured through tissue ingrowth.

## 4.4.6 Middle ear

There are two applications for implantable devices within the middle ear, those for the reconstruction of the ossicles and the implantable middle ear hearing aid. In the former case, the devices attempt to replace all or part of the ossicular chain, which can become damaged by trauma, or following infection, especially in cases of chronic otitis media. The technique is generally referred to as ossiculoplasty, the partial ossicular replacement (POR) being used when the stapes superstructure is still present, and the total ossicular replacement (TOR) used when both incus and stapes are missing. The devices are simple.

(a)

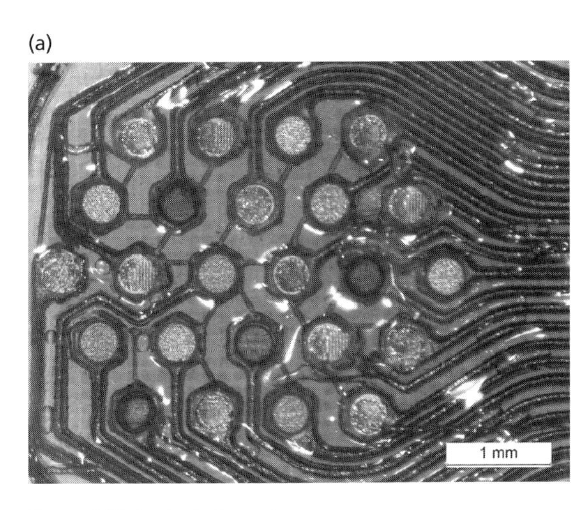


(b)

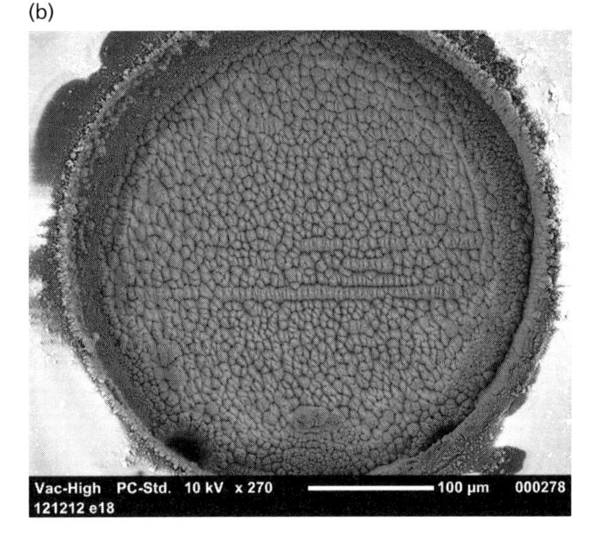


**Figure 4.30** Retinal electrodes.
**(a)** Stereoscopic view of a 24 channel suprachoroidal platinum-based electrode.
**(b)** Laser patterned surface that increases charge transfer area. Images courtesy of the bionic eye laboratory of Professor Nigel Lovell, Graduate School of Biomedical Engineering, University of New South Wales, Sydney, Australia, with special thanks to R.Green, C.Dodds, P.Matteucci and S.Suaning.

Preferred materials include cartilage, hydroxyapatite, titanium, porous PTFE and glass-ceramics. Depending on how success is defined, the effectiveness of the procedure is variable. The host response to the materials is often quite acceptable, but if success is measured by parameters of hearing restoration, the procedure does not appear that good. It is common to use the criterion of the development of a post-operative air-borne gap of 20 dB or more, and here success rates are usually no more than 50% at 5 years.

Conventional hearing aids, worn in the external ear (and not covered in this book) are very successful, especially with the significant improvements in miniaturization and appearance. There are some patients where the hearing improvements are not very significant and their condition is not appropriate for the use of the cochlear implants discussed in the next section, who may be treatable by an aid that is implanted in the middle ear. These devices typically use a piezoelectric or electromechanical transducer that is coupled to the ossicles, which stimulate the auditory pathway. These devices have yet to obtain widespread acceptance, largely because of significant technical issues with acoustic coupling, and also because the number of patients is not likely to be significantly large to support development costs.

## 4.4.7 Cochlear

The progression of sound through the middle ear results in transmission from the footplate of the stapes to the oval window of the cochlear. Movements of this membrane induce oscillations of pressure within the cochlear fluids that produce a wave of displacement along the basilar membrane that extends along the length of the cochlear. This is sensed by the hair cells, each of which contain fine rods of proteins that release chemical transmitters in response to their deflection as they sense this movement. These chemicals stimulate activity in the auditory neurons, changes in which are transmitted to the brain via the auditory nerve. The sensory hair cells are quite fragile and are easily damaged, leading to partial or complete deafness.

Cochlear implants have been designed to bypass the damaged, or even missing, hair cells through the direct stimulation of the neurons in the auditory nerve. The first implants produced in the 1980s were fairly elementary; they were able to restore hearing to the profoundly deaf, but without much quality. Since then, they have improved considerably in sophistication and offer very effective treatment, both for the congenitally deaf and for those who acquire cochlear-related deafness as a function of age, genetic diseases, infections, drug abuse and over-exposure to sound. It is not possible to quote success rates since in the vast majority of patients, some degree of hearing is restored, but there is variability in the quality, largely depending on individual conditions; producing or restoring sound transmission is not the only factor since the ability of the patient to interpret that sound is also very important. The use of a cochlear implant in an infant who has been deaf from birth has to be handled very delicately since they will have no previous perception of sound sensation.

The general principle of the cochlear implant is straightforward. As shown in Figure 4.31, the components include a type of microphone that senses the sound, a speech processor, a transcutaneous or percutaneous link to an implanted receiver/stimulator, and a cable that connects the latter to an array of electrodes. The electrodes may be placed in the scala tympani, one of the three cochlear chambers. The electrodes are usually made of platinum. Varying numbers are used, typically around 20, these being held in a carrier, usually of silicone elastomer. The implanted receiver is a small device, encased within a hermetically sealed container, in a very similar manner to other implantable electronic devices.

The performance of cochlear implants is largely dependent on the quality and sophistication of the speech processor and the electrode arrays. Improvements to the systems are continuously been made, not surprisingly, which target those patients who have residual difficulty in hearing, with emphases on the stimulation of both ears with bilateral implants and the use of combined electric and acoustic stimulation. Here the sound is separated into high

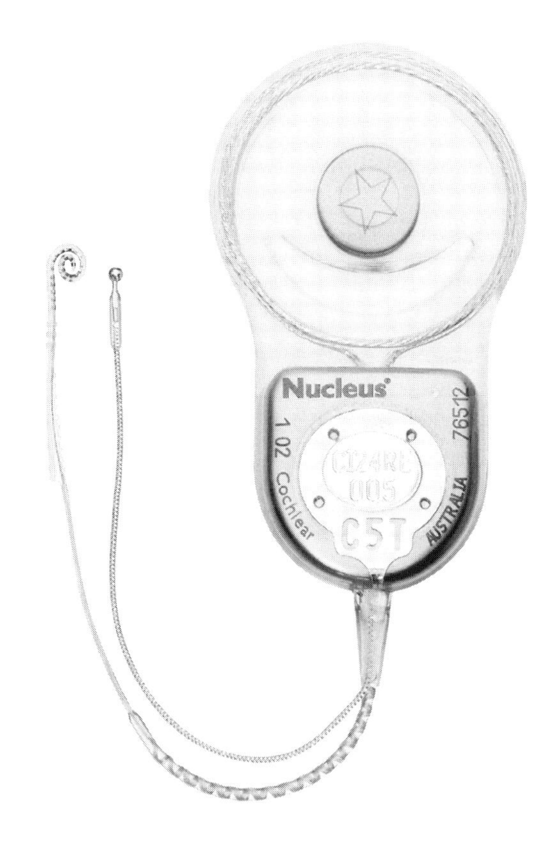


**Figure 4.31** The Nucleus® C124 cochlear implant. Image courtesy of Cochlear Europe Ltd, Surrey, UK.

and low frequency components, the former being processed in the normal cochlear implant and the latter processed acoustically via a ear mold. Clearly these developments have little to do with the materials of construction.

## 4.5 Central nervous system

Biomaterials are used in the central nervous system for a number of reasons. In some situations they are used to permit the delivery of drugs directly to the brain, overcoming the problems associated with the blood–brain barrier, discussed in Chapter 6. There is also current interest in biomaterial scaffolds for nerve tissue regeneration, which will be discussed in Chapter 5. We shall confine ourselves here to hydrocephalus shunts, brain stimulators and neurorecording electrodes.

## 4.5.1 Hydrocephalus shunts

Hydrocephalus is a condition that is frequently associated with spina bifida and involves an inability of cerebrospinal fluid (CSF) to pass from the ventricles of the brain, where it is produced, to the spinal cord, where it provides nourishment. This results in an accumulation of the CSF in the brain and a rise in intracranial pressure. This is a chronic condition, which necessitates a chronic, continuous solution, which is most frequently provided by the implantation of a tube, or shunt, which drains the CSF from the ventricles and subarachnoid space into either the peritoneal cavity or the right atrium of the heart, where it is reabsorbed. The shunt is universally made of the silicone elastomer polydimethylsiloxane, and contains a one-way valve to prevent reverse flow. The device (Figure 4.32) is life-saving since excessive intracranial pressure is eventually fatal, but it is not without problems, especially obstruction and infection. The obstruction may be caused by any of many factors, usually through the ingrowth of choroid plexus or glial tissue into openings at the proximal end of the catheter or the deposition of particulate debris around the valve. Infections can occur soon after the insertion of the catheter, or at any late stage – this is especially prevalent in spina bifida cases where urinary tract infections often arise and which can spread systemically. Attempts have been made to coat catheters with non-fouling or anti-bacterial substances, although they do not give any guarantee of successful elimination of infection risk.

## 4.5.2 Deep brain stimulator

Deep brain stimulator (DBS) technology has been available for a number of years for the treatment of pain, epilepsy and some movement disorders and, in more recent years, for cerebral palsy, Tourette's syndrome and depression. However, it has been the application of this technology to the movement disorders of Parkinson's disease, essential tremor and dystonia, and the development of the implantable DBS, that has brought this technology to the forefront.

Parkinson's disease involves the loss of dopaminergic cells in the substantia nigra, which leads to reduced dopamine production and increased inhibitory control from the basal ganglia, producing a variety of symptoms including bradykinesia and rigidity. When the pharmacological methods cease to give optimal results, as they do in many cases, surgical therapy using DBS may be considered, as noted in Chapter 1.

The DBS system includes arrays of insulated electrodes and a subcutaneously placed pulse generator (Figure 4.33). The design of the brain–electrode interface has evolved over the years in order to minimize tissue adhesion. Usually platinum–iridium is used as the electrode material, with a polyurethane insulation. Materials for the pulse generator are similar to those for a cardiac pacemaker. Targeting the relevant deep brain structure is clearly very important, these structures being mapped and defined relative to a stereotactic frame system or other type of device. Preoperative MRI or CT allows this mapping relative to the target sites, which could be the globus pallidus pars interna, the subthalamic nucleus or the ventral intermediate nucleus. After

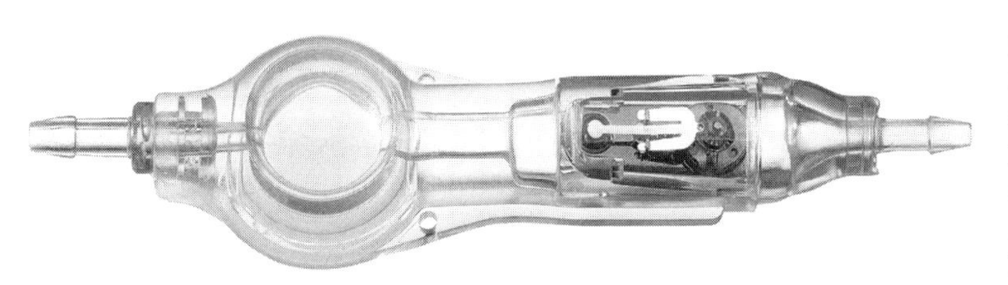

**Figure 4.32** Hydrocephalus shunt. The Codman Hakin Programmable Valve System used for the treatment of hydrocephalus. Image courtesy of the Codman Neuro Division, DePuySynthes, a Johnson & Johnson Company, Raynham, Massachusetts, USA.

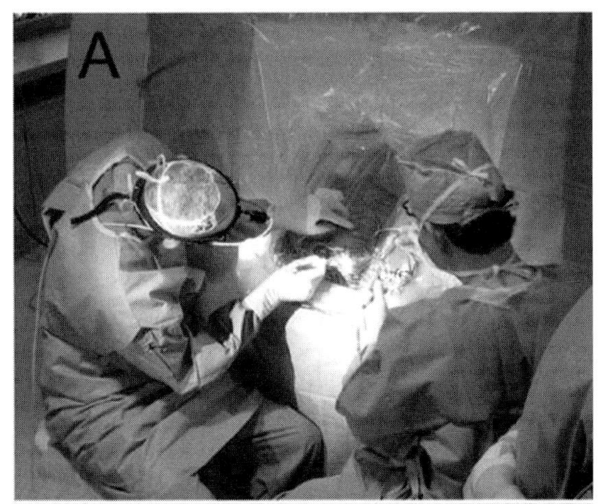


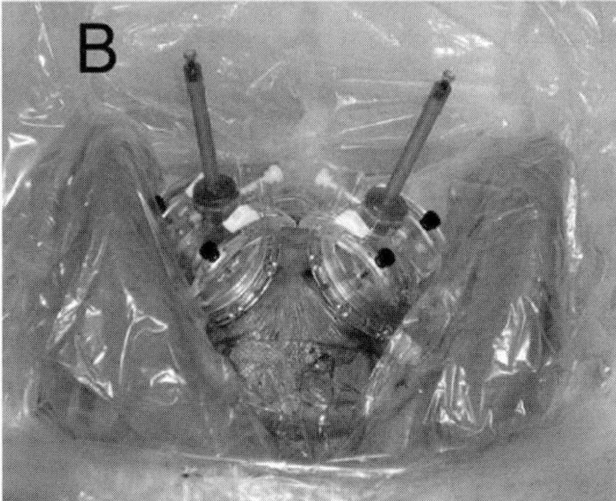


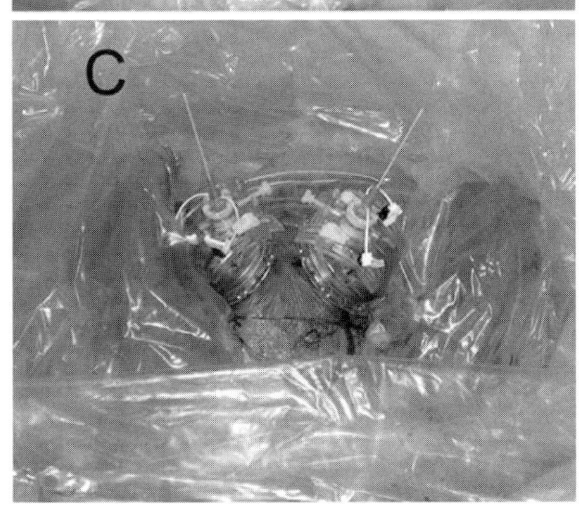


**Figure 4.33** Deep brain stimulator (see also Figure 1.2). Intraoperative photographs demonstrating surgical draping and trajectory guides.

(A) Patient's head is shown at the back of the MR bore, with a sterile drape.
(B) Trajectory guides with alignment stems.
(C) Trajectory guides with multilumen insert and peel-away sheath prior to advancing the sheath into the brain. The flexible radiofrequency receiving coils are covered with sterile blue towels. Reprinted from *The Journal of Neurosurgery*, 112, Starr P A *et al.*, Subthalamic nucleus deep brain stimulator placement using high-field interventional magnetic resonance imaging and a skull-mounted aiming device, 479–90, © 2010, with permission of JNSPG and Rockwater Inc. See plate section for color version.

successful target calculation, bur holes are placed in the skull and the electrodes inserted. Stimulation is performed intra-operatively to assess their effects before the electrodes are secured and connected to the generator. There is a risk of temporary morbidity, including hematoma formation, infection, confusion and seizures, and in the longer term there may be battery or electrode failures that necessitate revisions. There are some target-specific risks associated with spread of the stimulus, leading to dizziness, diplopia and severe depression.

## 4.5.3 Neural recording electrodes

One area which has received very significant attention but which has yet to produce the desired level of performance is that of neural recording electrodes. One of the concepts here is to obtain signals from an array of neurons and relay these into a computerized device that will allow paralyzed patients (for example) to control devices such as prosthetic limbs or wheelchairs.

The principal function of recording electrodes is to transpose the ionic current at the relevant tissue site into an electron current, which is different to the stimulating electrode in the deep brain stimulator discussed above which has to deliver electrical charge to the brain tissue in order to provoke excitable cells and tissues. Electrodes require good biocompatibility in order to prevent local host responses that detract from charge transfer. The electrode–tissue impedance has to be stable in order to give long-term reliability and reproducibility. Some material systems work well as recording electrodes but poorly as stimulating electrodes and vice versa. Gold can be used for recording but not stimulating, while iridium is the opposite.

Many materials have been investigated but few have been able to give consistent performance for very long because of the inflammatory response and, especially glial scarring, which increase electrode impedance. Microwire electrodes consisting of 20–50 μm diameter wires of conducting metals such as Pt, Au and Ir, insulated along their length except at the tips with PTFE or polyimide, give good performance as far as signal transport is concerned, but only for a matter of a few weeks. Silicon micromachined microprobes have also been developed but have similar limitations. Attempts are being made to control the inflammation through the local delivery of anti-inflammatory or anti-proliferative drugs and the use of alternative surface coatings, but success has been elusive. Some systems have been developed with polymer-based thin film electrodes, using, for example, polyimide, parylene or silicone substrates and Au, Cr, Pt, TiN electrode material deposited onto the substrate.

## 4.6 Genito-urinary system

The genito-urinary system is complex; it suffers from many diseases and conditions, of widely varying etiologies and consequences, which are generally not very amenable to solutions involving biomaterials. It is one area that has considerable potential for therapies of regenerative medicine, discussed in the next chapter. We discuss here implantable devices that are used in attempts to control urinary flow and devices that are intended to improve sexual performance.

### 4.6.1 The bladder

The bladder, a little like the heart, is quite a simple organ which we normally take for granted, but once it goes wrong, the quality of life markedly suffers. The word "simple" here is not an over-exaggeration. Urine is produced in the kidneys by a type of filtration process, and it drips down into this muscle-bound, expansible, container. In the healthy, urologically intact person, this is uneventful and a point is reached when he or she recognizes that the bladder is getting rather full and, in a purely voluntary fashion and without any urgency, chooses the time and place where to relieve the bladder of its contents, when the urine flows down the urethra to the outside world. We discuss below those situations, which unfortunately often occur in individuals who are least able to help themselves, that is the pediatric and geriatric cases, where medical technology has a role to play.

### 4.6.2 The overactive bladder in children

An overactive bladder is an alteration of the filling phase of the lower urinary tract, which results in the condition of urgency, when there is little control over the urination process. Once the problem has been identified in a child, it is usual to try so-called urotherapy methods, in which there is planned control of drinking and voiding; these do not usually work. There are also drug regimes, involving anti-muscarinic drugs, which may relieve symptoms but do not resolve the underlying condition. Electrical nerve stimulation for the treatment of intractable cases of this condition has been used for a number of years. This could be achieved through the use of implanted peri-urethral electrodes, electrodes placed in various genital or anal areas or through transcutaneous approaches. The results from clinical trials are good but not outstanding, with resolution rates between 30 and 70%.

### 4.6.3 Ureteral patency

Ureteral obstruction occurs relatively frequently, caused by calcifications, strictures, tumors and other factors. There are several treatment modalities, one of the more commonly used being the insertion of a ureteral stent. These have some features in common with arterial stents but some differences. Ureteral stents are primarily temporary devices, usually left in place for a few months. Their presence tends to promote ureteral healing and reduce inflammation, mainly by achieving better alignment of the ureteral wall and reduction of extravasation around the ureter. They may be constructed of either polymers or metals. Polyurethane and silicone elastomers are the more commonly used polymers. Metallic stents can be balloon-expandable or self-expandable using a shape memory alloy. Many have geometrical features, including the so-called double pigtail form. These stents can give good performance over these few months, but there are complications. Urothelial hyperplasia is a significant problem with metal stents, the hyperplastic tissue reintroducing stenosis. Encrustation is also common, and the major limiting factor as far as the length of time the stent remains *in situ* is concerned. Migration of stents may occur, as can urinary stasis associated with zones of poor peristalsis. Irritation on voiding is common. Overall discomfort, pain and incontinence may occur in over 50% of cases.

### 4.6.4 Urinary incontinence

The inability to exert effective control over urination, known as urinary incontinence, is a major problem, both in males and, especially, in females. There are several causes of incontinence and it can be manifested in a number of ways. It is usual to consider separately stress urinary incontinence, urge incontinence, overflow incontinence and functional incontinence. Stress urinary incontinence refers to those situations where the patient may have control under most circumstances but leaks urine on coughing, sneezing, lifting objects, etc. Urge incontinence also involves periods of control but with a very rapid onset of the need to urinate, with little warning. Overflow incontinence is often associated with strictures, stones, prostate conditions, etc., where the bladder essentially remains full and has to overflow involuntarily at frequent intervals. Functional incontinence occurs when some physical or mental disability prevents the person from dealing with the need to urinate in a socially acceptable manner.

Stress urinary incontinence (SUI) is the more prevalent form, particularly in women. There are several treatment modalities, which have varied and somewhat unpredictable outcomes and are also controversial. It is common to differentiate different forms of SUI: if the incontinence is associated with raised intra-abdominal pressure (as measured through urodynamic testing) in the absence of bladder contraction, it is usually referred to as genuine stress urinary incontinence. Treatment can include modification of lifestyle and pelvic muscle training, some medications, including estrogen, vaginal pessaries and surgical intervention.

Two main forms of intervention are available, involving the use of bulking agents and the use of devices such as slings and tapes, with a minor usage of artificial urinary sphincters.

#### Bulking agents

Bulking agents are materials that are injected into the urethral submucosa in order to augment the tissue and improve urethral coaptation. The injection may be placed at the bladder neck or in the midurethral region. Injections can be made by transurethral or periurethral routes under local anesthesia. Several types of material have been or are being used, with variable effects. As elsewhere in the body where tissue augmentation is required, collagen injections provide one possibility, usually delivered periurethrally. Again, as elsewhere, antigenicity is possible and skin testing prior to injection is usually performed. Similarly, it is not durable and re-injection may be required a few years later. The most commonly used synthetic material to be used is polydimethylsiloxane, which can also be injected in patients with decreased uretheral mobility. These injections are well accepted without significant adverse effects but success rates in terms of curing SUI may only be between 40 and 80% at one year.

Other materials to be used include ethylene vinyl alcohol (which was not very successful), carbon spheres, hydroxyapatite and hyaluronic acid-based polymers. Some of these are difficult to inject and some induce granulomas, although most are more durable than the collagen. SUI is less common in men than women, although rates have increased because of rises in prostate cancer and the use of prostatectomy procedures. Collagen injections have been used but are widely considered to have only modest success rates.

### Slings and tapes

The surgical treatment of stress urinary incontinence is rapidly growing in frequency but is not without controversy. There are two principal procedures, albeit with several variations in each. The first is retropubic urethropexy, sometimes known as colposuspension. A small suprapubic incision is used to place permanent sutures at the bladder neck, which are then attached to either Cooper's ligament or the periosteum of the pubic symphysis. Successful outcomes are common, with 10-year cure rates of over 50%. The second procedure is the pubovaginal sling. These procedures are very popular. They involve the insertion of tissue (such as autologous or cadaver-derived fascia) or synthetic mesh to support the bladder neck or the mid-urethra. The principle here is to raise the urethra by the use of a type of sling in order to control urodynamic pressure, but without placing undue stress on uretheral tissues; one trend has involved techniques that minimize tension within the system. Different forms of sling and different insertion procedures have been developed. They can give good performance but complications may arise. Infection is a significant factor as is the potential erosion of the urethral tissues and migration of the sling. Most slings are made of polypropylene and there has been much inconclusive discussion as to the ideal form of the polypropylene filaments and mesh geometry. Similar types of device are being used in the treatment of pelvic organ prolapse, where complications can also occur, as discussed in Section 4.6.5.

### Artificial urinary sphincters

The artificial urinary sphincter has been in use, in males and females, for several decades, primarily for neurologically derived incontinence, or post-prostatectomy (but not irradiated) males. A typical device consists of a urethral cuff, a pressure regulating balloon and a control pump; most parts are made of silicone elastomer. The pump is implanted in the scrotum of males or labia of females. Under normal, resting conditions the pressure of the fluid in the balloon holds the cuff closed and the pump is activated by hand to allow urination. Success rates are high and complications mild. Similar types of device are sometimes used for the treatment of severe fecal incontinence.

## 4.6.5 Pelvic organ prolapse

This is a very distressing situation that affects a surprising number of females, especially those who have had children. In this, female genito-urinary organs, including the bladder, cervix, bowel and rectum, descend within the pelvis and either distort the vaginal wall or, worse, bulge outside of the vagina. Vaginal delivery of multiple children coupled with obesity and advancing age are prominent risk factors. Protrusion of the bladder is called a cystocele, of the rectum a rectocele, and of the bowel an enterocele.

Many women who have this condition are relatively symptomless but for many this problem has become a huge quality-of-life issue. There are no drugs available apart from hormones in a few cases. Physical therapy in the early stages, including exercise and electrical stimulation, may be used but surgery is usually inevitable. There are three major surgical methods for managing this condition, using transvaginal, transabdominal or laparascopic approaches. The general term used to describe vaginal wall repair is colporrhaphy or vaginal vault suspension. In concept the process is simple; the surgeon attempts to move the displaced tissue back to the correct anatomical site and then use suturing techniques to secure and support the tissue between the vagina and the relevant organ. If the condition is accompanied by incontinence, the sling procedures may be applied at the same time. Difficulties occur with maintaining the functional quality of the tissue in order to prevent a recurrence and also controlling the tendency to infection. Some surgeons use a mesh, quite similar to that used in hernia repair of mid-urethral incontinence procedures, although these do sometimes have their own complications. Obviously infection is a risk in view of the potentially contaminated vaginal area. Dyspareunia may also be an undesirable complication, this being painful intercourse. There may be general pain, possibly associated with the fibrotic reaction to the mesh and also there is a finite risk of "erosion" a poorly defined response of the host tissue leading to lack of integration of the mesh and possible migration.

### 4.6.6 Penile reconstruction

Penile insufficiency or absence is quite rare but may have devastating consequences. It may occur in patients who have bladder dysfunction and are required to empty their bladder by catheterization, which may result in long-term damage, or following trauma or surgical resection of a tumor. Such patients have a need for some type of penile reconstruction, often referred to as phalloplasty. Female-to-male transsexuals may also require this type of procedure. The most common approach is to use a free flap usually obtained from the radial forearm, supplemented by a penile implant that increases stiffness. Several different types of implant are available, the most common being an inflatable device, with a reservoir placed within the abdomen, a pump placed in the scrotum and a pair of cylinders placed within the penis, the patient operating the pump in order to transfer fluid to the cylinders in order to achieve an erection. Materials are, once again, reliant on silicone elastomers, with cylinders often having surface treatment to reduce infection risk and wear. Success rates vary, with failures being due to infection, erosion, leakage and dysfunction. Total failure rates may be 30–40% after a few years.

### 4.6.7 Testicular implants

Testicular absence may follow an undescended or ectopic testis and testicular loss may result from surgical removal (orchidectomy) of testes following torsion, infection or malignancy. In all of these situations, but especially in testicular loss in adult men, this absence may have significant psychological consequences, resulting in a need for testicular implants. A similar need may also exist in female-to-male transsexuals undergoing gender re-alignment surgery. Although several materials have been used for the construction of these implants, most commercial varieties used silicone gel or saline-filled silicone elastomers, just as with the silicone breast implants discussed below. Some have a suture loop for attachment to the scrotum to prevent migration. Few problems appear to arise with these implants.

## 4.7 Gastrointestinal system

There are several reasons why interventions within the gastrointestinal system that involve medical devices have been introduced in recent years. They may be grouped into a small number of objectives, which are attempts to treat obstructions, attempts to treat or obviate reflux, or to reverse flow of matter within this system, and attempts to prevent absorption of food products. Regenerative procedures may also be considered, and this is done in the next chapter.

### 4.7.1 Gastric bands

Here we are dealing with chronic obesity. Until recently there were few available remedies, primarily based on dietary, pharmacological or cognitive interventions, which were used with varying degrees of success. Major surgical procedures were introduced at the end of the twentieth century, known as gastroplasty or gastric bypass, which were reasonably successful but carried a high risk of morbidity or even mortality. Laparoscopic techniques with adjustable bands were introduced in the 1990s that were aimed at reducing these risks and are now used in tens of thousands of patients annually worldwide, usually for patients who have BMI (body mass index) of 30–40. This area is often referred to as bariatric surgery.

The device is an adjustable inflatable band of silicone elastomer that is inserted by minimally invasive methods towards the top of the stomach. It is wrapped around the stomach leaving a pouch at the top. This allows for only a small amount of food to enter the pouch, which gives the sensation that the stomach is full and the patient stops eating. As the patient loses weight, so the tightness of the band can be adjusted; the band has attached to it a silicone tube which runs to a sub-dermal site and saline may be injected or withdrawn from the tube in order to adjust the tension.

### 4.7.2 Anti-reflux devices

Gastroesophageal reflux disease is a condition that allows periodic reverse flow of food from the stomach into the esophagus. This can be distinctly uncomfortable but is not life-threatening. Pharmacological methods are available, principally using proton-pump inhibitor drugs, which are effective in many patients but do have their limitations. In recent years various minimally invasive techniques have been introduced to provide alternative therapies. One of these involves the use of a catheter, an inflatable balloon and electrodes to deliver radiofrequency energy to the lower part of the esophagus and the gastric cardia, which is that part of the stomach attached to the esophagus. Several systems use biomaterials that are placed within the tissue in order to modify the esophagastric junction and control the reflux. One uses ethylene-vinyl-alcohol copolymer that is injected into the muscle of the cardia, where it hydrates and expands in order to enlarge the tissue. Another uses hydrogel cylinders that are placed submucosally and they also hydrate and expand. Significant long-term studies are not yet available to allow effective analyses of safety and efficacy and some adverse effects have been reported.

### 4.7.3 Gastrointestinal stents

Tumors can occur at various points along the gastrointestinal system. The current treatment of choice for the palliation of malignant obstructions here is that of the endoscopic insertion of a stent, similar in principle to those used in the vascular system. For example, dysphagia, which is a difficulty in swallowing, is a frequent complication of esophageal tumors, and this may be treated with a self-expanding metal stent. Often the stent is covered with a silicone sheet in order to prevent the overgrowth of tumor tissue. Malignant obstruction of the gastric outlet is a critical problem causing a wide variety of complications and again a polymer-covered metal stent provides an alternative to surgical intervention. The bile duct also suffers from malignant obstruction. Many different designs of biliary metallic stent are available. These tend to have added features to enhance performance, including some polymer coverings but also drainage side holes in order to avoid obstruction of the intrahepatic side ducts.

## 4.8 Reproductive system

Materials and devices used within the reproductive system are intended for contraceptive purposes. For males, there has been some interest in implantable devices for reversible vasectomies. The so-called IVD, or intra vas device, has a pair of silicone plugs placed within the vas deferens, which block the passage of sperm, but which can be removed should the individual decide he wishes to have a family at a later date. These are not common. Far greater use of contraceptive medical devices is seen with women.

### 4.8.1 Intravaginal devices

These are devices that provide a local delivery of contraceptive steroids directly into the vagina and are usually referred to as contraceptive vaginal rings (CVRs). In general these do not require health care professional involvement for insertion and have the advantage of being easily applied and removed so that the individual can control the menstrual cycle by the timing of insertion and removal. One common ring is made of ethylene vinyl acetate polymer that delivers the estrogen ethinyl estradiol and the progestogen etonogestrel. Failure rates are very low, although there may be some cases of expulsion and foreign body sensation. Other devices are made of silicone elastomer that may be used to deliver, continuously, a variety of agents.

### 4.8.2 Intrauterine devices

Devices have been placed in the uterus for contraceptive purposes for many years; they are referred to as intrauterine devices (IUDs). The first types of IUD were usually ring or coiled shaped and relied solely on the presence of the foreign material in the uterus to provide an intrauterine environment that inhibited conception. At a later stage it was realized that such devices could incorporate some active agent in order to enhance this effect. Two main forms of medicated intrauterine systems (IUS) are those that involve the release of copper or the delivery of a progestin. Copper IUDs are typically made of a T-shaped polymer frame (polyethylene and/or ethylene vinyl acetate, usually containing barium sulfate for ultrasound or radiological detection) to which is attached a fine copper wire. The devices usually have between 200 and 400 $mm^2$ Cu surface. The Cu is released continuously, with effective lifetimes of over 10 years, and shows both spermicidal activity and effects on the endometrium that inhibit conception. These devices are very effective and widely used. There are side effects, which include bleeding and pain, but could involve more serious consequences such as perforation and pelvic infection.

Progestin-releasing devices include those that deliver progesterone or levonorgestrel, again usually employing ethylene vinyl acetate as the polymeric carrier. The evidence suggests that these devices function by preventing fertilization rather than implantation. They perform very well, providing slightly better results than for copper IUDs but generally for a shorter period of time, i.e., 5–10 years.

## 4.9 Dentistry, maxillofacial and craniofacial tissues

As noted in Chapter 1, this book does not address the materials aspects of restorative/conservative/ prosthodontic/periodontic/orthodontic dentistry. There are a few situations within the broader fields of maxillofacial or craniofacial surgery where biomaterials are used and which have to be discussed here. These primarily include oral cancer, craniofacial reconstruction and dental implants.

### 4.9.1 Oral cancer

Oral cancer is the sixth most common cancer, although the prevalence varies considerably with variations in both genetics and lifestyle. In some parts of the world, the prevalence of oral cancer is high because of chewing tobacco products or other carcinogenic plants. Surgery in conjunction with radiotherapy is the most widely used treatment strategy, chemotherapy not being particularly effective. Oral cancers include those of the tongue, lips, floor of the mouth, salivary glands, pharynx, larynx and the bone of maxilla and mandible. In the majority of cases involving the soft tissues, reconstructive procedures after surgery involve repositioning of skin and possibly muscle, for example radial forearm free flap, but not the implantation of any biomaterial. Loss of bone through surgery is particularly devastating, potentially affecting appearance, speech, eating, drinking, etc. If surgery is required to remove a segment of the mandible, free flap reconstruction using bone from the fibula, the iliac crest, the scapula or the radial forearm may be used. It will often be necessary to harvest bone, muscle and skin for these procedures. This grafting of the patient's own tissue is usually very successful. Various implantable devices have been used but the free flaps give better outcomes unless the extent of the surgical removal of tissue is so large that biomaterials have to be incorporated in order to give shape and structure. This is rare, provided the cancer is detected early enough. Of course these free flaps are able to reconstruct the bone but not the dentition, and, once the bone has healed, most surgeons will use dental implants to complete the reconstruction, as discussed in the next section.

### 4.9.2 Dental implants

Dental implants are devices that are secured to oral tissues in order that they may support artificial teeth. They may be used in cases of single tooth loss as an alternative to a conventional bridge or in situations where there is multiple tooth loss, arising, for example, from the effects of periodontal disease. The history of dental implants goes back many years, although for several decades they were not very successful, largely arising from the difficulty of maintaining a device that was secured to submucosa tissue but which penetrated the mucosa to support a tooth intra-orally. Control of infection and, more importantly, control of the soft tissue at the point of passage through the mucosa were difficult challenges. In cases of multiple tooth loss, attempts were made to create major metallic structures that could be placed subperiosteally in maxilla or mandible, but a trend towards smaller individual implants became very popular in the 1980s. Several crucial factors emerged that produced considerable success with these so-called endosseous dental implants. The first was the need for meticulous preparation of the underlying bone, primarily through drilling and taping a hole, such that an implant could be placed and secured in this prepared hole. The second was the need to use a material with the highest degree of biocompatibility, and third the need to protect the implant–bone interface from mechanical loading during the healing phase. The latter characteristic was achieved by using a two-stage procedure in which the implant root was first placed into the prepared hole and the mucosa closed over it so that it was submerged and not subjected to any occlusal forces. This could be maintained for the length of time needed for bone apposition to the implant surface, following which the mucosa would be opened up and a post placed on the root, to which could be attached a single tooth crown or a multiple tooth structure. This process of bone adaptation to the biomaterial surface was called osseointegration, as discussed in Chapter 3; the material used for these implants was titanium.

In recent years much attention has been focused on the kinetics of osseointegration and the search for optimal surfaces that would lead to faster anchorage of the bone and the use of minimal time between the procedures. There is considerable debate as to this optimal nature of the implant surface, with both the surface chemistry and the topography appearing to play a role. Although bioactive surfaces on dental implants have been used, the recent trend has been away from these and back to titanium. Surfaces of the titanium may be sand- or grit-blasted, anodized or etched. It is widely agreed that some degree of roughness provided by these types of technique give better results than smooth surfaces. Some systems are using zirconia as the implant material, partly because of better aesthetics at the gingival margin. With respect to the need for a two-stage procedure, some dental surgeons believe that, all other things being equal, better results are obtained this way, but the trend is moving towards the immediate implant loading technique, as much influenced by patient demand as anything else. Both methods should give adequate results provided normal clinical practices are followed.

### 4.9.3 Craniofacial reconstruction

Reconstruction of craniofacial bone may be required following trauma or tumor surgery or in the case of congenital defects. Sometimes, the same type of free or pedicle bone flaps discussed in Section 4.9.1 may be used, but it is common to use either synthetic biomaterials or bone grafts for some of these procedures. Bone grafts may be either autologous (usually calvarian) or allogeneic (again usually banked calvarian). Biomaterials used for this reconstruction include polymers such as polymethylmethacrylate and porous polyethylene, and titanium mesh. Hydroxyapatite is commonly used, mostly as porous blocks or injectable cement. Quite often customized mixtures of materials, such as titanium mesh with hydroxyapatite or autologous bone, are used. Overall good results may be achieved, with minor rates of complications, including infections (more common with biomaterials compared to autologous bone), erosion, and too-rapid resorption of bone grafts. A great deal depends on the location of the reconstruction and its extent. For example, if the defect communicates with the frontal sinuses, better success and lower infection rates may be achieved with autologous bone. There is an increasing use of stimulatory factors such as BMPs and TGF-β to improve success rates.

A major challenge with craniofacial reconstruction involves the need to reproduce the required anatomical shape with good fidelity and accuracy. The ability to do this has been enhanced by the introduction of imaging technology coupled with personalized implant design and construction. For example, improvements in computerized tomography have allowed these techniques to be used for the detailed imaging of craniofacial tissues and software may be used to translate this information into stereolithography procedures for implant design.

## 4.10 General soft tissue repair, replacement and augmentation

### 4.10.1 Hernia

A hernia occurs when inner layers of abdominal muscle become weakened. The lining of the abdomen then bulges out into a small sac, and part of the intestine or abdominal tissue may enter the sac. Disturbances to collagen metabolism contribute to the etiology of hernias, which are also affected by genetic factors and lifestyle factors such as smoking. Hernias occur most commonly in the groin (inguinal hernia), the navel (umbilical hernia), and at the incision site of a previous surgery. A hiatal, or diaphragmatic, hernia occurs when the lower part of the esophagus and a portion of the stomach slide up through the esophageal hiatus.

Although individuals with hernias may remain relatively free from symptoms, a hernia can cause severe pain and other potentially serious problems such as infection and bowel obstruction. A hernia does not resolve on its own and surgery is the only way to resolve these serious cases. This may be achieved either by open surgery or laparoscopic procedures. The hernia sac is opened and the intestine or other tissue is placed back inside the abdomen. The weakened area is repaired and reinforced with a synthetic mesh and/or sutures.

The meshes currently used are derived from four polymers, polypropylene, polyethylene terephthalate, polytetrafluoroethylene and polyvinylidene fluoride. Polypropylene tends to be preferred although the mesh is stiffer than others. The meshes also vary in their characteristics. The material can be used as either monofilaments or multifilaments, and the pore size varies; generally large pore size is preferred, and some products have anisotropic distribution of pores.

The use of these meshes is not without complications and there may be both adverse acute and chronic events. Recurrence of the hernia is one of the more common adverse events. Infection is a major problem, with an incidence that varies with the site of the hernia and, whilst often occurring with a frequency of less than 1–2%, can be over 10%. If the mesh is placed in the peritoneal cavity, as with laparoscopic repair techniques, there is a finite risk of adhesions, which can be a serious complication. Chronic pain may occur in significant numbers of patients, possibly associated with the inflammatory response.

## 4.10.2 Breast reconstruction

It is fitting to conclude this chapter on implantable devices with a discussion of one of the most controversial of all devices, the breast implants. We shall return to this topic in Chapter 9 since it has been the subject of immense litigation battles, but we will confine ourselves here to the scientific and clinical aspects.

Breast implants are used in women for one of two reasons, either to reconstruct the breast after mastectomy (so-called reconstructive mammoplasty procedures) or to enlarge breasts in otherwise healthy and normal women (augmentation procedures). Although these situations have quite different psychological implications and, indeed, quite different anatomical and physiological conditions, the requirements and performance of the devices can be considered to be equivalent. At first glance, the functional requirements may seem quite trivial: the implant has to replicate some of the features of the tissues of the female breast. We are not trying to replicate the mammary function, just the space filling and consistency of the breast tissue. However, this is not so easy as it seems since nature has provided within breast tissue a characteristic that is extremely difficult to replicate in a synthetic material. This tissue is a composite of fatty material, muscle, glands and channels, which should have a balance of softness, firmness and compliance. There are no synthetic materials that intrinsically have these qualities, especially if we add the rider that such a material has to provide both shape and volume.

There are some materials that come close, and these are the hydrogels (or other gels) that are discussed in several places in this book. Even the softest, most compliant of synthetic elastomers, which could give both volume and shape, are of no value since they are still too rigid. The value of a hydrogel is that it could give the required compliance and volume. The difficulty is that, by itself, it could not give the required shape as a hydrogel, when unconstrained, would flow into surrounding tissues.

The solution to this dilemma emerged in the 1970s, at a time when augmentation mammoplasty was becoming popular, especially in the USA, and this involved encapsulating a gel inside an elastomeric shell. At that time, the preferred medical elastomer was polydimethylsiloxane, otherwise known as silicone elastomer or silicone rubber. Fortuitously this polymer was also available in the form of a gel and so the first significant breast implants were made from a silicone elastomer shell that contained a silicone gel. These implants became very popular and certainly met the expectations of patients and surgeons. There were many variations in design details, mostly concerned with the thickness of the silicone shell and the viscosity of the gel, both of which affected the overall deformability of the implant. This was important both for the ease of insertion and subsequent mechanical compatibility with the breast tissue. There were different opportunities for placement site, essentially reflecting the depth within the tissue, and this did depend to a significant extent on whether the implant was being used for augmentation or reconstruction after major mastectomy. There were also variations in manufacturing methods since it was necessary to fill the shell with fluid and then seal this in place with an adhesive patch. An example of a silicone gel breast implant is shown in Figure 4.34.

In spite of the apparent success, a few problems started to emerge. The first was that in a significant number of cases, the fibrous capsule that formed around the implant became thicker and thicker, and as this thickness increased, so the capsule contracted. This so-called capsular contraction, or constrictive fibrosis was troublesome for many patients since it could cause both distortion and pain. Since around 20% of patients suffered some degree of contraction, this became a really significant issue. The second problem was that it became obvious to anyone who handled a prosthesis that traces of the gel could be felt on the outer surface. This phenomenon became known as gel bleed. The significance was not clear, but it was apparent that some of the silicone gel could be released into the environment of the implant, i.e., the women's breast. Of greater concern

**Figure 4.34** Breast implant.

to many people was the fact that occasionally the silicone shell could rupture, potentially releasing much larger amounts of gel. The combination of these potentially clinically relevant problems led manufacturers to make many design modifications intended to ameliorate the effects.

We can summarize here the major technical issues, but leave most of this discussion until Chapter 9 since the time-course of breast implants has been played out in litigation courtrooms rather than in scientific circles.

Although very few in this industry acknowledged it, the reason for the constrictive fibrosis was relative movement between these implants (which were amongst the largest of all implantable devices) and the surrounding tissue. In other words, this was due to mechanotransduction and the mechanical stimulation of fibroblasts, consistent with the discussions of biocompatibility pathways of the previous chapter. There was no way of predicting susceptibility to this condition.

It was appreciated by some people that this problem could be ameliorated if the movement of the implant was minimized, and in some devices this was achieved by covering the silicone elastomer with a porous layer. A type of polyurethane was chosen for this and it did allow some degree of tissue ingrowth and prosthesis stabilization. Unfortunately it was not the optimal form of polyurethane and it experienced degradation; this was not a clinical problem since the stabilization was achieved in the early phases, but it became part of the litigation process since some degradation products of that type of isocyanate-based polyurethane were known to be potential carcinogens. Naturally this fact became very significant, since there was an extreme irony if the implants used for post-mastectomy reconstruction could be shown to be carcinogenic themselves.

The gel bleed issue caused manufacturers to adjust wall thickness of the silicone shell in attempts to produce "low bleed" devices. This was not particularly effective since bleed per se was never shown to be a clinical problem and since the diffusion process, with low molecular weight components of the gel diffusing through the shell, would result in an equilibrium position, with a hydrophobic gel-rich outer layer effectively closing down the concentration gradient; long-term release of silicone components in intact prostheses could never be demonstrated, in spite of sophisticated systemic studies.

Rupture was an important issue. It was not caused by any decrease in mechanical properties of the silicone shell over time, as claimed by some, but was a random event related to transient excursions of mechanical stress above safe levels associated with everyday activities. The fact remained that silicone elastomers do have a low tear strength and are notch sensitive, so that a finite rate of rupture is inevitable – damage to the device on implantation was a major consideration since instrument-induced nicks in the shell could eventually propagate. Manufacturers also tried to make thicker, tear-resistant shells, but these did not give the consistency that was required.

The result of all of these individual problems, coupled with the fact that the implants were used in many women, numbered in the millions in the USA alone, and the sensitivity to commercial opportunities making profits out of the female breast, led to massive controversy and regulatory opprobrium, and this part of the industry was essentially closed down. There were alternatives but none so functionally relevant as the silicone gel–silicone shell devices. These alternatives, and the consequences of their introduction into this controversial market are discussed in Chapter 9.

## 4.11 Commentary on implantable devices

In this chapter, a very wide range of clinical applications has been described, covering many medical conditions and different forms of technology. It is over 50 years since implantable devices were first used seriously and now they are provided for millions of patients on a global basis. In many of these situations, success rates are very high, perhaps over 90% after 20 years in patients. It is worthwhile reflecting on these achievements and the lessons learnt.

- In most situations, the selection of materials for long-term implantable devices has followed an increasingly narrow pathway towards a small group of well-characterized materials that can meet most of the required specifications. We can include titanium alloys, cobalt-based alloys, platinum group metals, shape memory nickel–titanium, alumina, various forms of carbon, some forms of calcium phosphate, polyethylene, expanded PTFE, PEEK, polyester textiles, silicone elastomers, some acrylics, a few polyurethanes and polypropylene in this list. In many, if not most, situations, it is unnecessary and may be unwise to look further.
- There are some situations where different properties are required, usually related to the need to be compliant with imaging modalities and further developments are needed here.
- In spite of the generally good success rates there are still several unresolved problems, largely related to uncertainties in the host tissue response, as we discussed in the previous chapter. These include wear debris related osteolysis, the response of endothelium to intravascular devices, the role of innate and acquired immunity in biocompatibility and the role of released active biomolecules. There are some clinical conditions where the issues of the biocompatibility of the materials that are used are not fully understood and success rates should be higher.
- In many of these clinical areas, success rates with the "average" patient are so high that clinicians are tempted to use devices in more and more difficult situations. This is laudable since innovations in medical technologies should be utilized in as wide a spectrum of patients as is possible and reasonable, but only up to a point. Contra-indications, in relation to co-morbidities, obesity, high blood pressure, psychiatric state and so on should always be considered.
- In the context of the above comment, the role of patient and clinical skills variables discussed in the previous chapter are especially important.

# Summary and learning objectives

1. This chapter has covered permanently implanted devices, short-term implants, invasive but removable devices and artificial organs and assist devices that are attached to the body. The description of these devices and the materials used in their construction covers the medical conditions themselves, the objectives and purposes of the devices and the clinical outcomes. We should always place these uses of medical technology in the context of alternative treatment options; the availability of medical devices does not imply that they

should be used as a first-line treatment and full consideration should be given to overall performance and potential complications.

2. In orthopedics, considerable success can be achieved with the replacement of many of the joints of the body, including hips, knees, ankles, elbows, shoulders and some joints of the fingers and toes. Some issues are still unresolved, including the eventual loosening of prostheses associated with osteolysis and immunological responses. Less success is achieved in tendon and ligament repair because of the difficult biomechanical environment and poorer intrinsic healing capability. In the spine, it is possible to correct deformities and treatment of degenerating discs can be effected by either fusion or replacement techniques. Bone fractures can be assisted in their repair by several types of device. Some areas remain problematic, especially spinal cord injury and cartilage lesions; these areas will necessitate regeneration rather than replacement methods.
3. Within the cardiovascular system, defective heart valves may be replaced by either mechanical or bioprosthetic devices, where potential adverse events such as thrombo-embolic complications and calcification have to be carefully managed. Attention has been turning towards less invasive methods of implantation. Although considerable success can be achieved, the treatment of diseased arteries still remains a challenge, primarily because of a lack of detailed knowledge of healing processes of the vascular endothelium. The use of angioplasty and intravascular stents has had a major impact here. Arrhythmias may be effectively treated by implanted pacemakers or defibrillators. Congenital heart defects may be surgically corrected using a variety of devices, but the treatment of myocardial infarction is a major issue and, again, will require developments in regenerative medicine rather than implanted devices.
4. Treatment of failing major organs has come a long way, and several technologies are available but this remains a daunting task. Kidney dialysis is routine, and lung oxygenation can be achieved in the short term. Replacement of liver and pancreas function is still very difficult.
5. Within sensory organs, very effective methods are available to facilitate the transmission of sound and light, through various forms of hearing devices and lenses and these are in widespread use. At the neurological end of these systems, the cochlear implant can provide hearing for those who are profoundly deaf and some early stage developments can be seen with implantable devices in the retina and artificial vision. These are very interesting areas that require interfacing biomaterials with electrophysiology.
6. Miscellaneous devices are used within other systems of the body, including the central nervous, genito-urinary, gastrointestinal and reproductive systems. These applications are rather diverse and have to be described on an individual basis. They often address significant quality-of-life issues such as incontinence and impotence; as such the specifications are not so easy to determine and success rates are more varied and even difficult to quantify. Similar arguments apply to soft tissue repair and augmentation, such as hernia repair, pelvic floor reconstruction and surgery of the female breast.
7. In general terms, although success rates do vary, massive progress has been made over the last half-century, and indeed these successes with implantable devices rank as some of the best medical achievements during this time. From a biomaterials perspective, experience has shown that we can concentrate on a small number of well-characterized materials for the construction of these devices, with only the occasional need to seek alternatives, and then usually for some

specialized property. It is necessary to consider the role of clinical skill and patient compliance in evaluating outcomes, and the use of these devices, however successful, must match patient needs in the light of alternative, and often less invasive, therapies.

## Questions

1. Discuss the various mechanisms that are available to attach joint replacement prostheses to bone.
2. Compare the advantages and disadvantages of metal-on-metal, metal-on-plastic and ceramic-on-ceramic combinations for total hip replacements.
3. Give the reasons why both cartilage and ligaments are difficult to repair or replace.
4. Compare and contrast the use of metallic and biodegradable polymeric internal bone fracture fixation devices.
5. Discuss the various technologies available for producing spinal fusion.
6. Describe the materials that are currently available for bioprosthetic heart valves. What are their main limitations?
7. The performance of vascular grafts for arteries in the abdomen and legs depends on the response of the tissue to the prostheses. Discuss this response in relation to the anastomotic and mid-graft regions.
8. What are the main similarities between implanted pacemakers and defibrillators and what are the main differences?
9. Discuss the biomaterial specifications for hemodialysis and oxygenation membranes.
10. Describe the materials that may be used for contact lenses and intraocular lenses.
11. Discuss the mechanisms of sound transmission in the ear and explain how this is assisted by devices in the middle ear.
12. Describe the condition of Parkinson's disease and explain how a deep brain stimulator can assist patients.
13. Discuss the factors that lead to urinary incontinence and the types of therapy that are available to treat this condition.
14. If you had to choose one metallic, one ceramic and one polymeric material as being the most versatile biomaterials, which ones would you choose and why?

# Recommended reading

## Orthopedics


Bohsall K I, Wirth M A, Rockwood C A, Complications of total shoulder arthroplasty, *Journal of Bone and Joint Surgery* 2006;88A:2279–90. *Discussion of clinical experiences and outcomes of total shoulder prostheses.*

Browne J A, Bechtold C D, Berry D J, Hanssen A D, Lewallen D G, Failed metal-on-metal hip arthroplasties, *Clinical Orthopedics and Related Research* 2010;468:2313–20. *Analysis of failure mechanisms of metal-on-metal hips.*

Chen J, Xu J, Wang A, Zheng M, Scaffolds for tendon and ligament repair: review of the efficacy of commercial products, *Expert Reviews of Medical Devices* 2009;6:61–73. *A review of devices and relevant clinical trials associated with tendon and ligament repair.*

Gamradt S C, Wang J C, Lumbar disc arthroplasty, *The Spine Journal* 2005;5:95–103. *Discussion of features that control the success of total disc arthroplasty.*

Goins M L, Wimberley D W, Yuan P S, Fitzhenry L N, Vaccaro A R, Nucleus pulposis replacement: an emerging technology, *The Spine Journal* 2005;5:317S–24S. *Description of surgical methods to relieve recalcitrant symptoms of disc degeneration using partial removal of the disc.*

Hailer N P, Garellick G, Karrholm J, Uncemented and cemented primary total hip arthroplasty in the Swedish Hip Arthroplasty Register: evaluation of 170,413 operations, *Acta Orthopaedica* 2010;81:34–41. *Major review of the outcomes of total hip replacement procedures.*

Heisel C, Kinkel S, Bernd L, Ewerbeck V, Megaprostheses for the treatment of malignant bone tumours of the lower limb, *International Orthopedics* 2006;30:452–7. *Discussion of prostheses used in reconstruction of the lower limb after cancer surgery.*

Hosman A H, Mason R B, Hobbs T, Rothwell A G, A New Zealand national joint registry review of 202 total ankle replacements followed for 6 years, *Acta Orthopaedica* 2007;78:584–91. *Detailed review of outcomes of total ankle replacement.*

Lee K, Goodman S B, Current state and future of joint replacements in the hip and knee, *Expert Reviews in Medical Devices* 2008;5:383–93. *Review of the performance of hip and knee replacement surgery.*

Morshed S, Bozic K J, Ries M D, Malchau H, Colford J M, Comparison of cemented and uncemented fixation of total hip replacement, *Acta Orthopaedica* 2007;78:315–26. *Detailed analysis of the outcomes of cemented and uncemented hip procedures.*

Prosser G H, Yates P J, Wood D A, *et al.*, Outcome of primary resurfacing hip replacement: evaluation of risk factors for early revision in 1203 replacements from the Australian Joint Registry, *Acta Orthopaedica* 2010;81:66–71. *Review of risk factors in metal-on-metal hip procedures as revealed by analysis of registry data.*

Rahaman M N, Yao A, Bal B S, Garino J P, Ries M D, Ceramics for prosthetic hip and knee replacement, *Journal of the American Ceramics Society.* 2007;40:1965–88. *Review of the use of ceramics in total joint replacement.*

Van den Eerenbeemt K D, Ostelo R W, van Royen B J, Peul W C, van Tulder M W, Total disc replacement surgery for symptomatic degenerative lumbar disc disease: a systematic review of the literature, *European Spine Journal* 2010;19:1262–80. *An evaluation of the effectiveness and safety of total disc replacement surgery, with comparison to spinal fusion.*

Ziran B H, Smith W R, Anglen J O, Tournetall P, External Fixation: how to make it work, *Journal of Bone and Joint Surgery* 2007;89A:1620–32. *Instructional lecture on the principles and practices of external fixation.*

Zywiel M G, Sayeed S A, Johnson A J, Schmalzried T P, Mont M A, Survival of hard-on-hard bearings in total hip arthroplasty, *Clinical Orthopedics and Related Research* 2011;469:1536–46. *A systematic review of the performance of metal-on-metal and ceramic-on-ceramic hip replacements.*


## The cardiovascular system


Borek P P, Wilkoff B L, Pacemaker and ICD leads: strategies for long-term management, *Journal of Interventional Cardiac Electrophysiology* 2008;23:59–72. *Review of practical aspects of managing patients with implanted arrhythmia control devices.*

Carabello B A, Mitral valve repair in the treatment of mitral regurgitation, *Current Treatment Options in Cardiovascular Medicine* 2009;11:419–25. *Describes techniques and outcomes for mitral regurgitation, including repair methods.*

Coeytaux R R, Williams J W, Gray R N, Wang A, Percutaneous heart valve replacement for aortic stenosis: state of the evidence, *Annals of Internal Medicine* 2010;153:314–24. *Discussion of the outcomes of clinical trials on percutaneous heart valve replacement.*

Epstein A E, Benefits of the implantable cardioverter-defibrillator, *Journal of the American College of Cardiology* 2008;52:1122–7. *Essay on the advantages and disadvantages of the ICD.*

Jorapur V, Cano-Gomez A, Condie C A, Should saphenous vein grafts be the conduits of last resort for coronary artery by-pass surgery, *Cardiology in Review* 2009;17:235–42. *A critical analysis of the respective roles of saphenous veins and internal mammary arteries in therapy for coronary artery disease.*

Kapadia M R, Popowich D A, Kibbe M R, Modified prosthetic vascular conduits, *Circulation* 2008;117:1873–82. *A review of approaches to modify vascular graft materials intended to improve patency rates.*

Lip G Y H, Tse H-F, Management of atrial fibrillation, *The Lancet* 2007;370:604–18. *An overview of pharmaceutical and non-pharmaceutical approaches to atrial fibrillation.*

Mahadevan V S, Gomperts N, Haberer K, *et al.*, Transcatheter closure of atrial septal defects with multiple devices in adults: procedure and clinical outcomes, *International Journal of Cardiology* 2009;133:359–63. *Describes the transcatheter technology for the closure of atrial septal defects.*

Pibarot P, Dumesnil J G, Prosthetic heart valves: selection of the optimal prosthesis and long term management, *Circulation* 2009;119:1034–49. *Review of criteria for heart valve prosthesis selection.*

Rahimtoola S H, Choice of prosthetic heart valve in adults, *Journal of the American College of Cardiology* 2010;55:2413–26. *Consideration of published data on life-time performance of mechanical and bioprosthetic valves and presentation of an algorithm for valve choice.*

Rayt H S, Sutton A J, London N J M, Sayers R D, Bown M J, A systematic review and meta-analysis of endovascular repair for ruptured abdominal aortic aneurysm, *European Journal of Vascular and Endovascular Surgery.* 2008;36:536–44. *Review of the literature dealing with outcomes of abdominal aortic aneurysm repair.*

Schoder M, Lammer J, Czerny M, Endovascular aortic arch repair: hopes and certainties, *European Journal of Vascular and Endovascular Surgery* 2009; 38:255–61. *Discusses the devices and techniques used in the treatment of aneurysms and dissections of the aortic arch and difficulties with the complex geometry of the area when implanting stent grafts.*

Schoen F J, Evolving concepts of cardiac valve dynamics: the continuum of development, functional structure, pathobiology and tissue engineering, *Circulation* 2008;118:1864–80. *Discussion of the role of valve dynamics on repair and regeneration philosophies.*

Schoen F J, Levy R J, Tissue heart valves: current challenges and future research perspectives, *Journal of Biomedical Materials Research* 1999;47:439–65. *Detailed perspective on the future of tissue heart valves – a little old but still valid.*

Schoen F J, Levy R J, Calcification of tissue heart valve substitutes: progress towards understanding and prevention. *Annals of Thoracic Surgery* 2005;79:1072–80. *A description of the mechanisms of heart valve leaflet calcification.*

Tung R, Zimetbaum P, Josephson M E, A critical appraisal of implantable cardioverter-defibrillator therapy for the prevention of sudden cardiac death, *Journal of the American College of Cardiology* 2008;52:1111–21. *Analysis of the advantages and disadvantages of the ICD.*

Wilson S R, Givertz M M, Stewart G C, Mudge G H, Ventricular assist devices, *Journal of the American College of Cardiology* 2009;54:1647–59. *A review of the devices available for mechanical assistance of the failing heart, their performance and the management of patients.*

Yasuda S, Shimokawa H, Acute myocardial infarction: the enduring challenge for cardiac protection and survival, *Circulation Journal* 2009;73:2000–8. *Essay on the therapies available for the treatment of myocardial infarction.*

Zarins C K, Arko F R, Crabtree T, *et al.*, Explant analysis of AneuRx stent grafts: relationship between structural findings and clinical outcome, *Journal of Vascular Surgery* 2004;40:1–11. *Discussion of the performance of stent grafts and their dependence on materials and designs.*

## The major organs

Cengiz E, Sherr JL, Weinzimer SA, Tamborlane WV, New generation diabetes management: glucose sensor augmented insulin pump therapy, *Expert Review of Medical Devices* 2011;8:449–58. *Perspective on the future technologies for insulin pumps.*

Humes HD, Fissell WH, Tiranathanagul K, The future of hemodialysis membranes, *Kidney International* 2006;69:1115–19. *Essay of the future developments for hemodialysis.*

## Eyes and ears

Backous DD, Duke W, Implantable middle ear hearing devices: current state of technology and market challenges, *Current Opinion in Otolaryngology and Head and Neck Surgery* 2006;14:314–18. *The technology of devices in the middle ear.*

McGlinchey SM, McCoy CP, Gorman SP, Jones DS, Key biological issues on contact lens development, *Expert Review in Medical Devices* 2008;5(5): 581–90. *Summary of key developments in contact lenses.*

Rao K, Ahmed I, Blake DA, Ayyala RS, New devices in glaucoma surgery, *Expert Review in Ophthalmology.* 2009;4(5):491–504. *Review of developments of therapies for glaucoma, especially implantable devices.*

Renner G. Lane RV. Auricular reconstruction: an update, *Current Opinion in Otolaryngology and Head and Neck Surgery.* 2004;12:277–80. *Review of techniques for ear reconstruction.*

Walter P, Implants for artificial vision, *Expert Review in Ophthalmology.* 2009;4(5):515–23. *Review of the technologies under development for artificial vision.*

Werner L, Biocompatibility of intraocular lens materials, *Current Opinions in Ophthalmology* 2008; 19(1):41–9. *Review of the factors that control the host response to intraocular lens materials.*

Wilson BS, Dorman MF, Cochlear implants: current designs and future possibilities, *Journal of Rehabilitation and Restorative Development.* 2008;45 (5):695–731. *Extensive review of the technology of cochlear implants.*

Yung M, Vowler SL, Long term results in ossiculoplasty; an analysis of prognostic factors, *Otology Neurology* 2006;27:874–81. *A review of the prognosis of treatment modalities in middle ear surgery.*

## The central nervous system

Collins KL, Lehmann EM, Patil PG, Deep brain stimulation for movement disorders, *Neurobiology of Disease* 2010;38:338–45. *Description of indications for deep brain stimulation, the technology and the outcomes.*

Myllymaa S, Myllymaa K, Lappalainen R, Flexible implantable thin film neural electrodes, *Recent Advances in Biomedical Engineering* 2009:Ch.9:165–89. *Extensive review of microfabrication techniques for recording and stimulating neural electrodes.*

Zhong Y, Bellamkonda RV, Biomaterials for the central nervous system, *Journal of the Royal Society Interface* 2008;5:957–75. *Review of the requirements for materials placed within the brain.*

## The genito-urinary system

Barroso U, Lordelo P, Electrical nerve stimulation for overactive bladder in children, *Nature Reviews Urology* 2011;8:402–7. *Discussion of the use of electrical nerve stimulation in children with a focus on mechanisms of action and clinical results.*

Bodiwala D, Summerton DJ, Terry TR, Testicular prostheses: development and modern usage, *Annals Royal College of Surgeons of England* 2007;89:349–53. *Review of controversies concerning biomaterials used for testicular prostheses.*

Kotb AF, Campeau L, Corcos J, Urethral bulking agents: techniques and outcomes, *Current Urology Reports* 2009;10:396–400. *Discussion of bulking agents for women suffering with stress urinary incontinence.*

Liatsikos E, Kallidonis P, Stolzenburg J-U, Karnabatidis D, Ureteral stents: past, present and future. *Expert Reviews in Medical Devices* 2009;6:313–24. *Discussion of materials most commonly used for ureteral stents and their complications.*

McIntyre M, Goudeelocke C, Rovner ES, An update on surgery for pelvic organ prolapse, *Current Opinion in Urology* 2010;20:490–4. *A review of the surgical management of pelvic organ prolapse with a commentary of the benefits and limitations of current techniques.*

Norton P, Brubaker L, Urinary incontinence in women, *Lancet* 2006;367:57–67. *Seminar type review of causes and treatment of incontinence, including the use of biomaterials and implanted devices.*

Sandhu JS, Treatment options for male stress urinary incontinence, *Nature Reviews in Urology* 2010;7:222–8. *Review of options for the management of male stress urinary incontinence, including artificial sphincters, bulking agents and slings.*

## Gastrointestinal system

Favretti F, Ashton D, Busetto L, Segato G, De Luca M, The gastric band: first choice procedure for obesity surgery, *World Journal of Surgery* 2009; 33:2039–48. *Review of the results of laparoscopic adjustable gastric banding and comparison with other techniques of obesity surgery.*

## The reproductive system

Benagiano G, Gabelnick H, Farris M, Contraceptive devices: intravaginal and intrauterine delivery systems, *Expert Reviews of Medical Devices* 2008;5:639–54. *Discusses the availability and performance of vaginal rings and intrauterine devices.*

## Maxillofacial and craniofacial surgery

Ehrenfest D M D, Coelho P G, Kang B-S, Sul Y-T, Albrektsson T, Classification of osseointegrated implant surfaces: materials, chemistry and topography, *Trends in Biotechnology* 2009;28:198–206. *Discussion of the role of surface features in the mechanism of osseointegration.*

Neovius E, Engstrand T, Craniofacial reconstruction with bone and biomaterials: review over the last 11 years, *Journal of Plastic, Reconstructive and Aesthetic Surgery* 2010;63:1615–23. *Analysis of the performance of biomaterials and bone grafts in the reconstruction of craniofacial bones in congenital defects.*

Orentlicher G, Goldsmith D, Horowitz A, Applications of 3-dimensional virtual computerized tomography technology in oral and maxillofacial surgery, *Journal of Oral and Maxillofacial Surgery* 2010;68:1933–59. *Description of technologies for imaging of maxillofacial defects and their reconstruction.*

Shah J P, Gil Z, Current concepts in management of oral cancer – surgery, *Oral Oncology* 2009;45:394–401. *A review of the options for the treatment of malignancies of the oral cavity.*

## Plastic and reconstructive surgery

Binnebosel M, von Trotha K T, Jansen P L, *et al.*, Biocompatibility of prosthetic meshes in abdominal surgery, *Seminars in Immunopathology* 2011;33:235–43. *Review of the factors that control the biocompatibility and clinical performance of meshes used in hernia repair.*

Spear S L, Jespersen M R, Breast implants: saline or silicone? *Aesthetic Surgery Journal* 2010;30:557–70. *Editorial on the merits of different types of breast implant.*

# 5 Regenerative medicine and tissue engineering

In this chapter the major alternative non-pharmacological, non-synthetic replacement methods to treat disease and injury are introduced. These are various methods of regenerative medicine, which incorporate cell therapies, gene therapies and tissue engineering. The first two of these are mentioned briefly since they involve biomaterials only minimally. Tissue engineering concerns the regeneration of tissues or organs through the stimulation of cells so that they recapitulate the power that they have during tissue development, but which they substantially lose once tissues become mature. This recapitulation may be achieved by either molecular and mechanical signaling, or usually by both. The conventional tissue engineering paradigm involves harvesting cells and stimulating them *ex vivo*, and implanting the resulting construct into the patient at the appropriate time. Other techniques attempt to use the environment of the patient's own body as the location for regeneration. We discuss here the types of materials that are used in the so-called scaffolds and matrices that form the template of the regenerated tissue and the interactions between these materials and the target cells. Tissue engineering processes have not yet become used in routine clinical practice and cannot at this stage be defined and classified as clearly as with implantable medical devices, but we do discuss their scientific and clinical status in a wide variety of situations.

## 5.1 Introduction to regenerative medicine

There are two stark conclusions that arise from the discussion of implantable medical devices given in the last chapter. First, although they can give very good performance, they will always be limited to situations involving mechanical or physical functions and will not, by themselves, be able to deal with conditions that require biological solutions. Secondly, even in those situations where the performance is good, it will always be less than 100% effective because so many variables impose themselves on the process, especially in the context of biocompatibility and the influence of clinical skills and patient compliance. Outcomes are variable and often unpredictable.

Biomaterials scientists, while working to optimize the applications and performance of their materials, have to recognize the implicit limitations to the concepts and practice of the long-term replacement of tissues and organs by synthetic materials. Put quite simply, a natural hip joint does not look like, or behave like, a combination of metals, polymers and ceramics. Natural tissues are usually heterogeneous, anisotropic, hydrated living substances; synthetic materials are not.

One of the alternatives to tissue replacement by synthetic biomaterials/medical devices involves the use of transplanted tissues, which are living structures derived from a human donor site. These could be small pieces of tissue (for example, a cornea or piece of skin) or major organs (such as the heart, kidney or liver). They may be derived from a recently deceased donor, a living, matched donor such as a close relative, or from a site within the patient themselves, as with bone, skin or nerve grafts.

Such transplants or grafts have the major advantage of being living, natural structures. Disadvantages include the logistics of donor supply, which is extremely limited, and the possibility of immunological rejection. In view of these difficulties, transplantation remains a minor option.

In between these two extremes of synthetic non-living replacements and living transplants are some more recent options for therapies that rely on the use of various methods and tools that result in the regeneration of the patient's own tissue, either physically or functionally. This area of clinical medicine has been termed "regenerative medicine," the essence of which is summarized in Box 5.1. There are currently three main strands of regenerative medicine. The first, usually called "cell therapy," involves the use of groups of cells, derived from the patient or elsewhere, that can be injected or otherwise placed at the site of disease or injury in the expectation that they will facilitate the spontaneous regeneration of the required tissue. Stem cells are

Box 5.1 **The essence of regenerative medicine**

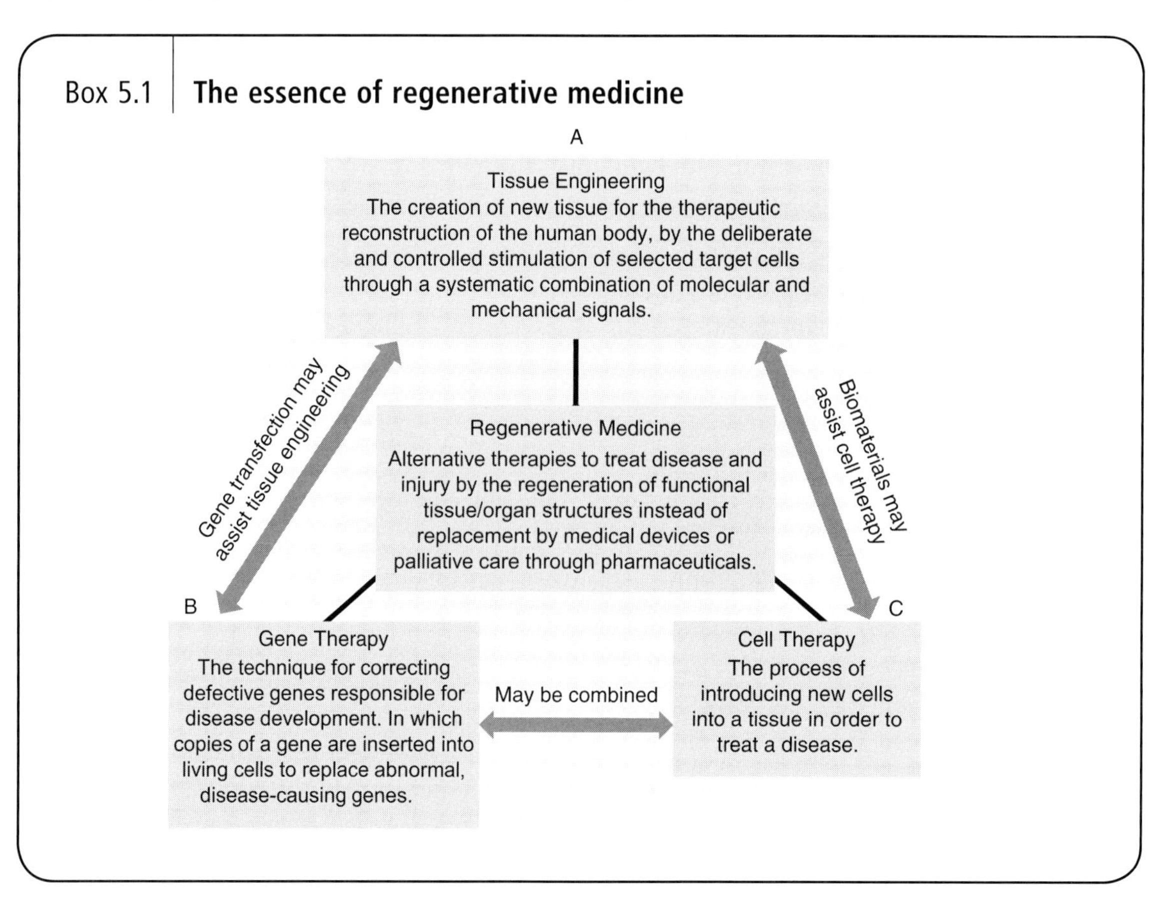

often discussed in this context. Potential applications include the use of dopamine-producing cells in the treatment of Parkinson's disease, chondrocytes in the treatment of cartilage lesions and cardiomyocytes in the treatment of myocardial infarction. At its conceptually simplest level, cell therapy does not involve conventional biomaterials. We therefore do not discuss cell therapies in any detail in this book, but they will be referred to where appropriate in order to place biomaterials-based tissue engineering processes in perspective.

The second strand is that of gene therapy where specific genes are inserted into specific cells in order to correct deficiencies in those cells. This may give rise to tissue regeneration or solve some other medical problem associated with the presence of a genetic defect. The process of gene transfection is not straightforward but can involve a variety of biomaterials, which are discussed in the following chapter.

The third strand is that of tissue engineering, which is now a major factor in the development of new biomaterials. I define tissue engineering as "the creation of new tissue for the therapeutic reconstruction of the human body, by the deliberate and controlled stimulation of selected target cells through a systematic combination of molecular and mechanical signals." The simple concept underpinning tissue engineering is based on the fact that, as noted in Chapter 1, adult humans have a very limited capacity to regenerate damaged tissue. This is largely limited to skin, bone, liver and peripheral nerve, and even then only under certain conditions and with incomplete results. Clearly, as a fetus, embryo or infant we have the capacity for new tissue growth but much of this capacity is lost once we reach maturity. It is possible, however, to switch these mechanisms back on under some circumstances, these processes involving delivery of the right signals to the affected tissue. The two most important forms of signals are those involving active biomolecules such as growth factors and those involving mechanical stimuli.

Molecular and mechanical signals do not directly imply the use of biomaterials, and indeed tissue engineering may in theory be carried out in the absence of biomaterials. Three factors, however, determine that biomaterials are most likely to be involved in tissue engineering processes. The first is that regenerated tissue needs form and structure, and injected cells by themselves are unlikely to provide this – a material template is very useful here, especially if it is biodegradable. Secondly, molecular signals are not easy to deliver with the appropriate spatial and temporal characteristics; a material template that contains and delivers such signals to the required cells in the requisite format would also be very beneficial. Thirdly, mechanical signals may be equally difficult to deliver, and again it is likely that better delivery could be sustained via a material construct.

In this chapter the essential principles of tissue engineering, which are summarized in Box 5.2, are discussed first. There are three main factors in tissue engineering products and processes, that is the cells which will express new tissue, the biomolecules that will provide the appropriate chemical signals for this to happen, and the biomaterial support. Each of these three components will be introduced before considering the progress that has been made in translating this triad of components into practical tissue engineering solutions to medical problems.

It will become obvious that this subject is far less advanced than the medical device technologies discussed in the previous chapter. Descriptions have to be less detailed, prognoses less certain and outlooks far more tentative. The specifications for the biomaterials required for tissue engineering are still largely unknown and choices of such materials have been more driven by pragmatism than logic.

## 5.2 The basic tissue engineering paradigms

The reason why there is the word "engineering" in the term "tissue engineering" is not really obvious. The vast majority of definitions of engineering

## Box 5.2 The essence of conventional tissue engineering

The starting point

The creation of new tissue for the therapeutic reconstruction of the human body, by the deliberate and controlled stimulation of selected target cells through a systematic combination of molecular and mechanical signals.

Step one

Obtain suitable cells, either stem cells or fully differentiated cells.
Manipulate cells to optimize cell number and phenotype.

Step two

Prepare suitable biomaterial support.
Arrange appropriate morphology and architecture of support.
Arrange appropriate surface characteristics.
Seed cells into biomaterial support.

Step three

Prepare suitable culture medium.
Arrange appropriate growth factors and other molecular signaling agents.
Arrange appropriate bioreactor conditions.

Step four

Incorporate resulting tissue engineered construct into tissues of patient.

invoke the use of scientific knowledge to solve practical problems and/or the systematic analysis of data to yield useful end products. Although not entirely unconnected, neither of these concepts is readily translated into the paradigms that are now represented by tissue engineering. Tissue engineering does have practical end products but the underlying science is far more related to cell, molecular and developmental biology than to the physical sciences that normally underpin classical engineering. Another meaning of engineering, however, which is best appreciated when we consider that the origin of the term is the Latin *ingenium*, is that it is ingenuity or creativeness that is really at the heart of the subject. This is not a matter of semantics but of immense importance in both the philosophy and practical development of tissue engineering, and the broader area of regenerative medicine.

Tissues and organs suffer from a wide variety of diseases and injuries, as a result of which they lose some degree of function. Primarily these conditions are associated with acute injury or chronic degenerative changes. Without any medical intervention, the response of the body is quite limited and mainly restricted to repair processes. Repair may lead to the restoration of continuity in the affected part by the synthesis of scar tissue, which is essentially collagenous and not reminiscent of the indigenous damaged tissue. This may be an effective front-line response to injury but does not lead to the restoration of normal structure and function and may, if uncontrolled, lead to detrimental effects in the patient.

The logical conclusion to the discussions that emphasize that repair is not an effective outcome, and that replacement has serious limitations with respect to logistics and lack of biological functionality, is to consider tissue regeneration as the only possible alternative, aimed at restoring normal structure and function through the production of new tissue that does replicate exactly that which has been lost. As noted above, adult mammals do not spontaneously regenerate many of their organs that are damaged and have only limited ability to regenerate certain tissues. If we wish to persuade the human adult to regenerate whole organs or tissues that do not spontaneously regenerate, then we have to give them some cues or signals, and superimpose on them a mechanism that is not the natural response to those conditions. Induced regeneration is the essence of tissue engineering, which is, of course, very different to either repair or replacement of tissues. Tissue engineering is, therefore, a matter of the creation of new tissue and to engineer here is, quite simply, to create.

Clearly it is not a trivial process persuading cells to produce new tissue under circumstances in which they do not normally do so. Moreover, it is of the utmost importance that, during this process, exactly the right type of tissue is generated, that the signals given to the cells can be switched off when the process is complete, and that the resulting tissue is fully functional. The process of tissue engineering starts with the sourcing of the relevant cells and ends with the full incorporation of the functional regenerated tissue into the host. The pathway between these two points can take many forms, but is essentially represented by the central tissue engineering paradigm, shown in Box 5.3, and its derivatives. The types of cells include those derived from autologous, allogeneic or, possibly, xenogeneic sources, and they may be fully differentiated cells or stem/progenitor cells. The degree of cell manipulation will depend on the origin of the cells and the complexity of the tissue, and may be dependent on gene transfer in order to optimize processes of, for example, cell expansion, or to control phenotype under these abnormal circumstances. Normally the cells will require some supporting structure, either a scaffold, a matrix or a membrane, within or on which they will express the new tissue. They will be persuaded to do so by molecular signals provided by relevant cytokines, growth factors or other molecules, and by mechanical signals, transmitted via the support and the fluid medium. The environment in which this takes place is usually described as a bioreactor. The tissue that forms, often referred to as a construct, will, if generated *ex vivo*, have to be placed within the host. There it has to be fully and functionally incorporated, taking into account the responses that should be avoided, such as excessive inflammation, immune responses and carcinogenicity or teratogenicity, and also the responses that may be required, such as vascularization and innervation, and indeed the further development and maturation of the tissue itself. It should be borne in mind that this paradigm does not have to be rigidly followed, and many tissue engineering processes are evolving with, for example, much of the regeneration actually occurring *in vivo* rather than *ex vivo*.

Having set out the framework of the generic tissue engineering approach, we have to identify the scientific and infrastructure factors that control the development of tissue engineering. It has to be recognized here that tissue engineering processes are complex and, as yet, have not been effectively translated into clinically acceptable procedures, or indeed commercial successes. There are several reasons for this but probably the most important is the difficulty of integrating all of these components into a coherent system that is able to accommodate the requirements and specifications for each phase of this paradigm into an efficient and cost-effective process within a quality-validated, clinically oriented environment, and which takes into account the impositions of regulatory, ethical and reimbursement schemes. A systems engineering approach to regenerative medicine appears to be an essential element of future developments with respect to this integration,

Box 5.3 **The central tissue engineering paradigms**

Tissue engineering is evolving with two distinct versions. The first is the conventional *ex vivo* approach involving bioreactors. The second, often referred to as "using the human body as the bioreactor" recruits resident cells into a template placed inside the host. The simplicity, lower costs and increased safety of the latter are obvious.

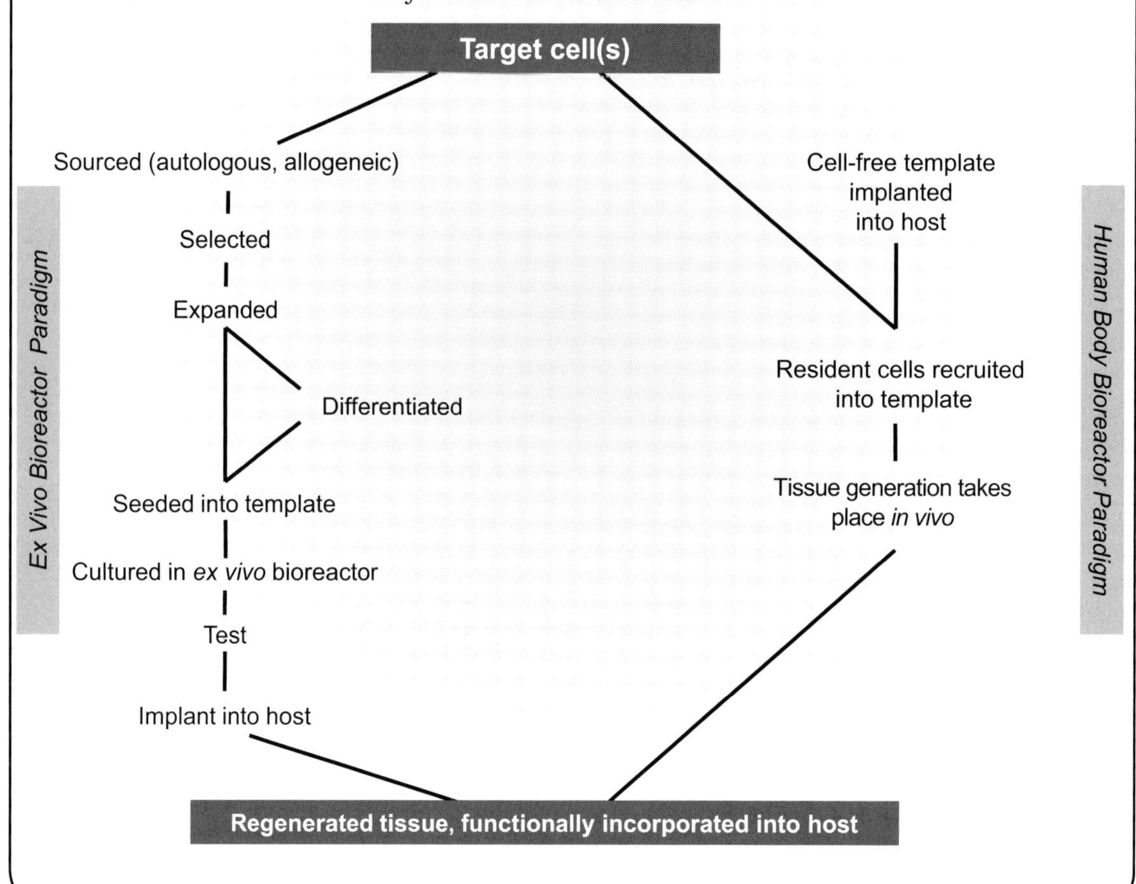


and it is possible that this will also require some elements of systems biology with respect to the underlying science. It has become clear that it is the dynamic interactions of molecules and cells that give rise to biological function, and that knowledge about individual biological components, from genes through to proteins, sub-cellular components, cells, tissues, organs and whole organisms does not in itself lead to an understanding of cell and organ function. It is rather the understanding of the inter- and intra-cellular processes that will do this, leading, for example, to a far greater appreciation of disease causation and drug design. So it is within tissue engineering and regenerative medicine.
The paradigm discussed in this chapter is not hierarchical but temporal, based on the practical transition from cell derivation to tissue construct integration.

## 5.3 Cells for tissue engineering

### 5.3.1 Cell sources

Different tissues of the human body have different capacities to repair themselves. Cartilage has extremely limited capacity, as do tendons and ligaments. Bone can regenerate provided the area of defective tissue is not too large. Skin can regenerate to heal an incisional wound but is very limited in producing effective repair after larger surface areas are lost, for example in burn wounds. Nerves in the peripheral nervous system can regenerate after some injuries but not normally the spine. By definition, tissues that suffer from a degenerative disease are not spontaneously regenerating.

In order for tissue regenerating processes to be effective, either the cells which are normally present in the tissue have to be stimulated into expressing new tissue or quite different cells have to be enticed to do this work, either *in situ* or elsewhere. This gives rise to the two main types of cell source (see Box 5.4). In the first case, fully differentiated cells of the right phenotype have to be obtained and appropriately stimulated. For cartilage regeneration, chondrocytes have to be recruited; for bone it will be the osteoblasts.

**Box 5.4 The main types of cell source for tissue engineering**

There are two general types of cell source for tissue engineering, fully differentiated cells and stem (SC) or progenitor (PC) cells. Some indication of the potential for such cells in the treatment of selective diseases are given here

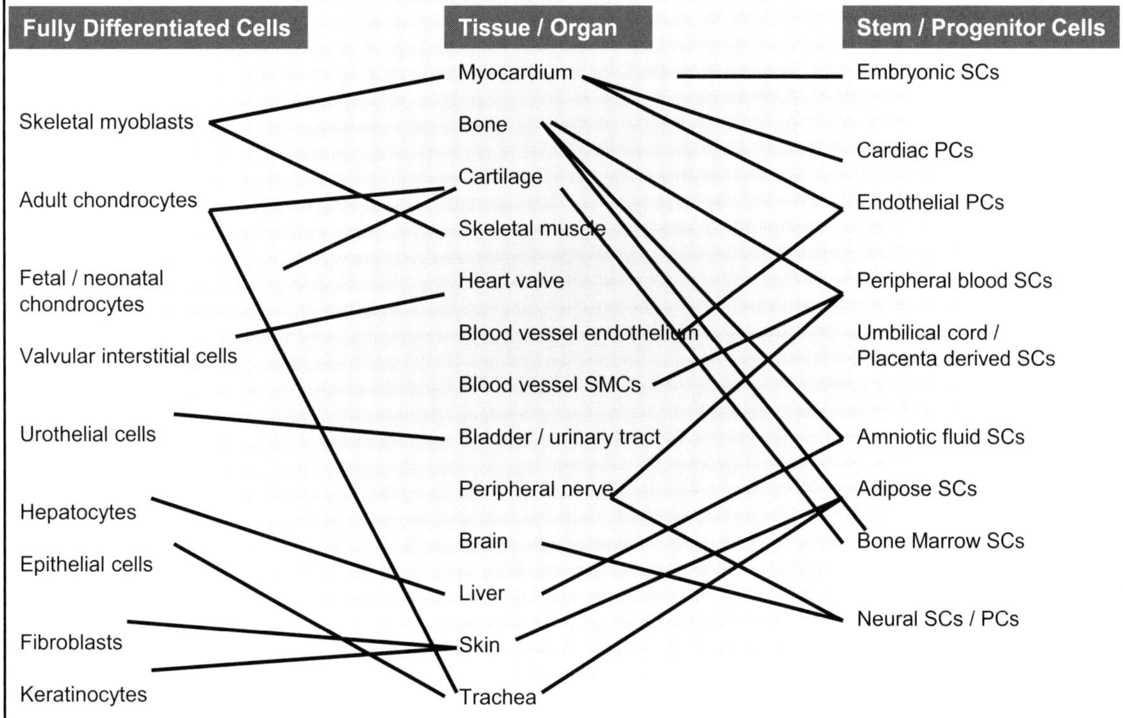

**Fully Differentiated Cells**

Skeletal myoblasts
Adult chondrocytes
Fetal / neonatal chondrocytes
Valvular interstitial cells
Urothelial cells
Hepatocytes
Epithelial cells
Fibroblasts
Keratinocytes

**Tissue / Organ**

Myocardium
Bone
Cartilage
Skeletal muscle
Heart valve
Blood vessel endothelium
Blood vessel SMCs
Bladder / urinary tract
Peripheral nerve
Brain
Liver
Skin
Trachea

**Stem / Progenitor Cells**

Embryonic SCs
Cardiac PCs
Endothelial PCs
Peripheral blood SCs
Umbilical cord / Placenta derived SCs
Amniotic fluid SCs
Adipose SCs
Bone Marrow SCs
Neural SCs / PCs

In some situations, indeed perhaps in most situations, at least two different cell types are required, smooth muscle cells and urothelial cells for regeneration of the urethra, fibroblasts and keratinocytes for the skin, and so on. In the second case it will be undifferentiated cells that are obtained and then persuaded to differentiate into the required phenotype. For the present purposes we shall call these stem cells, although we shall see below that this description needs some refinement. Within these two main classes of tissue-engineering cell there are several variations. With the fully differentiated cells they could be autologous, allogeneic or, in principle, xenogeneic. With stem cells they could be adult stem cells (sometimes called somatic stem cells) that are derived from sources such as peripheral blood, amniotic fluid or adipose tissue, or they could be embryo-derived.

There are some major differences between these types of cell, some quite obvious, some not so. The fully differentiated cells may have the power to express new tissue but they have to be given the right conditions; a major question is just how far they have to be manipulated in order to perform these functions, following on from which questions of safety naturally arise. With adult stem cells, if they are autologous, safety issues are minimal but questions arise about the numbers that can be sourced and what has to be done to expand the population in order to achieve a clinically relevant population. Embryonic stem cells should have the best potential for generating new tissue, because that is their natural function. Leaving aside the ethical issues of using embryos in this way, safety concerns over teratogenicity and immunogenicity have to be addressed. We will discuss stem cells here and fully differentiated cells will be covered in the tissue specific areas later. It should be remembered here that the discussion is not a full treatise on cell biology but an introduction to this complex subject that is of sufficient depth to understand the mechanisms by which these cells interact with biomaterials.

By way of introduction, we should emphasize a few basic definitions. Somatic cells are those cells of the body apart from sperm and egg cells, the cells from which these arise (the gametocytes) and undifferentiated stem cells. Somatic cells contain a full set of chromosomes, whereas reproductive cells contain only half. Stem cells are unspecialized cells that are capable of renewing themselves through cell division. In addition, under certain physiologic or experimental conditions, they can be induced to become specialized tissue- or organ-specific cells. In some organs, such as the bone marrow, stem cells regularly divide to repair damaged tissues but in most organs stem cells are infrequent and only divide under special conditions. When a stem cell divides, each new cell has the potential either to remain a stem cell or become another type of cell such as a myoblast, a red blood cell, or a nerve cell.

Regenerative medicine is not concerned with the process of reproduction, nor with the performance of germ cells. Clearly the use of embryonic stem cells encroaches upon the subject of reproduction, but is confined to the extraction of these stem cells from the reproductive process and their use for generating new tissue and not new organisms. It is also necessary to mention the subject of cloning here, which some might argue lies within the frame of regenerative medicine. In mammals, cloning allows almost identical genetic individuals (clones) to be produced. Somatic cell nuclear transfer is one method by which this can be achieved and involves removing the nucleus from a somatic cell. This nucleus may then be injected into an ovum of the same species which has had its own genetic material removed. This ovum no longer needs to be fertilized since it contains the correct amount of genetic material, but can be implanted into the uterus of an individual of the same species and permitted to develop. The reason why the clone is said to be only nearly identical is that small differences may arise because of some retained mitochondrial DNA.

## Stem cells

In the early stages of embryo development, at around 5–6 days, the stem cells that are present are pluripotent; they have the ability to become any cell type in the adult body. As they develop further, the

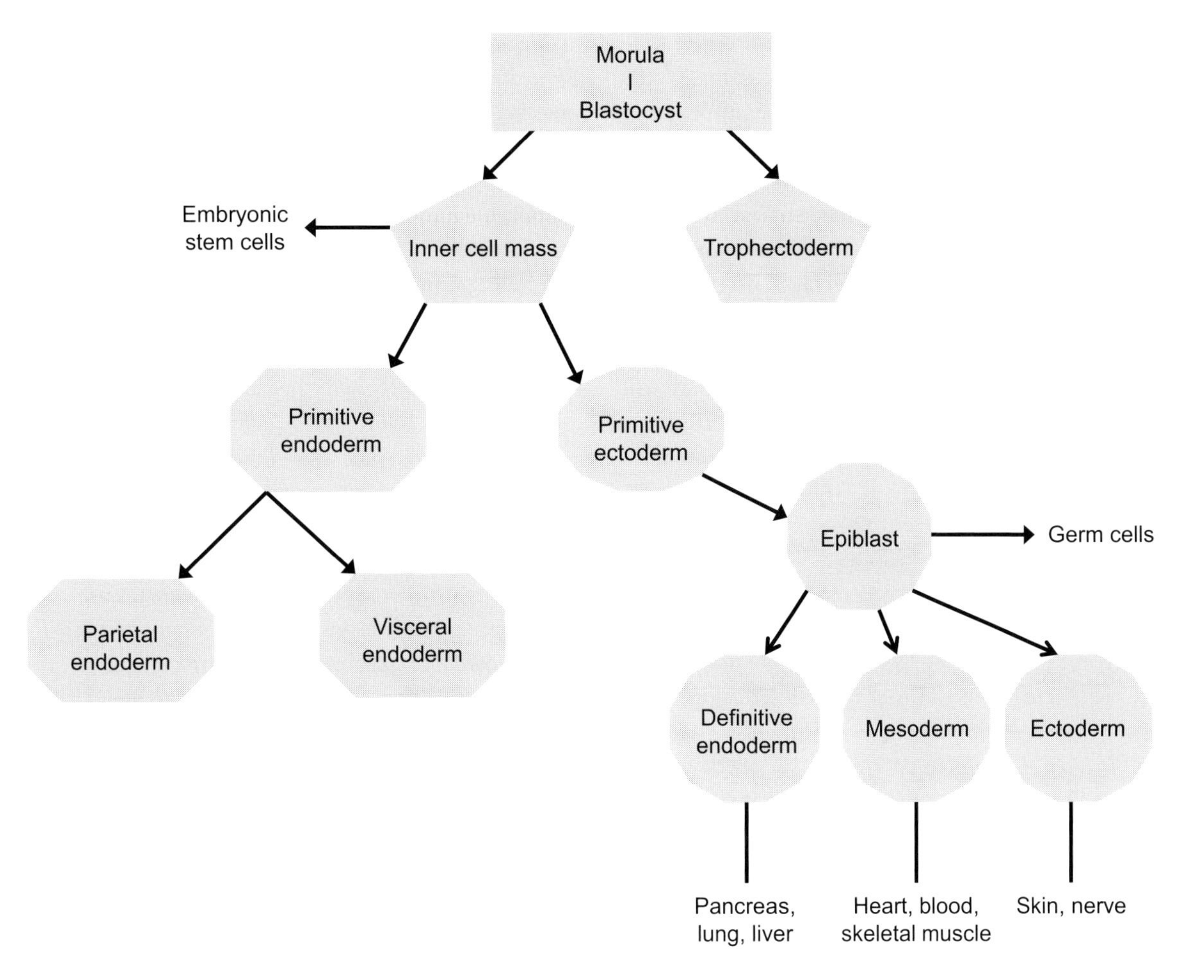


**Figure 5.1** Embryo development and stem cells. The morula is the spherical embryonic mass of cells, formed after the division of a fertilized ovum. This develops into the blastocyst, a thin-walled hollow structure, which has two parts. The trophectoderm gives rise to a peripheral layer of trophoblasts that attach the mass to the uterine wall. The inner cell mass is responsible for the deveopment of the fetus. Cells isolated from the inner cell mass and grown in culture are called embryonic stem cells. Derived from the inner cell mass are the primitive endoderm, the inner germ layer, which gives rise to parietal (cavity walls) and visceral (internal organs) endoderm, and the primitive ectoderm. The epiblast is the outer germ layer of the growing embryo, from which are derived germ cells. It is also possible to obtain and culture pluripotent stem cells from this layer. Arising from the epiblast are the three definitive germ layers of the embryo, which are responsible for tissues such as epithelial tissues from the endoderm, mesenchymal tissues from the mesoderm and nerve tissue from the ectoderm. Small numbers of stem cells will be found in these tissues throughout development and into adulthood.

stem cells commit to lineages that limit their potential to differentiate into specific cell types, becoming more multipotent than pluripotent, as depicted in Figure 5.1. As we shall see, by the time an individual has matured into an adult, stem cells are still present in tissue, but these adult stem cells can only replenish mature cells within that tissue.

The self-renewal of stem cells arises from cell division that is controlled by the features of the cell microenvironment. This is often referred to as the stem cell niche. Within this niche, the stem cell number is maintained constant by balancing those that self-renew and those that become progenitor cells. Also with adult stem cell niches, an organizational structure is maintained through the relationship between the stem cells and somatic cells, and topographical features of the niche influence the stem cell behavior and functionality. The cells are exposed within the niche to spatially and temporally controlled mixtures of molecules, both soluble and insoluble, which are able to provide the signals that determine features such as survival, self-renewal, migration, proliferation and differentiation. Critical niche features include growth factors, cell–cell contacts and cell–matrix adhesions. Their interplay will be important to determine whether any specific stem cell response can be sufficiently robust and resistant to *in vivo* perturbations to allow for their safe and efficient therapeutic use.

The ECM obviously mediates cell attachment and the presentation of signals to cells, but also is involved in the distribution of the molecules that carry out the signaling, by, for example, binding growth factors and controlling their diffusion. This gives some indication of the importance of engineered matrices in controlling the therapeutic use of the cells. Adhesion of stem cells to the surrounding matrix, and to other cells in their vicinity, is essential for their viability since individual cells do not thrive in suspension. One goal in stem cell-based tissue engineering is to develop nonhuman niche cells and matrices or matrix products in order to replicate these functions. In 3D cultures, a variety of hydrogels has proved suitable for the support of ESC growth in undifferentiated masses, that is in embryoid bodies. As we shall see below, these embryoid bodies can also be manipulated within biomaterial microwells in order to recapitulate these niches.

As hinted above, growth factors secreted by stem cells and other niche cells may have powerful effects on cell fate. In embryonic development, these growth factors are subject to tight spatial and temporal regulation. When stem cells are cultured, it is necessary to control their artificial niche with respect to interactions between the ECM equivalents and both secreted and added growth factors. Attempts to do this usually involve microfluidic devices that allow for the same type of spatiotemporal control of concentrations of active factors and of their interactions with ECM components on biomaterial surfaces. This has become an important issue since in standard culture, cells move semi-randomly, and it is necessary to exert dynamic control of the relationship between stem cells and gradients of growth factors. It is also becoming clear that different cell types might need to come into contact before they respond to locally secreted factors. Therefore, important as they are, growth factors have to be handled very carefully if they are to have the optimal effect on stem cell behavior, which has implications on substrate materials and microfluidics design.

The third major factor concerns mechanical forces. We have seen in Chapter 3 that the processes of mechanotransduction have powerful influences on cell behavior in many situations, and this is as true for stem cells as with fully differentiated cells. It is known, for example, that the first stages of cell differentiation in embryogenesis are blocked if force-generating myosin molecules are removed. Mechanical strains are common in solid tissues and substrate strains of the order of 5% can influence stem cell differentiation. Fluids that flow through the tissue also generate forces and can initiate the differentiation of some stem cells. Stem cells appear to have a variety of force-coupled signaling pathways that assist them to adapt to their microenvironments that involve these fluids, with varying fluid shear stresses, and strained solid tissues of varying elasticity. When stem cells are grown on gels of different elasticity and firmness, they may be able to differentiate down different pathways. Growth factors can either augment or oppose these effects.

Obviously the substrate materials on which stem cells are grown have a significant influence on the cell behavior, mediated by many of the factors discussed in Chapter 3, including mechanotransduction and the delivery of signaling molecules, but also the biophysical effects that control cell adhesion. This is clearly a systems phenomenon, where all of the prevailing factors are having supporting or opposing effects. Some of the challenges involved in the clinical translation of stem cell transplantation include the development and use of materials systems that create these specialized niches for the cells, providing adhesion for interacting cells and the control of presentation of adhesion sites such as integrins, thereby improving cell survival and tissue regeneration. As noted above, the biomaterial architecture can be used for patterning of the tissue structure that arises from interacting stem cells. These effects should be enhanced by the localized delivery of growth factors and/or cytokines from the biomaterial, or the broader delivery of trophic factors that can encourage more widespread or regional regeneration.

We shall now discuss how biomaterials relate to these signaling processes in embryonic, induced pluripotent and adult stem cells.

**Embryonic stem cells** Embryonic stem (ES) cells are derived from early embryos, and, as noted above, are capable of indefinite self-renewal *in vitro* while maintaining the potential to develop into all cell types of the body – they are, therefore, pluripotent.

The origin of the understanding of ES cells arose in the 1960s with observations on teratocarcinomas, which are malignant germ cell tumors that are characterized by an undifferentiated embryonal carcinoma (EC) component and differentiated derivatives that can include all three germ layers. It was demonstrated that a single EC cell was capable of both self-renewal and multilineage differentiation, and this first demonstration of a pluripotent stem cell provided the background for work on embryonic stem cells. Most early work on ES cells, in the 1970s, was carried out on mice. Cells of these EC have similar antigen and protein expression to the cells present in the inner cell mass (ICM) of the embryo, which suggested that they are comparable to pluripotent cells present in the ICM. However, there are limitations in the application of EC cells to regenerative medicine and other leads had to be followed.

In mammals, after fertilization, as the single-cell embryo migrates down the oviduct, it undergoes a series of divisions resulting in a morula, defined as a globular solid mass of blastomeres, which are any of the cells resulting from the cleavage of the fertilized ovum. Blastocyst formation then follows, a blastocyst consisting of the ICM, an internal cavity and an outer layer of cells. This outer cell layer of the morula delaminates from the rest of the embryo to form the trophectoderm. The ICM of the blastocyst gives rise to all the fetal tissues, that is the ectoderm, mesoderm and endoderm, and the trophectoderm gives rise to the trophoblast.

The pluripotent cells of the embryo usually have a transient existence as they quickly give rise to non-pluripotent cells through the normal developmental program. They receive signals that direct different compartments to differentiate into the definitive endoderm, ectoderm and mesoderm lineages. Once the pluripotency has been exploited to generate these lineages, that pluripotency is irretrievably lost. For these cells to be useful for regenerative processes, culture conditions were required that would allow the *in vitro* derivation of pluripotent stem cells directly from the embryo. Again this was first achieved with the mouse. It was found in the 1980s that ES cell cultures derived from a single cell could differentiate into a wide variety of cell types. However, it was found that they could also form teratocarcinomas when injected into mice, an issue we will return to shortly.

ES cell lines are usually obtained from the culture of the ICM. In addition mouse ES cells have also been derived from isolated primitive ectoderm, morula-stage embryos and individual blastomeres. It should be noted that, since no pluripotent cell in the intact embryo undergoes long-term self-renewal, ESCs are special artifacts of the culture conditions.

Although there were simultaneous attempts to derive human ES cells, species differences and difficulties with human embryo culture delayed progress. In the 1990s, ES cell lines were derived from non-human primates and this experience together with the introduction and development of human *in vitro* fertilization (IVF) treatment resulted in the derivation of human ES cell lines. These human ES cells (hES cells) had normal karyotypes and maintained the potential to contribute to advanced derivatives of all three germ layers. Obviously these developments

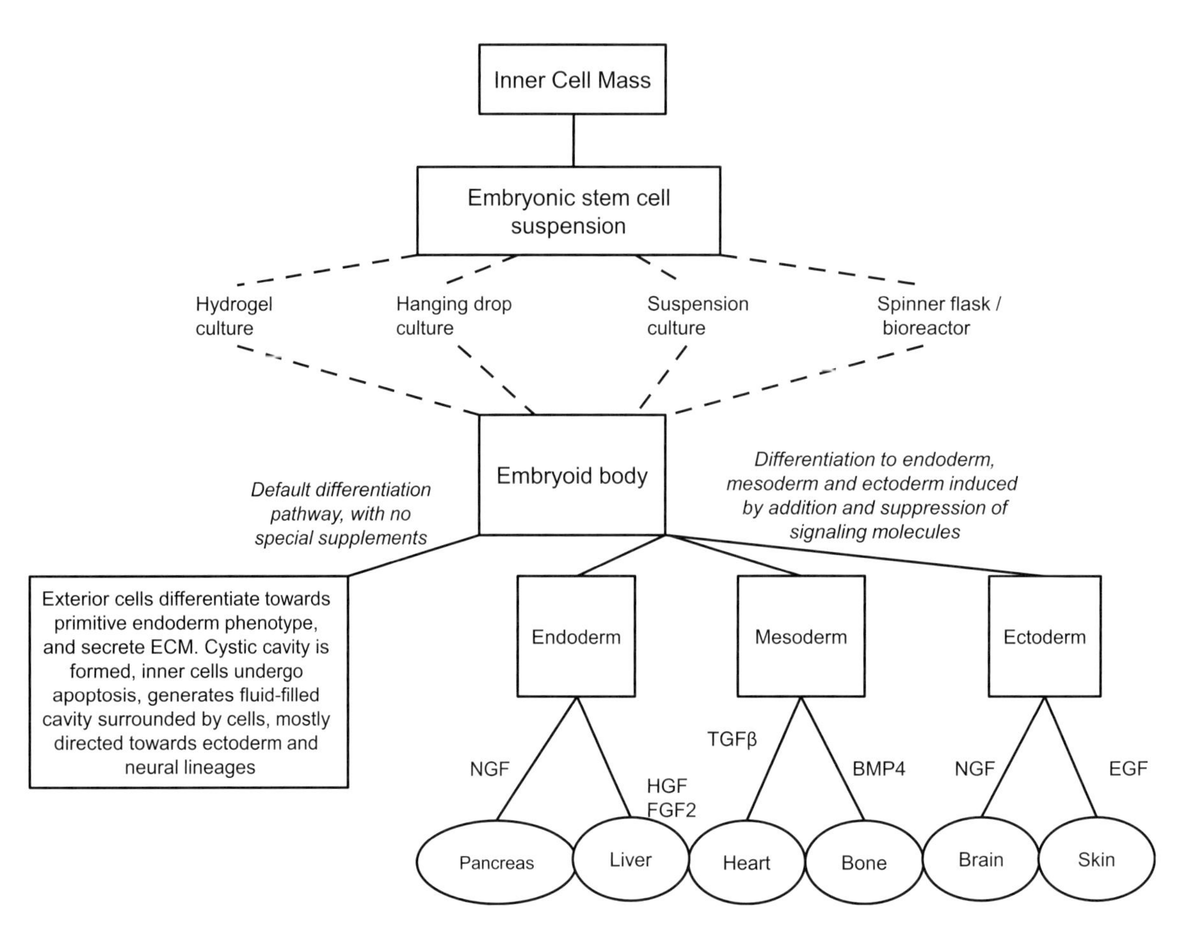


**Figure 5.2** Embryonic stem cells (ESCs) for regenerative medicine. ESCs are derived from the inner cell mass and cultured in suspension. They are transformed into embryoid bodies (EBs) by one of several culture/bioreactor techniques. If no special supplements are used, there is a default differentiation pathway which tends to favor the ectoderm tissues. By addition or suppression of a variety of signaling molecules, the differentiation can be directed towards endoderm, mesoderm or ectoderm paths, with generation of specific tissues by the use of different factors.

have been impeded by ethical and political interventions, discussed in Chapter 9, but progress has been made such that hundreds of hESC lines have been established worldwide.

To obtain physiologically functional cells from ES cells suitable for transplantation, the differentiation procedures must be controlled. Induction of differentiation generally occurs through an intermediate step of formation of embryoid bodies (EBs), which are complex three-dimensional cell aggregates, shown in Figure 5.2. This occurs in two phases. Within the first 2–4 days of suspension culture, endoderm forms on the surface of EBs, giving rise to structures known as simple embryoid bodies. Subsequently, around day 4, cystic EBs develop, forming a central cavity and the differentiation of a columnar epithelium with a basal lamina. *In vitro* culturing of EBs commences with the development of cells indicative of the ectoderm, mesoderm, and endoderm germ lineages. Further culturing can result in more differentiated cell types, including insulin-producing cells, neuronal cells and hematopoietic cells. The optimum basic medium requirements for culturing EBs are still a matter of debate. Glucose is very important for sustaining EB growth and differentiation. Media used in ES cell culture and differentiation are often supplemented with basic fibroblast growth factor (FGF2), which has a role in stimulating cell proliferation, and their maintenance in an undifferentiated state.

Of critical importance in directing ES cells down specific pathways to the required cell type is the signaling that they receive during these early stages. The endoderm and mesoderm lineages have a common progenitor, the mesendoderm, which is formed from the ES cell by both Wnt and TGFβ signaling. High levels of TGFβ then direct the differentiation towards the definitive endoderm, while BMP or Wnt signaling directs it towards the mesoderm. Suppression of both Wnt and TGFβ signaling leads to differentiation towards the ectoderm. Once committed to any of these three types of lineage, further signaling causes differentiation down the required specific pathway. Cells that are committed to the ectoderm may be directed to a neural fate by FGF4 or a BMP agonist. BMP4 elicits epidermal differentiation. Definitive endoderm progenitors can be directed towards a hepatic fate by BMP4 and FGF2, or to a pancreatic fate by retinoic acid and FGF10. A very large number of processes have now been identified that allow these pluripotent cells to be specifically and reliably directed towards many discrete cell types for therapeutic purposes. So far, biomaterials have not played any significant role here.

**Teratoma formation** It was noted above that knowledge of stem cell biology started with observations on teratocarcinomas and also that ES cells when injected into mice could form these carcinomas under some circumstances. There has, therefore, been a concern about tumor formation following the use of ES cells in any therapy. The major issue that has arisen is not really with teratocarcinomas but with teratomas. These are non-cancerous tumors, but they are tumors nevertheless and should be avoided in clinical practice. It would appear that susceptibility to teratoma formation varies with species and with the tissue or organ that is involved. It has been shown, for example, that injection of undifferentiated cells into the myocardium of mice frequently results in teratoma formation. In addition, cultured hES cells can acquire genetic and epigenetic changes that make them vulnerable to transformation, and some lines possess some features of neoplastic progression, including a high proliferative capacity, growth-factor independence, a major increase in frequency of tumor-initiating cells and aberrant lineage specification. Although there is no suggestion of malignancy at this stage, this is a matter of concern, which has to be borne in mind.

**Immunogenicity** Since, by definition, hES cells are not host-derived, it might be expected that they would be recognized by the host immune system as foreign, leading to some form of rejection. In principle, transplanted hES cells will possess both major and minor histocompatibility antigens (discussed in Chapter 3), suggesting that this is a real possibility. However, it is also possible for such cells to possess immune privilege properties, in line with the normal privilege situation at the fetus–placenta interface. Generally undifferentiated hES express low levels of MHC Class I and no MHC Class II; there is also an absence of co-stimulatory factors that are normally associated with allogeneic graft rejection. On the other hand, the differentiated derivatives of hES cells may not always be endowed with this immune privilege and immune responses and rejection have been noted in some cases where such cells have been used therapeutically.

**Induced pluripotent stem cells (iPS cells)** iPS cells have the potential to make a significant impact in cell therapies and regenerative medicine. Their inclusion in this section may be considered premature since there are no biomaterials-related applications or knowledge at this stage, but they certainly will have in the future so that a brief description is in order.

As noted above, stem cells progressively lose their potency to differentiate into many different cell types, proceeding from the pluripotent state that can give rise to any cell type, to multipotent that can give rise to cells within a tissue family, and then to terminally differentiated, which are locked into one specific identity. Under normal conditions, and leaving aside the behavior of cancer cells for the moment, it is not possible for cells to reverse this process, that is go back to a more primitive, less specialized type. The only way backwards would be by an artificially engineered process of cell reprogramming. The cloning experience mentioned in a previous section is an example of reprogramming. At the time that somatic cell nuclear transfer was developed, attempts were made to reprogram late stage embryonic stem cells to make them pluripotent, but these were not successful.

One solution to this difficulty has been found by turning adult cells directly into pluripotent cells without the use of either eggs or embryos, producing what have been termed induced pluripotent stem cells. Instead of introducing adult genetic material directly into an egg, the genes that are normally active only in eggs can be introduced into normal adult cells in order to reprogram them into an embryo-like state. In the first experiments with these processes, which were carried out on adult skin cells, cocktails of several dozens of genes were used, but it was soon found that only four genes were required to produce iPS cells. These are *Oct4*, *Sox2*, *Klf4* and *c-Myc* (see Table 5.1). The initial introduction of these genes into cells required the use of viral vectors, which makes it difficult to translate the technology into clinical practice, but the stage has been set for a major expansion of work to overcome these problems. Given that iPS cells were only announced in 2006, by Yamanaki, in Japan, there has been a remarkably rapid development in this technology.

The initial work was performed with mouse skin fibroblasts. The iPS cells produced showed that they had the characteristics of mouse ESCs, including:

- they showed ES cell morphology in culture;
- they have similar expansion rates to ES cells;
- they express key ES cell genes;
- their gene profiles resemble those of ES cells;
- they have similar epigenetic profiles of genes involved in ES cell function;
- they form embryoid bodies in culture;
- they can differentiate into cells of all germ layers in embryoid bodies in culture;
- they can form teratomas *in vivo* containing tissues of all germ layers.

Table 5.1 **Factors used in the formation of induced pluripotent stem cells in humans**

| Cell | Factors |
|---|---|
| Fibroblast | OKSM<br>OSLN<br>OKS |
| Peripheral blood cells | OKSM |
| Cord blood endothelial cells | OSLN |
| Cord blood stem cells | OKSM |
| Adipose-derived stem cells | OKS |
| Hepatocytes | OKSM |
| Keratinocytes | OKSM |
| Neural stem cells | O |
| Amniotic fluid-derived stem cells | OKSM<br>OSN |

O = *Oct4*; K = *Klf4*; S = *Sox2*; M = *c-Myc*; L = *Lin28*; N = *Nanog*

One of the very important factors that contrasts iPS cell technology with cloning is the potential to overcome the shortened lifespan and other abnormalities found with the latter. The lifespan of cells involved in regenerative medicine processes is governed by telomeres and telomerase activity. A significant number of nuclear-transfer-derived mammals die during or soon after gestation or have shortened lifespan – this happened with the now-famous first cloned mammal, Dolly the Sheep. This is mainly attributed to imprecise nuclear reprogramming, especially leading to shortened mean telomere restriction fragment lengths. Ensuring that iPS cells have appropriate telomere length and telomerase activity has become one of the main concerns with their potential to be used in regenerative processes.

An even more important question is whether iPS cells can be guided into relevant differentiation pathways using the same types of protocols used for ES cells. Several clinical targets have been identified, including neuronal differentiation for the treatment of Parkinson's disease. It has already been shown in rodents that iPS cell-derived dopamine neurons can be produced and implanted into animal models with the disease, with improved behavior. In the mouse, iPS cells have been differentiated *in vitro* into skeletal muscle stem/progenitor cells which, on transplantation into models of Duchenne muscular dystrophy, has led to sustained myogenic lineage differentiation. In addition iPS cells can become fully functional hepatocytes, opening up the potential to develop a variety of therapies for liver diseases.

Although considerable progress is being made on translating iPS cell science into clinically relevant therapies, a number of questions remain. The use of the viral vectors is an important issue and much effort is directed towards the delivery of the defined factors without resorting to viral vector integration. It may be possible to use these vectors transiently without integration, or removing vectors at a later stage, or completely circumventing vectors by delivery of recombinant protein-based factors or other molecules that are able to induce pluripotency.

**Adult stem cells** Adult stem cells are those undifferentiated cells that exist in various tissues and organs after embryonic development. They are sometimes referred to as somatic stem cells. As true stem cells, they have the capacity for self-renewal and differentiation into cells that are able to replenish dying cells and repair damaged tissue, but they are usually multipotent rather than pluripotent. In most tissues they are present in only very small numbers. The richest source of adult stem cells is the bone marrow, from which mesenchymal, hematopoietic and endothelial stem cells can be derived. As we shall see there are often multiple sources of individual adult stem cells and there may be a degree of plasticity between them, this being the transfer from one type to another. The main sources of stem cells are discussed here, but several others (for example, those derived from urine) may become more important in the future.

**Hematopoietic stem cells** Hematopoietic stem cells (HSCs), primarily reside in bone marrow and have the functions of maintaining blood formation and replenishing themselves throughout the adult's life, shown in Figure 5.3. In adult homeostasis, HSCs are relatively dormant but can extensively proliferate under conditions of regenerative stresses. Although relatively rare, constituting around 0.02% of the bone marrow cells, they are able to repopulate all of the hematopoietic lineages.

As far as the stem cell niche is concerned, in adult humans, these cells reside in the bone marrow cavity but closely associated with surrounding stromal cells. They tend to localize at the periosteum–endosteum border of the inner surface in close contact with osteoblasts. The osteoblasts are considered to provide essential cues for HSCs, suggesting that osteoblasts provide a direct physical niche for HSCs that maintains their self-renewing capacity through cell surface molecules.

During adult hematopoiesis, they generate both lymphoid and myeloid cells. The former are comprised mainly of T-cells, B cells, and natural killer cells, while the latter include granulocytes, macrophages and erythrocytes. The process is one of gradual differentiation that begins with HSCs and ends with the terminally differentiated lineages and several differentiation intermediates. There are two significant populations within bone marrow, the short-term HSCs and the multipotent progenitors. Transplantation of HSCs has long been used to treat patients with hematopoietic diseases.

**Mesenchymal stem cells** As noted above, bone marrow is a major source of mesenchymal stem cells (MSCs) and indeed it was observations on the

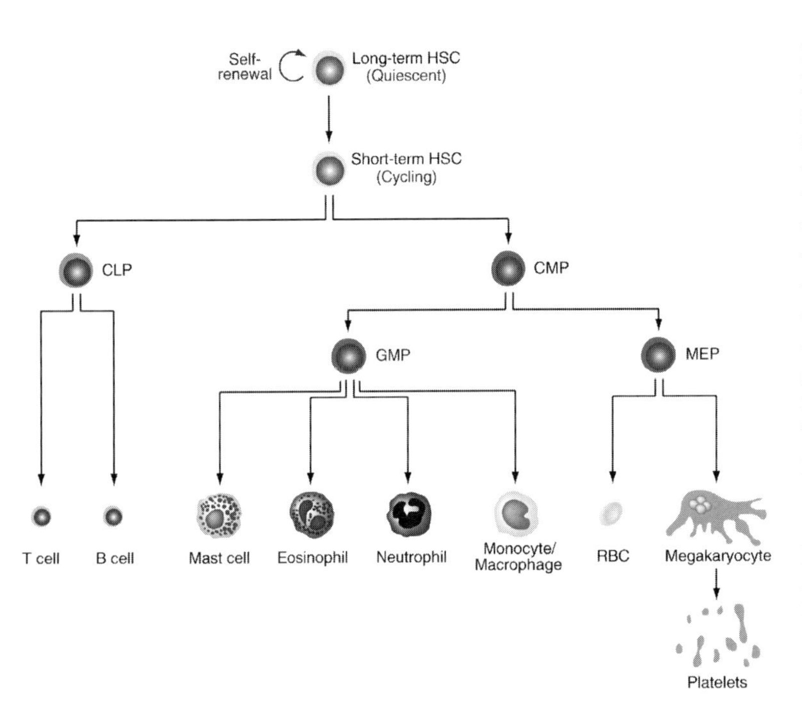


**Figure 5.3** Hematopoietic development. Hematopoietic stem cells (HSCs) give rise to progenitor cells that become increasingly lineage restricted and ultimately differentiate into mature blood cells of all lineages. Most long-term HSCs are quiescent and they enter the cell cycle only infrequently to undergo self-renewal. The transition from long-term to short-term HSCs is characterized by cell proliferation and subsequent differentiation. Common lymphoid progenitors (CLP) give rise to T cells and B cells. Common myeloid progenitors (CMP) give rise to granulocyte–macrophage progenitors (GMP) and megakaryocyte–erythroid progenitors (MEP). Reprinted from *Biochimica et Biophysica Acta*, 1830, Matsumoto A and Nakayama K I, Role of key regulators of the cell cycle in maintenance of hematopoietic stem cells, 2335–44 © 2013, with permission from Elsevier.

behavior of marrow-derived cells in culture that led to the first description of these cells. It was observed that fibroblast-like cells that were strongly adherent to tissue culture plastic existed in the bone marrow and that these tended to form colonies when seeded on these plates. These cells were originally called stromal cells, and were quickly considered as having potential for regenerative purposes if they could be cultured appropriately. Several names such as MSCs, mesenchymal progenitors, and stromal stem cells were used to describe them. Although scientifically many would consider a more appropriate name would be multipotent mesenchymal stromal cells, they are usually referred to simply as MSCs. The definition of MSCs is largely pragmatic, being based on these early observations, the general concept of a plastic-adherent type of cell that can be directed to differentiate *in vitro* into cells of lineages such as osteogenic, chondrogenic, adipogenic, and myogenic being satisfactory for most purposes (Figure 5.4). They proliferate and give rise to daughter cells that have the same pattern of gene expression and phenotype of the original cells; these characteristics are often grouped in order to categorize their so-called stemness.

On the other hand, there is much uncertainty about MSCs. Self-renewal and differentiation potential are present, by definition, but are mostly seen after *in vitro* manipulation, and evidence of these characteristics in non-manipulated MSCs *in vivo* is not so readily observed. Even more importantly, and in contrast to the hematopoietic stem cells, MSCs do not have a unique marker and they may undergo significant alterations during culture. In the absence of clear markers, cultured MSCs have been characterized by using cell surface antigens or their differentiation potential. It should be noted here that the CD nomenclature is widely used to characterize these cells. CD (cluster of differentiation) refers to the protocol for identifying cell surface molecules (often receptors), which leads to the process of immunophenotyping of cells.

Some minimal criteria are used to define human MSCs, which include:

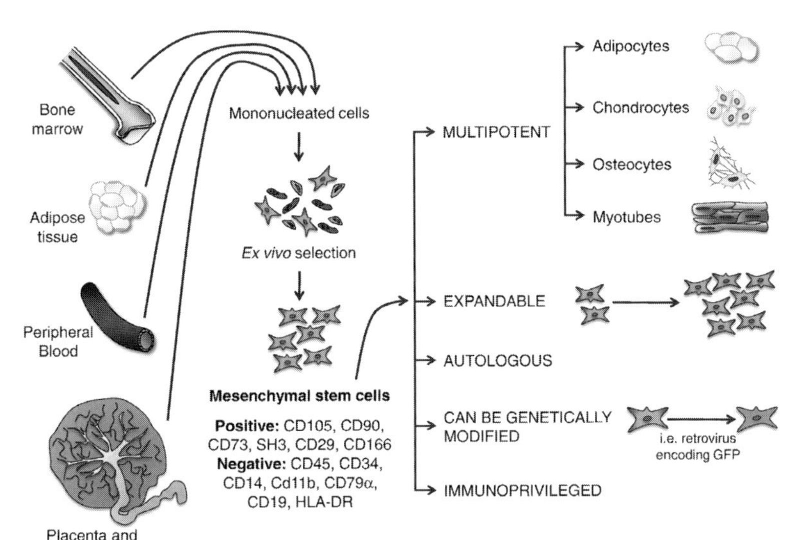


**Figure 5.4** Mesenchymal stem cells (MSCs) can be isolated from the bone marrow, adipose tissue, placenta and umbilical cord blood. Some investigators also described MSC in the peripheral blood. MSC can be expanded *ex vivo*, are multipotent and possess other favorable characteristics that make them suitable for cell therapy and myocardial repair. Reprinted from *Vascular Pharmacology*, 57, Gnecchi M *et al.*, Mesenchymal stem cell therapy for heart disease, 48–55, © 2012, with permission from Elsevier.

- they must be plastic-adherent when maintained in standard culture conditions and form colonies;
- they must differentiate to osteoblasts, adipocytes, and chondroblasts *in vitro*, and
- they must express CD105, CD73 and CD90 markers, and lack expression of CD45, CD34, CD14 or CD11b, CD79α or CD19, and HLA-DR surface molecules.

As indicated, the bone marrow is not the only source of MSCs and many tissues in the body contain a limited number of them. Moreover, there are several shortcomings of obtaining bone marrow MSCs, including pain, morbidity, and low cell numbers.

MSC-like cells are found circulating within the blood of young individuals, and in adults they appear to be resident in many tissues, where they can be stimulated to proliferate and differentiate to assist in repair processes. MSCs or MSC-like cells are found in adipose tissue, dermal tissue, tooth pulp, peripheral blood, amniotic fluid, human placenta and in umbilical cord blood (UCB).

MSCs from bone marrow, adipose tissue, and umbilical cord blood are morphologically and immunophenotypically very similar, although not quite identical. UCB-derived MSCs form the fewest colonies and show the highest proliferative capacity, whereas adipose tissue-derived MSCs form the greatest number of colonies. Bone marrow MSCs in fact have the lowest proliferative capacity of those mentioned. MSCs from adipose tissues and UCB may become more popular since these tissues are versatile and they are discarded if not used.

Some of these alternatives to bone marrow as MSC sources are discussed briefly and separately below.

**Adipose tissue-derived stem cells** Adipose (i.e., fat) tissue contains a readily available and abundant source of adult stem cells that can be directed towards several different lineages. They may be obtained during liposuction procedures, which can yield up to 3 L of tissue. One gram of adipose tissue yields around 5000 stem cells, which is much greater than the number isolated from an equivalent amount of bone marrow. The tissue is minced and washed and then dissociated by an enzyme such as collagenase. After centrifugation, a pelleted stromal vascular fraction, which also contains circulating blood cells, fibroblasts, pericytes and other cells, is derived and the adipose stem cells (ASCs) are selected and enriched based on adherence to tissue culture plastic. The ASC immunophenotype, for which there are numerous markers such as CD73 and CD90, largely resembles that seen with other adult MSCs such as bone marrow-derived and skeletal muscle stem cells. The cells secrete many cytokines such as VEGF, TGFβ, PDGF and BFGF, the cytokine profile being described as angiogenic and immunosuppressive. The anatomic site from which fat is harvested and patient characteristics such as age, gender and medical co-morbidities all influence ASC yield, plasticity and potency. White adipose tissue, found subcutaneously in abdominal and gluteal locations, is highly vascularized, and especially rich in ASCs. There have been several clinical trials performed, largely in areas of soft tissue augmentation.

**Amniotic fluid-derived stem cells** Amniotic fluid forms a liquid protective zone around the fetus, being rich in the nutrients required for fetal growth. The fluid contains a heterogeneous mixture, in small numbers, of fetal-derived cells, amniocytes, that originate from all three germ layers. These include a stem cell population that is characterized by highly specific morphology and antigen expression. These stem cells (AFSs) may be isolated using a number of techniques, giving diverse sub-populations with different self-renewal capacities, growth rates and multilineage differentiation potential.

Two approaches have been used to exploit AFSs in regenerative medicine. In the first, undifferentiated cells are used in situations where they receive tissue-specific signals in order to proliferate into the progeny appropriate to that tissue. In the second, specific cell types are differentiated *in vitro* prior to transplantation. As with other stem cell sources, AFSs have the potential to generate tissue constructs *in vitro* for tissue replacement and to be used for cellular therapy for *in vivo* repair. The former is an attractive option for the treatment of congenital abnormalities where the stem cells derived during amniocentesis can be used *in vitro*, parallel to gestation, for surgical implantation *in utero* or just after birth. There is potential here for the repair of cartilage or tendon tissues. For more generalized regenerative medicine approaches, AFSs are good candidates because the source material is associated with minimal ethical issues; moreover, the cells share marker expression with embryonic and somatic stem cells but lack the tumorigenicity potential of the former.

Typically a multipotential sub-population of amniotic fluid progenitor cells is isolated using selection of cells expressing the membrane receptor c-kit. These cells maintain a round shape for some time but then change morphology and become substrate adherent; they have high self-renewal capacity and do not require feeder layers. Depending on the culture conditions, they may be differentiated towards endoderm cells, such as hepatocytes, ectoderm cells such as neuronal cells and mesodermal cells, including myocytes, endothelial cells and osteocytes. There is no human experience at this stage but animal studies have shown some potential in nerve, kidney, myocardium, bone and cartilage regeneration, either as cell therapies or tissue engineering processes.

**Tooth-derived stem cells** Dental tissues are specialized tissues that do not undergo continuous remodeling as in bony tissue and it should be expected, therefore, that dental-tissue-derived stem or progenitor cells would be more restricted in their differentiation potential compared to bone marrow MSCs. However, dental mesenchyme has early interactions with the neural crest, so that dental stem cells may be expected to possess characteristics similar to those of neural crest cells. The neural crest consists of a group of embryonic cells that are separated during the formation of the neural tube (the precursor to the spinal cord) but do not remain as part of the central nervous system; neural crest cells generate a large number of differentiated cell types. In practice, the dental/periodontal complex is quite a rich source of stem cells and there are several different options. Three may be mentioned here, the dental pulp, stem cells from human exfoliated deciduous teeth (referred to as SHED) and the periodontal ligament. All of these are multipotent, usually with osteogenic, odontogenic, adipogenic, chondrogenic, myogenic and neurogenic potential.

The SHED source is particularly interesting having considerable proliferative capacity and, by definition, being readily and harmlessly obtained.

Since the normal function of dental pulp stem cells is concerned with the development of hard, mineralized tissues, it might be expected that they should be capable of being differentiated into cells of high osteogenic capacity. This would appear to be the case and c-kit$^+$/CD34$^+$ pulp stem cells may be differentiated into CD44$^+$/RUNX-2$^+$ osteoblast precursors and then to osteoblasts themselves. Woven bone-like tissue has been formed *in vitro* from such a system. Because of their similarity to neural crest cells, there have been hopes that tooth-derived stem cells could be used in neural regeneration but although there have been good signs of neural differentiation capacity *in vitro*, clinical applications are still far away.

**Peripheral blood stem cells** Peripheral blood, that is circulating blood that transiently passes through the blood vessels at the peripheral sites of the body, especially the arms and legs, is readily accessible. It can be removed from the circulation in significant volumes without detriment to the host. It also contains stem cells. It is therefore, *de facto*, an obvious source of autologous stem cells for use in regenerative medicine processes. It is known that the bone marrow is the source of stem cells in peripheral blood. HSCs, as we have seen, reside in adult bone marrow and serve to replenish lymphoid, megakaryocytic, erythroid, and myeloid hematopoietic lineages throughout adulthood. MSCs also reside in bone marrow. Many studies have indicated that there is a constant exchange of cells between the bone marrow and peripheral blood.

It is particularly important to note that peripheral blood may be used to derive stem cells that could be re-infused into patients following chemotherapy. Stem cells are mobilized using an agent such as granulocyte colony stimulating factor and collection is made during aphaeresis (the *ex vivo* separation of blood into its different components), and their infusion results in bone marrow homing.

## 5.3.2 Principles of cell manipulation

Having decided on the most appropriate source of cells, they have to be taken from this source and processed in some way in order for them to participate in the regeneration of the required tissue. This implies some form(s) of manipulation. The relevant principles are discussed here, and then the most important of these for biomaterials-promoted tissue engineering are discussed in more detail. We should mention here that some similarities will be seen for general cell therapy and tissue engineering as far as manipulation techniques are concerned, with the obvious exception that some templates are used in the latter in order to give the new tissue shape and form.

The degree of manipulation may be an important issue from several perspectives. There is a wide spectrum of possibilities. At one extreme in cell therapy, cells may be rapidly isolated from their source, and applied to the patient with minimal manipulation, perhaps without either the cells or the patient leaving the operating theatre. This has considerable advantages of low cost and minimal oversight by regulatory bodies. On the other hand, there is limited opportunity to induce the changes in these cells during this process, implying limited therapeutic efficiency in many cases. A common alternative involves isolation of the cells, their transportation to the appropriate laboratory, extensive manipulation, return to the clinical facility and re-implantation. This is obviously likely to be much more expensive and requires formidable regulatory control and compliance with the relevant current good manufacturing protocols. The latter technique is expanded into tissue engineering by infiltrating the cells into the biomaterial template as far as the manipulation is concerned.

Results with the first of these techniques, and indeed with many examples of the second, have confirmed that transplanted cells may have the capacity to assist in repair processes, but in general, very few transplanted cells survive (usually less than 5%) and there is very poor engraftment. With the increasing interest in using stem cells as the primary source for cell therapy, attention has to be focused on the methods of their differentiation into the required lineage.

Bearing in mind the tissue engineering paradigms discussed earlier, we can identify the various specific aspects of cell manipulation that may be required in order to express new tissue. These are given in Box 5.5. Many of these steps are discussed in the later sections.

## 5.3.3 Cell patterning

Although histological sections of most types of tissue would suggest that the distribution of cells within their tissue environment is rather random, in many tissues the physical juxtaposition of cells and their spatial patterns are very important for their functionality. Within tissue engineering processes, the ability of cells to function in a regenerative mode may equally depend on their spatial distribution. The patterning of cells, in both 2D and 3D, may therefore be a very important factor in the efficiency of tissue engineering processes.

Many patterning methods share a common approach. Cell-adhesive regions, i.e., the patterns, on the surface are separated by cell-repellent, non-adhesive areas. The adhesive patterns are usually created using ECM proteins or their alternatives such as cell-interactive peptides. There are numerous practical ways to achieve this, some of these methods having been in use in biology for some time. These include micro-contact printing (mCP), switchable substrates, elastomeric stencils, microwells, optical tweezers, electrophoresis and dielectrophoresis. Some of these specifically depend on fluid mechanics for their function, the details of which will be discussed in a later section. Because there may be different specific objectives to the patterning, there are several capabilities to consider. We may require patterning cells with single-cell resolution or with large numbers of cells. We may wish to allow the patterned cells room to grow and move, or not. Some cell types are very sensitive and require exceptionally gentle handling, whilst others are more robust. Each technique necessarily has some limitations and they have to be chosen carefully. Cell patterning techniques impose specific requirements on the substrate, such as specific surface chemistry or optical transparency.

Some examples of cell patterning are shown in Figure 5.5.

## 5.3.4 Cell sheet engineering

At this point, an alternative concept of harvesting cells and using them for regenerative purposes should be mentioned. This does not use biomaterials as templates within which cells grow and express new tissue, but rather uses biomaterials as interim supporting structures for the preparation of assemblies of cells. Since this process initially involved the generation of 2D arrays of cells, it has been given the name of cell sheet engineering. In this process, cell types that secrete significant amounts of ECM proteins are cultured in simple reactors to create sheet-like structures. However, the surfaces on which the cells attach are made of temperature-responsive polymers.

The principle of temperature-responsive materials was discussed in Chapter 2; most applications in cell sheet engineering utilize the polymer poly(N-isopropylacrylamide) (PIPAAm). This polymer can be covalently immobilized onto conventional culture surfaces at nanometer thickness. Changes in the surface properties of the polymer can be controlled by varying the temperature in the culture. Above PIPAAm's lower critical solution temperature (LCST) of 32°C, the surface is slightly hydrophobic, allowing cells to attach, spread and proliferate, in a similar manner to their behavior on normal tissue culture polystyrene. However, by reducing the temperature below the LCST, the PIPAAm grafted surfaces undergo a spontaneous change and become hydrophilic. A hydration layer

## Box 5.5 Principles of cell manipulation

Cell manipulation within tissue engineering involves phases of cell harvest, selection, expansion and differentiation.

Cell harvesting is usually performed by bone marrow aspiration, by tissue biopsies and retrieval from explanted or harvested tissues or organs.

- Bone marrow aspiration is normally obtained from the iliac crest under local anesthesia. The first part of the aspirate contains adipocytes, hematopoietic cells, endothelial progenitor cells and osteoprogenitor cells.
- Punch biopsies for obtaining autologous cells can be used for skin, cartilage, lungs, kidney, heart and liver. Sometimes this is done as part of an diagnostic procedures (e.g., knee cartilage); with organs, fluoroscopic control may be necessary.
- Allogeneic cells are usually obtained from organs that have been harvested for transplantation but found unsuitable, but can be digested and treated for cell isolation.

Cell selection is usually necessary because harvested cells will consist of several different types. There are several ways of selecting and separating cells.

- Density gradient centrifugation is often performed with Percoll medium, which consists of PVP-coated colloidal silica particles of 15–30 nm diameter. Cells are loaded onto the gradient and centrifuged, and separated either on the basis of size or density.
- Selective adhesion is based on the difference in adhesion properties of cells on surfaces such as fibronectin-coated plastics.
- Antibody techniques use the detection of antigens on cell surfaces by selective antibodies, which are linked to a fluorescent dye and separated by fluorescence activated cell sorting (FACS).

Cell expansion is required in order to achieve the appropriate number of cells.

- Some cells may be expanded in suspension culture, using a semi-solid medium such as agar or collagen where they proliferate under the right media conditions.
- Anchorage dependent cells may be expanded in monolayer culture, where they adhere to the substrate, which could be conventional tissue culture plastic, or a surface coated with laminin, fibronectin or collagen, where they proliferate.
- This process may be enhanced by using microcarriers to increase the surface area.
- The process may also be enhanced by using feeder cells, typically mouse embryonic fibroblasts, which have several supportive mechanisms for cell proliferation.

Cell differentiation very much depends on the cell system in question. Mechanical signaling, as discussed in Chapter 3, may be the most important process.

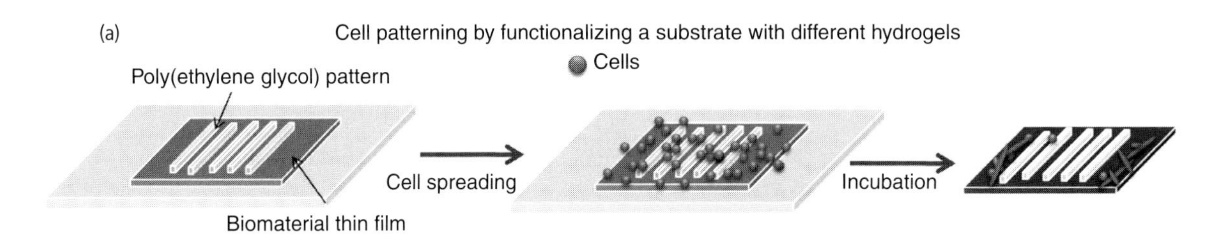


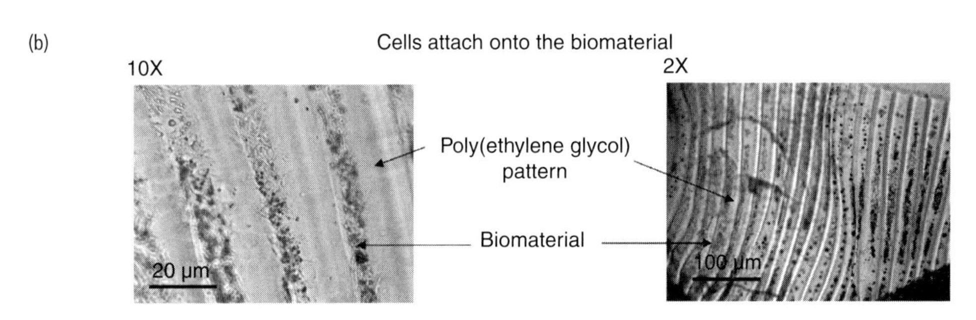


**Figure 5.5** Cell patterning.
**(a)** A glass slide is coated with a thin layer of photo-cross-linkable, gelatin-based hydrogel (GelMA), and subsequently patterned with a PEG pattern. Primary neonatal cardiomyocytes are dispensed onto this substrate and incubated for two weeks.
**(b)** Typical phase contrast images of cardiomyocytes proliferating on the GelMA layer along lines created by the PEG pattern. Images courtesy of Ali Khademhosseini Ph.D, Harvard-MIT Health Sciences & Technology, Boston, MA, USA.

forms between the culture surface and the attached cells, causing the cells to collectively detach from the surface, allowing for harvest of confluently cultured cells as intact sheets. Since these surfaces facilitate spontaneous cell detachment, the use of conventional proteolytic enzymes can be avoided, thereby preserving cell surface proteins such as growth factor receptors, ion channels and cell-to-cell junctions. Due to the presence of deposited ECM that is produced during incubation, the cell sheets can be easily transferred and attached to other surfaces such as culture dishes and host tissues. The process is depicted in Figure 5.6.

The maintenance of differentiated functions of several cell types, including epidermal keratinocytes, vascular endothelial cells, renal epithelial cells, periodontal ligaments and cardiomyocytes, has been demonstrated after this low-temperature cell sheet harvesting procedure. It is not so straightforward to use cell sheet engineering techniques with 3D structural tissues that are cell-sparse and ECM rich, such as bone and cartilage, but it is proving possible to generate some degree of thickness of the regenerated tissue by building up layers of cell sheets.

## 5.4 Biomolecules and nutrients for tissue engineering

Cells are grown under *in vitro* conditions for many reasons, in dishes, tubes, reactors and so on, and a great deal of expertise has been established over decades with the optimization of the culture conditions commensurate with the specific objectives. In many forms of tissue engineering, cells also have to be handled under *in vitro* conditions, either to produce the right quantity of the right cells for later *in vivo* transplantation or for the *in vitro*, bioreactor-based generation of new tissue.

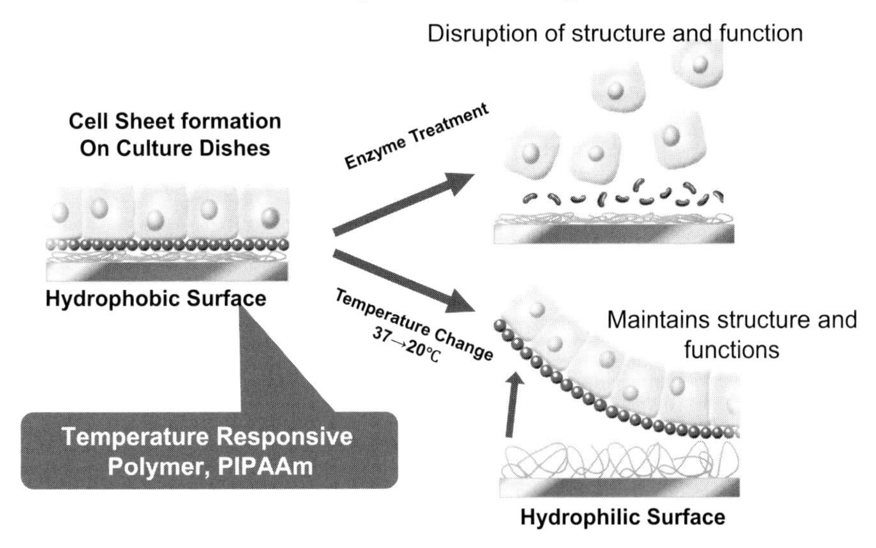


**Figure 5.6** Principle of cell sheet engineering. Conventional techniques to release cells from biomaterials surfaces using enzymes result in disruption and loss of some functions of the cells. Techniques of cell sheet engineering, involving the change of hydrophobic–hydrophilic balance of a thermally responsive polymer such as poly(N-isopropylacrylamide), PIPAAm, yields the release of sheets of cells without disruption and loss of function. Courtesy of Professor Teruo Okano, Institute of Advanced Biomedical Engineering and Science, TWIns, Tokyo Women's Medical University, Tokyo, Japan.

In either case we have to consider what are the optimal conditions for the performance of the cells in their specific situations. This largely revolves around the conditions under which the cells perform their normal function, with good survival, and with either maintenance of their original phenotype or differentiation to a different but specified phenotype and then the specific biomolecules that may have to be added in order for the tissue generation processes that are essential to tissue engineering to take place optimally. In the latter situation we have to consider growth factors and genes, while in the former we have to consider nutrients and oxygen for cell survival and growth. We shall deal with these only briefly since they may actually have little to do with the biomaterials used in tissue engineering, but there are several important connections. These are summarized in Box 5.6.

### 5.4.1 Culture media and oxygenation

#### Media and nutrients

When cells are used for *in vitro* tissue engineering processes, they have to be grown in culture conditions, as noted above. The culture medium may resemble normal cell culture conditions, but there have been some variations, both in concept and practical details, in recent years. It was considered necessary to supplement the standard, basal culture medium, of which there have been many commercially available varieties, with animal serum, for the purposes of enhancing cell growth and proliferation. The most widely used sera were of bovine origin especially derived from fetal tissue (fetal bovine serum, FBS). This is not fully of defined composition but can be used universally with quite good effects. There are several disadvantages, including the fact that it is ill defined,

## Box 5.6 Biomolecules and nutrients for tissue engineering

The culture medium used in *in vitro/ex vivo* tissue engineering is based on relatively standard solutions with some specific additions depending on individual requirements. Culture media are either simple, complex or chemically defined. Simple, or balanced, salt solutions are ionic buffers with energy and protein additions. Complex media are supplemented with amino acids, vitamins and some other nutrients. Chemically defined media, or serum-free media, is comprised solely of biochemically defined low molecular weight constituents, entirely free of animal-derived substances.

Inorganic salts usually include NaCl (6.5–7.5 g/L), $NaHCO_3$ (1.2–3.8 g/L), KCl (0.4 g/L), $NaH_2PO_4$ (0.1–0.8 g/L) and $MgSO_4$ (<0.1 g/L).

Amino acids usually include L-alanine (0.009 g/L), L-arginine (0.1–0.2 g/L), L-asparagine (0.01–0.05 g/L), L-cystine, (0.02–0.06 g/L), L-glutamic acid (0.01–0.02 g/L), L-glutamine (0.1–0.6 g/L), glycine (0.01–0.03 g/L), L-histidine (0.01–0.04 g/L), L-isoleucine (0.003–0.1 g/L), L-leucine (0.01–0.1 g/L), L-lysine (0.03–0.15 g/L), L-methionine (0.005–0.03 g/L), L-phenylalanine (0.005–0.07g/L), L-prolene (0.01–0.04 g/L), L-serine (0.01–0.04 g/L), L-trytophan (0.005–0.02 g/L) and L-tyrosine (0.002–0.05 g/L).

Vitamins frequently include folic acid (0.001 g/L), niacinamide (0.001 g/L), D-pantothenic acid (0.001 g/L), riboflavin (0.0001 g/L), thiamin (0.001 g/L), vitamin B-12 (0.001 g/L).

Growth factors are given in Table 5.3.

and it is of animal origin, so that the description of some cell therapy and tissue engineering products as being autologous is difficult to sustain. In many situations, other animal products were used to enhance the required cellular processes. An important example here is the use of a so-called mouse feeder layer, which is a collection of mouse embryonic skin cells that can be added to a culture dish/reactor before the target cells in order to provide an attachment surface for those cells, increasing proliferation, especially of stem cells.

For many reasons, including concerns of reproducibility and contamination, and the general factors of ethical positions on using animal products and their cost, there has been a move away from the use of these substances. This is not straightforward since the serum has so many, varied, functions in the medium, including the provision of hormonal factors that stimulate cell growth and proliferation. It also provides transport proteins, attachment factors and enzymes that act to stabilize the medium. All these have to be effectively replaced in alternatives, which may be simply described as serum-free medium, or, if the composition is really well known and characterized, as chemically defined medium. Characteristics of relevant tissue culture media are given in Table 5.2.

### Oxygen levels

Since oxygen, and oxygen gradients, play a significant role in normal tissue development, it should not be surprising that they are important factors in tissue engineering. It has been known for a long time, of course, that oxygen is involved in energy production in cells and is a major regulator of cellular

Table 5.2 **Characteristics of culture media for *ex vivo* tissue engineering with human MSCs. Table gives typical components used in culture media**

| **Osteogenic differentiation** | |
|---|---|
| Fetal calf serum | 10% |
| Dexamethasone | 10–100 nM |
| Ascorbic acid | 50–500 μM |
| B-glycerophosphate | 5–10 mM |
| BMP-2 | 20 nM |
| Vitamin D3 | 1–10 nM |
| TGFβ | 200 pM |
| **Chondrogenic differentiation** | |
| Dexamethasone | 10–100 nM |
| Ascorbic acid | 100–200 μM |
| TGFβ | 10–100 ng $ml^{-1}$ |
| Bovine serum albumin | 1–2 mg $ml^{-1}$ |
| BMP-6 | 500 ng $ml^{-1}$ |
| **Adipogenic differentiation** | |
| Fetal calf serum | 10% |
| Dexamethasone | 100 nM–1 mM |
| Insulin | 100 nM |
| 3-isobutyl-1-methylxanthine | 500 μM |
| Transferrin | 10 μg $ml^{-1}$ |
| **Smooth muscle cell differentiation** | |
| Fetal calf serum | 10% |
| TGFβ1 | 1–5 ng $ml^{-1}$ |
| PDGF | 10 ng $ml^{-1}$ |

metabolism. Disrupted oxygen availability plays a critical role in diseases such as stroke and many forms of cancer. On the other hand, it is also now clear that signal transduction pathways involving oxygen may control cell proliferation and morphogenesis during tissue development and that different effects may be seen when oxygen tensions rise and fall; it is not always the case that the more oxygen there is the better. Oxygen levels vary considerably, with blood showing levels of 13% in arteries and 5% in veins, and levels in cells under normal physiological conditions ranging from 1 to 5%, these conditions being described as normoxic. In many tissue engineering situations, these normoxic levels may be quite appropriate to support tissue regeneration with fully differentiated adult cells.

However, when it comes to the direction of stem cell differentiation, the situation may be quite different, and conditions of low oxygen tension, known as hypoxia, where pericellular oxygen levels can be below 0.5%, may promote the optimal stem cell behavior. A major controlling molecule here is the transcription complex HIF, hypoxia-inducible factor. This is a heterodimer with a number of variants that contain certain functional domains. HIF-1α is particularly important since at low oxygen levels this facilitates some signal transduction pathways that promote self-renewal and inhibit apoptosis mechanisms. Conversely, increasing oxygen levels tend to degrade HIF-1α, promoting apoptosis. Several tissues, and several cell types, have been shown to be particularly sensitive to oxygen levels with respect to tissue regeneration. Neural stem cells provide a good example. Low oxygen promotes the survival and differentiation into dopaminergic neurons. High levels of oxygen provide a source of cytotoxic reactive oxygen species that promote apoptosis of several CNS derivatives. Adipose-derived stem cells are also particularly sensitive to oxygen. It should also be noted that some tissues are naturally of low oxygen tension, their survival being based on adaptation to this condition. In many situations that involve cartilage, for example, chondrocytes may survive very well under low oxygen conditions, so that chondrogenesis should be favored by hypoxic conditions.

These factors indicate that control of oxygen levels may be a very important factor in tissue engineering processes. It should be expected that this is largely under the control of bioreactor culture conditions *ex vivo* rather than the template biomaterials. Generally there is little control over oxygen levels *in vivo*. In view of this, attempts have been made to exert some influence over oxygen generation, using a variety of so-called oxygen generating polymers.

## 5.4.2 Growth factors

Tissue engineering has been defined earlier as a process that involves combinations of molecular and mechanical signaling of cells. Signaling molecules are generally grouped into three types, those that stimulate cell division (mitogens), those that control the generation of tissue form (morphogens) and those that control migration, differentiation and proliferation (growth factors). We shall concentrate on the latter here.

Growth factors are soluble polypeptides that exist naturally in the ECM and which modulate cell behavior through binding to transmembrane receptors on their own target cells. A summary of the most relevant growth factors is provided in Table 5.3. The list includes VEGF, which controls blood vessel formation through effects on endothelial cells, BMPs that control bone and cartilage formation via effects on osteoblasts, NGF that controls neural cells affecting the nerve, spine and brain, and various FGFs that influence bone, skin and nerve. Transduction of the growth factor binding signal from the membrane to the nucleus involves many processes within the cytoskeleton, similar to those discussed in Chapter 3.

Within the ECM, growth factors have very short ranges of diffusion because of slow rates and short half-lives. In addition, the same growth factor can give different instructions to cells depending on the receptor to which it binds, and the same receptor can translate different messages depending on the

Table 5.3 **Growth factors in tissue engineering. Table gives some of the more prominent growth factors that are used in tissue engineering**

| Growth factor | Tissues |
|---|---|
| Angiopoietin 1 & 2 (Ang-1 and Ang-2) | Blood vessels, heart |
| Fibroblast growth factor 2 (FGF-2) | Blood vessels, bone, skin, nerve, muscle |
| Bone morphogenetic protein 2 (BMP-2) | Bone, cartilage |
| Bone morphogenetic protein 7 (BMP-7) | Cartilage, kidney |
| Epidermal growth factor (EGF) | Skin, nerve |
| Hepatocyte growth factor (HGF) | Bone, liver, muscle |
| Insulin-like growth factor 1 (IGF-1) | Bone, cartilage, lung, kidney, nerve, skin |
| Nerve growth factor (NGF) | Nerve, brain |
| Platelet-derived growth factor (PDGF) | Blood vessels, muscle, skin, cartilage |
| Transforming growth factor α (TGFα) | Brain, skin |
| Transforming growth factor β (TGFβ) | Bone, cartilage |
| Vascular endothelial growth factor (VEGF) | Blood vessels |

available intracellular transduction pathways. Therefore, although many growth factors possess names that imply one specific target cell and one type of biological response, in reality they have several different possible effects, which we have to bear in mind when considering the intentional use of growth factors in tissue engineering.

Growth factors have been used, although not often very successfully, as pharmaceutical agents outside of tissue engineering. Success here depends on the interaction of the administered growth factor with the natural ECM, since the latter will control the spatial presentation of the former through the presence of specific binding domains. The integrins, which we have seen in Chapter 3 are a family of cell surface ECM receptors, regulate many cell signaling processes in the activity of the growth factors. This regulation is very complex and failure to control all of the variables has led to disappointing clinical trials with respect to the therapeutic use of growth factors. The formulation of the factor, the required dose level and the method of administration have been difficult to optimize, so that highly promising applications, such as the use of intracoronary injections of growth factors to stimulate cardiac function and revascularize infarcted myocardium, have not usually produced reproducible results in clinical trials. VEGF, a very important and relevant factor here, has a half-life of less than 30 minutes, and the need for massive doses and multiple injections leads to unacceptable side effects.

The reason why this history of therapeutic uses is so important is that these unsatisfactory outcomes have led to the need to develop the technology for better growth factor delivery using biomaterials, which impacts on their use in tissue engineering. In general, two methods may be considered, in which the growth factor may be immobilized in or on the biomaterial such that they are effective if and when the target cells infiltrate the material, or alternatively, the growth factors are physically encapsulated within, and released from, the material.

Since many of the details are the same as for more conventional drug delivery systems, to be discussed in the next chapter, we need not pre-empt much of that general discussion here, but a few points should be emphasized. With respect to immobilization, either covalent or non-covalent methods may be used, but in either case, the immobilized factor has to be available to the cells that come into contact with the material. It is possible for them to be active in the bound state or become activated following cleavage from the material. Direct covalent attachment has many attractive features and can be achieved by conjugating the factors to functional groups on a polymer. Non-covalent incorporation can be obtained via absorption through secondary interactions or by the use of small proteins that can mimic the key fragments of larger proteins and act as intermediates.

The physical encapsulation of growth factors within polymer scaffolds is also appealing and may be achieved by phase separation and solvent casting techniques. As discussed elsewhere, there are many biodegradable synthetic polymers that can be used. The formulation and processing of such polymers can be arranged for the incorporation of multiple growth factors and their sequential release. Similarly, many of the natural polymers used for scaffolds can encapsulate growth factors, especially when they are in gel form and can be cross-linked to assist in the factor entrapment.

In addition to the normal diffusion or degradation-controlled release, there are systems that could allow release on demand. These include pH or temperature triggered release and the dependence on proteolytic initiated cleavage of growth factors from substrates, for example by the catalytic activity of matrix metalloproteinases; these are particularly effective with hydrogel matrices.

### 5.4.3 Gene transfer in tissue engineering

Gene therapy involves the introduction of genetic material into cells with the intention of altering cellular function or structure. This type of technique has immense clinical importance, and the role of biomaterials in the delivery of gene therapy in general is discussed in detail in the next chapter. If we consider, however, the general purpose of tissue engineering, which is the generation of new tissue through the stimulation of target cells, it becomes rather obvious that there are generic similarities between gene therapy and tissue engineering and that, moreover, there may be strong possibilities that the two techniques could be combined synergistically in order to provide more effective tissue regeneration. The simultaneous delivery of DNA to the target cells in tissue engineering alongside the other molecular and mechanical signals has been an attractive option for some time.

In Chapter 6, we shall see that there are two main methods for delivery of genes into cells, in which substances known as vectors are combined with the gene, which is then inserted into the cell, either *in vitro* or *in vivo*. These two methods involve, on the one hand a type of modified virus and, on the other hand, a chemical agent, this agent constituting a type of biomaterial. It is also possible for DNA to be transferred through physical processes, although these tend to be relatively non-specific in their action.

Assuming for a moment that non-viral transfer is preferred, there are, in principle, two ways in which we can accomplish this DNA insertion within a tissue engineering process. One involves the separate transfection of the target cells in which the biomaterial-conjugated gene is added to the culture medium containing the cells, which are then seeded into the tissue engineering scaffold or template for *in vitro* transfection. The second involves the incorporation of the biomaterial-conjugated gene into the biomaterial template surface, where it may influence cell behavior over a period of time during the regeneration process, either *in vitro* or *in vivo*. We may see the potential for gene transfer in tissue engineering by reference to a few different types of tissue.

Cartilage is a good place to start since it is composed of a single cell type, the chondrocyte, and has specific characteristics that provide unusual regeneration conditions. As noted later, it has proven particularly difficult to regenerate significant volumes of cartilage, largely because the tissue is avascular, with limited prospects of diffusion of molecules and nutrients from adjacent tissue. Because of the dense ECM that surrounds chondrocytes within cartilage, direct *in vivo* gene delivery is not possible. Moreover, autologous chondrocytes lose their chondrocytic phenotype when grown *in vitro* and readily assume a fibroblastic phenotype. Growth factor plasmids such as IGF-1 and TGF-$\beta$1 have been used to increase proliferation rates and the expression of type II collagen and aggrecan, while cells transduced with certain BMPs show improved healing of full thickness defects. With bone, BMP genes, such as BMP-2 can also enhance healing. Periosteal cells may also be transduced with BMP-7 for enhanced effects. There is much interest in transducing MSCs that are seeded into scaffolds with transcription factors such as the osteoblastic factor Runx2/Cbfa1. In nerve tissue engineering, it is essential that the cells present are able to secrete neurotrophic factors such as brain-derived growth factor, BDGF, and gene therapy has been used to assist in the controlled gene expression here. In skin, EGF, epidermal growth factor is an important molecule and it has been shown that delivery of the EGF gene can give a dramatic increase in the EGF protein concentration, with accelerated rates of healing.

## 5.5 Biomaterials for regenerative medicine and tissue engineering processes

### 5.5.1 Background

We now come to the main theme of this chapter, that is the biomaterials that are used in the processes of regenerative medicine, and especially, tissue engineering. It is tempting here to provide a list of those materials that have been used over the course of the last several decades for these processes, with a rationale for their selection and a discussion of their known performance. However, that would not be too helpful as a list of previously used materials does not at all provide a good perspective on the specifications for those biomaterials that are likely to be successful in the future. The selection of so-called scaffolds and matrices for tissue engineering processes was initially pragmatic, and understandably so. As tissue engineering products and processes were developed, it was natural that consideration had to be given to eventual regulatory approval for their use in human patients. This was already a formidable task when just considering the issues involving the manipulation of cells outside of the body, let alone the use of radically new materials to assist in that manipulation. Early stage products were almost wholly based on biomaterials that had prior regulatory approval in other non-tissue engineering applications. Regulatory approval procedures and the regulatory bodies such as the Food and Drugs Administration (FDA) in the USA will be discussed in Chapter 9, but it can be appreciated here that prior FDA approval became the principal specification for tissue engineering scaffolds at an early stage.

With hindsight, this was unfortunate and wrong. It was assumed that, since regulatory approval with respect to biomaterials-based devices is, to a large extent, concerned with the issues of safety and it was assumed that data on safety that had been clearly demonstrated in one non-tissue engineering application could be transferred to specific tissue engineering applications. However, biological safety is just a component of biocompatibility and we have seen that the performance of biomaterials with respect to biocompatibility varies from one application to another. The predicate biomaterials for tissue engineering scaffolds were assumed to be those materials (especially biodegradable materials) that had been safely used in implantable medical devices or drug delivery systems. These materials, however, had been designed to perform some non-biologic function in those devices without having any undesirable effect on the host, which usually meant having no biological activity at all.

We have already seen that in tissue engineering processes, certain cells, recruited for this purpose, have to be directed to express new tissue, and they require a combination of molecular and mechanical cues in order to do so. The biomaterial is the support or template within which these processes take place. We cannot expect, therefore, that these processes will be optimized, or even take place at all, if we use materials that cannot take part in any biological activity. Instead, the biomaterials that we use must be designed to facilitate cell performance and fully participate in the cell signaling and tissue expression processes.

### 5.5.2 The concept of tissue engineering templates

In this section we will now define the objectives of these biomaterials, set out a series of specifications for them, and describe the future array of biomaterials rather than the past catalog. One of our problems here is that some terminology has already been established that attempts to describe these materials and structures, where the words do not adequately express what is happening. The two most common are scaffold and matrix. It would not make any sense to ignore these since they are in common usage but that does not imply that we cannot be more creative with our descriptions. In addition, it is immensely important to recognize that the performance of the structures that we use in tissue engineering processes depends on both the material chemistry and its physical form, and indeed the latter may be the more important. The description of a "scaffold material" therefore falls far short of the characterization and specification that are required. The use of the composite noun "injectable scaffold" shows just how difficult the terminology discussion has become.

We therefore use the overarching term "template" to describe the biomaterials-based constructs that are used in tissue engineering. This embraces the external shape or form, the internal architecture,

**Box 5.7 Tissue engineering templates**

- Tissue engineering involves cells that are stimulated to generate new tissue by a combination of molecular and mechanical signals. Biomaterials are used to facilitate the delivery of these signals.
- Conventionally the material structures have been described as scaffolds, but they should be more than mechanical supports for the target cells.
- The biomaterial component is better considered as a template, which incorporates concepts of shape and volume, the chemistry and architecture necessary to support cell function, and the mechanical characteristics that provide both mechanotransduction, elasticity and strength.
- The template may be prefabricated for use in a bioreactor or for direct implantation, or may be injectable with appropriate cross-linking, gelation or self-assembly.
- The template must be conducive to the control of nutrient, oxygen and biomolecules supply to the cells and generating tissue.
- The template should have appropriate degradation profiles, with the biocompatibility necessary to support the processes of tissue generation without significant inflammation or immune responses.
- The template should be conducive to incorporation of the generated tissue into the host, with vascularization and innervation.

the material of construction, any surface-bound molecules or structures and any added biomolecules. This term subsumes the descriptors of scaffolds, matrices, gels, membranes and their variants, shown in Box 5.7.

### 5.5.3 The objectives of tissue engineering templates

Let us consider what we are really trying to do with our tissue engineering biomaterials. We start with a single cell that, in its pre-existing state, does not have the ability to express any ECM molecules. We need to change the internal characteristics of that cell and/or its environmental characteristics in order for it to recapitulate or gain that ability. Obviously we are unlikely to work with just a single cell and so we turn our attention to a population of identical cells and try to change their collective characteristics so that together they express multiple ECM molecules or other molecules that are of interest to us in a regenerative therapy. It may be possible to do this with the cells in suspension in culture, and, if the molecules they are releasing are the sole players in the therapy, all we have to do is deliver this suspension to the requisite site. This is the basis of cell therapy and could relate, for example, to the delivery of dopamine-producing cells to the brain for treatment of Parkinson's disease or stem cells that facilitate axon regeneration in the treatment of spinal cord injury. Templates and biomaterials may not have any role here.

However, in most situations it will be necessary to have spatial, and probably temporal, control over the behavior of these cells, and this is the control that the template should exercise. The following are the components of this control:

- There will need to be control over the volume and, in many cases, shape, of the region in which the cells are operating. This may simply be to confine the activity to the region of the lesion or target tissue or to give shape to the tissue that is being generated.
- There may need to be careful control of the mechanical signaling to the cells, which is unlikely to happen in most situations where they are simply injected into tissue.
- Cell-to-cell contact and signaling may be required in order to optimize their performance, again not readily achieved following free injection.
- Under *ex vivo* conditions, the supply of nutrients, including oxygen, needs to be regulated.
- Equally, molecular signaling through growth factors, or alteration of cell phenotype through gene transfer may be required, and this may need spatiotemporal control.
- In cases of complex tissue regeneration, more than one cell type is required for the expression of different types of ECM and this will definitely need spatiotemporal control – this becomes even more complex when we consider whole organ printing.
- In some situations, allogeneic or even xenogeneic cells may be used and they need protection from the host immune system through the use of some membrane template.
- The behavior of cells, whether stem cells or fully differentiated adult cells, and whether individually or in colonies, may have to be optimized through control over their interactions with substrates, usually best achieved through carefully engineered substrates, often at the nanoscale.

These, then, are the generic objectives of tissue engineering templates. We will now consider the form that such templates can take in practice. We do so by first producing a classification of tissue engineering templates based on the current portfolio of experimental and commercial structures. This classification refers to the characteristics of the templates themselves and not just the materials of their construction. We then extract information from the performance of these existing templates in order to characterize optimal template materials.

### 5.5.4 Classification of tissue engineering templates

As a starting point in the discussion of individual materials, we may look at the existing profile of those that have been used in templates experimentally and clinically. As we have implied, these templates may take several different forms. They may be simple homogeneously porous structures, into which cells are seeded under *ex vivo* culture conditions, and from which implantable constructs arise. They may be gels that attempt to mimic the environment of cells or they may be derived from naturally occurring ECM tissues. They may be used solely *ex vivo* to direct cell behavior or may provide *in vivo* protection to cells. The following system represents a broad classification of these templates:

Class I Homogeneous, isotropic, unmodified, porous structures: for example, a tricalcium phosphate block of uniform pore distribution; this could be implanted in a critical size defect in bone in order to facilitate osteoconduction, possibly assisted by prior cell seeding, Figure 5.7(a).

Class II Homogeneous, isotropic porous structures with chemically modified surfaces: for example, a similar construct to that in Class I but with BMP-2 attached to the pore surface, Figure 5.7(b).

Class III Homogeneous, anisotropic, unmodified, porous structures: for example, an electrospun homopolymer with oriented fibers; this could be fabricated as a tube to be used in blood vessel regeneration, Figure 5.7(c).

Class IV Homogeneous, anisotropic, porous structures with chemically modified surfaces: for example, a RGD modified electrospun homopolymer with oriented fibers, Figure 5.7(d).

Class V Homogeneous, isotropic porous structures with biomolecule delivery: for example, a porous polymer made by salt leaching that incorporates VEGF, Figure 5.7(e).

(a)

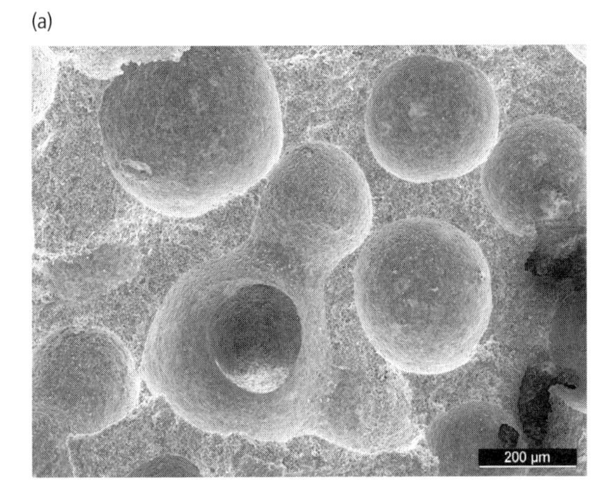


(b)

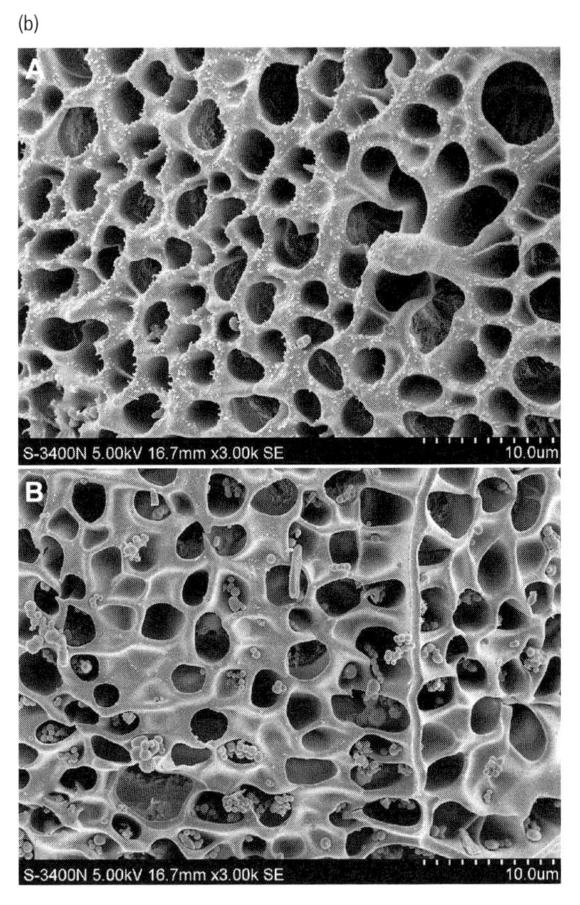


**Figure 5.7** Examples of tissue engineering scaffolds.
**(a)** Porous tricalcium phosphate, image courtesy of Dr. Marc Bohner, Skeletal Substitutes Group, RMS Foundation, Bettlach, Switzerland.
**(b)** A homogeneously porous glycidyl methacrylated dextran/gelatin scaffold with surfaces modified by growth factor loaded microparticles. Reprinted from *Biomaterials* 30, Chen F-M *et al.*, *In vitro* cellular responses to scaffolds containing two microencapsulated growth factors, 5215–24, © 2009, with permission from Elsevier.

Class VI Homogeneous, anisotropic porous structures with biomolecule delivery: for example, a polymer with aligned porosity that incorporates a neurotrophic factor for delivery to a nerve, Figure 5.7(f).

Class VII Heterogeneous, multiphase, unmodified, porous structures: for example, a blend of silk and elastin fibers, Figure 5.7(g).

Class VIII Heterogeneous, multiphase, porous structures with chemically modified surfaces: for example, an RGD-functionalized chitosan–gelatin porous solid, Figure 5.7(h).

Class IX Heterogeneous, multiphase, porous structures with biomolecule delivery: for example, a porous PCL-PHB structure containing growth factors, Figure 5.7(i).

Class X Derivatized ECM: for example, decellularized small intestine submucosa, Figure 5.7(j).

Class XI Injectable, *in situ* polymerizing or setting substance: for example, a photopolymerizable methacrylate modified Dextran hydrogel, Figure 5.7(k).

Class XII Non-injectable hydrogel: for example, a thermoresponsive polymer for cell sheet engineering, Figure 5.7(l).

Class XIII Membranes for cell encapsulation: for example, alginate microcapsules, Figure 5.7(m).

Class XIV Micro-engineered, patterned surfaces for cell culture: for example, microfabricated platform of PEG microwell arrays for ES culture, Figure 5.7(n).

### 5.5.5 Specifications for template materials

Having defined the generic structural characteristics of the templates, we now turn our attention to the

(c)

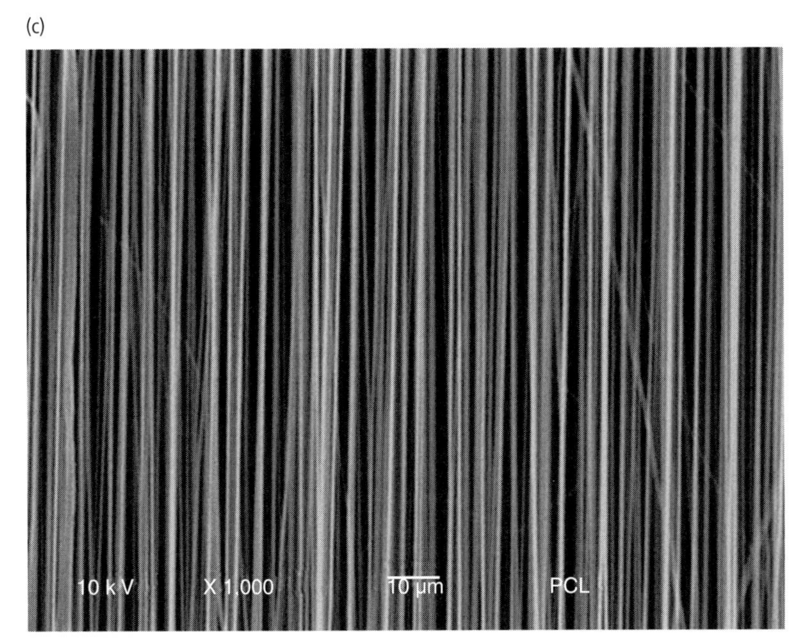


(d)

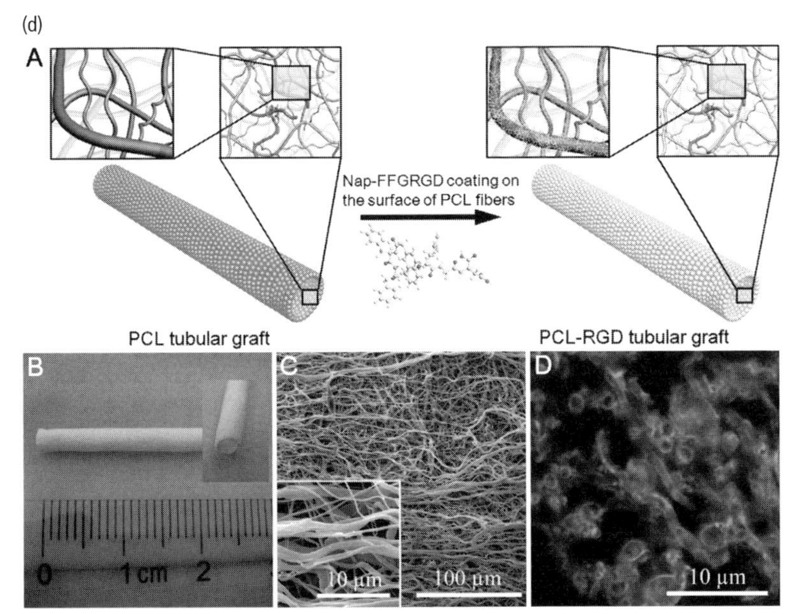


**Figure 5.7** *continued*
**(c)** Field emission scanning electron microscope image of electrospun polycaprolactone (Mw 80 000) oriented nanofibers (fiber diameter 480 ± 15 nm; solvent: methanol/chloroform (1:3): 9% w/v). Courtesy Professor Seeram Ramakrishna, National University of Singapore.
**(d)** RGD modified PCL tubular grafts: (A) schematic illustration of the structures of PCL and PCL-RGD grafts; (B) optical image of electrospun tubular PCL grafts; (C) scanning electron microscopy images of the luminal surface of PCL-RGD graft; (D) fluorescence microscopy image of the cross-section of tubular graft for FITC-labeled PCL-RGD. Images courtesy of Dr. Dan Ding and Professor Deiling Kong, The Key Laboratory of Bioactive materials, Nankai University, Tianjin, China.

materials that may be used in these templates, with the initial discussion centering on their requirements and specifications. Let us remind ourselves what we are attempting to do. The template should facilitate, even optimize, the delivery of molecular and mechanical signals to the target cells. It should accommodate those cells within an environment that replicates, as far as possible, the environment in which those cells normally reside, as with the stem cell niche, in order to maximize the chance of those

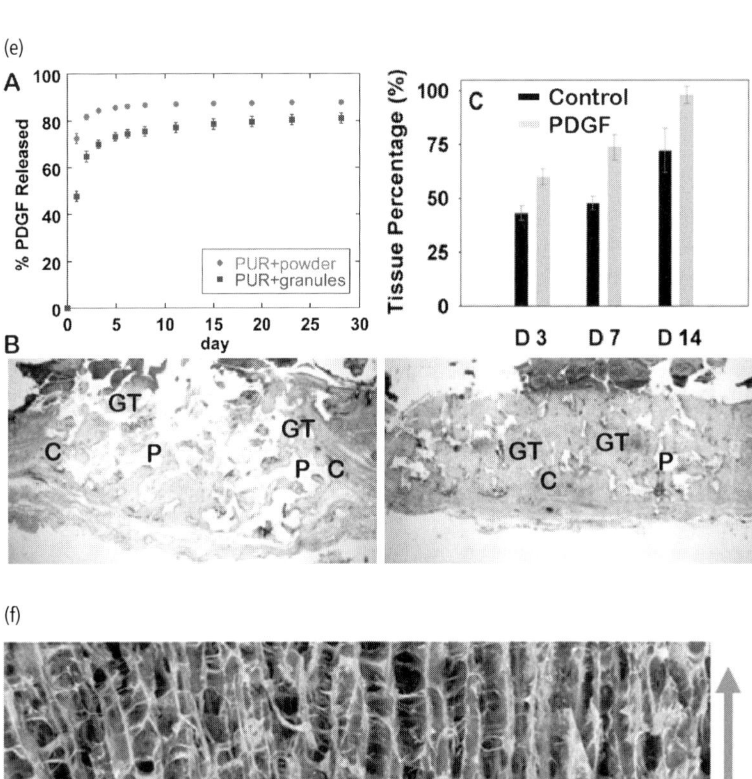


**Figure 5.7** *continued*
**(e)** Local delivery of growth factor incorporated into a porous polymer scaffold. These images show the effects of local delivery of the recombinant platelet-derived growth factor, rhPDGF-BB on healing of excisional wounds in rats. (A) rhPDGF-BB was added to resorbable polyurethane scaffolds as either a labile powder or bound to heparin (Hp)-modified PLGA microspheres embedded in gelatin granules. Binding rhPDGF to Hp resulted in a more sustained release compared to the diffusion-controlled release observed for the labile powder. (B) Histological sections of scaffolds augmented with either 0 (left) or 1.8 mg (11.5 mg/ml scaffold) rhPDGF-BB powder (right) implanted in 6-mm excisional wounds in rats at day 7. Local delivery of rhPDGF-BB resulted in accelerated polymer (P) degradation, as well as enhanced infiltration of granulation tissue (GT) and deposition of collagen (C). By day 14, wounds treated with scaffolds augmented with rhPDGF-BB had re-epithelialized while those treated with control scaffolds had not. Images courtesy of S.A. Guelcher, Vanderbilt University, USA. See plate section for color version.
**(f)** Structurally aligned collagen-GAG scaffold composed of longitudinally aligned, ellipsoidal pores. Fabricated via lyophilization using a directional solidification approach. Inset: stereology determined best-fit ellipse representation of mean pore shape. Images courtesy of Steven R. Caliari and Brendan Harley, Chemical and Biomolecular Engineering, Institute for Genomic Biology, University of Illinois at Urbana-Champaign, IL, USA.

cells expressing new tissue. The template should be responsive and adaptable to time sensitive changes in the environment. It should allow for optimal flow of nutrients and gases. It should not have any deleterious chemical effects on the cells. It should usually give shape and volume to the new tissue construct. It should have an architecture that facilitates innervation and vascularization. It will normally be expected to degrade during and after tissue regeneration without stimulating significant inflammation or other undesirable effects within the tissue.

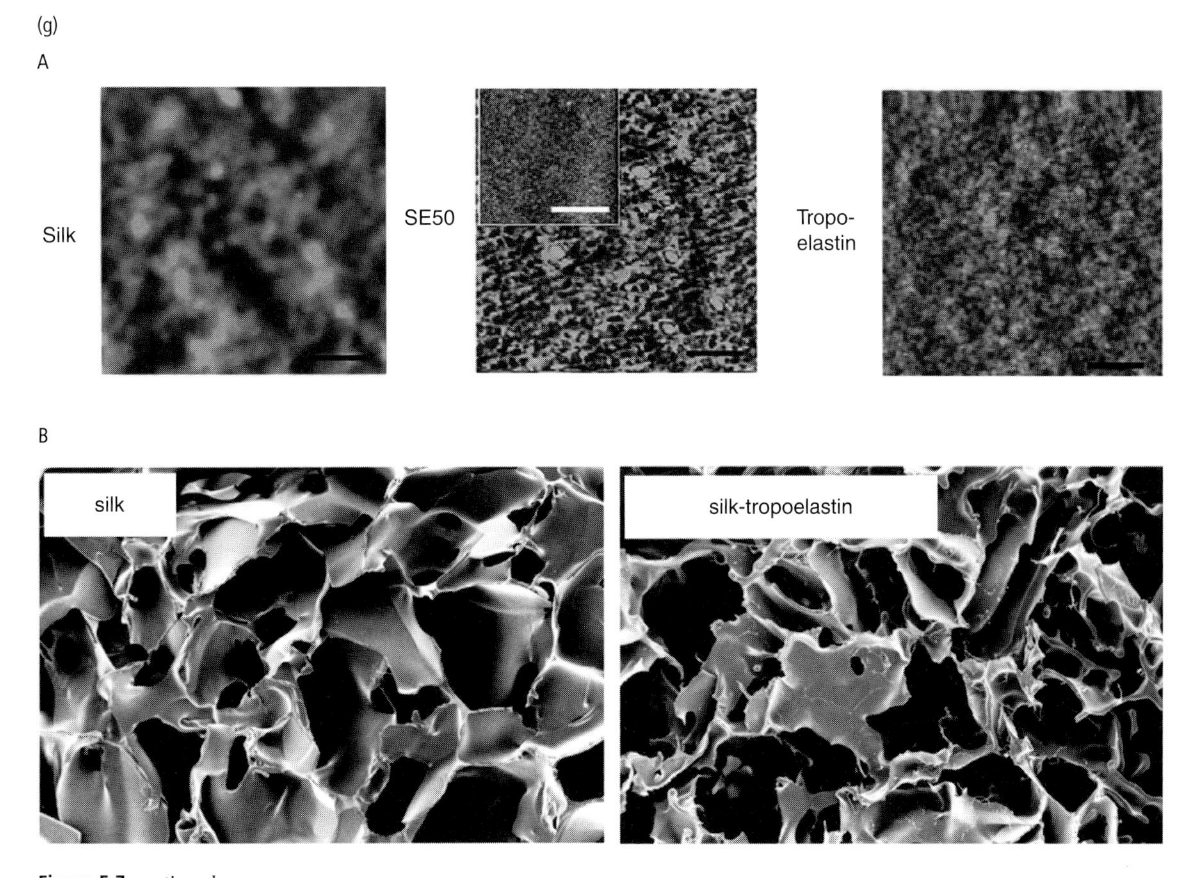


**Figure 5.7** *continued*
(g) Silk–tropoelastin blends. (A) AFM images: tropoelastin, silk, and silk–tropoelastin blend (50:50) in water, 0.2 wt%, cast on mica. Scale bars = 200 nm. (B) Porous protein matrices. Silk (left, 5% wt/v silk solution) and silk-tropoelastin blend (right, 5% wt/v silk solution with 5% wt/v tropoelastin solution, 70:30 ratio by volume), prepared by freezing at −20°C and lyophilization, followed by autoclaving at 121°C for 30 min to induce beta-sheet formation in the silk and stabilize the silk and tropoelastin. Scale bars = 100 um. In both cases, tropoelastin was supplied by Dr. Tony Weiss, University of Sydney, and images taken in the laboratory of Dr. David Kaplan, Tufts University, Boston, (A) by Xiao Hu and (B) by Jelena Rnjak-Kovacina.

This is a formidable list of requirements for the template. These are translated into the following specification for tissue engineering biomaterials:

### Mandatory

- The material should be capable of recapitulating the architecture of the niche of the target cells.
- Moreover, since the cell niche is changeable over time, the material should be capable of adapting to the constantly changing microenvironment.
- The material should have elastic properties, particularly stiffness, that favor mechanical signaling to the target cells in order to optimize differentiation, proliferation and gene expression.
- The material should have optimal surface or interfacial energy characteristics to facilitate cell adhesion and function.
- The material should be capable of orchestrating molecular signaling to the target cells, either by directing endogenous molecules or delivering exogenous molecules.

(h)

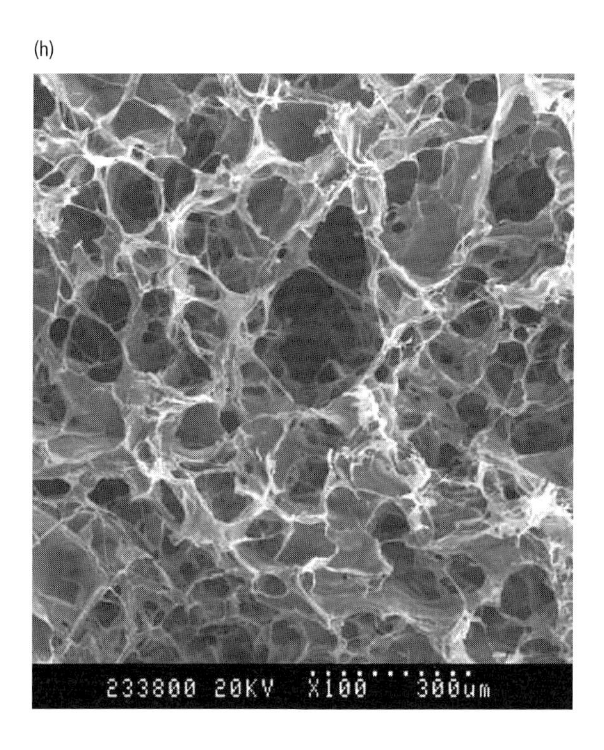


(i)

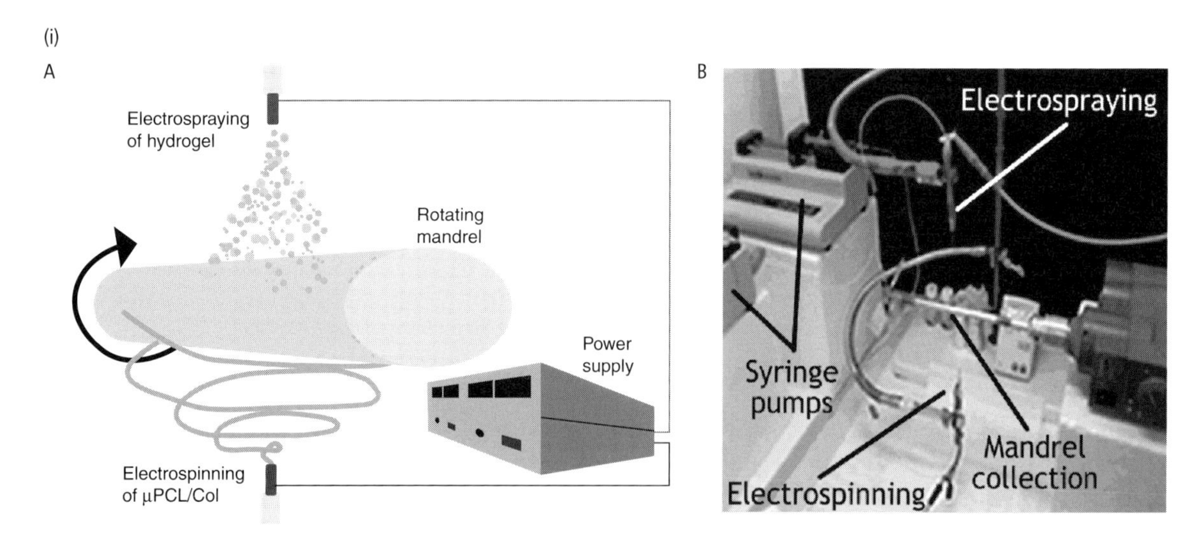


**Figure 5.7** *continued*

**(h)** Multiphase porous structure with chemically modified surface. SEM of gene-activated matrix, involving DNA-incorporated chitosan-gelatin. Image courtesy of Professor J Zhang, Nanjing University, China.

**(i)** A hybrid mesh of poly (3-caprolactone)-collagen blend (PCL/Col) and hyaluronic acid (HA) hydrogel created via a dual electrodeposition system. Simultaneous spraying of HA and spinning of PCL/Col allowed the dual loading of two angiogenic growth factors VEGF and PDGF-BB over a period of five weeks *in vitro*. Image courtesy of Professor Dietmar Hutmacher, Institute of Health and Biomedical Innovation, Queensland University of Technology, Brisbane, Queensland, Australia.

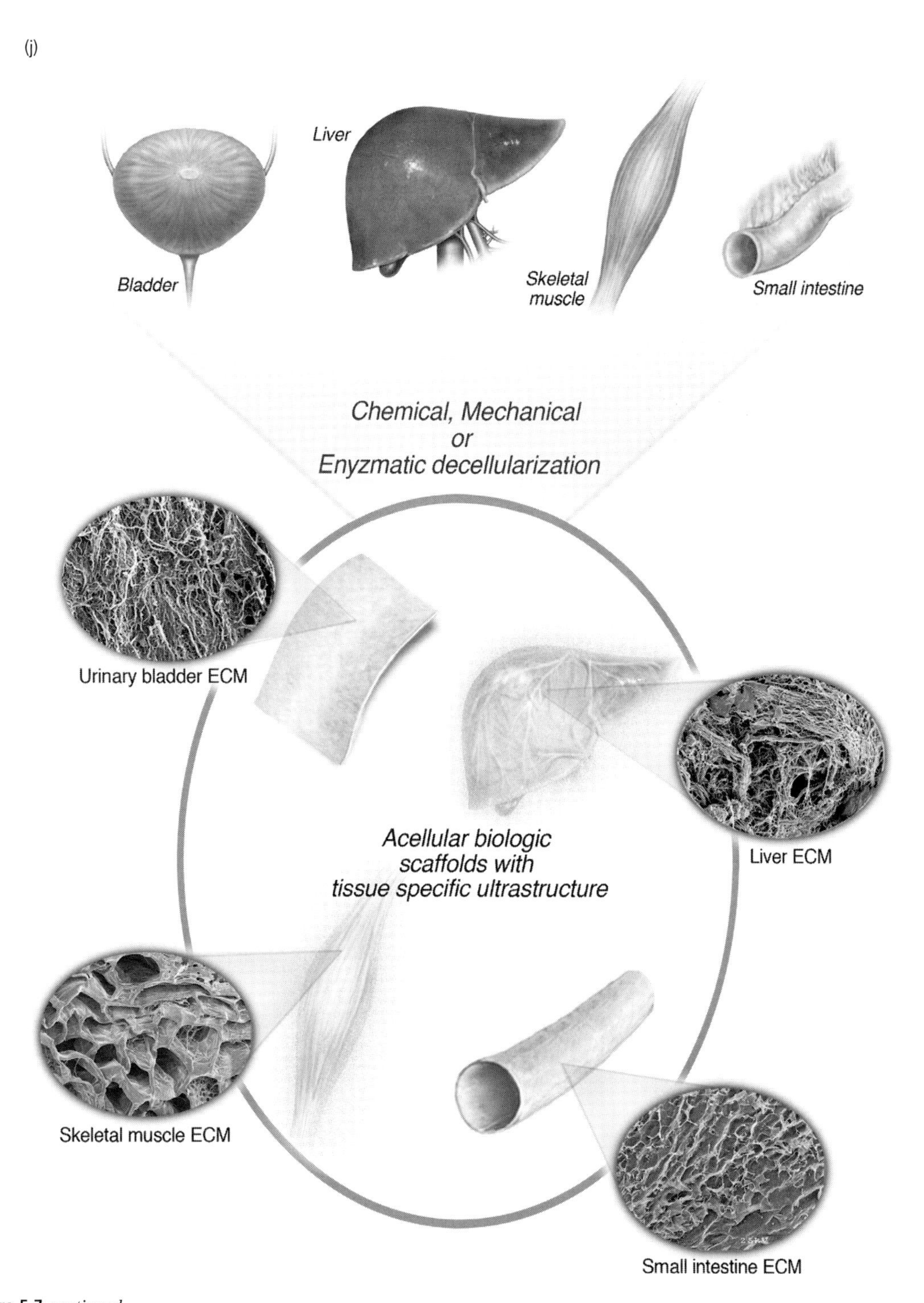


**Figure 5.7** *continued*
(j) Schematic showing the derivation of acellular biological scaffolds. Image courtesy of Dr. Steve Badylak, McGowan Institute for Regenerative Medicine, Department of Surgery, University of Pittsburgh, PA, USA.

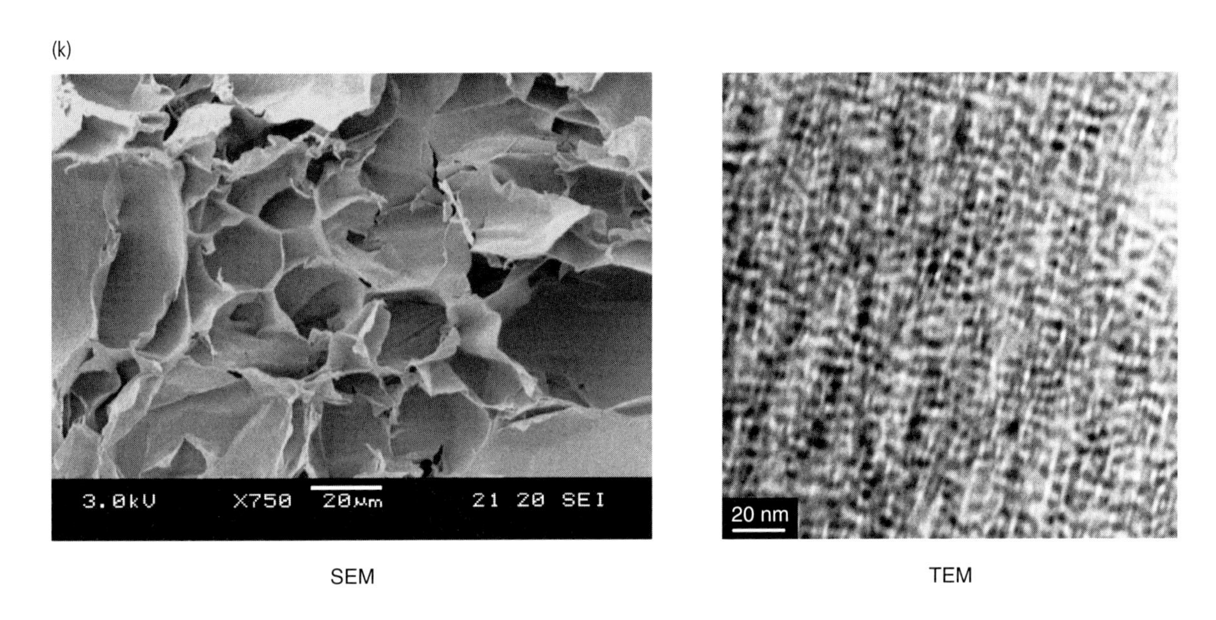


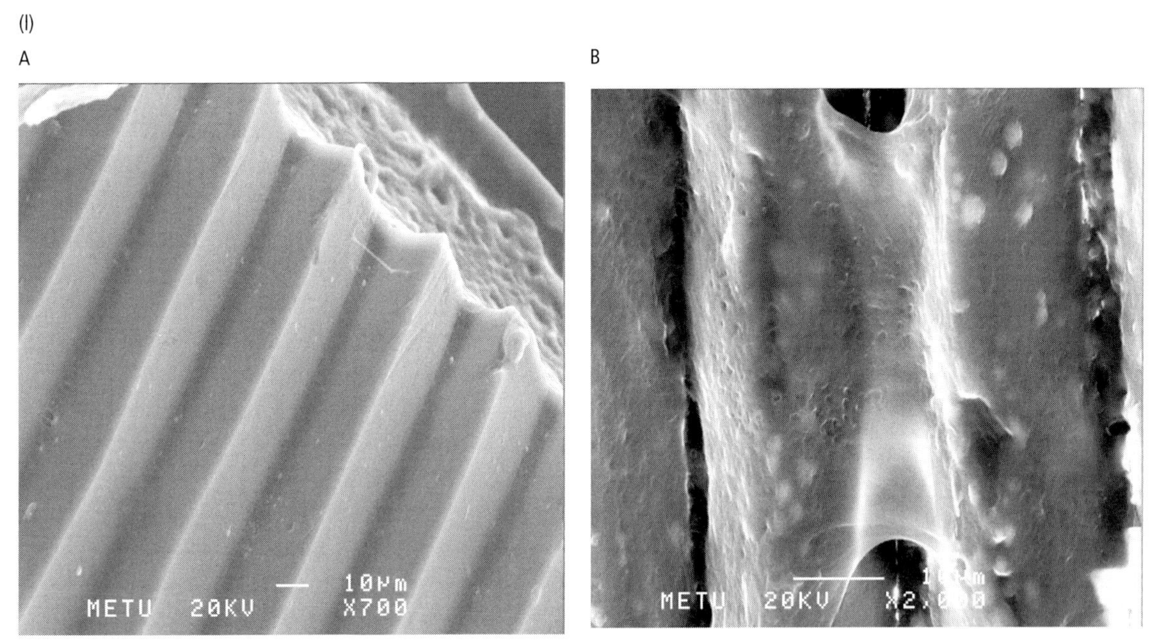


**Figure 5.7** *continued*

**(k)** Injectable hydrogel. Scanning and transmission electron micrographs of a pH-triggered injectable hydrogel of N-palmitoyl chitosan, Courtesy of Professor Hsing-Wen Sung, National Tsing Hua University, Hsinchu, Taiwan.

**(l)** Micropatterned hydrogels. (A) Pristine pNIPAM-pHEMA micropatterned film. (B) PolyNIPAM micropatterned film seeded with rat bone marrow mesenchymal stem cells cultured under dynamic conditions. Original SEMs from N. Ozturk's M. Sc. Thesis, courtesy of supervisor V. Hasirci, and co-supervisor G.T. Kose. Middle East Technical University, Ankara, Turkey.

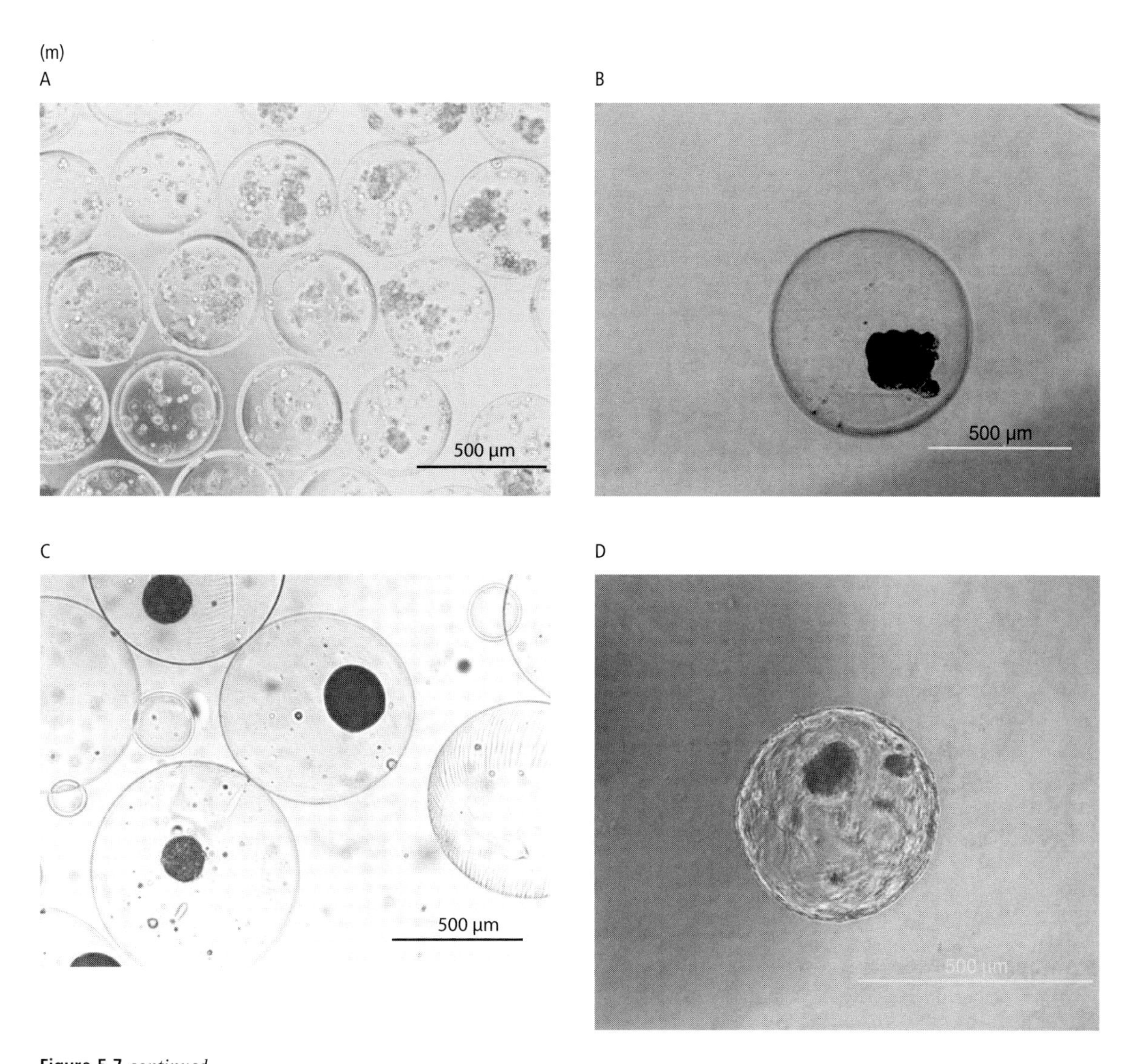


**Figure 5.7** *continued*
**(m)** Alginate encapsulated cells. Images courtesy of the Opara Lab at Wake Forest Institute of Regenerative Medicine, NC, USA. (A) Bone marrow-derived mesenchymal stem cells, encapsulated with 1.25% ultra pure alginate using an electrospraying technique, then coated with poly-L-lysine and 0.25% ultra pure alginate. (B) Rat islets, encapsulated using a microfluidic device, stained with the dye dithizone, indicating the presence of beta cells. (C) Encapsulated fetal porcine islets. (D) Encapsulated human islets.

- The material should be of a physical form that provides appropriate shape and size to the regenerated tissue.
- The material should be capable of forming into an architecture that optimizes cell, nutrient, gas and biomolecule transport, either *ex vivo* or *in vivo* or both, and facilitates blood vessel and nerve development.
- The material should be intrinsically non-cytotoxic and non-immunogenic, and minimally pro-inflammatory.

### Optional

- The material should be degradable if that is desired, with appropriate degradation kinetics and

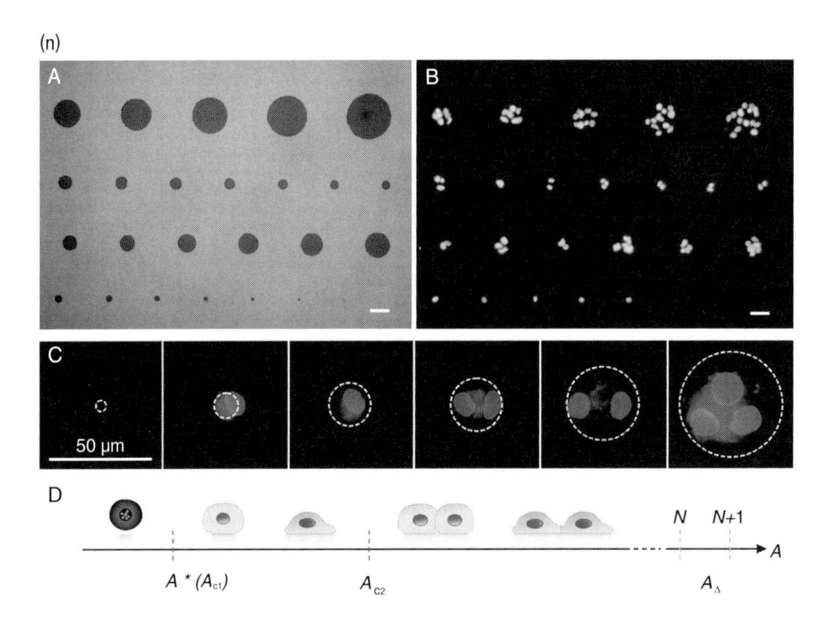


**Figure 5.7** *continued*

**(n)** Cell patterning and critical areas of cell adhesion on micropatterned surfaces with adhesion contrast. (A) Optical micrographs of micropatterns with RGD-peptide-grafted gold microislands of varying diameters on PEG hydrogel. (B) Fluorescence micrograph of MC3T3-E1 cells, a pre-osteoblast cell line, on a micropatterned surface with cellular nuclei labeled by DAPI. (C) Fluorescence micrograph of adherent cells with cellular nuclei labeled in blue and F-actin in red – the dashed circles indicate the contours of the underlying RGD-grafted microislands. (D) Schematic presentation of three characteristic areas for cells adhering on adhesive microislands on a cell-resistant background: $A_{c1}$, also named $A^*$, the critical area from apoptosis to survival; $A_{c2}$, the critical area from single-cell adhesion to multi-cell adhesion; $A_{\Delta}$, the characteristic area for one more cell to adhere. Courtesy of Professor Jiandong Ding and Dr. Ce Yan of Fudan University, China. See plate section for color version.

appropriate morphological and chemical degradation profiles.

- The material should be injectable if that is desired, with the appropriate rheological characteristics and transformation mechanisms and kinetics.
- Where necessary, the material should be compatible with the processing techniques that simultaneously pattern both the material and living cells.
- Where multiple cell types are involved, the material properties should be tunable in order to accommodate variable cellular requirements, with spatio-temporal control as appropriate.
- When used in a significantly stressed *in vivo* environment, the material must have sufficient strength and toughness.
- In those situations where the biomaterial encapsulates cells, optimal diffusion characteristics concerning key molecules are required.

## 5.5.6 Specific types of template material

In this section, we do not intend to discuss in detail each material that has been or could be used for these templates, since most of them are described in other parts of this book, and are summarized in Chapter 8, which gives a comprehensive guide and classification of biomaterials used in the different sectors of medical technology. Rather, we discuss the concepts that these materials represent so that we can have an overview of the landscape of materials for regenerative medicine.

If we distill the essential features of template materials from the long lists of template classes, requirements and specification, we arrive at a simple hierarchy of structures. There are, in fact, three groups of template biomaterials, the porous solids, the hydrogels and the decellularized tissues. There is some overlap between these groups since some polymers (either synthetic or natural) can be presented as a porous solid or a gel, and there are significant subdivisions within them, for example pre-formed and injectable gels, and soft and hard porous solids. Nevertheless, these three groups have sufficient distinctive features that allow these groupings to be robust and scientifically meaningful.

### Porous solids

Many of the first attempts to use biomaterials in tissue engineering involved therapies for skin regeneration and cartilage repair. In the former case the therapies were targeted towards burns and ulcers and the material was required to be in the form of a simple porous flat sheet. Cartilage lesions tended to be a little different and required more of a cylindrical shape. The tissues being addressed were soft and it was reasonably logical to turn to porous polymers as a basis for cell seeding and regeneration of tissue at the affected site; in both situations it was thought desirable for the polymer to degrade as the new tissue was formed. Moreover, it was assumed that since several biodegradable polymers had already been used successfully in medical devices, with regulatory approval, these would be ideal for the so-called tissue engineering scaffolds.

Thus tissue engineering products emerged with various porous polyester formulations, including PLLA and PGA and their copolymers and PCL. It is fair to say that these scaffold materials have not been very successful, although it is also true that many more factors other than the material have been at fault.

Considering the mandatory specifications, porous synthetic biodegradable polymers do not resemble the cell niche, either physically, chemically or mechanically. There have been some commendable modifications to the porous architecture, including the development of nanoarchitecture using techniques such as electrospinning as noted later, but it is essentially impossible to replicate that niche with manufactured synthetic fibers or open foam structures. It is hard to replicate tissue elasticity with such porous structures and very difficult to achieve the optimal hydrophilicity. These mean that the delivery of molecular and mechanical signaling is not optimal with these materials and structures.

Having said that, some success in experimental systems with such polymers has been obtained. The range of polymers has been extended to include most of the degradable systems discussed in Chapters 2 and 8. Some of the deficiencies can be addressed by surface modification of the polymers, for example by the attachment of RGD sequences to fiber surfaces or plasma treatments, and a wide variety of copolymerization techniques have been used in order to tune degradation profiles. Within hard tissues, limited success has been achieved with porous calcium phosphate materials, discussed later in this chapter and in Chapter 8, although it is recognized that bone is able to heal spontaneously in certain situations, so that the interpretation of "success" has to be qualified.

In many ways, a far better option than using synthetic degradable polymers in soft tissue engineering involves the use of natural biopolymers. There has been a significant move towards proteins (collagen, elastin, silk, etc.) and polysaccharides (alginates, chitosan, hyaluronan, etc.) all of which may be prepared in solid porous form. Several of these are clearly involved in signaling processes in their normal biological state and they may be associated directly with the cell niche. The use of copolymers and blends allow for fine-tuning of many of the properties, including elasticity and biodegradation. They may partially hydrate within a bioreactor or in tissues, which facilitates cell function, and many of them are capable of incorporating and releasing active molecules. As we shall see in the next section, they may also be formulated as gels.

A summary of the status of porous solids in templates is provided in Box 5.8.

## Box 5.8 **Porous solids in tissue engineering templates**

A variety of porous solids, including polymeric, ceramic and composite materials, have been used, mostly experimentally, in tissue engineering processes.

| **Natural polymers** | |
|---|---|
| Chitin/chitosan | Freeze-drying, wet spinning, thermally induced phase separation (TIPS)<br>Cartilage, osteochondral, bone, skin, nerve |
| Cellulose (bacterial) | Lyophilization<br>Bone, cartilage |
| Collagen | Freeze-drying, electrospinning, solid freeform fabrication (SFF), solvent evaporation<br>Cartilage, skin, cardiovascular |
| Silk | Native/regenerated/recombinant: freeze-drying, electrospinning<br>Bone, cartilage, skin, ligament |
| Blends | Silk–elastin, collagen–chitosan and other blends for various uses |
| **Synthetic polymers** | |
| Poly(lactic acid) | Up to 5 years degradation time<br>Electrospinning, TIPS<br>Skin, cartilage, ligament, tendon, bone |
| Poly(glycolic acid) | Up to 1 year degradation time<br>Electrospinning, TIPS<br>Nerve, urinary tract |
| Poly(lactic-co-glycolic acid) | Typically 6 months degradation time<br>Electrospinning<br>Skin, nerve, cartilage |
| Polycaprolactone | Up to 2 years degradation time<br>Electrospinning<br>Skin, cartilage, blood vessels, nerve, tendon |
| Polyanhydrides | Degradation within 1 month<br>Casting and porogen leaching<br>Bone |
| Poly(propylene fumarate) | Degradation time from 6 months to 3 years<br>Cross-linking with porogen leaching<br>Bone, blood vessels |
| Degradable polyurethane | Variable degradation times<br>Solvent casting and porogen leaching<br>Vessels |

| | |
|---|---|
| Also many copolymers and blends of the above and additional monomers | |
| **Ceramics** | |
| Hydroxyapatite | Sintering, SFF<br>Bone |
| Tricalcium phosphate | Sintering, SFF<br>Bone |
| Biphasic calcium phosphates | Sintering, SFF<br>Bone |
| **Composites** | |
| PLLA–hydroxyapatite | TIPS<br>Bone, vascular |
| PLA–PCL–hydroxyapatite | Electrospinning<br>Bone |

## Hydrogels

An indication of the potential for gels to facilitate the culturing of cells, including stem cells, can be seen with the ubiquitous use of the material Matrigel™ in laboratories. This is a commercially available gel that allows cells to replicate many of their normal complex behavior patterns under culture conditions, yielding phenomena that are not replicated with synthetic polymer substrates. The material is a gel derived from a tissue cultured mouse sarcoma line and contains laminin, type IV collagen, entactin, heparan sulfate proteoglycans and growth factors. It cannot be used in clinical applications in regenerative medicine in view of the sarcoma cell origin, but its performance characteristics indicate the potential for such naturally derived gel matrices.

This section focuses on gels, and more specifically hydrogels, the essential features of which have been described in Chapter 2. Hydrogels are 3D networks formed from hydrophilic polymers, or macromers that are cross-linked to form insoluble matrices. These polymers are usually soft and elastic with strong thermodynamic compatibility with water. The cross-linked structure is characterized by junctions formed from strong chemical bonds, physical entanglements, microcrystallite formation or weak interactions. The network morphology can be amorphous, semicrystalline, supramolecular, or colloidal aggregates, the physical structure and characteristics depending on the initial monomers and macromers. The polymers can be combined in the form of blends, copolymers, and interpenetrating networks. Gels based on blends may be prepared by processes such as the freeze–thaw process where the polymer is repeatedly frozen and thawed in cycles to form a network. Interpenetrating networks (IPNs) may be synthesized by sequentially polymerizing and cross-linking a monomer in the presence of another cross-linked network. Hydrogels can be neutral, cationic, anionic or ampholytic, determined by the pendant groups.

The network structure may be characterized by the polymer volume fraction in the swollen state, the average molecular weight between cross-links and a measure of distance between cross-links. These parameters are derived from equilibrium swelling theory and rubber elasticity theory. The former describes swelling by consideration of how the cross-linked polymers reach equilibrium within a fluid in terms of the thermodynamic force required to reduce entropy, taking into account synthesis and fabrication methods, solvent conditions and mechanical loading history. The latter reflects the fact that the hydrogels can exhibit a range of responses to mechanical stress from rapid, elastic recovery to a time-dependent viscous type of recovery. This behavior is dependent on the $T_g$ of the material. Also important are the mass transport characteristics since they will determine the distribution of nutrients, gases, waste products and bioactive agents. Diffusion is usually considered to be the driving force for this, which will itself be determined by parameters such as mesh size, pH and temperature.

Hydrogels used in regenerative medicine as scaffolds have also been used as drug or growth factor depots, as cell encapsulants and as adhesives. They are potentially attractive materials since they may be tailored to mimic properties of the ECM such as mechanical properties. Cells may be adherent to or suspended within the gel network. One strategy to effect cellular attachment involves functionalization with the RGD adhesion peptide sequence. A particular consideration with the use of scaffold hydrogels is the degradation mechanism and the relationship between this process and tissue regeneration and healing processes.

With respect to biocompatibility, it is recognized that the chemicals used in the preparation of hydrogels may display some toxicity and care has to be taken if the degree of conversion is not 100%; initiators, organic solvents, stabilizers, emulsifiers, unreacted monomers, cross-linking agents and other substances have to be considered in this light and such chemicals may need to be removed. Synthesis should be typically followed by purification processes, such as solvent washing or dialysis. When any of the materials are derived from natural sources, they may carry the risk of batch-to-batch variation, which also has to be taken into account.

When discussing individual types of hydrogel used in tissue engineering, we should bear in mind that most, if not all, of them have concurrent use in other medical applications. They are therefore discussed elsewhere in this book, and especially details are collated in Chapter 8, so that we need not dwell on individual characteristics here, except to point out their salient features with respect to these specific applications.

Synthetic hydrogels have some important advantages, including easier large-scale production and tunable and consistent properties. However, many are made using harsh synthetic chemistry, which, as noted above, requires care to ensure that contaminants and unreacted reagents present during synthesis are then removed. Reaction schemes for these hydrogels often rely on multifunctional cross-linking agents. Free radical polymerization is widely used with many tissue engineering systems that require *in situ* formation. Acrylate-based derivatives are common functional groups that can be polymerized through thermal or photoinitiated free radical initiators. Conjugate addition reactions may also be used, for example involving mixing an acrylated macromer with a thiolated macromer. Click chemistry has become attractive for cross-linking macromers (see Chapter 2) providing mild reaction conditions and good chemical selectivity, with high yields.

Among the more common synthetic hydrogels are poly(2-hydroxethyl methacrylate) (PHEMA), poly (ethylene glycol) (PEG) and poly(vinyl alcohol) (PVA), but only PEG has significant potential in tissue engineering. PEG homopolymer is a polyether that can be polymerized from ethylene oxide, the resulting chains possessing terminal hydroxyl groups. These may be derivatized to make PEG macromers for use in a wide variety of reaction schemes. Photocurable PEG-based gels are used to encapsulate cells within scaffolds. These scaffolds may also incorporate bioactive peptides attached to PEG chains, thereby influencing cellular behavior.

The biopolymer, or naturally derived, hydrogels include those based on hyaluronic acid, alginates, collagen, fibrin and peptides. They tend to be considered as superior to synthetic gels with respect to biocompatibility since they may offer better molecular and morphological cues to cells. Hyaluronic acid, or hyaluranon (HA), is a high molecular weight glycosaminoglycan with repeating disaccharide units composed of D-glucuronic acid and N-acetyl-D-glucosamine which is found in several soft connective tissues, including skin, umbilical cord, synovial fluid and vitreous humor. As a gel it has high viscoelasticity, a major factor in its use in ophthalmic surgery and in therapies for osteoarthritis. It can be degraded by reactive oxygen intermediates: unmodified HA is rapidly degraded and cleared from the site of administration. In order to reduce the rate of degradation, cross-links may be introduced. This polysaccharide offers multiple sites for this, using its carboxyl and hydroxyl groups. It may also be modified with peptides to enhance cell attachment, spreading and proliferation. For example, thiol-modified HA can be functionalized with the RGD sequence. These peptide functionalized gels may also be used as *in situ* gelling injectable constructs for *in vivo* tissue engineering.

Alginate is a linear block copolymer of D-mannuronic acid (M) and L-guluronic acid (G) residues that has been widely used in medical applications including wound healing and cell encapsulation. Commercially available alginate is extracted from brown seaweed algae; different sources of the algae yield different amount of M and G blocks and different M-G sequences. The viscosity of alginates and the stiffness after gelling depend on the concentration of the polymer and its molecular weight distribution. Cross-linking between polymer chains can be arranged through multivalent cations and with carboxylic acid groups in the sugars. Alginates have generally good biocompatibility and can be prepared as an injectable ionic solution. However, they have poorly controlled degradation and variable cell adhesion characteristics. Gamma irradiation may be used to break high molecular weight chains to allow faster degradation and clearance *in vivo*, while partial oxidation, for example with sodium periodate, increases susceptibility to hydrolysis.

Collagen may be prepared in various forms of gel for tissue engineering applications, including those with *in situ* formation capability. Many of these applications involve unmodified collagen; chemical cross-linkers can be used to inhibit degradation and resorption when necessary. As discussed in Chapter 8, collagen exists naturally in many different types.

The status of hydrogels in current tissue engineering templates is given in Box 5.9.

An increasingly important class of hydrogels for regenerative medicine are those made from self-assembled peptides (Figure 5.8). These are polypeptides that assemble under specific conditions to form nanoscale structures. One prominent example here is the class of self-assembled peptides made from amphiphilic molecules, derived from polypeptides linked to a polycarbon chain. The polypeptide region is typically hydrophilic while the hydrocarbon chain is hydrophobic. They can self-assemble into rod structures due to the arrangement of the hydrophobic regions as well as the charge shielding of the hydrophilic end groups by ionic molecules in the solution. These molecules can be decorated with functional groups to facilitate cellular adhesion and signaling. A number of other self-assembling peptides have been produced with advantages such as the ease of forming gels and functionalization. They do tend to be mechanically weak.

The significance of these engineered peptide hydrogels is that they epitomize this direction towards materials that can replicate cell niches, referred to above as the most important specification for tissue engineering templates. Considering the stem cell niche in particular, these contain ECM components, such as laminin and hyaluronan, which present cell-adhesion ligands, and also soluble factors such as cytokines and growth factors, with a constantly replenished supply of differentiation cues.

## Box 5.9 Hydrogels as tissue engineering templates

A variety of hydrogels, both natural and synthetic, have been used, mostly experimentally, in tissue engineering processes. The following gives a summary of the principal examples.

PEG
- As template with seeded fibroblasts for bone tissue engineering (t.e.).
- Encapsulation of stem cells in cardiovascular t.e.
- Encapsulation of chondrocytes or stem cells in cartilage t.e.
- Encapsulation of Islets in pancreatic t.e.
- Template or encapsulation of MSCs or smooth muscle cells in vascular t.e.

PEG–polyesters
- Template with seeded osteoblasts for bone t.e.
- Encapsulation of chondrocytes or stem cells in cartilage t.e.
- Encapsulation of Islets in pancreatic t.e.

Dextran
- Encapsulation of stem cells in cardiovascular t.e.

Alginates
- Encapsulation of chondrocytes in cartilage t.e.
- Template with chondrocytes for facial reconstruction.

Hyaluronic acid
- Template with fibroblasts for facial reconstruction.
- Template with fibroblasts for skin t.e.

Hyaluronic acid/gelatin
- Template for vocal cord regeneration.
- Template with fibroblasts for soft tissue reconstruction.

Collagen
- Template for neural cells for spinal cord regeneration.
- Encapsulation of chondrocytes or stem cells in cartilage t.e.

Peptide materials can be designed at a molecular level to give combined structural and biological activity features that start to address these niche characteristics. These engineered, self-assembled peptides contain relatively short chains of amino acids. Through the careful choice of amino acid monomer sequences, the peptides can fold into secondary structures such as β sheets, which themselves self-assemble into hierarchical structures such as fibers and micelles. These fibrous hydrogels

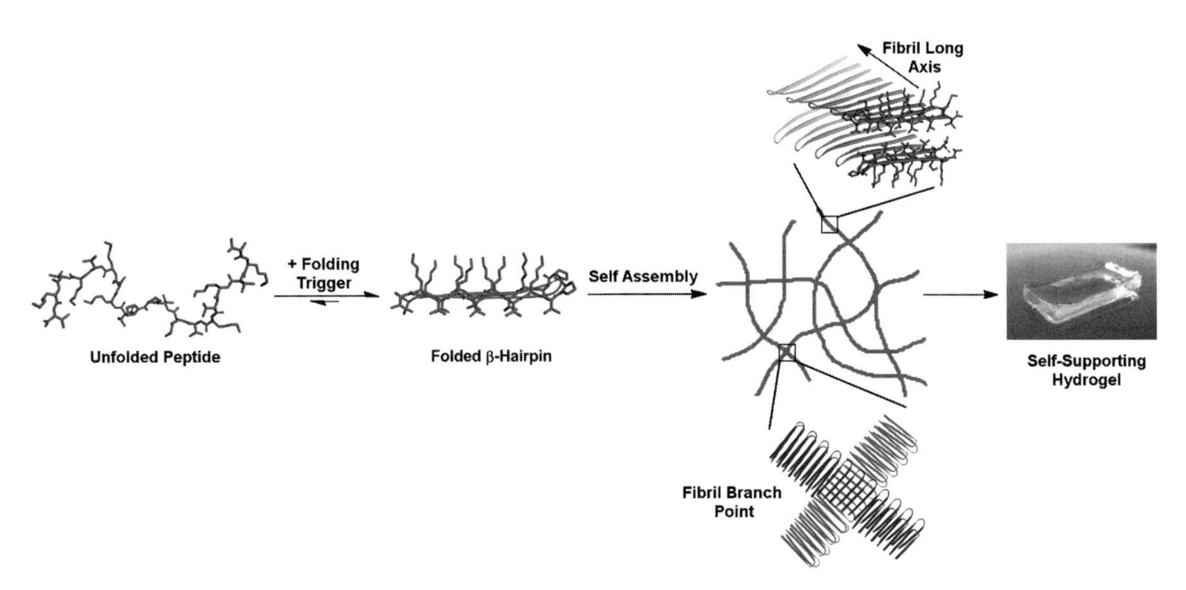


**Figure 5.8** Preparation of a peptide hydrogel. Self-assembling peptide hydrogels can be used to encapsulate and release macromolecules. One group of such materials involves unfolded peptides that undergo hydrogelation under physiological conditions to produce an amphiphilic β-hairpin structure that can self-assemble. This process can lead to a variety of fibrillar networks. Image courtesy of Dr. Joel Schneider, figure prepared by Daniel Smith in the Schneider Laboratory, National Cancer Institute, NIH.

far more replicate the required cell niches than other materials. It is possible that such structures may be reinforced by other nanoscale features in order to give robust templates.

## ECM-derived materials

The ECM, which we discussed in Chapter 3 in the context of biocompatibility, is, in principle, an obvious staring point for the development of a tissue-engineering template. The ECM, after all, comprises the structure in which cells reside in the various tissues and organs of the body and stands a good chance of providing the right type of environmental cues that drive cells into the regeneration process. The ECM in fact is in a state of dynamic equilibrium with the cells of that particular tissue, where it provides the best structural environment for its cells, which in turn provide the biochemical and molecular environment to support the matrix.

The ECM has a complex structure and composition, which varies to some extent from tissue to tissue, and from species to species. There are some common major features, however, that control the normal, natural homeostasis mechanisms in the tissue and should play some role if that ECM were to be used as a template in tissue engineering. The most structurally important and indeed most abundant protein in the ECM is collagen. Type I collagen is present in the largest quantities but many of the other 20 or so types will be present to some extent in each type of ECM. The presence of these "minor" collagens is important since they modulate both mechanical properties and ligand-mediated interactions with cells. Fibronectin is the second most abundant protein, being especially prevalent in submucosal structures, interstitial tissues and basement membranes. Fibronectin has multiple ligands to enable adhesion to many cell types. Since it is a dominant protein in developing embryos, we would expect it to play an important part in any regenerative process. Similarly, laminin is an adhesion protein with an important role in tissue development. In addition to these and other proteins, various types of glycosaminoglycans (GAGs) are present in ECM, their nature and abundance varying quite considerably from tissue to tissue. The main GAGs present in ECM are hyaluronic acid, heparan sulfate, heparin and some forms of chondroitin sulfate.

It is not surprising that many of these substances found in the ECM have been used separately as biomaterials in various forms of tissue engineering template, as we noted earlier in this section. One very fundamental question here is whether they are better used individually or collectively. One advantage of the individual use is the greater uniformity with which they can be manufactured, and the greater control that can be exerted on their biological properties, including immunogenicity. On the other hand, the totality of the ECM has a much better chance of recapitulating the ideal environment for tissue regeneration. One of the reasons for this is that the ECM also contains growth factors and cytokines, the activity of which should powerfully assist the cells that are effecting the regeneration. These growth factors include vascular endothelial cell growth factor (VEGF), platelet-derived growth factor (PDGF) and bone morphogenetic proteins (BMPs).

## 5.5.7 Fluid mechanics and templates

In many of the situations in which templates are used in practice, and in particular where they are used under *in vitro* conditions to control cell behavior, the characteristics of fluid mechanics associated with the templates and their immediate surroundings are very important. We examine a few of these situations here.

***In vitro* bioreactors** Bioreactors in tissue engineering are used to control the physicochemical parameters during the *ex vivo* processes of cultivation of cells and the generation of tissues. In many current situations when tissue engineering processes are still largely experimental, bioreactors are being used to develop protocols and establish the parameters that control these processes. The ultimate objective is to use bioreactors for standardized, automated, validated procedures for the routine manufacturing of tissue engineering products.

Bioreactors are used in several different phases of tissue engineering processes and with several different functions. The main functions are the control of cell seeding, the facilitation of mass transport and the physical conditioning of the tissues that are being generated. With cell seeding, it is important that cells are distributed with the appropriate uniformity, efficiency and viability, and the ease with which this is achieved will vary with different types of scaffold or matrix. The principal design parameter for bioreactors is the need to control the culture environment during the mechanical stimulation of the cells. The culture medium is usually agitated by stirring, rotation or perfusion, shown in Figure 5.9. The agitation rate will control the shear forces experienced by the target cells and, obviously, process variables such as the pH, oxygen concentration and metabolite levels have to be precisely controlled. The beneficial effects of bioreactors vary with cell source and the type of tissue involved. These factors are largely unrelated to the actual biomaterials in use, although of course the material architecture and elasticity should play an important role in controlling the mechanical stimulation.

### Microfluidics and microarrays

It has been noted earlier that the differentiation of ES cells may be directed by highly specific protocols *in vitro*. These protocols obviously involve the control of nutrients and growth factor concentrations, but also include the mechanics of the fluid environment. Under these conditions, ES cell differentiation involves the formation of embryoid bodies (EBs), the structure of which attempts to recapitulate the features of early embryonic development *in utero*. There are some conventional ways of forming EBs in suspension culture but they do not usually provide the optimal microenvironment, nor do they reproducibly yield EBs with similar characteristics, including differentiation potential. More sophisticated bioreactors have been designed in order to produce greater control over EB activity than is possible in traditional culture systems. In particular, systems that use microfabrication techniques and

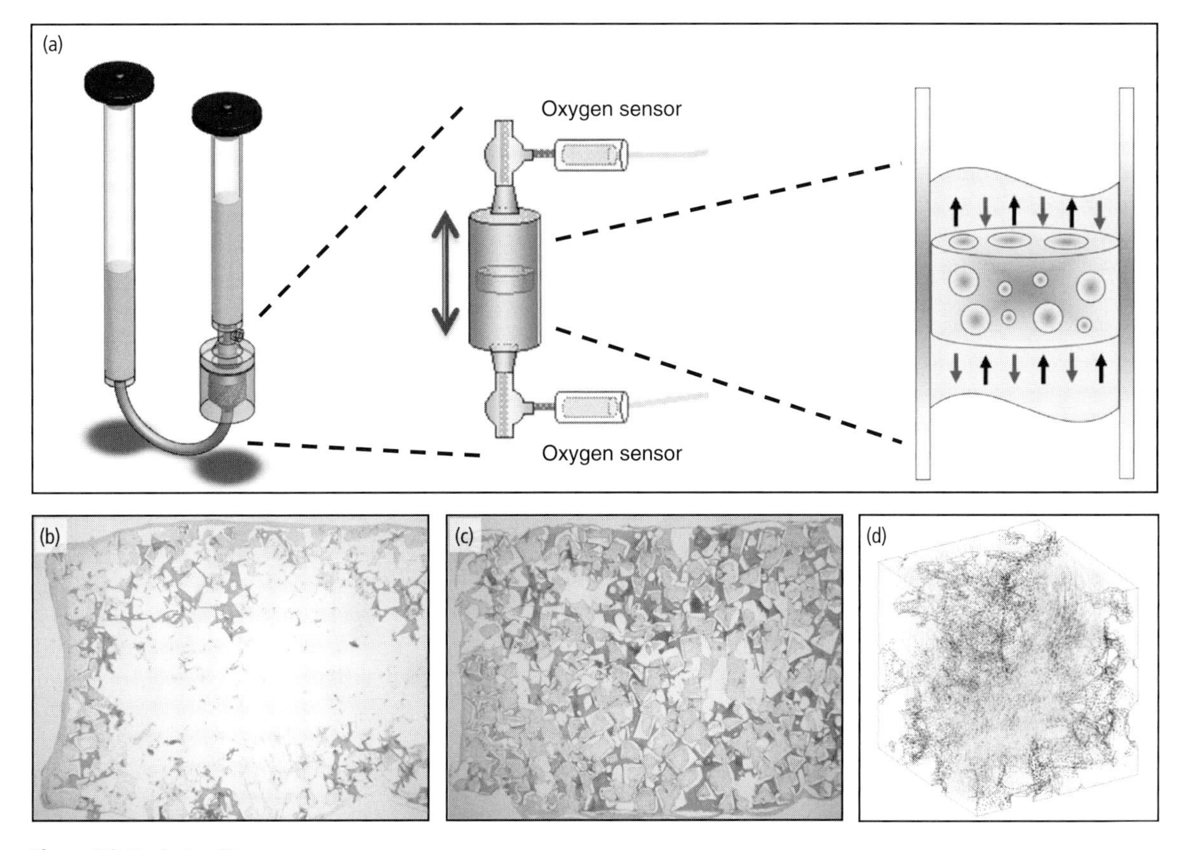


**Figure 5.9** Perfusion bioreactors.
**(a)** Perfusion bioreactor for engineering three-dimensional cell-scaffold constructs. Cells and culture medium are perfused directly through the pores of a 3D scaffold. The displayed configuration has been designed to prevent cell settling and allow uniform and efficient cell seeding into the scaffold pores by perfusion flow. Micro flow-through sensors can be integrated into the bioreactor system to monitor culture parameters (e.g., oxygen or pH) and to help establish defined and controlled culture conditions.
**(b)** Statically cultured construct. Due to mass transport limitations, cell necrosis is typically observed within the center of 3D constructs engineered under conventional static culture conditions.
**(c)** Perfusion cultured construct. Perfusion bioreactors may mitigate mass transport limitations within 3D constructs, producing constructs with a homogeneous distribution of viable cells and extracellular matrix.
**(d)** Computational fluid dynamics simulation of shear stress within construct. In conjunction with hypothesis driven experiments and sensorized bioreactors, computational modeling of shear stresses and mass transport of specific nutrients (e.g., oxygen) within the pores of a perfused scaffold can help to understand cell behavior in 3D scaffolds. Images courtesy of David Wendt, Elia Piccinini, and Ivan Martin, University of Basel, Switzerland. See plate section for color version.

microwell array designs have been introduced that allow, through control over the fluid microenvironment, the ability to regulate size, shape and homogeneity of EB populations (Figure 5.10).

The cell differentiation is directed by environmental stimuli, especially those derived from active biomolecules, which are mediated through the ECM, cell–cell interactions and physical stimuli. Within an EB there may be variations in these stimuli such that, for example, cells on the EB surface are maximally exposed to growth factors while those in the interior are subject to the effects of diffusion gradients. Mechanical forces exerted by the substrate and fluid shear stresses also influence the behavior. The control of the cellular microenvironment may be achieved by highly structured,

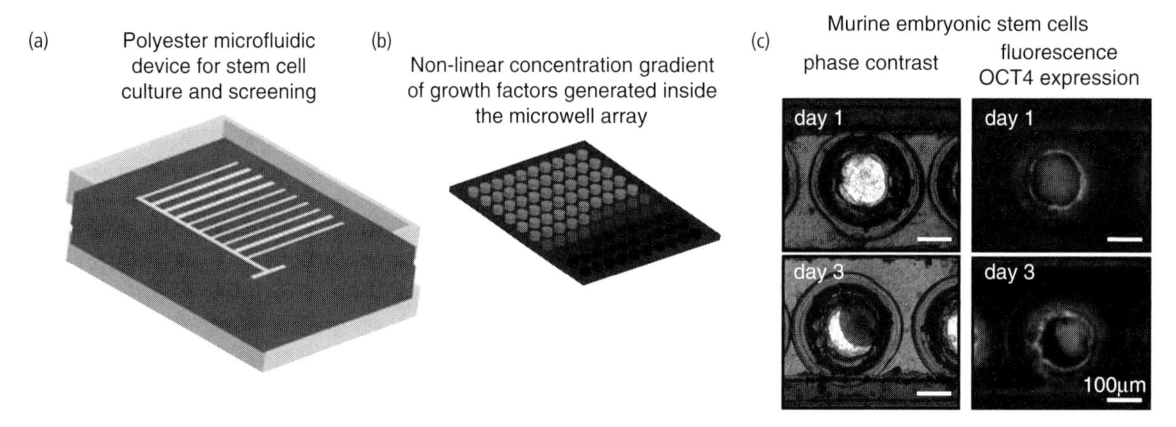


**Figure 5.10** Microfluidic device for culture of embryoid bodies.
**(a)** Sketch of the polyester microfluidic device.
**(b)** Simulation of a non-linear concentration gradient of rhodamine-labeled growth factors generated across the microwell array.
**(c)** Phase contrast and fluorescence images of typical murine embryonic stem cell aggregates on days 1 and 3 of perfusion culture. The cell cluster transforms by day 3 into an embryonic body. Courtesy of Ali Khademhosseini Ph.D, Harvard-MIT Health Sciences & Technology, Boston, MA, USA. See plate section for color version.

patterned surfaces, where reproducible EBs can form within arrays of microwells, protected from unwanted shear stresses if necessary, coupled with protection from unwanted non-specific adhesion to those surfaces by careful choice of the surface material.

It is not only stem cells that can have their behavior influenced by micropatterned surfaces. There is now an abundance of data to show that under culture conditions, the phenotype of cells can be influenced by the geometrical features of the surfaces. This can be very important since the influence of surface geometry can counterbalance the harmful effects of long-term culture of cells *in vitro* in the absence of their *in vivo* niche. For example, cells such as chondrocytes can lose the expression of marker molecules after several weeks in culture, essentially de-differentiating. It really becomes necessary to recapitulate at least some of the characteristics of their niche if they are to maintain their phenotype. This may be facilitated by growth factors and the control of the oxygen level, but the quasi-3D structure of a micropatterned surface also plays a part. This occurs because the relationship between cell phenotype and morphology may be lost in suspension culture but can be recovered if the cells are forced into a more desirable morphology by the geometrical characteristics of grooves and other features.

## 5.5.8 Fabrication routes for solid templates

Over the last decade or so, a wide variety of techniques has been used for the preparation of these templates, and especially for porous scaffolds. We review the most relevant in this section.

### Electrospinning and equivalent operations

Electrospinning, sometimes referred to as electrostatic spinning, is a relatively old process in industries such as textile fabrication. It utilizes electrostatic forces to generate polymer fibers and deposit them on a substrate with specified morphology and architecture. Since this is a variation on even older processes of electrospraying, in which droplets rather than fibers are deposited, the processing parameters have had to be changed and optimized to generate fibers with the required form. An electrospinning apparatus (Figure 5.11) consists of a capillary through which the liquid, typically a polymer in suitable solvent, is forced. A high-voltage

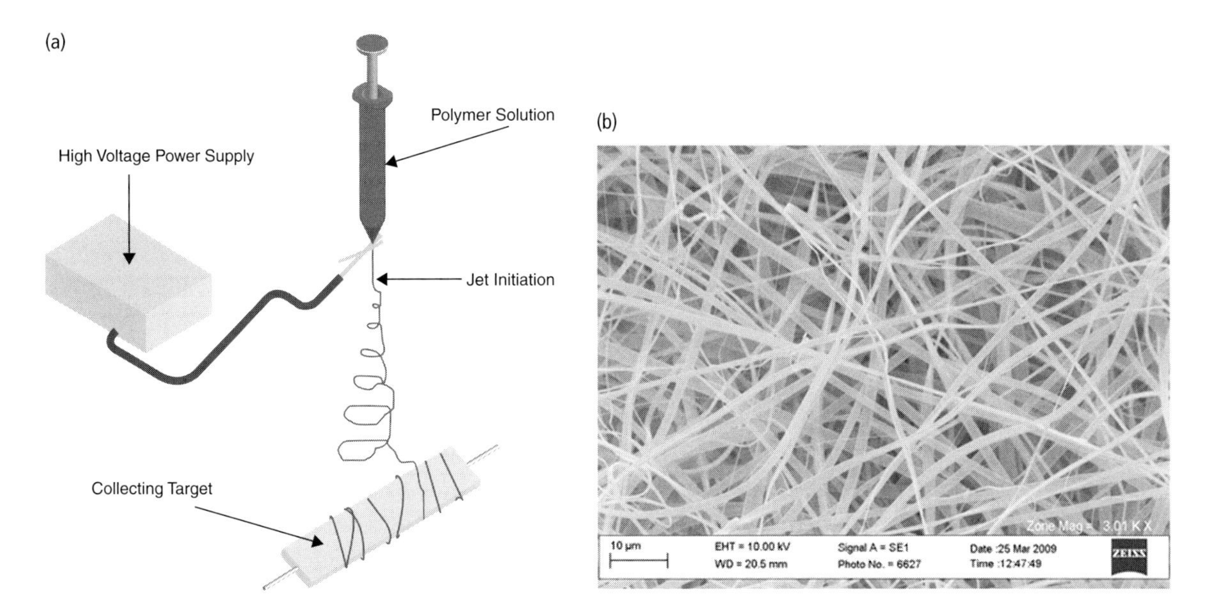


**Figure 5.11** Electrospinning of scaffolds.
**(a)** Basic electrospinning setup consisting of a polymer solution, high-voltage power supply, and rotating collecting target.
**(b)** Scanning electron micrograph of an electrospun silk fibroin scaffold created using such a system (×3000 magnification). Courtesy of Dr. Gary Bowlin, Department of Biomedical Engineering, Virginia Commonwealth University, USA.

source injects charge into the liquid and there is a grounded collector. Increasing the electric field strength causes repulsive interactions between like charges in the liquid. The attractive forces between the oppositely charged liquid and collector exert tensile forces on the liquid, elongating the pendant drop at the tip of the capillary. As the electric field strength is increased further, a point will be reached at which the electrostatic forces balance out the surface tension of the liquid leading to the development of a fiber jet, which will be ejected and accelerated toward the grounded collector.

As the fiber jet is accelerated through the atmosphere inside the apparatus it becomes unstable, thereby increasing the transit time and the path length to the collector. The fiber then thins and the solvent starts to evaporate. Essentially solid polymer fibers are deposited onto a grounded collector, which can be configured in a number of ways; for example, it may be a stationary plate or a rotating mandrel. A stationary plate collector will usually yield a randomly oriented fiber mat.

A rotating mandrel usually gives greater versatility since it can generate structures with fibers that are aligned or otherwise deliberately oriented. The mandrel rotation speed plays a significant role in determining the degree of anisotropy, while the overall morphology will also be controlled by a number of variables, including the voltage, the mandrel conductivity, the nature of the solvent, the flow rate, the polymer concentration, the nozzle configuration and so on.

Electrospinning is now widely used for tissue engineering templates, and is adaptable to many kinds of material. Adjustment of processing parameters allows for complex architectures, including multimodal fiber sizes and spaces, with both microscale and nanoscale structures. These are primarily suited to Class III templates but obviously can be used as the basis for modified structures of other classes.

## Solid freeform fabrication (SFF)

SFF techniques are computerized methods that quickly and conveniently produce complex 3D physical objects using data generated by systems such as computer-aided design and computer-based medical imaging. Consistent with the trends in biomaterials development in general which, as discussed in Chapter 1, are seeking bottom-up rather than top-down processes, and unlike conventional computerized machining processes which involve the removal of material from stock, these SFF methods involve building up 3D objects by the processing of sheet, liquid or powder material stocks using layer-by-layer techniques. These methods allow utilization of computer-generated models as inputs for scaffold fabrication, permitting complex scaffold designs to be realized, where individual patient-specific data can be incorporated into the scaffold design. The use of automated methods result in high throughput production. The high resolution of SFF and the ability to define and control individual process parameters yields highly accurate and consistent pore morphologies with, where necessary, anisotropic microstructures. Moreover, some SFF techniques allow pharmaceutical and biological agents to be incorporated into the templates during fabrication.

There are several different formats of SFF. Selective laser sintering (SLS) employs a laser beam to selectively sinter polymer or composite powders to form material layers (Figure 5.12(a)). The laser beam is directed onto the powder bed by a high precision laser scanning system. The fusion of material layers that are stacked on top of one another replicates the object's height. During fabrication, the object is supported and embedded by the surrounding unprocessed powders and has to be extracted from the powder bed after fabrication. Since the powders are subjected to low compaction forces during their deposition to form new layers, SLS-fabricated objects are usually porous. The porosity of SLS-fabricated objects can be controlled by adjusting the process parameters. Advantages of the technique include independent control of the level of porosity and pore size and the ability to develop fully interconnected porosity. It is not amenable to some low-temperature materials and produces relatively small pores.

In fused deposition modeling (FDM), filament material stock, typically a thermoplastic polymer, is melted inside a liquifier head and subsequently extruded through a nozzle with a small orifice. Successive layers of material are deposited, the

(a)

**Figure 5.12**

**(a)** Selective laser sintering. Visualization of a scaffold for SLS fabrication. CAD design (left), SLS fabricated scaffold (middle), and μCT visualization of the SLS fabricated scaffold (right). The structure measures approximately 20 mm diameter and 20 mm height with a 7.5 mm bore. Images courtesy of Dr. Stefan Lohfeld, National University of Ireland, Galway.

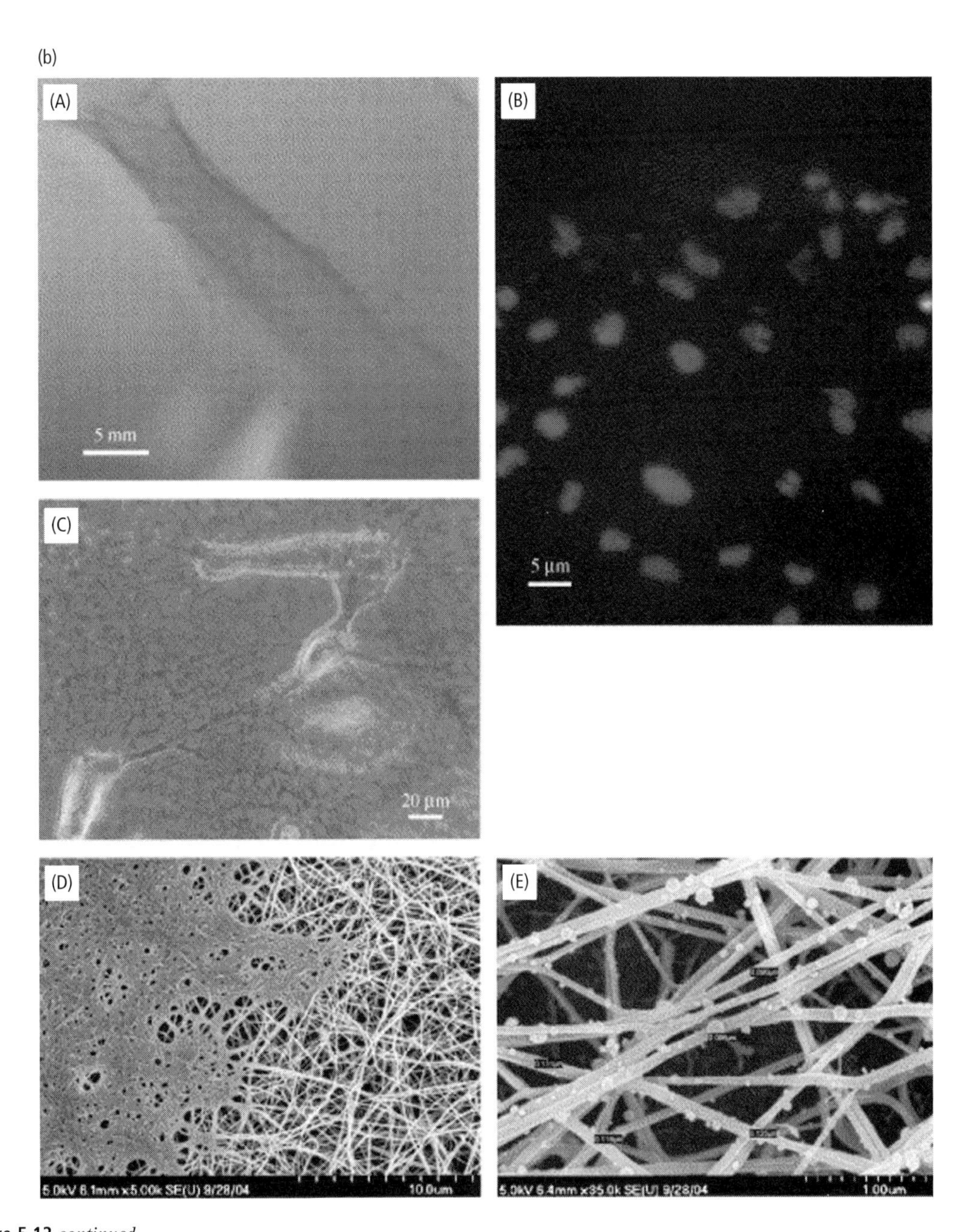


**Figure 5.12** *continued*

**(b)** Inkjet printing was first used for simple scaffolds and then for simultaneous printing of scaffolds and cells. In this example 3D neural sheets were fabricated by alternately printing fibrin gels and NT2 cells. (A) Printed neural 3D sheet 25 mm × 5 mm × 1 mm in the culture medium at day 1. (B) DAPI staining of NT2 cells within the printed sheet after 15 days of culture. (C) NT2 cells within the printed neural 3D sheet. NT2 cells after 12 days of culture. (D) The attachment of an individual NT2 neuron onto the fibrin fibers (E) SEMs of the fibrin scaffolds of the neural sheets exhibited the morphologies of fibrin fibers and porosities of the fibrin gel. Reprinted from *Biomaterials*, 27, Xu T *et al.*, Viability and electrophysiology of neural cell structures generated by the inkjet printing method, 3580–8, © 2006, with permission from Elsevier. See plate section for color version.

direction of material deposition being changeable. The spacing between the material strips can be controlled, giving highly uniform internal honeycomb-like structures with controllable pore morphology and complete pore interconnectivity. Removable supporting structures may be deposited alongside the scaffold to give additional support features. Again this technique only supports high-temperature-resistant polymers.

Three-dimensional printing (3D-P) techniques include inkjet printing technology for processing powder materials including polymers, metals and ceramics, as shown in Figure 5.12(b). A printer head is used to print a liquid binder onto thin layers of powder following the computer-generated profile. Sequential material layers are stacked and printed to give the full structure of the desired object. This technique and several sophisticated variations, are widely used, with good control over porosity and applicability to many materials. These are not restricted by thermal sensitivity but the process does involve solvents.

#### Phase manipulation

Multi-component polymer systems may be amenable to the manipulation of different phases when they show variations in thermodynamic stability. The most important process here is that of phase separation. A solution of two polymers may be manipulated such that the solution separates into two phases, a polymer-rich phase and a phase with low polymer concentration. When the solvent is removed, the polymer-rich phase solidifies. Depending on the conditions and the ease of removal of the second phase, this can result in an open pore foam, very suitable for Class I type templates. In a variation on this theme, thermally induced phase separation (TIPS) has been used to prepare porous PLLA structures for tissue engineering scaffolds. This method involves two thermodynamic processes, the nucleation and growth in the metastable state, and spinodal decomposition in the unstable region. The polymer solution is quenched from high temperature to a lower temperature, when the thermodynamic state of the solution at the time it is quenched determines the morphology of the resulting material. If the solution is quenched whilst in metastable state, an irregular pore size and poorly interconnected structure will be induced, whereas if quenching occurs in the latter state, an open, porous, and 3D structure will be induced. The open pore morphology can be controlled by parameters such as the quenching temperature, quenching rate and period, polymer concentration, ratio of solvent to non-solvent and the molecular structure. For both conventional phase separation and TIPS, a porogen may be added (see below) in order to give more complex porous structures.

#### Solvent casting and porogen leaching

One of the simplest and oldest methods to produce porous polymers involves mixing a porogen, which is an easily extractable substance, with the polymer, dissolving them in a suitable solvent and then casting the solution. On solidification, the porogen particles can be removed, leaving an open pore structure. A porogen such as sodium chloride can be removed easily in water. In some cases the porogen is a low melting point polymer that can be removed by heat. This technique is widely used although it obviously has limitations with respect to the ability to remove all of the porogen and to ensure continuity between the porogen particles. Some fairly sophisticated manufacturing systems are available for good control over the porosity.

### 5.5.9 Bioprinting and organ printing

Much of the discussion in this chapter has been based on the concept that an engineered template is required to give shape to regenerating 3D tissue, and to facilitate both mechanical and molecular signaling. This concept has led to the practical realization of 3D templates that are prepared by sophisticated processes, including those based on rapid prototype principles such as inkjet printing, and the subsequent seeding of cells into this template for their proliferation, migration and expression of ECM.

It should not be surprising to learn that alternative procedures have been developed using printing or other technologies in which template structures are prepared simultaneously with cell seeding, or where the template contains no exogenous material at all but is composed only of cells and the matrix they secrete. These methods are often called bio-printing and in the latter case are said to be scaffold-free tissue engineering processes.

The general bioprinting approach aims at building 3D functional tissues or organs, layer-by-layer, from the bottom up. In practice, jet-based methods have been taken from other non-biological areas such as metal transfer and adapted to biological material such as peptides, DNA and cells. For example, the laser-induced forward transfer (LIFT) technique was modified to print biological material, and named BioLP (biological laser printer) or LAB (laser-assisted bioprinter). These typically have three components: a pulsed laser source, a target (referred to as a ribbon) from which a biological material is printed, and a receiving substrate that collects the printed material. A suitable laser source would be a solid Nd:YAG crystal laser. The ribbon is often made of a thin absorbing layer of metal such as gold or titanium which is coated onto a laser-transparent support such as glass. The desired molecules or cells are prepared in a liquid solution (such as culture media), and deposited at the metal film surface. The laser pulse induces vaporization of the metal film and the production of a jet of solution, which is deposited onto the facing substrate. The cell fraction of each printed droplet should be close to 100%. The overall resolution of printing, that is the size and proximity of the droplets, depends on parameters such as the thickness of the biological ink layer coated onto the ribbon, the surface tension and the viscosity of the bioink, the wettability of the substrate, the laser fluence, the air gap between the ribbon and the substrate.

For scaffold-free tissue engineering processes, rapid prototyping technologies have been used. These are based on automated, computer-aided deposition of multicellular spheroids, sometimes referred to as bioinks since they are analogous to the particles of ink in inkjet printing, into a structure, usually collagen based, that is the equivalent to the paper in the printing world, sometimes referred to as biopaper. At first sight this appears to be a 2D process, but 3D tissue structures may be formed through the post-printing fusion of the bioink particles. This can be considered to be equivalent to the self-assembly phenomena in early morphogenesis.

There are, of course, many features of this conceptually simple process that are far from trivial in their realization. The collagen gelation is critical for the smooth deposition of the spheroids and the collagen concentration has to be carefully controlled to ensure fusion of the spheroids during the post-printing phase. Uneven gelation of successive collagen sheets in the layer-by-layer construction leads to progressive distortion of the construct beyond the first few layers. Refinements of these techniques have included attempts to eliminate the collagen biopaper altogether.

## 5.6 Tissue engineering and clinical reality

We now have to consider the question of how far tissue engineering has progressed towards clinically acceptable and effective techniques. This will be done by discussing some significant areas where there is potential and has been some progress, always using comparisons with alternative therapies.

### 5.6.1 Bone tissue engineering

Bone does have considerable ability to regenerate following injury, and the regenerated bone may be indistinguishable from normal healthy bone. The mechanisms of healing of bone injuries and defects are well understood and in most clinical situations the healing process can be easily managed. It may be that, as we have seen in the previous chapter, some form of internal or external support is required to assist the healing processes achieve optimal functional and aesthetic outcomes.

### Critical size defects: natural and accelerated bone healing

However, problems do arise when the injury is very extensive or complex, when there is underlying disease or when there are congenital defects that need restoration. It is under these types of situation that regenerative medicine approaches may be appropriate. A very important factor here is the limit to the amount of bone regeneration that can take place naturally, without assistance, and conversely, the extent of new bone formation that certainly requires specific therapeutic assistance. This is usually considered in terms of the "critical size defect." Intuitively this defines that limit expressed in the previous sentence, but in practice it is rather difficult to define. It is usually considered to be the smallest size intraosseous wound in a particular bone and species of animal that does not spontaneously heal, even over the lifetime of the animal. The very fact that this will depend on the nature of the bone in question, and on the species involved, makes it a rather inconsistent baseline, but it is important since so many experimental studies aimed at developing bone tissue engineering solutions focus on specific animal models; there is clearly no point in developing such solutions for sub-critical size defects since these should heal spontaneously without complex therapies.

One other issue should be addressed here, and that concerns the speed of natural bone repair/regeneration and the perceived need to accelerate this. In most situations, bone heals at a biologically defined rate; there is a temporal sequence of events that has to be followed and in general this cannot be manipulated or accelerated. It is not the purpose of bone tissue engineering to try to accelerate a natural process. There are a few situations in which it might be more convenient for patients if the healing process was faster, for example after some surgical procedures such as the insertion of dental implants, but these are exceptional and are best accommodated by optimization of conventional processes rather than by tissue engineering solutions. However, there are patients where bone healing is very slow or ineffective, as in delayed union after fracture, and indeed non-union, and regenerative medicine solutions may be necessary. There have already been many attempts to introduce new procedures for this, for example by using electrical stimulation, but these outcomes are very erratic and unpredictable.

Situations in which we cannot expect bone regeneration to take place, even in otherwise healthy individuals, are those in which there is no solid stable matrix on which new bone can form. In small defects, a hematoma will form after the injury, cell signaling processes then attracting inflammatory and osteoprogenitor cells and growth factors. Immature bone can form within this area, which is then mineralized. This bone forming process occurs by a combination of growth from pre-existing surfaces (osteoconduction) and also spontaneously within the defect as a result of these signaling processes (osteoinduction). If the defect is too large, osteoconduction cannot occur because there is a limit to the extent the bone can grow in this way, and osteoinduction is difficult since there is no stable support for the new solid tissue that would be formed. The classical approach to the treatment of large defects involves the use of bone grafts, either autologous grafts from iliac bone or ribs, or fresh or frozen allogeneic or xenogeneic products. There are limitations to all of these, including the lack of volume of autologous grafts and the donor site morbidity. Very large grafts may need the support of some biomaterial structure.

The need for regenerative medicine solutions with respect to bone are therefore:

- in situations in which bone should heal in a normal healthy patient but where, for some reason, it does not do so, or does not do so in a reasonable time;
- where there are congenital defects of such an extent that conventional autologous bone grafting could not provide sufficient material;
- similarly, where traumatic events have resulted in very significant bone loss;
- again, in a similar manner, where tumor resection involves substantial amount of bone loss with associated functional and/or aesthetic deficits.

**Box 5.10 Summary of bone tissue experiences**

For the treatment of long bone non-unions:

- Good experiences with biological enhancement of internal or external fixation through the use of BMP-7 augmented autologous bone graft.
- Good experiences with bone marrow-derived MSCs, with or without platelet-rich plasma (PRP) and some good results with allogeneic osteoblasts.

For enhancement of or alternatives to autologous bone grafts for bone defects:

- Autograft bone is very effective but there may be insufficient quantities available and there is the potential for donor site morbidity.
- Bone marrow aspirate may be used to supplement the autograft, but without high success.
- Allografts, either frozen or freeze-dried cancellous or cortical bone, may be used. This includes demineralized bone matrix, DBM. In spite of there being no osteogenesis and only limited osteoinduction, good success may be achieved.
- Limited success may be obtained with BMP and PRP supplementation.
- Limited success may also be achieved with MSC supplementation.
- Some success with various calcium phosphate ceramics, especially injectable materials.

Osteoarthritis, tumor resection, craniofacial reconstruction:
No routine clinical applications yet.

A summary of bone tissue engineering experiences is given in Box 5.10.

## 5.6.2 Cartilage tissue engineering

Sporting injuries to cartilage were mentioned in Chapter 1 and the possibilities to either repair or replace parts of this tissue were briefly mentioned in Chapter 4. We discuss here the progress that has been made in the regeneration of cartilage; this was one of the early targets of tissue engineering since injuries are common and the lesions often quite small. However, translating this hope into clinical reality has been difficult.

Concentrating on the articular cartilage that is found within joints, the main function of this tissue is to transmit loads through the joints, absorbing impact energy but providing low friction movement within the joint. To achieve this, cartilage has developed as a highly specialized tissue, which is avascular, without nerves and lymph vessels. It is very resilient but has extremely limited capacity for repair. There is a very slow natural turnover of tissue though degradation and synthesis, but once injured it loses structure and function.

The nature of cartilage does vary with anatomical location, both from joint to joint and across the various depths within one joint. In most synovial joints, the hyaline cartilage that is present is composed of the sole cell type, the chondrocyte, that is surrounded by ECM, which contains between 60 and 85% water and both collagenous and proteoglycans structures. The chondrocytes are rather few in number, each cell being responsible for a large volume of tissue. Most cartilage contains a majority of type II collagen, with small amounts of types VI, IX and XI. The orientation and distribution of these collagen molecules, and cross-linking between them, control the important mechanical properties, including compressive and tensile properties and shear stress resistance. Proteoglycans components contribute to shock-absorbing characteristics, provided by the hydrophilic environment associated with the polyionic aggregates formed by sulfated aggrecans and hyaluronic acid.

Upon injury, cartilage is usually unable to repair itself or regenerate effectively, the healing that occurs being achieved through scar tissue, which is essentially fibrocartilage. This is not a very effective substitute for hyaline cartilage and gradually loses structure and normal functions.

It is important to place potential tissue engineering solutions into the perspective of existing treatments. There are several palliative options, including conservative, non-interventional procedures which may alleviate symptoms but do not result in healing. The site may be debrided, that is the removal of debris produced by joint damage, or given a lavage treatment, but neither alter disease progression. Some reparative procedures are directed towards the initiation of bleeding in the subchondral bone, which allows the migration of bone marrow stem cells into the site. The most widely used procedure here is that of microfracture, where the subchondral bone is exposed after debridement and a pattern of V-shaped holes produced on this surface. The outcome again is fibrocartilage-based repair, which is moderately successful in young patients but is not an ideal permanent solution.

As noted in the previous chapter, there are also joint replacement strategies to deal with extensive joint damage. The challenge for regenerative medicine procedures is to be able to induce regeneration of cartilage in small defects, either caused by injury or disease, before they get too large or too advanced, with better and more sustainable results than these palliative or reparative techniques. Osteochondral transplantation, or grafting, has already been introduced in Chapter 1. Usually referred to as mosaicplasty, autologous osteochondral transplantation, involves removal of damaged fragments of cartilage, and the creation of a series of holes in the subchondral bone, into which osteochondral grafts taken from an adjacent but non-weight bearing part of the joint are placed. This works well for small focal lesions. It is also possible to consider allograft osteochondral grafting for large defects, using cadaveric or fresh grafts.

### Autologous chondrocyte implantation

The main tissue engineering solution is based on autologous chondrocyte implantation (ACI), which is a two-stage procedure. Chondrocytes are harvested arthroscopically from the patient, typically from the femoral interchondylar notch. These cells are cultured for around 6 weeks. The site of the lesion is debrided and covered with a periosteal flap, sealed in place with fibrin glue, and the cultured chondrocytes are injected under the flap. This technique can work well, producing a hyaline-like tissue that is functional for many years. However, there are disadvantages, including considerable variation in outcome due to chondrocyte leakage, dedifferentiation of the cells and hypertrophy of the periosteum. More reproducible results can be obtained if the cells can be characterized and selected during culture to allow the use of chondrocytes that have better potential for producing hyaline cartilage.

### Cells, templates, biomolecules and bioreactors

In order to see how this situation can be improved, to give both better, more consistent, performance and make it viable for larger lesions, we have to consider the basic tissue engineering paradigm. Questions arise concerning the nature of the cells we are using, whether they need a biomaterial support, whether they need better molecular signaling and whether the two-stage, *ex vivo* bioreactor approach is the best. With respect to the cells, the two possibilities are staying with autologous chondrocytes or changing to some stem cells, especially MSCs. Chondrocytes have a powerful advantage, especially as they can be harvested and expanded readily, but they have their own difficulties. In particular, in monolayer culture, chondrocytes are prone to dedifferentiate, with significant changes to the cell's gene expression profile and ability to produce cartilage matrix. There are differences in performance with the origin of the chondrocytes, including the depth, or zone, of the cartilage from which they are derived and their anatomical location. The performance is also mediated by growth factors and gene transfer, discussed below. MSCs can be obtained from a variety of sources, including bone marrow and adipose tissue. Bone marrow-derived MSCs are particularly popular in experimental approaches and much work has been performed to optimize culture conditions for inducing differentiation into pure chondrocytes.

The promotion of differentiation of MSCs into chondrocytes and of the expansion and maintenance of phenotype of chondrocytes themselves may be achieved with a variety of active molecules, notably those of the BMP and TGFβ families. Depending on the circumstances, BMP -2, -4, -6 and -7, and $TGF\beta_1$, $TGF\beta_2$ and $TGF\beta_3$, and various combinations of these have shown to be effective to some extent. These same molecules may also be considered candidates for gene transfer and several have been shown to be effective in experimental systems. Also there are several transcription factors that are essential for cartilage formation and which may be considered as targets for gene transfection, including Sox-9, Sox-5 and Sox-6.

With respect to biomaterials, the limited success with ACI has prompted the development of techniques that incorporate the chondrocytes into some form of matrix or scaffold in order for them to have a better chance of survival. This is sometimes described as matrix-assisted ACI. The requirements here are similar to some other scaffolds, in that the material should support chondrocyte function and have suitable mechanical properties, ultimately degrading, or being incorporated into the tissue in a non-inflammatory way. Various forms of collagen and gelatin have been extensively evaluated, as have hyaluronan, chitosan, alginate, fibrin and silk. Achieving good mechanical properties with these biopolymers has been a challenge and some approaches have involved hybrid structures in which cells are encapsulated in a biopolymer hydrogel that is then incorporated into a porous synthetic scaffold. At this stage, reasonable clinical success has been achieved in focal defects with hyaluronic acid derivatives and collagen I/III-based products.

In order to obtain success with larger defects or more generalized disease, variations in techniques may be necessary, with emphasis on the methods used in culture. While it is clinically and economically attractive to use single stage procedures, reducing or eliminating *ex vivo* culture time, this may not be a practical approach when significant volumes of cartilage are required, and more sophisticated bioreactors are necessary. It has been shown that simple static cultures have serious limitations with respect to the volume and homogeneity of cartilage constructs that can be produced. Far better results are achieved with large, automated, closed system perfusion bioreactors, and volumes of homogeneous cartilage can be generated around hyaluronic acid-based scaffolds in two weeks.

A summary of cartilage tissue engineering experiences is given in Box 5.11.

### 5.6.3 Skin tissue engineering

Skin tissue engineering received more attention than any other area of regenerative medicine in its early years, in the 1990s, partly because therapies for skin are non-invasive and easy to monitor, and partly because skin does have some powers of healing, which, in a suitable therapy, had to be reorganized and enhanced. Although there are some products in extensive clinical practice or clinical trial, this area has not developed as initially thought and hoped. There are several reasons for this, some scientific and some not. One main issue is that many skin wounds or injuries can heal themselves, possibly with inexpensive care, and there is little room for costly regenerative techniques. Some more complex injuries probably could benefit from skin tissue engineering methods but in the really significant areas where there is an urgent need for radically new and effective solutions, the tissue engineering processes have yet to succeed in a clinically and economically

**Box 5.11 Summary of cartilage tissue engineering experiences**

- Little success has been achieved with the regeneration of cartilage in association with degenerative disease.
- Most effects have been directed to the treatment of cartilage trauma, especially sporting injuries and the regeneration of cartilage-based tissues for cosmetic reasons.
- For trauma, the most successful regenerative medicine treatment is autologous chondrocyte implantation. This may be assisted by some biomaterial, in matrix-assisted ACI. In spite of vast amounts of experimental and pre-clinical work, with good results, no other biomaterials-related techniques have yet been used routinely for cartilage repair.
- Some success has been achieved with the generation of cartilage-like tissue for tubular structures such as the trachea and complex but cosmetic structures such as the external ear. These can be derived from adult chondrocytes (e.g., nasal septum chondrocytes) which are cultured on synthetic or biopolymer scaffolds.

effective manner. These two areas are burn wounds and chronic wounds such as venous ulcers and diabetic foot ulcers. We should note here that there have been many commercial products brought to the clinic, not always successful and often carrying different names. Many are simply called skin substitutes, or skin equivalents, sometimes qualified by engineered or bioengineered skin substitutes and equivalents. Some products are described as "living skin equivalents."

### Burn injuries

Burn wounds represent very difficult clinical challenges. Fortunately life-threatening skin burns are rare, although this does mean that the market for commercial products is inevitably small. Following the burn, a complex series of events is initiated, with inflammation and some regeneration, and then recruitment of fibroblasts and keratinocytes, revascularization and re-epithelialization, all of these events naturally being controlled by secreted biomolecules such as cytokines. However, this reparative sequence of events is restricted to small areas of tissue damage. With large areas, and especially where they are full thickness wounds that involve the destruction of regenerative capacity in the dermis, this healing process will not take place efficiently, the result usually being functionally and aesthetically compromised fibrous scar tissue.

Autologous skin transplantation is the standard-of-care for full thickness skin wounds, where split skin grafts, that incorporate epidermis and part of the dermis, are harvested from an uninjured area and applied to the damaged area. These techniques are very successful, especially with excellent control of fluids and infection, although they can result in hypertrophic scars and there may be insufficient donor tissue to cover the wounds. There is therefore a significant need for skin substitutes.

In theory, skin substitutes may be cell-free or they may be seeded with cells. For burn wounds, autologous cells are clearly inappropriate since treatment is required urgently. Dermal substitutes are therefore either cell-free constructs or are substances that are seeded with allogeneic cells, usually fibroblasts derived from human foreskin. Many of the cell-free products are derived from xenogeneic or allogeneic sources such as acellular human dermis, cross-linked porcine dermal collagen or acellular porcine small intestine submucosa.

### Diabetic foot ulcers

Diabetic foot ulcers are also problematic, but here the incidence is much greater. About 15% of patients with diabetes mellitus will develop a foot ulcer in their lifetime, and about 15% of these will go on to leg amputation because of the inability to treat this effectively. With some 250 million people affected by diabetes worldwide, this is clearly a major issue. There are palliative methods of treatment, with simple products and great care, but the condition not only affects the skin but also the underlying tissue, and regeneration of the underlying ECM and microvasculature is as important as the dermis and epidermis. Different forms of wound dressing have been developed which improve outcomes, and also negative pressure wound therapy, providing sub-atmospheric pressure around the wound may be beneficial. A few so-called living skin equivalent products have been used clinically in these situations. These have either synthetic or xenogeneic collagen scaffolds, with allogeneic cells (keratinocytes and fibroblasts); these cells appear to secrete cytokines for protracted periods of time after application. These ulcers are very difficult to heal, but, compared to conventional, conservative treatments, they do appear to give better chances of long-term performance without re-ulceration. The use of stem cells, either by themselves or in such products, has been considered for the treatment of these ulcers.

### Skin substitute products

As noted earlier, several different types of skin substitute have been used this far, and they are not easy to categorize since they contain different combinations of components. Roughly they fall into the following groups. Some examples are shown in Figure 5.13.

**Acellular xenogeneic graft** These are typically derived from porcine dermis or segments of the intestine. They are usually considered only as temporary covering for partial thickness wounds.

**Acellular allogeneic grafts** These are usually taken from cadaveric donors, being decellularized and cryopreserved, and have been used in both partial and full thickness wounds.

**Donor-cellularized collagen grafts** These may be either xenogeneic or allogeneic collagen scaffolds that are seeded with allogeneic fibroblasts, used for either partial or full thickness wounds. The product may have a synthetic protective outer layer.

**Autologous epidermal graft** A product that consists of a synthetic membrane, such as a hyaluronic acid derivative, that is cultured with autologous keratinocytes. This may be used for full thickness wounds in cases where there is insufficient skin for conventional grafting, although the 4 week culture time is not helpful, or for ulcers. Variations on this theme include co-culture with fibroblasts.

**Composite structures** Since skin is essentially bi-layered, it has been a natural evolution to make substitutes that have two layers. One product has incorporated a bovine collagen gel layer combined with dermal fibroblasts as the dermal layer with an epidermal layer of allogeneic keratinocytes. Another has a dermal layer of bovine collagen incorporating glycosaminoglycans, this being covered by an outer layer of silicone elastomer.

Although skin tissue engineering has been around for 20 or so years, there is still some way to go before these products and these techniques become routinely and widely used.

## 5.6.4 Skeletal muscle tissue engineering

Skeletal muscle is clearly a very important tissue. It comprises over 40% of the human body mass and suffers many types of injury and disease. There is a significant demand for techniques of muscle regeneration in cases of sports and other traumatic injuries, tumor ablation, various myopathies and congenital defects. For many common skeletal muscle injuries resulting from physical trauma but

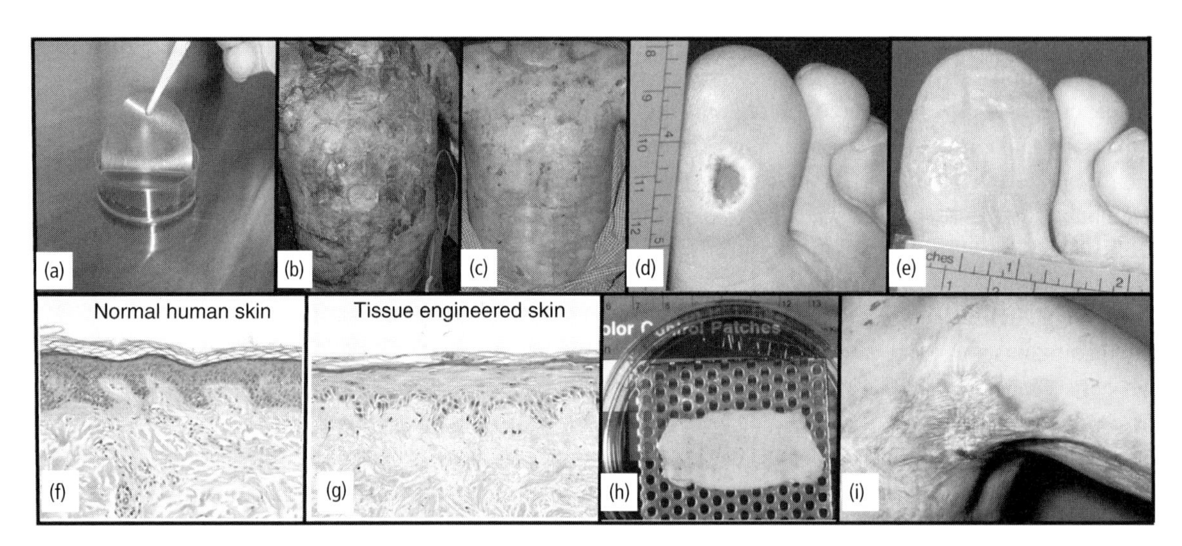


**Figure 5.13** Clinical applications of tissue engineered skin. For many applications autologous keratinocytes are expanded in the laboratory and then delivered to patients' wound beds to assist in the regeneration of a barrier layer for burns patients or to stimulate chronic wounds to heal.
(a) Carrier disc of silicone covered with a thin layer of acrylic acid using plasma polymerization, onto which laboratory expanded autologous keratinocytes are seeded (product developed as Myskin™).
(b) Patient with extensive full thickness burns who is being treated with wide mesh autograft over which Myskin™ dressings containing his own laboratory expanded cells have been placed.
(c) The same patient some 10 days later with an almost intact epithelium regenerated.
(d) Diabetic foot ulcer on a patient which had remained unhealed for 3 years.
(e) The same ulcer which healed after repeated applications of his own keratinocyte expanded in the laboratory and then delivered to this ulcer in outpatient clinics – eight applications of cells were required to stimulate the wound to heal. In many instances defects are full thickness and require tissue engineered skin containing both epidermal and dermal layers.
(f) Section through normal skin and (g) a similar section through tissue engineered skin. This skin was produced by taking cadaveric skin, stripping out all of the donor cells and sterilizing the skin with a mixture of glycerol and ethylene oxide. A small skin biopsy from the patient was used to culture and expand keratinocytes and fibroblasts, and then introducing both to the decellularized allodermis.
(g) Shows a well-attached, multi-layered differentiated epithelium with a normal looking dermis containing patient's fibroblasts.
(h) A piece of this skin grown in the laboratory at an air–liquid interface prior to surgical use.
(i) Patient who had suffered extreme skin contractures due to earlier burns injuries. Such contractures are managed by surgical release of the contracted area and then use of a split thickness skin graft sutured into place. Image shows the appearance of the tissue engineered skin (composed of patient's keratinocytes and fibroblasts and donor decellularized sterilized dermis) 2 weeks post transplantation. Images provided by Professor Sheila MacNeil, University of Sheffield. All patients gave consent for images to be used for research and educational purposes. See plate section for color version.

without significant tissue loss, the muscle has a good ability to regenerate because of a significant presence of mononuclear myogenic satellite cells. However, when injuries result in the loss of muscle tissue, the repair process is not so efficient. If the loss is greater than 20% of the muscle, the result is likely to be the formation of scar tissue, de-nervation of muscle distal to the defect and, in consequence, a loss of function. Currently available treatments involve surgical intervention with autologous muscle transplantation and transposition. These techniques may restore some function, but do not really regenerate the lost muscle and the results are often disappointing. It is appropriate to consider the potential role of regenerative medicine here, and, as with all sections here, we have to look at the cells, the biomolecules and the biomaterials, individually and collectively in order to examine this potential.

As with the response to injury in other tissues, the events following injury to skeletal muscle involve inflammatory, reparative and remodeling phases. The damaged myofibers expose their intracellular contents to the extracellular environment. Myofibrils are rapidly disintegrated following activation of calcium-dependent proteases. The chemotactic recruitment of neutrophils and macrophages takes place, which digest necrotic myofibers and cellular debris and release cytokines that amplify the inflammatory response and recruit muscle satellite cells to the injury. During the repair process, nerves, blood vessels and muscle cells infiltrate the area. Satellite cells differentiate into myoblasts that fuse with other myoblasts or with existing myofibers to form new skeletal muscle. It is also possible that other muscle-derived stem cell groups and circulating progenitor cells may take part in this process of new muscle formation either through myogenic differentiation or by secretion of paracrine factors. Scar tissue will also reconnect the remaining functional muscle fibers and preserve the transduction of force along the muscle. The problem is that in severe injuries with significant tissue loss, the scar tissue rapidly proliferates and blocks regenerating muscle fibers from bridging the wound. In the remodeling phase, new myofibers formed by invading progenitor cells subsequently mature and form attachments to the surrounding ECM with marked reorganization of the muscle tissue, such that the ends may form a new myotendinous junction with the scar tissue, remarkably affecting function.

Any regenerative therapy must clearly involve myogenic progenitor cells, which could occur through the isolation, expansion and injection of autologous cells or by the recruitment of resident or circulating progenitor cells *in situ*. Specifically, muscle regeneration requires cells that are capable of sustained proliferation, self-renewal and myogenesis. Muscle satellite cells reside between the plasma membrane and basement membrane of individual muscle fibers. They have the capacity for self-renewal but are relatively scarce within skeletal muscle. They remain quiescent until activated by injury when they can migrate within and between myofibers. They are unlikely to be recruited in high enough numbers to facilitate *in situ* repair for large injuries. They may be isolated and expanded *in vitro* although they may lose their regenerative capacity during such procedures. Both muscle-derived stem cell and adipose-derived stem cells have myogenic differentiation potential and have been shown in animal models to cause some regeneration. Several types of molecules could be considered as essential regulators of the muscle repair or regeneration processes, in particular controlling the proliferation and differentiation of muscle progenitor cells. Important growth factors include vascular endothelial growth factor (VEGF), hepatocyte growth factor (HGF), basic fibroblast growth factor (FGF), platelet-derived growth factors and especially insulin-like growth factor-1 (IGF-1). The situation is rather complex for, in addition to growth factors that stimulate repair, some, such as transforming growth factor β1 (TGFβ1), inhibit this process and determination of the optimal combination of growth factors that could significantly enhance muscle regeneration is not a trivial task.

Also the ECM itself is a regulator of muscle repair, type IV collagen, laminin and heparan sulfate proteoglycans being particularly important. Following skeletal muscle injury, invading macrophages and satellite cells release matrix metalloproteinases such as MMP-2 and MMP-9, which stimulate myoblast and progenitor cell migration, proliferation and differentiation.

With these comments about cells and growth factors in mind, a few options for regenerative medicine approaches can be seen. Cell therapy has been tried clinically, but only with limited success, primarily because of low retention and survival of the cells. Although growth factors can improve outcomes in experimental models, this is not yet feasible on a clinical scale.

With tissue engineering solutions, there are *ex vivo* and *in vivo* possibilities, shown in Figure 5.14.

(a)

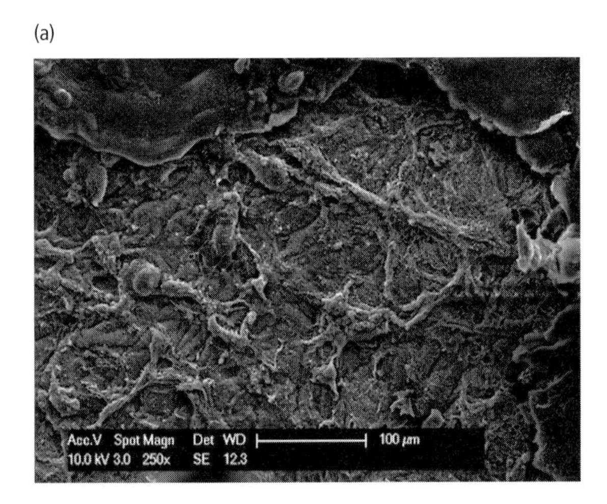


(b)

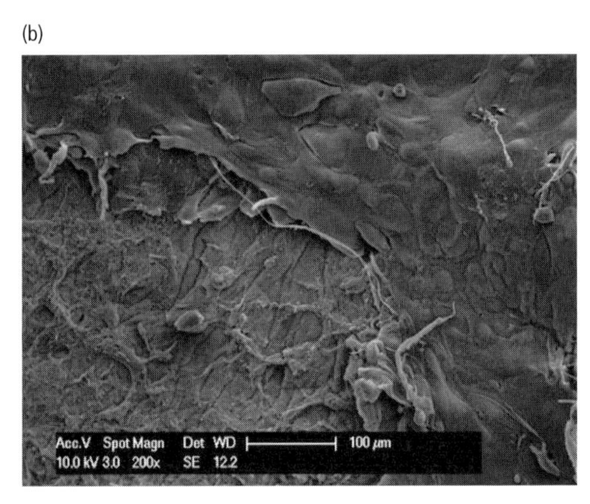


**Figure 5.14** Scaffolds and smooth muscle cells. SEM micrographs of a compressed collagen sheet incorporating human smooth muscles cells building a network inside the sheet **(a)** and human urothelial cells seeded on the same sheet's surface **(b)**, after 14 days in culture. Images courtesy of Professor Peter Frey, University Hospital of Lausanne, Switzerland.

Although there have been some significant advances in the formation of skeletal muscle *in vitro*, this approach has serious limitations. In functional muscle there are parallel alignments of muscle fibers, which has been difficult to reproduce. On implantation, it is necessary that functional vascular beds and innervation of the tissue take place with full integration in order to produce directed forces. *In vivo* tissue engineering implies that some template material, with or without seeded cells, could be placed at the site of injury where it would facilitate regeneration. The trend here has been towards using ECM-derived templates, either as implantable constructs such as decellularized ECM, or as injectable preparations that incorporate collagen, laminin and so on, much depending on the volume of tissue that needs to be formed, and its location. Under these circumstances, it would appear that seeding the template with cells before implantation is of no additional benefit since circulating progenitor cells migrate into the area.

## 5.6.5 Vascular tissue engineering

Cardiovascular disease and the implantable medical device solutions have been discussed in the previous chapter, where it will have been noted that good success can be achieved in the peripheral vasculature, although with some limitations concerning the long-term resistance to occlusion of the implanted vascular grafts, while no synthetic grafts are available for the coronary arteries. Blood vessels have therefore been a significant target for tissue engineering for some time.

### Endothelialization of synthetic materials

It could be argued that tissue engineering in the vascular system started with the engineering of the endothelium. The vascular grafts mentioned above have given excellent mechanical and physical performance in most respects in areas such as femoropopliteal bypass for many years; however, the patency limitations have been associated with the lack of a really appropriate endothelium. The overall performance could therefore be improved by coupling the physical attributes of textiles with a natural endothelial lining. Techniques of *in vitro* endothelialization of grafts were introduced in the late 1980s. This initially involved dissecting segments of either the external jugular or the cephalic veins and treating these with collagenase in order to harvest the endothelial cells. These were cultured and the cell suspension was applied to fibrin-coated inner

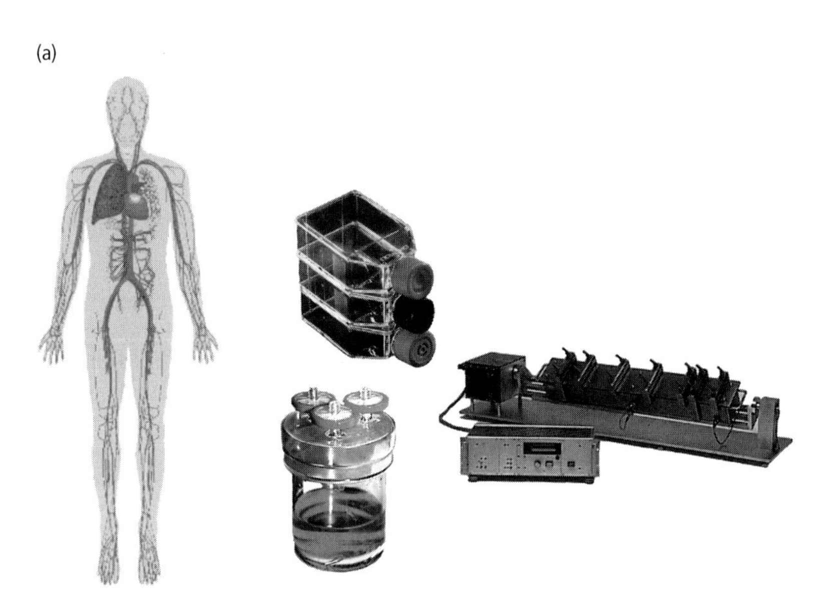


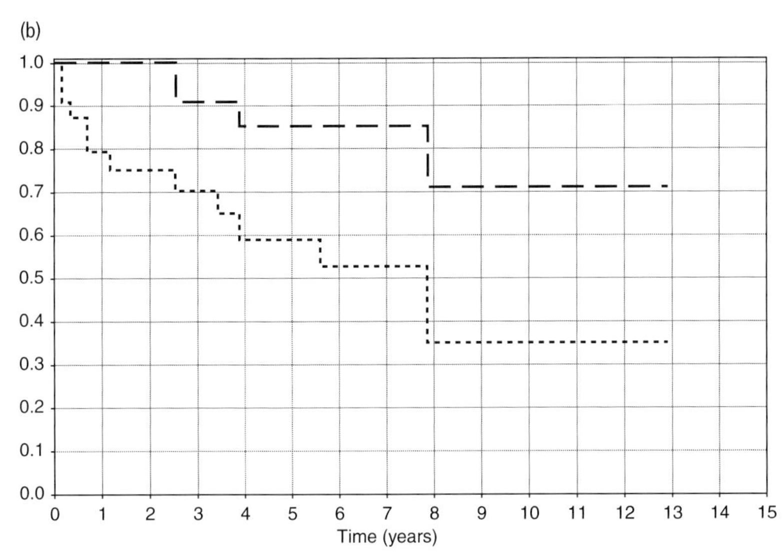


**Figure 5.15** Endothelialization of vascular grafts. This was one of the first applications of tissue engineering. The sequence is seen in **(a)** where endothelial cells are harvested and cultured and then seeded onto a vascular graft before that is implanted back into the same patient. **(b)** Shows the improved performance in the upper line compared to control patients, without endothelial cells, in the lower line. Images courtesy of Professor Peter Zilla, Cardiothoracic Research Unit, University of Cape Town, South Africa.

walls of grafts within a rotating seeding device, at over $10^7$ cells per graft. This endothelialization resulted in highly significant improvements in long-term patency rates (Figure 5.15).

The need for *in vitro* cell seeding gives a clinical and logistics disadvantage and many attempts have been made to develop techniques for *in vivo* endothelialization. Most of these involve surface modification of the graft material (still expanded PTFE and polyesters in most clinical applications) through plasma treatment, fibronectin coating and adhesion peptides such as RGD. It has been difficult to control inflammatory and thrombogenic responses to these surfaces.

### Tissue engineered blood vessels

Total tissue engineered blood vessels as opposed to endothelialized vascular grafts have proved difficult to develop. In principle these could involve either *in vitro* or *in vivo* approaches, but in both cases the development of adequate mechanical properties, especially burst strength, has been problematic. The *in vitro* techniques stand a much better chance of achieving this and most attention has been paid to them. Three specific approaches should be mentioned, involving biodegradable synthetic polymers, biopolymers, and scaffold-free techniques. Attempts with biodegradable polymers have mostly utilized polyglycolic acid and polycaprolactone-based systems, seeded with smooth muscle cells and endothelial cells. In the majority of situations with the scaffold-based methods, long-term patency has been reasonably successful but not the achievement of mechanical properties suitable for the pressures in the adult human circulation. This largely arises from the limited proliferative capacity of human smooth muscle cells within the porosity of the scaffolds.

Much attention has therefore been paid to cell sourcing and manipulation. Bone marrow mononuclear cells are particularly important. These are a heterogeneous population of progenitor cells that can develop into various vascular phenotypes during *in vitro* culture, creating layers of both endothelial cells and smooth muscle cells over a few weeks. It is also possible to implant bone marrow mononuclear cells directly, without pre-differentiation, although the mechanisms by which these generate blood vessels is not clear.

One situation where these techniques have been successful clinically is that of treating some congenital heart defects. For example, children may be borne with a single ventricle and require some type of cavo-pulmonary conduit for survival. This could be accomplished by a synthetic vascular graft, but these do not grow with the child, and have a significant number of problems. In 2001, the first tissue-engineered blood vessel was used clinically in such a situation. The scaffold utilized a combination of polyglycolic acid and polycaprolactone. Mononuclear cells derived from autologous bone marrow were seeded into the scaffolds, which were later implanted into these very young patients, where they functioned very successfully (Figure 5.16(a)).

Another important potential clinical application involves hemodialysis access, that is providing an arteriovenous shunt in patients who routinely require dialysis and in whom the natural blood vessels have deteriorated. Clinical studies have taken place in patients with end-stage renal disease that had had hemodialysis failure. The grafts were scaffold free and were obtained by wrapping sheets of autologous fibroblasts around a mandrel and culturing them for 10 weeks, followed by endothelialization of the inner layers (Figure 5.16(b)). These grafts have survived in some patients for over 12 months. Of course this does require an extensive *in vitro* culture period. Other possibilities include using scaffolds, for example decellularized allogeneic vessels with allogeneic smooth muscle cells.

Achieving appropriate mechanical properties and patency for long periods of time when exposed to adult arterial conditions have made the development of tissue-engineered blood vessels particularly difficult. There are indications of successful solutions that involve long *in vitro* culture times and/or non-mainstream clinical applications; these need to be enhanced in order to make them applicable to the major clinical areas such as peripheral vascular disease.

## 5.6.6 Cardiac tissue engineering: the myocardium

The heart has very limited powers of regeneration. Cardiomyocytes, the cells that populate the striated muscle of the myocardium die over time, with little opportunity for their natural replacement. Several heart disorders that result in chronic cardiac overload cause a slow but continual loss of cells, while a major myocardial infarction, or heart attack, can cause the loss of 25% of cardiomyocytes in the left ventricle. The massive increase in the incidence of heart disease globally provides a considerable incentive to develop strategies to assist in heart regeneration.

### Cell therapy

The strategies that have been under consideration include cell therapies, gene therapies,

(a)

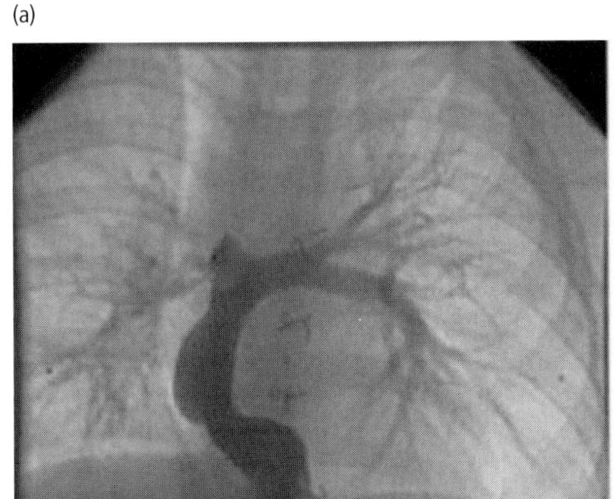


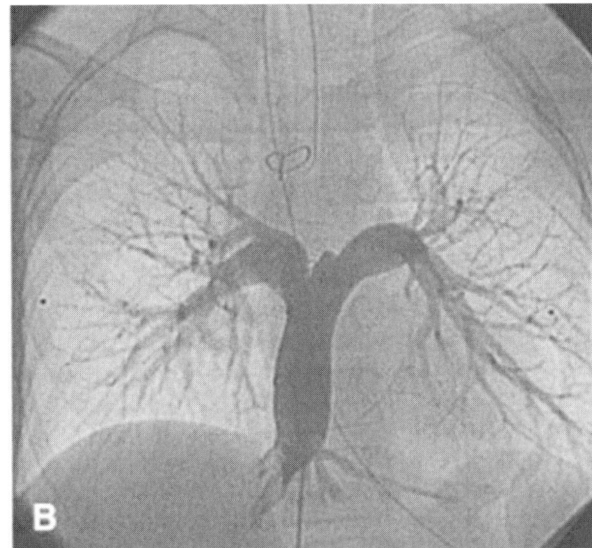


(b)

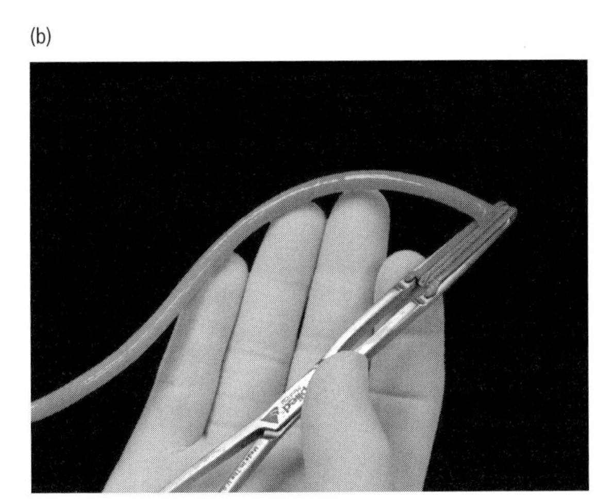

**Figure 5.16** Tissue engineered vascular grafts.
**(a)** Tissue engineered vascular graft used as a conduit in young patients with single ventricle physiology. Grafts were formed using a biodegradable polyester tubular scaffold seeded with mononuclear cells derived from autologous bone marrow. Angiographs 5 years after implantation in one patient (A) and after 4 years in another (B). In neither case was there stenosis, aneurysm formation nor ectopic calcification. Reprinted from *Journal of Thoracic and Cardiovascular Surgery*, 139(2), Hibino N *et al.*, Late-term results of tissue-engineered vascular grafts in humans, 431–6, © 2010, with permission from Elsevier.
**(b)** The first completely biological, tissue-engineered human blood vessel with clinically relevant mechanical strength for high-pressure arterial circulation of a human. Courtesy of Nicolas L'Heureux, Cytograft Tissue Engineering.

tissue-engineered cell-scaffold constructs, cell sheet engineering and hydrogel implantation. With cell therapies there are several types of cell source to consider. Some methods have involved cardiac progenitor cells, CPCs, where certain populations of autologous CPCs have been infused into patients after an infarction. Bone marrow-derived MSCs have also been used. There is some discussion about whether these actually transdifferentiate into cardiomyocytes or whether they exert effects in a paracrine manner by secreting cytokines. It is clear that the vast majority of MSCs that are transplanted into an infarcted myocardium will be lost within a few days but their positive effects may be seen over prolonged periods of time. It is possible that once the precise signaling pathways of their effects are understood, it should be possible to develop cell-free therapies using the identified proteins and small molecules (Figure 5.17(a)). It is also possible that either autologous

(a)

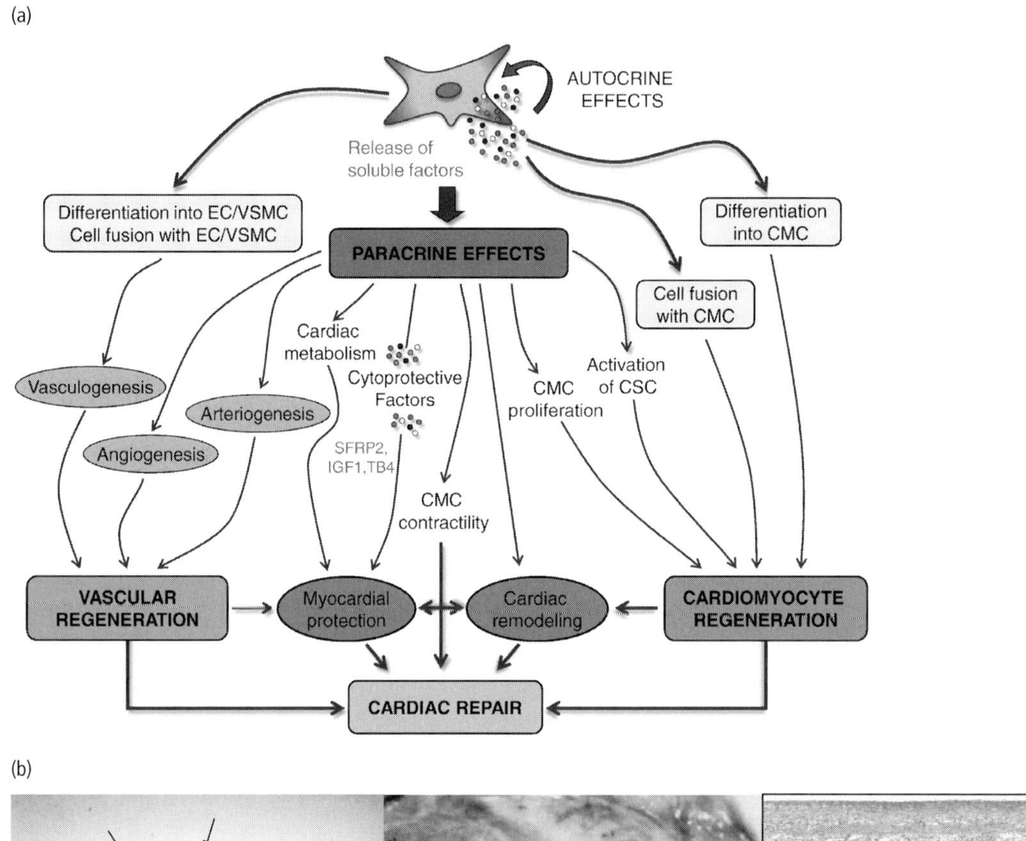


(b)

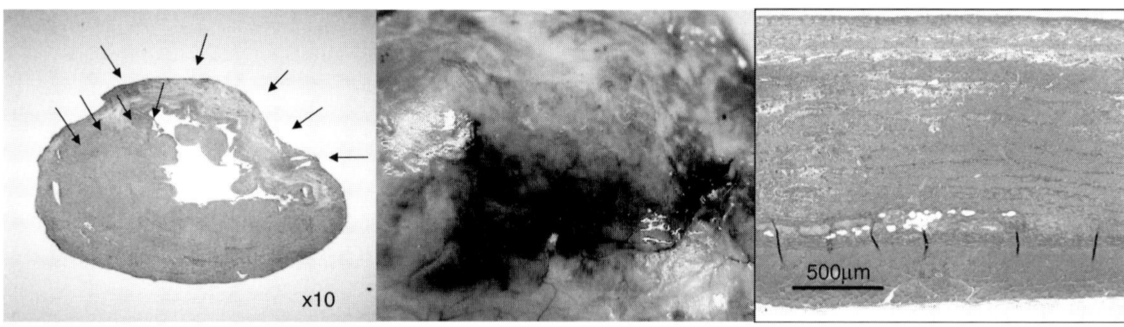


**Figure 5.17** Regenerative approaches to myocardial infarction. **(a)** Mesenchymal stem cell therapy for myocardial infarction. Transplanted mesenchymal stem cells may be able to differentiate into cardiomyocytes and vascular cells. This scheme demonstrates the mechanism by which cardiac regeneration and repair occurs after administration of these cells. Reprinted from *Vascular Pathology*, 57, Gnecchi M *et al.*, Mesenchymal stem cell therapy for heart disease, 48–55, © 2012, with permission from Elsevier. **(b)** The use of cell sheet engineering for generating myocardial tissue graft. The area of the myocardial infarction is seen on the left, the tissue graft is in the center, on the right is a section of a graft prepared from 3 sheets of cells at one day intervals. Courtesy of Professor Teruo Okano, Institute of Advanced Biomedical Engineering and Science, TWIns, Tokyo Women's Medical University, Tokyo, Japan. See plate section for color version.

or allogeneic MSCs can be administered intravenously or via an intracoronary route. Much work has been performed on both ES and iPS cells with respect to cardiomyocyte generation. This is technically possible although the cells produced tend to be rather immature, lacking the functional characteristics of adult ventricular cardiomyocytes.

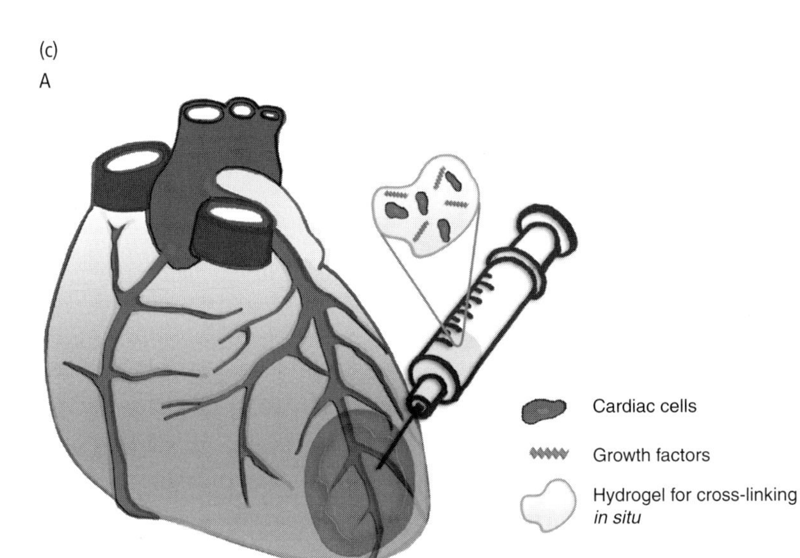


**Figure 5.17** *continued*
(c) *In situ* polymerization of hydrogel cell carriers for cardiac restoration following myocardial infarction. (A) Shows the most prevalent approach where a pre-polymer/cell solution is injected and cross-linked *in situ.* (B) An alternative strategy involves pre-polymerized microbeads of 100 μm diameter directly injected into the arterial circulation, where cross-linking can be controlled with more precision and the approach to the infarct through the arterial circulation can be more effective in delivering the cell graft to the intended target. Image courtesy of Alexandra Berdichevski and Professor Dror Seliktar of The Faculty of Biomedical Engineering, Technion: Israel Institute of Technology.

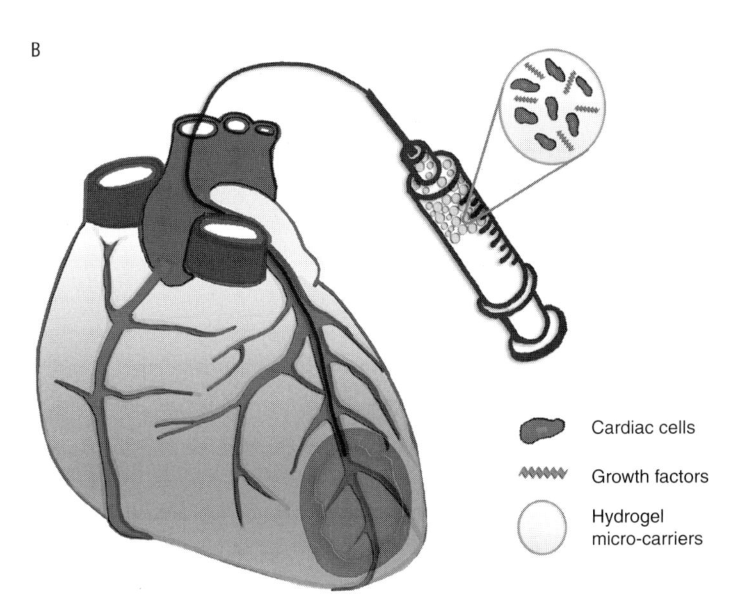


## Cell sheet engineering

The principle of cell sheet engineering has been introduced in Section 5.3.4. One of the most significant applications of this technology is in the treatment of myocardial infarction, where clinical studies have taken place. The same PIPAAm thermoresponsive polymers are used as the substrate for cell culture. It has been possible to layer cardiac cell sheets to obtain cell-rich 3D cardiac tissues, where gap junctions form between the cells. There is rapid establishment of electrical communication between cell sheets, giving synchronously pulsatile tissue. These cardiac patches can be transplanted onto the infarcted areas, with improvement to ventricular contraction and ejection fraction. This is shown in Figure 5.17(b).

### Protein and gene therapy

A variety of proteins and other biomolecules are being considered as therapeutic agents for ischemic heart disease. These include heat shock proteins, angiogenic growth factors (VEGF and FGF) and anti-apoptosis fusion proteins. It is likely that the co-delivery of several proteins will be required in order to optimize the benefits. As an alternative, gene therapy, and especially angiogenic gene therapy, may be considered. It is not clear just how long transgene expression would be required to sustain benefits in the infarcted heart but even short periods of administration of plasmid VEGF can increase blood flow and cardiac function in models of ischemic heart disease. Angiogenic therapy does have drawbacks, including safety issues concerned with unregulated expression. There are alternatives, including antioxidant and endothelial nitric oxide synthase gene therapy.

### Hydrogel therapy

The heart undergoes a process of ventricular remodeling in response to an infarct, in which the ventricular volume increases due to increased compliance of the tissue, which preserves the cardiac output but increases the overall stress on the muscle wall, eventually leading to cardiac failure unless treated. In view of this, approaches have been made to use biomaterial injections into the affected area in order to counter the increased stress. It is clear that the injection of non-contractile hydrogels into the wall can reduce this elevated stress in the myofibers (Figure 5.17(c)) using synthetic gels such as PEG or biological gels based on fibrin, collagen and alginates. The extent of the effect does vary with the stiffness of the gel and the timing and location of the injection. The use of these hydrogels also allows them to be functionalized with active molecules in order to combine the mechanical and biological effects. PEG is particularly effective here when combined with therapeutic proteins.

## 5.6.7 Cardiac tissue engineering: heart valves

The use of implantable prosthetic heart valves to treat valvular disease was discussed in the previous chapter. The good success of the long-term implantable devices suggests that there is little need for a tissue engineering approach to this problem. Nevertheless, great efforts have been made in this field, largely in order to address one specific issue. Although the vast majority of patients who require heart valve treatment are elderly, a significant number of children suffer from congenital heart disease, or acquire heart valve conditions early in life, for example as a consequence of rheumatic fever, leading to the need to replace their valves. Unfortunately, heart valve prostheses do not grow as the child ages, and so they are likely to need multiple surgeries during childhood and adolescence. Tissue engineering offers the prospect of preparing valves for these children that will respond dynamically within the growing individual.

As with other target tissues discussed in this chapter, a few different approaches have been tried. The first of these is the *in vitro* approach of seeding cells into scaffolds, maturing the developing scaffold-based valve in a bioreactor and then implanting the construct in the patient. As usual, the scaffold could be synthetic or natural. The alternative type of approach is the implantation of a structure that is remodeled *in vivo* through endogenous cells. Neither of these approaches is simple, and there are several significant obstacles to be overcome. The major issue here is that it is too simplistic to assume that a tissue engineered valve, produced by any paradigm, will be able to replicate the complex, dynamic structure and performance of the original valve. The remodeling which has to take place may not follow the characteristics of the valve. These functions depend on the quality of the ECM and of the performance of the relevant cells, the superficial valvular endothelial cells and the deep valvular interstitial cells. An example is seen in Figure 5.18.

With respect to the scaffold/template materials, there is a need to combine significant mechanical performance with biological properties associated

**Figure 5.18** Tissue engineered heart valves. A stented tissue engineered heart valve based on mature vascular-derived cells prior to *in vivo* implantation. Image courtesy of Dr. Benedikt Weber and Professor Simon Hoerstrup, Swiss Center for Regenerative Medicine, University Hospital Zurich, Switzerland.

with blood compatibility and biodegradation as well as stimulation of these target cells. PHB materials have received much attention. Fibrin-based materials have also been investigated. As with other areas, decellularized xenogeneic tissue is also a strong candidate and it has been shown that recellularization can take place in the valve position *in vivo*. Immunological problems associated with these xenogeneic tissues remain a concern. The source of cells is also an important factor. Adipose-derived stem cells, circulating endothelial progenitor cells and umbilical cord-derived cells are strong possibilities.

## 5.6.8 Nerve tissue engineering

It is necessary to consider separately the uses of regenerative medicine in the central and peripheral nervous systems (CNS and PNS respectively). This is partly because it has traditionally been thought that, in mammals, axons in the injured CNS do not regenerate whereas they do have some regenerative capacity in the PNS. The situation may not be as simple as this but it is still better to separate out these two compartments. It is not only traumatic injuries that affect the CNS but also a number of degenerative diseases have a major impact on health care and the prospects for regenerative medicine.

### Central nervous system: spinal cord injury

Severe spinal cord injury (SCI) is not very common, but the occurrence is usually devastating. Even in the best health care environments many patients suffering such injuries die before arriving at hospital or soon after. Quite a lot is known about the cellular and molecular events that occur following SCI, but difficulties do arise when studying SCI in animal models because lesions induced in animals do not really mimic those that occur in human patients. Full severance of the cord is not the usual manifestation of injury but rather it is contusions, crush injuries and lacerations that are more common. Most injuries occur in positions where the spine is most flexible, particularly the C5–C7 in the neck and in the lower back. There is usually significant vascular damage and hemorrhage, which initiate a number of secondary processes, especially as spinal neurons rapidly undergo apoptosis and necrosis. Axons do not regenerate to any significant extent in the central nervous system. Indeed, there is a loss of axonal connections and an inhibition of regeneration through the activity of glycoproteins and other inhibitory molecules released from reactive astrocytes. Glial scarring tends to form around the injured cord; this is a significant factor in attempts to devise therapies for these injuries since this profound scarring will have to be addressed before regenerative processes can be induced.

The use of regenerative methods for spinal cord injury remains a profound challenge. As elsewhere in this chapter, we see significant efforts to involve cell-based techniques, biomolecular therapies, scaffold-based technologies and combinations of these. In the first case, Schwann cells have been extensively investigated but although they can enhance functional recovery to some extent, primarily by releasing a number of growth factors that support the growth of axons, there is very poor survival of these cells in the medium term. Any biomaterials that are used in attempts to address spinal cord injury have to lessen glial scar as well as guiding regenerating axons across the injury site. Among the scaffold materials that have been investigated are collagen and chitosan, and synthetic polymers such as polylactic acid, polycaprolactone and PHB.

### Peripheral nerve repair

Peripheral nerve injury is common and may be seriously disabling, with possibilities of life-long disturbance of function and constant neuropathic pain. The effects naturally vary with the severity of the injury ranging from nerve crush, which has a good prognosis, to severance or destruction of the nerve, which is more difficult to treat. Generally, if a nerve is severed, and the gap is not too long (perhaps a few millimeters), it is usually possible to repair it by end-to-end tension free suturing. With longer gaps, such suturing produces tension, which affects axon regeneration. It has been known for a long time that it is possible to obtain a functional repair by using autologous peripheral nerve grafts. However, there are still limitations with serious nerve injury. The nerve graft can be augmented with the prolonged delivery of growth factors, especially glial cell-line-derived neurotrophic factor, or by physical processes such as electrical stimulation and phototherapy.

Alternatives to autologous nerve include autologous non-nerve tissue and allogeneic or xenogeneic grafts, although these are not yet very successful. For several decades attempts have been made to develop biomaterial-based conduits for encouraging nerve regeneration when autologous nerve grafting is inappropriate. The options are a non-degradable conduit, a synthetic biodegradable polymer and biopolymers. In these cases, both proximal and distal nerve stumps are inserted into the open ends of the conduit and fixed in place with epineural sutures. With non-degradable conduits, silicones and PTFE have been tried, with some limited clinical experience with the former. Both polycaprolactone and polyglycolic acid have been used in clinically approved conduits, as have bovine collagen tubes. Although reasonably successful in some situations, full sensory and motor function restoration is not guaranteed. The response to the initial injury is complex and it would seem that better therapeutic options would not be developed until there is a better understanding of these neurobiological mechanisms. There have been experimental attempts to involve stem cells and growth factors in association with biomaterial conduits with the objective of modulating nerve reorganization and amplifying adaptive responses of the system. Examples are seen in Figure 5.19.

## 5.6.9 Urological tissue engineering

### The bladder

There is a significant need for procedures to repair damaged bladders, especially in children. There are wide ranging congenital abnormalities, including myelomeningocele (a common form of spina bifida which usually involves loss of bladder function) and exstrophy (in which the bladder is essentially formed inside-out), which have severe implications for the bladder, as can a number of injuries. Drug treatment is not usually very effective and the use of reconstructive or regenerative procedures, referred to as cystoplasty, could be very beneficial. Over several decades, many groups tried to fashion bladders from synthetic materials, such as silicones, PTFE and collagen, but the difficult biomechanical environment has not allowed any significant success. As a result, cell-based approaches were considered in the 1990s and slow but good progress was made. The first tissue engineered autologous bladder to be used clinically was reported by Atala and colleagues in 2006. This involved a bladder biopsy obtained from the dome through a suprapubic incision. Muscle fragments were plated on culture dishes, and expanded using Dulbecco's Modified Eagles Medium and 10% fetal bovine serum. Separately, urothelial cells were expanded with keratinocyte growth medium. In this clinical series the first patients were treated with a product based on a homologous decellularized bladder submucosa scaffold, and the next group with a

(a)

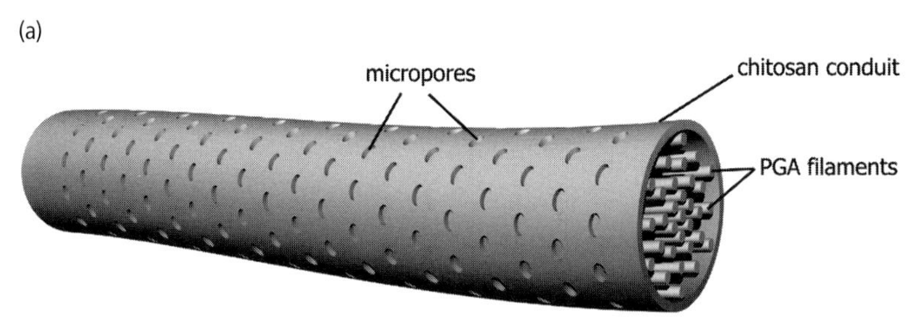


(b)

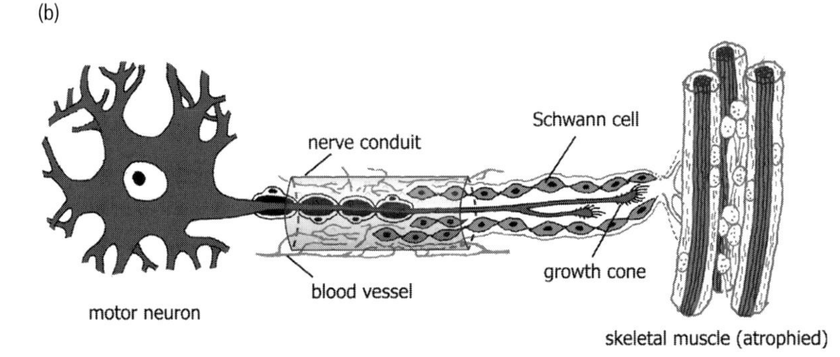


(c)

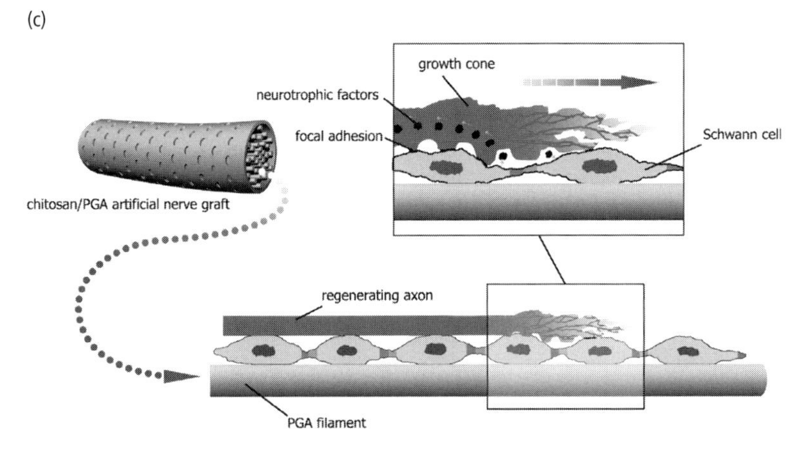


(d)

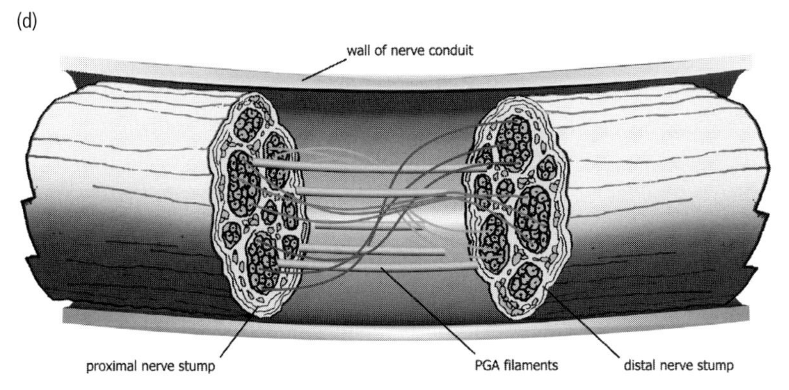


**Figure 5.19** Peripheral nerve tissue engineering.
**(a)** Schematic diagram showing the chitosan/PGA artificial nerve graft. The graft is made of a microporous conduit of chitosan, and filled with longitudinally aligned filaments of polyglycolic acid (PGA) within the conduit lumen.
**(b)** Schematic diagram showing guided nerve regeneration by the chitosan/PGA artificial nerve graft. The chitosan microporous conduit allowed vascularization from the stumps as well as trans-wall blood vessels infiltration, so as to provide nutrition for Schwann cell migration and axonal extension.
**(c)** Schematic diagram showing guiding role of the chitosan/PGA artificial nerve graft in Schwann cell migration and axonal extension.
**(d)** Schematic diagram showing preferential re-innervation of different basal lamina tubes in the distal stump by corresponding motor, sensory and sympathetic axons. All images courtesy of Professor Xiaosong Gu, Director, Key Laboratory of Neuroregeneration, Nantong University, China.

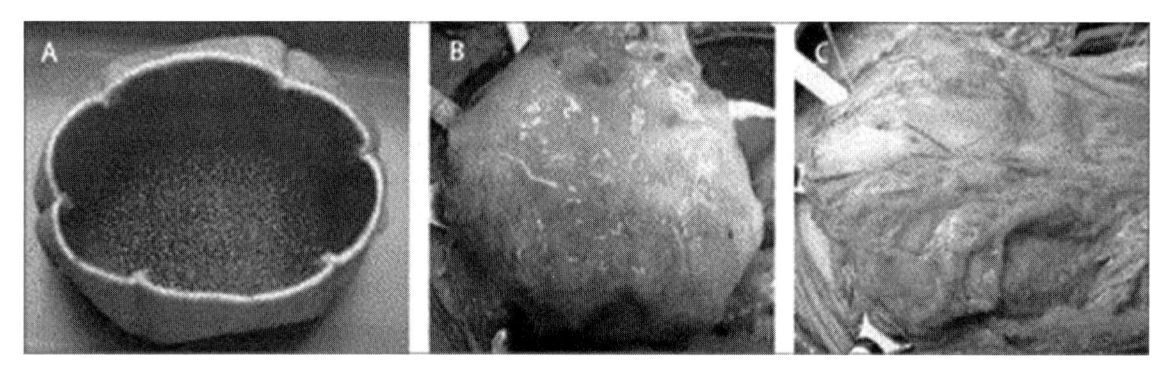


**Figure 5.20** Bladder tissue engineering. This was the first clinical study of bladder tissue engineering, in which 7 patients with the congenital abnormality myelomeningocele were treated and followed for up to 61 months. A bladder biopsy was obtained from the patients and urothelial and muscle cells derived from these were cultured and seeded into either collagen or collagen-polyglycolic acid scaffolds. Seven weeks after the biopsy, the autologous constructs were implanted into the patients. **(A)** shows the scaffold seeded with cells, **(B)** shows the anastomosis of the construct to the native bladder and **(C)** shows the construct covered with omentum. Reprinted from *The Lancet*, 367, Atala A *et al.*, Tissue-engineered autologous bladders for patients needing cystoplasty, 1241–6, © 2006, with permission from Elsevier.

composite scaffold of collagen and polyglycolic acid (Figure 5.20). In some patients these scaffolds were supplemented by a wrap of omental tissue that improved vascularization. The time between bladder biopsy and engineered bladder implantation was around 7–8 weeks. The reconstructed bladders showed improved function, which was maintained for the five year follow-up. In this situation, the combination of collagen and PGA provided an appropriate level of structural and cellular support, with assistance in revascularization from the omentum.

### The urethra

The urethra can be damaged by injury and disease. Short urethral defects may be repairable through end-to-end anastomosis, but this is difficult with long gaps and some form of tissue graft may be attempted. However, these are not very successful and tissue engineered urethral constructs may provide better solutions. In a similar manner to bladder regeneration, autologous cell-seeded scaffolds provides one approach. Tubular scaffolds may be made of degradable materials such as PLGA-PGA with or without collagen. Muscle cells and urothelial cells can be prepared from biopsy samples, and after suitable culture periods, the former may be seeded onto the outer tubular surface and the latter onto the luminal surface. The tissue-engineered construct should have a similar phenotypic and histological pattern as the native urethral tissue and may remain patent, avoiding the strictures that are often associated with conventional tissue grafts, over the long term.

## 5.6.10 Airway tissue engineering

### The trachea

The loss of airway tissue is devastating, but there are very few methods available for treatment. Resection of the trachea following severe benign or malignant diseases may be necessary but primary repair is very difficult. Only short segments may be removed and repaired without some form of functional replacement but autologous and synthetic grafts have proved relatively unsuccessful. The airway represents an interface between internal and external environments, which is unusual for transplantation procedures, where the mucosa has immunologically active cells which often cause acute graft rejection. This is one area where tissue engineering could make a significant contribution because

(a)

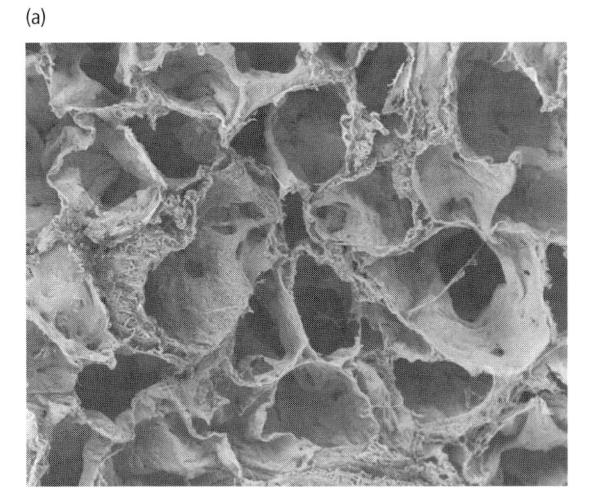

(b)

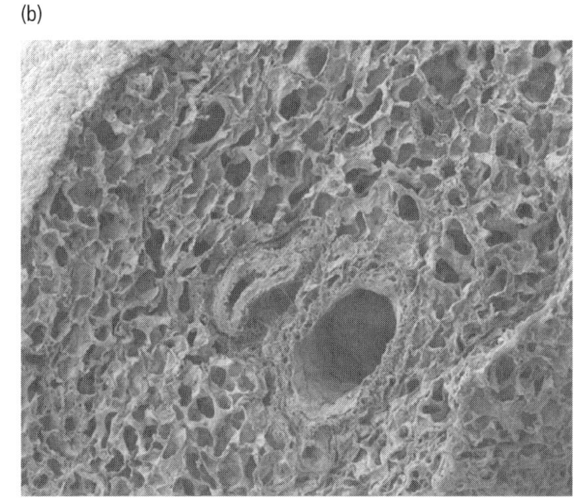

**Figure 5.21** Airway tissue engineering. The regeneration of functioning tissue in the airways represents a significant challenge. Shown here are two areas of rat alveoli that demonstrate their complexity, which will have to be replicated. Images courtesy of Panagiotis Maghsoudlou and Paolo De Coppi, University College London Institute of Child Health and Great Ormond Street Hospital, London, UK.

of this profound so far un-met clinical need. At a relatively early stage of development, some successes have been achieved with this form of airway tissue engineering. The main approach has been to use decellularized donor trachea as the template, which is populated by autologous epithelial and mesenchymal stem cell-derived chondrocytes (Figure 5.21).

The donor-derived tissue is rinsed with PBS and antibiotics and then decellularized over a 6-week period, with frequent sampling of the tissue to quantify remaining cells. Autologous epithelial cells can be obtained from bronchoscopic biopsy samples and prepared in culture that avoids the presence of fibroblasts. Autologous chondrocytes can be prepared from bone-marrow aspirate. Within a bioreactor, the tubular decellularized matrix is seeded on the external surface with the chondrocytes and on the inner surface with the epithelial cells. Typically the bioreactor will be slowly rotating, with different culture conditions for the two different cell types on inner and outer surfaces. The reactor alternates cells between liquid (the culture medium) and air phases; a total bioreactor time of around 96 hours should be sufficient. Depending on the precise location of the lesion or obstruction, the tissue engineered construct may be anastomosed end-to-end proximally and distally to the remaining trachea/bronchus. Obviously with limited data at this stage, it is difficult to predict success rates and ultimate outcomes. However, cytological analysis of the luminal surface at 4 days shows that we may expect abundant epithelial cells that retain their phenotype, and good lung function may be achieved.

## The lungs

Lung disease is far more common and is a significant cause of morbidity and mortality worldwide. Chronic obstructive pulmonary disorder (COPD) involves the anatomical narrowing and blocking of channels. In addition, restrictive pulmonary diseases, which includes pulmonary fibrosis, involve extensive scarring. In both cases there is limited respiration and oxygenation, where drug therapy, oxygen therapy and rehabilitation are the main forms of treatment. Physical replacement of the complex lung architecture is not really feasible, hence the significant interest in the tissue-engineered lung. So far, very limited progress has been made, again mainly because of the complexity of the organ. There are over 40 distinct cell types in the respiratory tract. The alveolar wall has a narrow connective tissue core with fibroblasts, myofibroblasts, capillary endothelial cells and ECM components. The epithelium itself has two cell types, types I and II pneumocytes. Clearly decisions have to be made as to what is the tissue engineering target and which cell types should be involved. The use of mature lung lineage cells derived from somatic lung progenitor cells is attractive from the autologous perspective.

Several materials have been used in experimental lung tissue engineering projects, including collagen, gelatin and biodegradable polyesters such as PLGA. One of the main problems has been concerned with the physical form of the construct: volumes of scaffold cultured with somatic lung progenitor cells have been shown to survive for some time when implanted in certain animal models, for example when implanted into the lobe of a lung, but support of lung epithelium is not readily achieved. The organization of regenerated tissue into identifiable pulmonary structures remains an important objective.

## 5.7 Tissue engineering of constructs for drug testing and tumor models

We finish this chapter with a mention of potential non-therapeutic uses of tissue engineering, primarily in the area of drug testing. The screening of new pharmaceutical molecules is traditionally undertaken with cell culture studies and animal testing. These are valuable but not totally effective, a fact which is demonstrated frequently by the failure of new drugs in clinical trials, either because of safety issues or for a lack of efficacy. There is therefore a strong impetus to develop alternative *in vitro* systems for new drug evaluation, and indeed for the study of disease mechanisms. Particularly important here is the need for *in vitro* three-dimensional models of tumors for a better understanding of tumor progression and the assessment of chemotherapeutic agents and imaging systems.

There have been several approaches. One of these involves the use of microfluidics technology to develop so-called “organs-on-chips,” which attempts to recapitulate the spatiotemporal chemical gradients and dynamic mechanical microenvironments of specific organs within micro-engineered systems. Microfluidic liver and kidney chips have been

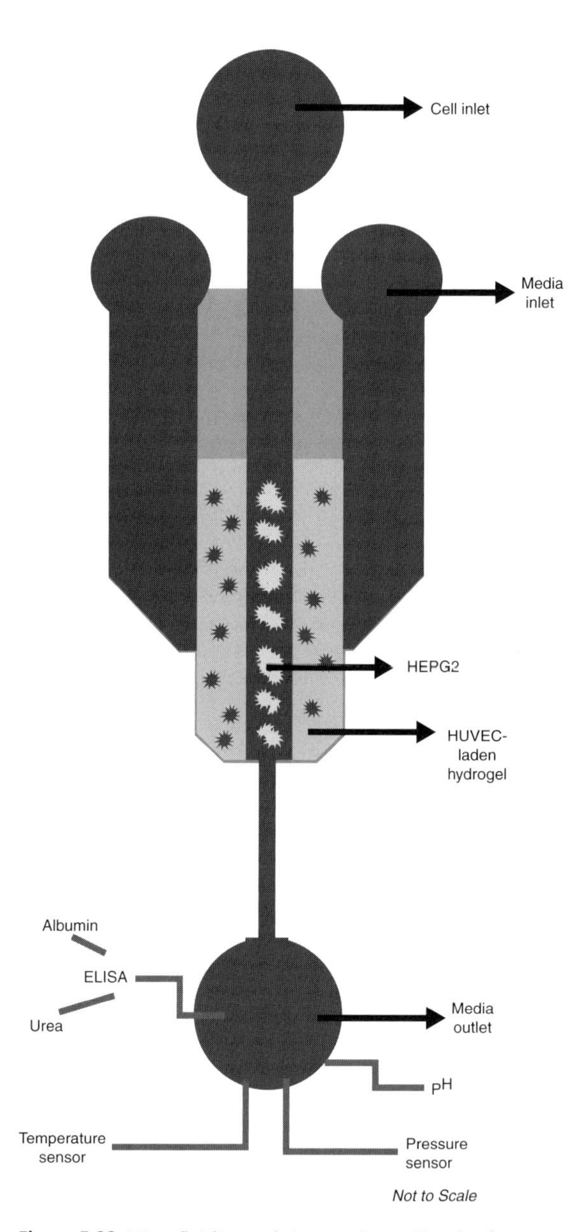


**Figure 5.22** Microfluidics and tissue culture. Sketch of a microfluidic device for culture of liver tissue. Image courtesy of Ali Khademhosseini Ph.D, Harvard-MIT Health Sciences & Technology, Boston, MA, USA.

developed, as shown in Figure 5.22. In the former case, flow chambers that are connected by baffles are able to separate cultured hepatocytes from blood flow, thus recreating the endothelial-hepatocyte interface. Similarly, the kidney chip recreates the interface between epithelium from flowing urine. In these systems, the biomaterial component is usually silicone elastomer. As these technologies develop, it will be necessary to incorporate more sophisticated biomaterial structures; for example, a microfluidic cornea model has been prepared with collagen membranes as a substrate for corneal epithelial cells.

Moving towards truly 3D models, several systems work with sandwich cultures, in which, for example, a monolayer of hepatocytes is placed between two ECM layers, the hepatocytes appearing to retain functional stability that allows studies of drug metabolism. Polymer scaffolds have been used in attempts to produce 3D mimics of most organs and tissues. With respect to tumor tissue models, many studies have been performed with multicellular spheroids, which try to mimic the heterogeneous structure of the tumor microenvironment. Other developments include the use of cell seeding within hollow polymer fibers (e.g., of polyvinylidine fluoride) and of multicellular layer models, where layers of tumor cells are cultured on collagen-coated microporous polymer membranes that allow the assessment of drug flux through the tumor tissue.

## Summary and learning objectives

1. This chapter has focused on the potential that is offered by the technologies of regenerative medicine and the philosophical and practical differences to those technologies of implantable devices and artificial organs that are intended to replace tissues and their functions. Regenerative medicine aims to regenerate natural structures rather than replace them with synthetic forms.
2. The basic tissue engineering paradigm has been introduced. This concerns the principles that have been developed in order to translate the objectives of persuading cells to express new tissue under conditions in which they would not normally do so. This is primarily achieved by combinations of molecular and mechanical signaling. In some circumstances this may be possible without the assistance of biomaterials, as in cell therapies, but usually a biomaterial structure is necessary. Such a structure is called a template. At the heart of this chapter is the discussion of the requirements and specifications of the materials for such templates. We discern the mandatory specifications for such template materials, which are mainly oriented towards the need for these structures to recapitulate the niche of the target cells.
3. In practice, although many different, and widely diverse, materials have been used for these templates, usually previously described as scaffolds, there are three groups that form the current state-of-the art. These are porous solids, hydrogels and decellularized ECM materials.
4. In order to understand the performance of these materials in tissue engineering, we have discussed, in a fairly elementary way, the principles of the target cells, especially stem cells, that may be used. This has included descriptions of embryonic stem cells and induced pluripotent stem cells, as well as the various forms and sources of mesenchymal stem cells that form the basis of much clinically oriented work. This chapter also addresses the technical issues of culturing and handling these cells, including bioreactors, cell patterning and cell sheet technologies, and the role of various nutrients and signaling molecules.
5. We have made it clear that these processes have not yet developed to the extent that they are used

routinely in clinical practice. Much of the chapter has therefore been devoted to a description of the progress that has been made in several of the main clinical areas to develop and apply these techniques. This has covered bone, cartilage, skin, muscle, cardiovascular, nerve, urological and pulmonary areas. There are many common approaches to the scientific and technical challenges faced.

## Questions

1. Discuss the rationale for regenerative medicine approaches to disease, comparing this to the alternative medical device technologies.
2. Describe the basic tissue engineering paradigm and discuss the variations to this paradigm that have emerged over time.
3. Discuss the characteristics of embryonic stem cells and describe the pros and cons of using these cells in regenerative medicine.
4. Describe the essential features of mesenchymal stem cells and compare the performance of such cells that are derived from bone marrow and those obtained from adipose tissue.
5. Give your opinions on the role of oxygen in *in vitro* tissue engineering processes.
6. Describe the processes involved in various printing techniques for cell patterning in relation to regenerative medicine.
7. Discuss the essential requirements for tissue engineering templates.
8. One of the principal specifications for templates is the ability to replicate the niche of the target cells. Discuss and compare how porous polymers and hydrogels are able to satisfy this requirement.
9. Describe the healing processes in bone following major resection of diseased tissue and speculate how tissue engineering solutions will be able to address those situations where full healing is not possible.
10. Discuss the potential role of tissue engineering in the treatment of heart valve disease.
11. Compare the therapies available for urinary tract and airway diseases, and the role that tissue engineering may play.
12. Discuss the potential for the development of tissue engineering constructs as models for drug testing.

# Recommended reading

Ahmed T A E, Hincke T, Strategies for articular cartilage lesion repair and functional restoration, *Tissue Engineering Part B* 2010;16:305–29. *Review of techniques, from conservative, reparative to regenerative, in treatment of cartilage lesions.*

Atala A, Bauer S B, Soker S, Yoo J J, Retik A B, Tissue engineered autologous bladders for patients needing cystoplasty, *The Lancet* 2006;367 (9518): 1241–6. *Description of tissue engineering techniques for bladder regeneration.*

* Atala A, Lanza R, Thomson J A, Nerem R, *Principles of Regenerative Medicine*, Academic Press, London, 2nd edition 2011.

Bleiziffer O, Eriksson E, Yao F, Horch R E, Kneser U, Gene transfer strategies in tissue engineering, *Journal of Cell and Molecular Medicine* 2007; 11:206–23. *Review of situations where gene transfer techniques may be used to enhance tissue regeneration.*

Deumens R, Bozkurt A, Meek M F, *et al.*, Repairing injured peripheral nerves: bridging the gap, *Progress in Neurobiology* 2010;92:245–76. *Discussion of experimental and clinical observations on peripheral nerve repair.*

Di Maggio N D, Piccinini E, Jaworski M, *et al.*, Toward modeling the bone marrow niche using scaffold-based 3D culture systems, *Biomaterials* 2011;32:321–9. *Discussion of the performance of 3D perfusion culture system that attempts to replicate the bone marrow niche.*

Discher D E, Mooney D J, Zandstra P, Growth factors, matrices and forces combine to control stem cells, *Science* 2009;324:1673–7. *Essay on the factors that control stem cell behavior within the stem cell niche.*

George A, Ravindram S, Protein templates for hard tissue engineering, *Nano Today* 2010;5:254–66. *Essay on nanofibrous protein structures used for bone tissue engineering.*

Groeber F, Holeiter M, Hampel M, Hinderer S, Schenke-Layland K, Skin tissue engineering *in vivo* and *in vitro* applications, *Advanced Drug Delivery Reviews* 2011; 128:352–66. *Review of tissue engineered skin substitutes and relevant biomaterials.*

Guillotin B, Souquet A, Catros S, *et al.*, Laser assisted bioprinting of engineered tissue with high cell density and microscale organization, *Biomaterials* 2010;31:725–6. *Description of a laser-controlled cell printing process that prints tissue with high cell density and micro-scale organization.*

Huang G T-J, Induced pluripotent stem cells – a new foundation in medicine. *Journal of Experimental and Clinical Medicine* 2010;2:202–17. *Review of the derivation and characterization of iPScells, and their potential for cell-based therapies.*

Huang G T-J, Gronthos S, Shi S, Mesenchymal stem cells from dental tissues versus those from other sources, *Journal of Dental Research* 2009;88:792–808. *Detailed discussion of the regenerative potential of MSCs derived from dental tissues.*

Huh, D, Torisawa Y, Hamilton G A, Kim H J, Ingber D E, Microengineered physiological biomimicry: organs-on-chips, *Lab on a Chip* 2012;12:2156–64. *Review of microfluidic approaches to* in vitro *organ mimicry for drug testing.*

Krawiec J T, Vorp D A, Adult stem cell-based tissue engineered blood vessels: a review, *Biomaterials* 2012;33:3388–400. *Review of the use in bone marrow mononuclear and other stem cells in the development of tissue engineered blood vessels.*

Laflamme M A, Murry C E, Heart regeneration, *Nature* 2011;473:326–35. *Review of the biology of heart regeneration and the potential therapies for heart failure.*

Lampe K J, Heilshorn S C, Building stem cell niches from the molecule up through engineered peptide materials, *Neuroscience Letters* 2012;519:138–46. *Review of self-assembled peptide and protein structures used in attempts to replicate the stem cell niche.*

Lee K, Silva E A, Mooney, D J, Growth factor delivery-based tissue engineering: general approaches and a review of recent developments, *Journal of the Royal Society Interface* 2011;8:153–70. *Overview of the methods for growth factor delivery in tissue engineering.*

Leong K F, Cheah C M, Chua C K, Solid freeform fabrication of three dimensional scaffolds for engineered replacement of tissues and organs, *Biomaterials* 2003;24:2363–78. *Review of the principles of SFF, with detailed discussion and comparisons of 3D printing, selective laser sintering and fused deposition modeling.*

Macchiarini P, Jungebluth P, Go T, *et al.*, Clinical transplantation of a tissue engineered airway, *The Lancet* 2008;372:2023–30. *Description of the first clinically successful use of a tissue engineered trachea, derived from decellularized donor tracheal tissue and autologous epithelial cells and chondrocytes.*

Moeller H-C, Mian M K, Shrivastava S, Ching B G, Khademhosseini A, A microwell array system for stem cell culture, *Biomaterials* 2008;29:752–63. *Detailed description of the design, fabrication and validation of microarray systems for ES cell culture.*

Nichols J E, Cortiella J, Engineering of a complex organ; progress toward development of a tissue-engineered lung, *Proceedings of the American Thoracic Society* 2008; 5:723–30. *Review of the factors that face the development of a tissue-engineered lung.*

Norotte C, Marga F S, Niklason L E, Forgacs G, Scaffold-free vascular tissue engineering using bioprinting, *Biomaterials* 2009; 30:5910–17. *Report on a fully biological self-assembly approach to rapid prototyping bioprinting for scaffold-free small diameter vascular reconstruction.*

Raya-Rivera A, Esquilliano D R E, Yoo J J, *et al.*, Tissue-engineered autologous urethras for patients who needed reconstruction: an observational study, *The Lancet* 2011;377:1175–82.*Discussion of procedures for the preparation and implantation of autologous tissue-engineered urethras used in cases of trauma in pediatric patients.*

Roh J D, Sawh-Martinez R, Brennan M P, *et al.*, Tissue engineered vascular grafts transform into mature blood vessels via an inflammation-mediated process of vascular remodeling, *Proceedings of the National Academy of Sciences* 2010;107:4669–74. *Discussion of techniques to produced tissue engineered vascular grafts based on degradable polymer scaffold and bone marrow mononuclear cells.*

Rosenthal A, Macdonald A, Voldman J, Cell patterning chip for controlling the stem cell microenvironment, *Biomaterials* 2007;28:3208–16. *Description of a microfabricated polymer chip, of thousands of microwells sized to trap down to a single stem cell.*

Sahni V, Kessler J A, Stem cell therapies for spinal cord injury, *Nature Reviews Neurology* 2010;6:363–72. *Consideration of the pathology of SCI and the role of stem cell therapy in its management.*

Santoro, R, Olivares A L, Brans G, *et al.*, Bioreactor based engineering of large-scale human cartilage grafts for joint resurfacing, *Biomaterials* 2010;31:8946–52. *Demonstration of techniques to form clinically relevant volumes of cartilage, using chondrocytes, hyaluronic acid scaffolds and perfusion bioreactors.*

Schoen F J, Heart valve tissue engineering: quo vadis? *Current Opinion in Biotechnology* 2011;22:698–705. *Authoritative opinion on the challenges and future possibilities for pediatric heart valve tissue engineering.*

Seekine H, Shimizu T, Okano T, Myocardial tissue engineering: towards a bioartificial pump, *Cell Tissue Research* 2012;347:775–82. *Review of technical and clinical aspects of cell sheet engineering in the myocardium.*

Sill T J, von Recum H A, Electrospinning: applications in drug delivery and tissue engineering, *Biomaterials* 2008;29: 1989–2006. *Comprehensive review of the technology and applications of electrospinning.*

Takahashi K, Yamanaka S, Induction of pluripotent stem cells from mouse embryonic and adult fibroblast cultures by defined factors, *Cell* 2006;126:663–76. *First description of iPSCs, produced by gene transfer.*

Turner NJ, Badylak S F, Regeneration of skeletal muscle, *Cell and Tissue Research* 2012;347:759–74. *Excellent review of approach to regeneration of skeletal muscle and therapeutic possibilities.*

* Van Blitterswijk C, Thomsen P, Lindahl A, *et al.* (eds), *Tissue Engineering*, Elsevier, Amsterdam 2008.

Werbowetski-Ogilvie T E, Bosse M. Stewart M, *et al.*, Characterization of human embryonic stem cells with features of neoplastic progression, *Nature Biotechnology* 2009;27:91–7. *Presentation and discussion of experimental work on teratoma formation following the use of injected ESCs.*

Woolfson D N, Mahmoud Z N, More than just bare scaffolds; towards multi-component and decorated fibrous biomaterials, *Chemical Society Review* 2010;39:3464–79. *Discussion of the evolution of fibrous biomaterials scaffolds, especially those based on peptides and with biologically active surfaces.*

Zhu J, Li J, Wang B, *et al.*, The regulation of phenotype of cultured tenocytes by microgrooved surface structure, *Biomaterials* 2010;31:6952–8. *Description of experiments that confirm the ability to control the phenotype of cells when cultured on surfaces with microgrooves.*

# 6 Drug and gene delivery

This chapter will introduce the concepts underlying the use of biomaterials to deliver various types of active molecules to the body for therapeutic or preventive purposes. This starts with the premise that many traditional pharmaceutical agents have been demonstrated to be effective in certain situations but their efficacy, specificity and safety is often compromised by poor bioavailability and high systemic toxicity when delivered through the conventional oral or intravenous routes. The chapter therefore concentrates on the technologies that can be used to produce more precise, controlled and targeted delivery of these active molecules. It covers drugs, genes, vaccines and other active molecules and deals with mechanical, physical, chemical and other techniques to produce the delivery mechanism. Many of these processes are based on the details of nanobiotechnology.

## 6.1 Introduction to active molecule delivery

Pharmaceutical agents, or drugs, are chemical substances that are applied to the body where they are metabolized and have, or are expected to have, a beneficial effect in treating, mitigating or preventing disease or discomfort, facilitating repair of injury or otherwise beneficially altering human physiological performance. Drugs have been the mainstay of many branches of medicine for centuries and have a powerful impact on both health status and health economics around the world.

The ability of a drug to achieve its desired effect efficiently and safely depends on many factors, including the ease with which it is administered to, and adsorbed by, the body, its mechanism of action in the body, and the way in which the body responds to the presence of the drug, taking into account the characteristics of distribution and metabolic fate. As drugs have become more complex, and their potential to do harm as well as good becomes more evident, attention has been increasingly drawn to the need for greater selectivity with respect to exactly where the drug is targeted in the body, to the preciseness with which the drug molecule attaches to the required site, and the relationship between the level of a drug in the body and time.

Traditionally and most easily, drugs are delivered to a patient by mouth, as in swallowing a tablet, or by inhalation as with an asthma nebulizer. If the drug is not easily absorbed by either of these routes, then alternatives such as intramuscular injection or intravenous infusion may be used, albeit with much lower levels of convenience. At the very simplest we can see the inherent problem with conventional delivery regimes by considering a patient with a headache or toothache who requires "long-lasting pain relief." By taking a tablet, by mouth, at the required frequency, say every four hours, the whole body experiences a peak load of the medication soon after a tablet is taken, which then decreases substantially as the drug is metabolized and cleared from the body, until the next dose. At times the level in the patient is likely to be higher than is strictly necessary, and at times it may well be lower than the threshold at which it has an effect. More importantly, the whole body and not just the relevant affected part of the head or tooth are experiencing these levels. This is hardly a well-controlled delivery process, although it is often quite effective and is usually very convenient as far as the patient is concerned.

This may seem a trivial issue and with many low-cost safe drugs available, it is. But now consider those drugs that are used for more serious conditions, where side effects are potentially much greater, and where costs may be astronomically high. Cancer therapy comes to mind. Everyone knows of the readily visible side effect of most forms of chemotherapy as patients lose their hair, but the effects of whole body doses of powerful anti-cancer drugs can be even more devastating. As the technology has been advancing, with, for example, immunotherapy added to the chemotherapy, so it has become necessary to address exactly where the drug molecules go, to which structures in the body they attach, and the mechanisms and kinetics of how the molecules and their metabolic products are distributed and cleared in and from the body.

The process of greater selectivity in drug delivery may be referred to as controlled drug delivery, controlled drug release or targeted drug delivery. It may be that the delivery mechanism is quite different, as with an implantable micropump to deliver morphine or insulin, or with transdermal or transmucosal patches. More often we find that the best results are obtained when the drug is combined with a suitable biomaterial. This could involve a physical dispersion of a drug in a biodegradable polymer, which is injected or implanted into the patient, where the drug is released by combinations of diffusion, erosion, osmotic and degradation processes. In addition, many new drugs do not have the required solubility in the body and biomaterials may assist in this process. It could also involve the chemical coupling of a drug to a soluble polymer, where the linkage between the two is broken under specific *in vivo* conditions.

This chapter deals with the technology of the controlled delivery and targeting of active molecules to the body and related matters. We start with drug delivery to patients, as introduced above. We also cover the situation where genes rather than drugs are involved, as anticipated in the previous chapter. It is also necessary to address those situations in regenerative medicine, and especially *ex vivo* tissue engineering, where the active molecule delivery takes place in the context of cells in culture rather than direct infusion into the patient. Before discussing the specific technologies and applications, however, it is useful to place these into the context of the reasons why we have to develop these technologies in order to get the best out of our drugs, and we do so by a logical and systematic overview given in the next section.

It is necessary to emphasize here that progress in drug delivery technologies has been rapid since the beginning of the twenty-first century. A vast amount of experimental work and pre-clinical studies have been undertaken. So far, relatively few have reached full regulatory approval for widespread clinical use. Quite a few are in various stages of clinical trial at the time of writing. It must not be assumed, however, that all ideas and concepts discussed in this chapter are or will be clinically successful. They are discussed here in order to demonstrate the emerging principles of these new technologies.

### 6.1.1 Rationale for drug delivery technologies

The starting point of this discussion is the "conventional" delivery of "conventional" drugs. This refers to the normal, routine methods by which everyday pharmaceuticals are delivered to patients. These

## Box 6.1 Glossary of terms

**anionic** characterized by an active anion, which is a negatively charged ion.

**antibacterial** a substance that is able to destroy or inhibit the growth or reproduction of bacteria.

**antibiotic** a chemical substance, derived from microorganisms, that has the capacity to selectively destroy or inhibit the growth of harmful microorganisms.

**antimicrobial** capable of killing microorganisms.

**antiviral** destroying viruses or inhibiting their replication.

**biodistribution** the static and dynamic distribution of substances within biological systems or within an organism.

**cationic** characterized by an active cation, which is a positively charged ion.

**chemotherapy** the treatment of disease, especially cancer, by chemical substances.

**clathrin-mediated endocytosis** the uptake of a material into a cell from the cell surface using clathrin-coated vesicles.

**conjugate** a covalently linked complex of two or more molecules.

**controlled release** regulation of the rate of drug delivery to provide a specified profile of release over the lifetime of the product.

**dendrimer** a polymer in which atoms are arranged with a branched structure, radiating out from a central core.

**dose** a specified quantity of a therapeutic agent.

**drug targeting** a strategy aimed at delivering a specific drug to a particular part of the body or to a specific feature within tissues.

**electroporation** the action of using electrical pulses to briefly open up pores in cell membranes that allow the take-up of drugs or other molecules.

**endocytic pathway** any one of the pathways by which macromolecules that attach to a cell surface become internalized.

**erosion** the process of gradual removal of material from the surface of an object.

**gene vector** any agent that is used to carry genetic material to target cells.

**hypodermic** relating to the area just under the skin.

**immunotherapy** the treatment of disease by inducing or enhancing an immune response.

**intrathecal** within the space under the arachnoid membrane, which covers the brain and spinal cord.

**intravenous** within or administered into a vein.

**lipofection** technique used to inject genetic material into a cell by means of liposomes.

**lipophilicity** having an affinity for, or being combined with, or dissolving in, lipids.

**lipoplex** a complex of plasmids and lipids for use in gene delivery.

**liposome** a vesicle formed by a lipid bilayer enclosing an aqueous compartment.

**micelle** a charged particle formed by an aggregate of molecules, usually polymeric, in certain colloidal solutions.

**multidrug resistance** the resistance of tumor cells to more than one type of chemotherapeutic agent.

**nanomedicine** the application of nanotechnology to medical therapies and diagnosis.

**nebulizer** a device that converts a liquid to a fine spray.

**oligonucleotide** a short nucleic acid chain, usually consisting of up to 20 nucleotides.

**osmosis** a process by which molecules of a solvent pass through a semi-permeable membrane from a less concentrated solution into a more concentrated one, equalizing the concentrations on each side.

**overexpression** the excessive expression of a gene.

**pharmacodynamics** the study of the effects of drugs and the mechanisms of their actions.

**pharmacokinetics** the process by which a drug is absorbed, distributed, metabolized and eliminated by the body.

**polymer therapeutics** nanoscale polymer-based pharmaceuticals.

**polyplex** a complex of a polymer and genetic material.

**theranostics** treatment strategies that combine diagnostics and therapeutics.

**transdermal** the application of a drug through the skin.

**transmucosal** the application of a drug through a mucosal surface.

**vaccine** a preparation of weakened or killed pathogens that may be administered in order to produce or increase immunity to a particular disease.

**zero-order release** release of a molecules of a drug from an agent that is independent of concentration.

**zwitterionic** characterized by a neutral compound that has delocalized electrical charges of opposite sign.

methods are not discussed in this book because they do not involve biomaterials, but they do provide a reference point, and especially allow us to visualize why these methods do not work with all drugs, either at all or only very inefficiently, and also to identify what new biomaterials-based technologies have been required in order to make them work. It is important to recognize here that these systems may involve biomaterials in their classical setting, as with simple synthetic polymers used to contain a drug for eventual delivery, and also new forms or concepts of biomaterials, discussed in Chapter 1, for example polymer–drug conjugates, engineered viral vectors or engineered vaccines.

Thus, in Box 6.2 we see this starting point at the top, in Group A. Included here are pain medicines and anti-inflammatory agents taken by mouth, the subcutaneous injection of insulin for diabetes therapy, the topical preparations such as creams, ointments, pessaries and eye drops which give a superficial local effect, and the anti-asthma agents inhaled with the use of a nebulizer. Note that the formulation of these drugs is straightforward; they have not been modified or engineered in order to

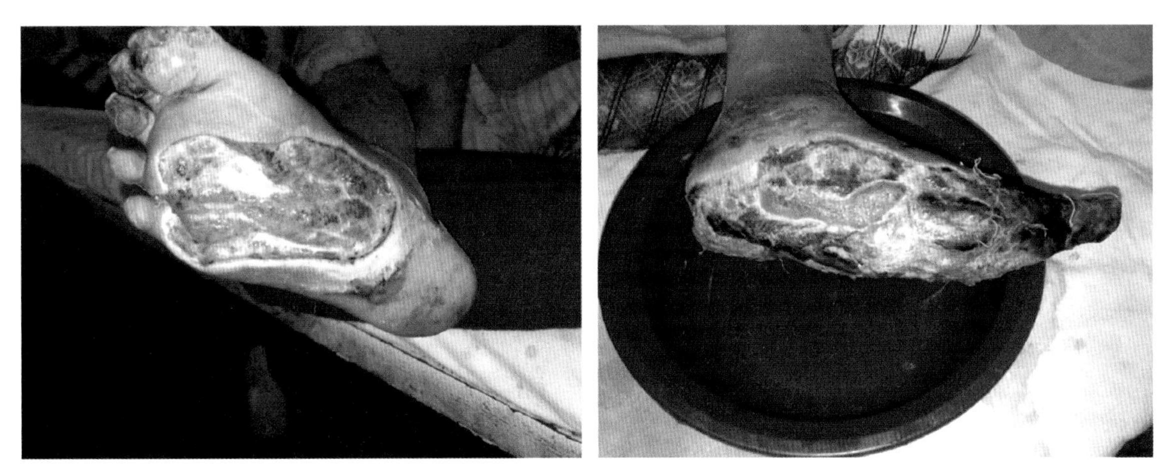

**Figure 1.6** Diabetic foot ulcer. One of the more severe consequences of diabetes is a compromised blood supply to the lower limbs, which results in the progressive development of ulcers. Examples here are severe, and may lead to limb amputation. Ulcers are very difficult to treat once they become established. Simple dressings may no longer suffice. Attempts are being made to develop techniques to regenerate the skin and underlying tissue, as discussed in Chapter 5. Reprinted from *The Foot*, 22(1), Zubair M *et al.*, Incidence, risk factors for amputation among patients with diabetic foot ulcer in a Northern India tertiary care hospital, 24, © 2012, with permission from Elsevier.

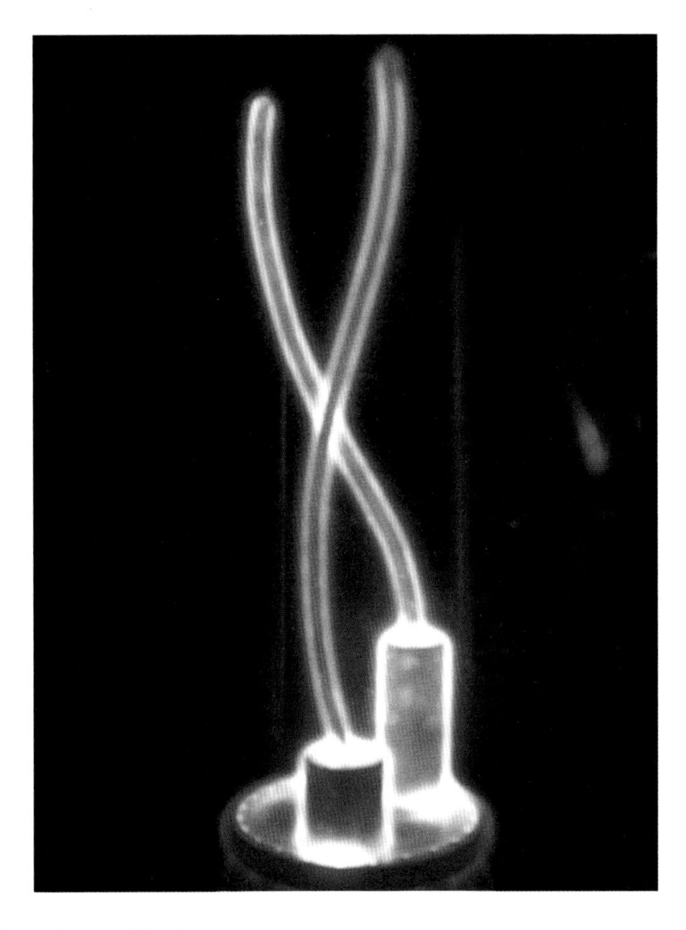

**Figure 2.12** Surface modifications.
3D plasma treatment of scoliosis correction rod made of NiTi to mitigate leaching of Ni ions into tissues. Image courtesy of Professor Paul Chu, Plasma Laboratory of City University of Hong Kong.

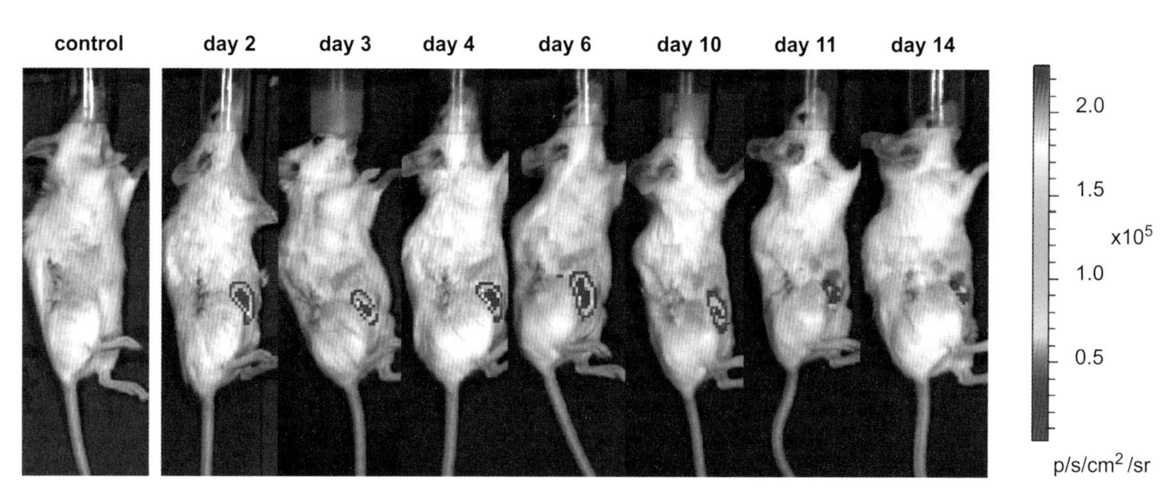


**Figure 2.16** Bioluminescence and the monitoring of biomaterial-related infection

A pseudo-color representation of the bioluminescent radiance from a surgical mesh, pre-contaminated with *S. aureus* Xen29 bacteria implanted subcutaneously in mice during a 15-day follow-up. The images show that the amount of viable bacteria contributing to the bioluminescence signal decreases in time during the course of infection. Bioluminescence was imaged with a CCD camera 2 days following implantation of the mesh and subsequently on selected days. Image was produced by Seyedmojtaba Daghighi in the laboratory of Dr. Jelmer Sjollema and Dr. Henk Busscher, University Medical Center Groningen, Netherlands.

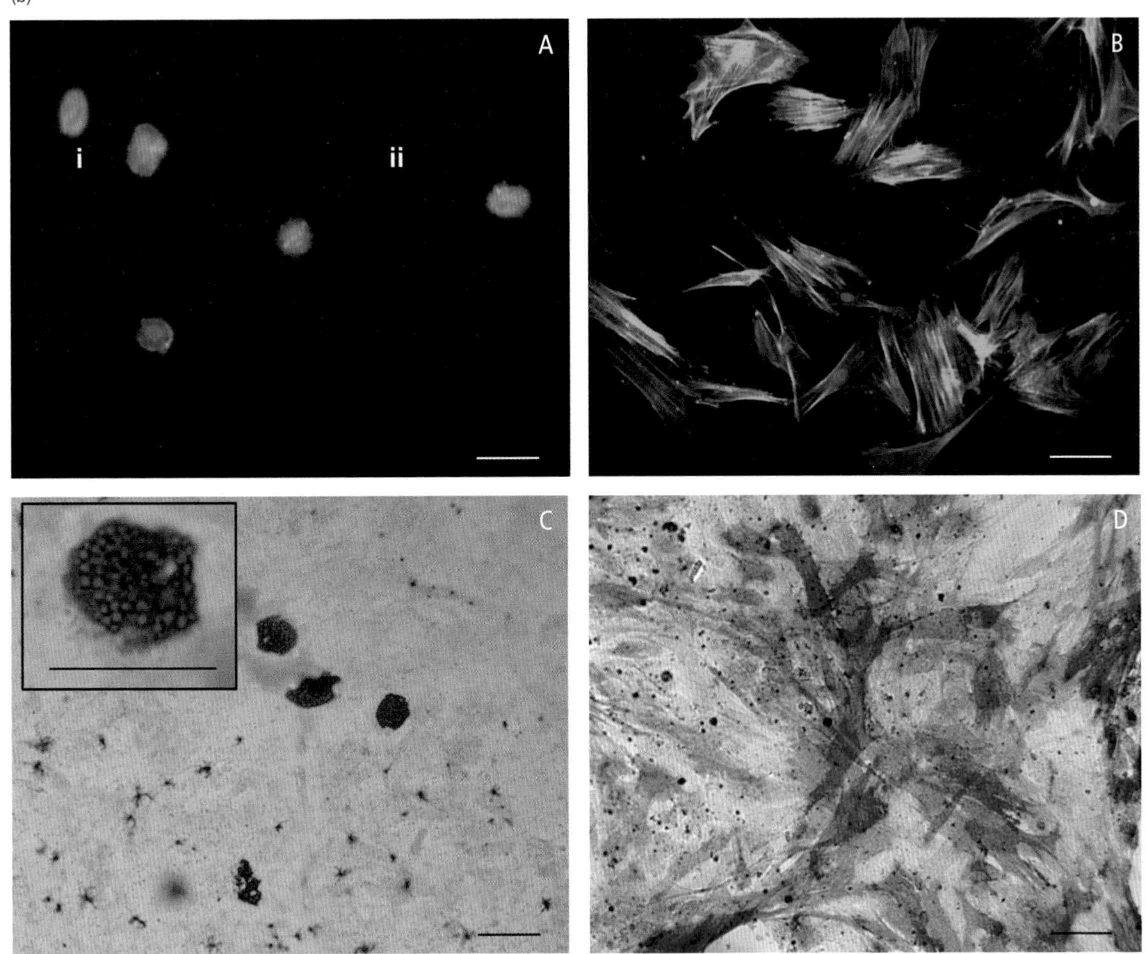


**Figure 3.1** Examples of biocompatibility phenomena.

**(b)** The effect of substrate stiffness on stem cell behavior. Fluorescent images of human mesenchymal stem cells cultured on (A) soft 3 kPa or (B) stiff 30 kPa hyaluronic acid-based hydrogels for 14 days in mixed adipogenic/osteogenic induction media and stained for F-actin (green) and nuclei (blue). Corresponding brightfield images (C) soft and (D) stiff of hMSCs stained for lipid droplets (red, adipogenic marker) and alkaline phosphatase (blue, osteogenic marker). Scale bars are 100 microns and 50 microns (inset). Image courtesy of Murat Guvendiren, PhD and Jason A. Burdick, PhD Department of Bioengineering University of Pennsylvania, USA.

(e)

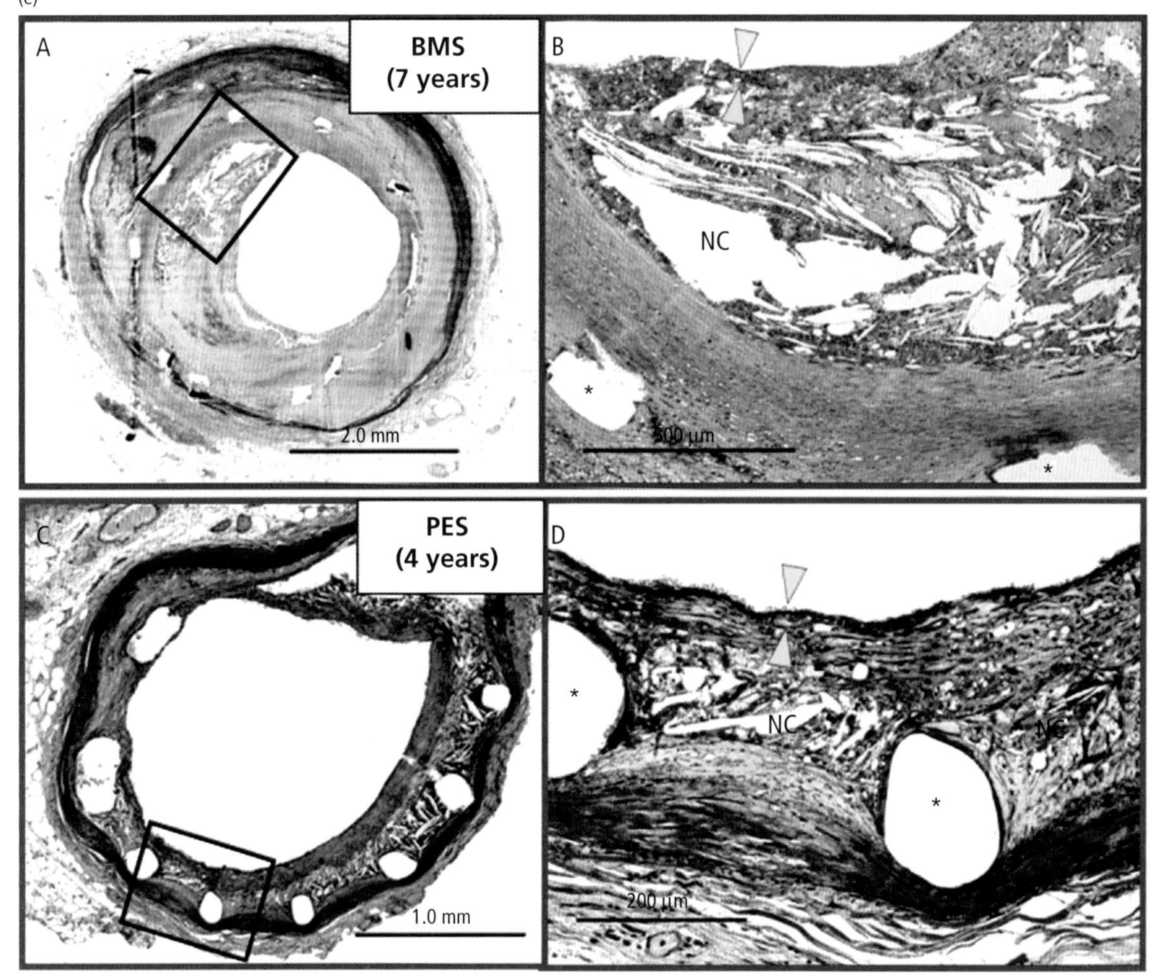


**Figure 3.1** *continued*

**(e)** In-stent restenosis. Intravascular stents, used to open blocked arteries, may themselves become blocked again in a process known as restenosis. (A) Cross-sectional histology of bare-metal stent (BMS) implanted in the coronary artery for 7 years. (B) High-power image of the box in A (×100). A large necrotic core (NC) containing cholesterol crystals is identified within the neointima. The fibrous cap overlying the NC is infiltrated by numerous foamy macrophages and is markedly thinned (yellow arrowheads point to thinnest portion), which resembles vulnerable plaque encountered in native coronary arteries. The asterisks represent metal struts. (C) Cross-sectional histology of paclitaxel-eluting stent (PES) implanted in the coronary artery. (D) High-power image of the box in C (×200). A relatively small NC containing cholesterol crystals is formed around metal struts (asterisk). The fibrous cap is infiltrated by numerous foamy macrophages and is markedly thinned (yellow arrowheads point to thinnest portion). Reprinted from *Journal of the American College of Cardiology*, 59, Park S-J *et al.*, In-stent neoatherosclerosis, 2051–7, © 2012, with permission of Elsevier.

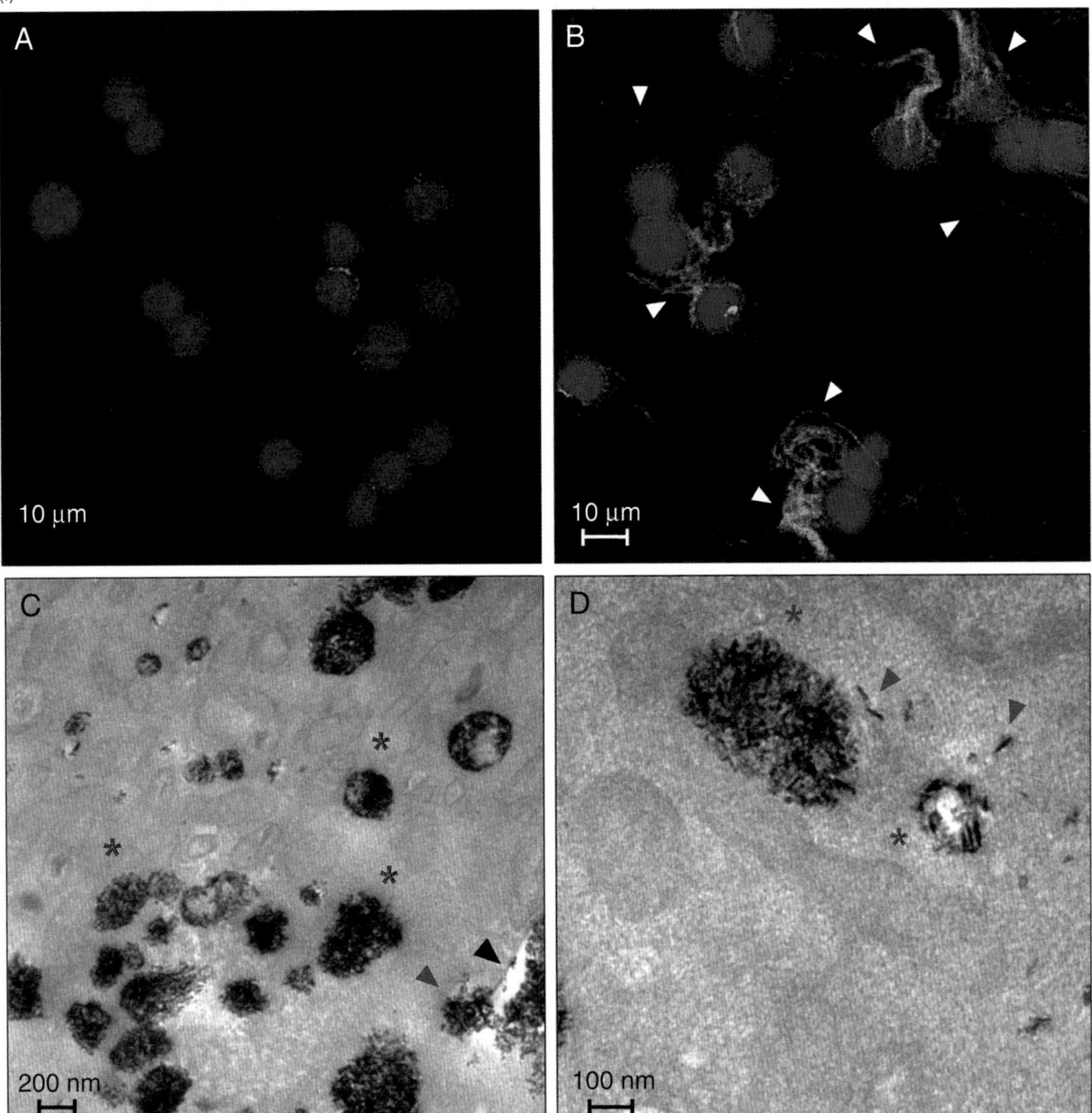


**Figure 3.1** *continued*
**(f)** Biocompatibility of contrast agents. Cultured mouse hepatocytes (NCTC 1469) were treated with a MRI contrast agent (ferucarbotran, Resovist), superparamagnetic iron oxide nanoparticle (SPION, 0.2 mM) alone (A) or concomitantly treated with the SPION and static magnetic field (SMF, 0.4 Tesla) for 1 h. In (B), the fibrillar SPION aggregates (green, arrowheads) were notably observed by confocal fluorescence microscopy in SMF-exposed cells, which were stained with anti-dextran antibody. Nuclei (blue) were stained with DAPI. The electron micrograph (C) of internalized SPION in the SMF-exposed cells show at the red arrowhead, an emerging phagocytic vesicle encapsulating SPION aggregates and at the surface of the cell membrane (black arrowhead). The asterisks show some of the trafficking vesicles containing SPION aggregates. The electron micrograph (D) show cytosolic SPION spilled by trafficking SPION-harboring vesicle(s) (asterisks). Note the spike-formed SPION aggregates in the vesicles and the cytosol. Image courtesy of Professor Kwon-Seok Chae, Kyungpook National University and Professor Myung-Jin Moon, Dankook University, Republic of Korea.

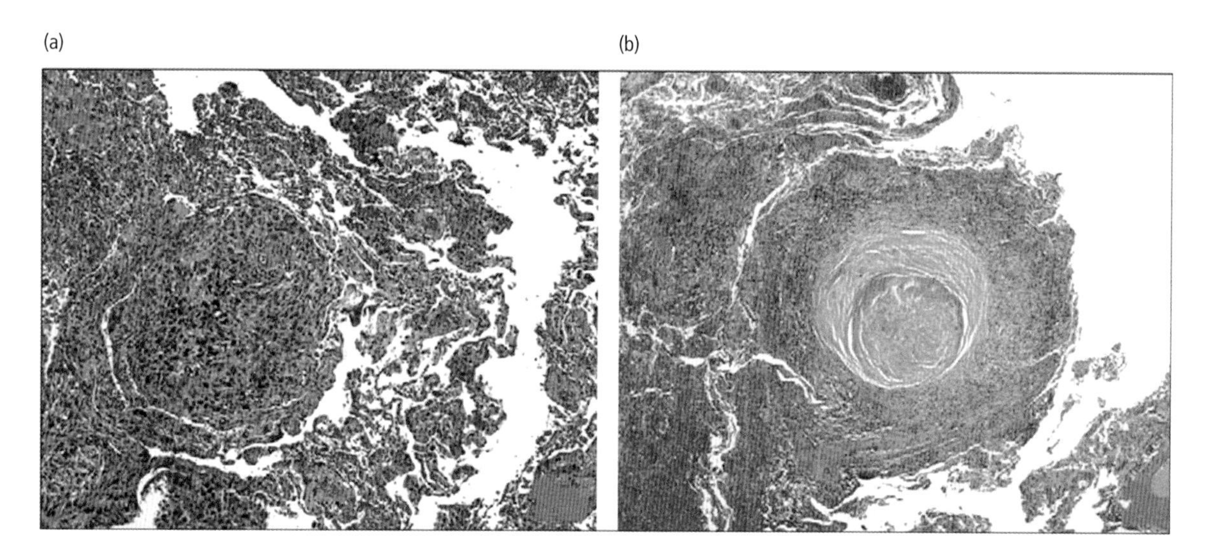


**Figure 3.27** Fibrosis of the lung. Histological sections of lung with silicotic lesions:
**(a)** early silicotic lesion as cellular nodule of dust-laden macrophages (×100).
**(b)** Chronic silicotic nodule with concentric fibrosis in the center and peripheral dust-laden macrophages (×40)
Reprinted from *The Lancet*, 379, Leung C C, Silicosis, 2008-18, 2012, with permission from Elsevier.

(a)

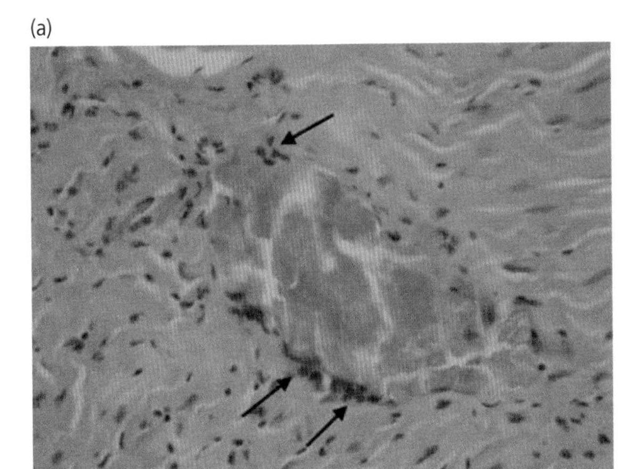

(b)

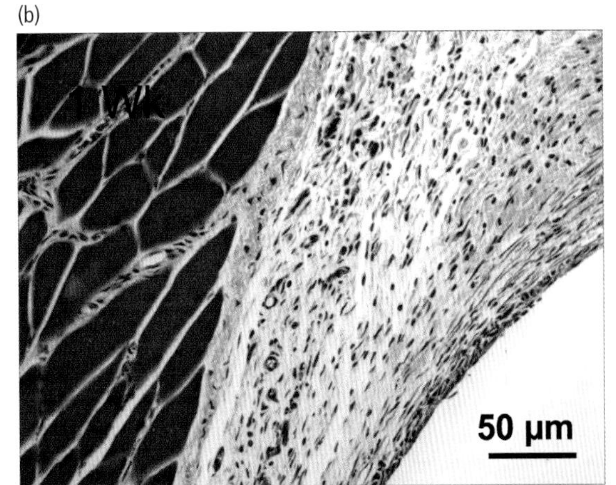


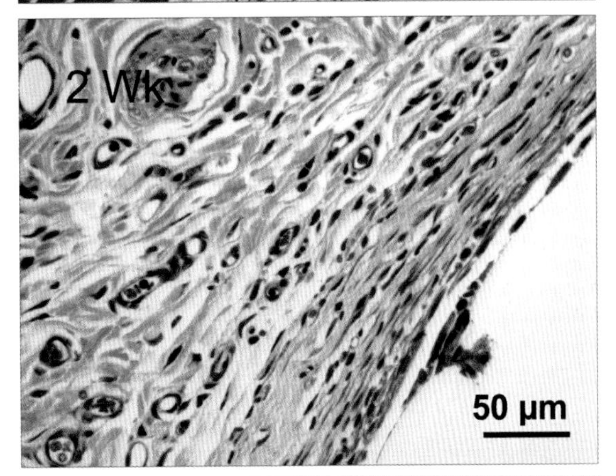


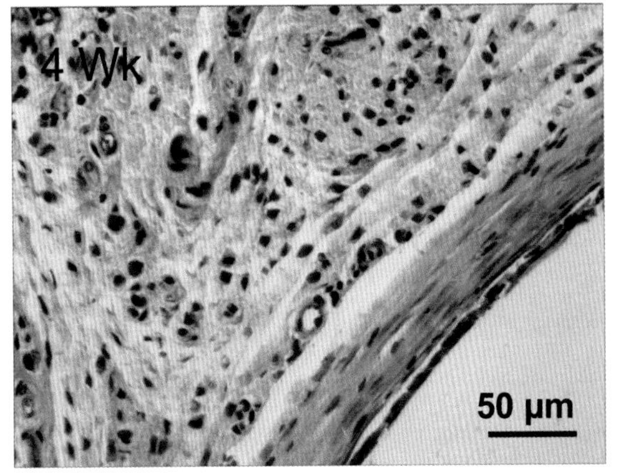

**Figure 3.29**

**(a)** Phagocytosis of corrosion products. Histological section showing dense fibrous tissue surrounding a corrosion deposit, demarcated by foreign body giant cells. Reprinted from *Acta Biomaterialia*, 5, Huber M *et al.*, Presence of corrosion products and hypersensitivity-associated reactions in periprosthetic tissue after aseptic loosening of total hip replacements with metal bearing surfaces, 172–80, © 2009, with permission of Elsevier.

**(b)** The initial tissue response to polymers is usually seen in terms of an inflammatory response and fibrotic tissue adjacent to the polymer. When polymer particles of any form are released into the tissue they may be associated with focal points of inflammatory cells. These sections show the tissue at 1, 2 and 4 weeks after the intramuscular implantation of elastomer cylinders. Images courtesy of Dr. Brian Amsden, Chemical Engineering, Queens University, Kingston, Ontario, Canada.

(b)

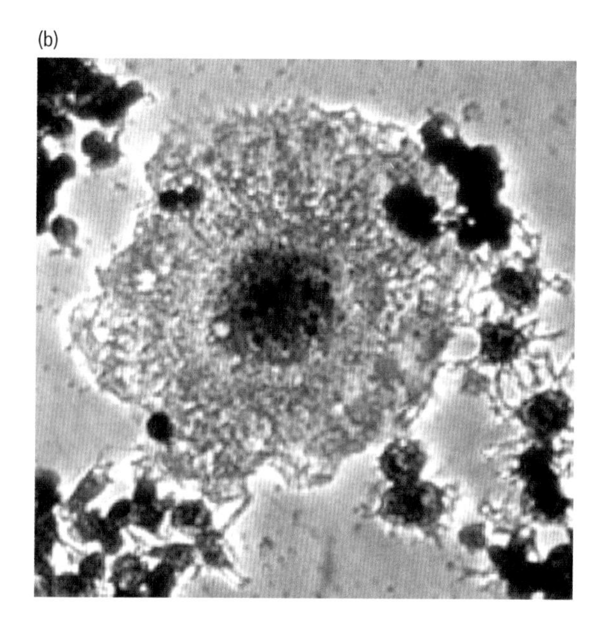

**Figure 3.30** Osteolysis.

**(b)** Osteoclasts. This image shows a multinucleated osteoclast, observed with tartrate-resistant acid phosphatase (TRAP) staining, that has been stimulated by endogenous macrophage-colony stimulating factor (MCSF) and RANKL. Reprinted from *Pharmaceutical Research*, 28(5), Newa, M *et al.*, Antibody-mediated "universal" osteoclast targeting platform using calcitonin as a model drug, 1131–43, © 2011, with kind permission from Springer Science and Business Media.

(a)

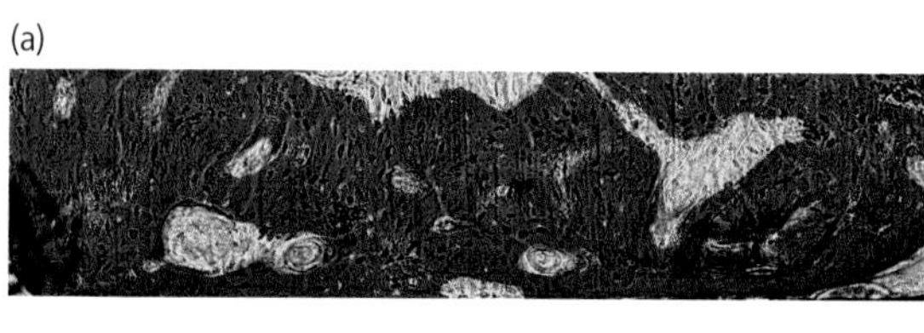

(b)

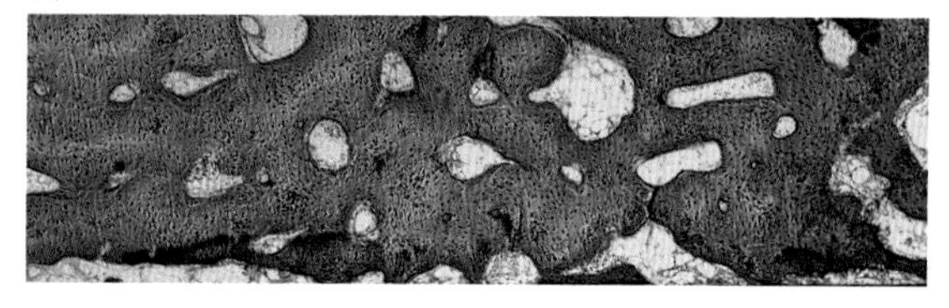

**Figure 3.32** Effects of nanotopography on bone attachment. Digital photographs of stained (methylene blue/basic fuchsine) histological sections at 4 weeks after implantation into minipig skull.

**(a)** Implant is titanium with a coating of 70 nm $TiO_2$ nanotubes.

**(b)** Implant is machined titanium. Images courtesy of Professor Zhenting Zhang, Capital Medical University, Beijing, China.

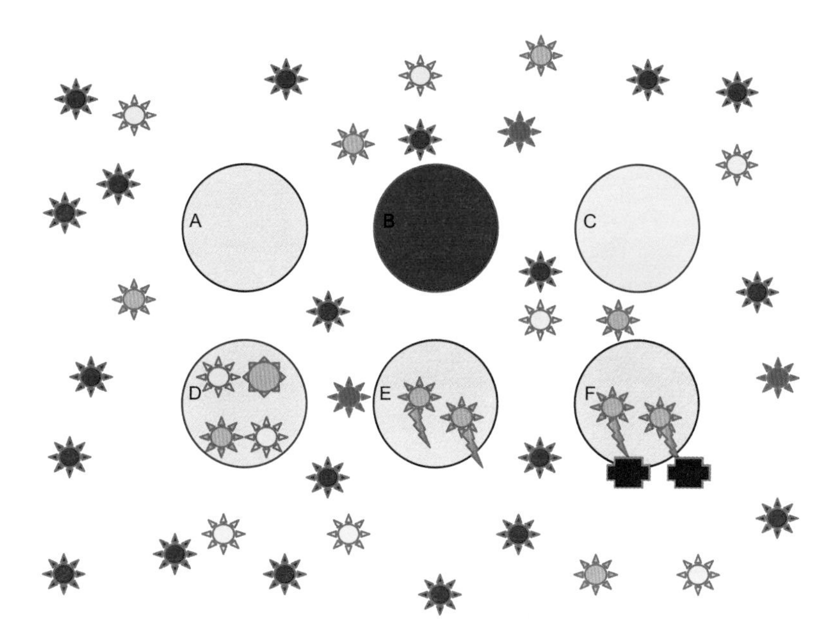


**Figure 3.33** The nanoparticle environment. Nanoparticles are associated with a protein corona once exposed to a biological environment. In A, the nanoparticle is introduced into the protein-rich environment, different protein species being colored red, yellow, blue and green. Protein red is the most abundant and will be initially adsorbed onto the nanoparticle surface preferentially (B). However, protein yellow has much higher affinity and may displace protein red (C). Protein blue has little affinity for the nanoparticle surface but does have affinity for protein yellow, and may attach to parts of that protein layer (D). Note that protein green is in low concentration and has little affinity and takes no part in the corona. In E we see that some epitopes on protein blue are exposed, which could combine with antibodies and other molecules. Based on diagrams in Monopoli *et al.*, *Nature Nanotechnology*, 2012, 7, 779–86.

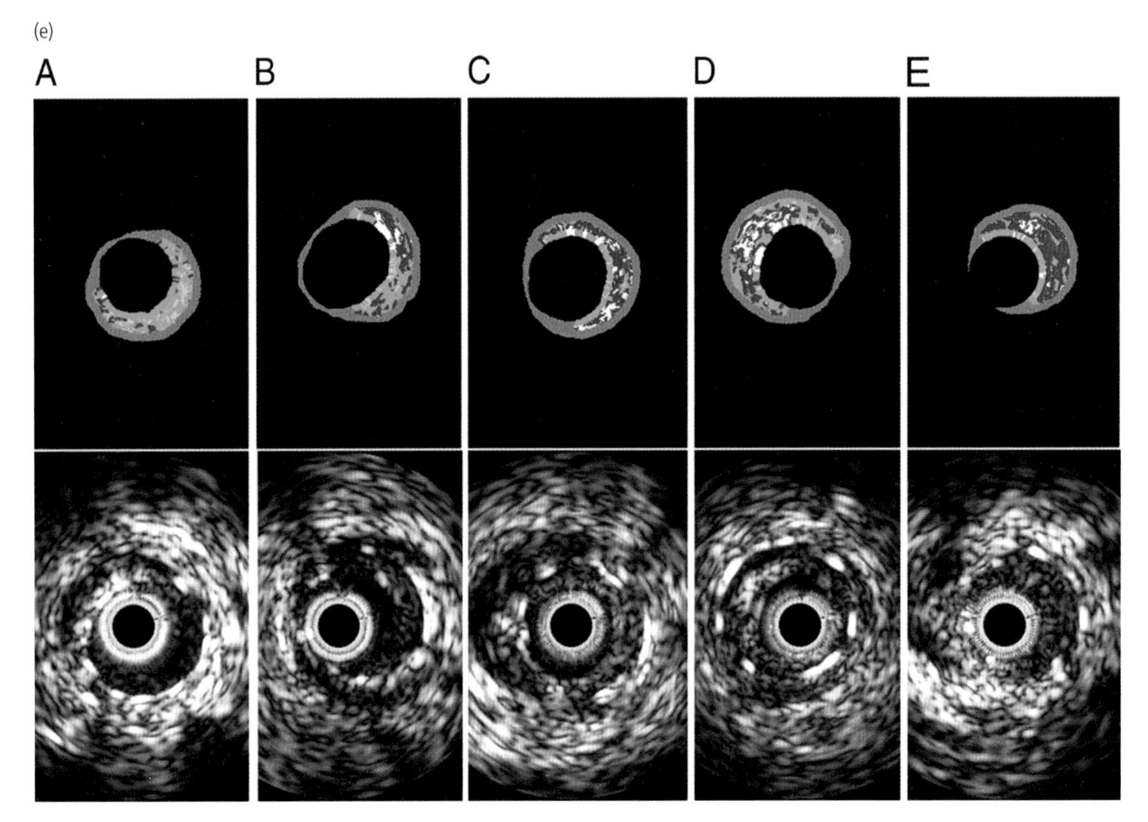


**Figure 3.36** Atherosclerosis and stenosis.
**(e)** Appearance of cross-section of arteries after atherosclerosis and stenting, using virtual histology intravascular ultrasonography. (A) Six-month follow-up of paclitaxel-eluting stent (percent necrotic core [%NC] 10%, percent dense calcium [%DC] 2%); (B) 9-month follow-up of paclitaxel-eluting stent (%NC 28%, %DC 8%); (C) 22-month follow-up of paclitaxel-eluting stent (%NC 39%, %DC 20%); (D) 48-month follow-up of bare-metal stent (%NC 40%, %DC 25%); and (E) 57-month follow-up of bare-metal stent (%NC 57%, %DC 15%). Reprinted from *Journal of the American College of Cardiology*, 59, Park S-J *et al.*, In-stent neoatherosclerosis, 2051–7, © 2012, with permission of Elsevier.

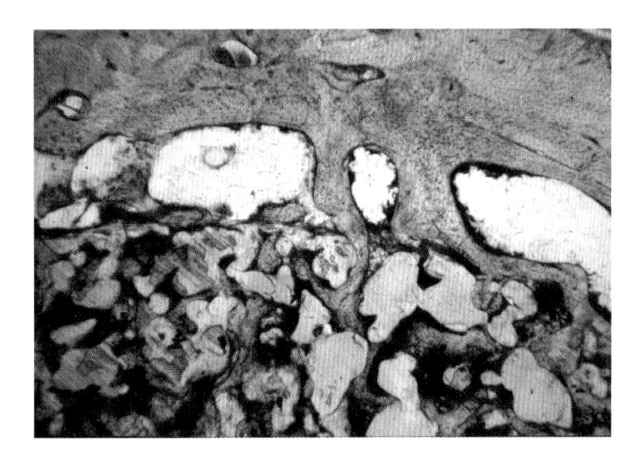

**Figure 4.1** Bone ingrowth into porous surfaces. Bone can grow into pores on the surfaces of most biomaterials, including metallic systems, both degradable and non-degradable bioceramics, and polymers, especially thermoplastics. In this example bone has invaded the pores of a high density polyethylene. The size of pores and of the interconnections between them are important determinants of the rate, extent and quality of the ingrown bone.

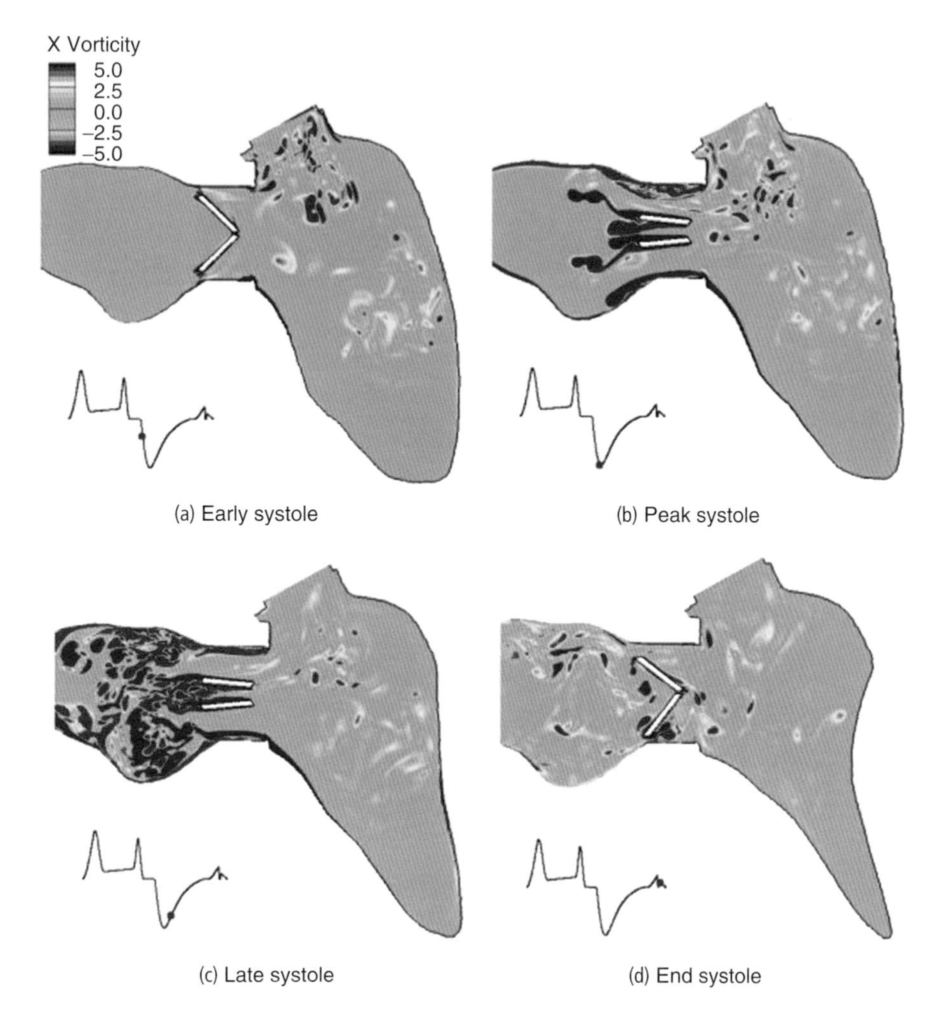


**Figure 4.21** Blood flow in heart valve prostheses. Blood flow through valves may be simulated in a kinematic model. In (a) at the early phase of systole, there is reasonably coherent flow. As the bileaflet valve opens at peak systole (b), an unstable shear layer forms at the leaflet surfaces, then complex 3D flow structures form at the end of systole at the aortic root (c). At closure (d) the leaflets vary slightly in closure velocity and there is some leakage back through the valve. Reprinted from *Journal of Computational Physics*, 244, 41–62, Bao Le *et al.*, Fluid–structure interaction of an aortic heart valve prosthesis driven by an animated anatomic left ventricle, 2013, with permission from Elsevier.

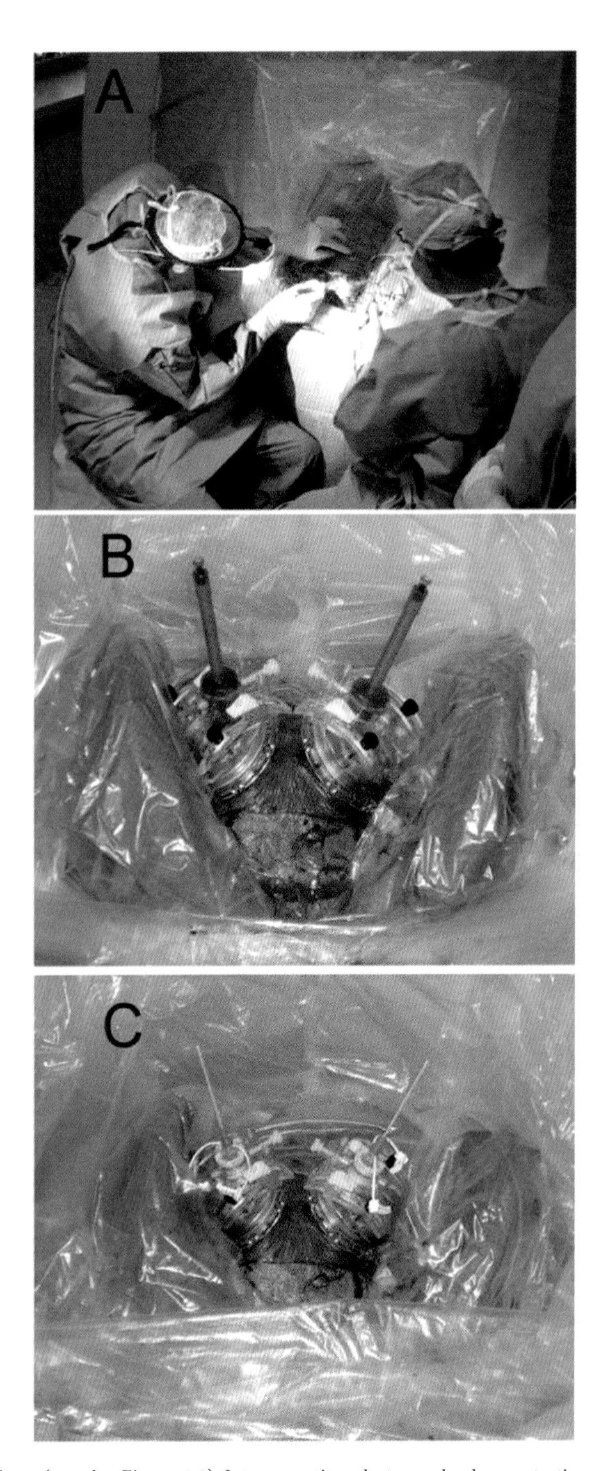


**Figure 4.33** Deep brain stimulator (see also Figure 1.2). Intraoperative photographs demonstrating surgical draping and trajectory guides. (A) Patient's head is shown at the back of the MR bore, with a sterile drape. (B) Trajectory guides with alignment stems. (C) Trajectory guides with multilumen insert and peel-away sheath prior to advancing the sheath into the brain. The flexible radiofrequency receiving coils are covered with sterile blue towels. Reprinted from *The Journal of Neurosurgery*, 112, Starr P A *et al.*, Subthalamic nucleus deep brain stimulator placement using high-field interventional magnetic resonance imaging and a skull-mounted aiming device, 479–90, © 2010, with permission of JNSPG and Rockwater Inc.

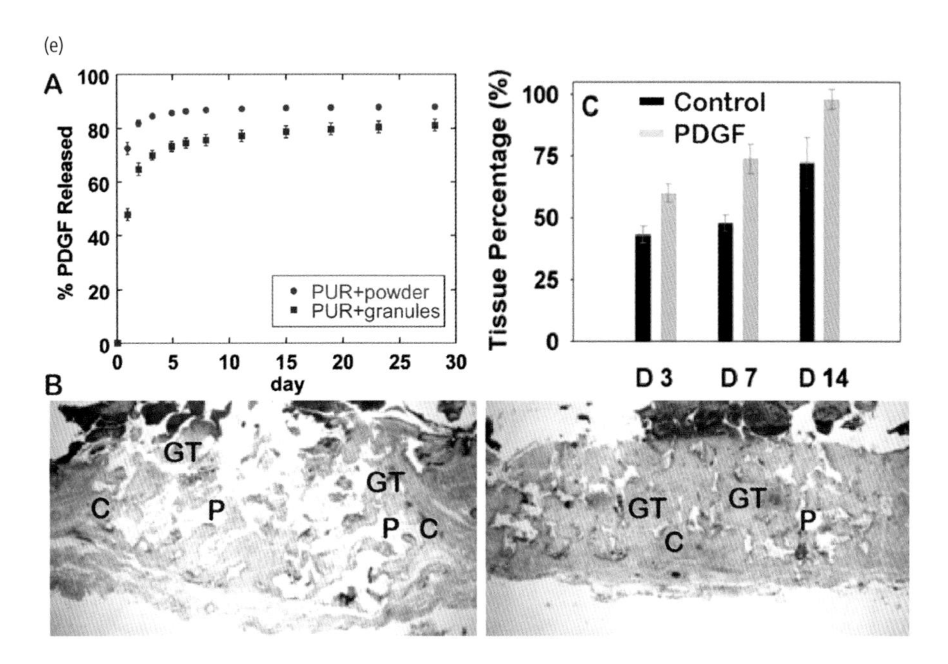


**Figure 5.7** Examples of tissue engineering scaffolds.

(e) Local delivery of growth factor incorporated into a porous polymer scaffold. These images show the effects of local delivery of the recombinant platelet-derived growth factor, rhPDGF-BB on healing of excisional wounds in rats. (A) rhPDGF-BB was added to resorbable polyurethane scaffolds as either a labile powder or bound to heparin (Hp)-modified PLGA microspheres embedded in gelatin granules. Binding rhPDGF to Hp resulted in a more sustained release compared to the diffusion-controlled release observed for the labile powder. (B) Histological sections of scaffolds augmented with either 0 (left) or 1.8 mg (11.5 mg/ml scaffold) rhPDGF-BB powder (right) implanted in 6-mm excisional wounds in rats at day 7. Local delivery of rhPDGF-BB resulted in accelerated polymer (P) degradation, as well as enhanced infiltration of granulation tissue (GT) and deposition of collagen (C). By day 14, wounds treated with scaffolds augmented with rhPDGF-BB had re-epithelialized while those treated with control scaffolds had not. Images courtesy of S.A. Guelcher, Vanderbilt University, USA.

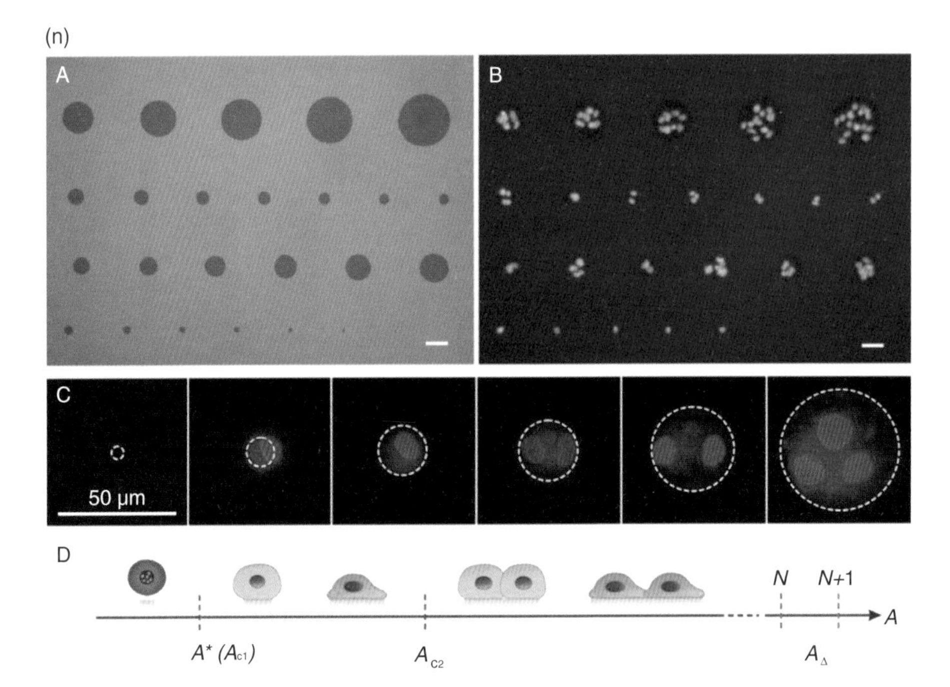


**Figure 5.7** *continued*

**(n)** Cell patterning and critical areas of cell adhesion on micropatterned surfaces with adhesion contrast. (A) Optical micrographs of micropatterns with RGD-peptide-grafted gold microislands of varying diameters on PEG hydrogel. (B) Fluorescence micrograph of MC3T3-E1 cells, a pre-osteoblast cell line, on a micropatterned surface with cellular nuclei labeled by DAPI. (C) Fluorescence micrograph of adherent cells with cellular nuclei labeled in blue and F-actin in red – the dashed circles indicate the contours of the underlying RGD-grafted microislands. (D) Schematic presentation of three characteristic areas for cells adhering on adhesive microislands on a cell-resistant background: $A_{c1}$, also named $A^*$, the critical area from apoptosis to survival; $A_{c2}$, the critical area from single-cell adhesion to multi-cell adhesion; $A_{\Delta}$, the characteristic area for one more cell to adhere. Courtesy of Professor Jiandong Ding and Dr. Ce Yan of Fudan University, China.

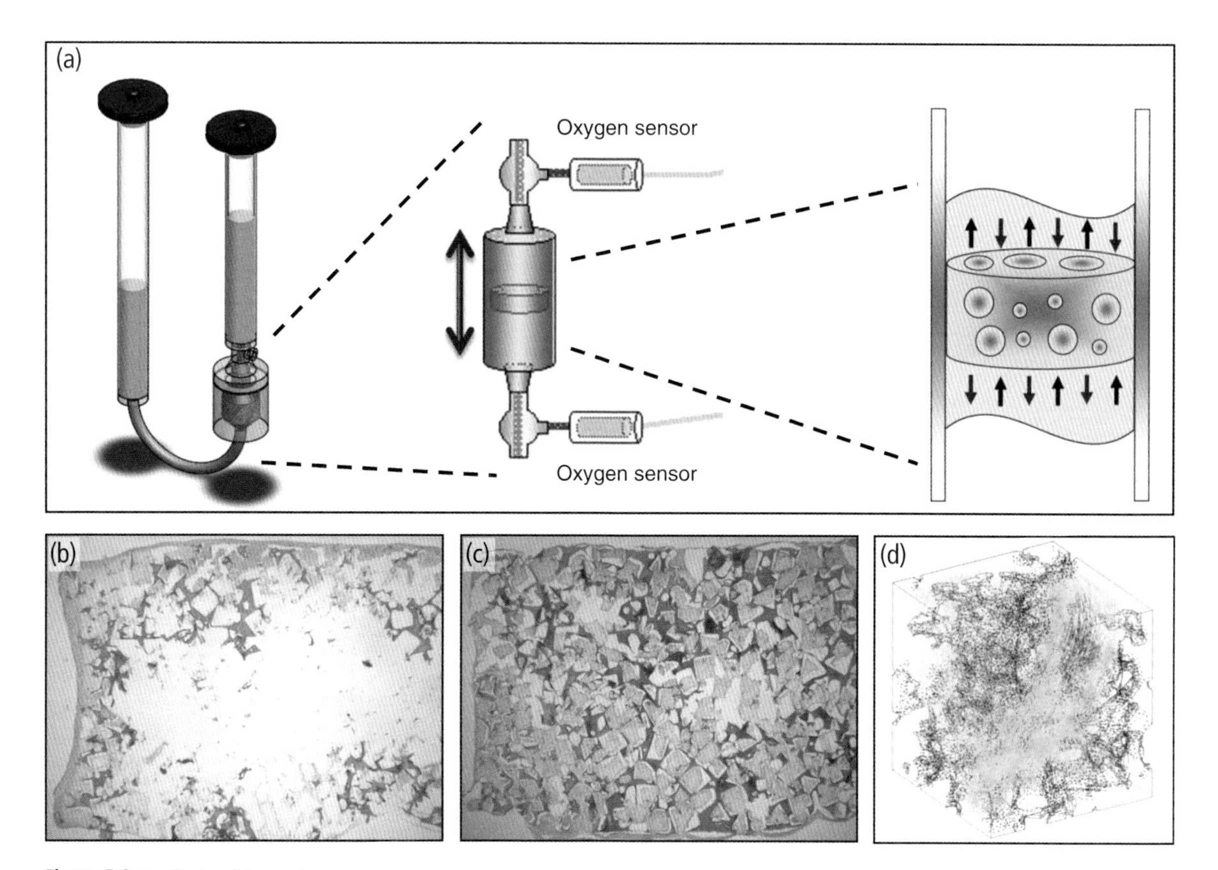


**Figure 5.9** Perfusion bioreactors.
**(a)** Perfusion bioreactor for engineering three-dimensional cell-scaffold constructs. Cells and culture medium are perfused directly through the pores of a 3D scaffold. The displayed configuration has been designed to prevent cell settling and allow uniform and efficient cell seeding into the scaffold pores by perfusion flow. Micro flow-through sensors can be integrated into the bioreactor system to monitor culture parameters (e.g., oxygen or pH) and to help establish defined and controlled culture conditions.
**(b)** Statically cultured construct. Due to mass transport limitations, cell necrosis is typically observed within the center of 3D constructs engineered under conventional static culture conditions.
**(c)** Perfusion cultured construct. Perfusion bioreactors may mitigate mass transport limitations within 3D constructs, producing constructs with a homogeneous distribution of viable cells and extracellular matrix.
**(d)** Computational fluid dynamics simulation of shear stress within construct. In conjunction with hypothesis driven experiments and sensorized bioreactors, computational modeling of shear stresses and mass transport of specific nutrients (e.g., oxygen) within the pores of a perfused scaffold can help to understand cell behavior in 3D scaffolds. Images courtesy of David Wendt, Elia Piccinini, and Ivan Martin, University of Basel, Switzerland.

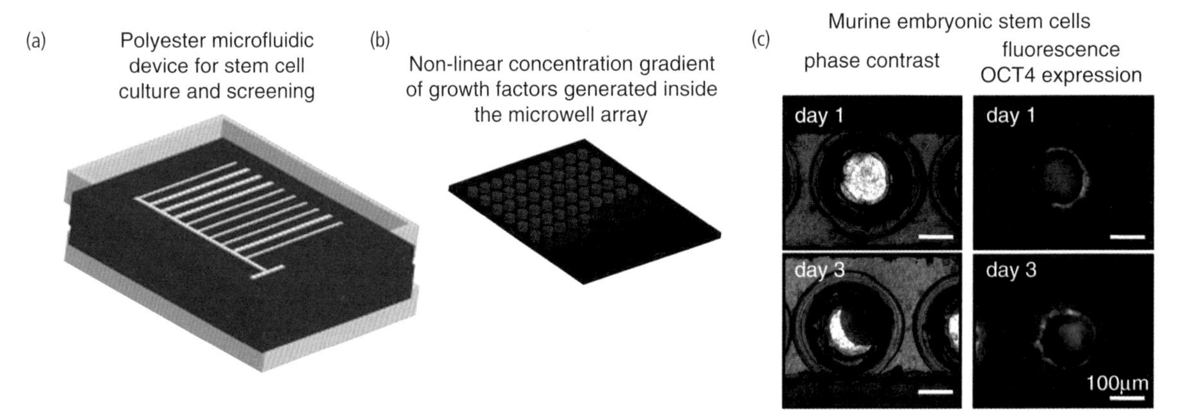


**Figure 5.10** Microfluidic device for culture of embryoid bodies.
**(a)** Sketch of the polyester microfluidic device.
**(b)** Simulation of a non-linear concentration gradient of rhodamine-labeled growth factors generated across the microwell array.
**(c)** Phase contrast and fluorescence images of typical murine embryonic stem cell aggregates on days 1 and 3 of perfusion culture. The cell cluster transforms by day 3 into an embryonic body. Courtesy of Ali Khademhosseini Ph.D, Harvard-MIT Health Sciences & Technology, Boston, MA, USA.

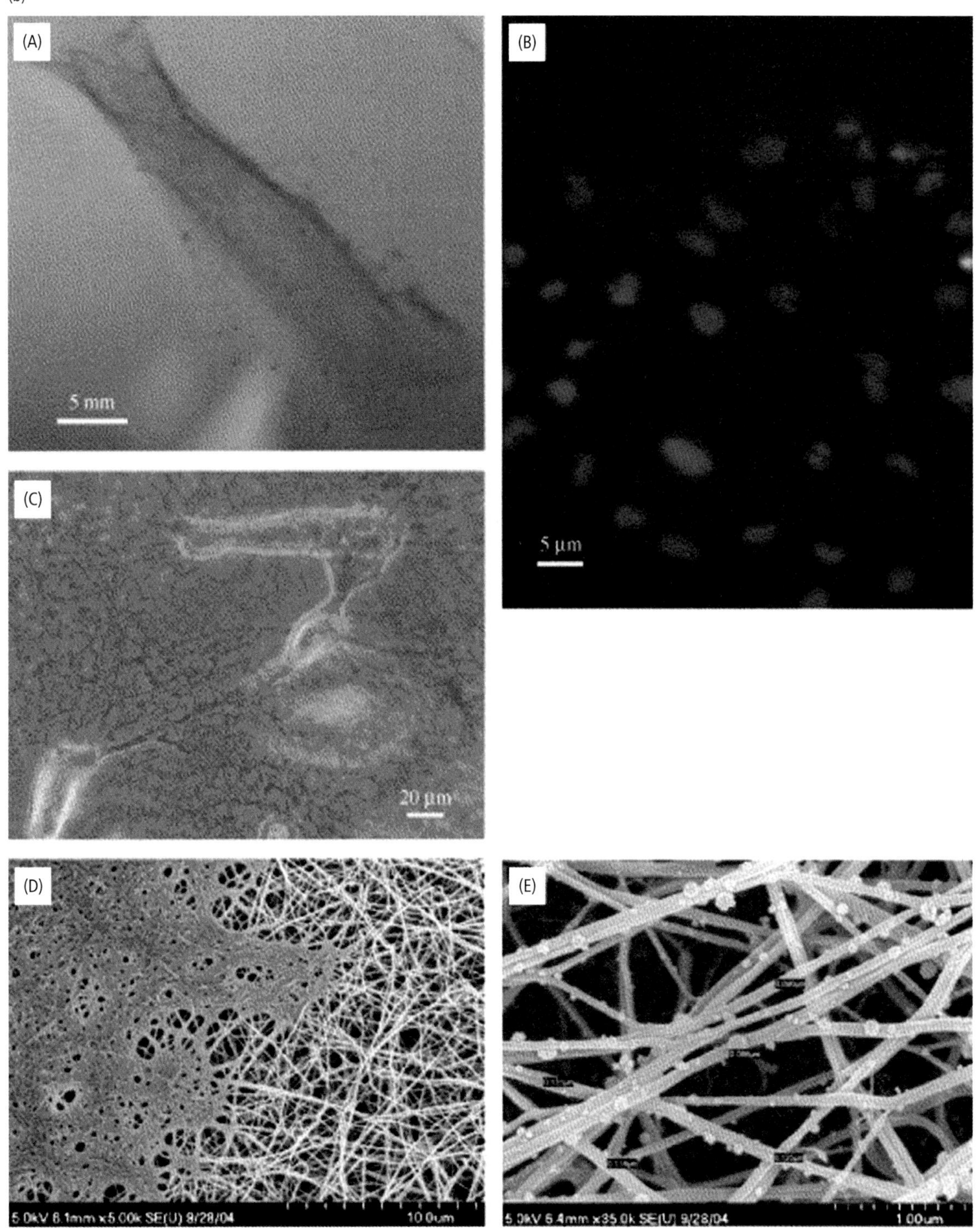


**Figure 5.12**

**(b)** Inkjet printing was first used for simple scaffolds and then for simultaneous printing of scaffolds and cells. In this example 3D neural sheets were fabricated by alternately printing fibrin gels and NT2 cells. (A) Printed neural 3D sheet 25 mm × 5 mm × 1 mm in the culture medium at day 1. (B) DAPI staining of NT2 cells within the printed sheet after 15 days of culture. (C) NT2 cells within the printed neural 3D sheet. NT2 cells after 12 days of culture. (D) The attachment of an individual NT2 neuron onto the fibrin fibers (E) SEMs of the fibrin scaffolds of the neural sheets exhibited the morphologies of fibrin fibers and porosities of the fibrin gel. Reprinted from *Biomaterials*, 27, Xu T *et al.*, Viability and electrophysiology of neural cell structures generated by the inkjet printing method, 3580–8, © 2006, with permission from Elsevier.

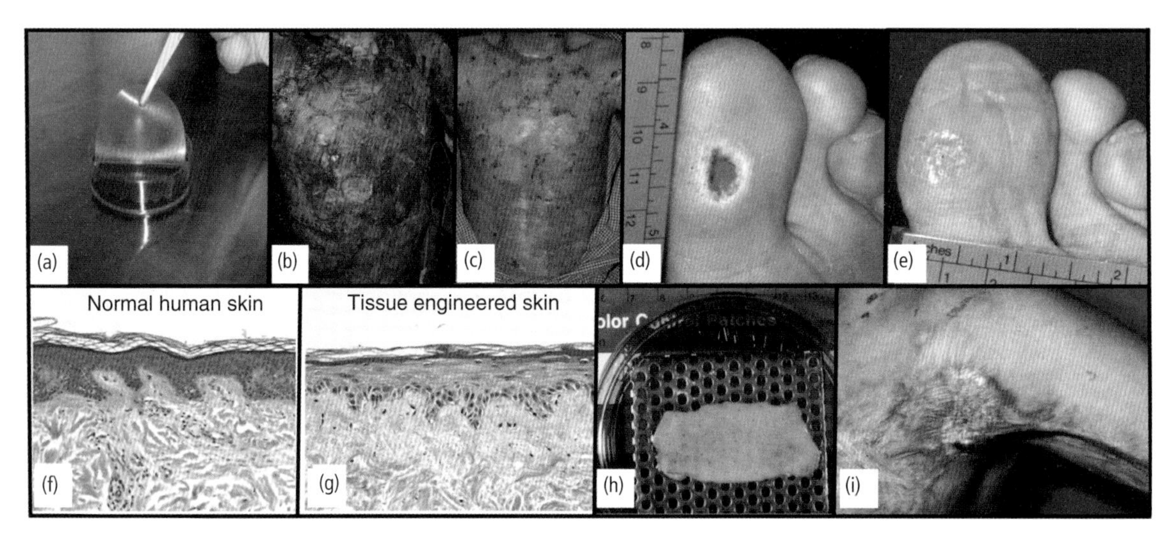


**Figure 5.13** Clinical applications of tissue engineered skin. For many applications autologous keratinocytes are expanded in the laboratory and then delivered to patients' wound beds to assist in the regeneration of a barrier layer for burns patients or to stimulate chronic wounds to heal.

**(a)** Carrier disc of silicone covered with a thin layer of acrylic acid using plasma polymerization, onto which laboratory expanded autologous keratinocytes are seeded (product developed as Myskin™).

**(b)** Patient with extensive full thickness burns who is being treated with wide mesh autograft over which Myskin™ dressings containing his own laboratory expanded cells have been placed.

**(c)** The same patient some 10 days later with an almost intact epithelium regenerated.

**(d)** Diabetic foot ulcer on a patient which had remained unhealed for 3 years.

**(e)** The same ulcer which healed after repeated applications of his own keratinocyte expanded in the laboratory and then delivered to this ulcer in outpatient clinics – eight applications of cells were required to stimulate the wound to heal. In many instances defects are full thickness and require tissue engineered skin containing both epidermal and dermal layers.

**(f)** Section through normal skin and (g) a similar section through tissue engineered skin. This skin was produced by taking cadaveric skin, stripping out all of the donor cells and sterilizing the skin with a mixture of glycerol and ethylene oxide. A small skin biopsy from the patient was used to culture and expand keratinocytes and fibroblasts, and then introducing both to the decellularized allodermis.

**(g)** Shows a well-attached, multi-layered differentiated epithelium with a normal looking dermis containing patient's fibroblasts.

**(h)** A piece of this skin grown in the laboratory at an air–liquid interface prior to surgical use.

**(i)** Patient who had suffered extreme skin contractures due to earlier burns injuries. Such contractures are managed by surgical release of the contracted area and then use of a split thickness skin graft sutured into place. Image shows the appearance of the tissue engineered skin (composed of patient's keratinocytes and fibroblasts and donor decellularized sterilized dermis) 2 weeks post transplantation. Images provided by Professor Sheila MacNeil, University of Sheffield. All patients gave consent for images to be used for research and educational purposes.

(b)

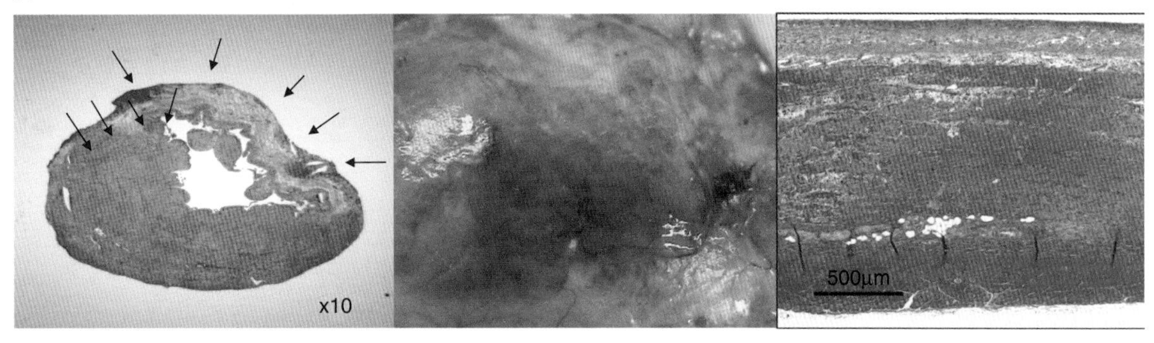


**Figure 5.17**
**(b)** The use of cell sheet engineering for generating myocardial tissue graft. The area of the myocardial infarction is seen on the left, the tissue graft is in the center, on the right is a section of a graft prepared from 3 sheets of cells at one day intervals. Courtesy of Professor Teruo Okano, Institute of Advanced Biomedical Engineering and Science, TWIns, Tokyo Women's Medical University, Tokyo, Japan.

**Figure 6.5** Drug release from solid polymer. The image shows a wafer of a polyanhydride that incorporates the drug BCNU, known as carmustine. The wafer is surgically implanted in the brain of patients affected by malignant gliomas. The polymer degrades over a few weeks. Image courtesy of Professor Robert Langer of the Massachusetts Institute of Technology, Cambridge, USA.

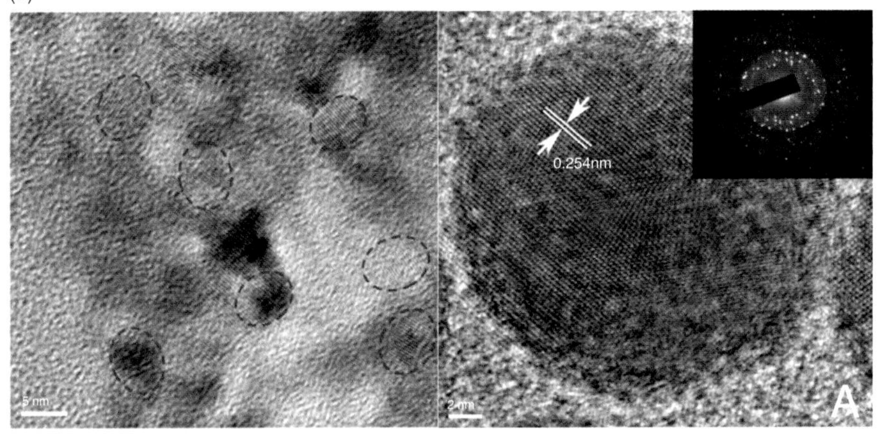


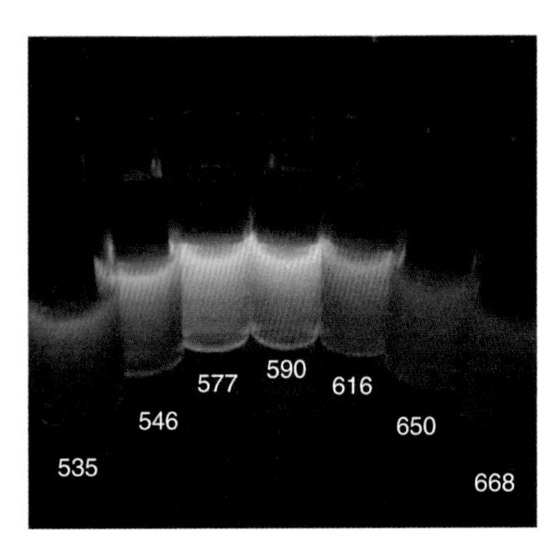


**Figure 6.16** Theranostic agents.
(a) Herceptin–quantum dot conjugates. Upper, high-resolution transmission electron micrographs of HER2–Quantum Dots. The HER2 monoclonal antibody is conjugated to CdTe quantum dots via molecules of the enzyme ribonuclease A. Depending on the size of the resulting nanoparticles, they emit light at different frequencies, shown in the lower panel. Images courtesy of Professor Daxiang Cui, Department of Bio-Nano Science and Engineering, Shanghai Jiao Tong University, China.

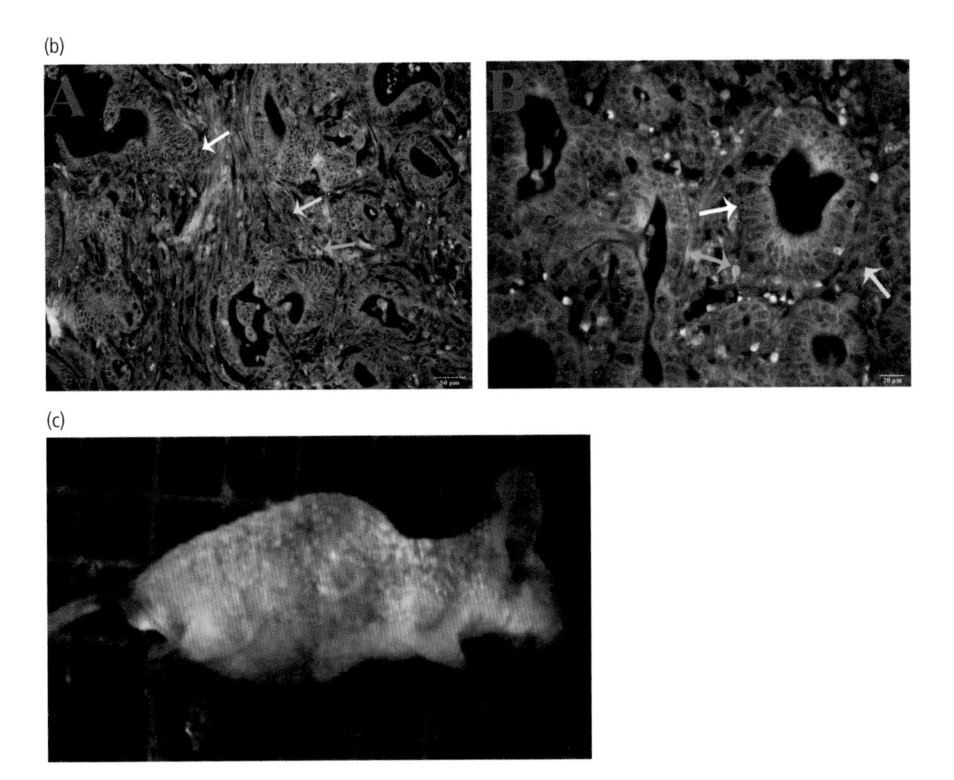

**Figure 7.7** Quantum dots for imaging.
**(b)** An image taken from diagnostic slides using QDs, scale bar 50 μm. On the left (A), the diagnostic detection of gastric cancer from the perspective of tumor microenvironment by QDs-based multiplexed molecular imaging. The tumor microenvironment varies during cancer progression, which could be used for GC diagnosis by simultaneously detecting the essential features (macrophages infiltrating, tumor angiogenesis and ECM remodeling). The tumor microenvironment is rich in infiltrating macrophages, which locate mainly in the invasion front (green arrow). Tumor angiogenesis, another essential event during cancer progression, is represented by the neovessel in the stromal, where are remodeled by infiltrating macrophages and Lysine oxidase (Lox). Lox is a major molecule accounting for ECM remodeling. Lox is observed both in cancer cells (white arrow) and stromal cells (yellow arrow). On the right (B), a high-power field image to show the infiltrating macrophages, neovessels and Lox. Red arrow: neovessels; green arrows: macrophages; yellow arrows: Lox in stromal; white arrows: Lox in cancer cells; scale bar 20 μm.
**(c)** *In vivo* imaging of subcutaneous tumor by quantum dots. Human hepatocellular carcinoma cell line HCCLM9 was inoculated subcutaneously into nude mice, and the tumor was imaged by CdSe/ZnS quantum dots-labeled alpha fetoprotein monoclonal antibody. The brown subcutaneous tumor was visible under the background of green auto-fluorescence of skin. All images courtesy of Professor Yan Li, Department of Oncology, Zhongnan Hospital of Wuhan University, Wuhan, China.

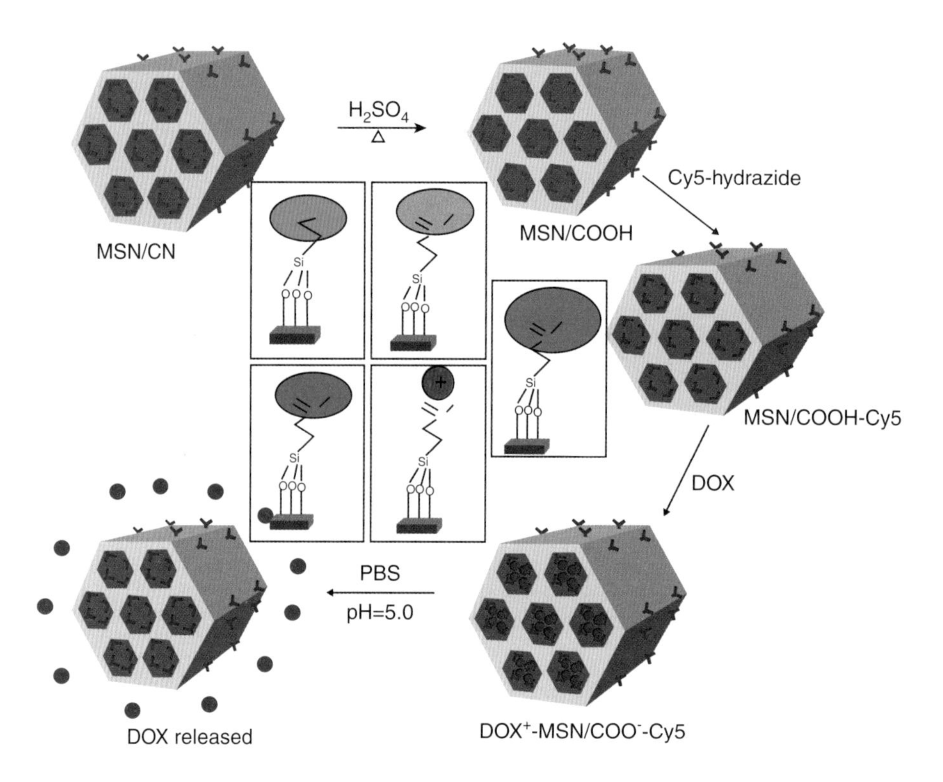


**Figure 8.9** Mesoporous silica nanoparticles (MSN). Illustration of the synthesis and chemical modification of MSN with COOH and Cy5, a near-infrared fluorescent dye, with doxorubicin loading, and pH dependent drug release. Reprinted from *Journal of Colloid and Interface Science*, 395, Xie M *et al.*, Hydrid nanoparticles for drug delivery and bioimaging: mesoporous silica nanoparticles functionalized with carboxyl groups and a near-infrared fluorescent dye, 306–14, © 2013, with permission from Elsevier.

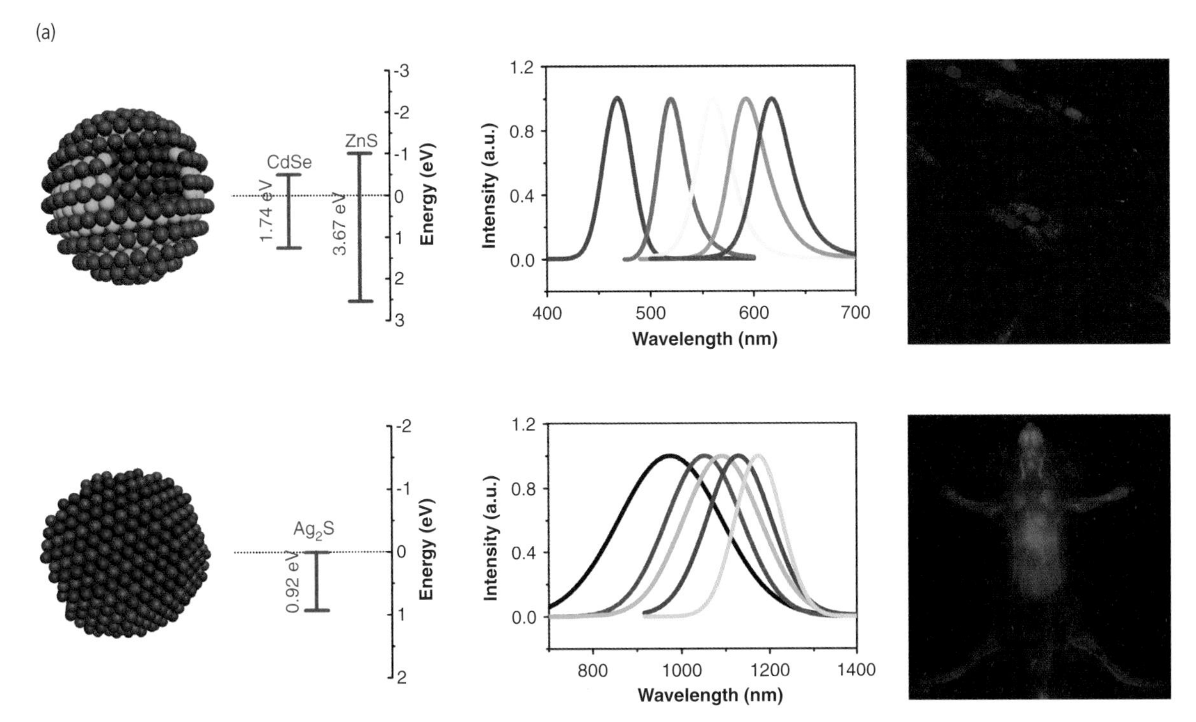


**Figure 8.13** Quantum dots and upconversion nanoparticles.
**(a)** Top: schematic illustration of CdSe@ZnS core-shell quantum dot and its energy band diagram, tunable visible photoluminescence spectra and cellular imaging. Bottom: Ag2S quantum dot with energy band diagram, tunable UV photoluminescence spectra and *in vivo* imaging. Images courtesy of Dr. Qiangbin Wang, Suzhou Institute of Nan-Tech and Nano-Bionics, Chinese Academy of Sciences, Suzhou, China.

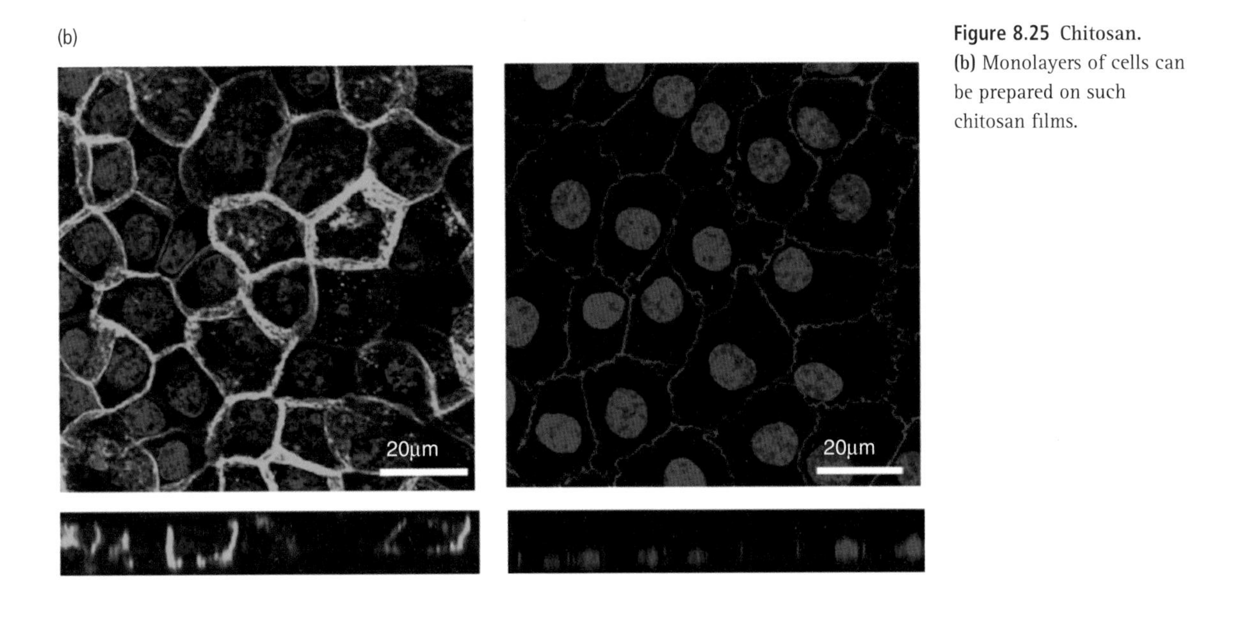


**Figure 8.25** Chitosan.
**(b)** Monolayers of cells can be prepared on such chitosan films.

Hematoxylin and Eosin

Masson Trichrome

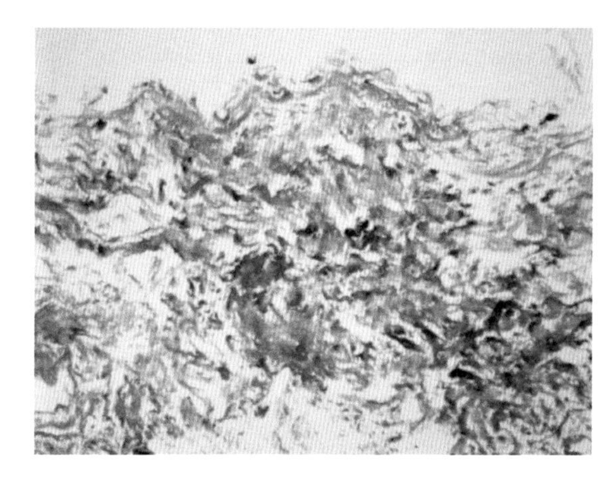

**Figure 8.37** Decellularized tissues. Microscope section of decellularized small intestine submucos (SIS) with peracetic acid (PAA) treatment. Fresh porcine-derived SIS was washed in distilled water on a rotary shaker at 200 RPM and 4°C for 3 days, oxidized by soaking in 5% PAA for 4 hrs, treated with 1% Triton X-100 solution for 2 days and then washed with distilled water for 2 more days. Hematoxylin and eosin, Masson Trichrome and DAPI staining showed that cellular DNA contents were almost completely removed. SIS matrix became visibly more porous, and the intra- and inter-fascicular space was increased in the lamina propria that appear more extended after decellularization procedures. Images courtesy of Dr. Yuanyuan Zhang, Wake Forest Institute of Regenerative Medicine, North Carolina, USA.

### Box 6.2 Rationale for drug delivery systems

These groups represent the major reasons for introducing different technologies to give more effective delivery of drugs.

Group A: Conventional drugs given by conventional routes, such as medication given orally, or by subcutaneous and intravenous injection or by topical application.

Group B: Orally delivered medication where tablets are modified to improve efficiency of delivery, such as pH-sensitive coatings to control absorption in GI tract.

Group C: Techniques to enhance non-oral delivery by increasing efficiency of passage through barriers, such as electrical charge or microneedles to assist transdermal delivery.

Group D: Technologies to deliver drugs to specific tissue sites, such as infusion pumps and implantable depots.

Group E: Devices that facilitate prolonged and sustained delivery such as depots placed in the eye, and contraceptive devices.

Group F: Systems that allow highly incompatible agents, such as toxic chemotherapeutic drugs, to be targeted to cells, for example by complexation with polymers.

Group G: Systems with a high degree of cellular and sub-cellular targeting, for example with highly specific targeting of biomarkers on cells in immunotherapy.

Group H: Combination products where drugs combined with devices improve the performance of the devices.

Group I: Combined therapeutic and diagnostic systems.

Group J: Delivery systems for *ex vivo* tissue engineering.

change their characteristics. The dosage has been determined by assessments of their pharmacokinetic and pharmacodynamic properties. Pharmacokinetics refers to the processes of uptake, distribution and elimination of the drug molecules and pharmacodynamics refers to the physiological effects of these molecules. These characteristics are determined in pre-clinical studies and in the various phases of pre-market drug testing.

The first group of drug delivery systems that we will discuss, although only briefly, in Group B, are the formulations that are modified in some way in order to improve the efficiency of delivery when using the same types of drugs and the same delivery routes. This primarily refers to the oral route of delivery where tablets can be coated with pH-sensitive or mucoadhesive polymers, or be designed as osmotic or swelling systems. Again note that these are conventional drugs that are delivered in minimally engineered systems solely to optimize their entry into the body.

The next group (C) is closely related to this, but involves more complex engineering or formulation changes that greatly enhance non-oral delivery. Included here are the systems such as electroporation and microneedles for transdermal delivery, polymer-based systems for transmucosal delivery and intravenous and intrathecal delivery systems. Some of these utilize chemicals that facilitate or enhance transport across membranes, whilst others use electrical or mechanical methods.

Following this is the group of devices (D) that deliver drugs to specific anatomical sites in the body for their local action. These site-specific delivery systems include continuous infusion pumps for the treatment of some forms of cancer and systems that deliver chronic pain management therapies to the spinal cord. They also include micro- or nanoparticulate formulations that can be delivered to surgical sites. Implantable depots of chemotherapeutic agents are also included; a good example here is the use of a bioerodible polymer that delivers chemotherapy to brain tumors, the blood–brain barrier making it difficult for intravenously injected agents to penetrate the tumor.

Devices and materials in Group E are intended to deliver drugs over a prolonged period of time, the so-called sustained delivery systems. One of the earliest, and more profound, situations where sustained delivery proved very effective was in the use of ganciclovir implants placed in the eye for the treatment of AIDS-related cytomegalovirus retinitis; the drug was released over a period lasting more than 6 months and obviated daily intravenous injections. Sustained release has become an important component of monoclonal antibody therapeutics, where biomaterial-based injectable systems have been developed.

Next are those systems that provide for the delivery of drugs that are intrinsically incompatible with human physiological systems (Group F). This is either because they are insoluble or intensively cytotoxic; they require significant protection. It is here that we see the powerful impact of biomaterials at the nanoscale on drug delivery and the growing interest in polymer therapeutics. Difficult but profoundly effective drugs can be conjugated with polymers or contained within nanoparticles to allow for their absorption and delivery, with some specific targeting.

Closely related are those drugs that require a high degree of cellular and sub-cellular targeting (Group G). Included here are those drugs that provide immunotherapy in cancer patients and have to be targeted to certain cells that display specific biomarkers, for example in the use of herceptin for HER2 positive breast cancer patients, which targets the HER2 receptor on the cancer cells, as discussed in Chapter 1. Also included here are gene therapy products, where, for example, genes are attached to some vector that is able to target relevant cells, pass through the cell membrane and gain access to the nucleus for delivery of its DNA payload.

One further group (Group H) involves the use of so-called combination products, where a drug is combined with a medical device in order for the drug to enhance the performance of the device or minimize the biological risks, for example, infection. The drugs here are not meant to have any other function, and a major factor in their development has been the need to maximize the beneficial effect whilst suppressing any risks that they themselves bring. Examples are the incorporation of antibiotics into materials used in joint replacement, the release of bone growth factors from bone fusion systems and the release of anti-proliferative agents from intravascular stents.

Other combinations have been developed for conjoined diagnostic and therapeutic functions (Group I), including agents that are able to image cancer cells and deliver drugs directly to them.

Finally, and with some similarity to those in Group H, are the systems of tissue engineering and cell therapy, discussed in the previous chapter, in which drugs or active biomolecules are required to facilitate cellular action (Group J). These include systems for the use of growth factors and gene vectors, and they may operate *ex vivo* or *in vivo*.

These groups, A–J, provide a convenient summary of the rationale for the development of drug delivery systems. Of course, for many products, more than one of these reasons may apply. Moreover, similar materials or mechanisms may apply to several of these areas. Because of these factors, it is more appropriate now to discuss active molecule delivery on the basis of the mechanisms and materials. We then conclude with some sections on specific areas that have their own features in addition to those of the generic mechanisms.

## 6.1.2 Oral drug delivery

As noted above, and mentioned with respect to Group B, it may be necessary to formulate orally delivered drugs with features that control the release of the drug at different sites or at different times in the passage of the tablet through the gastrointestinal system. The basic principles and mechanisms are discussed briefly here.

Orally delivered drugs are usually of low molecular weight, being based on synthetic organic molecules. The effectiveness of these drugs is primarily dependent on their hydrophilicity (i.e., solubility in the aqueous media of the GI tract) and lipophilicity (which controls permeability across epithelial membranes). Orally delivered drugs are often classified on the basis of these characteristics:

- Class I – high solubility and permeability.
- Class II – low solubility and high permeability.
- Class III – high solubility and low permeability.
- Class IV – low solubility and permeability.

With respect to solubility, the drug is released primarily by water penetration into the tablet, followed by its swelling and dissolution, often with the help of osmosis.

The GI tract is shown schematically in Figure 6.1. For the purposes of drug release, we may consider this to have three sections, the stomach, the small intestine and the large intestine. The stomach has a resting pH of around 4–5, but this drops to 1–2 during and immediately after eating; it is for this reason that instructions about whether to take medicines with or without food are important since the

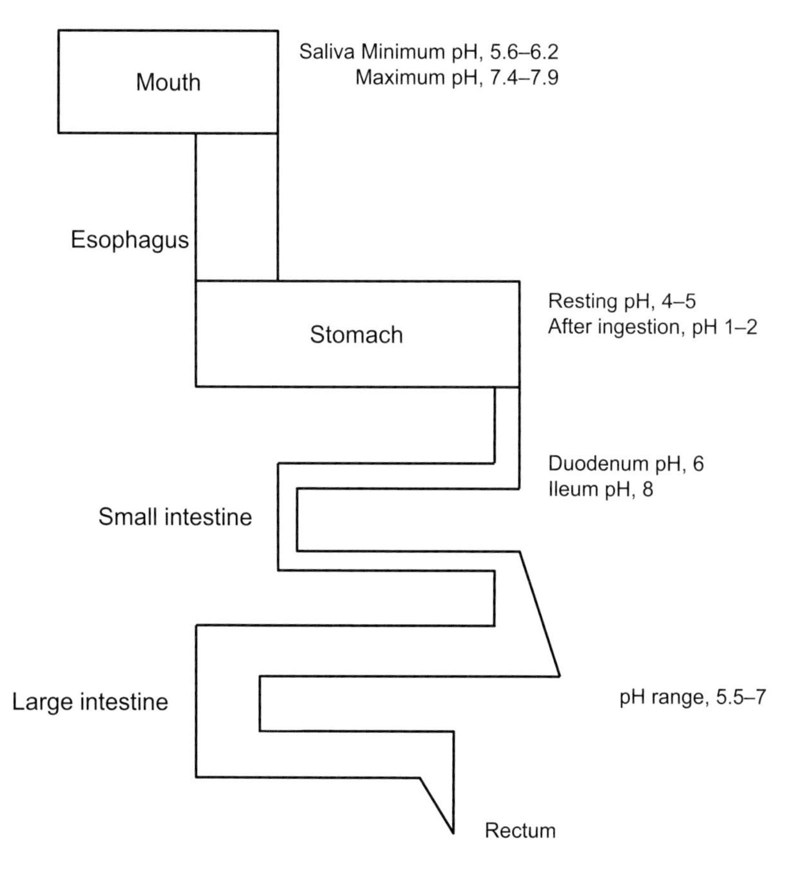


**Figure 6.1** The gastrointestinal tract. When medication is taken by mouth, very different environments are encountered within the fluids of the GI tract. While the resting pH of the stomach is usually between 4 and 5, it drops to a very acidic 1–2 after food. pH rises in the intestines. Different strategies are used to optimize delivery of drugs from tablets in different parts of the system.

drug release will vary considerably between these levels. Activities in the stomach are referred to as "gastric." The small intestine, which contains the duodenum and jejunum, has a much higher pH, usually between 7 and 9; the activity here is labeled "enteric." The large intestine has an intermediate pH of 5.5–7. This section mostly consists of the colon and the activity is "colonic."

Transit times of foods and tablets will vary, but typically will be less than 3 hours gastric, 3–5 hours enteric and 20 hours colonic. With some drugs it is preferable that they are released in the lower part of the GI tract rather than in the stomach. For example, aspirin and some non-steroidal anti-inflammatory agents can irritate the stomach so that gastric release should be avoided. The tablets of these formulations are usually given an enteric coating, in which a pH-sensitive polymer is used to coat the tablet. This should not dissolve in the acidic gastric conditions, delaying the release until the intestines are reached. Polymers that are protonated and insoluble under gastric conditions, but which ionize and swell under enteric conditions, are ideal. In other situations, the reverse may be required in order to accelerate gastric release. Materials for enteric coatings include cellulose acetate phthalate and poly(vinyl acetate phthalate).

A range of technologies has been developed to increase the gastric retention time of a drug. These include coating of the tablet with mucoadhesive microspheres that adhere to the stomach wall and the intragastric floating systems that allow the tablet to float at the top of the stomach contents and avoid gastric emptying. Tablets may contain a low density liquid or gas, or an effervescent substance.

It is sometimes preferable for the drug only to be released in the colon. More complex strategies have to be used, for example using an enteric coating that prevents dissolution in the stomach, which overlies a cationic polymer that prevents absorption in the small intestine. Some formulations may also include a substance that rapidly degrades in the cecal microflora in order to assist in the colonic release. Finally, the required delivery profile may be achieved with the assistance of ion-exchange and osmotic effects.

## 6.2 Non-oral delivery of conventional drugs

In this section we are primarily concerned with Group C discussed above and the technology that is available to deliver conventional drugs to the body via routes other than through the GI tract.

### 6.2.1 Transdermal delivery

Although the skin is a very efficient barrier to the penetration of the human body by microorganisms, it can be penetrated by chemicals under some circumstances. This is taken advantage of in transdermal delivery systems. The outermost layer of the skin is the stratum corneum that is up to 20 μm thick, underneath which is the epidermis, up to 100 μm thick, and then the dermis, up to 2 mm thick. This is a very effective barrier, especially the stratum corneum, but some chemicals can diffuse through the keratin and lipid regions of this layer, with somewhat different routes for hydrophilic and lipophilic molecules. Such penetration of the skin by pharmaceutical agents is not inherently efficient, and not many drugs are amenable to this route, so that the technologies here are aimed at enhancing diffusion or introducing new methods to persuade these molecules to pass through the skin barrier.

#### Transdermal patches

Transdermal patches were the first type of device to be used for transdermal delivery, following the use of scopolamine patches to control motion sickness and then the widespread use of nicotine patches as part of the strategies to minimize tobacco smoking. As shown in Figure 6.2, the drug, which may be in solid or gel form, is typically stored in a reservoir system surrounded by a number of layers, including an impermeable backing membrane, a rate-limiting

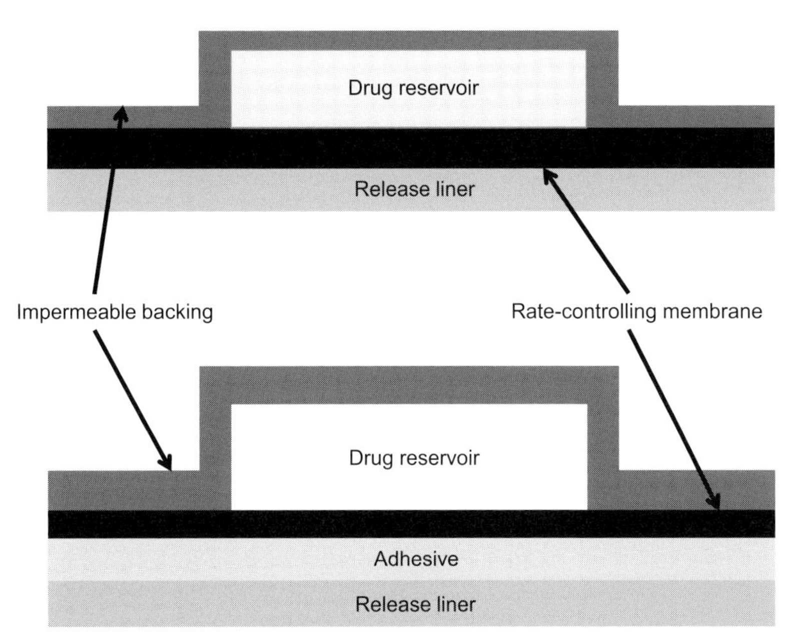


**Figure 6.2** Transdermal patches. The drug, in solid or gel form, is stored within a reservoir, with varying forms of impermeable backing membranes, rate-limiting semi-permeable membranes and adhesive layers.

semi-permeable barrier and an adhesive layer. In addition to the early uses mentioned above, transdermal patches are used for the delivery of nitroglycerine for the treatment of angina and also for pain relief medication. The patches are not without problems: the opioid antagonist fentanyl has been used for a number of years with transdermal delivery in a number of commercial products, but several of these had to be withdrawn because of a variable drug delivery rate which led to overdoses of this powerful drug.

## Chemical enhancers

The unpredictable and often rather slow rate of delivery of drugs from transdermal patches has been addressed by the use of chemical enhancers, which disrupt the structure of the stratum corneum, primarily by creating nanometer size defects in the lipid regions that facilitate transport through the layer. Great care has to be taken with the choice of the chemical since, not surprisingly, they have a tendency to affect and indeed irritate other compartments of the skin.

## Iontophoresis

Iontophoresis is a technique that uses an electrical driving force to transport drugs across skin or mucosal surfaces. It is non-invasive, as an electrode carrying the drug is placed on the external surface. The electrode is arranged to have the same charge as the drug, while a ground electrode of opposite charge is placed somewhere else on the body surface, the drug acting as a conductor of current as it passes through the tissue. The drug is delivered in proportion to the current and so this provides for an adjustable, even programmable, delivery rate, although there is a maximum rate that is determined by the irritation and pain associated with the current. The technique can be used to deliver local anesthesia and pain relief medication. One area of significant interest is ocular iontophoresis, where drugs can be delivered into the anterior chamber of the eye for localized effects. The drug can be contained in a hydrogel or sponge, such as a polyacetal, contained within a silicone elastomer shell that incorporates a silver–silver chloride electrode linked to the dose-controller.

### Electroporation

Electroporation uses a much higher voltage pulse for a short period of time. The pulses range from 10 V to 300 V, with single or multiple applications of 0.1 to 100 ms. These pulses again disrupt the lipid bilayer structure such that active and passive transport of the drug can take place over several hours. This technology is not widely used for drug delivery since extensive engineering is required to match microelectrode design with the control of pain and muscle stimulation. However, there is much interest in its use for the delivery of DNA vaccines and gene delivery more generally. Here the needles used to inject the preparation also act as conducting electrodes. Some systems use conventional hypodermic needles, but others use parylene-coated tungsten.

### Ultrasound

The use of ultrasound to deliver drugs through the skin is known as sonophoresis or phonophoresis. There are two main mechanisms by which ultrasound exerts its effect, one which is observed at high frequency and one at low frequency. The latter, which involves acoustic cavitation, is the more significant, and so these methods are usually referred to as non-cavitational and cavitational ultrasound drug delivery processes.

With non-cavitational ultrasound the drugs are driven through the skin by convection, mechanical pressure, thermal effects and disruption of the lipid components. This occurs at frequencies greater than 1 MHz. It is limited to small lipophilic compounds and care has to be taken with the effects of tissue heating. Cavitational effects are seen at low frequency, typically between 20 and 100 kHz. The acoustic cavitation effects involve the pulsation and growth of microbubbles that are present in the liquid phase of the skin, especially at its surface, and the continuous oscillation and collapse of bubbles. The resulting shock waves and microjets disrupt the lipid structure, increasing skin permeability. Systems using this technology have been used clinically for the delivery of local anesthetic agents such as lidocaine; such a system may have a power and control unit, a hand-held ultrasonic probe, a return electrode and a coupling medium disposable cartridge. There are no unusual material issues here and the probe is in contact with the skin for a very short time.

### Microneedles

The conventional hypodermic needle delivers a bolus of a drug solution to tissues underneath the skin layer. It is often perceived as painful because of the needle size and depth of penetration. Often, a good mode of delivery can be achieved without complete penetration of the skin through the use of multiple short needles. These are referred to as microneedles. Several approaches have been adopted. Arrays of solid microneedles may be used, which create an array of holes that can be covered by a drug-containing patch, seen in Figure 6.3(a). Alternatively, the microneedle array can be coated with the drug before placing in the skin. The needle arrays have been made from silicon wafers or thin metal films using lithography or etching techniques. A further refinement of this technology involves the use of so-called dissolving microneedles, made, for example, from carboxymethylcellulose or amylopectin, which dissolve quickly once inserted into the stratum corneum (Figure 6.3(b)).

### Ablation/abrasion

Thermal ablation may be used to facilitate transdermal delivery, during which the skin surface is vaporized. The thermal exposure is very short and the underlying tissue of the epidermis and dermis are unaffected. This technology is not widely used clinically as yet. A device suitable for the application involves a microchamber that heats and ejects super-heated steam, and masks that control the energy transfer to the skin. Standard materials can be used, for example tungsten and titanium for different parts of the mask and polymethylmethacrylate and polyethylene terephthalate for the chamber.

In a similar type of technique, the cosmetically used method of skin abrasion, known as

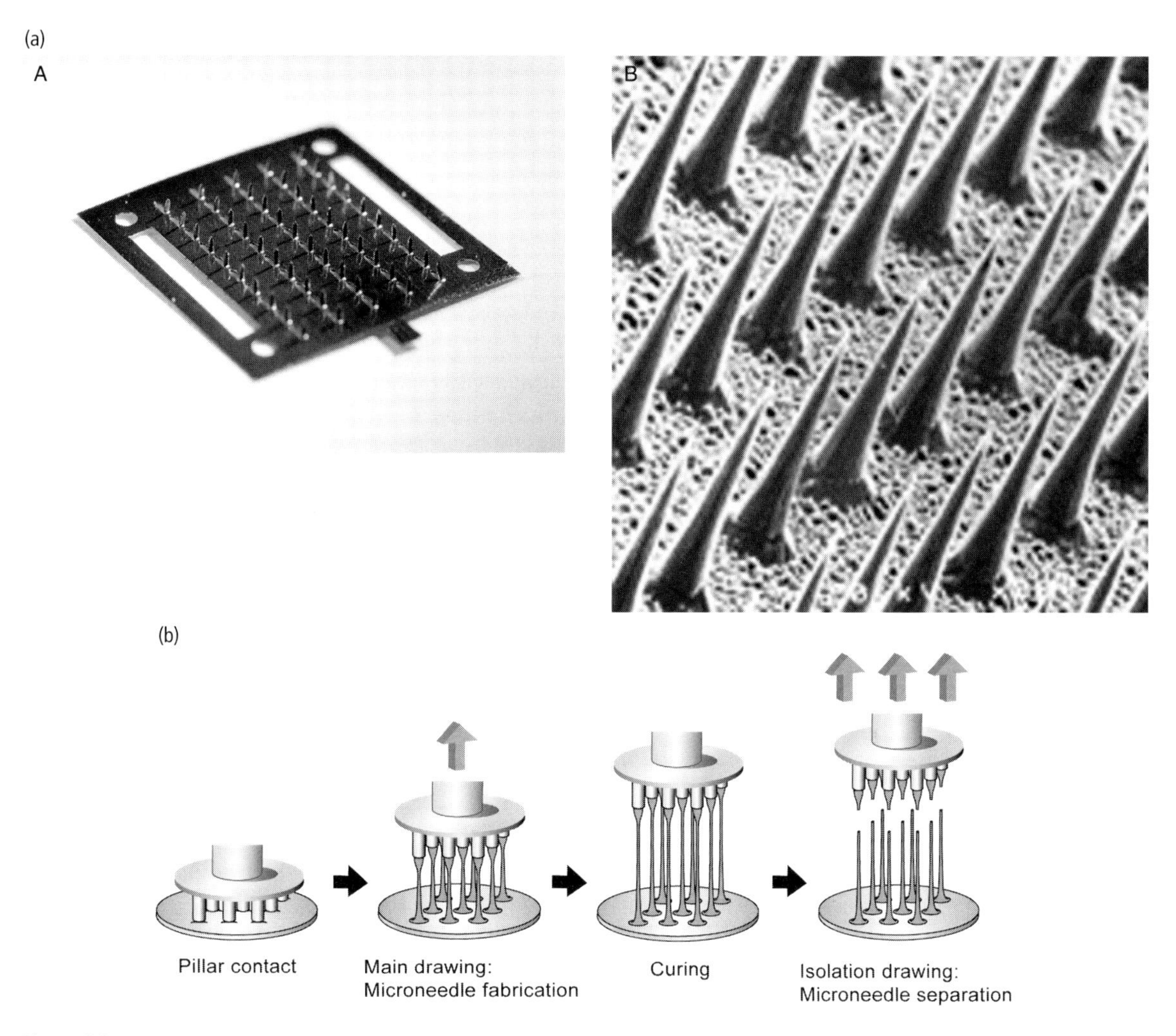


**Figure 6.3** Microneedles.

**(a)** Solid microneedles for drug delivery. Images courtesy of Professor Mark Prausnitz, Laboratory for Drug Delivery, Georgia Tech, USA. (A) 150 μm tall silicon microneedles (credit: Sebastien Henry, Georgia Tech). (B) A patch containing 50 metal microneedles (credit: Georgia Tech).

**(b)** Dissolving microneedles. The scheme shows the process of drawing lithography to produce a 3D dissolving microneedle. The matrix was spin coated and the main drawing was performed after the matrix contacted the patterned pillars, giving the appearance of an extended conical-shaped bridge between the matrix and pillar. The microneedles are cured to generate a rigid structure. Image courtesy of Professor Hyungil Jung, Department of Biotechnology, Yonsei University, Seoul, Korea.

microdermabrasion, may be used to modify the stratum corneum in order to enhance drug molecule transfer. Such systems normally use a high-pressure stream of alumina particles that remove and disrupt the outer layer.

## Needle-free injections

Most of the methods of enhancing transdermal delivery of drugs require skilled or trained personnel or relatively complex devices. In many situations, these techniques are not appropriate for the delivery of drugs or agents where frequent, self-administered doses are required but where oral delivery is not feasible. Inexpensive, contamination-free, pain-free simple procedures are often required, especially in developing nations where vaccinations and infection-control medications are so important. There have been several attempts to develop systems for these situations using the so-called needle-free injections. These require a simple inexpensive disposable device that can be held to the skin and deliver a precise controlled amount of drug without any part of the device penetrating the skin. These usually utilize the injection of a fluid or powder drug under extremely high pressure through the skin. These are hand-held and have some trigger mechanism, based on a spring or compressed gas, to activate the release, the drug passing through the skin at very high speed. This technology has been difficult to perfect since there are significant cost limitations (the cost of a disposable injection system has to be compared to the cost of the drug or vaccine) and reliability is an important factor – if a device is triggered but fails to deliver the payload it may or may not be obvious to the user. Also the ability of the drug bolus to pass through the skin varies from patient to patient and skin site to skin site.

### 6.2.2 Implantable electromechanical systems

The use of electromechanical systems for drug delivery intuitively appears to be a complex and expensive option; this is true and is a major reason why clinical use has so far been rather limited. With implantable systems it is also invasive. The real interest here, which is linked to some long-term goals, is in the way in which it points to individualized approaches to medicine, where drug delivery regimes can be tailored to specific and precise needs, the release profiles being controlled by sensitive, programmable systems.

There are broadly two mechanisms that can be used. The first involves microscale pumps; some pumps have been used for insulin delivery and pain management in cancer patients for a number of years. Recent developments have been directed towards the delivery of viscous solutions with low but constant flow rates with chronic reproducible performance and minimal power consumption.

Probably of greater interest are the devices that use chip technology. For example, reservoirs may be created on the surface of silicon microchips surrounding which are seals that can be opened and closed on command in order to release the drug in a programmable way. Alternatively, a solid drug-matrix formulation can be applied to a chip, which is electrically stimulated such that the rate of drug release is controlled by the applied current. One experimental system, for example, uses a conducting polymer film such as polypyrrole that contains the drug, and which can be stimulated by an array of gold microelectrodes set on a silicon chip.

## 6.3 Drug release from solid polymers in monolithic devices

In this section we deal with drugs that are combined with or contained within a solid polymer that, when placed within or on the body, releases the drug according to a planned schedule. In this situation the polymer is associated with a macroscopic, monolithic structure; the situation with particulate systems, both at the micro- and nanoscale will be covered in the next section. The mechanisms by which drugs are released will be discussed, but before that we need to cover a few general principles.

The release profile may take several different forms (Figure 6.4(a)). A linear relationship between the amount released and time, called zero-order release, may be considered ideal under many circumstances but is not always achieved. Instead, bi-phasic or especially tri-phasic profiles are more common. With a tri-phasic profile, there may be an initial burst effect, when drug molecules at the surface escape easily on the beginning of polymer hydration or find their way through surface cracks and pores. The second phase is usually slow, and may approximate to zero-order, as the drug is released through the

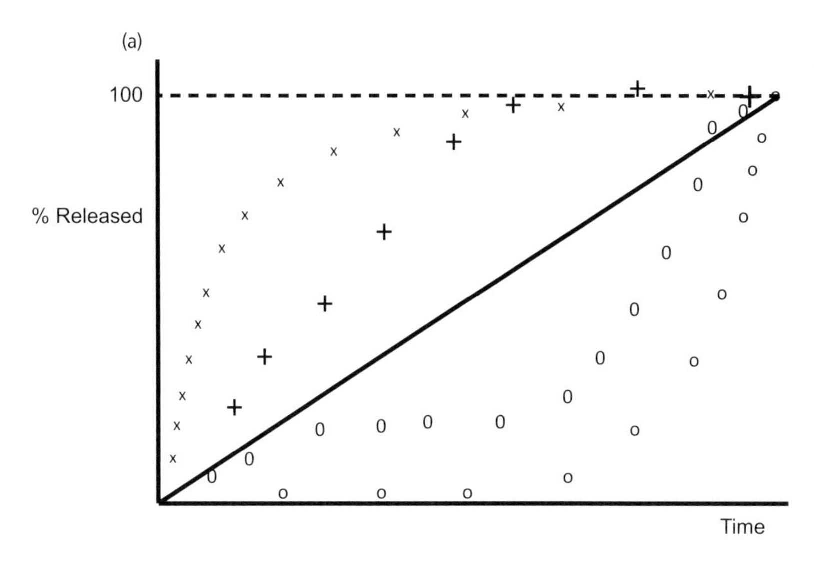


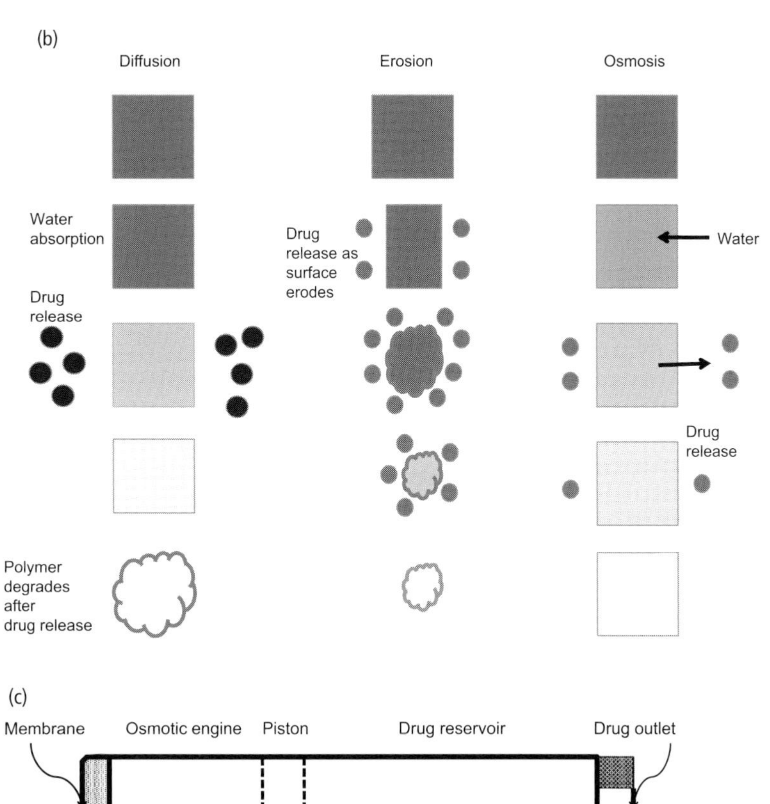


**Figure 6.4** Drug release from solid polymers.
**(a)** Release profiles of drugs from solid polymers. The straight line shows a theoretical constant release, referred to as zero-order release. In practice, the rate of release varies considerably, depending on the dominant release mechanism at any one particular time. There may be an initial burst effect, followed by a considerable slow down, or a very low release in early stages because of a low absorption of water, but a strong increase as the absorbed water is able to cause rapid diffusion.
**(b)** The main mechanisms are diffusion, erosion and osmosis. Diffusion involves absorption of water into the polymer and subsequent outward diffusion of drug into the tissues. The polymer will usually degrade after the drug has been released. In erosion, the drug, which is initially uniformly distributed in the polymer matrix, is released form the surface which gradually erodes within the aqueous environment. The third mechanism is osmosis, in which water enters the drug reservoir and the drug is forced out of the depot by osmotic pressure (see (c)).
**(c)** Implantable osmotic system. A polymer membrane covers one end of the reservoir. The osmotic engine (i.e., the draw solution) occupies a portion of the cylinder behind the membrane. An elastomeric piston separates the draw solution from the drug formulation in the drug reservoir. Upon diffusion of water into the osmotic engine, the piston is pushed and the drug is released through the drug outlet orifice. Redrawn from *Journal of Membrane Science*, 281, Cath T Y *et al.*, Forward osmosis: principles, applications, and recent developments, 70–87, © 2006, with permission from Elsevier.

predominant mechanism appropriate to the polymer–drug combination. Phase three is usually faster and follows the final stages of disappearance or collapse, by whatever mechanism, of the polymer. These profiles can vary. They will be determined, in detail, by polymer characteristics such as molecular weight and crystallinity, the nature of the drug and additives or excipients, the delivery system features such as porosity and shape, and the environmental conditions such as the host response to an implanted depot.

The mechanisms that we shall discuss concern the physical and chemical processes that allow for the drug release. These may involve several different stages, for example the initial absorption of water into the polymer, then the effect of the absorbed water on the polymer, for example a change to the crystallinity or polymer chain mobility, and then the diffusion of the drug outwards. Not all of these processes take place with the same kinetics and so in order to establish what happens in practice, the slowest or more sensitive mechanism has to be identified. This is the rate-controlling release mechanism. The processes of polymer change and drug diffusion may be fast, but if the absorption of water is slow, it is this that will control the overall rate of drug release.

### 6.3.1 Mechanisms of release

In the light of the above comments, it is important to identify the true processes by which a drug can be released from a polymer, and the polymer characteristics that can influence these processes. There are three true release mechanisms, diffusion, erosion and osmosis, shown in Figure 6.4(b). It is often stated that biodegradation is a release mechanism but this is not so. In most cases, especially with implantable drug depots or reservoirs, biodegradation takes place but this is mainly because it is required that the depot disappears after the drug has been released and only partly because it may assist the principal mechanisms. Diffusion may take place through the polymer itself, which is usually slow and limited to small molecule drugs, or through water-filled pores in the polymer. Osmosis occurs following water absorption, when the osmotic pressure forces the drug outwards. Erosion does not involve any transport of the drug within the polymer; instead the drug is released from the surface as the polymer at the surface disintegrates or collapses on a microscopic scale. Obviously the degradation of the polymer will influence diffusion as it should contribute to the opening up of the porosity, and hence faster transit of drug molecules, and it should influence erosion as it contributes to the process of polymer matrix collapse, but the degradation does not intrinsically cause the release.

#### Diffusion

Diffusion controlled systems rely on drug diffusivity through a membrane or through a polymer matrix; the membrane or matrix can be porous or non-porous and may or may not be degradable. The system may take the form of an orally delivered tablet that is coated with a hydrophilic layer. When this is hydrated, a gel barrier is formed through which the drug diffuses. Alternatively, and as we have seen in Section 6.2, it make be in the form of a patch such as a transdermal or ocular patch. The drug solution may be placed between a drug-impermeable backing layer and a skin contacting membrane. Alternatively, the drug is incorporated directly into a polymer matrix, which is attached to the skin by an adhesive. Many different polymers are used in the various parts of drug release systems, including natural polymers such as gelatin, starch, cellulosic materials, and synthetic polymers and elastomers including polyvinyl alcohol, silicones and polyurethanes.

If we concentrate on solid polymers that contain a drug (rather than a polymer membrane which allows for diffusion from one side to the other), we can see that the diffusion process may occur through the water-filled pores or through the polymer itself. The former is usually the more prominent and is clearly dependent on the pore structure of the polymer. The pores may establish a complete and uniform network or exist as discrete pores with interconnectivity between them. The tortuosity of the porous network obviously affects diffusion pathways. The initial diffusion coefficient will be dominated by the diffusion coefficient in the fluid within the pores. With sufficiently sized pores and interconnectivity, all or most of the drug could be released before any erosion or degradation has taken place; this certainly controls the initial burst effect. These effects contribute to the later pore forming or pore enhancing effects, which may determine a sigmoid release curve where the burst effect gives a rapid initial release, followed by a period of zero-order release and then a further faster release as erosion and degradation contribute to larger pore formation. It should be anticipated that diffusion through the aqueous phase in pores is the most important mechanism for the release of hydrophilic drugs from most polymers.

For small hydrophobic drugs, diffusion through the polymer itself may be possible. Diffusion is usually more effective in lower molecular weight polymers because of greater molecular flexibility and also in polymers that are in their rubbery rather than glassy state.

### Erosion

The erosion mechanism does not require diffusion of the drug through the polymer, either within solid matrices or pores. If we assume that the drug is homogeneously distributed throughout the polymer, the release rate of the drug will be identical to the rate of erosion. As the name implies, erosion is a surface-dominated process. As the polymer surface becomes quickly hydrated, drug molecules right at the surface are released. In most systems, erosion will occur simultaneously with hydration, as the surface regions are replenished with drug molecules moving outwards from the interior. As the polymer molecules themselves are released by the erosion process, so the surface profile changes. Since hydration is usually much faster than erosion, the water diffusion is likely to be rate-controlling, with a contribution from the erosion.

### Osmosis

At their simplest, osmotic systems consist of a reservoir constructed of a polymeric semi-permeable membrane that is permeable to water but not to the drug. This membrane usually has an orifice Figure 6.4(c). A concentrated solution of the drug is placed in the reservoir. Water passes through the membrane due to osmotic pressure and the drug solution is released through the orifice. The membrane is usually cellulose acetate. Various drugs can be delivered in this way, at relatively high rates.

In practice there are several variations on this simple theme. Most systems are used for oral drug delivery. The devices are grouped as single chamber (including the simple version above), multiple chambers, and various modified forms. The push–pull osmotic pump is a good example of the multiple chambers type; it is a bilayer tablet coated with the semi-permeable membrane. An elastic diaphragm separates the two compartments. One chamber contains the drug and has a very small delivery orifice to the outside. A polymeric osmotic agent is present in the other chamber, the swelling of which deforms the diaphragm that causes ejection of the drug solution.

Some osmotic systems have been developed for prolonged drug delivery following implantation, typically just under the skin. It may deliver the drug locally or to a specific site by an attached catheter. Typically such a system will be shaped like a narrow tube, made for example of titanium, with an orifice at one end and a semi-permeable membrane at the other. A piston separates the drug at the orifice end from an osmotic agent, for example salt, at the membrane end, and the osmotic pressure drives the piston towards the orifice, displacing the drug.

### Biodegradation

As noted above, the biodegradation of drug releasing polymer systems is mainly designed to facilitate the eventual disappearance of the polymer rather than as a mechanism for drug release, and in most systems, the drug will have been substantially released before significant degradation has taken place. Nevertheless, the polymer degradation process contributes to diffusion mechanisms through pore enlargement and to erosion, and so is an important part of the drug release phenomenon.

## 6.3.2 Solid polymers for drug release

Most of the polymers used in these systems are described in Chapter 8; we pick out a few examples here to indicate the special features that are required in order to optimize delivery. Some of the degradable systems may be used in either monolithic, microparticle or nanoparticle form; this following discussion concerns the generic issues.

### Degradable polyesters: poly(lactic-co-glycolic) acid

PLGA materials are probably the most widely used polymers in drug delivery systems, partly because of their generally good overall biocompatibility and partly because of relatively easy avenues to regulatory approval on account of their long track record. There are several commercial sources of the polymers, with quite a range of compositions and physico-chemical properties being available. Both molecular weight and lactide:glycolide ratios may be tuned in order to control release rates.

The release of drugs from PLGA follows the general principles outlined above. Water is absorbed quite rapidly. The absorbed water tends to form in minute pockets, which can be considered as types of pores; they eventually coalesce to provide an interconnected network, which controls the release process through diffusion within the porous network. There is minimal diffusion of drug within the polymer itself, especially at early times. Erosion, the surface-initiated mass loss of the polymer, also starts very early. This process rapidly extends into the interior as degradation starts around the edges of the water-filled pores.

The degradation process obviously influences the erosion process and hence diffusion, but there are many variables and characteristics to take into account. A very important factor is the acidic nature of the degradation products; this has several consequences including the potentially autocatalytic nature of the process as the more acidic products are produced the greater the rate of degradation, and the need to protect acid-sensitive drugs. These products also tend to plasticize the polymer, which increases the rate of water absorption. They may also affect the crystallinity of the polymer. Clearly the release rate will be profoundly affected if the material shows cracks or collapses. In reality, a number of structural changes simultaneously take place in the PLGA as erosion and degradation proceed, which influences the diffusion rate. The result is that different release profiles can be obtained, as shown in Figure 6.4. These processes are usually triphasic, sometimes perhaps approaching biphasic. There is often a burst effect, in which a significant amount of the drug is released early, primarily associated with the rapid release of drug particles at or near the surface. Most profiles have a slow, relatively constant rate, where the factors outlined above tend to balance each other. The final phase often has a higher rate, associated with the higher erosion and autocatalytic degradation.

### Polyanhydrides

Polyanhydrides appear to be very suitable vehicles for drug delivery because of the manner in which they degrade by surface erosion and the high water lability of the anhydride bond. They exist in a variety of structures, with either homo- or hetero- form depending on whether they are derived from aliphatic, aromatic or heterocyclic precursors.

These polymers have hydrophobic domains connected by the anhydride bonds. The hydrophobicity determines that degradation is largely controlled by surface erosion since it is difficult for water to penetrate the bulk of the polymer, which is different to the PLGA discussed in the previous section. There are several factors that control the actual rate of this erosion, especially the composition and crystallinity. Many forms of polyanhydrides used for drug delivery are copolymers involving both aliphatic and aromatic components. In most of these, the aliphatic chain is polysebacic acid. Other chains that may be used include isophthalic acid, fumaric acid and terephthalic acid. There are also fatty acid-derived polyanhydrides, based for example on oleic acid and linoleic acid. The hydrophobicity is profoundly controlled by the aliphatic/aromatic balance; purely aliphatic polymers can degrade in a matter of days, while fully aromatic polymers may take several years to degrade.

Perhaps the most important application of polyanhydrides in drug therapy involves the delivery of the chemotherapeutic agent BCNU to brain tumors. This is an FDA-regulated commercial product in which polymer wafers are surgically implanted into the brains of patients, as in Figure 6.5. The preferred form of polyanhydride has been a 20:80 molar ratio polymer of poly[*bis*(p-carboxyphenoxy) propane and sebacic acid, each wafer containing close to 200 mg of polymer and just under 8 mg BCNU. The survival time of patients with glioblastoma may be significantly increased by this technique. The drug appears to be released over several weeks with no appreciable increase of risk of toxicity.

**Figure 6.5** Drug release from solid polymer. The image shows a wafer of a polyanhydride that incorporates the drug BCNU, known as carmustine. The wafer is surgically implanted in the brain of patients affected by malignant gliomas. The polymer degrades over a few weeks. Image courtesy of Professor Robert Langer of the Massachusetts Institute of Technology, Cambridge, USA. See plate section for color version.

### Silicones

The above two sections referred to degradable solid polymers, degradation being an obvious requirement for invasive delivery systems. There are other ways in which drugs can be delivered, as we have seen earlier, and non-degradable polymer systems may be considered in these situations. The most widely used material here is polydimethylsiloxane, or silicone elastomer. This has well-established biocompatibility characteristics and appropriate flexibility for membrane-like depots. As elastomers, these have a helical type of molecular structure, with silicon–oxygen bond angles, both of which allow for the presence of considerable free volume within the structure. This allows for high gas permeability, as discussed in Chapter 4, but also good permeability for many substances, including several drugs.

Silicones have been used in devices in contact with epithelial surfaces in several situations, including intravaginal drug delivery, mainly for contraceptive agents. There are also transdermal and subdermal devices with applications ranging from anti-inflammatory treatment of acne, transdermal delivery of nitroglycerine for treatment of angina and motion sickness, and the use of reservoirs to deliver opioid analgesics.

## 6.4 Drug release from microparticulate and nanoparticulate systems

In the previous section, the release of drugs from monolithic solids was discussed. Obviously the size of the system is an important variable. This determines how the device can be applied to the body; with sizes more than a few millimeters, this implies skilled intervention, often with a surgical procedure, rather than self-administration. The size also determines diffusion pathways and degradation or erosion profiles. For many reasons, the possibility of using much smaller depots, but many more of them, has been attractive. This implies the use of particulate systems where individual particles can be of microscale dimensions or less, and where collective volumes of particles offer the possibility of injections and not surgical implantation.

It has to be said that this is not a trivial matter and the introduction of such systems into clinical practice has not been easy. A couple of disadvantages immediately present themselves. First, an injection of a mass of small particles within a carrier fluid is irreversible. The particles cannot be retrieved so that if they cause a local irritation or, worse, an allergic reaction, there is little that can be done to retrieve the situation. We should bear in mind here that, perhaps, a whole month's supply of the drug, intended to be released over that time will be an inevitable source of irritation until it is gone. Secondly, although the system may have been designed and engineered such that the particles remain locally, they, or their degradation products, may be distributed systemically, with potential adverse consequences during systemic transport or following deposition at remote sites.

The definition of the nanoscale is discussed elsewhere in this book. As far as drug delivery is concerned, it is possible to consider the size range as a continuum, from a millimeter down to a nanometer, smoothly and uniformly. It is more sensible, however, to introduce some discrimination in the description of size in view of the differing properties of and responses to particles of different size. The unit of the micron is a useful parameter for such discrimination. Most microparticles used in the context of drug delivery will be in the range of 1–500 μm. Nanoparticles used in drug delivery systems tend to be in the range 5–500 nm, although usually within sight of the 100 nm technical upper limit of the nanoscale. The significance is that the surface area to volume ratio is inversely proportional to the radius of the particle, so that the smaller the particle the greater will be the (theoretical) surface area from which a unit volume of drug solution can be released. In addition, the pathways for particle movement or translocation, and ability to penetrate cells, are different at the micro- and nanoscale.

Microparticles injected into a mass of tissue, for example muscle, or even into a cavity such as the peritoneal cavity, will stay in that location. Nanoparticles are much more likely to take part in systemic transport, and injections into the peritoneal cavity will be dispersed rapidly. As far as cell entry is concerned, even at the small end of the microscale, particles cannot enter most cells, with the exception of some of the larger professional phagocytic cells. However, nanoparticles can usually readily enter cells by passage through cell membranes, which can be very beneficial if it is intended to target the drug to intracellular locations. It should be recalled that particles have a tendency to aggregate, so that the effective size in practice may be different to the estimated size of individual particles. Furthermore samples of particles are rarely monodisperse but will have a distribution of particle size.

All of the above factors have to be taken into account when designing microparticulate or nanoparticulate drug delivery systems, especially in relation to the characteristics of the drugs and the intended site of their delivery. In general, these formulations will be designed to:

- be injected into and remain in one site for local action;
- be injected into a site and remains as a depot for release and subsequent systemic distribution;
- be injected systemically for systemic distribution and activity;
- be injected systemically but with targeting capability for distribution to their intended locus of activity.

### 6.4.1 Microparticles and drug delivery

As implied above, there are many potential advantages of microparticulate systems but they have not found significant clinical applications because of the difficulties already mentioned. In addition the advantages offered by nanoparticles are more profound, especially in relation to multifunctionality, the ability to carry more complex payloads and for

### Box 6.3 Vaccine delivery with nanoparticles and microparticles

Several forms of nanoparticle and microparticle offer considerable potential for enhanced vaccine delivery. Some of the main contenders are listed below.

- *Liposomes*: in clinical trials for delivery of vaccine against non-small cell lung cancer. Good biodegradability and biocompatibility but quickly cleared and not good for water-soluble proteins.
- *Virus-like particles*: one product marketed for vaccination against human papillomavirus, where the HPV major caspid protein L1 self-assembles as the virus-like nanoparticle. Some general safety concerns and only applicable to cancers caused by viruses.
- *PLGA*: good anti-tumor effects seen in mice. Good biocompatibility and biodegradation. Good manufacturing and scale-up.
- *Gold nanoparticles*: may be coated with DNA, evaluated for vaccination against melanoma.
- *Magnetite*: stimulates tumor-specific T cell activity but non-biodegradable. Has been in clinical trials for melanoma treatment.
- *Gelatin*: readily available, easily manufactured, with functional groups for immunoglobulin binding, active tumor targeting and may be pegylated for better opsonization.
- *Nanoemulsions*: they can have long circulating times and good take-up by antigen presenting cells, possibly used for protection against Hepatitis B, HIV and influenza by a mucosal route.

highly specific targeting, such that most attention has been transferred to this modality.

One area where there is still significant interest is that of vaccine delivery. Developments with biodegradable microparticles for vaccines have continued over many years, and many additional challenges have arisen. These included the instability of antigens within polymeric nanoparticles, difficulties with consistent quality manufacturing, and high costs. Many of these difficulties have been resolved and some microparticulate vaccine systems are on the market or in clinical trial. Materials used in these systems include degradable polymers such as poly(lactide-co-glycolide) and chitosan. Vaccine delivery using nanoparticles and microparticles are summarized in Box 6.3.

## 6.4.2 Nanoparticles and drug delivery

The advances made in nanotechnology in recent years, and especially in the biological and medical applications of nanotechnology, often referred to as bionanotechnology, have had a much more significant impact than developments at the microscale. This has not happened simply by reducing the size of the drug-containing particles, but by introducing other nanoscale functionalities into these systems. We usually continue to use the word nanoparticle here but the systems are often far more complex and sophisticated than implied by the use of the simple concept of a small particle. Included within the conceptual framework of nanoparticles are carbon nanotubes, nanofibers, DNA nanomachines and self-assembled polymers. It should be noted at this point that much of the apparent excitement about this impact of nanotechnology on drug delivery has not yet been substantially translated into robust clinical procedures and many scientific challenges associated with the *in vivo* performance of nanoparticulate systems have still to be resolved.

In this section, the general principles of pharmacology at the nanoscale, often referred to as nanomedicine, are discussed, giving both advantages and challenges. Different types of system are in use or under consideration for use, in various types of application, the details of which are covered in later sections. There are several different words used to describe the individual units of these systems, but in general, from a scientific or clinical perspective, it does not help to be too prescriptive. There are nanoparticles, nanospheres, nanorods, nanocapsules, nanotubes and so on, the geometric form of which should be fairly obvious, but not necessarily their function. For our purposes in drug delivery systems, we have to recognize that the nanoparticle (in its overarching general sense) may be either a matrix in which the drug(s) is physically dispersed or a vesicle-like object in which the drug is contained in a cavity surrounded by a material membrane. In the former case, the matrix could be a solid polymer, a hydrogel, supramolecular assemblies such as micelles, dendrimers or fullerenes, solid magnetic ceramics or quantum dots. In the latter case the membrane could be a polymer (giving a hollow shell nanoparticle) or a carbon tube or a lipid bilayer (giving a liposome).

The main advantage of nanoparticulate systems lies with the relationship between their size and their biological fate. Nanoparticles can extravasate through the endothelium and epithelium so that they may be able to penetrate microcapillaries, inflammatory sites and tumors. The targeting of nanomedicines should be placed in context. There is both active and passive targeting, shown in Figure 6.6. With passive targeting, the carrier system is taken up in organs by virtue of size and surface charge. In active targeting there is an interaction between surface ligands and specific receptors. One of the factors that determines the efficiency of targeting is the degree to which nanoparticles intentionally functionalized to target receptors actually get taken up in non-target cells by non-specific interactions.

The small size of nanoparticles also means that they should be capable of intravenous delivery without fear of embolization in the capillaries. It is also possible for some nanoparticle preparations to cross the blood–brain barrier. Nanoparticles are taken up by cells far more readily than microparticles, which means that cell targeting is more efficient and more widely applicable. The fate of the nanoparticles, including the speed of clearance, is also dependent on their hydrophobicity; prolongation of the circulating time may be achieved by surface coating of the particles with hydrophilic polymers, such as PEG. As far as cell targeting is concerned, it is necessary for the surface ligand to engage with the receptor; the chances of such engagement may be quite small if the attractive forces are only exerted over a very small distance.

### Fate of nanoparticle formulations: clearance and endocytosis

Spherical particles less than 5 nm in diameter tend to be rapidly removed through renal clearance and extravasation. Above 10 nm accumulation may occur through storage in the spleen, liver and bone marrow; uptake by individual cells is strongly dependent on the cell type. Cells of the reticuloendothelial system and Kupffer cells in the liver are particularly effective in nanoparticle removal. The removal process itself is influenced by the surface modification that takes place as soon as the particle comes into contact with plasma. This modification occurs through protein adsorption by a mechanism very similar to the opsonization of bacteria, where antibodies attach themselves to bacterial surfaces in order to facilitate engulfment by phagocytic cells; in fact the term opsonization is often used to describe this phenomenon with nanoparticles. Immunoglobulin and complement proteins are important here. One of the key issues facing the use of nanoparticulate drug formulations is the need to avoid

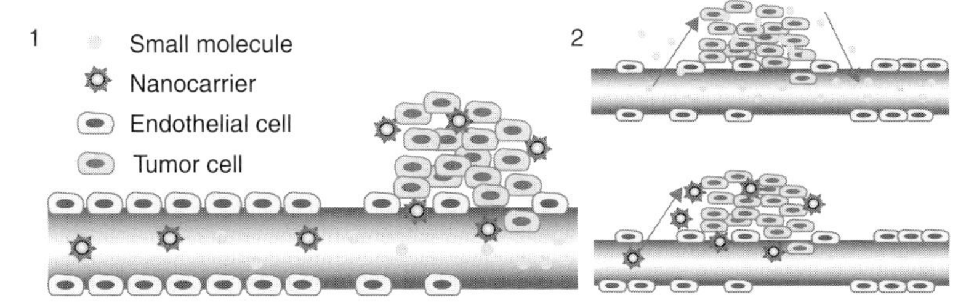


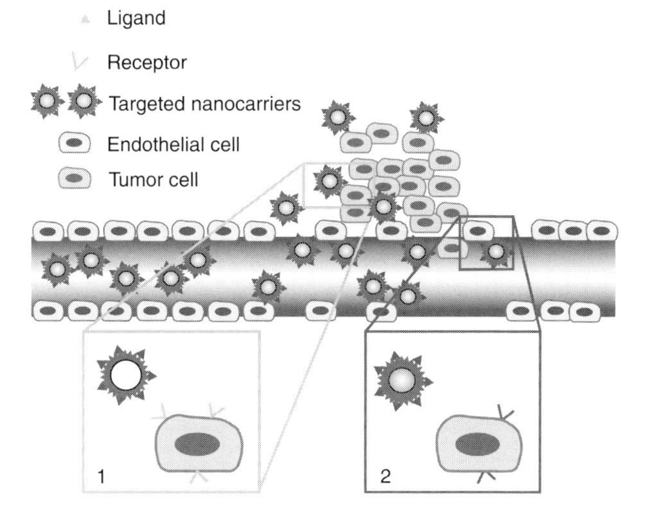


**Figure 6.6** Active and passive targeting of nanoparticles to tumor cells. This scheme shows (A) passive and (B) active targeting. In the former, drug nanocarriers reach tumors selectively through leaky vasculature surrounding the tumors (1). Drugs by themselves diffuse freely in and out of the tumor blood vessels and their concentration does not increase (2). Drug-loaded nanocarriers cannot diffuse back into the bloodstream because of their size, resulting in progressive accumulation, known as the Enhanced Permeability and Retention (EPR) effect. In (B), active targeting strategies are summarized as ligands are grafted at the surface of nanocarriers bind to receptors over expressed by (1) cancer cells or (2) angiogenic endothelial cells. Reprinted from *Journal of Controlled Release*, 148, Danhier F *et al.*, To exploit the tumor microenvironment: pressure and active tumor targeting of nanocarriers for anti-cancer drug delivery, 135–46, © 2010, with permission from Elsevier.

rapid take up and clearance by these cells but at the same time facilitate the entry into the target cells.

Nanoparticles are taken up by cells through the various processes of endocytosis, whereby a particle that contacts the exterior membrane of the cell can be actively transported through the membrane. Endocytosis is the general name for the passage of objects through the cell membrane; we have seen in Chapter 3 that "larger" objects gain cellular entry by phagocytosis. The small particles are internalized through the mechanism of pinocytosis. There are several stages, including the particle engulfment by membrane invaginations, which break away to form membrane-bound vesicles, which then deliver their cargo to specialized vesicular structures, which direct the cargo to various intracellular locations. There are several different intracellular pathways of endocytosis which, with respect to pinocytosis are classified on the basis of the endocytic proteins that opsonize the particles (Figure 6.7)

The so-called classical route for entry (known as CME or clathrin-mediated endocytosis) is dependent on clathrin, a fibrous protein of molecular weight 180 kDa, which forms a polyhedral coat around the particles. The process of endocytosis here involves engulfment in a pit in the membrane coated with clathrin, and the action of some assembly proteins to form vesicles that are pinched off by enzymatic activity. Some nanoparticle formulations follow this pathway, including the entry of cationic chitosan nanoparticles into the respiratory endothelium and silica nanotubes into cancer cells. The second main pathway is caveolae-mediated endocytosis, or clathrin-independent endocytosis, where caveolin, a low molecular weight protein, is found in the caveolae, which are lipid-based rafts in cell

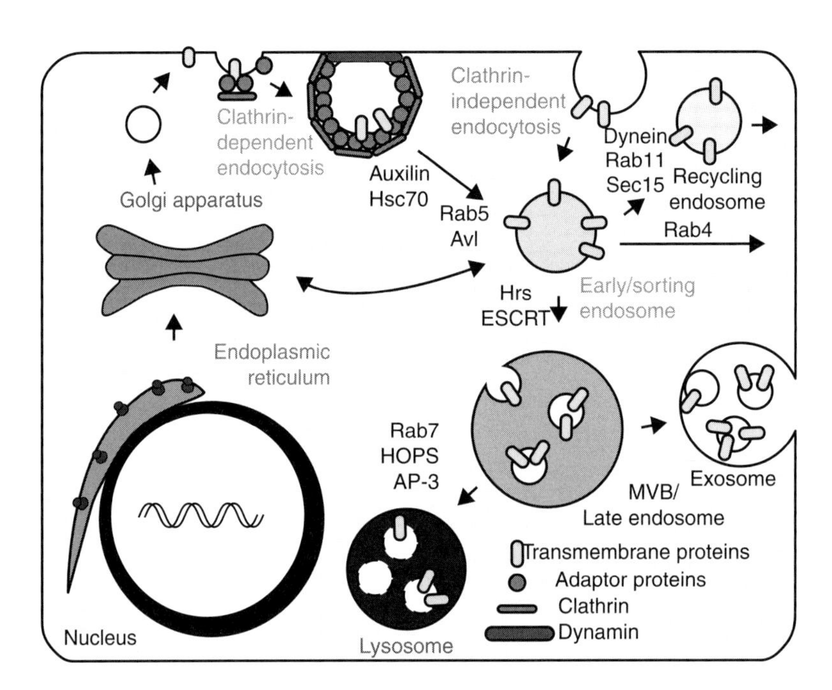


**Figure 6.7** Endocytosis. Endocytosis is a major process by which nanoparticles can be internalized; this mechanism is normally involved in the internalization of proteins. A wide variety of signaling pathways may be involved. Transmembrane proteins are made in the endoplasmic reticulum and traffic through the Golgi apparatus to reach the plasma membrane. From the cell surface, these proteins can re-enter the cell via various endocytosis pathways. In clathrin-dependent endocytosis, clathrin adaptor proteins, such as the AP-2 complex, recruit clathrin and cargo transmembrane proteins to the site of endocytosis. Endocytosis can also occur without clathrin referred to as "clathrin-independent endocytosis." After endocytosis, small GTPase Rab5 and SNARE protein Avalanche (Avl) mediate the fusion of endocytic vesicles with the early endosome. Endocytosed substances can recycle back to the plasma membrane directly or return back to the Golgi or travel to the late endosome and lysosome for degradation. Proteins or other substances destined for degradation are sorted into Rab7-positive late endosome or multivesicular bodies (MVB). In certain cell contexts, MVB can secrete their contents to extracellular regions. These secreted MVBs are referred to as exosomes. Finally, through HOPS and AP-3 complexes, MVB/late endosomes fuse with the lysosome and transmembrane proteins are degraded by proteases and acid hydrolases. Reprinted from *Current Topics in Developmental Biology*, 92, Yamamoto S *et al.*, Endocytosis and intracellular trafficking of notch and its ligands, 165–200, © 2010, with permission of Elsevier.

membranes. There are several isoforms of caveolin, including caveolin-1, which is found in the membranes of most cells.

## Liposomes

Liposomes were the first type of nanocarrier to be investigated and used clinically. They are artificial phospholipid vesicles of size ranging from 50 to 1000 nm, which may be loaded with a variety of drugs. In general they have low toxicity and immunogenicity, with good biocompatibility and biodegradability. They do, intrinsically, have the propensity to be rapidly cleared within the reticuloendothelial system, but, as implied earlier, their surface modification with PEG, and specifically the incorporation of PEG–lipid conjugates within the bilayer membrane, sterically stabilizes them against clearance. Whilst this does increase circulation times, the PEGylated liposomes may have difficulty in releasing the drug and mechanisms have been introduced to allow the PEG to remain during circulation but to detach from the lipid under intracellular conditions. Long circulating liposomes may be prepared with payloads of chemotherapeutic agents such as cisplatin, adriamycin and doxorubicin. They are also used for carrying amphotericin B in the treatment of fungal or parasitic infections such as cryptococcal meningitis and visceral leishmaniasis.

More accurate targeting of liposomes would be very helpful. Targeting agents may be directly bound to lipid anchors on the bilayer or attached via a linker. The targeting agents could be monoclonal antibodies, peptides, growth factors or glycoproteins and these may be attached via long chain PEG molecules. Much work has been undertaken here to enhance the delivery of the chemotherapeutic agents mentioned above. Antibodies against transferring receptors facilitate targeting tumors in which this receptor is over-expressed. It is far from clear that this type of targeting will have widespread use since the rate-limiting step for tumor localization is likely to be extravasation from the tumor vasculature and antibody–antigen couples would not be expected to influence this process.

In view of this, there has been more emphasis on engineering liposomes to be stimuli-sensitive, which can either respond to features that are characteristic of the targeted tissue, or be stimulated by some externally applied energy, for example ultrasound or electromagnetic fields. pH is the most common variable to be exploited here, since many target tissues, such as tumors and inflamed or infarcted areas are acidic relative to normal tissues. There are many examples of pH-sensitive liposomes that have been shown to release their contents in tissues or within cells after internalization. Both cisplatin and doxorubicin have been encapsulated in liposomes composed of the fusogenic lipid dioleoylphosphatidylethanolamine (DOPE) coupled with a pH-titratable carboxylate group. A low pH causes a neutralization of the excess negative charge of the carboxylate groups, resulting in a collapse of the lipid and release of these agents. The problem has been to increase circulation times and achieve a bioavailability that is consistent with clinical requirements. Better results are expected from liposomes that combine targeting with pH sensitivity, for example folate or transferring targets.

Temperature sensitive liposomes are the basis of hyperthermia therapy for tumors. Many of these liposomes contain dipalmitoylphosphatidylcholine (DPPC), which becomes leaky at a phase transition around 41°C. Alternatively the liposome may incorporate temperature sensitive polymers that have a lower critical solution temperature (LCST) that is just above normal body temperature, which can release a drug once the temperature rises above the LCST in tumor tissue. Again, achieving adequate bioavailability has been difficult and there are problems with the fact that such polymers are non-biodegradable. The same is true for magnetically sensitive liposomes; these can incorporate superparamagnetic iron oxide nanoparticles, which can be directed to tumors through the use of an extracorporeal magnet.

Examples of drug delivery with liposomes are given in Box 6.4.

## 6.5 Polymer therapeutics

The discussions in the previous sections have been based on the concept that the drug and the biomaterial are separate chemical entities, where the biomaterial contains and then releases the drug without there being any chemical change to the drug; the biomaterial is essentially a physiologically and pharmaceutically inert carrier. We now turn our attention to situations where this is not the case, and where the polymer actively plays a role in the chemistry of drug delivery. The materials involved, which include polymeric drugs, polymer–drug and polymer–protein conjugates, and polymer micelles covalently attached to drugs, are now collectively described as "polymer therapeutics." One very attractive feature here is that the clinical application of the developments of macromolecular drugs, including proteins and peptides, has been limited by the difficulty with effective delivery systems; polymer therapeutics offers massive potential for innovative delivery systems for these new drugs, especially those related to cancer therapy.

## Box 6.4 Liposome-based systems

A number of liposomal drug delivery systems have been introduced or tested in recent years. Some examples are listed below.

- *Morphine*: morphine sulfate may be encapsulated in liposomes and delivered through epidural injection for relief of post-surgical pain.
- *Amphotericin B*: several commercial formulations of liposomal amphotericin B are available for the treatment of fungal infections, particularly with some very resistant yeasts.
- *Cytarabine*: intrathecally delivered liposomal cytarabine is far more effective that cytarabine itself for the treatment of malignant lymphomatous meningitis.
- *Daunorubicin*: liposomal daunorubicin is given by infusion to treat patients with AIDs-related Karposi's sarcoma.
- *Doxorubicin*: pegylated liposomes containing the chemotherapeutic agent doxorubicin has been used for the treatment of tumors, where the liposomes release the drug after the application of heat following radiofrequency ablation.
- *Verteporfin*: age-related macular degeneration and other ophthalmological conditions may be treated by liposomal verteporfin.
- *Cisplatin*: the long-established chemotherapeutic drug cisplatin may be encapsulated within liposomes (lipoplatin) for the treatment of ovarian, bladder and testicular cancer.

A key feature of these polymers is that they are water-soluble; this immediately distinguishes them from traditional polymeric biomaterials and leads to both attractive features associated with conjugate chemistry and molecular morphology, and to new concepts in biocompatibility and toxicology. Typically the conjugates that have been developed have three domains, the polymer chain, the active component and a linker, shown in Figure 6.8(a). However, there are very many variations possible here, which include multiple active components and the incorporation of specific cell-targeting moieties and defined architectures. An alternative type of formulation involves the polymer chain itself having pharmacological activity. Some of these possibilities are discussed in relation to structures and properties and the clinical potential.

### 6.5.1 Polymeric drugs

Several polymeric materials have intrinsic pharmacological activity, including a variety of polysaccharides and synthetic polypeptides.

The intention here is the utilization of the high molecular weight and functional characteristics of polymers to selectively recognize, sequester and remove low molecular weight, disease-causing substances in the body, especially in intestinal fluid. They will not themselves be systemically absorbed, giving long-term safety profiles better than traditional small molecule drugs. They may also have multiple functional groups allowing polyvalent binding interactions. A powerful new class of therapeutic agents that can selectively bind and remove detrimental species from the gastrointestinal tract has emerged; this requires

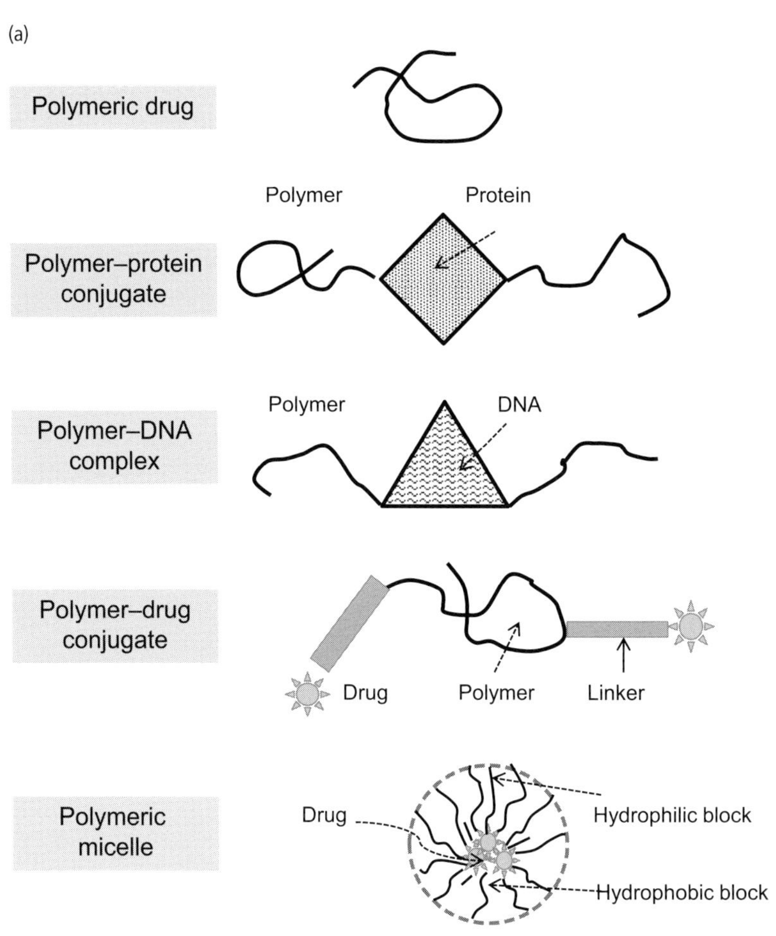


**Figure 6.8** Polymers as drugs.
**(a)** Polymer therapeutics. Representation of main situations where polymers and polymer–conjugates have pharmaceutical properties; these are known as polymer therapeutics. The molecule may be a polymeric drug, or the polymer can be conjugated with proteins, DNA or a drug. The polymer may have a micelle structure that contains a drug. Diagram based on publications of Dr. Ruth Duncan, e.g., *Advanced Drug Delivery Reviews 2013*, 65, 60–70.

a unique strategy which depends not only on the chemical nature of the target, but the location, concentration and quantity of the target to be removed at a therapeutically acceptable dose.

One early example of a clinically used polymer drug, sevelamer hydrochloride has been used in the treatment of hyperphosphatemia with effective, long-term control of serum phosphate levels in renal failure patients without exposure to toxic metal ions associated with other drugs. Bile acid sequestrants are cross-linked polymeric cationic gels that bind anionic bile acids in the GI tract and subsequently eliminate them from the body. They have a good safety record but poor patient compliance because of the high daily dose required.

Perhaps the most significant of the polymer drugs is that which is based on glatiramer acetate, derived from four amino acids, L-glutamic acid, L-alanine, L-tyrosine and L-lysine, which has a molecular weight in the range 5–9000 daltons. It is used in the treatment of multiple sclerosis.

## 6.5.2 Polymer–active agent conjugates

The main interest in polymer therapeutics has been with the conjugation of active molecules to polymers. The conjugates can be administered to the patient in a form which is inactive but the drug segment can be released under defined conditions in the body, when it is able to exert its pharmacological effect. These products have, as a minimum,

(b)

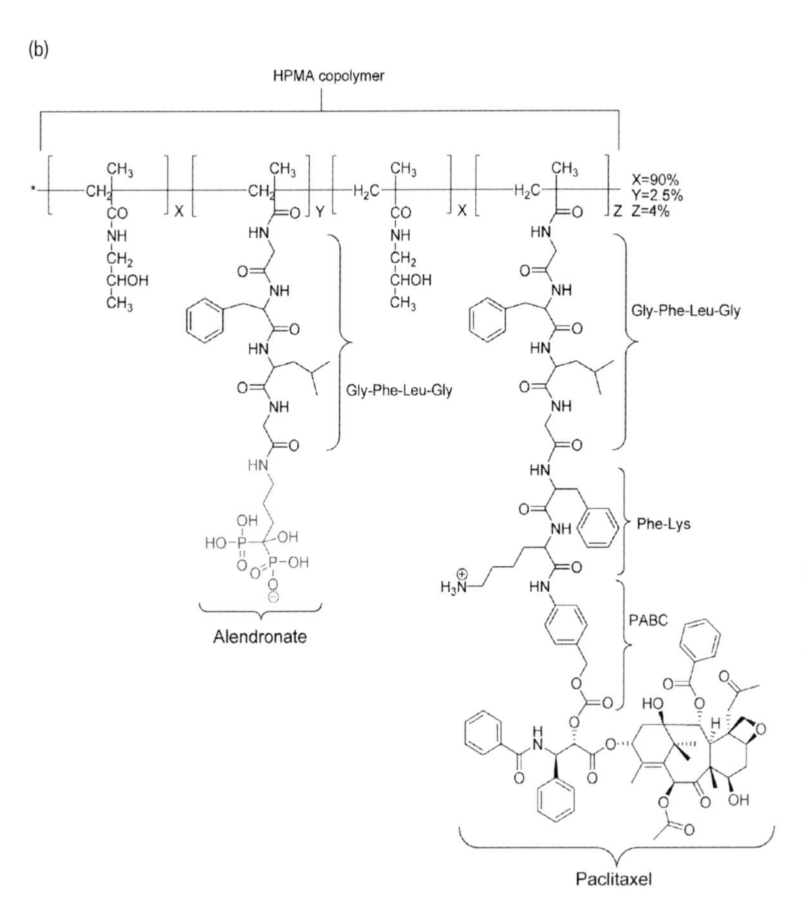


(c)

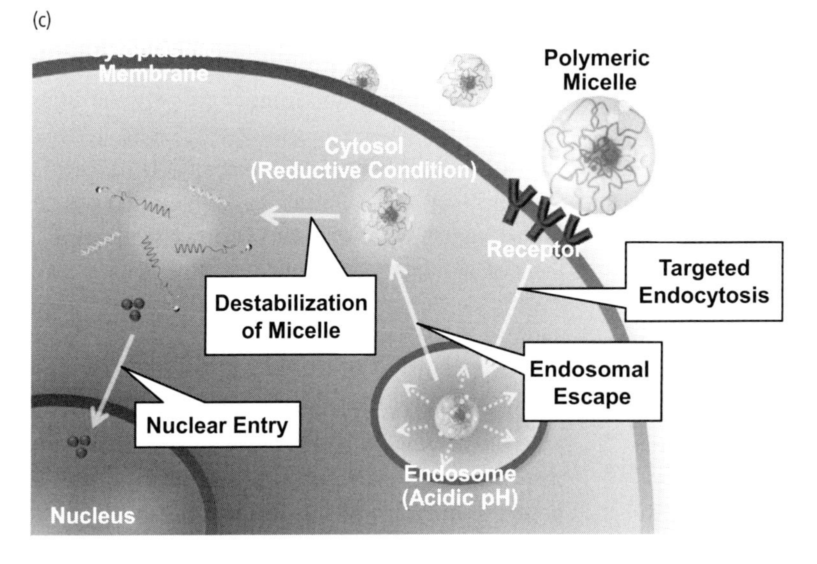


**Figure 6.8** *continued*
**(b)** Conjugation of chemotherapeutic agent to a polymer for targeted therapy This scheme shows the conjugation of a bone targeting moiety, aminobisphosphonate alendronate and the chemotherapeutic agent Paclitaxel to N-(2-hydroxypropyl)methacrylamide copolymer (HPMA), which can be targeted to prostate and breast cancer metastases. Reprinted from *Advanced Drug Delivery Reviews*, 61, Segal E and Satchi-Fainaro R, Design and development of polymer conjugates as anti-angiogenic agents, 1159–76, © 2009, with permission of Elsevier. **(c)** The use of polymeric micelles in drug and gene delivery and their passage through the cell (see previous discussion in Chapter 3). Image courtesy of Professor Kazunori Kataoka, University of Tokyo, Japan.

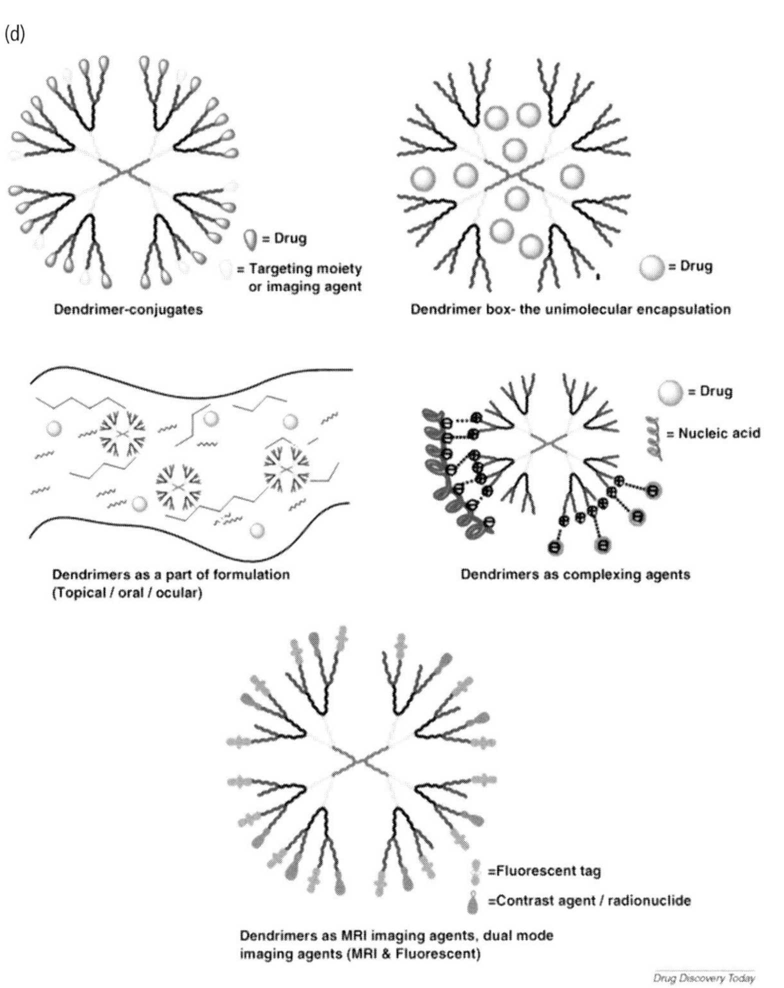


**Figure 6.8** *continued*
**(d)** Applications of dendrimers, including dendrimer-drug conjugates, drug encapsulation within the dendrimer interior and as carriers for imaging agents. Reprinted from *Drug Discovery Today*, 15, Menjoge A R *et al.*, Dendrimer-based drug and imaging conjugates, 171–85, © 2010, with permission of Elsevier.

the polymer, the drug payload and the linker, although other entities could be present, including imaging agents as will be discussed later in the chapter.

As noted above, there have been significant developments in the provision of macromolecular drugs, including those based on peptides, proteins and antibodies. They do have a few intrinsic problems, including immunogenicity in some cases, and poor stability in plasma. The first, and very successful, application of conjugation technology in this situation involved conjugation of proteins to poly(ethylene glycol), in the process that is now known as pegylation. This increases the solubility and stability of proteins. It is important that protein cross-linking does not take place during the conjugation, and various modifications to the pegylation process have been introduced in order to minimize this risk and to increase stability. Pegylation of proteins reduces immunogenicity and prolongs plasma half-life by preventing rapid renal clearance. Several of these conjugates are used clinically, including the use of interferon in the treatment of hepatitis C and PEG-conjugated to anti-TNF Fab antibodies in the treatment of rheumatoid arthritis and Crohn's disease. Other polymers used in conjugation include N-(2-hydroxypropyl)methacrylamide copolymers (HPMA), polyglutamic acid, dextrans and chitosans. The nature of the different polymer architectures that are possible are discussed below.

The linkers used in these systems are usually pH-sensitive molecules or peptidyl segments that can be cleaved by enzymes, including proteases found within tumors.

With low molecular weight drugs, the rationale for conjugation to polymer chains has been related to the need for specific cell or location targeting. This requires the careful design of the linker between polymer and drug, since this has to be stable at every stage of preparation and application, including transport in the body, up to the point where the conjugate arrives at the target site, when it has to be cleaved in order to provide an acceptable rate of release of the drug. This is especially attractive for the treatment of cancer and the potential for highly specific tumor cell targeting (Figure 6.8(b)). One of the most widely used polymers in this area is the HPMA mentioned above, often using a tetrapeptide, hydrazone or acetal linkage. HPMA–tetrapeptide systems have been used to conjugate chemotherapeutic agents such as doxorubicin and paclitaxel. Although the obvious advantage of such conjugates should be the reduction in toxicity associated with the free chemotherapeutic agent, the toxicity and general biocompatibility of the cleaved, non-biodegradable polymer has to be considered.

## 6.5.3 Polymer architecture

The available polymer architectures include linear, branched, hyperbranched and dendritic forms, these structures controlling the dynamic 3D shape of the molecule in solution. The simplest architectures may be below 20 nm in size, whereas some dendritic forms may be in excess of 100 nm. The latter are of classic nanoparticle dimension but they should be considered as supramolecular structures and not particles.

### Micelles

Of particular importance are micelle and dendrimer structures, details of which are given elsewhere in the book. Briefly, a micelle may be formed when amphiphilic molecules are dispersed in a liquid, when the hydrophilic head faces outwards of an essentially spherical structure and the hydrophobic tail stays in the central core (Figure 6.8(c)). Frequently the hydrophilic component is PEG, which tends to form hydrogen bonds with the surrounding medium, effectively protecting the core. The process of micelle formation in solution depends on concentration, and the critical micelle concentration (CMC) has to be exceeded. Even though this CMC is quite low for many systems, the concentration in blood after infusion may be below the threshold, leading to fast dissociation. Stabilization of micelles, for example by cross-linking of the core or the hydrophilic corona, is an important factor.

In addition to PEG, polymers such as poly(acryloylmorpholine), poly(trimethylene carbonate) and poly(vinylpyrrolidone) may be used for the corona. Block copolymers such as PEO-poly(amino acids) can be used to provide functional groups that are derivatized for targeting mechanisms. For the hydrophobic core there are several alternatives such as the biodegradable polyesters, poly(amino acids) and Pluronics. The polyesters include PGA, various PLAs and PLGAs, and polycaprolactone. Poly(amino acids) include poly(L-aspartate) and poly(L-glutamate). Pluronics include PEO-PPO-PEO block copolymers.

Drug loading of micelles may be achieved either by physical entrapment or chemical conjugation; the nature of the polymers used in the micelles should be adjusted to the hydrophilicity and solubility of the drug in question. For chemically conjugated drugs, the release of the drug from the micelle tends to occur through erosion or degradation of the polymer. Diffusion of the drug is the main mechanism of release from physically entrapped molecules. The relative sizes of the corona and core molecular segments and the presence of cross-links control the release rates. When used in chemotherapy, unmodified micelle drug conjugates will tend to accumulate in the tumor tissue, with drug release outside of the tumor cells, and passive diffusion into the cells. It may be possible to modify the surface of the micelles, for example with folate or antibodies, to enhance the possibility of receptor-mediated endocytosis.

#### Dendrimers

There are several similarities between micelles and dendrimers, but also some differences. Dendrimers have a well-defined topographical structure that has the form of a tree. In general, dendrimers are macromolecules that have three sections, a central core with two or more reactive groups, repeated units that are covalently attached to the central core and which are organized into radially homocentric layers, and functional groups on the outside. As with many micelles, PEG has been widely used as the core polymer and polyesters used for the tree structure (Figure 6.8(d)). Functionalization of the surface allows chemotherapeutic agents such as doxorubicin to be attached to the dendrimer. Dendrimers have the potential to be excellent drug delivery systems, and indeed some are in clinical use, because they can be produced with very high repeatability, well-established pharmacokinetics and pharmacodynamics, and versatile targeting opportunities. However, they can pose considerably toxicity profiles, which has limited their human use to some extent.

## 6.6 Cancer: chemotherapy and immunotherapy

The pharmaceutical treatment of cancer represents an excellent example of the difficulties of conventional drug delivery and the potential for the role of biomaterials in more effective therapies. Systemic chemotherapy has been the most important strategy for the treatment of cancer worldwide for several decades. However, this intravenous administration of potent anti-cancer drugs, especially at the optimal therapeutic/maximum tolerable dose, may lead to severe toxic effects. Moreover, it is quite an inefficient method since there are many physiological barriers to the delivery of these agents to solid tumors and through the interstitial spaces within the tumors. The main chemotherapeutic agents and the challenges with their delivery are shown in Box 6.5.

Many different strategies to improve the efficacy, safety and patient compliance of the drug treatment of cancer have been developed, many of which have been translated into clinical practice. These are indicated in Figure 6.9. In principle these involve the localized delivery of the unmodified drug into the tumor from a depot, minimizing the systemic distribution, or the targeting of a systemically administered but modified drug to the tumors cells.

### 6.6.1 Localized delivery of chemotherapeutic agents

We have already seen an example of the localized delivery process with the BCNU-polyanhydride wafers in the treatment of glioblastoma in Section 6.3.2. This approach offers some potential in other areas, and experimental work has involved injecting preparations into other tumors, but these have not been very successful as yet. They avoid excessive systemic distribution, but there is little control over the delivery characteristics other than by diffusion from a known drug concentration in the depot to the general area of the tumor.

### 6.6.2 Targeting chemotherapeutic agents to tumors

On the other hand, there has been major interest and success in the direct targeting of drugs to tumors. The techniques here involve either passive targeting, the active targeting to tumor cells, the targeting of the critical endothelial cells that control oxygen and nutrient supply to the tumor, or drug delivery to a tumor triggered by an external stimulus, as shown schematically in Figure 6.10.

## Box 6.5 Some existing chemotherapeutic agents

There are very many chemotherapeutic agents in use. Some of these are listed here according to their mode of action; see also Figure 6.9, which describes modes of action in terms of effects in the cell cycle.

- *Alkylating agents*: these include chlorambucil, given by oral delivery for the treatment of non-Hodgkin's lymphoma. Significant side effects on bone marrow suppression.
- *Anthracyclines*: includes some front-line agents such as daunorubicin and doxorubicin, the latter being given intravenously for many cancers, including hematological malignancies. Mechanism of action involves intercalation of DNA. Potential serious side effects of heart damage.
- *Cytoskeletal disruptors*: otherwise known as taxanes, being derived from the genus *Taxus*, or yew trees. Includes paclitaxel and docetaxel. Mechanism of action involves interference with the breakdown of microtubules during cell division.
- *Kinase inhibitors*: many examples, particularly imatinib, which inhibits the receptor tyrosine kinase, giving greater specificity. Effective against many cancers including chronic myelogenous leukemia.
- *Monoclonal antibodies*: also many examples, includes cetuximab, an inhibitor of epidermal growth factor receptors. Given intravenously for metastatic colorectal cancers and several others.
- *Nucleotide analogs*: includes fluorouracil, 5-FU, used for over 50 years in pancreatic and colorectal cancer, operates as thymidylate synthase inhibitor.
- *Peptide antibiotics*: includes bleomycin, which induces strand DNA breaks, and actinomycin which binds to DNA and inhibits transcription. Wide-ranging action, but side effects include lung damage.
- *Platinum agents*: include carboplatin, cisplatin and oxaliplatin, operate by cross-linking DNA. Serious side effects especially neurotoxicity.

### Passive targeting of tumors

This mechanism is largely associated with the so-called enhanced permeability and retention effect (EPR). Several clinically approved chemotherapeutic agents rely on this effect, including albumin-linked paclitaxel, pegylated and non-pegylated doxorubicin and micelle-based paclitaxel. EPR depends on the fact that solid tumors have a poorly developed and differentiated vasculature and they lack a functional lymphatic system. These two characteristics allow for a noticeable accumulation of certain drug formulations in the tumor, which is usually associated with nanoscale vehicles including the liposomes and micelles. There are some difficulties with this approach, mainly because of the considerable variation, from patient-to-patient and tumor-to-tumor, of the "leakiness" of the tumors.

### Active targeting of tumor cells

It would seem intuitively obvious that a preferential target for drugs would be the tumor cells themselves. The objective here should be to increase either or

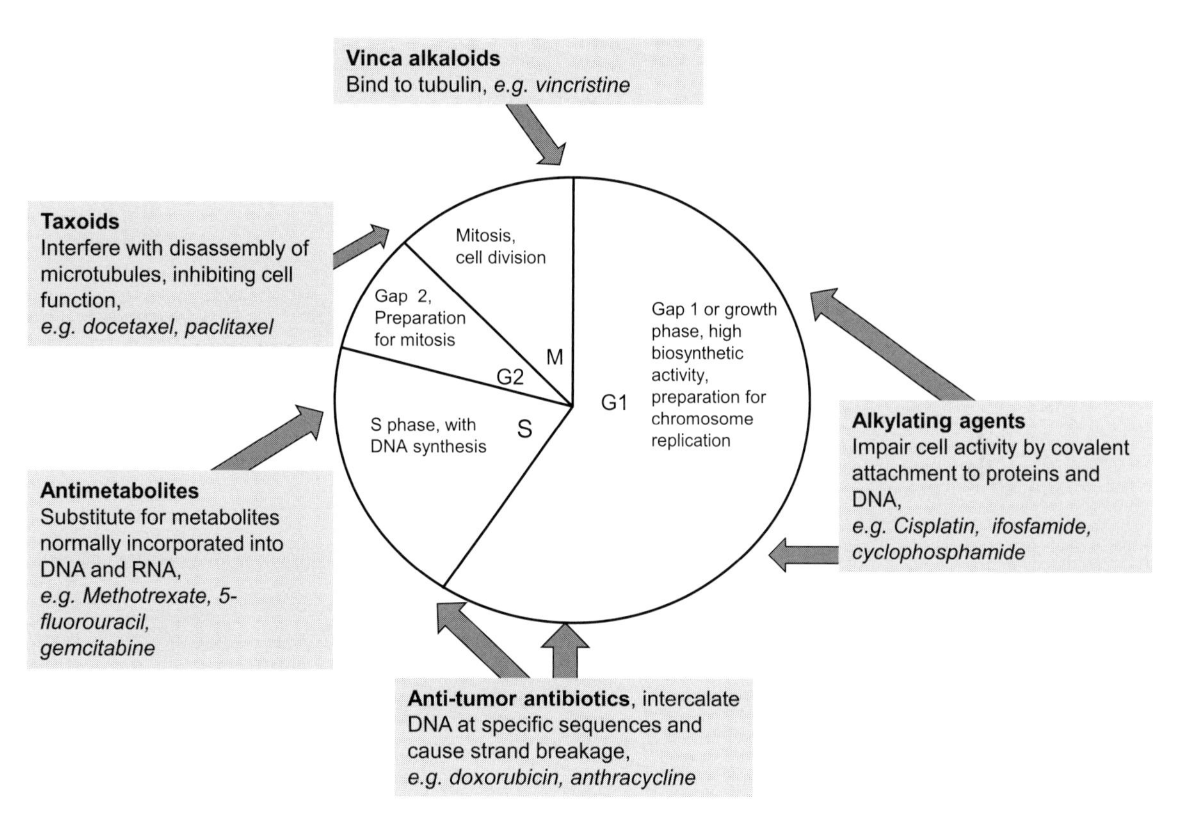


**Figure 6.9** Principles of chemotherapy. Chemotherapeutic agents may be classified on the basis of a number of features. A simple and effective method is to identify the effects of the drug on different parts of the cell cycle.

both the recognition of the target tumor cells by the drug entity and the uptake of the drug by these cells. Generally these would involve either peptides or antibodies that specifically bind to those receptors that are over-expressed by these cells. Ligands on the tumor cells that could be targeted include transferrin and folate. Although the principle is very sound, there are several barriers to entry of antibody- or peptide-targeted entities into the tumor cells, and such formulations are not yet widely used.

### Active targeting of endothelial cells

These barriers to finding and entering the cancer cells, largely associated with the high interstitial fluid pressure and high cell density within the tumor which both limit penetration, suggest that the endothelial cell may be a more realistic target. Some success has been achieved with this approach where antibody-modified low molecular weight drugs can bind to tumor blood vessels and be released into the tumor vasculature.

### Externally triggered drug delivery

The concept here is to create nanoscale structures that are themselves non-biologically active, but which can release their chemotherapeutic drug upon activation by an external stimulus. Such stimuli could include light, heat, magnetic fields and ultrasound. It is important that the trigger operates on a locoregional basis, meaning that the payload will only be released within the targeted area in the tumor.

### Nanomedicines for targeted delivery

All of the concepts outlined above require that the drugs are incorporated in some way within

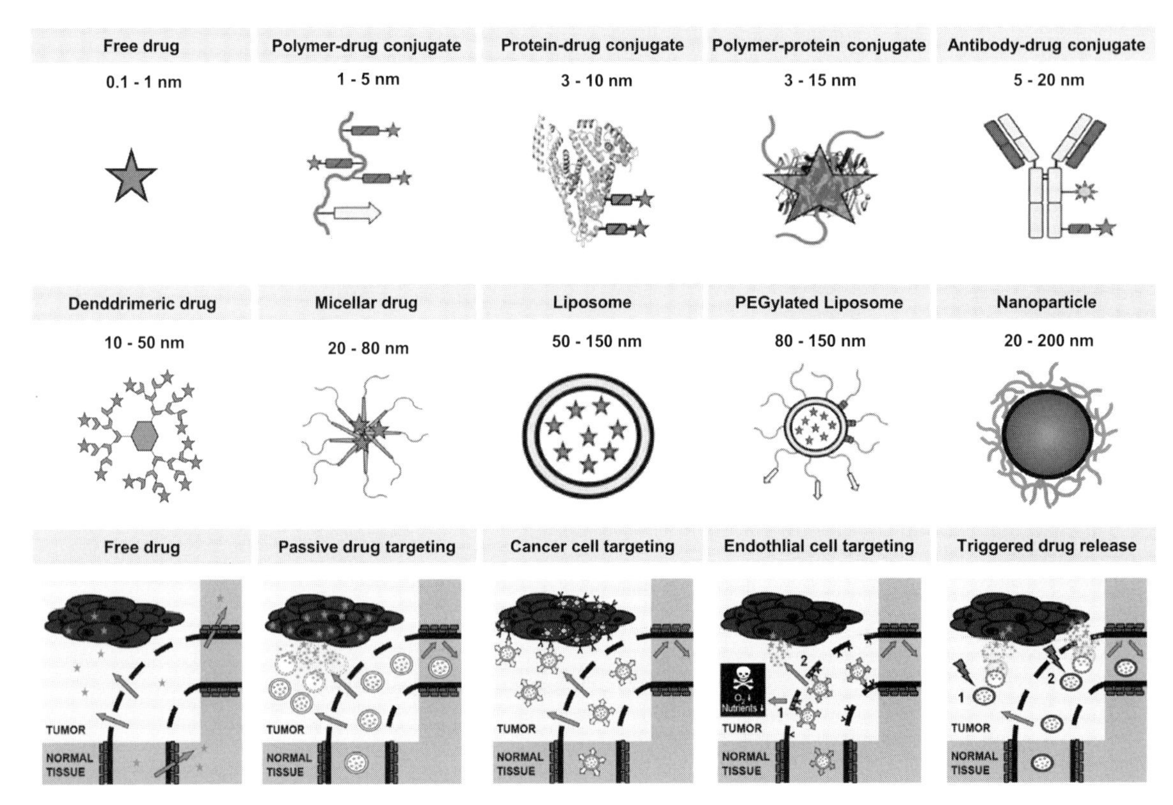


**Figure 6.10** Methods of tumor targeting. Schematic representation of ten forms of drug delivery systems and five methods by which they can be targeted to tumors. Image courtesy of Dr. Twan Lammers, Utrecht University, Netherlands.

a so-called nanomedicine formulation. The main categories of such nanomedicines are liposomes, polymer–drug conjugates, micelles, nanoparticles and antibody-linked drugs. Liposomes are probably the more advanced example here, with clinical use of a number of products, including doxorubicin-containing liposomes. A number of polymer-conjugated drugs are also in use, usually with PEG as the polymer. Paclitaxel has been prepared in micelle formulations and there are nanoparticle-based nucleic acid delivery systems.

## Cancer multidrug resistance

One unfortunate aspect of chemotherapy is that of multidrug resistance. Since tumors usually consist of mixed populations of malignant cells they will often show variations in drug sensitivity. Some cancer cells will be very sensitive to a drug, while others may be quite resistant. As the ratio of resistant to sensitive cells increases during treatment, the tumor may begin to grow again and the chemotherapy may fail to kill the most resistant. This resistance has a number of components. One of these concerns the molecular pumps in tumor cell membranes that can actively expel the drugs, allowing the cells to avoid the toxic effects within the cytoplasm and nucleus. A small number of proteins are associated with these molecular pumps, including P-glycoprotein, Pgp, and the so-called multi-drug resistance protein MRP, these being over-expressed in certain cells leading to the prevention of drug access. The fact that these proteins are only partially selective in terms of the drug molecules they affect does mean that any particular tumor may be resistant to multiple chemotherapeutic drugs.

As with other problems that require more effective delivery techniques, multidrug resistance has been tackled with a series of biomaterials-related potential solutions. Significant effort has been directed towards drug formulations that either evade the efflux effect or suppress the actual function of Pgp. Several Pgp inhibitor molecules have been developed, but they tend to lack the necessary specificity and efficacy when co-administered with the drugs.

Polymeric micelles provide an attractive approach since they may be able to bypass the Pgp-mediated efflux. Amphiphilic copolymers self-assembled in aqueous medium to form core-shell micelles may have the hydrophobic drugs contained in the hydrophobic core. These are potential carriers of water-insoluble drugs because of high solubilization capacity and stability, sustained release, prolonged circulation, and tumor localization by the EPR effect. Diblock copolymers such as PEG-b-PLA may be used to solubilize the hydrophobic anticancer agents paclitaxel and doxorubicin within micelles. Pluronic tri-block copolymers containing PEO and polypropylene oxide, PEO-PPO-PEO, also have lower cytotoxicity and weaker immunogenicity after topical and systemic administration. They are sterically stabilized and undergo limited opsonization and uptake by the reticuloendothelial system, allowing longer circulation in the blood. Nanoparticles also constitute a useful approach to overcome MDR. Carbon nanotubes enhance cellular accumulation of anticancer drugs by bypassing Pgp via endocytosis.

### Immunotherapy

One of the success stories in the treatment of some cancers concerns the use of immunotherapy, especially in breast cancer. This is not really a biomaterials story but one based on monoclonal antibodies. The humanized anti-HER2/neu monoclonal antibody trastuzumab may be used for the treatment of breast cancers that overexpress HER/neu. The drug is known as Herceptin and is very effective, with or without combination with chemotherapy.

Nevertheless there is a cohort of patients in whom there are serious side effects, and some develop drug resistance which renders the treatment ultimately ineffective. There is therefore a need to develop enhanced versions that allow high specific tumor targeting and accumulation in tumor cells without side effects. There have been attempts to use dendrimer conjugates for this purpose and also some biopolymers such as polymaleic acid that are able to deliver antisense oligonucleotides, which suppresses the proliferation of HER2/neu positive breast cancer cell lines.

### Photodynamic therapy

Photodynamic therapy, PDT, involves the use of an injectable photosensitizing agent that circulates in the body but is selectively retained in tumor tissue. At defined periods of time after injection, when the drug has been substantially cleared from normal tissues but remains in the tumor, the tumor is exposed to light of a specified wavelength. As the tumor absorbs the light energy it generates reactive oxygen species, which kill tumor cells. Since the light has a limited ability to pass through tissues, its use is confined to superficial tumors or those on the lining of internal organs or cavities.

There is currently a limited selectivity associated with PDT. There have been attempts, therefore to produce a more targeted version of PDT (TPDT), with well-defined targets. Some of these are passive and some are active methods. As seen elsewhere in this chapter, a passive approach involves the use of physicochemical characteristics of carriers, while active approaches involve some form of molecular recognition. In the former case, attention has been paid to both degradable and non-degradable biomaterials. Both liposomes and block copolymer micelles are included in the first group, while silica and gold nanoparticles are promising candidates in the latter group, with added advantages of being able to contribute to ROS formation. For active targeting, ligands that bind to receptors that are expressed within the target tumor are an obvious choice. Photosensitizers linked to peptides that have a high affinity for these receptors, aptamer-based nucleic acids and folic acid conjugated photosensitizers are under investigation.

## 6.7 Gene therapy and transfer

### 6.7.1 The concept of gene therapy

Genes are the basic physical and functional units of heredity and are carried on chromosomes. They are specific sequences of bases that encode instructions on how to make proteins. If genes are altered so that the encoded proteins are unable to carry out their normal functions, genetic disorders can result. Gene therapy is a technique for correcting defective genes responsible for disease development.

The most common approach to gene therapy involves the insertion of a normal gene into a non-specific location in order to replace a defective or non-functional gene. Alternatively, an abnormal gene could be replaced by a normal gene through homologous recombination, or it could be repaired through a process of selective reverse mutation, or it could be silenced. If a normal gene is inserted into the genome to replace an abnormal, disease-causing gene, it has to be delivered to the patient's target cells. The main problem here is that free oligonucleotides and DNA are rapidly degraded by serum nucleases if injected intravenously and indirect methods have had to be developed. Most attempts to do this involve a carrier molecule, which is called a vector, that unloads its protected genetic material containing the therapeutic human gene into the target cell. Here the therapeutic gene restores the target cell to a normal state.

Among the important properties required of a gene therapy vector, we can identify the following:

- The vector should be capable of targeting specific cells, especially when the relevant cells are widely dispersed throughout the body or part of a heterogeneous population. Ideally the vectors should totally avoid cells that mediate the immune response.
- There should be site-specific integration of the gene into the chromosome of the target cell or location of the gene in the nucleus such that it divides and segregates on cell division. It would also be helpful, however, if the vectors were capable of transfecting non-dividing cells such as neurons.
- Coding sequences of therapeutic genes vary considerably in size, and vectors should have no limit to the size which they can deliver.
- The vector should not elicit an immune response after delivery; this is particularly important where sustained high-level expression of the transgene is required.

There are two classes of gene therapy vector. By far the greater amount of attention has been paid to the use of viruses as vectors, which we shall discuss first. The alternatives are characterized as non-viral vectors and since these are more relevant to biomaterials science they will be dealt with in more detail. In addition there are a few methods for the direct introduction of therapeutic DNA into cells that try to overcome the degradation.

### 6.7.2 Viral vectors

Viruses have a genetic component that is required for their propagation. If these genetic components can be replaced by a therapeutic gene, then the modified, or engineered, virus can be used as a vector to deliver that gene to the target cells. It may be possible to achieve this in either *in vivo* or *ex vivo* procedures. In the former case the vector delivery can take place by either intravenous injection or localized implantation, most examples involving bone morphogenetic proteins as the target gene. Such methods in general tend to have relatively low transduction efficiencies and targeting difficulties, and risks of immune or inflammatory responses are quite high. With *ex vivo* approaches, specific populations of cells may be harvested from patients, genetic modification taking place *in vitro* and subsequently returned to a specific target site. Target cells could be stem cells (e.g., adipose-derived cells or bone marrow stromal cells) or fully differentiated cells such as fibroblasts or skeletal myoblasts. These techniques obviously allow for expansion of target cell populations and offer greater clinical safety. There are several different types of viruses that have been used, or are being considered, for gene therapy, summarized in Box 6.6.

### Retroviral vectors

A retrovirus is an RNA virus, meaning that it is composed of RNA not DNA. They have an enzyme, reverse transcriptase, that gives them the property of transcribing their RNA into DNA, this retroviral DNA then being capable of random integration into the chromosomal DNA of a host cell. Some retroviral vectors have been extensively characterized and used in clinical trials; these include the Moloney Murine Leukemia Virus. These have good, sustained transgene expression, and are relatively non-immunogenic, but they have poor specificity and cannot infect non-dividing cells. It is also known that they may integrate into some cells in a non-random manner, with a higher frequency in chromosomes that could lead to mutagenicity.

### Box 6.6 Summary of viral vectors

Viruses that have been considered as candidates for carriers in gene therapy and transfection can be grouped into a small number of categories.

- *Retroviruses*: viruses that can create double-stranded DNA copies of their RNA genomes, which can be integrated into the chromosomes of host cells. Human immunodeficiency virus (HIV) is a retrovirus. They usually only infect dividing cells, so that most applications are in *ex vivo* gene transfer. Direct administration to humans is usually ineffective. Some success has been achieved in the treatment of cancer-targeting T cell receptors in melanoma to give complete tumor regression. The subset of lentiviruses can transduce non-dividing cells but there are safety issues.
- *Adenoviruses*: viruses with double-stranded DNA genomes that cause respiratory, intestinal and eye infections in humans. Although they have a number of theoretical advantages, and can result in high levels of systemic gene transfer, experiences in humans are problematic, and they can result in severe toxicity at the levels required for clinical efficiency.
- *Adeno-associated viruses*: small, single-stranded DNA viruses that can insert their genetic material at a specific site on chromosome 19. These are easily prepared and purified but are very limited because of the small insertion site.
- *Herpes simplex viruses*: double-stranded DNA viruses that infect neurons. Herpes simplex virus type 1 is a common human pathogen that causes cold sores. They have the advantage of being able to infect a very wide range of cells, both dividing and non-dividing. They do have significant immunogenicity and toxicity problems. It is possible to engineer these viruses such that they can specifically infect tumor cells – giving oncolytic HSVs – which have been used in clinical trials for recurrent breast cancer, gliomas and liver metastases from colon cancer.

### Lentiviral vectors

Lentiviruses constitute a class of retrovirus that is able to infect non-dividing cells, which is very important for gene transfer since it means that they can transduce hematopoietic stem cells, neurons, hepatocytes and macrophages. They have a smaller risk of oncogene activation than the retroviruses discussed above since integration sites are more limited, but they still have safety risks as recombination events can lead to replication-competent viruses. The human immunodeficiency virus type 1 (HIV-1) is a lentivirus.

### Adenoviral vectors

These are capable of infecting both dividing and non-dividing cells. Unlike retroviruses, the genetic sequence that they inject into cells does not integrate into the host cell genome; thus they infect cells transiently and not constitutively, making them far less prone to insertional mutagenesis. On the other hand, there is a higher risk of host immunogenicity associated with the viral capsid proteins, which leads to rapid clearance by immune cells. There is usually, therefore, only brief transgene expression, which limits their applications.

### Adeno-associated vectors

Adeno-associated viruses are non-pathogenic single-stranded DNA viruses, which require extra genes to replicate. They are relatively non-immunogenic and are able to infect a wide variety of dividing and non-dividing cells.

### General comments on viral vectors

As noted, the rationale for using viruses as vehicles to introduce therapeutic genes into cells is very attractive. However, there are still significant practical problems, where many variables are difficult to control, especially in scale-up for routine commercial/clinical use, and there are several concerns for safety, especially insertional mutagenicity. Under the right circumstances, very high transfection efficiencies can be obtained, and hence interest in this approach is still strong, but the search for alternatives, with equally high efficiency and improved safety, is of prime importance.

## 6.7.3 Non-viral vectors

Non-viral vectors have been investigated as a group of alternative systems to address the dilemma with viral vectors. These involve a variety of chemical species that are able to interact with DNA and transport it to the relevant location in the target cell. There is one fundamental problem with this approach: whereas viruses have evolved strategies to overcome cellular barriers and immune defense mechanisms, non-viral vectors do not inherently have these capabilities, and hence there are major problems with accurate and efficient delivery. The concept behind non-viral vectors is centered on the negatively charged phosphate backbone of DNA and the requirement to develop cationic compounds that can complex with the DNA, leading to charge neutralization and compaction of the nucleotide fragment. The nature of the complex will naturally depend on the characteristics of the cationic compound; among the classes of compounds that have received extensive examination here are lipids, polymers (including dendrimers), polypeptides and nanoparticles.

Gene transfection may be achieved *in vitro*, where the objectives are to deliver the DNA–vector complex to the cell membrane, to facilitate the passage of the vector across that membrane and then through the cytosol to the nucleus. The complexes first attach to the cell membrane by electrostatic interactions with surface anionic proteoglycans and are then internalized, shown in Figure 6.11. The passage through the cell is determined by the various endocytic routes. Generally the endocytic pathway for the absorption of molecules and complexes into cells involves three main components. Early endosomes are vesicles that are located near the periphery of the cell and have a mildly acidic pH. These transfer internalized matter into the late endosomes, where the pH is around 5.5; here molecules and particles are sorted and passed into the lysosomes. These are 1–2 μm vacuoles at pH 4.8 and contain active lysosomal hydrolases and

(a)

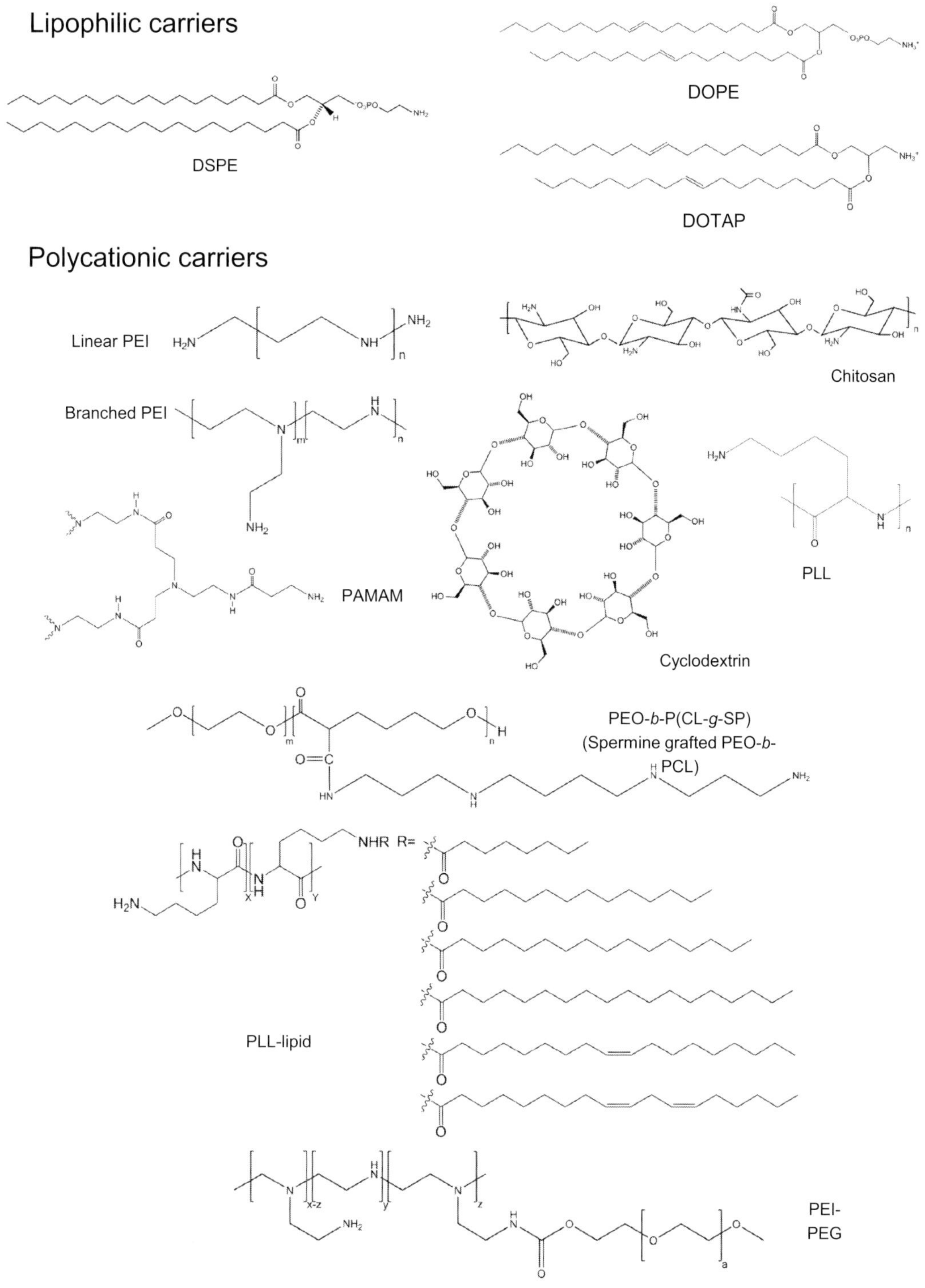

Lipophilic carriers

DSPE

DOPE

DOTAP

Polycationic carriers

Linear PEI

Chitosan

Branched PEI

PAMAM

Cyclodextrin

PLL

PEO-*b*-P(CL-*g*-SP)
(Spermine grafted PEO-*b*-PCL)

PLL-lipid

PEI-PEG

**Figure 6.11** Non-viral gene vector.

**(a)** A representative range of carriers used in delivery of gene-based medicines derived from DNA and RNA molecules. The carriers have been used on their own or as the base of further formulations with improved delivery functions. Abbreviations are: DOPE: Dioleoyl phosphatidylethanolamine; DOTAP: 1,2-dioleoyl-3-trimethylammonium-propane; DSPE: 1,2-distearoyl-sn-glycero-3-phosphoethanolamine; PAMAM: Poly(amidoamine); PEI: Polyethylenimine; PEO-*b*-PCL: poly(ethylene oxide)-block-poly(ε-caprolactone); PLL: Poly-L-lysine. PLL-lipid: lipid-substituted PLL; PEI-PEG: Poly(ethylene oxide)-substituted PEI. Image courtesy of Dr. Hamidreza M. Aliabadi and Dr. Hasan Uludag, University of Alberta, Canada.

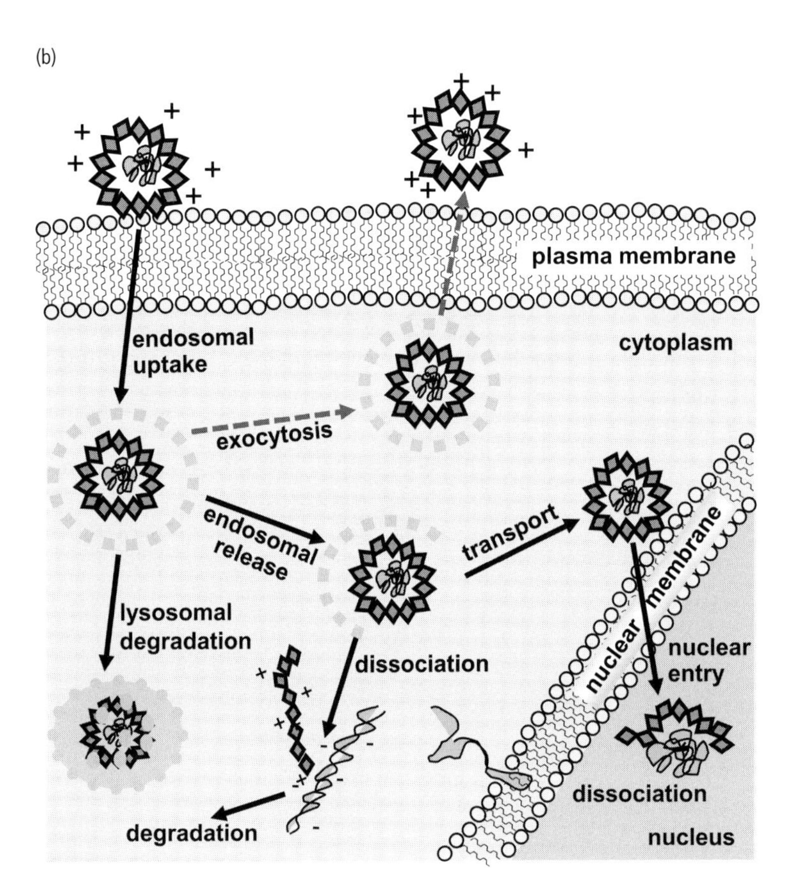


**Figure 6.11** *continued*
**(b)** Schematic of main trafficking pathways for polyplexes. Image courtesy of Dr. Vanessa Incani and Dr. Hasan Uludag, University of Alberta, Canada.

other enzymes. The normal operation of the endocytic pathway is directed towards the breakdown of various macromolecules and returning them to the cytoplasm. This, of course, is precisely what is not required with a DNA–vector complex that has to deliver its payload to the nucleus. Various methods of facilitating so-called "endosomal escape" to prevent the degradation of therapeutic agents during this process have been developed. For example, a protein sponge effect occurs when agents with a high buffering capacity and the ability to swell when protonated are incorporated into the delivery system. The protonation induces an inflow of water and ions into the endosomal environment, which ruptures the membrane and releases the molecules trapped inside. Poly(amidoamine) polymers are able to do this because of the protonated amine groups in their structure. Once in the cytoplasm, the vectors have to direct the DNA towards the nucleus, where degradation and disintegration are still possible.

Obviously to gain access to the transcriptional process in the nucleus, the DNA must cross the nuclear membrane. Trafficking across this membrane takes place through pore complexes within the membrane, involving passive diffusion for very small agents, and either ATP-dependent processes or nuclear localization for larger species.

## Lipid-based vectors

One of the earliest, and indeed one of the most widely explored, groups of non-viral vectors were those based on lipids. The lipid N-[1-(2,3-dioleyloxy) propyl]-N,N,N-trimethylammonium chloride (DOTMA) was shown capable of delivering genes to cells in the 1980s. We have already discussed liposomes in relation to drug delivery, and it should not be surprising that liposomes have played a part in gene vector development. They were particularly attractive since liposomes could fuse with cell membranes, and are capable of delivering DNA, chromosomes and mRNA to certain cells. This process of liposome-mediated gene delivery became known as "lipofection," and there are a number of commercially available agents.

Liposomes can have an anionic, cationic, zwitterionic and non-ionic character depending on the molecular structure of the polar head-groups. As expected, entry of the polyanionic-based gene vectors through the negatively charged cell membranes is inefficient, but cationic liposomes can form an overall positively charged complex with negatively charged DNA, the resulting lipid–DNA complexes, usually known as lipoplexes, being endocytosed by the cell plasma membrane. Cationic liposomes also protect DNA from attack by DNAses, and, being designed to compact the DNA, there is favorable intracellular movement and destabilization of endosomes. The mechanism of gene transfer with lipoplexes appears to follow the general mechanism outlined above, with incorporation into the cell by endocytosis and destabilization of the endosomal membrane. This involves reorganization of the phospholipids, which neutralizes the lipoplex and causes the DNA to dissociate into the cytoplasm.

The cationic liposomes have three structural domains, the cationic head-group itself, a hydrophobic domain and a linking domain, all of which play a part in the transfection efficiency and are capable of chemical manipulation (Figure 6.12). For example, simply replacing the ammonium group of monovalent cationic lipids can increase efficiency quite considerably. Multivalent head-groups may also be beneficial compared to monovalent ones, as can the use of branched and hyperbranched groups, and dendritic amphiphiles. Increased efficiency appears to be associated with higher surface charge density and improved buffering capacity. The length and type of aliphatic chain on the hydrophobic tail group also affects the performance of lipoplexes, generally gene transfer increasing with reduced chain length. Other manipulations include the introduction of alternative hydrophobic groups in place of the aliphatic chains. Molecules such as various cholesterol derivatives may alter the rigidity, biodegradability and fusogenic character, which may influence gene transfer. The linkers used in lipoplexes include amides, esters and carbamates, which vary in their stability; this is important for controlling the release of DNA after endocytosis.

There is a degree of cytotoxicity associated with cationic lipids, which is largely controlled by the nature of the polar and hydrophobic domains. Of the main hydrophobic moieties, aliphatic chains and

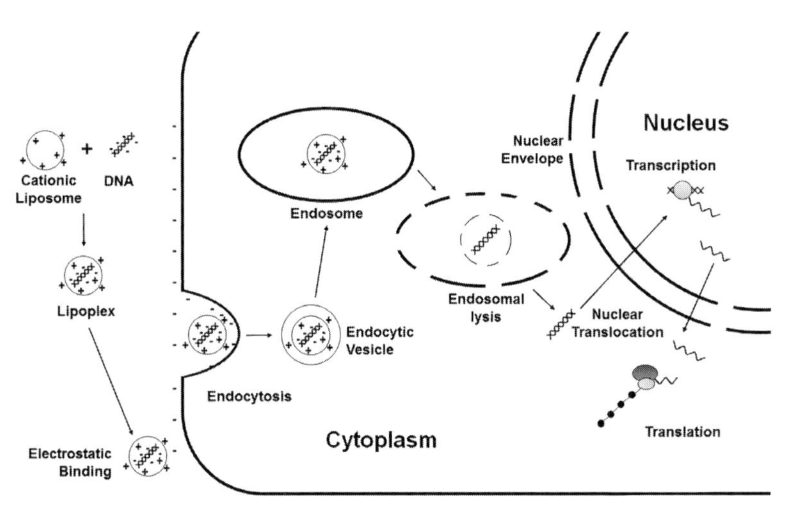


**Figure 6.12** Mechanism of cationic liposome-mediated gene delivery. The scheme shows the binding of a cationic liposome with DNA through electrostatic interactions to form a lipoplex, which is internalized by the process of endocytosis. This follows a route within the cytoplasm as an endocytic vesicle with endosomal escape via lysis and translocation through the nuclear envelope and transcription within the nucleus. Image courtesy of Bryant C. Yung and Robert J. Lee in the College of Pharmacy, The Ohio State University.

cholesterol derivatives, the latter are frequently inhibitors of protein kinase C, which gives rise to toxicity, while the former may also display toxicity, especially those with single-tailed lipids. With the polar domains, the head-groups are often ammonium salts, which tend to control toxicity, which is especially seen with quaternary ammonium amphiphiles. The degree of toxicity is related to the charge ratio between the cationic lipid and the nucleic acid and the particle size of the complexes. In general, the toxicity is manifested as an inflammatory response associated with the interaction of the liposome with immune cells, the activation of reactive oxygen species and interference with apoptosis and the caspase pathway. Also problematic is the relatively short-term expression, and applications that require sustained levels of transgene expression over many months would seem too difficult. There have been attempts to combine liposomes with biomaterial scaffolds including collagens, in order to prolong the duration of expression.

### Polypeptide vectors

Conjugates involving peptides and oligonucleotides are very attractive vectors that can have high cell specificity. The mechanism by which they operate utilizes short sequences of basic amino acid residues that are readily able to cross cell membranes. The peptide is usually covalently linked to the oligonucleotide rather than electrostatically complexed. It is also possible to complex with DNA and siRNA, the former by electrostatic methods and the latter by either. There are several types of amino acid sequence that can be used, these generally falling into the category of lysine-rich or arginine-rich peptides.

One prominent example of the lysine-rich class is the MPG peptide; multiple peptide molecules may interact with a single oligonucleotide strand to form aggregates. The hydrophobic regions of the peptides fold into ß-sheet structures, which facilitate cellular uptake by forming a pore-like structure in the cell membrane. A major example of the arginine-rich peptide is the TAT peptide, a well-known cell penetrating peptide. The efficiency achieved with the TAT peptide has not been high, partly because it has much lower capability with endosomal escape mechanisms than with cell membrane penetration. As a result, attention has turned to ternary complexes involving both TAT and liposomes conjugated with DNA.

### Polymeric vectors

A large number of polymers have been explored for their gene carrying ability; a few of these are now well characterized and used, but additions are made to the list on a regular basis. We shall discuss some representative examples here, with an overall summary given in Box 6.7. The complexes formed with nucleic acids are generally referred to as "polyplexes." The complexation with nucleic acids is again achieved electrostatically and polymeric gene vectors are cationic. These polymers do not readily form discrete sub-classes, but they do have a number of features in common. Since they do not contain hydrophobic moieties as with the cationic lipids, they are usually completely water-soluble. They are able to condense DNA molecules into a small size, which is beneficial for transfection efficiency. They are also easily manipulated and modified, with respect, for example, to molecular weight, degree of branching and ligand attachment, so that wide variations in properties are possible. Transfection efficiency is variable but not so good as achieved with viral vectors, and cytotoxicity is still a major concern.

**Polyethyleneimine (PEI)** PEI was one of the first cationic polymers to be investigated and is commercially available for gene transfer applications. Polyethyleneimine exists in both linear and branched forms, the former finding their way to commercial application before the latter. The transfection efficiency of PEI, which is relatively high for polymer vectors, is due to the proton sponge effect mentioned earlier, the buffering capacity associated with its very high level of protonation allowing the PEI to avoid lysosomal trafficking. The efficiency of PEI may be

## Box 6.7 Polymeric gene vectors

This list gives the main forms of non-viral vectors that have use or potential use in gene transfer or therapy.

- *Lipid-based vectors*: includes those where there are cationic head-group manipulations, those with hydrophobic tail-group manipulations, those with manipulations of linking groups. Also includes solid lipid nanoparticles.
- *Polymeric vectors*: includes polyethylenimine, poly(L-lysine), polymethacrylates, carbohydrate-based polymers such as cyclodextrin, chitosan and dextran, linear poly (amido-amine) and biodegradable polymers.
- *Dendrimer-based vectors*: such as polyamidoamine, poly(propylenimine) and poly(L-lysine) dendrimers.
- *Polypeptide vectors*
- *Nanoparticles*: includes quantum dots, gold nanoparticles, silica, carbon nanotubes and hydrogels.

increased by binding it to poly(ethylene glycol), which creates a hydrophilic exterior and reduces interactions with plasma proteins, and by the introduction of selective amine functionalization.

PEI vectors do have cytotoxicity issues. It is generally considered that two types of cytotoxicity mechanism are involved. The first is immediate toxicity associated with free PEI, mediated through membrane destabilization. Delayed toxicity occurs once the PEI is released during intracellular processing of the PEI/DNA complex. Attempts have been made to reduce the latter effect by incorporating biodegradable elements into the PEI chain, for example polycaprolactone segments or other acid-labile linkers.

**Poly(L-lysine) (PLL)** PLL was also an early contender for an effective cationic polymeric gene carrier. PLL molecules with a molecular weight above 3000 daltons efficiently condense DNA to form polyplexes. Although the efficiency increases with molecular weight, so does the toxicity. PLL structures also have a problem with endosomal escape since all the amino groups are already protonated and there is no buffering capacity. Various biodegradable polylysine conjugates have been developed in an attempt to improve efficiency and decrease toxicity.

**Polymethacrylates** Poly[2-(dimethylamino) ethyl methacrylate] (PDMAEMA) shows good transfection efficiency and acceptable cytotoxicity. The polymer is able to destabilize endosomes and to dissociate easily from the plasmid. However, the process of transfection is interfered with under a number of clinically relevant scenarios, being negatively affected by hyaluronic acid and albumin, and having a tendency to accumulate in the lungs. Various modifications to the polymeric structure may alleviate these effects, by, for example, the incorporation of additional tertiary amino groups to promote the protein sponge effect and balancing different functionalities to improve endosomal escape. Biodegradable varieties have been prepared with hydrolysable cationic side chains, including those with carbamate functionality. Improved efficiency and reduced toxicity may be achieved.

**Dendrimers** Polyamidoamine dendrimers (PAMAM) have some of the highest transfection rates of all non-viral vectors. The dendrimers have several features that promote both cellular delivery and endosomal release of DNA. The dendrimers induce lipid mixing associated with the ability of their spherical structure to alter the shape of the anionic membrane, promoting cellular penetration. In addition, it is possible to conjugate the structure to dextran through disulfide linkages that significantly improve endosomal swelling and lysis. Surface modification with arginine can also improve gene transfection efficiency and cell-specific uptake has been achieved by incorporating residues such as mannose or galactose onto cyclodextrin-PAMAM conjugates. Dendrimers based on polypropylenimine and poly(L-lysine) have also been investigated.

**Carbohydrate polymers** Several carbohydrate polymers have good potential as gene carriers, with most attention given to dextran and chitosan. Various derivatives of dextran have shown high efficiency coupled with low toxicity, including quaternary derivatives of dextran–spermine conjugates. Chitosan polyplexes have varying efficiency depending on molecular weight and degree of deacetylation (see Chapter 8). Generally higher molecular weight polymers entrap DNA better than low molecular weight polymers because of chain entanglement effects. Higher degrees of deacetylation give better transfection through higher stability associated with increased charge density. The performance of chitosan derivatives varies from cell to cell type depending on the presence of chitosan degrading enzymes within endosomal compartments. Low toxicity is usually seen with chitosan derivatives although this can vary; higher efficiency may be seen with trimethylchitosan derivatives but at the expense of higher toxicity. Chitosan may also be conjugated with cell-targeting ligands to allow delivery to specific cell types.

**Nanoparticles**
Several different types of nanoparticle have been used for gene transfer. Among the early candidates were gold nanoparticles: these can be reacted with cationic molecules to facilitate cellular entry via the endocytic pathway. Conjugation with PEI produces high efficiency. DNA–gold nanoplexes may be prepared with labile bonds that are cleaved when the complex is in the cytosol, for example by using photolabile thiol ligands. Colloidal silica nanoparticles are also attractive as they are easy to manipulate and have low toxicity. More recently, much attention has been given to quantum dots and carbon nanotubes. Several types of quantum dot, including CdSe/ZnS dots, can be covalently conjugated to plasmid DNA using a variety of encapsulating agents, which can be readily internalized and localized around cell nuclei, where they release the DNA. Both single-walled and multi-walled carbon nanotubes may be used as gene carriers. They are inherently insoluble but can be functionalized, either covalently or non-covalently to render them soluble. Ammonium functionalized nanotubes form cationic structures that bind to DNA by the usual electrostatic mechanisms. As well as DNA transfection, these carbon nanotubes may be functionalized to complex with siRNA, with good success at inhibition of target genes in certain cells, showing considerable potential for anticancer therapy.

## 6.8 Vaccines

The innate and adaptive immune systems have been discussed in Chapter 3. It will be recalled that there are both B-cells and T-cells in the adaptive immune system, which give antigen specificity and memory to the immune system. B-cells produce antibodies upon antigen recognition while T-cells recognize antigen-derived peptides of the antigen-presenting cells. Once activated the T-cells differentiate into effector T cells, their immunological role being supplemented by the effects of other white blood cells. Both B-cells and T-cells are able to remember previously encountered antigens, facilitating a rapid, robust response upon further exposure, illustrated in Figure 6.13. This phenomenon is utilized in the process of vaccination where deliberate exposure to weakened pathogens or purified components of antigens, provides immunity in the case of future exposure to that antigen.

Although highly successful in many ways, vaccination is not always effective, and not always safe. There are many highly dangerous pathogens, including the HIV viruses and many of the causative factors in tropical diseases that are difficult to safely vaccinate against. Recombinant vaccines have been produced in cases where live vaccines are too risky but they usually have low immunogenicity. There is therefore a need to develop safer and more effective vaccines, and here attention has been paid to biomaterials-enhanced systems.

A critical factor concerns the role of dendritic cells, also mentioned in Chapter 3. Dendritic cells are able to prime and modulate the effector T cell response. An important strategy in improving the performance of vaccines involves enhancing the ability of dendritic cells to do this. Substances that enhance or modulate the intrinsic immunogenicity of an antigen are known as adjuvants. The few adjuvants that have been used clinically are only marginally effective; part of the strategy therefore focuses on increasing the effectiveness of adjuvants through their interaction with dendritic cells, either by directly activating them or increasing antigen uptake and presentation by these cells. In addition, increased effectiveness may be produced by delivery of antigens within nano- or micro-particulate systems.

In general, particles within the size range between 50 nm and 5 μm can be internalized by dendritic cells. The selection of particles for this purpose follows the same types of pattern as seen with other delivery systems. The main interest has been in liposomes, emulsions, polymeric carriers, virus-like particles and some minerals.

## 6.9 Antimicrobial agents

A wide variety of chemical agents are used to control the effects of disease-causing microorganisms on humans. Because there are several different types of microorganism, which have their own distinct characteristics, there are several different forms of these agents. In general we can describe these as antimicrobial agents. When the pathogen is bacterial, the agents are collectively described as antibacterial, other categories primarily including antifungals and antivirals.

### 6.9.1 Antibacterial agents and antibiotics

All antibacterial agents interfere with the growth and reproduction of bacteria. In general, non-specific antibacterials are generally used as environmental disinfectants aimed at reducing the potential for bacteria to encounter humans. These may include consumer products such as those used in skincare, but not (usually) as medicines. Clinically used antibacterials are generally described as antibiotics.

For over 80 years hundreds of antibiotics have been used to control infections. While it is fairly straightforward to find chemicals that kill bacteria, it is much more difficult to find those substances that kill bacteria whilst having no adverse effects on the host. After all, bacteria are cells and human tissue contains cells, so we should expect there to be potential overlap between their effects. It is true that there are significant differences between bacterial cells, which are prokaryotic, not containing a nucleus, and mammalian tissue cells, which are eukaryotic; a major factor in the development of effective pharmaceutical preparations has been the ability to exploit these differences. Particularly antibiotics have been directed towards attacking metabolic pathways found in bacteria but not in humans. Table 6.1 gives a list of common antibiotics.

The majority of antibiotics have not needed complex delivery systems for their clinical use. However, in recent years a significant problem has arisen with

**I Subviral particle as vaccines**

Single viral protein
(low immunogenicity)

Subviral particle
(high immunogenicity)

Stabilized subunit particle
(high immunogenicity)

**II Polyvalent complexes as vaccines**

Fusion of two
dimeric viral proteins
(low immunogenicity)

Linear polyvalent complex
(High immunogenicity)

Fusion of three
dimeric viral proteins
(low immunogenicity)

Network polyvalent complex
(High immunogenicity)

**III Subviral particles as vaccine platforms**

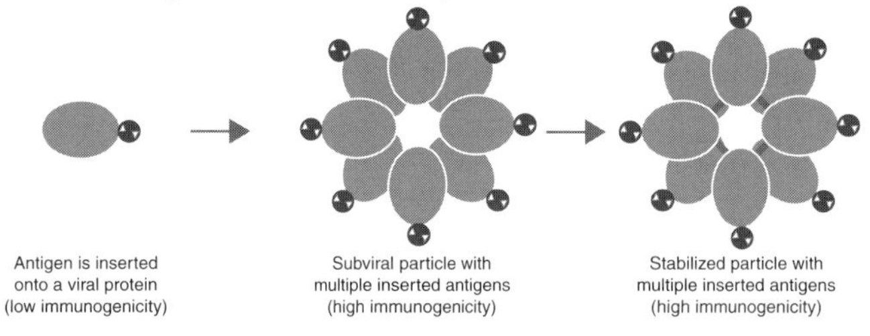


**IV Polyvalent complexes as vaccines**

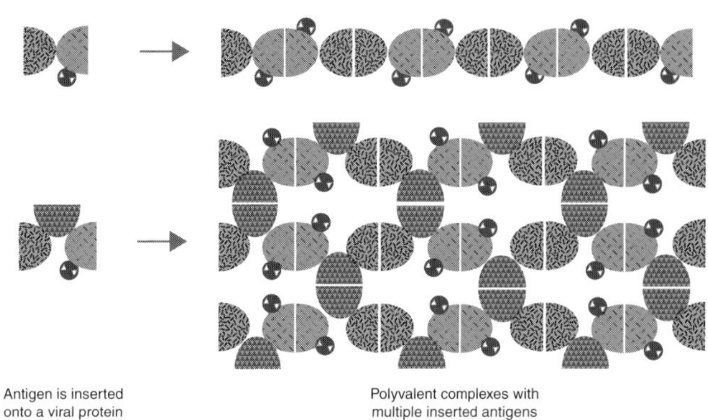


**Figure 6.13** Biomaterials and vaccines. Schematic illustration of biomaterials used in vaccine preparations. (I) Subviral particles as vaccines. Many viral structural proteins form subviral particles spontaneously when they are produced *in vitro*, including virus-like particles, capsid-like particles or other polyvalent particles or complexes. Such particles can be further stabilized by modification of the viral protein for increased intermolecular interactions (indicated by the bars). (II) Polyvalent complexes comprising viral antigens. Many truncated viral structural proteins can interact homotypically, forming dimers. When two or three such proteins are fused into one molecule, they will form linear or network polyvalent complexes through intermolecular dimerization. All polyvalent particles or complexes exhibit significantly higher immunogenicity than that of the monovalent viral protein, which is an important feature to be a vaccine. (III and IV) The subviral particles and the polyvalent complexes can be utilized as vaccine platforms for antigen presentation. An antigen (dark grey ball) can be inserted onto a viral protein. Together with the viral protein, this will become polyvalent in the assembled particles (III) and linear/network complexes (IV), by which immunogenicity of the inserted antigen and the viral protein will be significantly increased compared with the free antigen and the viral protein. Scheme courtesy of Dr. Ming Tan, Cincinnati Children's Hospital Medical Center.

Table 6.1 **Antibiotics. The main antibiotics are listed on the basis of their mechanism of action**

| Mechanism | Antibiotic | Notes |
|---|---|---|
| Cell wall synthesis inhibition | Penicillin, e.g., penicillin G and V aminopenicillins, methicillin | Effective against *Streptococcus pyogenes* and *agalactiae*. Problems with hypersensitivity reactions |
| | Cephalosporins, e.g., cefazolin and cefotaxime | Effective against *Staphylococcus aureus* and *epidermidis* and some gram-negative strains |
| | Vancomycin | Effective against MRSA |
| | Beta-lactamase inhibitors | Effective against *E.coli* |
| | Carbapenems | Very broad activity |
| | Bacitracin | Effective with topical gram-positive infections |
| Protein synthesis inhibition at 30s ribosomal subunit | Gentamicin, neomycin | Effective against aerobic gram-negative strains Prone to nephrotoxicity |
| | Tetracyclins | Effective in respiratory, urinary and genital tract infections |
| Protein synthesis inhibition at 50s ribosomal subunit | Macrolides, e.g., erythromycin | Used for bronchitis, diphtheria, Legionnaires' disease, pneumonia, rheumatic fever and ear infections |
| | Lincosamide, clindamycin | Effective for coagulase-negative Staph and Strep infections |
| DNA synthesis inhibition | Fluoroquinolones, e.g., ciprofloxacin and moxifloxacin | Effective against Strep infections, *Pseudomonas* and many gram-positive strains |
| RNA synthesis inhibition | Rifampin | Used to treat tuberculosis and meningitis. Some hepatotoxicity |
| Folic acid synthesis inhibition | Sulfonamides | Used to treat urinary tract infections and diarrhea |

so-called antimicrobial resistance, similar in some ways to the cancer multidrug resistance mentioned before. This mainly occurs when individual patients fail to comply with the instructions that they should complete each prescribed course of treatment, but then stop taking the drugs when they appear to be healthier. In such a scenario, many bacteria are killed, but the more robust bacteria survive, leading to the emergence of stronger and stronger surviving strains. Collectively this means that some bacterial strains become more resistant to the drugs, and in many cases some previously standard antibiotics now have limited power. One widely discussed situation here involves MRSA (methicillin-resistant *Staphylococcus aureus*), a bacterium that is highly infective, with initial infections being skin-based but rapidly invading other parts of the body, which is now resistant to antibiotics such as penicillin, amoxicillin and oxacillin.

While the problem is one of drug compliance, attempts to mitigate the situation have involved biomaterials-based formulations that provide extended release profiles such that reliance on patients taking tablets regularly is much less important. At its simplest this involves modifying the matrix of the dosage form to prolong residence time in the intestines (for example, using mucoadhesive formulations) or modifying the antibiotic molecule to increase serum half-life. The more complex systems involve microparticles, nanoparticles, coatings and polymer–drug conjugates. Ampicillin may be contained within microspheres. Nanoparticle systems include porous PLGA-rifampicin, oleoyl chitosan-doxorubicin, PEG/PLGA-teicoplanin and TAT/PEG/cholesterol micelles-ciprofloxacin. The development of the nanoparticle systems shows much promise but there are significant challenges with obtaining the optimal drug load and acceptable production costs.

We now give a few examples of specific biomaterials-based antibacterial agents.

### Antimicrobial peptides

Antibacterial hydrogels can be used to deliver small molecule antibiotics to tissue. Alternatively the material itself can be designed to be the antibacterial agent since surfaces can be endowed with antibacterial properties. This may be accomplished by covalently immobilizing known antibiotics on, or attaching silver nanoparticles or quaternary ammonium groups to, their surfaces. Materials that have polycationic surfaces may also be active against a broad spectrum of both gram-positive and gram-negative bacteria through a mechanism that involves bacterial membrane disruption.

Of importance here are the antibacterial hydrogels prepared from lysine-rich, self-assembling β-hairpin peptides, shown in Figure 6.14. These peptides assemble into a polycationic fibrillar network capable of killing bacteria by cell lysis. The fibrillar networks can be varied by modulating the amino acid composition of the peptide monomer. Antimicrobial peptides, AMPs, are small, water-soluble peptides that fold into amphiphilic conformations such as helices and β-sheets, which display opposing hydrophobic and polycationic surfaces. The cationic face engages the negatively charged surface of the bacteria's membrane via hydrogen bonding and electrostatic interactions. Once bound to the outer-leaflet of the membrane, the hydrophobic face of the peptide facilitates its insertion into the lipid part of the membrane, eventually disrupting the membrane and causing cell death. The polycationic surfaces usually contain many arginine residues whose side chain guanidinium groups can establish strong hydrogen bonds and salt bridges with the membrane surface of bacteria.

### Liposomes

Liposomal antibiotics have been developed in order to improve the pharmacokinetics and biodistribution of the active molecules, to decrease their toxicity to the host and to enhance and target activity towards intracellular pathogens. Liposomes are usually recognized as foreign antigens and taken up in the mononuclear phagocyte system, leading to low circulation times and rapid blood clearance. This uptake is dependent on liposome characteristics such as size, charge, rigidity and fluidity. A coating of the liposomes by PEG will give a hydrophilic surface layer which results in much longer circulation times, and

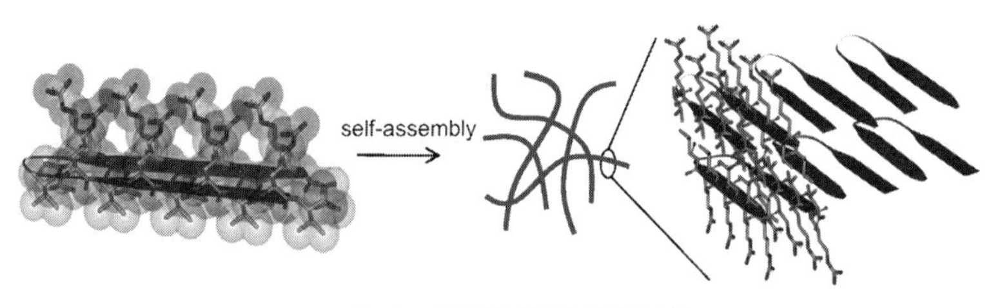


**Figure 6.14** Antimicrobial peptides. This shows the amphiphilic structure of the PEP8R hairpin peptide that self-assembles into β sheet rich fibrils that are incorporated into an injectable hydrogel. The PEP8R sequence is shown at the bottom. The hydrogel has significant antibacterial properties when injected *in vivo*. Reprinted from *Biomaterials*, 33, Veiga A S *et al.*, Arginine-rich self-assembling peptides as potent antibacterial gels, 8907–16, © 2012, with permission of Elsevier.

when these liposomes are stearically stabilized there may be selective accumulation in specific tissue sites.

Although of considerable potential, the clinical applications of liposomal antibiotics have been slow to emerge. In particular, these preparations tend to have a short shelf life and also there may be significant drug leakage from the lipid vesicles. Sterilization of products is also an important issue since liposomes are sensitive to heat, irradiation and chemical agents. Among the more successful liposomes are those that contain gentamicin, which is active against *Escherichia coli*, triclosan, which is active against oral streptococci, and vancomycin, which is active against *Staphylococcus epidermidis* and *Staphylococcus aureus*, both very common pathogens.

### Drug carriers for tuberculosis

Tuberculosis (TB) is a good example of an infectious disease that is readily curable but is a major problem because of poor patient compliance; it is a prominent target for more efficient delivery systems. It is very prevalent and highly contagious, especially in South East Asia and Africa; once considered to be virtually eradicated through vaccines and antibiotics, it now affects over ten million people worldwide. It involves infection with *Mycobacterium tuberculosis*, where the bacilli readily spread through the vascular and lymphatic systems. For those infected, there is a range of antimicrobials available, often delivered in combination. First-line agents include rifampicin and aminoglycosides, which inhibit bacterial RNA and protein synthesis.

The effectiveness of these agents may be improved by passive carriers such as liposomes, degradable polymers, microparticles and nanoparticles. Lecithin liposomes containing streptomycin or gentamicin are amongst those that have been investigated. PLGA microparticles have been loaded with rifampicin and poly(n-butylcyanoacrylate) nanoparticles with streptomycin. Rifampicin has also been incorporated in a variety of micelles, including those of PLLA and PDLA. As with other situations in this hugely important area of biomaterials-based systems to address antibiotic resistance, it is taking time to deal with all of the efficacy and safety issues, this obviously being necessary before widespread clinical acceptance.

### Nanosilver

It has been known for centuries that silver, either solid or in compound form, has antimicrobial activity. This has been used to good effect in many consumer/environmental situations and also in health care, for example with the coating of indwelling catheters. There has always been a discussion about the balance between antimicrobial effects and toxicity to human cells.

The effects of silver on bacteria are thought to be associated with the disruption of ATP production and DNA replication following uptake of free silver ions, by the generation of reactive oxygen species by silver and by direct damage to bacterial cell membranes. Interest in silver has increased significantly with the introduction of nanoscale formulations, primarily nanoparticles of metallic silver of size range 5–50 nm, the products being colloquially known as nanosilver, shown in Figure 6.15. Although many uses of nanosilver have been discussed and pursued, there are still concerns about general silver toxicity and also about the bioavailability of the effective silver ions because of their association with chloride ions within many media, and also complexation with proteins such as albumin. The silver may be stabilized, for example with dendrimers or within zeolites. Nanosilver products have been found effective against *E.coli*, *S. aureus*, *Pseudomonas aeruginosa* and many other bacterial strains and also against a number of fungi and algae. There is also some evidence of effects on the replication of some viruses.

## 6.10 Combinations of systems

We consider here a few situations where health care products involve a combination of components, one or more of which are drugs. The main groups here are

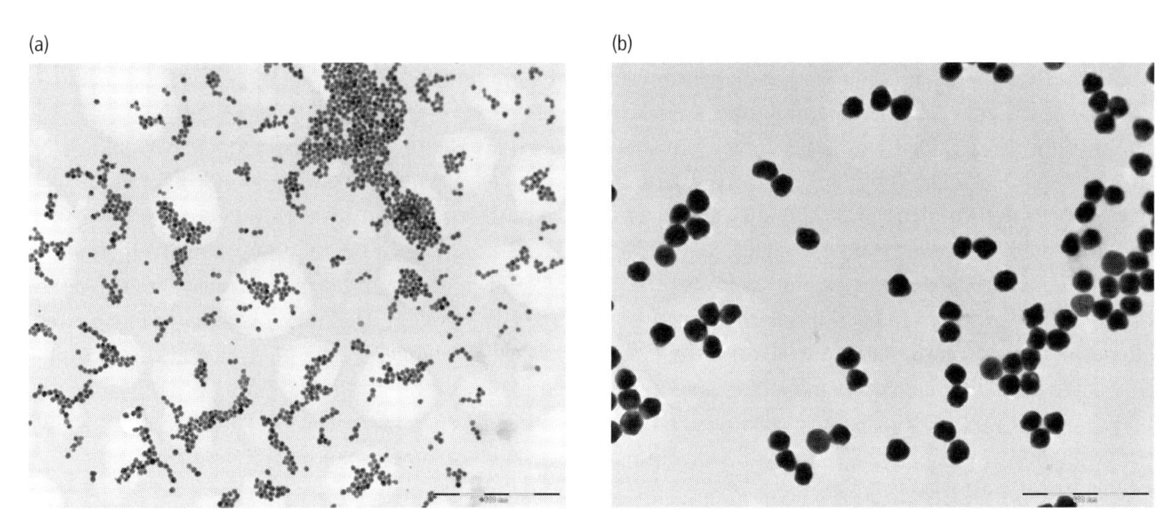


**Figure 6.15** Silver nanoparticles (nanosilver). Transmission electron micrographs of (a) 20 nm and (b) 80 nm nanosilver, showing single and loosely agglomerated nanoparticles. The nanoparticles were obtained from nanoComposix, San Diego,CA, USA, images courtesy of HM Braakhuis, MVDZ Pak, J-A Post, and Dr. Wim H De Jong. Laboratory for Health Protection Research, National Institute for Public Health and the Environment, Bilthoven, The Netherlands.

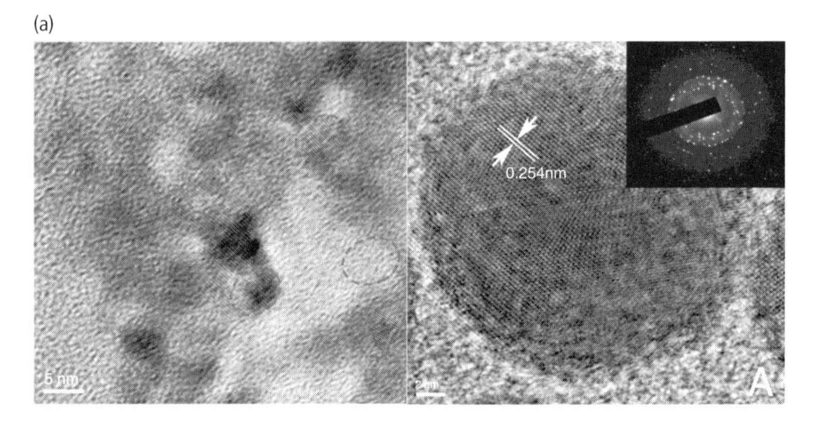


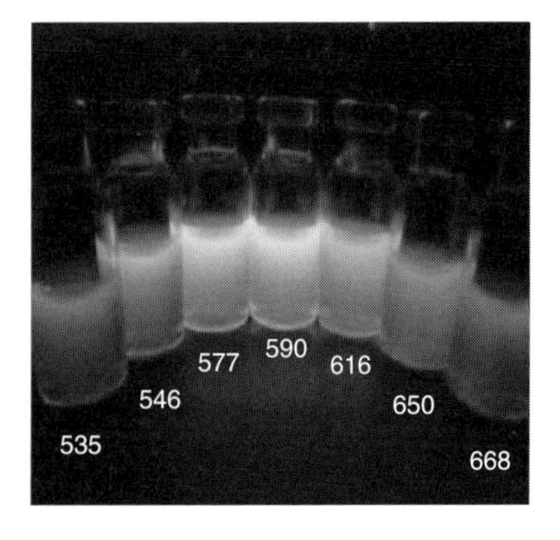


**Figure 6.16** Theranostic agents. **(a)** Herceptin–quantum dot conjugates. Upper, high-resolution transmission electron micrographs of HER2–Quantum Dots. The HER2 monoclonal antibody is conjugated to CdTe quantum dots via molecules of the enzyme ribonuclease A. Depending on the size of the resulting nanoparticles, they emit light at different frequencies, shown in the lower panel. Images courtesy of Professor Daxiang Cui, Department of Bio-Nano Science and Engineering, Shanghai Jiao Tong University, China. See plate section for color version.

the so-called theranostics and combinations of drugs with medical devices.

## 6.10.1 Theranostic agents

Theranostics, a word generated by a combination of therapy and diagnostics, means just that, a product that incorporates both a drug delivery mechanism and an imaging/diagnostic agent. This concept moves us closer to personalized medicine where therapies are directed towards individual patient features that are themselves made known to the physician by an integrated imaging agent. Not surprisingly, this possibility largely depends on manipulation of components at the nanoscale. Much of this work has been directed towards cancer. Cancers are very heterogeneous so that chemotherapeutic agents are effective only for a subset of patients, and then only at selective stages of the disease. If imaging agents can be associated with the chemotherapeutic agent, then it may be possible to use signals from the former to control the release of the latter. We shall only discuss this briefly here since we have not yet discussed imaging systems, which are covered in the following chapter.

Key factors here are the ability to match the activity of both components within the same structure, ensuring that sufficient amounts can be accumulated in the tumor, and the type of attachment between the two parts and the release of the drug when desired. Attachment may be achieved by direct

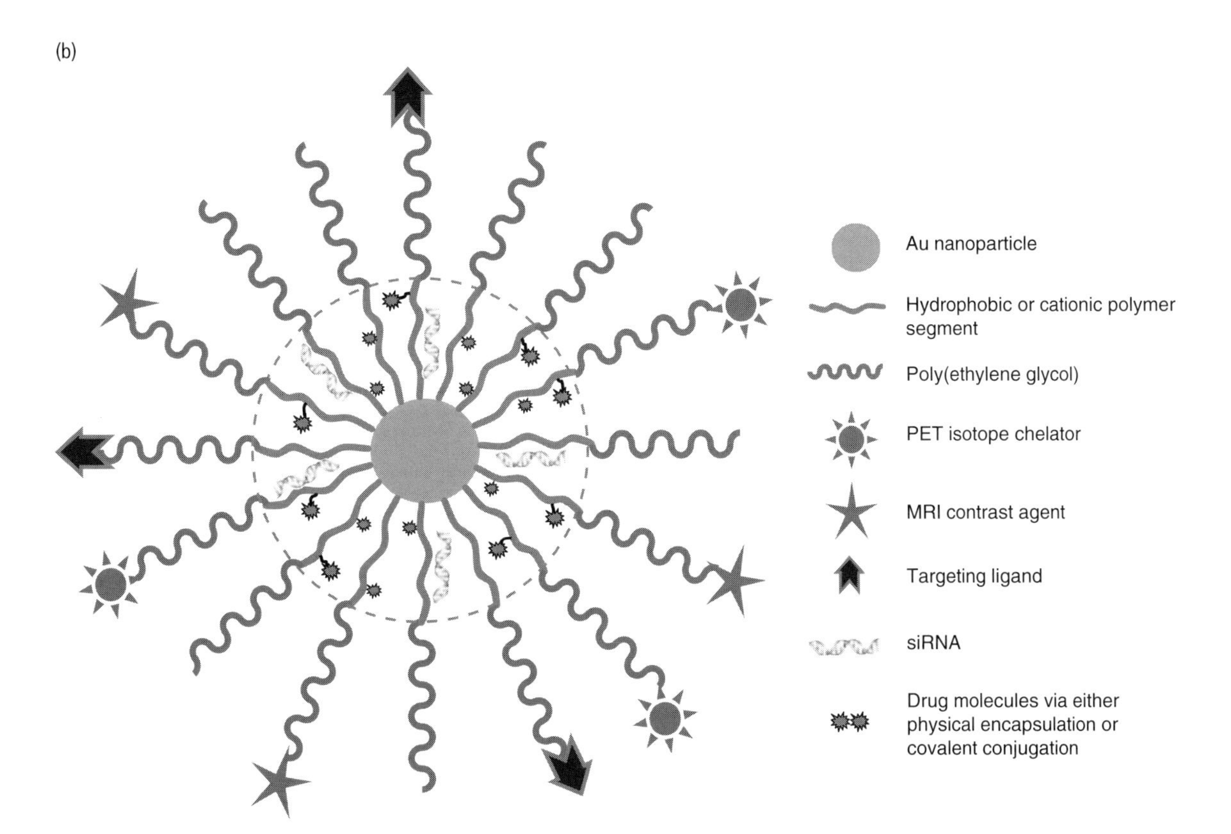


**Figure 6.16** *continued*
**(b)** Multifunctional theranostic gold nanoparticles. Gold nanoparticles used for targeted drug/siRNA delivery and imaging. Image courtesy of Virendra Gajbhiye and Shaoqin Gong, University of Wisconsin-Madison.

bioconjugation or by electrostatic interactions. A small number of nanoscale structures have been investigated.

#### Quantum dot-based nanostructured theranostic agents

Quantum dots (QDs) have been discussed in other parts of this book. They have considerable potential here, although there are significant challenges with the cytotoxicity of many QDs and their intrinsic water-insolubility. One of the attractive options is that of QD–Herceptin conjugates for breast cancers, Figure 6.16(a): QDs rendered water soluble through attachment to a PEG-based molecule may be coupled to Herceptin and a tumor-homing peptide. Alternatively, QD-aptamer-doxorubicin has also been used experimentally.

#### Gold nanoparticle-based agents

The production of gold nanoparticles is well advanced and their performance in a number of imaging systems is known to depend on a variety of physico-chemical characteristics. Both functional entities and therapeutic agents can be loaded into these nanoparticles. For example, with gold at the core, a nanoparticle can have a hydrophilic folate-conjugated outer shell for tumor targeting and a payload of doxorubicin conjugated to the hydrophobic inner shell with an acid-cleavable linkage (Figure 6.16(b)).

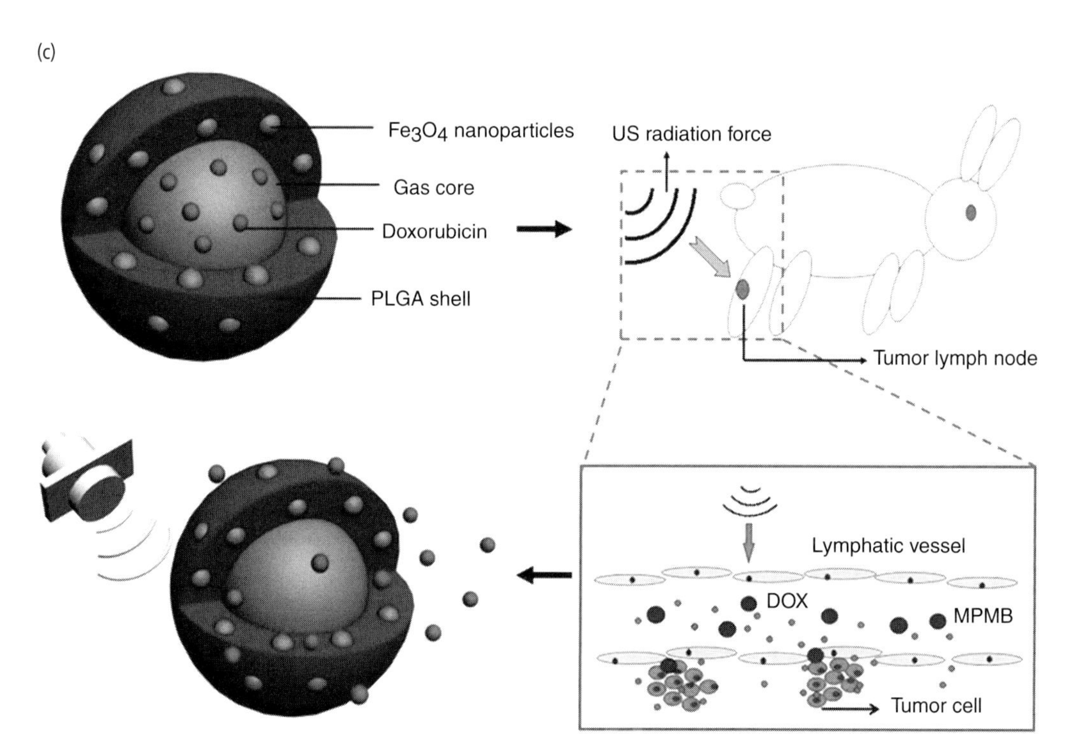


**Figure 6.16** *continued*
(c) Theranostics: doxorubicin loaded multifunctional imaging agents. This scheme describes a PLGA shell that contains a perfluorocarbon gas microbubble, the chemotherapeutic agent doxorubicin and iron oxide nanoparticles intended to provide dual-mode ultrasound/magnetic resonance imaging and drug therapy within lymph nodes. Reprinted from *Biomaterials*, 34(9), Niu, C *et al.*, Doxorubicin loaded superparamagnetic PLGA-iron oxide multifunctional microbubbles for dual-mode US/MR imaging and therapy of metastases in lymph nodes, 2307–17, © 2013, by permission from Elsevier.

#### Iron oxide nanoparticle-based agents

We will see elsewhere that iron oxide nanoparticles are nanocrystals made from magnetite or hematite. These have good magnetic properties and generally good biocompatibility and they have found a number of imaging applications. These nanoparticles can be coupled to many drugs (Figure 6.16(c)). Chemotherapeutic agents such as paclitaxel and methotrexate have been attached to gold nanoparticles through aminating the surface or coupling via phosphodiester groups. Targeting of the nanotheranostic agents to tumors, for example breast cancers, can be achieved, with partial success, by external magnetic fields.

### 6.10.2 Drug–device combinations

This is mentioned here for completeness; we have mentioned medical devices that may have their performance enhanced by the release of active pharmaceutical agents in both Chapters 3 and 4. There is no particular pattern that covers these uses; they are highly specific in terms of objectives and mechanisms.

## Summary and learning objectives

1. The ability of a drug to achieve its desired effects depends on many factors, including the ease with which it is administered to, and adsorbed by, the body, its mechanism of action in the body, and the way in which the body responds to its presence. Traditionally drugs are delivered by mouth or by inhalation as with an asthma nebulizer. If the drug is not easily absorbed by either of these routes, intramuscular injection or intravenous infusion may be used. Whatever the route, the level of the drug in the patient is likely to be higher than is strictly necessary, and at times it may well be lower than the threshold at which it has an effect. Also the whole body and not just the part that needs the drug will experience these levels. This is hardly a well-controlled delivery process, although it is often quite effective and is usually very convenient as far as the patient is concerned. In this chapter, we have discussed those techniques that are being developed in order to produce far greater control.
2. In contrast to biomaterials applications in medical devices, and similar to those in tissue engineering, these technologies are not yet in widespread clinical use. The scientific base is very strong, but the processes are complex and success is difficult to achieve. Targeting drugs precisely to individual cells requires highly specific *in vivo* transport routes and the evasion of defense systems; achieving efficacy and safety is not a trivial issue. In this chapter, the actual and potential technologies are discussed, with an indication of the achievements and hurdles yet to be overcome.
3. At its simplest, it may be necessary to formulate orally delivered drugs with features that control the release of the drug at different sites or at different times in the passage of the tablet through the gastrointestinal system. For example, a range of technologies has been developed to increase the gastric retention time of a drug, including mucoadhesive coatings of tablets and intragastric floating systems that allow the tablet to stay at the top of the stomach contents to avoid gastric emptying. It may be better for the drug only to be released in the colon, where more complex strategies have to be used, with an enteric coating that prevents dissolution in the stomach and a cationic polymer that prevents absorption in the small intestine.
4. We then discuss the technologies that are available to deliver conventional drugs to the body via routes other than through the GI tract. These systems are either transdermal, including transdermal patches, ablation and abrasion, the electrical methods of iontophoresis and electroporation, ultrasound, microneedles and so-called needle-free injections, or implantable

electromechanical systems such as pumps and microchip-based devices.

5. Drug release from solid polymers in monolithic devices is considered next, covering the mechanisms of release, that is diffusion, erosion and osmosis. The materials used are either degradable polyesters such as poly(lactic-co-glycolic) acid and polyanhydrides, or non-degradable silicones. A few products based on these principles and materials have proven very effective clinically, including the delivery of chemotherapeutic drugs to brain tumors. The release of drugs from these solid materials is naturally dependent on their placement and the control is not very sophisticated. Greater control, in theory, should be obtained by the use of microparticulate and nanoparticulate systems. We especially focus here on the fate of nanoparticles, with discussions of clearance and endocytosis. Liposomes are prominent examples and are used as a basis for discussion.
6. The above discussions have been based on situations where the drug and the biomaterial are separate chemical entities; the biomaterial is essentially a physiologically and pharmaceutically inert carrier. In recent years attention has been turned to situations where this is not the case, and where the polymer actively plays a role in the chemistry of drug delivery. The materials involved, which include polymeric drugs, polymer–drug and polymer–protein conjugates, and polymer micelles covalently attached to drugs, are now collectively described as polymer therapeutics. Particular attention is paid to polymer architecture with micelles and dendrimers. We discuss the essential details of the mechanisms of these potentially major innovations with respect to precise delivery, although few are yet clinically proven.
7. One of the most important clinical applications of controlled delivery is that of the treatment of cancer. The pharmaceutical treatment of cancer is a good example of the difficulties of conventional drug delivery and the potential for biomaterials to give more effective therapies. Systemic chemotherapy is the most important strategy for the treatment of cancer but this intravenous administration of potent anti-cancer drugs may lead to severe toxic effects. It is also quite an inefficient method in view of the physiological barriers to the delivery to solid tumors and through the interstitial spaces within them. The chapter considers targeting agents through passive methods but concentrates on the active targeting of tumor cells and endothelial cells. Special attention is given to cancer multidrug resistance and to improved targeting of immunotherapy and photodynamic therapy.
8. The delivery systems discussed in this chapter are not only applicable to conventional drugs. We discuss in some detail the special areas of gene therapy, vaccines and antimicrobial agents. Although details vary, there are common themes, based on the problems encountered with the more standard methods and the promise of biomaterials-based solutions to improve efficacy, specificity and safety. Once again, clinical successes have so far proved elusive in view of the significant biological challenges, but the potential is profound.

## Questions

1. Describe the generic problems with the conventional oral delivery of drugs and summarize the different approaches that may be used to address these problems.
2. Discuss the principles of polymer therapeutics. Give your views on whether polymer–drug conjugates will ever be clinically valuable.

3. Compare and contrast the various methods that have been developed to enhance delivery of drugs across the skin.
4. Polymer architecture is an important determinant of the ability to store and deliver drugs. Discuss the approaches that have been used to exploit the different forms of such architecture.
5. Describe the reasons why antibiotic resistance has emerged and discuss the role of biomaterials in the solution to these problems.
6. Liposomes play a large role in the development of nanoscale delivery systems. Discuss the scientific principles that underlie both the advantages and disadvantages of liposomes.
7. Describe the mechanisms whereby nanoparticles enter target cells and how they are subsequently processed by cells.
8. Discuss the philosophical aspects of the choice between viral and non-viral gene vectors.
9. Describe the principles of photodynamic therapy and the reasons why cancers are still difficult to treat by this technology.
10. Discuss the role of pH in the performance of biomaterials-based polymeric drug delivery systems.

## Recommended reading


Blanchette J, Kavimandan N, Peppas N A, Principles of transmucosal delivery of therapeutic agents, *Biomedicine and Pharmacotherapy* 2004;58:142–51. *Discussion of the use of hydrogels to increase the bioavailability of orally administered therapeutic agents.*

Bugaj A M, Targeted photodynamic therapy – a promising strategy of tumor treatment, *Photochemical and Photobiological Sciences* 2011;10:1097–1109. *Discussion of the principles of PDT and the potential benefits of targeting this treatment modality for cancer.*

De Koker S, Lambrecht B N, Willart M A, *et al.*, Designing polymeric particles for antigen delivery, *Chemical Society Reviews* 2011;40:320–39. *Detailed analysis of the chemistry of polymers used to enhance vaccine effectiveness.*

De Vry J, Martinez-Martinez P, Losen M, *et al.*, In vivo electroporation of the central nervous system: a non-viral approach for targeted gene delivery, *Progress in Neurobiology* 2010;92:227–44. *A discussion of the mechanisms of electroporation and the delivery of DNA into cells.*

Drulis-Kawa Z, Dorotkiewicz-Jach A, Liposomes as delivery systems for antibiotics, *International Journal of Pharmaceutics* 2010;387:187–98. *Review of the experimental and clinical work on liposomal antibiotics, including advantages and disadvantages.*

Duncan R, The dawning era of polymer therapeutics, *Nature Reviews Drug Discovery* 2003; 2:347–60. *The first paper to introduce the concept of polymer therapeutics.*

Eljarrat-Binstock E, Domb A J, Iontophoresis: a non-invasive ocular drug delivery, *Journal of Controlled Release* 2006;110:479–89. *Overview of the approaches to the development of ocular iontophoretic devices.*

Fredenberg S, Wahlgren M, Reslow M, Axelsson A, The mechanisms of drug release in poly(lactic-co-glycolic acid)-based drug delivery systems – a review, *International Journal of Pharmaceutics* 2011, 415:34–52. *A review of the mechanisms by which PLGA is able to release drugs.*

Giudice E L, Campbell J D, Needle-free vaccine delivery, *Advanced Drug Delivery Reviews* 2006;58:68–89.

*Review of the technologies for and physiological implications of needle-free vaccine delivery.*

Hubbell J A, Chilkoti A, Nanomaterials for drug delivery, *Science* 2012;337:303–5. *A short review of the principles of nanomedicines.*

Kohane D S, Microparticles and nanoparticles for drug delivery, *Biotechnology and Bioengineering* 2007; 96:203–9. *An essay on the general principles of particulate drug delivery systems and opportunities for future developments.*

Krishnamachari Y, Geary S M, Lemke C D, Salem A K, Nanoparticle delivery systems in cancer vaccines, *Pharmacological Research* 2011;28:215–36. *A discussion of the development of cancer vaccines based on nanoparticles.*

Kulkarni M, Greiser U, O'Brien T, Pandit A, Liposomal gene delivery mediated by tissue-engineered scaffolds, *Trends in Biotechnology* 2010, 28, 28–36. *A review of the mechanisms by which liposomes can deliver genes within tissue engineering scaffolds.*

Lammers T, Kiessling F, Hennink W E, Storm G, Drug targeting to tumors: principles, pitfalls and pre-clinical progress, *Journal of Controlled Release* 2012;161:175–87. *Significant review of the concepts and reality of targeting chemotherapeutic agents to tumors.*

Marambio-Jones C, Hoek E M V, A review of the antibacterial effects of silver nanomaterials and potential implications for human health and the environment, *Journal of Nanoparticle Research* 2010;12:1531–51. *Extensive discussion of the antibacterial activity of nanosilver.*

Mintzer M A, Simanek E E, Nonviral vectors for gene delivery, *Chemical Reviews* 2009;109:259–302. *An essay on the principles of non-viral delivery systems.*

Petros R A, DeSimone J M, Strategies in the design of nanoparticles for therapeutic applications, *Nature Reviews Drug Discovery* 2010;9:615–26. *Discusses the reasons why nanoparticles have difficulty in surmounting the biological barriers of drug delivery.*

Phillips J E, Gersbach C A, Garcia A J, Virus-based gene therapy strategies for bone regeneration, *Biomaterials* 2007; 28: 211–29. *Discussion of viral vectors in bone tissue engineering.*

Polat B E, Hart D, Langer R, Blankschtein D, Ultrasound-mediated transdermal drug delivery: mechanisms, scope and emerging trends, *Journal of Controlled Release* 2011;152:330–48. *A review of the use of low and high frequency ultrasound to increase skin penetration of drugs.*

Pradeep K, Dahl S, Holmes-Farley R, Huval C C, Jozefeak T H, Polymers as drugs, *Advances in Polymer Science* 2006;192:9–58. *Extensive review of the role of polymers in therapeutic systems.*

Prausnitz M R, Microneedles for transdermal drug delivery, *Advanced Drug Delivery Reviews* 2004;56:581–7. *Review of uses and mechanisms of microneedles.*

Prausnitz M R, Langer R, Transdermal drug delivery, *Nature Biotechnology* 2008;26:1261–8. *A discussion of various technologies for transdermal delivery, including chemical enhancers, ultrasound and iontophoresis, microneedles and electroporation.*

Ruenraroengsak P, Cook J M, Florence A T, Nanosystem drug targeting: facing up to complex realities, *Journal of Controlled Release* 2010;141:265–76. *Review of the limitations of nanomedicines, including nanoparticle instability and variations in nanoparticle transport and targeting.*

Sahay G, Alakhova D Y, Kabanov A V, Endocytosis of nanomedicines, *Journal of Controlled Release* 2010;145:182–95. *A review that classifies mechanisms of nanomedicines and discusses the effects of nanoparticle characteristics on these processes.*

Sawant R R, Torchilin V P, Liposomes as smart pharmaceutical nanocarriers, *Soft Matter* 2009;6:4026–44. *Discusses the evolution of liposomes, with an emphasis on stimuli-sensitive liposome systems.*

Singh R, Lillard J W, Nanoparticle-based targeted drug delivery, *Experimental and Molecular Pathology* 2009;86:215–23. *Review of nanomedicines including the formulations being developed for vaccines, therapeutic proteins and cancer therapeutics.*

Somia N, Verma I M, Gene therapy: trials and tribulations, *Nature Reviews Genetics* 2000;1:91–9. *Discussion of serious errors in early gene therapy trials and the lessons learnt.*

Stachowiak J C, Li T H, Arora A, Mitragotri S, Fletcher D A, Dynamic control of needle-free injection, *Journal of Controlled Release* 2009;135:104–12. *Discussion of the mechanisms of needle-free injections.*

Stamatialis D F, Papenburg B J, Girones M, *et al.*, Medical applications of membranes: drug delivery, artificial organs and tissue engineering, *Journal of Membrane Science* 2008;308:1–34. *Review of membrane materials in drug delivery.*

Staples M, Microchips and controlled-release drug reservoirs, *WIREs Nanomedicine and Nanobiotechnology* 2010;2:400–17. *Review of electromechanical devices in drug delivery systems.*

Xie J, Lee S, Chen X, Nanoparticle-based theranostic agents, *Advanced Drug Delivery Reviews* 2010;62:1064–79. *Introduction to the principles of theranostics.*

Yoo J-W, Irvine D J, Discher D E, Mitragotri S, Bio-inspired, bioengineered and biomimetic drug delivery carriers, *Nature Reviews: Drug Discovery* 2011, 10, 521–35. *Wide-ranging discussion of the role of biomimetics in drug delivery.*

# 7 Imaging and diagnostic systems

In this chapter we introduce some of the latest applications of biomaterials in medical technology. Previous chapters have concentrated on therapeutic processes, either by pharmaceutical components, regenerative medicine processes or implantable devices. All clinically successful therapeutic measures require good diagnosis of the patient's condition. Although the doctor's intrinsic clinical skills are extremely important, the need for very accurate, instrument-assisted, diagnosis, with good spatio-temporal resolution, is becoming increasingly significant. This is because early and precise diagnosis is essential if the doctors are able to use the most effective treatments for cancer, degenerative disease and other critical issues. Medical technologies have always played some role in diagnosis, but have rarely, until recently, involved the use of biomaterials. Most techniques are concerned with the application of some physical energy to the affected tissues and organs, using X-rays and other ionizing radiations, ultrasound and light, for example, and detecting the responses of the tissues to that energy. Conventional methods often provide relatively poor contrast between different types of tissues (and any lesions they contain), usually with restricted spatial resolution. They usually produce anatomic rather than functional information. The essential rationale for the use of biomaterials in these imaging systems is to enhance the contrast that can be seen, especially by accentuating differences between the response of different types of tissues and between different disease states. Most of these biomaterials are used in nanoparticulate form. Their applicability is controlled by their differential responses to the applied energy and by their handling within the tissues of the body.

## 7.1 Introduction

### 7.1.1 Anatomical and functional imaging

The majority of this book so far has been concerned with the therapeutic methods that may be used to treat diseases and conditions of the human body, where those methods significantly rely on the use of biomaterials. Before any therapy can be considered, it is necessary for the clinicians involved to be aware, as accurately as possible, of the nature of the disease or condition that is affecting the patient in question. For very many years, this process of identifying what is wrong, known as diagnosis, was informed by the skilled observations of the clinicians and the measurement of some relevant physiological parameters such as temperature, heart rate and blood pressure.

The instruments used to acquire this rather basic data were not sophisticated, and certainly did not involve biomaterials. Moreover, observations of the patient were largely confined to the outer surfaces and some orifices of the body. A significant step forward was made with the discovery of X-rays at the end of the nineteenth century and the development of X-ray equipment to allow physicians to acquire some images of the interior of the body, albeit with limited contrast in the early days. These first attempts at anatomical imaging, of course, still did not involve biomaterials since the technique involved the passage of very short wavelength electromagnetic radiation through the tissue and a visual representation of the attenuation of that radiation by use of a sensitive film. These attempts at anatomical visualization of some parts of the body could be enhanced by the use of radiopaque dyes, the first form of contrast agent. These were simple substances, such as a suspension of barium sulfate, and no-one considered these to be biomaterials, although there were some similarities to those biomaterials-based contrast agents that we will discuss in this chapter.

Anatomical imaging rapidly developed as a diagnostic modality some 50 years ago, with the introduction of alternative methods of visualization. Ultrasound was introduced, in which high frequency sound waves were directed into the body and their reflection back to the surface could be monitored. This was very ineffective until computerized techniques were developed in order to analyze the reflected waveforms, providing an image of structures, soon to be made available in real time in order to monitor, for example, blood flow, as well as the anatomical features controlling that blood flow. Computerized techniques also revolutionized X-ray diagnosis with the emergence of computerized tomography (CT), in which localized information from slices of tissue on which the X-ray beam is focused could be used to create highly specific anatomical images. Many variations of these techniques were developed, some of which also used contrast agents.

Successful, and helpful, as they were, these methods of anatomical imaging had significant drawbacks as far as diagnosis was concerned. Often the contrast obtained was insufficient for the purposes of really accurate diagnosis and prognosis and, moreover, even good images of anatomical structure do not give information about the functional characteristics. New methods were therefore required in order to improve contrast and provide data on physiological function. The history of magnetic resonance imaging (MRI) provides a good summary of the driving forces for these developments. MRI was introduced into clinical practice in the 1970s. As we shall see later, this imaging modality is dependent on the behavior of certain molecules within the body on the application of a high magnetic field. MRI has traditionally relied on the behavior of the hydrogen atoms in water, and good anatomical images can usually be obtained in the conventional mode. However, the contrast that is obtained is often inadequate for sensitive diagnoses. For this reason, a number of contrast agents have been developed which, after intravenous injection, are able to localize within relevant tissue structures, give much greater contrast and extend the mode to functional or molecular imaging at the same time.

Since such contrast agents are materials that affect diagnostic techniques through their interactions within the human body, they may be characterized as biomaterials. This chapter covers some of the major developments in contrast agents and their performance. We will cover MRI, CT and ultrasound, already mentioned above, as well as positron emission tomography (PET), single-photon emission computed tomography (SPECT) and various forms of optical imaging. Some of the basic physical principles involved have been discussed in Chapter 2. It is not intended to give any detailed coverage of the imaging systems here since they are beyond the scope of the book; we will just concentrate on the agents themselves. It should also be borne in mind that many of these systems are emerging rapidly and, as with a few previous chapters, clinical applications may not yet be validated.

## Box 7.1 Glossary of terms

**amperometry** the measurement of current at a single applied potential.

**aptamer** sequences of nucleic acid or protein molecules that bind to a specific target molecule.

**autofluorescence** self-induced fluorescence of substances, distinct from that of added fluorophores.

**biotinylation** the process of covalently attaching biotin, a crystalline, water-soluble vitamin of the vitamin B complex to a protein, nucleic acid or other molecule.

**cell penetrating peptide** short peptides that have the ability to enter cells by crossing the plasma membrane.

**contrast agent** a substance used to enhance the contrast of structures or fluids within the body in imaging.

**doping** adding an impurity to a material (especially a semiconductor) to produce a desired electrical characteristic.

***in situ* hybridization** a technique that localizes specific nucleic acid sequences within intact chromosomes or eukaryotic cells through the use of specific nucleic acid-labeled probes.

**lanthanide** any of the series of 15 metallic elements from lanthanum to lutetium in the Periodic Table.

**molecular imaging** the visualization, characterization, and measurement of biological processes at the molecular and cellular levels in humans and other living systems.

**near infrared** a subdivision in the infrared band between 800 and 2500 nm wavelengths.

**photobleaching** the photochemical destruction of a fluorophore.

**photon** a quantum of electromagnetic radiation.

**quantum yield** the ratio of the number of emitted photoelectrons to the number of incident photons.

**radionuclides** nuclides, which are atoms or ions characterized by the contents of their nucleus, that exhibit radioactivity.

**relaxation** the process where atoms, whose nuclei have first been aligned along a static magnetic field and then excited to a higher energy state by a radiofrequency signal, return to a lower energy equilibrium state.

**tomography** a method of producing a 3D image of the internal structures of the human body by recording the differences in the effects on the passage of waves of energy impinging on those structures.

**upconversion** the process where nanoparticles absorb energy in the near-infrared region and emit higher energy visible light.

### 7.1.2 Principles of molecular imaging

In line with the above comments, we should introduce the overarching term "molecular imaging" at this point. This is defined in terms of the methods that may be used to visualize and quantitatively measure the function of biological and cellular processes *in vivo*. Although applicable to many disease states, it is in the diagnosis and treatment planning of cancer that molecular imaging plays its greatest role. Referring to the various methods to treat cancer through active molecules discussed in the previous chapter, the ability to identify and quantify the molecular marker profile, especially related to growth factor receptor types on tumor cells, is an essential pre-requisite for targeted chemotherapy or immunotherapy.

At their simplest, contrast agents need two active components (Figure 7.1). The first is an agent that can bind to the target molecule: this could be an antibody, a peptide, a nucleic acid or some small molecule. The second is a label that can be detected by the imaging modality: for MRI this will be a substance, usually metal-based, that has good magnetic properties; for optical imaging it may be fluorescent dye; for CT it will be radiopaque, such as iodine; and so on. There will, of course, need to be a mechanism of binding these two principal agents together.

There are two significant variations on this theme. The first involves the theranostic approach discussed in the previous chapter, in which a third independent component is added, which is a pharmaceutically active molecule that can be released under circumstances determined by the activity of the imaging component. The second is seen in multi-modal imaging in which two or more imaging components are simultaneously employed within the same system.

Before considering the merits of each imaging modality, and the specific agents used in them, it is worth looking at the broad picture of what is currently available and the general merits and deficiencies of these systems, given in Box 7.2.

## 7.2 Magnetic resonance imaging

Nuclear magnetic resonance (NMR), first described in the 1940s, is a spectroscopy technique that uses the magnetic properties of the nucleus of atoms to determine the chemical composition of substances. Atoms with an odd number or protons and neutrons (including hydrogen) have a quantum property known as spin angular momentum. The atoms are able to occupy multiple different states of spin, but are usually randomly distributed between them. In the presence of an external magnetic field the atoms will tend to align their spin states either with or against the external magnetic field depending on their energy state. The object inside the magnetic field then possesses a very weak magnetic charge. Atoms can jump between high and low energy states by the application of radiofrequency energy. This transition between energy states results in a change

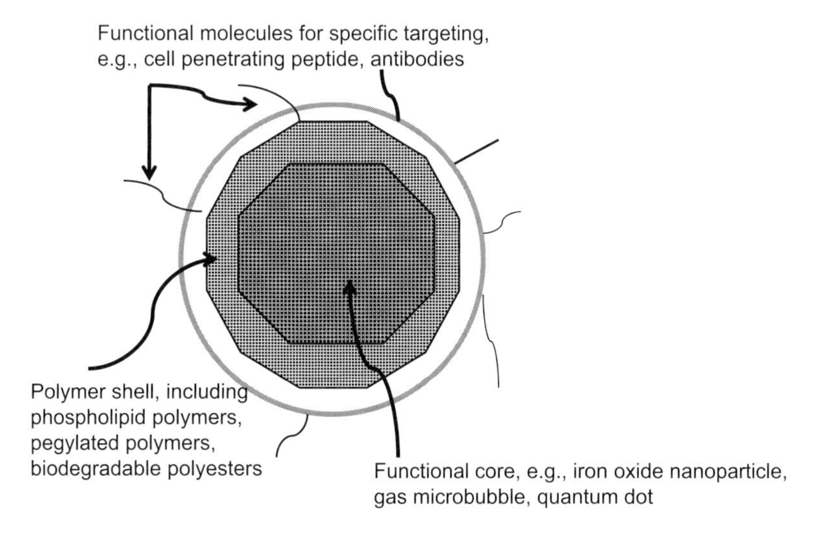


**Figure 7.1** Contrast agents. There are several types of contrast agent. They operate on the same general principle. There is a functional component that interacts with the imposed energy, such as a bubble that responds to ultrasound or a paramagnetic nanoparticle that responds to a magnetic field. The targeting of the agent to the requisite tissue components is achieved by one of several types of functional molecule on the surface, such as a cell penetrating peptide. The functional surface is attached to the functional core by a shell, typically of a polymer, which can have varying degrees of functionality itself.

### Box 7.2 Imaging nodalities

Imaging techniques have been continuously developed and improved over the last 50 years. The characteristics of the main modalities, with advantages and disadvantages, are as follows:

Magnetic resonance imaging

Targets: anatomical, physiological and molecular information
Main imaging agents: gadolinium, iron oxides, manganese oxide
Spatial resolution: 10–100 μm
Advantages: no radiation, no depth limit, good soft tissue contrast, quantitative
Disadvantages: expensive, long acquisition time, limited sensitivity

Computed tomography

Targets: anatomical and physiological information
Main imaging agents: iodine
Spatial resolution: 50 μm
Advantages: no depth limit, high resolution, short acquisition time, quantitative
Disadvantages: moderately expensive, some radiation, poor soft tissue contrast

Positive emission tomography

Targets: physiological and molecular information
Main imaging agents: $^{18}$F and $^{11}$C labeled compounds
Spatial resolution: 1–2 mm
Advantages: no depth limit, quantitative, high sensitivity
Disadvantages: expensive, long acquisition time

Ultrasound

Targets: physiological and anatomical information
Main imaging agents: microbubbles
Spatial resolution: 50 μm
Advantages: low cost and ease of use, quantitative, high resolution
Disadvantages: limited imaging depth, confined to vasculature

Optical imaging

Targets: physiological and molecular information
Main imaging agents: fluorophores
Spatial resolution: 1–5 mm
Advantages: high sensitivity, no radiation
Disadvantages: limited imaging depth, whole body imaging not possible

in this magnetic charge, which can be detected and measured by voltage induced in a surrounding coil, according to Faraday's Law. The radiofrequency required to cause the transition is dependent on intrinsic properties of the atom and the strength of the external magnetic field. This relationship is characteristic of the chemical composition, hence the use of NMR in chemical analysis.

The principles of NMR were used to create images of tissues in the 1970s. The magnetic properties of tissues, with 70% water content, are dominated by the hydrogen atom of the water. Bursts of specific radiofrequencies can excite hydrogen atoms into their higher energy states and change the magnetic state. An image can be created on the basis of the return of the magnetic state back to its original orientation. This is known as relaxation, and is fundamental to the generation and interpretation of magnetic resonance images. It is the speed of this process that allows MRI to differentiate between the tissues in the body.

There are two relaxation constants. T1 relaxation is the time constant for the return to alignment with the external magnetic field. T2 relaxation is the time constant to leave the plane perpendicular to the external magnetic field. These constants are material dependent, the water content being a major factor. Hard tissues have very fast relaxation constants while high water content soft tissues have long relaxation constants.

MRI can be used for characterization of tissues simply on the basis of the hydrogen in the water, and indeed the majority of investigations are carried out this way. However, about one-third of all MRI scans now make use of contrast agents, which allow for better tissue characterization and specificity and greater amounts of functional information. Most contrast agents are substances that include paramagnetic metal ions that alter relaxation constants of the tissues, influencing either T1 (positive agents) or T2 (negative agents). In addition, some other agents operate with quite different mechanisms. Examples of these contrast agents are given in Figure 7.2(a).

### 7.2.1 T1 agents

As we have seen, the T1 relaxation process involves the equilibration of net magnetization after a radiofrequency pulse as a consequence of energy transfer between proton spins and nearby molecules. The presence of paramagnetic species enhances this relaxation. The highly paramagnetic metal ions of several transition and lanthanide metals that have a large number of unpaired electrons exhibit very effective relaxation. The major advantage of such T1 contrast agents is their signal-enhancing positive contrast ability, allowing them to show anatomic details at high spatial resolution without the loss of signal. Most T1 contrast agents are based on gadolinium, and specifically $Gd^{3+}$, since $Gd^{3+}$ has seven unpaired electrons. Free $Gd^{3+}$ ions are, however, extremely toxic, there being no natural molecular systems based on this ion in the human body. These T1 contrast agents are therefore prepared in the form of paramagnetic metallochelates, composed of the metal ions and chelating ligands, which give thermodynamic and kinetic stability and make them less toxic.

In addition to gadolinium systems, manganese-based T1 agents are attractive since $Mn^{2+}$ has five unpaired electrons, with accompanying paramagnetic behavior. They do not have widespread applications since manganese is an essential trace element and $Mn^{2+}$ ions may be easily sequestered, markedly lowering stability. Some $Mn^{2+}$ complexes are preferentially taken up by hepatocytes, making these agents suitable for imaging liver diseases.

### 7.2.2 T2 agents

Agents that cause a shortening of T2 are known as T2 or negative agents, or sometimes as T2 susceptibility agents. T2 agents are superparamagnetic, the obvious examples being the oxides of iron. Particles of superparamagnetic iron oxide (SPIO) and their nanoscale derivatives (USPIO or SPION) induce large distortions of externally applied magnetic fields, leading to reduced intensity of T2-weighted images. This gives negative contrast since areas that contain

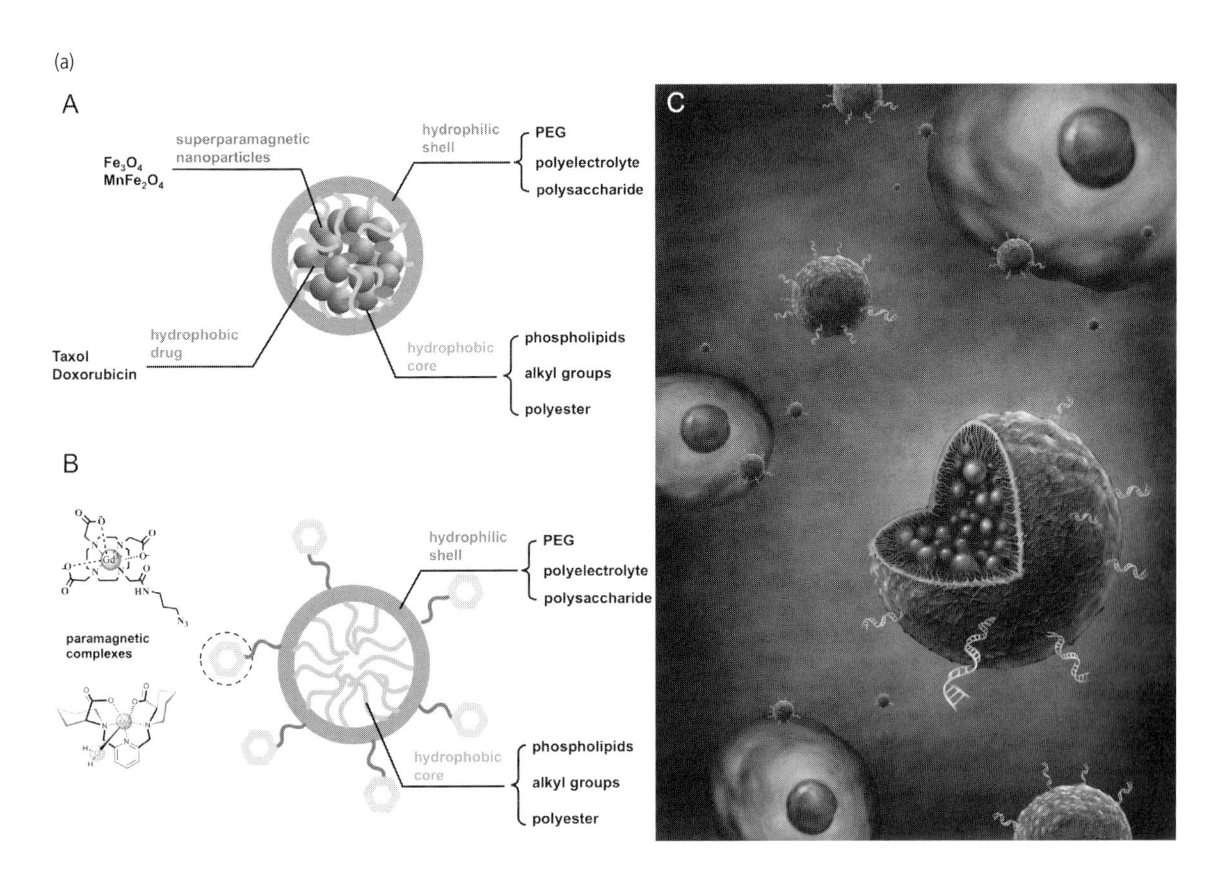


**Figure 7.2** Contrast agents for magnetic resonance imaging.
(a) The design structure of a T2 MRI probe (A). The key components of the probe include polymeric micelle and superparamagnetic iron oxide (SPIO) nanoparticles (e.g., $Fe_3O_4$, $MnFe_2O_4$). Polymeric micelles made of amphiphilic polymers with hydrophilic corona (e.g., poly(ethylene glycol), polyelectrolytes, and polysaccharides) and hydrophobic core (commonly used materials include: biodegradable polyesters, phospholipids, alkyl groups). The SPIO nanoparticles are closely packed inside the micelle, forming a clustering structure, which is beneficial for improving the probe's sensitivity through enhancing the T2 relaxivity. Hydrophobic small molecule drugs such as doxorubicin or paclitaxel can be loaded into micelles through physical entrapment. (B) Shows the design structure of a T1 MRI probe. The key components of the probe include polymeric micelle and paramagnetic metal ion (e.g., Mn, Gd) ligand complexes. (C) Illustrates a T2 MRI probe used to carry siRNA for cellular delivery.

the particles display fast transverse relaxation rates and low signal intensity.

Iron oxides are not as toxic as the Gd compounds, but care has to be taken with their distribution in the body. Usually they are prepared as colloids with dextran or similar substances. Their clearance from the body is very much dependent on size. The smaller size particles (USPIO), at less than 50 nm, have longer half-lives since their uptake by macrophages is more difficult than the larger SPIO particles, at 100–200 nm.

### 7.2.3 CEST and PARACEST

CEST stands for chemical exchange saturation transfer. CEST agents are exogenous mobile molecules that facilitate the transfer of saturation magnetization from tissue mobile protons to bulk water during radiofrequency irradiation. They do so by enhancing selective radiofrequency irradiation of the sharper NMR signal associated with their exchangeable protons, contained for example in NH or OH functional groups. These agents are either

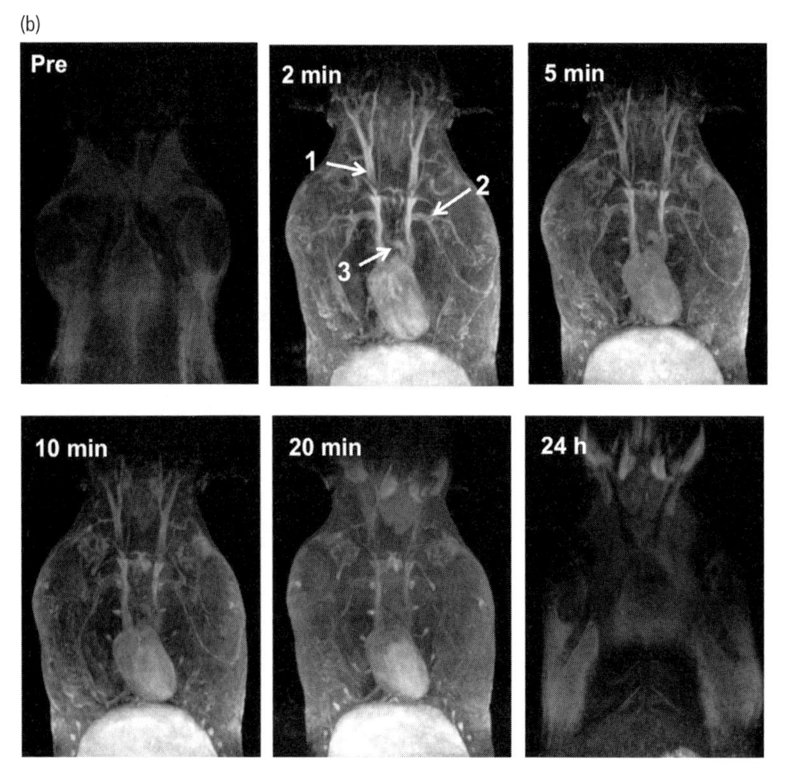


**Figure 7.2** *continued*
**(b)** T1-weighted MR image (Philips Medical System, Intera 3 T, TR = 5 ms, TE = 2 ms) of a rat at pre-injection, 2 min, 5 min, 10 min, 20 min and 24 h after administration of MnL complex (0.5 mmol $kg^{-1}$). MR signal intensity in the jugular vein (1), subclavian vein (2), and aortic arch (3) was enhanced significantly. All images courtesy of Professor Hua Ai, National Engineering Research Center for Biomaterials, Sichuan University, Chengdu, China.

diamagnetic or paramagnetic. Particularly important are the paramagnetic complexes (referred to as PARACEST) that contain the biologically important transition metal ion, $Fe^{2+}$ and also those based on $Eu^{3+}$ and $Tb^{3+}$. Complexes may be formed with tetra-amide derivatives of 1,4,7,10-tetraazacyclododecane-1,4,7,10-tetraacetic acid, known as DOTA. These are stable under physiological conditions and contain multiple protons for exchange with bulk water. Alternatively there are metal-free exogenous agents that consist of diamagnetic molecules that contain exchangeable protons, such as –OH and –NH, including sugars, amino acids, polypeptides and dendrimers, such as poly-L-lysine (PLL) and polyamidoamine dendrimer.

### 7.2.4 $^{19}F$ agents

It was noted earlier that MRI scans are usually based on the detection of hydrogen in water. This system does tend to have low sensitivity, and many efforts have been made to improve this. One promising alternative is the use of $^{19}F$ atoms instead of hydrogen nuclei. Fluorine naturally has a high intrinsic sensitivity and may be used to label and track individual cells. In practical terms, these agents are usually prepared as emulsions containing perfluorocarbons (PFCs), where individual droplets of PFC are surrounded by a lipid surfactant. The PFCs used for cell tracking include perfluorooctyl bromide and linear perfluoropolyethers.

Examples of images obtained by contrast enhanced MRI are shown in Figure 7.2(b).

## 7.3 CT imaging

As noted in the introduction to this chapter, computerized tomography (CT) involves the collection of localized information from slices of tissue on which an X-ray beam is focused, which is used to create highly specific anatomical images. The images are acquired with an X-ray tube, which passes a rotating fan beam of X-rays through the patient, measuring the transmission at thousands of points. The appearance of the tissue on a CT scan differentiates between bone, which is white, air, which is black, fat, which is dark gray and soft tissue, which is light gray. This technique has high spatial resolution but relatively poor contrast resolution compared to MRI.

For this reason, intravenously administered contrast agents may be used with CT in order to enhance the contrast. Contrast enhanced CT (CE-CT) has been introduced in recent years, and can give improved lesion detection and enhanced resolution of the vascular system and organs. It also has very good temporal resolution, allowing an effective dynamic version of CT imaging.

Because of the high radiopacity of the element iodine, most CT contrast agents are based on this element. In particular, the incorporation of iodine into a benzene ring gives a core molecule, 2,4,6-triiodinated benzene, from which most agents are derived. However, most regular CT contrast agents are relatively small molecules and are quickly excreted via the kidneys. A recent emphasis on the development of nanoparticle-based agents has been associated with the need to have longer circulation times through increased difficulty of clearance in this way, although they still have to avoid fast uptake in the liver. Gold nanoparticle and iodinated micelles based on PEG are possible alternative choices. There have also been attempts to develop targeting options for CT using antibodies, peptides and proteins. Gold nanoparticles can be functionalized with antibodies to target and image structures such as lymph nodes. Examples of the agents and images are shown in Figure 7.3. There is also a move towards multicolor CT imaging in which two different agents, for example gold nanoparticles and iodine compounds, are injected into the host. Since different materials have their own characteristic X-ray absorption profile, passage of a beam through tissues will allow distribution maps of the different materials to be obtained, giving simultaneous visualization of different pathologies.

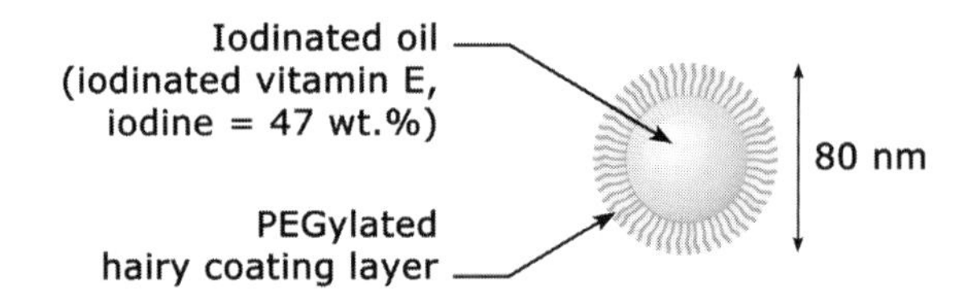


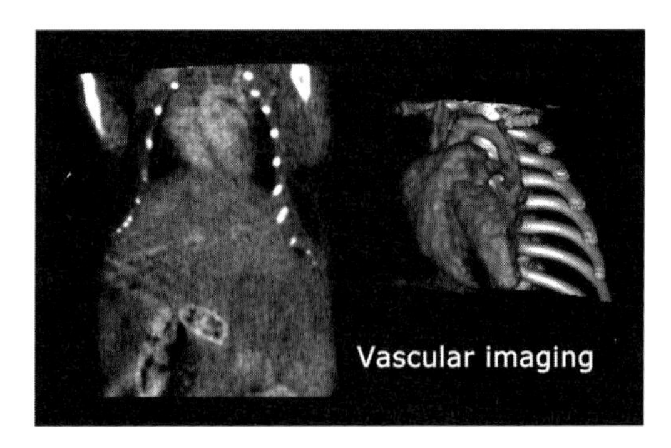


**Figure 7.3** Nanoemulsions as contrast agents. A PEG-coated iodinated oil nanoparticle used for contrast enhanced CT imaging of the vasculature. Image courtesy of Dr. Nicolas Anton, Faculté de Pharmacie, Université de Strasbourg, France.

## 7.4 PET and SPECT imaging

MRI systems generally have a spatial resolution in the range 10–100 μm, which is very good for most imaging requirements. However, the sensitivity in terms of the amount of the target substance that has to be present is in the region of $10^{-3}$ to $10^{-5}$ mole $L^{-1}$, which is not particularly good. MRI, and indeed CT, techniques are, therefore, generally used when high-resolution structural images are required. PET and SPECT imaging techniques may be used for functional imaging with lower resolution but much better sensitivity. In both techniques, the spatial resolution is of the order of a few millimeters but the sensitivity is in the region of $10^{-10}$ to $10^{-12}$ mole $L^{-1}$.

PET is positron emission tomography. This imaging modality utilizes radionuclides that decay within the tissue through the emission of positrons, which are positively charged particles within the nucleus. The radionuclides form part of the imaging agent. These positrons can travel just a short distance within the surrounding tissue before they combine with an electron and are annihilated. On annihilation, the combined masses are converted into energy that produces two 511 keV $\gamma$ rays. These are emitted simultaneously but in opposite directions. These $\gamma$ ray pairs are picked up by surrounding detectors, and the acquisition of information from these detectors allows for the construction of an image. This represents the spatial distribution of the radioactivity as a function of time; the distance traveled by a positron, known as the positron range, is an important factor in this functional imaging as is the energy of the positrons, which determines this range.

A number of nuclides can be incorporated into biomolecules and have potential for PET imaging. They are usually selected on the basis that their primary mode of decay is that of positron emission, the isotopes being chosen for half-lives that are comparable to that of the process to be imaged. The most frequently used radionuclides are shown in Table 7.1; $^{11}C$ and $^{18}F$ are the most common of these.

Table 7.1 **Radionuclides used in positron emission tomography**

| Nuclide | Half-life | Clinical application |
|---|---|---|
| $^{11}C$ | 20 minutes | EGF receptors<br>Amyloid plaque<br>Dopamine receptor |
| $^{13}N$ | 10 minutes | Blood flow |
| $^{15}O$ | 2 minutes | Blood flow<br>Oxygenation |
| $^{18}F$ | 110 minutes | Glucose metabolism<br>DNA proliferation |
| $^{64}Cu$ | 760 minutes | Hypoxia<br>Tumor antigens |
| 68 Ga | 68 minutes | Blood constituents |
| $^{76}Br$ | 970 minutes | Cancer cells |

Carbon is particularly advantageous since it is found in all tissues and the replacement of the normal $^{12}C$ by $^{11}C$ allows the imaging agent to be essentially indistinguishable from the nonradioactive component. $^{11}C$ does, however, have a short half-life, meaning that it is more suitable for radiopharmaceuticals that have short biological half-lives. Studies in patients can be repeated at short intervals but this also implies that the compounds can only be used in facilities with a cyclotron. $^{11}C$ is produced by proton bombardment of natural nitrogen. A gas mixture of 2% oxygen in nitrogen will produce radioactive carbon dioxide ($^{11}CO_2$) and 5% hydrogen in nitrogen will produce methane ($^{11}CH_4$). For the $^{11}C$-radiolabeling, a commonly used method involves N-methylation using $^{11}C$-methyl iodide ($^{11}CH_3I$). The short half-life of this radioisotope ($t_{1/2}$=20 min) imposes some constraints on labeling strategies.

Fluorine is not normally found in tissue biomolecules. However, F and H atoms are of similar size and F atoms may substitute for H atoms or hydroxyl groups to give a bioisosteric replacement. $^{18}F$ has many advantages. Probes may be made with high potency because of the strong fluorine electronegativity and the strength of the C-F bond means that the probes should be highly stable. The half-life is around 110 minutes. Radiofluorination is usually achieved via 2-[$^{18}F$]fluoro-2-deoxy-D-glucose ($^{18}FDG$) or 1-(3-[$^{18}F$]fluoro-2-hydroxypropyl)-2-nitroimidazole ($^{18}FMISO$). Labeling of both compounds depends on nucleophilic substitution reaction of an aminopolyether potassium complex.

The $^{18}FDG$ compound has been the most widely used PET tracer for many years, being very versatile. It is primarily associated with the monitoring of glucose metabolism, and is used in the detection of myocardial energy changes, many neurological diseases, tumor characterization and the monitoring of glucose management. It does have some disadvantages, including a rather unspecific mode of take-up so that some benign conditions such as inflammation may be difficult to distinguish from tumors. The $^{11}C$ probes may be used to target amyloid plaque in

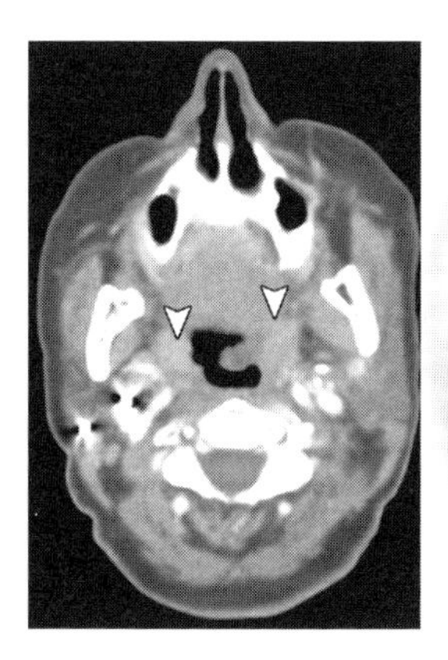

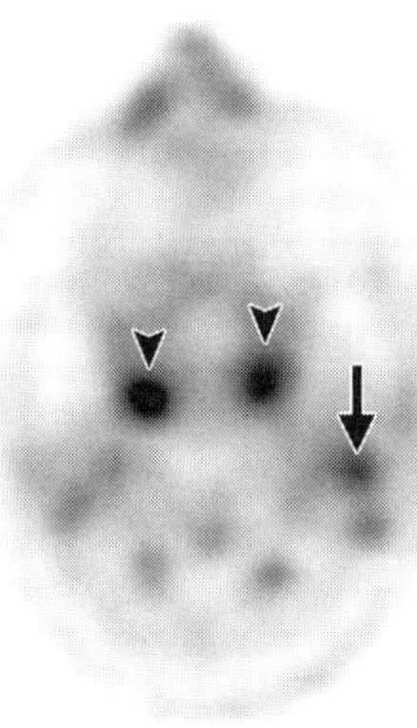

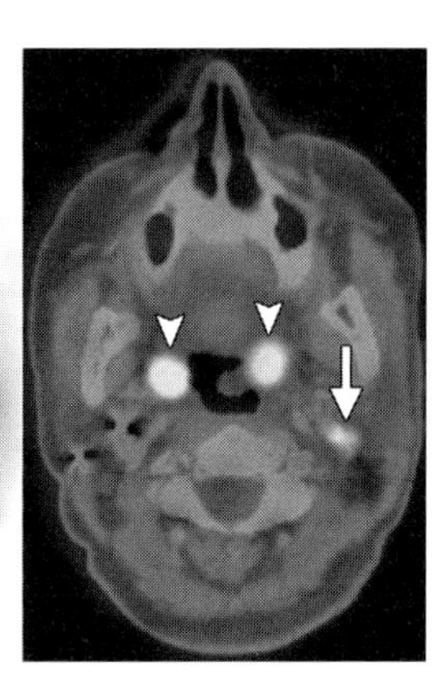

**Figure 7.4** Contrast-enhanced imaging. Axial contrast enhanced CT (left), FDG PET (center) and fused PET/CT (right) images. Reprinted from *Radiographics*, 3-, King K G *et al.* Cancers of the oral cavity and orophagus with contrast-enhanced CT in the post-treatment setting, 355–73, © 2011 The Radiological Society.

Alzheimer's disease, dopamine receptors in Parkinson's disease and EGF receptors in some cancers.

Examples of PET images obtained using these agents are shown in Figure 7.4.

## 7.5 Ultrasound imaging

Ultrasonic waves are sound waves of frequencies that are above the range that is audible to humans. As they travel through tissues, they are reflected back at points where there is a change in tissue density, the reflected echoes being analyzed with respect to their intensity and the position of the tissues that give rise to the echoes. Images are generated from this information that can be displayed in static form, or as real-time moving images through the use of rapid multiple scans.

Most ultrasound techniques have been entirely non-invasive, without any device or material being placed within the body; this, along with the fact that no ionizing radiation is involved, constitute the attractive characteristics of ultrasound with respect to patient safety. However, solely using the reflection of sound waves from tissues and fluids, without any assistance from exogenous agents, has limited the resolution and effectiveness of ultrasound scans. Contrast agent-free ultrasound methods are usually adequate in power or color Doppler mode for situations such as major intravascular investigations and some abdominal soft tissue examinations. They are, however, insufficient for the examination of tumors and the microvascularity.

In some applications of ultrasound imaging, significant improvements have been made by the use of contrast agents; with very few exceptions, these contrast agents have been based on microbubbles. Infusion of microbubble preparations into the vascular system allows them to behave as blood pool markers since they cannot pass through the vascular endothelium. The microbubbles are typically 1–4 μm in diameter, shown in Figure 7.5. They consist of a stabilizing shell and the internal gas. The first types of agents used had an albumin shell and contained air. These were followed by perfluoro gas–phospholipid membranes and some agents made of galactose microcrystals, on the surface of which air bubbles become attached when they are suspended in water and which are then stabilized by traces of the surfactant palmitic acid. The development of these microbubble agents allowed ultrasound to be used for the examination of the myocardium and heart function, and to diagnose conditions such as vesicoureteral reflux.

The principles of microbubble contrast agent performance are based on the fact that the change in density at the surface of a bubble in plasma provides a significant impedance mismatch, producing a major increase in echogenicity. In addition, the greater compressibility of the gases compared to soft tissue allow them to undergo alternate contraction and expansion, with profound vibration at their

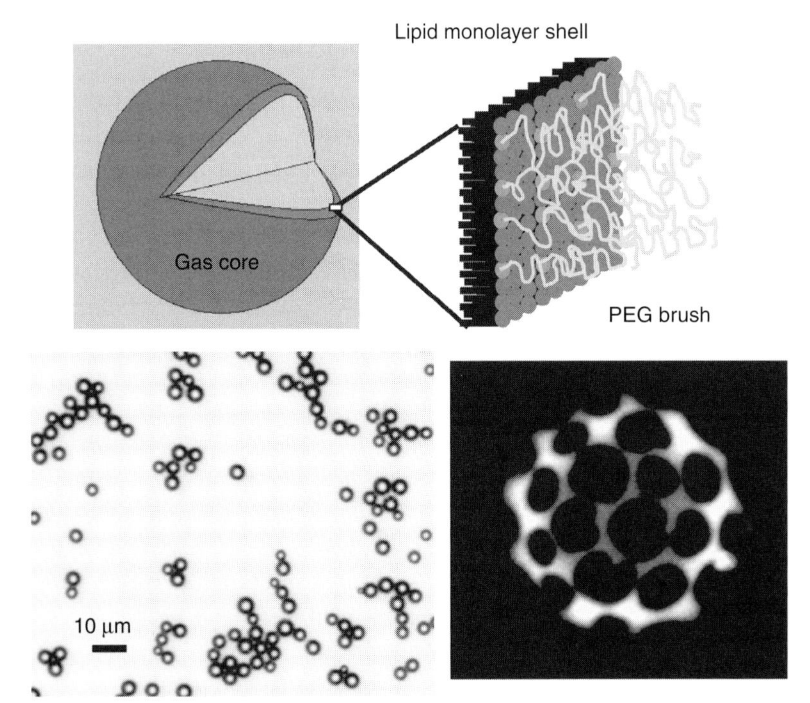


**Figure 7.5** Microbubbles. Ultrasound contrast agents are typically gas-filled colloidal particles (microbubbles) with diameters less than 10 μm. The surface comprises amphiphilic phospholipids self-assembled to form a lipid monolayer shell. Microbubbles can provide a sensitive acoustic response when detected using ultrasound because of the compressible gas core. Poly (ethylene glycol) chains can be incorporated into the shell of microbubbles in order to form a steric barrier against coalescence and adsorption of macromolecules. Owing to their small sizes, these agents can pass through the pulmonary vasculature and may exhibit contrast persistence longer than 10 min *in vivo*. Image courtesy of Dr. Mark Borden, Nicholas Rome Faculty Fellow, Department of Mechanical Engineering University of Colorado, Boulder, CO, USA.

resonance frequency. These frequencies for microbubbles of around 7 μm correspond to the frequencies used in diagnostic ultrasound, giving extremely high contrast. This can be enhanced even further by use of harmonics at high acoustic power. As this power increases, so the expansion and contraction phases become unequal, the response being nonlinear, giving rise to the harmonics with greater contrast.

## 7.6 Optical imaging

Optical imaging is increasing in importance as a group of diagnostic technologies, offering a number of advantages over more traditional radiological techniques. Primarily, optical imaging methods use non-ionizing radiation, significantly reducing patient radiation exposure and extending the time during which observations may be made. They offer the possibility of differentiating soft tissues, which may sometimes be enhanced through the use of endogenous or exogenous contrast media, with different photon absorption or scattering profiles at different wavelengths, yielding potential capabilities for capturing both functional and molecular level activities. Thirdly, optical imaging is generally amenable to multimodal imaging, often being readily combined with other, complementary imaging techniques.

These techniques are able to integrate the principles of microscopy with state-of-the-art optics technology. Some of the techniques recently introduced include spectral imaging, two-photon fluorescence correlation and optical coherence tomography. The latter, OCT, is a good example of the technologies that may be now used for performing high-resolution cross-sectional imaging. It is analogous to ultrasound imaging but uses light instead of sound. It can provide cross-sectional images of tissues at the micron scale *in situ* and in real time. Used in conjunction with catheters and endoscopes it can function as a type of optical biopsy, without the need for removal of a tissue specimen and processing for microscopic examination. This is advantageous where excisional biopsy is hazardous or impossible, and may be used to guide interventional procedures.

As interesting and important as this is, we can see that not all optical imaging techniques require the use or support of biomaterials, the contrast required for the imaging usually being derived from the differing photon absorption and/or characteristics of different tissues. There are some systems, however, which fundamentally depend on the use of exogenous agents in order to provide the contrast. Some of the principles here have been introduced in Chapter 2 in relation to the optoelectronic properties of materials.

The discussion in Chapter 2 concentrated on fluorescence, which is the emission of light by substances that have absorbed electromagnetic radiation of a different wavelength. Fluorescence is used as a non-destructive way of tracking or imaging biological molecules by means of the fluorescent emission at a specific frequency, particularly where there is no background from the excitation light, relatively few cellular components being naturally fluorescent. A protein or other component can be labeled with an extrinsic fluorescent dye, known as a fluorophore.

Conventional fluorophores may be grouped as either organic dyes or biological molecules. The small size of some organic dyes, including fluorescein, a synthetic organic dye and its derivatives such as fluorescein isothiocyanate, means that they can be linked to macromolecules such as antibodies without interfering with their biological function. The use of biological fluorophores was initiated when a green fluorescent protein, known as GFP, was synthesized from jellyfish and applied as a gene expression reporter. Several other biological fluorophores have been developed and they may be used in cells or whole organisms to study biological processes.

There were some deficiencies with both organic dyes and biological fluorophores so that other agents, were introduced, and it is these that are discussed in this chapter. These primarily include quantum dots and upconversion nanoparticles.

### 7.6.1 Upconversion nanoparticles

When discussing the optical properties of materials, it is useful to distinguish between linear and non-linear phenomena. With linear optics, the optical properties are independent of the intensity of the incident light. Non-linear optics covers situations where this is not the case, and where the optical properties depend on the radiant flux density of the exciting light. One of the more important non-linear phenomena is photon upconversion. In this there is conversion of long wavelength radiation, especially infrared or near infrared, to short wavelength radiation, usually in the visible range. There are several mechanisms by which this occurs, a detailed discussion of which is beyond the scope of this book, but they are all associated with the sequential absorption of two or more photons by long-lived but metastable energy states. This leads to the population of a highly excited state from which upconversion emission occurs.

There are several important requirements for this photon upconversion. These include a ladder-like arrangement, with similar spacing, of the energy levels, and long lifetimes of the excited states. Certain groups of elements provide these characteristics. It will be recalled from Chapter 2 that transition elements may be defined as the elements whose atoms or simple ions contain partially filled d-orbitals. The two series of inner transition elements, the lanthanoids and actinoids, known as rare earth elements, are particularly important here. A number of these transition metal ions show upconversion. Some of these, such as $Ti^{2+}$, $Ni^{2+}$, $Mo^{3+}$ and $Os^{4+}$, may do so, but have limitations concerning the low temperatures necessary to see the effect and the quality of the optical properties. Much better efficiencies are seen with the lanthanides, particularly with $Yb^{3+}$ and $Er^{3+}$. Structures of upconversion nanoparticles are shown in Figure 7.6(a).

These rare earth metals are not used by themselves, but they are used to dope other substances, which are ideally prepared as nanoparticles. The really critical features are the nature of the host lattice and the type and concentration of the dopant. The host lattice

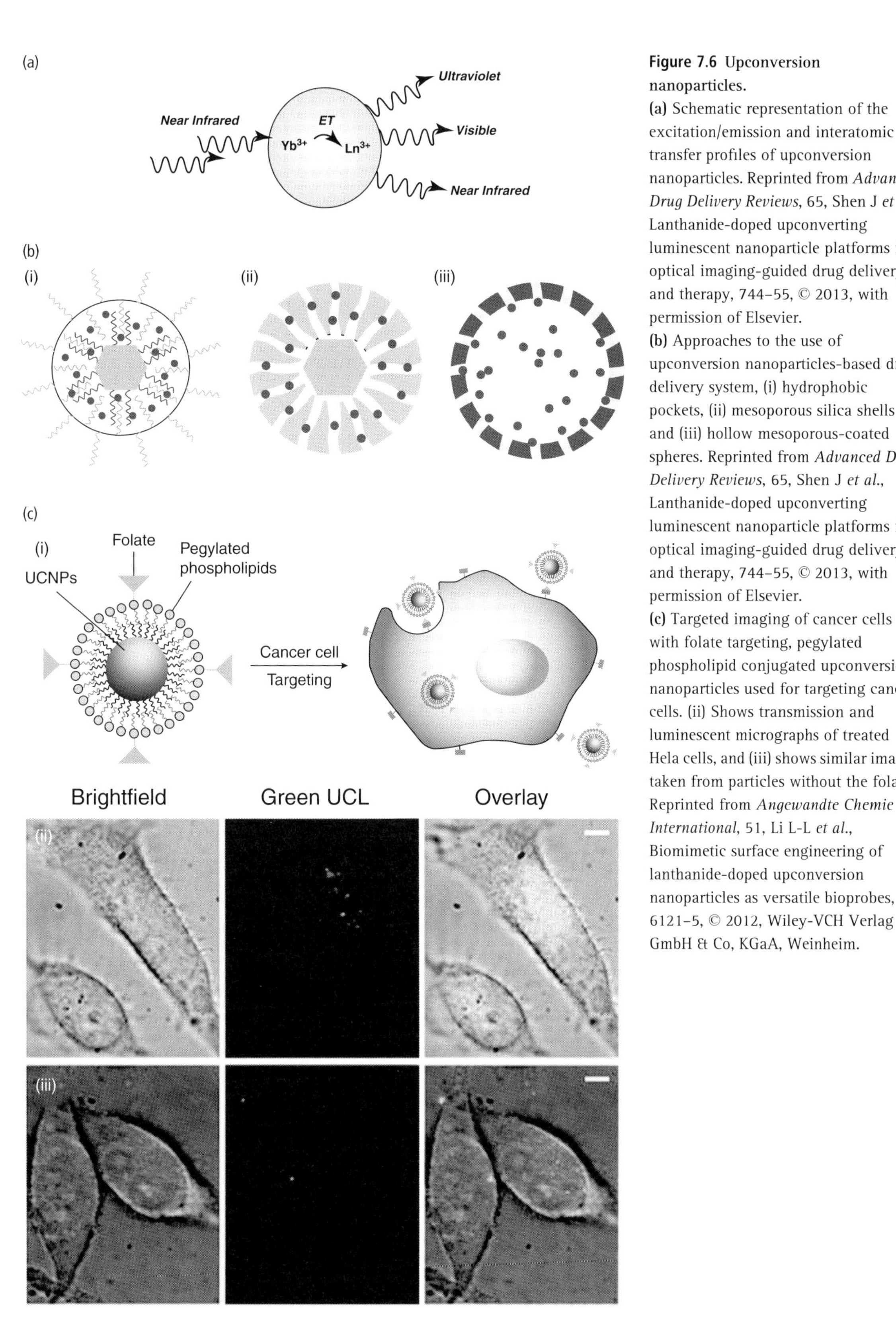


**Figure 7.6** Upconversion nanoparticles.
**(a)** Schematic representation of the excitation/emission and interatomic transfer profiles of upconversion nanoparticles. Reprinted from *Advanced Drug Delivery Reviews*, 65, Shen J *et al.*, Lanthanide-doped upconverting luminescent nanoparticle platforms for optical imaging-guided drug delivery and therapy, 744–55, © 2013, with permission of Elsevier.
**(b)** Approaches to the use of upconversion nanoparticles-based drug delivery system, (i) hydrophobic pockets, (ii) mesoporous silica shells and (iii) hollow mesoporous-coated spheres. Reprinted from *Advanced Drug Delivery Reviews*, 65, Shen J *et al.*, Lanthanide-doped upconverting luminescent nanoparticle platforms for optical imaging-guided drug delivery and therapy, 744–55, © 2013, with permission of Elsevier.
**(c)** Targeted imaging of cancer cells with folate targeting, pegylated phospholipid conjugated upconversion nanoparticles used for targeting cancer cells. (ii) Shows transmission and luminescent micrographs of treated Hela cells, and (iii) shows similar images taken from particles without the folate. Reprinted from *Angewandte Chemie International*, 51, Li L-L *et al.*, Biomimetic surface engineering of lanthanide-doped upconversion nanoparticles as versatile bioprobes, 6121–5, © 2012, Wiley-VCH Verlag GmbH & Co, KGaA, Weinheim.

provides an appropriate crystal structure while the dopant provides the luminescence centers, the spatial arrangement of which controls the efficiency and performance.

### The host lattice

The host lattice should have good transparency within the relevant wavelength range, with low phonon energy and a high threshold for optical damage. There should be close matching of the lattice parameters between the host and the dopant since the excited energies of the dopant ions may be absorbed by the host material through lattice vibrations, which will depend on these parameters. The general characteristics of the crystal structure are important since they will determine the overall optical properties of the nanoparticles. A great deal depends on the route of synthesis of the nanoparticles; the process must yield highly crystalline structures if efficient emissions are to be produced. It is also important for the nanoparticles to have a small particle size and high dispersity in order for them to be integrated into biological tissues.

Commonly used host crystals are given in Table 7.2. The element yttrium (Y) features very prominently in this list. Some early forms of Y-based crystals include $Y_20_2S$ and $Y_20_3$, but the structure $NaYF_4$ appears to be the most favored host lattice at this time.

Table 7.2 **Host lattices used in upconversion nanoparticles**

| Host lattice | Absorbing ion | Emitter ion | Emission color |
|---|---|---|---|
| $Y_20_2S$ | Ytterbium | Erbium<br>Thulium | Green<br>Blue |
| $Gd_20_2S$ | Ytterbium | Erbium | Red |
| YOF | Ytterbium | Thulium | Blue |
| $Y_30Cl_7$ | Ytterbium | Terbium | Green |
| $YF_3$ | Ytterbium | Erbium | Red |
| $NaYF_4$ | Ytterbium | Thulium | Blue |
| $Y_3Ga_50_{12}$ | Ytterbium | Erbium | Green |
| $YSi_20_5$ | Ytterbium | Holmium | Green |

### The dopant

As noted above, a small number of the lanthanide rare earth elements form the basis of the best dopants in upconversion nanoparticles. The vast majority of those used and investigated involve ytterbium (Yb), erbium (Eb), thulium (Tm) or a combination of these. Thus typical nanoparticles are $NaMF4{:}Yb3^{+}/Ln^{3+}$, where M is yttrium or gadolinium, and Ln is erbium, thulium or ytterbium.

### Optical properties

The absorption and emission spectra of the lanthanide ions arise from transitions of 4f electrons, and provided that the host lattice has been matched carefully, as discussed above, there should be few interactions between these electrons and that host lattice. The spectra show sharp lines. The resulting narrow absorption profile normally imposes significant constraints on the excitation source but laser systems are available at around 980 nm wavelength, which matches the absorption of the $Yb^{3+.}$ The emission peaks are also essentially independent of the host material and the emission colors are determined by the dopant concentration and emission wavelength.

Since the emission arising from these electron transitions do not cause any chemical bond breaks, the nanoparticles are largely immune to photobleaching or photodegradation, which are serious concerns for many fluorophores. The transitions also give long luminescence times, which allows for time-resolved luminescence detection, minimal interference from biological tissues and enhanced signal-to-noise ratios.

### Cytotoxicity and biocompatibility

Clearly there will always be concerns about possible adverse effects of injected nanoparticles. In general, and in contrast to many other imaging agents, these rare earth nanoparticles appear to show minimal cytotoxicity. Although few comprehensive studies have been performed, little or no cytotoxicity is observed with standard *in vitro* tests performed on these nanoparticles. Biodistribution studies tend to show few, if any, gross changes in the health of experimental animals. Intravenously injected nanoparticles are excreted through urine and feces fairly quickly, with a rapid accumulation and subsequent depletion in the lungs. Although many of the usual target organs will show some accumulation of the ions, especially Y, there is no sustained presence after a few days and no long-term associated pathology.

### Surface modification and functionalization

Upconversion nanoparticles, prepared by conventional nanotechnology synthesis routes, have no functional organic groups on their surfaces and they will have no intrinsic aqueous solubility. Their size will usually be controlled by capping ligands, which also stabilize against aggregation in solution. These characteristics suggest that the nanoparticles are intrinsically unreactive, which may account for good biocompatibility, but also that there is no biological functionality. As with many nanoparticles used in medical technologies, some biological functionality is usually beneficial.

Methods to functionalize the nanoparticles tend to follow four processes (Figure 7.6(b)):

- Ligand engineering through exchange processes: the hydrophobic ligands that naturally form on the nanoparticle surface may be replaced by hydrophilic organic molecules.
- Ligand engineering through attraction processes: this involves the absorption of amphiphilic polymeric molecules.
- Surface polymerization: the condensation of small monomers onto the nanoparticle core to give a dense cross-linked shell.
- Layer-by-layer assembly: this involves the electrostatic absorption of alternately charged polyions on the surface.

In addition, there are some direct coating methods such as with silica, which can give very stable surfaces. Silica can be deposited using silanes such as 3-aminopropyltriethoxysilane, the surface amino groups being used for the attachment of other biological molecules. These molecules include proteins, antibodies, peptides and nucleotides.

### Immunohistochemistry and other pathological techniques

Upconversion nanoparticles were first used in medical applications in the area of immunohistochemistry. We will not dwell on this topic for long since these nanoparticles, being used *ex vivo* do not really qualify as biomaterials. It should be noted, however, that their advantages became fairly obvious with respect to the lack of autofluorescence due to the very precise and narrow 980 nm excitation peak, the much improved contrast and the lack of photodecomposition over many months.

### *In vivo* imaging

The move from laboratory diagnostic techniques towards *in vivo* imaging started with the use of upconversion nanoparticles for cellular imaging *in vitro* and then imaging in small animals. It proved possible to track live cells *in vitro* by attaching the nanoparticles to cancer cells and imaging the particle internalization and the effects on those cells. Injection of nanoparticle-loaded cells into the vasculature of mice allows monitoring of those cells within the bloodstream with high signal-to-noise ratios. Examples are shown in Figure 7.6(c).

## 7.6.2 Quantum dots

The general principles of quantum dots have been discussed in Chapter 2 and their position within the overall classification of biomaterials will be given in Chapter 8. QDs are semiconductor nanocrystals (1–10 nm diameter) that show luminescent properties with absorption and emission bands that are defined by the band gap energy of the parent material and their size. They are generally derived from combinations of Periodic Table groups II and VI elements, such as CdSe, CdTe, CdS, CdHg and ZnS, or combinations of groups III and V elements, such as InAs, InP and GaAs, are possible.

QDs have been conventionally manufactured by injection of liquid precursors into organic solvents, where crystal growth occurs under controlled conditions. The process usually proceeds with the epitaxial growth of a layer of a substance such as ZnS around a core, which increases quantum yield and protects the nanocrystal from oxidation and leaching. Other techniques involve the injection of elemental precursors into hot surfactants, which causes the nucleation of very small nanocrystals, which are then allowed to grow by dropping the temperature. The production process can be designed with tight tolerances so that the emission spectra are similarly tight and symmetrical. They are, therefore, extremely bright and highly tunable.

QD crystals are essentially spherical. The larger the QD, the smaller the band gap which results in the emission of red light. Smaller QDs emit blue light because of their higher energy. They tend to have long fluorescence lifetimes (up to 40 ns) compared to the few nanoseconds for organic dyes. As with upconversion nanoparticles, they also show resistance to photobleaching. Typical structures of QDs are shown in Figure 7.7(a).

For biological applications, further manipulation steps are required because these QDs are not water-soluble but are likely to be cytotoxic. Attempts to address the solubility issues have been made by exchanging non-polar ligands on the surface with polar ones. Initially this was achieved with thiol functional groups but these tended to quench the fluorescence, reduce the brightness and decrease stability. Other steps involved cross-linking small ligands onto the surface, with conjugation to PEG, cysteine, polyallylamine and lysine being used. Care has to be taken not to increase the size of the QDs by surface modification; ideally they should still be under 10 nm after coating. Silanol- and methoxy-based molecules are quite good in this respect. Encapsulating the hydrophobic QDs with amphiphilic molecules such as phospholipids has also provided a good strategy. In this way, the brightness that is associated with non-polar native ligands is preserved, the amphiphilic polymer effectively surrounding the QDs with polar, chemically reactive groups on the outer surface.

As with other medically employed nanoparticles, there is a further need to functionalize the QDs in order to provide specific targeting or biological activity. QDs may be decorated with peptides, proteins, nucleic acids and other biomolecules, typically through conjugation with reactive functional groups such as primary amines, thiols and hydroxyl and carboxyl groups. For example, a carboxylic acid terminating group on the QD surface may be linked to a terminal amine group on a biomolecule using the agent 1-ethyl-3-(3-dimethylaminopropyl) carbodiimide. Also the traditional streptavidin–biotin conjugation reaction, used extensively in biotechnology and immunology, may be used, conjugating the avidin/streptavidin on the QD to biotin in the biological molecule.

Antibody–QD conjugation is the most widely used. As noted above, antibodies may be biotinylated and used with streptavidin-coated QDs. Antibody-conjugated QDs have been used experimentally to target prostate-specific membrane antigens in the detection of prostate cancer and to image HER2 breast cancer cells. Peptide-functionalized QDs can target some cellular proteins, including integrins and growth factor receptors. One regularly encountered problem is the non-specific cellular binding of QDs.

### Immunohistochemistry and *in situ* hybridization

As with upconversion nanoparticles, the first bioimaging uses of QDs were in the pathological environment. Streptavidin-coated QDs used with biotinylated antibodies may be used as biomarkers in immunohistochemistry, particularly for the detection of tumors through the targeting of tumor molecules such as E-cadherin and vimentin. *In situ* hybridization is also a technique that has benefited from QDs. Conventional fluorescence *in situ* hybridization (FISH) is quite effective in cancer detection, but with the use of organic fluorophores is limited by quick photobleaching and spectral overlap. QDs have

(a)

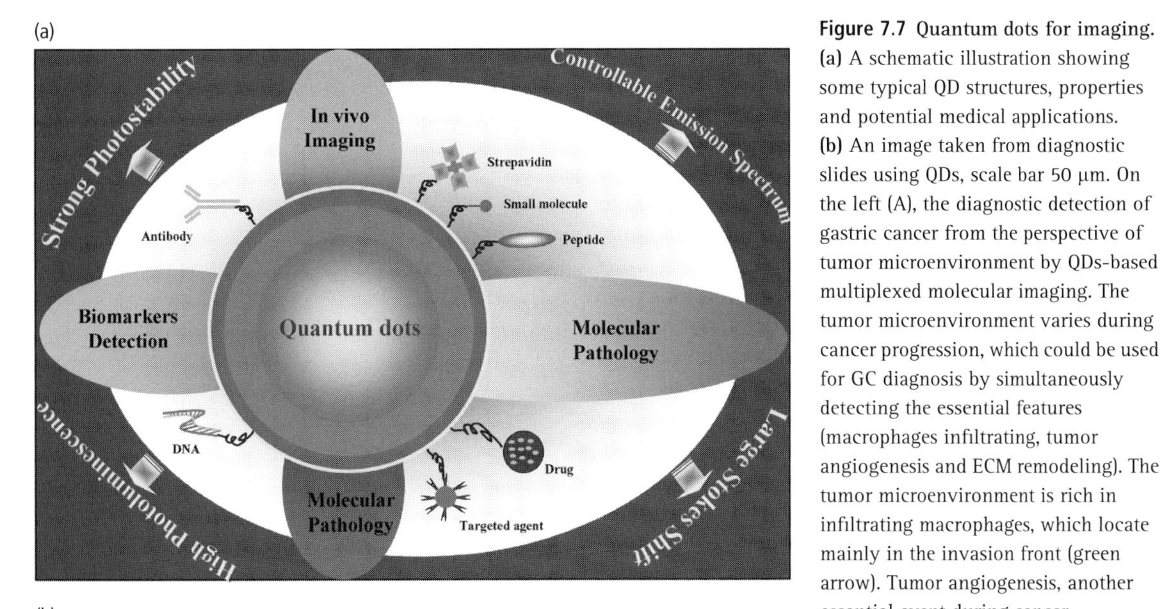


(b)

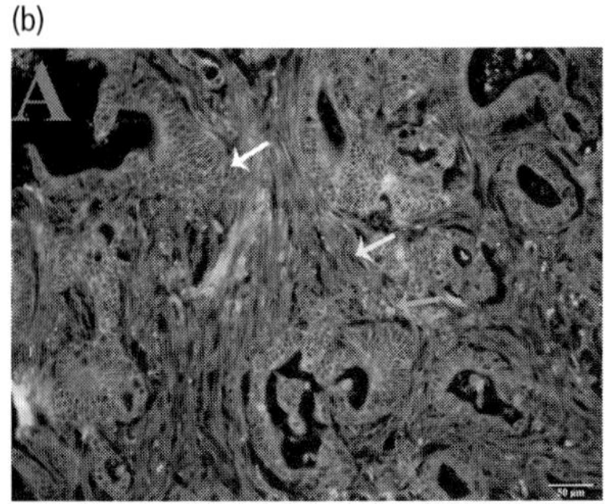


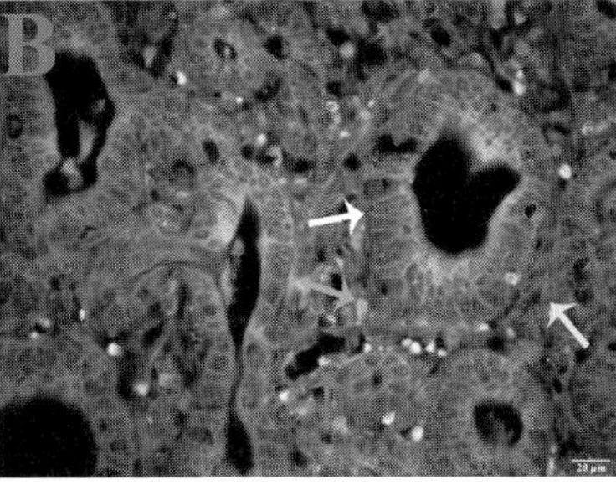


(c)

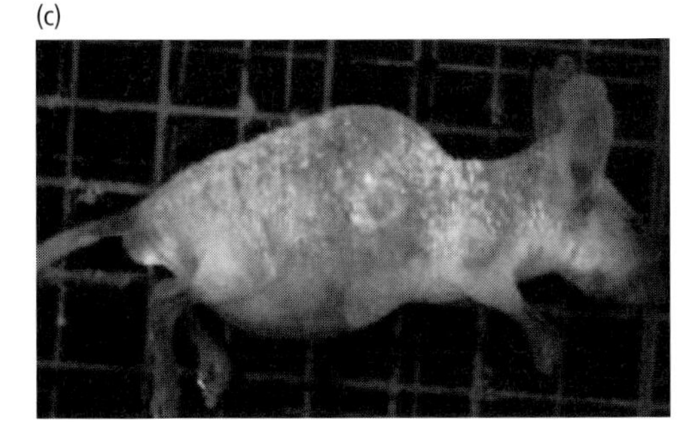

**Figure 7.7** Quantum dots for imaging. **(a)** A schematic illustration showing some typical QD structures, properties and potential medical applications. **(b)** An image taken from diagnostic slides using QDs, scale bar 50 μm. On the left (A), the diagnostic detection of gastric cancer from the perspective of tumor microenvironment by QDs-based multiplexed molecular imaging. The tumor microenvironment varies during cancer progression, which could be used for GC diagnosis by simultaneously detecting the essential features (macrophages infiltrating, tumor angiogenesis and ECM remodeling). The tumor microenvironment is rich in infiltrating macrophages, which locate mainly in the invasion front (green arrow). Tumor angiogenesis, another essential event during cancer progression, is represented by the neovessel in the stromal, which are remodeled by infiltrating macrophages and Lysine oxidase (Lox). Lox is a major molecule accounting for ECM remodeling. Lox is observed both in cancer cells (white arrow) and stromal cells (yellow arrow). On the right (B), a high-power field image to show the infiltrating macrophages, neovessels and Lox. Red arrow: neovessels; green arrows: macrophages; yellow arrows: Lox in stromal; white arrows: Lox in cancer cells; scale bar 20 μm. See plate section for color version. **(c)** *In vivo* imaging of subcutaneous tumor by quantum dots. Human hepatocellular carcinoma cell line HCCLM9 was inoculated subcutaneously into nude mice, and the tumor was imaged by CdSe/ZnS quantum dots-labeled alpha fetoprotein monoclonal antibody. The brown subcutaneous tumor was visible under the background of green auto-fluorescence of skin. See plate section for color version. All images courtesy of Professor Yan Li, Department of Oncology, Zhongnan Hospital of Wuhan University, Wuhan, China.

enhanced this technique, and QD-conjugates have been shown to be very effective in the labeling of membrane-associated HER2 receptor and of nuclear antigens in breast cancer cells. Attention is being paid to minimizing non-specific binding of QD probes to biomolecules, which reduces signal-to-noise ratio and sensitivity. Many QD probes are pegylated for this purpose. Examples of images obtained in these techniques are shown in Figure 7.7(b).

## Live cell and *in vivo* imaging

There is obviously a great deal of interest in extending the use of QDs to the imaging of live systems. This is not a trivial issue since QDs do not easily cross intact cell membranes and because QDs would be expected to have significant toxicity profiles in view of their metal content. Intracellular delivery has been addressed through engineered peptides and small biomolecules. Highly cationic cell penetrating peptides are partly effective but there are issues with lysosomal sequestration once the QDs are in the cells.

*In vivo* imaging is a particularly attractive option, potentially offering intraoperative image guidance, molecular targeting and real-time cell tracking. QD fluorescence imaging offers potential advantages over MRI, CT and PET with respect to sensitivity and spatial resolution. One particularly important area is in vascular imaging. The injection of QDs into the vasculature can allow the determination of blood circulation dynamics and their introduction into soft tissue can allow monitoring of the lymphatic system and transport mechanisms. It is important here that there is good stability and extended circulation times, with protective but non-fouling surfaces on the QDs. Some of the systems under investigation employ degradable polymer coatings and/or pegylation. Examples of *in vivo* images obtained with QDs are shown in Figure 7.7(c).

Further comments on these systems are made in the following sections on multi-modal imaging and theranostic systems.

## 7.7 Multi-modal imaging

We discussed in previous sections the use of individual imaging techniques. The advantages and disadvantages of each of these have been summarized in Box 7.2. When conventional equipment is used for these methods without contrast agents, they have to be considered as separate entities and clinicians have to select those that are most appropriate for the patient. With the advent of nanoparticle contrast agents, which may be designed to have different functionalities, it has become possible to combine imaging techniques in order to provide more information about that patient. This leads to so-called multi-modal imaging.

An important concept here is the use of techniques that combine the anatomical resolution of one mode with the quantification of functional information from another. The first multi-modal systems combined PET with either CT or MRI, where the PET provides the high sensitivity functional information and the CT or MRI give anatomical information. Other systems incorporating optical imaging are also very attractive.

The nanoparticles that may be used with multifunctional capability are usually combinations of organic structures, especially polymeric micelles, dendrimers or liposomes, with inorganic nanoparticles such as those based on gold, iron, carbon, silica and QDs. The organic components may carry radionuclides or near infrared dyes. Biological polymers may be preferred to synthetic polymers because of better general biocompatibility. With the inorganic nanoparticles, iron oxide is popular, again because of the lack of significant toxicity mentioned above and their good performance in MRI. An example could be an iron oxide-based MRI/optical probe where the nanoparticle is coated with cross-linked dextran and conjugated to a near infrared dye. Combined fluorescence optical and CT imaging may be achieved with gold nanoparticles, which have functionality in both systems.

## 7.8 Theranostics

The above discussion on multi-modality can be extended to combined imaging and therapy. This has been discussed in Chapter 6 and the term theranostics was introduced at that time. Once again it has been the emergence of functionalized nanoparticles that has prompted these developments. The concept involves the imaging of tissues before, during and after drug delivery through the use of a combined agent, in sufficient amounts, to the diseased area. These developments are at an early stage in terms of translation into clinical use.

Again, the preferred systems are based on iron oxide, gold and silica nanoparticles, carbon nanotubes and quantum dots. Iron oxide nanoparticles may be coated with suitable linkers and coupled to various drug molecules. For example, the surface may be aminated and then coupled to chemotherapeutic agents such as paclitaxel, methotrexate or doxorubicin. The particles may also be coupled to Herceptin antibody molecules. In addition to coupling of drug molecules to particle surfaces, these molecules may also be co-encapsulated with the nanoparticles in polymer matrices or contained within hollow nanoparticles. Cisplatin has been incorporated within porous iron oxide nanoparticles.

With gold nanoparticles, thiol–gold conjugation is the preferred route to functionalization and a number of drugs have been attached to them. Paclitaxel has been covalently coupled to 4-mercaptophenol modified gold nanoparticles. Chitosan is also an effective coating agent which, being highly positively charged, is able to bind to drugs through electrostatic interactions. In addition, these nanoparticles are good candidates for the delivery of genes. The nanoparticles may be functionalized, for example with alkylated quaternary ammonium, to which plasmid DNA may be attached.

Silica is also of interest. Silica nanostructures do not have intrinsic characteristics for imaging but they may provide a platform onto which small iron oxide nanoparticles or organic dyes may be attached, and into which drugs may be loaded. This possibility has been enhanced by the use of mesoporous silica, which has very small nanopores that can be loaded with significant quantities of drug molecules.

## 7.9 Biosensors

### 7.9.1 The principles of sensing chemical and biological agents in tissues

Detection of chemical and biological agents plays a major role in many aspects of modern life, including biomedical technologies. The development of highly sensitive, cost-effective, appropriately sized sensors requires input from many sciences and advanced technologies within the spheres of chemistry, physics, biology and material sciences. Sensors generally feature two functional components. First there has to be a recognition element that provides selective and specific binding with the target analytes. Secondly, there is a transducer component that produces signals that correspond to the binding event. The performance of both of these components control the recognition process in terms of response time, signal-to-noise ratio, selectivity, and limits of detection. In many situations the key feature that has improved performance in these areas has been the development of nanoscale materials.

There are several ways of classifying chemical/biological sensors depending on whether the emphasis is on the recognition element or the transducing components. In medical technology, the recognition element is usually of a biological nature and the vast majority of sensors used in diagnosis are usually referred to as biosensors. These elements could be immobilized enzymes, immunological agents, DNA, whole cells, cell surface receptors and so on. There are several types of transducer, which could be piezoelectric or mechanical, but are usually considered to be optical or electrochemical. The general working scheme for biosensors is depicted in Figure 7.8.

We discuss here the essential principles of these types of biosensor and the biomaterials used in their construction, and on which the performance depends.

### 7.9.2 Enzyme-based biosensors

Enzyme biosensors are probes that have a thin layer of an enzyme immobilized onto the surface of a working electrode. Since enzymes are highly substrate-specific, these sensors should be selective, the enzyme catalyzing the formation of an electroactive product. This can be detected by amperometry, where the current that is produced is measured in response to a constant applied voltage. Although these biosensors have many advantages, including

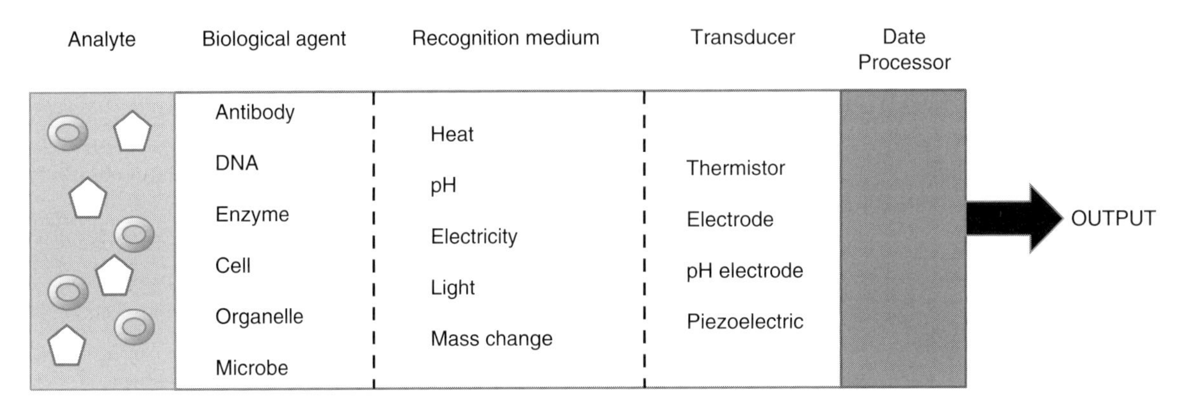


**Figure 7.8** Biosensors. This general scheme shows the pathways from exposure of a detecting biological agent to a sensor surface through to transduction of the resulting signal and data processing to give the output.

low cost and reusability, there are not too many analytical systems where enzyme activity is the most appropriate entity to analyze. Glucose monitoring has the longest history and most widespread use and constitutes a good basis for discussion. In the early forms, the enzyme was glucose oxidase (GOx), this being inexpensive, readily available and stable. Oxygen that was present in the fluids being analyzed acted as a co-substrate for optimal function of the enzyme. Generally the need for oxygen can be eliminated by the use of so-called artificial redox mediators, which are small soluble molecules that act as electron shuttles in these sensors. A great deal depends on the circumstances in which the electrode is used. The mediators are very effective in short-term monitoring situations, but the leaching of the mediator and toxicity limit their use in implantable configurations.

The continuous monitoring of glucose for closed-loop diabetes management has represented a challenge for biosensors. Implantable, *in vivo*, glucose biosensors are key components of these closed-loop systems since they cannot be directly accessed and must operate reliably and continuously over a prolonged period of time. The majority of glucose sensors are amperometric electrodes where GOx catalyzes the oxidation of glucose to gluconic acid and hydrogen peroxide, the latter being reduced to oxygen, hydrogen and electrons, which are detected. A typical device is a subcutaneously implanted needle-type electrode that will have a Ag-AgCl reference electrode. A Pt–Ir wire is the working electrode that is coated with PTFE, except for the small area of the sensing cavity, where the enzyme is incorporated into cellulose acetate/polyurethane layers. The main problem with these configurations is the deposition of proteins and other macromolecules on the surface of the materials, which inhibits the response. Various coating materials have been used to minimize this biofouling effect, including phospholipids, PEG and nitric oxide-releasing polymers.

### 7.9.3 Immunosensors

Many diseases may be assessed by the presence in tissues of certain cellular markers. Techniques that permit the detection, and preferably the sophisticated quantification, of these markers, are essential. This is particularly true for cancer, where the detection of tumor markers is of immense importance. Immunosensors are widely believed to provide the best options here. These techniques combine the high specificity of immunological reactions with the high sensitivity of the various methods that are available for the detection of the products of these reactions. In line with comments made above, these detection systems fall into one of the following types:

- electrochemical systems, including amperometric and potentiometric methods;

- optical systems, including fluorescence and luminescence;
- gravimetric;
- thermometric; and
- miscellaneous systems.

Among the various biomaterials that have been introduced into immunosensor technology recently are some nanotechnology-based materials. Most prominent here are carbon nanotubes, gold and iron oxide nanoparticles and nanowires. Immobilization of the relevant biological species on a substrate is essential for detection, and some nanostructures are able to provide protection and stabilization to these species. Single walled carbon nanotubes are particularly attractive since they are able to promote the electron transfer reactions that are necessary for measurement of these reactions. Their very high aspect ratio also provides a high surface area for them. The type of device that has been proposed here could include a microarray of platinum metal electrodes that are surface modified by carbon nanotubes, to which monoclonal antibodies to the relevant antigen are covalently attached. These could be, for example, monoclonal antibodies against total prostate-specific antigens, which may be attached to the nanotubes by ester linkers.

Gold nanoparticles are also of interest: tumor marker antibodies can be covalently attached to these nanoparticles following modification by monolayers of glutathione, and the nanoparticles themselves immobilized on gold electrodes through copolymerization with suitable agents. These, and other similar systems, are on the verge of routine clinical use. There are still issues of non-specific adsorption of biomolecules and a need for improved signal amplification. It is also possible for aptamers to be used in place of antibodies and for microfluidic devices to improve control over the reactions. A further alternative involves the use of so-called protein chips, which consist of immobilized biomolecules on planar substrates such as glass. These molecules, which could be proteins, peptides or oligonucleotides, are arranged in arrays on the substrate, which are probed with substances such as cellular extracts that monitor molecular recognition events.

## 7.10 General comments on nanoparticle toxicity in imaging systems

There are two aspects to consider with respect to toxicity and imaging contrast agents. The first concerns the actual adverse events that have been seen in patients. The second refers to the mechanisms by which agents can, theoretically and actually, interact with biological systems such as cells and DNA. We shall concentrate on the first of these here. We discussed biocompatibility at the nanoscale in Chapter 3 and we shall refer to just a few issues here that concern specific nanoparticle formulations.

### 7.10.1 Clinically observed effects

There is a wide range of general adverse effects that have been seen in patients who have been administered contrast agents. As usual when any chemicals, whether therapeutic or diagnostic, are administered intravenously, finite incidences of mild conditions such as nausea, vomiting, headaches and skin conditions will occur. Moderate effects include hypotension and bronchospasm and severe effects can include convulsions, cardiac and pulmonary collapse, pulmonary edema and unconsciousness. Some delayed adverse reactions such as flu-like and allergy-like reactions may also be seen. There are also some quite specific adverse effects. Intravascular iodinated agents can cause acute kidney injury, referred to as contrast induced nephropathy (CIN), which is associated with a marked increase in serum creatinine and a decline in glomerular filtration rate. Incidences appear to range from 10 to 20%, the level depending on the exact formulation used. Pre-exisiting renal insufficiency is, not surprisingly, a significant risk factor.

In MRI, as we have seen, Gd-based contrast agents are prominently used. These are not without problems, although at a lower incidence, and again it is the kidney that is primarily affected. A number of cases have been reported of a potentially severe, even fatal, reaction, specifically nephrogenic systemic fibrosis (NSF). This is a very rare condition that affects the skin, accompanied by pain and loss of mobility, but with systemic involvement of other organs particularly the kidneys. However, with patients who are on dialysis because of end-stage renal failure and undergo MRI using these contrast agents, about 3–6% may suffer severe NSF.

### 7.10.2 Nanoparticle toxicity and biodistribution

The biological effects of gold nanoparticles as used in imaging are dependent on size, dose and surface chemistry. Generally these particles are distributed in all tissues following intravenous injection, especially liver, lungs and spleen. Particles less than 15 nm are the most widely distributed and there is far less organ accumulation with particles over 200 nm. It is possible for particles less than 50 nm to cross the blood-brain barrier, this being dependent on surface functionalization. Small particles are also able to be absorbed in the gut. A large number of animal studies have monitored this distribution but only few noticeable effects have been seen. There are occasional observations of effects on gene expression and generalized adverse effects at high doses, but no serious hematological or pathological effects have emerged.

Iron oxides, including magnetite $Fe_3O_4$ and maghemite $Fe_2O_3$, are widely used and have good records for biocompatibility. Iron ions derived from the nanoparticles are incorporated into the body's iron pools and eventually incorporated into hemoglobin.

Greater concerns have been expressed with those nanoparticles based on more aggressive metals. As we have seen, the more effective QDs include cadmium. With CdSe QDs cytotoxic effects can be correlated with the release of free $Cd^{2+}$ ions. Unfortunately, the release of these ions is exacerbated by the degradation of the QDs under the acidic conditions found in the GI tract. The QDs themselves tend to target the liver, spleen and lymphatic systems, the circulation and clearance rates being controlled by size and surface coatings. Toxic effects are strongly related to the intracellular QDs amounts which localize in endosomes, where they generate reactive oxygen species, interfere with cell cytoskeleton and leach ions due to dissolution, resulting in increased toxicity and impeded QD fluorescence. Partitioning of QDs upon recurrent cell division results in a rapid loss of particles in live cells, limiting the use of QDs for long-term imaging.

The potential toxicity of Cd-based QDs has led to a search for non-Cd QDs. Included here are those based on silver and silicon.

## Summary and learning objectives

1. Before any therapy can be considered, clinicians have to be aware of the nature of the condition that is affecting the patient. This has led to the development of imaging techniques. The first attempts at anatomical imaging did not involve biomaterials since they involved the passage electromagnetic radiation through the tissue and a visual representation of the attenuation of the radiation by a sensitive film. These methods may be enhanced by the use of radiopaque dyes, the first form of contrast agent. Simple contrast agents were limited in terms of accuracy and did not give information about the functional characteristics of tissues. New methods were therefore required in order to improve contrast and provide data on physiological function. This chapter is concerned with the development of new biomaterials for a variety of imaging and sensing techniques.

2. Contrast agents need two active components. The first is an agent, such as an antibody, a peptide or some small molecule that can bind to the target molecule. The second is a label that can be detected by the imaging modality, using magnetic, optical or other methods. There will need to be a mechanism of binding these two principal agents together.
3. The magnetic properties of tissues are dominated by the hydrogen atom of water. Bursts of specific radiofrequencies can excite H atoms into higher energy states and change the magnetic state. An image can be created on the basis of the return of the magnetic state back to its original orientation. This is known as relaxation, and is fundamental to the generation and interpretation of magnetic resonance images. Relaxation constants, which are time constants concerned with relaxation processes, are material dependent and control the nature of the images obtained. About one-third of MRI scans now make use of contrast agents, which allow for better tissue characterization and specificity and greater amounts of functional information. These are substances that include paramagnetic metal ions that alter relaxation constants of the tissues, influencing either T1 or T2 agents. Most T1 agents are based on $Gd^{3+}$, which has seven unpaired electrons. T2 agents are superparamagnetic, especially the oxides of iron, which induce large distortions of externally applied magnetic fields and give negative contrast since areas that contain the particles display fast transverse relaxation rates.
4. CEST is chemical exchange saturation transfer. CEST agents are exogenous mobile molecules that facilitate the transfer of saturation magnetization from tissue mobile protons to bulk water during radiofrequency irradiation. These agents are either diamagnetic or paramagnetic. Particularly important are the paramagnetic complexes (referred to as PARACEST) that contain the biologically important transition metal ion $Fe^{2+}$ and also those based on $Eu^{3+}$ and $Tb^{3+}$.
5. Computerized tomography involves the collection of localized information from slices of tissue on which an X-ray beam is focused, which is used to create highly specific anatomical images. This technique has high spatial resolution but relatively poor contrast resolution. Intravenously administered contrast agents may be used in order to enhance the contrast. Because of its high radiopacity, most CT contrast agents are based on iodine. The incorporation of iodine into a benzene ring gives a core molecule, 2,4,6-triiodinated benzene, from which most agents are derived.
6. PET is positron emission tomography which utilizes radionuclides that decay within the tissue through the emission of positrons. The radionuclides form part of the imaging agent; $^{11}C$ and $^{18}F$ are the most common.
7. Ultrasound techniques are normally non-invasive, without any device or material being placed within the body. However, using the reflection of sound waves from tissues and fluids, without any assistance from exogenous agents limits the resolution and effectiveness of ultrasound scans. Significant improvements have been made by the use of contrast agents based on microbubbles. The microbubbles are typically 1–4 μm in diameter, consisting of a stabilizing shell and an internal gas, such perfluoro gas–phospholipid membranes.
8. Optical imaging methods use non-ionizing radiation, which offer the possibility of differentiating soft tissues. These may be enhanced by the use of endogenous or exogenous contrast media, with different photon absorption or scattering profiles at different wavelengths. Both organic dyes and biological fluorophores are used but they have limitations and several types of nanoparticles have been introduced. There are two main forms, upconversion nanoparticles and quantum dots. Nanoparticles for the former are typically

$NaMF4:Yb3^{+}/Ln^{3+}$, where M is yttrium or gadolinium, and Ln is erbium, thulium or ytterbium. The most widely used QDs are based on CdSe.

9. With both of these forms of nanoparticle, their effectiveness is enhanced by surface modification, which allows for targeting to specific biological features. There are concerns about the toxicity of most of these particles, especially Gd- and Cd-based systems, and these dangers have limited their clinical use. This has also led to the development of nanoparticles of reduced toxicity and more effective biodistribution.
10. The discussion of these nanoparticle-based imaging systems concludes with messages concerning combinations. More than one type of nanoparticle, or different functionalities within the same nanoparticle system, may be combined to give multi-modal imaging. Combinations of imaging enhanced nanoparticles with drug elution techniques has led to the introduction of theranostics, the merging of therapies with imaging.
11. Finally, we discuss biosensors, which have two functional components, a recognition element that provides binding with the target analytes and a transducer that produces signals that correspond to the binding event, allowing the detection of critical substances in body fluids, such as glucose, and the change in concentrations in real time. Critical biomaterials issues here involve the prevention of biomolecule accumulation on the sensing electrode surfaces, which impedes monitoring of the target substance.

## Questions

1. Discuss the essential principles that underpin the introduction of biomaterials-based contrast agents into molecular imaging techniques.
2. Describe the structure of T1 and T2 MRI contrast agents and discuss how these can improve the quality of MRI images.
3. Compare and contrast the clinical applications of PET, SPECT and CT imaging modalities and the contrast agents that may be used.
4. Discuss the mechanisms by which microbubbles can enhance ultrasound imaging.
5. Describe the phenomenon of upconversion as it relates to optical imaging. What are the features that determine the suitability of nanoparticles for this mode of image?
6. Discuss the mechanisms by which optical imaging nanoparticles can have toxicological effects on patients.
7. Discuss the influence of protein adsorption on the performance of implantable biosensors and the options for controlling the effects of this adsorption.
8. Explain the principles of both multi-modal imaging and theranostics and the contribution that nanoparticles can make to these processes and mechanisms.

# Recommended reading


Byers R J, Hitchman E R, Quantum dots brighten biological imaging, *Progress in Histochemistry and Cytochemistry* 2011;**45**:201–37. *Review of the uses of quantum dots in histopathology and in vivo imaging,*

Cao T, Yang Y, Gao Y, *et al.*, High-quality water-soluble and surface functionalized upconversion nanocrystals as luminescent probes for bioimaging, *Biomaterials* 2011;32:2959–68. *Discussion of the use of functionalization of upconversion luminescent nanoparticles.*

Chatterjee D K, Gnanasammandhan M K, Zhang Y, Small upconverting fluorescent nanoparticles for biomedical applications, *Small* 2010; 24:2781–95. *Introduction to upconverting fluorescent nanoparticles, their surface modification and functionalization and their prospects for biomedical imaging applications.*

Chen H, Jiang C, Yu, C, *et al.*, Protein chips and nanomaterials for application in tumor marker immunoassays, *Biosensors and Bioelectronics* 2009;24:3399–411. *Review of the role of nanomaterials in the diagnosis and understanding of cancer progression and etiology.*

Cheng L, Yang K, Li Y, *et al.*, Multifunctional nanoparticles for upconversion luminescence/MR multimodal imaging and magnetically targeted photothermal therapy, *Biomaterials* 2012;33:2215–22. *Experimental paper dealing with the preparation of upconversion nanoparticles and their application in multi-modal imaging.*

Dastru W, Longo D, Silvio A, Contrast agents and mechanisms, *Drug Discovery Today – Technologies* 2011; 8:e109–115. *Commentary on the status of contrast agents.*

Dawson P, Functional imaging in CT, *European Journal of Radiology* 2006;**60**:331–40. *The theoretical basis and applications of computed tomography.*

Haase M, Schafer H, Upconverting nanoparticles, *Angewandte Chemie International Edition* 2011;**50**:5808–23. *Review of the scientific basis of lanthanide-doped upconversion nanoparticles and their potential for medical imaging.*

Hahn M A, Singh A K, Sharma P, Brown S C, Moudgil B M, Nanoparticles as contrast agents for *in vivo* bioimaging: current status and future perspectives, *Annals of Bioanalytical Chemistry* 2011;**27**:3–27. *Extensive review of the role of nanotechnology and nanoparticles in in vivo imaging methods.*

Hasebrook K M, Serkova N J, Toxicity of MRI and Ct contrast agents, *Expert Opinion on Drug Metabolism and Toxicity* 2009;5:403–16. *Discussion of the major adverse events that can be associated with the clinical use of contrast agents.*

Kimmel D W, LeBlanc G, Meschievitz M E, Cliffel D E, Electrochemical sensors and biosensors, *Analytical Chemistry* 2012;**84**:685–707. *Review of electrochemical and biochemical sensors, especially potentiometric, voltammetric and immunosensors.*

Lee D-E, Koo H, Sun I-C, *et al.*, Multifunctional nanoparticles for multimodal imaging and theranostics, *Chemical Society Review* 2012;**41**:2656–72. *A tutorial review that highlights advances in the development of multifunctional nanoparticles for use in medicine.*

Li L-L, Zhang R, Yin L, *et al.*, Biomimetic surface engineering of lanthanide-doped upconversion nanoparticles as versatile bioprobes, *Angewandte Chemie International Edition* 2012; 51: 6121–5. *Discussion of the surface coating of upconversion nanoparticles by phospholipids and other agents.*

Li Y-F, Chen C, Fate and toxicity of metallic and metal-containing nanoparticles for biomedical applications, *Small* 2011;**7**:2965–80. *Explanation of uptake, trafficking, pharmacokinetics and clearance of nanomaterials in biological systems.*

Maurer-Jones M A, Bantz K C, Love S A, Marquis B J, Haynes C L, Toxicity of therapeutic nanoparticles, *Nanomedicine* 2009;4:219–41. *A perspective on the toxicology of nanoparticle formulations in both therapy and imaging.*

Na H B, Song I C, Hyeon T, Inorganic nanoparticles for MRI contrast agents, *Advanced Materials* 2009; 21;2133–48. *A review of the structure of contrast agents used in MRI.*

Pimlott S L, Sutherland A, Molecular tracers for the PET and SPECT imaging of disease, *Chemical Society Reviews* 2011; **40**:149–62. *A chemically oriented tutorial review of the molecular probes used for positron emission tomography and single photon emission computed tomography, especially for imaging cancer and neurological disease.*

Pysz MA, Gambhir SS, Willmann JK, Molecular imaging: current status and emerging strategies, *Clinical Radiology* 2010;**65**:500–16. *Review of recent developments with molecular imaging and effects on clinical strategies.*

Ronkainen NJ, Halsall HB, Heineman WR, Electrochemical biosensors, *Chemical Society Reviews* 2010;**39**:1747–63. *A critical review of biocatalytic devices and affinity sensors.*

Rosenthal SJ, Chang JC, Kovtun O, McBride JR, Tomlinson ID, Biocompatible quantum dots for biological applications, *Chemistry and Biology* 2011;**18**:10–24. *Tutorial style review of the biocompatibility of quantum dots.*

Wadsak W, Mitterhauser M, Basics and principles of radiopharmaceuticals for PET/CT, *European Journal of Radiology* 2010;**73**:461–9. *Brief review of the details of material specifications for radiopharmaceuticals.*

Wang F, Banerjee D, Liu Y, Chen X, Liu X, Upconversion nanoparticles in biological labeling, imaging and therapy, *Analyst* 2010;**135**:1839–54. *Review of optical biolabeling and imaging with upconversion nanoparticles.*

Xie J, Lee S, Chen X, Nanoparticle-based theranostic agents, *Advanced Drug Delivery Reviews* 2010;**62**:1064–79. *Review of the development of nanoparticles used in combined imaging and drug delivery processes.*

Xing H, Bu W, Zhang S, *et al.*, Multifunctional nanoprobes for upconversion fluorescence, MR and CT trimodal imaging, *Biomaterials* 2012;**33**:1079–89.

Zrazhevskiy P, Sena M, Gao X, Designing multifunctional quantum dots for bioimaging, detection, and drug delivery, *Chemical Society Reviews* 2010;**39**:4326–54. *Critical review of the design principles governing the engineering of QD probes.*

# 8 Contemporary and future biomaterials

This chapter provides a new way of classifying biomaterials and gives extensive information about the wide range of biomaterials that are either in current clinical use or showing considerable potential for clinical applications in the near future. There are six primary classes of biomaterials – metallic, polymeric and ceramics systems, carbons, composites and engineered tissues. In each of these classes you will see how real biomaterials are based on the principles of materials science, biology and biocompatibility given in the early chapters but also how they are adapted and modified to suit the specific requirements of the various clinical disciplines and medical technologies.

In this chapter, we bring together the details of all currently used biomaterials and those that appear to have considerable potential for the future. This is an extensive although not exhaustive list of the properties and the applications for each material and gives a generic classification of biomaterials, with an indication of their advantages and disadvantages.

Although this list of contemporary and future biomaterials covers the majority of clinical areas, the specific fields of restorative dentistry (filling materials, dental adhesives, prosthodontic materials, etc.) and orthodontics have been excluded, since these form a coherent group of materials that are not conventionally considered as biomaterials, and are better discussed separately. There are a few other areas where the inclusion of some materials as "biomaterials" may be considered questionable; with any classification system there are bound to be some borderline areas. These include some materials that may be considered as commodity materials that are used in a variety of health care products but are ancillary to the real tissue-interfacing biomaterials. Mostly these are excluded but there are a few, such as low density polyethylene and plasticized polyvinylchloride, which make such a vast difference to medical practice by the sheer number and range of their applications, and where in some cases their interactions with tissues are important, that are included. In other examples, there may be just a few situations where a material has a real or potential impact as a biomaterial but where the major medical uses are more mundane, as an excipient or inert additive in a pharmaceutical preparation, so they are included here for completeness.

It should also be made clear at this stage that this list has steered away from the use of trademarks and proprietary or commercial names, largely because such names are often ephemeral and it is wrong to describe a class of biomaterial on the basis of names that are past their expiry date. Nevertheless, it has been impossible to avoid mention of a few proprietary names since they have become synonymous with the type of product, just as the words hoover and nylon became adopted over time as descriptors of certain classes of product as opposed to brand names; this has been done cautiously and only when beneficial to our understanding of the provenance of some groups of biomaterials.

## 8.1 The classification of biomaterials

At the outset, we should discuss the options for the classification of biomaterials. There have been very few attempts to do this, and the few classifications that do exist do not make much sense today and have little value. Some classifications are based on biological properties, e.g., bioinert, bioactive and bioresorbable biomaterials. This is of no practical value since the meaning of these words, as we have seen, is vague and many individual groups of materials may have quite variable behavior with respect to these properties. There have been attempts to define biomaterials on the basis of the clinical areas where they are used, as with orthopedic, cardiovascular, ophthalmological and dental materials. With so much crossover between applications, however, this has no real value apart from splitting the large subject into smaller parts. We should also dismiss the concepts of generations of biomaterials. In several clinical areas the development of devices and materials has been marked by the apparent introduction of first, second, third and more generations – the third generation of intravascular stents, the fourth generation of bioactive glasses and so on. This does not make sense either: the term generation obviously comes from the descent of offspring from ancestors. This is fine for individual well-defined families, but soon loses usefulness when large groups of people are involved, with intermarriage, multiple partners and widely varying ages at which women have their children. At least these generations have one thing in common, the people themselves, but what is it about a material that constitutes a new generation, and who makes that decision – this type of classification has no place in biomaterials science.

It is obvious that the classification of biomaterials has to be based on objective, scientifically based, criteria, and this can only be rooted in the structure and composition of the material itself. Generally we can forget where the material might be used and what its properties are; it is what it is that is important. In the hierarchy of the classification, of course, some allowance has to be made for some highly specific biomedical characteristics. It is difficult, for example, to avoid a subdivision of polymers which is concerned with synthetic biodegradable polymers, but these are exceptions rather than the rules themselves.

Not surprisingly, the classification that is set out in this chapter follows the general themes of materials structures set out in the basic materials model of Chapter 2. Thus we have classes that are concerned with metallic systems, ceramic systems, polymeric

**Box 8.1 The first level of the classification of biomaterials**

Class 1 The metallic systems
Class 2 The ceramics systems
Class 3 The polymeric systems
Class 4 Carbon materials
Class 5 Composite materials
Class 6 Engineered biological materials

systems, carbon materials and composites. These broad classes might also be seen with generic classifications of electronic materials, aerospace materials, nuclear materials and so on, indicating that this is not a nonsensical approach. We do have one addition here, however, and that concerns engineered (or bioengineered) biological components, consistent with the discussions in previous chapters. It should also be recognized that some materials do not always fit into structure- or composition-based classes, but have to be identified by their form or morphology, for example as coatings or fabrics.

The first level in the hierarchy of the classification of biomaterials is therefore as given in Box 8.1.

## 8.2 Class 1 The metallic systems

Now let us consider each of the six classes in turn. Class 1, the metallic systems, is probably the easiest to sub-divide as seen in Box 8.2. Useful metallic materials are occasionally pure metals but the vast majority are alloys. All sub-classes in this section are based on individual elements, which we find are either used in their reasonably pure state or in alloys where the principal component is that element. Thus, Class 1.1 includes materials that are essentially pure titanium (usually referred to as commercially pure titanium) or alloys in which the parent element is titanium and where alloying elements have been added (such as aluminum, vanadium, niobium or zirconium) in order to confer specific structures and properties. In some of these classes, there are no biomedical uses of the pure metal and we are only concerned with alloys, such as those based on cobalt (Class 1.3), nickel (Class 1.4) or one of the platinum group of elements (Class 1.7). Steels are a little different since, as we shall see below, they are defined by the presence of two elements, iron and carbon (Class 1.2).

Each of these classes may be further sub-divided, usually on the basis of structural and compositional features, which define a coherent set of properties. The precise basis of this sub-division will vary from class to class. Thus titanium alloys are usually classified according to the relative presence and/or stability of the two main crystallographic phases of titanium, the alpha and beta phases. The alloys of the platinum group of metals are classified according to which of the elements platinum, palladium, rhodium, iridium, ruthenium and osmium are present. In some systems, the advent of nanostructured products has had an effect on the classification, such that we include nanostructured silver and gold, which appear to have distinct biological properties, in their own groups. Each of these third level metallic systems is discussed separately below.

### 8.2.1 Class 1.1 Titanium and titanium alloys

Titanium (Ti) was first introduced into implantable devices in the 1960s. It had become known as a very

### Box 8.2 The second level of the classification of biomaterials: metallic systems

Class 1.1 Titanium and titanium alloys
Class 1.1.1 Commercially pure titanium
Class 1.1.2 Alpha and near alpha titanium alloys
Class 1.1.3 Alpha–beta alloys
Class 1.1.3.1 Titanium–6% aluminum–4% vanadium
Class 1.1.3.2 Titanium–6% aluminum–7% niobium
Class 1.1.4 Beta titanium alloys
Class 1.2 Iron and steels
Class 1.2.1 Austenitic stainless steels
Class 1.2.1.1 ASTM 316 and 316L austenitic stainless steel
Class 1.2.1.2 High nitrogen/low nickel austenitic stainless steel
Class 1.2.2 Ferritic and duplex steels
Class 1.2.3 Iron nanowires
Class 1.3 Cobalt-based alloys
Class 1.3.1 Cobalt–chromium–molybdenum alloys
Class 1.3.2 Cobalt–chromium–tungsten–nickel alloys
Class 1.3.3 Cobalt–chromium–iron–nickel–molybdenum
Class 1.3.4 Cobalt–nickel–chromium–molybdenum alloys
Class 1.4 Nickel-based alloys
Class 1.4.1 Nickel–titanium shape memory alloy
Class 1.5 Tantalum and zirconium alloys
Class 1.5.1 Porous unalloyed tantalum
Class 1.5.2 Zirconium–niobium alloys and oxidized zirconium alloys
Class 1.6 Silver
Class 1.6.1 Silver coatings
Class 1.6.2 Silver electrodes
Class 1.6.3 Nanocrystalline silver and silver nanoparticles
Class 1.7 Platinum group metals and alloys
Class 1.7.1 Platinum and its alloys
Class 1.7.1.1 Platinum–iridium alloys
Class 1.7.2 Palladium-based alloys
Class 1.7.3 Iridium films
Class 1.8 Gold
Class 1.8.1 Metallic gold
Class 1.8.2 Gold nanoparticles
Class 1.9 Magnesium and its alloys

useful element in many engineering applications, especially those that required a high strength-to-weight ratio coupled with high corrosion resistance. Its density is 4.506 g $cm^{-3}$, making it much lighter than alternative metallic biomaterials. It has a Young's modulus of 116 GPa, which again is much less than that of steels and cobalt alloys, and much closer to that of bone.

Ti is a polymorphic transition element, of atomic number 22, exhibiting a close packed crystal structure (the α phase) up to 882.5°C and a body centered structure (the ß phase) from 882.5°C to the melting point at 1678°C. The mechanical properties of pure Ti are not good and so many Ti products are based on alloys. The alloying elements fall into three categories, those that stabilize the α phase, those which stabilize the ß phase and those which stabilize neither phase. The most important α stabilizer is Al although small non-metallic elements including O, N and C also have this effect. ß stabilizers include Mo, V, Nb, Ta, Fe, W, Cr, Si, Ni, Co and Mn. Neutral elements include Zr. The effect of different elements on titanium metallurgy can be seen in the generic binary phase diagram of Figure 8.1. On the basis of the elements added, and their amounts, several distinctive alloy microstructures can be generated, principally the α alloys, the near α alloys, the α + ß alloys and the ß alloys. As we see below these have different mechanical properties and, to a lesser extent, different corrosion resistance and biocompatibility. Ti alloys are defined by a numbering system developed by ASTM (American Society for Testing and Materials), currently ranging from Grade 1 to Grade 38, and also by the ASTM Standards. The standard specifications and main properties of Ti-based biomaterials are shown in Table 8.1.

### Class 1.1.1 Commercially pure titanium

For many years, a group of titanium materials, comprising ASTM Grades 1, 2, 3 and 4, were collectively known as commercially pure titanium, containing close to 99.5% titanium and small amounts of C, Fe, H, N and O. Today several more essential pure Ti materials have been added, incorporating small amounts of some other metallic elements. The non-metallic elements go into interstitial solid solution in the Ti and have a strengthening effect because of lattice distortion. O is probably the most important,

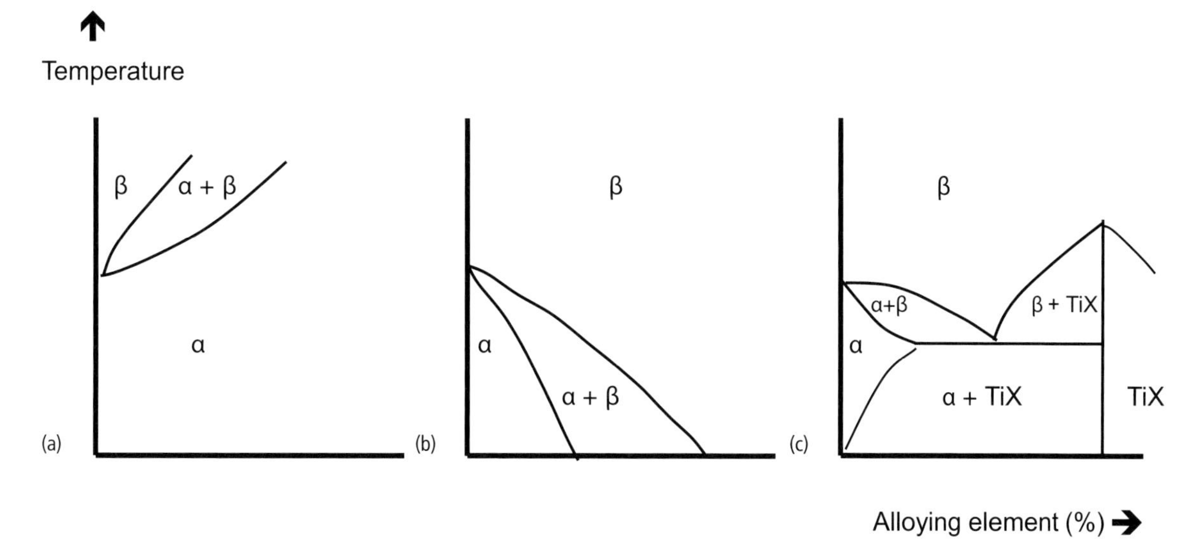


**Figure 8.1** Titanium alloys. Phase diagrams and alloy structures for titanium alloys depend on whether the alloying element is an α or β stabilizer: (a) the generic form of the titanium-rich end of a phase diagram for an α-stabilizing element; (b) generic diagram for a β-stabilizing element; and (c) generic diagram for an alloy that forms a eutectoid phase TiX.

Table 8.1 **The standard specifications and main properties of titanium and titanium alloys used as biomaterials. This table provides representatives specifications and average properties as given in several standards documents (* negligible)**

| Characteristic | CP Ti<br>Grade 1 ELI | CP Ti<br>Grade 1 | CP Ti<br>Grade 4 | Ti–Al–V | Ti–Mo–Zr–Al |
|---|---|---|---|---|---|
| Composition (%) | Ti balance<br>Al*<br>V*<br>Mo*<br>Zr*<br>Fe 0.10<br>N 0.012<br>C 0.03<br>H 0.012<br>O 0.10 | Ti balance<br>Al*<br>V*<br>Mo*<br>Zr*<br>Fe 0.20<br>N 0.03<br>C 0.10<br>H 0.125<br>O 0.18 | Ti balance<br>Al*<br>V*<br>Mo*<br>Zr*<br>Fe 0.50<br>N 0.05<br>C 0.10<br>H 0.0125<br>O 0.40 | Ti balance<br>Al 5.5–6.75<br>V 3.5 – 4.5<br>Mo*<br>Zr*<br>Fe < 0.3<br>N < 0.05<br>C < 0.08<br>H < 0.015<br>O < 0.2 | Ti balance<br>Al 2.5–3.5<br>V*<br>Mo 14.0–16.0<br>Zr 4.5 – 5.5<br>Fe < 0.3<br>N < 0.05<br>C < 0.08<br>H < 0.02<br>O < 0.20 |
| Tensile strength, minimum (MPa) | 200 | 240 | 550 annealed<br>680 cold work | 860 | 900 |
| Elongation (%) | 30 | 24 | 15 annealed<br>10 cold work | 10 | 12 |

it's level increasing from 0.12% in Grade 1 to 0.4% in Grade 4. Fe varies from 0.2 to 0.5% and N from 0.03 to 0.05%. Carbon is usually held at 0.1% to avoid inappropriate carbide formation and H is never above 0.015% to avoid hydrogen embrittlement. The UTS of annealed materials increases from around 240 MPa with Grade 1 to 550 MPa with Grade 5. As usual in these situations there is a concomitant loss of ductility, the elongation at break decreasing from around 24% to 15%. The other grades of commercially pure titanium incorporate Pd, 0.2% maximum in Grade 7 and Mo (0.3%) and Ni (0.8%) in Grade 12. The corrosion resistance of these materials is excellent, as is the general biocompatibility.

### Class 1.1.2 Alpha and near alpha titanium alloys

The α alloys are single phase but are strengthened by substitutional rather than interstitial solid solution effects. Most of these alloys contain Al as the α stabilizer, with small amounts of other elements such as Sn and Fe. Grade 6 alloy has 5%Al and 2.5%Sn, which results in a UTS of 950 MPa and elongation of 15%. Near α alloys have a small amount of a ß phase, produced by small levels of ß stabilizers, such as Mo or Si.

### Class 1.1.3 Alpha–beta alloys

The best mechanical properties in Ti alloys are achieved in the two-phase α + ß alloys, produced by the addition of a strong α stabilizer and some ß stabilizers. Al is almost universally chosen for the former and any one, or any combination, of the earlier mentioned ß stabilizers can be used to generate optimal microstructures.

**Class 1.1.3.1 Titanium–6% aluminum–4% vanadium** The most important, and most widely used, alloy here in the Ti-6Al-4V alloy (ASTM Grade 5). The ß phase gives both strength and ductility; if the alloy is given a solution heat treatment in the α + ß field and then rapidly cooled, a very fine dispersion of ß phase in the α can be produced, enhancing mechanical properties. Depending on working and heat treatments, a UTS of over 1100 MPa can be achieved, with a ductility over 10%. The alloy has good fatigue resistance and is amenable to many manufacturing operations.

**Class 1.1.3.2 Titanium–6% aluminum–7% niobium** An alternative α–β alloy has Nb instead of V, possessing very similar properties.

#### Class 1.1.4 Beta titanium alloys

The β phase is normally stabilized in β alloys by Mo, V, Nb, Cr, Zr, Ta and Fe. These alloys often contain a small amount of α stabilizer to control some second phase hardening. The β phase is ductile which means that the alloys can be cold-worked to a considerable extent and can achieve mechanical properties a little higher than the best α–β alloys. They are not used extensively as biomaterials at this stage.

### 8.2.2 Class 1.2 Iron and steels

Iron (Fe) is a polymorphic transition metal, atomic number 26 and Young's modulus 210 GPa. A few very distinctive features of Fe account for its versatility, but also its limitations. First, it is ferromagnetic, with a Curie Point at 777°C, which means that it is strongly magnetic at virtually all temperatures of interest. Secondly, it exists in three allotropic forms: up to 910°C it has a bcc crystal structure (α iron), at which temperature it transforms to an fcc structure (γ iron), which itself reverts to bcc (δ iron) at 1403°C. The melting point is 1535°C. As we have seen, a multiplicity of allotropes usually provides for extensive opportunities for alloying, which is very important since the mechanical properties of pure Fe are not good. The third point is that Fe normally has very poor environmental resistance, typified by the rusting that readily occurs with iron objects. Alloying procedures are again necessary if useful Fe-based products are required. The massive worldwide use of such products is based on the fact that alloying Fe with C gives very high strength steels, and alloying with Cr gives exceptional corrosion resistance.

By definition, steels are alloys that contain Fe and C, and any other elements that can confer special properties. The metallurgy of steels is very complex, but we need only cover a few of the principles here. The relevant part of the Fe–C phase diagram is shown in Figure 8.2. Carbon has very little solubility in Fe, since it is interstitial rather than substitutional. The limit of solubility of C in the α phase at room temperature is 0.008%, this phase being known as ferrite. Higher levels of C at this temperature cause the formation of a very brittle intermetallic phase, $Fe_3C$, known as cementite. The limit of solubility in the γ phase is much higher, at 0.8% just above the transition temperature, giving the austenite phase.

Many steels, for example mild steel or carbon steel, contain sufficient carbon to give a mixture of ferrite and cementite, the morphological distribution of the latter in the former controlling the mechanical properties. Even more profound effects are seen with manipulation of the austenite phase. As austenite, with its larger amount of C is cooled below the transition temperature, there should be a transition to ferrite plus cementite; indeed that can happen if the cooling takes place slowly enough for the atoms to move to new positions. If, however, the cooling takes place rapidly, in a process known as quenching, there will be insufficient time for these movements to take place by diffusion processes – if the steel is quenched into ice for example, the atoms will instantaneously lose thermal energy so that diffusion is impossible. Under these circumstances, a different transition takes place through shear rather than diffusion, the result being a metastable phase known as martensite, which is body centered tetragonal. This is immensely hard because of the lattice strains that are induced by the shear, but also very brittle. A martensitic steel is usually re-heated, in a process of tempering, to allow some lattice stress relief and softening. A wide variety of properties can be introduced in this way; many surgical instruments that require hardness and sharpness are made from martensitic steels.

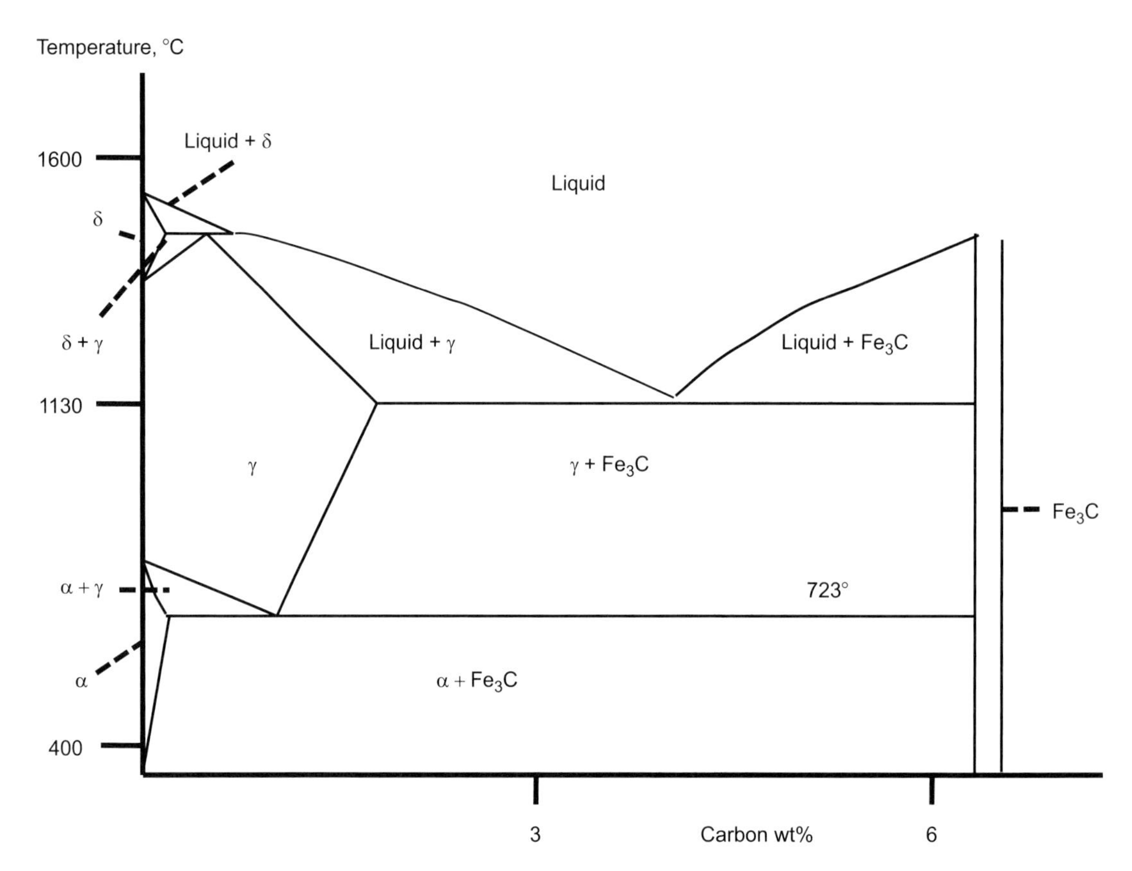


**Figure 8.2** The iron–carbon system. This is the iron-rich end of the phase diagram of iron and carbon. Carbon has very limited interstitial solubility in the iron lattice. At low temperature the stable phase is ferrite (α). At higher temperature the stable structure is austenite (γ). Solubility in the γ phase is higher than in the α phase. Above the limit of solubility an intermetallic compound $Fe_3C$, known as cementite forms. Steels are alloys of iron with carbon and many have equilibrium structures at room temperature of α + cementite. The presence of other alloying elements (e.g., Cr, Ni, Mo) and manipulation of phase transformations gives a wide range of properties to steels.

Also of considerable importance is the fact that the α–γ transition temperature can be altered by different alloying additions. In particular, elements such as Ni, Mn and N stabilize the γ phase such that austenitic steels may exist at room temperature.

### Class 1.2.1 Austenitic stainless steels

Austenitic steels are no more attractive than ferritic or martensitic steels in terms of mechanical properties, but they do provide one very important feature. If sufficient Ni is present to stabilize the austenite (around 12%), chromium and a few other elements may be added to confer corrosion resistance. Even though Cr is itself a ferrite stabilizer, if it is present in an austenitic steel at above 16%, if can form a sufficiently strong and continuous layer of chromium oxides on the surface to confer excellent corrosion resistance. Such alloys are known as austenitic stainless steels, which are used extensively in domestic and architectural applications. One particular form of this material has been used for surgical implants for many years.

**Class 1.2.1.1 ASTM 316 and 316L austenitic stainless steel** The ubiquitous biomaterial based on austenitic stainless steel is referred to by its ASTM

Table 8.2 **The standard specifications and main properties of stainless steel used as biomaterials. This table provides representatives specifications and average properties as given in several standards documents**

| Characteristic | 316L stainless steel | High nitrogen stainless steel |
|---|---|---|
| Composition (%) | Iron balance<br>Cr 17.0–19.0<br>Ni 13.0 - 15.0<br>Mo 2.25 - 3.0<br>Nb negligible<br>C < 0.030<br>Si < 1.0<br>Mn < 2.0<br>P < 0.025<br>S < 0.10<br>N < 0.10<br>Cu < 0.50 | Iron balance<br>Cr 19.5 - 22.0<br>Ni 9.0 - 11.0<br>Mo 2.0 - 3.0<br>Nb 0.25 - 0.8<br>C < 0.08<br>Si < 0.75<br>Mn 2.0 - 4.5<br>P < 0.025<br>S < 0.10<br>N 0.25–0.50<br>Cu < 0.25 |
| Tensile strength, annealed bars (MPa) | 490–690 | 750 |
| Tensile strength, cold drawn wire (MPa) | 1350–1600 | 1060 |
| Elongation, annealed bars (%) | 40 | 35 |
| Elongation, cold drawn wire (%) | < 5 | 12 |

designation of 316, details of which are given in Table 8.2. These steels have major alloying additions in the region of 16–18% Cr, 10–14% Ni and 2–3% Mo. The Ni maintains the austenitic structure. The Cr is the primary corrosion resistance additive but it is assisted by the small amount of Mo; without the Mo, the steel would be susceptible to corrosion in chloride media. There are low levels of other elements, including Mn (max 2%), Si (0.75%), P (max 0.045%), S (max 0.030%) and N (max 0.10%). Most of these are the type of impurities that are normally found in steels and very difficult to reduce to lower levels, but the Mn and the N do enhance the austenite stability. The maximum C allowed in 316 stainless steel is 0.08%.

The 316 steel typically has a UTS of 515 MPa and a ductility of 40%, which are very respectable values. Unlike many other steels, these materials are not heat treatable but their properties can be altered through cold work. Their corrosion resistance is very good. Usually implantable devices are passivated through an acid treatment in order to enhance general corrosion resistance. The corrosion resistance is, however, not as good as that for the Ti and Co alloys considered in this chapter, and oxide film breakdown can exceptionally occur under fretting and crevice conditions.

One further issue has to be addressed with 316 stainless steel. The maximum of 0.08% C should permit all of the C to remain in solution. If, however, the steel is kept at an elevated temperature during manufacturing, there is a possibility that some of the C can preferentially attach to Cr to form carbides such as $Cr_{23}C_6$. If they do this, then it is energetically favorable for them to do so at grain boundaries. The formation of these intergranular carbides effectively denudes the regions next to the boundaries with respect to Cr, which therefore lose their corrosion protection. This gives rise to intergranular corrosion and many cases of premature failure of 316 stainless steel orthopedic implants have occurred because of this problem. The solution to this has involved the introduction of a different formulation, known as 316L, or low carbon 316, for which the specification is the same except that the maximum C level is 0.03%.

**Class 1.2.1.2 High nitrogen/low nickel austenitic stainless steel** One biocompatibility issue that is sometimes raised with austenitic stainless steel is the potential for adverse effects arising from the Ni content; it is unclear how significant this is, but the possibility itself has led to the development of some alloys with reduced levels of Ni. These variously introduce greater levels on N and Mn, which are both austenite stabilizers and may be as effective as the Ni in this respect. Some of these alloys use N levels as high as 1% and Mn levels up to 15 or 20%.

#### Class 1.2.2 Ferritic and duplex steels

There is some interest in using ferritic rather than austenitic stainless steel; these usually contain 17–26% Cr but no Ni. They may have small amounts of Mo, Ti and/or Nb and very low C and N contents. They have exceptional corrosion resistance in chloride media but are prone to brittleness. Stainless steels with combined ferritic and austenitic phases, the former being the continuous phase, are also available. These are known as duplex steels; they have excellent strength but do not have either the ductility of austenitic steels nor the corrosion resistance of the 316 group.

#### Class 1.2.3 Iron nanowires

Nanowires are objects with one dimension that is very much longer than the other two, which are themselves very similar and of nanoscale dimension. Fe nanowires have attracted a great deal of attention in several arenas, including high density recording heads. They are made of α-Fe, most commonly by self-assembly during the decomposition of an appropriate perovskite mineral. Perovskites are naturally occurring compounds that, at their simplest, have compositions of $ABX_3$ where A and B are cations of widely different atomic radius and X is an anion such as oxygen. A composition involving lanthanum (La), strontium (Sr) and Fe in addition to oxygen decomposes to give an array of single crystal Fe nanorods embedded in a matrix of $LaSrFeO_4$. These have very high magnetic moments and have considerable potential for MRI contrast agents.

### 8.2.3 Class 1.3 Cobalt-based alloys

Cobalt (Co) is a group VIIIA transition element, with atomic number 27. It has a modulus of elasticity of 210 GPa and is ferromagnetic. At low temperatures it has an hcp structure and transforms to fcc at 417°C. Many useful alloys based on cobalt have been developed, their properties depending to a large extent on whether alloying additions favor the cubic structure (Ni, Mn, Fe, C) or the hexagonal structure (Cr, W, Mo). Considerable solid solution hardening effects are seen with additions of metals such as W, Ta and Mo, while precipitation hardening is achieved with the formation of carbide or carbonitride precipitates involving Cr, Zr, W, Nb and some other refractory metals. Cr and C together provide a very important platform for Co-based alloys, with excellent strength and hardness derived from carbides and corrosion resistance from the passivating effect of the Cr. For many years a family of Co–Cr alloys containing C and a little Ni and Mo, known as the Stellites because of their high corrosion resistance, were widely used in many engineering situations that required high-performance alloys, and these were introduced into clinical uses over 60 years ago. One of these alloys, Vitallium, was the principal surgical alloy for decades. The composition and properties of the main Co-based metallic biomaterials are shown in Table 8.3.

#### Class 1.3.1 Cobalt–chromium–molybdenum alloys

The Vitallium mentioned above is a Co–Cr–Mo alloy. In the classification scheme promoted by ASTM this type of alloy is referred to as F75. The compositional range is Cr 27–30%, Mo 5–7%, Ni 2.5% max and C 0.35%max, with traces of Fe, Mn and Si and the

Table 8.3 **The standard specifications and main properties of cobalt-based alloys used as biomaterials. This table provides representatives specifications and average properties as given in several standards documents**

| Characteristic | Wrought Co–Cr–Mo | Wrought Co–Ni–Cr–Mo–W–Fe |
|---|---|---|
| Composition | Co balance | Co balance |
| | Cr 26.0–30.0 | Cr 18.0–22.0 |
| | Ni < 1.0 | Ni 15.0–25.0 |
| | Mo 5.0–7.0 | Mo 3.0–4.0 |
| | W negligible | W 3.0–4.0 |
| | Fe < 0.75 | Fe 4.0–6.0 |
| | Ti negligible | Ti 0.5–3.50 |
| | Mn < 1.0 | Mn < 1.0 |
| | C 0.14 (low C alloy) | C < 0.05 |
| | C 0.15–0.35 (high C alloy) | |
| | Si < 1.0 | Si < 0.50 |
| Tensile strength, annealed (MPa) | 900 | 600 |
| Tensile strength, worked (MPa) | 1000 | 1300 |
| Elongation, annealed (%) | 20 | 50 |
| Elongation, worked (%) | 12 | 12 |

balance Co. The alloy forms with a hexagonal structure. In the early days of its use, the alloy could only be cast, which accounted for much use in dentistry, where intricate personalized shapes were required that could only be produced by casting. In the as-cast condition, the alloy was brittle with a maximum elongation of less than 10%. The yield strength had a maximum of just over 500 MPa and a UTS of 725 MPa. The fatigue strength was low, at around 250 MPa. The microstructure tends to be that of large grains, with precipitates of carbides, especially $Cr_{23}C_6$. Excessive carbide precipitates at grain boundaries reduced ductility. The properties of these alloys can be optimized by heat treatments, which can increase both the ultimate tensile strength and the ductility through control of grain size and carbide precipitates. Improvements in metallurgical processing conditions have allowed forged components to be made of F75, with particularly high strengths (UTS 1500 MPa, Fatigue 900 MPa) whilst maintaining good ductility through the use of cold-working techniques and reducing the dependence on carbide precipitates. Components of these alloys may also be prepared by powder metallurgical methods, for example hot isostatic pressing.

### Class 1.3.2 Cobalt–chromium–tungsten–nickel alloys

An alternative to F75 is F90. This has lower levels of both Cr (19–21%) and C (0.05–0.15%), no Mo, but substantial amounts of Ni (9–11%) and W (14–16%). The low levels of C here mean that the mechanical properties are not carbide dependent. The levels of Cr, Ni and W are such as to make both hexagonal and cubic phases potentially stable, and judicious use of cold-working and annealing cycles can result in a cubic matrix containing precipitated hexagonal platelets. At a maximum, the yield strength can be over 1300 MPa, the UTS over 1500 MPa, the fatigue strength around 600 MPa and the elongation 12%.

A fully annealed sample may have a ductility of 60% but a yield strength reduced to 350 MPa.

### Class 1.3.3 Cobalt–chromium–iron–nickel–molybdenum

Wrought cobalt alloys containing around 20 wt% Cr, 15% Ni, 7% Mo and 15% Fe have excellent mechanical properties and corrosion resistance and have been used in leads, electrodes and other components of implantable devices for many years. These alloys are known as Elgiloy alloys.

### Class 1.3.4 Cobalt–nickel–chromium–molybdenum alloys

Alloys of 35 wt% Co, 35% Ni, 20% Cr and 10% Mo are multiphase, and are generally known as MP35N alloys. These are wrought alloys with considerable capability of strengthening through work hardening and annealing so that they are tough and ductile; ultimate strengths in excess of 2000 MPa may be achieved. They have an fcc matrix that is strengthened by hexagonal precipitates. Of particular interest is their high fatigue strength, including high corrosion fatigue resistance, and they have been extensively used in implantable devices such as leads for pacemakers.

## 8.2.4 Class 1.4 Nickel-based alloys

Nickel (Ni) is a transition metal that has an fcc structure and many industrial uses, including applications in fuel cells, batteries and catalysis. It is very corrosion resistant and, as a pure metal, is widely used for coating other metals to give durability. There are many excellent Ni-based alloys, involving Fe, Cr and Mo, and several so-called nickel superalloys are used in many critical (especially high temperature) engineering applications. It is used, as we have seen, as an alloying addition in steels and cobalt alloys. In general there is little specific need for Ni alloys in medical technology, and the well-documented history of allergies to Ni has resulted in little enthusiasm to experiment with it in most of these situations. There is one exception, and that is the nickel–titanium alloys that exhibit shape memory effects; these are discussed below.

### Class 1.4.1 Nickel–titanium shape memory alloy

The basics of shape memory have been discussed in Chapter 2. Briefly, a shape memory alloy exhibits a phase transformation between an austenitic structure and a martensitic structure that is diffusionless in nature (i.e., atoms do not move to new positions by any diffusion process) but instead involves shear deformation, as with the martensitic transformation in steels discussed earlier. When the alloy is above the transition temperature it has an austenitic structure, and when cooled through the transition zone, undergoes a martensitic transformation. This is referred to as a zone here because the transformation takes place over a small range of temperature, starting at $M_s$, the start temperature, and completing it at the finish temperature, $M_f$. If the material, when fully martensitic, is deformed it will change shape. However, this change is fully recoverable on heating, as it goes through the austenite start temperature, $A_s$, and then the austenite finish temperature, $A_f$. This thermal shape recovery is the shape memory effect.

Although several alloy systems show shape memory, the one outstanding example is found in the binary Ni–Ti system, an alloy that is often referred to as Nitinol. The exact nature of the effect, including the transition zone, will depend on the precise Ni–Ti composition and the presence of impurities. The usual compositional range in which shape memory is seen is 48–52 atomic% Ni. A change of 0.1% in the Ni content can alter the transition temperatures by 10°C. Traces of Cr, Mn, Fe, V, Nb and Co tend to decrease these while traces of Au, Pt, Pd and Hf increase them. The presence of O and N can affect properties considerably since they tend to form complex oxides, nitrides and oxynitrides with either or both the Ni and Ti. In a typical medical product situation, the alloy is conditioned at 400–600°C to achieve the appropriate structure and properties and will be cooled to below body temperature where it is deformed to give it the pre-deployment shape. With the correct and optimal transition temperature, the

Table 8.4 **The specifications and properties of Ni–Ti alloys. This table provides representatives specifications and average properties as given in several standards documents**

| Characteristics | Nitinol type A | Nitinol type B |
|---|---|---|
| Composition (wt%) | Ti balance<br>Ni 54.5<br>0 < 0.05<br>C < 0.02<br>Inclusions < 2.8 | Ti balance<br>Ni 55.0<br>0 < 0.05<br>C < 0.02<br>Inclusions < 2.8 |
| Transformation temperature for finished product | 50–80°C | 30–50°C |
| Shape memory strain (%) | < 8.0 | < 8.0 |
| Tensile strength (MPa) | > 1070 | > 1070 |
| Total elongation (%) | > 10 | > 10 |

product may be inserted into the body, where the rise in temperature will cause it to attempt to recover its higher temperature shape. Table 8.4 gives the specifications and properties of Ni–Ti alloys.

Questions about the corrosion resistance of and potential allergies to these nickel-based alloys have been asked ever since the shape memory alloys were introduced clinically. A binary alloy involving around 50 atomic% Ni would be expected to generate significant problems with respect to allergies, or indeed even more direct and significant biological effects, but this has not happened. It would seem that the Ti in the alloy is in sufficient quantity to passivate the alloy and minimize the amount of Ni ions actually released.

The main critical clinical procedure that utilizes shape memory Ni–Ti is that of the catheter-delivered intravascular stent. Here, as discussed in Chapter 4, a cylindrical mesh structure of the alloy, which has been preconditioned in its ultimate shape, is cooled below the transition temperature and deformed such that it can be placed inside a catheter. The catheter is maneuvered inside a patient's blood vessel until it reaches the desired location where it is deployed. The rise in temperature and the release from the constraint of the catheter allows the stent to expand until it is compressing against the inner lumen of the vessel, where it remains. Another extensive use is found in orthodontics, where the forces associated with the shape recovery are used to direct tooth movement.

## 8.2.5 Class 1.5 Tantalum and zirconium alloys

Tantalum (Ta) and zirconium (Zr) are two similar transition elements that have some properties reminiscent of Ti. Ta is of atomic number 73 and Young's modulus 185 GPa. The pure metal has a UTS of 275 MPa and a ductility of 50%. Zr is of atomic number 40 and Young's modulus 88 GPa. Although both elements have been considered for implant use over many decades, largely on account of their good corrosion resistance and lack of toxicity these uses have not been of tremendous significance. Aneurysm clips and some stents are included here. However, two specific products have been gaining increasing interest in orthopedics.

### Class 1.5.1 Porous unalloyed tantalum

As noted elsewhere, many attempts have been made to develop materials with elastic moduli similar to bone. Ti is better than stainless steel or Co alloys but is still considerably more rigid than cortical bone, which is itself far more rigid than cancellous bone. One solution may involve porous metals. One proprietary porous metal that is used for this purpose is the so-called Trabecular Metal™, which is porous commercially pure Ta. It has a porosity in excess of 80%, and a Young's modulus of 3 GPa. The compressive strength is 50–80 MPa. This material is being used in joint replacement and spinal fusion products.

#### Class 1.5.2 Zirconium–niobium alloys and oxidized zirconium alloys

The alloy Zr–2.5% Nb is also of interest. This is used in a form where the surface is oxidized at very high temperatures, effectively giving a surface layer, around 5 microns thick, of zirconia. The proprietary product is Oxinium™, which has been used as the bearing surface of joint replacement prostheses.

### 8.2.6 Class 1.6 Silver

Silver (Ag) is a transition metal of atomic number 47; it has an fcc crystal structure. It is a soft and weak metal, with a maximum UTS of 350 MPa and a Young's modulus of 90 GPa. It is also very expensive. For these reasons, Ag is not a contender for structural applications. There are, however, some interesting and attractive properties that are relevant to biomaterials applications. The first is that Ag is an excellent electrical conductor and therefore may find use in electrodes, either for therapeutic or diagnostic purposes. The second is that Ag does have certain biological properties. The potential for Ag as a biomaterial is best realized when it is used as a biologically active coating or as an electrode. The introduction of Ag products at the nanoscale, and specifically "nanosilver," suggests that we should discuss these separately.

It is important at this stage to consider the biological activity since this does underpin the main interests, and controversy, over the medical applications of Ag. Crucially, Ag can have antibacterial properties, although the mechanism(s) by which this activity is derived are not entirely clear. As with other noble metals (see next sections) Ag binds avidly to many biological molecules; the silver ion $Ag^+$ has a strong affinity for electron donor groups such as those that contain sulfur, oxygen and nitrogen. Ag is therefore able to bind to plasma proteins such as albumin and structural proteins such as collagen. There is no doubt that Ag is bactericidal with a broad spectrum of effects in many types of bacteria. $Ag^+$ ions can readily pass through the bacterial cell wall and interfere with the intracellular DNA. This can happen when bacteria are exposed to various solutions, gels and creams. It is also clear that bacteria that contact metallic Ag are also affected, although this is probably also caused by free $Ag^+$ at the surface. Several commercial preparations of Ag antibacterial agents, especially silver sulfadiazine and silver nitrate, have been used for many years. The critical question then arises that if silver is toxic to bacterial cells, what is the effect on normal human cells? Experimental evidence suggests that Ag is indeed toxic to some cells under some conditions. A great deal depends on the availability of free $Ag^+$ in the immediate environment of the cells. In clinical circumstances there are many proteins and other macromolecules that are easily able to sequester these ions, for example albumin in the extracellular space and metallothioniens intracellularly, and reduce these risks of toxicity. There are very few indications of gross Ag toxicity in humans – the only significant manifestation of silver excess is the condition of argyria in which silver is deposited in certain tissues, which especially produces a characteristic color to the skin.

#### Class 1.6.1 Silver coatings

The potential for exploiting the antibacterial activity of Ag in medical technology is better realized by the use of coatings in order to minimize costs and obviate the poor mechanical properties. Several methods have been used for the deposition; the majority of these can be characterized as either chemical vapor deposition (CVD) or physical vapor deposition (PVD). CVD processes involve the use of reactive precursors that interact and deposit Ag on a surface. PVD techniques involve a physical process that directs Ag atoms or ions onto a surface under high vacuum conditions. Many PVD techniques are included under the general heading of sputtering, for example magnetron sputtering, or involve laser energy, such as with pulsed laser deposition. Two closely related PVD techniques used with Ag are ion implantation and ion beam assisted deposition (IBAD). The latter has been used on a number of implantable/invasive medical devices, including catheters, vascular grafts and heart valve sewing rings, shown in Figure 8.3. One commercially available process for such devices was known as Silzone. The heart valve in question had to be withdrawn from the market because of a very small number of clinical problems within a few months of implantation (see Chapter 9). Although some accusations were made about the role of silver toxicity in these cases, this was not proven and appears unlikely. It does emphasize the attention that has to be given to the differential effects of antibacterial agents on the bacteria themselves and the hosts.

### Class 1.6.2 Silver electrodes

Silver is extensively used in medical electrodes, for example for EEG recording. Nearly always these are sintered silver–silver chloride electrodes, which are durable, with low offset voltage and low noise.

### Class 1.6.3 Nanocrystalline silver and silver nanoparticles

Ag nanoparticles, often referred to colloquially as nanosilver, exist as clusters of Ag atoms with aggregate diameters ranging up to 100 nm. Small clusters of less than 10 nm may be produced by the photoreduction of aqueous $[Ag(NH_3)_2]^+$ in the presence of UV light, or by reduction of $AgNO_3$ in a chitosan/acetic acid solution under $\gamma$ irradiation. Clusters between 5 and 40 nm are prepared simply by applying an electric current between two silver wires in deionized water or, alternatively by the reduction of $AgNO_3$ in starch solutions. Other methods are also available with varying results in terms of size and purity. Ag nanoparticles are shown in Figure 6.15.

Although Ag nanoparticles, in common with other metallic nanoparticulate systems, do display some physical properties that are different from the bulk metal, these are not exploited medically. Instead, it is the biological characteristics that have attracted widespread interest, especially an enhanced antibacterial effect. Nanocrystalline Ag has varying antibacterial efficacy, the effects being very much strain dependent and also with some dependence on size and route of preparation. Nanocrystalline coatings may be incorporated into multilayer wound dressings and effects that are superior to the conventional silver sulfadiazine have been noted. Several types of polymeric catheter have been coated with nanocrystalline Ag in order to reduce the incidence of catheter-related infections. As with metallic Ag itself, there is still a lack of clarity on the borderline between antibacterial (or as has been implied anti-inflammatory) activity and toxicity to mammalian species.

## 8.2.7 Class 1.7 Platinum group metals and alloys

### Class 1.7.1 Platinum and its alloys

Platinum (Pt) is a transition metal of atomic number 78 with an fcc crystal structure. It is extremely corrosion resistant, resisting attack in many very hostile environments; it is considered as one of the noble metals, with a standard electrode potential of +1.2 V. The Young's modulus is 170 GPa. It is relatively soft and ductile, the UTS in the annealed state is 220–240 MPa and the elongation is usually around 40%. As such, Pt is not used in highly stressed situations; indeed even in the most important of its applications, in jewelry, it is not sufficiently strong and has to be alloyed. Most alloys will still have a Pt content of at least 50%, common alloying additions being Cu, Co and W.

In medical applications the high corrosion resistance of pure platinum should not be significantly compromised and the choice of alloying elements necessary to give adequate strength is predicated on this requirement. In most of these situations attention is given to a group of elements that have many similarities with Pt. These are collectively called the platinum group metals (PGMs), all of which are similar transition elements with their own good corrosion resistance. Next to platinum in the Periodic

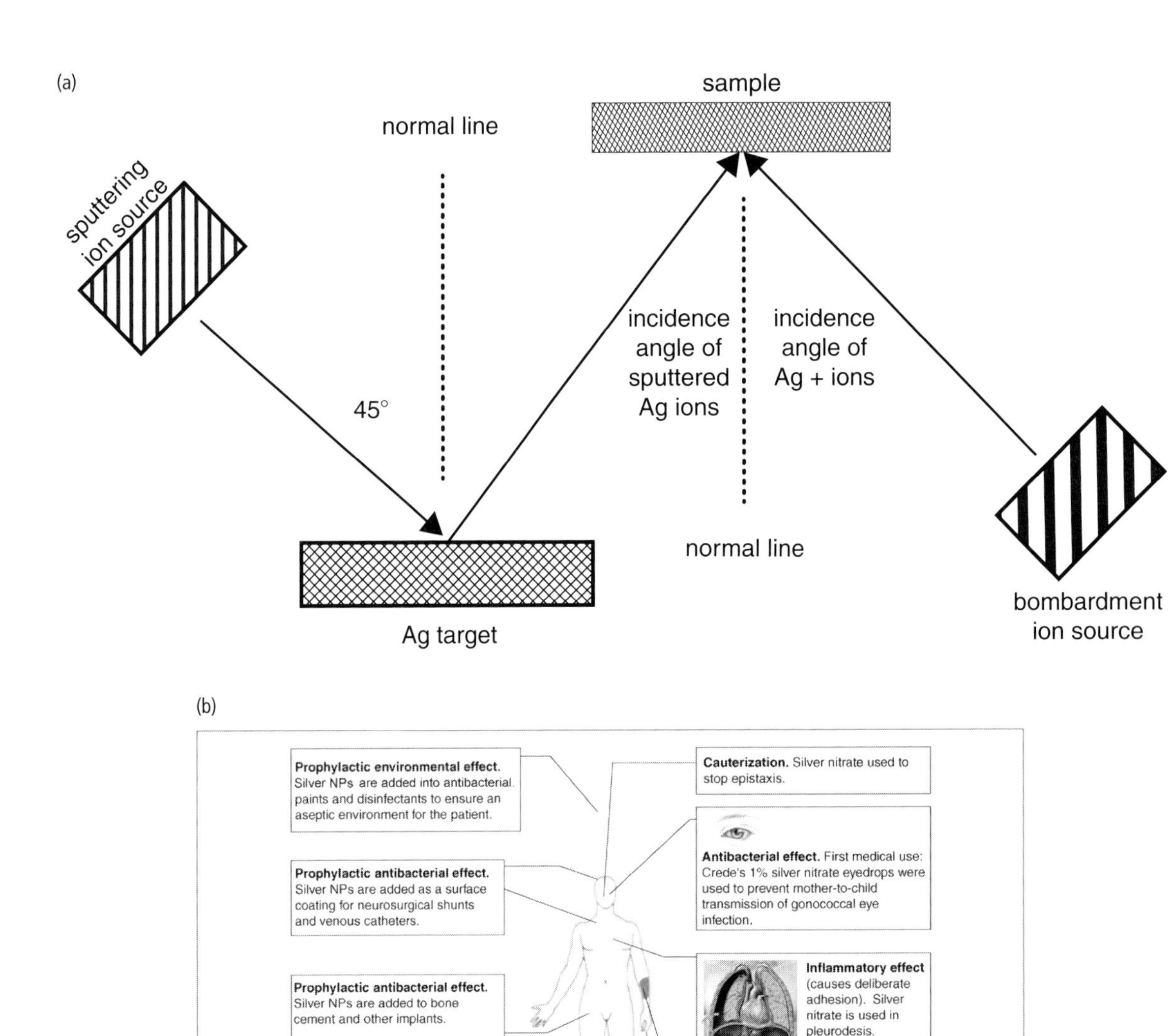


**Figure 8.3**
**(a)** Schematic diagram of the ion beam assisted deposition apparatus. Reprinted from *Applied Surface Science*, 254, Feng T *et al.*, Study of the orientation of silver films by ion-beam assisted deposition, 1565–8, © 2008, with permission from Elsevier.
**(b)** Silver nanoparticles. Traditional clinical applications of silver products (on the right) and current or potential applications of nanosilver (on the left). Reprinted from *Trends in Biotechnology*, 28, Chaloupka K *et al.*, Nanosilver as a new generation of nanoproduct in biomedical applications, 580–8, © 2010, with permission from Elsevier.

Table is iridium (Ir), atomic number 77 also with a fcc structure. This has the best corrosion resistance, and is the most rigid (Young's modulus 570 GPa) and strongest (UTS 2275 MPa). At atomic number 76 is osmium (Os), which is of hexagonal structure and of limited importance. Three further elements at atomic numbers 44, 45 and 46 respectively are ruthenium (Ru), rhodium (Rh) and palladium (Pd). The first is hexagonal and the other two are face centered cubic. Pd is the least rigid (Young's modulus 110) and weakest (UTS 200 MPa) of the group. Rh and Ru have intermediate properties.

The structure and properties of binary alloys involving combinations of these PGMs depends on their crystal structures. Four of these elements are fcc and are essentially mutually soluble in each other so that substitutional solid solutions are formed. With Pt-based alloys, the increase in strength gradually increases with addition element concentration and reasonable strengths are achieved with up to 20% of these elements. Complete solubility is, however, restricted to high temperatures. Figure 8.4 shows the phase diagram of Pt–Rh, which is representative of this group. The area below the dotted line refers to a "miscibility gap" in which the structure tends to occur as a mixture of platinum and palladium-rich areas rather than a complete range of solid solutions. Alloys involving the face centered cubic platinum with either of the hexagonal structures of Os and Ru are much more complex.

Before leaving the discussion of Pt, we must consider the biological properties of compounds of this

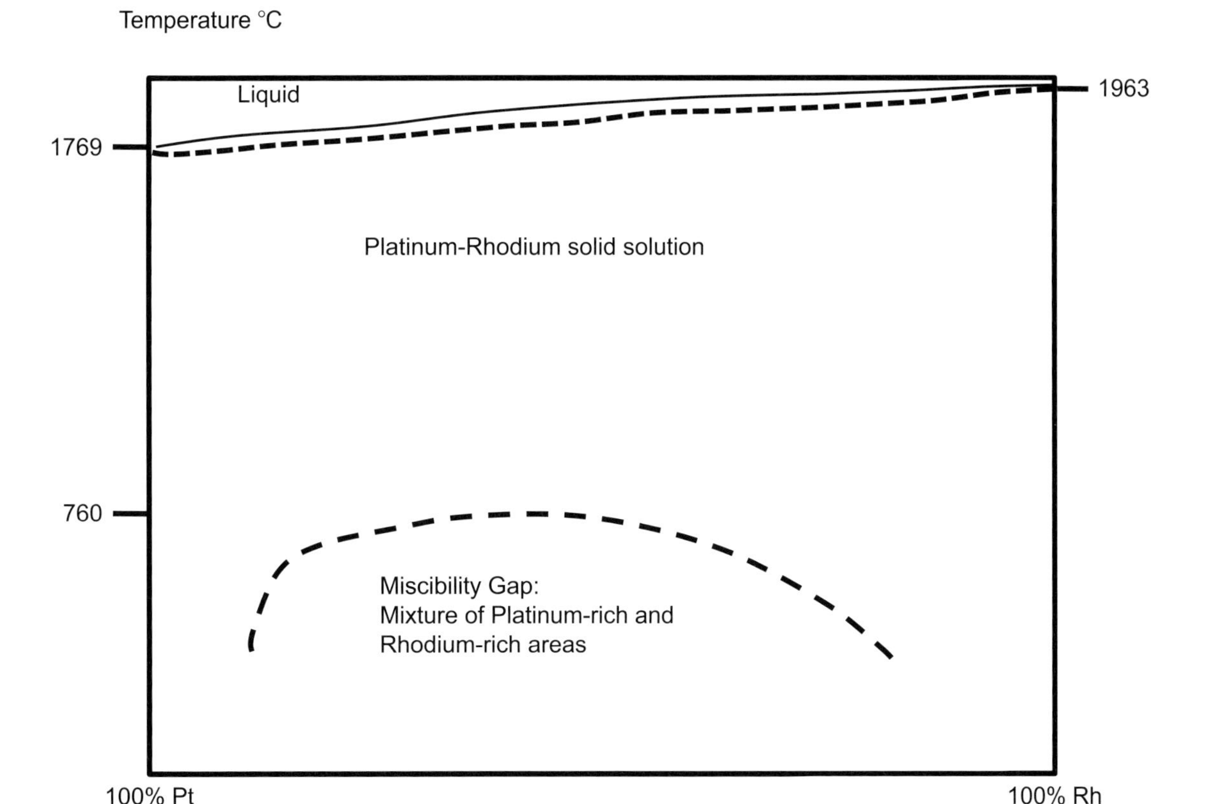


**Figure 8.4** Platinum alloys. Phase diagrams of platinum metal group alloys are very similar. This shows the Pt–Rh system. Generally there is good solubility but at low temperatures there may be problems with mutual miscibility such that the alloys consist of mixtures of Pt-rich and Rh-rich areas.

metal, just as we have done with Ag compounds above, and will do with gold compounds below. Pt is one of the most chemically inert of metallic elements but yet has an affinity for certain biological molecules, and especially DNA. This is especially seen with the molecule cis-diamminedichloroplatinum(II), commonly known as cisplatin, a widely used chemotherapeutic agent. This will cause death in all cells by binding to DNA at the guanine base, inhibiting DNA, protein and RNA synthesis. This occurs much faster with cells of a high turnover, such as tumor cells, but the ubiquity of its action is the cause of the serious side effects of this drug. It is only active in the cis-configuration, which is not charged.

**Class 1.7.1.1 Platinum–iridium alloys** The Pt–Ir system represents the most widely used group of alloys, in medical devices, jewelry and electrical contact applications. Compositions up to 20% Ir in Pt are used. The UTS can exceed 700 MPa. These alloys tend to be preferred for some forms of chronically implanted electrodes.

#### Class 1.7.2 Palladium-based alloys

In some situations it is possible to use Pd-based alloys, especially Pd–10% Re. The advantage of doing so is largely concerned with the much-reduced cost of Pd compared to Pt; the mechanical properties are essentially the same as for Pt alloys. Some medical devices, such as guidewires, utilize these alloys.

#### Class 1.7.3 Iridium films

It was noted above that Ir has some of the best properties of the PGMs. There is increasing interest in using unalloyed Ir in some microelectrodes and as coatings on Pt electrodes. It has high charge injection capability when used as a thin coating, with potential applications in stimulating electrodes.

### 8.2.8 Class 1.8 Gold

#### Class 1.8.1 Metallic gold

Gold (Au) is a noble metal, atomic number 79, with a very high density of 19.3 g/cm$^3$ and an fcc structure. Several features of gold severely limit its potential in biomedical applications. The first is that it has very poor mechanical properties, being both soft and weak; secondly it is very expensive. The mechanical properties can be improved by alloying, especially through the addition of copper, and these high strength alloys have been used very effectively in dentistry. Even here, however, the availability of high-quality, lower cost alternatives such as the tooth-colored, degradation-resistant, high-strength, oxide ceramics that can be prepared as crowns by CAD-CAM techniques has decreased the popularity of gold. There is little scope for gold as a bulk material in medical devices.

In such situations, two other routes to exploit some attractive properties of a material such as gold may be considered; these involve surface coatings and nanoparticles. The high corrosion resistance of gold has led many to believe that it should have good biocompatibility. In view of the need to improve the performance of metallic stents, and also to increase their radiopacity (see Chapter 4), attempts were made to coat stents with Au. This did not produce any better response, and indeed there were greater levels of restenosis, and Au coatings have not been utilized on any significant scale since then. There are some potential applications involving very small implantable devices in which the cost and mechanical properties of Au would not be too restrictive; one example here involves the minute implants used in middle ear surgery, such as stapes replacement prostheses, but the experience with surface coatings has been repeated as their clinical performance has been inferior to more conventional materials.

These comments on the biological performance of Au products led to the question of the overall biological properties of gold. As a transition metal, gold can exist in several oxidation states, and can be oxidized or reduced between these states depending on the conditions. Au (III) compounds are usually strong oxidizing agents, being reduced to Au (I), which may result in toxicity. This reduction is favored by the presence of naturally occurring reductants such as thiols, which are compounds that contain the functional group S-H. A number of gold thiolates exist, which have widely varying properties. They are able to undergo ligand exchange reactions with many biological groups such as the amino acid cysteine. As a consequence of this, a small number of gold thiolates have been used for their pharmacological activity, including sodium aurothiomalate and auranofin, both of which have anti-rheumatoid arthritis effects. One further positive effect of the gold-thiol interactions is that self-assembled thiol monolayers may be attached to gold surfaces, where they can have several forms of biological activity. This has best been demonstrated through Au nanoparticles, as discussed below.

### Class 1.8.2 Gold nanoparticles

Gold nanoparticles (GNPs) have several attractive features. They have exceptional optical and electronic properties and have the inherent ability to provide stable immobilization of biomolecules while retaining their biological activity. They can be readily prepared with a variety of sizes, from less than 10 nm to several hundred nm, and with a variety of shapes, including spheres, rods and cages.

The majority of GNP formulations exist as core-shell structures, with metallic gold cores and monolayer functional moieties at the surface as a shell. The nature of these structures is controlled by the conditions of the nanoparticle synthesis. For example, the nanoparticles can be prepared by the reduction of Au salts in solution in the presence of stabilizing agents that prevent aggregation. The core is established by generation of gold atoms during the reduction process and the addition of clusters of atoms to this atom through collision processes results in growth of the crystal. The size and shape of the resulting particles is controlled by the chemical composition of the medium. If the deposition of gold atoms onto the original atom is isotropic, then a spherical nanoparticle should be produced. Non-spherical particles can be prepared by the anisotropic growth of the particles, produced by certain surfactants. Au nanorods, for example, are produced through the use of the surfactant cetyltrimethylammonium bromide (CTAB), the concentration of the seed crystals, the temperature, the pH and other factors controlling the final shape of the rods. The crystal structure in the core of the particle may vary and does not necessarily follow the usual face centered pattern for bulk Au. Examples of gold nanoparticles are shown in Figure 8.5.

The shell of a GNP is usually a monolayer of a biologically active substance that can be anchored to the Au through thiol linkages. For example, the nanoparticle may be protected by a monolayer shell of p-mercaptobenzoic acid. The Au atoms may bind one or two sulfur atoms at the interface between core and shell. Functional groups may then be added to the shell, usually involving thiol, hydroxyl or phosphine groups. Thiol-modified oligonucleotides may be added to the solution and can selectively bind the complementary nucleic acids. The diverse functionalization of the nanoparticles permits a variety of applications in diagnostics and therapy. Moreover, several different functionalities can be generated by multiple ligand coating processes. A layer of a suitable polymer may stabilize the nanoparticle and target ligands, such as monoclonal antibodies, cell penetrating peptides or DNA sequences may be covalently attached to this layer in order to give selective recognition at different tissue or organ sites. As well as being used for drug and gene delivery systems based on these functionalities, the GNP may also be used in sensing and recognition processes using their special optical and electrochemical characteristics. Immunoassays have been developed based on the aggregation of antibody functionalized GNPs and their light scattering behavior. Au nanorods have been functionalized by peptides for the imaging of cancer cells, using differences in the effect of cancer and normal cells on Raman spectra after interaction with these nanorods. GNPs may be used as contrast agents for computed tomography in view of their good X-ray attenuation and long clearance times.

## 8.2.9 Class 1.9 Magnesium and its alloys

Magnesium (Mg) is an alkaline earth metal of atomic number 12. It is one of the lightest of metals, with a specific gravity of 1.74. It has an hcp crystal

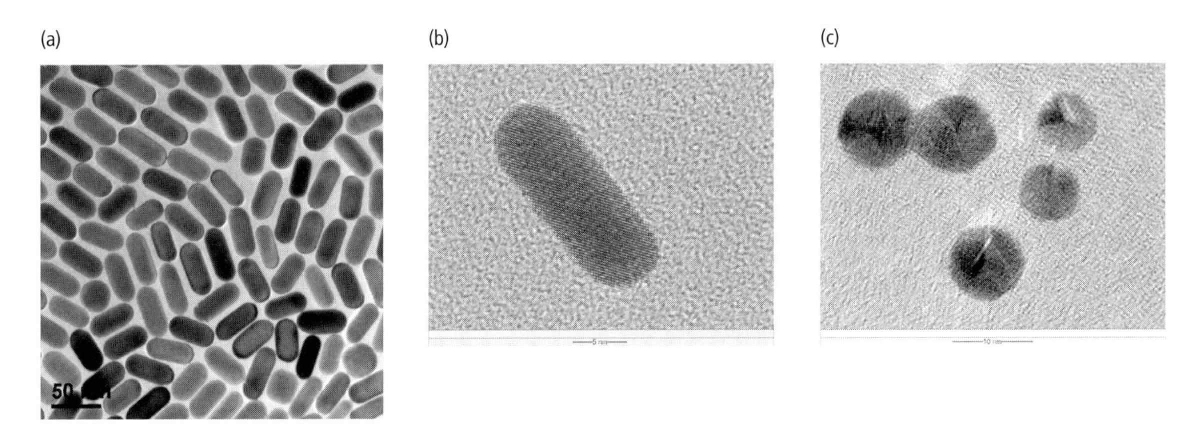


**Figure 8.5** Gold nanoparticles.
(a) Collection of gold nanorods.
(b) TEM image of aqueous monodispersed gold nanorods ($d$ = 50nm) with aspect ratio of 1:3 and cetyltrimethylammonium bromide as the shape controlling agent.
(c) High-resolution TEM image of 10 nm aqueous gold spherical nanoparticles from citrate reduction. The lattice patterns show the hexagonal close-packed (hcp) structure. Images courtesy of Dr. Lanry Yung Lin Yue, National University of Singapore.

structure. The attractiveness of magnesium in the context of biomaterials lies in its potential as a degradable material with minimal toxicity, since it is susceptible to corrosion in saline environments and is also an essential physiological element with an array of biological properties.

Commercially pure Mg has a minimum of 99.8% Mg, but with a UTS of only 20 MPa and an elongation at break of just 6% it does not find much use in structural applications. There are several types of Mg alloy, the common additions being Al, Mn and Zn but also with Zr, Ca and some rare earth metals occasionally being used. The alloys are conventionally described by a system that incorporates the first letters of the two major elements and two numbers that give their approximate percentage. Thus AZ31 incorporates 3% Al and 1% Zn; often a final letter such as A, B or C is added to distinguish small variations in content. The letter for Zr is K, for Fe it is F and for Cd it is D. Al is the most important alloying element. There is a maximum solubility of Al in Mg of around 12%; above this level an intermetallic compound $Mg_{17}Al_{12}$ is precipitated, which makes alloys much more brittle.

The most widely investigated alloys for medical use have been AZ21A (ASTM specification B275, 1.6–2.5% Al, 0.15% Mn, 0.8–1.6% Zn and 0.10–0.25% Ca), AZ31B (ASTM B107, 2.5–3.5% Al, 0.2–0.5% Mn and 0.6–1.4% Zn) and ZE41A (ASTM B93, 3.7–4.8% Zn, 4.75–5.5% Y). The alloys may be cast or wrought. A cast alloy can have a UTS of 250 MPa and elongation of 5%. Wrought alloys may have a UTS of 400 MPa and elongation of 20%. The Young's modulus is around 42 GPa.

Mg and its alloys corrode quite readily, especially in chloride solutions. The standard electrode potential is −2.37 V, indicating inherent vulnerability to corrosion; moreover, any oxide or hydroxide that forms on the surface is readily soluble in many media so there is little chance of passivation protection. Commercially pure Mg will corrode at a rate of 0.25 mm per year in seawater. With alloys, the rate of corrosion varies with composition. Al, Mn, Zr, Y and Sn have little influence on the corrosion rate at levels less than 5%. Cd, Zn, Ca and Ag have mild-moderate accelerating effects on the rate, while Fe, Ni, Cu and Co markedly increase the rate. An example of Mg corrosion is shown in Figure 8.6.

From the physiological perspective, Mg is a very important element. The adult human body contains about 30g Mg, mostly in muscle and bone. It is a

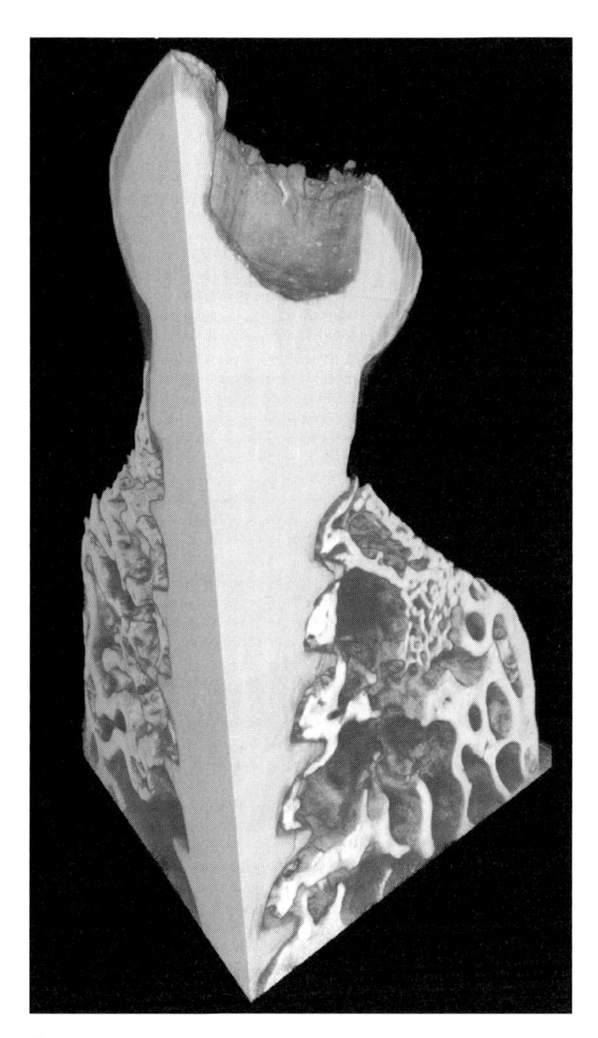

**Figure 8.6** Magnesium alloy screws. This is an image of an explanted AZ31 (Mg–3wt% Al–1wt% Zn) alloy screw, which had been inserted into the iliac crest of an adult sheep. The image was obtained using synchrotron-radiation micro-computed tomography. The screw has been inserted until all threads were buried into the bone to allow corrosion in two different environments: bone and muscle. The corrosion layer (dark grey) at the screw head is thicker than on the threads which are in direct contact to cancellous bone. Most of the threads are heavily corroded and/or cracked at 6 months after surgery. Image courtesy of Professor Frank Witte, Laboratory for Biomechanics and Biomaterials, Hannover Medical School, Germany.

cofactor for many enzymatic reactions and is involved in electrostatic binding to proteins and nucleic acids. It is essential for normal neurological and muscular function, and Mg depletion inhibits the functioning of osteoblasts and increases the activity of osteoclasts. Mg binds strongly to phosphates and therefore influences biomineralization processes. In the cardiovascular system, Mg depletion is associated with arrhythmias and vasoconstriction. Several Mg alloys have been shown to have good general biocompatibility, with little evidence of cytotoxicity.

In view of these biological properties, Mg alloys have been considered for use as biodegradable intravascular stents and bone fracture plates. Realization of the potential in these areas has not been easy, partly because of the relatively poor inherent mechanical properties and partly because of the difficulty of achieving a clinically relevant corrosion rate/dissolution time in alloys with acceptable biocompatibility.

## 8.3 Class 2 The ceramics systems

The biomaterials in Class 2 cover the ceramic systems. From a materials science perspective, ceramics are usually compounds involving both metallic and non-metallic elements for which the interatomic bonds are either totally ionic or predominantly ionic but having some covalent character. The word "usually" is included here since the boundaries around this class of material are a little vague, as are the ways in which these materials can be classified. We have a number of options here. For example, ceramics may be crystalline or they may be amorphous, in which state they are called glasses, or they may be somewhat in between, in which case they are glass-ceramics. They are often classified according to their uses, such as magnetic or semiconducting ceramics. As with the metallic systems, it is, for the most part, better to stay with the compositional approach, as indicated in Box 8.3.

The metallic element(s) in ceramics can be one of very many, but only a few non-metallic elements are involved, so it is preferable to class ceramics on the basis of the latter, either as a single non-metallic element (e.g., oxygen) or as combinations (e.g., silicon and oxygen). As such we find that the majority

**Box 8.3 The second level of the classification of biomaterials: ceramic systems**

Class 2.1 Oxides
Class 2.1.1 Aluminum oxide (alumina)
Class 2.1.2 Zirconium oxide (zirconia)
Class 2.1.2.1 Partially stabilized zirconia
Class 2.1.2.2 Stabilized zirconia: tetragonal zirconia polycrystals
Class 2.1.3 Alumina–zirconia ceramics
Class 2.1.4 Silicon oxides (silica)
Class 2.1.4.1 Crystalline and non-crystalline silica
Class 2.1.4.2 Mesoporous silica and silica-based nanoparticles
Class 2.1.5 Titanium oxides
Class 2.1.6 Iron oxide nanoparticles
Class 2.1.7 Iridium oxide
Class 2.1.8 Cerium oxide
Class 2.2 Phosphates
Class 2.2.1 Amorphous calcium phosphates
Class 2.2.2 Monocalcium phosphates
Class 2.2.3 Dicalcium phosphates
Class 2.2.4 Tricalcium phosphates
Class 2.2.5 Octacalcium phosphates
Class 2.2.6 Hydroxyapatite
Class 2.2.7 Biphasic calcium phosphates
Class 2.2.8 Calcium phosphate cements
Class 2.2.8.1 Apatite calcium phosphate cements
Class 2.2.8.2 Brushite calcium phosphate cements
Class 2.3 Sulfates
Class 2.3.1 Calcium sulfates
Class 2.4 Silicates and silica-based glasses
Class 2.4.1 Wollastonite
Class 2.4.2 Diopside and akermanite
Class 2.4.3 Zeolites: aluminosilicates
Class 2.4.4 Silica-based glasses: bioactive glasses
Class 2.5 Nitrides
Class 2.6 Carbides
Class 2.7 Titanates
Class 2.8 Optically active ceramic/metallic nanoparticles
Class 2.8.1 Semiconductor quantum dots
Class 2.8.2 Rare earth upconverting nanoparticles

of ceramic biomaterials are either oxides (Class 2.1), phosphates (Class 2.2), sulfates (Class 2.3), silicates (Class 2.4), nitrides (Class 2.5) or carbides (Class 2.6). Some are a little more complex than this, and we have a few discrete complex classes such as the semiconductor quantum dots (Class 2.7), although, as we shall see, there is a temptation to place these in other classes. We also resist the temptation to introduce classes based on biological properties or biomedical applications here, such as biologically active glasses or superparamagnetic oxides, but do, almost inevitably, have to introduce such features in lower levels of the hierarchy. It should also be noted that quite often (as indeed I myself have done in the past) authors include carbon materials within the class of ceramics. This no longer serves a useful purpose and they are considered in a separate class here.

## 8.3.1 Class 2.1 Oxides

Oxide ceramics may be considered to be amongst the simplest of ceramics and, for a long time, have been the major examples of ceramics as high-performance engineering materials. The simplicity depends, of course, on the chemistry. Equivalent valencies of metallic element and oxygen and perfect stoichiometry should provide that simplicity, but in reality this rarely happens, and multiple forms of crystal structures can arise when there are variations from this position. As we have seen in Chapter 2, the single most important disadvantage of ceramics in general, which is epitomized in structural oxides, is their brittleness and many of the materials science initiatives with these ceramics have been directed towards the reduction in brittleness, or more correctly the increase in toughness.

### Class 2.1.1 Aluminum oxide (alumina)

Aluminum oxide, $Al_2O_3$, is a relatively simple oxide ceramic that exists in several different forms. Its general usefulness is based on a high hardness and high melting point. It is quite easily prepared as large single crystals, for example as a synthetic version of the natural gem sapphire, and indeed has been used in this form to fabricate some implantable medical devices. However, the polycrystalline form offers much better performance. General industrial grades of alumina will have a purity between 85 and 99% $Al_2O_3$, with Young's modulus ranging from 220 to 350 GPa and compressive strength from 1800 to 2500 MPa over this compositional range. Materials from the higher purity end of this range were first used for implantable devices over 30 years ago but their mechanical properties were not sufficient for reliable quality, especially with a low toughness.

Gradually the technology of alumina ceramics has improved and the material is now used in highly stressed orthopedic applications, including alumina–alumina bearing surfaces in total hip replacements. It is α-alumina, of rhombohedral crystal structure with a minimum purity of 99.5%, which is used. The mechanical properties depend on the purity level and also on the porosity and grain size. Typical materials used in these products will have purities closer to 99.8%, a density of greater than 3.98 g/cm$^3$ and a grain size of 2–5 μm. The specifications and properties of alumina ceramics for implantable devices are summarized in Table 8.5. They have a compressive

Table 8.5 **Specifications and properties for alumina ceramics used as biomaterials. This table provides representatives specifications and average properties as given in several standards documents**

| Characteristic | Type A | Type B |
|---|---|---|
| $Al_2O_3$ mass fraction | > 99.7% | > 99.5% |
| MgO mass fraction | < 0.2% | < 0.2% |
| Impurities | < 0.1% | < 0.3% |
| Average bulk density, Kg/m$^3$ × 10$^3$ | > 3.94 | > 3.90 |
| Grain size, μm | < 2.5 | < 3.5 |
| Young's modulus, GPa | > 380 | > 370 |
| Biaxial flexural strength, MPa | > 300 | > 150 |
| Hardness, Vickers | > 1800 | > 1700 |

strength of 4500 MPa, with very good fracture toughness and fatigue resistance. It is noteworthy that this compressive strength falls off very quickly with increasing grain size, being 3500 MPa at 10 μm and 2500 at 20 μm. The wear rate of alumina–alumina bearing surfaces is lower than alumina–polymer and much lower than metal–polymer interfaces. The alumina, being a dense oxide ceramic, is very resistant to corrosion or degradation and is one of the most inert of biomaterials with respect to physiological fluids. Both soft and hard tissue responses appear to be minimal. Examples of products using alumina biomaterials are shown in Figure 8.7.

### Class 2.1.2 Zirconium oxide (zirconia)

Zirconium oxide, $ZrO_2$, is a simple oxide ceramic that has good mechanical properties and can be used for many applications in hostile environments. It is hard, although not as hard as alumina. Crucially, as we will see below, the toughness of zirconia can be significantly raised above that of alumina, making it an attractive possibility for structural applications in the body, possibly as the bearing surface for joint replacement prostheses. Since it essentially is a white opaque solid that can be colored by the selective use of dopants, it also became attractive for dental crowns. As with several important metallic biomaterials, the qualities of zirconia are dependent on the fact that it is polymorphic; that is, it exists with different crystal structures at different temperatures, the stability of which depends on the presence of some additives. Zirconia is monoclinic at room temperature. It transforms to a tetragonal structure at 1170°C, which itself transforms to a cubic phase at 2370°C. These transformation temperatures vary with the presence of small quantities of other substances, usually other oxides such as CaO, MgO, CeO and $Y_2O_3$, and the biomaterials story of zirconia can only be discussed in terms of these dopants and their influence on the immediate and long-term properties of the ceramic. This discussion should be read in conjunction with Section 2.8 on the transformation toughening of ceramics in Chapter 2.

**Class 2.1.2.1 Partially stabilized zirconia** Although the use of alumina as the bearing surface of joint prostheses was increasing in the late 1980s and 1990s, controversy still existed over the inherent brittleness of the alumina and the ever-present risk of brittle fracture. Zirconia was considered as a potential alternative since it was possible to raise the

**Figure 8.7** Ceramics in orthopedics. Examples of Biolox® alumina cups for hip replacement prostheses. Image courtesy of CeramTec International.

fracture toughness by the transformation toughening process. The tetragonal-monoclinic transformation should take place on cooling at around the 1170°C mark, but it is slow and usually is not complete until another 100°C lower. This transformation takes place with a volume expansion of 3–4%. In pure zirconia, the stresses generated by this expansion could cause cracking and disintegration. In order to overcome this, it was found that the stability of the crystal structures and the transformation between them, could be profoundly altered by addition of the stabilizing oxides mentioned above; in particular, cubic zirconia could be present as the major phase at room temperature, containing monoclinic or tetragonal minor phases. Crucially, any tetragonal phases present as a fine dispersion in the cubic matrix may transform into the monoclinic form on the application of stress. In particular, if a crack forms in the material, it may cause the tetragonal–monoclinic transformation to take place at the advancing crack tip. The stress field associated with the expansion that takes place during the transformation acts in opposition to the stress field that is promoting the crack propagation, thus dissipating the energy and blunting or arresting the crack. This produces a significant increase in the fracture toughness, and is shown in Figure 8.8(a)–(c). It has to be remembered that these structures exist in a metastable state, and other transformations to more stable states could occur if the energy conditions were favorable.

In practice, a zirconia material formed according to this principle is said to be partially stabilized (PSZ). Typically 8% MgO is added to the zirconia, which allows the formation of a fully cubic structure at around 1800°C. During controlled cooling and ageing, the metastable tetragonal phase forms by precipitation within the cubic matrix. Although several forms of PSZ were investigated for medical use, this was overtaken by the development of a different form of stabilized zirconia.

**Class 2.1.2.2 Stabilized zirconia: tetragonal zirconia polycrystals** The material widely known as TZP, tetragonal zirconia polycrystals, was developed to take this principle one step forward. This contains around 2–3% $Y_2O_3$, yttrium oxide, or yttria. This is constituted solely by sub-micron crystals of the tetragonal phase; again this is a metastable state. The properties of such a ceramic include an elastic modulus of 200 GPa, a flexural strength of 900 MPa and a fracture toughness, $K_{IC}$, of 17 MPa $m^{1/2}$.

TZP is not without problems. In the above paragraphs it has been emphasized that the transformation toughening process requires the presence of metastable phases. Metastability involves the existence of a phase under conditions where it should not theoretically be stable according to thermodynamic principles. The reason that it does exist under those conditions is usually because it cannot transform to the stable phase since the atoms do not have the ability to move to new lattice positions under those conditions, mainly because of poor atomic diffusion. However, metastable materials may transform under different conditions. With TZP this can happen in the presence of water or water vapor at relatively low temperatures. The transformation process can start in isolated grains near the surface, which then cascades to neighboring grains as volume expansion occurs, opening up microcracks and allowing the ingress of water into the bulk material. This process is often called low-temperature degradation (LTD) of zirconia and results in a catastrophic loss of properties. The resistance to LTD varies with the microstructure and thermomechanical history. This is of special relevance to the use of TZP in orthopedic devices: the material was used for femoral heads but some batches of these products received inappropriate treatment and large numbers of clinical failures ensued as the heads degraded.

### Class 2.1.3 Alumina–zirconia ceramics

The considerable attraction of both alumina and zirconia, coupled with the problems mentioned above with monolithic zirconia, have led to the development of ceramics involving both of these oxides. Some alumina–zirconia composites contain micro- or nanoscale dispersions of zirconia within an alumina matrix, as shown in Figure 8.8(d). Others, including one version which is widely used in clinical practice in hip replacements, are based on mixtures of grains of alumina, usually in the 1–2 μm range, and of zirconia, in the submicron range, at a ratio of 3:1, with small additions of other oxides, including chromium and strontium oxides, that improve mechanical properties.

### Class 2.1.4 Silicon oxides (silica)

**Class 2.1.4.1 Crystalline and non-crystalline silica** Silica, $SiO_2$, is a naturally occurring compound that exists in several different forms and is widely used as either the starting point for the manufacture of some high-performance materials or as additives in other substances. It can be crystalline or non-crystalline. In the crystalline form, pure silica exists as either quartz or cristobalite depending on

(a)

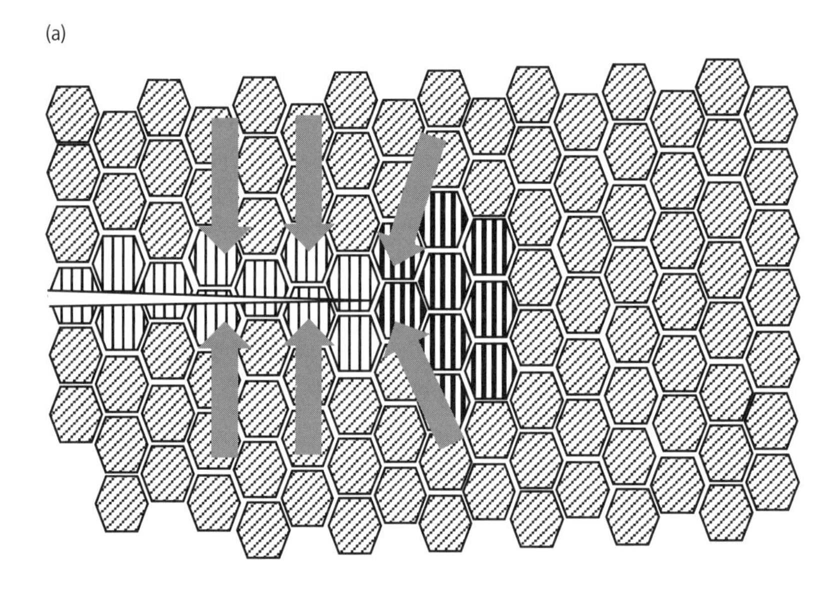

(b)

**Figure 8.8** Oxide ceramics.
**(a)** Phase transformation toughening mechanism. At the top, tensile stresses around a propagating crack trigger the phase transformation of tetragonal grains towards the monoclinic symmetry. The transformation, associated to a volume increase, leads to compressive stresses that hinder further crack propagation (arrows).
**(b)** and **(c)** illustrate phase transformations around a Vickers indentation (b) and a double torsion crack (c). Images courtesy of Jérôme Chevalier, Laurent Gremillard, Nicolas Courtois, Helen Reveron INSA-LYON, UMR CNRS 5510 (MATEIS), Villeurbanne, France.

the conditions, whereas less pure silica exists as tridymite. Non-crystalline silica, sometimes referred to as vitreous or amorphous silica, is essentially a super-cooled liquid. Both crystalline and amorphous silica are arranged as $SiO_4$ tetrahedra, from which short-range order is obtained. The long-range order which gives the crystalline form to the structure is determined by very precise Si–O–Si bond angles that join individual tetrahedral together. In the non-crystalline form, the bond angle randomly varies anywhere between 120° and 180° so that there is no equivalent long-range order, as discussed in Chapter 2.

There are no medical uses of monolithic silica. Silica particles are included as fillers in many polymeric biomaterials, including silicone elastomers and dental restorative materials. It should be noted that the biocompatibility and toxicology of silica is complex and controversial. Although amorphous silica appears to be readily cleared from human tissues, the

(c)

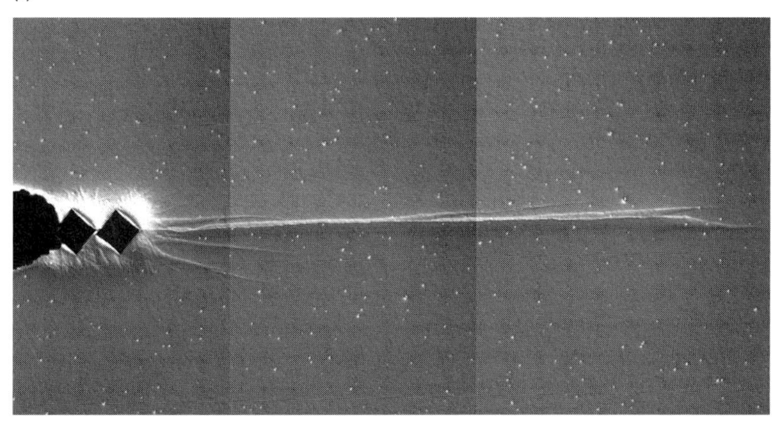

(d)

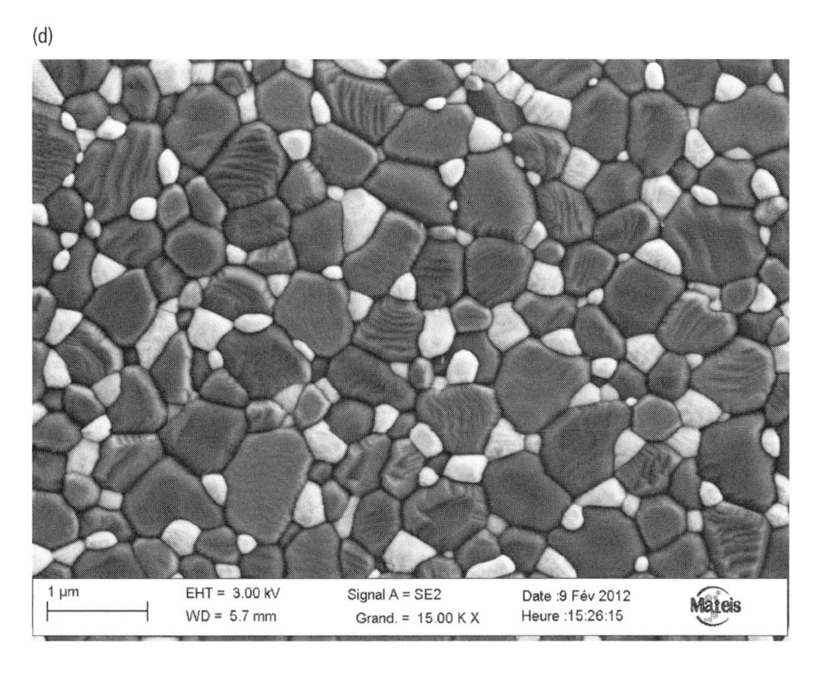


**Figure 8.8** *continued*
**(d)** Micrograph of an alumina–zirconia composite with 16 vol.% zirconia. Image courtesy of Jérôme Chevalier, Laurent Gremillard, Thierry Douillard, Katia Deuheuvels, INSA-LYON, UMR CNRS 5510 (MATEIS), Villeurbanne, France.

crystalline forms of silica have long residence times and may be associated with some pathologies, for example fibrosis of the lung.

**Class 2.1.4.2 Mesoporous silica and silica-based nanoparticles** In contrast to the situation with the monolithic silica, there is significant medical interest in nanostructured forms of silica, specifically mesoporous silica, mesoporous silica nanoparticles and silica-based core-shell nanoparticles. Mesoporous materials are defined as solids with pores of dimensions between 2 and 50 nm. These have important possibilities as molecular sieves and also in catalysis in view of their very high surface area, at greater than 1000 $m^2/g$. Silica is one of the most important materials from which mesoporous structures can be made. It is possible to produce mesoporous silica in the form of nanoparticles for targeted drug delivery, an example being shown in Figure 8.9. Also of considerable interest is the possibility of using core-shell silica nanoparticles as fluorescent labels (known as C-dots) in imaging and gene delivery. These nanoparticles have a core diameter around 60 nm and they may be functionalized with relevant agents, for example polyethylenimine for gene transfer.

### Class 2.1.5 Titanium oxides

The Ti–O phase diagram shows many stable phases, with a variety of crystal structures, ranging from $Ti_2O$ to $TiO_2$, and many of intermediate stoichiometry. Only titanium dioxide ($TiO_2$), sometimes referred to as titania, is relevant here. It exists in two major and several minor crystalline forms. Anastase and rutile both have tetragonal crystal structures but with quite different interatomic spacing. In both structures the basic units are slightly distorted octahedra with each Ti atom being surrounded by six O atoms. The elastic modulus ranges from 240 to 290 GPa. Various $TiO_2$ products are used as catalysts, in solar cells, in sensors and optical coatings, and, in the form of a white powder, as a pigment in paints and cosmetic products.

Titanium oxides are relevant to medical technology in a few respects. First, as noted elsewhere in this book, the surface of any metallic titanium product will consist of titanium oxide as a naturally derived

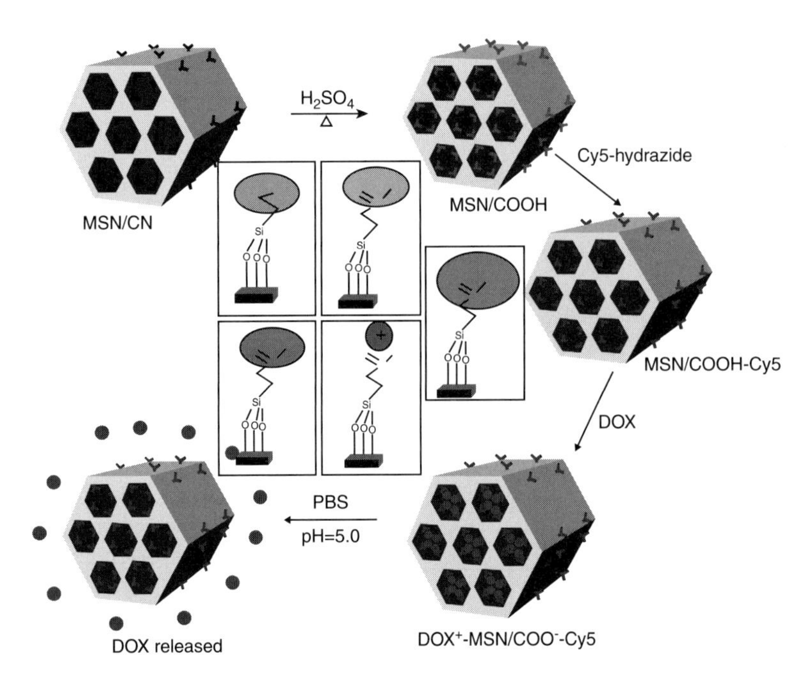


**Figure 8.9** Mesoporous silica nanoparticles (MSN). Illustration of the synthesis and chemical modification of MSN with COOH and Cy5, a near-infrared fluorescent dye, with doxorubicin loading, and pH dependent drug release. Reprinted from *Journal of Colloid and Interface Science*, 395, Xie M *et al.*, Hydrid nanoparticles for drug delivery and bioimaging: mesoporous silica nanoparticles functionalized with carboxyl groups and a near-infrared fluorescent dye, 306–14, © 2013, with permission from Elsevier. See plate section for color version.

**Figure 8.10** Titania nanotubes. SEM images of anodically produced nanotubes on titanium surface, formed at 20 V in 1% HF solution. Reprinted from *Journal of Crystal Growth*, 311, Lee J-H *et al.*, Fabrication of titania nanotubular film with metal nanoparticles, 638–41, © 2009, with permission from Elsevier.

result of oxidation of the surface. This has implications for the overall biocompatibility of the material, which is generally considered to be very good. The thickness of the oxide layer can be modulated by electrochemical processes, and substantial thicknesses can be produced by sol–gel techniques. It is also possible to specifically employ titania nanostructures to modify surfaces by, for example coating Ti with titania nanotubes, as shown in Figure 8.10. Titania exhibits a number of photochemical effects, and photofunctionalized surfaces have been considered for both enhanced bone biocompatibility and antibacterial effects.

Secondly, and perhaps of greater significance, $TiO_2$ nanoparticles have received considerable attention, both with respect to their use in sun-screens and in certain therapeutic systems; the possibility of enhanced biological effects, especially photobiological effects, may have both positive and negative effects. On the positive side, the significant photo-oxidative character of $TiO_2$ nanoparticles has led to considerable interest in their use in photodynamic therapy. On the other hand, the high mobility of charge carriers within the nanocrystals leads to high reactive oxygen species (ROS) generation, with potential consequences for toxicity. There are concerns that effects associated with ROS include interactions with DNA, leading to the possibility of carcinogenicity. Anastase nanoparticles seem to be much more toxic than those of rutile, although many commercial products contain mixtures of the two species.

### Class 2.1.6 Iron oxide nanoparticles

Iron oxides may be readily prepared as nanoparticles. A few different forms are of immense medical interest because of their special magnetic properties, which may be exploited in diagnostic and therapeutic applications. Of greatest relevance are those that have superparamagnetic properties. These nanoparticles are referred to as superparamagnetic iron oxide nanoparticles (SPION), or if they are smaller than 50 nm, as ultra-small superparamagnetic iron oxide nanoparticles (USPION), or similar variations on these acronyms. They may be based on magnetite, $Fe_3O_4$, or maghemite, $\gamma$-$Fe_2O_3$. They are usually co-precipitated from ferric or ferrous salts under alkaline conditions and coated with a dextran-based agent or a pegylated starch. Examples are shown in Figure 8.11.

The USPIONs are particularly effective as MRI contrast agents, having a long circulating half-life and being avidly taken up by a variety of cells including those in the liver and brain. Iron oxide

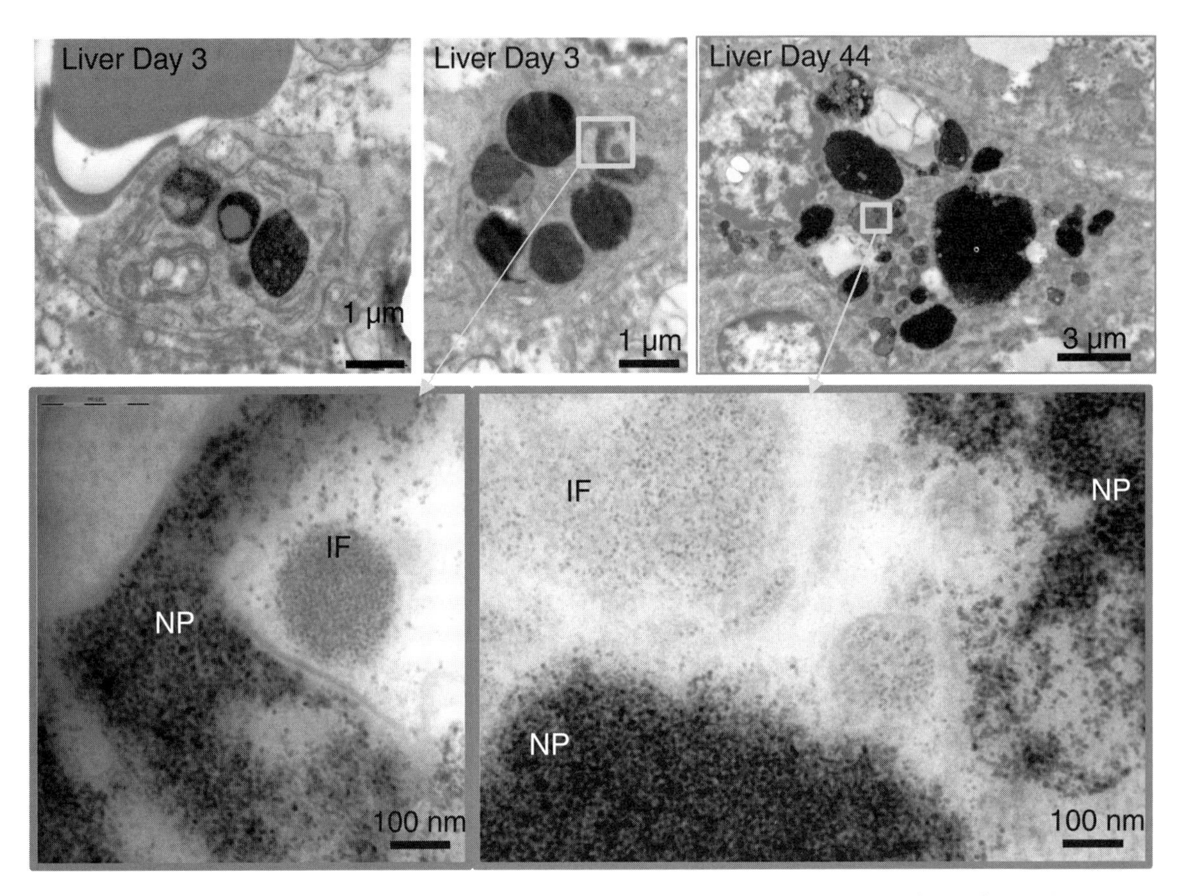


**Figure 8.11** Biotransformation of iron oxide nanoparticles. Iron oxide nanoparticles, 7–10 nm in size, designed as a MRI contrast agent were injected intravenously in mice. They are specifically taken up by macrophages of the reticulo-endothelial system (liver, spleen as well as inflammatory macrophages), in which they concentrate at very high concentration within phagolysosomes, as illustrated in TEM micrographs. Intracellular confinement persists over 44 days after NP injection, changing in turn the superparamagnetic properties of nanoparticles. Over time, the nanoparticles are locally degraded and coexist with iron-filled ferritin nanocavities, 4–6 nm in size. Image courtesy of Dr. F. Gazeau, Lab MSC, CNRS/Université Paris Diderot, Paris, France.

nanoparticles are internalized in cells and accumulate in lysosomes where the iron oxide is eventually reduced to Fe ions at the low pH, which are safely incorporated into hemoglobin. Several preparations have regulatory approval for use in the diagnosis of tumors in the central nervous system and other lesions.

## Class 2.1.7 Iridium oxide

Iridium oxide, $IrO_2$, may be used as a coating on neural stimulation electrodes; it allows the injection of charge into tissue while minimizing electrochemically irreversible processes at the interface. Common methods for applying a coating involve either deposition from iridium metal in an aqueous electrolyte by a process where electrochemical potential of the substrate metal is pulsed between negative and positive potential limits for electrolysis of water, or direct electrodeposition of iridium oxide. Substrate electrodes include Au, Pt and stainless steel.

## Class 2.1.8 Cerium oxide

Cerium (IV) oxide, $CeO_2$, usually referred to as ceria, is one of the oxides of the rare earth cerium (Ce). It is

Table 8.6 **Properties of calcium phosphate biomaterials. There are many calcium phosphate materials in clinical use. This table gives a brief summary of their performance**

| Characteristic | Porosity | Compressive strength (MPa) | Degradation |
|---|---|---|---|
| HA | 75% microporosity | 10 | None |
| HA | 85% macroporosity | 3 | Very slow |
| β-TCP | 75% macro-microporosity | 3 | 25 weeks |
| ACP, DCPD cement | 30–50% | 12 | 40 weeks |
| αTCP, HA, DCPD cement | 30–50% | 80 | 50 weeks |
| αTCP, CaCO3, MCPM cement | 30–50% | 50 | 50 weeks |

used as a catalyst, in photosensitive glass, and in solid fuel cells. When prepared as nanoparticles (CeNPs), it has catalytic antioxidant activity in some biological systems. CeNPs are able to catalyze the conversion of hydrogen peroxide to water and molecular oxygen. Essentially the redox state of Ce atoms at the surface of the nanoparticles modulates the reactivity of the material with respect to reactive oxygen species. The biological applications have not been fully explored as yet.

## 8.3.2 Class 2.2 Phosphates

We are solely concerned here with calcium phosphate compounds, which are widely used biomaterials. This use is based on the similarity between calcium phosphates and the mineral phases in musculoskeletal tissues, including bone, enamel and dentin. It is not surprising that these materials have generally good biocompatibility, especially within bone. Depending on their chemistry and formulation they may be used as manufactured solid components or injectable *in situ* setting cements and again depending on their chemistry and microstructure they exhibit varying degrees of biostability/degradation and resorption. There are several ways in which calcium phosphate biomaterials could be classified. We choose here a system based on their chemistry. The vast majority of commercial and experimental versions are calcium orthophosphates, that is they are based on the orthophosphate group $PO_4^{3-}$. The one material that is not an orthophosphate is calcium pyrophosphate, $Ca_2P_2O_7$; this is not discussed here. Within the various classes discussed below we will see that some materials are hydrated and some are not, while some products are obtained by precipitation in aqueous media and others are prepared by high-temperature processes. One of the most important variables with the calcium phosphates is the solubility/degradability. Generally those with a low Ca:P ratio have fast degradation rates, while those with high ratios are much more stable. The strength also increases with Ca:P ratio, reaching a maximum at the value of 1.67 for hydroxyapatite.

Table 8.6 summarizes calcium phosphate specifications and properties.

### Class 2.2.1 Amorphous calcium phosphates

Most calcium phosphate materials are crystalline. Amorphous calcium phosphate (ACP) may be prepared with a composition $Ca_3(PO_4)_2.nH_2O$, where n is between 3 and 4.5. It is not widely used by itself, but is a constituent of some calcium phosphate cements (see Class 2.2.8 below).

### Class 2.2.2 Monocalcium phosphates

Monocalcium phosphate, $Ca(H_2PO_4)_2$ exists in either hydrated (MCPM) or unhydrated (MCP) form. The Ca:P ratio is 0.50. MCPM, $Ca(H_2PO_4)_2.H_2O$, produced by low-temperature precipitation, is acidic and readily soluble, with generally poor biocompatibility when implanted by itself. MCP has rather similar properties but is used in some cement formulations.

### Class 2.2.3 Dicalcium phosphates

Dicalcium phosphate, $CaHPO_4$, DCP, is the mineral monetite. Dicalcium phosphate dihydrate, $CaHPO_4.2H_2O$, DCPD, is the mineral brushite. DCPD is metastable and may be converted to DCP or hydroxyapatite. DCP is reasonably stable and is also used in some cements.

### Class 2.2.4 Tricalcium phosphates

There are two forms of tricalcium phosphate, the α-tricalcium phosphate (α-TCP) and the ß-tricalcium phosphate (ß-TCP), both being $Ca_3(PO_4)_2$; both have a Ca:P ratio of 1.5. α-TCP has a monoclinic crystal structure, while ß-TCP is rhombohedral. α-TCP is the more soluble; it is used as a bone-substitute, in either granular, block or powder form. ß-TCP is more slowly degradable and can also be used as a bone filler, with some evidence that it is resorbed by osteoclast activity.

### Class 2.2.5 Octacalcium phosphate

Octacalcium phosphate (OCP) is $Ca_8H_2(PO_4)_6.5H_2O$; the Ca:P ratio is 1.33. OCP is metastable and not widely used.

### Class 2.2.6 Hydroxyapatite

Hydroxyapatite (HA) is $Ca_5(PO_4)_3.OH$, which has a Ca:P ratio of 1.67. It is highly crystalline and the most stable of the calcium phosphates. It is widely used as a bone substitute, both in solid and granular form, and as a coating on orthopedic and dental prostheses.

The emphasis on HA has been based on the similarity between its composition and structure with the main crystalline component of the mineral phase of bone, which is a calcium-deficient carbonate hydroxyapatite, with traces of magnesium, carbonate, hydroxyl, chloride, fluoride and citrate ions. The composition of materials used as bone substitute is critical to the properties and performance, as are the microstructure and method of synthesis and production. The stoichiometry is an important variable and has to be controlled with any thermal processing. The exact molar stoichiometric ratio of Ca to P is 1.67, but imbalances may lead to the formation of α or β calcium phosphates. Because of different views about the roles of various ions within the HA on biological properties, different formulations with ion substitution have been produced. Carbonate ions have been substituted for hydroxyl or phosphate groups, fluoride ions for hydroxyl groups, magnesium for calcium and silicon or silicate ions for phosphorus or phosphate groups.

### Class 2.2.7 Biphasic calcium phosphates

As the name implies, a biphasic calcium phosphate may contain two different forms of calcium phosphate; in practice the term tends to be reserved for those materials that consist of a combination of HA and ß-TCP. Most commercial products have these components in a ratio of around 60% HA:40% ß-TCP. The more soluble TCP degrades within the stable HA matrix and may be replaced by bone over time.

### Class 2.2.8 Calcium phosphate cements

Calcium phosphates have been developed as alternatives to implantable polymeric cements. The calcium phosphate cements (CPCs) are usually prepared by mixing water with one or several calcium phosphates, where the latter dissolve and interact, precipitating as less soluble forms of calcium phosphate. For example, a CPC may contain ß-TCP and MCPM to form DCPD. These setting reactions are only weakly exothermic. Their tensile strength is low, which restricts clinical applications. Although there are many potential formulations of CPCs, they all fit into either the class of apatite CPC or brushite CPC.

**Class 2.2.8.1 Apatite calcium phosphate cements** Most apatite CPCs have hydroxyapatite as the end product of the interaction, although in some formulations a small amount of carbonate is present, yielding a carbonate-apatite. The mechanical properties depend to a large extent on the formulation, the mixing process and the resulting porosity. A high phosphate powder to mixing liquid ratio favors low porosity and higher strengths. The resorption of these cements is variable but tends to be quite slow; however, the biocompatibility appears to be quite good.

**Class 2.2.8.2 Brushite calcium phosphate cements** The brushite CPCs have DCPD as the product of the setting reaction. Handling and mixing is generally similar to the apatite CPCs but they tend to degrade faster, may provoke an inflammatory response in association with this degradation, and they are usually a little weaker.

## 8.3.3 Class 2.3 Sulfates

Several sulfates are used extensively in medical applications, especially magnesium sulfate, which is used in drug formulations and topical agents (Epsom salts are based on this compound). However, only one sulfate, calcium sulfate, is used as a biomaterial.

### Class 2.3.1 Calcium sulfates

Calcium sulfate is a widely encountered mineral, generally known as gypsum. In its natural form it exists as the dihydrate, $CaSO_4.2H_2O$, but it loses water when heated to over 110°C, becoming the hemihydrate, $CaSO_4.1/2\ H_2O$, known as Plaster of Paris. This hemihydrate exists in two forms. The α-hemihydrate is hard and relatively insoluble and is used in dental technology as dental stone. The β-hemihydrate is much more soluble, the re-hydration process being convenient for setting under clinical conditions. Either pre-formed calcium sulfate shapes or powders/granules have been used as bone substitutes; the material resorbs more rapidly than most other bioceramics, often in a matter of weeks. The majority of clinical applications have been in dentistry and oral surgery, although some uses in orthopedic bone defects have been reported.

## 8.3.4 Class 2.4 Silicates and silica-based glasses

We have noted above that silica structures are based on the tetrahedral arrangements of four oxygen atoms around a silicon atom. This same unit forms the basis of a wide range of materials in which several other elements may be involved; these are known as silicates. In these structures, the tetrahedral units are anionic and have positively charged units to hold sheets of the tetrahedra together. The most common cations in silicates are those of Ca, Mg, K and Al. Well-known silicates include kaolin, $Al_2Si_2O_5(OH)_4$, and talc, $Mg_3Si_4O_{10}(OH)_2$. A few silicates are used in medical technology as shown below. Also included in this section are some silica-based glasses, which have commanded widespread interest in view of their apparent bioactivity.

### Class 2.4.1 Wollastonite

Wollastonite, $CaSiO_3$, is found naturally as one of the major forms of pyroxene rock-forming minerals and is widely used in the ceramic and cement industries. There are three forms, wollastonite TC (triclinic structure), wollastonite 2M (monoclinic structure) and pseudowollastonite (pseudo-orthorhombic structure). Wollastonite ceramics can be made by powder compaction and sintering processes. They are very brittle, which has severely limited their clinical applications, but they may be coated onto substrates such as Ti by plasma spraying. Such coatings will be primarily of triclinic structure, although some glassy phases may be present. The bond strength to Ti is higher than can be achieved with hydroxyapatite and the hardness is much higher. Wollastonite coatings appear to have good bioactivity with respect to bone formation.

### Class 2.4.2 Diopside and akermanite

Diopside, $CaMgSi_2O_6$, is also a pyroxene, specifically a mellite mineral of the sorosilicate group, where magnesium ions link the oxygen and calcium. It is again widely used in the ceramics industry and has higher strength than wollastonite. Akermanite, $CaMgSi_2O_7$, is very similar; it has a tetragonal structure. Containing Si, Ca and Mg, there are expectations that akermanite should have good bone bioactivity and some experimental evidence that this is so. Akermanite powders can be prepared by sol–gel procedures.

### Class 2.4.3 Zeolites: aluminosilicates

Zeolites are a family of chemically similar minerals based on crystalline hydrated aluminosilicates; there are many members of this family. One of the more important is clinoptilolite, $(Na_2,K_2,Ca)_3Al_6Si_{30}O_{72}.24H_2O$, which has considerable powers of absorption and can be used to soften water by ion exchange and in the treatment of radioactive waste. Many claims have been made about the biological properties and medical uses of this particular zeolite, including wound healing and components of anti-cancer agents and although most have not been confirmed, there is good evidence for its detoxification and absorption properties in biological environments.

### Class 2.4.4 Silica-based glasses: bioactive glasses

It has been known for many years that some types of glasses, ceramics and glass-ceramics are able to bond chemically to bone after implantation. This type of bonding is one form of bioactivity (Chapter 4) and such materials have become known as bioactive glasses (or -ceramics or -glass ceramics). The first examples of bioactive glasses, and still the most widely used, were some specific formulations containing $SiO_2$, $Na_2O$, CaO and $P_2O_5$. At first glance these appear rather similar to conventional glasses, but they have a higher $CaO/P_2O_5$ ratio, less than 60 mol% $SiO_2$ and higher $Na_2O$ and CaO levels. Formulations based on these principles are known as Bioglasses; the original was termed 45S5, which specifically has 45 wt% $SiO_2$, 6% $P_2O_5$, 24.5% CaO and 24.5% $Na_2O$. Bone-bonding is rather critically dependent on the composition; lower $CaO/P_2O_5$ ratios in particular result in loss of this ability.

The above formulation gives a glass. There are similar formulations that result in glass-ceramic structures. These tend to have slightly higher $SiO_2$ and significantly higher CaO levels, but no $P_2O_5$, much lower $Na_2O$ content and appreciable levels of $Ca(PO_3)_2$.

The dependence of bone-bonding on composition is depicted on the pseudo-ternary phase diagram shown in Figure 8.12. The mechanisms associated with the bonding process have been a matter of discussion for many years. Those materials that do bond to bone develop a large increase in surface area when exposed to physiological fluids through the generation of a nanoporous interface that arises from the formation of an active silica-rich gel. Hydroxyapatite crystals nucleate on this surface, which is followed by the incorporation of collagen, polysaccharides and glycoproteins into the structure. These events have to occur with the right kinetics; if the processes are too slow then a stable hydroxyapatite layer does not form, while if reactions are too fast, total dissolution of the glass may occur. These glasses tend to have low strength and toughness and cannot be used for monolithic implants in load-bearing situations; instead they are usually employed as coatings on tougher substrates.

## 8.3.5 Class 2.5 Nitrides

The vast majority of metals are able to form binary nitrides; these are exclusively synthetic since they are not found in nature. Some complex nitrides, including ternary and quaternary inorganic nitrides, are also possible. Many nitrides are very hard and have high melting points. Boron, aluminum and silicon nitrides are in commercial production with industrial applications. These materials are unlikely to be used in monolithic form, but they have been used experimentally and clinically either as nanoparticles or as coatings. Hexagonal boron nitride, and also silicon nitride, nanoparticles have been shown to be excellent deep UV light emitters in imaging applications and titanium nitride, or titanium-nitride-oxide, are used both as surface coatings on orthopedic devices and intravascular stents.

## 8.3.6 Class 2.6 Carbides

Carbides, which are compositionally simple compounds involving carbon and another element, are

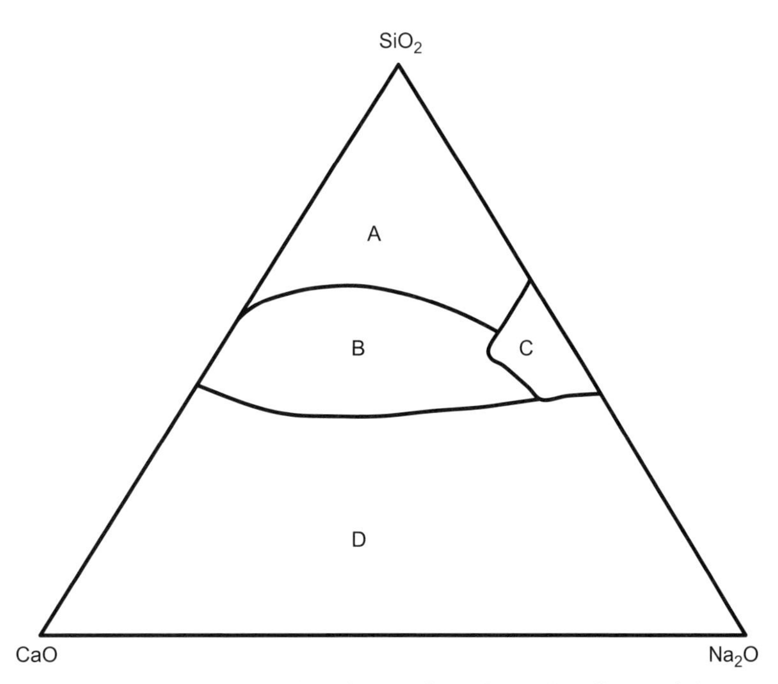


**Figure 8.12** Bioactive glasses. This is a quasi-ternary phase diagram that indicates the influence of glass composition on the performance in tissues. The three sides of the triangle represent binary compositions within the $SiO_2$–CaO, CaO–$Na_2O$ and $Na_2O$–$SiO_2$ systems. Points within the triangle represent the composition within $SiO_2$–$Na_2O$–CaO glasses with a constant addition of 6% $P_2O_5$. Glasses of composition within the area indicated by B show bone-bonding behavior. Those within C are too highly reactive and soluble. Those within A have a reactivity that is too low to allow bone-bonding. Compositions represented by area D are not glass-forming and do not bond to bone. The original Bioglass® composition was 45% $SiO_2$, 24.5% $Na_2O$, 24.5% CaO, 6% $P_2O_5$. This diagram is based on the work of Hench, see for example Hench L L, The story of bioglass, *Journal of Materials Science, Materials in Medicine*, 2006, 17, 967–78.

usually known and used because of their hardness and high-temperature properties. These are not normal characteristics of biomaterials and carbides have so far held little interest in medical technology apart from cutting tools used in dentistry and orthopedics. Some carbide coatings such as tungsten (WC) and silicon (SiC) have been used on medical devices, for example SiC as a coating on intravascular stents. SiC is an interesting material, being a covalently bound tetrahedral structure (these tetrahedra being considered as either $SiC_4$ or $CSi_4$), that can take on many different crystalline forms, mostly hexagonal or trigonal. These have quite different properties, including electrical/electronic properties. SiC is used in high-temperature, high-voltage semiconductor electronics and LEDs.

### 8.3.7 Class 2.7 Titanates

Titanates are compounds that display varying electrical properties that can be finely tuned based on composition. They may be used as simple compounds, or as mixtures since they are often able to form extensive solid solutions. The most common material here is barium titanate, $BaTiO_3$, which has an exceptionally high dielectric constant and both piezoelectric and ferroelectric properties. These features arise from the crystal structure that has corner-linked oxygen octahedral containing $Ti^{4+}$ ions at the octahedral sites, this structure giving high electrical polarizability. $BaTiO_3$ crystals have been widely used in medical ultrasound technology and in devices such as hearing aids. In nanocrystalline form they may be used as imaging probes because of the effectiveness in second harmonic emission. This arises from the non-centrosymmetric structure of $BaTiO_3$ nanocrystals which, when excited at a fundamental frequency, emit an optical signal with exactly doubled frequency, giving very efficient contrast between the target molecule and background.

## 8.3.8 Class 2.8 Optically active ceramic/metallic nanoparticles

The materials briefly discussed in this section are difficult to classify since they may have some characteristics of metals and some of ceramics. They may technically qualify as composites and many are polymer coated. They are included here as this seems the best location at this stage, but the field of application is growing fast and they may require a totally separate classification in the future.

### Class 2.8.1 Semiconductor quantum dots

The principles of semiconductors and quantum dots have been discussed in Chapters 2 and 7. QDs are semiconductor nanocrystals (1–10 nm diameter) that show luminescent properties with absorption and emission bands that are defined by the band gap energy of the parent material and their size. They have large absorption spectra but narrow and symmetric emission bands that can span from the UV to the IR, and can have very high quantum yield, with the ratio of emitted to absorbed photons being as high as 90%. In practice, although there is a wide range of compositions from which quantum dots can be generated, the majority is based on cadmium selenide (CdSe). Other combinations of Periodic Table groups II and VI elements, such as CdTe, CdS, CdHg and ZnS, or combinations of groups III and V elements, such as InAs, InP and GaAs are possible. QDs are usually manufactured by injection of liquid precursors into organic solvents, where crystal growth occurs under controlled conditions. This essentially produces a metallic core, which could be used by itself, but this would have a low quantum yield. Instead the process usually proceeds with the epitaxial growth of a layer of a substance such as ZnS around the core, which increases quantum yield and protects the nanocrystal from oxidation and leaching. Examples are shown in Figure 8.13(a). For biological applications, further steps are required because such QDs are not water-soluble but are likely to be cytotoxic. Addition of hydrophilic ligands addresses the solubility problem, but toxicity remains an issue. Functionalization of the surfaces with amine, carboxyl or mercapto groups allows direct conjugation to biomolecules. In the diagnostic laboratory, QDs have been used for immunohistochemistry and *in situ* hydridization. There is a great deal of interest in the use of QDs for live cell imaging and the detection of tumors *in vivo*. Whether or not the anticipated toxicity of cadmium-based QDs hinders such applications remains to be seen.

### Class 2.8.2 Rare earth upconverting nanoparticles

Upconverting nanoparticles have also been developed as part of the search for efficient but safe contrast agents for bioimaging. Although there are several forms available, the lanthanide-doped rare earth compounds serve as a good example. The rationale is that organic fluorophores are good contrast agents under some conditions, but they tend to have broad emission spectra, which makes them unsuitable for multiplex labeling. They often suffer photodegradation and autofluorescence. Upconversion nanoparticles, in contrast, have sharp emission bandwidth and tunable emission, with greater stability (including photostability) and longer lifetimes. They are also effective in the near infrared spectral region. These nanoparticles usually have an inorganic host that is doped with lanthanide ions, especially $Er^{3+}$. The host material requires adequate transparency within the relevant wavelength range, with lattice parameters that match those of the dopant ions in order to achieve high doping levels. With the lanthanide dopant ions, several compounds of inorganic rare earths, alkaline earths and transition metals provide suitable hosts, including $Y_2O_2S$ and $NaTF_4$. In contrast to regular luminescent

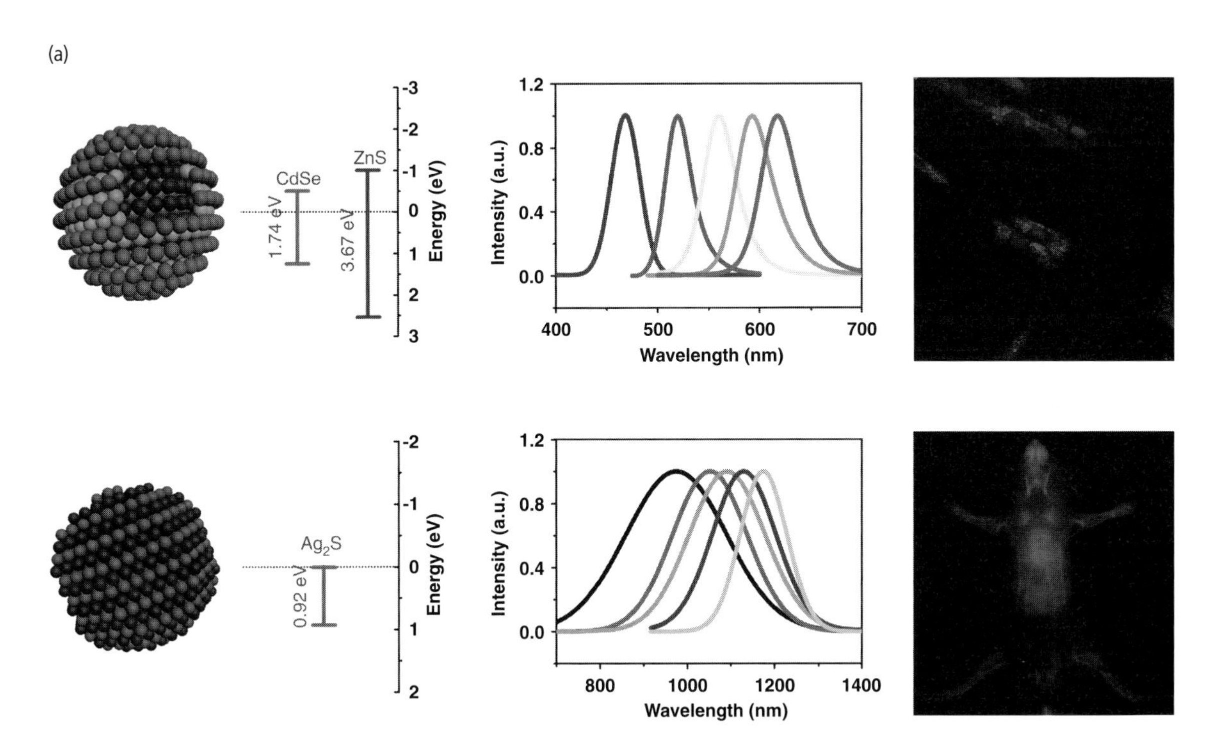


**Figure 8.13** Quantum dots and upconversion nanoparticles.
(a) Top: schematic illustration of CdSe@ZnS core-shell quantum dot and its energy band diagram, tunable visible photoluminescence spectra and cellular imaging. Bottom: Ag2S quantum dot with energy band diagram, tunable UV photoluminescence spectra and *in vivo* imaging. Images courtesy of Dr. Qiangbin Wang, Suzhou Institute of Nan-Tech and Nano-Bionics, Chinese Academy of Sciences, Suzhou, China. See plate section for color version.

materials that have a single ground state and a excited state, these nanoparticles have multiple intermediate states, which allow the accumulation of low-energy excitation photons. The nanoparticles are preferably highly crystalline to minimize energy loss associated with defects. Their surfaces are capped during synthesis to avoid excessive growth and aggregation. They may also be functionalized with pendant groups in order to enhance ligand targeting.

One of the potential advantages of the upconversion nanoparticles concerns their use in multimodal imaging, for example with simultaneous magnetic and optical imaging. The incorporation of gadolinium, as $Gd^{3+}$, may yield $NaGd_4$:Yb/Er nanoparticles for this purpose. They may also be used in conjunction with core-shell nanoparticle technology, where the upconversion nanoparticle acts as the core, surrounded by a porous silica shell that is impregnated with photosensitizers and surface functionalized with some cancer cell targeting moiety, an example being shown in Figure 8.13(b).

## 8.4 Class 3 The polymeric systems

The biomaterials of Class 3 cover polymeric systems. A polymer is generally considered to be a large molecule made up of chains or rings of linked monomer units. Since it is predominantly the four fold valency and covalent bonding associated with the carbon atom that allow such linking of monomers, polymers are generally organic, and they may be either synthetic or natural; of course it is not quite as simple as that. In this class, since there are an

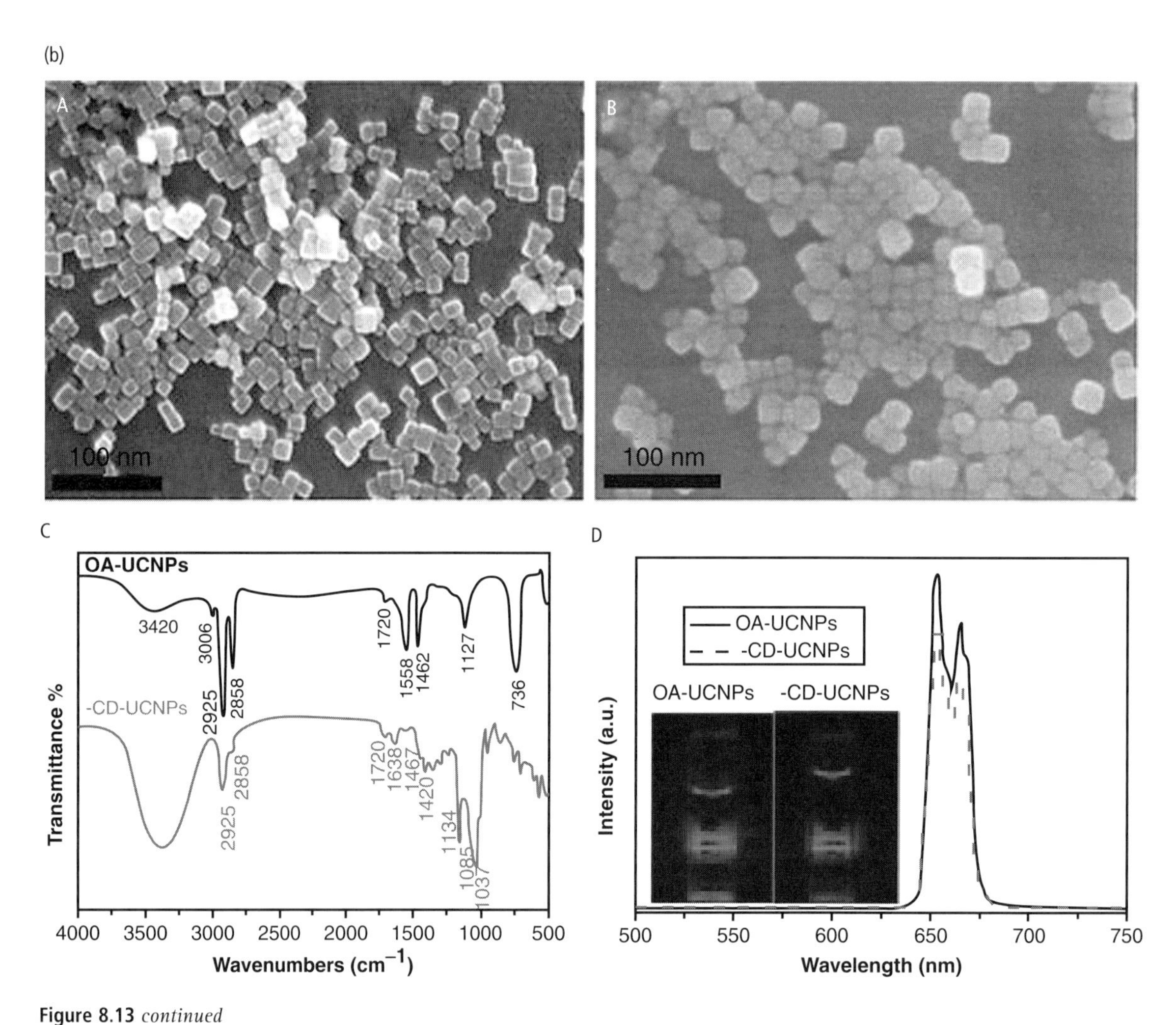


**Figure 8.13** *continued*
**(b)** Upconversion nanoparticles (UCNPs). SEM images of NaYF4:Yb/Er UCNPs (A) oleic-acid capped and (B) α-cyclodextrin capped. (C) FT-IR spectrum of UCNPs. (D) Room temperature upconversion emission of OA-UCNPs dispersed in cyclohexane (1 mg/mL, solid line) and α-CD-UCNPs dispersed in DI water (1 mg/mL, dashed line), respectively (inset: their corresponding luminescent photographs under 980-nm NIR excitation, power density: 0.5 W/cm$^2$). Reprinted from *Small*, 9, Tian G *et al.*, Red-emitting upconverting nanoparticles for photodynamic therapy in cancer cells under near-infrared excitation, 1929–38, Copyright © 2013 WILEY-VCH Verlag GmbH & Co. KGaA, Weinheim.

almost infinite number of combinations of C and other elements, in such a variety of configurations, it is too difficult to split them up into a small number of discrete sub-classes based on composition. Instead we use a system which is based on some key properties and especially the structure–composition characteristics; this second level tier in the hierarchy is shown in Box 8.4.

The first sub-division (Class 3.1) is that of the thermoplastic polymers, which are synthetic polymers that soften on heating, and are thus usually easily fabricated by techniques such as extrusion and molding. The second group (Class 3.2) is that of the thermosetting resins, which tend to decompose or melt on heating rather than soften; they cannot be molded or shaped when solid and are usually

## Box 8.4 The second level of the classification of biomaterials: polymeric systems

Class 3.1 Thermoplastic polymers
Class 3.1.1 Polyolefins
Class 3.1.1.1 Polyethylene
Class 3.1.1.1.1 Low density polyethylene
Class 3.1.1.1.2 High density polyethylene
Class 3.1.1.2 Polypropylene
Class 3.1.1.3 Polymethylpenetene (TPX)
Class 3.1.2 Fluorinated hydrocarbon (fluorocarbon) polymers
Class 3.1.2.1 Polytetrafluoroethylene
Class 3.1.2.2 Polyvinylidine fluoride
Class 3.1.2.3 Perfluorocarbons
Class 3.1.3 Acrylic polymers
Class 3.1.3.1 Acrylic acid-based materials
Class 3.1.3.2 Methacrylic acid-based materials
Class 3.1.4 Polyaryletherketones
Class 3.1.4.1 Poly(aryl-ether-ether-ketone) (PEEK)
Class 3.1.4.2 Carbon-fiber reinforced PEEK
Class 3.1.5 Polysulfones and polyethersulfones
Class 3.1.6 Polycarbonates
Class 3.1.7 Polyimides
Class 3.1.8 Polyurethanes
Class 3.1.9 Polyacetals
Class 3.2 Thermosetting resins
Class 3.2.1 Epoxy systems
Class 3.3 Synthetic polymeric sols and gels
Class 3.3.1 Poly(ethylene glycol)/poly(ethylene oxide)
Class 3.3.2 Pluronics
Class 3.3.3 Polyhydroxyethylmethacrylate
Class 3.3.4 Poly(vinyl alcohol)
Class 3.3.5 Polyglycerols
Class 3.3.6 Inverted colloid crystals
Class 3.4 Proteins and peptides
Class 3.4.1 Collagen derivatives
Class 3.4.1.1 Gelatin
Class 3.4.2 Elastin derivatives
Class 3.4.3 Resilin

Class 3.4.4 Fibrin derivatives
Class 3.4.5 Laminin derivatives
Class 3.4.6 Silk
Class 3.4.7 Keratins
Class 3.4.8 Zein
Class 3.4.9 Peptide nanomaterials
Class 3.4.10 Protein and peptide mimetics
Class 3.5 Polysaccharides
Class 3.5.1 Hyaluronan derivatives
Class 3.5.2 Alginates
Class 3.5.3 Chitin and its derivatives
Class 3.5.4 Pullulan
Class 3.5.5 Dextran polymers
Class 3.5.6 Cellulose
Class 3.5.6.1 Microbial cellulose
Class 3.5.6.2 Methylcellulose and carboxymethylcellulose
Class 3.6 Lipids
Class 3.6.1 Phospholipids
Class 3.6.2 Liposomes
Class 3.7 Biodegradable structural polymers
Class 3.7.1 The poly (α-hydroxy acids): polylactides and polyglycolides
Class 3.7.2 Polycaprolactone
Class 3.7.3 Polydioxanone
Class 3.7.4 Poly(ortho esters)
Class 3.7.5 Polyanhydrides
Class 3.7.6 Polyketals
Class 3.7.7 Sebacate polymers
Class 3.7.8 Fumarate polymers
Class 3.7.9 Cyanoacrylate polymers
Class 3.7.10 Degradable polyurethanes
Class 3.7.11 Polyhydroxyalkanoates
Class 3.8 Water-soluble polymers
Class 3.8.1 Polyethylenimine
Class 3.8.2 Hydroxypropyl methacrylamide
Class 3.8.3 Polyvinylpyrrolidone
Class 3.8.4 Polyamidoamines
Class 3.9 Polymers with ionizable or ionic groups
Class 3.9.1 Conducting polymers
Class 3.9.1.1 Polypyrrole

Class 3.9.2 Polyelectrolytes
Class 3.10 Elastomers
Class 3.10 1 Silicone elastomers
Class 3.10 2 Polyurethane elastomers
Class 3.10.3 Poly(styrene-block-isobutylene-block-styrene) (SIBS)
Class 3.10.4 Plasticized polyvinylchloride
Class 3.11 Fibers, fabrics and textiles
Class 3.11.1 Polyethylene terephthalate materials
Class 3.11.2 Microporous expanded polytetrafluoroethylene
Class 3.12 Environmentally responsive polymers
Class 3.12.1 Thermoresponsive polymers
Class 3.12.2 pH responsive polymers

fabricated *in situ* by some variation of cross-linking. The third group (Class 3.3) are those polymers which are sols or gels. Such structures may in general be synthetic or natural, but in this classification of biomaterials, the natural gel structures are considered separately and Class 3.3 is reserved for synthetic gels, and these are usually hydrogels. We then have a series of naturally occurring polymers, including proteins and peptides (Class 3.4), polysaccharides (Class 3.5) and lipids (Class 3.6). Finally there are some groups of polymers which do not fit easily into any of the above groups or which are especially relevant to biomaterials applications. These are synthetic biodegradable structural polymers (Class 3.7), water-soluble polymers (Class 3.8), polymers containing ionizable or ionic groups (Class 3.9), elastomers (Class 3.10), textiles (Class 3.11) and environmentally responsive polymers (Class 3.12). Most of these second tier classes of polymers may be further sub-divided, as noted in the following sections.

It will become clear that although this classification is logical on the basis of the structure–property considerations, there are many situations in which a very minor change to the molecular structure of a material can move it from one class to another. Such examples are highlighted where appropriate. Moreover, some polymers do not easily fit into any of these classes while others could readily fit into several of them.

## 8.4.1 Class 3.1 Thermoplastic polymers

The essential structure and properties of thermoplastic materials were discussed in Chapter 2. By definition, these are structural polymers that soften on heating, and may therefore be shaped by thermal processes. Many of these materials were among the first group of polymers to be used as biomaterials, and they feature in many types of permanently implantable devices.

### Class 3.1.1 Polyolefins

Polyolefins are based on olefin monomers, which are unsaturated aliphatic hydrocarbons, of general formulae $C_nH_{2n}$. These contain one carbon double bond, C=C, and are readily polymerized through the use of certain catalysts. These addition polymers therefore have an all-C backbone and are generally light, strong, flexible and environmentally stable. They are also easily processed by many techniques including various forms of molding and extrusion.

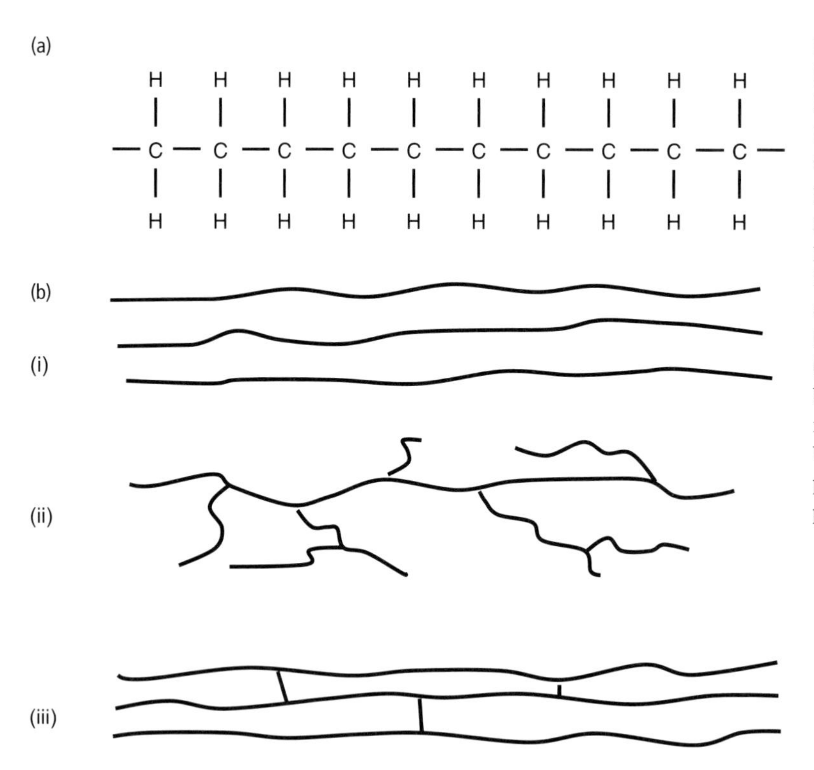


**Figure 8.14** Hydrocarbon polymers. **(a)** Molecular structure of polyethylene. **(b)** Branching and cross-linking. (i) unbranched, un-cross-linked structure, which is represented by close packing of molecules, high density and good crystallinity (e.g., high density polyethylene, HDPE), (ii) this molecule is branched, preventing close approximation of adjacent molecules (e.g., low density polyethylene, LDPE), (iii) cross-links have been introduced between adjacent molecules, improving mechanical properties, used with UHMWPE for improved wear performance in joint replacement prostheses.

The two simplest polyolefins, polyethylene and polypropylene are used extensively in health care products.

**Class 3.1.1.1 Polyethylene** Polyethylene is the simplest of all polymers, being derived from ethylene, $C_2H_4$, with the structure shown in Figure 8.14(a). The all carbon backbone gives stability. There are two main determinants of the properties of various forms of polyethylene. The first is the extent of chain branching, shown in Figure 8.14(b). Very little branching allows for higher density of packing of the long chain molecules. The second is the average length of the molecules, which determines molecular weight. Polyethylenes are normally defined by their density, with molecular weight being a secondary factor. Generally four varieties are available, high density (HDPE), medium density (MDPE), low density (LDPE) and linear low density (LLDPE) polyethylenes. As we shall see below, there are some sub-divisions within these.

**Class 3.1.1.1.1 Low density polyethylene** LDPE is probably the most widely known of all plastics – it is inexpensive and readily processed into many commercial and domestic items such as bags. It is mentioned only briefly here since it is not a classical biomaterial, but it is widely used in medical products. Some 10% of all disposable and non-disposable plastic medical products, including syringes, bottles and packaging materials are made from LDPE. The density of LDPE ranges from 0.91 to 0.94 g/cm$^3$.

**Class 3.1.1.1.2 High density polyethylene** HDPE, which is semi-crystalline, generally has a density of

(c)

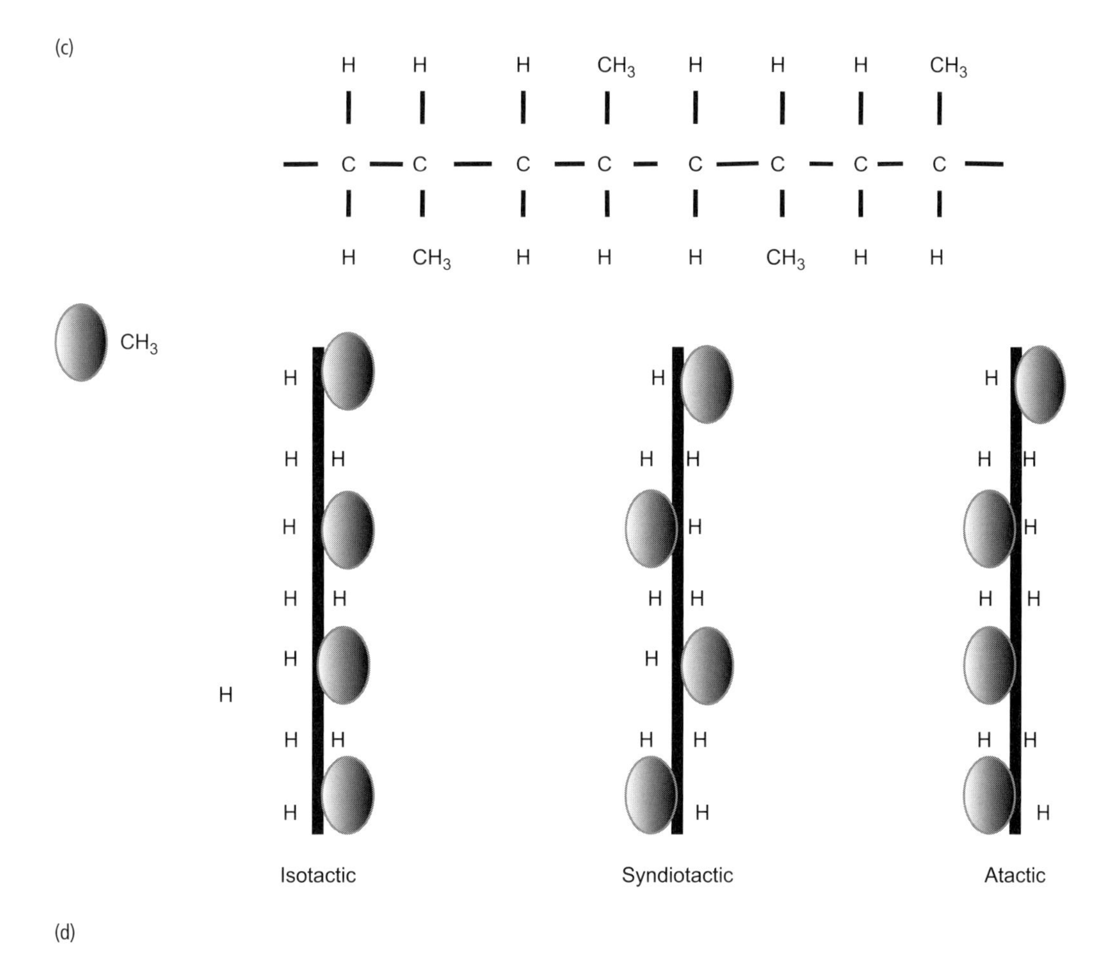


(d)

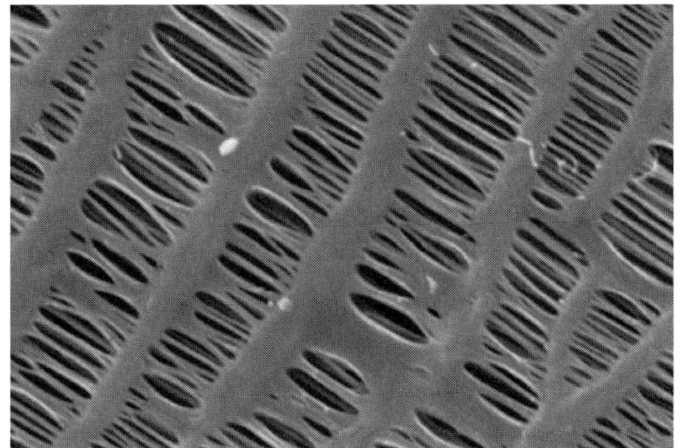

**Figure 8.14** *continued*
**(c)** Molecular structure of polypropylene, with representation of the different forms depending on the positioning of the $CH_3$ side groups.
**(d)** A microporous polypropylene membrane.

0.93–0.94 g/cm$^3$. It is widely used in general engineering but has special interest for implantable biomaterials when used in one particular form, known universally as UHMWPE, which is ultra-high molecular weight HDPE. UHMWPE is defined as having a molecular weight of at least 1 million daltons, although the forms used as biomaterials will normally have molecular weights

(e)

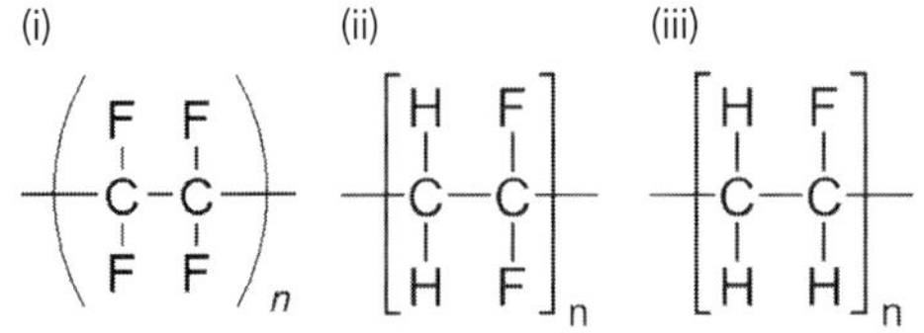

(i) (ii) (iii)

**Figure 8.14** *continued*
(e) Fluorocarbon polymers: (i) polytetrafluoroethylene, (ii) polyvinylidene fluoride, (iii) polyvinylfluoride.

between 2 and 6 million daltons. In the crystalline phase, the chains are folded into highly oriented lamellae, up to 50 nm thick, giving an orthorhombic structure. $T_g$ is around −150°C so that the material is well above this transition temperature in the body. The yield strength will be around 24 MPa, the UTS around 40 MPa and the elongation at break just under 400%.

The most profound medical use of UHMWPE is in total joint prostheses; as indicated in Chapter 4 the material is used as one of the bearing surfaces in the majority of total hip and knee replacements worldwide. It performs very well, although the eventual wear of the material, and the osteolytic response to the wear debris is a major cause of failure of these devices. Although structurally a very simple material, UHMWPE does have complex mechanical behavior such that the deformation characteristics that contribute to wear are highly dependent on the thermomechanical history, especially associated with processing, because of reorientation of the lamellae and changes in crystallinity. Of equal significance is the fact that oxidation of the polymer may occur under certain sterilization procedures, which can radically alter the mechanical properties. For a number of years, the components were sterilized by gamma irradiation, at a dose of 25 kGy, this being carried out in air. However, oxidative degradation takes place under these conditions since free radicals are formed by the combined effects of radiation and air. Hydroperoxides are formed in the polyethylene, which yield even more free radicals and cause the formation of carbonyl groups along the molecular backbone. This process continues even after the radiation cycle has ceased, which causes brittleness, a decrease in fracture toughness, reduced elongation and increased rigidity. This can result in sub-surface damage and delamination of the material at the surface, considerably increasing the effective wear rate.

Sterilization carried out in vacuum, or inert gas packaging addresses this problem to some extent, but much better performance has been obtained through the modification of the polyethylene by cross-linking. Several cross-linking processes have been developed. One method actually involves exposing the material to even higher (>50 kGy) doses of gamma irradiation followed by a post radiation thermal treatment; other methods involve repeated cycles of irradiation and annealing. Some processes include concurrent mechanical deformation, while others are based on the incorporation of vitamin E, a natural antioxidant, into the polymer. These different processes all reduce the wear rate but they have mixed effects on other mechanical properties and long-term clinical data on their performance is not yet available.

The composition, structure and properties of different forms of UHMWPE are given in Table 8.7.

**Class 3.1.1.2 Polypropylene** Polypropylene (PP) is the addition polymer derived from propylene. It exists with varying degrees of crystallinity depending on the regularity of the positioning of the $CH_3$ group along the backbone, as shown in Figure 8.14(c), giving isotactic, syndiotactic and

Table 8.7 **Characteristics of UHMWPE used for joint replacements. This table summarizes the main mechanical properties of the different routes to manufacture UHMWPE for joint prostheses. These properties are very similar. The main difference claimed for these materials are concerned with clinical wear, which depends on design**

| Characteristic | Conventional UHMWPE Radiation sterilization (25–40 kGy) | Radiation cross-linked (50–100 kGy) Melting, non-radiation sterilization | Radiation cross-linked (100 kGy) Annealing, radiation sterilization | Vitamin E diffused UHMWPE, homogenization radiation sterilization |
|---|---|---|---|---|
| Crystallinity (%) | 70 | 60–80 | 60–80 | 50–60 |
| Yield Strength (MPa) | 14 | 20 | 25 | 22 |
| Ultimate Strength (MPa) | 60 | 40 | 60 | 50 |
| Elongation (%) | 370 | 350 | 280 | 350 |

atatic forms. Like LDPE it is widely used in consumer products, indeed some 25% of all disposable medical products incorporate PP.

The most widespread use of PP in implantable devices is in surgical sutures: known as Prolene, this suture has one of the best performances in the long term. PP is also used as a mesh for tissue repair, especially for hernia and other peritoneal applications and, increasingly over the last decade for urological and urogynecological reconstruction procedures. As discussed in Chapter 5 these have not been without controversy, especially concerning the resistance to degradation. PP is known to oxidize under some conditions, particularly involving elevated temperatures. For this reason, the PP used in most engineering applications incorporates an antioxidant. PP will not normally degrade under implantation conditions although there is evidence that cracking can occur.

Microporous polypropylene is also used in hollow-fiber membrane oxygenators. These have channels that are approximately 1 micron in diameter, as shown in Figure 8.14(d).

**Class 3.1.1.3 Polymethylpentene (TPX)** In the above paragraph the use of PP in oxygenators was mentioned. In the last decade, an alternative polyolefin has been introduced into this application. This is poly (4-methyl-1-pentene) sometimes known as TPX or PMP. Its structure is $-CH_2-CH(CH_2CHCH_3CH_3)-$. It is a transparent polymer with several engineering uses. These oxygenators appear to offer some advantages, although these are both design and material related. Smaller priming volumes are required; they are also heparin coated.

### Class 3.1.2 Fluorinated hydrocarbon (fluorocarbon) polymers

Fluorocarbon polymers include polytetrafluoroethylene (PTFE), polyvinylfluoride, polyvinylidinefluoride and fluorinated ethylene-propylene copolymers. They are characterized by having a carbon backbone and fluorine, or fluorine plus other halogen side groups, as shown in Figure 8.14(e). The carbon–fluorine bond is exceptionally strong, giving some very stable polymers, particularly the PTFE, which is fully fluorinated.

**Class 3.1.2.1 Polytetrafluoroethylene** The high strength associated with the carbon–fluorine bond makes the fabrication of PTFE rather difficult; it is a thermoplastic polymer but most thermoplastic routes are impractical since it is extremely viscous, even at high temperature. The polymer is prepared by a free radical addition polymerization reaction in aqueous dispersion under pressure. Molecular weights tend to be between 500 000 and 5 000 000 and it is highly crystalline. The material is usually sintered, using pressures in excess of 50 MPa, and the resulting products are readily machined. It is resistant to most chemicals and can withstand extremes of temperature. The surface energy of PTFE is very low so that it is difficult for any other material to be bonded to it; for similar reasons, it has a low coefficient of friction. The main disadvantages with PTFE are associated with poor mechanical properties, with a compressive strength as low as 10 MPa and the tensile strength ranges from 15–35 MPa. Ductility may be in the region of 300%.

**Class 3.1.2.2 Polyvinylidinefluoride** Polyvinylidinefluoride (PVDF), with a backbone of $CH_2CF_2$ repeating units, is an interesting polymer that can have piezoelectric properties and is capable of being used in many formats, including fibers and textiles, and as a coating. When the polymer is synthesized and stretched with exposure to a corona discharge, crystal domains form where the fluorine and hydrogen atoms lie on opposite sides of the backbone, creating a strong molecular dipole. The charge can be varied dynamically with spatial reorientation of the dipoles, creating interesting electrical properties, which appear to be able to influence cell behavior, especially neuronal cells. The material has been used in monofilament form as sutures and as a coating on implantable meshes and filters.

**Class 3.1.2.3 Perfluorocarbons** Perfluorocarbons are perfluorinated aliphatic compounds that have structures such as those shown in Figure 8.14(e). Although these compounds are rather diverse in their molecular arrangements, it can be seen that the C–F covalent bond is dominant, which accounts for their usual characteristics of chemical stability. Of considerable significance, these substances, which are liquid, often prepared as emulsions, have very high oxygen and carbon dioxide solubility. Compared to $O_2$ and $CO_2$ solubility in water of 30 ml/l and 570 ml/l respectively, the figures for perfluoroctylbromide are 530 ml/l and 2100 ml/l, with similar figures for other compounds. They therefore have potential as oxygen carriers, for example during liquid ventilation. Although interatomic bonds are strong, intermolecular bonds are weak and the liquids have high volatility and they are cleared rapidly in the lung. They have very low surface tension so they travel freely through narrow airway spaces. Safety issues have prevented routine clinical use at this stage.

### Class 3.1.3 Acrylic polymers

Acrylic polymers (or resins) are the polymers of derivatives of acrylic acid or methacrylic acid. Some of these polymers were among the first commercial plastics, dating back to the 1940s, and their first clinical applications originated in dentistry. The majority of acrylic polymers used clinically today are derived from esters of these acids, and most of these applications are based on one of two specific attributes, their optical transparency and their ease of low-temperature synthesis. A few are derived from acrylic acid but the majority is based on methacrylic acid. The structure of the some acrylic monomers and the principle polymer are shown in Figure 8.15.

**Class 3.1.3.1 Acrylic acid-based materials**

Polyacrylic acid, $-CH_2-CHCOOH-$, is extensively used in dentistry as a reactive component used in the formulation of cements, such as the glass ionomer cements, but has not found significant use in medical applications. It is best known as a superabsorbent material for disposable diapers.

**Class 3.1.3.2 Methacrylic acid-based materials** The most common monomer used for methacrylic acid-based polymers is the methyl ester, although ethyl

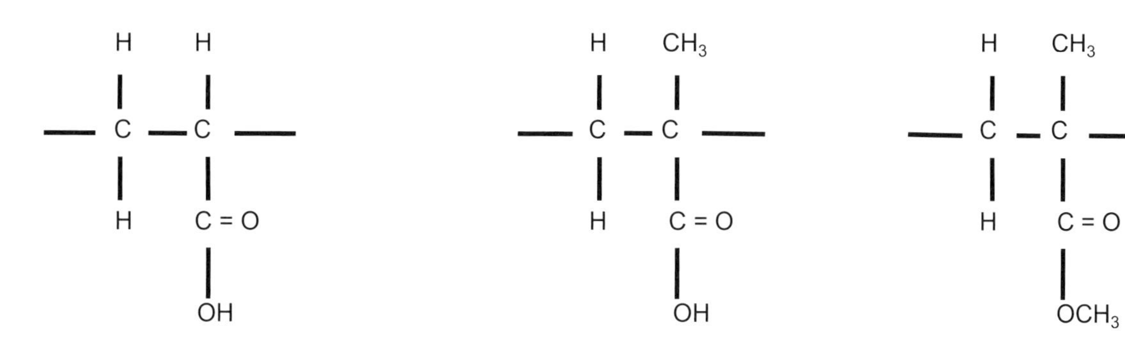


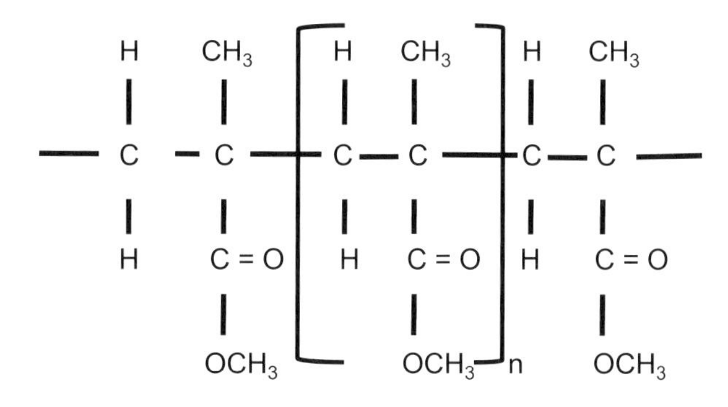


**Figure 8.15** Acrylic polymers. The main monomers used in acrylic polymers and the principle acrylic medical polymer, polymethylmethacrylate.

and butyl esters are sometimes used, often copolymerized with the methyl ester. In some situations, di-functional esters such as ethylene glycol dimethacrylate may be used, in particular when cross-linking of methacrylates is required. Polymerization of the methacrylates takes place by free radical initiated addition reactions. Most commonly, free radicals are generated by the decomposition of dibenzoylperoxide. This may be achieved by heating, which enables solid objects to be produced within a mold, or by reaction at room temperature with an accelerator such as N,N-dimethyl-p-toluidine, which allows for *in situ* curing. The latter process is of significance for the use of polymethylmethacrylate (PMMA) bone cements, which are prepared peri-operatively by mixing two components, a powder that consists of prepolymerized PMMA particles together with the peroxide initiator and a liquid, which is monomeric methylmethacrylate and the accelerator. Many additives may be present in this system, including those that reduce the exotherm, some stabilizers and, in some situations, radiopacifying substances and antibacterial agents. There are several important uses of PMMA in medical technology. The bone cement mentioned above has been in clinical use for over 40 years. For much of this time, the basic formulation has stayed essentially the same. The powder contains more than 90% of pre-polymerized PMMA, with 4–8% $BaSO_4$ and other additives and 1–2% dibenzoylperoxide. The liquid would consist of greater than 85% MMA and 10–15% of comonomers, with 1–2% N,N-dimethyl-p-toluidine and up to 100 ppm of a stabilizer such as hydroquinone. Whereas initially the mixing and application were carried out manually, some specific techniques have been introduced to achieve better consistency, with reduced porosity and, hopefully, improved strength. Vacuum mixing is usually used. Ingredients are pre-packed in the sealed mixing system. Delivery of the cement to the bone site is achieved with pressurization devices. The main change to the composition has been the inclusion, in certain products, of an antibiotic, typically gentamicin.

### Class 3.1.4 Polyaryletherketones

Polyaryletherketones comprise a family of high-performance thermoplastics that have found widespread application in many areas of engineering during the last few decades. Their mechanical and environmental properties have led to a number of medical applications, especially in orthopedic and spinal devices. The polymers have an aromatic backbone (Figure 8.16) with interconnections by ketone and ether functional groups. Several forms have been produced depending on the relative frequency of the ketone and ether groups, including poly(aryl-ether-ketone-ether-ketone-ketone) (commonly known as PEKEKK) and poly(aryl-ether-ether-ketone) (or PEEK). Only the latter variety is used as a biomaterial at this stage.

**Class 3.1.4.1 Poly(aryl-ether-ether-ketone) (PEEK)** Commercially available PEEK typically has a molecular weight of 85 000–115 000. $T_g$ is 143°C and $T_m$ is 343°C; the polymer is therefore in the glassy state at body temperature, with around 30–35% crystallinity. The elastic modulus is in the range 3–4 GPa. It may be processed by most normal methods, including injection and compression molding and extrusion. It has a tensile strength of 90 MPa and an elongation of 30–40%. The molecular structure of PEEK renders it very stable and it remains unattacked by virtually all solvents. It is not hydrolyzed in aqueous media and does not degrade in the body. It is stable on high doses of radiation and may be sterilized by gamma irradiation. There appear to be no biocompatibility concerns with respect to long-term implantation. The main medical uses involve spinal fixation devices, and specifically interbody fusion devices. It is also used for CAD-CAM produced craniofacial implants.

**Class 3.1.4.2 Carbon-fiber reinforced PEEK** Although PEEK has inherently good mechanical properties, there have been significant

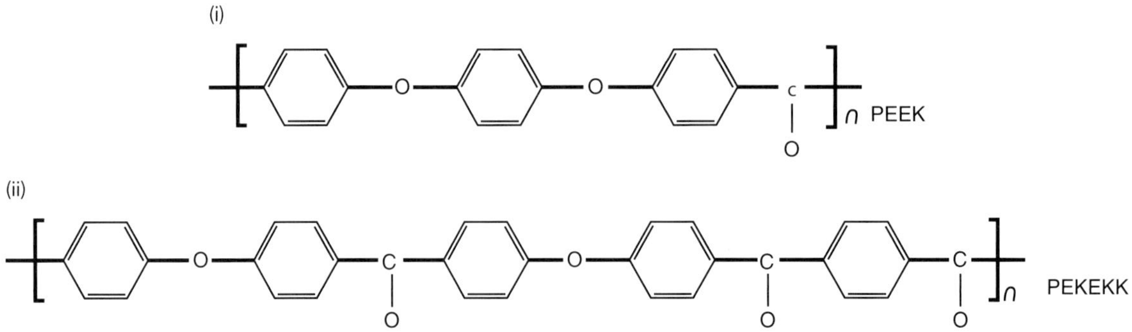

(i)

PEEK

(ii)

PEKEKK

**Figure 8.16** The molecular struture of polyaryletherketones: (i) polyetheretherketone (PEEK) and (ii) polyetherketoneetherketoneketone (PEKEKK). Reprinted from *Biomaterials*, 28, Kurtz S M and Devine J N, PEEK biomaterials in trauma, orthopedic and spinal implants, 4845–69, © 2007, with permission from Elsevier.

attempts to manipulate these properties by the use of carbon fiber composites. It is possible to produce both chopped fiber and continuous fiber reinforcement, in the former case typically with 30% fiber and close to 70% in the latter case. The elastic modulus can be arranged such that it is close to that of cortical bone and the strength may be significantly increased to 170 MPa with chopped fiber and over 2000 MPa with continuous fiber. These materials are being used as the bearing surface and stems of joint prostheses respectively, although long-term data is not yet available.

### Class 3.1.5 Polysulfones and polyethersulfones

These are two closely related families of polymers based on the diphenylsulfone group (Figure 8.17(a)). The repeating unit of polysulfone is shown in Figure 8.17(b) and that of polyethersulfone (PES) in Figure 8.17(c). They are both amorphous, transparent tough polymers that can be fabricated by a variety of conventional processing methods. Ultimate tensile strengths are of the order of 130 MPa and the Young's modulus 7000 MPa. For medical applications PES has the better properties, especially a greater resistance to water absorption and the ability to be repeatedly sterilized, although both are finding use in hollow fiber microfiltration and ultrafiltration membranes, including those in hemodialysis. A variety of phase separation methods have been developed for the fabrication of these membranes, including thermally induced phase separation.

### Class 3.1.6 Polycarbonates

Polycarbonates are very versatile engineering thermoplastics that can be used as homopolymers or as components of blends or copolymers. The polymers are based on the carbonate group (-$OCO_2$-), which may be formed during the reaction between a dihydroxy compound and a derivative of carbonic acid. The dihydroxy compound can be either aliphatic or aromatic, although the vast majority will use an aromatic compound, and especially bisphenol A, 2,2-bis(4-hydroxyphenyl)propane. These reactions usually lead to linear homopolymers with molecular weights up to 30 000, which is controlled by the use of phenol derivative chain terminators. The polymers are insoluble in water and alcohol. They

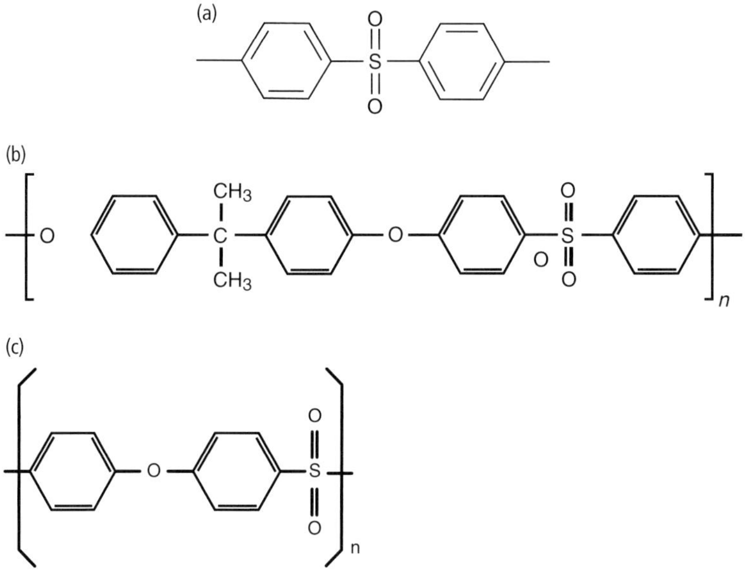

(a)

(b)

$CH_3$ C $CH_3$ O S O O O n

(c)

O S O O n

**Figure 8.17** Polysulfones: (a) the diphenylsulfone group; (b) polysulfone; (c) polyethersulfone.

are partially crystalline, but the degree of crystallinity is usually around 20% such that the materials are transparent. This transparency, coupled with high impact resistance, accounts for many of the industrial and domestic uses. Many of the medical uses of polycarbonates involve clinical sundries such as bottles and syringes. They have been used in some blood contacting applications but do not have excellent blood compatibility, which has been a main limitation. Some aliphatic polycarbonates have been developed as biodegradable tissue engineering scaffolds.

It will be recalled from Chapter 3 that there are some concerns about the safety of products derived from bisphenol A, especially in relation to genotoxicity and reproductive toxicity, and this has also been a limitation. There has therefore been more emphasis on aliphatic polycarbonates, for example poly(propylcarbonate). One main possibility here is with biodegradable polycarbonates based on tyrosine-derived polymers; these polymers include copolymers involving polycarbonate, polyarylates and poly(ethylene glycol), which have potential for drug delivery systems, for example in ophthalmology.

### Class 3.1.7 Polyimides

Polyimides constitute a family of polymers that have some very useful engineering properties. The polymers are based on the imide group; there are two general forms. The first, a linear polyimide is a simple addition polymer. The second is an aromatic heterocyclic structure where the imide group is found within a cyclic unit on the polymer chain; most commercially available polyimides belong to this latter group, for example poly(4,4'-oxydiphenylene-pyromellitimide). The molecular chains of these polymers can be arranged in a stack-like structure where the carbonyl groups of one chain are aligned next to the nitrogen of an adjacent chain, the charge transfer complexes that are established producing strong intermolecular bonds, resulting in very strong materials. The UTS may be 230 MPa and the Young's modulus 2.5 GPa.

Polyimides have a high electrical resistivity and high temperature resistance and they have found many uses as an insulator, especially at elevated temperatures. They have been used as an insulator within the circuitry of implantable defibrillators. Polyimides do absorb fluids, which has limited other implantable applications.

### Class 3.1.8 Polyurethanes

Polyurethanes (PUs) comprise a large group of polymers that contain the urethane group, -NHCOO-, as seen in Figure 8.18(a). This group can be contained within many different, and often complex, chemical structures; depending on the precise structure the materials may exist as rigid thermoplastics, thermoplastic elastomers, sealants, adhesives, foams and other forms. The urethane group is best considered as resulting from the reaction between an isocyanate and an alcohol, as shown in Figure 8.18(b). The versatility of this polymer family lies with the fact that R and R′ in the polymer structure may be substituted by a variety of groups. R is usually oligomeric, typically consisting of hydroxy-terminated polyethers or polyesters with a molecular weight ranging from a few hundred to 5000. R′ may be either aliphatic or aromatic.

For much of the history of PU in medical devices, attention was focused on three types, poly(ester urethane), poly(ether urethane) and poly(ester urethane urea). The attractive features here were those of elasticity and apparent biocompatibility and blood compatibility. One major problem surfaced when it was realized that, as a family, PUs are susceptible to a number of degradation mechanisms *in vivo*, with some forms being much more stable than others. Unfortunately the better mechanical properties tended to be found in the more degradable forms. This has caused a radical rethink of the formulation of PUs in order to optimize the balance between mechanical properties and biostability, especially with the development of poly(carbonate urethane) and urethane–siloxane copolymers. Since these efforts have almost wholly been directed towards

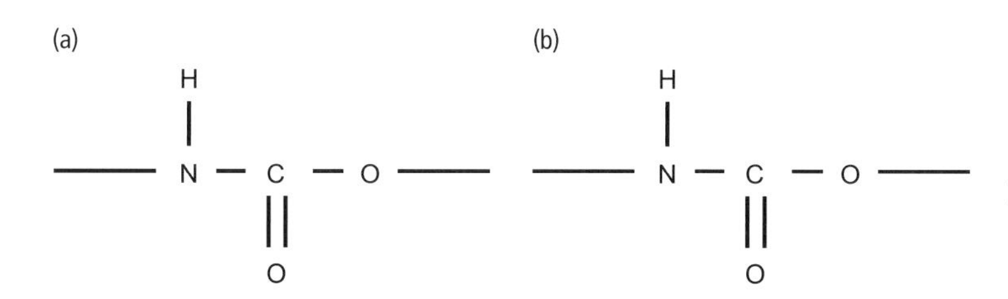


**Figure 8.18**
**(a)** The urethane group.
**(b)** The reaction between an isocyanate and an alcohol in the synthesis of polyurethanes.

polyurethane elastomers, this discussion will be reserved for section 8.4.10 below.

### Class 3.1.9 Polyacetals

Polyacetals are linear polymers with alternating carbon and oxygen atoms in their backbone. They are prepared by the addition polymerization of aldehydes. Polyoxymethylene, also called polyformaldehyde, is the simplest form. It may be copolymerized with trioxane and ethylene glycol. These polymers are readily molded and have widespread industrial uses. For most applications, the low but finite water absorption that occurs is of no consequence, but this is not the case for implantable devices. The homopolymer has been used in two types of device. It was used as the moving disc in prosthetic heart valves. Although giving good performance in many patients it was not ideal and was superseded by carbon-based materials. It was also used as the acetabular component of hip replacements, but the relatively poor wear resistance led to major clinical failures.

## 8.4.2 Class 3.2 Thermosetting resins

The principles of thermosetting resins have been discussed in Chapter 2. There are three main forms of thermosetting resins and a large variety of lesser known ones. The main forms are epoxy resins, phenol formaldehyde and polyurethanes. The others include saturated and unsaturated polyester resins and alkyd resins. There are relatively few situations in which thermosetting resins are used as biomaterials; epoxy resins are used in the majority of these.

### Class 3.2.1 Epoxy systems

The term "epoxy resin" embraces a wide variety of chemically distinct polymers. They are formed by reacting an epoxy monomer, which contains two or more reactive epoxide groups, and any one of a large number of curing agents. The most common type of monomer is that based on the reaction product of epichlorohydrin and bisphenol A. These products can be prepared with different molecular weights and viscosities. The curing agents include primary amines and aliphatic diamines and polyamides. Two problems have prevented epoxy resins from being used extensively in implantable devices, namely the water absorption and potential toxicity. Early forms of cardiac pacemaker were encapsulated in epoxy resins. They were adequate for short-term use, but as both power sources and pacing leads were improved to give performances much longer than a year, the epoxy resin proved problematic, as water was able to diffuse slowly into the interior of the devices. The properties have been improved over the years, and epoxy resins are used in a variety of non-implantable and short-term medical devices. In particular, one-part, fast-cure epoxy adhesives have been developed that bond well to most relevant substrates, giving an instant cure and very high bond strengths.

## 8.4.3 Class 3.3 Synthetic polymeric sols and gels

Hydrogels are water-swollen polymer networks. They may be of natural or synthetic origin. For many years the most clinically relevant hydrogels were cross-linked, covalently bonded synthetic materials, many of which have been based on the principles of the polymerization of hydroxyethylmethacrylate (see below). These polymers were developed in the 1970s for specific use in contact lenses. They may absorb up to two-thirds of their own weight of water to become elastic, permeable membranes. The amount of water absorbed by a hydrogel is governed by the nature of the hydrophilic monomer, the nature and density of cross-linking and external factors such as pH and temperature. The biomedical applications of hydrogels are largely based on this equilibrium water content, and the properties that are conferred on the material by the water, both in relation to the many similarities between hydrogels and soft tissues and the ability of hydrogels to contain and deliver active biomolecules.

Hydrogels may involve just one polymer (homopolymer hydrogels), two co-monomer units (copolymer hydrogels) or more (multipolymer hydrogels). They may also take the form of interpenetrating hydrogels, where a first network is established that is then swollen in a different monomer. Hydrogels may be semi-crystalline, in which there are some dense regions of ordered chains, or they may be amorphous, with no specific order to the chains. They may be electrically neutral, or they may be anionic, cationic or ampholytic. In this section we are dealing with synthetic hydrogels, recognizing that many of the medically useful gels are based on the natural products that are discussed in Sections 8.4.4 and 8.4.5 below.

For the synthetic hydrogels, it is usual to prepare a sol containing the monomer or non-cross-linked polymer of interest and then alter the conditions such that a sol–gel transition takes place. In many of the biomedical situations in which these materials are used, especially in tissue engineering and drug delivery, these conditions cannot be too harsh in order to avoid irreversible changes to drugs, cells or biomolecules. Photo-polymerization can be very effective here, as we shall see with examples such as poly(ethylene glycol) below. This is particularly important when the sol–gel transition is desired to take place *in situ*, for example with injectable materials.

### Class 3.3.1 Poly(ethylene glycol)/ polyethylene oxide

Poly(ethylene glycol), commonly abbreviated to PEG, and otherwise known as poly(ethylene oxide), is perhaps the most versatile of all synthetic hydrogels and is widely used in its own right in medical devices, drug delivery systems and tissue engineering, and also when combined with other molecules and materials systems. As shown in Figure 8.19, PEG is a diol with two hydroxy end groups, being formed by linking repeating units of ethylene glycol. The versatility arises from two facts: first it has linear and branched forms, and secondly these end groups can be readily functionalized, for example through carboxyl, amine, thiol or azide groups. These functional end groups can be symmetric or asymmetric, the latter option being very effective since it allows for simultaneous but different properties to be achieved. A very popular form of PEG is the 4-arm-PEG, but other geometries are possible.

The versatility is also facilitated by the possibility of cross-linking by different methods, under different conditions. Free radical polymerization, condensation reactions, enzymatic reactions and Click chemistry can all be used, but the most common cross-linking procedure is photo-polymerization. This allows conversion of liquid to solid state under ambient physiological conditions, *in situ*, with good spatial and temporal control and with the possibility of simultaneous incorporation of biological species or active agents. Best results are normally achieved through the use of acrylates, such as diacrylates or dimethacrylates, as the macromers. PEG is normally non-biodegradable and has very little intrinsic biological reactivity; one of the major uses of PEG is as a coating to minimize protein adsorption to surfaces. Unmodified PEG is unattractive for tissue engineering applications because of the lack of any support for cell function.

PEG itself, in low molecular weight, unmodified form, has a number of rather mundane medical uses,

(a)

$$H-O\left[\begin{array}{c}H\ \ H\\ |\ \ \ |\\ C-C-O\\ |\ \ \ |\\ H\ \ H\end{array}\right]_n H$$

(b)

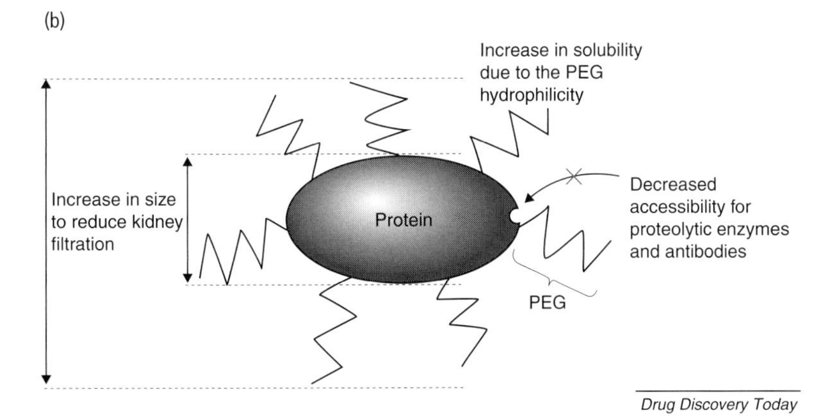


**Figure 8.19** Poly(ethylene glycol). **(a)** The basic PEG polymer. **(b)** Pegylation is the modification of protein, peptide or other molecules by the linking of one or more poly (ethylene glycol) chains. The scheme shows a polymer–protein conjugate, where the PEG shields the protein surface from degrading agents through steric hindrance. The increased size of the conjugate also decreases kidney clearance and increases residence times in the body, which are very important for protein drugs. Reprinted from *Drug Discovery Today*, 21, Veronese F M and Pasut G, Pegylation, successful approach to drug delivery, 1451–8, © 2005, with permission from Elsevier.

largely as OTC preparations such as laxatives and skin moisturizing agents. The real attraction of PEG as a biomaterial is associated with its combination with other molecules. There are two main scenarios here, either the PEG is used to surface modify a structure to provide or hide properties, or other molecules are used to modify PEG hydrogels in order to capitalize on the hydrogel characteristics but provide some specific biological activity. The first of these options is referred to as pegylation. There are many examples, ranging from pharmaceutical molecules to nanoparticles. It has been known for some time that pegylation of protein and polypeptide drugs can alter both their pharmacokinetic and pharmacodynamic properties, for example increasing water solubility, minimizing cytotoxicity and reducing renal clearance. At its simplest, the PEG makes the drug molecule larger: each ethylene glycol subunit is tightly associated with two or three water molecules that makes the pegylated molecules appear up to 10 times larger than the unmodified molecule. The PEG acts as a shield around the molecule protecting it from degradation and rapid clearance. The effect may be much more subtle, depending on the chemistry involved. For example, the bond between the PEG and drug molecule may be intentionally unstable in order to improve targeting, such as when the bond is cleavable by enzymes within the endosomal compartment of cells, which releases the peptide or protein molecule within the cell. Examples of pegylation are shown in Figure 8.19(b).

With respect to nanoparticles, as discussed in several places in this book, many attractive properties are associated with these systems, especially the physical properties of metal and semiconductor nanoparticles, but they often need surface coatings in order to obtain suitable biological properties. In many situations, PEG is ideal for this since it allows for the development of a hydrophilic outer shell and, with the use of PEG-conjugated ligands, the possibility of receptor-mediated drug and gene delivery and multifunctional cell labels.

Bioactive modification of PEG hydrogels for tissue engineering applications may facilitate cellular function. A variety of ECM protein-derived cell adhesive molecules may be incorporated into the hydrogel in order to increase cell adhesiveness. The hydrogels may be made degradable by the incorporation of hydrolytically or enzymatically susceptible segments. Growth factors that have been functionalized may be covalently attached to the PEG hydrogels, especially those which involve multithiol or multiacrylate groups.

### Class 3.3.2 Pluronics

Pluronic polymers, sometimes known as poloxamers are block copolymers that consist of hydrophilic poly(ethylene oxide) and hydrophobic poly (propylene oxide) blocks that are arranged in a triblock form, PEO–PPO–PEO. These may have different numbers of ethylene oxide and propylene oxide units, giving varying hydrophilic–lipophilic balances. Their amphiphilic character gives these polymers surfactant properties and in solution they can self-assemble into micelles. These micelles will be between 10–100 nm in diameter, the core consisting of the hydrophobic PPO blocks and the shell is of hydrated PEO. This shell allows the micelles to remain in a dispersed state while the core can accommodate significant amounts of water-insoluble drugs. Several Pluronics are commercially available and used in a variety of biological and medical situations, each being designated by alphanumeric notations such as Pluronic P105 or Pluronic L61.

### Class 3.3.3 Polyhydroxyethylmethacrylate

For many years polyhydroxyethylmethacrylate, polyHEMA, stood out as being the only material that had been developed (in the 1960s) specifically and solely for biomedical use, all other biomaterials being derived from materials that originated in other engineering spheres. The application here was the contact lens, which required a soft, optically transparent, tear-film wettable, high oxygen transmissible, non-fouling, biostable and inexpensive material that would also be easy to clean and disinfect. PolyHEMA, whose structure is shown in Figure 8.20 is a lightly cross-linked hydrogel that had, in original formulations, an equilibrium water content of around 40%. The oxygen transmission varies with the water content and, obviously, with thickness. Over the years, variations of polyHEMA that had higher water content and/or could be made with thinner dimensions, were introduced. This was achieved through copolymerization, for example with ethylene glycol dimethacrylate, methacrylic acid or vinylpyrrolidone. In order to improve properties and make use of the oxygen transmissibility of polydimethylsiloxane (PDMS), alternative contact lens materials have been developed that are essentially copolymers of the hydrophilic monomers used in these hydrogels and silicone macromers such as PDMS or tris(trimethylsiloxy)silanes.

### Class 3.3.4 Poly(vinyl alcohol)

Poly(vinyl alcohol) (PVA) is a hydrogel that is derived by the hydrolysis of polyvinyl acetate. Many of the repeat units have pendant hydroxyl groups, which allow for considerable variability in water content and functionality. The properties of the PVA

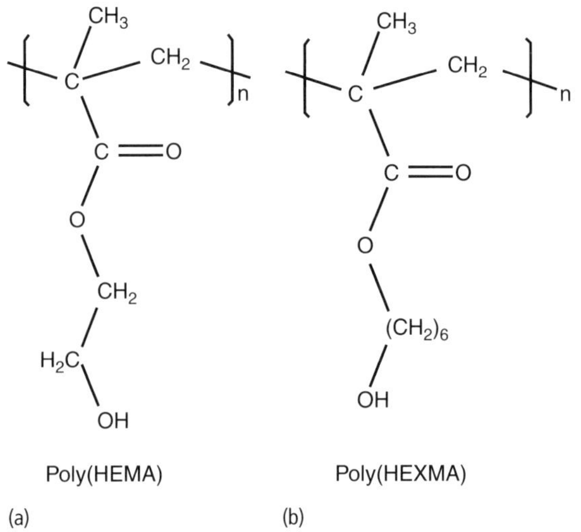

(a) Poly(HEMA) (b) Poly(HEXMA)

**Figure 8.20** Acrylic hydrogels: **(a)** polyhydroxyethylmethacrylate; **(b)** polyhydroxyhexylmethacrylate.

can be modified through cross-linking, especially photo-cross-linking. Different macromers, including those with acrylamide groups, may be incorporated into the gels for these purposes. PVA has been used as an alternative contact lens material and, experimentally, for cell encapsulation.

#### Class 3.3.5 Polyglycerols

Polyglycerols could be included in several classes of polymer, and it will be briefly mentioned in the section on water-soluble dendrimers. Polyglycerol hydrogels have many similarities to PEG. These are prepared as hyperbranched networks, with an inert polyether backbone and functional hydroxyl groups at the branch-ends. Glycidol is the usual precursor of the hydrogel. They may be functionalized, for example with glycidylmethacrylate, and, as with PEG, they may be photo-polymerized for *in situ* gelation. Polyglycerol derivatives, with neutral, cationic or anionic charges have been prepared with dendritic structures, having a defined aliphatic polyether background with multiple functional end groups.

#### Class 3.3.6 Inverted colloid crystals

Colloidal crystals are ordered arrays of colloidal particles, seen for example with the close packed array of silica particles in opal gems. It is possible to create equivalent colloidal crystals with synthetic polymers, for example with polystyrene spheres and then use these as templates to create inverted colloidal crystals, for example with a cross-linkable hydrogel, which gives a porous inverted replica of the original polymer, creating a finely tuned 3D scaffold that can be used to control cell behavior in culture.

### 8.4.4 Class 3.4 Proteins and peptides

Class 3.1, 3.2 and 3.3 biomaterials are based on synthetic, organic macromolecules. Most of these provide good functionality and acceptable safety in their chosen biomedical applications. They are not, however, directly analogous to the tissues they are interacting with, and we must now turn our attention to those materials that do possess such characteristics. We begin with materials based on proteins or protein-like molecules. It has to be stated at the very beginning that the attractiveness of using natural biomaterials with these characteristics is offset to some extent by a few disadvantages. First, while the inherent issues of toxicity that may arise with synthetic materials should not arise with natural materials (unless toxic species are introduced during processing), they may be replaced by issues of immunogenicity. Moreover, since materials such as proteins are, in their natural state, undergoing constant turnover/renewal, at varying rates, it is difficult to achieve the required level of biostability or biodegradability. In addition, natural materials are structurally far more complex than synthetic polymers, with inherent variability from source to source, so that the manufacturing technology of producing biomaterials of the required reproducibility and quality is not necessarily straightforward.

Notwithstanding the above challenges, a wide variety of proteins, and their derivatives, have been developed as biomaterials. We must bear in mind here that protein-based materials could be used in several different ways. The majority of proteins that are used as biomaterials are based on those found in mammalian tissue. These include the structural proteins collagen and elastin, and also some that are derived from plasma proteins, including fibrin and fibrinogen. The structural proteins here could be used in what is essentially their natural form, that is as the mammalian tissues themselves, with varying degrees of processing. Alternatively, they may be extracted from such tissues and subjected to some form of purification and reconstitution. Thirdly, because of the inherent variability of such products, it may be possible to prepare the materials by recombinant technologies. Of these options, the natural tissues themselves are discussed in Section 8.7. The others are discussed here.

There are some generic, and important, differences between the reconstituted and recombinant forms of protein biomaterials. In the former case, there is always a risk of contamination with prions and viruses, but this risk is not present with recombinant proteins. The latter are fully characterized, consistent and reproducible, whereas the former are dependent on source quality, and batch-to-batch variability may be high. From the commercial manufacturing perspective, recombinant proteins will usually be very expensive, but are amenable to proprietary processes that can be protected by patents and trademarks, which is not so readily applicable to the natural products.

### Class 3.4.1 Collagen derivatives

Collagen, in one form or another, has been used as a biomaterial for many years. Catgut sutures, now rarely used, were the mainstay of surgical wound closure for a very long time, and were made from the collagen of bovine intestines. More recently, purified collagen has been used in injectable form as a tissue filler, both for functional (vocal chords, urethra) and cosmetic (facial) effects. Some of these products are xenogeneic, being derived, for example, from bovine dermis, while others are allogeneic, being obtained from cultured human dermal fibroblasts, or from human cadavers. Formulations vary quite considerably. Some are simple suspensions (at less than 5%) in buffered saline. Others are prepared as micronized particles contained in syringes that are hydrated just prior to use. Depending on the source of the collagen, these products do have the potential to be immunogenic, with a small number of patients showing immunological intolerance. Procedures usually involve prior testing of sensitivity. The collagen filler does become incorporated into the patient's tissue but some volume loss will occur over time.

The precise biochemical characteristics of the collagens used in the above products are not very significant since the material is simply acting as a space filler. The applications of collagen in tissue engineering, wound healing and drug delivery are more dependent on these characteristics. The structural order of collagen occurs at several different levels and there are many different forms of the protein in mammalian tissue. Of relevance here is the need to balance the mechanical properties and biological activity of the products and this balance will change with the specific type of collagen.

At this stage, collagen types I and III, either alone or in combination, are most commonly used in products of regenerative medicine. Conduits for peripheral nerve repair are good examples. There are products made from type I or mixed types I and III in clinical use. These have somewhat different structures; some conduits are homogeneous fibrillar structures while others have heterogeneous structures with concentric cylindrical structures. The materials are processed in ways to minimize antigenicity, for example by cross-linking or enzymatic removal of antigenic non-helical telopeptides. Degradation rates vary from a few months to a year or so. In some situations the collagen may be copolymerized, for example with glycosaminoglycans molecules. Collagen type I is also used in a number of products in bone tissue engineering, although not on its own. It is usually combined with hydroxyapatite or tricalcium phosphate, either as a reinforced composite or as phosphate-coated collagen fibrils. A resume of collagen-based biomaterials is given in Table 8.8.

Because collagen is implicated in normal hemostasis processes, it has potential for use in hemostatic devices that control bleeding. It has been used as sheets, foams and gels both for topical hemostasis and to facilitate sealing of arterial puncture sites after percutaneous procedures.

It was noted earlier that recombinant collagen may have some advantages over animal- or human-derived materials. A number of methods are available for recombinant collagen production. This occurs in bioreactor-based eukaryotic systems, mammalian cell culture, insect cell culture and many other systems. However, best results, and the formation of collagens most appropriate for human regenerative medicine applications, are obtained with mammalian cells transfected with collagen genes, where hydroxylated full-length collagens are produced. Recombinant human collagen of types I, II and III can be reconstituted into fibrils that are processed into forms

Table 8.8 **Collagen biomaterials. Collagen is one of the most widely used biopolymers in medical technology. Most products are collagen-rich and represent mimics of natural tissues. In a few areas, products that are substantially made of collagen have been used**

| Clinical area | Collagen characteristics |
|---|---|
| Injectable dermal fillers | (a) Bovine collagen plus 3% lidocaine, resorbs in 3 months<br>(b) Human collagen plus 3% lidocaine, resorbs in 4–7 months |
| Nerve guides | (a) Human collagen, type I, resorbs in 4 years<br>(b) Human collagen, type II, resorbs in 8 months |
| Topical hemostats | (a) Microfibrillar collagen, absorbed in 8 weeks<br>(b) Gelatin foams, absorbed in 4–6 weeks |
| Tissue engineering scaffolds | (a) Injectable, with cross-linking by chemistry or UV radiation<br>(b) Membranes, produced by solvent casting, phase separation, electrospinning<br>(c) Sponges, produced by solid free-form, freeze-drying, phase separation<br>(d) Micro- and nanoparticles, produced by thermally induced phase separation, emulsification, electrostatic fields. |

such as fleeces, 3D gels and sponges, where it is anticipated that type I will be used in bone tissue engineering, type II in cartilage and type III for vascular tissues.

**Class 3.4.1.1 Gelatin** Gelatin is obtained by the partial hydrolysis of collagen obtained from bone and other connective tissues of animals. When obtained under acid conditions it is known as Type A gelatin and under alkaline conditions as Type B. It is used in foodstuffs and in many pharmaceutical formulations. It is also used as a gel to provide an initial seal within vascular grafts, in which situation it may also be a drug carrier, for example for the delivery of antibiotics.

### Class 3.4.2 Elastin derivatives

Elastin is the dominant protein of elastic tissue fibers and, as such, it is a very important component of the extracellular matrix of tissues, such as those of the lungs, skin and blood vessels, that depend on elasticity for their function. It is derived *in vivo* from the cross-linking of the tropoelastin monomer, and is essentially insoluble. The tropoelastin/elastin molecules have hydrophobic regions from which the resulting elasticity is derived and hydrophilic regions, which provide sites for amine-dependent cross-linking and biological signaling. The elastin possesses a number of peptide motifs that are able to influence cell behavior, including proliferation and differentiation. Such interactions take place through several cell-surface receptors, including the elastin-laminin receptor, often referred to as the elastin-binding protein (EBP). The signaling of cells in wound healing by elastin controls the relative activity of dermal fibroblasts and contractile myofibroblasts and hence determines the mechanical properties of the subsequent repaired skin.

The insolubility of elastin, which restricts processing, and its relatively poor strength, has limited the practical applications of this protein in biomedical applications. Elastin-like or elastin-based materials have therefore attracted more attention. There are solubilized forms of elastin, including α-elastin that is obtained under acid conditions and κ-elastin obtained under alkaline conditions. Recombinant techniques have allowed the preparation of materials that are able to mimic the important and desirable regions of the elastin. Synthetic analogs of elastin may also be prepared from the aqueous processing of replicas of tropoelastin. Many of these substances are amenable to scale-up manufacture, giving a range of elastin-derived products for biomedical uses. Some of these substances can be coated onto polymeric and metallic substrates; recombinant elastin constructs in particular may be robustly coated onto substrates. For tissue engineering applications, many of these derivatives may be processed by electrospinning or prepared as hydrogel matrices. It is also possible to prepare blends involving elastin derivatives and other proteins, for example silk and collagen.

### Class 3.4.3 Resilin

Resilin is a structural protein that is found in some parts of the bodies of arthropods where there are highly repetitive movements found in the hind legs of jumping insects such as fleas, the wings of dragonflies and the vibrating membranes of cicadas. It is naturally synthesized as proresilin, whose molecules are then cross-linked through di- and tri-tyrosine links. This results in a stable highly elastic, fatigue-resistant material that resides in the extra-cellular matrix, where it can be compressed to store energy for very rapid release. Resilin-mimetic materials have been produced, using DNA recombinant technology, for biomedical applications. For example, the first exon of the *Drosophila melanogaster* CG 15920 gene in *Escherichia coli*, resulting in the material described as rec1-resilin, can be cast into a rubber-like form with rapid photochemical cross-linking. Figure 8.21 depicts the major elastin and resilin-based biomaterials.

### Class 3.4.4 Fibrin derivatives

Fibrinogen, as we have seen in Chapter 3, is a soluble protein in blood that is converted to an insoluble fibrin network in the presence of thrombin during coagulation. The fibrinogen has three pairs of polypeptide chains that are joined by disulfide groups. The central domain contains fibrinopeptides A and B which are cleaved in the presence of thrombin to form the fibrin monomer. These monomeric units form two-stranded fibrils that undergo covalent cross-linking to form the fibrin network, a process that is facilitated by $CaCl_2$. This process can be recapitulated artificially in biomaterial products, generally known as fibrin glues or fibrin sealants, in which separate preparations containing fibrinogen and thrombin are mixed just prior to application at sites of surgical injury, for example during cardio-pulmonary bypass, to assist in the sealing of tissue defects such as fistulae and to facilitate tissue repair in peripheral nerves.

It is possible for fibrin glue to be derived both from autogenous or allogeneic blood, although the latter is more common in commercial products. Blood components are obtained from the blood and undergo various screening and purification procedures, especially where pooled blood is involved. Fibrinogen may be isolated from the blood using centrifugation and cryoprecipitation. Typically sodium citrate solution is added to the blood to anticoagulate it prior to centrifugation. The cryoprecipitation procedure may be carried out by a variety of regimes, usually using temperatures between −20 and −80°C. $CaCl_2$ and aprotinin are added to the solutions to optimize the gelation of the fibrinogen and thrombin solutions once mixed.

The role that fibrin plays in response to injury of vascularized tissue may also be recapitulated in the formation of tissue engineering scaffolds, which have been used in many experimental systems involving several different tissues. Fibrin hydrogels prepared, as above, from commercially purified allogeneic fibrinogen and thrombin does have some attractive features but is not ideal. Mechanically it is weak and not very stiff, and it also has a tendency to degrade very fast. The gels also suffer considerable shrinkage during formation. In addition, the fibrin is not particularly active biologically in this format. Several procedures may be used to modify the gel to obtain better performance. Shrinkage may be minimized by the incorporation of poly-L-lysine. Composite or copolymer scaffolds, using a variety of synthetic polymers or calcium phosphate additions, may improve the mechanical properties. Cross-linking agents may improve stability. In addition biologically active peptides or therapeutic proteins can be

(a)

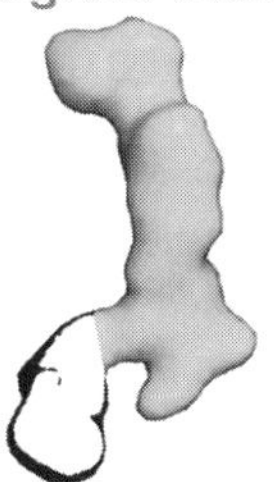


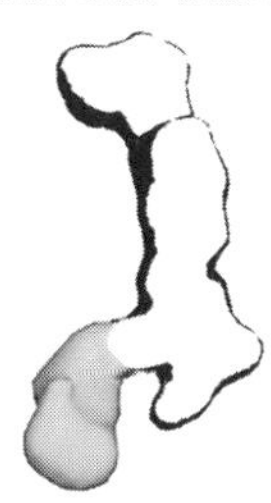


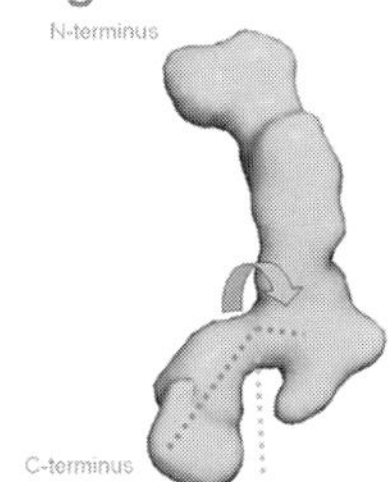

(b)

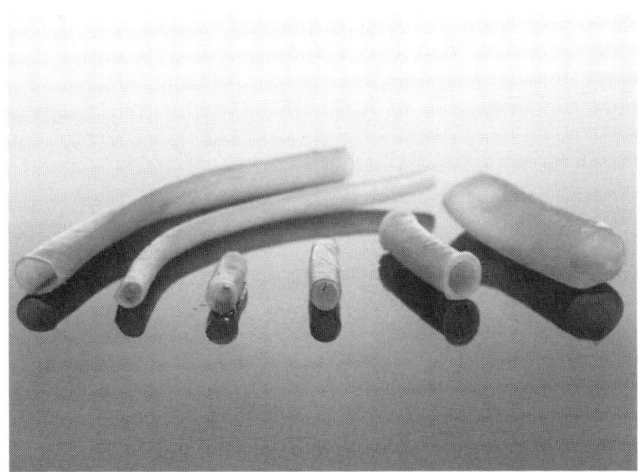

(c)

**Figure 8.21** Highly elastic proteins.
(a) Tropoelastin, the precursor of elastin. Solution structure of tropoelastin shows the major functional regions. Image courtesy of Professor A.S. Weiss and Dr. G. Yeo, University of Sydney, Australia.
(b) Bespoke vascular conduits made from synthetic human elastin. Image courtesy of Professor A.S. Weiss and Dr. S.M. Mithieux, University of Sydney, Australia.
(c) Resilin is a highly elastomeric natural polymer that provides for flying and jumping activity in many insects. A recombinant form of this protein has been made and is considerd for use in medical applications. The image shows a dragonfly (actual size: five centimetres long) and a UV-illuminated rod of cross-linked recombinant resilin (actual rod size: one millimetre in diameter). Credits: Resilin knot photograph by Dr. David Merritt, Univeristy of Queensland; dragonfly image by David McClenaghan. Layout by Dr. Nancy Liyou, Ted Hagemeijer. Images courtesy of Dr. Chris Elvin, Senior Principal Research Scientist, CSIRO Animal, Food and Health Sciences, Queensland Australia.

incorporated into the hydrogels, for example by the attachment of functionalized poly(ethylene glycol).

### Class 3.4.5 Laminin derivatives

Laminin is a major component of basement membranes that is particularly important in neural cell behavior. It is expressed in the central nervous system during development although not in adults, but does so in situations in the peripheral nervous system. Laminin will not form a monolithic biomaterial but can be used as a coating or surface modification. It can self-assemble *in vitro* and form mesh-like structures. Applied to scaffold substrates it can modulate neuroplasticity and positively influence both neural cell adhesion and neurite extension.

### Class 3.4.6 Silk

Silks are proteins that are synthesized by Lepidoptera larvae such as silkworms and spiders. They are biosynthesized in epithelial cells and secreted into the lumen of specialized glands, where they are stored and subsequently spun into fibers. The properties of silks vary considerably with their source, with different amino acid sequences and with mechanical properties that are tuned to their specific function. The most widely used and investigated silks with respect to biomaterials applications are derived from the domesticated silkworm, *Bombyx mori*, and from spiders such as *Nephila clavipes* and *Araneus diadematus*, examples being shown in Figure 8.22.

Silkworm silk is the most popular and has been used for medical devices such as sutures for many years, and in general textiles for much longer. This silk has two major fibroin proteins, light (25 kDa) and heavy (150 kDa) chains, where the core sequence repeats include alanine-glycine with serine or tyrosine. These core fibers are encased in the glue-like sericin protein. Spider silk proteins range from 70 to 700 kDa; many such silks are characterized by polyalanine and glycine regions. Spider silk is not easy to harvest and much emphasis has been placed on the use of genetic engineering techniques to produce synthetic versions. Cloning and expression of silks has been achieved in a number of host systems since the sequences of cDNA and genomic clones encoding spider silks show highly repetitive structures that can be used to construct genetically engineered spider silk-like proteins.

Silk fibers have significant hydrophobic regions and high crystallinity, with extensive hydrogen bonding, giving good environmental stability and mechanical properties. They are insoluble in most solvents including water. The crystallinity is due to the presence of very small $\beta$ sheets within the fibers. *B. mori* silk can have a UTS of 740 MPa, a Young's modulus of 10 GPa and a 20% strain at break. Spider silk may show values of 950 MPa, 12 GPa and 18% respectively.

The biocompatibility of silk products does vary, largely because of the varying levels of non-fibroin components such as the sericin. When used as a suture material, silk does elicit a greater inflammatory response than most synthetic polymers. Although it is classified as nonabsorbable, it does degrade slowly, silk fibers losing their tensile strength over a year or so. Proteases such as chymotrypsin can cleave proteins to peptides, especially in amorphous regions. Various silks have found utility in tissue engineering applications since they are often able to support cell growth.

### Class 3.4.7 Keratins

Keratins are structural proteins that are found in, and may be isolated from, a variety of tissues. There are several different forms; they are often divided into two types, the soft keratins that are found in epithelial tissues and the hard tissues found in protective tissues such as nails and hair. The latter have been considered as the source of biomaterials for a number of applications. These proteins have a high sulfur content, with extensive disulfide bonding; these can be denatured under either oxidative or reductive conditions (giving kerateines and keratoses respectively), extracted from the tissue and processed into gels, films or fibers. Hair is the more common source of keratin biomaterials. There are three forms. Alpha keratins have a relatively low sulfur content and an

(a)

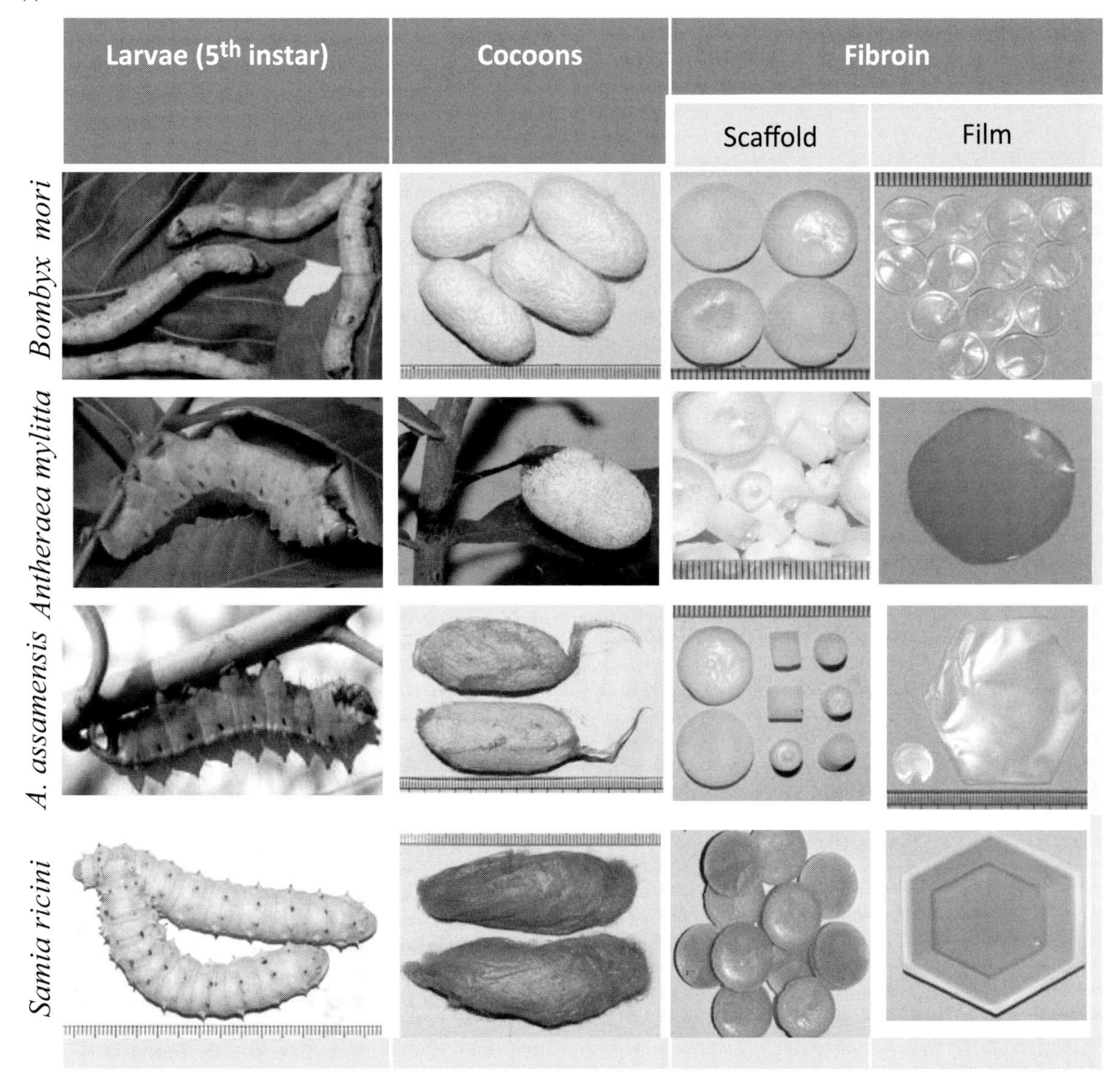


**Figure 8.22** Silk.
(a) Different silkworms and silk-derived biomaterials. Note that sericin can also be fabricated into different biomaterials such as scaffolds, films, hydrogels, nanofibers and particles.

alpha-helical tertiary structure. Beta keratins form the majority of the cuticle and are not considered very useful. Gamma keratins are globular and of high sulfur content. Keratin-based biomaterials have a range of properties depending on extraction and processing conditions. They generally have good biocompatibility, with potential applications in wound healing, hemostasis and nerve repair.

### Class 3.4.8 Zein

Zein is a storage protein found in corn. It has a helical wheel-shaped structure with nine homologous

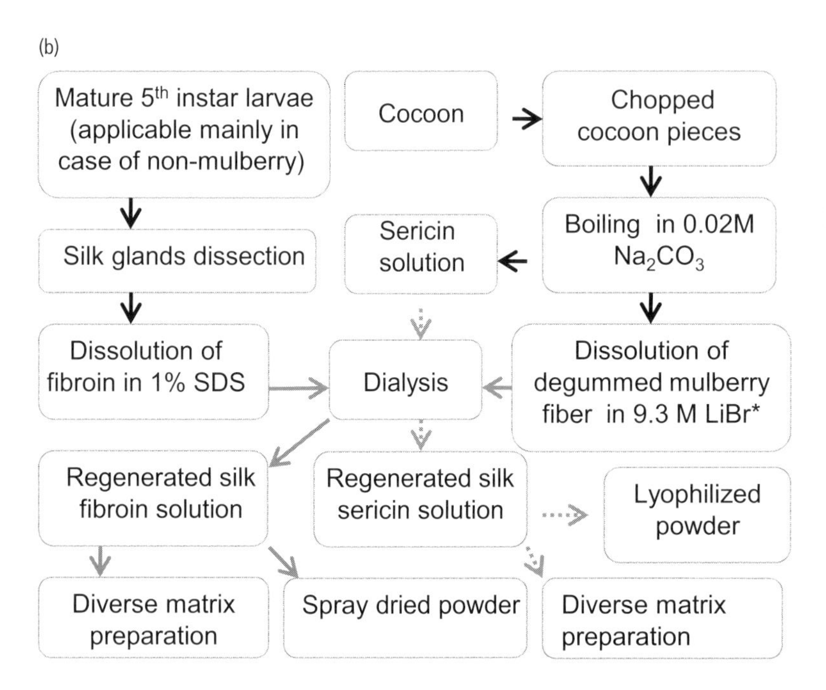


**Figure 8.22** *continued*
**(b)** Regeneration of mulberry and non-mulberry silk protein fibroins and sericins. Images courtesy of Dr. Deboki, Dr. Banani, and Professor Kundu, Indian Institute of Technology Kharagpur, India.

repeating units in a hydrogen-stabilized anti-parallel form. !t has a high proportion of non-polar amino acid residues and has some solubility characteristics that are unusual for proteins so that it can be formed into various shapes including coatings. From solution it can self-assemble into 3D forms such as sponges and spheres and can be applied as tough coatings, which may be patterned on surfaces for control of protein and cell attachment. There is interest in using zein proteins as tissue engineering scaffolds.

### Class 3.4.9 Peptide nanomaterials

The ability of structural features of polypeptides to encode supramolecular structure has led to the development of new forms of peptide-based materials that can be used in a variety of technologies. The range of amino acid segments available gives considerable versatility to the resulting structures, architectures and properties. The exact properties that can be produced depend on both molecular level encoding of the peptide blocks and the process of self-assembly of the supramolecular structure. It is becoming apparent that precise control of molecular, temporal and spatial events in self-assembly is necessary. Functionality of the supramolecular structure that is not present in the peptide blocks becomes possible with this control, giving new mechanical, electronic and optical properties, including the effects of bioactive ligands that may be present. Of particular significance here is the possibility of producing materials that combine biological functionality with biocompatibility and biodegradability. Some of these nanostructures have responsive surfaces that are electro-switchable: surface bound peptide ligands can be selectively linked to cell surface integrins, which can be switched between adhesive ("on") and non-adhesive ("off") states. It is also possible to peptide functionalize nanoparticles, such as the carbon nanotubes discussed later in this chapter, with potential for biosensing applications.

#### Class 3.4.10 Protein and peptide mimetics

It has occasionally been possible to develop synthetic substances that are inspired by natural proteins or peptides and which possess some critical functions of those natural products. It is difficult to provide a general discussion of these since structures and functions will be case specific. One example is that of a multifunctional polymer, polydopamine, that has been inspired by the adhesive proteins used by mussels to adhere to surfaces. Mussels are able to attach to a wide variety of organic, inorganic and metallic substrates, including some materials that virtually nothing else will adhere to. It has been found that proteins which dominate the mussel–substrate interface are rich in 3,4-dihydroxy-L-phenylalanine (DOPA). DOPA can be electropolymerized onto a variety of substrates, including noble metals, oxide and apatite ceramics and synthetic polymers, to produce multifunctional surfaces. These surfaces may mediate secondary reactions through specific chemical properties, including the control of cell behavior, in medical applications.

### 8.4.5 Class 3.5 Polysaccharides

In this section we deal with the main group of natural alternative materials to proteins, that is the polysaccharides. Generically, a polysaccharide is a complex carbohydrate that is composed of a chain of monosaccharide units linked together by glycosidic groups; we bear in mind here that carbohydrates are compounds (which include starches, sugars and cellulose) that only contain C, H and O. Sugars are monosaccharides or disaccharides; polysaccharides have much larger molecules. They clearly have some similarities, but structural differences between different types mean that it is difficult to find much common ground that distinguishes the group as a whole from other materials and they are best dealt with individually. We should note that some of those discussed here are glycosaminoglycans, that is mucopolysaccharides, which are long linear unbranched polysaccharides of repeating disaccharide units. Some, such as hyaluronan, are non-sulfated, but several, including heparan sulfate, dermatan sulfate and keratan sulfate are sulfated.

#### Class 3.5.1 Hyaluronan derivatives

Hyaluronan (HA), otherwise known as hyaluronic acid, is a linear glycosaminoglycan, shown in Figure 8.23, that has a molecular mass of between $10^6$ and $10^7$ Da, with very long molecules that consist of linear chains of repeating units of disaccharides of glucuronic acid and N-acetylglucosamine.

There are two main features of HA that have contributed to its attractiveness as a biomaterial. First, it is contained in, and contributes to, the properties of many tissues in the human body, suggesting the possibility to recapitulate some of these within therapeutic products. Secondly, it can be functionalized and chemically modified to present a range of physical characteristics, with wide-ranging solubility and mechanical properties. Among the specific biological properties of HA are its role in embryonic development and wound healing. It is present in high concentration in synovial fluid and in the extracellular matrix of cartilage so that it plays a significant role in the functioning of articulating joints. HA interacts with some cell surface receptors and is involved in angiogenesis, cell migration and motility, and tissue organization. It plays a role in inflammation and the stimulation of cytokine activity.

HA can be derived from a number of sources, one of the most common being rooster comb. It is highly soluble, especially at low pH, and has a high rate of turnover in human tissue. For the purposes of creating a practical biomaterial, HA is normally cross-linked, for which a number of methods are available and it may be modified with other substances.

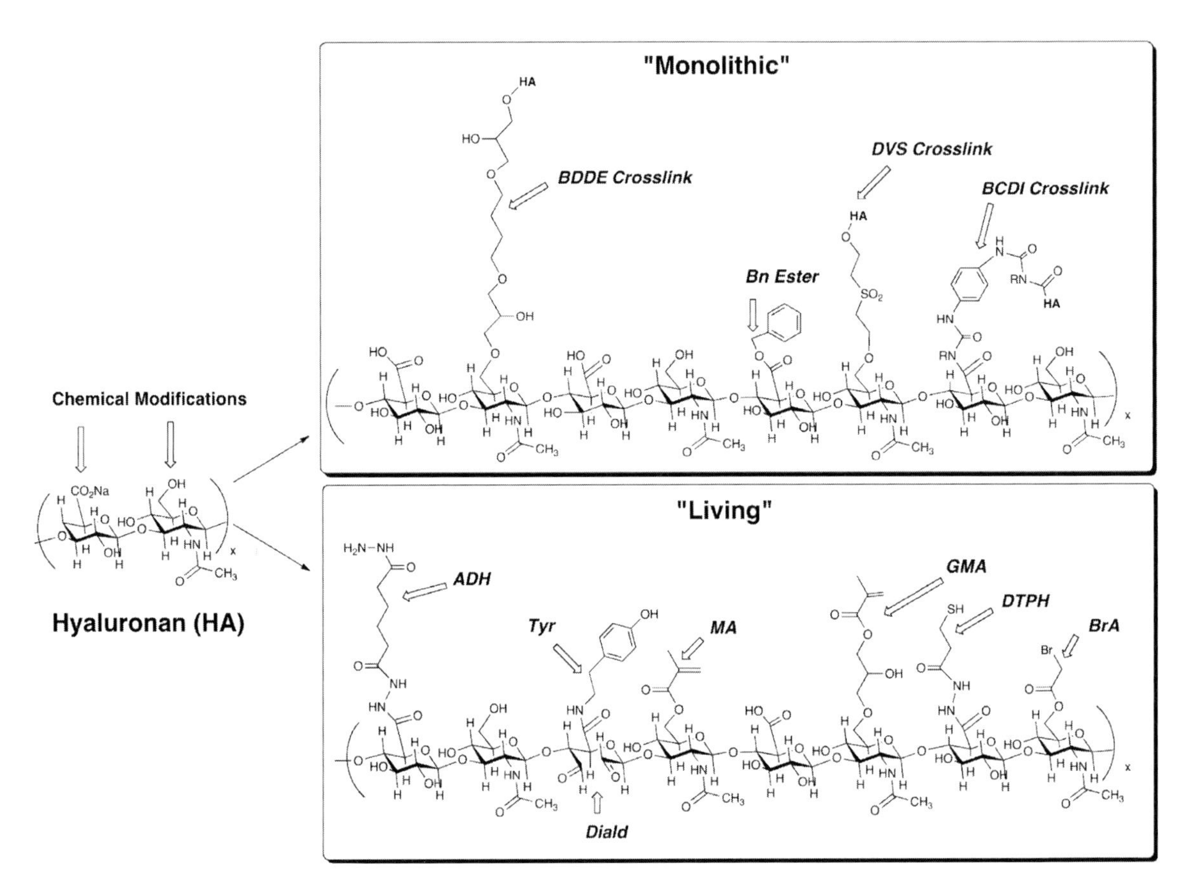


**Figure 8.23** Hyaluronic acid (hyaluronan). Hyaluronic acid is a naturally occurring polysaccharide, which may be modified in various ways to give a range of properties suitable for medical applications. These are representative chemical substructures of monolithic (top) and living (bottom) chemical modifications of hyaluronic acid. Key for top: BDDE, butane 1,4-diol diglycidyl ether (e.g., Restylane®, Juvederm®); Bn, benzyl ester (e.g., HYAFF®); DVS, divinylsulfone (e.g., Synvisc®); BCDI, biscarbodiimide (e.g., Incert®, Monovisc®). Key for bottom: ADH, adipic dihydrazide; Tyr, tyramide (used in CoreGel®), Diald, dialdehyde; MA, methyacrylate; GMA, glycidyl methacrylate; DTPH, dithiopropionyl hydrazide (used in Sentrx®, HyStem®); BrA, bromoacetate. Image courtesy of Professor Glenn Prestwich, Department of Medicinal Chemistry, University of Utah, USA.

Covalent cross-linking provides the opportunity to achieve hydrogels, sponges and other solid forms whilst maintaining biological functionality. Cross-linking may take place using water-soluble agents such as a carbodiimide or by the use of photo-cross-linking using glycidylmethacrylate or methacrylic anhydride. Several HA polymeric systems, referred to as the HYAFF family are derived from the esterification of HA. For example, HYAFF-11is obtained by the total esterification of 80–200 kDa sodium hyaluronate with benzyl alcohol; this material, which is readily prepared as a non-woven mesh, will typically degrade over a 10-week period. Total esterification is not necessary. HYAFF 11p75 is 75% esterified and breaks down much faster. The HA may be formulated together with collagen, chitosan or other materials for greater control over properties.

Some relatively simple HA preparations have been used for a number of years for viscosupplementation, in which regular injections can be made into joint spaces for the treatment of osteoarthritis, or into the eye to increase the viscosity of the vitreous humor. Some wound dressings contain HA in order to enhance healing. Much attention is being paid to the use of HA products in tissue engineering, where they possess many advantageous properties.

### Class 3.5.2 Alginates

Algae are living organisms, mostly found in water; they can easily be harvested and provide substances for many industrial uses. These uses are mainly based on their polysaccharide content, although they are often also rich in amino acids such as proline, glycine and lysine, which accounts for widespread use as food additives. Seaweeds constitute a very important source of harvested algae-based substances since their cell walls contain polysaccharides, which can be readily extracted. Some seaweeds produce agar (from red seaweed) and others produce carrageenans. Brown algae (Phaephyta) produce alginates and several other polysaccharides. The alginates are probably the most important seaweed-derived products, and certainly the most significant from a biomaterials perspective. These are harvested in their wild state, since cultivation is too expensive, which accounts for some variability in the extracted alginates since there is some species and seasonal dependence. Alginates are extracted from the seaweed using sodium carbonate and precipitated as either sodium or calcium alginate. This is treated with diluted HCl to produce alginic acid, which is purified and reconstituted into different ionic forms depending on the application.

Alginates are linear block copolymers of 1,4-linked β-D-mannuronic acid (M) and α-L-guluronic acid (G), shown in Figure 8.24. The properties of the material will depend on the M:G ratio. Alginates form gels and readily retain water. Different ionic forms have different solubilities; the transition from sol to gel can be achieved very easily by conversion between calcium and sodium forms, a process that was extensively used in dentistry to produce elastomeric impression materials. In the solid state, alginates can form films and fibers of good structural quality. When prepared as meshes of non-woven fibers, they make very useful wound dressings, as they are able to absorb large amounts of exudate from the wound bed. Alginate products are used as food additives, moisturizing components of cosmetics and as ingestible preparations for treating inflamed mucosal surfaces. Again it is in tissue engineering and cell therapies that most attention has been paid. It is particularly notable that alginates present one of the best options for cell encapsulation since the hydrated material will allow the diffusion of small molecules essential for metabolic activity but not the immunoglobulins that would attack the cells.

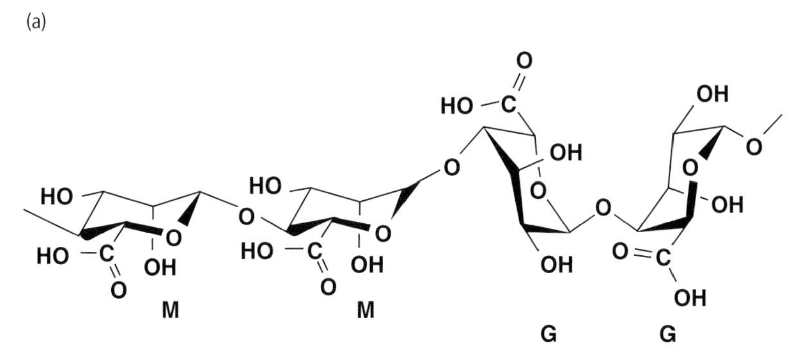


**Figure 8.24** Alginates. Representative alginate structure showing (a) chain conformation and (b) block distribution. Reprinted from *Biomaterials*, 33, Pawar S N and Edgar K J, Alginate derivarization: a review of chemistry, properties and applications, 3279–305, © 2012, with permission from Elsevier.

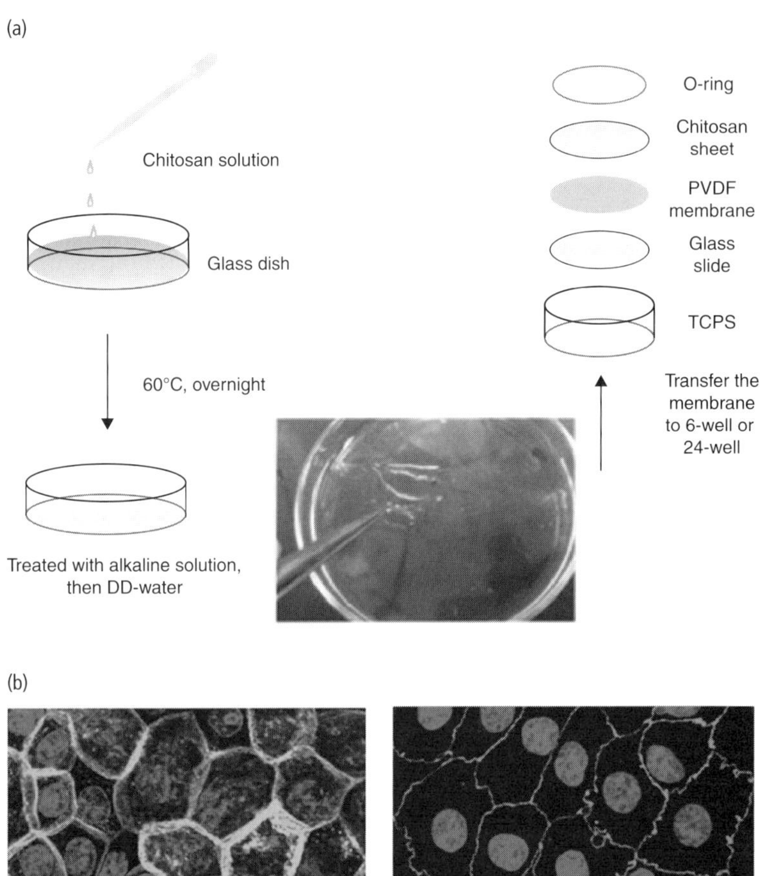


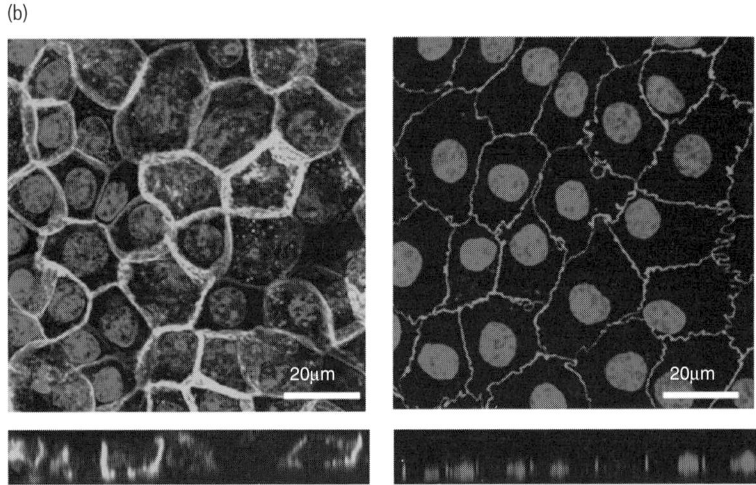


**Figure 8.25** Chitosan.
**(a)** Scheme for preparation of chitosan films.
**(b)** Monolayers of cells can be prepared on such chitosan films. See plate section for color version. Images courtesy of Professor Tai-Horng Young, Institute of Biomedical Engineering, National Taiwan University, Taipei, Taiwan.

## Class 3.5.3 Chitin and its derivatives

Chitin was first prepared from the cuticle of beetles, from which it derives its name. It is widely distributed in both animals and plants, being found in the shells of crustaceans and mollusks, the backbone of squid, the cell wall of many fungi, within marine diatoms and so on. Chitin is a linear polysaccharide of β-(1–4)-2-acetamido-2-deoxy-D-glucopyranose, where all the residues are comprised entirely of N-acetyl-glucosamine, that is it is fully acetylated. Chitosan is a derivative of chitin (Figure 8.25), which is a linear polymer of β-(1–4)-2-amino-2-deoxy-D-glucopyranose, where all the residues are comprised entirely of N-glucosamine, which is fully deacetylated. In nature, it is rare for the material to exist as either pure chitin or pure chitosan, and the natural biopolymer will be a copolymer of the two. Generally when the number of acetamide groups exceeds 50% the material is referred to as chitin, and the actual percentage is termed the degree of acetylation. Conversely, when the amino groups dominate, the material is referred to as chitosan.

The dry shells of animal sources such as crabs and lobsters contain 20–40% chitin, the remainder being proteins and calcium carbonate. Demineralization and deproteinization steps are used in the process to prepare raw chitin products. Chitosan can be prepared from the chitin by deacetylation methods involving sodium hydroxide.

Chitin may exist in three polymorphic states, the α, β and γ forms, the α-chitin being the most common. The biostability varies with the source of the material, the crystallinity and the degree of acetylation. Generally, chitosan is resistant to degradation in aqueous media but it is susceptible to enzymatic degradation. Within animal species, a variety of chitinases are able to break the chitin down into oligosaccharides, which can then be degraded by enzymes such as β-N-acetyl-glucosaminadase to yield N-acetyl-glucosamine. Similar mechanisms exist for the degradation of chitosan to N-glucosamine.

Medical applications have included wound dressings, drug delivery vehicles, tissue engineering scaffolds and gene vectors.

### Class 3.5.4 Pullulan

Several microorganisms produce a substance that attaches itself to the cell wall as an amorphous slime; these are known as exopolysaccharides. Although many bacterial exopolysaccharides have received attention, those derived from yeasts are also of interest here. In particular, the yeast-like fungus *Aureobasidium pullulans* produces a water-soluble random coil glucan, known as pullulan. There are variations on the structure, but they may be considered as polysaccharides based on repeating α-(1→6)-linked (1→4)-α-D-triglucosides. They have been used as food additives. Since they may be prepared as powders that can be processed by a number of methods to produce interesting solid materials, which are biodegradable, there has been interest in their use in drug delivery and tissue engineering.

### Class 3.5.5 Dextran polymers

Dextran is a bacterial-derived polysaccharide, consisting essentially of α-1,6 linked D-glucopyranose residues with some α-1,2-, α-1,3-, or α-1,4-linked side chains. These hydrogels have been used in drug delivery systems and as volume expanders in intravenous solutions.

### Class 3.5.6 Cellulose

Cellulose is a linear polymer of β-(1,4)-D-glucose units (Figure 8.26). It is the main component of the primary cell wall of plants. It forms as crystalline microfibrils, which encapsulate the cell with a mesh-like structure, and controls, along with hemicellulose, pectin and lignin, the mechanical properties of the plants. There is interest in using some derivatives of cellulose as biomaterials in, for example, scaffolds and drug delivery systems.

**Class 3.5.6.1 Microbial cellulose** Microbial cellulose is a polymer that is synthesized by *Acetobacter xylinum*, a simple gram-negative bacterium. During the synthesis, various carbon compounds are utilized by the bacteria, polymerized into single linear β-1,4-glucan chains and then secreted through pores to the cell exterior. These chains then self-assemble into sub-fibrils and then microfibrils and bundles, which yields a highly 3D crystalline structure with considerable mechanical strength. This nanostructure results in a large surface area that can hold a large amount of water. It can be prepared as a gelatinous membrane that is highly nanoporous. It may be treated with strong bases at elevated temperatures to remove the cells that are embedded in the cellulose net; it is free of lignin and hemicelluloses so that the final material should be non-pyrogenic, non-cytotoxic and non-immunogenic. It has shown considerable potential, with support from clinical evidence, for uses in wound healing and possibly as a tissue

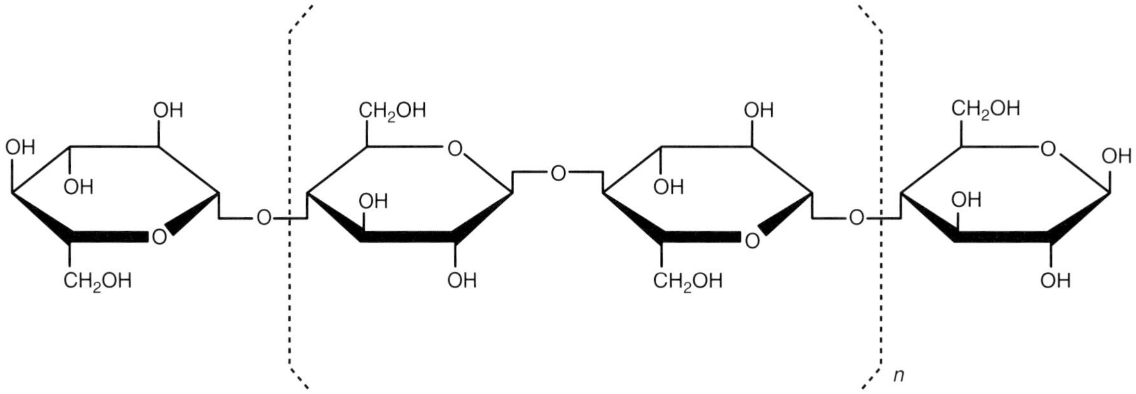

**Figure 8.26** General structure of cellulose.

engineering scaffold. Economic large-scale fermentation systems have been difficult to optimize, which has restricted commercial applications so far.

**Class 3.5.6.2 Methylcellulose and carboxymethylcellulose** Methylcellulose is a derivative of cellulose, being obtained by heating cellulose under alkaline conditions in the presence of methyl chloride. This is not a biomaterial, but is used clinically in the management of constipation. Carboxymethylcellulose is obtained by heating cellulose with chloroacetic acid. It can be prepared as gels with a range of properties, including viscosity, that depend on the extent of hydroxylation and other factors. It is used as a food additive and has been used, as a 0.5% solution, in artificial tears in the management of dry eye. The substance may also be cross-linked to form viscous gels with the intention of use as an injectable agent in soft tissue augmentation. It has also been used as an alternative to silicone gel as the filler in silicone elastomer shell breast implants.

## 8.4.6 Class 3.6 Lipids

Lipids constitute a wide-ranging group of hydrophobic molecules that form fats, waxes, oils, steroids and other substances. Fats are made from glycerol and fatty acids. They may be saturated, mono-unsaturated or poly-unsaturated. They do not have many applications within the biomaterials space, but we have to mention the groups of phospholipids and liposomes.

### Class 3.6.1 Phospholipids

A phospholipid is any lipid that contains phosphorus. The molecules, which may have a glycerol or a sphingosine backbone, have a hydrophilic head and two hydrophobic tails and in solution they tend to form bilayers with the tails directed towards the center (Figure 8.27). These bilayers are crucially involved in the construction of some biological membranes, especially cell membranes. In most situations, there is asymmetrical phospholipid structure, with negatively charged molecules found on the inner cytoplasmic side of the membrane and neutral, zwitterionic, lipid molecules on the outside; the membranes usually involve proteins for support. There are many naturally occurring phospholipids; their relevance to biomaterials science lies in the exploitation of phospholipid-based substances that mimic the structure and properties of these naturally occurring substances. The best examples are phosphorylcholines, such as phosphatidylcholine, which are prominent in the outer surface of red blood cells. The concept of phospholipid mimicry is based on the fact that such cells have very little interaction with plasma proteins by virtue of the biological inertness of this outer layer, which may be recapitulated in synthetic molecules that are structurally similar. There have been several attempts to do this, largely

(a)

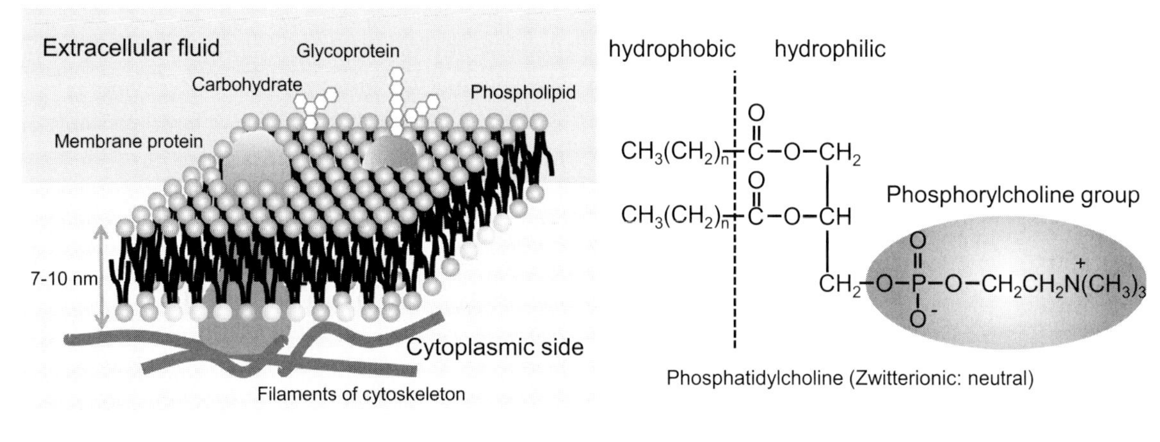


(b)

Phosphorylcholine group

$CH_2=C(CH_3)-C(=O)-OCH_2CH_2OP(=O)(O^-)OCH_2CH_2N^+(CH_3)_3$ + $CH_2=C(CH_3)-C(=O)-O(CH_2)_3CH_3$

2-methacryloyloxyethyl phosphorylcholine (MPC)

radical initiator, solvent →

$-(CH_2-C(CH_3)(C(=O)OCH_2CH_2OP(=O)(O^-)OCH_2CH_2N^+(CH_3)_3))_n-(CH_2-C(CH_3)(C(=O)O(CH_2)_3CH_3))_m-$

Poly(MPC-*co*-BMA)

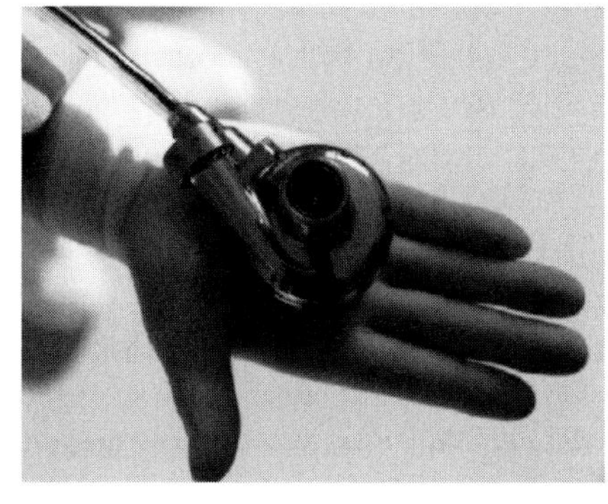

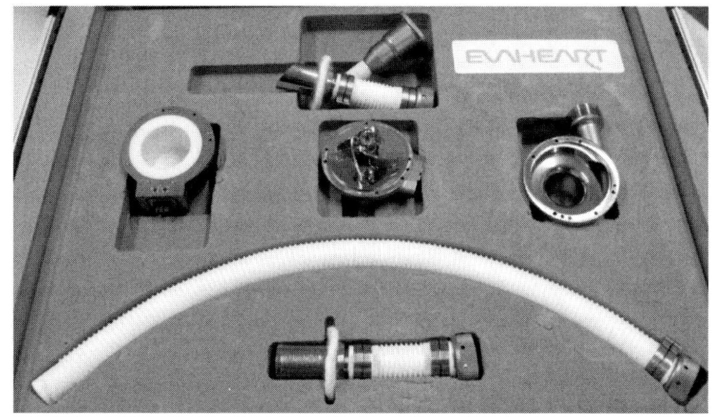


**Figure 8.27** Phospholipids.
**(a)** Schematic representation of the structure of cell membrane is indicated on the left-hand side. It is a molecular assembly with phospholipids, proteins and carbohydrates. Phospholipids construct double-layered membrane and membrane proteins are mostly embedded in the membrane through hydrophobic interactions. A representative chemical structure of the phospholipid having zwitterionic and neutral phosphorylcholine polar group, phosphatidylcholine, is shown in the right-hand side.
**(b)** One of the MPC polymers, poly(MPC-*co*-BMA) is coated on an implantable artificial heart, EVAHEART®, at every blood contacting surface, including cannula, to reduce risk of thrombus formation. The artificial heart has been implanted into more than 100 human patients since 2005. Images courtesy of Professor Kazuhiko Ishihara, Department of Materials Engineering, The University of Tokyo, Japan

by preparing polymeric analogs. For example, a methacrylate monomer with phospholipid polar groups can be prepared and then polymerized. The monomer is 2-methacryloyloxyethyl phosphorylcholine, MPC. Polymerization is achieved through the methacrylate group; the homopolymer is water-soluble. Quite a wide variety of polymers can be produced through the use of comonomers such as butyl methacrylate. It is clear that the coating of medical devices with layers of such phospholipid polymers results in lower protein adsorption in many situations and this technology has been especially used in short-term blood-contacting devices such as extracorporeal circulation and biosensors.

#### Class 3.6.2 Liposomes

Liposomes are artificial phospholipid vesicles (Figure 8.28), which range in diameter from 50 to 1000 nm. They have been extensively used in pharmaceutical technology since they are able to carry a wide range of drugs, are biodegradable and have low toxicity and immunogenicity. They can also be functionalized so that they may be targeted to specific sites. The structure itself is straightforward, as a closed spherical bilayer of phospholipid, with variations being derived by the molecules attached to the external surface. At its very simplest, the pharmaceutical liposome is a spherical shell that contains a water-soluble drug within its aqueous interior.
A water-insoluble drug could be incorporated into the phospholipid membrane. The biological performance may be altered by modifications to this membrane; for example, a layer of PEG could be attached in order to increase the circulation life. Antibodies may be attached covalently to increase targeting efficiency or pH/temperature-sensitive molecules may enable the liposomes to be stimuli-sensitive.

### 8.4.7 Class 3.7 Biodegradable structural polymers

As noted in several places in this book, the availability of polymers that are able to degrade in the body has opened up many biomaterials opportunities. As far as the choice of polymer is concerned, obviously they have to have mechanical and other functional properties consistent with their intended application, but also are required to degrade with the appropriate kinetics and to do so without inducing any adverse effects in the host. In this section we shall deal with solid polymers that have physical form and structural applications. We shall deal with soluble degradable polymers later, although in several of these classes, both structural and soluble forms are possible. There are no obvious hierarchical rules by which we can logically and consistently classify biodegradable polymers; their characteristics clearly depend on their molecular structure, but very small differences in these structures can have powerful effects on the resulting degradation characteristics such that ostensibly very similar polymers can have very dissimilar degradation profiles, whereas, conversely, polymers based on quite different monomers can have very similar profiles. Even more important is the fact that many commercially available and experimental degradable polymers are copolymers and blends and are not readily classified at all. The classification system given here is based on a rather pragmatic approach to constructing a list where major chemical similarities determine where in that list the various materials should be placed. We start with the aliphatic polyesters that have the longest tradition.

#### Class 3.7.1 The poly ($\alpha$-hydroxy acids): polylactides and polyglycolides

Polylactides, polyglycolides and their copolymers were developed for medical applications in the 1960s and remain the materials of choice in many areas where biodegradation and bioresorption are required. Lactic acid, otherwise known as 2-hydroxy propanoic acid, is a simple hydroxy acid, and has an asymmetric carbon atom (Figure 8.29). Of special interest is the fact that this molecule can exist in two different forms based on this symmetry, known as isomers. These have identical molecular formulae but different arrangements of atoms within the molecules. The isomerism that occurs with lactic acid is stereo-

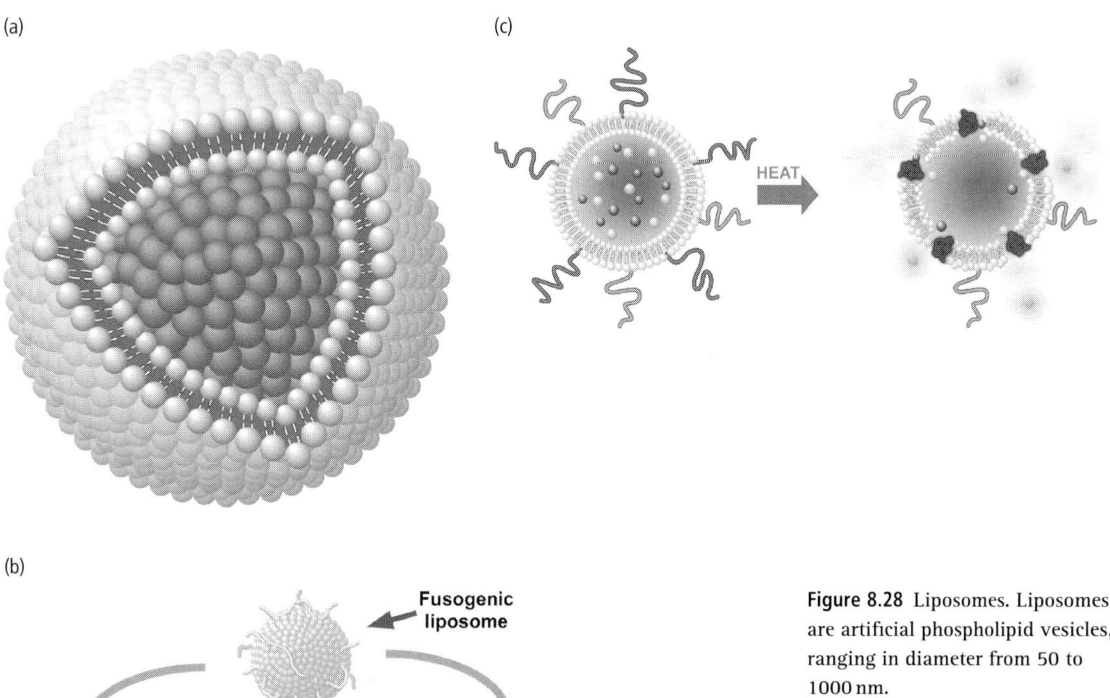


**Figure 8.28** Liposomes. Liposomes are artificial phospholipid vesicles, ranging in diameter from 50 to 1000 nm.
**(a)** Schematic of the structure with its bi-layer membrane structure and internal contents.
**(b)** Fusogenic-modified liposome which delivers its content into the cell either by fusing directly with the cell membrane or by fusing with endosome membranes after endocytosis.
**(c)** Multifunctional liposome with PEG attached to the phospholipid with temperature-responsive character, and a $Gd^{3+}$ bearing dendron lipid, allowing MR imaging and drug delivery. Images courtesy of Professor Kenji Kono, Osaka Prefecture University, Japan.

isomerism, which means that there are two different forms that are mirror images of each other. Molecules of this form are said to be chiral and the two different forms are called enantiomers. The two different enantiomers of lactic acid are L-lactic acid and D-lactic acid, as shown in Figure 8.29(a). These are optical isomers that can rotate the plane of polarization of plane-polarized light. One enantiomer rotates the polarized light clockwise and is the L (+) enantiomer; the other rotates it anticlockwise and is the D (–) enantiomer.

Lactic acid can be polymerized to polylactic acid. Generally the term polylactic acid is used when the polymer is produced by direct polymerization. It is, however, more common to produce the polymer via the formation of lactide, followed by ring opening

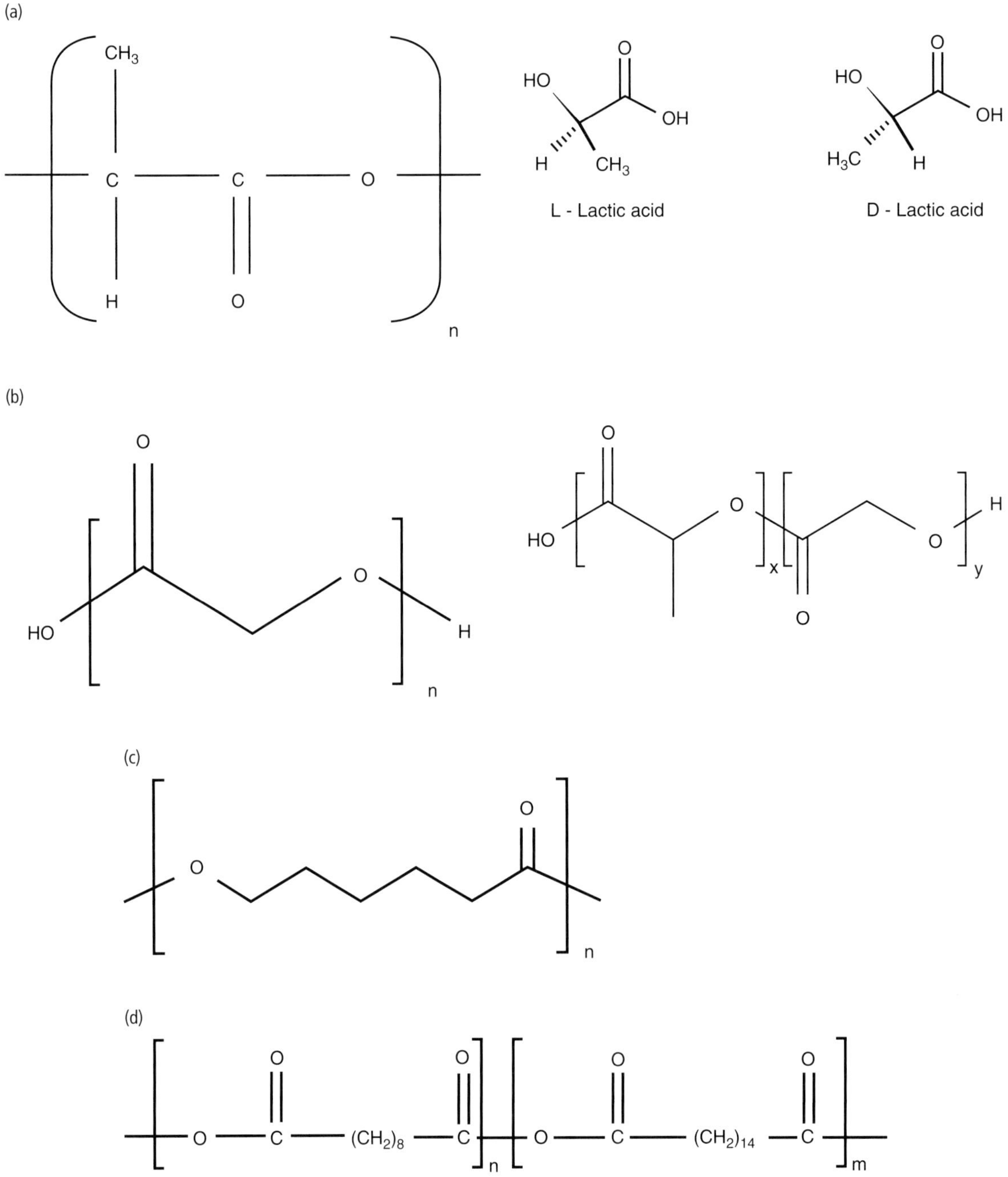

(a)

(b)

(c)

(d)

**Figure 8.29** Biodegradable polymers.
**(a)** Polylactic acid, on the left, with the two isomers L- and D-lactic acid on the right.
**(b)** Polyglycolic acid on the left and a poly(lactic acid-glycolic acid) copolymer on the right.
**(c)** Polycaprolactone; note the short-hand version of polymer structures with repeating $CH_2$ groups, in this case five successive groups, indicated in the zig-zag structure, within the repeating unit.
**(d)** Polyanhydrides; there are many variations, based on the repeating unit $-CO-(R)_n-CO-$.

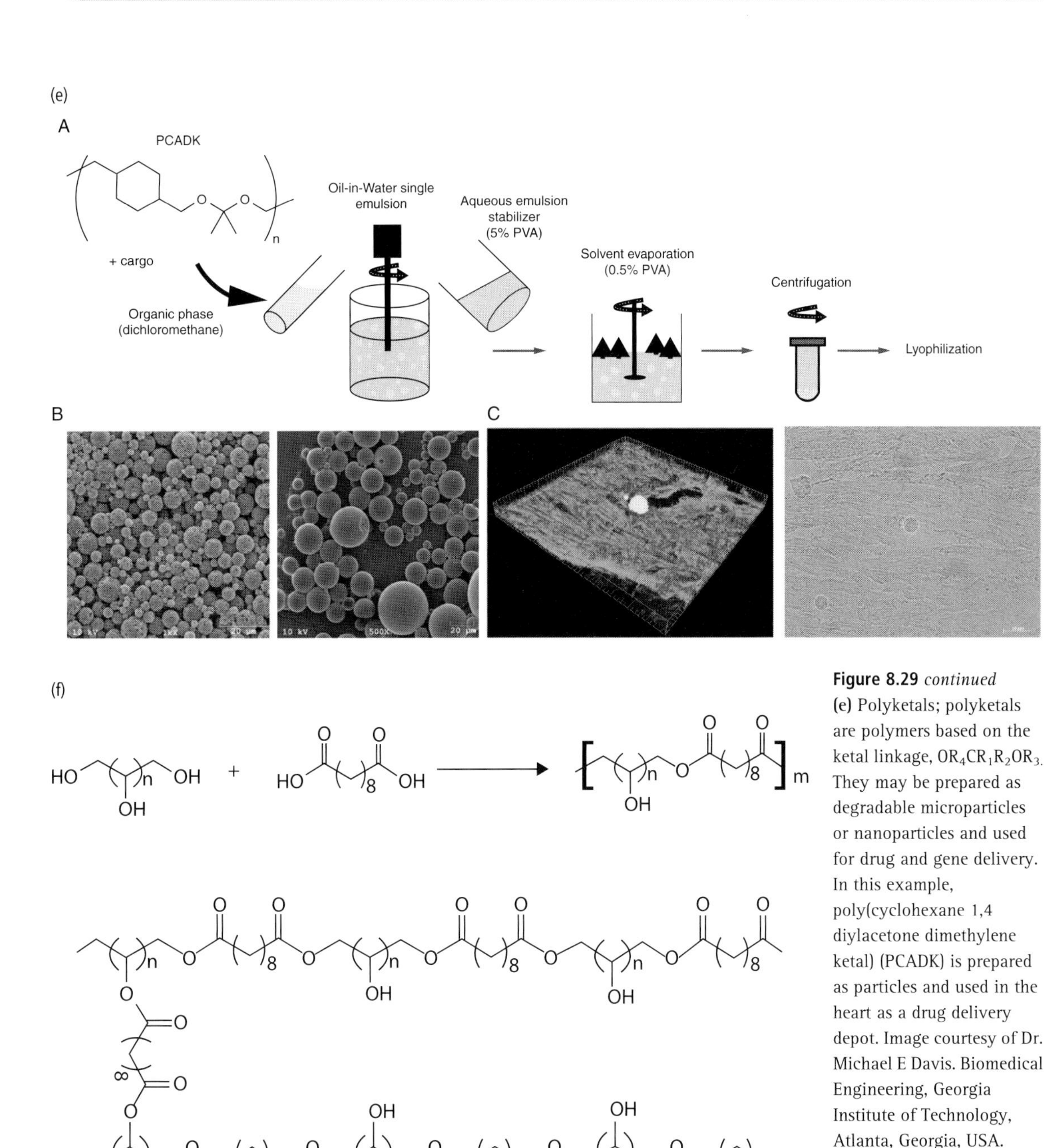


**Figure 8.29** *continued*
**(e)** Polyketals; polyketals are polymers based on the ketal linkage, $OR_4CR_1R_2OR_3$. They may be prepared as degradable microparticles or nanoparticles and used for drug and gene delivery. In this example, poly(cyclohexane 1,4 diylacetone dimethylene ketal) (PCADK) is prepared as particles and used in the heart as a drug delivery depot. Image courtesy of Dr. Michael E Davis. Biomedical Engineering, Georgia Institute of Technology, Atlanta, Georgia, USA.
**(f)** Poly (polyol sebacate)s (PPS) comprise a family of cross-linked polyester elastomers. They are synthesized from a polyol (an alcohol containing multiple hydroxyl groups) and sebacic acid, which is a dicarboxylic acid. The reaction, at the top, involves the formation of PPS chains, which are then cross-linked, at the bottom. The n can be 2, as in glycol, 3 as in glycerol, 5 as in xylitol, 6 as in mannitol, etc. The polyglycerol sebacate (PGS) is the most widely investigated for biomedical use. Reprinted from *Progress in Polymer Science* 38(3–4), Chen Q *et al.*, Elastomeric biomaterials for tissue engineering, 584–671, © 2013, with permission of Elsevier.

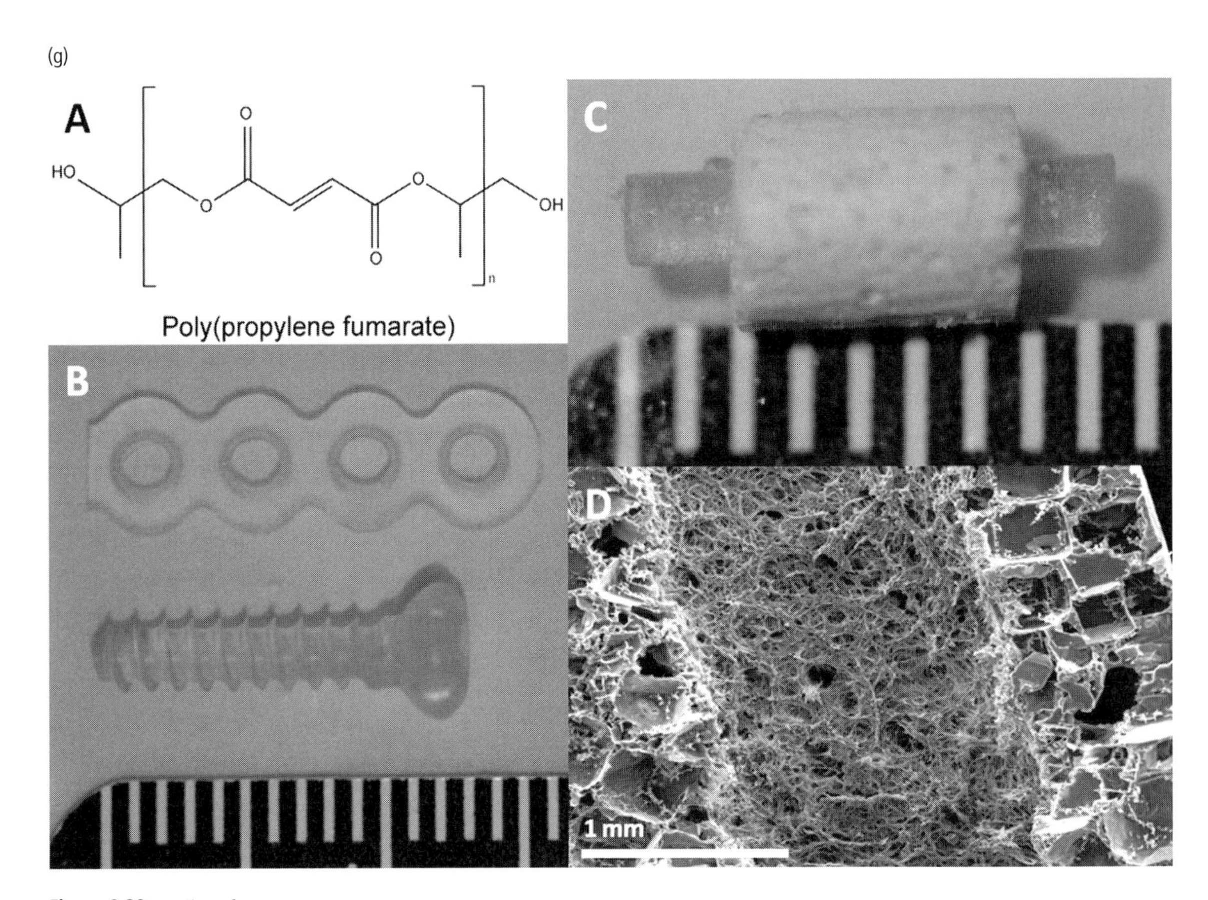


**Figure 8.29** *continued*
(g) Fumarate polymers. These illustrations describe poly(propylene fumarate) (PPF) and PPF-based implants. (A) Chemical structure of PPF. (B) Degradable orthopedic plate and screw implant geometries fabricated via photo-cross-linking of PPF through a transparent silicone mold. (C) A composite PPF tissue engineering scaffold comprising a solid PPF intramedullary rod and a porous PPF sleeve for mechanical stabilization and tissue ingrowth in segmental long bone defects. (D) Scanning electron micrograph of a cross-section of a porous PPF sleeve (as appears in (C)), where the porosity was introduced via a particulate leaching technique employing NaCl crystals (300–500 μm). Tick marks in (B) and (C) represent millimeters; scale bar in (D) represents 1 millimeter. Images courtesy of Dr. Kurtis Kasper and Professor Tony Mikos, Department of Bioengineering, Rice University, Houston, Texas.

polymerization, in which case it is usually referred to as polylactide. Lactic acid itself has some trouble taking part in chemical reactions because the hydroxyl group (-OH) is so close to the carboxylic group (-COOH). It may, however, form a dimer by the combination of two molecules, which forms a cyclic compound, called lactide. Glycolic acid is a even simpler hydroxyl acid (Figure 8.29(b)). It does not exist in isomeric forms and so the chemistry is simpler. As with the lactic acid polymers, we have the same terminology issues, with either polyglycolic acid or polyglycolide being used.

A wide range of polymers and copolymers based on the combinations of the different lactides and glycolides have been developed over the last 50 years. We are concerned here primarily with the poly (L-lactide)s, poly(D-lactide)s, poly(DL-lactide)s, polyglycolides, L-lactide/DL-lactide copolymers, and DL-lactide/glycolide copolymers. In addition, they may be used in conjunction with caprolactone,

(h)

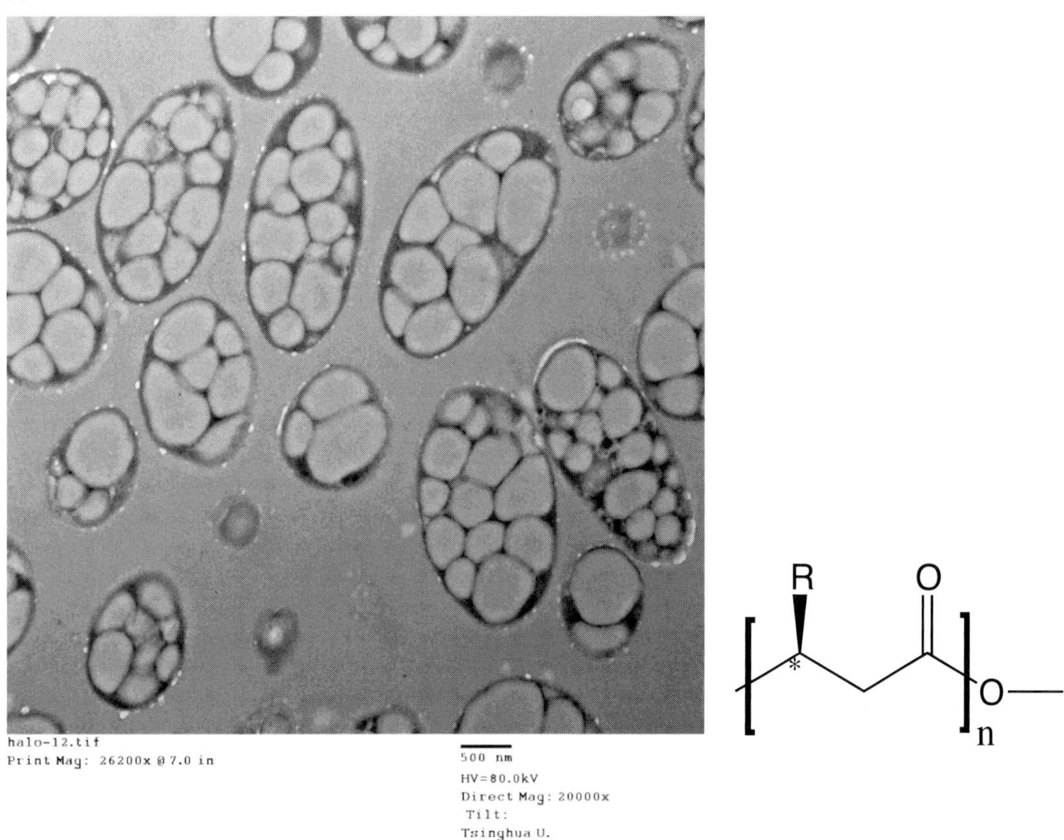


**Figure 8.29** *continued*

**(h)** Polyhydroxyalkanoates (PHA). These polymers are stored as intracellular granules by many bacteria under conditions of excess carbon source supply and limited nitrogen or phosphorous (upper picture, taken by Tan Dan of Lab of Microbiology at Tsinghua Univ, China). PHA have a common molecular structure as illustrated (lower picture drawn by Guo-Qiang Chen of Lab of Microbiology at Tsinghua Univ, China). The carbon atom with a star sign indicates the chiral center of the polymer. R is highly variable. So far, over 150 different PHA (with different R) have been reported.

discussed below, giving, for example, L-lactide/$\epsilon$-caprolactone copolymers, D-lactide/$\epsilon$-caprolactone copolymers, DL-lactide/$\epsilon$-caprolactone copolymers and glycolide/$\epsilon$-caprolactone copolymers. With the polylactide materials, they may be referred to by the PLA shorthand, with PLLA, PDLA being used for the homopolymers of the L and D forms respectively and PDLLA for polymers of mixed enantiomers. Enantiomerically pure PLAs are semi-crystalline polymers. PLLA is generally about 40% crystalline, $T_g$ being around 55–60°C and the melting temperature around 180°C. Racemic PDLLA, that is containing equal amounts of L and D forms, is a completely amorphous polymer. Because of this, PLLA tends to have higher tensile strength. However, the mechanical properties of both PLLA and PDLLA can be varied quite considerably by modifying the molecular weight and crystallinity, typically by annealing processes. Tensile strengths in the range 20–70 MPa can be achieved. The tensile modulus of the semi-crystalline PLA is in the region of 3GPa. Elongations are usually below 10%, which implies

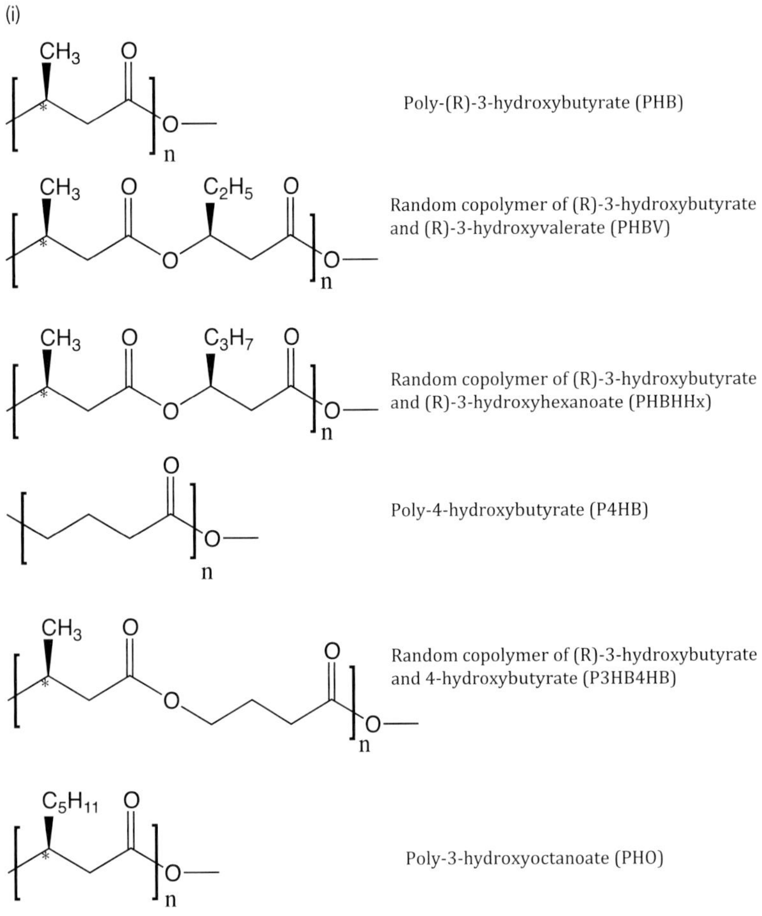

(i)

Poly-(R)-3-hydroxybutyrate (PHB)

Random copolymer of (R)-3-hydroxybutyrate and (R)-3-hydroxyvalerate (PHBV)

Random copolymer of (R)-3-hydroxybutyrate and (R)-3-hydroxyhexanoate (PHBHHx)

Poly-4-hydroxybutyrate (P4HB)

Random copolymer of (R)-3-hydroxybutyrate and 4-hydroxybutyrate (P3HB4HB)

Poly-3-hydroxyoctanoate (PHO)

**Figure 8.29** *continued*
**(i)** Structure of main forms of PHA that are of biomedical interest. Images for (h) and (i) courtesy of Guo-Qiang Chen of Lab of Microbiology at Tsinghua Univ, China.

that PLA is rather brittle. PLA is also hydrophobic, with a contact angle of around 80°. This does inhibit cell attachment and activity at the surface, which may be a disadvantage in some applications.

With respect to copolymers, there are several options. We may start with the copolymers derived from the two enantiomers of the lactides, with compositions usually in the range 70/30 L-lactide/DL -lactide to 80/20 L-lactide/DL-lactide. Copolymers of lactide/glycolide range from 10/90 L-lactide/glycolide to 85/15 L-lactide/glycolide, and from 50/50 DL-lactide/glycolide to 75/25 DL-lactide/glycolide. Copolymers with caprolactone are typically in the ratio of 70/30 L-lactide/caprolactone. Other compositions are possible, including 95/5 and 85/15, these intermediate ratios giving intermediate properties. The lactides/glycolide copolymers have tensile strengths in the range 40–70 MPa, with elongations usually below 6%. Tensile moduli are in the same range as the homopolymers, between 3 and 3.8 GPa. The lactides/caprolactone copolymers have much lower moduli, in the region of 0.03 GPa, and lower tensile strengths of 20 MPa, but elongations over 100%.

PLA degrades through the hydrolysis of backbone ester groups and the degradation rate depends on the degree of crystallinity, the molecular weight and its distribution, morphology, water diffusion rate into the polymer, and the stereoisomeric content. The amorphous parts of the polymer have a higher rate of water uptake and a higher rate of hydrolysis than the crystalline regions. The amorphous racemic PLA therefore degrades faster than PLLA. The degradation rates are rather slow and could lead to long residence times in the body, up to many years in some situations. One of the interesting points about PLA degradation is that the process is essentially random in terms of which ester groups are hydrolyzed at any one time, which means that the material suffers a gradual reduction in molecular weight over a long period of time, but without any appreciable mass loss for most of the time. However, with solid masses of polymer, it may be noticed that the inside degrades faster at the later stages of the process leading to a buildup of compounds containing carboxylic end groups. These cannot easily diffuse to the surface and contribute to an autocatalytic reaction through acid-catalyzed chain-end scission in the center.

Polyglycolide or polyglycolic acid (PGA) is the analogous polymer to PLA, and is the simplest of all aliphatic polyesters. It has a lower $T_g$ but a higher melting temperature and has crystallinity in the region of 45–55% depending on molecular weight. This gives the polymer good mechanical properties. The polymer degrades by random chain scission of the ester backbone and will lose strength over a few months and all mass within a year. Because of its initial strength and fast rate of degradation, coupled with very good fiber forming ability it has been used in surgical sutures; indeed the first FDA-approved synthetic surgical sutures were made from PGA.

Poly(lactic-co-glycolic acid), or PLGA, is probably the best known biodegradable platform, with widespread commercial uses. These polymers are usually prepared by the random co-ring-opening polymerization of lactide (either L or DL forms) and glycolide dimers, again using a variety of possible catalysts. The result is a copolymer with ester linkages between lactic and glycolic acid units. Most often it is the racemic DL mixture of PLA that is used, and the ratios of lactide to glycolide may vary from 10/90, 50/50, 70/30, 75/25 and up to 85/15. The variety of properties that are obtainable with PLGA gives this platform a strong position in the field of resorbable medical polymers; this is not surprising since they should be able to combine the better characteristics of the individual homopolymers, giving blends of strength and degradability suitable for many situations. In general, the 50/50 copolymers degrade over a couple of months, with the 75/25 doing so in 4–5 months and 85/15 in 5–6 months. Again the degradation is produced by bulk hydrolytic attack on the ester backbones.

As noted above, there are different methods of polymerization of the lactic acid. One is the direct polycondensation route in which lactic acid in aqueous solution is converted directly into polylactic acid, in the presence of a catalyst, through the elimination of water. The resulting polymers tend to have low molecular weight and low mechanical properties. They may be used for consumer goods but not, in general, for high-performance applications. The more popular method for biomaterials applications is that of ring opening polymerization. This involves the generation of the cyclic dimers and then the opening of the rings and their subsequent attachment to each other, under high temperature and low pressure, in the presence of a suitable catalyst. These methods produce high molecular weights, a high conversion rate (or yield) and very attractive properties. The choice of catalyst is important and many are available. For many years metal-derived catalysts were preferred, including those based on iron, tin and zinc. Developments in catalyst systems, especially with organocatalysts have become significant, including guanidines and amidines. Different catalysts and different variations of the ring opening polymerization mechanism give rise to different characteristics of the resulting polymer chain and the material properties. Among the more significant features are the molecular lengths and the nature of the groups that terminate the molecules. These end groups can include those of carboxylic, alkoxide, hydroxyalkyl or acidic nature.

### Class 3.7.2 Polycaprolactone

Caprolactone is a cyclic ester, $C_6H_{10}O_2$, with the structure given in Figure 8.29(c). There are different forms depending on the position of the oxygen atom in the ring, and these are described by Greek letters. The most widely encountered form, and that used for the synthesis of biodegradable polymers, is epsilon caprolactone ($\epsilon$-caprolactone). Polycaprolactone is also a semi-crystalline polymer that can be synthesized by ring opening polymerization. It has a very low $T_g$, at −60°C, which gives it a low tensile strength of around 20 MPa and an elongation in excess of 600%. The degradation mechanism is the same but is slow, typically taking more than a few years for complete mass loss to occur. For this reason, caprolactone is often copolymerized for use in medical devices. The caprolactone platform here is typified by a poly(L-lactide/caprolactone) at a ratio of 70/30. These polymers have faster degradation times and also much better mechanical properties.

### Class 3.7.3 Polydioxanone

Polydioxanone is very similar to other polylactones and was initially specifically developed for use in absorbable sutures. It is synthesized from the monomer paradioxanone via ring opening polymerization. It is partly crystalline and, because of the presence of the ether group in the backbone, is very flexible. Degradation takes place over a matter of months, somewhere between PGA and PCL.

### Class 3.7.4 Poly(ortho esters)

Poly(ortho esters) were initially designed as surface erodible polymers. They had no cleavable bonds in the backbone but had carboxylic acid groups that could interact with water at the material surface, undergo acid disassociation and slowly dissolve. There have been many forms of poly(ortho ester) developed over many years, very few of which have achieved success because of the difficulty of balancing mechanical properties with erosion/degradation rates. The ortho ester linkages are often acid sensitive and degradation may be difficult to control because of the autocatalytic effect associated with the generation of acidic by-products. Some poly(ortho esters) are copolymerized with lactic or glycolic acids, where the lactic or glycolic acid segments hydrolyze first, generating polymer fragments that contain a carboxylic acid end group which catalyze ortho ester hydrolysis. The polymers are rather hydrophobic and degradation takes place over a long time period.

### Class 3.7.5 Polyanhydrides

Polyanhydrides are synthesized from diacid monomers; they comprise a large family of degradable polymers (Figure 8.29(d)). They may be aromatic or aliphatic and the anhydride group may be either in the backbone or as side groups. The degradation profile of these polymers varies considerably with the structure. Generally the polymers of greatest value degrade by hydrolysis of anhydride groups in the backbone, with rates that are easily tuned by control of the molecular weight; the polymer may start degradation by surface erosion and then change to bulk erosion. The rate tends to be quite fast and care has to be taken over storage to avoid premature degradation. They provoke minimal tissue responses and are readily eliminated. They are also prepared by facile methods and can be compounded with an array of bioactive agents. The mechanical properties, however, are poor. Polyanhydrides are used for drug delivery applications in critical situations, for example the release of the chemotherapeutic agent BCNU to the brain for the treatment of tumors.

### Class 3.7.6 Polyketals

Polyketals are polymers that have ketal linkages in their backbone (Figure 8.29(e)). These linkages hydrolyze, with kinetics that are strongly pH dependent. Some polyketals have been specifically designed for drug delivery. Poly(1,4-phenylene acetone dimethylene ketal), or PPADK, degrades to benzene dimethanol. Poly(cyclohexane-

1,4-diyl acetone dimethylene ketal) degrades into acetone and 1,4-diylhexanedimethanol, both of which have excellent toxicity profiles. These polymers have attractive properties for intracellular drug delivery since the hydrolysis takes place much faster at the low phagosomal pH.

### Class 3.7.7 Sebacate polymers

There are several degradable polymers based on sebacic acid. Poly(glycerol sebacate) (Figure 8.29(f)) is a biodegradable thermosetting elastomer that degrades by surface erosion. Degradation rates and mechanical properties are very variable depending on the polymerization conditions and the nature of cross-linking. Polycondensation of sebacic acid and glycerol can produce a lightly cross-linked polymer that is very hydrophilic, degrades *in vivo* in a matter of months and has a tensile strength of around 1 MPa and an elongation of 300%. Poly(polyol sebacate)s are similar in many ways, being obtained by reacting sebacic acid with polyols in a polycondensation manner. Polyols that could be used include sorbitol, xylitol and mannitol. Randomly cross-linked networks are produced, whose properties are readily tunable. Young's modulus can vary from less than 1 to more than 300 MPa and $T_g$ can range from 5 to 45°C. Degradation occurs by hydrolysis, the end-products being potentially metabolized since the polyols are intermediates in carbohydrate metabolism and sebacic acid in fatty acid oxidation.

### Class 3.7.8 Fumarate polymers

Fumaric acid is a naturally occurring metabolite found in the Krebs cycle that can be used as the basis of a series of cross-linkable polymers (Figure 8.29(g)). It is usual to use this structure as part of a copolymer, good examples being polypropylene fumarate and poly(propylene fumarate-co-ethylene glycol). These polymers can be cross-linked in similar manner to methacrylates using benzoyl peroxide as an initiator and can be used as an injectable form with both thermal and light activation. Mechanical properties are not ideal but they can be improved with nanocomposite formulations, such as those involving alumina or carbon nanotubes. The degradation occurs by hydrolysis of the ester groups in the backbone. One interesting variation on these polymers is the oligo(poly(ethylene glycol)fumarate), which can be formulated as a hydrogel that supports cell adhesion and proliferation.

### Class 3.7.9 Cyanoacrylate polymers

Cyanoacrylate is the generic name for a group of acrylic polymers that have exceptional properties as adhesives, being widely used in domestic and industrial applications. Various forms are used, such as methyl-2-cyanoacrylate, ethyl-2-cyanoacrylate and 2-octyl cyanoacrylate. They have been employed as surgical adhesives, both in periodontal surgery and now as a fast-acting skin adhesive. These applications were first used several decades ago. The tissue reactions to the first products could be quite significant, but improvements to formulations have given better clinical outcomes.

### Class 3.7.10 Degradable polyurethanes

Polyurethanes have been discussed in several places in this chapter. In recent years, there have been some developments in intentionally degradable polyurethane formulations. We have noted before that thermoplastic polyurethanes are usually prepared from a diisocyanate, a chain extender and a macrodiol, which produce segmented copolymers with alternating soft and hard segments. Intentional degradation can be designed into polyurethane structures by manipulation of the soft segment chemistry. These segments can incorporate polycaprolactone, polylacides and polyglycolides and PEG. Since the overall properties of the polyurethanes depends on the nature and balance of soft and hard segments, it is necessary to carry out soft segment modifications with great care. Chain extenders with hydrolysable groups, for example with DL-lactic acid/PEG structures, represents an attractive option to achieve this balance,

### Class 3.7.11 Polyhydroxyalkanoates

Polyhydroxyalkanoates (PHAs) are polyesters that are involved in energy storage processes in bacteria. They may be extracted from bacterial cultures, purified and processed as thermoplastics. Several decades ago a great deal of interest was focused on these polymers as potential non-petroleum-based plastic materials. Widespread industrial uses have not followed, but one very attractive feature, that of controlled biodegradability, has ensured continued interest from a biomaterials perspective.

The general molecular structure of PHAs is shown in Figure 8.29(h). Many bacteria synthesize and accumulate PHAs for carbon and energy storage, this process being mediated by PHA synthesases, the precise structure produced being dependent on the bacteria, the culture conditions and the media composition. They are usually derived from hydroxyalkanoic acid monomers that have a variety of aliphatic or aromatic side groups. The figure shows a group of PHAs, including poly(3-hydroxybutyrate), poly(4-hydroxybutyrate), poly(3-hydroxyhexanoate) and poly(5-hydroxyvalerate). The molecular weight ranges from 200 000 to 3 million daltons. The polymers accumulate in bacterial cells as granules. The polyhydroxybutyrates (PHBs) are the most widely used; they typically have a $T_g$ of 15°C, a Young's modulus of 3.5 GPa, a tensile strength of 40 MPa and an elongation of 5%. They are quite highly crystalline.

If solid PHAs are buried in soil they will be broken down by bacteria; hence the interest in environmentally compatible biodegradable plastics. However, degradability is not guaranteed in any biological environment; indeed it requires specific conditions for PHAs to exhibit biodegradation. Simply placing a high molecular weight PHB in a subcutaneous or intramuscular site does not yield rapid degradation. It has been necessary to tune the molecular structure, molecular weight and crystallinity in order to produce desirable degradation kinetics for implantable and tissue engineering applications.

Although poly(3-hydroxybutyric acid) received most attention in early days, a small group of homopolymers and copolymers has emerged as possessing superior properties for these applications, as indicated in Table 8.9. Poly(4-hydroxybutyric acid) has far greater elasticity – up to 1000% – and faster degradation. Poly(3-hydroxybutyrate-co-4-hydroxybutyrate) is also more degradable and has an elasticity of more than 100%. Poly(3-hydroxybutyrate-co-valerate) at valerate levels of between 5 and 12% has very attractive rates of degradation. Poly(3-hydroxybutyrate-co-hydroxyhexanoate), known as PHBHHx, has good combinations of properties; it degrades over a period of 6–9 months *in vivo*, preferentially through amorphous regions, by both hydrolysis and enzyme-assisted processes.

## 8.4.8 Class 3.8 Water-soluble polymers

This is one of the most difficult classes to construct and define, but its existence is absolutely essential

Table 8.9 **Forms of polyhyroxybutyrates**

| Composition | $T_m$°C | Tensile strength (MPa) | Elongation (%) | Degradation |
|---|---|---|---|---|
| PHB | 177 | 43 | 5 | Slow, several years |
| P(HB-co-10% HV) | 150 | 25 | 20 | Medium, several months |
| P(HB-co-20% HV) | 135 | 20 | 100 | Faster, few months |
| P(HB-co-10% HHx) | 127 | 21 | 400 | Faster, few months |
| P(HB-co-17% HHx) | 120 | 20 | 850 | Fast, several weeks |

when we take into account many of the developments in polymeric biomaterials in recent years. To allocate a material to either a class that is called soluble or to a class that is called insoluble is fraught with danger as solubility almost always depends on the conditions; this is especially true within the physiological environment. Many general engineering and construction disciplines have utilized classical, simple, water-soluble polymers for a long time, in water-based paints, detergents, surfactants and so on. What we have in mind here, however, are those substances that are polymeric in nature and exist as water-soluble entities to facilitate drug or gene delivery or the dispersal of agents for imaging systems. These entities could be (as discussed in Chapter 6) polymeric drugs themselves, polymer–drug combinations, polymer–protein conjugates, micelles, dendrimers and polyplexes. These agents may often be considered to be at the interface between macromolecules and nanoparticles; this focuses our attention on the difficulty of characterizing a substance as water soluble since with materials at the nanoscale, the solubility becomes a magnitude which varies with the size and shape of the solid; the solubility of a compound in a solvent at a given temperature cannot be considered as a thermodynamic characteristic of the system. It is too early in the development of these polymers to give a detailed hierarchy, and just a few examples are provided. It should also be mentioned that several polymers that could be listed here are classified elsewhere, such as PEG.

### Class 3.8.1 Polyethylenimine

Polyethylenimines (PEIs) are highly water-soluble polymers based on the repeating unit of ethylamine (Figure 8.30(a)). They are basic, positively charged aliphatic polymers containing primary, secondary and tertiary amino groups; molecular weights can range from 700 Da to 1000 kDa and they exist in both linear and branched forms. There is great interest in the use of PEIs as gene carriers, as discussed in Chapter 6. As a non-viral vector, the polymer has the potential to carry genes with less toxicity and risk than viral vectors, although with lower efficiency. Even so the cationic nature of PEIs does carry some risk of dose-related toxicity and much of the development work has been involved with trying to balance efficiency and toxicity risk through manipulation of the chemistry. The emphasis has been on modification to the polymer backbone to reduce the positive charge. The linear form of PEI, obtained from polyisoxazoline precursors, have high transfection potency at a molecular weight around 22 kDa. With branched PEIs acetylation and succinylation may show good efficiency and reduced toxicity by screening the charge on the polymer. PEGylation may have the same effect. Generally the PEI products have been quite effective with plasmid-based gene delivery but less effective with siRNA-mediated gene silencing. In aqueous solution, PEI condenses DNA and the resulting positively charged complex interacts with the negatively charged cell membrane and is readily internalized. It has substantial buffering capacity inside the cell and within endosomes, facilitating endosomal escape.

### Class 3.8.2 Hydroxypropyl methacrylamide

Hydroxypropyl methacrylamide (HPMA) has been used for some time in various forms as water-soluble carriers for drugs and DNA. This use has centered on the conjugation of HPMA with anti-cancer drugs such as doxorubicin. The properties can be varied by incorporation of enzyme cleavable groups in the backbone of the copolymers. The material has also been used as a blood plasma expander.

### Class 3.8.3 Polyvinylpyrrolidone

Polyvinylpyrrolidone is the water-soluble polymer of N-vinylpyrrolidone (Figure 8.30(b)) that has also been used as a blood plasma expander. Also known as Povidone it has a number of ancillary medical uses. More recently it has been explored as a vehicle for the conjugation of drugs and has been found to be one of the best polymers with respect to increasing the residence time of drugs in the circulation.

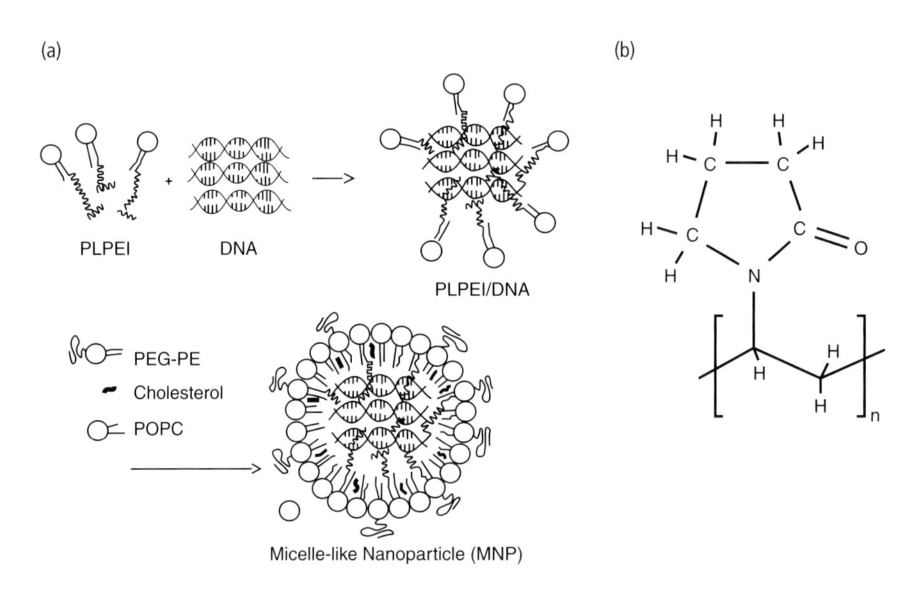
(a)
PLPEI
DNA
PLPEI/DNA
PEG-PE
Cholesterol
POPC
Micelle-like Nanoparticle (MNP)
(b)

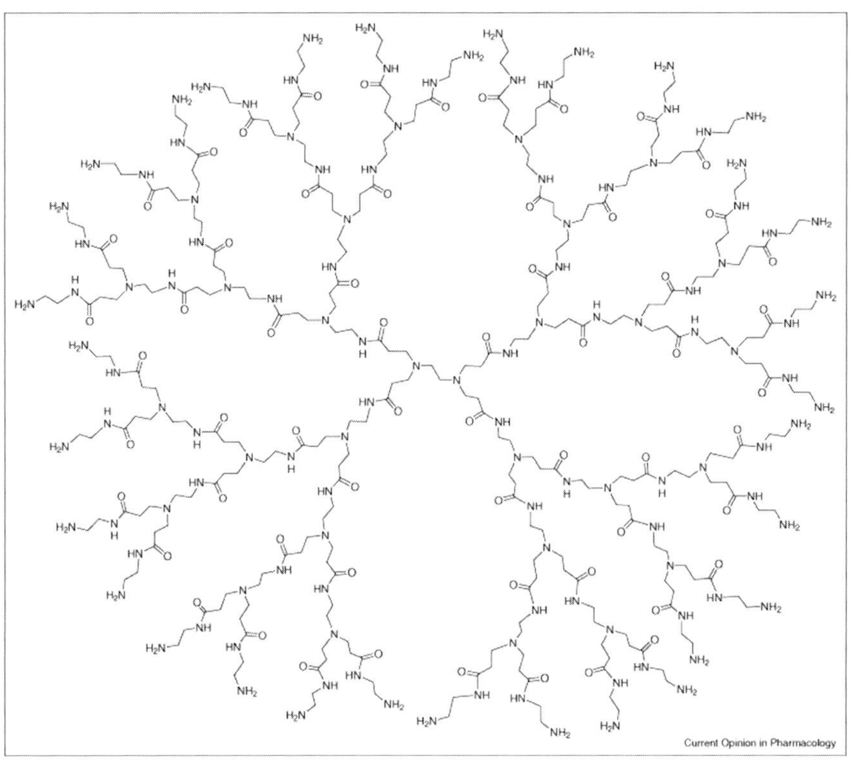
(c)
Current Opinion in Pharmacology

### Class 3.8.4 Polyamidoamines

Polyamidoamines (PAMAM) may be prepared as well-controlled dendrimers with precise numbers of terminal amino groups at the various branching levels (Figure 8.30(c)). PAMAM–DNA complexes can mediate relatively efficient transfer of plasmid DNA into various cell types *in vitro*.

## 8.4.9 Class 3.9 Polymers with ionizable or ionic groups

### Class 3.9.1 Conducting polymers

Although we noted in Chapter 2 that organic polymers generally have extremely low electrical conductivity, there is a small group of polymers that defy this position and have very interesting electrical properties. These are often called conducting polymers (CPs) or electroactive polymers. These polymers have chains of alternating double- and single-bonded $sp^2$ hybridized atoms, which are generated by an electron cloud overlap of p-orbitals to form $\pi$ orbitals, giving a so-called conjugated system, which imparts metal-like semiconducting properties to the polymers. It is particularly important for effective conductivity that the conformation of the double-bond system maximizes sideways overlap between these $\pi$ orbitals. A pure polymer of this type is a semiconductor rather than a conductor. Significant conductivity is induced if the band gap between conducting band and valence band is minimized and this is achieved through the process of doping.

Doping involves either oxidizing (p-doping) or reducing (n-doping) the neutral polymer and adding either a counter anion or cation. Electrons in one repeat unit of the chain are attracted to nuclei in adjacent chains, giving charge mobility both within and between chains and the formation of charge carriers. It is the movement of these charge carriers along the polymer backbone that endows the material with electrical conductivity. Generally one of these polymers without doping will have conductivity in the region of $10^{-10}$ S/cm, which, it will be recalled from Chapter 2, indicates that they are essentially insulators. After doping the conductivity can be raised to over $10^2$ S/cm, which places them, just, into the category of conductors.

One of the first CPs to be developed was polyacetylene, but it is not very stable. More success was achieved with cyclic polyenes, such as polyphenylene, where the aromatic groups give better stability. More recently, and with even better success, polyheterocyclic polymers such a polyaniline and, especially, polypyrrole, have become popular. These polymers can be synthesized either chemically or electrochemically, the latter generally being easier and capable of producing thin films. Dopants include iodine (as $I^-_3$ or $I^-_5$), $BF_4$, $ClO^-_4$ and $SO_3CF^-_3$, and also some large molecules such as hyaluronic acid or polystyrene sulfonate.

It has been shown that many CPs are able to modulate a variety of biological phenomena, especially in tissues and organs that are electrically

**Figure 8.30** Non-viral gene vectors.
(a) Self-assembly of micelle-like nanoparticles involving PEI. This example shows the formation of nanoparticle micelles by condensing plasmid DNA with a conjugate of phospholipid and polyethyleneimine (PLPEI) and coating these complexes with an envelope of a lipid monolayer (POPC) containing PEG, this layer providing *in vivo* stability, protecting the DNA payload. Reprinted from *Journal of Controlled Release*, 133, Ko Y T *et al.*, Self-assembling micelle-like nanoparticles based on phospholipid-polyethyleneimine conjugates for systemic gene delivery, 132–8, © 2009, with permission from Elsevier.
(b) Structure of polyvinylpyrrolidone.
(c) Structure of PAMAM dendrimers. Poly(amidoamine), or PAMAM, is the most well-known dendrimer. The core of PAMAM is a diamine, usually ethylenediamine, which is reacted with methyl acrylate, and another ethylenediamine to make the so-called generation-0 PAMAM. Successive reactions create higher generations, which tend to have different properties. Medium sized, G-3 shown here, or G-4 have internal space that is essentially separated from the outer shell of the dendrimer, this space being potentially suitable for holding drugs. Reprinted from *Current Opinion in Pharmacology*, 6, Najlah M and D'Emanuele A, Crossing cellular barriers using dendrimer nanotechnologies, 522–7, © 2006, with permission of Elsevier.

responsive. The potential to transfer charge from a physiological reaction, to store and release biologically active molecules according to physical stimuli, and to modulate DNA synthesis and protein secretion indicates significant potential in many medical technology applications, including biosensors, tissue engineering scaffolds and neural probes/electrodes. They may be employed in a variety of forms including thin film coatings and as hydrogels.

**Class 3.9.1.1 Polypyrrole** Polyprrole, or PPy, whose structure is shown in Figure 8.31 is the most widely investigated CP with respect to biomedical applications. Its conductivity ranges between 40 and 200 S/

| Conductive Electroactive Polymers (CEPs) and their bulk Conductivities (S cm-1) | Molecular Structure |
|---|---|
| Polyacetylene (PA)<br>1000 | 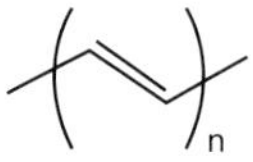 n |
| Poly(paraphenylene) (PP)<br>100-500 | 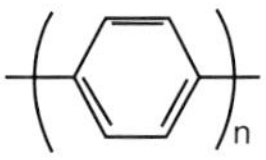 n |
| Poly(para-phenylene vinylene) (PPV)<br>3 | 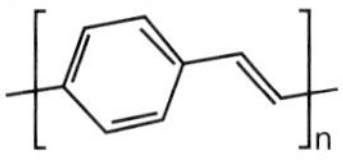 n |
| Poly(para-phenylene sulfide) (PPS)<br>1-100 | 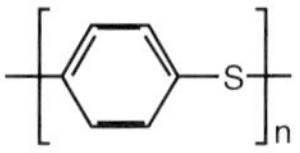 S n |
| Polyaniline (PAn)<br>1-100 | 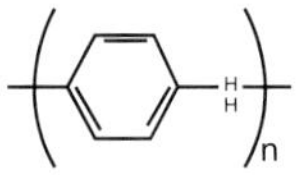 H H n |
| Polypymole<br>(PPy) | 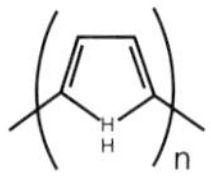 H H n |
| Polythiophene (PTh)<br>10-100 |  n |
| Polyazulene<br>0.1 | 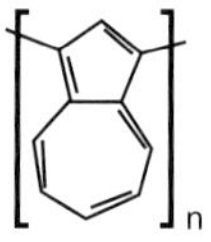 n |
| Poly(3,4-ethylenedioxythiophene) (PEDOT)<br>10-100 | 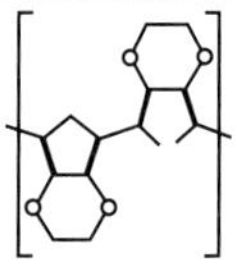 |

**Figure 8.31** Conductive electroactive polymers. The molecular structure of electroactive polymers of biomedical interest. Reprinted from *Biomaterials*, 31, Guiseppi-Elie A, Electroconductive hydrogels: synthesis, characterization and biomedical applications, 2701–16, © 2010, with permission of Elsevier.

cm and it is usually p-doped with $CF_3SO^-_3$ or $ClO^-_4$, giving a generic structure of $(C_4H_3N)X_{0.3}$, where X is the dopant. There is evidence concerning its performance in amperometric biosensors involving glucose oxidase, urease and cholesterol oxidase, where the biomolecule may be physically adsorbed or deposited by electropolymerization methods onto the sensor substrate. PPy supports cell adhesion and growth with several different cell types including mesenchymal stem cells. PPY doped with p-toluene sulfonate has good compatibility with many cells and it may be used to modulate cell behavior through electrical stimulation. One limitation with respect to tissue engineering applications is the lack of biodegradability. PPy is an attractive candidate for neurological devices because of its good biocompatibility coupled with the ability to promote ion exchange at electrode–tissue interfaces. PPy doped with sodium dodecylbenzenesulfonate is very effective under these conditions. It may also be doped with peptides and peptide fragments for greater effects.

### Class 3.9.2 Polyelectrolytes

These comprise a general group of polymers that carry either positively or negatively charged ionizable groups that have varying properties that depend on the fraction of dissociated groups, salt concentrations and dielectric constant. They have a wide range of commodity applications such as food additives. More recently they have taken on new significance as polyelectrolyte multilayers, which can be created by layer-by-layer techniques with alternating charged layers.

## 8.4.10 Class 3.10 Elastomers

The principles of elastic deformation, and the basis of elastomeric materials, have been discussed in Chapter 2. There are very many situations, including a number in medical technology, where the elastic deformability of a material is one of the most important functional specifications. The insertion of tubes into channels in the body, in the circulatory and urinary systems for example, provide good evidence for this. The provision of elastomeric biomaterials with the appropriate biocompatibility characteristics is not a trivial issue and only a few are used routinely in clinical practice.

### Class 3.10.1 Silicone elastomers

Silicone polymers differ from virtually all other polymers by virtue of having a backbone based on silicon rather than carbon. Specifically, the backbone consists of repeating Si–O bonds, the Si atoms also being bonded to organic groups (Figure 8.32). The repeating unit is known as a siloxane. By far the most common polymer involves two methyl groups attached to each Si atom, giving polydimethylsiloxane (PDMS), although many other organic groups can substitute for the methyl groups. PDMS has many excellent properties and may be formulated as elastomers, fluids, adhesives, resins and emulsions. The materials have good general biocompatibility, which accounts for much of the interest in them for medical technologies. They have very good stability in many types of environments and very good retention of properties at both extremes of temperature. In the form of elastomers, these materials are readily molded during curing, which accounts for their use as dental impression materials, as a potting compound in electronics and as a substrate for various forms of microprinting and patterning in biological environments.

The PDMS may be converted into the useful products mentioned above through a variety of cross-linking reactions. Silicone adhesives tend to be cured by condensation reactions in which the siloxane chains react with moisture upon release from a sealed tube or cartridge; this takes place at room temperature and the products are described as room temperature vulcanized (RTV). Tin-based catalysts are usually used in this process. Most silicone products, especially those that require high fidelity in casting or molding, are cross-linked by addition reactions, where vinyl end-blocked molecules are reacted with H–Si groups. These are catalyzed by platinum or rhodium complexes.

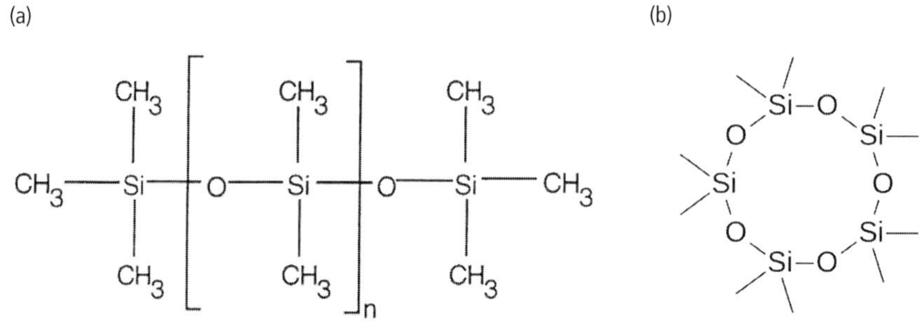

(a) (b)

**Figure 8.32** Silicone polymers.
**(a)** Polydimethylsiloxane, the most widely used silicone polymer. Depending on "n" and the degree of cross-linking, the polymer can exist as a fluid, a gel or an elastomer.
**(b)** Some cyclic oligomers may be present in polymers after synthesis. This is the cyclic oligomer D5. There have been some concerns over the biological activity of such oligomers.

Although PDMS elastomers possess exceptional elasticity, they do tend to have poor strength, including poor tear strength. This disadvantage is ameliorated to some extent by the incorporation of fillers. Fumed silica is a common filler, the reinforcement effect usually being enhanced by a silane coupling agent. In some situations, radio-paque fillers, such as barium sulfate may be added. Depending on the amount and nature of the fillers, such elastomers may have an elastic modulus of 1–5 MPa, a tensile strength of 5–10 MPa and an elongation of 200–1000%. The glass transition temperature is very low, below −100°C, which accounts for the high elasticity seen over a wide range of temperature. Silicones have a low surface tension, which assists in their adhesive properties since they are capable of wetting most surfaces. The molecular structure of the cross-linked polymer provides for a much higher free volume than typical organic polymers and this allows for high solubility and diffusion coefficients for gases, especially with a high permeability to oxygen and water vapor. These characteristics have been responsible, in part, for the use of silicones in gas permeable membranes; for example, with ophthalmological lenses and artificial respirators.

The use of silicones in breast implants is covered in some detail elsewhere in this book. Of relevance to this general discussion on the properties of silicones with respect to this application is the composition of the silicone gels that were used in many of these products. These cross-linked gels contain relatively low molecular weight PDMS, which, of course, give the product the viscosity and conformability that is required. There is no doubt that, depending on the thickness and characteristics of the silicone elastomer shell that contains the gel, the diffusion of gel through the elastomer is possible. This should be minimal and self-limiting although great care has to be taken with the formulation and processing of the gel, since lower molecular weight species are likely to diffuse at a higher rate. As well as PDMS molecules themselves, there may be low levels of oligomers and other species. One possible species is a cyclic oligomer, known as D5, Figure 8.32(b). It has been speculated that such an oligomer could have different biocompatibility characteristics than higher molecular weight species.

PDMS elastomers do have extremely good biostability and biocompatibility under most physiological circumstances, which is one of the main reasons for their very strong track record in medical devices. As with all biomaterials, and as discussed in Chapter 3, biocompatibility can never be guaranteed and will depend on individual circumstances. One possible cause of concern is the fact that although silicones are very hydrophobic, they are lipophilic to some extent, which means that in some physiological situations, PDMS may swell after absorption of lipids, and may crack or suffer a diminution of the mechanical properties.

### Class 3.10.2 Polyurethane elastomers

The family of polymers known as the polyurethanes is one of the most versatile and widely exploited group of materials of any class; they were invented in the 1930s, very early in the period of synthetic polymer development. The fundamental feature of all polyurethanes is the urethane group, but the overall molecular structure is almost infinitely variable. As noted earlier, based on highly variable chain stiffness, interchain attraction, crystallinity, branching and cross-linking, polyurethanes can exist as rigid or flexible foams, surface coatings, fibers, films, rigid plastics or elastomers. Although some other forms have been used medically, it is the elastomeric form that is by far the most important.

Some early forms of polyurethane were based on the reaction between diisocyanates and diols. One prominent example was the linear polyurethane produced by the reaction between 1,4-butane diol and hexamethylene diisocyanate. The NCO group, crucial to the creation of polyurethanes, which is present in the diisocyanate, may also react with compounds containing OH groups, giving rise to this vary large family. If the reactants are bifunctional, then linear polymers will be produced, but higher functionality leads to branched and cross-linked polymers. Increasingly with the development of polyurethane chemistry, the reactants moved from simple molecules such as the diisocyanate and diol to intermediates, which were often polymeric themselves, and which had terminal groups that were capable of further reactions. The effects of these changes included quite different cross-linked structures and different molecular flexibilities. These polyurethanes, which still contained the defining urethane group, therefore often had a majority of linkages that were non-urethane. If the reactions involved urea groups, then polyurethane ureas would form. Common examples of polymeric intermediates are polyesters and polyethers, in which case polyester urethanes and polyether urethanes are produced.

Urethane elastomers are essentially linear block copolymers, with the blocks providing quite different characteristics; because of this they are often called segmented urethanes. The segmented form can take many different forms but generally they can be considered as having three types of structure in the chain, as shown in Figure 8.33(a). First, there are the rigid isocyanate blocks. Secondly, there will be the diol flexible block, and thirdly there will be sections that consist of chain extenders, which can be of variable flexibility. The segmented urethane therefore has some hard, rigid segments, usually quite short, around 150 nm in length, and some soft, flexible segments, usually ten times longer. Cross-links are usually established between the hard segments. There are several alternative synthesis routes, for example there is a one-step process in which polyol, diisocyanate and chain extender are mixed in a reactor, together with a catalyst. On the other hand, a two-step process may be used where the isocyanate and polyol are reacted together to form a so-called pre-polymer, which is converted to the final material by reaction with the chain extender.

There are several possible isocyanates that could be used in the preparation of polyurethane elastomers, but just two types are of real significance here. The first are the toluene diisocyanates, either the 2,4- or 2,6- varieties, TDI. The second are the various forms of MDI, 4,4'diphenylmethane diisocyanate. There is a major difference between these aromatic and aliphatic isocyanates, since the former has a much more significant toxic profile. As noted above, most polyurethane elastomers utilize polymeric diols. Polyesters are usually prepared by the reaction between a dibasic acid such as adipic acid with a diol such as ethylene glycol, diethylene glycol or 1,2-propylene glycol. Polyethers include polypropylene glycol and polytetramethylene glycol. In general, polyester urethanes have poorer resistance to

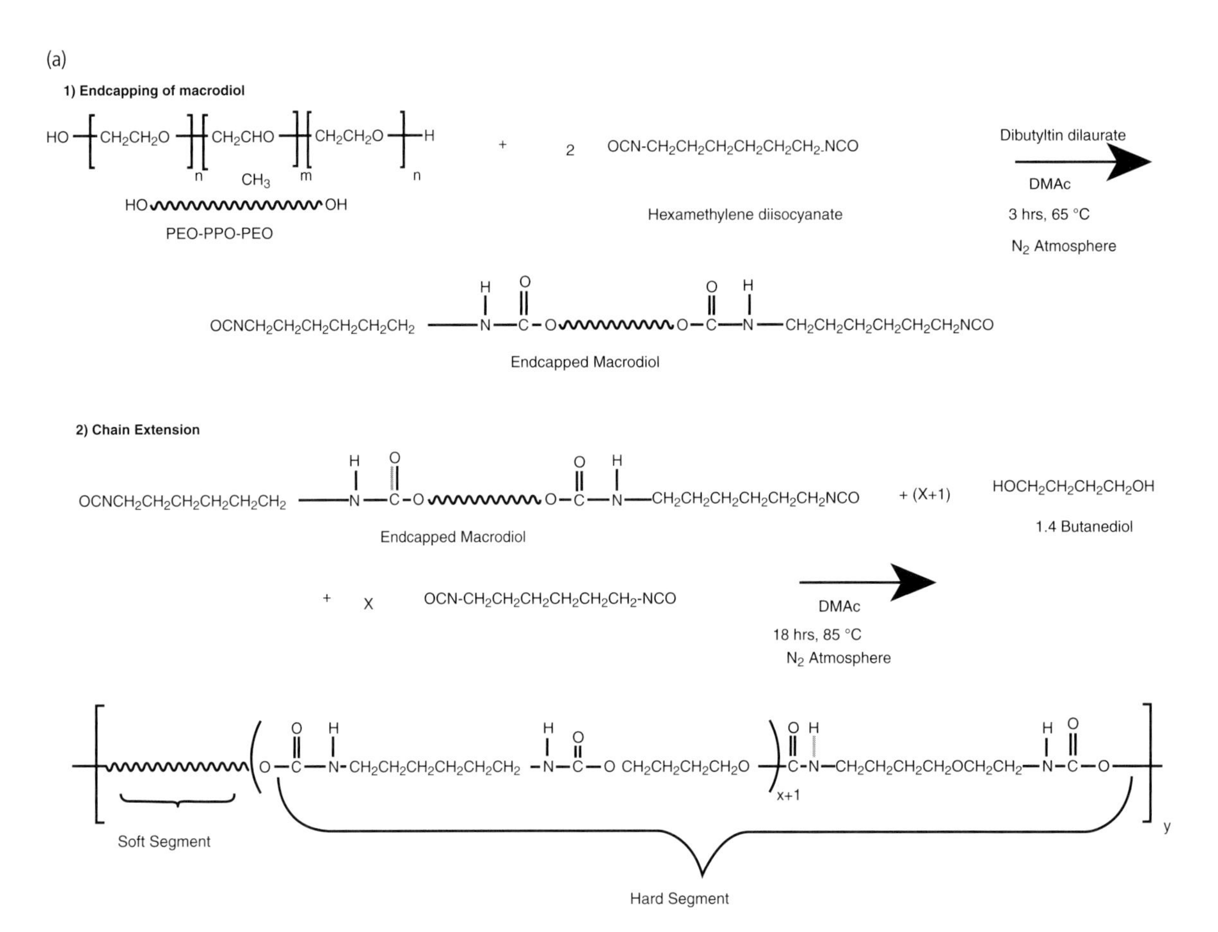


**Figure 8.33** Synthetic elastomers.
**(a)** Segmented polyurethanes. Segmented polyurethanes have soft and hard segments, the nature of which, and their morphological relationship, control properties. Polyurethanes form one important collection of semicrystalline segmented polymers. This scheme shows the two-step solution synthetic route to one high molecular weight variety of this material. Reprinted from *Polymer*, 47, Korley L T J *et al.*, Effect of the degree of soft and hard segment ordering on the morphology and mechanical behavior of semi-crystalline segmented polyurethane, 3073–82, © 2006, with permission of Elsevier.

degradation than polyether urethanes, which has significant implications for long-term implantation. There are many chain extenders in conventional polyurethanes, one of the most common being 1,4 butane diol. Other examples include some aromatic compounds such as diaminobenzene.

Clearly the mechanical properties of materials within this family are going to vary considerably on the basis of the components and their molar ratios. Typical values can be found within the ranges of 20–60 MPa for the tensile strength, elongation at break between 300 and 60% and a modulus of elasticity at 300% strain of 10–30 MPa.

The medical applications of polyurethane elastomers have followed a very interesting pattern with respect to the susceptibility of these materials to degradation and the search for biostable elastomers that can perform for many years in critical implantable situations. It should be of no surprise that polyurethanes do experience some degree of

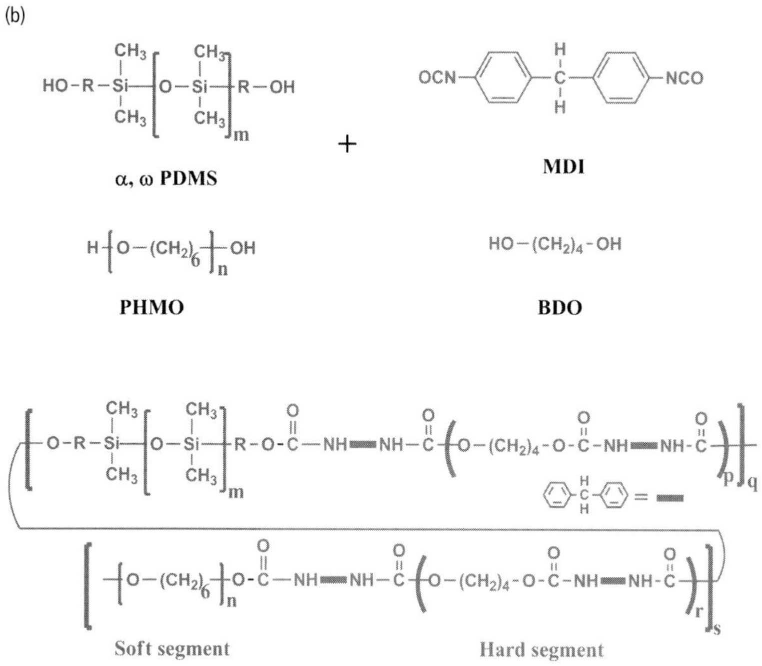

(b)

α, ω PDMS

MDI

PHMO

BDO

Soft segment

Hard segment

**Figure 8.33** *continued*
(b) Urethane–siloxane-based elastomers. This is an example of the synthesis of a highly biostable siloxane-based polyurethane (Elast-Eon™). MDI = diphenylmethane diisocyanate, PHMO = poly(hexamethylene oxide), BDO = 1,4-butanediol. Image courtesy of Dr. P Gunatillake, CSIRO, Melbourne, Australia.

degradation over the long term, in view of the plurality of hydrolysis – susceptible bonds in the varying molecular structures. The situation became very complex since multiple degradation mechanisms are operative in these polymers. The major problems arose with the use of elastomeric insulating materials for cardiac pacing leads; in this situation oxidative degradation, environmental stress cracking and metal ion accelerated degradation all appeared to contribute to the degradation. Because of concerns over degradation, two polyurethane materials, Pellethane and Biomer were removed from medical applications in the early 1990s. The search for more biostable polyurethanes has involved a number of strategies, the task being made more difficult by having to achieve greater stability without compromising the elastomeric properties, many avenues of increasing the resistance to cleavage of polymeric chains becoming closed by the increase of stiffness associated with the changes in the molecule backbone.

One of these strategies has involved the use of polycarbonate diols, giving the so-called polycarbonate urethanes. For example, the reaction between 1,6-hexanediol and ethylene carbonate yields poly(1,6-hexyl 1,2-ethyl carbonate) diol, which can be reacted with an aliphatic diisocyanate and 1,4 butane diol to give a polycarbonate urethane, Chronoflex. These polymers may have tensile strengths over 5000 MPa and an elongation of 600%. A second strategy involves the use of silicones within the structure, which could be considered as urethane–silicone copolymers. One procedure here has involved the use of polydimethylsiloxane as a macrodiol. For example, one family of such polymers incorporate mixed macrodiols of polydimcthylsiloxane and poly(-hexamethylene oxide), with MDI and 1,4-butanediol, known as Elast-Eon (Figure 8.33(b)). These materials, which can be made of different hardness and flexibility, certainly have much improved biostability.

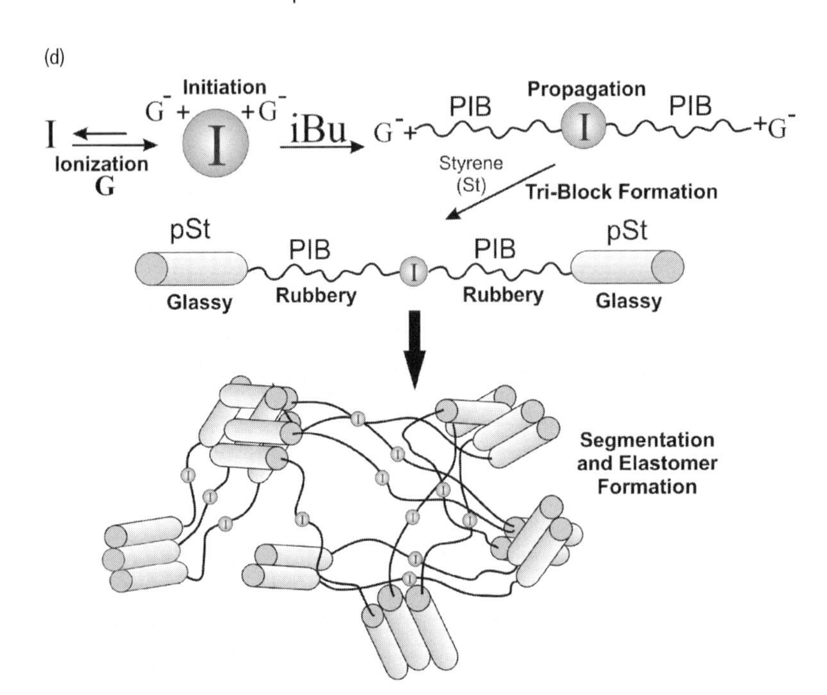


**Figure 8.33** *continued*
(c) and (d) SIBS. The chemical structure of SIBS is shown in (c). A schematic of the polymerization process is shown in (d), with initiation and propagation phases and the morphology of the resulting segmented elastomer. Images courtesy of Dr. Len Pinchuk, Innovia LLC, Florida, USA

## Class 3.10.3 Poly(styrene-block-isobutylene-block-styrene) (SIBS)

Polyisobutylene is a homochain polymer that has alternating secondary and quaternary carbon atoms in the backbone, with hydrogen and methyl side groups (Figure 8.33(c)). It is used in many industrial applications because of its elasticity/viscoelasticity, being synthesized inexpensively from isobutylene, but is not cross-linked and cannot be used to manufacture solid shape-retentive shapes. However, polyisobutylene chains can be bound to hard glassy polystyrene domains to yield a thermoplastic elastomer, poly(styrene-block-isobutylene-block-styrene), known as SIBS, which is essentially a self-assembled physically cross-linked polymer that is thermo- and solution-formable. It is soluble in a variety of non-polar solvents and can therefore be spray-coated or solvent cast. It has a UTS of 10–30 MPa, and an elongation of up to 1000%, the properties depending on the composition, which can range from 5–50 mole percent styrene. The best-known medical use of SIBS is in the drug-eluting coronary stent, where it has been employed as the carrier of paclitaxel. It shows, and indeed was designed to show, excellent long-term biostability, with good blood compatibility and general biocompatibility.

#### Class 3.10.4 Plasticized polyvinylchloride

Much controversy surrounds the use of polyvinylchloride (PVC) in medical technology. It could be argued that it is the most important of all polymers used in health care since it is the most widely used, in applications such as containers for blood and intravenous solutions, heart and lung bypass sets, containers for urine incontinence and ostomy products, catheters and cannulae and tubing for dialysis, endotracheal feeding and pressure monitoring. PVC is a very simple polymer, being produced by the addition polymerization of vinyl chloride. As a pure polymer it would have very good biocompatibility and biostability. The problem arises with the influence of the molecular structure on the mechanical properties. The presence of very small H atoms and much larger Cl atoms as the sole side groups on the molecular chains, where they exist in unequal number, means that it is very difficult for the polymer to crystallize, it is essentially amorphous and the molecular chains find it hard to slide over each other, which limits the flexibility. Although such products have good properties for many industrial uses, this is a severe limitation for the medical devices mentioned above, which require flexibility and suppleness. The solution to this problem has, for many years, been the inclusion of a plasticizer in the formulation, where the low molecular weight, short molecular length substance acts as a molecular lubricant, giving flexibility. There are many potential plasticizers; the most common and certainly the most effective for most situations, is di-(2-ethylhexyl) phthalate, DEHP. The vast majority of PVC medical products have used the DEHP plasticized form; it is very effective and inexpensive.

However, this has not really solved the problem. The very reason for using such a molecule as a plasticizer lies with its molecular mobility. Whilst this may give good flexibility, it is associated with a potential disadvantage: these molecules may diffuse through the polymer matrix because of their small size. Diffusion through the matrix implies that this could result in diffusion to the surface and, because of the inevitable concentration gradient, from the surfaces into any surrounding fluid environment. This could be the blood product in a blood bag, or the tissues of a patient. This is controversial because phthalates are associated with certain biological effects. It is certainly true that they can accumulate in animal tissues when administered to them, and can have some effects on reproductive organs. There is no evidence of any significant adverse effects in human patients but there are concerns that such effects could arise, especially in very young children. There have been many attempts to produce non-phthalate plasticizers, including some phosphates, trimellitates and some polyesters (see Figure 8.34), but it has proved difficult to produce ideal properties. The majority of medical device regulators believe that the clinical benefits of DEHP-PVC outweigh the potential but not demonstrated risks.

### 8.4.11 Class 3.11 Fibers, fabrics and textiles

This is a class of biomaterial that is governed more by the microstructure or morphology of the products than the chemistry of the material, but the applications are sufficiently profound and discrete that a separate classification is justified. The morphology in question concerns the formation of thin polymeric fibers into porous membranes or fabrics. The nearest common term that describes this type of material is "textile," and indeed such products are often termed "medical textiles." A textile is normally considered to be a collection of fibers, filaments or

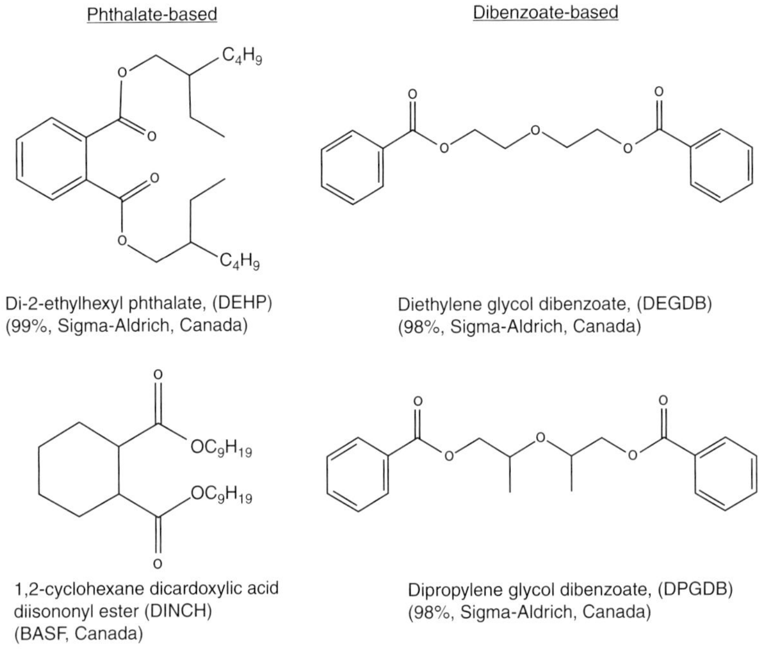

Phthalate-based

Dibenzoate-based

Di-2-ethylhexyl phthalate, (DEHP)
(99%, Sigma-Aldrich, Canada)

Diethylene glycol dibenzoate, (DEGDB)
(98%, Sigma-Aldrich, Canada)

1,2-cyclohexane dicardoxylic acid
diisononyl ester (DINCH)
(BASF, Canada)

Dipropylene glycol dibenzoate, (DPGDB)
(98%, Sigma-Aldrich, Canada)

**Figure 8.34** Polyvinylchloride (PVC) plasticizers. Examples of the main plasticizers used in PVC. DEHP at the top left has had the major impact in medical devices and has been controversial with respect to putative biological effects. Reprinted from *Science of the Total Environment*, 432, Kastner J *et al.*, Aqueous leaching of di-ethylhexyl phthalate and green plasticizers from poly(vinyl chloride), 357–64, © 2012, with permission of Elsevier.

yarns used in making cloth; we need not be too concerned about whether they involve fibers, filaments or yarns, since a medical textile may not, strictly speaking, consist of any of these, nor do we have to equate the products with a cloth. The description is useful, however, in conveying the overall impression of the materials, and in fact, the two main varieties of medical textiles are derivatives of well-known clothing materials, the synthetic polyesters used for shirts and similar garments, and the expanded polytetrafluoroethylene (ePTFE) used for waterproof clothing.

If we assume for the moment that the textile is made conventionally from fibers, there are several ways of forming such fibers into porous macroscopic objects, yielding forms known as wovens, nonwovens, braids and knits. A woven fabric has primary structural yarns that are oriented at 90° to each other, as shown in Figure 8.35(a). The machine direction is known as the warp direction and the cross direction is the weft. There are many different forms of weave depending on how the fibers cross each other. These structures tend to have rather isotropic properties, with high strength and low elongation. Nonwoven fabrics have fibers that are bonded or joined together, using adhesives, thermal effects or mechanical interlocking. Knitted structures are made by looping yarns in horizontal rows and vertical columns, as shown in Figure 8.35(b). This tends to produce a more open structure than found with the woven fabrics and therefore have greater porosity and permeability. Again there are different geometric forms, one of which, the velour form, has an extra yarn in the structure, which gives the fabric a surface texture as these fibers extend from the surface. Braiding involves the twisting of fibers around each other in a quasi-parallel fashion, as shown in Figure 8.35(c).

The fibers themselves can be produced by a variety of processes, generally called spinning processes, which include melt spinning, wet spinning and electrospinning. Many polymers, both synthetic and natural, are amenable to some form of spinning and fabric processing. Since the polymer molecules generally find themselves oriented more or less parallel to the fiber long axis, the strength of the fibers will largely depend on the covalent bonds of polymer chains, resulting in high strength. If these fibers

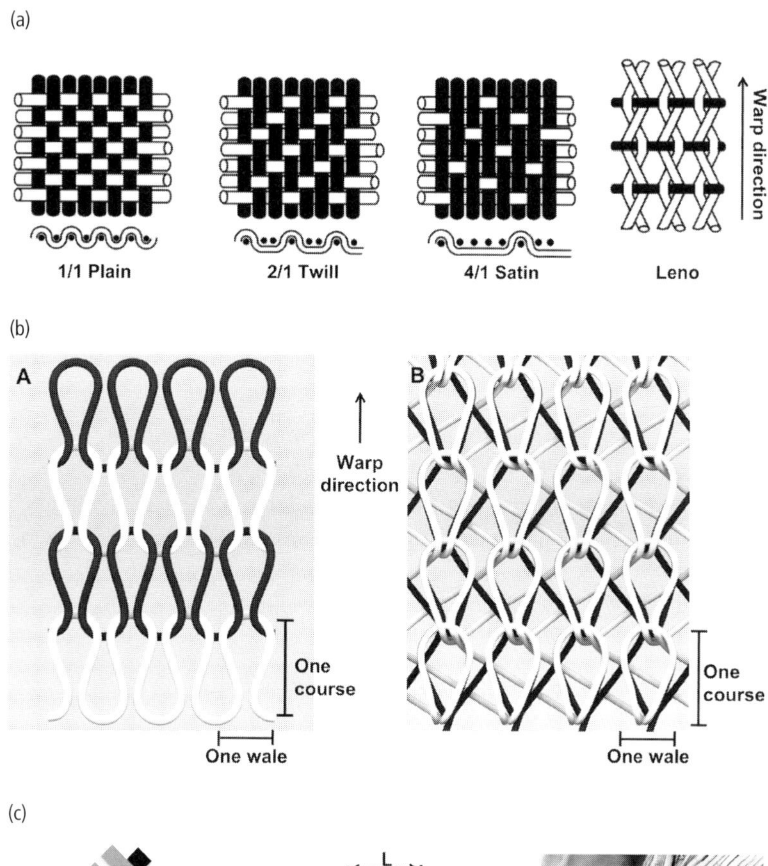


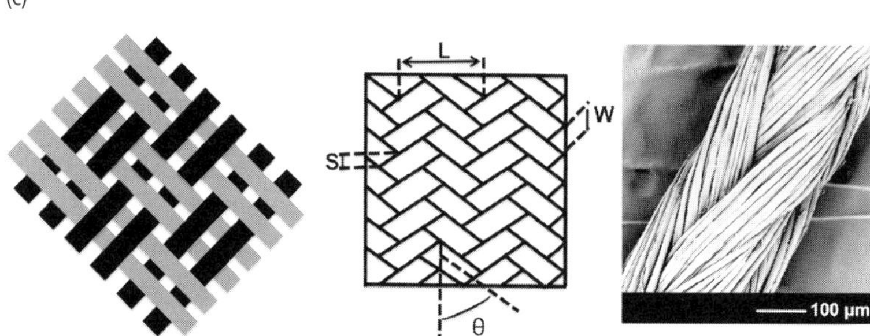


**Figure 8.35** Medical textiles.
**(a)** Diagram showing differences in woven structure and interlacement pattern between plain, twill, satin and leno weaves.
**(b)** Diagram showing differences in knitted structure and stitch pattern between (A) single jersey weft knitted fabric and (B) two bar locknit warp knitted fabric.
**(c)** Diagrams (left) showing the structure, braid angle and pick repeat distance of a 2/2 regular biaxial braided fabric. Scanning electron micrograph of a braided silk suture (right) showing yarns lying in both clockwise and counter-clockwise directions. Images prepared by Mike Freeman, Computer Operations, College of Textiles, North Carolina State University, USA.

themselves are aligned, very high strength materials are formed, including silk, which is discussed elsewhere in the chapter, and synthetic products such as nylon; surgical suture materials provide a good example of high strength spun fibers. The overall properties of the fabric will depend on the final geometrical form of the textile as well as the chemistry of the fibers.

Medical uses of textiles are manifold. Many do not qualify to be described as biomaterials since they are not actively and directly involved with patient care, but some do; these applications include wound dressings and implantable hernia repair meshes. The most important examples are the textiles used in prosthetic vascular grafts, as discussed in the following two sections.

### Class 3.11.1 Polyethylene terephthalate materials

Polyethylene terephthalate (PET) is an aromatic polyester. PET fabric is widely used in the clothing and other industries and is one of those materials that has come to be known by a common trade name, Dacron. The fabric can be made in different forms; in medical uses it is generally used as knitted or woven fabrics, using multifilament yarns. Weaving tends to result in a tighter structure with relatively small spaces between yarns and knitting allows looser patterns with larger spaces. These differences in fabric structure result in different biological performance, for example with respect to tissue ingrowth.

#### Class 3.11.2 Microporous expanded polytetrafluoroethylene

Expanded polytetrafluoroethylene (ePTFE) is also usually synonymous with a proprietary product, in this case Gore-Tex, which is itself based on a material that is best known by a trademarked product, Teflon. PTFE is a very simple polymer, which has attractive properties of inertness and inactivity, as noted elsewhere in this book. It has been used in monolithic form in medical devices, although this has been problematic when used in stress-bearing situations since the polymer, although heat and chemical resistant, is not mechanically strong. The expanded form is microporous, in which the structure consists of a dense meshwork of very fine fibrils, which meet up at so-called nodes (Figure 4.15).

### 8.4.11 Class 3.12 Environmentally responsive polymers

In the medical technology industry, as elsewhere, there has been an increasing demand for materials that can alter structure and properties depending on the conditions in which they are used. We have already seen with shape memory materials that some can do this on the basis of stress-induced structural changes that lead to changes in shape. More often it is environmental changes, especially temperature and pH, that are responsible for these effects. It has become fashionable for such materials to be called "smart" or "intelligent"; these words rather exaggerate what is possible since they imply to make changes to behavior under many conditions, including those that are unexpected. With responsive materials, the changes that may be seen will follow defined mechanisms that occur under defined, and anticipated, conditions. In most situations it is polymers that are used in responsive systems, and most commonly these are in the form of hydrogels, primarily because they can be functionalized and because reactivity can be facilitated by the water that they contain; this usually takes place by control over swelling and shrinking.

#### Class 3.12.1 Thermoresponsive polymers

Thermo-responsive, or temperature-sensitive, hydrogels are able to undergo a reversible volume-phase transition in response to a change in temperature. Such polymers have a transition temperature, known as lower critical solubility temperature (LCST). These hydrogels swell as the ambient temperature is lowered through the LCST and shrink or collapse when the temperature is raised higher than the LCST. One of the most common examples here is poly(N-isopropyl acrylamide), or PNIPAAm.

#### Class 3.12.2 pH responsive polymers

In hydrogel structures that have weak acid or base groups in side groups, branches and cross-links, any movement of water into or out of the network can result in changes to the ionization of these groups. This allows for a change in the degree of swelling on the basis of the pH of the solution in which the hydrogel is placed; the equilibrium degree of swelling for hydrogels containing weak acid pendant groups increases with the pH of the external solution, and vice-versa for hydrogels containing weak basic pendant groups. Examples of pH-sensitive hydrogels include poly(methacrylic acid), polyacrylic acid and polyacrylamide.

## 8.5 Class 4 Carbon materials

For much of the history of materials science, carbon was considered to have two significant allotropes, diamond and graphite. These had rather limited uses from an engineering perspective and were often discussed under the general heading of ceramics. In more recent times, and especially in the last 20 years, several other forms of carbon have been discovered and these materials have become far more important. They therefore warrant a separate classification, and many of these are directly applicable to medical and biological sciences, shown in Box 8.5. In the hierarchy, diamond constitutes one major sub-class (Class 4.1), although we shall see that it exists in several different forms as coatings. There are also various forms of graphitic carbon in use (Class 4.2) including carbon fibers, as well as glassy, vitreous or amorphous carbons (Class 4.3). Most of the new developments have been associated with advances in nanotechnology (Class 4.4), with the fullerenes, nanotubes and graphene. The varying properties of carbon-based materials arise from several factors including the electronic structure of the carbon atoms in these different forms. Carbon generally has the electronic structure $1s^2 2s^2 2p^2$; the differences between the types of carbon are based on the interaction between s and p orbitals.

## 8.5.1 Class 4.1 Diamond and diamond-like materials

Diamond is the form of carbon that is fully crystalline, with a cubic structure, in which each carbon atom is linked to four other carbon atoms arranged tetrahedrally around it (some forms are shown in Figure 8.36). Here a single s orbital and three p orbitals undergo a $sp^3$ hydridization, which gives rise to this tetrahedral relationship. We shall see below that graphite has a different hybridization character. Diamond is extremely hard but also extremely expensive. Although so-called industrial diamonds exist, and may be synthesized from graphite, they do not, in general, have properties relevant to medical technology. Interest in diamond, especially its extreme stability and electronic/optical properties, has led to the development of some analogs which do have medical relevance.

### Class 4.1.1 Diamond-like carbon

Diamond-like carbon is the name given to a variety of amorphous carbon materials, with a varying amount of hydrogen. They contain significant

**Box 8.5 The second level of the classification of biomaterials: carbon biomaterials**

Class 4.1 Diamond and diamond-like materials
Class 4.1.1 Diamond-like carbon
Class 4.1.2 Tetrahedral amorphous carbon
Class 4.1.3 Nanocrystalline and ultrananocrystalline diamond
Class 4.2 Graphitic materials
Class 4.2.1 Pyrolytic carbon
Class 4.2.2 Activated charcoal
Class 4.3 Glassy or vitreous carbon
Class 4.4 Hexagonally bonded carbon nanostructures
Class 4.4.1 Fullerenes
Class 4.4.2 Carbon nanotubes
Class 4.4.3 Graphene

(a)

A

B

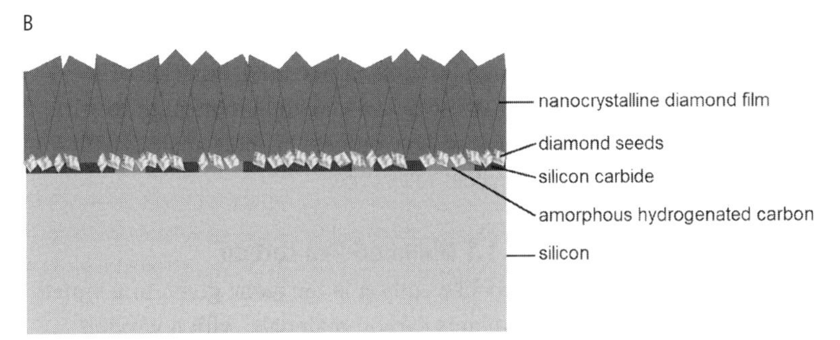


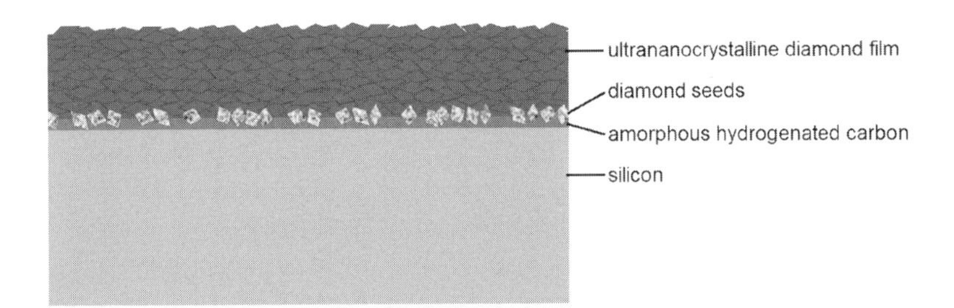


**Figure 8.36** Carbon materials. (a) Diamond. (A) Image of nanocrystalline diamond, with crystal size in the region of 100nm. (B) Nucleation mechanisms for nanocrystalline and ultrananocrystalline diamond, the latter having a much smaller crystal size. Images courtesy of Dr. Roger Narayan, North Carolina State University, USA.

amounts of $sp^3$ bonds, which accounts for some of their diamond-like properties. Those forms that contain low levels of hydrogen (around 1%), sometimes referred as a-C, contain the highest levels of $sp^3$ bonds, up to 85%, are considered to be non-hydrogenated carbon, or tetrahedral carbon, and are considered in the next section. Those forms that contain high levels of hydrogen (up to 50%) have lower levels of $sp^3$ bonds; these forms are referred to as a-C:H, or more commonly as diamond-like carbon (DLC). DLC is a metastable amorphous material, which may be considered as a random network of covalently bonded carbon atoms in the different hybridizations. The material can only be used as a thin film, which may be deposited by plasma-assisted chemical vapor deposition using a variety of precursors and hydrogen-containing environments. This is essentially a line-of-sight process. Non-line-of-sight processes such as plasma source ion implantation may also be used. The films have high hardness and high elastic modulus, but the deposition processes have a tendency to result in high internal stresses. As

(b)

(c)

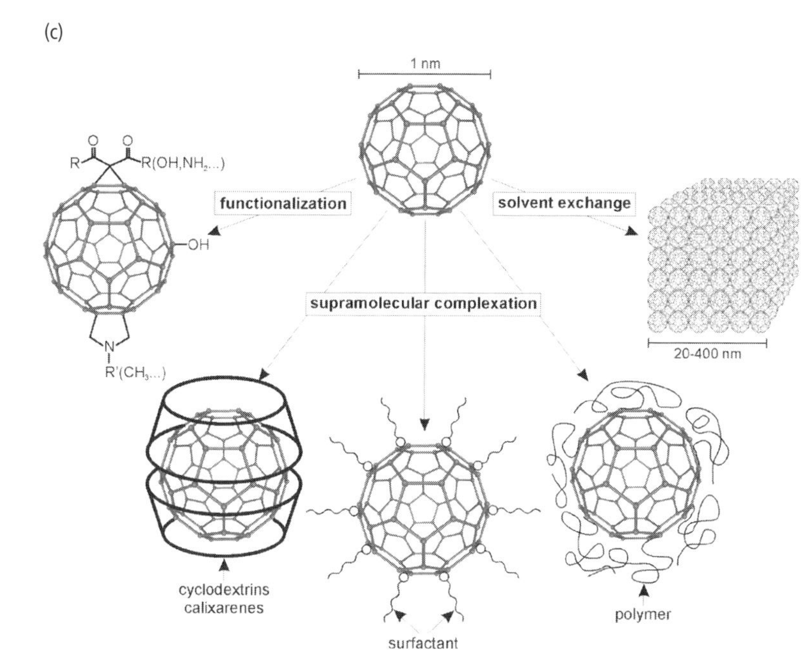


**Figure 8.36** *continued*
**(b)** Pyrolytic carbon. This is one of the oldest uses of carbon in medical technology and still very successful in devices such as mechanical heart valves, where the annulus and leaflets are made from pyrolytic carbon. Image courtesy of On-X Life Technologies Inc, Austin, Texas, USA.
**(c)** Fullerene nanoparticles, and methods of preparation; fullerenes represent an allotrope of carbon, in which the atoms are arranged in closed shells with the structure of a truncated icosahedron. These shells may be modified in a number of ways to give different properties and functions. Fullerene derivatives and supramolecular complexes, while depicted as single units for clarity, also form aggregates in water suspension. Images courtesy of Vladimir Trajkovic, School of Medicine, University of Belgrade and Zoran Markovic, Vinca Institute of Nuclear Sciences, University of Belgrade, Belgrade, Serbia.

metastable films, their structure may change over time, especially at high temperature. The films are chemically inert and water resistant, of high wear resistance and low coefficient of friction, low dielectric constant and high electrical resistivity, and infrared transparency. These properties have led to widespread use of DLC in many industrial sectors. Since numerous experiments have shown generally good biocompatibility, there have been attempts to use DLC in medical devices such as total joint replacements (because of high hardness and wear resistance), stents and guidewires. The introduction of DLC films into medical technology has been slow, however, mostly because of problems of adhesion, where spallation and delamination can occur as a result of the internal stresses in the films.

## Class 4.1.2 Tetrahedral amorphous carbon

We have seen that the carbon bonds in graphite have $sp^2$ hybridization, which gives 3-fold co-ordination for each atom. Diamond exhibits $sp^3$ hybridization, which gives 4-fold tetragonal or tetrahedral

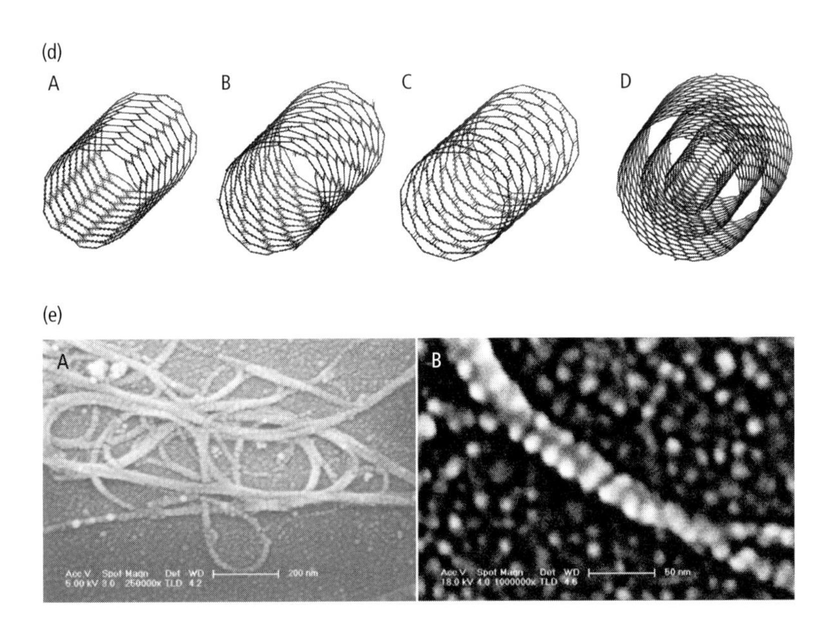


**Figure 8.36** *continued*
(d) and (e) Carbon nanotubes (CNTs).
**(d)** Schematic of CNTs: (A) Armchair 6,6 Single Wall Nanotube (SWNT), (B) Chiral 9.3 SWNT, (C) Zigzag 10,0 SWNT and (D) Multi-Walled Nanotubes.
**(e)** Scanning electron microscope images of amylose-coated SWNTs. Images courtesy of Dr. Qinghua Lu, Shanghai Jiaotong University, China

co-ordination. Amorphous carbons, which by definition have no long-range order, do have local bonding, and this is intermediate between these two. Tetrahedral amorphous carbon has much more, i.e., $> 75\%$ of the 4-fold type. This amorphous carbon is a thin-film material that has strength, modulus and hardness much closer to crystalline diamond. These films can be produced by pulsed vapor deposition onto appropriate substrates such as silicon. These films have good soft tissue biocompatibility.

#### Class 4.1.3 Nanocrystalline and ultrananocrystalline diamond

In a similar manner to amorphous DLC, thin films may be made from crystalline diamond with high hardness, chemical stability and exceptional electronic and optical properties. The better properties are obtained with very small crystal sizes, and two related forms, nanocrystalline (NCD) and ultrananocrystalline (UNCD), have been developed. NCD films may be synthesized using microwave plasma chemical vapor deposition with an argon-hydrogen-methane gas mixture. Crystal sizes of around 20 nm are produced. The high volume of crystal, or grain, boundaries and the grain size gives significantly improved properties since the carbon atoms are $\pi$-bonded in these regions. UNCD films may be produced by similar methods but slightly different conditions, yielding 3–5 nm crystallites with $sp^2$ bonds at the 0.2–0.3 nm grain boundaries, shown in Figure 8.36(a).

### 8.5.2 Class 4.2 Graphitic materials

Graphite is the second allotrope of carbon. Here one s orbital and two p orbitals interact to give $sp^2$ hybridization, which gives rise to a trigonal planar structure in which each carbon has three nearest neighbors within a graphite sheet. One p orbital is unaffected. Although there are strong interatomic forces within sheets, there are only weak forces between them. The combination of mechanical and electrical properties of graphite makes it suitable for several engineering applications, but not in medical engineering.

#### Class 4.2.1 Pyrolytic carbon

Pyrolytic carbon is a synthetic version of carbon, which does not occur naturally, but can be coated onto material substrates. It is formed from the thermal decomposition of hydrocarbons in the absence of oxygen; these hydrocarbons include methane, acetylene and propane. This is carried out in a fluidized bed reactor in which a vertical furnace is heated at temperatures up to 2000°C. The furnace contains small refractory ceramic particles, in which objects to be coated are placed. An inert gas is used to fluidize the bed and, at the appropriate temperature, the hydrocarbon gas is introduced. Pyrolysis takes place and, depending on process parameters such as the gas flow rate and the temperature, carbon is deposited on the target substrate. This carbon forms as a turbostratic structure where there is order within graphite-like layers but little interaction between these layers. Clearly because this is a high-temperature process there are significant restrictions on the nature of the substrate, usually being limited to refractory metals and graphite. A pure pyrolytic carbon, often termed low-temperature isotropic (LTI) carbon will have a Young's modulus of 30 GPa, and a flexural strength of around 500 MPa. The strain at failure is around 1.5%. In most medical applications, some silicon carbide is added to the mix, giving a 6% silicon content, in order to improve some properties, including the hardness. It is also possible to produce pyrolytic carbon by vapor deposition processes, this form being termed ultra-low-temperature isotropic carbon (ULTI). These materials do have a low fracture toughness, hence the limitation to their use as surface coatings, which can vary in thickness from 25 to 500 μm. The main biomedical use of pyrolytic carbon is in heart valves, where it is coated onto graphitic substrates for the manufacture of valve leaflets shown in Figure 8.36(b).

#### Class 4.2.2 Activated charcoal

Activated charcoal is made by heating carbon-based substances such as peat and coal to produce regular charcoal and then heating this in the presence of an oxidizing gas such as steam, when the charcoal becomes highly porous with activated surfaces within the pores. These are highly adsorbent and may be used for detoxification. This can be achieved following oral ingestion of tablets or powders. It has also been used in liver perfusion columns for the treatment of serious acute poisoning.

### 8.5.3 Class 4.3 Glassy or vitreous carbon

Vitreous carbon, also known as glassy carbon, is produced by slowly heating a solid polymeric substance in order to drive off volatile constituents, leaving carbon behind. It has no long-range order but does have some short-range order with bonds similar to those found in graphite. It has extremely good high-temperature resistance and is therefore useful for furnaces. Although biostability and biocompatibility appear favorable, and attempts were made to use the material in dental implants and percutaneous devices, medical applications have not been forthcoming because of the mechanical property limitations, including brittleness.

### 8.5.4 Class 4.4 Hexagonally bonded carbon nanostructures

Three different forms of hexagonally bonded carbon nanostructures have received massive attention in recent years, with many potential medical applications.

#### Class 4.4.1 Fullerenes

Fullerenes are large cage-like molecules of carbon, which may be considered as a 3D carbon allotrope. The most commonly investigated and used is buckminsterfullerene, $C_{60}$, which has 60 carbon atoms in an essentially spherical structure, shown in Figure 8.36(c). This is a truncated icosahedron, which has 12 pentagons and 20 hexagons, where every carbon atom forms bonds to three other adjacent atoms, either with single bonds in the pentagons or double bonds in the hexagons. These are, by definition, molecules, but with a diameter of 0.7 nm come very close to being characterized as nanoparticles. In fact, collections of fullerene molecules tend to aggregate within solutions such that their properties are more nanoparticle-like than molecular. Although there have been widely discussed potential applications of fullerenes in areas such as quantum computing and other nanoelectronic fields, such potential has not been extensively exploited, but much attention remains with biological and medical applications and with functionalized fullerenes. Examples of the relevant properties include the ability to act as anti-inflammatory agents via free-radical sponge behavior and there are potential uses in drug and gene delivery through the use of the space within the spheres. Some imaging and sensing systems utilize their electronic and optoelectronic properties. There are two main limitations at this stage, based first on the uncertainty over the toxicity of $C_{60}$-based structures and on the extremely low water solubility. Attempts have been made to improve water solubility by, for example, chemical modification of the carbon cage by attaching functional groups such as -OH, -COOH and $NH_2$ to facilitate formation of hydrophilic adducts or incorporation into water-soluble supramolecular structures.

One way forward for fullerenes is to develop these molecules into practical derivatives. For example, fullerene architecture can be extended to the macromolecular scale by polymerization. Fullerenes may be covalently incorporated into a polymer chain, giving side-chain polymers, main-chain polymers, star-shaped polymers and so on. Alternatively, it is possible to prepare all-carbon poly[60]fullerenes in which only $C_{60}$ molecules are connected to each other by covalent bonds without the use of any linking group.

### Class 4.4.2 Carbon nanotubes

Carbon nanotubes are cylindrical tubes based entirely on carbon atoms that typically have lengths greater than 100 nm and much smaller diameters. They may exist in the form of a single tube, in which case they are called single-walled carbon nanotubes (SWCNs), or of concentric cylinders, referred to as multi-walled carbon nanotubes (MWCNs), shown in Figure 8.36(d). Although other methods are available, carbon nanotubes are usually prepared by chemical vapor deposition, usually involving hydrocarbon feedstock at high temperatures in the presence of metal catalysts. These have to be purified since the residual catalyst may be cytotoxic. The morphology and geometry will depend on the reaction conditions; the resulting materials have varied and attractive electronic, thermal and biological properties. Because of these properties, the carbon nanotubes may be used in cell imaging and biological sensing, biomolecule delivery and tissue engineering. They possess optical transitions in the near infrared spectrum which is very useful in the optical imaging field, whilst they may be functionalized with gadolinium for potential in MRI. One advantage of tubular carbon species is that they can be heterogeneously functionalized so that ends and side walls have different properties, leading to dual functionality. As with fullerenes, and indeed with any such nanostructures, some questions remain unanswered about their toxicity.

### Class 4.4.3 Graphene

We discussed graphite in Section 8.5.2 above; it is an allotropic form of carbon composed of layers of $sp^2$-bonded carbon atoms in a honeycomb arrangement. Individual atom layers interact only weakly, primarily through van der Waals forces. There are several million such layers in a millimeter thick piece of graphite. Such layers, if produced as separate individual entities, constitute the new material, graphene, which is currently receiving widespread attention in many scientific scenes. There are expectations that graphene will be a successor of silicon in electronics and in many similar applications. At the present time, uniformed ordered growth of graphene on an insulating substrate is necessary. Deposition processes involving high vacuums or argon atmospheres have been used on silicon carbide substrates; in many situations this involves oxidized forms of graphene. So-called chemically derived graphene oxide (GO) is prepared by introducing oxygen functional groups during chemical exfoliation of graphite. This material is claimed to have significant biological properties, including antibacterial activities. Macroscopic products may be made from suspensions of GO sheets and, in particular, formed into sheets or antibacterial paper.

## 8.6 Class 5 Composite materials

With the first four classes of biomaterials, the majority of properties of individual materials within each group, both at first and second level classifications, can be anticipated from the general behavior of those groups. Thus we would expect all examples within the metallic system to be conductors of heat and electricity, and we would expect all those in the ceramics system, under normal circumstances, to display brittleness and relatively poor toughness. There will, of course, be variations in these properties at all levels – not all metals will have the same electrical conductivity nor all ceramics the same level of toughness – and we would expect to be able to modulate those properties by adjusting the composition and treatment; after all, that is the job of the materials scientist. In general, however, we can only go so far with this approach and still stay within the same class of material. If we add too much of a different component we may alter the characteristics of the substance such that it can no longer qualify for inclusion in any particular class. It is for this reason that we have a new major class of material (and hence biomaterial), which we shall designate Class 5 and call these materials composites.

The classical derivation of composites comes from the need to strengthen polymers and/or toughen ceramics. Polymers deform plastically too easily and they are relatively weak, but they can show prodigious ductility simply because they deform so readily. Ceramics have no intrinsic mechanism for plastic deformation and will much rather undergo brittle fracture than plastic deformation. However, their interatomic bonds are very strong so that very high levels of stress are required before the structure fractures, that is, it can be very strong, even if it has zero ductility. If, however, we take a ductile but weak polymer, and disperse ceramic particles within it, then the particles resist the polymer plasticity and the polymer matrix deflects or stops cracks in the ceramic, and the result is a strong *and* ductile material. Clearly, this is neither a polymer nor a ceramic system, but a composite, in this case a polymer–ceramic composite. Composites, by definition, are materials made from two entirely different structures where the best properties of each are retained, and the worse properties are lost. There are many forms of composite depending on the relationships between the phases. In theory, we could consider different sub-classes of composite depending on the two (or possibly more) phases that are present (perhaps polymer–ceramic, ceramic–carbon, carbon–polymer), but in this case such a classification principle does not really work, since the biomaterials applications of composites depend on the required properties. In Box 8.6 we see some of the possibilities.

**Box 8.6 The second level of the classification of biomaterials: composite biomaterials**

Class 5.1 Fiber reinforced thermoplastic polymers
Class 5.2 Fiber reinforced resins
Class 5.3 Ceramic microparticle/biostable polymers
Class 5.4 Ceramic microparticle/biodegradable polymers
Class 5.5 Nanocomposites

One of the problems with a classification of composites is that their development has been based on ad hoc opportunities, without an entirely logical set of generic specifications. Also, as far as biomaterials applications are concerned, not all composites have been successful and some general problems have become apparent. The first is that the mechanical properties of composites depend to a considerable extent on the interface between the components, and many interfacial regions are affected by the aqueous environment of the body. Secondly, the architecture of most composites is characterized by micro- and nanoscale features. Small particles released from these composites can have serious consequences with respect to biocompatibility, as discussed in Chapter 3, and this has reduced the possibilities for composites in many applications.

In this classification, we use generic combinations of components, and give examples, where they exist, within these classes. In some situations, the materials could equally well be designated as a sub-class of one or other of the major components, as we see in the next section.

### 8.6.1 Class 5.1 Fiber reinforced thermoplastic polymers

Fiber reinforced polymers usually involve fibers of considerable length and small diameter in order to optimize the mechanical property advantages. Most carbon fibers in traditional composites are usually around 7 microns in diameter. Since thermoplastic polymers cannot be cured in the same way as a resin, any composite involving fibers in such polymers has to be fabricated by thermoplastic routes, which means that it is not possible to use very long fibers nor achieve a high volume fraction. For this reason, the so-called short fiber composites have generally had limited use in industrial applications. Early attempts to incorporate short carbon fibers in polymeric biomaterials, for example as reinforcements in acrylic bone cements or the polyethylene of joint prostheses, were not successful because of adverse effects on biocompatibility; the host response is disadvantaged by the presence of, say, 20 microns long, 7 microns diameter hard fibers.

In spite of the generic limitations discussed above, one composite has now made an impact in biomaterials applications, this being the carbon-fiber reinforced PEEK. Since the properties are closely aligned to the parent polymer, but with improvements to mechanical properties, this material has already been discussed within Class 3.1.4.2.

### 8.6.2 Class 5.2 Fiber reinforced resins

Many of the early industrial applications of fiber reinforced polymers utilized long fibers, initially of glass and then carbon, which could be incorporated into thermosetting resins such as phenolic and epoxy resins. By using a resin, shapes may be prepared before the resin is cured, allowing very high volume percentages of fibers to be incorporated into the composite. The carbon fiber–epoxy systems have been used in some medical devices such as bone fracture plates and stems of joint prostheses, the main potential benefit being the close matching of the elastic modulus of these materials to that of bone, with an additional advantage of being amenable to significant anisotropy based on the orientation of the fibers. Nevertheless significant clinical advantages of these materials have not been fully realized.

### 8.6.3 Class 5.3 Ceramic microparticle reinforced biostable polymers

The mechanical properties of polymers can also be improved by small isotropic particles as well as by fibers. There have been few attempts to use such composites in implantable devices. One form of particulate reinforced resins is used very extensively in restorative dentistry, where light cured resins such as urethane dimethacrylate are heavily reinforced by ceramic particles, giving good wear resistance as well as high-quality aesthetics and ease of clinical use.

A great deal of effort has been devoted towards the incorporation of bioactive ceramics into thermoplastic polymers in order to improve overall biological performance. The rationale for much of this effort was based on the fact that bone material is essentially a composite of the ceramic hydroxyapatite within a polymer, collagen, and that therefore incorporating a bioceramic in a synthetic polymer should emulate the natural bone. Obviously this is a rather simplistic analogy since a synthetic polymer-based composite is non-vital and the hydroxyapatite contained within the matrix has little chance of operating in any "bioactive" manner. Such materials have been used clinically but not in any extensive manner.

### 8.6.4 Class 5.4 Ceramic microparticle reinforced biodegradable polymers

Similar types of argument have been made to justify the development of bioactive particle-biodegradable polymer composites. Certainly the mechanical properties of biodegradable polymers can be improved by appropriate use of particulates but whether this is a clinical necessity is not clear. Perhaps of greater potential practical value is the incorporation of bioactive particles into degradable polymers in the belief that their biological performance, for example in tissue engineering scaffolds, could be improved. This question has been discussed in Chapter 5.

### 8.6.5 Class 5.5 Nanocomposites

This is a potentially large category of biomaterials but we have to be careful with inclusion criteria. Strictly speaking, a nanocomposite is a composite material where one of the phases has dimensions in one, two or three directions of less than 100 nm. It is necessary, however, to add the caveat that this nanoscale phase makes a significant contribution to the properties of the material by virtue of its nanoscale features, not just because it is smaller than a traditional microscale phase. Many of those composite materials under consideration for biomedical use that would be included in Class 5.3 do have some particles with nanoscale dimensions, but this does not imply that they should now be described as nanocomposites.

There are also some difficulties with the terminology of nanoscale materials in drug delivery or imaging contrast agents, and especially with multifunctional systems. Questions arise when, for example, an iron oxide nanoparticle is conjugated to a polymeric micelle; should the term nanocomposite be used here and what are the consequences of doing so? This aspect of biomaterials science is too young to form definitive positions on these questions. Suffice it to say that a classification of 5.5, Nanocomposites is relevant, but we do not know what materials systems should be included.

## 8.7 Class 6 Engineered biological materials

Classes 1–5 above could appear, perhaps in somewhat different formats and with different details, in any general classification of materials. Class 6, which we shall call "engineered biological materials," will appear here in the classification of biomaterials and in no other general materials classification system (Box 8.7). It is here that we extend the conventional thinking of what is a biomaterial and move to recognize how biomaterials science is evolving in order to meet the current and future challenges of clinical medicine. These arguments have all been addressed in one way or another in earlier chapters of this book, but are brought together here in order to provide the complete rationale for this special class of biomaterial. However, as with the previous section on composites, this field is rather young and a fully rational classification has not really emerged.

One consequence of the early stage of development of engineered biological components, and the sensitive nature of the tissues and processes, is that procedures are often experimental or non-commercial and products do not always follow the pattern of classical synthetic materials. There are several ways in which these products and materials

**Box 8.7 The second level of the classification of biomaterials: engineered biological materials**

Class 6.1 Autologous tissues
Class 6.2 Allogeneic tissues
Class 6.2.1 Allograft bone
Class 6.2.2 Allograft cartilage
Class 6.2.3 Allograft dermis
Class 6.2.4 Allograft blood vessels
Class 6.2.5 Allograft amniotic membrane
Class 6.2.6 Allograft dura mater
Class 6.2.7 Allograft fascia lata
Class 6.3 Xenogeneic tissues
Class 6.3.1 Xenogeneic bone
Class 6.3.2 Xenogeneic small intestine submucosa
Class 6.3.3 Xenogeneic pericardium
Class 6.3.4 Xenogeneic aortic valve tissue
Class 6.3.5 Xenogeneic whole organs

could be categorized. The simple classification process introduced here is based on the desire to avoid a long list of closely related but slightly different products whilst keeping the criteria based on scientific principles. There is a temptation to classify the materials on the basis of clinical usage, but this does not really help. Instead we use a system based on the source of the biological components, with sub-classes based on methods of preparation. The inclusion of materials in this list does not imply that they have good clinical data or even regulatory approval. Indeed some of these materials may have had poor clinical outcomes; the reason that they are included is that the performance of many options needs to be considered carefully, as mechanisms for their incorporation into the human body are not fully understood and further developments may be anticipated.

There is an inevitable overlap between some of these materials and some products we have classified in either the polymeric systems (e.g., proteins and peptides) or the ceramics systems (e.g., calcium phosphates). The distinction used here is that such systems involve individual well-characterized components whereas in the current section the products bear significant resemblance to the source tissue, such as pericardium as distinct from pure collagen derived from pericardium.

Since the definition of biomaterials, as discussed in Chapter 1, includes many forms of engineered tissue, there has also been a temptation to include engineered cells, either mammalian or bacterial, and engineered viruses such as those used in gene transfer. This temptation has also been resisted because these fields are insufficiently developed to sustain a rigorous and mechanistic classification at this time.

### 8.7.1 General comments on tissue processing

The majority of engineered biological components are derived from donated or harvested tissues, of either human or animal sources. The extracellular matrix of animal tissues (including human tissue) is a rich source of potential biomaterials. Many different tissues, including bone, ligament, blood vessels, fascia lata, dura mater, dermis, pericardium and intestinal submucosa, have been explored for medical use, and many used clinically on a regular basis. The rationale for this use is based on the fact that these tissue derivatives may have very similar physical and mechanical properties to human tissues in which they may be placed or are intended to replace, and also because they may be able to offer biological activity that supports the intended function of the product. The ECM consists of the secreted products of the cells that reside in these tissues and organs and, as such, provide a natural basis for the presentation of signals and cues that affect cell behavior, such as migration, proliferation and differentiation. On the negative side, the natural origin of the tissues may equally pose biological risks, for example immunogenicity and infectivity. The use of ECM derivatives in regenerative medicine is therefore predicated on the ability to process the sourced tissue in order to eliminate any risks whilst maintaining structural features and biological cues.

The processing of these tissues is clearly a very important factor in determining the resulting properties of these biomaterials. The processing aims to remove unwanted components and, in the majority of situations, this will include the removal of cells. For the supply of tissues from animals for commercial products, processing will normally include mechanical and chemical steps, with rinsing, dehydration and sterilization phases. The mechanical steps include cutting to form sheets or grinding to yield powder. Chemical processing is usually directed towards decellularization, disinfection and stabilization. Removal of bacteria and viruses without profoundly altering the properties of the tissue is not straightforward. Glutaraldehyde is widely used for this purpose, which should assist in killing microorganisms and also cross-link the collagen, which assists in controlling immunogenicity and increasing biostability.

**8.7.1.1 Decellularization processes** Since, as implied above, residual cellular material may adversely affect tissue regeneration and remodeling, one of the most important steps in the processing is aimed at the complete removal of cells and cellular debris. The majority of decellularization processes involve either or both chemical or biological agents, although some physical processes may also be used. Included in the chemical agents are detergents, hypo- or hypertonic solutions, acids, bases and solvents. Biological agents primarily include enzymes and chelating agents. Physical agents are temperature and electroporation.

The use of detergents is based on the fact that, whether they are ionic, non-ionic or zwitterionic, they are able to solubilize cell membranes and dissociate DNA from proteins. Triton X-100 is a popular non-ionic detergent and sodium dodecylsulfate (SDS) a good ionic detergent. All detergents have the disadvantage of being able to disrupt proteins in the ECM itself. Optimal choice is based on the type of tissue involved since some are better in thick sections and others in thin ones, some better in highly cellular organs such as the liver and others in less cellular tissues such as tendons, and so on. Solvents include alcohols such as glycerol, isopropanol, ethanol and methanol and non-alcohol substances such as acetone and tributyl phosphate. Both acids and bases can cause hydrolytic degradation of key molecules, solubilization of cytoplasmic components, and denaturation of proteins and nucleic acids, common examples being peracetic acid and sodium or calcium hydroxide; these can also cause significant disruption of structural proteins such as collagen. Hypotonic solutions cause cell lysis by osmotic effects and hypertonic solutions dissociate DNA from proteins. Chelating agents are able to bind metallic ions, disrupting cell adhesion – EDTA is a good example. As expected, enzymes have high specificity for particular cellular components, and agents include trypsin, collagenase, nucleases and lipase. With respect to the use of temperature, freeze–thaw processes lyse cells, although by themselves do not result in the removal of intracellular components. Electroporation involves the use of pulsed electrical fields to disrupt cell membranes.

With many tissues, combinations of these agents and techniques may be preferable. Freeze–thaw is usually seen as the first phase, followed by variations of enzymes, osmotic solutions, detergents and acid/base solutions. At its simplest, these procedures may be used by immersion with agitation sequences. Inducing pressure gradients can help with preserving the ECM structure, especially with tubular organs, and better elimination of the cellular components may be achieved through the use of supercritical fluids. When whole organs are being decellularized, perfusion of the complete structure may also be beneficial in maintaining architecture. Some methods of decellularization are given in Box 8.8 and an example of a decellularized tissue is shown in Figure 8.37.

## 8.7.2 Class 6.1 Autologous tissues

If autologous tissue is used for the reconstruction of part of a patient, it is normal for the tissue to be harvested from the donor site and transplanted directly into the recipient site without any significant manipulation. As such these are considered to be viable tissue grafts and not biomaterials. In principle, it is possible to harvest tissue and use some manipulation technique in order to improve or optimize the graft for its intended application. This could simply involve preservation in order to extend the value of the tissue. Alternatively, as with the so-called Ross procedure, this could use tissue from one site and implant it in a more significant site, in this case taking a pulmonary heart valve and placing it in the more demanding aortic position, for which purpose the shape and supporting structures have to be modified.

These possibilities for engineered autologous tissue are rare at this stage, but may increase in the future as planned tissue engineering procedures are being developed.

## 8.7.3 Class 6.2 Allogeneic tissues

Allogeneic tissues, by definition, are derived from human donors. These are mostly obtained from cadavers, and are treated to remove potential infectivity and processed by a variety of techniques that are intended to optimize their performance in specific situations. The products are stored in tissue banks for commercial companies.

### Class 6.2.1 Allograft bone

Bone is one of the major targets here, allograft bone being used as an alternative to autograft bone in situations where harvesting sufficient bone from the patient is not possible. The allograft bone may be fresh frozen or freeze-dried. It may also be demineralized (known as demineralized bone matrix).

### Class 6.2.2 Allograft cartilage

In a very similar way, cryopreserved cartilage may be used in knee injuries; it is often used as a combined cartilage–bone construct.

### Class 6.2.3 Allograft dermis

Several commercially available allograft dermal matrix products have been used over the last few decades. Dermal tissue is separated from the epidermis and decellularized and prepared as solid implantable or micronized injectable forms. They have been used in skin repair and in cosmetic applications for eliminating wrinkles or enlarging penises.

### Class 6.2.4 Allograft blood vessels

Cryopreserved vein and artery allografts have been used for vascular access in the treatment of patients with hemodialysis, but not with great success.

### Class 6.2.5 Allograft amniotic membrane

The amniotic membrane is the innermost layer of the placenta, and consists of epithelial cells, basement membrane and an avascular connective tissue matrix. It can be removed from the remaining placental tissue, cleaned, dehydrated and sterilized and

## Box 8.8 Methods of decellularization

There are several generic mechanisms of action for decellularization and several agents that are used for each mechanism. There are also different techniques that are used to apply these agents to the tissue samples.

The groups of mechanisms involve chemical, biological or physical effects:

### Chemical effects

These include solubilization of cytoplasmic components including nucleic membranes and lipids, cell lysis by osmotic shock or dehydration, disruption of DNA–protein, protein–protein and lipid–protein interactions.

The main agents are acids and bases, hypo- and hypertonic solutions, non-ionic detergents such as Triton X-100, ionic detergents such as Triton X-200 and sodium dodecyl sulfate (SDS), zwitterionic detergents, solvents such as alcohols and acetone.

Many of these chemicals cause some disruption of the tissue ultrastructure and removal of GAGs and growth factors. Effectiveness usually depends on tissue thickness.

### Biological efects

These are usually enzymes. Effective enzymes include nucleases, which catalyze the hydrolysis of ribonucleotide and deoxyribonucleotide chains, trypsin, which cleaves specific peptide bonds, and dispase, which cleaves peptides such as fibronectin and some collagens. Chelating agents such as EDTA may be used in conjunction with enzymes.

Prolonged exposure to enzymes can disrupt the ECM ultrastructure and remove many of the constituents such as GAGs, fibronectin, laminin and elastin.

### Physical effects

Mechanical force can burst and eliminate cells, although it can also damage the ECM.

Freezing and thawing can produce ice crystals that disrupt cell membranes.

Hydrostatic pressure can burst cells, but again can damage the ECM.

Pulsed electrical fields can disrupt cell membranes.

### Application of agents

Generally agitation can help lyse cells and facilitate exposure of cells to the agent and remove the cellular material.

Supercritical fluids also lyses cells.

Perfusion, possibly assisted by pressure gradients, also facilitates exposure and removal.

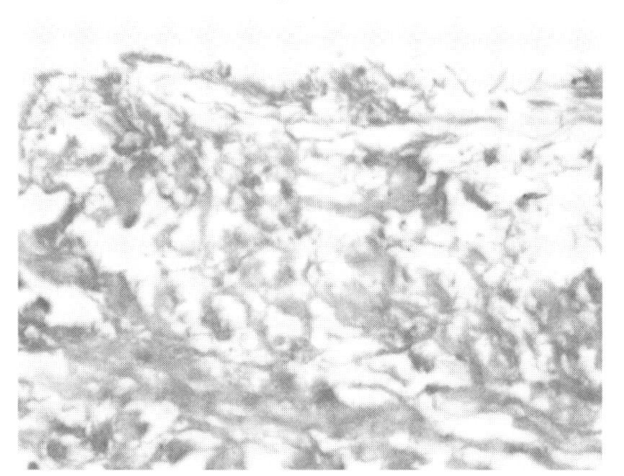

**Figure 8.37** Decellularized tissues. Microscope section of decellularized small intestine submucos (SIS) with peracetic acid (PAA) treatment. Fresh porcine-derived SIS was washed in distilled water on a rotary shaker at 200 RPM and 4°C for 3 days, oxidized by soaking in 5% PAA for 4 hrs, treated with 1% Triton X-100 solution for 2 days and then washed with distilled water for 2 more days. Hematoxylin and eosin, Masson Trichrome and DAPI staining showed that cellular DNA contents were almost completely removed. SIS matrix became visibly more porous, and the intra- and inter-fascicular space was increased in the lamina propria that appear more extended after decellularization procedures. Images courtesy of Dr. Yuanyuan Zhang, Wake Forest Institute of Regenerative Medicine, North Carolina, USA. See plate section for color version.

used as a membrane to assist in healing processes. This product is particularly successful in the treatment of damaged ocular surfaces.

#### Class 6.2.6 Allograft dura mater

The dura mater is the outermost of the three layers of the meninges that surrounds the brain and spinal cord. It is a tough, inelastic membrane and has been used in some forms of surgery, particularly in periodontal surgery. Questions naturally arise with respect to infectivity since brain tissue is one of the major risk factors in the transmission of prion diseases, which has reduced enthusiasm for the material.

#### Class 6.2.7 Allograft fascia lata

The fascia lata is also a tough membrane, derived from the thigh, for example in the iliotibial band that extends from the iliac crest to the tibia. This can be harvested and cleaned in similar ways and has been used in ligament regeneration.

### 8.7.4 Class 6.3 Xenogeneic tissues

These materials are derived from animals. Many species are potential sources, although the majority of products are either of porcine or bovine origin. Procedures are very similar to those used for allogeneic-sourced products

#### Class 6.3.1 Xenogeneic bone

Porcine and bovine bone products have been used in much the same way as human bone.

#### Class 6.3.2 Xenogeneic small intestine submucosa

One of the more interesting developments on xenogeneic tissues in recent years has involved the small intestine submucosa (SIS). The submucosa is found between the mucosa and muscle layers of the small intestine. It has a fibrous, porous structure that consists of collagens, proteoglycans, glycosaminoglycans, together with biomolecules such as growth factors. It is usually harvested from pigs and rigorously decellularized. It is in clinical use in a variety of situations to assist in reconstruction and repair processes, including those in the peritoneal cavity and genito-urinary locations. The uses of SIS have been discussed in full in Chapter 5.

### Class 6.3.3 Xenogeneic pericardium

The pericardium is the collagen-rich structure that surrounds the heart. Pericardium is extensively used in the construction of bioprosthetic heart valves. Bovine pericardium is preferred but other animals have been considered including more exotic species including the ostrich and yak. This has been discussed in previous chapters, including Chapters 3 and 4.

### Class 6.3.4 Xenogeneic aortic valve tissue

Similarly, bioprosthetic heart valves may be made from treated xenogeneic valves, especially obtained from pigs.

### Class 6.3.5 Xenogeneic whole organs

At the time of writing, there is much interest in extending the concept of decellularization of xenogeneic tissues to whole organs such as the liver in order to provide a template for tissue engineering. Some of these options have been mentioned in Chapter 5.

## Summary and learning objectives

1. This chapter has brought together all of the types of biomaterial discussed in previous chapters and stratified them in terms of a hierarchical classification. This is not a simple list of biomaterials but a discussion of the structure–property relationships for each type, with an indication of why and how their properties control current and future biomedical applications.
2. We first introduce the rationale for the classification. There are six first level classes, these being the metallic, ceramic and polymeric systems, carbons, composites and engineered tissues. It is the intention here to give the reader a guide to why each type of biomaterial has its own characteristics that are not randomly assigned but are dependent of their essential structure–property relationships, usually deliberately chosen and enhanced or modified for particular medical applications.
3. With metals, we discuss titanium and its alloys, steels, cobalt alloys, nickel alloys, tantalum and niobium, silver, platinum group metals, gold and magnesium alloys. The main learning objective here is to understand how basic metallurgical principles have been employed to optimize the mechanical and corrosion resistance properties for major structural medical devices and, in addition, how their biological properties at the nanoscale have been put to use in recent years in a wide range of pharmaceutical and imaging applications.
4. The section on ceramics and glasses logically classifies these materials as oxides, phosphates, sulfates, silicates, nitrides, carbides and titanates, with a special section on optically active nanoparticulate ceramic or ceramic-like materials. Again their properties range from the structural to the biologically active to the physically active.
5. The group of polymer-based materials constitute the most complex class, part of the complexity being due to the manner in which slight variations in composition and formulation can change the characteristics considerably, possibly changing their classification. Also it is noted that practical applications often involve copolymers, blends or conjugates that involve more than one class.
   A major learning objective here is to understand just how wide ranging these structures can be, but also how the classification can be dictated by the

intending applications. The main subclasses here are the thermoplastics, thermosetting resins, hydrogels, proteins and peptides, polysaccharides, lipids, degradable polymers, water-soluble polymers, ionizable polymers, elastomers, textiles and environmentally responsive polymers.

6. For many years, carbon-based materials constituted a minor class of (bio)material. Here we have some of the long-established carbon materials such as graphitic materials and glassy carbon. However, the current classification has to include those forms which have been introduced because of nanotechnology, giving the nanostructured diamond materials and the fullerenes, carbon nanotubes and graphene.
7. Composites are materials that are comprised of two or more different classes where all those components present add very specific features such that the properties of the composite combine the best of the different components. They are not simply mixtures of different substances. They have been developed on an ad hoc basis without any overall logical pattern, and each has been designed with specific objectives in mind.
8. With engineered biological components, the generation of a rigorous classification is difficult at this stage in the development of these materials. It is better to categorize them on the basis of the source, that is primarily allogeneic and xenogeneic sources, with consideration being given to the type of tissue and the treatment protocols that are used.
9. As indicated, the outcome of such a classification should not be a mere catalogue. Instead it should be a guide to how biomaterials may be selected for medical applications on the basis of their properties.

## Questions

1. Explain the rationale of a classification system for biomaterials that is based on the structure–property relationships of materials.
2. Discuss the difficulties that arise when using this type of classification system to categorize composite materials.
3. Describe how the properties of titanium alloys are controlled by their elemental composition and phase relationships.
4. Discuss the relative merits of titanium alloys, stainless steels and cobalt–chromium alloys with respect to orthopedic implants.
5. Certain noble metals have medical applications based on their inertness, yet they have derivatives with significant biological or pharmaceutical characteristics. Discuss this dichotomy in relation to gold, silver and platinum.
6. Discuss the variety of properties that are associated with oxide ceramics.
7. Calcium phosphates are often considered to have bioactive properties, especially in relation to bone bonding. Discuss how the different forms of calcium phosphate perform in relation to this bioactivity.

8. Give some reasons why PEEK is often preferred to polyethylene in current medical applications.
9. Polyurethanes have many different forms and quite different mechanical behavior. Discuss the relationships between the properties of polyurethanes and their molecular structure.
10. Compare and contrast the major forms of biodegradable structural polymers.
11. Explain the mechanisms by which some polymers show environmental responsiveness.
12. Discuss the essential features of nanocomposites and give two examples of such materials in medical applications.
13. Describe the relative merits of fiber- and particle-based composite materials.
14. Compare the risks inherent in xenogeneic and allogeneic engineered tissues.

## Recommended reading

### Class 1 materials: the metallic systems

Balasubramanian S K, Yang L, Yung L-Y L, Ong C-N. Ong W-Y, Characterization, purification and stability of gold nanoparticles, *Biomaterials* 2010;31: 9023–30. *Experimental demonstration of practical aspects of gold nanoparticle preparation and properties.*

*Brunette D M. Tengvall P, Textor M, Thomsen P (eds) *Titanium in Medicine*, Springer, Berlin, 2001. *Comprehensive textbook on titanium.*

Chaloupka K, Malam Y, Seifalian A M, Nanosilver as a new generation of nanoproduct in biomedical applications. *Trends in Biotechnology* 2010;28:580–8. *Review of the medical applications of silver nanoparticles.*

Gu X, Zheng Y, Cheng Y, Zhong S, Xi T, In vitro corrosion and biocompatibility of binary magnesium alloys, *Biomaterials* 2009;30:484–98. *Review of experimental evidence on magnesium corrosion and tissue responses.*

Levine B R, Sporer S, Poggie R A, Della Valle C J, Jacobs J J, Experimental and clinical performance of porous tantalum in orthopedic surgery, *Biomaterials* 2006; 27: 4671–81. *Review of clinical evidence concerning porous tantalum.*

Minoda Y, Kobayashi A, Iwaki H, *et al.*, Comparison of bone mineral density between porous tantalum and cemented tibial total knee arthroplasty, *Journal of Bone & Joint Surgery* 2010; 92A: 700–6. *Discusses advantages and disadvantages of porous metals in joint replacement.*

Nguyen D T, Kim D-J, Kim K-S, Controlled synthesis and biomolecular probe application of gold nanoparticles, *Micron* 2011;42:207–27. *Discusses the controlled synthesis of gold nanoparticles and demonstrates their application in the detection of proteins and nucleic acids.*

Rondelli G, Torricelli P, Fini M, Giardino R, In vitro corrosion study by EIS of a nickel-free stainless steel for orthopedic applications, *Biomaterials* 2005;26:739–44. *Experimental determination of the corrosion characteristics of nickel-free steels.*

Shabalovskaya S, Rondelli G C, Undisz A L, *et al.*, The electrochemical characteristics of Nitinol surfaces, *Biomaterials* 2009;30:3662–71. *Discusses the effects of surface features on the chemical and biological properties of nickel–titanium.*

### Class 2 materials: ceramics

Benzaid R, Chevalier J, Saadaoui M, *et al.* Fracture toughness, strength and slow crack growth in a ceria stabilized zirconia-alumina nanocomposite for medical applications, *Biomaterials* 2008;29:3636–41. *Explanation of the mechanical properties of nanostructured zirconia-alumina ceramics.*

Byers R J, Hitchman E R, Quantum dots brighten biological imaging, *Progress in Histochemistry and Cytochemistry* 2011; 45:201–37. *Review of imaging possibilities with quantum dots.*

Chevalier J, What future for zirconia as a biomaterial? *Biomaterials* 2006;27:535–43. *Places the technology of transformation toughened zirconia in perspective.*

Dorozhkin S V, Bioceramics of calcium orthophosphates, *Biomaterials* 2010;31:1465–85. *Extensive review of medical uses of the main group of calcium phosphates.*

Fuller J E, Zugates G T, Ferreira L S, *et al.*, Intracellular delivery of core-shell fluorescent silica nanoparticles, *Biomaterials* 2008;29:1526–32. *Discusses mechanisms of silica nanoparticle delivery to cells*

Gabriel G, Erill I, Caro J, *et al.*, Manufacturing and full characterization of silicon carbide-based multi-sensor micro-probes for biomedical applications, *Microelectronics Journal* 2007;38:406–15. *Examines the feasibility of using silicon carbide substrates for multi-sensor micro-probes for monitoring organs during transplantation.*

Hsieh C-L, Grange R, Psaltis, Bioconjugation of barium titanate nanocrystals with immunoglobulin G antibody for second harmonic radiation imaging probes, *Biomaterials* 2010;31:2272–7. *Discusses conjugation of nanostructured barium titanate with amine groups and antibodies for imaging probes.*

Liu X, Morra M, Carpi A, Li B, Bioactive calcium silicate ceramics and coatings, *Biomedicine and Pharmacotherapy* 2008;62:526–9. *Brief practical details of calcium silicate coatings.*

Thomas M V, Puleo D A, Calcium sulfate: properties and clinical applications, *Journal of Biomedical Materials Research, Part B: Applied Biomaterials* 2008;88B:597–610. *Review of properties, clinical applications and limitations of calcium sulfates.*

Wang F, Banerjee D, Liu Y, Chen X, Liu X, Upconversion nanoparticles in biological labeling, imaging and therapy, *Analyst* 2010;135:1839–54. *Review of optical bio-labeling and bio-imaging techniques using upconversion nanoparticles.*

Weinstein J S, Varallyay C G, Dosa E, *et al.*, Superparamagnetic iron oxide nanoparticles: diagnostic magnetic resonance imaging and potential therapeutic applications in neurooncology and central nervous system inflammatory pathologies: a review, *Journal of Cerebral Blood Flow & Metabolism* 2010;30:15–35. *Comprehensive review of the applications of iron oxide nanoparticles, especially in the central nervous system.*

Wu C, Chang J, Synthesis and apatite-formation ability of akermanite, *Materials Letters* 2004;58:2415–17. *Brief discussion of fabrication and biocompatibility of akermanite.*

### Class 3 materials: polymeric materials

Ahmed T A E, Dare E V, Hincke M, Fibrin: a versatile scaffold for tissue engineering applications, *Tissue Engineering Part B* 2008;14(2):199–210. *Reviews fibrin hydrogels, glues and scaffolds.*

Allison D D, Grande-Allen K J, Hyaluronan: a powerful tissue engineering tool, *Tissue Engineering* 2006;12:2131–40. *Review of the tissue engineering potential for hyaluronan and the stimulation of the endogenous production of HA by cells.*

Almine J F, Bax D V, Mithieux S M, *et al.*, Elastin-based materials, *Chemical Society Reviews* 2010;39:3371–9. *Review of the physical and biological properties of elastin-based materials.*

Altman G H, Diaz F, Jakuba C, *et al.*, Silk-based biomaterials, *Biomaterials* 2003;24:401–16. *Extensive discussion of the properties of silk-based biomaterials.*

Badylak S F, Xenogeneic extracellular matrix as a scaffold for tissue reconstruction, *Transplant Immunology* 2004;12:367–77. *Discussion of the principles of the ECM scaffold materials.*

Batrakova W V, Kabanov A V, Pluronic block copolymers: evolution of drug delivery concept from inert nanocarriers to biological response modifiers, *Journal of Controlled Release* 2008;130:98–106. *Review of the biological response modifying activities of Pluronics and their application in nanomedicine.*

Boland E D, Coleman B D, Barnes C P, *et al.*, Electrospinning polydioxanone for biomedical applications, *Acta Biomaterialia* 2005;1:115–23. *Discusses the electrospinning technology adapted for polydioxanone.*

Bruggeman J P, de Bruin B-J, Bettinger C J, Langer R, Biodegradable poly(polol sebacate) polymers, *Biomaterials* 2008;29:4726–35. *Describes development of biodegradable polymers based on structural units endogenous to human metabolism.*

Caldorera-Moore M, Peppas N A, Micro- and nanotechnologies for intelligent and responsive biomaterial-based medical systems, *Advance Drug Delivery Reviews* 2009;61:1391–1401. *Reviews microscale and nanoscale environmentally responsive biomaterials for sensors and drug delivery.*

Castro C I, Briceno J C, Perfluorocarbon-based oxygen carriers; review of products and trials, *Artificial Organs* 2010;34:622–34. *Presentation of commercial products based on perfluorocarbon materials and their medical applications.*

Chen G-Q, Wu Q, The application of polyhydroxyalkanoates as tissue engineering materials, *Biomaterials* 2005;26:6565–78. *Extensive review of the PHB family of biodegradable polymers.*

Crapo P M, Gilbert T W, Badylak S F, An overview of tissue and whole organ decellularization processes, *Biomaterials* 2011;32:3233–43. *Extensive review of procedures and outcomes of different decellularization techniques.*

Czaja W, Krystynowicz A, Bielecki S, Brown R M, Microbial cellulose – the natural power to heal wounds, *Biomaterials* 2006;27:145–51. *Essay on the potential of bacterial cellulose.*

Dimitrov I, Trzebicka B, Muller A H E, Dworal A, Tsvetanov C, Thermosensitive water-soluble copolymers with doubly responsive reversibly interacting entities, *Progress in Polymer Science* 2007;32:1275–43. *Demonstration of amphiphilic water-soluble, multi-sensitive copolymer systems with thermoresponsive properties.*

Dutta N K, Choudhury N R, Truong M Y, Kim M, Elvin C M, Physical approaches for fabrication of organized nanostructure of resilin-mimetic elastic protein rec1-resilin. *Biomaterials* 2009;30:4868–76. *Practical aspects of resilin mimetic elastic proteins.*

Falco E E, Patel M, Fisher J P, Recent developments in cyclic acetal biomaterials for tissue engineering, *Pharmaceutical Research* 2008;25:2348–56. *Review of cyclic acetal polymers, polyacetals and polyketals.*

Goda T, Ishihara K, Soft contact lens biomaterials from bioinspired phospholipid polymers, *Expert Review of Medical Devices* 2006;3:167–74. *A discussion of the use of phospholipid polymers that enhance the biocompatibility and anti-fouling properties of soft contact lenses.*

Gong S, Wang H, Sun Q, Xue S-T, Wang J-Y, Mechanical properties and in vitro biocompatibility of porous zein scaffolds, *Biomaterials* 2006;27:3793–9. *Examination of the biocompatibility of porous zein scaffolds.*

Gopferich A, Tessmar J, Polyanhydride degradation and erosion, *Advanced Drug Delivery Reviews* 2002;54:911–31. *Reviews the major classes of anhydrides and the mechanisms of degradation and erosion of polyanhydrides.*

Guimard N K, Gomez N, Schmidt C E, Conducting polymers in biomedical engineering, *Progress in Polymer Science* 2007;32:876–921. *Extensive review of the structure–property relationships in conducting polymers.*

Hakkarainen M, Migration of monomeric and polymeric PVC plasticizers, *Advances in Polymer Science* 2008;211:159–85. *Discussion of DEHP plasticizer and polymeric alternatives.*

Harris J M, Chess R B, Effect of pegylation on pharmaceuticals, *Nature Reviews Drug Discovery* 2003;2:214–21. *Review on the principles of regulation for drug delivery.*

Heller J, Barr J, Ng S Y, Abdellauoi K S, Gurny R, Poly(ortho esters): synthesis, characterization, properties and uses, *Advanced Drug Delivery Reviews* 2002;54;1015–39. *Review of history and optimization of poly(ortho esters).*

Hill P, Brantley H, van Dyke M, Some properties of keratin biomaterials: Kerateines. *Biomaterials* 2010;31:585–93. *Discussion of experimentally determined biocompatibility properties of keratin-based biomaterials.*

Ishihara K, Takai M, Bioinspired interface for nanobiodevices based on phospholipid polymer chemistry, *Journal of the Royal Society Interface* 2009;6:S279–91. *Review of phospholipid surfaces on medical devices.*

Iwasaki Y, Ishihara K, Phosphorylcholine-containing polymers for biomedical applications, *Analytical and Bioanalytical Chemistry* 2005;381:534–46. *Review of the chemistry and biochemistry of phosphorylcholine-based polymers.*

Khandare J, Mohr A, Calderon M, *et al.*, Structure-biocompatibility relationship of dendritic derivatives, *Biomaterials* 2010;31:4268–77. *Demonstrates that dendritic polyglycerols can be synthesized and used for delivery of therapeutic agents.*

Kumar N, Langer R S, Domb A J, Polyanhydrides: an overview, *Advanced Drug Delivery Reviews* 2002;54:889–910. *Review of structure of the various polyanhydrides.*

Kurtz S M, Devine J N, PEEK biomaterials in trauma, orthopedic and spinal implants, *Biomaterials* 2007;28:4845–69. *Extensive review of materials science, biocompatibility and clinical applications of PEEK-based materials.*

Lee H, Dellatore S M, Miller W M, Messersmith P B, Mussel-inspired surface chemistry for multi-functional coatings, *Science* 2007;318:426–30. *Report of procedures to coat a variety of materials with surface-adherent polydopamine films that mimic the adhesive proteins found in mussels.*

Lee J, Shanbhag S, Kotov N A, Inverted colloidal crystals as three-dimensional microenvironments for cellular co-culture. *Journal of Material Chemistry* 2006;16:3558–64. *Discussion of the principles of inverted colloidal crystals.*

Lee S, Yang S C, Heffernan M J, Taylor W R, Murthy N, Polyketal microparticles: a new delivery vehicle for superoxide dismutase, *Bioconjugate Chemistry* 2007;18:4–7. *Initial paper introducing polyketals for drug delivery.*

Letellier P, Mayaffre A, Turmine M, Solubility of nanoparticles; nonextensive thermodynamics approach, *Journal of Physics: Condensed Matter* 2007;19:4362–99. *Analysis of thermodynamics of nanoparticle solubility characteristics; complex but good introduction,*

Martens P, Blundo J, Nilasaroya A, *et al.*, Effect of poly (vinyl alcohol) macromer chemistry and chain interactions on hydrogel mechanical properties, *Chemistry of Materials* 2007;19:2641–8. *Detailed analysis of poly(vinyl alcohol) and the structure–property relationships in hydrogels.*

Martin D J, Poole Warren L A, *et al.*, Polydimethylsiloxane/polyether-mixed macrodiol-based polyurethane elastomers: biostability, *Biomaterials* 200;21:1021–9. *Description of the synthetic chemistry underpinning the production of biostable polyurethanes using polydimethylsiloxane macrodiol.*

Nguyen K F, West J L, Photopolymerizable hydrogels for tissue engineering applications, *Biomaterials* 2002;23:4307–14. *Reviews the technology of photopolymerization of hydrogels.*

Otsuka H, Nagasaki Y, Kataoka K, PEGylated nanoparticles for biological and pharmaceutical applications, *Advanced Drug Delivery Reviews* 2003;55:403–19. *Discusses the principles of pegylation and medical uses.*

Oudshoorn M H M, Rissmann R, Bouwstra J A, Hennink W E, Synthesis and characterization of hyperbranched polyglycerol hydrogels, *Biomaterials* 2006;27:5471–9. *Describes the polymer chemistry of polyglycerol hydrogels.*

Pinchuk L, Wilson G J, Barry J J, *et al.*, Medical applications of poly(styrene-block-isobutylene-block-stryrene) ("SIBS"), *Biomaterials* 2008;29:448–60. *Thorough review of SIBS applications.*

Qu X-H, Wu Q, Zhang K-Y, Chen G Q, In vivo studies of poly(3-hydroxybutyrate-co-3-hydroxyhexanoate) based polymers; biodegradation and tissue reactions, *Biomaterials* 2006;27:3540–8. *Discusses experimental evidence of the degradation of PHB-based materials.*

Rinaudo M, Main properties and current applications of some polysaccharides as biomaterials, *Polymer International* 2008;57:397–430. *Extensive background to the applications of polysaccharides in medical technology.*

Rubehn B, Stieglitz T, In vitro evaluation of the long term stability of polyimide as a material for neural implants, *Biomaterials* 2010;31:3449–58. *Examination of the biocompatibility of polyimide neural electrodes.*

Santerre J P, Woodhouse K, Laroche G, Labow R S, Understanding the biodegradation of polyurethanes: from classical implants to tissue engineering materials, *Biomaterials* 2005;26:7457–70. *Provides an analysis of the mechanisms of polyurethane degradation.*

Sawant R R, Torchilin V P, Liposomes as "smart" pharmaceutical nanocarriers, *Soft Matter* 2010; 6: 4026–44. *A thorough review of responsive liposomes.*

Singh R S, Saina G K, Kennedy J F, Pullulan: microbial sources, production and applications, *Carbohydrate Polymers* 2008;73:515–31. *Critical appraisal of fungal exopolysaccharides including pullulan.*

Sobieraj M C, Rimnac C M, Ultra high molecular weight polyethylene; mechanics, morphology and clinical behavior, *Journal of the Mechanical Behavior of Biomedical Materials* 2009;2:433–43. *Detailed discussion of the mechanical properties of UHMW polyethylene.*

Soon A S C, Stabenfeldt S E, Brown W E, Barker T H, Engineering fibrin matrices; the engagement of polymerization pockets through fibrin knob technology for the delivery and retention of therapeutic proteins, *Biomaterials* 2010;31:1944–54. *Presents the technology for the therapeutic uses of engineered fibrin.*

Tangpasuthadol V, Pendharkar P, Kohn J, Hydrolytic degradation of tyrosine-derived polycarbonates, a class of new biomaterials. Part 1: Study on model compounds, *Biomaterials* 2000;21:2371–8. *Discusses rationale for the development of biodegradable polycarbonate that has hydrolytically labile bonds in the repeating units of the polymer chain.*

Tatai L, Moore T G, Adhikari R, *et al.*, Thermoplastic biodegradable polyurethanes; the effect of chain extender structure on properties and in vitro degradation, *Biomaterials* 2007;28:5407–17. *Experimental analysis of the role of molecular structure on the properties of degradable polyurethanes.*

Teeri T T, Brumer H, Daniel G, Gatenholm P, Biomimetic engineering of cellulose-based materials, *Trends in Biotechnology* 2007;25:299–306. *Overview of the structures of plant cell walls and approaches to mimic these cellulosic materials.*

Torchilin V P, Micellar nanocarriers: pharmaceutical perspectives, *Pharmaceutical Research* 2007;24:1–8. *Reviews the structure of nanoscale micelles, micelle-forming compounds and amphiphilic copolymers for drug delivery.*

Vicennati P, Giuliano A, Ortaggi G, Masotti A, Polyethylenimine in medicinal chemistry, *Current Medicinal Chemistry* 2008;15:2826–39. *Detailed discussion of the chemistry of PEI.*

Wang Y, Ameer G A, Sheppard B J, Langer R, A tough biodegradable elastomer, *Nature Biotechnology* 2002;20:602–6. *Explains the rationale and synthetic chemistry of poly(glycerol-sebacate).*

Welle A, Kroger M, Doring M, *et al.*, Electrospun aliphatic polycarbonates as tailored tissue scaffold materials, *Biomaterials* 2007;28:2211–19. *Describes synthesis and electrospinning of aliphatic polycarbonates.*

Yang C, Hillas P J, Baez J A, *et al.*, The application of recombinant human collagen in tissue engineering, *BioDrugs* 2004;18:103–19. *Discusses collagens and the prospects for making individual collagen types by recombinant technologies.*

Zelzer M, Ulijn R V, Next generation peptide nanomaterials: molecular networks, interfaces and supramolecular functionality, *Chemical Society Reviews* 2010;39:3351–7. *Analysis of the functionality of peptide nanomaterials.*

Zhu J, Bioactive modification of poly(ethylene glycol) hydrogels for tissue engineering, *Biomaterials* 2010;31:4639–56. *Detailed and informative review of PEG hydrogels.*

### Class 4 materials: carbon materials

Giacalone F, Martin N, Fullerene polymers; synthesis and properties, *Chemical Reviews* 2006;106:5136–90. *Detailed discussion of chemistry of fullerene polymers.*

Grill A, Diamond-like carbon: state of the art, *Diamond and Related Materials* 1999;8:428–34. *Review of the structure and fabrication of DLC.*

Harrison B S, Atala A, Carbon nanotubes applications for tissue engineering, *Biomaterials* 2007;28:344–53. *Review of performance of carbon nanotubes and potential medical applications.*

Hu W, Peng C, Luo W, *et al.*, Graphene-based antibacterial paper. *ACS Nano* 2010;4:4317. *Discussion of the antibacterial properties of graphene.*

LaVan D A, Padera R F, Friedmann T A, *et al.*, In vivo evaluation of tetrahedral amorphous carbon, *Biomaterials* 2005;26:465–73. *Biocompatibility of tetrahedral amorphous carbon.*

Markovic Z. Trajkovic V, Biomedical potential of the reactive oxygen species generation and quenching by fullerenes (C60), *Biomaterials* 2008;29:3561–73. *Mechanisms and biological consequences of ROS generation by fullerenes.*

Roy R K, Lee K-R, Biomedical applications of diamond-like carbon coatings; a review. *Journal of Biomedical Materials Research, Part B Applied Biomaterials* 2007;83B:72–84. *Discussion of the advantages and disadvantages of DLC materials and coatings.*

### Class 5 materials: composite materials

Saito N, Aoki K, Shimizu M, *et al.*, Application of carbon fibers to biomaterials: a new era of nano-level control of carbon fibers after 30 years of development, *Chemical Society Reviews* 2011;40:3824–34. *Review of the history of carbon fiber applications to biomaterials and the role of carbon nanofibers in biomaterial composites.*

### Class 6 materials: engineered biological components

Crapo P M, Gilbert T W, Badylak S F, An overview of tissue and whole organ decellularization processes, *Biomaterials* 2011;32:3233–43. *Review of principles and practical aspects of decellularization.*

# 9 Infrastructure of the biomaterials industry

In this chapter we shall address many of the practices and procedural issues that affect and control the production, supply and clinical use of biomaterials and products that are based on these materials. These are not trivial matters. They should be considered as essentials, rather than as afterthoughts or "nice-to-haves." The reason why these issues are addressed towards the end of this book and not at the beginning is a reflection of the need for the reader to first understand the essential scientific, clinical and technological basis before coming to grips with the infrastructure issues, rather than any suggestion of lesser importance. This chapter covers the regulation of biomaterials-based health care products, the pre-clinical testing and clinical evaluation of materials and products, and the ethical, legal and economic matters that have a major influence over this industry.

## 9.1 Introduction

Although it could be argued that, globally, the biomaterials and medical technology industries are quite heterogeneous and fragmented, there are several facets of the industry that are highly structured. These deserve discussion in a textbook on the essentials of biomaterials science. The fundamental reason for this is that biomaterials can do harm as well as good and it is essential that they are used as safely and effectively as possible. This will not happen if each application of a medical device that incorporates biomaterials is developed in isolation, in ignorance of the relevant history of that type of device and without regard to the broader issues of patient (and doctor) safety and the ethical and legal aspects of clinical applications. That there are standards and codes of practice in industrial sectors where safety is a primary concern, such as in aerospace and nuclear industries, is no surprise. The sheer volume of medical devices used on a global scale, and the potential number of individuals that they can affect, for good or bad, suggests that there should be extensive control over these products as well.

There are numerous facets of this infrastructure. The first directly concerns patient safety and the existence of legally enforceable industry regulations. We shall therefore discuss regulatory control of medical technology through pre-market approval processes. Whenever industry regulations are written, they have to refer to standards and testing protocols, which again we discuss here. Once a product is on the market, new types of issues arise. In most cases, especially those whose performance is not life threatening, clinical trials are not required before widespread usage is authorized, mainly because of the impact on commercial viability of the significant time delays that would be involved. Records have to be made, however, of any notable safety issues; we discuss here the significance of both clinical trials and the post-market surveillance of products. Naturally, and inevitably, problems do arise and not all patients are satisfied with the outcomes of biomaterials-based therapies. In extreme cases, patients die or are severely injured through this use. Because in many countries there are legal requirements to produce devices that are safe and not defective, the occurrence of failures often leads to litigation, either through product liability (i.e., it is assumed to be the fault of the device) or medical malpractice (i.e., it appeared to be the doctor's fault). In many ways the outcome of legal proceedings has shaped the biomaterials industry over the years, and we take note of this here.

Three additional factors need to be considered. First are the ethical issues associated with these biomaterials-related therapies. Then there are the difficult questions of health economics, i.e., who can afford such technologies and who should pay, and how much. And finally there is the question of intellectual property: who owns the ideas and the technologies, and how does this impact on product availability.

## 9.2 Principles of medical technology regulation

It is widely accepted that certain groups of manufactured products or services have the power or potential to cause harm as well as good, and that some form of control over their use is essential. There are controls over the use of materials used in products that may catch fire, controls over the construction of buildings through various codes in order to provide for electrical safety, controls over the manufacture of aircraft, and controls over the construction of nuclear installations. These are all obvious and sensible systems. In medical technology, it has become clear over the last couple of decades that the regulation of products is also sensible since medical devices can be associated with adverse events as well as positive outcomes. The purpose of this regulation is to minimize the possibility of such adverse events. Regulation is a matter for governments, and therefore they have to be discussed on a regional basis.

There are some guiding principles that have controlled the development of the regulatory procedures. The first is that there have to be clear definitions of the different types of product. At a very simple level, health care products that require regulation may be divided into those that are deemed to be pharmaceutical products and those that are considered to be medical devices. Over time, this distinction has become blurred. We have, on the one hand, those products that relate to drug delivery and which could be considered as drug–device combinations, or devices that have ancillary drug action, or drugs that have ancillary device function. On the other hand, we have products that are neither classical drugs nor devices but contain live cells, as in tissue engineering or cell therapy products. We also have to consider the borderline between health care products and cosmetics, this borderline being influenced by the nature of the products themselves, the manufacturers' claims about their activity and purpose, and the marketing emphasis given to them. Different regulatory bodies have taken different positions on all of these definitions and criteria.

The second is that for health care products that contain biomaterials, the degree of regulatory control should be proportional to the level of risk. For this reason, all regulatory systems stratify products on the basis of this risk, which may be determined by the degree of invasiveness, the duration of contact with the body and so on. The classification of products that arises from the risk stratification leads to different required levels of pre-clinical testing and the monitoring of clinical performance.

The third is that the regulatory process should give an acceptable balance of risk and innovation. If a process is dominated by the mantra that the device should never do any harm, then there will be little or no innovation and, perversely, patients could be denied the benefits of new technologies because the regulatory pathway was overwhelmingly difficult. It has to be borne in mind that compliance with regulations is both time and resource consuming, such that the very difficult regulations would make it unacceptably expensive or impractical to bring some new developments to market.

This latter point leads to the dilemma of what actually constitutes a new development or a new product. Does a minor change to a device automatically mean that it is a new device that requires the full implementation of regulatory testing? Is it possible to use data on the performance of existing devices in order to justify a very light touch with the regulatory control over a similar but new device?

With the design and manufacturing of products, regulatory control is largely concerned with the establishment of, and adherence to, quality systems and quality control, the guiding principle being that having qualified a type of device as being "effective and safe," it is essential that every individual device that gets to a patient is of exactly the same quality.

The pre-clinical testing that is required, which means the testing that has to be performed on the device and its components before it is ever implanted in a human, includes establishing, as far as possible, the functional efficiency of the product and its biological safety. A series of standards, at both national and international levels, exist to guide manufacturers through these processes. It is important to note that such standards, which will be discussed briefly later, are not usually mandatory. However, manufacturers are well advised to use them, wherever possible, and to provide detailed justification when they chose to use alternative methods, or when they chose not to carry out some tests at all.

The major question now arises as to how a health care product is introduced into clinical practice. New drugs require extensive clinical trials and there are well-defined pathways towards the general release of drugs, following four phases of trial. Medical devices are quite different. For example, with drugs, large numbers of patients can be recruited to a trial since it may only involve taking tablets by mouth over a short period of time and monitoring outcomes and adverse events. Medical devices, and especially implantable medical devices, are used on an individual basis, where the outcomes may not be assessed effectively for many years. Moreover, as we have seen, these outcomes may be profoundly affected by the skills of the clinical staff. Clinical trials are therefore very expensive and may not yield helpful data for a long time. All too often, product development timetables produce newer versions of devices before the results of clinical trials are fully known. More and more, the medical specialties are turning to registries for the monitoring of implantable devices, which will be discussed later. Regulators usually require a system of post-market surveillance for the monitoring of adverse events.

### 9.2.1 Risk assessment and risk management

Most regulatory procedures are, as noted above, based on the interpretation of risk. Here two terms must be separated and understood, risk assessment and risk management, in addition to risk itself. All of these terms have slightly different meanings depending on the specific situations. With health care products, the questions about risks tend to follow those of exposure to chemical substances. Here *risk* is defined as a measure of the probability that a harmful event, such as death, injury or loss, ensuing from exposure to a chemical or physical agent may occur under specified conditions of use. *Risk assessment* involves the identification and quantification of risk resulting from a specified use or occurrence of a chemical or physical agent. *Risk management* is the decision-making process to select the optimal steps for reducing a risk to an acceptable level. In most spheres, risk management has to take into account scientific, technical, sociological and legal factors.

The assessment and management of risk are not usually trivial tasks when contemplating the use of new biomaterials and new health care products. With chemicals and medicinal products, the process of risk assessment will take into account dose and effect relationships, coupled to exposure assessments that consider both anticipated and accidental uses. This is not so easy to do with exposure to risk, whether chemical, physical or otherwise, for most biomaterials applications. This is especially true with nanoparticle formulations and biological products. It should be of no surprise that sometimes manufacturers and regulators get this risk–benefit analysis wrong.

We should also note that the manufacturer has considerable responsibility here. In the development of a new product, a procedure that identifies potential risks should be carried out. There are a number of procedural systems in use here, with the Failure Modes and Effects Analysis (FMEA) or its derivative Failure Modes and Effects Criticality Analysis (FMECA) being the most common. With new products requiring the highest level of regulatory scrutiny, the FMEA documentation should be included in the Device (or Design) History File, which gives a clear view of the risk assessment and the steps undertaken with risk management. The FMEA cycle is shown in Figure 9.1. Step 1 involves the detection of a potential failure mode. Step 2 is an assessment of the severity, graded on a scale of 1 to 10, should the event occur. Step 3 is an assessment of the probability of the event occurring, graded on a scale of 1 to 10. Step 4 is an assessment of the ease of

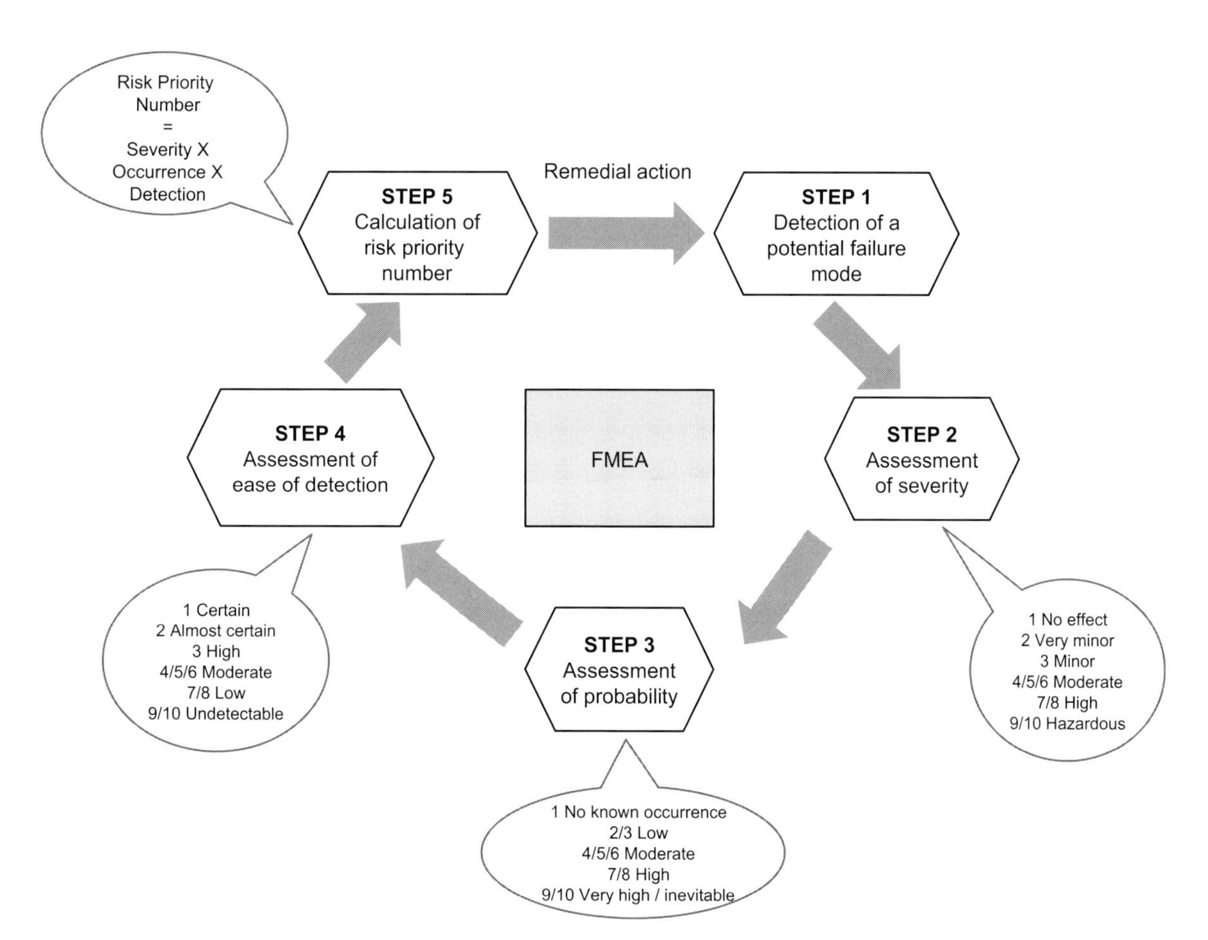


**Figure 9.1** The Failure Modes and Effects Analysis cycle. This cycle starts at Step 1, which determines that there is a potential failure mode, and then passes through steps to assess severity, probability and ease of detection, leading to the calculation of a risk priority number and decisions on any remedial action that is needed.

detection of the event (although this is not always included) and Step 5 is the calculation of the Risk Priority Number, which is the product of the assigned grades. The final step is the determination of the priority of risks and recommendations of the steps necessary to reduce the risk associated with each event.

As noted above, the focal point of risk assessment with medical devices is the product that is used in or on the patient. The considerations include, naturally, the nature of the material(s) used in the construction of the product, but also involve aspects of design, manufacturing, testing and clinical use. In all of these individual areas, manufacturers have a duty to comply with broader industry regulations. With materials, for example, there are many non-medical-specific regulations, where the highly focused medical device regulations do not necessarily preclude the application of broader requirements. Within Europe, recently introduced legislation, known as REACH, the Registration, Evaluation, Authorisation and Restriction on Chemical Substances, aims to protect human health, especially in the workplace environment, from inappropriate uses of chemicals. Whilst there are often some safeguards to protect highly strategic or specialized materials, including minimum amounts of manufactured materials that can be covered by a regulation, manufacturers of new biomaterials have to take these factors into account. It should also be noted that while some substances can be included in a list of Generally Recognized as Safe (GRAS) substances in order to avoid excessive and redundant testing, such that their use as food additives may be considered as acceptable, this rarely applies to health care products.

## 9.2.2 Regulation in the USA

Regulation of health care products in the USA is carried out by the Food and Drugs Administration (FDA). This federal agency was established in 1930 in order to provide some degree of regulation of foods and pharmaceutical products. Although the legislation that created the FDA included clauses that enabled medical devices to be subjected to such regulation, no details were included at that time and in reality no such regulation took place until the Medical Device Amendment Act was passed in May 1976, which set out the provisions for this regulation.

There are several offices within the FDA that have separate and diverse responsibilities. Drugs are regulated by CDER, the *Committee for Drug Evaluation and Research*. We shall not consider their activities here but their presence does have relevance to the regulation of biomaterials. Medical devices are regulated by CDRH, the *Committee on Devices and Radiological Health*. A group of products collectively known as Biologics is the subject of CBER, the *Committee of Biologics Evaluation and Research*. Some products contain elements of more than one type and come under the auspices of the *Office of Combination Products* (OCP).

### Definitions

A few definitions are in order here. From a statutory perspective, the US Food Drugs and Cosmetics Act determines that the term "*drug*" means:

(1) articles recognized in the official United States Pharmacopoeia, official Homoeopathic Pharmacopoeia of the United States, or official National Formulary, or any supplement to any of them; and
(2) articles intended for use in the diagnosis, cure, mitigation, treatment, or prevention of disease in man or other animals; and
(3) articles (other than food) intended to affect the structure or any function of the body of man or other animals; and
(4) articles intended for use as a component of any articles specified in clause (1), (2) or (3).

The same act provides that the term "*device*" means an instrument, apparatus, implement, machine, contrivance, implant, *in vitro* reagent, or other similar or related article, including any component, part, or accessory, which is:

(1) recognized in the official National Formulary, or the United States Pharmacopeia, or any supplement to them;

(2) intended for use in the diagnosis of disease or other conditions, or in the cure, mitigation, treatment, or prevention of disease, in man or other animals; or
(3) intended to affect the structure or any function of the body of man or other animals, and which does not achieve its primary intended purpose through chemical action within or on the body of man or other animals and which is not dependent upon being metabolized for the achievement of its primary intended purposes.

There are several definitions of biologics, or biological products, which is not surprising since they comprise a varied group of products that are often best characterized by analogies or examples. In practice, "*biologics*" includes a wide range of products such as vaccines, blood and blood components, allergenics, somatic cells, gene therapy agents, tissues and recombinant therapeutic proteins. Each of these groups has their own definitions.

The term "*combination product*" includes:

(1) a product comprised of two or more regulated components, i.e., drug/device, biologic/device, drug/biologic, or drug/device/biologic, that are physically, chemically, or otherwise combined or mixed and produced as a single entity;
(2) two or more separate products packaged together in a single package or as a unit and comprised of drug and device products, device and biological products, or biological and drug products;
(3) a drug, device, or biological product packaged separately that according to its investigational plan or proposed labeling is intended for use only with an approved individually specified drug, device, or biological product where both are required to achieve the intended use, indication, or effect and whereupon approval of the proposed product the labeling of the approved product would need to be changed, e.g., to reflect a change in intended use, dosage form, strength, route of administration, or significant change in dose; or
(4) any investigational drug, device, or biological product packaged separately that according to its proposed labeling is for use only with another individually specified investigational drug, device, or biological product where both are required to achieve the intended use, indication, or effect.

We shall consider each of these in turn but only in brief, concentrating on the principles. The subject is extremely important for legal and manufacturing perspectives but is only marginally related to biomaterials science. It is also constantly changing and latest details are best found on the FDA website, www.fda.gov.

## Medical devices

Dealing with medical devices first, these are classified into Class I, II and III. The FDA has established classifications for approximately 1700 different generic types of devices and grouped them into 16 medical specialties referred to as panels. Device classification depends on the intended use of the device and also upon indications for use. Classification is also risk-based, that is, the risk the device poses to the patient and/or the user is a major factor in the class it is assigned. Class I includes devices with the lowest risk and Class III includes those with the greatest risk.

From a practical perspective, the major feature of the classification is the determination whether a product can be marketed with a Premarket Notification, known as a 510(k), or whether it requires Premarket Approval (PMA). Most Class I and II devices require the 510(k) route in which a manufacturer submits evidence to the FDA that the product is substantially equivalent to a device(s) that is or has been legally marketed in the USA, which is called a predicate device. Products requiring PMAs are Class III devices that pose a significant risk of illness or injury, or devices found not substantially equivalent to Class I or II predicate devices through the 510(k) process. The PMA process is more involved and includes the submission of clinical data to support claims made for the device. An Investigational Device Exemption (IDE) allows a device to be used in a clinical study in order to obtain the necessary safety and effectiveness data required to support a PMA application. Studies with devices of significant risk must be approved by FDA and by an Institutional Review Board (IRB) before the study can start.

Obviously, the 510(k) route is simpler, faster and much less expensive, and manufacturers understandably seek evidence that supports a suitable predicate device for this route. On occasion this is problematic since substantial equivalence is not easily defined or shown, and the borderline between 510(k) and PMA may shift in order to recognize this.

The FDA procedures for medical devices are summarized in Box 9.1.

## Biologics

With biologics, the matter is far less well developed than medical devices and regulatory procedures do vary from one type of product to another. They have more similarities to drug regulation procedures than those for medial devices.

Human cells or tissue intended for implantation, transplantation, infusion, or transfer into a human recipient are regulated as a human cell, tissue, and cellular- and tissue-based product or HCT/P. Examples of such tissues are bone, skin, corneas, ligaments, tendons, dura mater, heart valves, hematopoietic stem/progenitor cells derived from peripheral and cord blood, oocytes and semen. CBER does not regulate the transplantation of vascularized human organ transplants such as kidney, liver, heart, lung or pancreas since the Health Resources Services Administration (HRSA) oversees the transplantation of vascularized human organs. Establishments that deal with human tissue are required to screen and test donors, and to follow strict procedures for the prevention of the spread of communicable disease. They register and list their HCT/P procedures with FDA. Current good tissue practices for HCT/Ps have also been established.

CBER also has the responsibility to regulate human gene therapy products, although the FDA had not approved any human gene therapy product for sale before 2011. In addition, CBER has the responsibility for oversight of xenotransplantation procedures that involves the transplantation, implantation or infusion into a human recipient of either live cells, tissues, or organs from a nonhuman animal source, or human body fluids, cells, tissues or organs that have had *ex vivo* contact with live nonhuman animal cells, tissues or organs.

## Combination products

We noted above that there are several different types of combination product. Examples of combination products where the components are physically, chemically or otherwise combined include:

- a monoclonal antibody combined with a therapeutic drug;
- a device coated or impregnated with a drug or biologic such as a drug-eluting stent, pacing lead with steroid-coated tip or a catheter with antimicrobial coating;
- transdermal patches;
- skin substitutes with cellular components;
- an orthopedic implant with growth factors;
- prefilled syringes;
- insulin injector pens.

Examples of combination products where the components are packaged together include drug or biologics packaged with a delivery device and surgical trays with surgical instruments. Combination products where the components are separately provided but labeled for use together include iontophoretic drug delivery patches with a controller.

A combination product is assigned to the Center that will have primary jurisdiction for its premarket review and regulation. This assignment for primary jurisdiction is based on a determination of the primary mode of action product, which is the most important therapeutic action of the combination product. In some situations, a combination product may have two independent modes of action, neither of which is subordinate to the other. Here the FDA follows a procedure to give an assignment based on consistency with other combination products raising similar types of safety and effectiveness questions, or

### Box 9.1 Summary of FDA procedures for medical device regulation

CDRH of the FDA is responsible for regulating firms who manufacture or import medical devices sold in the USA. Medical devices are classified into Class I, II and III, regulatory control increasing from Class I to III. Most Class I devices are exempt from Premarket Notification 510(k); most Class II devices require Premarket Notification 510(k); and most Class III devices require Premarket Approval. Manufacturers and importers of medical devices must register their establishments with the FDA and verified annually; foreign manufacturers must also designate a US agent.

#### Premarket Notification 510(k)

If the device requires the submission of a Premarket Notification 510(k), it cannot be commercially distributed until an authorizing letter of substantial equivalence is received from FDA. A 510(k) must demonstrate that the device is substantially equivalent to one legally in commercial distribution in the United States. Most Class I devices and some Class II devices are exempt from the Premarket Notification 510(k) submission, a list of exempt devices being published by the FDA.

#### Premarket Approval (PMA)

Products requiring PMAs are Class III devices that pose a significant risk of illness or injury, or devices found not substantially equivalent to Class I or II predicates through the 510(k) process. The PMA process is more involved and includes the submission of clinical data to support claims made for the device.

#### Investigational Device Exemption (IDE)

An Investigational Device Exemption (IDE) allows the investigational device to be used in a clinical study in order to collect safety and effectiveness data required to support a Premarket Approval (PMA) application or a Premarket Notification 510(k) submission to FDA. Clinical studies with devices of significant risk must be approved by FDA and by an Institutional Review Board (IRB) before the study can begin. Studies with devices of non-significant risk must be approved by the IRB only before the study can begin.

#### Quality System Regulation (QSR)/Good Manufacturing Practices (GMP)

The quality system regulation includes requirements related to the methods used in and the facilities and controls used for designing, purchasing, manufacturing, packaging, labeling, storing, installing and servicing of medical devices.

#### Labeling

Labeling includes labels on the device as well as descriptive and informational literature.

#### Medical Device Reporting

Incidents in which a device may have caused or contributed to a death or serious injury must to be reported to FDA under the Medical Device Reporting program.

to the center with the best expertise to evaluate the safety and effectiveness questions raised by the product.

### 9.2.3 Regulation in Europe

Procedures in Europe have tended to mirror the political changes that have occurred in this continent in recent decades. In the 1950s and 1960s a process of forming closer relationships between the different countries of Europe began, leading to the formation of the European Union (EU). Although each country that became a member of the EU retained their sovereign status, there has been an increasing level of common practices, and indeed common laws, promoted and policed from the European Commission, based in Brussels. Much of the drive for this harmonization was based on the need for greater collaboration between these countries, which had seen devastating conflict during World War II, and especially greater economic collaboration in order to speed up the post-war recovery. Central to this economic harmonization was the principle of freedom of trade between the countries and the simplification of regulation and trade tariffs that otherwise hindered the movement of goods. The Commission set up a framework of harmonization in which certain types of goods could be qualified as attaining a suitable mark of quality that would permit free movement within the Member States without recourse to further examination within each country. This process was known as the Communite Europeane marking scheme, where such qualified goods were awarded a CE mark. It was into this scheme that medical devices were drawn in the 1990s through the adoption of three Directives dealing with medical devices. Prior to this point, the ability to sell medical devices in any country in Europe had been a matter for countries to deal with separately according to their national laws. Many countries had only rudimentary schemes for examination of medical devices and often all that was required was the registration of the company selling or distributing the goods. Thus the process of regulation of medical devices in Europe was borne out of a desire to permit the cross-border movement of goods rather than the protection of the patient.

One main difference between the USA and European systems is that Europe has no equivalent of the combination product classification. Until very recently there was no equivalent of the biologics category but that has now changed. Drugs and devices are regulated very differently. As noted above, devices are regulated through the CE mark procedure but drugs are not – they are regulated through a very different set of rules, quite similar to the FDA's procedures, with an FDA equivalent central agency, the European Medicines Agency (EMA) in London, taking all responsibility.

#### Medical devices

The regulation of medical devices is governed by the legal framework of the Medical Device Directive, 93/42/EEC. This is under revision but sets out the principles well. The implementation of the Directive is assisted by a number of guidance documents, referred to as MEDDEV Guidance Documents. Again these are constantly updated and the different versions of these documents may be found at the Europa website, http://ec.europa.eu/health/medical-devices. It should be noted that Directive 93/42/EEC covers the main features of medical device regulation but that there are subsidiary directives dealing with active implantable devices and *in vitro* diagnostic systems.

Again we shall only deal with the principles of the regulatory framework here and not the details. The regulatory process is not centralized but is managed by a series of licensed commercial organizations called Notified Bodies. There are many of these all over Europe and a manufacturer may use any one of these so long as it is licensed for the type of product in question. Because Europe does not have a federal structure, each country (or Member State) does have the power to override a decision made by a Notified Body, but this power is rarely used. Each country has an arm or agency of government with this responsibility, for example the Medicines and Healthcare Products Regulatory Agency (MHRA) of the UK,

Agence Francoişe de Sécurité Sanitaire de Produits de Sante (Afssaps) of France and Bundesinstitut für Arzneimittel und Medizinprodukte (BfArM) of Germany.

Europe also has a classification scheme, but this does not involve a database such as at the FDA, where a manufacturer seeks the closest comparator for inclusion in a class, but rather a series of rules. Each product is considered in its own right and a decision tree is followed in order to place it in a class. There are four, not three classes here, I, IIa, IIb and III, again increasing in the level of risk. Risk is stratified according to the intended purpose of the device, the duration of contact with the patient, and the invasiveness of the device. There are 18 rules, including a set of rules for non-invasive devices, a set for invasive devices, a set for active devices and a set for special rules. For example, Rule 1 is the default position that says that all non-invasive devices are in Class 1 (the lowest risk and therefore commanding minimal regulation) unless they have a special feature that is covered by a higher rule. Thus stethoscopes and non-invasive electrodes are in Class 1. However, all non-invasive devices intended for modifying the biological or chemical composition of blood, other body liquids or other liquids intended for infusion into the body, such as hemodialysis concentrates, are in Class IIb, unless the treatment consists of filtration, centrifugation or exchange of gas or heat, in which case they are in Class IIa.

With implantable devices, which obviously concern most biomaterials applications, a transient use usually places a device in Class IIa, and a short-term use (less than 30 days) places it in Class IIb. Devices intended to be implanted for longer than 30 days are placed in IIb or II. Any device that is in direct contact with, or controls, the central nervous system or the central circulatory system are automatically in Class III. Each manufacturer of each product has to establish which rules apply to that product and make a submission.

### Advanced therapy medicinal products

The European Commission struggled for many years to devise regulations to deal with cell therapies and tissue engineering. At one stage it appeared that a Directive similar to that concerned with medical devices would be produced, but ultimately, after a great deal of political and religious arguments had taken place, a new Regulation was introduced in 2007. This Regulation set out specific rules concerning the authorization, supervision and pharmacovigilance of so-called advanced therapy medicinal products. The Commission determined that these products should be subject to a centralized authorization procedure, involving a single scientific evaluation of the quality, safety and efficacy of the product, which would be carried by the European Medicines Agency. In recognizing that the evaluation of advanced therapy medicinal products would often require very specific expertise that goes beyond that associated with the traditional pharmaceutical field the Commission created a Committee for Advanced Therapies, which would be responsible for preparing a draft opinion on the quality, safety and efficacy of each advanced therapy medicinal product for final approval by the Agency's Committee for Medicinal Products for Human Use.

A summary of the key aspects of this Regulation is given in Box 9.2.

## 9.2.4 Regulation in the Asia-Pacific region

The various countries of the Asia-Pacific region generally have their own regulations, which tend to combine some of the elements of European and American procedures.

### Australia

The regulation of medicines, devices, blood products and biologics is subject to oversight by the Therapeutic Goods Administration, TGA, and are based on the Therapeutic Goods Act of 1989. Although Australia has a high level of innovation in health care products, especially in sensory devices, ophthalmology and, recently, in cell therapies, the majority of products are imported and this Act has a main function of controlling

## Box 9.2 Summary of advanced therapy medicinal products regulation of the European Union

### Definitions

"Advanced therapy medicinal product" means any of the following:

- a gene therapy medicinal product;
- a somatic cell therapy medicinal product;
- a tissue engineered product.

"Tissue engineered product" means a product that:

- contains or consists of engineered cells or tissues, and
- is presented as having properties for, or is used in or administered to human beings with a view to regenerating, repairing or replacing a human tissue.

### Context

A tissue engineered product may contain cells or tissues of human or animal origin, or both. The cells or tissues may be viable or non-viable. It may also contain additional substances, such as cellular products, biomolecules, biomaterials, chemical substances, scaffolds or matrices.
Cells or tissues shall be considered "engineered" if they fulfill at least one of the following conditions:

- the cells or tissues have been subject to substantial manipulation, so that biological characteristics, physiological functions or structural properties relevant for the intended regeneration, repair or replacement are achieved;
- the cells or tissues are not intended to be used for the same essential function or functions in the recipient as in the donor.

"Combined advanced therapy medicinal product" means an advanced therapy medicinal product that fulfills the following conditions:

- it must incorporate one or more medical devices or one or more active implantable medical devices, and
- its cellular or tissue part must contain viable cells or tissues, or
- its cellular or tissue part containing non-viable cells or tissues must be liable to act upon the human body with action that can be considered as primary;
- where a product contains viable cells or tissues, the pharmacological, immunological or metabolic action of those cells or tissues shall be considered as the principal mode of action of the product.

Requirements for regulatory submission

- Composition of the product: (a) general description of the product, (b) qualitative and quantitative composition in terms of the active substances and other constituents of the product, (c) a detailed description of these cells or tissues and of their specific origin.
- Clinical particulars: (a) therapeutic indications, (b) instructions for use, (c) contra-indications, (d) special factors and warnings.
- Pharmaceutical form and pharmacological properties: (a) pharmacodynamic properties, (b) pharmacokinetic properties, (c) pre-clinical safety data.
- Quality particulars: (a) list of excipients, (b) incompatibilities, (c) shelf life, (d) special precautions.
- Marketing authorization details.

the quality of imported goods; essentially, any product for which therapeutic claims are made must be included in the Australian Register of Therapeutic Goods before it can be supplied in Australia.

The procedures by which the TGA regulates medicines and medical devices are quite different, this difference largely reflecting those seen in Europe. With medical devices, similar systems as used for the CE mark process are in place, although this is achieved through centralized action at the TGA not through third parties. In summary, the regulatory system utilises the following features:

- a rule-based classification scheme based on the risk posed in use;
- compliance with essential principles of quality, safety and performance of the medical devices, ensuring that the product performs the way the manufacturer intends and that the benefits outweigh the residual risk;
- application of a conformity assessment procedure encompassing quality management systems;
- options as to how compliance with the essential principles can be satisfied;
- the use of recognized standards to demonstrate compliance with the essential principles;
- a comprehensive post-market surveillance and adverse incident reporting program;
- appropriate regulatory controls for the manufacturing processes of medical devices.

Currently, therapeutic products that utilize cells and tissues are subject to differing regulatory requirements, based on the level of risk posed:

- whole organs for transplant are exempt from the legislation;
- tissue for implantation in the human body that is obtained, stored and supplied without any deliberate alteration to its biological or mechanical properties is exempt from the requirements for entry on the Register provided that the institution complies with the Code of GMP (cGMP) for Blood and Human Tissues;
- cell and tissue products that are custom made for a particular person are exempt from the requirements for pre-market approval;
- human tissue and cell extracts, whose principal therapeutic purposes are achieved through chemical, pharmacological, or metabolic actions that are regulated as medicines.

Blood, blood components and hematopoietic progenitor cells are generally regulated as medicines. However, since May 2011 human cell- and tissue-based products have been regulated as a distinct group of therapeutic goods called "biologicals," but with a distinct process that utilizes:

- a rule-based classification system that defines four classes of biologicals based on the risk posed by the product (see below);
- requirements for the risk management approach to be documented through all stages of the product's life, from concept or collection to release and clinical use;
- revision of the cGMP for Blood and Tissues (2000), which will result in a new cGMP for human blood, blood components, human tissues and human cellular therapy products;
- standards for minimization of infectious disease transmission;
- a comprehensive post-market surveillance and adverse event reporting program.

The four classes are:

- Class I means a biological that is declared in the Regulations as a Class 1 biological of low risk;
- Class II, also of low risk, means a biological that is both (i) processed using only one or more of the actions of minimal manipulation and (ii) is for homologous use, or is declared in the Regulations as a Class II biological;
- Class III, of medium risk, means a biological that is processed (i) using a method in addition to any of the actions of minimal manipulation and (ii) in a way that does not change an inherent biochemical, physiological or immunological property, or is declared in the Regulations as a Class III biological;
- Class IV, of high risk, means a biological that is processed (i) using a method in addition to any of the actions of minimal manipulation and (ii) in a way that changes an inherent biochemical, physiological or immunological property, or is declared in the Regulations as a Class IV biological.

### Japan

For many years Japan had a reputation of being a very difficult market to penetrate, especially for overseas and multinational companies. Because the Japanese government felt that this reputation, either real or imaginary, was affecting the availability of new drugs and devices for the population, changes were introduced in the early part of the twenty-first century. Some revisions to the Pharmaceutical Affairs Law were introduced over the period 2002–2005. The new laws first established so-called Market Authorization Holders (MAHs) who had to be established in Japan, which changed the concept that overseas companies had to be represented by a Japanese company to one where the overseas company could establish their own company in the country. A classification scheme was put in place in which Class I devices are of extremely low risk (General Medical Devices), Class II of low risk (Specially Controlled), Class III of medium risk (Highly Controlled) and Class IV of high risk (also Highly Controlled). Different procedures apply to different classes. The Pharmaceutical and Medical Devices Agency (PMDA) was established with primary responsibility for approval of health care products. Up-to-date information may be obtained from www.pmda.go.jp.

### China

China represents a very interesting case. Along with many other aspects of society in this country, there have been profound changes in the use of medical devices, including manufacture and regulatory control, during recent years. Until the late 1990s, the regulations were a little opaque and it was extremely difficult for any overseas manufacturer to gain access to their internal market. In 1998, changes were instituted, including the merger of the Ministry of Health's Department of Drug Administration with the State Pharmaceutical Administration of China to form the State Drug Administration (SDA). This was restructured in 2003 to become the SFDA, the State Food and Drug Administration, which was modeled on the FDA of the USA. This body has responsibility to draft laws and regulations on the management of the safety of food, health foods, cosmetics, drugs and medical devices (see www.sfda.com). Following a series of serious problems in some products that were exported, including some baby foods and contaminated heparin products, the SFDA (now CFDA) has tightened up procedures, which now appear to show equivalence with the USA and Europe.

Many of the definitions and classifications are similar to those discussed in other jurisdictions. There are, as in the USA, three classes of medical devices, although the wording is slightly different:

- Class I medical devices are those for which safety and effectiveness can be ensured through routine administration.
- Class II medical devices are those for which further control is required to ensure safety and effectiveness.
- Class III medical devices are those which are implanted into the human body, or used for life support or sustenance, or pose potential risk to the human body and thus must be strictly controlled in respect to safety and effectiveness.

Key issues in China include the encouragement of research and development into new medical devices and the fact that local rules apply in some provinces and autonomous regions. Medical institutions are allowed to develop their own devices to suit their particular needs and use them within their clinics under the guidance of some licensed doctors.

The CFDA have made considerable efforts to introduce regulations for tissue engineering and cell-based products. They rely heavily on regional centers of excellence for advice and testing, including some in Beijing, Shanghai, Chengdu and Xi'an, although a central body has been established for the purposes of quality control in cell-based products. A series of internal guidance documents has been published for the purposes of regulating a number of products and processes, including skin substitutes, the biomaterial hyaluronic acid, collagen type I, chitosan and alginate, and processes for tissue and cell processing. These developments are taking place rapidly.

## 9.3 Biomaterials and medical device standards and guidelines

Although the regulatory processes work fairly well on a regional basis, it has not been easy to produce a harmonized approach on a global basis, mainly because there are political and cultural differences on matters such as risk perceptions and benefit expectations around the world. What is consistent, however, is the underlying technology and there is reasonable global agreement that regulatory procedures have to be based on sound, consistent testing regimes and standards. These apply to systems for quality manufacturing and pre-clinical testing, and guidelines for materials and designs. One of the most important series of standards is concerned with the assessment of biological safety, which we shall consider separately in the following section. Here some of the other systems are briefly mentioned. Many of the procedures are set out in standard documents produced by the International Standards Organization (ISO), which are written by international panels of experts, and endorsed by national standards institutions.

Most regulatory processes are centered on quality systems. There are very many systems used throughout the world that pertain to quality manufacturing procedures. The most widely used is the ISO standard on quality management systems for medical devices (ISO 13485:2003). Any organization that manufactures medical devices that are subject to regulation for international trade are required to establish, document, implement and maintain a quality management system, in which criteria and methods needed to ensure the operation and control of all processes are appropriate and effective. Documentation is key to these procedures. Although documentation can be in any form, it should logically constitute a quality manual with strict rules for the handling and control of documents. The standard specifically states that top management has to be fully committed to quality systems.

ISO also produces standards for engineering design and biomaterials. These tend to be rather general documents that give guidance on basic engineering design and materials specification, but of course they cannot define fine details of medical device engineering. Most types of implantable device are governed by relevant standards. Take hip replacement procedures, for example (ISO 21535:2007). The standard specifies the minimum range of angular movement between the femoral and acetabular components. These ranges are at least 100° in flexion/extension, 60° in abduction/adduction and 90° in rotation; it is obvious that these figures give a great deal of latitude for manufacturers. There are general guidelines on tolerances, sphericity and surface finish and the thickness of polyethylene acetabular heads. There is little guidance on the materials to be used, with the exception that unalloyed titanium and titanium alloys should not be used in articulating surfaces "unless an appropriate surface treatment is undertaken and demonstrated to be suitable" (ISO 21535:2007). More importantly, and this is true for many standards, detailed guidance is given for the evaluation of the design parameters and pre-clinical functional testing, especially concerning mechanical performance such as endurance and wear testing.

In addition, ISO produces technical reports that offer suggestions for design and validation procedures, which are particularly valuable for some critical situations. For example, ISO/TR 22442-4:2010 provides advice on the elimination and/or activation of transmissible spongiform encephalopathy agents when using animal tissues in medical devices.

Outside of the ISO, there are other, usually national, organizations that play a major role in setting standards and guidance. The USA is the major contributor of such documents. For example, the American Society for Testing and Materials (ASTM) has a significant portfolio for standards in medical and surgical materials and devices. These are produced by the F04 Committee (www.astm.org/standard/committees/F04.html). The standards include those for terminology, implantable devices (arthroplasty, cardiovascular devices, implantable hearing devices, neurosurgical devices, osteosynthesis, plastic and reconstructive surgery, spinal devices, urological devices), tissue-engineering products, biocompatibility test methods, biomaterials (ceramic materials, metallurgical materials, polymeric materials), materials test methods, and surgical instruments.

The FDA also provides extensive documentation to assist in the filing of regulatory submissions. These are usually guidance documents, intended for both industry and their own staff. As such they contain non-binding recommendations but they are usually very detailed, thorough and extensive. For example, the draft guidance for the design, testing and manufacture of heart valves is extensively used by industry as a template for IDE or PMA submissions. This covers a range of pre-clinical *in vitro* assessment (durability, fatigue, corrosion, hemodynamic performance, etc.), pre-clinical animal studies (study design, data analysis, etc.) and clinical investigations (details of investigators and sites, data collection, control data, etc.).

### 9.3.1 Standards for the pre-clinical testing of biomaterials

The discussions in Chapter 3 made it clear that biocompatibility characteristics are crucial determinants of biomaterials and device performance. It follows that the regulatory process for biomaterials-based health care products should aim to determine that these characteristics are appropriate for effectiveness and patient safety. Since biocompatibility is complex and, indeed, refers to more than safety, the standard tests are not (or should not be) discussed in terms of their biocompatibility, but rather in terms of biological safety. ISO have produced a series of tests for this purpose – this series is referred to as ISO 10993. In 2009 there were already 20 different parts of this series, with more in development. The first part (ISO 10993-1:2009, *Biological evaluation of medical devices: evaluation and testing within a risk management process*), summarizes the underlying principles of these evaluations, and introduces the individual components of the series.

A summary of these different parts of ISO 10993 is provided in Box 9.3. These provide a framework for tests rather than a rigid set of rules. Any assessment of biological safety should take into account a number of issues, including the known physical and chemical characteristics of the materials and their previous history in medical

### Box 9.3 Parts of ISO 10993: *Biological evaluation of medical devices*

Note: Parts 1, 2, 7, 9,12, 17, 18 and 19 cover general issues such as animal welfare and reference materials.

Part 3 Genotoxicity, carcinogenicity, reproductive toxicity: a battery of tests employing mammalian or non-mammalian cell culture techniques to determine gene mutations, chromosomal changes and other DNA effects. If any of these tests are positive, *in vivo* mutagenicity tests are usually performed. If there is no information from other sources, carcinogenicity test may be considered (although rare). If the application of the device is relevant to reproductive potential of patients, reproductive and developmental tests may be performed.

Part 4 Interactions with blood: these tests include hemolysis, although this really assesses effects of materials on cell membranes (red blood cells) rather than blood compatibility itself. The main tests are concerned with complement activation and clotting times.

Part 5 *In vitro* cytotoxicity: these tests employ cell culture techniques to determine cell lysis, cell growth inhibition, colony formation, etc.

Part 6 Local effects after implantation: these tests determine local pathologic effects in suitable animal models, taking into account the intended route and duration of contact.

Part 10 Irritation and skin sensitization: includes tests for delayed-type hypersensitivity concerning contact sensitization and intracutaneous reactivity.

Part 11 Systemic toxicity: estimates the potential harmful effects arising from acute contact in a suitable animal model. In some situations these acute tests may be supplemented by subacute and subchronic toxicity.

Part 13 Degradation products from polymers; Part 14 Degradation products from ceramics; Part 15 Degradation products from metals and alloys: these tests are required if the device is intended to be biodegradable or if potentially toxic products are likely to be released over times greater than 30 days.

Part 16 Toxicokinetics of degradation products and leachable: tests to evaluate absorption, distribution, metabolism and excretion (ADME) of relevant chemicals, assessed on the basis of the composition of the product.

Part 20 Immunotoxicology: tests to be considered where the nature of the materials suggests there is a potential for immunotoxicological effects.

devices. The presence of intended additives, process contaminants and residues, of leachable substances and degradation products all have to be taken into account. The selection of tests for a particular device has to be based on the end-use application and the tests themselves undertaken on the final product, where that is possible, or on extracts from the final product.

The selection of tests to be performed according to these standards also takes into account the nature and duration of the intended contact between the device and the patients. With respect to the nature of the contact, devices are categorized according to whether there will be surface contact, external communication or implantation. In the first case there could be contact with skin, with mucosal membranes or breached/compromised surfaces. External communicating devices include indirect blood path contact, direct contact with circulating blood or tissue/bone/dentin surfaces. Implantation devices primarily include those in contact with bone, soft tissue or blood. Categorization by duration of contact places devices in either *limited exposure*, up to 24 hours, *prolonged exposure*, more than 24 hours but less than 30 days, and *permanent contact*, at greater than 30 days. The standard includes a table that indicates the panel of tests that are recommended for types of device that come into each of these categories. At the lowest level (limited surface contact up to prolonged mucosal membrane contact) three types of test are mandated, cytotoxicity, sensitization and irritation/intracutaneous reactivity. An implantable device in permanent contact with blood should be tested additionally for acute systemic toxicity, subchronic toxicity, genotoxicity, implantation and hemocompatibility.

These tests clearly have a significant impact on the commercialization of biomaterials and products based on them. In general they have been very valuable, although it is obvious that they cannot fully predict safety and performance in patients, and there have been many cases where devices have "passed" all relevant tests and been given regulatory approval, only to provide inadequate performance clinically because of unacceptable biocompatibility. This problem is likely to be aggravated as applications of biomaterials extend to situations that cannot easily be replicated by these standard tests. The biomaterials used in tissue engineering, nanoscale diagnostic systems and many drug delivery situations come into these categories.

## 9.4 Clinical trials and post-market surveillance

### 9.4.1 Clinical trials of health care products

Clinical trials are used to determine the overall performance of health care products and may be a requirement for regulatory approval. Procedures for clinical trials on pharmaceutical products have been established for many years. They are carried out in four phases:

- In Phase I trials, a drug is evaluated in a small group of people (20–80) for the first time to evaluate safety, determine a safe dosage range, and identify side effects.
- In Phase II trials, the drug is given to a larger group of people (100–300) to see if it is effective.
- In Phase III trials, the drug is given to an even larger group (1000–3000) to confirm its effectiveness, monitor side effects and compare it to commonly used treatments.
- Phase IV trials are effectively post-marketing studies that acquire additional information including the drug's risks, benefits and optimal use.

It will be obvious that such trials are rarely appropriate for medical devices, including implantable medical devices. The drug trials are hugely expensive but procedurally they are straightforward since they involve administration of the drug by the preferred route and monitoring the performance in patients. It is usually easy to carry out randomized double blind studies, where neither the patient nor the lead physician knows what treatment (drug versus control or placebo) is used on any one individual, and the patients are usually followed for a short time.

When the product is a complex implantable device rather than a tablet that has to be swallowed, numbers of patients may have to be severely restricted, and while randomization may be possible, it is usually very difficult (and possibly unethical) to mask the specific treatment from patient or doctor. In addition, it may be necessary to follow the patients for months or years in order to assess the performance. Quite different procedures have therefore had to be devised.

The main features of these procedures are:

**Permissions and oversight** The process of documenting a planned clinical trial and requesting permission to carry it out, is the responsibility of an Ethics Committee, EC, also usually known as an Institutional Review Board (IRB), established in the main institution where the trial is based, which is an independent body that protects the rights, safety and well-being of human subjects.

**Risk evaluation** The investigators and sponsors carry out a risk evaluation, which is included in EC documentation.

**Justification for design of the trial** The investigators and sponsors have to evaluate all pre-clinical data and accordingly justify the protocol for the trial. This includes the formulation of an investigator's brochure (IB) that makes it clear to all investigators what the sponsor's data and objectives are.

**Clinical investigation plan (CIP)** This is a detailed plan of the trial, which gives all factual information about sponsors, the identification and description of the investigational device, the anticipated clinical benefits and adverse effects and the risk–benefit rationale, the precise design of the investigation (including methods and end-points), inclusion and exclusion criteria for subject selection, criteria for discontinuation, the monitoring plan with detailed descriptions of all clinical-investigation-related procedures, statistical considerations and data management, allowable deviations, procedures for handling adverse events, statements of compliance and publication policy.

**Clinical investigation conduct** This is a detailed set of rules for documentation, decision-making and accountability.

**Suspension, termination and close-out** This sets out the agreed rules for suspension or termination of the trial by the sponsor, and subsequent responsibilities and reporting requirements.

**Responsibilities of sponsor and investigator** These include appropriate qualifications, compliance issues and safety reporting.

Obviously there cannot be one set of rules for the conduct of clinical trials covering all types of medical devices. Some of the ethical issues arising from these trials are covered later.

The situation with new health care products that are neither drugs nor devices is obviously complicating this position; included here are tissue engineering products and cell therapies. There are no universal rules for clinical trials here. As a general position, those products that are substantially cell-driven together with active molecules are most likely to be directed towards the pharmacological route of trials. Those that are more biomaterials-driven may well be able to use different types of clinical trial, more similar to the medical devices route. There are some real problems here since major tissue engineering processes are so resource intensive and directed at specific individuals that classical clinical trials of either type are not really relevant, For this reason, the pivotal developmental studies are carried out as clinical observational studies rather than clinical trials. It is through the publication of these studies that progress is being made, although they do not have the rigor and value of trials.

### 9.4.2 Post-market surveillance

Once a medical device has been given regulatory approval, which technically usually means that a company can place the device on the market, we cannot assume that it has been determined that the product carries no risk. In most cases the product will not have undergone a full clinical trial, and indeed a clinical trial may have been conducted with a relatively small number of subjects. Risk are always with us and the purpose of the ongoing risk–benefit analysis of any product is to ensure that these factors are monitored throughout the life cycle of the product. Most regulatory agencies ensure that some formal monitoring process is put in place so that any adverse events are recorded and reported and well-informed decisions can be made about the continued use of a product. This procedure is usually referred to as post-market surveillance. These procedures involve recording and analysis of customer complaints, recording and control of non-conforming products and materials and physician-reported failures. Companies are responsible for analysis of the data and, where significant problems emerge, for their reporting to regulatory agencies such as Medical Device Reporting and Vigilance Reporting.

These systems may be comprehensive but questions arise as to how effective they are. A "failure" of a device may be reported and described, but often without too much detail. A manufacturer will often trace a failed device through manufacturing steps and conclude that no irregularities were seen, apparently absolving the device itself from the failure process. In the vast majority of situations, the chain of custody for a "failed" device may be unclear (often the patient will retain a device following explantation) and is usually unavailable for any forensic type of analysis. It is also the case that adverse events are only reported in the context of a single product and no comparison with similar products can be made. Because of this, there has been an increasing emphasis on implant registries in recent years.

## 9.4.3 Implant registries

In order to overcome these deficiencies in single-product surveillance of problems, a number of organizations have established different types of registries of implants. The concept is simple; a professional, governmental or industry-based organization establishes a central database where details of all implants of one particular category (e.g., hip replacements, heart valves, cardiac pacemakers, breast implants) are recorded and routinely assessed. The database can be as simple or complex as one wishes. At one end of the spectrum, the data can simply include the name of the hospital where implantation is carried out, the hospital where explantation (or other revision procedure) is performed and the name or designation of the product. Such registries are anonymous as far as patient and surgeon's identity are concerned, the principal data produced being a rather crude "failure rate." At the other end of the spectrum are the databases that contain personal and clinical details. The latter group has the advantage of being useable in cases of a serious recall of a product, and has been used to trace patients with cardiovascular devices where significant risks involving revision have been identified.

Some significant successes have been achieved with registries, particularly in orthopedics and especially in Scandinavian countries. These have evolved in the last 40 years and are now very sophisticated. The Swedish hip registry contains details of all demographic data, frequencies of procedures, types of implant and fixation regime used and their survival. The database can be interrogated at national and regional levels, but not for individual units. These processes have been successful in identifying higher than normal failure rates within 1–2 years, where analysis of post-market surveillance data may not reveal any trends for many years.

There are several problems with registries, the two main ones being their cost and questions of privacy. With respect to the former, costs may be passed on to the manufacturers, hospitals or individuals, or may be borne centrally by governments or their agencies. Although the evidence of beneficial effects of registries is beyond doubt, several countries, including the USA and the UK have moved away from these practices on grounds of cost and political issues. The privacy issues remain in many sectors. Breast implants provide a good example here since while it may be straightforward in opening a file for individual patients, they may then disappear from all records since they usually do not want to part of an ongoing analysis of implant use and failure. It should also be noted that whilst registries are potentially useful for detecting early trends, they are less useful for quantifying "failure rates" at later stages since revision procedures are likely to be influenced by the publicity of early events.

## 9.5 Product liability litigation in medical technology

The influence of litigation on the availability of biomaterials and medical devices has been immense, especially in the USA, and this has shaped the industry more than any other factor. As implantable medical devices became more and more popular in the 1960s and 1970s, the attention of large corporations, especially pharmaceutical companies, was drawn to this potentially lucrative sector. Small medical device companies were acquired by these large companies, and products were marketed aggressively. At that time, regulation of products was cursory and often confined to conformance with simple testing procedures and quality control systems. A poorly regulated but potentially highly profitable sector is very attractive to risk takers. The consequences of failure were not so obvious. In the 1970s and 1980s, four important episodes of failure were initiated. The first case concerned an intrauterine contraceptive device (IUD). The second involved a very small implantable device used in the surgical treatment of temporomandibular joint dysfunction. The third concerned the design and clinical use of a prosthetic heart valve. The fourth, which became a massive cause célèbre in litigation, involved silicone gel breast implants. We shall look at these briefly, and also consider one further very recent case, in order to judge the implications and lessons.

### 9.5.1 The Dalkon Shield

The Dalkon Shield was an IUD that was introduced into the fertility control arena in the early 1970s. It was designed by a gynecologist and engineer in the USA in 1968 and the product was taken up by a small pharmaceutical company in 1970. That was at a time when contraception was a new and highly profitable medical sector and the IUD became popular because of concerns over the safety of the contraceptive pill. IUDs are medical devices that are inserted into the uterus where they interfere with conception mechanisms. The Dalkon Shield had a shield-like structure attached to the lower end of which was a nylon string. Whilst the shield itself was inserted into and remained within the uterus, the string passed through the cervix and protruded into the vagina. The string was made of multifilament nylon thread, which was surrounded by a nylon sheath (see Figure 9.2).

The problem with this design was that the uterus is essentially sterile whilst the vagina contains bacteria-laden fluids. The multifilamentous nature of the string allowed bacteria to wick up the thread and into the shield, where they were protected from normal defense mechanisms. The region between the uterus and vagina, the cervical plug, is normally a very effective antibacterial barrier but the nylon thread allowed bacteria to evade the normal processes. The result was that many patients contracted a uterine infection, often leading to the very serious

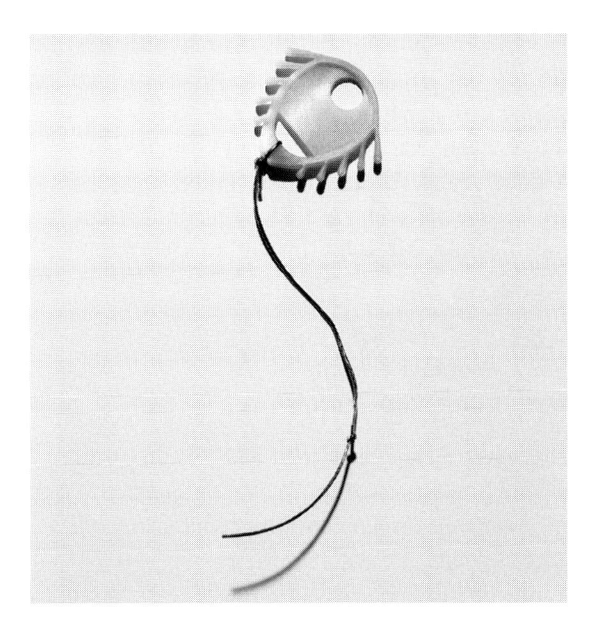

**Figure 9.2** The Dalkon Shield intrauterine contraceptive device.

condition of pelvic inflammatory disease, leading to severe pain, sterility and sometimes death. The manufacturers ignored the warning signs and accelerated the marketing of the device, soon becoming the market-leader. The increase in infections forced the company to withdraw the device from the US Market in 1974 but it was still sold abroad. At this time there was no effective regulatory control in the USA since the Medical Device Amendment Act was not passed until 1976. No other countries had laws that could have prevented importation of such devices.

The Dalkon Shield became infamous as a medical device that caused great harm. It also ushered in the beginnings of mass tort, that is where the various parties take legal action in order to obtain monetary compensation. Many individual claims were made against the company in 1975, and in some of these judges and juries awarded both compensatory and punitive damages, the first being a sum that compensates for actual monetary loss and the second being a figure assessed to punish the company for making such a gross error. At this point the company and their insurers became seriously concerned and the nature of the litigation took several new turns, including the folding of many cases into class actions, where one trial covered a whole group of similar cases, and the declaration of bankruptcy by the company. The whole process of compensation became tied up the rules of bankruptcy and the proceedings dragged on for many years until it was finally resolved in 1989, at a cost of several billion dollars.

## 9.5.2 Proplast and the temporomandibular joint

The temporomandibular joint (TMJ) connects the mandible to the rest of the facial skeleton. Its efficient functioning is essential for eating and talking. As with all joints, it can suffer disease; the slow destruction of the bone and cartilage leads to severe pain and difficulties with these essential functions. The condition is especially seen in middle age women, typically those with post-menopausal osteoporosis, and it is exacerbated by the phenomenon of bruxism, in which the patient continuously grinds the lower jaw against the upper one, especially when asleep. This clinical problem, dealt with by periodontists and oral surgeons, is a difficult one, and is usually treated conservatively. In the 1970s attempts were made to design prostheses to replace parts of this joint. The chosen material was known by the trademark Proplast, which consisted of a composite based on the polymer polytetrafluoroethylene (PTFE) filled with either alumina or carbon. The biomechanics of the TMJ were uncertain, with confusion over the forces that were exerted on the joint. The result was that, in many patients, the implants suffered mechanical wear and fragmentation, resulting in significant tissue destruction.

The issue to be discussed here was that the Proplast products were made by a small company, Vitek, in Texas, which was a spin-out from a major hospital. Lawyers for the patients involved in the use of these devices started to sue the manufacturer, the hospital and the surgeons. These, however, did not have unlimited financial resources or appropriate insurance, and so the attention was turned to the supplier of the PTFE. This was the Du Pont Company, one of the world's leading chemical companies, who evidently did have good financial resources. The problem was that Du Pont denied any liability since they did not knowingly supply the PTFE to this small company. The reality was that only very small amounts of material were used and these were obtained through intermediate suppliers, Du Pont having no knowledge of this use. Du Pont prevailed in court, but with significant defense costs. The company decided that they would not supply any material to any company manufacturing medical devices. This had a profound effect since Du Pont manufactured and supplied many polymeric materials, including polyurethanes and polyesters, to the medical device industry, and the possibility of large chemical companies being found financially responsible for any failures of health care products manufactured by other companies, had significant worldwide consequences. In the USA, a special law, the Biomaterials Access Act, was introduced in order to mitigate this problem, but it has only been in the last few years that chemical/materials companies have willingly targeted this particular area.

### 9.5.3 The Bjork–Shiley heart valve

As we have seen, prosthetic heart valves are generally very successful products, used safely in many patients worldwide. In the 1970s and 1980s it was seen as a lucrative market and some of the major pharmaceutical companies became involved, usually by acquiring small heart valve companies. One very successful type of mechanical valve was designed by the surgeon Bjork and manufactured by the small company Shiley. Several types of Bjork–Shiley valve were introduced and their success led to the acquisition of Shiley by Pfizer, a major pharmaceutical company who had built up quite a large portfolio of medical device subsidiaries.

A new version of the Bjork–Shiley valves, the convexo-concave tilting disc valves, were designed and introduced into clinical practice in the 1980s. These valves were constructed with a cobalt–chromium annular ring, to which were attached two struts. A pyrolytic carbon disc was inserted between these struts and the valve operated by the rotation of the disc from the fully closed position when it completely lay within the annulus, and the open position, when it lay at an angle, typically 60° to that annulus. The valve appeared to be well designed and tested and offered very good hemodynamic characteristics.

Sometime after the valves were introduced clinically, a problem emerged: one of the struts on some valves fractured, an event that was often fatal, as shown in Figure 9.3. It became clear that the failure was that of fatigue in the strut. Surprisingly, the strut that failed was on the inflow side of the valve; it had been assumed that this strut would suffer minimal stress on valve closure, whilst the outflow strut would experience higher stresses on valve opening. The valve was re-designed in order to minimize these

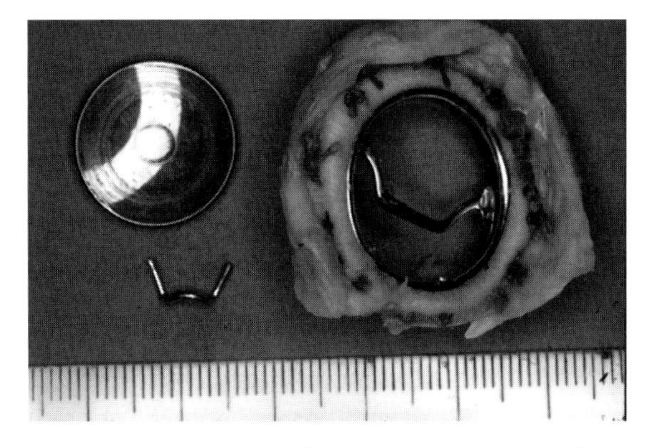

**Figure 9.3** The Bjork–Shiley heart valve. This mechanical heart valve had exceptionally good hemodynamic performance. However, one of the metallic struts was prone to fatigue failure. The lower image shows parts of a valve removed at post-mortem from a patient who died as a result of a strut fracture. The strut fragment travelled down the aorta, and the carbon disc remained loose within the heart. The valve itself, with sewing ring and metallic annulus was cut from the heart wall.

strut stresses but the damage had already been done – the number of clinical failures, and deaths were increasing. The valve had to be taken off the market, and the litigation started. However, a huge dilemma arose. Some 80 000 patients had received these valves on a worldwide basis. Within a few years of the problem being known, some 700 of these patients had experienced a fracture. Whilst clearly this is less than 1% of the patients, since it was a fatigue failure it could be anticipated that many patients would ultimately be vulnerable. The question then was whether all these patients should be re-operated in order to replace the valve. The difficulty was that statistically around 6% of all operations to replace a mechanical valve resulted in death of the patient. A full actuarial analysis was carried out on failures so that the most significant risk factors, such as age, gender and weight of the patient, whether it was the mitral or aortic valve, who and where was the operating surgeon, could be determined and those most at risk could be considered for re-operation. This was a massive undertaking, made more complex by claims that there were manufacturing faults associated with the welding of the struts to the annulus.

It was of no surprise that Pfizer, with very significant liabilities, divested themselves of the Shiley company as soon as they could, and also most of their medical device subsidiaries. Other drug companies, who had become attracted to the potentially very profitable medical device sector, soon changed their minds as well. The whole of the medical device industry underwent major changes at this time, as did the regulatory process in view of questions being asked about how the regulatory agencies allowed such a product to gain access to the market and be placed in patients. These points were significantly magnified by the issues arising from the next example.

### 9.5.4 The silicone breast implant story

Reconstructive surgery, sometimes called plastic surgery, became very popular in the 1970s and 1980s. One of the major areas here was that of breast enlargement. This was considered for post-mastectomy reconstruction and, especially, cosmetic breast augmentation. Several companies started to manufacture implants and, within the space of few years, hundreds of thousands of women, especially in the USA, received such implants.

The technical issues were not themselves too difficult. Breast tissue is a relatively soft mass of fat, glands and muscle. There is no synthetic material that has the same qualities and which can be manufactured into an object with defined shape and size. The solution was to take a substance that had these material qualities, typically a gel, and encase it within a flexible membrane. Almost universally, the medical device companies chose a silicone gel and encased this in a silicone elastomer shell. Once again the pharmaceutical companies of the time acquired the breast implant companies, and indeed manufacturers of the silicones became producers of the devices themselves. Very quickly some of the major companies in the USA, including Dow Corning, Baxter, 3M, General Electric and Bristol-Myers Squibb were all involved.

It was known fairly early on that not all procedures were successful, and about 15% of women reported a painful hardening of the tissue and distortion of the breast (shown in Chapter 3 in Figure 3.6), a condition usually known as constrictive fibrosis or capsular contracture. Some of these implants had to be removed or altered, but the problem was contained. The situation became substantially more difficult when some women reported rupture of the silicone elastomer and release of the silicone gel into the tissue. Manufacturers had, understandably, been taking note of patients comments about the un-natural feel of some prostheses, and had introduced new designs with thinner shells. Not surprisingly some of these did suffer tears and some silicone gel could get into the tissue. Again this matter could have been contained as long as the numbers were low, but questions started to be asked about the fate of the silicone gel. This then became a matter of major public concern, with gender issues and the treatment of susceptible females at the forefront, and the story was played out in the media and in political circles. The problem was that complaints were being made that the silicone could be the cause of auto-immune diseases. These include some very serious diseases such as scleroderma, lupus and rheumatoid arthritis. In the USA, the FDA reacted quickly and effectively banned silicone gel breast implants.
A class action lawsuit followed and, without any real evidence of causation, the main companies were collectively fined over US$4 billion. In the ensuing years, as individual lawsuits were brought against individual companies each seeking massive punitive damages, it became clearer that no such evidence of causation existed, as one after another, epidemiological studies showed no correlation between breast implants and auto-immunity. However, the damage had already been done: the large companies, having lost hundreds of millions of dollars defending themselves, withdrew from this sector. Millions of women who still required implants were left without any products.

Some companies, especially in Europe, tried to fill the void with non-silicone products, usually with far inferior performance and much greater potential damage to the patients. In most countries, saline-filled silicone elastomer shells were allowed, but to the vast majority of people, including patients and surgeons, these were inferior. The saline did not have the consistency of the silicone gel and could not give the same control over shape and feel. Also, the implants were far more likely to rupture since the silicone shell, receiving far less support from the saline than the gel, was more prone to developing creases and folds, with increased rupture rates.

The alternative fillers that were tried included cellulosic materials but these were mostly shown to have inferior properties and safety profiles. One particularly important case involved the use of a fatty substance, specifically a triglyceride. There were two unfortunate consequences of this choice. The first was that due attention had not been paid to the literature and the potential effects of lipid-like materials on the silicone polymer shell. The silicone elastomer appeared to gradually lose strength such that rupture and failure was more-or-less inevitable. Even more significantly, the triglyceride was itself prone to auto-oxidation, which may not have mattered if it stayed totally confined by the shell but once released into the tissue, this oxidation led to significant tissue responses. Indeed it was known that one of the by-products of auto-oxidation was a mutagen, which led to the perverse possibility that a woman who had a mastectomy following breast cancer could develop a further cancerous condition because a mutagenic substance had been used in her reconstruction prosthesis. This point was used publicly by one regulatory authority to justify withdrawal of these devices from the market.

## 9.5.5 Silzone heart valves

This case was resolved just before this chapter was written. The proceedings were concluded some 13 years after a putative problem with a medical device was identified. Although legal proceedings were initiated soon after this initial event, it eventually became the subject of a class action within the jurisdiction of the Superior Court of Justice in Ontario, Canada.

The device was a mechanical prosthetic heart valve manufactured by the US company St. Jude Medical. These valves were very similar to existing valves of the company, except they had a sewing ring that was coated with a layer of silver. The intention was that the silver, applied by a proprietary process producing a layer known as Silzone, would exert antibacterial activity in order to reduce the incidence of the potentially fatal bacterial endocarditis. After testing and process development, and subsequent regulatory approval in many jurisdictions, the valve started to be implanted in patients in many countries in 1997. In November 1999, the Medical Devices Agency in the UK issued an Advice Notice to physicians warning them of possible thromboembolic complications. There was evidence of a very small number of affected patients in one UK hospital, in Cardiff, Wales. A post-marketing multi-centered clinical trial initially found no other evidence of this effect but soon a few other incidents were found and some agencies around the world took action to withdraw regulatory approval. The company had no choice but to recall the valves, by which time some 36 000 had been sold.

The burden on the court (a single judge) was determined by a number of so-called common issues. The more important biomaterials-related issues concerned the defendant's conduct in designing, testing and marketing the Silzone valve, whether causation could be proved (i.e., could Silzone have an adverse effect on tissue healing) and whether the future risk of medical complications would be greater for Silzone patients than with other heart valve recipients.

There were many factors concerning the preclinical testing of the valves and especially whether they were sufficiently extensive and thorough (i.e., numbers of animals and duration of exposure) to have determined the level, if any, of additional risk. Scientifically, the issue revolved around whether a layer of silver on the fabric of the valve sewing ring could have affected the healing of valve tissue following valve implantation, It was contended that the valve complications, although few in number, were associated with delayed or compromised healing around the valve since silver has a degree of toxicity. The question of the levels of silver ions necessary to produce the intended antibacterial activity without inducing any significant degree of toxicity was of major significance. The judge eventually deduced that the weight of evidence showed that silver could not have caused any significant delay in healing, and that causation relating the deaths of patients and the presence of silver could not be demonstrated. A significant factor here was the long-term results of the clinical trial showed that although soon after implantation there was a slightly higher risk of valve-related complications, these diminished with time and indeed, after a few years, Silzone patients were less at risk.

This trial took over 2 years to complete and the written judgment was over 200 pages in length. It provides a landmark in medical device/biomaterials litigation. The biocompatibility of biomaterials is a complex issue, as we have seen in Chapter 3. Balancing risk and innovation in such cases is not a trivial process, which clearly has to be undertaken with caution and diligence.

### 9.5.6 General observations

The question arises as to whether there are any lessons in these examples. Obviously these cases span several decades and many things have changed. They are also, of course, USA-centric.

There are some themes that are common to many of the cases where devices have not performed to the required or expected level, both as seen in the above examples and in others not discussed here. The first is that many of them have been associated with the perceptions of big markets, and therefore big profits, in newly developed areas of medical technology, the originating inventors and small companies not having the experience and ability to carry out risk analyses. Quite frequently the small originating companies were acquired by much larger companies, where a surprising lack of effective due diligence became obvious. In the earlier days it was the large pharmaceutical companies that acquired small medical device start-ups; many have learned their lessons and divested themselves of these liabilities. Now it is usually the large multinational medical device companies that carry out the acquisition and they have more knowledge and experience. It has to be said, however, that these powerful lessons should not be forgotten as even newer technologies and materials are exposed to the medical arena.

One very important issue concerns the understanding of risk and the identification of warning signs. In most of the examples quoted, there were warning signs in both pre-clinical testing and early clinical studies. In some cases they were not considered important, in other cases deliberately ignored. This can never be a good policy. The requirements of regulatory approval, discussed earlier, are meant to rule out such events, but they can still take place.

As a final point here, we should qualify the term "failure." It is now common practice, and this is not limited to the USA, for those who perceive that the biomaterial or medical product that has been applied to them has "failed," believe that they deserve compensation. There is some justification for this in certain cases. However, so often the fact is that the procedure may have failed to give the level of satisfaction that was anticipated, but this does not mean that the device has failed. It will be obvious from previous chapters, that 100% success can never be guaranteed and there will always be a percentage of patients in whom the procedure failed, and this usually increases with time. The lack of success, or less than optimal degree of satisfaction, does not necessarily mean that the device has failed.

## 9.6 Health economics and social policy

Whatever the quality of health care products that are generated by biomaterials-related research and development, and whatever exquisite skills the clinicians bring to the operating theatre or diagnostic facility for their implementation, deserving patients will not receive these products unless the economic considerations are right, which implies the existence of appropriate social policies and overall political will. This is an enormous subject, which can only be touched on briefly here. It is, inevitably, a global issue, although most data comes from just a few countries and organizations. We shall mention three factors in this section, the global inequalities in health care and related differences in disease incidence, medical technology as a main driver of economies and the overall assessment of the effects of biomaterials-related technologies on the quality of life.

## 9.6.1 Global inequalities in health and health care

That inequalities in health care exist on a global basis should be of no surprise. Several factors are involved, although they are usually interrelated. First, the incidence and prevalence of disease vary considerable with geography and climate. In general, people in developed countries suffer more from degenerative diseases and conditions related to lifestyle issues (diet, exercise, tobacco, alcohol, etc.) including cardiovascular disease. In developing countries it is still infectious diseases associated with, for example, tropical insect-related and parasite-transmitted diseases such as malaria and schistosomiasis that dominate illness and disability patterns. The prevalence of arthritis in sub-Saharan Africa is, at 150 per 100 000 people, just one-tenth that in Western Europe. On the other hand, in some parts of central Africa, over 10% of individuals over the age of 50 are blind, the figure for high-income parts of the USA being less than 0.5%.

The second issue concerns wealth and its distribution. There is a significant, and interesting relationship between life expectancy and gross domestic product (GDP) per capita. At low levels of national income there is a profound relationship between GDP and life expectancy, measured as the expectation of longevity at birth; increasing incomes markedly increase the overall health of individuals and their longevity. Above a certain level, however, the relationship is far less notable, suggesting that there is a limit to how much money can benefit the delivery of health care. Many factors are involved, including the existence of social policies that mandate the effective use of the financial resources.

These two issues of disease prevalence and economy-related longevity are interacting in a major way. As economies improve in the developing world, so the population ages and lifestyles change. Those degenerative diseases that are age-related and rarely seen in people in the poorest countries in the past, suddenly become noticed in those countries when the population is now collectively older. This has significant implications for the demand for biomaterials-related products as the developing countries transitions towards the developed countries. One of the most significant changes is seen with age-related dementias. Recent estimates by the WHO indicate that there are over 35 million people worldwide with dementia, the treatment and care of these individuals already costing a staggering 1% of the world's cumulative GDP. While this was considered until recently to be a condition largely confined to Europe and North America, the highest rates of increase in prevalence are now seen in Southern Asia (mostly India) and Eastern Asia (mostly China). It is almost non-existent at this stage in parts of sub-Saharan Africa. Cancer also provides significant and interesting data. Currently the countries with the highest cancer mortality rates are the Netherlands (430 deaths per 100 000 people), several other European countries, the USA (310 per 100 000) and Australia (300 per 100 000). In China, the rate was 75 per 100 000 in 1970, which increased to 108 per 100 000 in 1990 and 140 per 100 000 by 2005.

Obesity and diabetes tend to go hand-in-hand, but regional differences are significant. There are large genetic and gender differences, as well as dietary and lifestyle factors. Pacific Island countries have highest rates of obesity: Papua-New Guinea have almost 80% of the population described as obese (with a body mass index above 30). In typical developed countries, the figures are between 10 and 30% (USA 30.6%, UK 23%, Finland 13%, France 9%), while in very poor countries (Nepal, Eritrea, Ethiopia) less than 1% will be obese. Most Eastern Asian countries have low figures, with both Japan and China showing around 2%. Females tend to be more obese than males (Saudi Arabia, 26% males, 44% females, South Africa 8% males, 27% females). Translating these figures into the incidence of diabetes, we note that in 2010, 285 million people were estimated to live with diabetes, which is projected to rise to 438 million by 2030. While the prevalence of diabetes follows obesity rates to some extent (highest in Pacific Islands, lowest in sub-Saharan Africa), this is not always the case. India and China which have low obesity rates now have the world's largest diabetes populations, 50 million in India and 43 million in China.

With some life-threatening diseases, the mortality figures reflect the lifestyle effects on prevalence and also the quality of medical services that are used to treat the diseases. Cardiovascular diseases provide an excellent example of this. The age-standardized death rates per 100 000 people for ischemic heart disease (heart attacks) and cerebrovascular disease (strokes) largely reflect the latter factor. In parts of the former Soviet Bloc and the Middle East (Ukraine, Turkmenistan, Afghanistan), the figures are typically 400 and 150 respectively. In the USA, with exceptional care services, the equivalent figures are 80 and 25, in Japan 31 and 36, and France 29 and 21.

### 9.6.2 Medical technology as a driver of economy

For individuals who look towards a biomaterials-based career, the state of the financial health of related industries is of considerable importance. There is no doubt that whatever the state of the global economy, which at this point in the twenty-first century is particularly volatile, the health care sector will remain relatively robust. It will be vulnerable to health care policies, especially the political interference in the private/public provision debate, but the attainment and maintenance of health is the most important issue facing most individuals and their families and communities, so that health care is in the lead when it comes to the good years and the last to be given up in the poor years. Since manufactured products form a major pillar of health care practices, this means of course that the medical technology industry is a key driver of many economies. The UK parliament repeatedly claims that this industry is a significant component of the UK economy, with, at the present time, 5000 companies, employing 55 000 people and annual sales equivalent to $10 billion. These figures are dwarfed by those of the USA, with $140 billion worth of manufactured medical technology goods, employing a workforce of 425 000 people. There is a wide variation from state to state, with some locations, particularly Minnesota, California and Massachusetts, taking greater stakes in this market.

Globally, some countries have made a specific and distinct effort to form medical technology clusters. Ireland is a good example, employing the highest number of employees per head of population (a total of 25 000) of any European country, with $10 billion worth of exports. Seventeen of the world's top 25 medical technology companies have invested significantly in Ireland, and continue to do so even during that country's difficult financial climate. After slow starts, both China and India now have significant medical industries. Interestingly, many medical device global companies are in the process of moving headquarters and manufacturing facilities to Singapore and China.

### 9.6.3 Medical technologies, cost-effectiveness and quality of life

A recent WHO report on global health (*Closing the Gap in a Generation*, published in 2008) posed the question "That the life of a slum dweller in Caracas is generally shorter and more brutish that the earthly span of a rich person in Cologne or Chicago is hardly surprising. But why do men in Carlton, a rough part of Glasgow (Scotland, UK), tend to die more than two decades sooner than men from the dormitory town of Lenzie a few miles away. Why do American Asian families live, on average, to 87, while the life expectancy of black males (in similar environments) is only 69." This critical question covers the philosophical matters raised in this section, and also raises questions of what determines the quality of life, how is it to be measured, and how health care practices influence this quality.

With certain diseases and therapies there are standard methods of scoring that give both individual and collective indications of patients' conditions before and after treatments. The New York Association Classification of Heart Failure (range I–IV) is used for assessing the condition of the heart. The Harris Hip Score is used to assess the condition of a hip (measured by scores for pain, limp, support required, distance walked, etc.) before and after replacement.

On a broader scale, there are procedures to measure the overall quality of life as a function of medical intervention, The best known is based on the concept of QALYs, which is the acronym for Quality-Adjusted Life Year. Developed and promoted by the UK's National Institute for Health & Clinical Excellence (NICE), this concept effectively measures a person's length of life weighted by a valuation of their health-related quality of life.

A QALY is determined on the basis of five facets of life, that is mobility, pain/discomfort, self-care, anxiety/depression and activity, with three categories in each, corresponding to no problems, some/moderate problems and severe issues. A person who has no problems in any of the facets has a utility score of 1.000. It would take one year of perfect health at this 1.000 utility score to generate one QALY. The greater number and severity of the problems leads to lower scores. A person with some mobility problems, unable to wash and dress, unable to perform usual activities, moderate pain, and moderate anxiety would have a utility score of 0.079.

One year for a patient in a health state associated with a utility score of 0.5 is equivalent to half a QALY. If a number of treatment options are available for the main condition of this patient, the effectiveness, and therefore the cost-effectiveness, can be determined from the change in utility score over a period of time. For example, it could be determined that one additional QALY could be generated by one method but two additional QALYs by another. The cost utility ratio, which is then used by health care providers and/or insurers, is determined by the differential cost of these treatments divided by the difference in QALYs achieved.

It will be readily appreciated that the use of the QALY concept, or indeed any other measure of the cost-effectiveness of medical therapies, is not an exact science, but as the practices of health economics become more refined and robust they will certainly play a greater role. This will be true for therapies that involve medical devices and biomaterials. Whether they will be of real practical value in determining the cost-benefits of a procedure that involves an implantable total artificial heart, with capital costs close to $250 000 and annual costs of $50 000, remains to be seen.

## 9.7 Intellectual property

The questions surrounding intellectual property and its protection are probably the most complex of all infrastructure issues in medical technology. They are supremely important since much of the industry's economic position discussed in the previous section, and indeed the competitive battlegrounds of this industry, focus on the rules of IP protection. We can only mention the main features briefly here. There are several forms of IP, including copyright and trade secrets, but we are primarily concerned with patents.

Patents provide the exclusive right, granted by a governmental body, to make use of an invention or process for a specified period of time. One of the reasons why this subject is complex is that patents are specific to individual jurisdictions. There is international agreement on many aspects of patent processing but there is no such thing as an international patent. Most relevant countries accept national phase patent applications based on the World Intellectual Property Organizations under the Patent Cooperation Treaty (PCT). However, a patent awarded by the government of a country is valid only within its territorial boundaries. It is therefore necessary for inventors to apply for patent protection in all countries where they wish to be active in either marketing or manufacturing a product. Such country-specific applications are usually based on the PCT application which gives a priority date, but grant of a patent in the home jurisdiction of a PCT application gives no guarantee of grant anywhere else. Countries within the European Union are covered by a single European Patent Office.

There are three essential ingredients of a patent, which are usefulness, novelty and non-obviousness. Usefulness is fairly straightforward and implies that the inventor is able to make practical use of the invention with respect to at least one of the claims. Novelty is crucial since it implies that there can have been no public announcement about the invention before the patent is filed or before the accepted priority date. There are some subtle differences in priority and filing dates that need not concern us here, as there are between different jurisdictions, but one important principle applies everywhere. Many patents are based on the work of individuals, who are often academics. If work in a laboratory is likely to lead to an invention, it is essential that the academics do not talk of their invention in public, even in a seminar or conference abstract, before the IP has been protected. Prior disclosure, usually inadvertent, is the main reason for patent invalidity. The non-obviousness part is also important. This is a complex subject, which is a major factor in attempts by competitors to invalidate a patent, since the invention should not have been obvious "to a person skilled in the art" at the time of the priority date. Of significant interest, the USA has just changed the law in this respect. Effective from early 2013, the law now recognizes the "first-to-file" as the critical event rather than "first-to-invent." Up until then, if a company could demonstrate that they had made a patentable invention at a certain time but chose to keep this secret rather than placing the information in the public domain through a patent filing, that date could take preference over a later patent priority date of a competitor; this is no longer the case.

Patent litigation is a major feature within the medical device industry. One company may sue a competing company for patent infringement, in which they claim that a product from that competitor actually copies features of the original patented product. The defendant usually attempts to disprove infringement and at the same time counter-sues on the grounds that the original patent should have not been granted, or is invalid. Again these suits are jurisdiction specific, which means that they are very costly, but the rewards of victory could be measured in millions of dollars and the costs of failure could mean the end of the company.

## 9.8 Ethical issues

If we believe that the one and only purpose of biomaterials science is to provide effective materials that can be used for the advancement of health care procedures to the benefit of patients, we can immediately identify a number of situations where difficult ethical decisions have to be made. These primarily concern clinical decision-making factors involved with the optimal delivery of care to individual patients and, in some situations, the sourcing of biological material that has to be used in these procedures. In reality, of course, we cannot consider the biomaterials/medical device sector independent from society in general and the ethical dimension is much broader than this. There is no point here in discussing those ethical matters that are equivalent to those that are met in other societal and commercial situations, but it is worth summarizing these ethical scenarios. We do so in a logical sequence starting with ideas, research and intellectual property and finishing with the marketing and full clinical uses of health care products.

Debates about ethical issues in health care in general incorporate a number of basic principles, including:

- avoiding harm, referred to as nonmaleficence;
- protection and defense of the rights of individuals;
- preventing harm and removing existing harm;
- fair opportunity and distribution of resources;
- human dignity.

### 9.8.1 Research data and publishing

There is nothing specific to the ethics of basic research in biomaterials science that does not apply to research in general, but there are some issues that relate to the dissemination of data in medical research, including (and perhaps especially including) research that relates to product development that should be mentioned here. These issues mostly revolve around the sponsorship of research and the relationship between sponsorship and right to publish, the questions that concern the publishing of negative results, and the preferred forums in which publication takes place.

Most medical research, especially basic research, is funded by governmental bodies or charitable organizations. These agencies do not normally have a say in the interpretation of data and publishing of conclusions. However, the commercial bodies that will eventually market those products fund a significant amount of research and development concerning products. Although many have argued that manufacturers should not sponsor extramural medical research (that is research that is not carried out within the companies themselves) because of the significant risk of conflicts of interest, the general consensus is that such sponsorship is entirely valid provided there are agreed procedures for the handling of the results and conclusions. This is a very difficult area and editors of scientific journals are well aware that attempts have been made to give a bias to submitted publications that favors a particular product. In some journals that deal with health care products, including those that involve biomaterials, it is easy to find presentations of papers that will support future marketing practices rather than reflecting the actual conclusions. Most good journals now have policies that are geared to avoid this and have publishable declarations of financial support and conflicts of interest.

The situation with clinical trials is particularly important. There is a natural tendency for those who sponsor clinical trials to try to publish those results that support the use of a product but to delay or even avoid the publication of unsupportive data. It is widely believed in the medical community that the majority of clinical trials should be published in the peer-reviewed literature whatever the outcomes. Some governments have introduced legislation that is intended to improve the public reporting of clinical trials results on the basis that a failure to disseminate the data disrupts scientific progress, leads to redundant efforts and generally fosters misconceptions about the nature of clinical evidence. This is most evident in the USA, as detailed in the website http://clinicaltrials.gov. Amongst other features, this legislation requires sponsors and principal investigators to register trials at their inception; without registration the results cannot later be published in the major journals.

As a final point here, it should be noted that there is considerable controversy over the optimal, and indeed ethical, processes of publishing, especially in the medical area. The traditional model has involved commercial publishers, who make profits from the process of producing scientific journals. The source of contention here rests with the fact that much research is publicly funded but the public has to pay again, for example through journal subscriptions, to see the papers that arise from the work. An alternative model, referred to as open access, allows the public, whoever and wherever they are, to have access to the scientific literature free of charge. There are many problems with this model, which need not be discussed here, and extensive and fair open access procedures have yet to evolve, although it should be noted that some high profile funders, such as NIH in the USA and the Wellcome Trust in the UK, require grant recipients to publish their work in some form of open access within, for example, one year of completion.

### 9.8.2 Intellectual property

Some crucial features on intellectual property law have been briefly discussed earlier. Since the process of patent prosecution, however complex, is governed by laws and international agreements and conventions, it would seem that there is little involvement of ethics here; you are either within or without the law, which is a matter of legal judgment and not ethical positions. There are two aspects, however, that we must briefly consider.

The first is very general, but highly relevant to medical devices. The whole purpose of patents is to facilitate placing know-how and invention details within the public domain, the reward for doing so being a monopoly for the commercial use of that information for a period of time, typically 20 years. This is particularly important in medicine, and especially health care products, since it is vital that such products are placed on the market for the benefit of patients. In practice, patent law can be used for the opposite effect, that of preventing inventions being used in order to secure a commercial advantage. The patent infringement process is a business in its own right, and is hugely expensive. It is not unknown for medical device companies to sue other companies simply to slow down the market opportunities of the latter, or even to force competitor companies out of business through the high cost of defending themselves. There are also some companies, referred to colloquially as patent trolls, whose sole intention is to buy patent portfolios, never to use them for the benefit of society, but to sue, for massive amounts of money, any company that they consider is infringing these otherwise unused patents. There are significant ethical as well as legal aspects of these matters.

The second is of major significance in the area of regenerative medicine, especially those procedures that involve, often in association with biomaterials, the use of stem cells. It is well known, as we shall discuss a little later, that there are significant controversies over the use of embryonic stem cells (ESCs); what is less well understood are the arguments over intellectual property rights associated with the use and handling of stem cells. Although the situation remains rather fluid, a divergence of opinion is emerging between the USA and Europe. In the USA, legal judgments have sided with the view that ESC techniques should be patentable. However, the European Court of Justice has determined, without recourse to appeal and affecting all countries of the European Union, that no procedure that involves ESCs can be patented, resulting in the loss of protection for companies that are involved in the very expensive research and development processes.

### 9.8.3 Regulatory submissions

Once again here it might be imagined that the plethora of legally binding regulations worldwide might lead to a limited scope for ethical interjections. However, when so much is at stake when regulatory submissions are made, and when so many of the rules, classifications and procedures are based on subjective judgments as well as hard facts, the temptations to dwell on the positive data and opinion and downplay the negatives are rather high. The following examples, which are real but have not necessarily been discussed in the public domain, demonstrate the extent of this situation.

**Demonstration of equivalence** In some jurisdictions, the regulatory process relies heavily on the demonstration that the product in question is equivalent to one that is already, legally, on the market. The USA FDA 510(k) route is a good example. Problems arise when there is a lack of clarity on exactly what features of the device are equivalent to the so-called predicate device. It could be the desired function or the biomaterials of construction, for example. There have been many cases of predicate devices that have the same function but not precisely the same construction, and vice versa, often being used to improve the chance of approval. Not surprisingly, these procedures have come under scrutiny in recent years.

**Biological safety tests** As noted earlier, there is a well-established series of international standards that set out tests that may be performed to establish biological safety of biomaterials. There are, however, several contentious issues here. The standards are permissive and not obligatory and manufacturers can use other tests provided they give appropriate justification. A manufacturer may perform a number of tests and only submit to the regulatory agency the reports that indicate a pass result. In the same manner, the pass–fail criteria are somewhat arbitrary and it is not unusual for the same test to be performed several times on the same material and for it to pass sometimes and fail in others. The manufacturer is more likely (and legally correct) to submit the pass results, sometimes showing that there was a technical error when the fail result was obtained.

**Submission of all data** In a similar manner a manufacturer may carry out many diverse tests during the development phase of a product. Although a manufacturer should include all tests and all data within a submission, for example within a Design History File, it is quite possible to pay minimal attention to a problematic result by reference to experimental difficulties, etc.

**Demarcation issues** As noted earlier, an increasing number of products fall within the gray boundary areas between different device classes. It is often perceived that regulatory approval is easier with medical devices compared to other types of product, although this is not necessarily the case. It is usually up to a manufacturer to indicate their preference for classification. While some may select the classification that would seem to be the easiest, this is not always the case. A manufacturer may select a different class or category because such products qualify for much higher reimbursement rates, potentially yielding much greater profits. Of paramount importance here is the equivalence of the claims that appear in a regulatory submission and those which appear in subsequent company marketing documents. On more than one occasion a manufacturer has indicated that the principal intended function of a product is purely mechanical in order to qualify for unambiguous medical device classification, only then to claim exceptional biological performance during marketing of the product.

### 9.8.4 Organ transplantation and organ assist

Although there are still some individuals and religious groups that oppose organ transplantation on what might be described as ethical grounds, the philosophy and practice of using organs from deceased individuals for the benefit of others has been largely accepted in many countries. What ethical issues remain are associated with some critical decision-making processes and health economics.

The main ethical issues about organ sourcing revolve around the so-called dead donor rule. In the early days of transplantation it was agreed that the surgeon and his/her team carrying out the transplantation was removed from the decision on death of the potential donor. Diagnosis of death had previously been based on the long-established criteria that the body was cold, blue and stiff but these are useless criteria if you wish to extract a viable organ from the cadaver. Rules were then written which defined donor death in terms of brain death, that is the potential donor had devastating and "irreversible" neurological injury. These rules appeared to be sound, but the ability to keep individuals alive with considerable metabolic activity even if there were no intracranial activity has caused doubts to be raised about these criteria, and in recent years, different conditions have been discussed, including criteria related to cardiac rather than brain death.

The availability of donor organs remains a crucial ethical dilemma. The main question is whether individuals have to opt-in or opt-out of organ donation. In the former case, the individual has to carry a donor card that specifies he or she agrees to be a donor in case of death. In the latter case, which is gaining favor in many countries because of the acute shortage of willing donors (assumed to be due to the inertia of individuals for never get around to acquiring a donor card), people have to carry cards or other forms of identification which specifically state that they do not give permission for their organs to be used. There have also been very controversial, and usually illegal, practices of individuals who donate a healthy kidney for a monetary award. Also one government backed a scheme for using organs from executed prisoners; the furor that this caused in international circles, with the boycott of the physicians involved, caused that government to desist from this practice.

The health economics arguments refer both to the direct costs of organ transplants and to the broader factors concerning the diversion of health care finance and indirect resources to the small number of recipients away from general medical care for the majority. As far as the direct costs are concerned, it is necessary to take into account the cost of pre-transplantation procedures, organ procurement, hospital costs, physicians and staff costs, post-transplant care and immunosuppression. In 2011 in the USA, the total costs of a kidney transplant were $250 000, a liver $500 000, a heart $1 million and both heart–lung and intestine transplants around $1 250 000. These considerable costs have encouraged so-called medical tourism, where patients travel to lower-cost countries for their transplants, which raises ethical issues of their own.

The direct and indirect costs of organ transplantation do not tell the whole story, of course, since these figures have to be weighed against the costs of alternative treatments, or the "costs" of waiting for someone to die. It is in this context we have to consider the costs, and associated ethical challenges, associated with medical device alternatives. The left ventricular assist device provides a good example. The LVAD was designed and intended for use as a so-called "bridge-to-transplant" for patients with chronic end-stage heart failure who were waiting for a suitable donor. However, one major clinical trial showed that these devices could produce longer benefits and, in the USA, approval was given for them to be used for the long term, without any planned transplant, a procedure now known as destination therapy. This has not met with universal approval. The costs may well approach $250 000 and the overall quality of life in the recipient is not always good. These devices can profoundly alter end-of-life trajectories because of repeated infections, device malfunction and serious neurological complications. The financial burden and psychological deficits for the next-of-kin/care-givers give great cause for concern over the extension of this use to patients.

## 9.8.5 Animal tissues

The vast majority of biomaterials are synthetic and non-living. However, as we have seen in the preceding chapter, there is now an increasing number of biomaterials that are based upon, or incorporate, some form of animal-derived tissues. The ethical dimensions to this position, both with respect to the rights of the patient receiving products that utilize these materials and the animals from which they are derived, deserve some consideration.

This is a large subject area since it subsumes the transplantation of living organs from animal donors to human patients, the use of relatively simple molecules, such as bovine serum albumin, and hyaluronic acid derived from rooster combs and the by-products of organisms such as silkworm-derived silk proteins. In the former case, major new considerations arise, while in the latter case, the sourcing and manufacturing processes may be no different to those found in, for example, food and clothing industries.

Dealing first with the live transplantation processes, these are described under the general heading of xenotransplantation. This in general is the inter-species transplantation of cells, tissues and organs or the *ex vivo* interspecies therapeutic use of cells, tissues and organs. There were some attempts at transplantation of organs from non-human primates to human patients towards the end of the twentieth century, but these were, predictably, unsuccessful and both ethical and technological/immunological difficulties have prevented progress in recent years. It is unlikely that xenotransplantation involving whole viable organs will play a major role in the future, with the possible exception of the use of genetically modified pigs as the source of liver transplants. It is also possible that xenogeneic extracorporeal liver perfusion may be beneficial in patients waiting for human liver transplantation. There are clearly some ethical issues here, both concerning the philosophical use of living animal tissues within humans and the possibility of disease transmission. Although a number of religions oppose xenotransplantation, the majority accepts the possibility in the context of the primacy of human over animal life.

Concerning animal-derived products within medical technology, in line with the last comment of the previous paragraph, society has come to accept, ethically and pragmatically, that we have to use these products if we are to optimize the technologies. In most jurisdictions, it is believed that existing legislation over the care of animals is sufficient to safeguard animals during medically directed tissue procurement. The question of disease transmission remains, but this has largely been subsumed within the practical aspects of viral and bacterial elimination. This whole question was brought into focus during the bovine spongiform encephalopathy crisis in Europe in the 1980s and 1990s. In order to minimize the risk of prion transmission from cows and a few other species to humans, in products ranging from food to pharmaceutical excipients and medical devices (such as catgut sutures), a wide-ranging series of precautions were introduced, including animal sourcing, veterinary control and tissue processing, with considerable effectiveness. There have to be some lingering doubts, for example over the possibility of transmission of porcine endogenous retroviruses to humans, which if it occurred could result in human diseases for which there was no resistance, but very effective risk management procedures have been put in place.

### 9.8.6 Stem cells and human-derived products

Ever since stem cell technology was developed to the point that it could conceivably be used for therapeutic interventions, debates have raged over the world about the ethical implications of such work. There is not, yet, any international consensus on the regulation of stem cell research, primarily concerning human embryonic stem cells (hESCs). Several countries (including Catholic-oriented countries in Europe) have produced legislation that is largely restrictive of such practices and others (such as Sweden, the UK, Switzerland and Korea) have tended towards permissive regimes. The USA is somewhere in the middle, and policies have been changing with political mood.

The issues with hESCs are those of embryo destruction and the ethical definition of "personhood," that is, when does a collection of cells constitute life. The dilemma is that, to most scientists working in this area, the value of stem cells varies with cellular potency, with differences between totipotency, pluripotency and multipotency, the less differentiated being more malleable but of highest ethical concern. It is no wonder that this position has come to be epitomized with the philosophy that the derivation of stem cells from those embryos that remain after infertility treatment is justifiable, but only if there are no less ethically problematic alternatives available.

These issues have become somewhat more complex with the development of techniques that are aimed at achieving pluripotency in non-stem cells. The first such technology was that of somatic cell nuclear transfer (often referred to as cloning, made famous through the creation of Dolly the sheep). The second is that of the nuclear reprogramming of fibroblasts through the retroviral transduction of four transcription factors into these cells, giving the induced pluripotent stem cells (iPS cells). Although not involving the methods to source hESCs, these processes have their own ethical issues, particularly the strong moral objection to cloning individual humans. The concerns with iPS cells are the future development of processes with demonstrably acceptable risk–benefit ratios and the possibility of using such cells in a highly unorthodox manner.

### 9.8.7 Commercialization of the human body

The commercialization of parts of the human body would appear to be an ethical nightmare. However, some consistent practices have emerged over several decades, which have essentially been transposed into laws in many countries.

We are mainly talking about those activities that involve profiting from the use of any parts or components of the human body. Two such uses gained prominence in the 1950s, one for research purposes and one for therapy. The first involved the development of an immortal human cell line derived from cervical cancer cells of a patient in the USA, Henrietta Lacks. These cells became known as HeLa cells. The scientist who developed the line, and the methods for handling them, gave materials and knowledge to anyone asking for them; they have been used, very successfully, in many lines of research, The patient, who died soon after the cells were removed, of course received no benefit, even though the cell lines were subsequently commercialized. Although this might seem unethical, it has been made clear in several courts of law that a patient's discarded tissue, whatever it might be, does not belong to the patient but to the doctor or his/her institution, who are free to commercialize the product. The ethical matter essentially stops there with respect to discarded tissue components.

The second involved the use of cadaver-derived growth hormone to treat hormone-deficient children. In the USA this practice was formalized by the creation of a special NIH body to control the procurement of this hormone from the pituitary glands. Informed consent was not required from the family of the deceased and no payment was made. This practice eventually ceased when several cases of the devastating neurological condition Creutzfeldt–Jakob disease arose from the transfer of the hormone from an infected but asymptomatic donor to a group of patients, the solution being the development of a recombinant hormone product.

These ethical and practical issues, although largely solved in principle, may never disappear entirely. The rapid development of cell therapy and tissue engineering techniques has resulted in significant commercialization activity based on cells sourced from human patients, ranging from foreskin fibroblasts obtained at circumcision to samples of cardiac fibroblasts obtained at surgery or biopsies. This has troubled a number of religious groups, some of whom fought an extensive campaign in Europe designed to prevent new legislation that would allow and regulate cell therapies and tissue engineering techniques.

### 9.8.8 Clinical trials

The procedures for clinical trials of medical devices and other health care products have been discussed previously. Obviously it is expected that such trials will be carried out according to the highest ethical standards, which are referred to in a number of international documents such as the Declaration of Helsinki. At the center of all considerations is the role and performance of the ethics committee (sometimes referred to as the IRB, or Institutional Review Board) which has the responsibility of receiving, reviewing and monitoring all relevant information from the sponsors. There are many aspects of this process, the main ones being:

**Compensation and inducement** Although patients may be reimbursed expenses and out-of-pocket costs, such compensation should not be so large as to unduly encourage subjects to participate. Arrangements have to be made to allow for additional costs should the patient suffer an adverse event.

**Vulnerable populations** Quite often new products are addressing the needs of individuals who have exceptional needs and may be considered vulnerable to unwarranted suggestions of major improvement to their health. There are detailed procedures sent out in standards documents and guidelines, which address this potentially serious situation.

**Informed consent** This is a major factor in the use of any medical device, but is of particular significance with respect to clinical trials. It is possible that adverse events will occur with health care products even if there were no indications of dangers from all the pre-clinical testing. It is immensely important that all subjects in a clinical trial are given full and complete data so that they can make an informed judgment on whether to proceed. It might be expected that the probability of serious side effects were greater with pharmaceutical products, as indeed witnessed in the UK in 2006 with the testing of a monoclonal antibody where a number of patients experienced multiple organ failure, but no-one can be complacent with any product, and full disclosure is necessary in the interests of all concerned. This is not a trivial point since not all subjects will have the knowledge or intelligence to assess the meaning and significance of pre-clinical data. For example, it is not unknown for a product to be tested in animal models in the pre-clinical phase, wherein one or more animals die. Animals do die during experimental procedures, often quite unrelated to the specifics of the product in question. Should this data, along with an explanation from the sponsor of why the death(s) had no relationship to the performance of the product, be given to subjects and families?

### 9.8.9 Marketing

There is little to say about the ethics of the marketing of medical devices that does not apply to other product sectors. Consistent with other sections in this chapter, the critical areas are:

- inducements to doctors and hospital purchasing staff for using the company's product;
- the use of internet marketing directly to patients in attempts to persuade implanting physicians to use their products in preference to their normal practice;
- wording in marketing material, including journal, newspaper and television advertising that is not fully in accordance with the evidence of device performance in the average patient.

### 9.8.10 Product recalls

If a problem arises with a medical device in use, the company is required to take some action, as implied above with post-market surveillance procedures. What actions are taken revolve around hard commercial decisions, compliance with the regulatory laws, pragmatic issues of product identification and tracking, and ethical considerations. Manufacturers will usually spend significant sums of money bringing a product to market and will, quite naturally, be reluctant to stop its use unless they are really sure that the device is both defective and a danger to health or unless sales are in significant decline. A manufacturer might not wish to take any action if the problem was obviously caused by poor clinical technique or wrong patient selection. Procedures are in place in most countries that mandate companies to take seriously, and report, incidences of adverse events. A company may or may not decide to recall a device, but the decision-making process should be transparent and agreeable to the agency concerned. There are different types of recall, ranging from cessation of marketing while some deficiency is corrected right up to total cessation and prevention of all further use. There are practical issues here since many companies are international and it may not be easy to track all sold devices in many different countries. If a manufacturer, on examining the evidence, does not issue a voluntary recall, the regulatory agency may force them to do so. These are highly contentious issues that often can only be fully resolved in courts of law, where of course, the ethical dimensions to the company's behavior may be taken into account.

### 9.8.11 Off-label use

The various regulatory processes that exist, although somewhat complicated, would appear to have definitive conclusions. A submission is made to an agency, which includes details of the device, the intended use of the device, the instructions for use and the labeling that will accompany it. After successful submission, the manufacturer should be free to market and sell the device within the geographic area of the agency. However, it is not quite that simple. What happens if a physician or surgeon likes a device so much, and gets very good results with it, such that he/she would like to use it in applications that were not covered in a regulatory submission? On the face of it, this would seem wrong since all of the data is concerned with the intended use and not any other. On the other hand, such a dogmatic interpretation of the rules might result in a lack of choice as far as the genuine wishes of an experienced doctor are concerned. Once a different application has been identified, it could take several years to progress that new application through regulatory approval.

The answer to this conundrum in many jurisdictions has been effectively answered by siding with the physician, who may be entitled to use a legally marketed device in any situation he/she wishes if it is used to treat patients in his/her care, where there is no better alternative. This is referred to as "off-label use." No further regulatory approval is required although the doctor may be wise, or even directed, to tell his/her institutional IRB or insurers that that is what he/she is doing. That would seem to make sense. The real difficulty comes in the "promotion" of off-label use. Again in most countries the manufacturers of devices are prohibited from advertising or promoting off-label use. This is a very controversial area, since it is difficult to define promotion (does word-of-mouth dissemination of information count?) and several high-profile cases, for example in the use of BMP products for spinal fusion in parts of the spine not approved, or the use of intravascular stents in vessels that were not approved, have drawn attention to these ethical problems. In the USA there is a tendency to be a little more lenient, by, for example, allowing manufacturers to distribute "scientific papers" that support off-label use.

### 9.8.12 Re-use of single-use items

Although one would assume that a product labeled "single-use only" would be used with one patient, and only one patient, the re-use of putative single-use devices is a major ethical and economical issue. In the majority of cases of single-use devices, their disposable nature should eliminate the need for cleaning, repair and/or reprocessing, ensuring greater patient safety and, in theory, reduced hospital costs associated with the elimination of these processes. In more complex devices, questions also arise as the effectiveness and overall performance of devices that have already been used in patients. Arguments to support the re-use of devices range from superior economy if the costs of cleaning and re-processing are lower than the costs of purchasing new products to the beneficial effects on the environment by reducing medical waste and to the humanitarian benefits if the pre-used devices are cleaned and sent to developing countries that are unable to afford the cost of new devices.

There are no simple solutions, although most arguments tend to support the avoidance of re-use. There are many examples of proven contaminated pre-used devices being used in new patients, with clear associated adverse events; the existence of just a few such cases should suggest major caution here. The humanitarian argument makes sense at first sight, since many patients have no access to new devices, and something may appear to be better than nothing. There are, however, strong ethical arguments against this on the grounds that rich countries should not be offering lower-cost, higher-risk devices to poorer developing countries and, if the preprocessing can be achieved with minimal increased risks, then it would be better for those devices to be used under proper regulatory control in the originating countries. There are better economic models for providing lower-cost devices to developing countries than those that involve re-use.

To reinforce the position that this is not a simple matter, we have to take into account that in many cases of "re-use," the device packet has been opened within a sterile field but has not been used or soiled in any way: the surgeon or physician could change his/her mind on device use mid-procedure, or may decide on a different size than that predicted pre-operatively, or indeed the patient could die intra-operatively. In these situations there may be no possibility of contamination. There are also some situations with complex implantable devices where powerful economic arguments could be used to encourage re-use; the clinical use of implantable pacemakers comes into this category. The longevity of a pacemaker, which is usually an expensive product, is largely dependent on the battery life, which may be 10 years. If a patient dies from unrelated causes just six months after implantation, it may be argued that nine and a half years of battery life will then be wasted. There are good grounds for allowing, as some jurisdictions do, the re-use of sealed pacemaker can, fully detached from the leads, in new patients following decontamination and re-sterilization.

### 9.8.13 Developing countries

The vast majority of readers of this book will live and study/work in countries that are considered to be "developed," where access to health care and medical technology is largely taken for granted. Conversely, much of the world's population live in areas that are considered to be "developing," where equivalent access may not be readily available. This is compounded by the fact that patients in these developing countries are often affected by diseases, especially infectious diseases, that are not experienced in the developed countries of North America, Western Europe and some parts of the Asian Pacific Rim. This means that the procedures necessary to prevent and treat these diseases do not normally receive any priority and often go unchecked. There are some high-profile diseases, especially HIV/Aids and malaria, which have received major attention in recent years but many others have not. This is a major ethical issue that governments, non-governmental organizations and the medical industry have to face. Affordable medical devices, including those that are based on biomaterials, figure prominently in this dilemma.

### 9.8.14 The overall relationships between doctors/medical associations and industry

One of the most difficult issues within the overall framework of ethics in the health care sector is the relationships between individual doctors and their professional association with the health care industry. These relationships partially subsume many of the individual ethical concerns already discussed in this section, and it seems relevant to bring some common themes together.

Dealing first with the professional medical associations, these of course play significant roles in bringing together physicians of the same specialty or sub-specialty for purposes of advanced medical education, the establishment of guidelines and practices and the promulgation of codes of practice. In some countries, the associations also have legislative powers over individual conduct. Obviously the operation of these associations may be costly, with large infrastructure commitments and questions arise as to who pays for this. They also potentially wield a great deal of power and prestige. Individual members pay dues but many would argue that is very onerous for doctors to personally pay the cost of the promotion and protection of health care. A few of these associations publish prestigious journals, which can be very lucrative. Where statutory duties are involved, governments will cover some costs, but often these associations struggle to finance themselves. This is where the potential conflict of interest arises, for the pharmaceutical and medical devices industries may well, through various means, subsidize the associations. They may sponsor conferences and other events; they may pay for special issues of the association journals and pay large fees to advertise their products in both events and publications. The imprimatur of a globally recognized medical association is a serious temptation which most companies find hard to ignore. There have been attempts in recent years to establish guidelines to minimize these risks from real or perceived conflicts of interest, especially in the differentiation between industry-sponsored education and marketing.

Naturally, many of these risks of conflicts at corporate and institutional levels are transposed to the level of the individual physician. Because of some high-profile cases, with very many lesser, almost routine, cases, the practices of industry-financed benefits to doctors in return for favorable treatment and use of products has been reduced considerably, but still clearly exists. These can take the form of individual payments and other benefits or sponsorship of an individual doctor's research, or donations to the doctor's hospital or university.

## Summary and learning objectives

1. Although appearing at the end of the book, the subject matter of this chapter is of immense importance. Without due attention to those matters included here, there will be no effective biomaterials industry and no safe delivery of products for patients. The topics discussed here are given in no particular logical order, they are all significant.
2. All products based on biomaterials that are used in the treatment or diagnosis of disease are regulated by government-controlled bodies. This is sensible, and in keeping with many industrial sectors where products or services may impose risk on humans, since it is necessary to provide some degree of protection to patients with respect to the risks to which they are exposed. Regulatory procedures vary from one jurisdiction to another and from one type of product to another. The purpose of this section has been to explain the fundamental principles of regulation and to give examples of the ways in which devices are classified and how the regulatory process is designed to reflect the level of risk to the patients.
3. Regulation could not be effective if there were no standards or guidelines. Standards, both international and national, exist in order to provide a framework for effective design and evaluation of materials and products. Many materials standards have been published, and are constantly revised on the basis of practical experience, so that manufacturers can benefit from prior successful materials applications and assist regulators in setting benchmarks for materials specifications. Test procedures, especially for the biological safety of biomaterials, constitute a major platform for standards and are used on a global basis. With some complex but critical life-saving devices, very detailed guidelines for product development and testing are provided by bodies such as the FDA.
4. The clinical testing of materials and devices is of critical importance but is also controversial. This section discusses the potential benefits but also the difficulties. At the one extreme we have the possibility of systems that allow unregulated use of devices in humans with no obligations for either clinical trials or clinical follow-up. This situation existed even in the USA before 1976. It is clearly inappropriate today. On the other hand, the other extreme could involve multi-centered clinical trials on every marketed device and lifetime follow-up. In the vast majority of circumstances, this would be prohibitively expensive and the life-cycle of products would mean that sufficient evidence of safety could not be produced before the natural end of the product cycle. It has become obvious that the use of implant registries is very important, although even here there are problems of cost and privacy.
5. Not every device, nor every application of a biomaterial in patients, will be totally successful. If products fail to meet expectations, which may not mean failure of the device itself, and may be more related to patient variables and clinical techniques than the materials, there is a chance of product liability litigation. In this section a number of examples, mostly taken from the author's own professional experience, are summarized, allowing an overview of the important factors to be generated.
6. The next section makes it abundantly clear that whatever the quality of devices and clinical skills,

there will be no implementation without appropriate economic and social infrastructure. Among the issues discussed here are the influence of global economic, political and social issues on health and health care, and their changing dynamics, the role of medical technology as a driver of local and national economies, and the measurement of the cost-effectiveness of health care.

7. We mention briefly the importance of intellectual property and its protection. This is a complex subject, made even more complex by the lack of common international positions. Under discussion here are issues of novelty of inventions and the fine line between patent infringement and invalidity.
8. Finally there is a section on ethical matters. These are difficult to summarize, as they extend from ethical issues associated with the use of stem cells, to the patenting of biological materials and omission/commission dilemmas in regulatory submissions. Each case has to be considered on its own merits; it is appropriate for this to constitute the last section of this book since ethical positions have to be the basis of everything a biomaterials scientist does.

## Questions

1. Discuss the essential reasons why regulatory approval processes are considered necessary and important for health care products.
2. Different regulatory bodies have different definitions of health care products such as medical devices and tissue engineering products. Give your own views of the important differences between the different types of product.
3. Discuss the main principles on which the ISO 10993 standard for the biological evaluation of biomaterials are based. Do you have any significant criticisms of this panel of tests?
4. Compare and contrast the main ground rules for conducting clinical trials on medical devices and pharmaceuticals.
5. Explain the concept of implant registries. Discuss the main advantages that they have over single product evaluation schemes and comment on the various difficulties that are encountered in the establishment and maintenance of registries.
6. Discuss the main drivers of product liability litigation. Do you believe that such litigation is a good method of holding medical device companies responsible for their products, or do the disadvantages outweigh any benefits?
7. Take any one major form of disease and discuss the global dimensions to the prevalence of the disease and the health care systems that are available to treat or contain this disease.
8. Of all the ethical issues that face an inventor-clinician in either medical device or regenerative medicine fields, chose one that you feel is among the most important, explaining the reasons for your choice.

## Recommended reading


CSDH, *Closing the Gap in a Generation: Health equity through action on the social determinants of health. Final report of the Commission on Social Determination of Health*, World Health Organization, Geneva, 2008. *Extensive analysis of the interplay between social and economic policy and health care.*

Haihn R W, Klovers K B, Singer H J, The need for greater price transparency in the medical device industry: an economic analysis, *Health Affairs* 2008;27:1554–9. *Discussion about medical device pricing and related economic issues.*

International Standards Organization, *Clinical Investigation of Medical Devices for Human Subjects – Good clinical practice*, ISO 14155, 2011. *Detailed standard for the conduct of human clinical trials on medical devices.*

Karrholm J, The Swedish Hip Arthroplasty Register, *Acta Orthopaedica* 2010, 81, 3–4. *A summary of the principles and workings of the first implant registry, established in Sweden in the 1970s.*

Klepinski R J, Off-label use of medical devices in the USA, *Journal of Medical Device Regulation* 2009;6:8–19. *Essay about the pros and cons of the off-label use of medical devices.*

Kramer D B, Xu, S, Kesselheim A S, Regulation of medical devices in the United States and Europe, *New England Journal of Medicine* 2012;336:845–55. *Discussion of the similarities and differences between regulation in the USA and Europe.*

Rizzieri, A G, Verheijde J L, Rady M Y, McGregor J L, Ethical challenges with the left ventricular assist device as a destination therapy, *Philosophy, Ethics and Humanities in Medicine* 2008;3:20. *A review of the value and difficulties of organ assist devices and the ethical issues that surround issues of quality of life for patients and care givers.*

Superior Court of Justice, Ontario, Canada, Anderson v St Jude Medical Inc, 2012, ONSC 3660, Court File 00-CV-195906CV, Toronto, 26 June 2012. *Written judgment on class action concerning silver-coated heart valves.*

World Health Organization, *Global Atlas on Cardiovascular Disease Prevention and Control*, Mendis S, Puska P, Norrving B (eds.). Geneva, 2011. *Detailed facts about the global incidence of cardiovascular disease.*

Zacharias D G, Nelson T J, Mueller P S, Hook C C, The science and ethics of induced pluripotency: what will become of embryonic stem cells, *Mayo Clinic Proceedings* 2011;86:634–40. *Discussion of the debates that concern the ethical issues associated with the use of embryonic stem cells and iPS cells.*

# INDEX